THE
HUTCHINSON
GUIDE TO THE
WORLD

EDITORS

Editorial director
Michael Upshall

Project editor
Gian Douglas Home

Text editors
Edith Harkness
Claire Jenkins

Update editors
Sue Croft
Frances Lass
Ingrid von Essen

Pronunciation editor
J C Wells MA PhD

Proofreaders
Jennifer Goss
Helen McCurdy

Researcher
Anna Farkas

Cartography
Swanston Graphics

Design
Terry Caven

Systems administrator
Graham Duncan

Production
Tony Ballsdon

CONTRIBUTORS

Ian D Derbyshire MA, PhD
J Denis Derbyshire BSc, PhD, FBIM
Dougal Dixon BSc, MSc
Bob Moore PhD
David Munro BSc, PhD

ACKNOWLEDGEMENTS

We are grateful to several readers, including Gerald
Lefebvre, who pointed out errors and omissions in the
first edition.

THE
HUTCHINSON
GUIDE TO THE
WORLD

Helicon

First published 1990
(as *The Hutchinson Paperback Guide
to the World*)

Second edition 1993
(*The Hutchinson Guide to the World*)

Revised and updated 1994
Copyright © Helicon Publishing Ltd 1994
Maps copyright © Helicon Publishing Ltd 1994

Helicon Publishing Limited
42 Hythe Bridge Street
Oxford OX1 2EP

Typeset by Aesthetex, Edinburgh

Printed and bound in Great Britain by The Bath Press Ltd, Bath, Avon

ISBN 0 09 177521 3 (cased)
ISBN 0 09 178251 1 (paper)

British Library Cataloguing in Publication Data

A catalogue record for this book is available
from the British Library

PREFACE

This second edition of the Hutchinson Guide to the World, first published in 1990, provides comprehensive coverage of the world's countries, regions, towns, and geographical features. In addition to extensive modifications and updates to existing entries, it contains over 500 new entries (including 18 new sovereign states created since 1991). Every attempt has been made to keep the reader abreast of political developments.

Editorial policy overall has been to include cities and towns with populations over 250,000 (excepting China), but to include also smaller places if justified on other grounds. Where relevant and of interest, historical and cultural data supplement information on geographical location, population, and industries.

Arrangement of entries

Entries are ordered alphabetically, as if there were no spaces between words. Thus, entries for words beginning 'New' follow the order:

New Amsterdam
Newark
New Bedford

Words beginning 'St' and 'Saint' are both treated as if they were spelt 'Saint'.

Foreign names

Names of foreign places are usually shown in their English form, except where the foreign name is more familiar; thus, there is an entry for Florence, not Firenze, but for Livorno not Leghorn.

Cross-references

Selective use has been made of cross-references, principally in entries for former names, foreign names, or alternative spellings that refer the reader directly to a main entry.

Chinese names

Pinyin, the preferred system for transcribing Chinese names of places, is generally used: thus, there is an entry at Beijing, not Peking.

Pronunciations

Pronunciations are given for the names of places, using a transcription which conforms to the International Phonetic Alphabet (IPA). If the name is from a foreign language, the pronunciation given is the nearest equivalent; this provides the English speaker with an intelligible pronunciation of the same name. A key to the pronunciation symbols is shown on page vi.

PRONUNCIATION KEY

ɑː	father/'fɑːðə/, start /stɑːt/		ɬ	Llanelli /ɬæ'neɬi/
aɪ	price /praɪs/, high /haɪ/		m	minimum /'mɪnɪməm/
aʊ	mouth /maʊθ/, how /haʊ/		n	nine /naɪn/
æ	trap /træp/, man /mæn/		ŋ	sing /sɪŋ/, uncle /'ʌŋkl/
b	baby /'beɪbɪ/		ɒ	lot /lɒt/, watch /wɒtʃ/
d	dead /ded/		ɔː	thought /θɔːt/, north /nɔːθ/
dʒ	judge /dʒʌdʒ/		ɔɪ	choice /tʃɔɪs/, boy /bɔɪ/
ð	this /ðɪs/, other /'ʌðə/		p	paper /'peɪpə/
e	dress /dres/, men /men/		r	red /red/, carry /'kærɪ/
eɪ	face /feɪs/, wait /weɪt/		s	space /speɪs/
eə	square /skweə, fair /feə/		ʃ	ship /ʃɪp/, motion /'məʊʃən/
ɜː	nurse /nɜːs/, pearl /pɜːl/		t	totter /'tɒtə/
ə	another /ə'nʌðə/		tʃ	church /tʃɜːtʃ/
əʊ	goat /gəʊt/, snow /snəʊ/		θ	thick /θɪk/, author /'ɔːθə/
f	fifty /'fɪftɪ/		uː	goose /'guːs/, soup /suːp/
g	giggle /'gɪgl/		u	influence /'ɪnfluəns/
h	hot /hɒt/		ʊ	foot /fʊt/, push /pʊʃ/
iː	fleece /fliːs/, sea /siː/		ʊə	poor /pʊə/, cure /kjʊə/
i	happy /hæpi/, glorious /'glɔːrɪəs/		v	vivid /'vɪvɪd/
ɪ	kit /kɪt/, tin /tɪn/		ʌ	strut /strʌt/, love /lʌv/
ɪə	near /nɪə/, idea /aɪ'dɪə/		w	west /west/
j	yellow /'jeləʊ/, few /fjuː/		x	loch /lɒx/
k	kick /kɪk/		z	zones /zəʊnz/
l	little /'lɪtl/		ʒ	pleasure /'pleʒə/

Consonants

p b t d k g tʃ dʒ f v θ ð s z ʃ ʒ m n ŋ r l w j ɬ x

Vowels and Diphthongs

iː ɪ e æ ɑː ɒ ɔː ʊ uː ʌ ɜː ə eɪ əʊ aɪ aʊ ɔɪ ɪə eə ʊə

Stress marks

' (primary word stress)   ˌ (secondary word stress)

Aachen /'ɑːxən/ (French *Aix-la-Chapelle*) German cathedral city and spa in the *Land* of North Rhine–Westphalia, 72 km/45 mi SW of Cologne; population (1988) 239,000. It has thriving electronic, glass, and rubber industries, and is one of Germany's principal railway junctions.

Aachen was the Roman *Aquisgranum*, and from the time of Charlemagne until 1531 the German emperors were crowned there.

Charlemagne was born and buried in Aachen, and founded the cathedral 796. The 14th-century town hall, containing the hall of the emperors, is built on the site of Charlemagne's palace.

Aalborg /'ɔːlbɔːɡ/ (Danish *Ålborg*) port in Denmark 32 km/20 mi inland from the Kattegat, on the south shore of the Limfjord; population (1990) 155,000. One of Denmark's oldest towns, it has a castle and the fine Budolfi church. It is the capital of Nordjylland county in Jylland (Jutland); the port is linked to Nørresundby on the north side of the fjord by a tunnel built 1969.

Aalst /ɑːlst/ (French *Alost*) industrial town (brewing, textiles) in East Flanders, Belgium, on the river Dender 24 km/15 mi NW of Brussels; population (1991) 76,400.

Aarhus /'ɔːhuːs/ (Danish *Århus*) second largest city of Denmark, on the east coast overlooking the Kattegat; population (1990) 261,400. It is the capital of Aarhus county in Jylland (Jutland) and a shipping and commercial centre.

Abadan /,æbə'dɑːn/ Iranian oil port on the east side of the Shatt-al-Arab waterway; population (1986) 294,000. Abadan is the chief refinery and shipping centre for Iran's oil industry, nationalized 1951.

Abakan /,æbə'kæn/ coal-mining city and capital of Khakass Autonomous Region, Krasnoyarsk Territory, in Siberian Russia; population (1987) 181,000.

Abbeville /'æbvɪl/ town in N France in the Somme *département*, 19 km/12 mi inland from the mouth of the river Somme; population (1990) 24,500.

Abeokuta /,æbiəu'kuːtə/ agricultural trade centre in Nigeria, W Africa, on the Ogun River, 103 km/64 mi N of Lagos; population (1983) 309,000.

Aberbrothock /,æbə'brɒθək/ another name for ▷Arbroath, a town in Scotland.

Aberdare /,æbə'deə/ (Welsh *Aberdâr*) town in Mid Glamorgan, Wales, formerly producing high-grade coal, but now with electrical and light engineering industries; population (1981) 36,621.

Aberdeen /,æbə'diːn/ city and seaport on the east coast of Scotland, administrative headquarters of Grampian Region; population (1991) 201,100. Industries include agricultural machinery, paper, and textiles; fishing; ship-building; granite-quarrying; and engineering. There are shore-based maintenance and service depots for the North Sea oil rigs. Aberdeen is Scotland's third largest city.

It is rich in historical interest and fine buildings, including the Municipal Buildings (1867); King's College (1494) and Marischal College (founded 1593, and housed in one of the largest granite buildings in the world 1836), which together form Aberdeen University; St Machar Cathedral (1378); and the Auld Brig o'Balgownie (1320).

Oil discoveries in the North Sea in the 1960s–70s transformed Aberdeen into the European 'offshore capital', with an airport and heliport linking the mainland to the rigs.

Aberdeenshire /,æbə'diːnʃə/ former county in E Scotland, merged 1975 into Grampian Region.

Aberfan /,æbə'væn/ mining village in Mid Glamorgan, Wales. Coal waste overwhelmed a school and houses 1966; of 144 dead, 116 were children.

Aberystwyth /,æbə'rɪstwɪθ/ resort town in Wales; population (1989 est) 15,000. It is the unofficial capital of the Welsh-speaking area of Wales. The University College of Wales 1872, Welsh Plant Breeding Station, and National Library of Wales are here.

Abidja'n /,æbiː'dʒɑːn/ port and former capital (until 1983) of the Republic of Ivory Coast, W Africa; population (1982) 1,850,000. Products include coffee, palm oil, cocoa, and timber (mahogany). Yamoussoukro became the new capital 1983, but was not internationally recognized as such until 1992.

Abilene /'æbəliːn/ city in central Texas, USA, southwest of Fort Worth; seat of Taylor County; population (1990) 106,650. It is a centre for oil-drilling equipment. Abilene was founded 1881 as the terminus for the Texas and Pacific Railroad.

Abilene /'æbəliːn/ town in Kansas, USA, on the Smoky Hill River; population (1990) 6,240. In the 1860s Abilene was the northern terminus of the Chisholm Trail cattle drive and the point from which the herds were shipped east by rail. Industries include the manufacture of aircraft and missile components and oil-field equipment.

Abingdon /'æbɪŋdən/ town in Oxfordshire, England, on the Thames 10 km/6 mi S of Oxford; population (1989 est) 28,600. The remains of the 7th-century abbey include Checker Hall, restored as an Elizabethan-type theatre. The 15th-century bridge was reconstructed 1929. There are light industries.

Abkhazia /æbˈkɑːziə/ autonomous republic in Georgia, situated on the Black Sea
capital Sukhumi
area 8,600 sq km/3,320 sq mi
products tin, fruit, and tobacco
population (1989) 526,000
history The region has been the scene of secessionist activity on the part of the minority Muslim Abkhazi community since 1989, culminating in the republic's declaration of independence 1992. Georgian troops invaded and took control Aug 1992, but secessionist guerrillas subsequently gained control of the northern half of the republic.

A Georgian kingdom from the 4th century, Abkhazia was inhabited traditionally by Abkhazis, an ethnic group converted from Christianity to Islam in the 17th century. By the 1980s some 17% of the population were Muslims and two-thirds were of Georgian origin. In March–April and July 1989, Abkhazis demanded secession from Georgia and reinstatement as a full Union republic, conflicting with Georgian nationalists' demands for the republic to be incorporated as part of Georgia. The dispute triggered civil unrest in Abkhazia and nationalist demonstrations throughout Georgia. In July 1992 the local parliament unilaterally declared Abkhazia's independence. The kidnapping of senior Georgian officials in Aug provoked an invasion of Abkhazia by Georgian troops, who seized the capital and set up an interim government. By early Sept Russia's president Boris Yeltsin had successfully brokered a cease-fire, but in a surprise Oct offensive, secessionist guerrillas reclaimed half of Abkhazia, gaining control of all of the region N of the capital, Sukhumi.

Åbo /ˈɔːbuː/ Swedish name for ▷Turku, a port in Finland.

Abomey /əˈbəumi/ town and port of Benin, W Africa; population (1982) 54,500. It was once the capital of the kingdom of Dahomey, which flourished in the 17th–19th centuries, and had a mud-built defence wall 10 km/6 mi in circumference.

Abruzzi /əˈbrutsi/ mountainous region of S central Italy, comprising the provinces of L'Aquila, Chieti, Pescara, and Teramo; area 10,800 sq km/4,169 sq mi; population (1990) 1,272,000; capital L'Aquila. Gran Sasso d'Italia, 2,914 m/9,564 ft, is the highest point of the Apennines.

Abu Dhabi /ˌæbuːˈdɑːbi/ sheikhdom in SW Asia, on the Persian Gulf, capital of the United Arab Emirates; area 67,350 sq km/26,000 sq mi; population (1982 est) 516,000. Formerly under British protection, it has been ruled since 1971 by Sheik Sultan Zayed bin al-Nahayan, who is also president of the Supreme Council of Rulers of the United Arab Emirates.

Abuja /əˈbuːdʒə/ capital of Nigeria (formally designated as such 1982, although not officially recognized until 1992); population (1991)
378,700 (federal capital territory). Shaped like a crescent, the city was designed by Japanese architect Kenzo Tange and began construction 1976 as a replacement for Lagos. The movement of administrative offices was still in progress late 1992.

Abu Musa /ˈaebuːˈmuːsɑː/ small island in the Persian Gulf. Formerly owned by the ruler of Sharjah, it was forcibly occupied by Iran 1971.

Abu Simbel /ˌæbuː ˈsɪmbəl/ former site of two ancient temples cut into the rock on the banks of the Nile in S Egypt during the reign of Ramses II, commemorating him and his wife Nefertari. The temples were moved, in sections, 1966–67 before the site was flooded by the Aswan High Dam.

Abydos /əˈbaɪdɒs/ ancient city in Upper Egypt; the Great Temple of Seti I dates from about 1300 BC.

Abyssinia /ˌæbɪˈsɪniə/ former name of ▷Ethiopia.

Acadia /əˈkeɪdiə/ (French *Acadie*) name given to ▷Nova Scotia by French settlers 1604, from which the term Cajun derives.

Acapulco /ˌækəˈpulkəu/ or *Acapulco de Juarez* port and holiday resort in S Mexico; population (1990) 592,200. There is deep-sea fishing, and tropical products are exported. Acapulco was founded 1550 and was Mexico's major Pacific coast port until about 1815.

Accra /əˈkrɑː/ capital and port of Ghana; population (1984) 964,800. The port trades in cacao, gold, and timber. Industries include engineering, brewing, and food processing. Osu (Christiansborg) Castle is the presidential residence.

Accrington /ˈækrɪŋtən/ industrial town (textiles, engineering) in Lancashire, England; population (1981) 36,000.

Achaea /əˈkiːə/ in ancient Greece, and also today, an area of the N Peloponnese. The *Achaeans* were the predominant society during the Mycenaean period and are said by Homer to have taken part in the siege of Troy.

Achill Island /ˈækɪl/ or *Eagle Island* largest of the Irish islands, off County Mayo; area 148 sq km/57 sq mi.

Aconcagua /ˌækənˈkægwə/ extinct volcano in the Argentine Andes, the highest peak in the Americas; 6,960 m/22,834 ft. It was first climbed by Edward Fitzgerald's expedition 1897.

Acre /ˈeɪkə/ or *'Akko* seaport in Israel; population (1983) 37,000. Taken by the Crusaders 1104, it was captured by Saladin 1187 and retaken by Richard I (the Lionheart) 1191. Napoleon failed in a siege 1799. British field marshal Allenby captured the port 1918. From being part of British mandated Palestine, it became part of Israel 1948.

ACT abbreviation for ▷*Australian Capital Territory*.

Adana /ˈædənə/ capital of Adana (Seyhan) province, S Turkey; population (1990) 916,150. It is a major cotton-growing centre and Turkey's fourth largest city.

Addis Ababa /'ædɪs 'æbəbə/ or *Adis Abeba* capital of Ethiopia; population (1984) 1,413,000. It was founded 1887 by Menelik, chief of Shoa, who ascended the throne of Ethiopia 1889. His former residence, Menelik Palace, is now occupied by the government.

The city is the headquarters of the Organization of African Unity (OAU).

Adelaide capital and industrial city of South Australia; population (1990) 1,049,100. Industries include oil refining, shipbuilding, and the manufacture of electrical goods and cars. Grain, wool, fruit, and wine are exported. Founded 1836, Adelaide was named after William IV's Queen.

It is a fine example of town planning, with residential districts separated from commercial areas by a 'green belt', one of the earliest examples of its kind, developed by William Light (1786–1839). The business district of the city is situated on the river Torrens, which is dammed to form Adelaide's water supply. Impressive streets include King William Street and North Terrace, and fine buildings include Parliament House, Government House, the Anglican cathedral of St Peter, the Roman Catholic cathedral, two universities, the state observatory, and the museum and art gallery.

Adélie Land /ə'deɪli/ (French *Terre Adélie*) region of Antarctica which is about 140 km/87 mi long, mountainous, covered in snow and ice, and inhabited only by a research team. It was claimed for France 1840.

Aden /'eɪdn/ (Arabic *'Adan*) main port and commercial centre of Yemen, on a rocky peninsula at the southwest corner of Arabia, commanding the entrance to the Red Sea; population (1984) 318,000. The city's economy is based on oil refining, fishing, and shipping. A British territory from 1839, Aden became part of independent South Yemen 1967; it was the capital of South Yemen until 1990.

It comprises the new administrative centre Madinet al-Sha'ab, the commercial and business quarters of Crater and Tawahi, and the harbour area of Ma'alla.

history After annexation by Britain, Aden and its immediately surrounding area (121 sq km/ 47 sq mi) were developed as a ship-refuelling station following the opening of the Suez Canal 1869. It was a colony 1937–63 and then, after a period of transitional violence among rival nationalist groups and British forces, was combined with the former Aden protectorate (290,000 sq km/ 112,000 sq mi) to create the Southern Yemen People's Republic 1967, which was renamed the People's Democratic Republic of Yemen 1970–90.

Adige /'ɑːdɪdʒeɪ/ second longest river (after the Po) in Italy, 410 km/255 mi in length. It crosses the Lombardy Plain and enters the Adriatic just N of the Po delta.

Adirondacks /ˌædə'rɒndæks/ mountainous area in NE New York State, USA, rising to 1,629 m/ 5,344 ft at Mount Marcy; the source of the Hudson and Ausable rivers. Adirondacks is named after a native American people, and is known for its scenery and sports facilities.

Admiralty Islands /'ædmərəlti/ group of small islands in the SW Pacific, part of Papua New Guinea; population (1980) 25,000. The main island is Manus. The islands became a German protectorate 1884 and an Australian mandate 1920.

Adowa /'ædəwɑː/ alternative form of ◊Aduwa, the former capital of Ethiopia.

Adriatic Sea /ˌeɪdri'ætɪk/ large arm of the Mediterranean Sea, lying NW to SE between the Italian and the Balkan peninsulas. The western shore is Italian; the eastern is Croatian, Yugoslav, and Albanian. The sea is about 805 km/500 mi long, and its area is 135,250 sq km/52,220 sq mi.

Aduwa /'ædəwɑː/ or *Adwa*, or *Adowa* former capital of Ethiopia, about 180 km/110 mi SW of Massawa at an altitude of 1,910 m/6,270 ft; population (1982) 27,000.

Aegean Islands /iː'dʒiːan/ islands of the Aegean Sea, but more specifically a region of Greece comprising the Dodecanese islands, the Cyclades islands, Lesvos, Samos, and Chios; area 9,122 sq km/ 3,523 sq mi; population (1991) 460,800.

Aegean Sea /iː'dʒiːən/ branch of the Mediterranean between Greece and Turkey; the Dardanelles connect it with the Sea of Marmara. The numerous islands in the Aegean Sea include Crete, the Cyclades, the Sporades, and the Dodecanese. There is political tension between Greece and Turkey over sea limits claimed by Greece around such islands as Lesvos, Chios, Samos, and Kos.

The Aegean Sea is named after the legendary Aegeus, who drowned himself in the belief that Theseus, his son, had been killed.

I still sigh for the Aegean. Shall you not always love its bluest of waves & brightest of all skies?

On the **Aegean Sea** Lord Byron,
letter of 1812

Aegina /iː'dʒaɪnə/ (Greek *Aíyna* or *Aíyina*) Greek island in the Gulf of Aegina about 32 km/20 mi SW of Piraeus; area 83 sq km/32 sq mi; population (1981) 11,100. In 1811 remarkable sculptures were recovered from a Doric temple in the NE, restored by Thorwaldsen, and taken to Munich.

Aeolian Islands /iː'əʊliən/ another name for the ◊Lipari Islands.

Afars and the Issas, French Territory of the /'æfɑːz, 'ɪsəz/ former French territory that became the Republic of ◊Djibouti 1977.

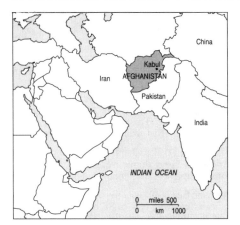

Afghanistan Republic of (*Jamhuria Afghanistan*)
area 652,090 sq km/251,707 sq mi
capital Kabul
towns Kandahar, Herat, Mazar-i-Sharif
physical mountainous in centre and NE, plains in N and SW
environment an estimated 95% of the urban population is without access to sanitation services
features Hindu Kush mountain range (Khyber and Salang passes, Wakhan salient and Panjshir Valley), Amu Darya (Oxus) River, Helmand River, Lake Saberi
head of state Burhanuddin Rabbani from 1992
head of government Gulbuddin Hekmatyar from 1993
political system emergent democracy
political parties Homeland Party (Hezb-i-Watan, formerly People's Democratic Party of Afghanistan, PDPA) Marxist-Leninist; Hezb-i-Islami and Jamiat-i-Islami, Islamic fundamentalist mujaheddin; National Liberation Front, moderate mujaheddin
exports dried fruit, natural gas, fresh fruits, carpets; small amounts of rare minerals, karakul lamb skins, and Afghan coats
currency afgháni
population (1989) 15,590,000 (more than 5 million became refugees after 1979); growth rate 0.6% p.a.
life expectancy (1986) men 43, women 41
languages Pushtu, Dari (Persian)
religion Muslim (80% Sunni, 20% Shi'ite)
literacy men 39%, women 8% (1985 est)
GNP $3.3 bn (1985); $275 per head
GDP $1,86 bn; $111 per head
chronology
1747 Afghanistan became an independent emirate.
1839–42 and 1878–80 Afghan Wars instigated by Britain to counter the threat to British India from expanding Russian influence in Afghanistan.
1919 Afghanistan recovered full independence following Third Afghan War.
1953 Lt-Gen Daud Khan became prime minister and introduced reform programme.
1963 Daud Khan forced to resign and constitutional monarchy established.
1973 Monarchy overthrown in coup by Daud Khan.
1978 Daud Khan ousted by Taraki and the PDPA.
1979 Taraki replaced by Hafizullah Amin; USSR entered country to prop up government; it installed Babrak Karmal in power. Amin executed.
1986 Replacement of Karmal as leader by Dr Najibullah Ahmadzai. Partial Soviet troop withdrawal.
1988 New non-Marxist constitution adopted.
1989 Complete withdrawal of Soviet troops; state of emergency imposed in response to intensification of civil war.
1991 US and Soviet military aid withdrawn. Mujaheddin began talks with Russians and Kabul government.
1992 April: Najibullah government overthrown. June: Burhanuddin Rabbani named interim head of state; Islamic law introduced. Sept: Hezb-i-Islami barred from government participation after shell attacks on Kabul. Dec: Rabbani elected president.
1993 Jan: renewed bombardment of Kabul by Hezb-i-Islami and other rebel forces. Interim parliament appointed by constituent assembly. March: peace agreement signed between Rabbani and dissident mujaheddin leader Gulbuddin Hekmatyar, under which Hekmatyar became prime minister.

Africa /ˈæfrɪkə/ second largest of the continents, three times the area of Europe
area 30,097,000 sq km/11,620,451 sq mi
largest cities (population over 1 million) Cairo, Algiers, Lagos, Kinshasa, Abidja'n, Cape Town, Nairobi, Casablanca, El Gîza, Addis Ababa, Luanda, Dar es Salaam, Ibadan, Douala, Mogadishu
features Great Rift Valley, containing most of the great lakes of E Africa (except Lake Victoria); Atlas Mountains in the NW; Drakensberg mountain range in the SE; Sahara Desert (world's largest desert) in the N; Namib, Kalahari, and Great Karoo deserts in the S; Nile, Zaïre, Niger, Zambezi, Limpopo, Volta, and Orange rivers; the Sahel is a narrow belt of savanna and scrub forest which covers 700 million hectares/1.7 billion acres in W and centre
highest points Mount Kilimanjaro 5,900 m/19,364 ft, and Mount Kenya 5,200 m/17,058 ft
lowest point Lac Assal in Djibouti –144 m/–471 ft
geographical extremities of mainland Cape Hafun in the E, Cape Almadies in the W, Ras Ben Sekka in the N, and Cape Agulhas in the S
physical dominated by a uniform central plateau comprising a southern tableland (mean altitude 1,070 m/3,000 ft) that falls northwards to a lower elevated plain (mean altitude 400 m/1,300 ft). Overall Africa has a mean altitude of 610 m/2,000 ft, two times greater than Europe. 75% of the continent lies within the tropics
products 11% of the world's crude petroleum, 58% of its cocoa (Ivory Coast, Ghana, Cameroon, Nigeria), 23% of its coffee (Uganda, Ivory Coast, Zaire, Ethiopia, Cameroon, Kenya), 20% of its groundnuts (Senegal, Nigeria, Sudan, Zaire), and 21% of its hardwood timber (Nigeria, Zaire, Tanzania, Kenya). Africa has 30% of the world's

minerals including diamonds (51%) and gold (47%)

population (1988) 610 million; more than double the 1960 population of 278 million, and rising to an estimated 900 million by 2000; annual growth rate 3% (10 times greater than Europe); 27% of the world's undernourished people live in sub-Saharan Africa, where an estimated 25 million are facing famine

languages over 1,000 languages spoken in Africa; Niger-Kordofanian languages including Mandinke, Kwa, Lingala, Bemba, and Bantu (Zulu, Swahili, Kikuyu), spoken over half of Africa from Mauritania in the W to South Africa; Nilo-Saharan languages, including Dinka, Shilluk, Nuer, and Masai, spoken in central Africa from the bend of the Niger River to the foothills of Ethiopia; Afro-Asiatic languages, including Arabic, Berber, Ethiopian, and Amharic, N of the equator; Khoisan languages with 'click' consonants spoken in the SW by Kung, Khoikhoi, and Nama people of Namibia

religion Islam in the N and on the east coast as far S as N Mozambique; animism below the Sahara, which survives alongside Christianity (both Catholic and Protestant) in many central and southern areas.

Africa, Horn of /'æfrɪkə/ projection constituted by Somalia and adjacent territories.

Agadir /ˌægə'dɪə/ resort and seaport in S Morocco, near the mouth of the river Sus; population (1984) 10,500. It was rebuilt after being destroyed by an earthquake 1960.

Agaña /ə'gɑːnjə/ capital of Guam, in the W Pacific; population (1980) 896. It is a US naval base.

The town was largely destroyed during World War II; growth has since been confined to the nearby business centres of Tamung and Piti.

Agra /'ɑːgrə/ city of Uttar Pradesh, India, on the river Jumna, 160 km/100 mi SE of Delhi; population (1981) 747,318. A commercial and university centre, it was the capital of the Mogul empire 1527–1628, from which period the Taj Mahal dates.

history Zahir ud-din Muhammad (known as 'Babur'), the first great Mogul ruler, made Agra his capital 1527. His grandson Akbar rebuilt the Red Fort of Salim Shah 1566, and is buried outside the city in the tomb at Sikandra. In the 17th century the buildings of Shah Jahan made Agra one of the most beautiful cities in the world. The Taj Mahal, erected as a tomb for the emperor's wife Mumtaz Mahal, was completed 1650. Agra's political importance dwindled from 1658, when Aurangzeb moved the capital back to Delhi. It was taken from the Marathas by Lord Lake 1803.

Agrigento /ˌægrɪ'dʒentəʊ/ town in Sicily, known for Greek temples; population (1981) 51,300.

The Roman *Agrigentum*, it was long called *Girgenti* until renamed Agrigento 1927 under the Fascist regime.

Aguascalientes /ˌægwəskæli'enteɪs/ city in central Mexico, and capital of a state of the same name; population (1990) 506,384. It has hot mineral springs.

Agulhas /ə'gʌləs/ southernmost cape in Africa. In 1852 the British troopship *Birkenhead* sank off the cape with the loss of over 400 lives.

Ahaggar /ə'hægə/ or *Hoggar* mountainous plateau of the central Sahara, Algeria, whose highest point, Tahat, at 2,918 m/9,576 ft, lies between Algiers and the mouth of the Niger. It is the home of the formerly nomadic Tuareg.

Ahmadnagar /ˌɑːməd'nʌgə/ city in Maharashtra, India, 195 km/120 mi E of Bombay, on the left bank of the river Sina; population (1981) 181,000. It is a centre of cotton trade and manufacture.

Ahmedabad /'amədəbɑːd/ or *Ahmadabad* capital of Gujarat, India; population (1981) 2,515,195. It is a cotton-manufacturing centre, and has many sacred buildings of the Hindu, Muslim, and Jain faiths.

Ahmedabad was founded in the reign of Ahmad Shah 1412, and came under the control of the East India Company 1818. In 1930 Gandhi marched to the sea from here to protest against the government salt monopoly.

Ahváz /ɑː'vɑːz/ industrial capital of the province of Khuzestan, on the river Karun, W Iran; population (1986) 590,000. The ancient city was rebuilt in the 3rd century AD; it became a prosperous city in the 20th century after the discovery of oil.

Ahvenanmaa Island /'ɑːvənənmɑː/ island in the Gulf of Bothnia, Finland; largest of the ▷Åland Islands.

Ailsa Craig /'eɪlsə 'kreɪg/ rocky islet in the Firth of Clyde, Scotland, about 16 km/10 mi off the coast of Strathclyde, opposite Girvan. Ailsa Craig rock is used in the manufacture of curling stones. It is a breeding ground for birds.

Ain /æn/ river of E central France, giving its name to a *département*; length 190 km/118 mi. Rising in the Jura Mountains, it flows S into the Rhône.

Aintab /aɪn'tɑːb/ Syrian name of ▷Gaziantep, a city in Turkey.

Aisne /eɪn/ river of N France, giving its name to a *département*; length 282 km/175 mi.

Aix-en-Provence town in the *département* of Bouches-du-Rhône, France, 29 km/18 mi N of Marseille; population (1990) 126,800. It is the capital of Provence and dates from Roman times.

Its Roman name was *Aquae Sextiae*. It has a Gothic cathedral and a university founded 1409. The painter Paul Cézanne was born here.

Aix-la-Chapelle /'eɪks læ ʃæ'pel/ French name of ▷Aachen, an ancient city in Germany.

Aix-les-Bains /'eɪks leɪ 'bæn/ spa with hot springs in the *département* of Savoie, France, near Lake Bourget, 13 km/8 mi N of Chambéry; population (1990) 24,800.

Ajaccio /æ'ʒæksiəʊ/ capital and second largest port of Corsica; population (1990) 59,300. Founded by the Genoese 1492, it was the birthplace of Napoleon; it has been French since 1768.

Ajman /'ædʒmɑːn/ smallest of the seven states that make up the ◊United Arab Emirates; area 250 sq km/96 sq mi; population (1985) 64,318.

Ajmer /ɑːdʒ'mɪə/ town in Rajasthan, India; population (1981) 376,000. Situated in a deep valley in the Aravalli Mountains, it is a commercial and industrial centre, notably of cotton manufacture. It has many ancient remains, including a Jain temple.

It was formerly the capital of the small state of Ajmer, which was merged with Rajasthan 1956.

AK abbreviation for ◊*Alaska*, a state of the USA.

Akaba /'ækəbə/ alternative transliteration of ◊Aqaba, a gulf of the Red Sea.

'Akko /ɑː'kəu/ Israeli name for the port of ◊Acre.

Akola /ə'kəulə/ town in Maharashtra, India, near the Purnar; population (1981) 176,000. It is a major cotton and grain centre.

Akron /'ækrən/ (Greek 'summit') city in Ohio, USA, on the Cuyahoga River, 56 km/35 mi SE of Cleveland; population (1990) 660,000. Known as the 'Rubber Capital of the World,' it is home to the headquarters of several major tyre and rubber companies, although production there had ended by 1982.

Akron was first settled 1807. Dr B F Goodrich established a rubber factory 1870, and the industry grew immensely with the rising demand for tyres from about 1910.

Akrotiri /ˌækrəu'tɪəri/ peninsula on the south coast of Cyprus; it has a British military base.

Aksai Chin /ˌæksaɪ/ part of Himalayan Kashmir lying to the E of the Karakoram range. It is occupied by China but claimed by India.

Aktyubinsk /æk'tjuːbɪnsk/ industrial city (chemicals, metals, electrical equipment) in Kazakhstan; population (1987) 248,000. Established 1869, it expanded after the opening of the Trans-Caspian railway 1905.

AL abbreviation for ◊*Alabama*, a state of the USA.

Alabama /ˌælə'bæmə/ state in southern USA; nickname Heart of Dixie/Cotton State
area 134,700 sq km/51,994 sq mi
capital Montgomery
towns Birmingham, Mobile, Huntsville, Tuscaloosa

Alabama

Montgomery

physical the state comprises the Cumberland Plateau in the N; the Black Belt, or Canebrake, which is excellent cotton-growing country, in the centre; and S of this, the coastal plain of Piny Woods. The Alabama River is the largest in the state
features Alabama and Tennessee rivers; Appalachian Mountains; George Washington Carver Museum at the Tuskegee Institute (a college founded for blacks by Booker T Washington); Helen Keller's birthplace at Tuscumbia
products cotton (still important though no longer prime crop); soya beans, peanuts, wood products, coal, iron, chemicals, textiles, paper
population (1990) 4,040,600
famous people Nat King Cole, Helen Keller, Joe Louis, Jesse Owens, Booker T Washington
history first settled by the French in the early 18th century, it was ceded to Britain 1763, passed to the USA 1783, and became a state 1819. It was one of the Confederate States in the Civil War.

Åland Islands /'ɔːlənd/ (Finnish *Ahvenanmaa* 'land of waters') group of some 6,000 islands in the Baltic Sea, at the southern extremity of the Gulf of Bothnia; area 1,481 sq km/572 sq mi; population (1990) 24,600. Only 80 are inhabited; the largest island has a small town, Mariehamn. The main sectors of the island economy are tourism, agriculture, and shipping.
history The islands were Swedish until 1809, when they came, (with Finland), under Russian control. The Swedes tried, unsuccessfully, to recover the islands at the time of the Russian Revolution 1917. In 1921 the League of Nations ruled that the islands remain under Finnish sovereignty, be demilitarized, and granted autonomous status. Although the islands' assembly voted for union with Sweden 1945, the 1921 declaration remains valid. The islands became a member of the Nordic Council 1970.

Alaska /ə'læskə/ largest state of the USA, on the northwest extremity of North America, separated from the lower 48 states by British Columbia; nickname Last Frontier
total area 1,530,700 sq km/591,004 sq mi
land area 1,478,457 sq km/570,833 sq mi
capital Juneau
towns Anchorage, Fairbanks, Fort Yukon, Holy Cross, Nome
physical much of Alaska is mountainous and includes Mount McKinley (Denali), 6,194 m/ 20,322 ft, the highest peak in North America, surrounded by Denali National Park. Caribou (descended from 2,000 reindeer imported from Siberia in the early 1900s) thrive in the Arctic tundra, and elsewhere there are extensive forests
features Yukon River; Rocky Mountains, including Mount McKinley and Mount Katmai, a volcano that erupted 1912 and formed the Valley of Ten Thousand Smokes (from which smoke and steam still escape and which is now a national monument); Arctic Wild Life Range, with the only large herd of indigenous North American caribou; Little Diomede Island, which is only 4 km/2.5 mi from Russian Big Diomede/Ratmanov

Alaska

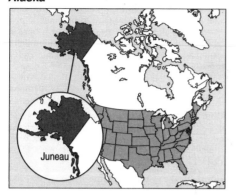

Island. The chief railway line runs from Seward to Fairbanks, which is linked by highway (via Canada) with Seattle; near Fairbanks is the University of Alaska
products oil, natural gas, coal, copper, iron, gold, tin, fur, salmon fisheries and canneries, lumber
population (1990) 550,000; including 9% American Indians, Aleuts, and Inuits
history various groups of Indians crossed the Bering land bridge 60,000–15,000 years ago; the Eskimo began to settle the Arctic coast from Siberia about 2000 BC; the Aleuts settled the Aleutian archipelago about 1000 BC. The first European to visit Alaska was Vitus Bering 1741. Alaska was a Russian colony from 1744 until purchased by the USA 1867 for $7,200,000; gold was discovered five years later. It became a state 1959.

A Congressional act 1980 gave environmental protection to 42 million ha/104 million acres. Valuable mineral resources have been exploited from 1968, especially in the Prudhoe Bay area to the SE of Point Barrow. An oil pipeline (1977) runs from Prudhoe Bay to the port of Valdez. Oilspill from a tanker in Prince William Sound caused great environmental damage 1989. Under construction is an underground natural-gas pipeline to Chicago and San Francisco.

Alaska Highway /ə,læskə 'haɪweɪ/ road that runs from Dawson Creek, British Columbia, to Fairbanks, Alaska (2,450 km/1,522 mi). It was built 1942 as a supply route for US military forces in Alaska.

The highway, which runs along the eastern edge of the Rocky Mountains, is paved in Alaska but mostly gravel-surfaced in Canada.

Alba /'ælbə/ Gaelic name for ◊Scotland.

Albacete /,ælbə'θeɪti/ market town in the province of the same name, SE Spain; population (1991) 134,600. Once famous for cutlery, it now produces clothes and footwear.

Alba Iulia /ælbe 'juːliːe/ (German **Karlsburg**) city on the river Mures, W central Romania; population (1985) 64,300. It was founded by the Romans in the 2nd century AD. The Romanian kings were crowned here, Michael the Brave (1601) and the kings of modern Romania until World War II.

Albania Republic of (*Republika e Shqipërisë*)
area 28,748 sq km/11,097 sq mi
capital Tiranë
towns Shkodër, Elbasan, Vlorë, chief port Durrës
physical mainly mountainous, with rivers flowing E–W, and a narrow coastal plain
features Dinaric Alps, with wild boar and wolves
head of state Sali Berisha from 1992
head of government Alexandr Meksi from 1992
political system emergent democracy
political parties Democratic Party of Albania (PSDS), moderate, market-oriented; Socialist Party of Albania (PSS), ex-communist; Human Rights Union (HMU), Greek minority party
exports crude oil, bitumen, chrome, iron ore, nickel, coal, copper wire, tobacco, fruit, vegetables
currency lek
population (1990 est) 3,270,000; growth rate 1.9% p.a.
life expectancy men 69, women 73
languages Albanian, Greek
religion Muslim 70%, although all religion banned 1967–90
literacy 75% (1986)
GNP $2.8 bn (1986 est); $900 per head
GDP $1.3 bn; $543 per head
chronology
c. 1468 Albania made part of the Ottoman Empire.
1912 Independence achieved from Turkey.
1925 Republic proclaimed.
1928–39 Monarchy of King Zog.
1939–44 Under Italian and then German rule.
1946 Communist republic proclaimed under the leadership of Enver Hoxha.
1949 Admitted into Comecon.
1961 Break with Khrushchev's USSR.
1967 Albania declared itself the 'first atheist state in the world'.
1978 Break with 'revisionist' China.
1985 Death of Hoxha.
1987 Normal diplomatic relations restored with Canada, Greece, and West Germany.

1990 One-party system abandoned; first opposition party formed.
1991 Party of Labour of Albania (PLA) won first multiparty elections; Ramiz Alia re-elected president; three successive governments formed. PLA renamed PSS.
1992 Presidential elections won by PSDS; Sali Berisha elected president. Alia and other former communist officials charged with corruption and abuse of power; totalitarian and communist parties banned.

---

*Land of Albania! where Iskander rose,/
Theme of the young, and beacon of the
wise ...*

On **Albania** Lord Byron *Childe Harold's
Pilgrimage* 1812

Albany /ˈɔːlbəni/ capital of New York State, USA, situated on the west bank of the Hudson River, about 225 km/140 mi N of New York City; population (1990) 101,000. With Schenectady and Troy it forms a metropolitan area, population (1980) 794,298.

Albany /ˈɔːlbəni/ city on the Flint River, Georgia, USA, SE of Columbus; seat of Dougherty County; population (1990) 78,100. It is a commercial centre for the production of pecans and peanuts, chemicals, and lumber.

Albany /ˈɔːlbəni/ port in Western Australia, population (1986) 14,100. It suffered from the initial development of Fremantle, but has grown with the greater exploitation of the surrounding area. The *Albany Doctor* is a cooling breeze from the sea, rising in the afternoon.

Alberta /ælˈbɜːtə/ province of W Canada
area 661,200 sq km/255,223 sq mi
capital Edmonton
towns Calgary, Lethbridge, Medicine Hat, Red Deer
physical Rocky Mountains; dry, treeless prairie in the centre and S; towards the N this merges into a zone of poplar, then mixed forest. The valley of the Peace River is the most northerly farming land in Canada (except for Inuit pastures)

Alberta

features Banff, Jasper, and Waterton Lake national parks; annual Calgary stampede; extensive dinosaur finds near Drumheller
products coal; wheat, barley, oats, sugar beet in the S; more than a million head of cattle; oil and natural gas
population (1991) 2,501,400
history in the 17th century much of its area was part of a grant to the Hudson's Bay Company for the fur trade. It became a province 1905.

Albert Canal /ˈælbət kəˈnæl/ canal designed as part of Belgium's frontier defences; it also links the industrial basin of Liège with the port of Antwerp. It was built 1930–39 and named after King Albert I.

Albert, Lake /ˈælbət/ former name of Lake ◊Mobutu in central Africa.

Albertville /ˌælbeəˈviːl/ resort town at the entrance to the Val d'Arly in the *département* of Savoie, SE France; population (1981) 17,500. It was the scene of the 1992 Winter Olympics.

Albi /ælˈbiː/ chief town in Tarn *département*, Midi-Pyrénées, SW France, on the river Tarn, 72 km/45 mi NE of Toulouse; population (1990) 48,700. It was the centre of the Albigensian heresy and the birthplace of the artist Toulouse-Lautrec. It has a 13th-century cathedral.

Ålborg /ˈɔːlbɔːg/ alternative form of ◊Aalborg, a port in Denmark.

Albufeira /ˌælbuˈfeərə/ fishing village and resort on the Algarve coast of S Portugal, 43 km/27 mi W of Faro; it is known as the St Tropez of the Algarve. There are Moorish remains.

Albuquerque /ˈælbəkɜːki/ largest city of New Mexico, USA, situated E of the Rio Grande, in the Pueblo district; population (1990) 384,750. Founded 1706, it was named after Afonso de Albuquerque. It is a resort and industrial centre, specializing in electronic products and aerospace equipment.

Albury-Wodonga /ˈɔːbəri wəˈdɒŋgə/ twin town on the New South Wales/Victoria border, Australia; population (1981) 54,214. It was planned to relieve overspill from Melbourne and Sydney, and produces car components.

Alcatraz /ˈælkətræz/ small island in San Francisco Bay, California, USA. Its fortress was a military prison 1886–1934 and then a federal penitentiary until closed 1963. The dangerous tides allowed few successful escapes.

Inmates included the gangster Al Capone and the 'Birdman of Alcatraz', a prisoner who used his time in solitary confinement to become an authority on caged birds. American Indian 'nationalists' briefly took over the island 1970 as a symbol of their lost heritage.

Aldabra /ælˈdæbrə/ high limestone island group in the Seychelles, some 420 km/260 mi NW of Madagascar; area 154 sq km/59 sq mi. A nature reserve since 1976, it has rare plants and animals, including the giant tortoise.

Aldeburgh /ˈɔːldbərə/ small town and coastal

resort in Suffolk, England; population (1989 est) 2,900. It is the site of an annual music festival founded by Benjamin Britten, and is also the home of the Britten–Pears School for Advanced Musical Studies.

Alderney /'ɔːldəni/ third largest of the Channel Islands, with its capital at St Anne's; area 8 sq km/ 3 sq mi; population (1980) 2,000. It gives its name to a breed of cattle, better known as the Guernsey.

Aldershot /'ɔːldəʃɒt/ town in Hampshire, England, SW of London; population (1985 est) 38,000. It has a military camp and barracks dating from 1854.

Aleksandrovsk /ˌælɪk'saːndrɒfsk/ former name (until 1921) of ▷Zaporozhye, a city in the Ukraine.

Alençon /ˌælɒn'sɒŋ/ capital of the Orne *département* of France, situated in a rich agricultural plain to the SE of Caen; population (1990) 31,100. Lace, now a declining industry, was once a major product.

Alentejo /ˌlian 'tezuː/ region of E central Portugal divided into the districts of Alto Alentejo and Baixo Alentejo. The chief towns are Evora, Neja, and Portalegre.

Aleppo /ə'lepəu/ (Syrian *Halab*) ancient city in NW Syria; population (1981) 977,000. There has been a settlement on the site for at least 4,000 years.

Alessandria /ˌælɪ'sændriə/ town in N Italy on the river Tanaro; population (1987) 101,000. It was founded 1168 by Pope Alexander III as a defence against Frederick I Barbarossa.

Aletsch /'aːletʃ/ most extensive glacier in Europe, 23.6 km/14.7 mi long, beginning on the southern slopes of the Jungfrau in the Bernese Alps, Switzerland.

Aleutian Islands /ə'luːʃn/ volcanic island chain in the N Pacific, stretching 1,200 mi/1,900 km SW of Alaska, of which it forms part; population 6,000 Aleuts (most of whom belong to the Orthodox Church, plus a large US defence establishment). There are 14 large and more than 100 small islands running along the Aleutian Trench. The islands are mountainous, barren, and treeless; they are ice-free all year but are often foggy, with only about 25 days of sunshine recorded annually. *history* The islands were settled by the Aleuts around 1000 BC and discovered by a Russian expedition 1741; they passed to the USA with the purchase of Alaska 1867. The Japanese occupied Attu and Kiska islands 1942–43; Attu was retaken in the only ground fighting on North American soil during World War II.

Alexandretta /ˌælɪgzaːn'dretə/ former name of ▷Iskenderun, a port in S Turkey.

Alexandria /ˌælɪg'zændriə/ city in central Louisiana, USA, on the Red River, NW of Baton Rouge; seat of Rapides parish; population (1990) 49,188. It is a livestock and meatpacking centre.

Alexandria /ˌælɪg'zaːndriə/ or *El Iskandariya*

city, chief port, and second largest city of Egypt, situated between the Mediterranean and Lake Maryut; population (1986) 5,000,000. It is linked by canal with the Nile and is an industrial city (oil refining, gas processing, and cotton and grain trading). Founded 331 BC by Alexander the Great, Alexandria was the capital of Egypt for over 1,000 years.

history The principal centre of Hellenistic culture, Alexandria has since the 4th century AD been the seat of a Christian patriarch. In 641 it was captured by the Muslim Arabs, and after the opening of the Cape route its trade rapidly declined. Early in the 19th century it began to recover its prosperity, and its growth was encouraged by its use as the main British naval base in the Mediterranean during both world wars. Of the large European community, most were expelled after the Suez Crisis 1956 and their property confiscated.

Few relics of antiquity remain. The Pharos, the first lighthouse and one of the seven wonders of the ancient world, has long since disappeared. The library, said to have contained 700,000 volumes, was destroyed by the caliph Omar 640. Pompey's Pillar is a column erected, as a landmark visible from the sea, by the emperor Diocletian. Two obelisks that once stood before the Caesarum temple are now in London (Cleopatra's Needle) and New York respectively.

The pallor of this city is the pallor of a hiatus – a pause in history, at the point where civilizations meet.

On **Alexandria** James Morris *Places* 1972

Algarve /æl'gaːv/ (Arabic *al-gharb* 'the west') ancient kingdom in S Portugal, the modern district of Faro, a popular holiday resort; population (1981) 323,500. Industries include agriculture, fishing, wine, mining, and tourism. The Algarve began to be wrested from the Moors in the 12th century and was united with Portugal as a kingdom 1253.

It includes the SW extremity of Europe, Cape St Vincent, where the British fleet defeated the Spanish 1797.

Algeciras /ˌældʒɪ'sɪərəs/ port in S Spain, to the W of Gibraltar across the Bay of Algeciras; population (1991) 101,400. Founded by the Moors 713, it was taken from them by Alfonso XI of Castile 1344.

Virtually destroyed in a fresh attack by the Moors, Algeciras was re-founded 1704 by Spanish refugees who had fled from Gibraltar after it had been captured by the British. Following a conference of European Powers held here 1906, France and Spain were given control of Morocco.

Algeria Democratic and Popular Republic of (*al-Jumhuriya al-Jazairiya ad-Dimuqratiya ash-Shabiya*)
area 2,381,741 sq km/919,352 sq mi
capital Algiers (al-Jazair)
towns Constantine/Qacentina; ports are Oran/Ouahran, Annaba

physical coastal plains backed by mountains in N; Sahara desert in S

features Atlas mountains, Barbary Coast, Chott Melrhir depression, Hoggar mountains

head of state Ali Kafi from 1992

head of government Redha Malek from 1993

political system semi-military rule

political parties National Liberation Front (FLN), nationalist socialist; Socialist Forces Front (FSS), Berber-based

exports oil, natural gas, iron, wine, olive oil

currency dinar

population (1990 est) 25,715,000 (83% Arab, 17% Berber); growth rate 3.0% p.a.

life expectancy men 59, women 62

languages Arabic (official); Berber, French

religion Sunni Muslim (state religion)

literacy men 63%, women 37% (1985 est)

GDP $64.6 bn; $2,796 per head

chronology

1954 War for independence from France led by the FLN.

1962 Independence achieved from France. Republic declared. Ahmed Ben Bella elected prime minister.

1963 Ben Bella elected Algeria's first president.

1965 Ben Bella deposed by military, led by Colonel Houari Boumédienne.

1976 New constitution approved.

1978 Death of Boumédienne.

1979 Benjedid Chadli elected president. Ben Bella released from house arrest. FLN adopted new party structure.

1981 Algeria helped secure release of US prisoners in Iran.

1983 Chadli re-elected.

1988 Riots in protest at government policies; 170 killed. Reform programme introduced. Diplomatic relations with Egypt restored.

1989 Constitutional changes proposed, leading to limited political pluralism.

1990 Fundamentalist Islamic Salvation Front (FIS) won Algerian municipal and provincial elections.

1991 Dec: FIS won first round of multiparty elections.

1992 Jan: Chadli resigned; military took control of government; Mohamed Boudiaf became

president; FIS leaders detained. Feb: State of emergency declared. March: FIS ordered to disband. June: Boudiaf assassinated; Ali Kafi chosen as new head of state.

1993 Worsening civil strife; assassinations of politicians and other public figures. Nov: killings of foreign workers began following expiry of FIS deadline for foreigners to leave the country.

Algiers /æl'dʒɪəz/ (Arabic *al-Jazair*; French *Alger*) capital of Algeria, situated on the narrow coastal plain between the Atlas Mountains and the Mediterranean; population (1984) 2,442,300. Founded by the Arabs AD 935, Algiers was taken by the Turks 1518 and by the French 1830. The old town is dominated by the Kasbah, the palace and prison of the Turkish rulers. The new town, constructed under French rule, is in European style.

Algoa Bay /æl'gəuə/ broad and shallow inlet in Cape Province, South Africa, where Bartholomeu Diaz landed after rounding the Cape 1488.

Alicante city and seaport in E Spain on the Mediterranean Sea, 123 km/77 mi S of Valencia; population (1986 est) 265,500. It is the commercial port for Madrid, exporting wine, olive oil, and fruit; there are manufacturing industries.

Believed to occupy site of the ancient Roman city of *Lucentum*, Alicante was captured by the Moors 713, retaken by James I of Aragon 1265, and besieged by the French 1709 and 1812.

Aligarh /ɑːlɪ'gɜː/ city in Uttar Pradesh, N central India; population (1981) 320,000. Industries include agricultural manufacturing and processing, engineering, and textiles. The city is also named Koil; Aligarh is the name of a nearby fort.

Allahabad /ˌæləhə'bɑːd/ ('city of god') historic city in Uttar Pradesh state, NE India, 580 km/360 mi SE of Delhi, on the Yamuna River where it meets the Ganges and the mythical Seraswati River; population (1981) 642,000. A Hindu religious event, the festival of the jar of nectar of immortality (Khumbh Mela), is held here every 12 years with the participants washing away sin and sickness by bathing in the rivers. Fifteen million people attended the festival Jan–March 1989.

Allegheny Mountains /ˌælɪ'geɪni/ range over 800 km/500 mi long extending from Pennsylvania to Virginia, USA, rising to more than 1,500 m/4,900 ft and averaging 750 m/2,500 ft. The mountains are a major source of timber, coal, iron, and limestone. They initially hindered western migration, the first settlement to the west being Marietta 1788.

Allen, Bog of /'ælən/ wetland E of the river Shannon in the Republic of Ireland, comprising some 96,000 ha/240,000 acres of the counties of Offaly, Leix, and Kildare; the country's main source of peat fuel.

Allen, Lough /'ælən/ lake in County Leitrim, Republic of Ireland, on the upper course of the river Shannon. It is 11 km/7 mi long and 5 km/3 mi broad.

Allentown /'æləntaun/ city in E Pennsylvania, USA, on the Lehigh River, just NW of Philadelphia; population (1990) 105,090. It is an industrial centre for textiles, machinery, and electronic equipment. During the American War of Independence it was a centre for munitions production.

Allier /,æli'eɪ/ river in central France, a tributary of the Loire; it is 565 km/350 mi long and gives its name to a *département*. Vichy is the chief town on the Allier.

Alma-Ata /æl'mɑː ɔ'tɑː/ (formerly until 1921 *Vernyi*) capital of Kazakhstan; population (1991) 1,151,300. Industries include engineering, printing, tobacco processing, textile manufacturing, and leather products.

Established 1854 as a military fortress and trading centre, the town was destroyed by an earthquake 1887.

Almaden /,ælmə'ðen/ mining town in Ciudad Real province, Castilla-La Mancha, central Spain; population (1981) 9,700. It has the world's largest supply of mercury, worked since the 4th century BC.

Almería /,ælme'riːə/ Spanish city, chief town of a province of the same name on the Mediterranean; population (1991) 157,800. The province is famous for its white grapes, and in the Sierra Nevada are rich mineral deposits.

Alost /ɑː'lɒst/ French name for the Belgian town of ▷ Aalst.

Alps /ælps/ mountain chain, the barrier between N Italy and France, Germany and Austria.

Famous peaks include *Mont Blanc*, the highest at 4,809 m/15,777 ft, first climbed by Jacques Balmat and Michel Paccard 1786; *Matterhorn* in the Pennine Alps, 4,479 m/14,694 ft, first climbed by Edward Whymper 1865 (four of the party of seven were killed when a rope broke during their descent); *Eiger* in the Bernese Alps/Oberland, 3,970 m/13,030 ft, with a near-vertical rock wall on the north face, first climbed 1858; *Jungfrau*, 4,166 m/13,673 ft; and *Finsteraarhorn* 4,275 m/14,027 ft.

Famous passes include *Brenner*, the lowest, Austria/Italy; *Great St Bernard*, one of the highest, 2,472 m/8,113 ft, Italy/Switzerland (by which Napoleon marched into Italy 1800); *Little St Bernard*, Italy/France (which Hannibal is thought to have used); and *St Gotthard*, S Switzerland, which Suvorov used when ordered by the tsar to withdraw his troops from Italy. All have been superseded by all-weather road/rail tunnels. The Alps extend down the Adriatic coast into Slovenia, Croatia, Bosnia-Herzegovina, Yugoslavia, and N Albania with the Julian and Dinaric Alps.

But lo! the Alps, ascending white in air/
Toy with the sun and glitter from afar.

On the **Alps** William Wordsworth
Descriptive Sketches 1792

Alps, Australian /ælps/ highest area of the E Highlands in Victoria/New South Wales, Australia, noted for winter sports. They include the *Snowy Mountains* and *Mount Kosciusko*, Australia's highest mountain, 2,229 m/7,316 ft, first noted by Polish-born Paul Strzelecki 1829 and named after a Polish hero.

Alps, Southern /ælps/ range of mountains running the entire length of South Island, New Zealand. They are forested to the west, with scanty scrub to the east. The highest point is Mount Cook, 3,764 m/12,349 ft. Scenic features include gorges, glaciers, lakes, and waterfalls. Among its lakes are those at the southern end of the range: Manapouri, Te Anau, and the largest, Wakatipu, 83 km/52 mi long, which lies about 300 m/1,000 ft above sea level and has a depth of 378 m/1,242 ft.

Alsace /æl'sæs/ region of France; area 8,300 sq km/3,204 sq mi; population (1986) 1,600,000. It consists of the *départements* of Bas-Rhin and Haut-Rhin, and its capital is Strasbourg.

Alsace-Lorraine /æl'sæs lɒ'reɪn/ area of NE France, lying west of the river Rhine. It forms the French regions of Alsace and Lorraine. The former iron and steel industries are being replaced by electronics, chemicals, and precision engineering. The German dialect spoken does not have equal rights with French, and there is autonomist sentiment.

history Alsace-Lorraine formed part of Celtic Gaul in Caesar's time, was invaded by the Alemanni and other Germanic tribes in the 4th century, and remained part of the German Empire until the 17th century. In 1648 part of the territory was ceded to France; in 1681 Louis XIV seized Strasbourg. The few remaining districts were seized by France after the French Revolution. Conquered by Germany 1870–71 (chiefly for its iron ores), it was regained by France 1919, then again annexed by Germany 1940–44, when it was liberated by the Allies.

Altai /ɑːl'taɪ/ territory of the Russian Federation, in SW Siberia; area 261,700 sq km/101,043 sq mi; capital Barnaul; population (1985) 2,744,000. Industries include mining, light engineering, chemicals, and timber. Altai was colonized by the Russians from the 18th century.

Altai Mountains /æl'taɪ/ mountain system of Kazakhstan, W Siberian Russia, W Mongolia, and N China.

It is divided into two parts, the Russian Altai, which includes the highest peak, Mount Belukha, 4,506 m/14,783 ft, and the Mongolian or Great Altai.

Altamira /,æltə'mɪərə/ Amazonian town in the state of Pará, NE Brazil, situated at the junction of the Trans-Amazonian Highway with the Xingu River, 700 km/400 mi SW of Belém; population (1991) 157,884.

In 1989 a protest by Brazilian Indians and environmentalists against the building of six dams focused world attention on the devastation of the Amazon rainforest.

Altdorf /'æltdɔːf/ capital of the Swiss canton Uri at the head of Lake Lucerne, Switzerland; population 9,000. It was the scene of the legendary exploits of William Tell.

Altiplano /ˌæltuˈplɑːnəu/ densely populated upland plateau of the Andes of South America, stretching from S Peru to NW Argentina. The height of the Altiplano is 3,000–4,000 m/10,000–13,000 ft.

Alton /'ɔːltən/ city in W Illinois, USA, on the Mississippi River, just NE of St Louis, Missouri; population (1980) 34,171. Alton is an industrial centre with flour mills and oil refineries.

Altoona /ælˈtuːnə/ city in S central Pennsylvania, USA, in the Allegheny Mountains, NW of Harrisburg; population (1980) 57,078. It is a railroad manufacturing and repair centre, and there is coal mining. The first steel railroad tracks in the USA were laid from Pittsburgh to Altoona.

Alwar /'ʌlwɑː/ city in Rajasthan, India, chief town of the district (formerly princely state) of the same name; population (1981) 146,000. It has fine palaces, temples, and tombs. Flour milling and trade in cotton goods and millet are major occupations.

Amagasaki /ˌæməgəˈsɑːki/ industrial city on the NW outskirts of Osaka, Honshu island, Japan; population (1990) 498,900.

Amalfi /əˈmælfi/ port 39 km/24 mi SE of Naples, Italy, situated at the foot of Monte Cerrato, on the Gulf of Salerno; population 7,000. For 700 years it was an independent republic. It is an ancient archiepiscopal see (seat of an archbishop) and has a Romanesque cathedral.

Amarillo /ˌæməˈrɪləu/ town in the Texas panhandle, USA; population (1980) 149,230. The centre of the world's largest cattle-producing area, it processes the live animal into frozen supermarket packets in a single continuous operation on an assembly line. It is also a centre for assembly of nuclear warheads.

Amazon /'æməzən/ (Indian *Amossona* 'destroyer of boats') South American river, the world's second longest, 6,570 km/4,080 mi, and the largest in volume of water. Its main headstreams, the Marañón and the Ucayali, rise in central Peru and unite to flow E across Brazil for about 4,000 km/2,500 mi. It has 48,280 km/30,000 mi of navigable waterways, draining 7,000,000 sq km/2,750,000 sq mi, nearly half the South American land mass. It reaches the Atlantic on the equator, its estuary 80 km/50 mi wide, discharging a volume of water so immense that 64 km/40 mi out to sea, fresh water remains at the surface. The Amazon basin covers 7.5 million sq km/3 million sq mi, of which 5 million sq km/2 million sq mi is tropical forest containing 30% of all known plant and animal species (80,000 known species of trees, 3,000 known species of land vertebrates, 2,000 freshwater fish). It is the wettest region on Earth; average rainfall 2.54 m/8.3 ft a year.

The opening up of the Amazon river basin to settlers from the overpopulated east coast has resulted in a massive burning of tropical forest to create both arable and pasture land. The problems of soil erosion, the disappearance of potentially useful plant and animal species, and the possible impact of large-scale forest clearance on global warming of the atmosphere have become environmental issues of international concern.

Brazil, with one third of the world's remaining tropical rainforest, has 55,000 species of flowering plant, half of which are only found in Brazilian Amazonia. In June 1990 the Brazilian Satellite Research Institute announced that 8% of the rainforest in the area had been destroyed by deforestation, amounting to 404,000 sq km/155,944 sq mi (nearly the size of Sweden).

Amazonia /ˌæməˈzəuniə/ those regions of Brazil, Colombia, Ecuador, Peru, and Bolivia lying within the basin of the Amazon River.

Ambala /əmˈbɑːlə/ or *Umballa* city and railway junction in N India, situated 176 km/110 mi NW of Delhi; population (1981) 121,200. Food processing, flour milling, and cotton ginning are the foremost industries. It is an archaeological site with prehistoric artefacts.

Amboina /æmˈbɔɪnə/ or *Ambon* small island in the Moluccas, Republic of Indonesia; population (1980) 209,000. The town of Amboina, formerly a historic centre of Dutch influence, has shipyards.

America /əˈmerɪkə/ western hemisphere of the Earth, containing the continents of North America and South America, with Central America in between. This great land mass extends from the Arctic to the Antarctic, from beyond 75° N to past 55° S. The area is about 42,000,000 sq km/16,000,000 sq mi, and the estimated population is over 500 million.

The name America is derived from Amerigo Vespucci, the Florentine navigator who was falsely supposed to have been the first European to reach the American mainland 1497. The name is also popularly used to refer to the United States of America, a usage which many Canadians, South Americans, and other non-US Americans dislike.

American Samoa see ⬦Samoa, American.

Amersfoort /'ɑːməzfɔːt/ town in the Netherlands, 19 km/12 mi NE of Utrecht; population (1991) 102,000. Industries include brewing, chemicals, and light engineering.

Amiens ancient city of NE France at the confluence of the rivers Somme and Avre; capital of Somme *département* and centre of a market-gardening region irrigated by canals; population (1990) 136,200. It has a magnificent Gothic cathedral with a spire 113 m/370 ft high and gave its name to the battles of Aug 1918, when British field marshal Douglas Haig launched his victorious offensive in World War I.

Amman /əˈmɑːn/ capital and chief industrial centre of Jordan; population (1986) 1,160,000. It is a major communications centre, linking historic trade routes across the Middle East.

Amman is built on the site of the Old Testament Rabbath-Ammon (Philadelphia), capital of the Ammonites.

Amoy /əˈmɔɪ/ ancient name for ⟩Xiamen, a port in SE China.

Amritsar /æmˈrɪtsə/ industrial city in the Punjab, India; population (1981) 595,000. It is the holy city of Sikhism, with the Guru Nanak University (named after the first Sikh guru) and the Golden Temple, from which armed demonstrators were evicted by the Indian army under General Dayal 1984, 325 being killed. Subsequently, Indian prime minister Indira Gandhi was assassinated in reprisal. In 1919 the city was the scene of the Amritsar Massacre.

Amsterdam /ˈæmstədæm/ capital of the Netherlands; population (1990) 695,100. Canals cut through the city link it with the North Sea and the Rhine, and as a Dutch port it is second only to Rotterdam. There is shipbuilding, printing, food processing, banking, and insurance.

Art galleries include the Rijksmuseum, Stedelijk, Vincent van Gogh Museum, and the Rembrandt house. Notable also are the Royal Palace 1655 and the Anne Frank house.

Amu Darya /æˈmuː dɑːriˈɑː/ (formerly *Oxus*) river in central Asia, flowing 2,530 km/1,578 mi from the Pamirs to the Aral Sea.

… rolls its stately burden down from a hoar antiquity through the legends and annals of the East.

On the **Amu Darya** George Nathaniel Curzon *Russia in Central Asia* 1889

Amur /əˈmuə/ river in E Asia. Formed by the Argun and Shilka rivers, the Amur enters the Sea of Okhotsk. At its mouth at Nikolevsk it is 16 km/10 mi wide. For much of its course of over 4,400 km/2,730 mi it forms, together with its tributary, the Ussuri, the boundary between Russia and China.

Under the treaties of Aigun 1858 and Peking 1860, 984,200 sq km/380,000 sq mi of territory N and E of the two rivers were ceded by China to the tsarist government. From 1963 China raised the question of its return and there have been border clashes.

Anaconda /ˌænəˈkɒndə/ town in Montana, USA, which had the world's largest copper smelter (closed 1980); population (1990) 10,278. The city was founded as Copperopolis 1883 by the Anaconda Copper Mining Company, and was incorporated as Anaconda 1888. The town is 1,615 m/5,300 ft above sea level and 42 km/26 mi NW of Butte.

Anaheim /ˈænəhaɪm/ city in SW California, USA, southeast of Los Angeles; industries include electronic and farm equipment and processed foods; population (1990) 266,406. Disneyland amusement park is here. Anaheim was settled by German immigrants 1858 as a wine-producing community.

Anatolia /ˌænəˈtəʊliə/ (Turkish *Anadolu*) Asian part of Turkey, consisting of a mountainous peninsula with the Black Sea to the N, the Aegean Sea to the W, and the Mediterranean Sea to the S.

Anchorage /ˈæŋkərɪdʒ/ port and largest city of Alaska, USA, at the head of Cook Inlet; population (1990) 226,340. Established 1918, Anchorage is an important centre of administration, communication, and commerce. Oil and gas extraction and fish canning are also important to the local economy.

Ancona /ænˈkəʊnə/ Italian town and naval base on the Adriatic Sea, capital of Marche region; population (1988) 104,000. It has a Romanesque cathedral and a former palace of the popes.

Andalusia /ˌændəˈluːsiə/ (Spanish *Andalucía*) fertile autonomous region of S Spain, including the provinces of Almería, Cádiz, Córdoba, Granada, Huelva, Jaén, Málaga, and Seville; area 87,300 sq km/33,698 sq mi; population (1986) 6,876,000. Málaga, Cádiz, and Algeciras are the chief ports and industrial centres. The *Costa del Sol* on the south coast has many tourist resorts, including Marbella and Torremolinos.

Andalusia has Moorish architecture, having been under Muslim rule from the 8th to the 15th centuries.

Andaman and Nicobar Islands /ˈændəmən, ˈnɪkəbɑː/ two groups of islands in the Bay of Bengal, between India and Myanmar, forming a Union Territory of the Republic of India; capital Port Blair; area 8,300 sq km/3,204 sq mi; population (1991) 278,000. The economy is based on fishing, timber, rubber, fruit, and rice.

The *Andamans* consist of five principal islands (forming the Great Andaman), the Little Andaman, and about 204 islets; area 6,340 sq km/2,447 sq mi; population (1981) 158,000. They were used as a penal settlement 1857–1942. The *Nicobars*, consisting of 19 islands (7 of which are uninhabited), are 120 km/75 mi S of Little Andaman; area 1,953 sq km/754 sq mi; population (1981) 30,500. The main items of trade are coconut and areca nut. The Nicobars were British 1869–1947.

Anderson /ˈændəsn/ city in E central Indiana, USA, northeast of Indianapolis; seat of Madison County; population (1990) 59,500. Industries include car accessories and paper products.

Andes /ˈændiːz/ great mountain system or *cordillera* that forms the western fringe of South America, extending through some 67° of latitude and the republics of Colombia, Venezuela, Ecuador, Peru, Bolivia, Chile, and Argentina. The mountains exceed 3,600 m/12,000 ft for half their length of 6,500 km/4,000 mi.

Geologically speaking, the Andes are new mountains, having attained their present height by vertical upheaval of the entire strip of the Earth's crust as recently as the latter part of the Tertiary era and the Quaternary. But they have been greatly affected by weathering; rivers have cut deep gorges, and glaciers have produced characteristic valleys. The majority of the individual mountains are volcanic; some are still active.

The whole system may be divided into two almost parallel ranges. The southernmost extremity is Cape Horn, but the range extends into the sea and forms islands. Among the highest peaks are Cotopaxi and Chimborazo in Ecuador, Cerro de Pasco and Misti in Peru, Illampu and Illimani in Bolivia, Aconcagua (the highest mountain in the New World) in Argentina, and Ojos del Salado in Chile.

Andean mineral resources include gold, silver, tin, tungsten, bismuth, vanadium, copper, and lead. Difficult communications make mining expensive. Transport for a long time was chiefly by pack animals, but air transport has greatly reduced difficulties of communications. Three railways cross the Andes from Valparaiso to Buenos Aires, Antofagasta to Salta, and Antofagasta via Uyuni to Asunción. New roads are being built, including the Pan-American Highway.

The majority of the sparse population is dependent on agriculture, the nature and products of which vary with the natural environment. Newcomers to the Andean plateau, which includes Lake Titicaca, suffer from *puna*, mountain sickness, but indigenous peoples have hearts and lungs adapted to altitude.

Andhra Pradesh /ˈændrə prɑːˈdeʃ/ state in E central India
area 276,700 sq km/106,845 sq mi
capital Hyderabad
towns Secunderabad
products rice, sugar cane, tobacco, groundnuts, cotton
population (1991) 66,304,900
languages Telugu, Urdu, Tamil
history formed 1953 from the Telegu-speaking areas of Madras, and enlarged 1956 from the former Hyderabad state.

Andhra Pradesh

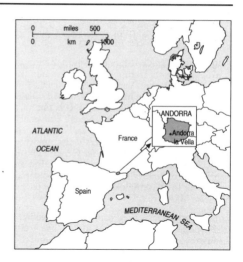

Andorra Principality of (*Principat d'Andorra*)
area 468 sq km/181 sq mi
capital Andorra-la-Vella
towns Les Escaldes
physical mountainous, with narrow valleys
features the E Pyrenees, Valira River
heads of state Joan Marti i Alanis (bishop of Urgel, Spain) and François Mitterrand (president of France)
head of government Oscar Riba Reig from 1989
political system semi-feudal co-principality
political party Democratic Party of Andorra
exports main industries tourism and tobacco
currency French franc and Spanish peseta
population (1990) 51,000 (30% Andorrans, 61% Spanish, 6% French)
languages Catalan (official); Spanish, French
religion Roman Catholic
literacy 100% (1987)
GDP $300 million (1985)
chronology
1278 Treaty signed making Spanish bishop and French count joint rulers of Andorra (through marriage the king of France later inherited the count's right).
1970 Extension of franchise to third-generation women and second-generation men.
1976 First political organization (Democratic Party of Andorra) formed.
1977 Franchise extended to first-generation Andorrans.
1981 First prime minister appointed by General Council.
1982 With the appointment of an Executive Council, executive and legislative powers were separated.
1991 Andorra's first constitution planned; links with European Community (EC) formalized.

It is too extraordinary a place to be invaded. Let it stand forever as it is, a museum piece.

On **Andorra** Napoleon

Aneto, Pico de /æ,netəu'piːkəu/ highest peak of the Pyrenees mountains, rising to 3,404 m/11,168 ft in the Spanish province of Huesca.

Angel Falls /'eɪndʒəl/ highest waterfalls in the world, on the river Caroní in the tropical rainforest of Bolívar Region, Venezuela; total height 978 m/3,210 ft. They were named after the aviator and prospector James Angel who flew over the falls and crash-landed nearby 1935.

Angers /,ɒn'ʒeɪ/ ancient French town, capital of Maine-et-Loire *département*, on the river Maine; population (1990) 146,100. Products include electrical machinery and Cointreau liqueur. It has a 12th–13th-century cathedral and castle and was formerly the capital of the duchy and province of Anjou.

Angkor /'æŋkɔː/ site of the ancient capital of the Khmer Empire in NW Cambodia, north of Tonle Sap. The remains date mainly from the 10th-12th centuries AD, and comprise temples originally dedicated to the Hindu gods, shrines associated with Theravada Buddhism, and royal palaces. Many are grouped within the enclosure called *Angkor Thom*, but the great temple of *Angkor Wat* (early 12th century) lies outside.

Angkor was abandoned in the 15th century, and the ruins were overgrown by jungle and not adequately described until 1863. Buildings on the site suffered damage during the civil war 1970–75.

Angkor is perhaps the greatest of Man's essays in rectangular architecture that has yet been brought to light...

On **Angkor** Arnold Toynbee
East to West 1958

Anglesey /'æŋgəlsi/ (Welsh *Ynys Môn*) island off the northwest coast of Wales; area 720 sq km/278 sq mi; population (1991) 67,800. It is separated from the mainland by the Menai Straits, which are crossed by the Britannia tubular railway bridge and Telford's suspension bridge, built 1819–26 but since rebuilt. It is a holiday resort with rich fauna (notably bird life) and flora, and many buildings and relics of historic interest. The ancient granary of Wales, Anglesey now has industries such as toy-making, electrical goods, and bromine extraction from the sea. Holyhead is the principal town and port; Beaumaris was the county town until the county of Anglesey was merged into Gwynedd 1974.

Angola People's Republic of (*República Popular de Angola*)
area 1,246,700 sq km/481,226 sq mi
capital and chief port Luanda
towns Lobito and Benguela, also ports; Huambo, Lubango
physical narrow coastal plain rises to vast interior plateau with rainforest in NW; desert in S
features Cuanza, Cuito, Cubango, and Cunene rivers; Cabinda enclave

head of state and government José Eduardo dos Santos from 1979
political system socialist republic
political parties People's Movement for the Liberation of Angola–Workers' Party (MPLA–PT), Marxist-Leninist; National Union for the Total Independence of Angola (UNITA); National Front for the Liberation of Angola (FNLA)
exports oil, coffee, diamonds, palm oil, sisal, iron ore, fish
currency kwanza
population (1989 est) 9,733,000 (largest ethnic group Ovimbundu); growth rate 2.5% p.a.
life expectancy men 40, women 44
languages Portuguese (official); Bantu dialects
religions Roman Catholic 68%, Protestant 20%, animist 12%
literacy 20%
GDP $2.7 bn; $432 per head
chronology
1951 Angola became an overseas territory of Portugal.
1956 First independence movement formed, the People's Movement for the Liberation of Angola (MPLA).
1961 Unsuccessful independence rebellion.
1962 Second nationalist movement formed, the National Front for the Liberation of Angola (FNLA).
1966 Third nationalist movement formed, the National Union for the Total Independence of Angola (UNITA).
1975 Independence achieved from Portugal. Transitional government of independence formed from representatives of MPLA, FNLA, UNITA, and Portuguese government. MPLA proclaimed People's Republic of Angola under the presidency of Dr Agostinho Neto. FNLA and UNITA proclaimed People's Democratic Republic of Angola.
1976 MPLA gained control of most of the country. South African troops withdrawn, but Cuban units remained.
1977 MPLA restructured to become the People's Movement for the Liberation of Angola–Workers' Party (MPLA–PT).

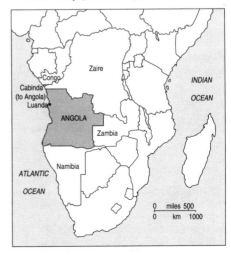

1979 Death of Neto, succeeded by José Eduardo dos Santos.
1980 UNITA guerrillas, aided by South Africa, continued raids against the Luanda government and bases of the South West Africa People's Organization (SWAPO) in Angola.
1984 South Africa promised to withdraw its forces if the Luanda government guaranteed that areas vacated would not by filled by Cuban or SWAPO units (the Lusaka Agreement).
1985 South African forces officially withdrawn.
1986 Further South African raids into Angola. UNITA continued to receive South African support.
1988 Peace treaty, providing for the withdrawal of all foreign troops, signed with South Africa and Cuba.
1989 Cease-fire agreed with UNITA broke down and guerrilla activity restarted.
1990 Peace offer by rebels. Return to multiparty politics promised.
1991 Peace agreement signed, civil war between MPLA–PT and UNITA officially ended. Amnesty for all political prisoners.
1992 MPLA–PT's general-election victory fiercely disputed by UNITA, plunging the country into renewed civil war. UNITA offered, and eventually accepted, seats in the new government, but fighting continued.
1993 Fighting continued. Key city fell to UNITA. MPLA–PT made power-sharing agreement with UNITA.

Angostura /ˌæŋgəˈstjʊərə/ former name of ◊Ciudad Bolívar, a port in Venezuela.

Angoulême /ˌɒŋguːˈleɪm/ French town, capital of the *département* of Charente, on the Charente River; population (1990) 46,100. It has a cathedral, and a castle and paper mills dating from the 16th century.

Anguilla /æŋˈgwɪlə/ island in the E Caribbean
area 160 sq km/62 sq mi
capital The Valley
features white coral-sand beaches; 80% of its coral reef has been lost through tourism (pollution and souvenir sales)
exports lobster, salt

Anguilla

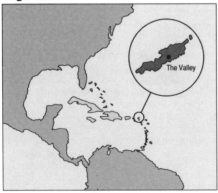

currency Eastern Caribbean dollar
population (1988) 7,000
languages English, Creole
government from 1982, governor, executive council, and legislative house of assembly
history a British colony from 1650, Anguilla was long associated with St Christopher–Nevis but revolted against alleged domination by the larger island and in 1969 declared itself a republic. A small British force restored order, and Anguilla retained a special position at its own request; since 1980 it has been a separate dependency of the UK.

Angus /ˈæŋgəs/ former county and modern district on the east coast of Scotland, merged 1975 into Tayside Region; population (1991) 92,900.

Anhui /ˌænˈhweɪ/ or **Anhwei** province of E China, watered by the Chang Jiang (Yangtze River)
area 139,900 sq km/54,000 sq mi
capital Hefei
products cereals in the N; cotton, rice, tea in the S
population (1990) 56,181,000.

Anhwei /ˌænˈhweɪ/ alternative spelling of ◊Anhui.

Ankara /ˈæŋkərə/ (formerly **Angora**) capital of Turkey; population (1990) 2,559,500. Industries include cement, textiles, and leather products. It replaced Istanbul (then in Allied occupation) as capital 1923.
It has the presidential palace and Grand National Assembly buildings; three universities, including a technical university to serve the whole Middle East; the Atatürk mausoleum on a nearby hilltop; and the largest mosque in Turkey at Kocatepe.

Annaba /ˈænəbə/ (formerly **Bône**) seaport in Algeria; population (1983) 348,000. The name means 'city of jujube trees'. There are metallurgical industries, and iron ore and phosphates are exported.

Annapolis /əˈnæpəlɪs/ seaport and capital of Maryland, USA; population (1984) 31,900. It was named after Princess (later Queen) Anne 1695.
It was in session here Nov 1783–June 1784 that Congress received George Washington's resignation as commander in chief 1783, and ratified the peace treaty of the War of American Independence. The US Naval Academy is here, and the naval hero John Paul Jones is buried in the chapel crypt.

Annapurna /ˌænəˈpɜːnə/ mountain 8,075 m/26,502 ft in the Himalayas, Nepal. The north face was first climbed by a French expedition (Maurice Herzog) 1950 and the south by a British team 1970.

Ann Arbor /ˌæn ˈɑːbə/ city in SE Michigan, USA, west of Dearborn and Detroit, on the Huron River; seat of Washtenaw county; population

(1990) 109,600. It is a centre for medical, aeronautical, nuclear, and chemical research, and the site of the University of Michigan (1837).

Annecy /æn'si:/ capital of the *département* of Haute-Savoie, SE France, at the northern end of Lake Annecy; population (1990) 51,100. It has some light industry, including precision instruments, and is a tourist resort.

It is like living in a picture postcard, especially when there is full sunshine.

On **Annecy** Arnold Bennett *Journal* 1928

Anniston /'ænɪstən/ city in E Alabama, USA, northeast of Birmingham; seat of Calhoun county; population (1990) 26,600. The site of iron mines, its industries include iron products as well as textiles and chemicals.

Annobón /ˌænəu'bɒn/ island in Equatorial Guinea; area 17 sq km/7 sq mi; population (1984 est) 3,000. Its inhabitants are descended from slaves of the Portuguese and still speak a form of that language. From 1973–79 the island was called *Pagalu*.

Anshan /ˌæn'ʃæn/ Chinese city in Liaoning province, 89 km/55 mi SE of Shenyang (Mukden); population (1986) 1,280,000. The iron and steel centre started here 1918, was expanded by the Japanese, dismantled by the Russians, and restored by the Communist government of China. It produces 6 million tonnes of steel annually.

Antakya /æn'tɑ:kjə/ or *Hatay* city in SE Turkey, site of the ancient Antioch, capital of Syria 3rd–1st century BC and an early Christian centre; population (1990) 378,200.

Antalya /æn'tɑ:ljə/ Mediterranean port on the west coast of Turkey and capital of a province of the same name; population (1990) 378,200. The port trades in agricultural and forest produce.

Antananarivo /ˌæntəˌnænə'ri:vəu/ (formerly *Tananarive*) capital of Madagascar, on the interior plateau, with a rail link to Tamatave; population (1986) 703,000. Industries include tobacco, food processing, leather goods, and clothing.

Antarctica /ænt'ɑ:ktɪkə/ continent surrounding the South Pole, arbitrarily defined as the region lying S of the Antarctic Circle. Occupying 10% of the world's surface, Antarctica contains 90% of the world's ice and 70% of its fresh water
area 13,900,000 sq km/5,400,000 sq mi (the size of Europe and the USA combined)
features Mount Erebus on Ross Island is the world's southernmost active volcano; the Ross Ice Shelf is formed by several glaciers coalescing in the Ross Sea
highest point Vinson Massif (5,139 m/16,866 ft)
physical formed of two blocs of rock with an area of about 8 million sq km/3 million sq mi, Antarctica is covered by a cap of ice that flows slowly towards its 22,400 km/14,000 mi coastline,

reaching the sea in high ice cliffs. The most southerly shores are near the 78th parallel in the Ross and Weddell seas. E Antarctica is a massive bloc of ancient rocks that surface in the Transantarctic Mountains of Victoria Land. Separated by a deep channel, W Antarctica is characterized by the mountainous regions of Graham Land, the Antarctic Peninsula, Palmer Land, and Ellsworth Land. Little more than 1% of the land is ice-free. With an estimated volume of 24 million cu m/5.9 million cu mi, the ice-cap has a mean thickness of 1,880 m/6,170 ft and in places reaches depths of 5,000 m/16,000 ft or more. Each annual layer of snow preserves a record of global conditions
climate winds are strong and temperatures are cold, particularly in the interior where temperatures can drop to −70°C/−100°F and below. Precipitation is largely in the form of snow or hoar-frost rather than rain which rarely exceeds 50 mm/2 in per year (less than the Sahara Desert)
flora and fauna the Antarctic ecosystem is characterized by large numbers of relatively few species of higher plants and animals, and a short food chain from tiny marine plants to whales, seals, penguins, and other sea birds
products cod, Antarctic icefish, and krill are fished in Antarctic waters. Whaling, which began in the early 20th century, ceased during the 1960s as a result of overfishing, although Norway and Iceland defied the ban 1992. Petroleum, coal, and minerals, such as palladium and platinum exist, but their exploitation is prevented by a 50-year ban on commercial mining agreed by 39 nations 1991
population no permanent residents; settlement limited to scientific research stations with maximum population of 2,000 to 3,000 during the summer months. Sectors of Antarctica are claimed by Argentina, Australia, Chile, France, the UK, Norway, and New Zealand.

... no other phenomenon on Earth demonstrates so clearly man's complete and utter impotence ...

On **Antarctica** Jules Dumont d'Urville 1838

Antarctic Ocean /ænt'ɑ:ktɪk/ popular name for the reaches of the Atlantic, Indian, and Pacific oceans extending S of the Antarctic Circle (66° 32′S). The term is not used by the International Hydrographic Bureau.

Antarctic Peninsula /ænt'ɑ:ktɪk/ mountainous peninsula of W Antarctica extending 1,930 km/1,200 mi N toward South America; originally named *Palmer Land* after a US navigator, Captain Nathaniel Palmer, who was the first to explore the region 1820. It was claimed by Britain 1832, Argentina 1940, and Chile 1942. Its name was changed to the Antarctic Peninsula 1964.

Antibes /ɒn'ti:b/ resort, which includes Juan les Pins, on the French Riviera, in the *département* of Alpes Maritimes; population (1990) 70,600. There is a Picasso collection in the 17th-century castle museum.

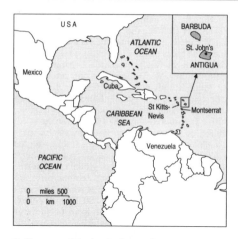

Antigua and Barbuda State of

area Antigua 280 sq km/108 sq mi, Barbuda 161 sq km/62 sq mi, plus Redonda 1 sq km/0.4 sq mi

capital and chief port St John's

towns Codrington (on Barbuda)

physical low-lying tropical islands of limestone and coral with some higher volcanic outcrops; no rivers and low rainfall result in frequent droughts and deforestation

features Antigua is the largest of the Leeward Islands; Redonda is an uninhabited island of volcanic rock rising to 305 m/1,000 ft

head of state Elizabeth II from 1981 represented by governor general

head of government Vere C Bird from 1981

political system liberal democracy

political parties Antigua Labour Party (ALP), moderate, left of centre; Progressive Labour Movement (PLM), left of centre

exports sea-island cotton, rum, lobsters

currency Eastern Caribbean dollar

population (1989) 83,500; growth rate 1.3% p.a.

life expectancy 70 years

language English

media no daily newspaper; weekly papers all owned by political parties

religion Christian (mostly Anglican)

literacy 90% (1985)

GDP $173 million (1985); $2,200 per head

chronology

1493 Antigua visited by Christopher Columbus.

1632 Antigua colonized by English settlers.

1667 Treaty of Breda formally ceded Antigua to Britain.

1871–1956 Antigua and Barbuda administered as part of the Leeward Islands federation.

1967 Antigua and Barbuda became an associated state within the Commonwealth, with full internal independence.

1971 PLM won the general election by defeating the ALP.

1976 PLM called for early independence, but ALP urged caution. ALP won the general election.

1981 Independence from Britain achieved.

1983 Assisted US invasion of Grenada.

1984 ALP won a decisive victory in the general election.

1985 ALP re-elected.

1989 Another sweeping general election victory for the ALP under Vere Bird.

1991 Bird remained in power despite calls for his resignation.

Anti-Lebanon /ˌænti'lebənən/ or *Antilibanus* mountain range on the Lebanese-Syrian border, including Mount Hermon, 2,800 m/9,200 ft. It is separated from the Lebanon Mountains by the Bekaa Valley.

Antilles /æn'tɪliːz/ group of West Indian islands, divided N–S into the *Greater Antilles* (Cuba, Jamaica, Haiti–Dominican Republic, Puerto Rico) and *Lesser Antilles*, subdivided into the Leeward Islands (Virgin Islands, St Christopher–Nevis, Antigua and Barbuda, Anguilla, Montserrat, and Guadeloupe) and the Windward Islands (Dominica, Martinique, St Lucia, St Vincent and the Grenadines, Barbados, and Grenada).

Antofagasta /ˌæntəfə'gæstə/ port of N Chile, capital of the region of Antofagasta; population (1990) 218,800. The area of the region is 125,300 sq km/48,366 sq mi; its population (1982) 341,000. Nitrates from the Atacama Desert are exported.

Antrim /'æntrɪm/ county of Northern Ireland

area 2,830 sq km/1,092 sq mi

towns Belfast (county town), Larne (port)

features Giant's Causeway of natural hexagonal basalt columns, which, in legend, was built to enable the giants to cross between Ireland and Scotland; Antrim borders Lough Neagh, and is separated from Scotland by the North Channel, 30 km/20 mi wide

products potatoes, oats, linen, synthetic textiles

population (1981) 642,000.

Antwerp /'æntwɜːp/ (Flemish *Antwerpen,* French *Anvers*) port in Belgium on the river Scheldt, capital of the province of Antwerp; population (1991) 467,500. One of the world's busiest ports, it has shipbuilding, oil-refining, petrochemical, textile, and diamond-cutting industries. The home of the artist Rubens is preserved, and many of his works are in the Gothic

Antilles

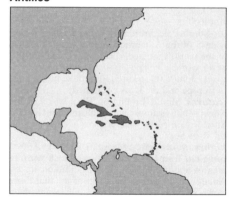

cathedral. The province of Antwerp has an area of 2,900 sq km/1,119 sq mi; population (1987) 1,588,000.

history It was not until the 15th century that Antwerp rose to prosperity; from 1500 to 1560 it was the richest port in N Europe. After this Antwerp was beset by religious troubles and the Netherlands' revolt against Spain. In 1648 the Treaty of Westphalia gave both shores of the Scheldt estuary to the United Provinces, which closed it to Antwerp trade. The Treaty of Paris 1814 opened the estuary to all nations on payment of a small toll to the Dutch, abandoned 1863. During World War I Antwerp was occupied by Germany Oct 1914–Nov 1918; during World War II, May 1940–Sept 1944.

Anuradhapura /ə'nuərədəpuərə/ ancient holy city in Sri Lanka; population (1981) 36,000. It was the capital of the Sinhalese kings of Sri Lanka 5th century BC–8th century AD; rediscovered in the mid-19th century. Sacred in Buddhism, it claims a Bo tree descended from the one under which the Buddha became enlightened.

Anvers /ɒŋ'veə/ French form of ⋄Antwerp, a province in N Belgium.

Anyang /ˌæn'jæŋ/ city in Henan province, E China; population (1986) 550,000. It was the capital of the Shang dynasty (13th–12th centuries BC). Rich archaeological remains have been uncovered since the 1930s.

Anzhero-Sudzhensk /æn'ʒeərəu 'suːdʒənsk/ town in W Siberian Russia, 80 km/50 mi N of Kemerovo in the Kuznetsk basin; population (1985) 110,000. Its chief industry is coal mining.

Aomori /'aumɒri/ port at the head of Mutsu Bay, on the north coast of Honshu Island, Japan; 40 km/25 mi NE of Hirosaki; population (1990) 287,800.

The port handles a large local trade in fish, rice, and timber.

Aosta /ɑː'ɒstə/ Italian city, 79 km/49 mi NW of Turin; population (1981) 37,200. It is the capital of Valle d'Aosta (French-speaking) autonomous region, and has extensive Roman remains.

Aotearoa /'æɒˌtɪəˌɒːə/ (Maori 'land of the long white cloud') Maori name for ⋄New Zealand.

Aouzu Strip /ɑː'uːzuː/ disputed territory 100 km/ 60 mi wide on the Chad–Libya frontier, occupied by Libya 1973. Lying to the N of the Tibesti massif, the area is rich in uranium and other minerals.

Apeldoorn /'ɑːpəldɔːn/ commercial city in Gelderland province, E central Netherlands; population (1991) 148,200. Het Loo, which is situated nearby, has been the summer residence of the Dutch royal family since the time of William of Orange.

Apennines /'æpənaɪnz/ chain of mountains stretching the length of the Italian peninsula. A continuation of the Maritime Alps, from Genoa it swings across the peninsula to Ancona on the east coast, and then back to the west coast and into

the 'toe' of Italy. The system is continued over the Strait of Messina along the N Sicilian coast, then across the Mediterranean Sea in a series of islands to the Atlas Mountains of N Africa. The highest peak is Gran Sasso d'Italia at 2,914 m/9,560 ft.

Apia /'ɑːpiə/ capital and port of Western Samoa, on the north coast of Upolu Island, in the W Pacific; population (1981) 33,000. It was the final home of the writer Robert Louis Stevenson from 1888–94.

Apo, Mount /'ɑːpəu/ active volcano and highest peak in the Philippines, rising to 2,954 m/9,692 ft on the island of Mindanao.

Appalachians /ˌæpə'leɪtʃənz/ mountain system of E North America, stretching about 2,400 km/ 1,500 mi from Alabama to Québec, composed of very ancient eroded rocks. The chain includes the Allegheny, Catskill, and Blue Ridge mountains, the last-named having the highest peak, Mount Mitchell, 2,045 m/6,712 ft. The eastern edge has a fall line to the coastal plain where Philadelphia, Baltimore, and Washington stand.

Appleton /'æpltən/ city in E central Wisconsin, USA, northwest of Oshkosh, on the Fox River; seat of Outagamie County; population (1990) 65,700. It is a manufacturing centre for paper products. Founded 1847, it claims to have the world's first hydroelectric plant, built 1882.

Apulia /'puːljə/ English form of ⋄Puglia, a region of Italy.

Aqaba, Gulf of /'ækəbə/ gulf extending for 160 km/100 mi between the Negev and the Red Sea; its coastline is uninhabited except at its head, where the frontiers of Israel, Egypt, Jordan, and Saudi Arabia converge. The two ports of Eilat (Israeli 'Elath') and Aqaba, Jordan's only port, are situated here.

Aquitaine /ˌækwɪ'teɪn/ region of SW France; capital Bordeaux; area 41,300 sq km/15,942 sq mi; population (1986) 2,718,000. It comprises the *départements* of Dordogne, Gironde, Landes, Lot-et-Garonne, and Pyrénées-Atlantiques. Red wines (Margaux, St Julien) are produced in the Médoc district, bordering the Gironde. Aquitaine was an English possession 1152–1452.

history Early human remains have been found in the Dordogne region. Aquitaine coincides roughly with the Roman province of Aquitania and the ancient French province of Aquitaine. Eleanor of Aquitaine married the future Henry II of England 1152 and brought it to him as her dowry; it remained in English hands until 1452.

AR abbreviation for ⋄*Arkansas*, a state of the USA.

Arab Emirates see ⋄United Arab Emirates.

Arabia /ə'reɪbiə/ peninsula between the Persian Gulf and the Red Sea, in SW Asia; area 2,600,000 sq km/1,000,000 sq mi. The peninsula contains the world's richest oil and gas reserves. It comprises the states of Bahrain, Kuwait, Oman, Qatar, Saudi Arabia, the United Arab Emirates, and Yemen.

physical A sandy coastal plain of varying width borders the Red Sea, behind which a mountain chain rises to about 2,000–2,500 m/ 6,600–8,200 ft. Behind this range is the plateau of the Nejd, averaging 1,000 m/3,300 ft. The interior comprises a vast desert area: part of the Al-Hamad (Syrian) Desert in the far north, Nafud in N Saudi Arabia, and Rub'al Khali in S Saudi Arabia.

history The Arabian civilization was revived by Muhammad during the 7th century, but in the new empire created by militant Islam, Arabia became a subordinate state, and its cities were eclipsed by Damascus, Baghdad, and Cairo. Colonialism only touched the fringe of Arabia in the 19th century, and until the 20th century the interior was unknown to Europeans. Nationalism began actively to emerge at the period of World War I (1914–18), and the oil discoveries from 1953 gave the peninsula significant economic power.

Arabian Gulf /ə'reɪbɪən/ another name for the ◊Persian Gulf.

Arabian Sea northwestern branch of the ◊Indian Ocean.

Arabistan /ˌærəbɪ'stɑːn/ former name of the Iranian province of Khuzestan, revived in the 1980s by the 2 million Sunni Arab inhabitants who demand autonomy. Unrest and sabotage 1979–80 led to a pledge of a degree of autonomy by Ayatollah Khomeini.

Arad /'æræd/ Romanian town on the river Mures, 160 km/100 mi NE of Belgrade; population (1985) 185,900. It is a major trading centre with many industries.

Arafura Sea /ˌærə'fuərə/ area of the Pacific Ocean between N Australia and Indonesia, bounded by the Timor Sea in the W and the Coral Sea in the E. It is 1,290 km/800 mi long and 560 km/ 350 mi wide.

Aragon /'ærəgən/ autonomous region of NE Spain including the provinces of Huesca, Teruel, and Zaragoza; area 47,700 sq km/18,412 sq mi; population (1986) 1,215,000. Its capital is Zaragoza, and products include almonds, figs, grapes, and olives. Aragon was an independent kingdom 1035–1479.

history A Roman province until taken in the 5th century by the Visigoths, who lost it to the Moors in the 8th century, it became a kingdom 1035. It was united with Castile 1479 under Ferdinand and Isabella.

The name alone is enough to fill a man with delight and to magnify him with the story of twelve hundred years.

On **Aragon** Hilaire Belloc *Many Cities* 1928

Arakan /ˌærə'kɑːn/ state of Myanmar (formerly Burma) on the Bay of Bengal coast, some 645 km/ 400 mi long and strewn with islands; population

(1983) 2,046,000. The chief town is Sittwe. It is bounded along its eastern side by the Arakan Yoma, a mountain range rising to 3,000 m/ 10,000 ft. The ancient kingdom of Arakan was conquered by Burma 1785.

Aral Sea /ɑːrəl/ inland sea divided between Kazakhstan and Uzbekistan, the world's fourth largest lake; former area 62,000 sq km/ 24,000 sq mi, but decreasing. Water from its tributaries, the Amu Darya and Syr Darya, has been diverted for irrigation and city use, and the sea is disappearing, with long-term consequences for the climate.

Between 1960 and 1990 the water level dropped 13 m/40 ft, reducing the lake to two-thirds of its original area. Between 1989 and 1991 alone, its area was cut by nearly half, from 63,800 sq km/ 24,626 sq mi to approximately 36,000 sq km/ 13,896 sq mi. All native fish species have disappeared, and winds drop 43 million tonnes of salt a year on the surrounding cropland.

Aran Islands /'æran/ three rocky islands (Inishmore, Inishmaan, Inisheer) in the mouth of Galway Bay, Republic of Ireland; population approximately 4,600. The capital is Kilronan. J M Synge used the language of the islands in his plays.

Aranjuez /ˌærəŋ'xweθ/ Spanish town on the river Tagus, 40 km/25 mi SE of Madrid; population (1991) 35,900. The palace was a royal residence for centuries.

Ararat /'ærəræt/ double-peaked mountain on the Turkish-Iranian border; the higher, Great Ararat, 5,137 m/16,854 ft, was the reputed resting place of Noah's Ark after the Flood.

Ararat /'ærəræt/ wheat and wool centre in NW Victoria, Australia; population (1986) 8,000. It is a former gold-mining town.

Arbil /'ɑːbɪl/ Kurdish town in a province of the same name in N Iraq; population (1985) 334,000. Occupied since Assyrian times, it was the site of a battle 331 BC at which Alexander the Great defeated the Persians under Darius III. In 1974 Arbil became the capital of a Kurdish autonomous region set up by the Iraqi government.

Arbroath /ɑː'brəuθ/ fishing town in Tayside, Scotland; population (1981) 24,100. In 1320 the Scottish Parliament asserted Scotland's independence here in a letter to the pope.

Arcadia /ɑː'keɪdɪə/ (Greek *Arkadhia*) central plateau of S Greece; area 4,419 sq km/ 1,706 sq mi; population (1981) 108,000. Tripolis is the capital town.

The English poet Philip Sidney idealized the life of shepherds here in antiquity.

Archangel /'ɑːkeɪndʒəl/ (Russian *Arkhangel'sk*) port in N Russia; population (1987) 416,000. Although blocked by ice during half the year, it is the chief timber-exporting port of Russia. Archangel was made an open port by Boris Godunov and was of prime importance until Peter the Great built St Petersburg.

It was used 1918–20 by the Allied interventionist armies in collaboration with the White Army in their effort to overthrow the newly established Soviet state. In World War II it was the receiving station for Anglo-American supplies. Plesetsk, to the S, is a launch site for crewed space flight.

Arctic, the /ɑːktɪk/ that part of the northern hemisphere surrounding the North Pole; arbitrarily defined as the region lying N of the Arctic Circle (66° 32′N) or N of the tree line. There is no Arctic continent; the greater part of the region comprises the Arctic Ocean, which is the world's smallest ocean. Arctic climate, fauna, and flora extend over the islands and northern edges of continental land masses that surround the Arctic Ocean (Svalbard, Iceland, Greenland, Siberia, Scandinavia, Alaska, and Canada)
area 36,000,000 sq km/14,000,000 sq mi
physical pack-ice floating on the Arctic Ocean occupies almost the entire region between the North Pole and the coasts of North America and Eurasia, covering an area that ranges in diameter from 3,000 km/1,900 mi to 4,000 km/2,500 mi. The pack-ice reaches a maximum extent in Feb when its outer limit (influenced by the cold Labrador Current and the warm Gulf Stream) varies from 50°N along the coast of Labrador to 75°N in the Barents Sea N of Scandinavia. In spring the pack-ice begins to break up into ice floes which are carried by the south-flowing Greenland Current to the Atlantic Ocean. Arctic ice is at its minimum area in Aug. The greatest concentration of icebergs in Arctic regions is found in Baffin Bay. They are derived from the glaciers of W Greenland, then carried along Baffin Bay and down into the N Atlantic where they melt off Labrador and Newfoundland.

The Bering Straits are icebound for more than six months each year, but the Barents Sea between Scandinavia and Svalbard is free of ice and is navigable throughout the year. Arctic coastlines, which have emerged from the sea since the last Ice Age, are characterized by deposits of gravel and disintegrated rock. The area covered by the Arctic icecap shrank 2% 1978–87
climate permanent ice sheets and year-round snow cover are found in regions where average monthly temperatures remain below 0°C/32°F, but on land areas where one or more summer months have average temperatures between freezing point and 10°C/50°F, a stunted, treeless tundra vegetation is found. Mean annual temperatures range from –23°C at the North Pole to –12°C on the coast of Alaska. In winter the Sun disappears below the horizon for a time, but the cold is less severe than in parts of inland Siberia or Antartica. During the short summer season there is a maximum of 24 hours of daylight at the summer solstice on the Arctic Circle and six months' constant light at the North Pole. Countries with Arctic coastlines established the International Arctic Sciences Committee 1987 to study ozone depletion and climatic change
flora and fauna the plants of the relatively infertile Arctic tundra (lichens, mosses, grasses, cushion plants, and low shrubs) spring to life during the short summer season and remain dormant

for the remaining ten months of the year. There are no annual plants, only perennials. Animal species include reindeer, caribou, musk ox, fox, hare, lemming, wolf, polar bear, seal, and walrus. There are few birds except in summer when insects, such as mosquitoes, are plentiful
natural resources coal (Svalbard, Russia), oil and natural gas (Alaska, Canadian Arctic, Russia), and mineral resources including gold, silver, copper, uranium, lead, zinc, nickel, and bauxite. Because of climatic conditions, the Arctic is not suited to navigation and the exploitation of these resources. Murmansk naval base on the Kola Peninsula is the largest in the world
population there are about 1 million aboriginal people including the Aleuts of Alaska, North American Indians, the Lapps of Scandinavia and Russia, the Yakuts, Samoyeds, Komi, Chukchi, Tungus, and Dolgany of Russia, and the Inuit of Siberian Russia, the Canadian Arctic, and Greenland.

... the North alone is silent and at peace. Give man time and he will spoil that too.

On the **Arctic** Stephen Leacock *My Discovery of the West* 1937

Arctic Ocean ocean surrounding the North Pole; area 14,000,000 sq km/5,400,000 sq mi. Because of the Siberian and North American rivers flowing into it, it has comparatively low salinity and freezes readily.

The ocean comprises:
Beaufort Sea off Canada/Alaska coast, named after British admiral Francis Beaufort; oil drilling allowed only in winter because the sea is the breeding and migration route of the bowhead whales, staple diet of the local Inuit people;
Greenland Sea between Greenland and Svalbard;
Norwegian Sea between Greenland and Norway.
From W to E along the north coast of Russia:
Barents Sea named after Willem Barents, which has oil and gas reserves and was strategically significant as the meeting point of the NATO and Warsaw Pact forces. The White Sea is its southernmost part;
Kara Sea renowned for bad weather and known as the 'great ice cellar';
Laptev Sea between Taimyr Peninsula and New Siberian Island;
East Siberian Sea and **Chukchi Sea** between Russia and the USA; the seminomadic Chukchi people of NE Siberia finally accepted Soviet rule in the 1930s.
The Arctic Ocean has the world's greatest concentration of nuclear submarines, but at the same time there is much scientific cooperation on exploration, especially since Russia needs Western aid to develop oil and gas in its areas.

Ardabil /ɑːdəˈbiːl/ town in NW Iran, near the frontier with Azerbaijan; population (1986) 282,000. Ardabil exports dried fruits, carpets, and rugs.

Ardèche /ɑːˈdeʃ/ river in SE France, a tributary of the Rhône. Near Vallon it flows under the Pont d'Arc, a natural bridge. It gives its name to a *département*.

Arden, Forest of /ˈɑːdn/ former forest region of N Warwickshire, England, the setting for William Shakespeare's play *As You Like It*.

Ardennes /ɑːˈden/ wooded plateau in NE France, SE Belgium, and N Luxembourg, cut through by the river Meuse; also a *département* of Champagne-Ardenne. There was heavy fighting here in World Wars I and II.

Arequipa /ˌærerˈkiːpə/ city in Peru at the base of the volcano El Misti; population (1990 est) 965,000. Founded by Pizarro 1540, it is the cultural focus of S Peru and a busy commercial centre (soap, textiles).

Arezzo /əˈretsəʊ/ town in the Tuscan region of Italy, 80 km/50 mi SE of Florence; population (1981) 92,100. The writers Petrarch and Aretino were born here. It is a mining town and also trades in textiles, olive oil, and antiques.

Argenteuil /ˌɑːʒɒnˈtɜːi/ northwestern suburb of Paris, France, on the river Seine; population (1982) 96,000.

Argentina Republic of (*República Argentina*)
area 2,780,092 sq km/1,073,116 sq mi
capital Buenos Aires (to move to Viedma)
towns Rosario, Córdoba, Tucumán, Mendoza, Santa Fe; ports are La Plata and Bahía Blanca
physical mountains in W, forest and savanna in N, pampas (treeless plains) in E central area, Patagonian plateau in S; rivers Colorado, Salado, Paraná, Uruguay, Río de la Plata estuary
territories part of Tierra del Fuego; disputed claims to S Atlantic islands and part of Antarctica
environment an estimated 20,000 sq km/ 7,700 sq mi of land has been swamped with salt water
features Andes mountains, with Aconcagua the highest peak in the W hemisphere; Iguaçú Falls
head of state and government Carlos Menem from 1989
political system emergent democratic federal republic
political parties Radical Civic Union Party (UCR), moderate centrist; Justicialist Party, right-wing Peronist
exports livestock products, cereals, wool, tannin, peanuts, linseed oil, minerals (coal, copper, molybdenum, gold, silver, lead, zinc, barium, uranium); the country has huge resources of oil, natural gas, hydroelectric power
currency peso = 10,000 australs
population (1990 est) 32,686,000 (mainly of Spanish or Italian origin, only about 30,000 American Indians surviving); growth rate 1.5% p.a.
life expectancy men 66, women 73
languages Spanish (official); English, Italian, German, French
religion Roman Catholic (state-supported)
literacy men 96%, women 95% (1985 est)
GDP $70.1 bn (1990); $2,162 per head

chronology
1816 Independence achieved from Spain, followed by civil wars.
1946 Juan Perón elected president, supported by his wife 'Evita'.
1952 'Evita' Perón died.
1955 Perón overthrown and civilian administration restored.
1966 Coup brought back military rule.
1973 A Peronist party won the presidential and congressional elections. Perón returned from exile in Spain as president, with his third wife, Isabel, as vice president.
1974 Perón died, succeeded by Isabel.
1976 Coup resulted in rule by a military junta led by Lt-Gen Jorge Videla. Congress dissolved, and hundreds of people, including Isabel Perón, detained.
1976–83 Ferocious campaign against left-wing elements, the 'dirty war'.
1978 Videla retired. Succeeded by General Roberto Viola, who promised a return to democracy.
1981 Viola died suddenly. Replaced by General Leopoldo Galtieri.
1982 With a deteriorating economy, Galtieri sought popular support by ordering an invasion of the British-held Falkland Islands. After losing the short war, Galtieri was removed and replaced by General Reynaldo Bignone.
1983 Amnesty law passed and 1853 democratic constitution revived. General elections won by Raúl Alfonsín and his party. Armed forces under scrutiny.
1984 National Commission on the Disappearance of Persons (CONADEP) reported on over 8,000 people who had disappeared during the 'dirty war' of 1976–83.
1985 A deteriorating economy forced Alfonsín to introduce an austerity programme.
1986 Unsuccessful attempt on Alfonsín's life.
1988 Unsuccessful army coup attempt.
1989 Carlos Menem, of the Justicialist Party, elected president.

1990 Full diplomatic relations with the UK restored. Menem elected Justicialist Party leader. Revolt by army officers thwarted. **1992** New currency introduced.

It is the United States of the Southern Hemisphere.

On **Argentina** James Bryce *South America: Observations and Impressions* 1912

Argyllshire /ɑːˈgaɪlʃə/ former county on the west coast of Scotland, including many of the Western Isles, which was for the most part merged in Strathclyde Region 1975, although a small area to the NW including Ballachulish, Ardgour, and Kingairloch went to Highland Region.

Århus /ˈɔːhuːs/ alternative form of ◊Aarhus, a port in Denmark.

Arica /əˈriːkə/ port in Chile; population (1990) 177,300. Much of Bolivia's trade passes through it, and there is contention over the use of Arica by Bolivia to allow access to the Pacific Ocean. It is Chile's northernmost city.

Arica is on a rainless coastline. It was several times devastated by earthquake, and was razed 1880 when captured by Chile from Peru.

Ariège /ˌæriˈeɪʒ/ river in S France, a tributary of the Garonne, which rises in the Pyrenees; length 170 km/106 mi. It gives its name to a *département*.

Arizona /ˌærɪˈzəʊnə/ state in southwestern USA; nickname Grand Canyon State
area 294,100 sq km/113,500 sq mi
capital Phoenix
towns Tucson, Scottsdale, Tempe, Mesa, Glendale, Flagstaff
physical Colorado Plateau in the N and E, desert basins and mountains in the S and W, Colorado River, Grand Canyon
features Grand Canyon National Park (the multicoloured-rock gorge through which the Colorado River flows, 6–29 km/4–18 mi wide, up to 1.7 km/1.1 mi deep, and 350 km/217 mi long); Organ Pipe Cactus National Monument Park; deserts: Painted (including the Petrified Forest of fossil trees), Gila, Sonoran; dams: Roosevelt, Hoover; old London Bridge (transported 1971 to the tourist resort of Lake Havasu City)

products cotton under irrigation, livestock, copper, molybdenum, silver, electronics, aircraft
population (1990) 3,665,000; including 4.5% American Indians (Navajo, Hopi, Apache), who by treaty own 25% of the state
famous people Cochise, Wyatt Earp, Geronimo, Barry Goldwater, Zane Grey, Percival Lowell, Frank Lloyd Wright
history part of New Spain 1715; part of Mexico 1824; passed to the USA after Mexican War 1848; territory 1863; statehood achieved 1912.

Arkansas /ˈɑːkənsɔː/ state in S central USA; nickname Wonder State/Land of Opportunity
area 137,800 sq km/53,191 sq mi
capital Little Rock
towns Fort Smith, Pine Bluff, Fayetteville
physical Ozark Mountains and plateau in the W, lowlands in the E; Arkansas River; many lakes
features Hot Springs National Park
products cotton, soya beans, rice, oil, natural gas, bauxite, timber, processed foods
population (1990) 2,350,700
famous people Johnny Cash, Bill Clinton, J William Fulbright, Douglas MacArthur, Winthrop Rockefeller
history explored by Hernando de Soto 1541; European settlers 1648, who traded with local Indians; part of Louisiana Purchase 1803; statehood achieved 1836.

The first European settlement was Arkansas Post, founded by some of the companions of the French explorer La Salle. After seceding from the Union 1861, it was readmitted 1868.

Arles /ɑːl/ town in Bouches-du-Rhône *département*, SE France, on the left bank of the Rhône; population (1990) 52,600. It is an important fruit- and vine-growing district. Roman relics include an amphitheatre for 25,000 spectators. The cathedral of St Trophime is a notable Romanesque structure. The painter Van Gogh lived here 1888–89, during which period he painted some of his major works.

Arlington /ˈɑːlɪŋtən/ county in Virginia, USA, and suburb of Washington, DC; population (1990) 170,900. It is the site of the National Cemetery for the dead of the US wars. The grounds were first used as a military cemetery 1864 during the American Civil War. By 1975, 165,142 military, naval, and civilian persons had been buried there and numbered, including the Unknown Soldier of both world wars, President John F Kennedy, and his brother Robert Kennedy.

Arizona

Arkansas

Arlington /ɑːlɪŋtən/ city in N Texas, USA, located between Dallas and Fort Worth; population (1990) 261,700. Industries include machinery, paper products, steel, car assembly, rubber, and chemicals.

Armagh /ɑːˈmɑː/ county of Northern Ireland
area 1,250 sq km/483 sq mi
towns Armagh (county town), Lurgan, Portadown, Keady
physical flat in the N, with many bogs; low hills in the S; Lough Neagh
features smallest county of Northern Ireland. There are crops in the better drained parts, especially flax. The chief rivers are the Bann and Blackwater, flowing into Lough Neagh, and the Callan tributary of the Blackwater
products chiefly agricultural: apples, potatoes, flax
population (1981) 119,000.

Armagh /ɑːˈmɑː/ county town of Armagh, Northern Ireland; population (1981) 13,000. It became the religious centre of Ireland in the 5th century when St Patrick was made archbishop. For 700 years it was the seat of the kings of Ulster. The Protestant archbishop of Armagh is nominally 'Primate of All Ireland'.

Armenia Republic of
area 29,800 sq km/11,500 sq mi
capital Yerevan
towns Kumayri (formerly Leninakan)
physical mainly mountainous (including Mount Ararat), wooded
features State Academia Theatre of Opera and Ballet; Yerevan Film Studio
head of state Levon Ter-Petrossian from 1990
head of government Gagik Arutyunyan from 1991
political system emergent democracy
products copper, molybdenum, cereals, cotton, silk
population (1991) 3,580,000 (90% Armenian, 5% Azeri, 2% Russian, 2% Kurd)
language Armenian
religion traditionally Armenian Christian
chronology
1918 Became an independent republic.
1920 Occupied by the Red Army.
1936 Became a constituent republic of the USSR.

1988 Feb: demonstrations in Yerevan called for transfer of Nagorno-Karabakh from Azerbaijan to Armenian control. Dec: earthquake claimed around 25,000 lives and caused extensive damage.
1989 Jan–Nov: strife-torn Nagorno-Karabakh placed under 'temporary' direct rule from Moscow. Pro-autonomy Armenian National Movement founded. Nov: civil war erupted with Azerbaijan over Nagorno-Karabakh.
1990 March: Armenia boycotted USSR constitutional referendum. Aug: nationalists secured control of Armenian supreme soviet; former dissident Levon Ter-Petrossian indirectly elected president; independence declared. Nakhichevan affected by Nagorno-Karabakh dispute.
1991 March: overwhelming support for independence in referendum. Dec: Armenia joined new Commonwealth of Independent States (CIS); Nagorno-Karabakh declared its independence; Armenia granted diplomatic recognition by USA.
1992 Jan: admitted into Conference on Security and Cooperation in Europe (CSCE). March: joined United Nations (UN). Conflict over Nagorno-Karabakh worsened.

Armentières /ˌɑːmɒntiˈeə/ town in N France on the Lys River; population (1990) 26,200. The song 'Mademoiselle from Armentières' originated during World War I, when the town was held by the British. The town was flattened by German bombardment 1918 and rebuilt.

Armidale /ˈɑːmɪdeɪl/ town in New South Wales, Australia; population (1985) 21,500. The University of New England is here, and mansions of the squatters (early settlers) survive.

Arnhem /ˈɑːnəm/ city in the Netherlands, on the Rhine SE of Utrecht; population (1991) 131,700. It produces salt, chemicals, and pharmaceuticals. In World War II, it was the scene of an airborne operation by the Allies, 17–26 Sept 1944, to secure a bridgehead over the Rhine and a possible early end to the war. The operation was only partly successful, with 7,600 casualties.

Arnhem Land /ˈɑːnəm/ plateau of the central peninsula in Northern Territory, Australia. It is named after a Dutch ship which dropped anchor there 1618. The chief town is Nhulunbuy. It is the largest of the Aboriginal reserves, and a traditional way of life is maintained, now threatened by mineral exploitation.

Arno /ˈɑːnəʊ/ Italian river 240 km/150 mi long, rising in the Apennines, and flowing westwards to the Mediterranean Sea. Florence and Pisa stand on its banks. A flood 1966 damaged virtually every Renaissance landmark in Florence.

Arran /ˈærən/ large mountainous island in the Firth of Clyde, Scotland, in Strathclyde; area 427 sq km/165 sq mi; population (1981) 4,726. It is popular as a holiday resort. The chief town is Brodick.

Arras /ˈærəs/ French town on the Scarpe River NE of Paris; population (1990) 42,700. It is the capital of Pas-de-Calais *département*, and was formerly known for tapestry.

It was the birthplace of the French revolutionary leader Robespierre.

Arthur's Pass /ˈɑːθəz ˈpɑːs/ road-rail link across the Southern Alps, New Zealand, at 926 m/ 3,038 ft, linking Christchurch with Greymouth.

Arthur's Seat /ˈɑːθəz ˈsiːt/ hill of volcanic origin, Edinburgh, Scotland; height 251 m/823 ft. It is only fancifully linked with King Arthur.

Aruba /əˈruːbə/ island in the Caribbean, the westernmost of the Lesser Antilles; an overseas part of the Netherlands
area 193 sq km/75 sq mi
population (1989) 62,400
history Aruba obtained separate status from the other Netherlands Antilles 1986 and has full internal autonomy.

Arunachal Pradesh /ˌɑːrəˈnɑːtʃəl prɑːˈdeʃ/ state of India, in the Himalayas on the borders of Tibet and Myanmar
area 83,600 sq km/32,270 sq mi
capital Itanagar
products rubber, coffee, spices, fruit, timber
population (1991) 858,400
languages 50 different dialects
history formerly nominally part of Assam, known as the renamed Arunachal Pradesh ('Hills of the Rising Sun'). It became a state 1986.

Arundel /ˈærəndl/ town in Sussex, England, on the river Arun; population (1981) 2,200. It has a magnificent castle (much restored and rebuilt), the seat for centuries of the earls of Arundel and dukes of Norfolk.

Arvand River /ɑːˈvɑːnd/ Iranian name for the ▷Shatt-al-Arab waterway.

Aryana /ˌeəriˈɑːnə/ ancient name of ▷Afghanistan.

Asante alternative form of ▷Ashanti, a region of Ghana.

Ascension /əˈsenʃən/ British island of volcanic origin in the S Atlantic, a dependency of St Helena since 1922; population (1982) 1,625. The chief settlement is Georgetown.

A Portuguese navigator landed there on Ascension Day 1501, but it remained uninhabited until occupied by Britain 1815. There are sea turtles and sooty terns. It is known for its role as a staging post to the Falkland Islands.

Ascot /ˈæskət/ village in Berkshire, England 9.5 km/6 mi SW of Windsor. Queen Anne established the racecourse on Ascot Heath 1711, and the Royal Ascot meeting is a social as well as a sporting event. Horse races include the Gold Cup, Ascot Stakes, Coventry Stakes, and King George VI and Queen Elizabeth Stakes.

Ashanti /əˈʃænti/ or *Asante* region of Ghana, W Africa; area 25,100 sq km/9,700 sq mi; population (1984) 2,089,683. Kumasi is the capital. Most Ashanti are cultivators and the main crop is cocoa, but the people also produce fine metalwork and textiles.

For more than 200 years Ashanti was an independent kingdom. During the 19th century the Ashanti and the British fought for control of trade in West Africa. The British sent four expeditions against the Ashanti and formally annexed their country 1901. Otomfuo Sir Osei Agyeman, nephew of the deposed king, Prempeh I, was made head of the re-established Ashanti confederation 1935 as Prempeh II. The Golden Stool (actually a chair), symbol of the Ashanti peoples since the 17th century, was returned to Kumasi 1935 (the rest of the Ashanti treasure is in the British Museum). The Asantahene (King of the Ashanti) still holds ceremonies in which this stool is ceremonially paraded.

Ashby-de-la-Zouch /ˈæʃbi də lə zuːʃ/ market town in Leicestershire, England; 26 km/16 mi NW of Leicester; population (1985) 11,906. It was named after the La Zouche family who built the castle, which was used to imprison Mary Queen of Scots 1569. The 15th-century castle features in Walter Scott's novel *Ivanhoe.*

Ashdod /ˈæʃdɒd/ deep-water port of Israel, on the Mediterranean 32 km/20 mi S of Tel-Aviv, which it superseded 1965; population (1982) 66,000. It stands on the site of the ancient Philistine stronghold of Askalon.

Asheville /ˈæʃvɪl/ textile town in the Blue Ridge Mountains of North Carolina, USA; population (1990) 61,600. Showplaces include the 19th-century Biltmore mansion, home of millionaire George W Vanderbilt, and the home of the writer Thomas Wolfe.

Ashford /ˈæʃfəd/ town in Kent, England, on the river Stour, SW of Canterbury; population (1989 est) 50,000. It expanded in the 1980s as a new commercial and industrial centre for SE England.

Ashkhabad /ˌæʃkəˈbæd/ capital of Turkmenistan; population (1989) 402,000. Industries include glass, carpets ('Bukhara' carpets are made here), cotton; the spectacular natural setting has been used by the film-making industry.

It was established 1881 as a military fort on the Persian frontier, occupying an oasis on the edge of the Kara-Kum Desert.

Ashland /ˈæʃlənd/ city in NE Kentucky, USA, on the Ohio River, E of Louisville; population (1990) 23,600. Industries include chemicals, coal, oil, limestone, coke, petroleum products, steel, clothing, and leather goods.

Ashmore and Cartier Islands /ˈæʃmɔː, ˈkɑːtieɪ/ group of uninhabited Australian islands comprising Middle, East, and West Islands (the Ashmores), and Cartier Island, in the Indian Ocean, about 190 km/120 mi off the northwest coast of Australia; area 5 sq km/2 sq mi. They were transferred to the authority of Australia by Britain 1931. Formerly administered as part of the Northern Territory, they became a separate territory 1978. West Ashmore has an automated weather station. Ashmore reef was declared a national nature reserve 1983.

Ashton under Lyne /ˈæʃtən ʌndə ˈlaɪn/ town in Greater Manchester, England; population (1987 est) 44,400. There are light industries, coal, and cotton.

Asia largest of the continents, occupying one-third of the total land surface of the world
area 44,000,000 sq km/17,000,000 sq mi
largest cities (population over 5 million) Tokyo, Shanghai, Osaka, Beijing, Seoul, Calcutta, Bombay, Jakarta, Bangkok, Tehran, Hong Kong, Delhi, Tianjin, Karachi
features rivers (over 3,200 km/2,000 mi) include Chiang Jiang (Yangtze), Huang He (Yellow River), Ob-Irtysh, Amur, Lena, Mekong, Yenisei; lakes (over 18,000 sq km/7,000 sq mi) include Caspian Sea (the largest lake in the world), Aral Sea, Baikal (largest freshwater lake in Eurasia), Balkhash; deserts include the Gobi, Takla Makan, Syrian Desert, Arabian Desert, Negev
highest point Mount Everest (8,872 m/29,118 ft), the world's highest mountain
lowest point Dead Sea (−394 m/−1,293 ft), the world's lowest point below sea level
geographical extremities of mainland Cape Chelyubinsk in N, Cape Piai at the southern tip of the Malay Peninsula in the S, Dezhneva Cape in the E, Cape Baba in Turkey in the W
physical lying in the eastern hemisphere, Asia extends from the Arctic Circle to just over 10° S of the equator. The Asian mainland, which forms the greater part of the Eurasian continent, lies entirely in the northern hemisphere

Containing the world's highest mountains and largest inland seas, Asia can be divided into five physical units:
1) at the heart of the continent, a central triangle of plateaus at varying altitudes (Tibetan Plateau, Tarim Basin, Gobi Desert), surrounded by huge mountain chains which spread in all directions (Himalayas, Karakoram, Hindu Kush, Pamirs, Kunlun, Tien Shan, Altai);
2) the western plateaus and ranges (Elburz, Zagros, Taurus, Great Caucasus mountains) of Afghanistan, Iran, N Iraq, Armenia, and Turkey;
3) the lowlands of Turkestan and Siberia which stretch N of the central mountains to the Arctic Ocean and include large areas in which the subsoil is permanently frozen;
4) the fertile and densely populated eastern lowlands and river plains of Korea, China, and Indochina, and the islands of the East Indies and Japan;
5) the southern plateaus of Arabia, and the Deccan, with the fertile alluvial plains of the Euphrates, Tigris, Indus, Ganges, Brahmaputra, and Irrawaddy rivers.

In Asiatic Russia are the largest areas of coniferous forest (taiga) in the world
climate showing great extremes and contrasts, the heart of the continent becoming bitterly cold in winter and extremely hot in summer. When the heated air over land rises, moisture-laden air from the surrounding seas flows in, bringing heavy monsoon rains to all SE Asia, China, and Japan between May and Oct
products 46% of the world's cereal crops (91% of its rice); mangoes (India); groundnuts (India, China); 84% of its copra (Philippines, Indonesia); 93% of its rubber (Indonesia, Malaysia,

Thailand); tobacco (China); flax (China, Russia); 95% of its jute (India, Bangladesh, China); cotton (China, India, Pakistan); silk (China, India); fish (Japan, China, Korea, Thailand); 55% of its tungsten (China); 45% of its tin (Malaysia, China, Indonesia); oil (Saudi Arabia is the world's largest producer)
population (1988) 2,996,000; the world's largest, though not the fastest growing population, amounting to more than half the total number of people in the world; between 1950 and 1990 the death rate and infant mortality were reduced by more than 60%; annual growth rate 1.7%; projected to increase to 3,550,000 by the year 2000
languages predominantly tonal languages (Chinese, Japanese) in the E, Indo-Iranian languages (Hindi, Urdu, Persian) in S Asia, Altaic languages (Mongolian, Turkish) in W and Central Asia, Semitic languages (Arabic, Hebrew) in the SW
religions the major religions of the world had their origins in Asia—Judaism and Christianity in the Middle East, Islam in Arabia, Buddhism, Hinduism, and Sikhism in India, Confucianism in China, and Shintoism in Japan.

Asia Minor /ˈeɪʃə ˈmaɪnə/ historical name for ◊Anatolia, the Asian part of Turkey.

Asian Republics, Central see ◊Central Asian Republics.

Asmara /æsˈmɑːrə/ or *Asmera* capital of Eritrea; 64 km/40 mi SW of Massawa on the Red Sea; population (1984) 275,385. Products include beer, clothes, and textiles. It has a naval school. In 1974 unrest here precipitated the end of the Ethiopian Empire.

Asnières /ɑːniˈeə/ northwestern suburb of Paris, France, on the left bank of the river Seine; population (1982) 71,220. It is a boating centre and pleasure resort.

Aspen /ˈæspən/ town in W central Colorado, on the Roaring Fork River, lying at an altitude of 2,417 m/7,930 ft; population (1991) 3,700. Established as a silver-mining town in the 1880s, Aspen is now a recreational (winter skiing and summer river rafting) and cultural centre.

Assam/æˈsæm/ state of NE India
area 78,400 sq km/30,262 sq mi
capital Dispur
towns Guwahati
products half India's tea is grown and half its oil produced here; rice, jute, sugar, cotton, coal
population (1991) 24,294,600, including 12 million Assamese (Hindus), 5 million Bengalis (chiefly Muslim immigrants from Bangladesh), Nepalis, and 2 million indigenous people (Christian and traditional religions)
language Assamese
history a thriving region from 1000 BC; Assam migrants came from China and Myanmar (Burma). After Burmese invasion 1826, Britain took control and made Assam a separate province 1874; it was included in the Dominion of India,

Assam

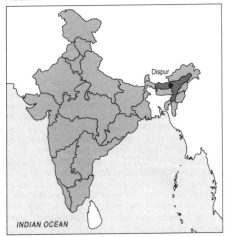

Dispur

INDIAN OCEAN

except for most of the Muslim district of Silhet, which went to Pakistan 1947. Ethnic unrest started in the 1960s when Assamese was declared the official language. After protests, the Gara, Khasi, and Jainitia tribal hill districts became the state of Meghalaya 1971; the Mizo hill district became the Union Territory of Mizoram 1972. There were massacres of Muslim Bengalis by Hindus 1983. In 1987 members of the Bodo ethnic group began fighting for a separate homeland. Direct rule was imposed by the Indian government Nov 1990 following separatist violence from the Marxist-militant United Liberation Front of Assam (ULFA), which had extorted payments from tea-exporting companies. In March 1991 it was reported that the ULFA, operating from the jungles of Myanmar, had been involved in 97 killings, mainly of Congress I politicians, since 27 Nov 1990.

Assisi /ə'siːzi/ town in Umbria, Italy, 19 km/ 12 mi SE of Perugia; population (1981) 25,000. St Francis was born here and is buried in the Franciscan monastery, completed 1253. The churches of St Francis are adorned with frescoes by Giotto, Cimabue, and others.

Assiut /æ'sjuːt/ alternative transliteration of ♢Asyut, a town in Egypt.

Assuan /æ'swɑːn/ alternative transliteration of ♢Aswan, a town in Egypt.

Assy /'æsi/ plateau in Haute-Savoie, E France, 1,000 m/3,280 ft above sea level. The area has numerous sanatoriums. The church of Nôtre Dame de Toute Grâce, begun 1937 and consecrated 1950, is adorned with works by Braque, Chagall, Matisse, Derain, Rouault, and other artists.

Asti /'æsti/ town in Piedmont, SE of Turin, Italy; population (1983) 76,439. Asti province is famed for its sparkling wine. Other products include chemicals, textiles, and glass.

Astrakhan /ˌæstrə'kɑːn/ city in Russia, on the delta of the river Volga, capital of Astrakhan region;

population (1989) 509,000. In ancient times a Tatar capital, it became Russian 1556. It is the chief port for the Caspian fisheries.

... the principal seat of our Asiatic commerce, and the general magazine of fish for the whole Russian empire in Europe ...

On **Astrakhan** P S Pallas 1793

Asturias /æ'stuəriəs/ autonomous region of N Spain; area 10,600 sq km/4,092 sq mi; population (1986) 1,114,000. Half of Spain's coal comes from the mines of Asturias. Agricultural produce includes maize, fruit, and livestock. Oviedo and Gijón are the main industrial towns.

It was once a separate kingdom, and the eldest son of a king of Spain is still called prince of Asturias.

The season for travelling in Asturias is spring, summer, or not at all.

On **Asturias** H O'Shea *A Guide to Spain* 1865

Asunción /æˌsuːnsi'ɒn/ capital and port of Paraguay, on the Paraguay River; population (1984) 729,000. It produces textiles, footwear, and food products. Founded 1537, it was the first Spanish settlement in the La Plata region.

Aswan /ˌæs'wɑːn/ winter resort town in Upper Egypt; population (1985) 183,000. It is near the High Dam, built 1960–70, which keeps the level of the Nile constant throughout the year without flooding. It produces steel and textiles.

Asyut /æs'juːt/ commercial centre in Upper Egypt, near the Nile, 322 km/200 mi S of Cairo; population (1985) 274,400. An ancient Graeco-Egyptian city, it has many tombs of 11th- and 12th-dynasty nobles.

Atacama Desert /ˌætə'kɑːmə/ desert in N Chile; area about 80,000 sq km/31,000 sq mi. There are mountains inland, and the coastal area is rainless and barren. The desert has silver and copper mines, and extensive nitrate deposits.

Atatürk Dam /'ætətɜːk/ dam on the river Euphrates, in the province of Gaziantep, S Turkey, completed 1989. The lake, 550 km/340 mi SE of Ankara, covers 815 sq km/315 sq mi (when full, it holds four times the annual flow of the Euphrates). In 1990 it was filled for the first time, submerging 25 villages, all of whose 55,000 inhabitants were relocated.

It is the world's fifth largest dam, and is part of the Great Anatolia Project (GAP), which is designed to harness the power of the Euphrates and Tigris rivers and provide irrigation water. The impact of the dam on river flow has caused tension with Iraq and Syria.

Athabasca /ˌæθə'bæskə/ lake and river in Alberta and Saskatchewan, Canada, with immense tarsand deposits (source of the hydrocarbon mixture 'heavy oil') to the SW of the lake.

Athens /'æθɪnz/ (Greek **Athinai**) capital city of Greece and of ancient Attica; population (1981) 885,000, metropolitan area (1991) 3,096,800. Situated 8 km/5 mi NE of its port of Piraeus on the Gulf of Aegina, it is built around the rocky hills of the Acropolis 169 m/555 ft and the Areopagus 112 m/368 ft, and is overlooked from the NE by the hill of Lycabettus, 277 m/909 ft high. It lies in the S of the central plain of Attica, watered by the mountain streams of Cephissus and Ilissus. It has less green space than any other European capital (4%) and severe air and noise pollution.
features The Acropolis dominates the city. Remains of ancient Greece include the Parthenon, the Erechtheum, and the temple of Athena Nike. Near the site of the ancient Agora (marketplace) stands the Theseum, and S of the Acropolis is the theatre of Dionysus. To the SE stand the gate of Hadrian and the columns of the temple of Olympian Zeus. Nearby is the marble stadium built about 330 BC and restored 1896.
history The site was first inhabited about 3000 BC, and Athens became the capital of a united Attica before 700 BC. Captured and sacked by the Persians 480 BC, subsequently under Pericles it was the first city of Greece in power and culture. After the death of Alexander the Great the city fell into comparative decline, but it flourished as an intellectual centre until AD 529, when the philosophical schools were closed by Justinian. In 1458 it was captured by the Turks who held it until 1833; it was chosen as the capital of Greece 1834. Among present-day buildings are the royal palace and several museums.

Athens /'æθɪnz/ city in NE Georgia, USA, on the Oconee River, NE of Atlanta; seat of Clarke County; population (1980) 42,549. The University of Georgia was established here 1801. Industries include cotton and electrical products.

Athos /'eɪθɒs/ mountainous peninsula on the Macedonian coast of Greece. Its peak is 2,033 m/6,672 ft high. The promontory is occupied by a community of 20 Basilian monasteries inhabited by some 3,000 monks and lay brothers.

Atlanta /ət'læntə/ capital and largest city of Georgia, USA; population (1990) 394,000, metropolitan area 2,010,000. It was founded 1837 and was partly destroyed by General Sherman 1864. There are Ford and Lockheed assembly plants, and it is the headquarters of Coca-Cola. In 1990 it was chosen as the host city for the 1996 summer Olympic Games.
Originally named **Terminus** 1837, it was renamed 1845, and was burned 1864 by General Sherman during the American Civil War. Nearby Stone Mountain Memorial shows the Confederate heroes Jefferson Davis, Robert E Lee, and Stonewall Jackson on horseback.

I heard it said that the 'architecture' of Atlanta is rococola.

On **Atlanta** John Gunther *Inside USA* 1947

Atlantic City /ət,læntɪk 'sɪti/ seaside resort in New Jersey; population (1990) 38,000. Formerly a family resort, Atlantic City has become a centre for casino gambling, which was legalized 1978.
It is known for its 'boardwalk' (a wooden pavement along the beach).

Atlantic Ocean /ət'læntɪk/ ocean lying between Europe and Africa to the E and the Americas to the W, probably named after the legendary island continent of Atlantis (said to have sunk about 9600 BC); area of basin 81,500,000 sq km/31,500,000 sq mi; including the Arctic Ocean and Antarctic seas, 106,200,000 sq km/41,000,000 sq mi. The average depth is 3 km/2 mi; greatest depth the Milwaukee Depth in the Puerto Rico Trench 8,648 m/28,374 ft. The **Mid-Atlantic Ridge**, of which the Azores, Ascension, St Helena, and Tristan da Cunha form part, divides it from N to S. Lava welling up from this central area annually increases the distance between South America and Africa. The N Atlantic is the saltiest of the main oceans, and has the largest tidal range.
In the 1960s–80s average wave heights increased by 25%, the largest from 12 m/39 ft to 18 m/59 ft.

Atlas Mountains /'ætləs/ mountain system of NW Africa, stretching 2,400 km/1,500 mi from the Atlantic coast of Morocco to the Gulf of Gabes, Tunisia, and lying between the Mediterranean on the N and the Sahara on the S. The highest peak is Mount Toubkal 4,167 m/13,670 ft.
Geologically, the Atlas Mountains compare with the Alps in age, but their structure is much less complex. They are recognized as the continuation of the great Tertiary fold mountain systems of Europe.

Attica (Greek **Attiki**) region of Greece comprising Athens and the district around it; area 3,381 sq km/1,305 sq mi. It is renowned for its language, art, and philosophical thought in Classical times. It is a prefecture of modern Greece with Athens as its capital.

Attleboro /'ætlbʌrə/ city in SE Massachusetts, USA, northwest of New Bedford; population (1990) 38,400. Industries include jewellery, tools, silver products, electronics, and paper goods.

Attock /ə'tɒk/ city in Punjab province, E Pakistan, near Rawalpindi; capital of Attock district; population (1981) 40,000. Under British rule, it was known as **Campbellpore**.

Aube /əub/ river of NE France, a tributary of the Seine, length 248 km/155 mi; it gives its name to a *département*.

Auburn /'ɔ:bən/ city in SW Maine, USA, on the Androscoggin River, W of Lewiston; seat of Androscoggin County; population (1990) 24,300. Industries include shoes, textiles, poultry, livestock, and bricks.

Aubusson /,əubju:'sɒn/ town in the *département* of Creuse, France; population (1982) 6,500. Its carpet and tapestry industry dates from the 15th century.

Auckland /'ɔ:klənd/ city in New Zealand, situated in N North Island; population (1991) 315,900. It fills the isthmus that separates its two harbours (Waitemata and Manukau), and its suburbs spread N across the Harbour Bridge. It is the country's chief port and leading industrial centre, having iron and steel plants, engineering, car assembly, textiles, food processing, sugar refining, and brewing.

There was a small whaling settlement on the site in the 1830s, and Auckland was officially founded as New Zealand's capital 1840, remaining so until 1865. The university was founded 1882.

Auckland Islands /'ɔ:klənd/ six uninhabited volcanic islands 480 km/300 mi S of South Island, New Zealand; area 60 sq km/23 sq mi.

Aude /əud/ river in SE France, 210 km/130 mi long; it gives its name to a *département*. Carcassonne is the main town through which it passes.

Audenarde /əud'nɑ:d/ French form of ▷Oudenaarde, a town in Belgium.

Augrabies Falls /ɔ:'xrɑ:biz/ waterfalls in the Orange River, NW Cape Province, South Africa; height 148 m/480 ft.

Augsburg /'auksbɜ:g/ industrial city in Bavaria, Germany, at the confluence of the Wertach and Lech rivers, 52 km/32 mi NW of Munich; population (1988) 246,000. It is named after the Roman emperor Augustus, who founded it 15 BC.

Augusta /ɔ:'gʌstə/ capital of Maine, USA, located in the southwestern part of the state, on the Kennebec River, NE of Lewiston and Auburn; population (1990) 21,300. Industries include cotton, timber, and textiles.

Augusta /ɔ:'gʌstə/ city in E central Georgia, USA, on the Savannah River on the South Carolina border; seat of Richmond County; population (1990) 44,600. It is a manufacturing city and terminal for river barges. Industries include textiles and other cotton products and building materials. Established 1736 as Fort Augusta, an Indian trading post, it was the site of several battles during the War of American Independence and served as Georgia's capital 1786–95.

Aurora /ə'rɔ:rə/ city in NE Illinois, USA, on the Fox River, W of Chicago; population (1990) 99,600. Industries include transportation equipment, glass, and chemicals. Aurora was a pioneer in the use of electric street lights.

Austin /'ɒstɪn/ capital of Texas, on the Colorado River; population (1990) 465,600. It is a centre for electronic and scientific research.

Australasia /ˌɒstrə'leɪziə/ loosely applied geographical term, usually meaning Australia, New Zealand, and neighbouring islands.

Australia /ɒs'treɪliə/ Commonwealth of
area 7,682,300 sq km/2,966,136 sq mi
capital Canberra
towns Adelaide, Alice Springs, Brisbane,

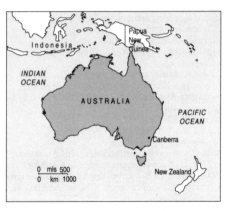

Darwin, Melbourne, Perth, Sydney, Hobart, Geelong, Newcastle, Townsville, Wollongong
physical the world's smallest, flattest, and driest continent (40% lies in the tropics, one-third is desert, and one-third is marginal grazing); Great Sandy Desert; Gibson Desert; Great Victoria Desert; Simpson Desert; the Great Barrier Reef (largest coral reef in the world, stretching 2,000 km/1,250 mi off E coast of Queensland); Great Dividing Range and Australian Alps in the E (Mount Kosciusko, 2,229 m/7,136 ft, Australia's highest peak). The fertile SE region is watered by the Darling, Lachlan, Murrumbridgee, and Murray rivers; rivers in the interior are seasonal. Lake Eyre basin and Nullarbor Plain in the S
territories Norfolk Island, Christmas Island, Cocos (Keeling) Islands, Ashmore and Cartier Islands, Coral Sea Islands, Heard Island and McDonald Islands, Australian Antarctic Territory
environment an estimated 75% of Australia's northern tropical rainforest has been cleared for agriculture or urban development since Europeans first settled there in the early 19th century
features Ayers Rock; Arnhem Land; Gulf of Carpentaria; Cape York Peninsula; Great Australian Bight; unique animal species include the kangaroo, koala, platypus, wombat, Tasmanian devil, and spiny anteater; of 800 species of bird, the budgerigar, cassowary, emu, kookaburra, lyre bird, and black swan are also unique as a result of Australia's long isolation from other continents
head of state Elizabeth II from 1952 represented by governor general
head of government Paul Keating from 1991
political system federal constitutional monarchy
political parties Australian Labor Party (ALP), moderate left of centre; Liberal Party of Australia, moderate, liberal, free enterprise; National Party of Australia (formerly Country Party), centrist non-metropolitan
exports world's largest exporter of sheep, wool, diamonds, alumina, coal, lead and refined zinc ores, and mineral sands; other exports include cereals, beef, veal, mutton, lamb, sugar, nickel (world's second largest producer), iron ore;

principal trade partners are Japan, the USA, and EC member states
currency Australian dollar
population (1990 est) 16,650,000; growth rate 1.5% p.a.
life expectancy men 75, women 80
languages English, Aboriginal languages
religions Anglican 26%, other Protestant 17%, Roman Catholic 26%
literacy 98.5.% (1988)
GDP $286.9 bn (1992)
chronology
1901 Creation of Commonwealth of Australia.
1927 Seat of government moved to Canberra.
1942 Statute of Westminster Adoption Act gave Australia autonomy from UK in internal and external affairs.
1944 Liberal Party founded by Robert Menzies.
1951 Australia joined New Zealand and the USA as a signatory to the ANZUS Pacific security treaty.
1966 Menzies resigned after being Liberal prime minister for 17 years, and was succeeded by Harold Holt.
1967 A referendum was passed giving Aborigines full citizenship rights.
1968 John Gorton became prime minister after Holt's death.
1971 Gorton succeeded by William McMahon, heading a Liberal–Country Party coalition.
1972 Gough Whitlam became prime minister, leading a Labor government.
1975 Senate blocked the government's financial legislation; Whitlam dismissed by the governor general, who invited Malcolm Fraser to form a Liberal–Country Party caretaker government. This action of the governor general, John Kerr, was widely criticized.
1978 Northern Territory attained self-government.
1983 Labor Party, returned to power under Bob Hawke, convened meeting of employers and

unions to seek consensus on economic policy to deal with growing unemployment.
1986 Australia Act passed by UK government, eliminating last vestiges of British legal authority in Australia.
1988 Labor foreign minister Bill Hayden appointed governor general designate. Free-trade agreement with New Zealand signed.
1990 Hawke won record fourth election victory, defeating Liberal Party by small majority.
1991 Paul Keating became new Labor Party leader and prime minister.
1992 Keating's popularity declined as economic problems continued. Oath of allegiance to British crown abandoned.
1993 Labor Party won general election, entering fifth term of office.

Earth is here so kind that just tickle her with a hoe and she laughs with a harvest.

On **Australia** Douglas William Jerrold
A Man Made of Money 1849

Australian Antarctic Territory islands and territories south of 60° S, between 160° E and 45° E longitude, excluding Adélie Land; area 6,044,000 sq km/2,332,984 sq mi of land and 75,800 sq km/29,259 sq mi of ice shelf. The population on the Antarctic continent is limited to research personnel.

There are scientific bases at Mawson (1954) in MacRobertson Land, named after the explorer; at Davis (1957) on the coast of Princess Elizabeth Land, named in honour of Mawson's second-in-command; at Casey (1969) in Wilkes Land, named after Lord Casey; and Macquarie Island (1948). The Australian Antarctic Territory came into being 1933, when established by a British Order in Council.

Australian Capital Territory territory ceded to Australia by New South Wales 1911 to provide the site of Canberra, with its port at Jervis Bay, ceded 1915; area 2,400 sq km/926 sq mi; population (1987) 261,000.

Austral Islands /ˈɒstrəl/ alternative name for ◊Tubuai Islands, part of French Polynesia.

Australia, Commonwealth of

state	capital	area in sq km
New South Wales	Sydney	801,600
Queensland	Brisbane	1,727,200
SouthAustralia	Adelaide	984,000
Tasmania	Hobart	67,800
Victoria	Melbourne	227,600
Western Australia	Perth	2,525,500
territories		
Northern Territory	Darwin	1,346,200
Capital Territory	Canberra	2,400
		7,682,300
external territories		
Ashmore and Cartier Islands		5
Australian Antarctic Territory		6,044,000
Christmas Island		135
Cocos (Keeling) Islands		14
Coral Sea Islands		1,000,000
Heard Island and McDonald Islands		410
Norfolk Island		40

Australian Capital Territory

Austria Republic of (*Republik Österreich*)
area 83,500 sq km/32,374 sq mi
capital Vienna
towns Graz, Linz, Salzburg, Innsbruck
physical landlocked mountainous state, with Alps in W and S and low relief in E where most of the population is concentrated
environment Hainburg, the largest primeval forest left in Europe, under threat from a dam project (suspended 1990)
features Austrian Alps (including Grossglockner and Brenner and Semmering passes); Lechtaler and Allgauer Alps N of river Inn; Carnic Alps on Italian border; river Danube
head of state Thomas Klestil from 1992
head of government Franz Vranitzky from 1986
political system democratic federal republic
political parties Socialist Party of Austria (SPÖ), democratic socialist; Austrian People's Party (ÖVP), progressive centrist; Freedom Party of Austria (FPÖ), moderate left of centre; United Green Party of Austria (VGÖ), conservative ecological; Green Alternative Party (ALV), radical ecological
exports lumber, textiles, clothing, iron and steel, paper, machinery and transport equipment, foodstuffs
currency schilling
population (1990 est) 7,595,000; growth rate 0.1% p.a.
life expectancy men 70, women 77
language German
religions Roman Catholic 85%, Protestant 6%

Austria: provinces

province	capital	area in sq km
Burgenland	Eisenstadt	4,000
Carinthia	Klagenfurt	9,500
Lower Austria	St Pölten	19,200
Salzburg	Salzburg	7,200
Styria	Graz	16,400
Tirol	Innsbruck	12,600
Upper Austria	Linz	12,000
Vienna	Vienna	420
Vorarlberg	Bregenz	2,600

literacy 98% (1983)
GDP $184.7 bn (1992)
chronology
1867 Emperor Franz Josef established dual monarchy of Austria–Hungary.
1914 Archduke Franz Ferdinand assassinated by a Serbian nationalist; Austria–Hungary invaded Serbia, precipitating World War I.
1918 Habsburg empire ended; republic proclaimed.
1938 Austria incorporated into German Third Reich by Hitler (the *Anschluss*).
1945 Under Allied occupation, constitution of 1920 reinstated and coalition government formed by the SPÖ and the ÖVP.
1955 Allied occupation ended, and the independence of Austria formally recognized.
1966 ÖVP in power with Josef Klaus as chancellor.
1970 SPÖ formed a minority government, with Dr Bruno Kreisky as chancellor.
1983 Kreisky resigned and was replaced by Dr Fred Sinowatz, leading a coalition.
1986 Dr Kurt Waldheim elected president. Sinowatz resigned, succeeded by Franz Vranitzky. No party won an overall majority; Vranitzky formed a coalition of the SPÖ and the ÖVP, with ÖVP leader, Dr Alois Mock, as vice chancellor.
1989 Austria sought European Community membership.
1990 Vranitzky re-elected.
1991 Bid for EC membership endorsed by the Community.
1992 Thomas Klestil elected president, replacing Waldheim.

The chief crop of provincial Austria is—scenery.

On **Austria** John Gunther *Inside Europe* 1938

Auvergne /əʊ'veən/ ancient province of central France and a modern region comprising the *départements* of Allier, Cantal, Haute-Loire, and Puy-de-Dôme
area 26,000 sq km/10,036 sq mi
population (1986) 1,334,000
capital Clermont-Ferrand
physical mountainous, composed chiefly of volcanic rocks in several masses
products cattle, wheat, wine, and cheese
history named after the ancient Gallic Avenni tribe whose leader, Vercingetorix, led a revolt against the Romans 52 BC. In the 14th century the Auvergne was divided into a duchy, dauphiny, and countship. The duchy and dauphiny were united by the dukes of Bourbon before being confiscated by Francis I 1527. The countship united with France 1615.

Auxerre /əʊ'seə/ capital of Yonne *département*, France, 170 km/106 mi SE of Paris, on the river Yonne; population (1990) 40,600. The Gothic cathedral, founded 1215, has exceptional sculptures and stained glass.

Avernus /ə'vɜːnəs/ circular lake, near Naples, Italy. Because it formerly gave off fumes that killed birds, it was thought by the Romans to be the entrance to the lower world.

Aviemore /ˌævɪ'mɔː/ winter sports centre, in the Highlands, Scotland, SE of Inverness among the Cairngorm Mountains.

Avignon /'æviːnjɒn/ city in Provence, France, capital of Vaucluse *département*, on the river Rhône NW of Marseilles; population (1990) 89,400. An important Gallic and Roman city, it has a 12th-century bridge (only half still standing), a 13th-century cathedral, 14th-century walls, and two palaces built during the residence here of the popes, Le Palais Vieux (1334–42) and Le Palais Nouveau (1342–52). Avignon was papal property 1348–1791.

Avila /'ævɪlə/ town in Spain, 90 km/56 mi NW of Madrid; population (1986) 45,000. It is the capital of a province of the same name. It has the remains of a Moorish castle, a Gothic cathedral, and the convent and church of St Teresa, who was born here. The medieval town walls are among the best preserved in Europe.

Avila is a miniature Jerusalem that must have been transported to Castile from Palestine on angels' wings.

On **Avila** Arnold Toynbee *Between Niger and Nile* 1965

Avon /'eɪvən/ county in SW England
area 1,340 sq km/517 sq mi
towns Bristol (administrative headquarters), Bath, Weston-super-Mare
features low-lying basin bordered by Cotswold Hills in NE, Mendip Hills in S; river Avon flows W into Severn estuary

Avon

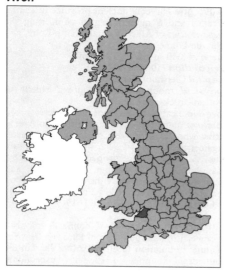

products aircraft and other engineering, tobacco, chemicals, printing, dairy products
population (1987) 919,800
famous people John Cabot, Thomas Chatterton, W G Grace
history formed 1974 from the city and county of Bristol, part of S Gloucestershire, and part of N Somerset.

Avon /'eɪvən/ any of several rivers in England and Scotland. The Avon in Warwickshire is associated with the playwright William Shakespeare.

The Upper or Warwickshire Avon, 154 km/96 mi, rises in the Northampton uplands near Naseby and joins the Severn at Tewkesbury. The Lower, or Bristol, Avon, 121 km/75 mi, rises in the Cotswolds and flows into the Bristol Channel at Avonmouth. The East, or Salisbury, Avon, 104 km/65 mi, rises S of the Marlborough Downs and flows into the English Channel at Christchurch.

Awash /'aːwaːʃ/ river that rises to the S of Addis Ababa in Ethiopia and flows NE to Lake Abba on the frontier with Djibouti. Although deep inside present-day Ethiopia, the Awash River was considered by Somalis to mark the eastern limit of Ethiopian sovereignty prior to the colonial division of Somaliland in the 19th century.

Awe /ɔː/ longest (37 km/23 mi) of the Scottish freshwater lochs, in Strathclyde, SE of Oban. It is drained by the river Awe into Loch Etive.

Axholme, Isle of /'ækshəʊm/ area of 2,000 ha/5,000 acres in Humberside, England, bounded by the Trent, Don, Idle, and Torne rivers, where a form of 'medieval' open-field strip farming is still practised. The largest village, Epworth, is the birthplace of the Methodist John Wesley.

Ayacucho /ˌaɪə'kuːtʃau/ capital of a province of the same name in the Andean Mountains of central Peru; population (1988) 94,200. The last great battle against Spanish troops in the war of independence was fought near here Dec 1824.

Aycliffe /'eɪklɪf/ town in Durham, England, on the river Skerne; population (1981) 36,825. It developed from 1947 as a new town.

Ayers Rock /eəz/ (Aboriginal *Uluru*) vast ovate mass of pinkish rock in Northern Territory, Australia; 335 m/1,110 ft high and 9 km/6 mi around. It is named after Henry Ayers, a premier of South Australia.

For the Aboriginals, whose paintings decorate its caves, it has magical significance.

Ayot St Lawrence /'eɪət sənt lɒrəns/ village in Hertfordshire, England, where Shaw's Corner (home of the playwright George Bernard Shaw) is preserved.

Ayr /eə/ town in Strathclyde, Scotland, at the mouth of the river Ayr; population (1985 est) 49,400. Auld Bridge was built in the 5th century, the New Bridge 1788 (rebuilt 1879). Ayr has associations with Robert Burns.

Ayrshire /'eəʃə/ former county of SW Scotland, with a 113 km/70 mi coastline on the Firth of Clyde. In 1975 the major part was merged into the region of Strathclyde.

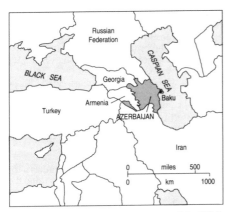

AZ abbreviation for *Arizona*, a state of the USA.

Azerbaijan Republic of
area 86,600 sq km/33,400 sq mi
capital Baku
towns Gyandzha (formerly Kirovabad), Sumgait
physical Caspian Sea; the country ranges from semidesert to the Caucasus mountains
head of state and government Albufaz Elchibey from 1992
political system emergent democracy
political parties Republican Democratic Party, ex-communist dominated; Popular Front, democratic nationalist; Islamic Party, fundamentalist
products oil, iron, copper, fruit, vines, cotton, silk, carpets
population (1990) 7,145,600 (83% Azeri, 6% Russian, 6% Armenian)
language Turkic
religion traditionally Shi'ite Muslim
chronology
1917–18 A member of the anti-Bolshevik Transcaucasian Federation.
1918 Became an independent republic.
1920 Occupied by the Red Army.
1922–36 Formed part of the Transcaucasian Federal Republic with Georgia and Armenia.
1936 Became a constituent republic of the USSR.
1988 Riots followed Nagorno-Karabakh's request for transfer to Armenia.
1989 Jan–Nov: strife-torn Nagorno-Karabakh placed under 'temporary' direct rule from Moscow. Azerbaijan Popular Front established. Nov: civil war erupted with Armenia.

1990 Jan: Soviet troops dispatched to Baku to restore order. Aug: communists won parliamentary elections. Nakhichevan affected by Nagorno-Karabakh dispute.
1991 Aug: Azeri leadership supported attempted anti-Gorbachev coup in Moscow; independence declared. Sept: former communist Ayaz Mutalibov elected president. Dec: joined new Commonwealth of Independent States (CIS); Nagorno-Karabakh declared independence.
1992 Jan: admitted into Conference on Security and Cooperation in Europe (CSCE); March: Mutalibov resigned; Azerbaijan became a member of the United Nations (UN); accorded diplomatic recognition by the USA. June: Albufaz Elchibey, leader of the Popular Front, elected president; renewed campaign against Armenia in the fight for Nagorno-Karabakh.

Azerbaijan, Iranian /ˌæzəbaɪˈdʒɑːn/ two provinces of NW Iran, *Eastern Azerbaijan* (capital Tabriz), population (1986) 4,114,000, and *Western Azerbaijan* (capital Orúmiyeh), population (1986) 1,972,000. Azeris in Iran, as in the Republic of Azerbaijan, are mainly Shi'ite Muslim ethnic Turks, descendants of followers of Khans from the Mongol Empire.

There are about 5 million in Azerbaijan, and 3 million distributed in the rest of the country, where they form a strong middle class. In 1946, with Soviet backing, they briefly established their own republic. Denied autonomy under the Shah, they rose 1979–80 against the supremacy of Ayatollah Khomeini and were forcibly repressed, although a degree of autonomy was promised.

Azores /əˈzɔːz/ group of nine islands in the N Atlantic, belonging to Portugal; area 2,247 sq km/867 sq mi; population (1987) 254,000. They are outlying peaks of the Mid-Atlantic Ridge and are volcanic in origin. The capital is Ponta Delgada on the main island, San Miguel.

Portuguese from 1430, the Azores were granted partial autonomy 1976, but remain a Portuguese overseas territory. The islands have a separatist movement. The Azores command the Western shipping lanes.

Azov /ˈeɪzɒv/ (Russian *Azovskoye More*) inland sea of Europe forming a gulf in the NE of the Black Sea, between Ukraine and Russia; area 37,555 sq km/14,500 sq mi. Principal ports include Rostov-on-Don, Kerch, and Taganrog. Azov is a good source of freshwater fish.

B

Baabda /'bɑːbdə/ capital of the province of Jebel Lubnan in central Lebanon, SE of Beirut. It is the site of the country's presidential palace.

Bab-el-Mandeb /'bæb el 'mændeb/ strait that joins the Red Sea and the Gulf of Aden, and separates Arabia and Africa. The name, meaning 'gate of tears', refers to its currents.

Babi Yar /'bɑːbi 'jɑː/ ravine near Kiev, Ukraine, where more than 100,000 people (80,000 Jews; the others were Poles, Russians, and Ukrainians) were killed by the Nazis 1941. The site was ignored until the Soviet poet Yevtushenko wrote a poem called 'Babi Yar' 1961 in protest at plans for a sports centre on the site.

Bacău /'bɑːkəu/ industrial city in Romania, 250 km/155 mi NNE of Bucharest, on the river Bistrita; population (1985) 175,300. It is the capital of Bacău county, a leading oil-producing region.

Badajoz /ˌbædə'xəuθ/ city in Extremadura, Spain, on the Portuguese frontier; population (1991) 129,700. It has a 13th-century cathedral and ruins of a Moorish castle. Badajoz has often been besieged and was stormed by the Duke of Wellington 1812 with the loss of 59,000 British troops.

Baden /'bɑːdn/ town in Aargau canton, Switzerland, near Zurich; at an altitude of 388 m/1,273 ft; population (1990) 14,780. Its hot sulphur springs and mineral waters have been visited since Roman times.

Baden-Baden /'bɑːdn 'bɑːdn/ Black Forest spa in Baden-Württemberg, Germany; population (1984) 49,000. A fashionable resort in the 19th century, it is now a conference centre.

*The prettiest town of all places where
Pleasure has set up her tents.*

On **Baden-Baden** William Makepiece
Thackeray *The Newcomes* 1853–55

Baden-Württemberg /'bɑːdn 'vuətəmbɜːg/ administrative region (German *Land*) of Germany

area 35,800 sq km/13,819 sq mi
capital Stuttgart
towns Mannheim, Karlsruhe, Freiburg, Heidelberg, Heilbronn, Pforzheim, Ulm
physical Black Forest; Rhine boundary S and W; source of the river Danube
products wine, jewellery, watches, clocks, musical instruments, textiles, chemicals, iron, steel, electrical equipment, surgical instruments
population (1988) 9,390,000
history formed 1952 (following a plebiscite) by the merger of the *Länder* Baden, Württemberg-Baden, and Württemberg-Hohenzollern.

Bad Godesburg /'bæd 'gəudəsbɜːg/ southeastern suburb of Bonn, Germany, formerly a spa, and the meeting place of Chamberlain and Hitler before the Munich Agreement 1938.

Baffin Island /'bæfɪn/ island in the Northwest Territories, Canada
area 507,450 sq km/195,875 sq mi
features largest island in the Canadian Arctic; mountains rise above 2,000 m/6,000 ft, and there are several large lakes. The northernmost part of the strait separating Baffin Island from Greenland forms Baffin Bay, the southern end is Davis Strait. It is named after William Baffin, who carried out research here 1614 during his search for the Northwest Passage.

Baghdad /ˌbæg'dæd/ historic city and capital of Iraq, on the river Tigris; population (1985) 4,649,000. Industries include oil refining, distilling, tanning, tobacco processing, and the manufacture of textiles and cement. Founded 762, it became Iraq's capital 1921. During the Gulf War 1991, the UN coalition forces bombed it in repeated air raids and destroyed much of the city.

To the SE, on the river Tigris, are the ruins of *Ctesiphon*, capital of Parthia about 250 BC–AD 226 and of the Sassanian Empire about 226–641; the remains of the Great Palace include the world's largest single-span brick arch 26 m/85 ft wide and 29 m/95 ft high.

A transportation hub from the earliest times, it was developed by the 8th-century caliph Harun al-Rashid, although little of the *Arabian Nights* city remains. It was overrun 1258 by the Mongols, who destroyed the irrigation system. In 1639 it was taken by the Turks. During World War I, Baghdad was captured by General Maude 1917.

A long flat city in a flat land.

On **Baghdad** Freya Stark, letter of 1929

Baguio /bæ'gwiːəu/ summer resort on Luzon island in the Philippines, 200 km/125 mi N of Manila, 1,370 m/4,500 ft above sea level; population (1980) 119,000. It is the official summer residence of the Philippine president.

Bahamas Commonwealth of the
area 13,864 sq km/5,352 sq mi
capital Nassau on New Providence
towns Alice Town, Andros Town, Hope Town,

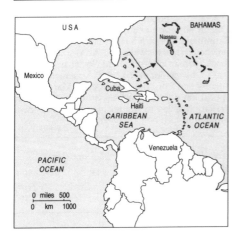

Spanish Wells, Freeport, Moss Town, George Town
physical comprises 700 tropical coral islands and about 1,000 cays
features desert islands: only 30 are inhabited; Blue Holes of Andros, the world's longest and deepest submarine caves; the Exumas are a narrow spine of 365 islands
principal islands Andros, Grand Bahama, Great Abaco, Eleuthera, New Providence, Berry Islands, Biminis, Great Inagua, Acklins, Exumas, Mayaguana, Crooked Island, Long Island, Cat Island, Rum Cay, Watling (San Salvador) Island
head of state Elizabeth II from 1973 represented by governor general
head of government Hubert Ingraham from 1992
political system constitutional monarchy
political parties Progressive Liberal Party (PLP), centrist; Free National Movement (FNM), centre-left
exports cement, pharmaceuticals, petroleum products, crawfish, salt, aragonite, rum, pulpwood; over half the islands' employment comes from tourism
currency Bahamian dollar
population (1990 est) 251,000; growth rate 1.8% p.a.
languages English and some Creole
media three independent daily newspapers
religions 29% Baptist, 23% Anglican, 22% Roman Catholic
literacy 95% (1986)
GDP $2.7 bn (1987); $11,261 per head
chronology
1964 Independence achieved from Britain.
1967 First national assembly elections.
1972 Constitutional conference to discuss full independence.
1973 Full independence achieved.
1983 Allegations of drug trafficking by government ministers.
1984 Deputy prime minister and two cabinet ministers resigned. Pindling denied any personal involvement and was endorsed as party leader.
1987 Pindling re-elected despite claims of frauds.
1992 FNM led by Hubert Ingraham won absolute majority in assembly elections.

Bahawalpur /bəˌhɑːwəlˈpuə/ city in the Punjab, Pakistan; population (1981) 178,000. Once the capital of a former state of Bahawalpur, it is now an industrial town producing textiles and soap. It has a university, established 1975.

Bahia /bəˈiːə/ state of E Brazil
area 561,026 sq km/216,556 sq mi
capital Salvador
physical low coastal plain rising to central plateau in W, crossed by São Francisco River
industries oil, chemicals, agriculture; industrial diamonds are mined
population (1986) 10,949,000.

Bahía Blanca /bəˈiːə ˈblæŋkə/ port in S Argentina, on the river Naposta, 5 km/3 mi from its mouth; population (1991) 271,500. It is a major distribution centre for wool and food processing. The naval base of Puerto Belgrano is here.

Bahrain State of (*Dawlat al Bahrayn*)
area 688 sq km/266 sq mi
capital Manama on the largest island (also called Bahrain)
towns Muharraq, Jidd Hafs, Isa Town; oil port Mina Sulman
physical 35 islands, composed largely of sand-covered limestone; generally poor and infertile soil; flat and hot
environment a wildlife park features the oryx on Bahrain; most of the south of the island is preserved for the ruling family's falconry
features causeway linking Bahrain to mainland Saudi Arabia; Sitra island is a communications centre for the lower Persian Gulf and has a satellite-tracking station
head of state and government Sheik Isa bin Sulman al-Khalifa (1933–) from 1961
political system absolute emirate
political parties none
exports oil, natural gas, aluminium, fish
currency Bahrain dinar
population (1990 est) 512,000 (two-thirds are nationals); growth rate 4.4% p.a.
life expectancy men 67, women 71
languages Arabic (official); Farsi, English, Urdu
religion 85% Muslim (Shi'ite 60%, Sunni 40%)
literacy men 79%, women 64% (1985 est)

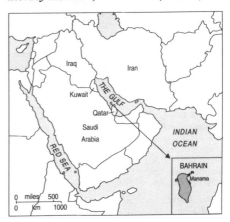

GDP $3.5 bn (1987); $7,772 per head
chronology
1861 Became British protectorate.
1968 Britain announced its intention to withdraw its forces. Bahrain formed, with Qatar and the Trucial States, the Federation of Arab Emirates.
1971 Qatar and the Trucial States withdrew from the federation and Bahrain became an independent state.
1973 New constitution adopted, with an elected national assembly.
1975 Prime minister resigned and national assembly dissolved. Emir and his family assumed virtually absolute power.
1986 Gulf University established in Bahrain. A causeway was opened linking the island with Saudi Arabia.
1988 Bahrain recognized Afghan rebel government.
1991 Bahrain joined United Nations coalition that ousted Iraq from its occupation of Kuwait.

Baikal /baɪ'kæl/ (Russian **Baykal Ozero**) freshwater lake in S Siberian Russia, the largest in Asia, and the eighth largest in the world (area 31,500 sq km/12,150 sq mi); also the deepest in the world (up to 1,640 m/5,700 ft). Fed by more than 300 rivers, it is drained only by the Lower Angara. It has sturgeon fisheries and is rich in fauna.

The lake has 1,155 species of animals and 1,085 species of plants—more than 1,000 of these are not found anywhere else in the world. The water at the bottom of the lake holds sufficient oxygen to allow animals to live at depths of over 1,600 m/5,200 ft.

Baile Atha Cliath /blɑː 'klɪə/ official Gaelic name of ◊ Dublin, capital of the Republic of Ireland, from 1922.

Baja California /'bɑːhɑː/ mountainous peninsula that forms the twin northwestern states of Lower (Spanish *baja*) California, Mexico; area 143,396 sq km/55,351 sq mi; population (1990) 1,736,200.
The northern state, *Baja California Norte*, includes the cities of Mexicali (its capital) and Tijuana. Bordering the USA, it contains many factories. The southern state, *Baja California Sur*, attracts many US tourists for fishing, whale-watching (grey whales winter here), swimming, and sunbathing. La Paz is the state capital; Cabo San Lucas is a fashionable resort.

Bakersfield /'beɪkəzfiːld/ city in S California, USA, northeast of Santa Barbara, on the Kern river; the seat of Kern County; population (1990) 174,800. It is known for its oil wells and oil products. Oil was first discovered 1899.

Bakhtaran /ˌbæktə'rɑːn/ (formerly until 1980 *Kermanshah*) capital of Bakhtaran province, NW Iran; population (1986) 561,000. The province (area 23,700 sq km/9,148 sq mi; population 1,463,000) is on the Iraqi border and is mainly inhabited by Kurds. Industries include oil refining, carpets, and textiles.

Baku /bɑː'kuː/ capital city of the Republic of Azerbaijan, industrial port (oil refining) on the

Caspian Sea; population (1987) 1,741,000. It is a major oil centre and is linked by pipelines with Batumi on the Black Sea. In Jan 1990 there were violent clashes between the Azeri majority and the Armenian minority, and Soviet troops were sent to the region; over 13,000 Armenians subsequently fled from the city. In early March 1992, opposition political forces sponsored protests in the city that led to the resignation of President Mutalibov.

New Baku may well be called the Paris of the Caspian.

On **Baku** Henry Morton Stanley *My Early Travels and Adventures* 1895

Bala Lake /'bælə/ (Welsh *Llyn Tegid*) lake in Gwynedd, N Wales, about 6.4 km/4 mi long and 1.6 km/1 mi wide. Bala Lake has a unique primitive species of fish, the gwyniad (a form of whitefish), a protected species from 1988.

Balaton /'bɒlətɒn/ lake in W Hungary; area 600 sq km/230 sq mi.

Bâle /bɑːl/ French form of Basle or ◊ Basel, a town in Switzerland.

Balearic Islands /ˌbæli'ærɪk/ (Spanish *Baleares*) group of Mediterranean islands forming an autonomous region of Spain; including Majorca, Minorca, Ibiza, Cabrera, and Formentera
area 5,000 sq km/1,930 sq mi
capital Palma de Mallorca
products figs, olives, oranges, wine, brandy, coal, iron, slate; tourism is crucial
population (1986) 755,000
history a Roman colony from 123 BC, the Balearic Islands were an independent Moorish kingdom 1009–1232; they were conquered by Aragon 1343.

Bali /'bɑːli/ island of Indonesia, E of Java, one of the Sunda Islands
area 5,800 sq km/2,240 sq mi
capital Denpasar
physical volcanic mountains
features Balinese dancing, music, drama; 1 million tourists a year (1990)
products gold and silver work, woodcarving, weaving, copra, salt, coffee
population (1989) 2,787,000
history Bali's Hindu culture goes back to the 7th century; the Dutch gained control of the island by 1908.

Balikesir /ˌbɑːlɪke'sɪə/ city in NW Turkey, capital of Aydin province; population (1990) 170,600. There are silver mines nearby.

Balikpapan /ˌbɑːlɪk'pɑːpən/ port in Indonesia, on the east coast of S Kalimantan, Borneo; population (1980) 280,900. It is an oil-refining centre.

Bali Strait /'bɑːli/ narrow strait between the two islands of Bali and Java, Indonesia. On 19–20 Feb 1942 it was the scene of a naval action between

Baltic Sea

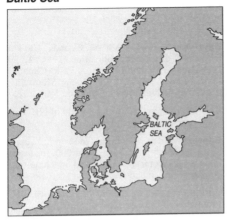

Japanese and Dutch forces that served to delay slightly the Japanese invasion of Java.

Balkans /'bɔːlkənz/ (Turkish 'mountains') peninsula of SE Europe, stretching into the Mediterranean Sea between the Adriatic and Aegean seas, comprising Albania, Bosnia-Herzegovina, Bulgaria, Croatia, Greece, Romania, Slovenia, the part of Turkey in Europe, and Yugoslavia. It is joined to the rest of Europe by an isthmus 1,200 km/750 mi wide between Rijeka on the W and the mouth of the Danube on the Black Sea to the E.

The ethnic diversity resulting from successive waves of invasion has made the Balkans a byword for political dissension. The Balkans' economy developed comparatively slowly until after World War II, largely because of the predominantly mountainous terrain, apart from the plains of the Save-Danube basin in the N. Political differences have remained strong, for example, the confrontation of Greece and Turkey over Cyprus, and the differing types of communism that prevailed until the early 1990s in the rest. More recently, ethnic interfighting has dominated the peninsula as first Slovenia and Croatia, and then Bosnia-Herzegovina, have battled to win independence from the Serb-dominated Yugoslav federation. Despite international recognition being awarded to all three republics early 1992, fierce fighting between Serb, Croat, and Muslim factions continued in Bosnia-Herzegovina. To '*Balkanize*' is to divide into small warring states.

Balkhash /bæl'xɑːʃ/ salt lake in Kazakhstan; area 17,300 sq km/6,678 sq mi. It is 600 km/375 mi long and receives several rivers, but has no outlet. It is very shallow and is frozen throughout the winter.

Balkhash /bæl'xɑːʃ/ town on the northern shore of Lake Balkhash in Kazakhstan; population (1985) 112,000. It was founded 1928. Chief industries include copper mining and salt extraction.

Ballarat town in Victoria, Australia; population (1986) 75,200. It was founded in the 1851 gold rush; the mining village and workings have been restored for tourists. The Eureka Stockade miners' revolt took place here 1854.

Ballinasloe /ˌbælɪnə'sləu/ town in Galway Bay, Republic of Ireland; population (1991) 5,800. The annual livestock fair every Oct is the largest in Ireland.

Baltic Sea /'bɔːltɪk/ large shallow arm of the North Sea, extending NE from the narrow Skagerrak and Kattegat, between Sweden and Denmark, to the Gulf of Bothnia between Sweden and Finland. Its coastline is 8,000 km/5,000 mi long, and its area, including the gulfs of Riga, Finland, and Bothnia, is 422,300 sq km/163,000 sq mi. Its shoreline is shared by Denmark, Germany, Poland, the Baltic States, Russia, Finland, and Sweden.

Many large rivers flow into it, including the Oder, Vistula, Niemen, Western Dvina, Narva, and Neva. Tides are hardly perceptible, salt content is low; weather is often stormy and navigation dangerous. Most ports are closed by ice from Dec until May. The Kiel canal links the Baltic and North seas; the Göta canal connects the two seas by way of the S Swedish lakes. Since 1975 the Baltic Sea has been linked by the St Petersburg–Belomorsk seaway with the White Sea.

Baltic States /'bɔːltɪk/ collective name for the states of ◊Estonia, ◊Latvia, and ◊Lithuania, former constituent republics of the USSR (from 1940). They regained independence Sept 1991.

Baltimore /'bɔːltɪmɔː/ industrial port and largest city in Maryland, USA, on the western shore of Chesapeake Bay, NE of Washington, DC; population (1990) 736,000. Industries include shipbuilding, oil refining, food processing, and the manufacture of steel, chemicals, and aerospace equipment.

Named after the founder of Maryland, Lord Baltimore (1606–1675), the city dates from 1729 and was incorporated 1797.

... the gastronomic metropolis of the Union.

On **Baltimore** Oliver Wendell Holmes
The Professor at the Breakfast-Table 1859

Baltistan /ˌbɔːltɪ'stɑːn/ region in the Karakoram range of NE Kashmir, held by Pakistan since 1949. It is the home of Balti Muslims of Tibetan origin. The chief town is Skardu, but Ghyari is of greater significance to Muslims as the site of a mosque built by Sayyid Ali Hamadani, a Persian who brought the Shia Muslim religion to Baltistan in the 14th century.

Baluchistan /bə,luːtʃɪ'stɑːn/ mountainous desert area, comprising a province of Pakistan, part of the Iranian province of Sistán and Balúchestan, and a small area of Afghanistan. The Pakistani province has an area of 347,200 sq km/134,019 sq mi and a population (1985) of 4,908,000; its capital is Quetta. Sistán and Balúchestan has an area of 181,600 sq km/70,098 sq mi and a population (1986) of

1,197,000; its capital is Zahedan. The port of Gwadar in Pakistan is strategically important, situated on the Indian Ocean and the Strait of Hormuz.

The common religion of the Baluch (or Baluchi) people is Islam, and they speak Baluchi, a member of the Iranian branch of the Indo-European language family. In the drier areas they make use of tents, moving when it becomes too arid. Although they practise nomadic pastoralism, many are settled agriculturalists.

history Originally a loose tribal confederation, Baluchistan was later divided into four principalities that were sometimes under Persian, sometimes under Afghan suzerainty. In the 19th century British troops tried to subdue the inhabitants until a treaty 1876 gave them autonomy in exchange for British army outposts along the Afghan border and strategic roads. On the partition of India 1947 the khan of Khalat declared Baluchistan independent; the insurrection was crushed by the new Pakistani army after eight months. Three rebellions followed, the last being from 1973 to 1977, when 3,300 Pakistani soldiers and some 6,000 Baluch were killed.

Bamako /ˌbæməˈkəu/ capital and port of Mali on the river Niger; population (1976) 400,000. It produces pharmaceuticals, chemicals, textiles, tobacco, and metal products.

Bamberg town in Bavaria, Germany, on the river Regnitz; population (1985) 70,400. The economy is based on engineering and the production of textiles, carpets, and electrical goods. It has an early 13th-century Romanesque cathedral.

Banaba /ˈbɑːnəbə/ (formerly *Ocean Island*) island in the Republic of ▷Kiribati.

Banaras /bəˈnɑːrəs/ alternative transliteration of ▷Varanasi, a holy Hindu city in Uttar Pradesh, India.

Banbury /ˈbænbəri/ town in Oxfordshire, England; population (1981) 35,800.

The *Banbury Cross* of the nursery rhyme was destroyed by the Puritans 1602, but replaced 1858. *Banbury cakes* are criss-cross pastry cases with a mince-pie-style filling.

Bandar Abbas /ˈbændər ˈæbəs/ port and winter resort in Iran on the Strait of Hormuz, Persian Gulf; population (1983) 175,000. Formerly called Gombroon, it was renamed and made prosperous by Shah Abbas I (1571–1629). It is a naval base.

Bandar Seri Begawan /ˈbændə ˈseri bəˈgɑːwən/ (formerly until 1970 *Brunei Town*) capital and largest town of Brunei, 14 km/9 mi from the mouth of the Brunei River; population (1987 est) 56,300.

Features include the Mesjid Sultan Omar Ali Saifuddin Mosque (1958) and Kampong Ayer water village. Since 1972 the town's main trade outlet has been the deepwater port of Muara at the mouth of the Brunei.

Bandung /ˈbændʊŋ/ commercial city and capital of Jawa Barat province on the island of Java, Indonesia; population (1980) 1,463,000. Bandung is the third largest city in Indonesia and was the administrative centre when the country was the Netherlands East Indies.

Banff /bænf/ resort in Alberta, Canada, 100 km/ 62 mi NW of Calgary; population (1984) 4,250. It is a centre for Banff National Park (Canada's first, founded 1885) in the Rocky Mountains. Industries include brewing and iron founding.

Banffshire /ˈbænfʃə/ former county of NE Scotland, now in Grampian region.

Bangalore /ˌbæŋgəˈlɔː/ capital of Karnataka state, S India; population (1981) 2,600,000. Industries include electronics, aircraft and machine-tools construction, and coffee.

Bangka /ˈbæŋkə/ or *Banka* or *Banca* island in Indonesia off the east coast of Sumatra
area 12,000 sq km/4,600 sq mi
capital Pangkalpinang
towns Mintok (port)
products tin (one of the world's largest producers)
population (1970) 300,000.

Bangkok /ˌbæŋˈkɒk/ capital and port of Thailand, on the river Chao Phraya; population (1990) 6,019,000. Products include paper, ceramics, cement, textiles, and aircraft. It is the headquarters of the Southeast Asia Treaty Organization (SEATO).

Bangkok was established as the capital by Phra Chao Tak 1769, after the Burmese had burned down the former capital, Avuthia, about 65 km/ 40 mi to the N. Features include the temple of the Emerald Buddha and the vast palace complex.

Bangkok is stuck as thick with pagodas as a duff with plums.

On **Bangkok** Crosbie Garstin *The Dragon and the Lotus* 1928

Bangladesh People's Republic of *(Gana Prajatantri Bangladesh)* (formerly *East Pakistan*)
area 144,000 sq km/55,585 sq mi
capital Dhaka (formerly Dacca)
towns ports Chittagong, Khulna
physical flat delta of rivers Ganges (Padma) and Brahmaputra (Jamuna), the largest estuarine delta in the world; annual rainfall of 2,540 mm/ 100 in; some 75% of the land is less than 3 m/ 10 ft above sea level and vulnerable to flooding and cyclones; hilly in extreme SE and NE
environment deforestation on the slopes of the Himalayas increases the threat of flooding in the coastal lowlands of Bangladesh, which are also subject to devastating monsoon storms. The building of India's Farakka Barrage has reduced the flow of the Ganges in Bangladesh and permitted salt water to intrude further inland. Increased salinity has destroyed fisheries, contaminated drinking water, and damaged forests
head of state Abdur Rahman Biswas from 1991

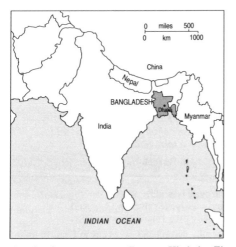

head of government Begum Khaleda Zia from 1991
political system emergent democratic republic
political parties Bangladesh Nationalist Party (BNP), Islamic, right of centre; Awami League, secular, moderate socialist; Jatiya Dal (National Party), Islamic nationalist
exports jute, tea, garments, fish products
currency taka
population (1991 est) 107,992,100; growth rate 2.17% p.a.; just over 1 million people live in small ethnic groups in the tropical Chittagong Hill Tracts, Mymensingh, and Sylhet districts
life expectancy men 50, women 52
language Bangla (Bengali)
religions Sunni Muslim 85%, Hindu 14%
literacy men 43%, women 22% (1985 est)
GDP $17.6 bn (1987); $172 per head
chronology
1947 Formed into eastern province of Pakistan on partition of British India.
1970 Half a million killed in flood.
1971 Bangladesh emerged as independent nation, under leadership of Sheik Mujibur Rahman, after civil war.
1975 Mujibur Rahman assassinated. Martial law imposed.
1976–77 Maj-Gen Zia ur-Rahman assumed power.
1978–79 Elections held and civilian rule restored.
1981 Assassination of Maj-Gen Zia.
1982 Lt-Gen Ershad assumed power in army coup. Martial law reimposed.
1986 Elections held but disputed. Martial law ended.
1987 State of emergency declared in response to opposition demonstrations.
1988 Assembly elections boycotted by main opposition parties. State of emergency lifted. Islam made state religion. Monsoon floods left 30 million homeless and thousands dead.
1989 Power devolved to Chittagong Hill Tracts to end 14-year conflict between local people and army-protected settlers.
1990 Following mass antigovernment protests, President Ershad resigned; Shahabuddin Ahmad became interim president.

1991 Feb: elections resulted in coalition government with BNP dominant. April: cyclone killed around 139,000 and left up to 10 million homeless. Sept: parliamentary government restored; Abdur Rahman Biswas elected president.

Bangor /'bæŋgə/ cathedral city in Gwynedd, N Wales; population (1981) 46,585. University College of the University of Wales is here. The cathedral was begun 1495. Industries include chemicals and electrical goods.

Bangui /bɒŋ'giː/ capital and port of the Central African Republic, on the river Ubangi; population (1988) 597,000. Industries include beer, cigarettes, office machinery, and timber and metal products.

Banjermasin /ˌbaːnjə'maːsɪn/ river port in Indonesia, on the island of Borneo; capital of Kalimantan Selatan province; population (1980) 381,300. It exports rubber, timber, and precious stones.

Banjul /bæn'dʒuːl/ capital and chief port of Gambia, on an island at the mouth of the river Gambia; population (1983) 44,536. Established 1816 as a settlement for freed slaves, it was known as Bathurst until 1973.

Banka alternative form of ◊Bangka, an Indonesian island.

Barbados
area 430 sq km/166 sq mi
capital Bridgetown
towns Speightstown, Holetown, Oistins
physical most easterly island of the West Indies; surrounded by coral reefs; subject to hurricanes June–Nov
features highest point Mount Hillaby 340 m/1,115 ft
head of state Elizabeth II from 1966 represented by governor general Hugh Springer from 1984
head of government prime minister Erskine Lloyd Sandiford from 1987
political system constitutional monarchy
political parties Barbados Labour Party (BLP), moderate, left of centre; Democratic Labour Party (DLP), moderate, left of centre; National Democratic Party (NDP), centre
exports sugar, rum, electronic components, clothing, cement
currency Barbados dollar
population (1990 est) 260,000; growth rate 0.5% p.a.
life expectancy men 70, women 75
languages English and Bajan (Barbadian English dialect)
media two independent daily newspapers
religions 70% Anglican, 9% Methodist, 4% Roman Catholic
literacy 99% (1984)
GDP $1.4 bn (1987); $5,449 per head
chronology
1627 Became British colony; developed as a sugar-plantation economy, initially on basis of slavery.
1834 Slaves freed.
1951 Universal adult suffrage introduced. BLP won general election.
1954 Ministerial government established.

of the Mediterranean. The city was devastated in the Catalonian Revolt 1652 and again during the War of the Spanish Succession 1714. At the forefront of the fight for regional autonomy during the Spanish Civil War, it suffered as a result of insurrections 1835, 1856, and 1909, and was held by the Republicans 1936–39.

... refuge of foreigners, school of chivalry, and epitome of all that a civilized and inquisitive taste could ask for.

On **Barcelona** Cervantes, letter of 1705

1961 Independence achieved from Britain. DLP, led by Errol Barrow, in power.
1966 Barbados achieved full independence within Commonwealth. Barrow became the new nation's first prime minister.
1972 Diplomatic relations with Cuba established.
1976 BLP, led by Tom Adams, returned to power.
1983 Barbados supported US invasion of Grenada.
1985 Adams died; Bernard St John became prime minister.
1986 DLP, led by Barrow, returned to power.
1987 Barrow died; Erskine Lloyd Sandiford became prime minister.
1989 New NDP opposition formed.
1991 DLP, under Erskine Sandiford, won general election.

Barbican, the /'bɑːbɪkən/ arts and residential complex in the City of London. The Barbican Arts Centre (1982) contains theatres, cinemas, and exhibition and concert halls. The architects were Powell, Chamberlin, and Bon.

Barbuda /bɑːˈbjuːdə/ one of the islands that form the state of ◊Antigua and Barbuda.

Barcelona /ˌbɑːsəˈləʊnə/ capital, industrial city (textiles, engineering, chemicals), and port of Catalonia, NE Spain; population (1991) 1,653,200. As the chief centre of anarchism and Catalonian nationalism, it was prominent in the overthrow of the monarchy 1931 and was the last city of the republic to surrender to Franco 1939. In 1992 the city hosted the Summer Olympics.
features The Ramblas, tree-lined promenades leading from the Plaza de Cataluña, the largest square in Spain; Gaudí's unfinished church of the Holy Family 1883; the Pueblo Español 1929, with specimens of Spanish architecture; a replica of Columbus's flagship the *Santa María*, in the Maritime Museum; a large collection of art by Picasso.
history Founded in the 3rd century BC, Barcelona was ruled independently by the Counts of Barcelona from the 9th century, becoming a commercial centre for Aragon and Catalonia in the 13th–14th centuries and one of the leading ports

Bardsey Island /'bɑːdzi/ (Welsh *Ynys Enlli*) former pilgrimage centre in Gwynedd, Wales, with a 6th-century ruined abbey.

Bareilly /bəˈreɪli/ industrial city in Uttar Pradesh, India; population (1981) 438,000. It was a Mogul capital 1657 and at the centre of the Indian Mutiny 1857.

Barents Sea /'bærənts/ section of the E ◊Arctic Ocean. It has oil and gas reserves.

Bari /'bɑːri/ capital of Puglia region, S Italy, and industrial port on the Adriatic Sea; population (1988) 359,000. It is the site of Italy's first nuclear power station; the part of the town known as Tecnopolis is the Italian equivalent of California's Silicon Valley.

Barisal /ˌbʌrɪ'sɑːl/ river port and capital city of Barisal region, S Bangladesh; population (1981) 142,000. It trades in jute, rice, fish, and oilseed.

Barking and Dagenham /'bɑːkɪŋ, 'dægənəm/ borough of E Greater London
products Ford motor industry at Dagenham
population (1991) 139,900.

Barkly Tableland /'bɑːkli/ large-scale, open-range, cattle-raising area in Northern Territory and Queensland, Australia.

Barletta /bɑːˈletə/ industrial port on the Adriatic Sea, Italy; population (1981) 83,800. It produces chemicals and soap; as an agriculture centre it trades in wine and fruit. There is a Romanesque cathedral built 1150 and a 13th-century castle.

Barnaul /ˌbɑːnɑːˈuːl/ industrial city in S central Siberian Russia, capital of Altai Territory; population (1987) 596,000. Industries include engineering, textiles, and timber. Founded 1738, it developed alongside the Turkestan-Siberian railway.

Barnet /'bɑːnɪt/ borough of NW Greater London
features site of the Battle of Barnet 1471 in one of the Wars of the Roses; Hadley Woods; Hampstead Garden Suburb; department for newspapers and periodicals of the British Library at Colindale; residential district of *Hendon*, which includes the Metropolitan Police Detective Training and Motor Driving schools and the Royal Air Force Battle of Britain and Bomber Command museums
population (1991) 283,000.

Barnsley /'bɑːnzli/ town in South Yorkshire, England; population (1985 est) 225,800. It is an industrial town (iron and steel, glass, paper, carpets, clothing) on one of Britain's richest coalfields.

Baroda /bə'rəudə/ former name of ⟩Vadodara, a town in Gujarat, India.

Barossa Valley /bə'rɒsə/ wine-growing area in the Lofty mountain ranges, South Australia.

Barquisimeto /bɑːˌkiːsɪ'meɪtəu/ capital of Lara state, NW Venezuela; population (1989) 764,200.

Barra /'bærə/ southernmost island of the larger Outer Hebrides, Scotland; area 90 sq km/ 35 sq mi; population (1981) 1,340. It is separated from South Uist by the Sound of Barra. The main town is Castlebay.

Barrancabermeja /bəræŋkəbəmeɪxə/ port and oil-refining centre on the Magdalena River in the department of Santander, NE Colombia; population (1980 est) 70,000
It is a major outlet for oil from the De Mares fields, which are linked by pipeline to Cartagena on the Caribbean coast.

Barranquilla /ˌbærən'kiːljə/ seaport in N Colombia, on the Magdalena River; population (1985) 1,120,900. Products include chemicals, tobacco, textiles, furniture, and footwear. It is Colombia's chief port on the Caribbean and the site of Latin America's first air terminal 1919.

Barren Lands/Grounds tundra region of Canada, W of Hudson Bay.

Barrow /'bærəu/ most northerly town in the USA, at Point Barrow, Alaska; the world's largest Inuit settlement. There is oil at nearby Prudhoe Bay.

Barrow-in-Furness /'bærəu ɪn 'fɜːnɪs/ port in Cumbria, England; population (1985) 72,600. Industries include shipbuilding and the manufacture of nuclear submarines.

Barry /'bæri/ (Welsh *Y Barri*) port in S Glamorgan, Wales; population (1981) 44,000. With *Barry Island*, it is a holiday resort.

Basel /'bɑːzəl/ or *Basle* (French *Bâle*) financial, commercial, and industrial city (dyes, vitamins, agrochemicals, dietary products, genetic products) in Switzerland; population (1990) 171,000. Basel was a strong military station under the Romans. In 1501 it joined the Swiss confederation and later developed as a centre for the Reformation.
It has the chemical firms Hoffman–La Roche, Sandoz, and Ciba-Geigy. There are trade fairs, and it is the headquarters of the Bank for International Settlements. There is an 11th-century cathedral (rebuilt after an earthquake 1356), a 16th-century town hall, and a university dating from the 15th century.

Bashkir /bæʃ'kɪə/ autonomous republic of Russia, with the Ural Mountains on the east
area 143,600 sq km/55,430 sq mi

capital Ufa
products minerals, oil, natural gas
population (1982) 3,876,000
languages Russian, Bashkir (about 25%)
history annexed by Russia 1557; became the first Soviet autonomous republic 1919. Since 1989 Bashkirs have demanded greater independence.

Basildon /'bæzldən/ industrial new town in Essex, England; population (1991) 157,500. It was designated as a new town 1949 from several townships. Industries include chemicals, clothing, printing, and engineering.

Basilicata /bəˌzɪlɪ'kɑːtə/ mountainous region of S Italy, comprising the provinces of Potenza and Matera; area 10,000 sq km/3,860 sq mi; population (1990) 624,500. Its capital is Potenza. It was the Roman province of Lucania.

Basingstoke /'beɪzɪŋstəuk/ industrial town in Hampshire, England, 72 km/45 mi WSW of London; population (1984 est) 80,000. It is the headquarters of the UK Civil Service Commission.

Basle /bɑːl/ alternative form of ⟩Basel, a city in Switzerland.

Basque Country /bæsk/ (Basque *Euskal Herria*) homeland of the Basque people in the W Pyrenees, divided by the Franco-Spanish border. The Spanish Basque Country (Spanish *País Vasco*) is an autonomous region (created 1979) of central N Spain, comprising the provinces of Vizcaya, Alava, and Guipúzcoa (Basque *Bizkaia, Araba,* and *Gipuzkoa*); area 7,300 sq km/ 2,818 sq mi; population (1988) 2,176,790. The French Basque Country (French *Pays Basque*) comprises the *département* of Pyrénées-Atlantiques, including the arrondissements of Labourd, Basse-Navarre, and Soule (Basque *Lapurdi, Nafarroa Beherea,* and *Zuberoa*); area 7,633 sq km/4770 sq mi; population (1981) 555,700. To Basque nationalists *Euskal Herria* also includes the autonomous Spanish province of Navarre.

Basque Provinces /bæsk/ see Basque Country.

Basra /'bæzrə/ (Arabic *al-Basrah*) principal port in Iraq, in the Shatt-al-Arab delta, 97 km/ 60 mi from the Persian Gulf, founded in the 7th century; population (1977) 1.5 million (1991) 850,000. Exports include wool, oil, cereal, and dates. Aerial bombing during the 1991 Gulf War destroyed bridges, factories, power stations, water-treatment plants, sewage-treatment plants, and the port. A Shi'ite rebellion March 1991 was crushed by the Iraqi army, causing further death and destruction.

Basra was about the hottest place on earth in the summer ... the place where the fable of frying eggs on the street came true.

On **Basra** Cedric Belfrage
Away From It All 1936

Bassein /bɑː'seɪn/ port in Myanmar (Burma), in the Irrawaddy delta, 125 km/78 mi from the sea; population (1983) 355,588. Bassein was founded in the 13th century.

Basse-Normandie /'bæs ˌnɔːmɒn'diː/ or *Lower Normandy* coastal region of NW France lying between Haute-Normandie and Brittany (Bretagne). It includes the *départements* of Calvados, Manche, and Orne; area 17,600 sq km/6,794 sq mi; population (1986) 1,373,000. Its capital is Caen. Apart from stock farming, dairy farming, and textiles, the area produces Calvados (apple brandy).

The invasion of Europe by Allied forces began June 1944 when troops landed on the beaches of Calvados.

Basseterre /'bæs 'teə/ capital and port of St Christopher–Nevis, in the Leeward Islands; population (1980) 14,000. Industries include data processing, rum, clothes, and electrical components.

Basse-Terre /'bæs 'teə/ main island of the French West Indian island group of Guadeloupe; area 848 sq km/327 sq mi; population (1982) 141,300. It has an active volcano, Grande Soufrière, rising to 1,484 m/4,870 ft.

Basse-Terre /'bæs 'teə/ port on Basse-Terre, one of the Leeward Islands; population (1982) 13,600. It is the capital of the French overseas *département* of Guadeloupe.

Bass Rock /bæs/ islet in the Firth of Forth, Scotland, about 107 m/350 ft high, with a lighthouse.

Bass Strait /bæs/ channel between Australia and Tasmania, named after British explorer George Bass; oil was discovered here in the 1960s.

Bastia /'bæstiə/ (Italian *bastiglia* 'fortress') port and commercial centre in NE Corsica, France; capital of the *département* of Haute-Corse; population (1990) 38,700. Founded by the Genoese 1380, it was the capital of Corsica until 1811. There are several fine churches.

Basutoland /bə'suːtəʊlænd/ former name for ◊Lesotho, a kingdom in southern Africa.

Bataan /bə'tɑːn/ peninsula in Luzon, the Philippines, which was defended against the Japanese in World War II by US and Filipino troops under General MacArthur 1 Jan–9 April 1942. MacArthur was evacuated, but some 67,000 Allied prisoners died on the *Bataan Death March* to camps in the interior.

Despite Bataan being in an earthquake zone, the Marcos government built a nuclear power station here, near a dormant volcano. It has never generated any electricity, but in 1989 cost the country $350,000 a week in interest payments.

Batavia /bə'teɪviə/ former name until 1949 for ◊Jakarta, the capital of Indonesia on the island of Java.

Bath /bɑːθ/ historic city in Avon, England, situated SE of Bristol on the river Avon; population (1991) 79,900.

features Hot springs; the ruins of the baths after which it is named, as well as a great temple, are the finest Roman remains in Britain. Excavations 1979 revealed thousands of coins and 'curses', offered at a place which was thought to be the link between the upper and lower worlds. The Gothic Bath Abbey has an unusually decorated west front and fan vaulting. There is much 18th-century architecture, notably the Royal Crescent by John Wood. The Assembly Rooms 1771 were destroyed in an air raid 1942 but reconstructed 1963. The University of Technology was established 1966. The Bath Festival Orchestra is based here

history The Roman city of Aquae Sulis ('waters of Sul'—the British goddess of wisdom) was built in the first 20 years after the Roman invasion. In medieval times the hot springs were crown property, administered by the church, but the city was transformed in the 18th century to a fashionable spa, presided over by 'Beau' Nash. At his home here the astronomer William Herschel discovered Uranus 1781. Visitors included the novelists Tobias Smollett, Henry Fielding, and Jane Austen.

Bathurst /'bæθɜːst/ town in New South Wales, on the Macquarie River, Australia; population (1981) 19,600. It dates from the 1851 gold rush.

Bathurst /'bæθɜːst/ port in New Brunswick, Canada; population (1981) 19,500. Industries include copper and zinc mining; products include paper and timber.

Bathurst /'bæθɜːst/ former name (until 1973) of ◊Banjul, the capital of the Gambia.

Baton Rouge /'bætn 'ruːʒ/ deepwater port on the Mississippi River, USA, the capital of Louisiana; population (1990) 219,530. Industries include oil refining, petrochemicals, and iron.

The bronze and marble state capitol was built by Governor Huey Long.

Battersea /'bætəsi/ district of the Inner London borough of Wandsworth on the south bank of the Thames. It has a park (including a funfair 1951–74), a classically styled power station, now disused, and Battersea Dogs' Home (opened 1860) for strays.

Battle /'bætl/ town in Sussex, England, named after the Battle of Hastings, which took place here; population (1985 est) 5,200. There are remains of an abbey founded by William the Conqueror.

Battle Creek /'bætl kriːk/ city in S Michigan, USA, directly E of Kalamazoo; population (1990) 53,540. It became known as the cereal capital of the world after J H Kellogg, W K Kellogg, and C W Post established dry cereal and grain factories here. Battle Creek was also a station on the Underground Railroad.

Batumi /bə'tuːmi/ Black Sea port and capital of the autonomous republic of Adzhar, in Georgia; population (1987) 135,000. Main industries include oil refining, food canning, and engineering.

Bavaria /bə'veəriə/ (German *Bayern*) administrative region (German *Land*) of Germany

area 70,600 sq km/27,252 sq mi
capital Munich
towns Nuremberg, Augsburg, Würzburg, Regensburg
features largest of the German *Länder*; forms the Danube basin; festivals at Bayreuth and Oberammergau
products beer, electronics, electrical engineering, optics, cars, aerospace, chemicals, plastics, oil refining, textiles, glass, toys
population (1988) 11,000,000
famous people Lucas Cranach, Adolf Hitler, Franz Josef Strauss, Richard Strauss
religion 70% Roman Catholic, 26% Protestant
history the last king, Ludwig III, abdicated 1918, and Bavaria declared itself a republic.

The original Bavarians were Teutonic invaders from Bohemia who occupied the country at the end of the 5th century. From about 555 to 788 Bavaria was ruled by Frankish dukes of the Agilolfing family. In the 7th and 8th centuries the region was christianized by Irish and Scottish monks. In 788 Charlemagne deposed the last of the Agilolfing dukes and incorporated Bavaria into the Carolingian Empire, and in the 10th century it became part of the Holy Roman Empire. The house of Wittelsbach ruled parts or all of Bavaria 1181–1918; Napoleon made the ruler a king 1806. In 1871 Bavaria became a state of the German Empire.

Bavarians say that the difference between a rich farmer and a poor farmer is that the poor farmer cleans his Mercedes himself.

On **Bavaria** J W Murray in *Observer* 17 June 1979

Bay City /'beɪ 'sɪti/ industrial city in Michigan, USA; population (1990) 38,900. Industries include shipbuilding and engineering.

Bayern /'baɪən/ German name for ♢Bavaria, a region of Germany.

Bayeux /baɪ'ɜː/ town in N France; population (1982) 15,200. Its museum houses the Bayeux Tapestry. There is a 13th-century Gothic cathedral. Bayeux was the first town in W Europe to be liberated by the Allies in World War II, 8 June 1944.

Bay of Pigs inlet on the south coast of Cuba about 145 km/90 mi SW of Havana. It was the site of an unsuccessful invasion attempt by 1,500 US-sponsored Cuban exiles 17–20 April 1961; 1,173 were taken prisoner.

Bayonne /baɪ'ɒn/ river port in SW France; population (1990) 41,800. It trades in timber, steel, fertiliser, and brandy. It is a centre of Basque life. The bayonet was invented here.

Bayreuth /baɪ'rɔɪt/ town in Bavaria, S Germany, where opera festivals are held every summer; population (1983) 71,000. It was the home of composer Richard Wagner, and the Wagner theatre was established 1876.

Bayview /'beɪvjuː/ hamlet near Brownsville, Texas, USA, the site of a tent city, erected 1989, for the detention of Central American immigrants filing invalid claims for political asylum.

Beachy Head /'biːtʃi/ (French *Béveziers*) loftiest headland (162 m/532 ft high) on the south coast of England, between Seaford and Eastbourne in Sussex, the eastern termination of the South Downs. The lighthouse off the shore is 38 m/125 ft high.

Beaconsfield /'bekənzfiːld/ town in Buckinghamshire, England; 37 km/23 mi WNW of London; population (1981) 10,900. It has associations with Benjamin Disraeli (whose title was Earl of Beaconsfield), political theorist Edmund Burke, and the poet Edmund Waller.

Beagle Channel /'biːgl/ channel to the S of Tierra del Fuego, South America, named after the ship of Charles Darwin's voyage. Three islands at its eastern end, with krill and oil reserves within their 322 km/200 mi territorial waters, and the dependent sector of the Antarctic with its resources, were disputed between Argentina and Chile and awarded to Chile 1985.

Beas /'biːəs/ river in Himachal Pradesh, India, an upper tributary of the Sutlej, which in turn joins the Indus. It is one of the five rivers that give the Punjab its name. The ancient Hyphasis, it marked the limit of the invasion of India by Alexander the Great.

Beaufort Sea /'bəufət/ section of the Arctic Ocean off Alaska and Canada, named after Francis Beaufort. Oil drilling is allowed only in the winter months because the sea is the breeding and migration route of bowhead whales, the staple diet of the local Inuit people.

Beaulieu /'bjuːli/ village in Hampshire, England; 9 km/6 mi SW of Southampton; population (1985) 1,200. The former abbey is the home of Lord Montagu of Beaulieu and has the Montagu Museum of vintage cars.

Beauly Firth /'bjuːli/ arm of the North Sea cutting into Scotland N of Inverness, spanned by Kessock Bridge 1982.

Beaumont /'bəumɒnt/ city and port in SE Texas, USA, on the Neches River, NE of Houston; seat of Jefferson County; population (1990) 114,320. It is an oil-processing centre for the surrounding oil fields and a shipping point via the Sabine–Neches canal to the Gulf of Mexico; other industries include shipbuilding and paper production.

Beaune /bəun/ town SW of Dijon, France; population (1990) 22,100. It is the centre of the Burgundian wine trade, and has a wine museum. Other products include agricultural equipment and mustard.

Beauvais /bəu'veɪ/ town 76 km/47 mi NW of Paris, France; population (1990) 56,300. It is a market town trading in fruit, dairy produce, and agricultural machinery. Beauvais has a Gothic cathedral, the tallest in France (68 m/223 ft), and

Bedfordshire

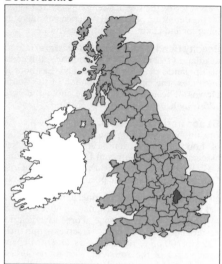

is renowned for tapestries (which are now made at the Gobelins factory, Paris).

The cathedral was planned to be the greatest church ever built. Its choir, built from 1250, was the tallest of any Gothic cathedral (48 m/158 ft), but it collapsed 1284, and the tower over the crossing followed it 1573.

Bebington town on Merseyside, England; population (1981) 64,150. Industries include oil and chemicals. There is a model housing estate originally built 1888 for Unilever workers, Port Sunlight.

Bechuanaland /ˌbetʃuˈɑːnəlænd/ former name (until 1966) of ⟡Botswana.

Bedford /ˈbedfəd/ administrative headquarters of Bedfordshire, England; population (1983) 89,200. Industries include agricultural machinery and airships. John Bunyan wrote *The Pilgrim's Progress* (1678) while imprisoned here.

Bedfordshire /ˈbedfədʃə/ county in S central England
area 1,240 sq km/479 sq mi
towns Bedford (administrative headquarters), Luton, Dunstable
features Whipsnade Zoo 1931, near Dunstable, a zoological park (200 hectares/500 acres) belonging to the London Zoological Society; Woburn Abbey, seat of the duke of Bedford
products cereals, vegetables, agricultural machinery, electrical goods
population (1991) 514,200
famous people John Bunyan, John Howard, Joseph Paxton.

Beds abbreviation for *Bedfordshire*.

Beersheba /bɪəˈʃiːbə/ industrial town in Israel; population (1987) 115,000. It is the chief centre of the Negev Desert and has been a settlement from the Stone Age.

Beijing /ˈbeɪˈdʒɪŋ/ or *Peking* capital of China; part of its northeast border is formed by the Great Wall of China; population (1989) 6,800,000. The municipality of Beijing has an area of 17,800 sq km/6,871 sq mi and a population (1990) of 10,819,000. Industries include textiles, petrochemicals, steel, and engineering.
features Tiananmen Gate (Gate of Heavenly Peace) and Tiananmen Square, where, in 1989, Chinese troops massacred over 1,000 students and civilians demonstrating for greater freedom and democracy; the Forbidden City, built between 1406 and 1420 as Gu Gong (Imperial Palace) of the Ming Emperors, where there were 9,000 ladies-in-waiting and 10,000 eunuchs in service (it is now the seat of the government); the Great Hall of the People 1959 (used for official banquets); museums of Chinese history and of the Chinese revolution; Chairman Mao Memorial Hall 1977 (shared from 1983 with Zhou Enlai, Zhu De, and Liu Shaoqi); the Summer Palace built by the dowager empress Zi Xi (damaged by European powers 1900, but restored 1903); Temple of Heaven (Tiantan); and Ming tombs 50 km/30 mi to the NW.
history Beijing, founded 2,000 years ago, was the 13th-century capital of the Mongol emperor Kublai Khan. Later replaced by Nanking, it was again capital from 1421, except from 1928 to 1949, when it was renamed Peiping. Beijing was held by Japan 1937–45.

It is lovely, but eerie.

On **Beijing** James Cameron *Witness* 1966

Beira /ˈbaɪrə/ port at the mouth of the River Pungwe, Mozambique; population (1986) 270,000. It exports minerals, cotton, and food products. A railway through the *Beira Corridor* links the port with Zimbabwe.

Beirut /ˌbeɪˈruːt/ or *Beyrouth* capital and port of Lebanon, devastated by civil war in the 1970s and 1980s, when it was occupied by armies of neighbouring countries; population (1988 est) 1,500,000.
history Beirut dates back to at least 1400 BC. Before the civil war 1975–90, Beirut was an international financial and educational centre, with four universities (Lebanese, Arab, French, and US); it was also a centre of espionage. Subsequent struggles for power among Christian and Muslim factions caused widespread destruction. From July–Sept 1982 the city was besieged and sections were virtually destroyed by the Israeli army to enforce the withdrawal of the forces of the Palestinian Liberation Organization (PLO). After the cease-fire, 500 Palestinians were massacred in the Sabra–Shatila camps 16–18 Sept, 1982, by dissident Phalangist and Maronite troops, with alleged Israeli complicity. Civil disturbances continued, characterized by sporadic street fighting and hostage taking. In 1987 Syrian troops entered the city and remained. Intensive fighting broke out between Christian and Syrian troops in Beirut, and by 1990 the strength of Syrian

military force in greater Beirut and E Lebanon was estimated at 42,000. In Oct 1990 President Elias Hwari formally invited Syrian troops to remove the Maronite Christian leader General Michel Aoun from his E Beirut stronghold; the troops then went on to dismantle the 'Green Line' separating Muslim western and Christian eastern Beirut. The Syrian-backed 'Greater Beirut Security Plan' was subsequently implemented by the Lebanese government, enforcing the withdrawal of all militias from greater Beirut.

A profound carelessness pervaded the life of Beirut, an empirical, striated disorder and randomness.

On **Beirut** John Knowles *Double Vision: American Thoughts Abroad* 1964

Bejaia /bɪˈdʒaɪə/ (formerly *Bougie*) port in Algeria, 193 km/120 mi E of Algiers; population (1983) 124,100. It is linked by pipeline with oil wells at Hassi Messaoud. It exports wood and hides.

Bekka, the /beˈkaː/ or *El Beqa'a* governorate of E Lebanon separated from Syria by the Anti-Lebanon Mountains. Zahlé and the ancient city of Baalbek are the chief towns. The Bekka Valley was of strategic importance in the Syrian struggle for control of N Lebanon. In the early 1980s the valley was penetrated by Shia Muslims who established an extremist Hezbollah stronghold with the support of Iranian Revolutionary Guards.

Belarus Republic of
area 207,600 sq km/80,100 sq mi
capital Minsk (Mensk)
towns Gomel, Vitebsk, Mogilev, Bobruisk, Grodno, Brest
physical more than 25% forested; rivers W Dvina, Dnieper and its tributaries, including the Pripet and Beresina; the Pripet Marshes in the E; mild and damp climate
environment large areas contaminated by fallout
features Belovezhskaya Pushcha (scenic forest reserve)
head of state Stanislav Shushkevich from 1991
head of government Vyacheslav Kebich from 1990
political system emergent democracy
products peat, agricultural machinery, fertilizers, glass, textiles, leather, salt, electrical goods, meat, dairy produce
currency rouble and dukat
population (1990) 10,200,000 (77% Byelorussian 'Eastern Slavs', 13% Russian, 4% Polish, 1% Jewish)
languages Byelorussian, Russian
religions Roman Catholic, Russian Orthodox, with Baptist and Muslim minorities
chronology
1918–19 Briefly independent from Russia.
1937–41 More than 100,000 people were shot in mass executions ordered by Stalin.
1941–44 Occupied by Nazi Germany.
1945 Became a founding member of the United Nations.
1986 April: fallout from the Chernobyl nuclear reactor in Ukraine contaminated a large area.
1989 Byelorussian Popular Front established as well as a more extreme nationalist organization, the Tolaka group.
1990 Sept: Byelorussian established as state language and republican sovereignty declared.
1991 April: Minsk hit by nationalist-backed general strike, calling for disbandment of Communist Party (CP) workplace cells. Aug: declared independence from Soviet Union in wake of failed anti-Gorbachev coup; suspended CP. Sept: Shushkevich elected president. Dec: Commonwealth of Independent States (CIS) formed in Minsk; Belarus accorded diplomatic recognition by USA.
1992 Jan: admitted into Conference on Security and Cooperation in Europe (CSCE). May: protocols signed with USA agreeing to honour START treaty.

Belau, Republic of /bəˈlaʊ/ (formerly *Palau*) self-governing island group in Micronesia
area 500 sq km/193 sq mi
capital Koror
features 26 larger islands (eight inhabited) and about 300 islets
population (1990) 15,100
history Spain held the islands from about 1600, and sold them to Germany 1899. Japan seized them in World War I, administered them by League of Nations mandate, and used them as a naval base during World War II. They were captured by the USA 1944, and became part of the US Trust Territory of the Pacific Islands three years later. Belau became internally self-governing 1980. It is the only remaining member of the Trust Territory.

Belém /bəˈlem/ port and naval base in N Brazil; population (1991) 1,235,600. The chief trade centre of the Amazon basin, it is also known as Pará, the name of the state of which it is capital. It was founded about 1615 as Santa Maria de Belém do Grās Pará.

Belfast /ˌbelˈfaːst/ industrial port (shipbuilding, engineering, electronics, textiles, tobacco) and capital of Northern Ireland since 1920; population (1985) 300,000 (Protestants form the majority in E Belfast, Catholics in the W). Since 1968 the city has been heavily damaged by civil disturbances.

history Belfast grew up around a castle built 1177 by John de Courcy. With the settlement of English and Scots, Belfast became a centre of Irish Protestantism in the 17th century. An influx of Huguenots after 1685 extended the linen industry, and the 1800 Act of Union with England resulted in the promotion of Belfast as an industrial centre. It was created a city 1888, with a lord mayor from 1892.

The Lord in His Mercy be good to Belfast/ The grief of the exile she soothed as he passed.

On **Belfast** old Irish ballad

Belfort /bel'fɔ:/ town in NE France; population (1990) 51,900. It is in the strategic *Belfort Gap* between the Vosges and Jura mountains and is the capital of the *département* of Territoire de Belfort. Industries include chemicals, engineering, plastics, and textiles.

Belgaum /bel'gɔ:m/ city in Karnataka, S India; population (1981) 300,000. The main industry is cotton manufacture. It is known for its Jain temples.

Belgian Congo /ˌbeldʒən 'kɒŋgəʊ/ former name (1908–60) of ◊Zaire.

Belgium Kingdom of (French *Royaume de Belgique*, Flemish *Koninkrijk België*)
area 30,510 sq km/11,784 sq mi
capital Brussels
towns Ghent, Liège, Charleroi, Bruges, Mons, Namur, Leuven; ports are Antwerp, Ostend, Zeebrugge
physical fertile coastal plain in NW, central rolling hills rise eastwards, hills and forest in SE
environment a 1989 government report judged the drinking water in Flanders to be 'seriously substandard' and more than half the rivers and canals in that region to be in a 'very bad' condition
features Ardennes Forest; rivers Scheldt and Meuse
head of state King Albert from 1993
head of government Jean-Luc Dehaene from 1992
political system liberal democracy
political parties Flemish Christian Social Party (CVP), centre-left; French Social Christian Party (PSC), centre-left; Flemish Socialist Party (SP), left of centre; French Socialist Party (PS), left of centre; Flemish Liberal Party (PVV), moderate centrist; French Liberal Reform Party (PRL), moderate centrist; Flemish People's Party (VU), federalist; Flemish Green Party (Agalev); French Green Party (Ecolo)
exports iron, steel, textiles, manufactured goods, petrochemicals, plastics, vehicles, diamonds
currency Belgian franc

population (1990 est) 9,895,000 (comprising Flemings and Walloons); growth rate 0.1% p.a.
life expectancy men 72, women 78
languages in the N (Flanders) Flemish (a Dutch dialect, known as *Vlaams*) 55%; in the S (Wallonia) Walloon (a French dialect) 32%; bilingual 11%; German (E border) 0.6%; all are official
religion Roman Catholic 75%
literacy 98% (1984)
GDP $218.7 bn (1992)
chronology
1830 Belgium became an independent kingdom.
1914 Invaded by Germany.
1940 Again invaded by Germany.
1948 Belgium became founding member of Benelux Customs Union.
1949 Belgium became founding member of Council of Europe and NATO.
1951 Leopold III abdicated in favour of his son Baudouin.
1952 Belgium became founding member of European Coal and Steel Community.
1957 Belgium became founding member of the European Economic Community.
1971 Steps towards regional autonomy taken.
1972 German-speaking members included in the cabinet for the first time.
1973 Linguistic parity achieved in government appointments.
1974 Leo Tindemans became prime minister. Separate regional councils and ministerial committees established.
1978 Wilfried Martens succeeded Tindemans.
1980 Open violence over language divisions. Regional assemblies for Flanders and Wallonia and a three-member executive for Brussels created.
1981 Short-lived coalition led by Mark Eyskens was followed by the return of Martens.
1987 Martens head of caretaker government after break-up of coalition.
1988 Following a general election, Martens formed a new CVP–PS–SP–PSC–VU coalition.
1992 Martens-led coalition collapsed; Jean-Luc Dehaene formed a new CVP-led coalition.

1993 Federal system adopted, based on Flanders, Wallonia, and Brussels. King Baudouin died and was succeeded by his brother Albert.

Belgium suffers severely from linguistic indigestion.

On **Belgium** R W G Penn in *Geographical Magazine* March 1980

Belgrade /ˌbelˈɡreɪd/ (Serbo-Croatian *Beograd*) capital of Yugoslavia and Serbia, and Danube river port linked with the port of Bar on the Adriatic Sea; population (1981) 1,470,000. Industries include light engineering, food processing, textiles, pharmaceuticals, and electrical goods.

Belgravia /belˈɡreɪvɪə/ district of London, laid out in squares by Thomas Cubitt (1788–1855) 1825–30, and bounded to the N by Knightsbridge.

Belitung /bɪˈliːtuŋ/ alternative name for the Indonesian island of ◊Billiton.

Belize (formerly *British Honduras*)
area 22,963 sq km/8,864 sq mi
capital Belmopan
towns ports Belize City, Dangriga, Punta Gorda; Orange Walk, Corozal
physical tropical swampy coastal plain, Maya Mountains in S; over 90% forested
environment since 1981 Belize has developed an extensive system of national parks and reserves to protect large areas of tropical forest, coastal mangrove, and offshore islands. Forestry has been replaced by agriculture and ecotourism, which are now the most important sectors of the economy; world's first jaguar reserve created 1986 in the Cockscomb Mountains
features world's second longest barrier reef; Maya ruins
head of state Elizabeth II from 1981, represented by governor general
head of government George Price from 1989
political system constitutional monarchy
political parties People's United Party (PUP), left of centre; United Democratic Party (UDP), moderate conservative
exports sugar, citrus fruits, rice, fish products, bananas
currency Belize dollar
population (1990 est) 180,400 (including Mayan minority in the interior); growth rate 2.5% p.a.
life expectancy (1988) 60 years
languages English (official); Spanish (widely spoken), native Creole dialects
media no daily newspaper; several independent weekly tabloids
religions Roman Catholic 60%, Protestant 35%
literacy 93% (1988)
GDP $247 million (1988); $1,220 per head
chronology
1862 Belize became a British colony.
1954 Constitution adopted, providing for limited internal self-government. General election won by George Price.

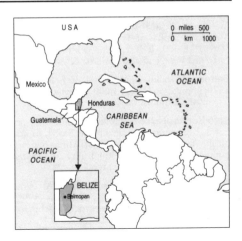

1964 Self-government achieved from the UK (universal adult suffrage introduced).
1965 Two-chamber national assembly introduced, with Price as prime minister.
1970 Capital moved from Belize City to Belmopan.
1973 British Honduras became Belize.
1975 British troops sent to defend the disputed frontier with Guatemala.
1980 United Nations called for full independence.
1981 Full independence achieved. Price became prime minister.
1984 Price defeated in general election. Manuel Esquivel formed the government. The UK reaffirmed its undertaking to defend the frontier.
1989 Price and the PUP won the general election.
1991 Diplomatic relations with Guatemala established.

If the world had any ends British Honduras would certainly be one of them.

On **Belize** (formerly *British Honduras*) Aldous Huxley *Beyond the Mexique Bay* 1934

Belize City /bɪˈliːz/ chief port of Belize, and capital until 1970; population (1991) 46,000. After the city was destroyed by a hurricane 1961 it was decided to move the capital inland, to Belmopan.

Belleville /ˈbelvɪl/ city in SW Illinois, USA, southeast of East St Louis; seat of St Clair County; population (1990) 42,800. Industries include coal, beer, furnaces and boilers, and clothing.

Bellingham /ˈbelɪŋhæm/ city and port in NW Washington, USA, just S of the Canadian border, on Bellingham Bay in the Strait of Georgia; population (1990) 52,180. It is a port of entry for the logging and paper industry; there are also shipbuilding and food processing industries.

Bellingshausen Sea section of the S Pacific off the Antarctic coast. It is named after the Russian explorer Fabian Gottlieb von Bellingshausen.

Bellinzona /ˌbelɪnt'səunə/ town in Switzerland on the river Ticino, 16 km/10 mi from Lago Maggiore; capital of Ticino canton; population (1990) 16,900. It is a traffic centre for the St Gotthard Pass, and also a tourist centre.

Belmopan /ˌbelmə'pæn/ capital of Belize from 1970; population (1991) 4,000. It replaced Belize City as the administrative centre of the country.

Belo Horizonte /'beləu ˌhɒrɪ'zɒnteɪ/ industrial city (steel, engineering, textiles) in SE Brazil, capital of the fast-developing state of Minas Gerais; population (1991) 2,103,300. Built in the 1890s, it was Brazil's first planned modern city.

Beloit /bə'lɔɪt/ city in SE Wisconsin, USA, on the Rock River, SE of Madison; population (1990) 35,600. Industries include electrical machinery, shoes, generators, and diesel engines.

Belorussia /ˌbelə'rʌʃə/ alternative form of ◊Belarus, a country in E central Europe.

Benares /bɪ'nɑːrɪz/ alternative transliteration of ◊Varanasi, a holy city in India.

Bendigo /'bendɪgəu/ city in Victoria, Australia, about 120 km/75 mi NNW of Melbourne; population (1986) 62,400. Founded 1851 at the start of a gold rush, the town takes its name from the pugilist William Thompson (1811–1889), known as 'Bendigo'.

Benevento /ˌbenɪ'ventəu/ historic town in Campania, S Italy; population (1981) 62,500. It is known for the production of Strega liqueur.

Bengal, Bay of /ˌbeŋ'gɔːl/ part of the Indian Ocean lying between the east coast of India and the west coast of Myanmar (Burma) and the Malay Peninsula.
  The Irrawaddy, Ganges, and Brahmaputra rivers flow into the bay. The principal islands are to be found in the Andaman and Nicobar groups.

Benghazi /ben'gɑːzi/ or **Banghazi** historic city and industrial port in N Libya on the Gulf of Sirte; population (1982) 650,000. It was controlled by Turkey between the 16th century and 1911, and by Italy 1911–42; it was a major naval supply base during World War II.
  Colonized by the Greeks in the 7th century BC (as **Euhesperides**), Benghazi was taken by Rome in the 1st century BC (renamed **Berenice**) and by the Vandals in the 5th century AD. It became Arab in the 7th century. With Tripoli, it was co-capital of Libya 1951–72.

Benguela /ben'gweɪlə/ port in Angola, SW Africa; population (1970) 41,000. It was founded 1617. Its railway runs inland to the copper mines of Zaire and Zambia.

Benidorm /'benɪdɔːm/ fishing village and popular tourist resort on the Mediterranean Costa Blanca, in Alicante province, E Spain; population (1991) 74,900.

Benin People's Republic of (*République Populaire du Bénin*)
area 112,622 sq km/43,472 sq mi

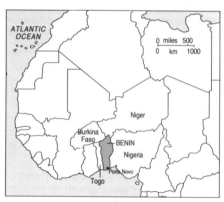

capital Porto Novo (official), Cotonou (de facto)
towns Abomey, Natitingou, Parakou; chief port Cotonou
physical flat to undulating terrain; hot and humid in S; semiarid in N
features coastal lagoons with fishing villages on stilts; Niger River in NE
head of state and government Nicéphore Soglo from 1991
political system socialist pluralist republic
political parties Party of the People's Revolution of Benin (PRPB); other parties from 1990
exports cocoa, peanuts, cotton, palm oil, petroleum, cement, sea products
currency CFA franc
population (1990 est) 4,840,000; growth rate 3% p.a.
life expectancy men 42, women 46
languages French (official); Fon 47% and Yoruba 9% in south; six major tribal languages in north
religions animist 65%, Christian 17%, Muslim 13%
literacy men 37%, women 16% (1985 est)
GDP $1.6 bn (1987); $365 per head
chronology
1851 Under French control.
1958 Became self-governing dominion within the French Community.
1960 Independence achieved from France.
1960–72 Acute political instability, with switches from civilian to military rule.
1972 Military regime established by General Mathieu Kerekou.
1974 Kerekou announced that the country would follow a path of 'scientific socialism'.
1975 Name of country changed from Dahomey to Benin.
1977 Return to civilian rule under a new constitution.
1980 Kerekou formally elected president by the national revolutionary assembly.
1989 Marxist-Leninism dropped as official ideology. Strikes and protests against Kerekou's rule mounted; demonstrations banned and army deployed against protesters.
1990 Referendum support for multiparty politics.
1991 Multiparty elections held. Kerekou defeated in presidential elections by Nicéphore Soglo.

Benin /be'niːn/ former African kingdom 1200–1897, now a province of Nigeria. It reached the height of its power in the 14th–17th centuries when it ruled the area between the Niger Delta and Lagos.

Benin traded in spices, ivory, palm oil, and slaves until its decline and eventual incorporation into Nigeria. The oba (ruler) of Benin continues to rule his people as a divine monarch. The present oba is considered an enlightened leader and one who is helping his people to become part of modern Nigeria.

Artworks honouring the Oba of Benin were looted by a British military expedition 1897. They included cast bronzes and carved ivories and have since found their way into museums and into the hands of collectors worldwide.

Ben Nevis /ben 'nevɪs/ highest mountain in the British Isles (1,343 m/4,406 ft), in the Grampian Mountains, Scotland.

Read me a lesson, muse, and speak it loud/
Upon the top of Nevis blind in Mist!

On **Ben Nevis** John Keats
'On Ben Nevis' 1818

Benoni /bɪ'nəuni/ city in the Transvaal, South Africa, 27 km/17 mi E of Johannesburg; population (1980) 207,000. It was founded 1903 as a gold-mining centre.

Bentiu /'bentiuː/ oil-rich region to the W of the White Nile, in the Upper Nile province of S Sudan.

Benton Harbor /ˌbentən 'hɑːbə/ city in SW Michigan, USA, northeast of Chicago, Illinois, which is across Lake Michigan; population (1990) 12,820. Industries include iron and other metal products and food processing. The religious sect, House of David, was established here 1903.

Benue /'benueɪ/ river in Nigeria, largest tributary of the river Niger; it is navigable for most of its length of 1,400 km/870 mi.

Beograd /'beɪəugræd/ Serbo-Croatian form of ▷Belgrade, the capital of Yugoslavia.

Berbera /'bɜːbərə/ seaport in Somalia, with the only sheltered harbour on the south side of the Gulf of Aden; population (1982) 55,000. It is in a strategic position on the oil route and has a deep-sea port completed 1969. It was under British control 1884–1960.

Berdichev /bɪə'diːtʃef/ town in W Ukraine, 48 km/30 mi S of Zhitomir; population (1980) 60,000. Industries include engineering and food processing.

Berdyansk /bɪə'djænsk/ city and port on the Berdyansk Gulf of the Sea of Azov, in SE Ukraine; population (1985) 130,000.

Berezniki /bɪˌreznɪ'kiː/ city in Russia, on the Kama River north of Perm; population (1987) 200,000. It was formed 1932 by the amalgamation of several older towns. Industries include chemicals and paper.

Bergamo /'beəgəməu/ city in Lombardy, Italy; 48 km/30 mi NE of Milan; population (1989) 117,600. Industries include silk and metal. The Academia Carrara holds a fine collection of paintings.

Bergen /'beəgən/ industrial port (shipbuilding, engineering, fishing) in SW Norway; population (1991) 213,300. Founded 1070, Bergen was a member of the Hanseatic League.

Bergen-op-Zoom /'beəxən ɒp 'zəum/ fishing port in the SW Netherlands; population (1991) 46,900. It produces chemicals, cigarettes, and precision goods.

Bergisch Gladbach /'beəgɪʃ 'glædbæx/ industrial city (paper, metal products) in North Rhine-Westphalia, Germany; population (1988) 102,000.

Bering Sea /'beərɪŋ/ section of the N Pacific between Alaska and Siberia, from the Aleutian Islands north to the Bering Strait.

Bering Strait /'beərɪŋ/ strait between Alaska and Siberia, linking the N Pacific and Arctic oceans.

Berkeley /'bɜːkli/ city on San Francisco Bay in California; population (1990) 102,700. It is the site of an acclaimed branch of the University of California, noted for its nuclear research at the Lawrence Berkeley Laboratory.

Berks abbreviation for *Berkshire*.

Berkshire /'bɑːkʃə/ or *Royal Berkshire* county in S central England
area 1,260 sq km/486 sq mi
towns Reading (administrative headquarters), Eton, Slough, Maidenhead, Ascot, Bracknell, Newbury, Windsor

Berkshire

features rivers Thames and Kennet; Inkpen Beacon, 297 m/975 ft; Bagshot Heath; Ridgeway Path, walkers' path (partly prehistoric) running from Wiltshire across the Berkshire Downs into Hertfordshire; Windsor Forest and Windsor Castle; Eton College; Royal Military Academy at Sandhurst; atomic-weapons research establishment at Aldermaston; the former main UK base for US cruise missiles at Greenham Common, Newbury
products general agricultural and horticultural goods, electronics, plastics, pharmaceuticals
population (1991) 716,500
famous people Jethro Tull, William Laud, Stanley Spencer.

Berlin /bɜːˈlɪn/ industrial city (machine tools, electrical goods, paper, printing) and capital of the Federal Republic of Germany; population (1990) 3,102,500.
features Unter den Linden, the tree-lined avenue once the whole city's focal point, has been restored in what was formerly East Berlin. The fashionable Kurfürstendamm and the residential Hansa quarter form part of what was once West Berlin. Prominent buildings include the Reichstag (former parliament building); Schloss Bellevue (Berlin residence of the president); Schloss Charlottenburg (housing several museums); Congress Hall; restored 18th-century State Opera; Dahlem picture gallery. The environs of Berlin include the Grünewald forest and Wannsee lake.
history First mentioned about 1230, the city grew out of a fishing village, joined the Hanseatic League in the 15th century, became the permanent seat of the Hohenzollerns, and was capital of the Brandenburg electorate 1486–1701, of the kingdom of Prussia 1701–1871, and of united Germany 1871–1945. From the middle of the 18th century it developed into a commercial and cultural centre. In World War II air raids and conquest by the Soviet army 23 April–2 May 1945, destroyed much of the city. After the war, Berlin was divided into four sectors—British, US,

French, and Soviet—and until 1948 was under quadripartite government by the Allies. Following the Berlin blockade the city was divided, with the USSR creating a separate municipal government in its sector. The other three sectors (West Berlin) were made a *Land* of the Federal Republic May 1949, and in Oct 1949 East Berlin was proclaimed capital of East Germany. The Berlin Wall divided the city from 1961 to 1989, but in Oct 1990 Berlin became the capital of a unified Germany, once more with East and West Berlin reunited as the 16th *Land* (state) of the Federal Republic.

All free men, wherever they live, are citizens of Berlin, and therefore, as a free man, I take pride in the words 'Ich bin ein Berliner.'

On **Berlin** John F Kennedy 1963

Bermuda /bəˈmjuːdə/ British colony in the NW Atlantic Ocean
area 54 sq km/21 sq mi
capital and chief port Hamilton
features consists of about 150 small islands, of which 20 are inhabited, linked by bridges and causeways; Britain's oldest colony
products Easter lilies, pharmaceuticals; tourism and banking are important
currency Bermuda dollar
population (1988) 58,100
language English
religion Christian
government under the constitution of 1968, Bermuda is a fully self-governing British colony, with a governor (Lord Waddington from 1992), senate, and elected House of Assembly (premier from 1982 John Swan, United Bermuda Party)
history the islands were named after Juan de Bermudez, who visited them 1515, and were settled by British colonists 1609. Indian and African slaves were transported from 1616, and soon outnumbered the white settlers. Racial violence 1977 led to intervention, at the request of the government, by British troops.

Bermuda

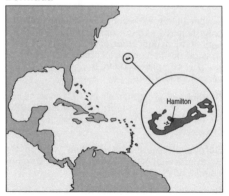

Berlin 1945–89

⧄ French sector	⧄ US sector
☐ British sector	▨ Soviet sector

Bermuda is essentially a small town in a glamorous setting.

On **Bermuda** David Shelley Nicholl, letter to the *Times* 17 Feb 1978

Bermuda Triangle /bə'mjuːdə/ sea area bounded by Bermuda, Florida, and Puerto Rico, which gained the nickname 'Deadly Bermuda Triangle' 1964 when it was suggested that unexplained disappearances of ships and aircraft were exceptionally frequent there. Analysis of the data has not confirmed the idea.

Bern /beən/ (French *Berne*) capital of Switzerland and of Bern canton, in W Switzerland on the Aare River; population (1990) 134,600; canton 945,600. It joined the Swiss confederation 1353 and became the capital 1848. Industries include textiles, chocolate, pharmaceuticals, light metal and electrical goods.

Berne was founded 1191 and made a free imperial city by Frederick II 1218. Its name is derived from the bear in its coat of arms, and there has been a bear pit in the city since the 16th century. The minster was begun 1421, the town hall 1406, and the university 1834. It is the seat of the Universal Postal Union.

Bernese Oberland /'bɜːniːz 'əubəlænd/ or **Bernese Alps** mountainous area in the S of Berne canton. It includes the Jungfrau, Eiger, and Finsteraarhorn peaks. Interlaken is the chief town.

Berwickshire /'berɪkʃə/ former county of SE Scotland, a district of Borders region from 1975.

Berwick-upon-Tweed /'berɪk əpɒn 'twiːd/ town in NE England, at the mouth of the Tweed, Northumberland, 5 km/3 mi S of the Scottish border; population (1981) 26,230. It is a fishing port. Other industries include iron foundries and shipbuilding.
features Three bridges cross the Tweed: the Old Bridge 1611–34 with 15 arches, the Royal Border railway bridge 1850 constructed by Robert Stephenson, and the Royal Tweed Bridge 1928.
history Held alternately by England and Scotland for centuries, Berwick was made a neutral town 1551; it was attached to Northumberland 1885.

Besançon /bə'zɒnsɒn/ town on the river Doubs, France; population (1990) 119,200. It is the capital of Franche-Comté. The first factory to produce artificial fibres was established here 1890. Industries include textiles and clockmaking. It has fortifications by Vauban, Roman remains, and a Gothic cathedral. The writer Victor Hugo and the Lumière brothers, inventors of cinematography, were born here.

Bessarabia /ˌbesə'reɪbiə/ region in SE Europe, divided between Moldova and Ukraine. Bessarabia was annexed by Russia 1812, but broke away at the Russian Revolution to join Romania. The cession was confirmed by the Allies, but not by Russia, in a Paris treaty of 1920; the USSR reoccupied it 1940 and divided it between the Moldavian and Ukrainian republics (now independent Moldova and Ukraine). Romania recognized the position in the 1947 peace treaty.

Bethlehem /'beθlɪhem/ (Hebrew *Beit-Lahm*) town on the west bank of the river Jordan, S of Jerusalem; population (1980) 14,000. It was occupied by Israel 1967. In the Bible it is mentioned as the birthplace of King David and Jesus.

Bethlehem /'beθlɪhem/ city in E Pennsylvania, USA; population (1990) 71,428. Its former steel industry has been replaced by high technology.

Béthune /beɪ'tjuːn/ city in N France, W of Lille; population (1990) 25,200. Industries include textiles, machinery, and tyres.

Betws-y-coed /'betus ə 'kɔɪd/ village in Gwynedd, N Wales; population (1981) 750. It is a tourist centre. There are waterfalls nearby.

Beverly Hills /'bevəli/ residential city and a part of greater Los Angeles, California, USA; population (1990) 31,900. It is known as the home of Hollywood film stars.

Beverly Hills ... is a caricature of a stockbroker's suburb, enlivened by illusions of greater grandeur.

On **Beverly Hills** James Morris *Coast to Coast* 1956

Bexhill-on-Sea /ˌbeks'hɪl/ seaside resort in E Sussex, England; population (1989 est) 38,000.

Bexley borough of SE Greater London
features 16th-century Hall Palace; Red House, home of William Morris 1860–65
population (1991) 211,200.

Béziers /bez'jeɪ/ city in Languedoc-Roussillon, S France; population (1990) 72,300. It is a centre of the wine trade. It was once a Roman station and was the site of a massacre 1209 in the Albigensian Crusade.

Bhagalpur /'bɑːglpuə/ town in N India, on the river Ganges; population (1981) 225,000. It manufactures silk and textiles. Several Jain temples are here.

Bhamo /bə'məu/ town in Myanmar (Burma), near the Chinese frontier, on the Irrawaddy River. It is the inland limit of steam navigation and is mainly a trading centre.

Bharat /'bʌrət/ Hindi name for ◊India.

Bhatgaon /bɑː't'gɑːɒn/ or *Bhadgaon* or *Bhaktapur* town in Nepal, 11 km/7 mi SE of Katmandu; population (1981) 48,500. It has been a religious centre since the 9th century; there is a palace.

Bhavnagar /bau'nʌgə/ port in Gujarat, NW India, in the Kathiawar peninsula; population (1981)

308,000. It is a centre for textile industry. It was capital of the former Rajput princely state of Bhavnagar.

Bhopal /bəu'pɑːl/ industrial city (textiles, chemicals, electrical goods, jewellery) and capital of Madhya Pradesh, central India; population (1981) 672,000. Nearby Bhimbetka Caves, discovered 1973, have the world's largest collection of prehistoric paintings, about 10,000 years old. In 1984 some 2,600 people died from an escape of the poisonous gas methyl isocyanate from a factory owned by US company Union Carbide; another 300,000 suffer from long-term health problems.

The city was capital of the former princely state of Bhopal, founded 1723, which became allied to Britain 1817. It was merged with Madhya Pradesh 1956.

Bhubaneswar /ˌbuvə'neɪʃwə/ city in NE India; capital of Orissa state; population (1981) 219,200. Utkal University was founded 1843. A place of pilgrimage and centre of Siva worship, it has temples of the 6th–12th centuries.

It was capital of the Kesaris (Lion) dynasty of Orissa 474–950.

Bhutan Kingdom of (*Druk-yul*)
area 46,500 sq km/17,954 sq mi
capital Thimbu (Thimphu)
towns Paro, Punakha, Mongar
physical occupies southern slopes of the Himalayas; cut by valleys formed by tributaries of the Brahmaputra; thick forests in S
features Gangkar Punsum (7,529 m/24,700 ft) is one of the world's highest unclimbed peaks
head of state and government Jigme Singye Wangchuk from 1972
political system absolute monarchy
political parties none officially; illegal Bhutan People's Party (BPP)
exports timber, talc, fruit and vegetables, cement, distilled spirits, calcium carbide
currency ngultrum; also Indian currency
population (1990 est) 1,566,000 (75% Ngalops and Sharchops, 25% Nepalese); growth rate 2% p.a.
life expectancy men 44, women 43
languages Dzongkha (official, a Tibetan dialect), Sharchop, Bumthap, Nepali, and English
religions 75% Lamaistic Buddhist (state religion), 25% Hindu
literacy 5%
GDP $250 million (1987); $170 per head
chronology
1865 Trade treaty with Britain signed.
1907 First hereditary monarch installed.
1910 Anglo-Bhutanese Treaty signed.
1949 Indo-Bhutan Treaty of Friendship signed.
1952 King Jigme Dorji Wangchuk installed.
1953 National assembly established.
1959 4,000 Tibetan refugees given asylum.
1968 King established first cabinet.
1972 King died and was succeeded by his son Jigme Singye Wangchuk.
1979 Tibetan refugees told to take Bhutanese citizenship or leave; most stayed.
1983 Bhutan became a founding member

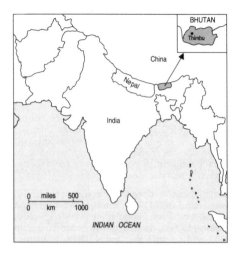

of the South Asian Association for Regional Cooperation organization (SAARC).
1988 King imposed 'code of conduct' suppressing Nepalese customs.
1990 Hundreds of people allegedly killed during prodemocracy demonstrations.

Biafra, Bight of /bi'æfrə/ name until 1975 of the Bight of ⟐Bonny, W Africa.

Białystok /bjæ'wɪstɒk/ city in E Poland; capital of Białystok region; population (1990) 270,600. Industries include textiles, chemicals, and tools. Founded 1310, the city belonged to Prussia 1795–1807 and to Russia 1807–1919.

Biarritz /bɪə'rɪts/ town on the Bay of Biscay, France, near the Spanish border; population (1990) 28,900. A seaside resort and spa town, it was popularized by Queen Victoria and Edward VII.

Biel /biːl/ (French *Bienne*) town in NW Switzerland; population (1990) 52,700. Its main industries include engineering, scientific instruments, and watch-making.

Bihar

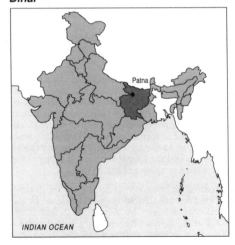

Bielefeld /'biːləfeld/ city in North Rhine-Westphalia, Germany, 55 km/34 mi E of Münster; population (1988) 299,000. Industries include textiles, drinks, chemicals, machinery, and motorcycles.

Bielostok /ˌbjelə'stɒk/ Russian form of ▷Białystok, a city in Poland.

Bienne /bjen/ French form of ▷Biel, a town in Switzerland.

Bihar /bɪ'hɑː/ or **Behar** state of NE India
area 173,900 sq km/67,125 sq mi
capital Patna
features river Ganges in the N, Rajmahal Hills in the S
products copper, iron, coal, rice, jute, sugar cane, grain, oilseed
population (1991) 86,338,900
languages Hindi, Bihari
famous people Chandragupta, Asoka
history the ancient kingdom of Magadha roughly corresponded to central and S Bihar. Many Bihari people were massacred as a result of their protest at the establishment of Bangladesh 1971.

Bikaner /ˌbɪkə'nɪə/ city in Rajasthan, N India; population (1981) 280,000. Once capital of the Rajput state of Bikaner, it is now a centre for carpet-weaving.

Bikini /bɪ'kiːni/ atoll in the Marshall Islands, W Pacific, where the USA carried out 23 atomic- and hydrogen-bomb tests (some underwater) 1946–58.

The islanders were relocated by the USA before 1946. Some returned after Bikini was declared safe for habitation 1969, but they were again removed in the late 1970s because of continuing harmful levels of radiation. In 1990 a US plan was announced to remove radioactive topsoil, allowing 800 islanders to return home.

Bilbao /bɪl'bau/ industrial port (iron and steel, chemicals, cement, food) in N Spain, capital of Biscay province; population (1991) 372,200.

Billings /'bɪlɪŋz/ city in S central Montana, USA, on the north shore of the Yellowstone river; seat of Yellowstone County; population (1990) 81,100. It is a centre for transporting livestock and animal products and for vegetable and grain processing. Nearby, is Big Horn Indian reservation where the Battle of Little Big Horn took place 1876.

Billingsgate /'bɪlɪŋzgeɪt/ chief London wholesale fish market, formerly (from the 9th century) near London Bridge. It re-opened 1982 at the new Billingsgate market, West India Dock, Isle of Dogs.

Billiton /'bɪlɪtɒn/ Indonesian island in the Java Sea, between Borneo and Sumatra, one of the Sunda Islands; area 4,830 sq km/1,860 sq mi. The chief port is Tanjungpandan. Tin mining is the chief industry.

Biloxi /bɪ'lɒksi/ port in Mississippi, USA; population (1990) 46,300. Chief occupations include tourism and seafood canning. Named after a local Indian people, Biloxi was founded 1719 by the French.

Binghamton /'bɪŋəmtən/ city in S central New York State, USA, where the Chenango River meets the Susquehanna River; population (1990) 53,000. Industries include electronic, computer, and camera equipment and textiles. Johnson City and Endicott, directly to the W on the Susquehanna River, form the Triple Cities with Binghamton.

Bío-Bío /'biːəu 'biːəu/ longest river in Chile; length 370 km/230 mi from its source in the Andes to its mouth on the Pacific. The name is an Araucanian term meaning 'much water'.

Bioko island in the Bight of Bonny, W Africa, part of Equatorial Guinea; area 2,017 sq km/786 sq mi; products include coffee, cacao, and copra; population (1983) 57,190. Formerly a Spanish possession, as **Fernando Po**, it was known 1973–79 as **Macías Nguema Bijogo**.

Birkenhead /'bɜːkənhed/ seaport in Merseyside, England, on the Mersey estuary opposite Liverpool; population (1981) 123,884. Chief industries include shipbuilding and engineering. The rail Mersey Tunnel 1886 and road Queensway Tunnel 1934 link Birkenhead with Liverpool.

history The first settlement grew up round a Benedictine priory, and Birkenhead was still a small village when William Laird established a small shipbuilding yard, the forerunner of the immense Cammell Laird yards. In 1829 the first iron vessel in the UK was built at Birkenhead. Wallasey dock, the first of the series, was opened 1847.

Birmingham /'bɜːmɪŋəm/ industrial city in the West Midlands, second largest city of the UK; population (1991 est) 934,900, metropolitan area 2,632,000. Industries include motor vehicles, machine tools, aerospace control systems, plastics, chemicals, and food.

features The National Exhibition Centre and Sports Arena; Aston University, linked to a science park; a school of music and symphony orchestra; the City Art Gallery containing a Pre-Raphaelite collection; the repertory theatre founded 1913 by Sir Barry Jackson (1897–1961) (since 1990 it has been the home of the Royal Ballet); Symphony Hall (holding over 4,000) opened 1991; Sutton Park, in the residential suburb of Sutton Coldfield, has been a public country recreational area since the 16th century.

history A medieval township, Birmingham supplied large quantities of weapons to the Parliamentarians during the Civil War. Its location on the edge of the S Staffordshire coalfields and its reputation for producing small arms allowed it to develop rapidly during the 18th–19th centuries (much of the city was rebuilt 1875–82). It has continued to expand in the 20th century due to a gradual change from heavy to high-tech industries. Lawn tennis was invented here.

What wilt thou doe, black Vulcans noysey Towne,/Old Bremigham? lowd Fame to thee affords/A title from the Make not Use of Swords.

On **Birmingham** Abraham Cowley
'The Civil War' c. 1643

Birmingham /ˈbɜːmɪŋhæm/ commercial and industrial city (iron, steel, chemicals, building materials, computers, cotton textiles) and largest city in Alabama, USA; population (1990) 266,000.

Birobijan /ˌbɪrəbɪˈdʒɑːn/ town in Khabarovsk Territory, E Russia, near the Chinese border; population (1989) 82,000. Industries include sawmills and clothing. It was capital of the Jewish Autonomous Region 1928–51 (sometimes also called Birobijan).

Biscay, Bay of /ˈbɪskeɪ/ bay of the Atlantic Ocean between N Spain and W France, known for rough seas and exceptionally high tides.

Bishkek /bɪʃˈkek/ (formerly *Pishpek* until 1926, and *Frunze* 1926–92) capital of Kyrgyzstan; population (1987) 632,000. It produces textiles, farm machinery, metal goods, and tobacco.

Biskra /ˈbɪskrɑː/ oasis town in Algeria on the edge of the Sahara Desert; population (1982) 123,100.

Bismarck /ˈbɪzmɑːk/ capital of North Dakota, USA, on the Missouri river in Burleigh County, in the southern part of the state; population (1990) 49,200. It is a shipping point for the region's agricultural and livestock products from surrounding farms and for oil products from nearby oil wells.

Serving as the capital of the Dakota Territory from 1883, it remained the capital when North Dakota became a state 1889. Named for German chancellor Otto von Bismarck, it was the terminus of the heavily German-funded Northern Pacific Railroad.

Bismarck Archipelago /ˈbɪzmɑːk/ group of over 200 islands in the SW Pacific Ocean, part of Papua New Guinea; area 49,660 sq km/19,200 sq mi. The largest island is New Britain.

Bissau /bɪˈsaʊ/ capital and chief port of Guinea-Bissau, on an island at the mouth of the Geba River; population (1988) 125,000. Originally a fortified slave-trading centre, Bissau became a free port 1869.

Bismarck Archipelago

Black Sea

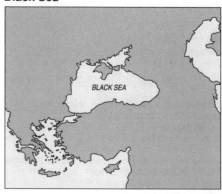

Bitolj /ˈbiːtɒl/ or *Bitola* town in the Former Yugoslav Republic of Macedonia, 32 km/20 mi N of the Greek border; population (1981) 137,800. *history* Held by the Turks (under whom it was known as Monastir) from 1382, it was taken by the Serbs 1912 during the First Balkan War. Retaken by Bulgaria 1915, it was again taken by the Allies Nov 1916.

Bizerta /bɪˈzɜːtə/ or *Bizerte* port in Tunisia, N Africa; population (1984) 94,500. Industries include fishing, oil refining, and metal works.

Björneborg /ˌbjɜːnəˈbɔːri/ Swedish name of ◊Pori, a town in Finland.

Blackburn /ˈblækbɜːn/ industrial town (engineering) in Lancashire, England, 32 km/20 mi NW of Manchester; population (1989 est) 140,000.

It was pre-eminently a cotton-weaving town until World War II.

Black Country central area of England, around and to the N of Birmingham. Heavily industrialized, it gained its name in the 19th century from its belching chimneys, but antipollution laws have changed its aspect.

Black Forest /blæk/ (German *Schwarzwald*) mountainous region of coniferous forest in Baden-Württemberg, W Germany. Bounded to the W and S by the Rhine, which separates it from the Vosges, it has an area of 4,660 sq km/1,800 sq mi and rises to 1,493 m/4,905 ft in the Feldberg. Parts of the forest have recently been affected by acid rain.

Blackheath /ˌblækˈhiːθ/ English common that gives its name to a residential suburb of London partly in Greenwich, partly in Lewisham. Wat Tyler encamped on Blackheath in the 1381 Peasants' Revolt.

Black Hills mountains in the Dakotas and Wyoming, USA. They occupy about 15,500 sq km/6,000 sq mi and rise to 2,207 m/7,242 ft at Harney Peak, South Dakota.

Black Mountain ridge of hills in the Brecon Beacons National Park in Dyfed and Powys, S Wales, stretching 19 km/12 m N from Swansea. The hills are made of limestone and red sandstone.

Black Mountains upland massif with cliffs and steep-sided valleys in Powys and Gwent, SE Wales, lying to the W of Offa's Dyke. The highest peak is Waun Fach (811 m/2,660 ft).

Blackpool /'blækpuːl/ seaside resort in Lancashire, England, 45 km/28 mi N of Liverpool; population (1987 est) 144,100.

The largest holiday resort in N England, the amusement facilities include 11 km/7 mi of promenades, known for their 'illuminations' of coloured lights, funfairs, and a tower 152 m/ 500 ft high. Political party conferences are often held here.

Black Sea (Russian *Chernoye More*) inland sea in SE Europe, linked with the seas of Azov and Marmara, and via the Dardanelles strait with the Mediterranean. Uranium deposits beneath it are among the world's largest. About 90% of the water is polluted, mainly by agricultural fertilizers.

Blantyre-Limbe /'blæntaɪə 'lɪmbeɪ/ chief industrial and commercial centre of Malawi, in the Shire highlands; population (1987) 331,600. It produces tea, coffee, rubber, tobacco, and textiles.

It was formed by the union of the towns of Blantyre (named after the explorer David Livingstone's birthplace) and Limbe 1959.

Blarney /'blɑːni/ small town in County Cork, Republic of Ireland, possessing, inset in the wall of the 15th-century castle, the *Blarney Stone*, reputed to give persuasive speech to those kissing it.

Blenheim /'blenɪm/ centre of a sheep-grazing area in the NE of South Island, New Zealand; population (1986) 18,300.

Bloemfontein /'bluːmfənteɪn/ capital of the Orange Free State and judicial capital of the Republic of South Africa; population (1985) 204,000. Founded 1846, the city produces canned fruit, glassware, furniture, and plastics.

Blois /blwɑː/ town on the river Loire in central France; population (1990) 51,500. It has a château partly dating from the 13th century.

Bloomington /'bluːmɪŋtən/ city in S central Indiana, USA, southwest of Indianapolis; seat of Monroe County; population (1990) 42,100. It is an exporter of limestone from nearby quarries. It is also a centre for the manufacture of electrical products and lifts.

Bloomington /'bluːmɪŋtən/ city in central Illinois, USA, southeast of Peoria; seat of McLean County; population (1990) 42,200. It is in the middle of a rich farming and livestock raising area.

Bloomsbury area in Camden, London. It contains the London University headquarters, the British Museum, and the Royal Academy of Dramatic Arts. Between the world wars it was the home of the Bloomsbury Group of writers and artists.

Bluefields /'bluːfiːldz/ one of three major port facilities on the east coast of Nicaragua, situated on an inlet of the Caribbean Sea; population (1990 est) 18,000.

Blue Mountains part of the Great Dividing Range, New South Wales, Australia, ranging 600–1,100 m/2,000–3,600 ft and blocking Sydney from the interior until the crossing 1813 by surveyor William Lawson, Gregory Blaxland, and William Wentworth.

Blue Nile /,bluː 'naɪl/ (Arabic *Bahr el Azraq*) river rising in the mountains of Ethiopia. Flowing W then N for 2,000 km/1,250 mi, it eventually meets the White Nile at Khartoum.

The river is dammed at Roseires where a hydroelectric scheme produces 70% of Sudan's electricity.

Blue Ridge Mountains range extending from West Virginia to Georgia, USA, and including Mount Mitchell 2,045 m/6,712 ft; part of the Appalachians.

Bobruisk /bə'bruːɪsk/ town in Belarus, on the Beresina River; population (1987) 232,000. Industries include timber, machinery, tyres, and chemicals.

Boca Raton /,bəʊkə rə'təʊn/ city in SE Florida, USA, on the Atlantic Ocean, N of Miami; population (1990) 61,500. Although it is mainly a resort and residential area, there is some light industry.

Bochum /'bəʊxʊm/ town in the Ruhr district, Germany; population (1988) 381,000. Industries include metallurgy, vehicles, and chemicals.

Bodensee /'bəʊdnzeɪ/ German name for Lake ◊Constance, N of the Alps.

Bodmin /'bɒdmɪn/ market town in Cornwall, England, 48 km/30 m from Plymouth; population (1984) 15,000. *Bodmin Moor* to the NE is a granite upland, culminating in Brown Willy 419 m/1,375 ft.

Bognor Regis /'bɒgnə 'riːdʒɪs/ seaside resort in West Sussex, England, 105 km/66 mi SW of London; population (1988 est) 20,000. It owes the Regis part of its name to the convalescent visit by King George V 1929.

Bogotá /,bɒgə'tɑː/ capital of Colombia, South America; 2,640 m/8,660 ft above sea level on the edge of the plateau of the E Cordillera; population (1985) 4,185,000. It was founded 1538.

Bohemia /bəʊ'hiːmɪə/ area of the Czech Republic, a kingdom of central Europe from the 9th century. It was under Habsburg rule 1526-1918, when it was included in Czechoslovakia. The name Bohemia derives from the Celtic Boii, its earliest known inhabitants.

history It became part of the Holy Roman Empire as the result of Charlemagne's establishment of a protectorate over the Celtic, Germanic, and Slav tribes settled in this area. Christianity was introduced in the 9th century, the See of Prague being established 975, and feudalism was introduced by King Ottaker I of Bohemia (1197–1230). From the 12th century onward, mining attracted large numbers of German settlers, leading to a strong Germanic influence in culture and society. In 1310, John of Luxemburg (died 1346) founded a German-Czech royal dynasty that lasted until 1437. His son, Charles IV, became Holy Roman Emperor 1355, and during his reign the See of Prague was elevated to an archbishopric and a university was founded here. During the 15th century, divisions within the nobility and religious conflicts culminating in the Hussite Wars (1420–36) led to its decline.

He who holds Bohemia holds mid-Europe.

On **Bohemia** Prince Otto von Bismarck

Boise /'bɔɪzi/ capital of Idaho, USA, located in the western part of the state, on the Boise River in the western foothills of the Rocky Mountains; population (1990) 125,700. It serves as a centre for the farm and livestock products of the region and has meatpacking and food-processing industries; steel and lumber products are also manufactured.

It was founded during the Idaho gold rush of 1862 and served as territorial capital 1864–90, when Idaho became a state.

Bois-le-Duc /'bwɑ: lə 'dju:k/ French form of ◊'s Hertogenbosch, a town in North Brabant, the Netherlands.

Bokhara /bɒ'kɑːrə/ another form of ◊ Bukhara, a city in Uzbekistan.

Bolivia Republic of (*República de Bolivia*)
area 1,098,581 sq km/424,052 sq mi
capital La Paz (seat of government), Sucre (legal capital and seat of judiciary)
towns Santa Cruz, Cochabamba, Oruro, Potosí
physical high plateau (Altiplano) between mountain ridges (cordilleras); forest and lowlands (llano) in the E
features Andes, lakes Titicaca (the world's highest navigable lake, 3,800 m/12,500 ft) and Poopó; La Paz is the world's highest capital city (3,600 m/11,800 ft)
head of state and government Jaime Paz Zamora from 1989
political system emergent democratic republic
political parties National Revolutionary Movement (MNR), centre-right; Nationalist Democratic Action Party (ADN), extreme right-wing; Movement of the Revolutionary Left (MIR), left of centre; Solidarity Civil Union (UCS), radical free market
exports tin, antimony (second largest world producer), other nonferrous metals, oil, gas (piped to Argentina), agricultural products, coffee, sugar, cotton
currency boliviano
population (1990 est) 6,730,000; (Quechua 25%, Aymara 17%, mestizo (mixed) 30%, European 14%); growth rate 2.7% p.a.
life expectancy men 51, women 54
languages Spanish, Aymara, Quechua (all official)
religion Roman Catholic 95% (state-recognized)
literacy men 84%, women 65% (1985 est)

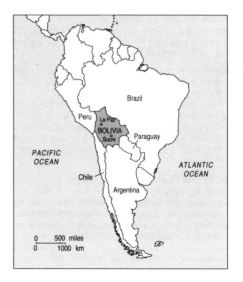

GDP $4.2 bn (1987); $617 per head
chronology
1825 Liberated from Spanish rule by Simón Bolívar; independence achieved (formerly known as Upper Peru).
1952 Dr Víctor Paz Estenssoro elected president.
1956 Dr Hernán Siles Zuazo became president.
1960 Estenssoro returned to power.
1964 Army coup led by vice president.
1966 Gen René Barrientos became president.
1967 Uprising, led by 'Che' Guevara, put down with US help.
1969 Barrientos killed in plane crash, replaced by Vice President Siles Salinas. Army coup deposed him.
1970 Army coup put General Juan Torres González in power.
1971 Torres replaced by Col Hugo Banzer Suárez.
1973 Banzer promised a return to democratic government.
1974 Attempted coup prompted Banzer to postpone elections and ban political and trade-union activity.
1978 Elections declared invalid after allegations of fraud.
1980 More inconclusive elections followed by another coup, led by General Luis García. Allegations of corruption and drug trafficking led to cancellation of US and EC aid.
1981 García forced to resign. Replaced by General Celso Torrelio Villa.
1982 Torrelio resigned. Replaced by military junta led by General Guido Vildoso. Because of worsening economy, Vildoso asked congress to install a civilian administration. Dr Siles Zuazo chosen as president.
1983 Economic aid from USA and Europe resumed.
1984 New coalition government formed by Siles. Abduction of president by right-wing officers. The president undertook a five-day hunger strike as an example to the nation.
1985 President Siles resigned. Election result

inconclusive; Dr Paz Estenssoro, at the age of 77, chosen by congress as president.

1989 Jaime Paz Zamora (MIR) elected president in power-sharing arrangement with Hugo Banzer Suárez, pledged to maintain fiscal and monetary discipline and preserve free-market policies.

1992 The new UCS party gained support.

Bologna /bə'lɒnjə/ industrial city and capital of Emilia-Romagna, Italy, 80 km/50 mi N of Florence; population (1988) 427,000. It was the site of an Etruscan town, later of a Roman colony, and became a republic in the 12th century. It came under papal rule 1506 and was united with Italy 1860.

The city has a cathedral and medieval towers, and the university, which dates from the 11th century, laid the foundations of the study of anatomy and was attended by the poets Dante, Petrarch, and Tasso, and the astronomer Copernicus. An annual international children's book fair is held here.

Bolton /'bəultən/ city in Greater Manchester, England, 18 km/11 mi NW of Manchester; population (1991) 253,300. Industries include chemicals and textiles.

Bolzano /bɒlt'sɑːnəu/ (German *Bozen*) town in Italy, in Trentino-Alto Adige region on the river Isarco in the Alps; population (1988) 101,000. Bolzano belonged to Austria until 1919. Its inhabitants are mostly German-speaking.

Boma /'bəumə/ port in Zaire, on the estuary of the river Zäire 88 km/55 mi from the Atlantic; population (1976) 93,965. The oldest European settlement in Zaire, it was a centre of the slave trade, and capital of the Belgian Congo until 1927.

Bombay /ˌbɒm'beɪ/ industrial port (textiles, engineering, pharmaceuticals, diamonds), commercial centre, and capital of Maharashtra, W India; population (1981) 8,227,000. It is the centre of the Hindi film industry.

features World Trade Centre 1975; National Centre for the Performing Arts 1969.

history Bombay was founded in the 13th century, came under Mogul rule, was occupied by Portugal 1530, and passed to Britain 1662 as part of Catherine of Braganza's dowry. It was the headquarters of the East India Company 1685–1708. The city expanded rapidly with the development of the cotton trade and the railway in the 1860s.

The huge city which the West had built and abandoned with a gesture of despair.

On **Bombay** E M Forster *A Passage to India* 1924

Bône /bəun/ (or *Bohn*) former name of ◊Annaba, a port in Algeria.

Bonin and Volcano islands /'bəunɪn/ Japanese islands in the Pacific, N of the Marianas and

1,300 km/800 mi E of the Ryukyu Islands. They were under US control 1945–68. The *Bonin Islands* (Japanese *Ogasawara Gunto*) number 27 (in three groups), the largest being Chichijima: area 104 sq km/40 sq mi, population (1991) 2,430. The *Volcano Islands* (Japanese *Kazan Retto*) number three, including Iwo Jima, scene of some of the fiercest fighting of World War II; total area 28 sq km/11 sq mi. They have no civilian population, but a 200-strong maritime self-defence force and 100-strong air self-defence force are stationed here.

Bonn /bɒn/ industrial city (chemicals, textiles, plastics, aluminium) and seat of government of the Federal Republic of Germany, 18 km/15 mi SSE of Cologne, on the left bank of the Rhine; population (1988) 292,000.

Once a Roman outpost, Bonn was captured by the French 1794, annexed 1801, and was allotted to Prussia 1815. Beethoven was born here. It was capital of West Germany 1949–90.

Bonneville Salt Flats /'bɒnəvɪl/ bed of a prehistoric lake in Utah, USA, of which the Great Salt Lake is the surviving remnant. The flats, near the Nevada border, have been used to set many land speed records.

Bonny, Bight of /'bɒni/ name since 1975 of the former Bight of Biafra, an area of sea off the coasts of Nigeria and Cameroon.

Bootle /'buːtl/ port in Merseyside, England, adjoining Liverpool; population (1981) 62,463. The National Girobank headquarters are here.

Bophuthatswana /bəuˌpuːtət'swɑːnə/ Republic of; self-governing black 'homeland' within South Africa
area 40,330 sq km/15,571 sq mi
capital Mmbatho or Sun City, a casino resort frequented by many white South Africans
features divided into six 'blocks'
exports platinum, chromium, vanadium, asbestos, manganese
currency South African rand
population (1985) 1,627,000
languages Setswana, English
religion Christian
government executive president elected by the Assembly: Chief Lucas Mangope
recent history first 'independent' Black National State from 1977, but not recognized by any country other than South Africa.

Bora-Bora /ˌbɔːrə'bɔːrə/ one of the 14 Society Islands of French Polynesia; situated 225 km/140 mi NW of Tahiti; area 39 sq km/15 sq mi; population (1977) 2,500. Exports include mother-of-pearl, fruit, and tobacco.

Borås /bu'rɔːs/ town in SW Sweden; population (1990) 101,766. Chief industries include textiles and engineering.

Bordeaux /bɔː'dəu/ port on the river Garonne, capital of Aquitaine, SW France, a centre for the wine trade, oil refining, and aeronautics and space industries; population (1990) 213,300. Bordeaux was under the English crown for three centuries until 1453. In 1870, 1914, and 1940 the French government was moved here because of German invasion.

Bordeaux is ... dedicated to the worship of Bacchus in the most discreet form.

On **Bordeaux** Henry James *A Little Tour in France* 1882

Borders /'bɔːdəz/ region of Scotland
area 4,700 sq km/1,815 sq mi
towns Newtown St Boswells (administrative headquarters), Hawick, Jedburgh
features river Tweed; Lammermuir, Moorfoot, and Pentland hills; home of the novelist Walter Scott at Abbotsford; Dryburgh Abbey, burial place of Field Marshal Haig and Scott; ruins of 12th-century Melrose Abbey
products knitted goods, tweed, electronics, timber
population (1991) 102,600
famous people Duns Scotus, James Murray, Mungo Park.

This queer compromise between fairyland and battlefield which is the Border.

On the **Borders** Henry Vollam Morton *In Search of Scotland* 1929

Borneo /'bɔːniəʊ/ third-largest island in the world, one of the Sunda Islands in the W Pacific; area 754,000 sq km/290,000 sq mi. It comprises the Malaysian territories of *Sabah* and *Sarawak*; *Brunei*; and, occupying by far the largest part, the Indonesian territory of *Kalimantan*. It is mountainous and densely forested. In coastal areas the people of Borneo are mainly of Malaysian

Borders

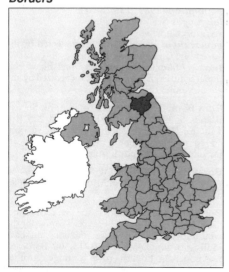

origin, with a few Chinese, and the interior is inhabited by the indigenous Dyaks. It was formerly under both Dutch and British colonial influence until Sarawak was formed 1841.

Bornholm /ˌbɔːn'həum/ Danish island in the Baltic Sea, 35 km/22 mi SE of the nearest point of the Swedish coast; it constitutes a county of the same name
area 587 sq km/227 sq mi
capital Rönne
population (1985) 46,000.

Borobudur site of a Buddhist shrine near Yogyakarta, Indonesia.

Bosnia-Herzegovina Republic of
area 51,129 sq km/19,745 sq mi
capital Sarajevo
towns Banja Luka, Mostar, Prijedor, Tuzla, Zenica
physical barren, mountainous country
features part of the Dinaric Alps, limestone gorges
population (1990) 4,300,000 including 44% Muslims, 33% Serbs, 17% Croats; a complex patchwork of ethnically mixed communities
head of state Alija Izetbegović from 1990
head of government Mile Akmadzic from 1992
political system emergent democracy
political parties Party of Democratic Action (SDA), Muslim-oriented; Serbian Democratic Party (SDS), Serbian nationalist; Christian Democratic Union (HDS), centrist; League of Communists, left-wing
products citrus fruits and vegetables; iron, steel, and leather goods; textiles
language Serbian variant of Serbo-Croatian
religions Sunni Muslim, Serbian Orthodox, Roman Catholic
chronology
1918 Incorporated in the future Yugoslavia.
1941 Occupied by Nazi Germany.
1945 Became republic within Yugoslav Socialist Federation.
1980 Upsurge in Islamic nationalism.
1990 Ethnic violence erupted between Muslims and Serbs. Nov–Dec: communists defeated in multiparty elections; coalition formed by Serb, Muslim, and Croatian parties.
1991 May: Serbia–Croatia conflict spread disorder into Bosnia-Herzegovina. Aug: Serbia revealed plans to annex the SE part of the republic. Sept: Serbian enclaves established by force. Oct: 'sovereignty' declared. Nov: plebiscite by Serbs favoured remaining within Yugoslavia; Serbs and Croats established autonomous communities.
1992 Feb–March: Muslims and Croats voted overwhelmingly in favour of independence; referendum boycotted by Serbs. April: USA and EC recognized Bosnian independence. Ethnic hostilities escalated, with Serb forces occupying E and Croatian forces much of W; state of emergency declared; all-out civil war. May: admitted to United Nations. June: UN forces drafted into Sarajevo to break three-month siege of city by Serbs. Accusations of 'ethnic cleansing'

being practised, particularly by Serbs. Oct: UN ban on military flights over Bosnia-Herzegovina. First British troops deployed.
1993 UN–EC peace plan introduced, failed. USA began airdrops of food and medical supplies. Croat-Serb partition plan proposed, rejected by Muslims. UN relief convoys attacked.

Bosporus /'bɒspərəs/ (Turkish **Karadeniz Boğazi**) strait 27 km/17 mi long, joining the Black Sea with the Sea of Marmara and forming part of the water division between Europe and Asia; its name may be derived from the Greek legend of Io. Istanbul stands on its west side. The **Bosporus Bridge** 1973, 1,621 m/5,320 ft, links Istanbul and Anatolia (the Asian part of Turkey). In 1988 a second bridge across the straits was opened, linking Asia and Europe.

Boston /'bɒstən/ industrial and commercial centre, capital of Massachusetts, USA; population (1990) 574,300; metropolitan area 4,171,600. It is a publishing centre and industrial port on Massachusetts Bay, but the economy is dominated by financial and health services and government. Boston and Northeastern universities are here. A centre of opposition to British trade restrictions, it was the scene of the Boston Tea Party 1773.

Boston is a state of mind.

On **Boston** attributed to Mark Twain and others

Boston /'bɒstən/ seaport in Lincolnshire, England, on the river Witham; population (1985 est) 27,000. St Botolph's is England's largest parish church, and its tower 'Boston stump' is a landmark for sailors.

Botany Bay inlet on the east coast of Australia, 8 km/5 mi S of Sydney, New South Wales. Chosen 1787 as the site for a penal colony, it proved unsuitable. Sydney now stands on the site of the former settlement. The name Botany Bay continued to be popularly used for any convict settlement in Australia.

Botswana Republic of
area 582,000 sq km/225,000 sq mi
capital Gaborone
towns Mahalpye, Serowe, Tutume, Francistown
physical desert in SW, plains in E, fertile lands and swamp in N
environment the Okavango Swamp is threatened by plans to develop the area for mining and agriculture
features Kalahari Desert in SW; Okavango Swamp in N, remarkable for its wildlife; Makgadikgadi salt pans in E; diamonds mined at Orapa and Jwaneng in partnership with De Beers of South Africa
head of state and government Quett Ketamile Joni Masire from 1980
political system democratic republic
political parties Botswana Democratic Party (BDP), moderate centrist; Botswana National Front (BNF), moderate, left of centre
exports diamonds (third largest producer in world), copper, nickel, meat products, textiles
currency pula
population (1990 est) 1,218,000 (Bamangwato 80%, Bangwaketse 20%); growth rate 3.5% p.a.
life expectancy (1988) 59 years
languages English (official), Setswana (national)
religions Christian 50%, animist 50%
literacy (1988) 84%
GDP $2.0 bn (1988); $1,611 per head
chronology
1885 Became a British protectorate.
1960 New constitution created a legislative council.
1963 End of rule by High Commission.
1965 Capital transferred from Mafeking to Gaborone. Internal self-government achieved. Sir Seretse Khama elected head of government.
1966 Independence achieved from Britain. New constitution came into effect; name changed from Bechuanaland to Botswana; Seretse Khama elected president.
1980 Seretse Khama died; succeeded by Vice President Quett Masire.
1984 Masire re-elected.

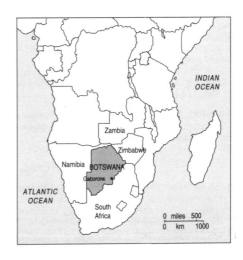

1985 South African raid on Gaborone.
1987 Joint permanent commission with Mozambique established, to improve relations.
1989 The BDP and Masire re-elected.

Bottrop /'bɒtrɒp/ city in North Rhine–Westphalia, Germany; population (1988) 112,000.

Bougainville /'buːgənvɪl/ island province of Papua New Guinea; largest of the Solomon Islands archipelago
area 10,620 sq km/4,100 sq mi
capital Kieta
environment waste from one of the world's largest copper mines, owned by Australian company CRA Minerals, devastated the island's environment (according to rebel government sources) 1992, silting up the streams of the Jaba River system and contaminating the water with dangerous heavy metals
products copper, gold, and silver
population (1989) 128,000
history named after the French navigator Bougainville who arrived 768. In 1976 Bougainville became a province (with substantial autonomy) of Papua New Guinea. A state of emergency was declared 1989 after secessionist violence. In 1990 the secessionist Bougainville Revolutionary Army took control of the island, declaring it independent; government troops regained control 1992.

Bougie /buː'ʒiː/ former name (until 1962) of ⟩Bejaia, a port in Algeria.

Bou Kraa /'buːkraː/ principal phosphate-mining centre of Western Sahara, linked by conveyor belt to the Atlantic coast near Laâyoune.

Boulder /'bəʊldə/ city in N central Colorado, USA, northwest of Denver, in the eastern foothills of the Rocky Mountains; population (1990) 83,300. A centre of scientific research, especially space research, it also has agriculture, mining, and tourism.

Boulogne-sur-Mer /buː'lɔɪn sjuə 'meə/ town on the English Channel in the *département* of Pas-de-Calais, France; population (1990) 44,200. Industries include oil refining, food processing, and fishing. It is also a ferry port (connecting with Dover and Folkestone) and seaside resort. Boulogne was a medieval countship, but became part of France 1477.
In World War II it was evacuated by the British 23 May 1940 and recaptured by Canadian forces 22 Sept 1944.

Boundary Peak highest mountain in Nevada, USA, rising to 4,006 m/13,143 ft on the Nevada–California frontier.

Bourbon /'buəbən/ name 1649–1815 of the French island of ⟩Réunion, in the Indian Ocean.

Bourges /buəʒ/ city in central France, 200 km/125 mi S of Paris; population (1990) 78,800. Industries include aircraft, engineering, and tyres. It has a 13th-century Gothic cathedral and notable art collections.

Bourgogne /'buə'gɔθn/ French name of ⟩Burgundy, a region of E France.

Bournemouth /'bɔːnməθ/ seaside resort in Dorset, England; population (1991) 154,400.

Bouvet Island /'buːveɪ/ uninhabited island in the S Atlantic Ocean, a dependency of Norway since 1930; area 48 sq km/19 sq mi. Discovered by the French captain Jacques Bouvet 1738, it was made the subject of a claim by Britain 1825, but this was waived in Norway's favour 1928.

Boyne /bɔɪn/ river in the Republic of Ireland. Rising in the Bog of Allen in County Kildare, it flows 110 km/69 mi NE to the Irish Sea near Drogheda. The Battle of the Boyne was fought at Oldbridge near the mouth of the river 1690.

Boyoma Falls /bɔɪ'əumə/ series of seven cataracts in under 100 km/60 mi in the Lualaba (upper Zaïre River) above Kisangani, central Africa. They have a total drop of over 60 m/200 ft.

Bozen /'bəutsən/ German form of ⟩Bolzano, a town in Italy.

Brabant /brə'bænt/ (Flemish **Braband**) former duchy of W Europe, comprising the Dutch province of North Brabant and the Belgian provinces of Brabant and Antwerp. They were divided when Belgium became independent 1830. The present-day Belgian province of Brabant has an area of 3,400 sq km/1,312 sq mi and a population (1987) of 2,245,900.
history During the Middle Ages Brabant was an independent duchy, and after passing to Burgundy, and thence to the Spanish crown, was divided during the Dutch War of Independence. The southern portion was Spanish until 1713, then Austrian until 1815, when the whole area was included in the Netherlands. In 1830 the French-speaking part of the population in the S Netherlands rebelled, and when Belgium was recognized 1839, S Brabant was included in it.

Bracknell /'bræknəl/ new town in Berkshire, England, founded 1949; population (1985 est) 51,500. The headquarters of the Meteorological Office is here, and Bracknell (with Washington, DC) is one of the two global area forecasting centres (of upper-level winds and temperatures) for the world's airlines.

Bradenton /'breɪdntən/ city in W Florida, USA, on the southern shores of Tampa Bay, SW of Tampa; seat of Manatee County; population (1990) 43,800. It is a resort centre during the winter months. Some travertine, a sparkling sheet of calcium carbonate formed on cave walls and floors, is quarried here.

Bradford /'brædfəd/ industrial city (engineering, machine tools, electronics, printing) in West Yorkshire, England, 14 km/9 mi W of Leeds; population (1989 est) 466,000.
features A 15th-century cathedral; Cartwright Hall art gallery; the National Museum of Photography, Film, and Television 1983 (with Britain's largest cinema screen 14 x 20 m); and the Alhambra, built as a music hall and restored for ballet, plays, and pantomime.

history From the 13th century, Bradford developed as a great wool- and, later, cloth-manufacturing centre, but the industry declined from the 1970s with competition from developing countries and the European Economic Community (EEC). The city has received a succession of immigrants, Irish in the 1840s, German merchants in the mid-19th century, then Poles and Ukrainians, and more recently West Indians and Asians.

Braga /ˈbrɑːgə/ city in N Portugal 48 km/30 mi NNE of Pôrto; population (1981) 63,800. Industries include textiles, electrical goods, and vehicle manufacture. It has a 12th-century cathedral, and the archbishop is primate of the Iberian peninsula. As *Bracara Augusta* it was capital of the Roman province Lusitania.

Bragança /brəˈgænsə/ capital of a province of the same name in NE Portugal, 176 km/110 mi NE of Pôrto; population (1981) 13,900. It was the original family seat of the House of Braganza, which ruled Portugal 1640–1910.

Brahmaputra /ˌbrɑːməˈpuːtrə/ river in Asia 2,900 km/1,800 mi long, a tributary of the Ganges.
It rises in the Himalayan glaciers as Zangbo and runs E through Tibet, to the mountain mass of Namcha Barwa. Turning S, as the Dihang, it enters India and flows into the Assam Valley near Sadiya, where it is now known as the Brahmaputra. It flows generally W until, shortly after reaching Bangladesh, it turns S and divides into the Brahmaputra proper, without much water, and the main stream, the Jamuna, which joins the Padma arm of the Ganges. The river is navigable for 1,285 km/800 mi from the sea.

Brăila /brəˈiːlə/ port in Romania on the river Danube; 170 km/106 mi from its mouth; population (1983) 226,000. It is a naval base. Industries include the manufacture of artificial fibres, iron and steel, machinery, and paper. It was controlled by the Ottoman Empire 1544–1828.

Brandenburg /ˈbrændənbɜːg/ administrative *Land* (state) of Germany
area 25,000 sq km/10,000 sq mi
capital Potsdam
towns Cottbus, Brandenburg, Frankfurt-on-Oder
products iron and steel, paper, pulp, metal products, semiconductors
population (1990) 2,700,000
history the Hohenzollern rulers who took control of Brandenburg 1415 later acquired the powerful duchy of Prussia and became emperors of Germany. At the end of World War II, Brandenburg lost over 12,950 sq km/5,000 sq mi of territory when Poland advanced its frontier to the line of the Oder and Neisse rivers. The remainder, which became a region of East Germany, was divided 1952 into the districts of Frankfurt-on-Oder, Potsdam, and Cottbus. When Germany was reunited 1990, Brandenburg reappeared as a state of the Federal Republic.

Brandenburg /ˈbrændənbɜːg/ town in Germany, on the river Havel; 60 km/36 mi W of Berlin; population (1981) 94,700. Industries include textiles, cars, and aircraft. It has a 12th-century cathedral.

Brasília /brəˈzɪliə/ capital of Brazil from 1960, 1,000 m/3,000 ft above sea level; population (1991) 1,841,000. It was designed by Lucio Costa (1902–1963), with Oscar Niemeyer as chief architect, as a completely new city to bring life to the interior.

Brasov /brɑːˈsɒv/ (Hungarian *Brassó*, German *Krondstadt*) industrial city (machine tools, industrial equipment, chemicals, cement, woollens) in central Romania at the foot of the Transylvanian Alps; population (1985) 347,000. It belonged to Hungary until 1920.

Bratislava (German *Pressburg*) industrial port (engineering, chemicals, oil refining) and capital of the Slovak Republic, on the river Danube; population (1991) 441,500. It was the capital of Hungary 1526–1784 and capital of the Slovak Federative Republic within Czechoslovakia until 1993.

Braunschweig /ˈbraʊnʃvaɪk/ German form of ◊Brunswick, a city in Lower Saxony, Germany.

Brazil Federative Republic of (*República Federativa do Brasil*)
area 8,511,965 sq km/3,285,618 sq mi
capital Brasília
towns Sao Paulo, Belo Horizonte, Curitiba, Manaus, Fortaleza; ports are Rio de Janeiro, Belém, Recife, Pôrto Alegre, Salvador
physical the densely forested Amazon basin covers the northern half of the country with a network of rivers; the south is fertile; enormous energy resources, both hydroelectric (Itaipú dam on the Paraná, and Tucuruí on the Tocantins) and nuclear (uranium ores)
environment Brazil has one-third of the world's tropical rainforest. It contains 55,000 species of flowering plants (the greatest variety in the world)

and 20% of all the world's bird species. During the 1980s at least 7% of the Amazon rainforest was destroyed by settlers who cleared the land for cultivation and grazing
features Mount Roraima, Xingu National Park; Amazon delta; Rio harbour
head of state and government Itamar Franco from 1992
political system emergent democratic federal republic
political parties Social Democratic Party (PDS), moderate, left of centre; Brazilian Democratic Movement Party (PMDB), centre-left; Liberal Front Party (PFL), moderate, left of centre; Workers' Party (PT), left of centre; National Reconstruction Party (PRN), centre-right
exports coffee, sugar, soya beans, cotton, textiles, timber, motor vehicles, iron, chrome, manganese, tungsten and other ores, as well as quartz crystals, industrial diamonds, gemstones; the world's sixth largest arms exporter
currency cruzado ; inflation 1990 was 1,795%
population (1990 est) 153,770,000 (including 200,000 Indians, survivors of 5 million, especially in Rondônia and Mato Grosso, mostly living on reservations); growth rate 2.2% p.a.
life expectancy men 61, women 66
languages Portuguese (official); 120 Indian languages
religions Roman Catholic 89%; Indian faiths
literacy men 79%, women 76% (1985 est)
GDP $352 bn (1988); $2,434 per head
chronology
1822 Independence achieved from Portugal; ruled by Dom Pedro, son of the refugee King John VI of Portugal.
1889 Monarchy abolished and republic established.
1891 Constitution for a federal state adopted.
1930 Dr Getúlio Vargas became president.
1945 Vargas deposed by the military.
1946 New constitution adopted.
1951 Vargas returned to office.
1954 Vargas committed suicide.
1956 Juscelino Kubitschek became president.
1960 Capital moved to Brasília.
1961 João Goulart became president.
1964 Bloodless coup made General Castelo Branco president; he assumed dictatorial powers, abolishing free political parties.
1967 New constitution adopted. Branco succeeded by Marshal da Costa e Silva.
1969 Da Costa e Silva resigned and a military junta took over.
1974 General Ernesto Geisel became president.
1978 General Baptista de Figueiredo became president.
1979 Political parties legalized again.
1984 Mass calls for a return to fully democratic government.
1985 Tancredo Neves became first civilian president in 21 years. Neves died and was succeeded by the vice president, José Sarney.
1988 New constitution approved, transferring power from the president to the congress. Measures announced to halt large-scale burning of Amazonian rainforest for cattle grazing.
1989 Forest Protection Service and Ministry

for Land Reform abolished. International concern over extent of burning of Amazon forest. Fernando Collor (PRN) elected president Dec, pledging free-market economic policies.
1990 Government won the general election offset by mass abstentions.
1992 Earth Summit, global conference on the environment, held in Rio de Janeiro. Collor impeached for corruption and forced to resign. Replaced by vice president Itamar Franco.

The country has been nicely described as a continent with its colonies inside.

On **Brazil** John Gunther *Inside South America* 1967

Brazzaville /'bræzəvɪl/ capital of the Congo, industrial port (foundries, railway repairs, shipbuilding, shoes, soap, furniture, bricks) on the river Zaïre, opposite Kinshasa; population (1984) 595,000. There is a cathedral 1892 and the Pasteur Institute 1908. It stands on Pool Malebo (Stanley Pool).
Brazzaville was founded by the Italian Count Pierre Savorgnan de Brazza (1852–1905), employed in African expeditions by the French government. It was the African headquarters of the Free (later Fighting) French during World War II.

Brecknockshire /'breknɒkʃə/ or **Breconshire** (Welsh *Sir Frycheiniog*) former county of Wales, merged into Powys 1974.

Breda /breɪ'dɑː/ town in North Brabant, the Netherlands; population (1991) 124,800. It was here that Charles II of England made the declaration that paved the way for his restoration 1660.

Breizh Celtic name for ◊Brittany, a region of France.

Bremen /'breɪmən/ administrative region (German *Land*) of Germany, consisting of the cities of Bremen and Bremerhaven; area 400 sq km/154 sq mi; population (1988) 652,000.

Bremen /'breɪmən/ industrial port (iron, steel, oil refining, chemicals, aircraft, shipbuilding, cars) in Germany, on the river Weser 69 km/43 mi from the open sea; population (1988) 522,000.
Bremen was a member of the Hanseatic League, and a free imperial city from 1646. It became a member of the North German Confederation 1867 and of the German Empire 1871.

Bremerhaven /ˌbreɪmə'hɑːfən/ (formerly until 1947 *Wesermünde*) port at the mouth of the river Weser, Germany; population (1988) 132,000. Industries include fishing and shipbuilding. It serves as an outport for Bremen.

Bremerton /'bremətən/ city in W Washington, SW of Seattle and NW of Tacoma, on an inlet of Puget Sound; population (1990) 38,100. It serves as a port for the area's fish, dairy, and lumber products. Tourism is also important to the economy.

Brenner Pass /ˈbrenə/ lowest of the Alpine passes, 1,370 m/4,495 ft; it leads from Trentino–Alto Adige, Italy, to the Austrian Tirol, and is 19 km/ 12 mi long.

Brent borough of NW Greater London
features Wembley Stadium
population (1991) 226,100.

Brescia /ˈbreʃə/ (ancient *Brixia*) historic and industrial city (textiles, engineering, firearms, metal products) in N Italy, 84 km/52 mi E of Milan; population (1988) 199,000. It has medieval walls and two cathedrals (12th and 17th century).

Breslau /ˈbreslau/ German name of ◊ Wrocław, a town in Poland.

Brest /brest/ naval base and industrial port (electronics, engineering, chemicals) on *Rade de Brest* (Brest Roads), a great bay at the western extremity of Brittany, France; population (1983) 201,000. Occupied as a U-boat base by the Germans 1940–44, the town was destroyed by Allied bombing and rebuilt.

Brest /brest/ town in Belarus, on the river Bug and the Polish frontier; population (1990) 153,000. It was in Poland (*Brześć nad Bugiem*) until 1795 and again 1921–39. The *Treaty of Brest-Litovsk* (an older Russian name of the town) was signed here.

Bretagne /brəˈtœɲ/ French form of ◊ Brittany, a region of NW France.

Bridgeport /ˈbrɪdʒpɔːt/ city in Connecticut, USA, on Long Island Sound; population (1990) 141,700. Industries include metal goods, electrical appliances, and aircraft, but many factories closed in the 1970s. The university was established 1927.

Bridgeton /ˈbrɪdʒtən/ city in SW New Jersey, USA, at the head of the Cohansey River; seat of Cumberland County; population (1990) 18,900. Industries include food processing, glassmaking, clothing, and dairy products.

Bridgetown /ˈbrɪdʒtaun/ port and capital of Barbados; population (1987) 8,000. Sugar is exported through the nearby deep-water port. Bridgetown was founded 1628.

Bridgwater /ˈbrɪdʒwɔːtə/ port in Somerset, England, on the river Parret; population (1981) 26,000. Industries include plastics and electrical goods.

Brighton /ˈbraɪtn/ resort on the E Sussex coast, England; population (1981) 146,000. It has Regency architecture and the Royal Pavilion 1782 in Oriental style. There are two piers and an aquarium. The University of Sussex was founded 1963.

Originally a fishing village called Brighthelmstone, it became known as Brighton at the beginning of the 19th century, when it was already a fashionable health resort patronized by the Prince Regent, afterwards George IV. In 1990 the Royal Pavilion reopened after nine years of restoration.

... Brighton, that always looks brisk, gay and gaudy, like a harlequin's jacket ...

On **Brighton** William Makepiece Thackeray
Vanity Fair 1847

Brindisi /ˈbrɪndɪzi/ (ancient *Brundisium*) port and naval base on the Adriatic Sea, in Puglia, on the heel of Italy; population (1981) 90,000. Industries include food processing and petrochemicals. It is one of the oldest Mediterranean ports, at the end of the Appian Way from Rome. The poet Virgil died here 19 BC.

Brisbane /ˈbrɪzbən/ industrial port (brewing, engineering, tanning, tobacco, shoes; oil pipeline from Moonie), capital of Queensland, E Australia, near the mouth of Brisbane River, dredged to carry ocean-going ships; population (1990) 1,301,700.

Bristol /ˈbrɪstəl/ industrial port (aircraft engines, engineering, microelectronics, tobacco, chemicals, paper, printing) and administrative headquarters of Avon, SW England; population (1991 est) 370,300. The old docks have been redeveloped for housing, industry, yachting facilities, and the National Lifeboat Museum. Further developments include a new city centre, with Brunel's Temple Meads railway station at its focus, and a weir across the Avon nearby to improve the waterside environment.
features 12th-century cathedral; 14th-century St Mary Redcliffe; 16th-century Acton Court, built by Sir Nicholas Poynz, a courtier of Henry VIII; Georgian residential area of Clifton; the Clifton Suspension Bridge designed by Brunel, and the *SS Great Britain*, which is being restored in dry dock.
history John Cabot sailed from here 1497 to Newfoundland, and there was a great trade with the American colonies and the West Indies in the 17th–18th centuries, including slaves. The poet Chatterton was born here.

Bristol /ˈbrɪstl/ city in Virginia and Tennessee (the border runs through the centre of the city), USA, northeast of Knoxville; population (1990) 23,400 (Tennessee), 18,400 (Virginia). This dual city is divided politically, but one unit economically. Industries include lumber, steel, office machines, pharmaceuticals, missile parts, and textiles.

Bristol /ˈbrɪstl/ city in central Connecticut, USA, southwest of Hartford; population (1990) 60,600. Known as the clock-making capital of the USA, its products also include tools and machinery parts.

Britain /ˈbrɪtn/ another name for ◊ Great Britain.

British Antarctic Territory colony created 1962 and comprising all British territories south of latitude 60° S: the South Orkney Islands, the South Shetland Islands, the Antarctic Peninsula

British Colombia

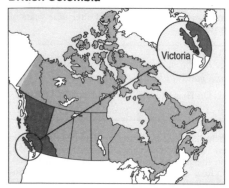

and all adjacent lands, and Coats Land, extending to the South Pole; total land area 660,000 sq km/170,874 sq mi; population (exclusively scientific personnel) about 300.

British Columbia /kə'lʌmbiə/ province of Canada on the Pacific Ocean
area 947,800 sq km/365,851 sq mi
capital Victoria
towns Vancouver, Prince George, Kamloops, Kelowna
physical Rocky Mountains and Coast Range; deeply indented coast; rivers include the Fraser and Columbia; over 80 lakes; more than half the land is forested
products fruit and vegetables; timber and wood products; fish; coal, copper, iron, lead; oil and natural gas; hydroelectricity
population (1991) 3,185,900
history Captain Cook explored the coast 1778; a British colony was founded on Vancouver Island 1849, and the gold rush of 1858 extended settlement to the mainland; it became a province 1871. In 1885 the Canadian Pacific Railroad linking British Columbia to the east coast was completed.

Such a land is good for an energetic man. It is also not so bad for the loafer.

On **British Columbia** Rudyard Kipling 1908

British Honduras /hɒn'djuərəs/ former name (until 1973) of ◊Belize.

British Indian Ocean Territory British colony in the Indian Ocean directly administered by the Foreign and Commonwealth Office. It consists of the Chagos Archipelago some 1,900 km/1,200 mi NE of Mauritius
area 60 sq km/23 sq mi
features lagoons; US naval and air base on Diego Garcia
products copra, salt fish, tortoiseshell
population (1982) 3,000
history purchased 1965 for $3 million by Britain from Mauritius to provide a joint US/UK base. The islands of Aldabra, Farquhar, and Desroches, some 485 km/300 mi N of Madagascar, originally formed part of the British

Indian Ocean Territory but were returned to the administration of the Seychelles 1976.

British Isles group of islands off the northwest coast of Europe, consisting of Great Britain (England, Wales, and Scotland), Ireland, the Channel Islands, the Orkney and Shetland islands, the Isle of Man, and many other islands that are included in various counties, such as the Isle of Wight, Scilly Isles, Lundy Island, and the Inner and Outer Hebrides. The islands are divided from Europe by the North Sea, Strait of Dover, and the English Channel, and face the Atlantic to the W.

British Virgin Islands part of the ◊Virgin Islands group in the West Indies.

Brittany /'brɪtənɪ/ (French **Bretagne**, Breton **Breiz**) region of NW France in the Breton peninsula between the Bay of Biscay and the English Channel; area 27,200 sq km/10,499 sq mi; capital Rennes; population (1987) 2,767,000. A farming region, it includes the *départements* of Côtes-du-Nord, Finistère, Ille-et-Vilaine, and Morbihan.
history Brittany was the Gallo-Roman province of Armorica after being conquered by Julius Caesar 56 BC. It was devastated by Norsemen after the Roman withdrawal. Established under the name of Brittany in the 5th century AD by Celts fleeing the Anglo-Saxon invasion of Britain, it became a strong, expansionist state that maintained its cultural and political independence, despite pressure from the Carolingians, Normans, and Capetians. In 1171, the duchy of Brittany was inherited by Geoffrey, son of Henry II of England, and remained in the Angevin dynasty's possession until 1203, when Geoffrey's son Arthur was murdered by King John, and the title passed to the Capetian Peter of Dreux. Under the Angevins, feudalism was introduced, and French influence increased under the Capetians. By 1547 it had been formally annexed by France, and the Breton language was banned in education. A separatist movement developed after World War II, and there has been guerrilla activity.

Brno /'bɜːnəʊ/ industrial city (chemicals, arms, textiles, machinery) in the Czech Republic; population (1991) 388,000. Now the second largest city in the Czech Republic, Brno was formerly the capital of the Austrian crown land of Moravia.

Brno is a very nice city, but we didn't get a chance to spnd mch tme thre ...

On **Brno** Art Buchwald *More Caviar* 1958

Broads, Norfolk /brɔːdz/ area of navigable lakes and rivers in England; see ◊Norfolk Broads.

Brocken /'brɒkən/ highest peak of the Harz Mountains (1,142 m/3,746 ft) in Germany. On 1 May (Walpurgis night), witches are said to gather here.

The **Brocken Spectre** is a phenomenon of mountainous areas, so named because it was first scientifically observed at Brocken 1780. The greatly enlarged shadow of the observer, accompanied by coloured rings, is cast by a low sun upon a cloud bank.

Brockton /'brɒktən/ city in SW Massachusetts, USA, south of Boston; population (1990) 92,800. Industries include footwear, tools, and electronic equipment.

Broken Hill /'brəukən/ mining town in New South Wales, Australia; population (1981) 27,000. It is the base of the Royal Flying Doctor Service.

Broken Hill /'brəukən/ former name (until 1967) of ◊Kabwe, a town in Zambia.

Bromberg /'brɒmbɜːg/ German name of ◊Bydgoszcz, a port in Poland.

Bromley borough of SE Greater London
features 17th-century Bromley College; 47 High Street was the birthplace of H G Wells 1866
population (1991) 281,700.

Bronx, the /brɒŋks/ borough of New York City, USA, northeast of Harlem River; area 109 sq km/ 42 sq mi; population (1990) 1,169,000. Largely residential, it is named after an early Dutch settler, James Bronck. The New York Zoological Society and Gardens are here popularly called the Bronx Zoo and the Bronx Botanical Gardens.

The bronx? No thonx.

On the **Bronx** Ogden Nash
in *New Yorker* 1931

Brooklyn /'bruklɪn/ borough of New York City, USA, occupying the southwest end of Long Island. It is linked to Manhattan Island by the Brooklyn-Battery Tunnel, the Brooklyn Bridge 1883, the Williamsburg and the Manhattan bridges, and to Staten Island by the Verrazano-Narrows Bridge 1964. There are more than 60 parks of which Prospect is the largest. There is also a museum, botanical garden, and a beach and amusement area at Coney Island.

Browns Ferry /ˌbraunz 'feri/ site of a nuclear power station on the Alabama River, central Alabama, USA. A nuclear accident 1975 resulted in the closure of the plant for 18 months. This incident marked the beginning of widespread disenchantment with nuclear power in the USA.

Brownsville /'braunzvɪl/ city in S Texas, USA, on the Rio Grande just before it flows into the Gulf of Mexico, S of Corpus Christi and N of Matamoros, Mexico; seat of Cameron County; population (1990) 99,000. It is a port of entry to the USA; industries include chemicals and food products; tourism is also important. Originally Fort Taylor 1846, it was an important Confederate port during the Civil War.

Bruges /bruːʒ/ (Flemish *Brugge*) historic city in NW Belgium; capital of W Flanders province, 16 km/10 mi from the North Sea, with which it is connected by canal; population (1991) 117,100.

Bruges was the capital of medieval Flanders and was the chief European wool manufacturing town as well as its chief market. The contemporary port handles coal, iron ore, oil, and fish; local industries include lace, textiles, paint, steel, beer, furniture, and motors.
features Among many fine buildings are the 14th-century cathedral, the church of Nôtre Dame with a Michelangelo statue of the Virgin and Child, the Gothic town hall and market hall; there are remarkable art collections. It was named for its many bridges. The College of Europe is the oldest centre of European studies.

Brugge /bruːxə/ Flemish form of ◊Bruges, a town in Belgium.

Brunei Islamic Sultanate of (*Negara Brunei Darussalam*)
area 5,765 sq km/2,225 sq mi
capital Bandar Seri Begawan
towns Tutong, Seria, Kuala Belait
physical flat coastal plain with hilly lowland in W and mountains in E; 75% of the area is forested; the Limbang valley splits Brunei in two, and its cession to Sarawak 1890 is disputed by Brunei
features Temburong, Tutong, and Belait rivers; Mount Pagon (1,850 m/6,070 ft)
head of state and of government HM Muda Hassanal Bolkiah Mu'izzaddin Waddaulah, Sultan of Brunei, from 1968
political system absolute monarchy
political party Brunei National United Party (BNUP)
exports liquefied natural gas (world's largest producer) and oil, both expected to be exhausted by the year 2000
currency Brunei dollar
population (1990 est) 372,000 (65% Malay, 20% Chinese—few Chinese granted citizenship); growth rate 12% p.a.
life expectancy 74 years
languages Malay (official), Chinese (Hokkien), English
religion 60% Muslim (official)
literacy 95%

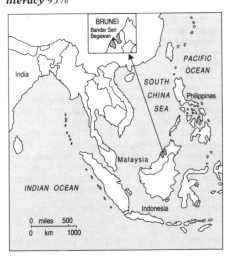

GDP $3.4 bn (1985); $20,000 per head
chronology
1888 Brunei became a British protectorate.
1941–45 Occupied by Japan.
1959 Written constitution made Britain responsible for defence and external affairs.
1962 Sultan began rule by decree.
1967 Sultan abdicated in favour of his son, Hassanal Bolkiah.
1971 Brunei given internal self-government.
1975 United Nations resolution called for independence for Brunei.
1984 Independence achieved from Britain, with Britain maintaining a small force to protect the oil and gas fields.
1985 A 'loyal and reliable' political party, the Brunei National Democratic Party (BNDP), legalized.
1986 Death of former sultan, Sir Omar. Formation of multiethnic BNUP.
1988 BNDP banned.

Brunei Town /'bruːnaɪ/ former name (until 1970) of ◊Bandar Seri Begawan, the capital of Brunei.

Brünn /brʊn/ German form of ◊Brno, a town in the Czech Republic.

Brunswick /'brʌnzwɪk/ (German *Braunschweig*) former independent duchy, a republic from 1918, which is now part of Lower Saxony, Germany.

Brunswick /'brʌnzwɪk/ (German *Braunschweig*) industrial city (chemical engineering, precision engineering, food processing) in Lower Saxony, Germany; population (1988) 248,000. It was one of the chief cities of N Germany in the Middle Ages and a member of the Hanseatic League. It was capital of the duchy of Brunswick from 1671.

Brusa /bruːˈsɑː/ alternative form of ◊Bursa, a town in Turkey.

Brussels /'brʌsəlz/ (Flemish *Brussel*, French *Bruxelles*) capital of Belgium, industrial city (lace, textiles, machinery, and chemicals); population (1987) 974,000 (80% French-speaking, the suburbs Flemish-speaking). It is the headquarters of the European Economic Community (EEC) and since 1967 of the international secretariat of NATO. First settled in the 6th century, and a city from 1312, Brussels became the capital of the Spanish Netherlands 1530 and of Belgium 1830.
features It has fine buildings including the 13th-century church of Sainte Gudule; the Hôtel de Ville, Maison du Roi, and others in the Grande Place; the royal palace. The Musées Royaux des Beaux-Arts de Belgique hold a large art collection. The bronze fountain statue of a tiny naked boy urinating, the Manneken Pis (1388) is to be found here. Brussels is also the site of the world's largest cinema complex, with 24 film theatres.

Bruxelles French form of ◊Brussels, the capital of Belgium.

Bryan /'braɪən/ city in E Texas, USA, northwest of Houston; seat of Brazos County; population

Buckinghamshire

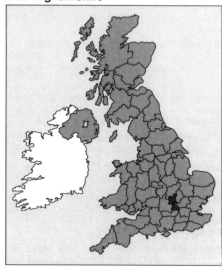

(1990) 55,000. Its industries include the manufacture of cotton gins.

Bryansk /bri'ænsk/ city in W central Russia, SW of Moscow on the river Desna; population (1987) 445,000. Industries include sawmills, textiles, and steel.

Brześć nad Bugiem /ˌbʒeʃtʃ nad 'buːgjem/ Polish name of ◊Brest, a town in Belarus.

Bubiyan /ˌbuːbɪ'jɑn/ uninhabited island off Kuwait, occupied by Iraq 1990; area about 1,000 sq km/380 sq mi. On 28 Feb 1991, following Allied success in the Gulf War, Iraqi troops were withdrawn.

Bucaramanga /buˌkɑːrə'mæŋgə/ industrial and commercial city (coffee, tobacco, cacao, cotton) in N central Colombia; population (1985) 493,929. It was founded by the Spanish 1622.

Bucharest /ˌbuːkə'rest/ (Romanian *Bucuresti*) capital and largest city of Romania; population (1985) 1,976,000, the conurbation of Bucharest district having an area of 1,520 sq km/587 sq mi and a population of 2,273,000. It was originally a citadel built by Vlad the Impaler to stop the advance of the Ottoman invasion in the 14th century. Bucharest became the capital of the princes of Wallachia 1698 and of Romania 1861. Savage fighting took place in the city during Romania's 1989 revolution.

Bucharest is the town of one street, one church, and one idea.

On **Bucharest** Romanian proverb

Buckingham market town in Buckinghamshire, England, on the river Ouse; population (1981) 6,600. University College was established 1974, and was given a royal charter as the University of Buckingham 1983.

Buckinghamshire /'bʌkɪŋəmʃə/ county in SE central England
area 1,880 sq km/726 sq mi
towns Aylesbury (administrative headquarters), Buckingham, High Wycombe, Beaconsfield, Olney, Milton Keynes
features Chequers (country seat of the prime minister); Burnham Beeches and the church of the poet Gray's 'Elegy' at Stoke Poges; Cliveden, a country house designed by Charles Barry (now a hotel, it was used by the newspaper-owning Astors for house parties); Bletchley Park, home of World War II code-breaking activities, now used as a training post for GCHQ (Britain's electronic surveillance centre); Open University at Walton Hall; homes of the poets William Cowper at Olney and John Milton at Chalfont St Giles, and of the Tory prime minister Disraeli at Hughenden; Stowe gardens
products furniture, chiefly beech; agricultural goods
population (1991) 619,500
famous people William Herschel, George Gilbert Scott, Edmund Waller, John Hampden, Ben Nicholson.

Bucks abbreviation for ◊*Buckinghamshire.*

Budapest /ˌbjuːdə'pest/ capital of Hungary, industrial city (chemicals, textiles) on the river Danube; population (1989) 2,115,000. Buda, on the right bank of the Danube, became the Hungarian capital 1867 and was joined with Pest, on the left bank, 1872.
  Budapest saw fighting between German and Soviet troops in World War II 1944–45 and between the Hungarians and Soviet troops in the uprising of 1956.

Budějovice see ◊České Budějovice, a town in the Czech Republic.

Budweis /'budvaɪs/ German form of ◊České Budějovice, a town in the Czech Republic.

Buenos Aires /ˌbweɪnɒs 'aɪrɪz/ capital and industrial city of Argentina, on the south bank of the Río de la Plata; population (1991) 2,961,000, metropolitan area 7,950,400. It was founded 1536, and became the capital 1853.
features Congress building; the cathedral and presidential palace (known as the Pink House), both on the Plaza de Mayo; university 1821.

Buenos Aires is at first glance, and for days afterwards, a most civilized ant-hill.

On **Buenos Aires** Paul Theroux
The Old Patagonian Express 1979

Buffalo /'bʌfələu/ industrial port in New York State, USA, at the east end of Lake Erie; population (1990) 328,100. It is linked with New York City by the New York State Barge Canal.

Bug /buːg/ name of two rivers in E Europe: the *West Bug* rises in SW Ukraine and flows to the Vistula, and the *South Bug* rises in W Ukraine and flows to the Black Sea.

Buganda /buː'gændə/ either of two provinces (North and South Buganda) of Uganda, home of the Baganda people and formerly a kingdom from the 17th century. The *kabaka* or king, Edward Mutesa II (1924–1969), was the first president of independent Uganda 1962–66, and his son Ronald Mutebi (1955–   ) is *sabataka* (head of the Baganda clans).

Bujumbura /ˌbuːdʒum'buərə/ capital of Burundi; population (1986) 272,600. Formerly called *Usumbura* (until 1962), it was founded 1899 by German colonists. The university was established 1960.

Bukavu /buː'kɑːvuː/ port in E Zaire, on Lake Kivu; population (1984) 171,100. Mining is the chief industry. Called *Costermansville* until 1966, it is the capital of Itivu region.

Bukhara /buˈxɑːrə/ or *Bokhara* central city in Uzbekistan; population (1987) 220,000. It is the capital of Bukhara region, which has given its name to carpets (made in Ashkhabad). It is an Islamic centre, with a Muslim theological training centre. An ancient city in central Asia, it was formerly the capital of the independent emirate of Bukhara, annexed to Russia 1868. It was included in Bukhara region 1924.

Bukharest alternative form of ◊Bucharest, the capital of Romania.

Bukovina /ˌbukə'viːnə/ region in SE Europe, divided between the Ukraine and Romania. It covers 10,500 sq km/4,050 sq mi.
history Part of Moldavia during the Turkish regime, it was ceded by the Ottoman Empire to Austria 1777, becoming a duchy of the Dual Monarchy 1867–1918; then it was included in Romania. N Bukovina was ceded to the USSR 1940 and included in Ukraine as the region of Chernovtsy; the cession was confirmed by the peace treaty 1947, but the question of its return has been raised by Romania. The part of Bukovina remaining in Romania became the district of Suceava.

Bulawayo /ˌbulə'weɪəu/ industrial city and railway junction in Zimbabwe; population (1982) 415,000. It lies at an altitude of 1,355 m/4,450 ft on the river Matsheumlope, a tributary of the Zambezi, and was founded on the site of the kraal (enclosed village), burned down 1893, of the Matabele chief, Lobenguela. It produces agricultural and electrical equipment. The former capital of Matabeleland, Bulawayo developed with the exploitation of gold mines in the neighbourhood.

Bulgaria Republic of (*Republika Bulgaria*)
area 110,912 sq km/42,812 sq mi
capital Sofia
towns Plovdiv, Ruse; Black Sea ports Burgas and Varna
physical lowland plains in N and SE separated by mountains that cover three-quarters of the country
environment pollution has virtually eliminated all species of fish once caught in the Black Sea. Vehicle-exhaust emissions in Sofia have

led to dust concentrations more than twice the medically accepted level

features key position on land route from Europe to Asia; Black Sea coast; Balkan and Rhodope mountains; Danube River in N

head of state Zhelyu Zhelev from 1990

head of government Lyuben Berov from 1992

political system emergent democratic republic

political parties Union of Democratic Forces (UDF), right of centre; Bulgarian Socialist Party (BSP), left-wing, ex-communist; Movement for Rights and Freedoms (MRF), centrist

exports textiles, leather, chemicals, nonferrous metals, timber, machinery, tobacco, cigarettes

currency lev

population (1990 est) 8,978,000 (including 900,000–1,500,000 ethnic Turks, concentrated in S and NE); growth rate 0.1% p.a.

life expectancy men 69, women 74

languages Bulgarian, Turkish

religions Eastern Orthodox Christian 90%, Sunni Muslim 10%

literacy 98%

GDP $25.4 bn (1987); $2,836 per head

chronology

1908 Bulgaria became a kingdom independent of Turkish rule.

1944 Soviet invasion of German-occupied Bulgaria.

1946 Monarchy abolished and communist-dominated people's republic proclaimed.

1947 Soviet-style constitution adopted.

1949 Death of Georgi Dimitrov, the communist government leader.

1954 Election of Todor Zhivkov as Communist Party general secretary; made nation a loyal satellite of USSR.

1971 Constitution modified; Zhivkov elected president.

1985–89 Large administrative and personnel changes made haphazardly under Soviet stimulus.

1987 New electoral law introduced multi-candidate elections.

1989 Programme of 'Bulgarianization' resulted in mass exodus of Turks to Turkey. Nov: Zhivkov

ousted by Petar Mladenov. Dec: opposition parties allowed to form.

1990 April: BCP renamed Bulgarian Socialist Party (BSP). Aug: Dr Zhelyu Zhelev elected president. Nov: government headed by Andrei Lukanov resigned, replaced Dec by coalition led by Dimitur Popov.

1991 July: new constitution adopted. Oct: UDF beat BSP in general election by narrow margin; formation of first noncommunist, UDF-minority government under Filip Dimitrov.

1992 Zhelev became Bulgaria's first directly elected president. Relations with West greatly improved. Dimitrov resigned after vote of no confidence; replaced by Lyuben Berov.

Burgas /'buəgəs/ Black Sea port and resort in Bulgaria; population (1987) 198,000.

Burgenland /'buəgənlænd/ federal state of SE Austria, extending S from the Danube along the western border of the Hungarian plain; area 4,000 sq km/1,544 sq mi; population (1989) 267,200. It is a largely agricultural region adjoining the Neusiedler See, and produces timber, fruit, sugar, wine, lignite, antimony, and limestone. Its capital is Eisenstadt.

Burgess Shale Site /'bɜːdʒɪs/ site of unique fossil-bearing rock formations created 530 million years ago by a mud slide, in Yoho National Park, British Columbia, Canada. The shales in this corner of the Rocky Mountains contain more than 120 species of marine invertebrate fossils. Although discovered 1909 by US geologist Charles Walcott, the Burgess Shales have only recently been used as evidence in the debate concerning the evolution of life. In *Wonderful Life* 1990 Stephen Jay Gould drew attention to a body of scientific opinion interpreting the fossil finds as evidence of parallel early evolutionary trends extinguished by chance rather than natural selection.

Burgos /'buəgɒs/ city in Castilla-León, Spain, 217 km/135 mi N of Madrid; population (1991) 169,300. It produces textiles, motor parts, and chemicals. It was capital of the old kingdom of Castile, and the national hero El Cid is buried in the cathedral, built 1221–1567 and regarded as one of the most important examples of Gothic architecture in Europe.

Burgos tipifies what is another characteristic of Spain, and that is the endless fertility of architectural adventure: the flowering of stone.

On **Burgos** Hilaire Belloc *Places* 1942

Burgundy /'bɜːgəndi/ (French *Bourgogne*) modern region and former duchy of France that includes the *départements* of Côte-d'Or, Nièvre, Sâone-et-Loire, and Yonne; area 31,600 sq km/12,198 sq mi; population (1986) 1,607,000. Its capital is Dijon.

Burgundy is renowned for its wines, such as Chablis and Nuits-Saint-Georges, and for its cattle (the Charolais herd-book is maintained at Nevers). A duchy from the 9th century, it was part of an independent medieval kingdom and was incorporated into France 1477.

Burkina Faso The People's Democratic Republic of (formerly *Upper Volta*)
area 274,122 sq km/105,811 sq mi
capital Ouagadougou
towns Bobo-Dioulasso, Koudougou
physical landlocked plateau with hills in W and SE; headwaters of the river Volta; semiarid in N, forest and farmland in S
environment tropical savanna subject to overgrazing and deforestation
features linked by rail to Abidjan in Ivory Coast, Burkina Faso's only outlet to the sea
head of state and government Blaise Compaoré from 1987
political system transitional
political parties Organization for Popular Democracy–Workers' Movement (ODP–MT), nationalist left-wing; FP–Popular Front
exports cotton, groundnuts, livestock, hides, skins, sesame, cereals
currency CFA franc
population (1990 est) 8,941,000; growth rate 2.4% p.a.
life expectancy men 44, women 47
languages French (official); about 50 native Sudanic languages spoken by 90% of population
religions animist 53%, Sunni Muslim 36%, Roman Catholic 11%
literacy men 21%, women 6% (1985 est)
GDP $1.6 bn (1987); $188 per head
chronology
1958 Became a self-governing republic within the French Community.
1960 Independence from France, with Maurice Yaméogo as the first president.
1966 Military coup led by Col Lamizana. Constitution suspended, political activities banned, and a supreme council of the armed forces established.
1969 Ban on political activities lifted.
1970 Referendum approved a new constitution leading to a return to civilian rule.
1974 After experimenting with a mixture of military and civilian rule, Lamizana reassumed full power.
1977 Ban on political activities removed.

Referendum approved a new constitution based on civilian rule.
1978 Lamizana elected president.
1980 Lamizana overthrown in bloodless coup led by Col Zerbo.
1982 Zerbo ousted in a coup by junior officers. Major Ouédraogo became president and Thomas Sankara prime minister.
1983 Sankara seized complete power.
1984 Upper Volta renamed Burkina Faso, 'land of upright men'.
1987 Sankara killed in coup led by Blaise Compaoré.
1989 New government party ODP–MT formed by merger of other pro-government parties. Coup against Compaoré foiled.
1991 New constitution approved. Compaoré re-elected president.
1992 Multiparty elections won by FP–Popular Front.

Burlington /ˈbɜːlɪŋtən/ city in N central North Carolina, USA, northwest of Durham; population (1990) 39,500. Industries include textiles, chemicals, furniture, and agricultural products.

Burlington /ˈbɜːlɪŋtən/ city in NW Vermont, USA, on the east shore of Lake Champlain; seat of Chittenden County; population (1990) 39,100. It is a port of entry to the USA; industries include computer parts, steel, marble, lumber, dairy products, and tourism.

Burma /ˈbɜːmə/ former name (to 1989) of ◊Myanmar.

Burnley /ˈbɜːnli/ town in Lancashire, England, 19 km/12 mi NE of Blackburn; population (1989 est) 85,400. It was formerly a cotton-manufacturing town.

Bursa /ˈbɜːsə/ city in NW Turkey, with a port at Mudania; population (1990) 834,600. It was the capital of the Ottoman Empire 1326–1423.

Burton upon Trent /ˈbɜːtn əpɒn ˈtrent/ town in Staffordshire, England, NE of Birmingham; population (1988 est) 57,700. Industries include brewing, tyres, and engineering.

Say, for what were hop-yards meant/Or why was Burton built on Trent?

On **Burton-upon-Trent** A E Housman
A Shropshire Lad 1896

Burundi Republic of (*Republika y'Uburundi*)
area 27,834 sq km/10,744 sq mi
capital Bujumbura
towns Gitega, Bururi, Ngozi, Muyinga
physical landlocked grassy highland straddling watershed of Nile and Congo
features Lake Tanganyika, Great Rift Valley
head of state and government transitional military rule from 1993
political system one-party military republic
political party Union for National Progress (UPRONA), nationalist socialist
exports coffee, cotton, tea, nickel, hides,

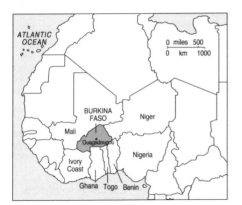

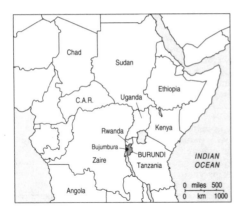

Bury /'beri/ town in Greater Manchester, England, on the river Irwell, 16 km/10 mi N of central Manchester; population (1986) 173,600. Industries include cotton, chemicals, and engineering.

Buryat /ˌburi'aːt/ autonomous republic of Russia, in East Siberia
area 351,300 sq km/135,600 sq mi
capital Ulan-Udé
physical bounded S by Mongolia, W by Lake Baikal; mountainous and forested
products coal, timber, building materials, fish, sheep, cattle
population (1986) 1,014,000
history settled by Russians 17th century; annexed from China by treaties 1689 and 1727.

livestock, cigarettes, beer, soft drinks; there are 500 million tonnes of peat reserves in the basin of the Akanyaru River
currency Burundi franc
population (1990 est) 5,647,000 (of whom 15% are the Nilotic Tutsi, still holding most of the land and political power, 1% are Pygmy Twa, and the remainder Bantu Hutu); growth rate 2.8% p.a.
life expectancy men 45, women 48
languages Kirundi (a Bantu language) and French (both official), Kiswahili
religions Roman Catholic 62%, Protestant 5%, Muslim 1%, animist 32%
literacy men 43%, women 26% (1985)
GDP $1.1 bn (1987); $230 per head
chronology
1962 Separated from Ruanda-Urundi, as Burundi, and given independence as a monarchy under King Mwambutsa IV.
1966 King deposed by his son Charles, who became Ntare V; he was in turn deposed by his prime minister, Capt Michel Micombero, who declared Burundi a republic.
1972 Ntare V killed, allegedly by the Hutu ethnic group. Massacres of 150,000 Hutus by the rival Tutsi ethnic group, of which Micombero was a member.
1973 Micombero made president and prime minister.
1974 UPRONA declared the only permitted political party, with the president as its secretary general.
1976 Army coup deposed Micombero. Col Jean-Baptiste Bagaza appointed president by the Supreme Revolutionary Council.
1981 New constitution adopted, providing for a national assembly.
1984 Bagaza elected president as sole candidate.
1987 Bagaza deposed in coup Sept. Maj Pierre Buyoya headed new Military Council for National Redemption.
1988 Some 24,000 majority Hutus killed by Tutsis.
1992 Referendum approved new constitution.
1993 June: Melchior Ndadaye, a Hutu, elected president. Oct: Ndadaye and other government figures assassinated in military coup. Intertribal massacres followed.

Bury St Edmunds /'beri/ market town in Suffolk, England, on the river Lark; population (1989 est) 32,000. It was named after St Edmund, and there are remains of a large Benedictine abbey founded 1020.

... it is the neatest place that ever was seen.

On **Bury St Edmunds** William Cobbett 1830

Bute /bjuːt/ island and resort in the Firth of Clyde, Scotland; area 120 sq km/46 sq mi. The chief town is Rothesay. It is separated from the mainland in the N by a winding channel, the *Kyles of Bute*. With Arran and the adjacent islands it comprised the former county of Bute, merged 1975 into the region of Strathclyde.

Butte /bjuːt/ mining town in Montana, USA, in the Rocky Mountains; population (1990) 33,900. Butte was founded 1864 during a rush for gold, soon exhausted; copper was found some 20 years later on what was called 'the richest hill on earth'.

... built on the 'richest hill on earth', and generally described as the greatest mining camp ever known ...

On **Butte** John Gunther *Inside USA* 1947

Buxton /'bʌkstən/ spa town in Derbyshire, England; population (1981) 21,000. Known from Roman times for its hot springs, it is today a source for bottled mineral water. It has a restored Edwardian opera house and an annual opera festival.

Bydgoszcz /'bɪdgɒʃtʃ/ industrial river port in N Poland, 105 km/65 mi NE of Poznań on the river Warta; population (1990) 381,500. As *Bromberg* it was under Prussian control 1772–1919.

Byelorussia alternative form of ◊Belarus, a country in E central Europe.

CA abbreviation for ◊*California*, a state of the USA.

Cabinda /kəˈbɪndə/ or **Kabinda** African coastal enclave, a province of Angola; area 7,770 sq km/ 3,000 sq mi; population (1980) 81,300. The capital is Cabinda. There are oil reserves. Attached to Angola 1886, the enclave has made claims to independence.

Cadiz /kəˈdɪz/ Spanish city and naval base, capital and seaport of the province of Cadiz, standing on Cadiz Bay, an inlet of the Atlantic, 103 km/ 64 mi S of Seville; population (1991) 156,600. After the discovery of the Americas 1492, Cadiz became one of Europe's most vital trade ports. The English adventurer Francis Drake burned a Spanish fleet here 1587 to prevent the sailing of the Armada.

Probably founded by the Phoenicians about 1100 BC, it was a centre for the tin trade with Cornwall, England. It was recaptured from the Moors by the king of Castile 1262. Development was restricted by its peninsular location until a bridge to the opposite shore of Cadiz Bay was completed 1969.

Caen /kɑːn/ capital of Calvados *département*, France, on the river Orne; population (1990) 115,600. It is a business centre, with ironworks and electric and electronic industries; Caen building stone has a fine reputation. The town is linked by canal with the nearby English Channel to the northeast.

The church of St Etienne was founded by William the Conqueror, and the university by Henry VI of England 1432. Caen was captured by British forces in World War II on 9 July 1944 after five weeks' fighting, during which the town was badly damaged.

Caerleon /kɑːˈliːən/ small town in Gwent, Wales, on the river Usk, 5 km/3 mi NE of Newport; population (1981) 6,711. It stands on the site of the Roman fortress of Isca. There is a Legionary Museum and remains of an amphitheatre.

Caernarvon /kəˈnɑːvn/ or **Caernarfon** administrative headquarters of Gwynedd, N Wales, situated on the southwest shore of the Menai Strait; population (1981) 10,000. Formerly a Roman station, it is now a market town and port.

The first Prince of Wales (later Edward II) was born in Caernarvon Castle; Edward VIII was invested here 1911 and Prince Charles 1969. The Earl of Snowdon became constable of the castle 1963.

Caernarvonshire /kəˈnɑːvənʃə/ (Welsh *Sir Gaernarfon*) former county of N Wales, merged in ◊Gwynedd 1974.

Caerphilly /kəˈfɪli/ (Welsh *Caerffili*) market town in Mid Glamorgan, Wales, 11 km/7 mi N of Cardiff; population (1981) 42,700. The castle was built by Edward I. The town gives its name to mild Caerphilly cheese.

Caesarea Mazaca /ˌsiːzəˈrə məˈzɑːkə/ ancient name for the Turkish city of ◊Kayseri.

Cagliari /kælˈjɑːri/ capital and port of Sardinia, Italy, on the Gulf of Cagliari; population (1988) 222,000.

Cagnes-sur-Mer /ˈkæn sjuə ˈmeə/ town SW of Nice, France; capital of the *département* of Alpes-Maritimes; population (1990) 41,300. The château (13th–17th century) contains mementos of the Impressionist painter Renoir, who lived here 1900–19.

Cahora Bassa /ˈkɑːhɒɪrɑ ˈbæsə/ largest hydroelectric project in Africa, created as a result of the damming of the Zambezi River to form a reservoir 230 km/144 mi long in W Mozambique.

Cairngorm Mountains /ˈkeəngɔːm/ mountain group in Scotland, northern part of the Grampian Mountains, the highest peak being Ben Macdhui 1,309 m/4,296 ft.

Aviemore (Britain's first complete holiday and sports centre) was opened 1966, and 11 km/7 mi to the S is the Highland Wildlife Park at Kincraig.

Cairns /keənz/ seaport of NE Queensland, Australia; population (1984) 38,700.

Its chief industry is the export of sugar and timber; tourism is important. Tours to the Great Barrier Reef start here.

Cairo /ˈkaɪrəu/ (Arabic *El Qahira*) capital of Egypt, on the east bank of the river Nile 13 km/ 8 mi above the apex of the delta and 160 km/ 100 mi from the Mediterranean; the largest city in Africa and in the Middle East; population (1985) 6,205,000; metropolitan area (1987) 13,300,000. An earthquake in a suburb of the city Oct 1992 left over 500 dead.

history El Fustat (Old Cairo) was founded by Arabs about AD 642, Al Qahira about 1000 by the Fatimid ruler Gowhar. Cairo was the capital of the Ayyubid dynasty, one of whose sultans, Saladin, built the Citadel in the late 1100s.

Under the Mamelukes 1250–1517 the city prospered, but declined in the 16th century after conquest by the Turks. It became the capital of the virtually autonomous kingdom of Egypt established by Mehmet Ali 1805. During World War II it was the headquarters of the Allied forces in N Africa.

features Cairo is the site of the mosque that houses the El Azhar university (972). The

Mosque of Amr dates from 643; the 12th-century Citadel contains the impressive 19th-century Muhammad Ali mosque. The city is 32 km/20 mi N of the site of the ancient Egyptian centre of Memphis. The Great Pyramids and Sphinx are at nearby Gîza.

The government and business quarters reflect Cairo's position as a leading administrative and commercial centre, and the semi-official newspaper *al Ahram* is an influential voice in the Arab world. Cairo's industries include the manufacture of textiles, cement, vegetable oils, and beer. At Helwan, 24 km/15 mi to the S, an industrial centre is developing, with iron and steelworks powered by electricity from the Aswan High Dam. There are two secular universities: Cairo University (1908) and Ein Shams (1950).

Calabar /'kæləbɑː/ port and capital of Cross River State, SE Nigeria, on the Cross River, 64 km/40 mi from the Atlantic Ocean; population (1983) 126,000. Rubber, timber, and vegetable oils are exported. It was a centre of the slave trade in the 18th and 19th centuries.

Calabria /kə'læbriə/ mountainous earthquake region occupying the 'toe' of Italy, comprising the provinces of Catanzaro, Cosenza, and Reggio; capital Catanzaro; area 15,100 sq km/5,829 sq mi; population (1990) 2,153,700. Reggio is the industrial centre.

Calais /'kæleɪ/ port in Pas-de-Calais *département*, N France; population (1990) 75,800. Taken by England's Edward III 1347, it was saved from destruction by the personal surrender of the Burghers of Calais, commemorated in Auguste Rodin's sculpture; the French retook it 1558. In World War II, following German occupation May 1940–Oct 1944, it surrendered to the Canadians.

When I am dead and opened, you shall find 'Calais' lying in my heart.

On **Calais** Mary Tudor about 1558

Calais, Pas de /'pɑ də 'kæleɪ/ French name for the Strait of ⇨Dover.

Calcutta /kæl'kʌtə/ largest city of India, on the river Hooghly, the westernmost mouth of the river Ganges, some 130 km/80 mi N of the Bay of Bengal. It is the capital of West Bengal; population (1981) 9,166,000. It is chiefly a commercial and industrial centre (engineering, shipbuilding, jute, and other textiles). Calcutta was the seat of government of British India 1773–1912. There is severe air pollution.
features Buildings include a magnificent Jain temple, the palaces of former Indian princes; and the Law Courts and Government House, survivals of the British Raj. Across the river is the city of Howrah, and between Calcutta and the sea there is a new bulk cargo port, Haldia, which is the focus of oil refineries, petrochemical plants, and fertilizer factories. There is a fine museum; educational institutions include the

University of Calcutta (1857), oldest of several universities; the Visva Bharati at Santiniketan, founded by Rabindranath Tagore; and the Bose Research Institute.
history Calcutta was founded 1686–90 by Job Charnock of the East India Company as a trading post. Captured by Suraj-ud-Dowlah 1756, during the Anglo-French wars in India, it was retaken by Robert Clive 1757.

It is quite impossible to forget or ignore Calcutta's imperial past, for the city has been pickled in its origins.

On **Calcutta** Geoffrey Moorhouse
Calcutta 1971

Caldey Island /'kɔːldi/ or *Caldy Island* (Welsh *Ynys Bŷr*) island in Carmarthen Bay off the coast of Dyfed, Wales, near Tenby, separated from the mainland by Caldey Sound. It has been inhabited by Celtic monks since the 6th century. A monastery, built by Anglican Benedictines 1906, is now occupied by Belgian Trappist monks. There is a small village and a lighthouse.

Caledonian Canal /ˌkælɪ'dəʊniən/ waterway in NW Scotland, 98 km/61 mi long, linking the Atlantic and the North Sea. Of its 98 km/61 mi length only a 37 km/23 mi stretch is artificial, the rest being composed of lochs Lochy, Oich, and Ness. The canal was built by Scottish civil engineer Thomas Telford 1803–23.

Calgary /'kælgəri/ city in Alberta, Canada, on the Bow River, in the foothills of the Rocky Mountains; at 1,048 m/3,440 ft it is one of the highest Canadian towns; population (1986) 671,000. It is the centre of a large agricultural region and is the oil and financial centre of Alberta and W Canada. The 1988 Winter Olympic Games were held here.

Founded as Fort Calgary by the North West Mounted Police 1875, it was reached by the Canadian Pacific Railway 1885 and developed rapidly after the discovery of oil 1914.

Cali /kæ'liː/ city in SW Colombia, in the Cauca Valley 975 m/3,200 ft above sea level; population (1985) 1,398,276. Cali was founded 1536. It has textile, sugar, and engineering industries.

California /ˌkælɪ'fɔːniə/ Pacific-coast state of the USA; nicknamed the Golden State (originally because of its gold mines, more recently because of its orange groves and sunshine)
area 411,100 sq km/158,685 sq mi
capital Sacramento
cities Los Angeles, San Diego, San Francisco, San José, Fresno
physical Sierra Nevada, including Yosemite and Sequoia national parks, Lake Tahoe, Mount Whitney (4,418 m/14,500 ft, the highest mountain in the lower 48 states); the Coast Range; Death Valley (86 m/282 ft below sea level, the lowest point in the western hemisphere); Colorado and Mojave deserts; Monterey Peninsula; Salton Sea; the San Andreas fault; huge,

California

offshore underwater volcanoes with tops 5 mi/
8 km across
features California Institute of Technology
(Caltech); Lawrence Berkeley and Lawrence
Livermore laboratories of the University of
California, which share particle physics and
nuclear weapons research with Los Alamos;
Stanford University, which has the Hoover Institute and is the powerhouse of Silicon Valley; Paul
Getty art museum at Malibu, built in the style of
a Roman villa; Hollywood
products leading agricultural state with fruit
(peaches, citrus, grapes in the valley of the San
Joaquin and Sacramento rivers), nuts, wheat,
vegetables, cotton, and rice, all mostly grown by
irrigation, the water being carried by immense
concrete-lined canals to the Central and Imperial
valleys; beef cattle; timber; fish; oil; natural
gas; aerospace technology; electronics (Silicon
Valley); food processing; films and television programmes; great reserves of energy (geothermal) in
the hot water that lies beneath much of the state
population (1990) 29,760,000, the most populous state of the USA (69.9% white, 25.8%
Hispanic, 9.6% Asian and Pacific islander,
including many Vietnamese, 7.4% black, 0.8%
American Indian)
famous people Luther Burbank, Walt Disney,
William Randolph Hearst, Jack London, Marilyn
Monroe, Richard Nixon, Ronald Reagan, John
Steinbeck
history colonized by Spain 1769; ceded to the
USA after the Mexican War 1848; became a state
1850. The discovery of gold in the Sierra Nevada
Jan 1848 was followed by the gold rush 1849–56.

---

*California, where the twentieth century
is a burning and a shining neon light,
and where, anyway, cows are rarely seen
out-of-doors nowadays.*

On **California** Malcolm Muggeridge
Observer April 1965

California, Lower /ˌkælɪˈfɔːniə/ English name for
◊Baja California.
Callao /kaɪˈau/ chief commercial and fishing port
of Peru, 12 km/7 mi SW of Lima; population
(1988) 318,000. Founded 1537, it was destroyed
by an earthquake 1746. It is Peru's main naval
base, and produces fertilizers.

Calpe /ˈkælpi/ name of ◊Gibraltar in ancient
Phoenician and Carthaginian times.

Caltanissetta /ˌkæltənɪˈsetə/ town in Sicily, Italy,
96 km/60 mi SE of Palermo; population (1981)
61,146. It is the centre of the island's sulphur
industry. It has a Baroque cathedral.

Camagüey /ˌkæməˈgweɪ/ city in Cuba; population
(1989) 283,000. It is the capital of Camagüey
province in the centre of the island. Founded
about 1514, it was the capital of the Spanish
West Indies during the 19th century. It has a
17th-century cathedral.

Camargue /kæˈmɑːg/ marshy area of the Rhône
delta, S of Arles, France; about 780 sq km/
300 sq mi. Black bulls and white horses are bred
here, and the nature reserve, which is known for
its bird life, forms the southern part.

Cambodia State of (formerly **Khmer Republic**
1970–76, **Democratic Kampuchea** 1976–79,
People's Republic of Kampuchea 1979–89)
area 181,035 sq km/69,880 sq mi
capital Phnom Penh
towns Battambang, the seaport Kompong Som
physical mostly flat forested plains with
mountains in SW and N; Mekong River runs N–S
features ruins of ancient capital Angkor; Lake
Tonle Sap
head of state Prince Norodom Sihanouk
from 1991
head of government Hun Sen from 1985
political system transitional
political parties Cambodian People's Party
(CPP), reform socialist (formerly the communist Kampuchean People's Revolutionary Party
(KPRP)); Party of Democratic Kampuchea
(Khmer Rouge), ultranationalist communist;
Khmer People's National Liberation Front
(KPNLF), anticommunist
exports rubber, rice, pepper, wood, cattle
currency Cambodian riel
population (1990 est) 6,993,000; growth rate
2.2% p.a.
life expectancy men 42, women 45

languages Khmer (official), French
religion Theravāda Buddhist 95%
literacy men 78%, women 39% (1980 est)
GDP $592 mn (1987); $83 per head
chronology
1863–1941 French protectorate.
1941–45 Occupied by Japan.
1946 Recaptured by France.
1953 Independence achieved from France.
1970 Prince Sihanouk overthrown by US-backed Lon Nol.
1975 Lon Nol overthrown by Khmer Rouge.
1978–79 Vietnamese invasion and installation of Heng Samrin government.
1982 The three main anti-Vietnamese resistance groups formed an alliance under Prince Sihanouk.
1987 Vietnamese troop withdrawal began.
1989 Sept: completion of Vietnamese withdrawal. Nov: United Nations peace proposal rejected by Phnom Penh government.
1991 Oct: Peace agreement signed in Paris, providing for a UN Transitional Authority in Cambodia (UNTAC) to administer country in transition period in conjunction with all-party Supreme National Council; communism abandoned. Nov: Sihanouk returned as head of state.
1992 Political prisoners released; freedom of speech and party formation restored. Oct: Khmer Rouge refused to disarm in accordance with peace process.

Before the fall of Sihanouk it was the last paradise, the last paradise.

On **Cambodia** helicopter pilot

Camborne-Redruth /ˈkæmbɔːn, redˈruːθ/ town in Cornwall, 16 km/10 mi SW of Truro, England; population (1985) 18,500. It has tin mines and there is a School of Metalliferous Mining.

Cambrai /kɒmˈbreɪ/ chief town of Nord *département*, France, on the river Escaut (Scheldt); population (1990) 34,200. Industries include light textiles (cambric is named after the town) and confectionery. The Peace of Cambrai or Ladies' Peace (1529) was concluded on behalf of Francis I of France by his mother Louise of Savoy and on behalf of Charles V by his aunt Margaret of Austria.

Cambridge /ˈkeɪmbrɪdʒ/ city in England, on the river Cam (a river sometimes called by its earlier name, Granta), 80 km/50 mi N of London; population (1989) 101,000. It is the administrative headquarters of Cambridgeshire. The city is centred on Cambridge University (founded 12th century), whose outstanding buildings, including Kings College Chapel, back onto the river. Present-day industries include the manufacture of scientific instruments, radio, electronics, paper, flour milling, and fertilizers.

As early as 100 BC, a Roman settlement grew up on a slight rise in the low-lying plain, commanding a ford over the river.

Apart from those of Cambridge University, fine buildings include St Benet's church, the oldest building in Cambridge, the round church of the Holy Sepulchre, and the Guildhall (1939). The Fitzwilliam Museum (1816) houses a fine art collection.

University colleges include Peterhouse, founded 1284, the oldest college; King's College 1441; Queen's College 1448; Jesus College 1496; St John's College 1511; and Trinity College 1546, the largest college. The university library is one of the oldest in the country and its treasures include the first book ever printed in English.

...My Mother Cambridge, whom as
with a Crowne/He doth adorne, and is
adorn'd of it/With many a gentle Muse,
and many a learned wit.

On **Cambridge** Edmund Spenser
The Faerie Queene 1596

Cambridge /ˈkeɪmbrɪdʒ/ city in Massachusetts, USA; population (1991) 101,000. Industries include paper and publishing. Harvard University 1636 (the oldest educational institution in the USA, named after John Harvard 1607–1638, who bequeathed his library to it along with half his estate), Massachusetts Institute of Technology (1861), and the John F Kennedy School of Government are here, as well as a park named after him.

Cambridgeshire /ˈkeɪmbrɪdʒʃə/ county in E England
area 3,410 sq km/1,316 sq mi
towns Cambridge (administrative headquarters), Ely, Huntingdon, Peterborough
features rivers: Ouse, Cam, Nene; Isle of Ely; Cambridge University; at RAF Molesworth, near Huntingdon, Britain's second cruise missile base was deactivated Jan 1989

Cambridgeshire

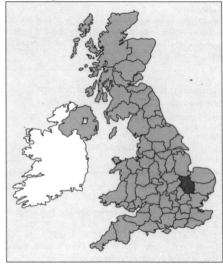

products mainly agricultural
population (1991) 640,700
famous people Oliver Cromwell, Octavia Hill,
John Maynard Keynes.

Cambs abbreviation for *Cambridgeshire*, an
English county.

Camden /ˈkæmdən/ industrial city of New Jersey,
USA, on the Delaware River; population (1990)
87,500. The city is linked with Philadelphia,
Pennsylvania, by the Benjamin Franklin suspension bridge (1926). The Walt Whitman House,
where the poet lived 1884–92, is now a museum.

Camden /ˈkæmdən/ inner borough of NW Greater
London
It includes the districts of:
Bloomsbury, site of London University, Royal
Academy of Dramatic Art (RADA), and the
British Museum; home between World Wars I
and II of writers and artists including Leonard and
Virginia Woolf and Lytton Strachey;
Fitzrovia, W of Tottenham Court Road, with the
Telecom Tower and Fitzroy Square as its focus;
Hampstead, site of Primrose Hill, Hampstead
Heath, and nearby Kenwood House; the poet
John Keats's home, now a museum; the church-
yard where the painter Constable is buried; and
Hampstead Garden Suburb;
Highgate, with the burial site of George Eliot,
Michael Faraday, and Karl Marx;
Holborn, with the Inns of Court (Lincoln's
Inn and Gray's Inn); Hatton Garden (diamond
dealers), the London Silver Vaults;
Somers Town, between Euston and King's
Cross railway stations
population (1991) 170,500.

Camembert /ˈkæməmbeə/ village in Normandy,
France, which gives its name to a soft cheese that
originated here.

Cameroon Republic of (*République du Cameroun*)
area 475,440 sq km/183,638 sq mi
capital Yaoundé
towns chief port Douala; Nkongsamba, Garova
physical desert in far north in the Lake Chad
basin, mountains in W, dry savanna plateau in the
intermediate area, and dense tropical rainforest
in the south
environment the Korup National Park preserves
1,300 sq km/500 sq mi of Africa's fast-
disappearing tropical rainforest. Scientists have
identified nearly 100 potentially useful chemical
substances produced naturally by the plants of
this forest
features Mount Cameroon 4,070 m/13,358 ft,
an active volcano on the coast, W of the Adamawa
Mountains
head of state and of government Paul Biya
from 1982
political system emergent democratic republic
political parties Democratic Assembly of the
Cameroon People (RDPC), nationalist, left of
centre; Social Democratic Front (SDF), centre left; Social Movement for New Democracy
(MSND), left of centre; Union of the Peoples of
Cameroon (UPC), left of centre

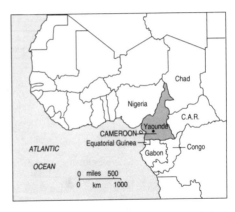

exports cocoa, coffee, bananas, cotton, timber,
rubber, groundnuts, gold, aluminium, crude oil
currency CFA franc
population (1990 est) 11,109,000; growth rate
2.7% p.a.
life expectancy men 49, women 53
languages French and English in pidgin variations (official); there has been some discontent
with the emphasis on French—there are 163
indigenous peoples with their own African
languages
media heavy government censorship
religions Roman Catholic 35%, animist 25%,
Muslim 22%, Protestant 18%
literacy men 68%, women 45% (1985 est)
GDP $12.7 bn (1987); $1,170 per head
chronology
1884 Treaty signed establishing German rule.
1916 Captured by Allied forces in World
War I.
1922 Divided between Britain and France.
1946 French Cameroon and British Cameroons
made UN trust territories.
1960 French Cameroon became the independent
Republic of Cameroon. Ahmadou Ahidjo elected
president.
1961 Northern part of British Cameroon merged
with Nigeria and southern part joined the Republic of Cameroon to become the Federal Republic
of Cameroon.
1966 One-party regime introduced.
1972 New constitution made Cameroon a unitary
state, the United Republic of Cameroon.
1973 New national assembly elected.
1982 Ahidjo resigned and was succeeded by
Paul Biya.
1983 Biya began to remove his predecessor's
supporters; accused by Ahidjo of trying to create a police state. Ahidjo went into exile in
France.
1984 Biya re-elected; defeated a plot to overthrow
him. Country's name changed to Republic of
Cameroon.
1988 Biya re-elected.
1990 Widespread public disorder. Biya granted
amnesty to political prisoners.
1991 Constitutional changes made.
1992 Ruling RDPC won in first multiparty
elections in 28 years. Biya's presidential victory
challenged by opposition.

Canada: provinces

province	capital	area in sq km
Alberta	Edmonton	661,200
British Columbia	Victoria	947,800
Manitoba	Winnipeg	650,000
New Brunswick	Fredericton	73,400
Newfoundland	St John's	405,700
Nova Scotia	Halifax	55,500
Ontario	Toronto	1,068,600
Prince Edward Island	Charlottetown	5,700
Québec	Québec	1,540,700
Saskatchewan	Regina	652,300

territory	capital	area in sq km
Northwest Territories	Yellowknife	3,426,300
Yukon Territory	Whitehorse	483,500

Campagna Romana /kæm'pænjə rəu'mɑːnə/ lowland stretch of the Italian peninsula, including and surrounding the city of Rome. Lying between the Tyrrhenian Sea and the Sabine Hills to the NE, and the Alban Hills to the SE, it is drained by the lower course of the river Tiber and a number of small streams, most of which dry up in the summer. Prosperous in Roman times, it later became virtually derelict through overgrazing, lack of water, and the arrival in the area of the malaria-carrying *Anopheles* mosquito. Extensive land reclamation and drainage in the 19th and 20th centuries restored its usefulness.

Campania /kæm'pænjə/ agricultural region (wheat, citrus, wine, vegetables, tobacco) of S Italy, including the volcano Vesuvius; area 13,600 sq km/5,250 sq mi; population (1990) 5,853,900. The capital is Naples; industrial centres include Benevento, Caserta, and Salerno. There are ancient sites at Pompeii, Herculaneum, and Paestum.

Campeche /kæm'petʃi/ port on the Bay of Campeche, Mexico; population (1984) 120,000. It is the capital of Campeche state. Timber and fish are exported, and there is a university, established 1756.

Campeche, Bay of /kæm'petʃi/ southwestern area of the Gulf of Mexico, site of a major oil-pollution disaster from the field off Yucatán peninsula 1979.

Campinas /kæm'piːnəs/ city of Sao Paulo, Brazil, situated on the central plateau; population (1980) 566,700. It is a coffee-trading centre. There are also metallurgical and food industries.

Campobasso /ˌkæmpəu'bæsəu/ capital of Molise region, Italy, about 190 km/120 mi SE of Rome; population (1981) 48,300. It has a high reputation for its cutlery.

Cam Ranh /'kæm 'ræn/ port in S Vietnam; population (1989) 114,000. In the Vietnam War it was a US base; it later became a staging complex for the Soviet Pacific fleet.

Canada
area 9,970,610 sq km/3,849,674 sq mi
capital Ottawa
towns Toronto, Montréal, Vancouver, Edmonton, Calgary, Winnipeg, Quebec, Hamilton, Saskatoon, Halifax
physical mountains in W, with low-lying plains in interior and rolling hills in E. Climate varies from temperate in S to arctic in N
environment sugar maples are dying in E Canada as a result of increasing soil acidification; nine rivers in Nova Scotia are now too acid to support salmon or trout reproduction
features St Lawrence Seaway, Mackenzie River; Great Lakes; Arctic Archipelago; Rocky Mountains; Great Plains or Prairies; Canadian Shield; Niagara Falls; the world's second largest country
head of state Elizabeth II from 1952, represented by governor general
head of government Jean Chretien from 1993
political system federal constitutional monarchy
political parties Progressive Conservative Party, free-enterprise, right of centre; Liberal Party, nationalist, centrist; New Democratic Party (NDP), moderate left of centre
exports wheat, timber, pulp, newsprint, fish (salmon), furs (ranched fox and mink exceed the value of wild furs), oil, natural gas, aluminium, asbestos (world's second largest producer), coal, copper, iron, zinc, nickel (world's largest producer), uranium (world's largest producer), motor vehicles and parts, industrial and agricultural machinery, fertilizers, chemicals
currency Canadian dollar
population (1990 est) 26,527,000 – including 300,000 North American Indians, of whom 75% live on more than 2,000 reservations in Ontario and the four Western provinces; some 300,000 Métis (people of mixed race) and 19,000 Inuit of whom 75% live in the Northwest Territories. Over half Canada's population lives in Ontario and Québec. Growth rate 1.1% p.a.
life expectancy men 72, women 79
languages English, French (both official; about 70% speak English, 20% French, and the rest are bilingual); there are also North American Indian languages and the Inuit Inuktitut
religion Roman Catholic 46%, Protestant 35%
literacy 99%
GDP $562 bn (1992)
chronology
1867 Dominion of Canada founded.
1949 Newfoundland joined Canada.
1957 Progressive Conservatives returned to power after 22 years in opposition.
1961 NDP formed.
1963 Liberals elected under Lester Pearson.
1968 Pearson succeeded by Pierre Trudeau.
1979 Joe Clark, leader of the Progressive Conservatives, formed a minority government; defeated on budget proposals.
1980 Liberals under Trudeau returned with a large majority. Québec referendum rejected demand for independence.
1982 Canada Act removed Britain's last legal control over Canadian affairs; 'patriation' of Canada's constitution.

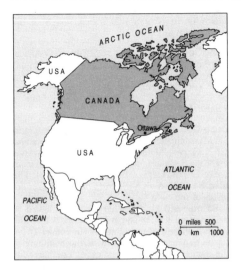

is no moisture, no artificial light pollution, and little natural airglow.

The Organization of African Unity (OAU) supports an independent Guanch Republic (so called from the indigenous islanders, a branch of the N African Berbers) and revival of the Guanch language.

Canberra /'kænbərə/ capital of Australia (since 1908), situated in the Australian Capital Territory, enclosed within New South Wales, on a tributary of the Murrumbidgee River; area (Australian Capital Territory including the port at Jervis Bay) 2,432 sq km/939 sq mi; population (1988) 297,300.

It contains the Parliament House (first used by the Commonwealth Parliament 1927), the Australian National University (1946), the Canberra School of Music (1965), and the National War Memorial.

Cancún /kæŋ'kuːn/ Caribbean resort in Mexico where, in 1981, a North–South summit was held to discuss the widening gap between the industrialized countries and the Third World.

Candia /'kændiə/ Italian name for the Greek island of ◊Crete. It was also formerly the name of Crete's largest city, Iráklion, founded about AD 824.

Canea /'kɑːniə/ (Greek **Khaniá**) capital and administrative centre of Crete, on the NW coast; population (1981) 47,338. It was founded 1252 by the Venetians and is surrounded by a wall. Vegetable oils, soap, and leather are exported.

Heavy fighting took place here during World War II, after the landing of German parachutists May 1941.

Cannes /kæn/ resort in Alpes-Maritimes *département*, S France; population (1990) 69,400. A prestigious film festival is held here annually. Formerly only a small seaport, in 1834 it attracted the patronage of Lord Brougham and other dis-

1983 Clark replaced as leader of the Progressive Conservatives by Brian Mulroney.
1984 Trudeau retired and was succeeded as Liberal leader and prime minister by John Turner. Progressive Conservatives won the federal election with a large majority, and Mulroney became prime minister.
1988 Conservatives re-elected with reduced majority on platform of free trade with the USA.
1989 Free-trade agreement signed. Turner resigned as Liberal Party leader, and Ed Broadbent as NDP leader.
1990 Collapse of Meech Lake accord. Canada joined the coalition opposing Iraq's invasion of Kuwait.
1992 Gradual withdrawal of Canadian forces in Europe announced. Self-governing homeland for Inuit approved. Constitutional reform package, the Charlottetown Accord, rejected in national referendum.
1993 Feb: Mulroney resigned leadership of Conservative Party. June: Kim Campbell, the new party leader, became prime minister. Oct: Conservatives defeated in general election. Liberal leader Jean Chretien became prime minister.

Canary Islands /kə'neəri/ (Spanish *Canarias*) group of volcanic islands 100 km/60 mi off the NW coast of Africa, forming the Spanish provinces of Las Palmas and Santa Cruz de Tenerife; area 7,300 sq km/2,818 sq mi; population (1986) 1,615,000.
features The chief centres are Santa Cruz on Tenerife (which also has the highest peak in extracontinental Spain, Pico de Teide, 3,713 m/12,186 ft), and Las Palmas on Gran Canaria. The province of Santa Cruz comprises Tenerife, Palma, Gomera, and Hierro; the province of Las Palmas comprises Gran Canaria, Lanzarote, and Fuerteventura. There are also six uninhabited islets.

The Northern Hemisphere Observatory (1981) is on the island of La Palma. Observation conditions are exceptionally good because there

Canary Islands

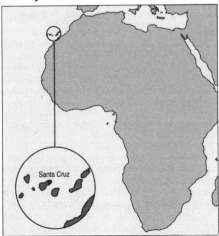

tinguished visitors and soon became a fashionable holiday resort. A new town (La Bocca) grew up facing the Mediterranean.

Cantabria /kæn'tæbriə/ autonomous region of N Spain; area 5,300 sq km/2,046 sq mi; population (1986) 525,000. The capital is Santander. From the coastline on the Bay of Biscay it rises to the Cantabrian Mountains. Mining is the major industry.

Cantabrian Mountains /kæn'tæbriən/ (Spanish *Cordillera Cantabrica*) mountain range running along the north coast of Spain, reaching 2,648 m/8,688 ft in the Picos de Europa massif. The mountains contain coal and iron deposits.

Cantal /kɒn'taːl/ volcanic mountain range in central France, which gives its name to Cantal *département*. The highest point is the Plomb du Cantal, 1,858 m/6,096 ft.

Canterbury /'kæntəbəri/ historic cathedral city in Kent, England, on the river Stour, 100 km/62 mi SE of London; population (1989 est) 127,000. In 597 King Ethelbert welcomed Augustine's mission to England here, and the city has since been the metropolis of the Anglican Communion and seat of the archbishop of Canterbury.

The Roman *Durovernum*, Canterbury was the Saxon capital of Kent. The present name derives from *Cantwarabyrig* (Old English 'fortress of the men of Kent'). In the Middle Ages it was a centre of pilgrimage to the tomb of St Thomas à Becket, murdered in the cathedral 1170, but the shrine was destroyed by Henry VIII.

Canterbury Plains /'kæntəbəri/ area of rich grassland between the mountains and the sea on the east coast of South Island, New Zealand; area 10,000 sq km/4,000 sq mi. Canterbury lamb is produced here.

Canton /'kæntən/ city in NE Ohio, USA, SE of Akron; seat of Stark County; population (1990) 84,161. Its products include office equipment, ceramics, and steel. The home of President William McKinley is here, as is the Football Hall of Fame.

Canton /ˌkæn'tɒn/ alternative spelling of Kwangchow or ⟡Guangzhou, a city in China.

Cape Breton /'keɪp 'bretn/ island forming the northern part of the province of Nova Scotia, Canada; area 10,282 sq km/3,970 sq mi; population (1988) 170,000. Bisected by a waterway, it has road and rail links with the mainland across the Strait of Canso. It has coal resources and steelworks, and there has been substantial development in the strait area, with docks, oil refineries, and newsprint production from local timber. In the N, the surface rises to 550 m/1,800 ft at North Cape, and the coast has many fine harbours. There are cod fisheries. The climate is mild and very moist. The chief towns are Sydney and Glace Bay.

history The first British colony was established 1629 but was driven out by the French. In 1763 Cape Breton was ceded to Britain and attached to Nova Scotia 1763–84 and from 1820.

Cape Byron /'keɪp 'baɪrən/ eastern extremity of Australia, in New South Wales, just S of the border with Queensland.

Cape Coast /'keɪp 'kəʊst/ port of Ghana, W Africa, 130 km/80 mi W of Accra; population (1982) 73,000. It has been superseded as the main port since 1962 by Tema. The town, first established by the Portuguese in the 16th century, is built on a natural breakwater, adjoining the castle.

Cape Cod /'keɪp 'kɒd/ hook-shaped peninsula in SE Massachusetts, USA, 100 km/60 mi long and 1.6–32 km/1–20 mi wide. Its beaches and woods make it a popular tourist area. It is separated from the rest of the state by the Cape Cod Canal. The islands of Martha's Vineyard and Nantucket are just S of the cape.

Basque and Norse fisherfolk are believed to have visited Cape Cod many years before the English Pilgrims landed at Provincetown 1620. It was named after the cod which were caught in the dangerous shoals of the cape. The Kennedy family home is at the resort of Hyannis Port.

Cape Horn /'keɪp hɔːn/ southernmost point of South America, in the Chilean part of the archipelago of Tierra del Fuego; notorious for gales and heavy seas. It was named 1616 by Dutch explorer Willem Schouten (1580–1625) after his birthplace (Hoorn).

Cape of Good Hope /'keɪp əv ɡʊd 'həʊp/ South African headland forming a peninsula between Table Bay and False Bay, Cape Town. The first European to sail around it was Bartholomew Diaz 1488. Formerly named Cape of Storms, it was given its present name by King John II of Portugal.

> *... the fairest Cape we saw in the whole circumference of the Earth.*
>
> On the **Cape of Good Hope** Francis Drake
> 1598–1600

Cape Province /'keɪp 'prɒvɪns/ (Afrikaans *Kaapprovinsie*) largest province of the Republic of South Africa, named after the Cape of Good Hope
area 641,379 sq km/247,638 sq mi, excluding Walvis Bay
capital Cape Town
towns Port Elizabeth, East London, Kimberley, Grahamstown, Stellenbosch
physical Orange River, Drakensberg, Table Mountain (highest point Maclear's Beacon, 1,087 m/3,567 ft); Great Karoo Plateau; Walvis Bay
products fruit, vegetables, wine; meat, ostrich feathers; diamonds, copper, asbestos, manganese
population (1985) 5,041,000; officially including 44% coloured; 31% black; 25% white; 0.6% Asian
history Dutch traders established the first European settlement on the Cape 1652, but it was taken by the British 1795, after the French Revo-

lutionary armies had occupied the Netherlands, and was sold to Britain for £6 million 1814. The Cape achieved self-government 1872. It was an original province of the Union 1910. The Orange River was proclaimed the northern boundary 1825. Griqualand West (1880) and the southern part of Bechuanaland (1895) were later incorporated. In 1991 South Africa and Namibia agreed to joint administration of Walvis Bay (solely under South African control from 1922); a joint administrative body was set up 1992.

Cape Town /'keɪptaun/ (Afrikaans **Kaapstad**) port and oldest town (founded 1652) in South Africa, situated in the SW on Table Bay; population (1985) 776,617. Industries include horticulture and trade in wool, wine, fruit, grain, and oil. It is the legislative capital of the Republic of South Africa and capital of Cape Province.

The Houses of Parliament, City Hall, Cape Town Castle 1666, and Groote Schuur ('great barn'), the estate of Cecil Rhodes (he designated the house as the home of the premier, and a university and the National Botanical Gardens occupy part of the grounds) are here. The naval base of **Simonstown** is to the SE; in 1975 Britain's use of its facilities was ended by the Labour government in disapproval of South Africa's racial policies.

... there is perhaps not a place in the known World that can equal this in affording refreshments of all kinds to Shipping.

On **Cape Town** Capt James Cook *Journal* 1771

Cape Verde Republic of (*República de Cabo Verde*)
area 4,033 sq km/1,557 sq mi
capital Praia
towns Mindelo, Sal-Rei, Porto Novo
physical archipelago of ten volcanic islands 565 km/350 mi W of Senegal; the windward (Barlavento) group includes Santo Antão, São Vicente, Santa Luzia, São Nicolau, Sal, and Boa Vista; the leeward (Sotovento) group comprises Maio, São Tiago, Fogo, and Brava; all but Santa Luzia are inhabited
features strategic importance guaranteed by its domination of western shipping lanes; Sal, Boa Vista, and Maio lack water supplies but have fine beaches
head of state Mascarenhas Monteiro from 1991
head of government Carlos Viega from 1991
political system socialist pluralist state
political parties African Party for the Independence of Cape Verde (PAICV), African nationalist; Movement for Democracy (MPD)
exports bananas, salt, fish
currency Cape Verde escudo
population (1990 est) 375,000 (including 100,000 Angolan refugees); growth rate 1.9% p.a.
life expectancy men 57, women 61
language Creole dialect of Portuguese
religion Roman Catholic 80%
literacy men 61%, women 39% (1985)

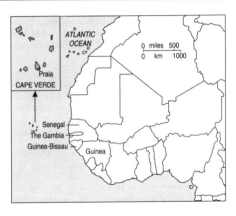

GDP $158 million (1987); $454 per head
chronology
15th century First settled by Portuguese.
1951–74 Ruled as an overseas territory by Portugal.
1974 Moved towards independence through a transitional Portuguese–Cape Verde government.
1975 Independence achieved from Portugal. National people's assembly elected. Aristides Pereira became the first president.
1980 Constitution adopted providing for eventual union with Guinea-Bissau.
1981 Union with Guinea-Bissau abandoned and the constitution amended; became one-party state.
1991 First multiparty elections held. New party, MPD, won majority in assembly. Pereira replaced by Mascarenhas Monteiro.

Cape Wrath /'keɪp rɑːθ/ (Norse *huaf* 'point of turning') headland at the northwest extremity of Scotland, extending 159 m/523 ft into the Atlantic Ocean. Its lighthouse dates from 1828.

Cape York /'keɪp jɔːk/ peninsula, the northernmost point (10° 41′ S) of the Australian mainland, named by Capt James Cook 1770. The peninsula is about 800 km/500 mi long and 640 km/400 mi wide at its junction with the mainland. Its barrenness deterred early Dutch explorers, although the S is being developed for cattle (Brahmin type). In the N there are large bauxite deposits.

Caporetto /ˌkæpəˈretəu/ former name of ◊Kobarid, a village in Slovenia.

Capri /kəˈpriː/ Italian island at the southern entrance of the Bay of Naples; 32 km/20 mi S of Naples; area 13 sq km/5 sq mi. It has two towns, Capri and Anacapri, a profusion of flowers, beautiful scenery, and an ideal climate. The Blue Grotto on the north coast is an important tourist attraction.

Caprivi Strip /kəˈpriːvi/ northeastern part of Namibia, a narrow strip between Angola and Botswana, giving the country access to the Zambezi River.

Capua /'kæpjuə/ Italian town in Caserta province on the Volturno River, in a fertile plain

N of Naples; population (1981) 18,000. There was heavy fighting here 1943 during World War II, which almost destroyed the Romanesque cathedral.

Caracas /kə'rækəs/ chief city and capital of Venezuela, situated on the slopes of the Andes Mountains, 13 km/8 mi S of its port La Guaira on the Caribbean coast; population of metropolitan area (1989) 3,373,100. Founded 1567, it is now a large industrial and commercial centre, notably for oil companies.

It is the birthplace of the South American liberator Simón Bolívar. It has many fine buildings, including Venezuela University (1725); there are also two modern universities. The city has suffered several severe earthquakes.

Carcassonne /,ka:ka'son/ city in SW France, capital of Aude *département*, on the river Aude, which divides it into the ancient and modern town; population (1990) 45,000. Its medieval fortifications (restored) are the finest in France.

Cardiff /'ka:dɪf/ (Welsh *Caerdydd*) capital of Wales (from 1955) and administrative headquarters of South and Mid Glamorgan, at the mouth of the Taff, Rhymney, and Ely rivers; population (1991) 272,600. Besides steelworks, there are car-component, flour-milling, paper, cigar, and other industries.

The city dates from Roman times, the later town being built around a Norman castle. The castle was the residence of the earls and marquesses of Bute from the 18th century and was given to the city 1947 by the fifth marquess. Coal was exported until the 1920s. As coal declined, iron and steel exports continued to grow, and an import trade in timber, grain and flour, tobacco, meat, and citrus fruit developed. The docks on the Bristol Channel were opened 1839 and greatly extended by the second marquess of Bute (1793–1848). They have now been redeveloped for industry.

In Cathays Park is a group of public buildings including the Law Courts, City Hall, the National Museum of Wales, the Welsh Office (established 1964), a major part of the University of Wales (Institute of Science and Technology, National School of Medicine, and University College of South Wales), and the Temple of Peace and Health. Llandaff, on the right bank of the river Taff, was included in Cardiff 1922; its cathedral, virtually rebuilt in the 19th century and restored 1948–57 after air-raid damage in World War II, has Jacob Epstein's sculpture Christ in Majesty. At St Fagan's is the Welsh National Folk Museum, containing small, rebuilt historical buildings from rural Wales in which crafts are demonstrated. The city is the headquarters of the Welsh National Opera.

Cardiganshire /'ka:dɪgənʃə/ (Welsh *Ceredigion* or *Sir Aberteifi*) former county of Wales. It was merged, together with Pembrokeshire and Carmarthenshire, into Dyfed 1974.

Caribbean Sea /,kærɪ'bi:ən/ western part of the Atlantic Ocean between the southern coast of North America and the northern coasts of South

America. Central America is to the W and the West Indies are the islands within the sea, which is about 2,740 km/1,700 mi long and 650–1,500 km/400-900 mi wide. It is from here that the Gulf Stream turns towards Europe.

The Caribbean Sea was named after the Carib Indians who inhabited the area when it was discovered by Spanish explorers, beginning with Christopher Columbus 1492. It is a major maritime trade route for oil, other raw materials, seafood, and tropical agricultural products. It is heavily polluted by untreated sewage, which destroys mangrove forests and coral reefs.

Carinthia /kə'rɪnθiə/ (German *Kärnten*) federal province of Alpine SE Austria, bordering Italy and Slovenia in the S; area 9,500 sq km/3,667 sq mi; population (1987) 542,000. The capital is Klagenfurt. It was an independent duchy from 976 and a possession of the Habsburg dynasty 1276–1918.

Carisbrooke /'kærɪzbruk/ village SW of Newport, Isle of Wight, England. Charles I was imprisoned in its castle 1647–48.

Carlisle /ka:'laɪl/ city in Cumbria, NW England, situated on the river Eden at the western end of Hadrian's Wall, administrative centre of the county; population (1991) 99,800. It is a leading railway centre; textiles, engineering, and biscuit making are the chief industries. There is a Norman cathedral and a castle. The bishopric dates from 1133.

Carlisle /'ka:laɪl/ city in S Pennsylvania, USA, in the Cumberland Valley, SW of Harrisburg; seat of Cumberland County; population (1990) 18,400. Industries include electronics and steel products. The US Army War College and Dickinson College are here.

Carlow /'ka:ləu/ county in the Republic of Ireland, in the province of Leinster; county town Carlow; area 900 sq km/347 sq mi; population (1991) 40,900. Mostly flat except for mountains in the S, the land is fertile, and well suited to dairy farming.

Carlsbad /'ka:lzbæd/ German name of ◊Karlovy Vary, a spa town in W Bohemia, Czech Republic.

Carmarthenshire /kə'ma:ðənʃə/ (Welsh *Sir Gaerfyrddin*) former county of S Wales, part of Dyfed from 1974. The county town was Carmarthen.

Carnac /'ka:næk/ Megalithic site in Brittany, France, where remains of tombs and stone alignments of the period 2000–1500 BC have been found. The largest of the latter has 1,000 stones up to 4 m/13 ft high arranged in 11 rows, with a circle at the western end. Named after the village of Carnac; population about 4,000.

Carnarvon alternative spelling of ◊Caernarvon, a town in Wales.

Carnarvon Range /kə'na:vən/ section of the Great Divide mountain range, Queensland, Australia, with peaks up to 1,310 m/4,300 ft high. There are many Aboriginal paintings in the sandstone caves along its 160 km/100 mi length.

Carnatic region of SE India, in Madras state. It is situated between the Eastern Ghats and the Coromandel coast and was formerly a leading trading centre.

Carniola /,kɑːniˈəʊlə/ former crownland and duchy of Austria, most of which was included in Slovenia, part of the kingdom of the Serbs, Croats, and Slovenes (later Yugoslavia), 1919. The western districts of Idrija and Postojna, then allocated to Italy, were transferred to Yugoslavia 1947.

Carolina /,kærəˈlaɪnə/ either of two separate states of the USA; see ◊North Carolina and ◊South Carolina.

Carolines /ˈkærəlaɪnz/ scattered archipelago in Micronesia, Pacific Ocean, consisting of over 500 coral islets; area 1,200 sq km/463 sq mi. The chief islands are Ponape, Kusai, and Truk in the eastern group, and Yap and Belau in the western group.

The Carolines are well watered and productive. Occupied by Germany 1899, and Japan 1914, and mandated by the League of Nations to Japan 1919, they were fortified, contrary to the terms of the mandate. Under Allied air attack in World War II they remained unconquered. They were part of the US Trust Territory of the Pacific Islands 1947–90.

Carpathian Mountains /kɑːˈpeɪθɪən/ central European mountain system, forming a semicircle through Slovakia – Poland – Ukraine – Romania, 1,450 km/900 mi long. The central *Tatra Mountains* on the Slovak–Polish frontier include the highest peak, Gerlachovka, 2,663 m/8,737 ft.

Carpentaria, Gulf of /,kɑːpənˈteərɪə/ shallow gulf opening out of the Arafura Sea, N of Australia. It was discovered by the Dutch navigator Abel Tasman 1606 and named 1623 in honour of Pieter Carpentier, governor general of the Dutch East Indies.

Carrara /kəˈrɑːrə/ town in Tuscany, Italy, 60 km/37 mi NW of Livorno; population (1981) 66,000. It is known for its quarries of fine white marble, which were worked by the Romans, abandoned in the 5th century, and came into use again with the revival of sculpture and architecture in the 12th century.

Carrickfergus /,kærɪkˈfɜːgəs/ seaport on Belfast Lough, County Antrim, Northern Ireland; population (1989 est) 34,000. There is some light industry. The remains of the castle, built 1180, house two museums.

Carse of Gowrie /ˈkɑːsəv ˈgaʊri/ fertile lowland plain bordering the Firth of Tay, Scotland. It is 24 km/15 mi long, and is one of the country's most productive agricultural areas. William III landed here before the Battle of the Boyne, 1690.

Carson City /ˈkɑːsən ˈsɪti/ capital of Nevada, USA; population (1990) 40,400. Settled as a trading post 1851, it was named after the frontier guide Kit Carson 1858. It flourished as a boom town after the discovery of the nearby Comstock silver-ore lode 1859.

Cartagena /,kɑːtəˈdʒiːnə/ city in the province of Murcia, Spain, on the Mediterranean coast; population (1991) 172,200. It is a seaport and naval base. It was founded as *Carthago Nova* about 225 BC by the Carthaginian Hasdrubal, son-in-law of Hamilcar Barca. It continued to flourish under the Romans and the Moors and was conquered by the Spanish 1269. It has a 13th-century cathedral and Roman remains.

Cartagena /,kɑːtəˈdʒiːnə/ or *Cartagena de los Indes* port, industrial centre, and capital of the department of Bolívar, NW Colombia; population (1985) 531,000. Plastics and chemicals are produced here.

It was founded 1533 and taken by the English buccaneer Francis Drake 1586. A pipeline brings petroleum to the city from the De Manes oilfields.

Casablanca /,kæsəˈblæŋkə/ (Arabic *Dar el-Beida*) port, commercial and industrial centre on the Atlantic coast of Morocco; population (1982) 2,139,000. It trades in fish, phosphates, and manganese. The Great Hassan II Mosque, completed 1989, is the world's largest; it is built on a platform (40,000 sq m/430,000 sq ft) jutting out over the Atlantic, with walls 60 m/200 ft high, topped by a hydraulic sliding roof, and a minaret 175 m/574 ft high.

Cascade Range /kæˈskeɪd/ volcanic mountains in the western USA and Canada, extending 1,120 km/700 mi from N California through Oregon and Washington to the Fraser River. They include Mount St Helens and Mount Rainier (the highest peak, 4,392 m/14,408 ft), which is noteworthy for its glaciers. The mountains are the most active in the USA, excluding Alaska and Hawaii.

Cascais /kəʃˈkaɪʃ/ fishing port and resort town on the Costa do Sol, 25 km/16 mi W of Lisbon, Portugal; population (1981) 12,500. The 17th-century citadel is the summer palace of the president. There is local fishing and canning; tourism is important.

Caserta /kəˈzɜːtə/ town in S Italy, 33 km/21 mi NE of Naples; population (1981) 66,318. It trades in chemicals, olive oil, wine, and grain. The base for Garibaldi's campaigns in the 19th century, it was the Allied headquarters in Italy 1943–45, and the German forces surrendered to Field Marshal Alexander here 1945.

Casper /ˈkæspə/ city in E central Wyoming, USA, on the North Platte River, NW of Cheyenne; seat of Natrona County; population (1990) 46,700. The largest city in Wyoming, it serves as the marketing centre for the region's livestock and petroleum products.

Caspian Sea /ˈkæspɪən/ world's largest inland sea, divided between Iran, Azerbaijan, Russia, Kazakhstan, and Turkmenistan; area about 400,000 sq km/155,000 sq mi, with a maximum depth of 1,000 m/3,250 ft. The chief ports are Astrakhan and Baku. Drainage in the N and damming of the Volga and Ural rivers for hydroelectric power left the sea approximately

28 m/90 ft below sea level. In June 1991 opening of sluices in the dams caused the water level to rise dramatically, threatening towns and industrial areas.

An underwater ridge divides it into two halves, of which the shallow northern half is almost saltfree. There are no tides. The damming has led to shrinkage over the last 50 years, and the growth of industry along its shores has caused pollution and damaged the Russian and Iranian caviar industries. 40% of Russia's polluted waste water flows into the sea.

Cassel /'kæsəl/ alternative spelling of ◊Kassel, an industrial town in Germany.

Cassino /kæ'siːnəu/ town in S Italy, 80 km/50 mi NW of Naples, at the foot of Monte Cassino; population (1981) 31,139. It was the scene of heavy fighting during World War II 1944, when most of the town was destroyed. It was rebuilt 1.5 km/1 mi to the N. The abbey on the summit of Monte Cassino, founded by St Benedict 529, was rebuilt 1956.

Castel Gandolfo /kæs'tel gæn'dɒlfəu/ village in Italy 24 km/15 mi SE of Rome, named after a castle that once stood on the site. There is a 17th-century palace, built by Pope Urban VIII, which is still used by the pope as a summer residence.

Castellón de la Plana port in Spain, facing the Mediterranean to the E; population (1991) 137,500. It is the capital of Castellón province and is the centre of an orange-growing district.

Castilla–La Mancha /kæ'stiːljə laː 'mæntʃə/ autonomous region of central Spain; area 79,200 sq km/30,571 sq mi; population (1986) 1,665,000. It includes the provinces of Albacete, Ciudad Real, Cuenca, Guadalajara, and Toledo. Irrigated land produces grain and chickpeas, and merino sheep graze here.

Castilla–León /kæ'stiːljə leɪ'ɒn/ autonomous region of central Spain; area 94,100 sq km/36,323 sq mi; population (1986) 2,600,000. It includes the provinces of Ávila, Burgos, León, Palencia, Salamanca, Segovia, Soria, Valladolid, and Zamora. Irrigated land produces wheat and rye. Cattle, sheep, and fighting bulls are bred in the uplands.

Castleford /'kɑːsəlfəd/ town in West Yorkshire, England, at the confluence of the rivers Aire and Calder, 15 km/10 mi SE of Leeds; population (1981) 39,400. Industries include chemicals, glass, and earthenware products. The town stands on the site of the Roman *Lagentium*.

Castries /kæ'striːz/ port and capital of St Lucia, on the northwest coast of the island in the Caribbean; population (1988) 53,000. It produces textiles, chemicals, tobacco, and wood and rubber products. The town was rebuilt after destruction by fire 1948.

Catalonia /ˌkætə'ləuniə/ (Spanish *Cataluña*, Catalan *Catalunya*) autonomous region of NE Spain; area 31,900 sq km/12,313 sq mi; population (1986) 5,977,000. It includes Barcelona (the capital), Gerona, Lérida, and Tarragona. Industries include wool and cotton textiles; hydroelectric power is produced.

The N is mountainous, and the Ebro basin breaks through the Castellón Mountains in the S. The soil is fertile, but the climate in the interior is arid. Catalonia leads Spain in industrial development. Tourist resorts have developed along the Costa Brava.

history The region has a long tradition of independence. It enjoyed autonomy 1932–39 but lost its privileges for supporting the republican cause in the Spanish Civil War. Autonomy and official use of the Catalan language were restored 1980.

Cataluña /ˌkætə'luːnjə/ Spanish name for ◊Catalonia.

Catalunya /ˌkætə'luːnjə/ Catalan name for ◊Catalonia.

Catania /kə'tɑːniə/ industrial port in Sicily, just S of Mount Etna; population (1988) 372,000. It exports local sulphur; there are also shipbuilding and textile industries.

Catskills /'kætskɪlz/ US mountain range, mainly in SE New York State; the highest point is Slide Mountain, 1,281 m/4,204 ft.

Catterick /'kætərɪk/ village near Richmond in North Yorkshire, England, where there is a large military camp.

Caucasus /'kɔːkəsəs/ series of mountain ranges between the Caspian and Black seas, in the republics of Russia, Georgia, Armenia, and Azerbaijan; 1,200 km/750 mi long. The highest peak is Elbruz, 5,633 m/18,480 ft.

Arabian thoroughbreds are raised at Tersk farm in the northern foothills.

Oh who can hold a fire in his hand/
By thinking on the frostie Caucasus?

On the **Caucasus** William Shakespeare
Richard II 1593–96

Cauvery /'kɔːvəri/ or **Kaveri** river of S India, rising in the W Ghats and flowing 765 km/475 mi SE to meet the Bay of Bengal in a wide delta. It has been a major source of hydroelectric power since 1902 when India's first hydropower plant was built on the river.

Cavan /'kævən/ agricultural county of the Republic of Ireland, in the province of Ulster; area 1,890 sq km/730 sq mi; population (1991) 52,800.

The river Erne divides it into a narrow, mostly low-lying peninsula, 30 km/20 mi long, between Leitrim and Fermanagh, and an eastern section of wild and bare hill country. The soil is generally poor and the climate moist and cold. The chief towns are Cavan, the capital, population about 3,000; Kilmore, seat of Roman Catholic and Protestant bishoprics; and Virginia.

Cavite /kə'viːti/ town and port of the Philippine Republic; 13 km/8 mi S of Manila; population

Cayman Islands

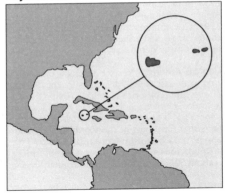

(1980) 88,000. It is the capital of Cavite province, Luzon. It was in Japanese hands Dec 1941–Feb 1945. After the Philippines achieved independence 1946, the US Seventh Fleet continued to use the naval base.

Cawnpore /,kɔːnˈpɔː/ former spelling of ⟶Kanpur, a city in India.

Cayenne /keɪˈen/ capital and chief port of French Guiana, on Cayenne Island, NE South America, at the mouth of the river Cayenne; population (1990) 41,700. It was founded 1634 by the French, and used as a penal settlement 1854–1946.

Cayman Islands /ˈkeɪmən/ British island group in the West Indies
area 260 sq km/100 sq mi
features comprises three low-lying islands: Grand Cayman, Cayman Brac, and Little Cayman
government governor, executive council, and legislative assembly
exports seawhip coral, a source of prostaglandins; shrimps; honey; jewellery
currency CI dollar
population (1988) 22,000
language English
history discovered by Chrisopher Columbus 1503; acquired by Britain following the Treaty of Madrid 1670; a dependency of Jamaica from 1863. In 1962 the islands became a separate colony, although the inhabitants chose to remain British. From that date, changes in legislation attracted foreign banks and the Caymans are now an international financial centre and tax haven as well as a tourist resort.

Cebu /seɪˈbuː/ chief city and port of the island of Cebu in the Philippines; population (1990) 610,400; area of the island 5,086 sq km/1,964 sq mi. The oldest city of the Philippines, Cebu was founded as San Miguel 1565 and became the capital of the Spanish Philippines.

Cedar Falls /ˌsiːdə ˈfɔːlz/ city in NE Iowa, USA, on the Cedar River, W of Waterloo; population (1990) 35,000. Industries include farm and other heavy equipment, rotary pumps, and tools.

Central

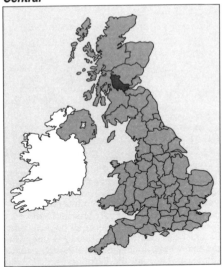

Cedar Rapids /ˌsiːdə ˈræpɪdz/ city in E Iowa, USA; population (1990) 108,800. It produces communications equipment, construction machinery, and processed foods.

Celebes /səˈliːbɪz/ English name for ⟶Sulawesi, an island of Indonesia.

Celtic Sea /ˈkeltɪk/ sea area bounded by Wales, Ireland, and SW England; the name is commonly used by workers in the oil industry to avoid nationalist significance. The Celtic Sea is separated from the Irish Sea by St George's Channel.

Central region of Scotland, formed 1975 from the counties of Stirling, S Perthshire, and West Lothian
area 2,600 sq km/1,004 sq mi
towns Stirling (administrative headquarters), Falkirk, Alloa, Grangemouth
features Stirling Castle; field of Bannockburn; Loch Lomond; the Trossachs
products agriculture; industries including brewing and distilling, engineering, electronics
population (1991) 268,000
famous people William Alexander (founder of Nova Scotia), Rob Roy Macgregor.

Central African Republic (*République Centrafricaine*)
area 622,436 sq km/240,260 sq mi
capital Bangui
towns Berbérati, Bouar, Bossangoa
physical landlocked flat plateau, with rivers flowing N and S, and hills in NE and SW; dry in N, rainforest in SW
environment an estimated 87% of the urban population is without access to safe drinking water
features Kotto and Mbali river falls; the Oubangui River rises 6 m/20 ft at Bangui during the wet season (June–Nov)
head of state and government André Kolingba from 1981

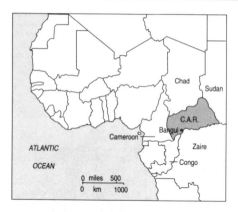

political system one-party military republic
political parties Central African Democratic Assembly (RDC), nationalist; all political activity has been banned since the 1981 coup, but the main opposition groups, although passive, still exist. They are the Patriotic Front Ubangi Workers' Party (FPO-PT), the Central African Movement for National Liberation (MCLN), and the Movement for the Liberation of the Central African People (MPLC)
exports diamonds, uranium, coffee, cotton, timber, tobacco
currency CFA franc
population (1990 est) 2,879,000 (more than 80 ethnic groups); growth rate 2.3% p.a.
life expectancy men 41, women 45
languages Sangho (national), French (official), Arabic, Hunsa, and Swahili
religions 25% Protestant; 25% Roman Catholic; 10% Muslim; 10% animist
literacy men 53%, women 29% (1985 est)
GDP $1 bn (1987); $374 per head
chronology
1960 Central African Republic achieved independence from France; David Dacko elected president.
1962 The republic made a one-party state.
1965 Dacko ousted in military coup led by Col Bokassa.
1966 Constitution rescinded and national assembly dissolved.
1972 Bokassa declared himself president for life.
1977 Bokassa made himself emperor of the Central African Empire.
1979 Bokassa deposed by Dacko following violent repressive measures by the self-styled emperor, who went into exile.
1981 Dacko deposed in a bloodless coup, led by General André Kolingba, and an all-military government established.
1983 Clandestine opposition movement formed.
1984 Amnesty for all political party leaders announced. President Mitterrand of France paid a state visit.
1985 New constitution promised, with some civilians in the government.
1986 Bokassa returned from France, expecting to return to power; he was imprisoned and his trial started. General Kolingba re-elected. New

constitution approved by referendum.
1988 Bokassa found guilty and received death sentence, later commuted to life imprisonment.
1991 Government announced that a national conference would be held in response to demands for a return to democracy.

Central America /'sentrəl ə'merɪkə/ part of the Americas that links Mexico with the Isthmus of Panama, comprising Belize, Costa Rica, El Salvador, Guatemala, Honduras, Nicaragua, and Panama. It is also an isthmus, crossed by mountains that form part of the Cordilleras, rising to a maximum height of 4,220 m/13,845 ft. There are numerous active volcanoes. Central America is about 523,000 sq km/200,000 sq mi in area and has a population (1980) estimated at 22,700,000, mostly Indians or mestizos (of mixed white-Indian ancestry). Tropical agricultural products and other basic commodities and raw materials are exported.

Much of Central America formed part of the Maya civilization. It was discovered by Christopher Columbus 1502. Spanish settlers married indigenous women, and the area remained out of the mainstream of Spanish Empire history. When the Spanish Empire collapsed in the early 1800s, the area formed the Central American Federation, with a constitution based on that of the USA. The federation disintegrated 1840, however. Completion of the Panama Canal 1914 enhanced the region's position as a strategic international crossroads. Demand for cash crops (bananas, coffee, cotton), especially from the USA, created a strong landowning class controlling a serflike peasantry by military means. There has been US military intervention in the area—for example, in Nicaragua, where the dynasty of General Anastasio Somoza was founded. President Carter officially reversed support for such regimes, but in the 1980s the Reagan and Bush administrations again favoured military and financial aid to selected political groups, including the Contras in Nicaragua and the ruling regime in El Salvador. Continuing US interest was underscored by its invasion of Panama Dec 1989.

To understand European politics, one should read the history of Central America ... Central America, being just Europe in miniature and with the lid off, is the ideal laboratory in which to study the behaviour of the Great Powers.

On **Central America** Aldous Huxley
Beyond the Mexique Bay 1934

Central Asian Republics group of five republics: Kazakhstan, Kyrgyzstan, Tajikistan, Turkmenistan, and Uzbekistan. Formerly part of the Soviet Union, their independence was recognized 1991. They comprise a large part of the geographical region of Turkestan and are the home of large numbers of Muslims.

The area was conquered by Russia as recently as 1866–73 and until 1917 was divided into the khanate of Khiva, the emirate of Bokhara, and the governor-generalship of Turkestan. The Soviet government became firmly established 1919, and in 1920 the Khan of Khiva and the Emir of Bokhara were overthrown and People's Republics set up. Turkestan became an Autonomous Soviet Socialist Republic 1921. Boundaries were redistributed 1925 along nationalist lines, and Uzbekistan, Tajikistan, and Turkmenistan became republics of the USSR, along with Bokhara and Khiva. The area populated by Kazakhs was united with Kazakhstan, which became a Union Republic 1936, the same year as Kyrgyzstan.

Central Lowlands one of the three geographical divisions of Scotland, occupying the fertile and densely populated plain that lies between two geological fault lines, which run nearly parallel NE–SW across Scotland from Stonehaven to Dumbarton and from Dunbar to Girvan.

Central Mount Stuart /ˈstjuːət/ flat-topped mountain 844 m/2,770 ft high, at approximately the central point of Australia.

It was originally named 1860 by explorer J McDouall Stuart after another explorer, Charles Sturt—Central Mount Sturt—but later became known by his own name.

Central Provinces and Berar /beɪˈrɑː/ former British province of India, now part of Madhya Pradesh.

Centre /sɒntr/ region of N central France; area 39,200 sq km/15,131 sq mi; population (1986) 2,324,000. Centre includes the *départements* of Cher, Eure-et-Loire, Indre, Indre-et-Loire, Loire-et-Cher, and Loiret. Its capital is Orléans.

Centre, the /ˈsentə/ region of central Australia, including the tourist area between the Musgrave and MacDonnell ranges which contains Ayers Rock and Lake Amadeus.

Cephalonia /ˌsefəˈləʊniə/ English form of ◊Kefallinia, the largest of the Ionian Islands, off the west coast of Greece.

Ceram /səˈræm/ or **Seram** Indonesian island in the Moluccas island group; area 17,142 sq km/6,621 sq mi. The principal town is Ambon.

Cernăuţi /ˌtʃeənəˈuts/ Romanian form of ◊Chernovtsy, a city in Ukraine.

České Budějovice /ˈtʃeskeɪ ˈbuːdʒəuˌviːtseɪ/ (German *Budweis*) town in the Czech Republic, on the river Vltava; population (1991) 123,400. It is a commercial and industrial centre for S Bohemia, producing beer, timber, and metal products.

Cetinje /ˈtsetiːnjeɪ/ town in Montenegro, Yugoslavia, 19 km/12 mi SE of Kotor; population (1981) 20,213. Founded 1484 by Ivan the Black, it was capital of Montenegro until 1918. It has a palace built by Nicholas, the last king of Montenegro.

Ceuta /ˈsjuːtə/ Spanish seaport and military base in Morocco, N Africa, 27 km/17 mi S of Gibraltar and overlooking the Mediterranean approaches to the Straits of Gibraltar; area 18 sq

km/7 sq mi; population (1986) 71,000. It trades in tobacco and petrol products.

Cévennes /seˈven/ series of mountain ranges on the southern, southeastern, and eastern borders of the Central Plateau of France. The highest peak is Mount Mézenc, 1,754 m/5,755 ft.

... that undecipherable labyrinth of hills ...

On the **Cévennes** Robert Louis Stevenson
Travels with a Donkey in the Cévennes 1879

Ceylon /sɪˈlɒn/ former name (until 1972) of ◊Sri Lanka.

Chaco /ˈtʃɑːkəʊ/ province of Argentina; area 99,633 sq km/38,458 sq mi; population (1991) 838,300. Its capital is Resistencia, in the SE. The chief crop is cotton, and there is forestry.

It includes many lakes, swamps, and forests, producing timber and quebracho (a type of wood used in tanning). Until 1951 it was a territory, part of Gran Chaco, a great zone, mostly level, stretching into Paraguay and Bolivia.

Chad Republic of (*République du Tchad*)
area 1,284,000 sq km/495,624 sq mi
capital Ndjamena (formerly Fort Lamy)
towns Sarh, Moundou, Abéché
physical landlocked state with mountains and part of Sahara Desert in N; moist savanna in S; rivers in S flow NW to Lake Chad
head of state Idriss Deby from 1990
head of government Fuidel Mounyar (interim) from 1993
political system emergent democratic republic
political parties National Union for Independence and Revolution (UNIR), nationalist; Alliance for Democracy and Progress (RDP), centre-left; Union for Democracy and Progress (UPDT), centre-left
exports cotton, meat, livestock, hides, skins
currency CFA franc
population (1990 est) 5,064,000; growth rate 2.3% p.a. Nomadic tribes move N–S seasonally in search of water
life expectancy men 42, women 45
languages French, Arabic (both official), over 100 African languages spoken

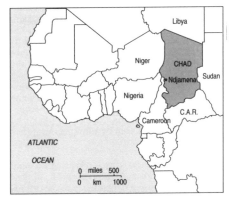

religions Muslim 44% (N), Christian 33%, animist 23% (S)
literacy men 40%, women 11% (1985 est)
GDP $980 million (1986); $186 per head
chronology
1960 Independence achieved from France, with François Tombalbaye as president.
1963 Violent opposition in the Muslim north, led by the Chadian National Liberation Front (Frolinat), backed by Libya.
1968 Revolt quelled with France's help.
1975 Tombalbaye killed in military coup led by Félix Malloum. Frolinat continued its resistance.
1978 Malloum tried to find a political solution by bringing the former Frolinat leader Hissène Habré into his government but they were unable to work together.
1979 Malloum forced to leave the country; an interim government was set up under General Goukouni. Habré continued his opposition with his Army of the North (FAN).
1981 Habré now in control of half the country. Goukouni fled and set up a 'government in exile'.
1983 Habré's regime recognized by the Organization for African Unity (OAU), but in the north Goukouni's supporters, with Libya's help, fought on. Eventually a cease-fire was agreed, with latitude 16°N dividing the country.
1984 Libya and France agreed to a withdrawal of forces.
1985 Fighting between Libyan-backed and French-backed forces intensified.
1987 Chad, France, and Libya agreed on cease-fire proposed by OAU.
1988 Full diplomatic relations with Libya restored.
1989 Libyan troop movements reported on border; Habré re-elected, amended constitution.
1990 President Habré ousted in coup led by Idriss Deby. New constitution adopted.
1991 Several anti-government coups foiled.
1992 Anti-government coup foiled. Two new opposition parties approved.

Who holds Chad holds Africa.

On **Chad** French maxim

Chad, Lake /tʃæd/ lake on the northeastern boundary of Nigeria. It once varied in extent between rainy and dry seasons from 50,000 sq km/20,000 sq mi to 20,000 sq km/7,000 sq mi, but a series of droughts 1979–89 reduced its area by 80%. The S Chad irrigation project used the lake waters to irrigate the surrounding desert, but the 4,000 km/2,500 mi of canals dug for the project are now permanently dry because of the shrinking size of the lake. The Lake Chad basin is being jointly developed for oil and natron by Cameroon, Chad, Niger, and Nigeria.
The lake was first seen by European explorers 1823.

Chagos Archipelago /'tʃɑːgəs ˌɑːkɪ'peləgəu/ island group in the Indian Ocean; area 60 sq km/23 sq mi. Formerly a dependency of Mauritius,

it now forms the British Indian Ocean Territory. The chief island is Diego Garcia, now a US-British strategic base.

Chalatenango /tʃəˌlætɪ'næŋgəu/ department on the northern frontier of El Salvador; area 2,507 sq km/968 sq mi; population (1981) 235,700. The capital is Chalatenango.

Châlons-sur-Marne /ʃɑː'lɒn sjuə 'mɑːn/ capital of the *département* of Marne, NE France; population (1990) 51,500. It is a market town and trades mainly in champagne. Tradition has it that Attila was defeated in his attempt to invade France, at the *Battle of Châlons* 451 by the Roman general Aëtius and the Visigoth Theodoric.

Chalon-sur-Saône /ʃɑː'lɒn sjuə 'səun/ town in the *département* of Saône-et-Loire, France, on the river Saône and the Canal du Centre; population (1990) 56,300. It has mechanical and electrical engineering and chemical industries.

Chambéry /ˌʃɒmbe'riː/ former capital of Savoy, now capital of Savoie *département*, France; population (1990) 55,600. It is the seat of an archbishopric and has some industry; it is also a holiday and health resort. The town gives its name to a French vermouth.

Chamonix /'ʃæməni/ holiday resort at the foot of Mont Blanc, in the French Alps; population (1982) 9,255. It was the site of the first Winter Olympics 1924.

This place is altogether the Paradise of Wilderness.

On **Chamonix** Lord Byron, letter of 1816

Champagne-Ardenne /ʃæmˌpeɪn ɑː'den/ region of NE France; area 25,600 sq km/9,882 sq mi; population (1986) 1,353,000. Its capital is Reims, and it comprises the *départements* of Ardennes, Aube, Marne, and Haute-Marne. It has sheep and dairy farming and vineyards.
It forms the plains E of the Paris basin. Its chief towns are Epernay, Troyes, and Chaumont. The capital of the ancient province of Champagne was Troyes.

Champaign /ʃæm'peɪn/ city in E central Illinois, USA, directly W of Urbana; population (1990) 63,500. Industries include electronic equipment, academic clothing, and air-conditioning equipment. Together with Urbana, it is the site of the University of Illinois.

Champlain, Lake /ʃæm'pleɪn/ lake in northeastern USA (extending some 10 km/6 mi into Canada) on the New York–Vermont border; length 201 km/125 mi; area 692 sq km/430 sq mi. It is linked by canal to the St Lawrence and Hudson rivers.
The largest city on its shores is Burlington, Vermont. Lake Champlain is named after explorer Samuel de Champlain, who saw it 1609. It was the scene of a US naval victory over the British 1814.

Channel Islands

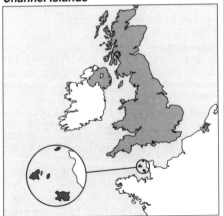

Chandernagore /ˌtʃʌndənəˈɡɔː/ ('city of sandalwood') city on the river Hooghly, India, in the state of West Bengal; population (1981) 102,000. Formerly a French settlement, it was ceded to India by treaty 1952.

Chandigarh /ˌtʃʌndɪˈɡɑː/ city of N India, in the foothills of the Himalayas; population (1981) 421,000. It is also a Union Territory; area 114 sq km/44 sq mi; population (1991) 640,725.

Planned by the architect Le Corbusier, the city was inaugurated 1953 to replace Lahore (capital of British Punjab), which went to Pakistan under partition 1947. Since 1966, when Chandigarh became a Union Territory, it has been the capital city of both Haryana and Punjab, until a new capital is built for the former.

Changchiakow /ˈtʃæŋ ˌtʃɪə ˈkau/ alternative transcription of ⮑Zhangjiakou, a trading centre in China.

Changchun /ˌtʃæŋˈtʃun/ industrial city and capital of Jilin province, China; population (1989) 2,020,000. Machinery and motor vehicles are manufactured. It is also the centre of an agricultural district.

As Hsingking ('new capital') it was the capital of Manchukuo 1932–45 during Japanese occupation.

Chang Jiang /ˈtʃæŋ dʒiˈæŋ/ or **Yangtze Kiang** longest river of China, flowing about 6,300 km/3,900 mi from Tibet to the Yellow Sea. It is a main commercial waterway.

It has 204 km/127 mi of gorges, below which is Gezhou Ba, the first dam to harness the river. The entire length of the river was first navigated 1986.

Changsha /ˌtʃæŋˈʃɑː/ port on the river Chang Jiang, capital of Hunan province, China; population (1989) 1,300,000. It trades in rice, tea, timber, and nonferrous metals; works antimony, lead, and silver; and produces chemicals, electronics, porcelain, and embroidery.

Channel Country /ˈtʃænl/ area of SW Queensland, Australia, in which channels such as Cooper's Creek (where explorers Robert Burke and William Wills died 1861) are cut by intermittent rivers. Summer rains supply rich grass for cattle, and there are the 'beef roads', down which herds are taken in linked trucks for slaughter.

Channel, English see ⮑English Channel.

Channel Islands /ˈtʃænl/ group of islands in the English Channel, off the northwest coast of France; they are a possession of the British crown. They comprise the islands of Jersey, Guernsey, Alderney, Great and Little Sark, with the lesser Herm, Brechou, Jethou, and Lihou.
area 194 sq km/75 sq mi
features very mild climate, productive soil; financially the islands are a tax haven
exports flowers, early potatoes, tomatoes, butterflies
currency English pound, also local coinage
population (1981) 128,900
languages official language French (Norman French) but English more widely used
religion chiefly Anglican
famous people Lillie Langtry
government the main islands have their own parliaments and laws. Unless specially signified, the Channel Islands are not bound by British acts of Parliament, though the British government is responsible for defence and external relations
history originally under the duchy of Normandy, they are the only part still held by Britain. The islands came under the same rule as England 1066, and are dependent territories of the British crown. Germany occupied the islands during World War II June 1940–May 1945.

Channel Tunnel tunnel built beneath the English Channel, linking Britain with mainland Europe. It comprises twin rail tunnels, 50 km/31 mi long and 7.3 m/24 ft in diameter, located 40 m/130 ft beneath the seabed. Specially designed shuttle trains carrying cars and lorries will run between terminals at Folkestone, Kent, and Sangatte, W of Calais, France. It was begun 1986 and is scheduled to be operational 1993. The French and English sections were linked Dec 1990.

Surely John Bull will not endanger his birthright, his liberty, his property simply in order that men and women may cross between England and France without running the risk of sea-sickness.

On the **Channel Tunnel** proposals of 1882
—Field Marshal Viscount Wolseley

Chantilly /ʃænˈtɪli/ town in Oise *département*, France, NE of Paris; population (1982) 10,208. It is the centre of French horseracing and was the headquarters of the French military chief Joseph Joffre 1914–17. It formerly produced lace and porcelain.

Chao Phraya /ˈtʃau prəˈjɑː/ chief river (formerly Menam) of Thailand, flowing 1,200 km/750 mi into the Bight of Bangkok, an inlet of the Gulf of Thailand.

Charente /ʃæ'rɒnt/ French river, rising in Haute-Vienne *département* and flowing past Angoulême and Cognac into the Bay of Biscay below Rochefort. It is 360 km/225 mi long. Its wide estuary is much silted up. It gives its name to two *départements*, Charente and Charente-Maritime (formerly Charente-Inférieure).

Charing Cross /'tʃeərɪŋ 'krɒs/ district in Westminster, London, around Charing Cross railway station. It derives its name from the site of the last of 12 stone crosses erected by Edward I 1290 at the resting-places of the coffin of his queen, Eleanor. The present cross was designed by A S Barry 1865.

Charleroi /ʃɑːlə'rwɑː/ town in Belgium on the river Sambre, Hainault province; population (1991) 206,200. Its coal industry declined in the 1970s; other industries include steel, electronics and electrical engineering, pharmaceuticals, glass, and cement.

Charleston /'tʃɑːlstən/ capital and chief city of West Virginia, USA, on the Kanawha River; population (1990) 57,300. It is the centre of a region that produces coal, natural gas, salt, clay, timber, and oil, and it is an important chemical-producing centre. Charleston developed from a fort built 1788.

Charleston /'tʃɑːlstən/ main port and city of South Carolina, USA; population (1990) 80,400. Industries include textiles, clothing, and paper products. A nuclear-submarine naval base and an air-force base are nearby. The city dates from 1670.

Fort Sumter, in the sheltered harbour of Charleston, was bombarded by Confederate batteries April 12–13, 1861, thus beginning the Civil War. There are many historic houses and fine gardens. Charleston was hard hit by Hurricane Hugo 1989.

Charlotte /'ʃɑːlət/ city in North Carolina, USA, on the border with South Carolina; population (1990) 395,900. Industries include data processing, textiles, chemicals, machinery, and food products. It was the gold-mining centre of the country until gold was discovered in California 1849.

The University of North Carolina–Charlotte is here, and the Mint Museum of Arts has paintings, sculpture, and ceramics. Settled around 1750, Charlotte enjoyed rapid growth in the 1970s and is the largest city in the state.

Charlotte Amalie /'ʃɑːlət ə'mɑːljə/ capital, tourist resort, and free port of the US Virgin Islands, on the island of St Thomas; population (1980) 11,756. Boatbuilding and rum distilling are among the economic activities. It was founded 1672 by the Danish West India Company.

Charlotte Amalie was formerly (1921–37) called *St Thomas*. The College of the Virgin Islands and the Museum of the Virgin Islands are here.

Charlottesville /'ʃɑːlətsvɪl/ city in central Virginia, USA, in the Blue Ridge mountain foothills, NW of Richmond, on the Rivanna River; seat of Albemarle County; population (1990) 40,300. Tourism is important, and some textiles are manufactured.

Charlottetown /'ʃɑːləttaun/ capital of Prince Edward Island, Canada; population (1986) 16,000. The city trades in textiles, fish, timber, vegetables, and dairy produce. It was founded by French settlers in the 1720s.

Chartres /'ʃɑːtrə/ capital of the *département* of Eure-et-Loir, NW France, 96 km/59 mi SW of Paris, on the river Eure; population (1990) 41,850. The city is an agricultural centre for the fertile Plaine de la Beauce. Its cathedral of Nôtre Dame, completed about 1240, is a masterpiece of Gothic architecture.

Chatham /'tʃætəm/ town in Kent, England; population (1983) 146,200. The Royal Dockyard 1588–1984 was from 1985 converted to an industrial area, marina, and museum as a focus of revival for the whole Medway area.

Chatham Islands /'tʃætəm/ two Pacific islands (Chatham and Pitt), forming a county of South Island, New Zealand; area 960 sq km/371 sq mi; population (1981) 750. The chief settlement is Waitangi.

Chattanooga /ˌtʃætə'nuːgə/ city in Tennessee, USA, on the Tennessee River; population (1990) 152,500. It is the focus of the Tennessee Valley Authority area. Developed as a salt-trading centre after 1835, it now produces chemicals, textiles, and metal products.

Cheapside /'tʃiːpsaɪd/ street running from St Paul's Cathedral to Poultry, in the City of London, England. It was the scene of the 13th-century 'Cheap', a permanent fair and general market. Christopher Wren's church of St Mary-le-Bow in Cheapside has the Bow Bells.

Checheno-Ingush /tʃɪ'tʃenəu ɪŋ'guːʃ/ autonomous republic in southern Russia, on the northern slopes of the Caucasus Mountains; area 19,000 km/7,350 sq mi; population (1986) 1,230,000.

It was conquered in the 1850s, and has a large oilfield. The capital is Grozny. The population includes Chechens (53%) and Ingushes (12%). In Oct 1991 the region declared its independence. After a brief, unsuccessful attempt to quell the rebellion, Moscow agreed to enter into negotiations over the republic's future. In March 1992 Checheno-Ingush refused to be party to a federal treaty signed by 18 of Russia's 20 main political subdivisions. By the end of the month at least five people had been killed in clashes between anti- and pro-separatist factions, forcing the republic's parliament to declare a state of emergency.

Cheddar /'tʃedə/ village in Somerset, England where Cheddar cheese was first produced. Nearby are a limestone gorge and caves with stalactites and stalagmites. In 1962 excavation revealed the site of a Saxon palace.

Chefoo /ˌtʃiː'fuː/ former name of ◊Yantai, a port in China.

Chekiang /ˌtʃekiˈæŋ/ alternative transcription of ▷Zhejiang, a province of China.

Chelmsford /ˈtʃelmzfəd/ town in Essex, England, 48 km/30 mi NE of London; population (1985 est) 144,000. It is the administrative headquarters of the county, and a market town with radio, electrical, engineering, and agricultural machinery industries.

Chelsea /ˈtʃelsi/ historic area of the Royal Borough of Kensington and Chelsea, London, immediately N of the Thames where it is crossed by the Albert and Chelsea bridges.

The Royal Hospital was founded 1682 by Charles II for old and disabled soldiers, 'Chelsea Pensioners', and the National Army Museum, founded 1960, covers campaigns 1485–1914. The Physic Garden for botanical research was established in the 17th century; the home of the essayist Thomas Carlyle in Cheyne Row is a museum. The Chelsea Flower Show is held annually by the Royal Horticultural Society in the grounds of Royal Hospital. Ranelagh Gardens 1742–1804 and Cremorne Gardens 1845–77 were popular places of entertainment.

Cheltenham /ˈtʃeltənəm/ spa town at the foot of the Cotswold Hills, Gloucestershire, England; population (1988 est) 86,400. There are annual literary and music festivals, a racecourse (the Cheltenham Gold Cup is held annually), and Cheltenham College (founded 1854).

The home of the composer Gustav Holst is now a museum, and to the SW is Prinknash Abbey, a Benedictine house that produces pottery. Cheltenham is also the centre of the British government's electronic surveillance operations (GCHQ). The Universities' Central Council on Admissions (UCCA) 1963 and the Polytechnics and Colleges Admissions System (PCAS) 1985 are here.

Chelyabinsk /ˌtʃeliˈæbɪnsk/ industrial town and capital of Chelyabinsk region, W Siberia, Russia; population (1987) 1,119,000. It has iron and engineering works and makes chemicals, motor vehicles, and aircraft.

It lies E of the Ural Mountains, 240 km/150 mi SE of Ekaterinburg (Sverdlovsk). It was founded 1736 as a Russian frontier post.

Chemnitz /ˈkemnɪts/ industrial city (engineering, textiles, chemicals) in the state of Saxony, Federal Republic of Germany, on the river Chemnitz, 65 km/40 mi SSE of Leipzig; population (1990) 310,000. As a former district capital of East Germany it was named *Karl-Marx-Stadt* 1953–90.

Chemulpo /ˌtʃemulˈpəu/ former name for ▷Inchon, a port and summer resort on the west coast of South Korea.

Chenab /tʃɪˈnæb/ tributary of the river ▷Indus, in India and Pakistan.

Chengchow /ˌtʃeŋˈtʃau/ alternative transcription of ▷Zhengzhou, the capital of Henan province, China.

Chengde /ˌtʃeŋˈdeɪ/ or *Chengteh* town in Hebei province, China, NE of Beijing; population (1984) 325,800. It is a market town for agricultural and forestry products. It was the summer residence of the Manchu rulers and has an 18th-century palace and temples. As *Jehol*, it was capital of a former province of the same name.

Chengdu /ˌtʃeŋˈduː/ or *Chengtu* ancient city, capital of Sichuan province, China; population (1989) 2,780,000. It is a busy rail junction and has railway workshops, and textile, electronics, and engineering industries. It has well-preserved temples.

Chengteh /ˌtʃeŋˈdɜː/ alternative transliteration of ▷Chengde, a town in China.

Chengtu /ˌtʃeŋˈduː/ alternative transliteration of ▷Chengdu, a city in China.

Chepstow /ˈtʃepstəu/ (Welsh *Casgwent*) market town in Gwent, Wales, on the river Wye; population (1984) 12,500. The high tides, sometimes 15 m/50 ft above low level, are the highest in Britain. There is a Norman castle, and the ruins of Tintern Abbey are 6.5 km/4 mi to the N.

Cher /ʃeə/ French river that rises in Creuse *département* and flows into the river Loire below Tours, length 355 km/220 mi. It gives its name to a *département*.

Cherbourg /ˈʃeəbuəg/ French port and naval station at the northern end of the Cotentin peninsula, in Manche *département*; population (1990) 28,800. There is an institute for studies in nuclear warfare, and Cherbourg has large shipbuilding yards.

During World War II, Cherbourg was captured June 1944 by the Allies, who thus gained their first large port of entry into France. It was severely damaged; restoration of the harbour was completed 1952. There is a nuclear processing plant at nearby Cap la Hague. There are ferry links to England (Southampton, Weymouth, and Rosslare).

Cherepovets /ˌtʃerɪpəˈvets/ iron and steel city in W Russia, on the Volga-Baltic waterway; population (1985) 299,000. It was originally a tax-exempt settlement.

Chernigov /tʃəˈnɪgɒf/ town and port on the river Desna in N Ukraine; population (1987) 291,000. Lumbering, textiles, chemicals, distilling, and food-canning are among its industries. It has an 11th-century cathedral.

Chernobyl /tʃəˈnəubəl/ town in central Ukraine; site of a nuclear power station. In April 1986 a leak, caused by overheating, occurred in a non-pressurized boiling-water nuclear reactor. The resulting clouds of radioactive isotopes were traced as far away as Sweden; over 250 people were killed, and thousands of square miles contaminated.

Chernovtsy /ˌtʃəːnɒftˈsiː/ city in W Ukraine; population (1987) 254,000. Industries include textiles, clothing, and machinery. It has formerly been called *Czernowitz* (before 1918), *Cernăuţi* (1918–1940, when it was part of Romania), and *Chrenovitsy* (1940–44).

Cheshire

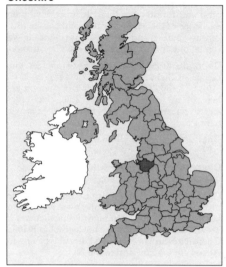

Ches. abbreviation for ◊ *Cheshire* .

Chesapeake Bay /ˈtʃesəpiːk/ largest of the inlets on the Atlantic coast of the USA, bordered by Maryland and Virginia. It is about 320 km/200 mi in length and 6–64 km/4–40 mi in width.

Among the rivers that flow into the bay are the James, York, Potomac, Rappahannock, Patuxent, and Susquehanna. Deepwater ports on the bay are Newport News, Norfolk, Portsmouth, and Baltimore. The Chesapeake and Delaware Canal links the bay to the Delaware River and the Wilmington-Philadelphia port area. Pollution has greatly diminished the once plentiful shellfish in the bay.

Cheshire /ˈtʃeʃə/ county in NW England
area 2,320 sq km/896 sq mi
towns Chester (administrative headquarters), Warrington, Crewe, Widnes, Macclesfield, Congleton
physical chiefly a fertile plain; rivers: Mersey, Dee, Weaver
features salt mines and geologically rich former copper workings at Alderley Edge (in use from Roman times until the 1920s); Little Moreton Hall; discovery of Lindow Man, the first 'bogman' to be found in mainland Britain, dating from around 500 BC; Quarry Bank Mill at Styal is a cotton-industry museum
products textiles, chemicals, dairy products
population (1991) 937,300
famous people Charles Dodgson (Lewis Carroll); the novelist Mrs Gaskell lived at Knutsford (the locale of *Cranford*).

Chesil Bank /ˈtʃezəl/ shingle bank extending 19 km/11 mi along the coast of Dorset, England, from Abbotsbury to the Isle of Portland.

Chester /ˈtʃestə/ city in Cheshire, England, on the river Dee 26 km/16 mi S of Liverpool; population (1991) 115,000. It is the administrative headquarters of Cheshire. Industries include engineering and the manufacture of car components. Its name derives from the Roman *Castra Devana* ('the camp on the Dee'), and there are many Roman and later remains. It is the only English city to retain its city walls (2 mi/3 km long) intact. The cathedral dates from the 11th century but was restored 1876. The church of St John the Baptist is a well-known example of early Norman architecture. The 'Rows' are covered arcades dating from the Middle Ages.

From 1070 to the reign of Henry III, Chester was the seat of a county palatine (a county whose lord exercised some of the roles usually reserved for the monarch). The town hall dates from 1869. Although the silting-up of the Dee destroyed Chester's importance as a port, navigation has been greatly improved by dredging.

Chesterfield /ˈtʃestəfiːld/ market town of Derbyshire, England; 40 km/25 mi N of Derby, on the Rother River; population (1988 est) 72,200. Industries include coal-mining, engineering, and glass. It is the burial place of the engineer George Stephenson. All Saints' Church is renowned for its crooked spire.

Cheviot Hills /ˈtʃiːviət/ range of hills 56 km/35 mi long, mainly in Northumberland, forming the border between England and Scotland for some 48 km/30 mi. The highest point is the Cheviot, 816 m/2,676 ft. For centuries the area was a battleground between the English and the Scots. It gives its name to a breed of sheep.

Cheyenne /ʃaɪˈæn/ capital of Wyoming, USA, located in the southeastern part of the state, just N of the Colorado border in the foothills of the Laramie Mountains; population (1990) 50,000. An agricultural and transportation centre, its industries include oil refining, fertilizers, electronics, restaurant equipment, and ceramics. Tourism is also important to the economy.

Chiang Mai alternative name for ◊ Chiengmai, a town in Thailand.

Chiba /ˈtʃiːbə/ industrial city (paper, steel, textiles) in Kanton region, E Honshu Island, Japan, 40 km/25 mi W of Tokyo; population (1990) 829,500.

Chicago /ʃɪˈkɑːgəʊ/ financial and industrial city in Illinois, USA, on Lake Michigan. It is the third largest US city; population (1990) 2,783,700, metropolitan area 8,065,000. Industries include iron, steel, chemicals, electrical goods, machinery, meatpacking and food processing, publishing, and fabricated metals. The once famous stockyards are now closed.
features The world's first skyscraper was built here 1885 and some of the world's tallest skyscrapers, including the tallest, the Sears Tower at 443 m/1,454 ft, are in Chicago. The Museum of Science and Industry, opened 1893, has 'hands-on' exhibits including a coal mine, a World War II U-boat, an Apollo spacecraft and lunar module, and exhibits by industrial firms. The Chicago River cuts the city into three 'sides'. Chicago is known as the Windy City, so called from the breezes of Lake Michigan, as well as from its citizens' (and, allegedly, politicians')

voluble talk; the lake shore ('the Gold Coast') is occupied by luxury apartment blocks. It has a symphony orchestra, an art institute, the University of Chicago (site of the first controlled nuclear reaction), DePaul and Loyola universities, a campus of the University of Illinois, and the Illinois Institute of Technology. Chicago-O'Hare International Airport is the nation's busiest. The Board of Trade, Mercantile Exchange, and Options Exchange are among the world's largest commodity markets.

history The site of Chicago was visited by Jesuit missionaries 1673, and Fort Dearborn, then a frontier fort, was built here 1803. The original layout was a rectangular grid, but many outer boulevards have been constructed on less rigid lines. As late as 1831 Chicago was still a village, but railroads from the east coast reached it by 1852, and by 1871, when it suffered a disastrous fire, it was a city of more than 300,000 inhabitants.

Rapid development began again in the 1920s, and during the years of Prohibition 1919–33, the city became notorious for its gangsters. The opening of the St Lawrence Seaway 1959 brought Atlantic shipping to its docks.

She is always a novelty; for she is never the Chicago you saw when you passed through the last time.

On **Chicago** Mark Twain *Life on the Mississippi* 1883

Chichester /ˈtʃɪtʃɪstə/ city and market town in Sussex; 111 km/69 mi SW of London, near Chichester Harbour; population (1989 est) 25,000. It is the administrative headquarters of West Sussex. It was a Roman township, and the remains of the Roman palace built around AD 80 at nearby Fishbourne are unique outside Italy. There is a cathedral consecrated 1108, later much rebuilt and restored, and the Chichester Festival Theatre (1962).

Chiclayo /tʃiˈklaɪəʊ/ capital of Lambayeque department, NW Peru; population (1988) 395,000. At the centre of an agricultural area, its industries include rice milling, cotton ginning, brewing, and tanning.

Chico /ˈtʃiːkəʊ/ city in N California, USA, NW of Sacramento; population (1990) 40,000. Situated in the fertile Sacramento Valley, a farming region, its industries include food processing and lumber products.

Chiengmai /dʒiˈeŋ ˈmaɪ/ or **Chiang Mai** town in N Thailand; population (1990) 164,900. There is a trade in teak and lac (as shellac, a resin used in varnishes and polishes) and many handicraft industries. It is the former capital of the Lan Na Thai kingdom.

Chihuahua /tʃɪˈwɑːwə/ capital of Chihuahua state, Mexico, 1,285 km/800 mi NW of Mexico City; population (1984) 375,000. Founded 1707, it is the centre of a mining district and has textile mills.

Chile Republic of (*República de Chile*)
area 756,950 sq km/292,257 sq mi
capital Santiago
towns Concepción, Viña del Mar, Temuco; ports Valparaíso, Antofagasta, Arica, Iquique, Punta Arenas
physical Andes mountains along E border, Atacama Desert in N, fertile central valley, grazing land and forest in S
territories Easter Island, Juan Fernández Islands, part of Tierra del Fuego, claim to part of Antarctica
features Atacama Desert is one of the driest regions in the world
head of state and government Patricio Aylwin from 1990
political system emergent democratic republic
political parties Christian Democratic Party (PDC), moderate centrist; National Renewal Party (RN), right-wing
exports copper (world's leading producer), iron, molybdenum (world's second largest producer), nitrate, pulp and paper, steel products, fishmeal, fruit
currency peso
population (1990 est) 13,000,000 (the majority are of European origin or are mestizos, of mixed American Indian and Spanish descent); growth rate 1.6% p.a.
life expectancy men 64, women 73
language Spanish
religion Roman Catholic 89%
literacy 94% (1988)
GDP $18.9 bn (1987); $6,512 per head
chronology
1818 Achieved independence from Spain.
1964 PDC government under Eduardo Frei.
1970 Dr Salvador Allende became the first democratically elected Marxist president; he embarked on a programme of social reform.
1973 Government overthrown by the CIA-backed military, led by General Augusto Pinochet. Allende killed. Repression began.
1983 Growing opposition to the regime from all sides, with outbreaks of violence.

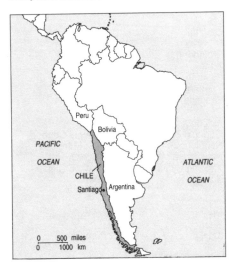

1988 Referendum on whether Pinochet should serve a further term resulted in a clear 'No' vote. *1989* President Pinochet agreed to constitutional changes to allow pluralist politics. Patricio Aylwin (PDC) elected president (his term would begin 1990); Pinochet remained as army commander in chief. *1990* Aylwin reached accord on end to military junta government. Pinochet censured by president. *1992* Future US–Chilean free-trade agreement announced.

Chile is God's mechanism for keeping Argentina from the Pacific.

On **Chile** Santiago saying

Chiltern Hills /'tʃɪltən/ range of chalk hills extending for some 72 km/45 mi in a curve from a point N of Reading to the Suffolk border. Coombe Hill, near Wendover, 260 m/852 ft high, is the highest point.

Chimbote /tʃɪm'bəuti/ largest fishing port in Peru; population (1981) 216,000. Sugar and fish products are exported; other industries include iron and steel.

China People's Republic of (*Zhonghua Renmin Gonghe Guo*)
area 9,596,960 sq km/3,599,975 sq mi
capital Beijing (Peking)
towns Chongqing (Chungking), Shenyang (Mukden), Wuhan, Nanjing (Nanking), Harbin; ports Tianjin (Tientsin), Shanghai, Qingdao (Tsingtao), Lüda (Lü-ta), Guangzhou (Canton)
physical two-thirds of China is mountains or desert (N and W); the low-lying E is irrigated by rivers Huang He (Yellow River), Chang Jiang (Yangtze-Kiang), Xi Jiang (Si Kiang)
features Great Wall of China; Gezhouba Dam; Ming Tombs; Terracotta Warriors (Xi'ain); Gobi Desert; world's most populous country
head of state Yang Shangkun from 1988
head of government Li Peng from 1987
political system communist republic
political party Chinese Communist Party (CCP), Marxist-Leninist-Maoist
exports tea, livestock and animal products, silk, cotton, oil, minerals (China is the world's largest producer of tungsten and antimony), chemicals, light industrial goods
currency yuan

China: provinces

province	alternative transcription	capital	area in sq km
Anhui	Anhwei	Hefei	139,900
Fujian	Fukien	Fuzhou	123,100
Gansu	Kansu	Lanzhou	530,000
Guangdong	Kwantung	Guangzhou	231,400
Guizhou	Kweichow	Guiyang	174,000
Hainan		Haikou	34,000
Hebei	Hopei	Shijiazhuang	202,700
Heilongjiang	Heilungkiang	Harbin	463,600
Henan	Honan	Zhengzhou	167,000
Hubei	Hupei	Wuhan	187,500
Hunan		Changsha	210,500
Jiangsu	Kiangsu	Nanjing	102,200
Jiangxi	Kiangsi	Nanchang	164,800
Jilin	Kirin	Changchun	187,000
Liaoning		Shenyang	151,000
Qinghai	Tsinghai	Xining	721,000
Shaanxi	Shensi	Xian	195,800
Shandong	Shantung	Jinan	153,300
Shanxi	Shansi	Taiyuan	157,100
Sichuan	Szechwan	Chengdu	569,000
Yunnan		Kunming	436,200
Zhejiang	Chekiang	Hangzhou	101,800
autonomous region			
Guangxi Zhuang	Kwangsi Chuang	Nanning	220,400
Nei Mongol	Inner Mongolia	Hohhot	450,000
Ningxia Hui	Ninghsia-Hui	Yinchuan	170,000
Xinjiang Uygur	Sinkiang Uighur	Urumqi	1,646,800
Xizang	Tibet	Lhasa	1,221,600
municipality			
Beijing	Peking		17,800
Shanghai			5,800
Tianjin	Tientsin		4,000
		TOTAL	9,139,300

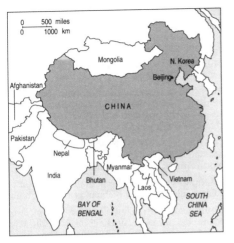

population (1990 est) 1,130,065,000 (the majority are Han or ethnic Chinese; the 67 million of other ethnic groups, including Tibetan, Uigur, and Zhuang, live in border areas). The number of people of Chinese origin outside China, Taiwan, and Hong Kong is estimated at 15–24 million. Growth rate 1.2% p.a.
life expectancy men 67, women 69
languages Chinese, including Mandarin (official), Cantonese, and other dialects
religions officially atheist, but traditionally Taoist, Confucianist, and Buddhist; Muslim 13 million; Catholic 3–6 million (divided between the 'patriotic' church established 1958 and the 'loyal' church subject to Rome); Protestant 3 million
literacy men 82%, women 66% (1985 est)
GDP $293.4 bn (1987); $274 per head
chronology
1949 People's Republic of China proclaimed by Mao Zedong.
1954 Soviet-style constitution adopted.
1956–57 Hundred Flowers Movement encouraged criticism of the government.
1958–60 Great Leap Forward commune experiment to achieve 'true communism'.
1960 Withdrawal of Soviet technical advisers.
1962 Sino-Indian border war.
1962–65 Economic recovery programme under Liu Shaoqi; Maoist 'socialist education movement' rectification campaign.
1966–69 Great Proletarian Cultural Revolution; Liu Shaoqi overthrown.
1969 Ussuri River border clashes with USSR.
1970–76 Reconstruction under Mao and Zhou Enlai.
1971 Entry into United Nations.
1972 US president Nixon visited Beijing.
1975 New state constitution. Unveiling of Zhou's 'Four Modernizations' programme.
1976 Deaths of Zhou Enlai and Mao Zedong; appointment of Hua Guofeng as prime minister and Communist Party chair. Vice Premier Deng Xiaoping in hiding. Gang of Four arrested.
1977 Rehabilitation of Deng Xiaoping.
1979 Economic reforms introduced. Diplomatic relations opened with USA. Punitive invasion of Vietnam.

1980 Zhao Ziyang appointed prime minister.
1981 Hu Yaobang succeeded Hua Guofeng as party chair. Imprisonment of Gang of Four.
1982 New state constitution adopted.
1984 'Enterprise management' reforms for industrial sector.
1986 Student prodemocracy demonstrations.
1987 Hu was replaced as party leader by Zhao, with Li Peng as prime minister. Deng left Politburo but remained influential.
1988 Yang Shangkun replaced Li Xiannian as state president. Economic reforms encountered increasing problems; inflation rocketed.
1989 Over 2,000 killed in prodemocracy student demonstrations in Tiananmen Square; international sanctions imposed.
1991 March: European Community (EC) and Japanese sanctions lifted. May: normal relations with USSR resumed. Sept: UK prime minister John Major visited Beijing. Nov: relations with Vietnam normalized.
1992 China promised to sign 1968 Nuclear Non-Proliferation Treaty. Historic visit by Japan's emperor.

China is one of the few countries an American President can visit without having to face demonstrations.

On **China** Gough Whitlam 1971

China Sea /ˈtʃaɪnə/ area of the Pacific Ocean bordered by China, Vietnam, Borneo, the Philippines, and Japan. Various groups of small islands and shoals, including the Paracels, 500 km/300 mi E of Vietnam, have been disputed by China and other powers because they lie in oil-rich areas.

N of Taiwan it is known as the **East China Sea** and to the S as the **South China Sea**.

Chinghai /ˌtʃɪŋˈhaɪ/ alternative transcription of ◊Qinghai, a province of China.

Chişinău /kiʃiˈnəʊ/ (Russian **Kishinev**) capital of Moldova, situated in a rich agricultural area; population (1989) 565,000. It is a commercial and cultural centre; industries include cement, food processing, tobacco, and textiles.

Founded 1436, it became Russian 1812. It was taken by Romania 1918, by the USSR 1940, and by Germany 1941, when it was totally destroyed. The USSR recaptured the site 1944, and rebuilding soon began. Nationalist demonstrations were held in the city 1989, prior to Moldova gaining independence 1991.

Chita /tʃiːˈtɑː/ town in E Siberia, on the Chita River; population (1987) 349,000. It is on the Trans-Siberian railway, and has chemical and engineering works and coal mines.

Chittagong /ˈtʃɪtəɡɒŋ/ city and port in Bangladesh, 16 km/10 mi from the mouth of the Karnaphuli River, on the Bay of Bengal; population (1981) 1,388,476. Industries include steel, engineering, chemicals, and textiles.

Chkalov /ˈxkɑːlɒv/ former name 1938–57 of ◊Orenburg, a town in Russia.

Chongjin /ˌtʃʊŋˈdʒɪn/ port and capital of North Hamgyong province on the northeast coast of North Korea; population (1984) 754,000. Timber, iron, and textiles are exported; there is steel, pig-iron, and fish processing.

Chongqing /ˌtʃʊŋˈtʃɪŋ/ or **Chungking**, also known as **Pahsien** city in Sichuan province, China, that stands at the confluence of the Chang Jiang and Jialing Jiang rivers; population (1984) 2,733,700. Industries include iron, steel, chemicals, synthetic rubber, and textiles.

For over 4,000 years it has been a major commercial centre in one of the most remote and economically deprived regions of China. It was opened to foreign trade 1891, and remains a focal point of road, river, and rail transport. When both Beijing and Nanjing were occupied by the Japanese, it was the capital of China 1938–46.

Christchurch /ˈkraɪstʃɜːtʃ/ town in Dorset, S England, adjoining Bournemouth at the junction of the Stour and Avon rivers; population (1983) 40,300. Light industries include plastics and electronics. There is a Norman and Early English priory church.

Christchurch /ˈkraɪstʃɜːtʃ/ city on South Island, New Zealand, 11 km/7 mi from the mouth of the Avon River; population (1991) 292,500, urban area 306,900. It is the principal city of the Canterbury plains and the seat of the University of Canterbury. Industries include fertilizers and chemicals, canning and meat processing, rail workshops, and shoes.

Christchurch uses as its port a bay in the sheltered Lyttelton Harbour on the northern shore of the Banks Peninsula, which forms a denuded volcanic mass. Land has been reclaimed for service facilities, and rail and road tunnels (1867 and 1964 respectively) link Christchurch with Lyttelton.

It represents a dream carried 13,000 miles, down into the Antipodes, and subtly changing as soon as it began to be realized.

On **Christchurch** (New Zealand)
J B Priestley *A Visit to New Zealand* 1974

Christiania /ˌkrɪstiˈɑːniə/ former name (1624–1924) of the Norwegian capital of ⬦Oslo, after King Christian IV who replanned it after a fire 1624.

Christmas Island /ˈkrɪsməs/ island in the Indian Ocean, 360 km/224 mi S of Java; area 140 sq km/54 sq mi; population (1986) 2,000. It has phosphate deposits. Found to be uninhabited when reached by Capt W Mynars on Christmas Day 1643, it was annexed by Britain 1888, occupied by Japan 1942–45, and transferred to Australia 1958.

After a referendum 1984, it was included in Northern Territory.

Christmas Island former name of ⬦Kiritimati, an island in Kiribati, in the central Pacific.

Chubu /ˈtʃuːbuː/ mountainous coastal region of central Honshu Island, Japan; area 66,774 sq km/25,791 sq mi; population (1986) 20,694,000. The chief city is Nagoya.

Chufu /ˈtʃuːˈfuː/ alternative transcription of ⬦Qufu, a town in Shandong province, China.

Chugoku /tʃuːˈgəʊkuː/ southwestern region of Honshu Island, Japan; area 31,881 sq km/12,314 sq mi; population (1986) 7,764,000. The chief city is Hiroshima.

Chukchi Sea /ˈtʃʊktʃiː/ part of the Arctic Ocean, situated N of the Bering Strait between Asia and North America.

Chungking /ˌtʃʊŋˈkɪŋ/ alternative transcription of ⬦Chongqing, a city in Sichuan province, China.

Churchill /ˈtʃɜːtʃɪl/ town in the province of Manitoba, Canada, situated on Hudson Bay; population (1986) 1,217. Although the port is ice-free only three months a year, Churchill handles about 500,000 tonnes of grain annually, as well as fuel oil and bulk cargo. It is a major centre for Arctic research, health programmes, and education. The Hudson's Bay Company established a post here and named it for Lord Churchill (later 1st duke of Marlborough).

Chuvash /ˈtʃuːvæʃ/ autonomous republic of Russia, lying W of the Volga River, 560 km/350 mi E of Moscow; area 18,300 sq km/7,100 sq mi; population (1986) 1,320,000. The capital is Cheboksary, population (1985) 389,000. The economy is based on lumbering and grain growing and there are phosphate and limestone deposits and electrical and engineering industries.

Cienfuegos /ˌsiːenˈfweɪɡɒs/ port and naval base in Cuba; population (1985) 124,600. It trades in sugar, fruit, and tobacco. It was founded 1819, destroyed by a storm 1825, and rebuilt.

Cilicia /saɪˈlɪsiə/ ancient region of Asia Minor, now forming part of Turkey, situated between the Taurus Mountains and the Mediterranean. Access from the N across the Taurus range is through the *Cilician Gates*, a strategic pass that has been used for centuries as part of a trade route linking Europe and the Middle East.

history Successively conquered by the Persians, Alexander the Great, and the Romans under Pompey, Cilicia became an independent Armenian principality 1080 and a kingdom 1198. Sometimes referred to as Lesser Armenia, it was absorbed into the Ottoman Empire during the 15th century.

Cincinnati /ˌsɪnsɪˈnæti/ city and port in Ohio, USA, on the Ohio River; population (1990) 364,000. Chief industries include machinery, clothing, furniture making, wine, chemicals, and meatpacking. Founded 1788, Cincinnati became a city 1819. It attracted large numbers of European immigrants, particularly Germans, during the 19th century.

Circassia /səˈkæsiə/ former name of an area of the N Caucasus, ceded to Russia by Turkey 1829 and now part of the Karachai-Cherkess region of Russia.

Cirencester /'saɪrən,sestə/ market town in Gloucestershire, England; population (1989 est) 17,000. It is the 'capital' of the Cotswolds. Industries include engineering and the manufacture of electrical goods. It was the second largest town in Roman Britain, and has an amphitheatre which seated 8,000, and the Corinium Museum. The Royal Agricultural College is based here.

Ciskei, Republic of /ˌsɪs'kaɪ/ Bantu homeland in South Africa, which became independent 1981, although this is not recognized by any other country
area 7,700 sq km/2,974 sq mi
capital Bisho
features one of the two homelands of the Xhosa people created by South Africa (the other is Transkei)
products wheat, sorghum, sunflower, vegetables, timber, metal products, leather, textiles
population (1985) 925,000
language Xhosa
government president (Brig Oupa Gqozo from 1990), with legislative and executive councils
recent history In Sept 1992 Ciskei troops fired on African National Congress (ANC) demonstrators demanding the ousting of the territory's military leader, Brig Gqozo; 28 people were slain and about 200 injured. Brig Gqozo had assumed power in a military takeover 1990. Initially sympathetic to the aims of the ANC, he had pledged to restore civilian rule and called for the reincorporation of Ciskei into South Africa. Subsequently, he turned against the organization and has been accused of repressing its supporters.

Citlaltépetl /ˌsɪtlæl'tepek/ (Aztec 'star mountain') dormant volcano, the highest mountain in Mexico at 5,700 m/18,700 ft, N of the city of Orizaba (after which it is sometimes named). It last erupted 1687.

Ciudad Bolívar /sju:'ðɑːð bɒ'liːvɑː/ former name (1824–49) *Angostura* capital of Bolívar state, SE Venezuela, on the river Orinoco, 400 km/250 mi from its mouth; population (1989) 277,000. Gold is mined in the vicinity. The city is linked with Soledad across the river by the Angostura Bridge (1967), the first to span the Orinoco.

Ciudad Guayana /sju:'ðɑːð gwaɪ'ɑːnə/ city in Venezuela, on the south bank of the river Orinoco, population (1989) 516,500. Main industries include iron and steel. The city was formed by the union of Puerto Ordaz and San Felix, and has been opened to ocean-going ships by dredging.

Ciudad Juárez /sju:'ðɑːð 'xwɑːres/ city on the Rio Grande, in Chihuahua, N Mexico, on the US border; population (1990) 797,650. It is a centre for cotton.

Ciudad Real /sju:'ðɑːð reɪ'æl/ city of central Spain, 170 km/105 mi S of Madrid; population (1991) 59,400. It is the capital of Ciudad Real province. It trades in livestock and produces textiles and pharmaceuticals. Its chief feature is its huge Gothic cathedral.

Ciudad Trujillo /sju:'ðɑːð tru:'xiːəu/ former name (1936–61) of ⟡Santo Domingo, capital city and seaport of the Dominican Republic.

Civitavecchia /ˌtʃiːvɪtə'vekjə/ ancient port on the west coast of Italy, in Lazio region, 64 km/40 mi NW of Rome; population (1981) 43,550. Industries include fishing and the manufacture of cement and calcium carbide.

Clackmannanshire /klæk'mænənʃə/ former county (the smallest) in Scotland, bordering the Firth of Forth. It became a district of Central Region 1975. The administrative centre is Alloa.

Clacton-on-Sea /'klæktən ɒn 'siː/ seaside resort in Essex, England; 19 km/12 mi SE of Colchester; population (1981) 43,600. The 16th-century St Osyth's priory is nearby.

Clare /kleə/ county on the west coast of the Republic of Ireland, in the province of Munster; area 3,190 sq km/1,231 sq mi; population (1991) 90,800. Shannon airport is here.
The coastline is rocky and dangerous, and inland Clare is an undulating plain, with mountains on the E, W, and NW, the chief range being the Slieve Bernagh Mountains in the SE rising to over 518 m/1,700 ft. The principal rivers are the Shannon and its tributary, the Fergus. There are over 100 lakes in the county, including Lough Derg on the eastern border. The county town is Ennis. At Ardnachusha, 5 km/3 mi N of Limerick, is the main power station of the Shannon hydroelectric installations. The county is said to be named after Thomas de Clare, an Anglo-Norman settler to whom this area was granted 1276.

Clarksville /'klɑːksvɪl/ city in N Tennessee, USA, at the confluence of the Cumberland and Red rivers, NW of Nashville; population (1980) 54,777. Industries include tobacco products, clothing, air-conditioning and heating equipment, rubber, and cheese.

Clearwater /'klɪəwɔːtə/ city in W central Florida, USA, on the Gulf of Mexico, NW of St Petersburg; seat of Pinellas County; population (1990) 98,800. Industries include tourism, citrus fruits, fishing, electronics, and flowers.

Cleethorpes /'kliːθɔːps/ seaside resort in Humberside, NE England, on the Humber estuary; population (1987 est) 35,500. Fishing is important to the local economy.

Clermont-Ferrand /'kleəmɒn fe'rɒn/ city, capital of Puy-de-Dôme *département*, in the Auvergne region of France; population (1990) 140,200. It is a centre for agriculture, and its rubber industry is the largest in France.
Car tyres are manufactured here; other products include chemicals, preserves, foodstuffs, and clothing. The Gothic cathedral is 13th century. Urban II ordered the First Crusade at a council here 1095. The 17th-century writer Blaise Pascal was born here.

Cleveland /'kliːvlənd/ county in NE England
area 580 sq km/224 sq mi
towns Middlesbrough (administrative headquarters), Stockton on Tees, Billingham, Hartlepool

Cleveland

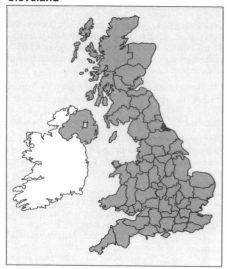

features river Tees, with Seal Sands wildfowl refuge at its mouth; North Yorkshire Moors National Park; Teesside, the industrial area at the mouth of the Tees, has Europe's largest steel complex (at Redcar) and chemical site (ICI, using gas and local potash), as well as an oil-fuel terminal at Seal Sands
products steel, chemicals
population (1987) 555,000
famous people Capt James Cook, Thomas Sheraton, Compton Mackenzie.

Cleveland /'kliːvlənd/ largest city of Ohio, USA, on Lake Erie at the mouth of the river Cuyahoga; population (1990) 505,600, metropolitan area 2,759,800. Its chief industries are iron and steel and petroleum refining.

Iron ore from the Lake Superior region and coal from Ohio and Pennsylvania mines are brought here.

Cluj /kluːʒ/ (German *Klausenberg*) city in Transylvania, Romania, located on the river Somes; population (1985) 310,000. It is a communications centre for Romania and the Hungarian plain. Industries include machine tools, furniture, and knitwear.

There is a 14th-century cathedral, and Romanian (1872) and Hungarian (1945) universities.

Cluny /'kluːni/ town in Saône-et-Loire *département*, France, on the river Grosne; population (1982) 4,500. Its abbey, now in ruins, was the foundation house 910–1790 of the Cluniac order, originally a reformed branch of the Benedictines. Cluny was once a lace-making centre; it has a large cattle market.

Clutha /'kluːθə/ longest river in South Island, New Zealand, 322 km/201 mi long. It rises in the Southern Alps, has hydroelectric installations, and flows to meet the sea near Kaitangata.

Clwyd /'kluːɪd/ county in N Wales
area 2,420 sq km/934 sq mi
towns Mold (administrative headquarters), Flint, Denbigh, Wrexham; seaside resorts: Colwyn Bay, Rhyl, Prestatyn
physical rivers: Dee, Clwyd; Clwydian range of mountains with Offa's Dyke along the main ridge
features Chirk, Denbigh, Flint, and Rhuddlan castles; Greenfield Valley, NW of Flint, was in the forefront of the Industrial Revolution before the advent of steam and now has a museum of industrial archaeology
products dairy and meat products, optical glass, chemicals, limestone, microprocessors, plastics
population (1991) 400,500
languages 19% Welsh, English
famous people George Jeffreys, Henry Morton Stanley.

Clyde /klaɪd/ river in Strathclyde, Scotland; 170 km/103 mi long. The Firth of Clyde and Firth of Forth are linked by the Forth and Clyde Canal, 56 km/35 mi long. The shipbuilding yards have declined in recent years.

The nuclear-submarine bases of Faslane (Polaris) and Holy Loch (USA Poseidon) are here.

The Clyde made Glasgow and Glasgow made the Clyde.

On the river **Clyde** old Scottish saying

Clydebank /'klaɪdbæŋk/ town on the river Clyde, Strathclyde, Scotland, 10 km/6 mi NW of Glasgow; population (1991) 44,600. At the John Brown yard, liners such as the *Queen Elizabeth II* were built.

CO abbreviation for ◊*Colorado*, a state of the USA.

Coatbridge /'kəʊtbrɪdʒ/ town in Strathclyde, Scotland, 13 km/8 mi E of Glasgow; population (1986 est) 46,100. Coal and iron are mined nearby.

Industries include iron, ore, steel, and engineering.

Cobh /kəʊv/ seaport and market town on Great Island, Republic of Ireland, in the estuary of the Lee, County Cork; population (1991) 6,200. Cobh was formerly a port of call for transatlantic steamers. The town was known as Cove of Cork until 1849 and Queenstown until 1922.

Coblenz /'kəʊblents/ alternative spelling of the German city ◊Koblenz.

Coburg /'kəʊbɜːg/ town in Bavaria, Germany, on the river Itz, 80 km/50 mi SE of Gotha; population (1984) 44,500. Industries include machinery, toys, and porcelain. Formerly the capital of the duchy of Coburg, it was part of Saxe-Coburg-Gotha 1826–1918, and a residence of its dukes.

Cochabamba /ˌkɒtʃəˈbæmbə/ city in central Bolivia, SE of La Paz; population (1988)

Clwyd

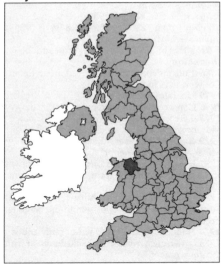

377,200. Its altitude is 2,550 m/8,370 ft; it is a centre of agricultural trading and oil refining.

Its refinery is linked by pipeline with the Camiri oilfields. It is the third largest city in Bolivia.

Cochin /'kəutʃɪn/ present-day region and former princely state lying W of the Anamalai hills in S India. It was part of Travancore-Cochin from 1949 until merged into Kerala 1956.

Cochin /'kəutʃɪn/ seaport in Kerala state, India, on the Malabar coast; population (1983) 686,000. It is a fishing port, naval training base, and an industrial centre with oil refineries; ropes and clothing are also manufactured here. It exports coir, copra, tea, and spices. Vasco da Gama established a Portuguese factory at Cochin 1502, and St Francis Xavier made it a missionary centre 1530. The Dutch held Cochin from 1663 to 1795, when it was taken by the English.

Cochin-China /'kɒtʃɪn 'tʃaɪnə/ region of SE Asia. With Cambodia it formed part of the ancient Khmer empire. In the 17th–18th centuries it was conquered by Annam. Together with Cambodia it became, 1863–67, the first part of the Indochinese peninsula to be occupied by France. Since 1949 it has been part of Vietnam.

Cocos Islands /'kəukɒs/ or *Keeling Islands* group of 27 small coral islands in the Indian Ocean, about 2,770 km/1,720 mi NW of Perth, Australia; area 14 sq km/5.5 sq mi; population (1986) 616. They are owned by Australia.

Discovered by William Keeling 1609, they were uninhabited until 1826, annexed by Britain 1857, and transferred to Australia as the Territory of Cocos (Keeling) Islands 1955. The Australian government purchased them from John Clunies-Ross 1978. In 1984 the islanders voted to become part of Australia.

Cody /'kəudi/ resort town in NW Wyoming, USA, on the Shoshone River, situated at the entrance to Yellowstone National Park; population (1990)

7,897. It was founded by William F Cody ('Buffalo Bill') 1901. There is a Museum of the Old West and the 'Buffalo Bill' Historical Center.

Cognac /'kɒnjæk/ town in Charente *département*, France, 40 km/25 mi W of Angoulême; population (1982) 21,000. Situated in a vine-growing district, Cognac has given its name to a brandy.

Bottles, corks, barrels, and crates are manufactured here.

Coimbatore /kəuˌɪmbə'tɔː/ city in Tamil Nadu, S India, on the Noyil River; population (1981) 917,000. It has textile industries and the Indian Air Force Administrative College.

Coimbra /kəu'ɪmbrə/ city in Portugal, on the Mondego River, 32 km/19 mi from the sea; population (1981) 71,800. It produces fabrics, paper, pottery, and biscuits. There is a 12th-century Romanesque cathedral incorporating part of an older mosque, and a university, founded in Lisbon 1290 and transferred to Coimbra 1537. Coimbra was the capital of Portugal 1139–1385.

Colchester /'kəultʃɪstə/ town and river port in England, on the river Colne, Essex; 80 km/50 mi NE of London; population (1989 est) 87,500. In an agricultural area, it is a market centre with clothing manufacture and engineering and printing works. The University of Essex (1961) is at Wivenhoe to the SE.

history Claiming to be the oldest town in England (Latin **Camulodunum**), Colchester dates from the time of Cymbeline (*c.* AD 10–43). It became a colony of Roman ex-soldiers AD 50, and one of the most prosperous towns in Roman Britain despite its burning by Boudicca (Boadicea) 61. Most of the Roman walls remain, as well as ruins of the Norman castle, and St Botolph's priory. Holly Tree Mansion (1718) is a museum of 18th- and 19th-century social life.

College Station /ˌkɒlɪdʒ 'steɪʃn/ city in E central Texas, USA, NW of Houston, adjoining the city of Bryan; population (1990) 52,500. Texas A & M University is here.

Colmar /'kɒlmɑː/ capital of Haut-Rhin *département*, France, between the river Rhine and the Vosges Mountains; population (1990) 64,900. It is the centre of a wine-growing and market-gardening area. Industries include engineering, food processing, and textiles. The church of St Martin is 13th–14th century, and the former Dominican monastery, now the Unterlinden Museum, contains a Grünewald altarpiece.

Cologne /kə'ləun/ (German **Köln**) industrial and commercial port in North Rhine–Westphalia, Germany, on the left bank of the Rhine, 35 km/22 mi SE of Düsseldorf; population (1988) 914,000. To the N is the Ruhr coalfield, on which many of Cologne's industries are based. They include motor vehicles, railway wagons, chemicals, and machine tools. Cologne is an important transshipment centre.

Founded by the Romans 38 BC and made a colony AD 50 under the name **Colonia Claudia Arae Agrippinensis**, it became a leading

Frankish city and during the Middle Ages was ruled by its archbishops. It was a free imperial city from 1288 until the Napoleonic age. In 1815 it passed to Prussia. The great Gothic cathedral was begun in the 13th century, but its towers were not built until the 19th century (completed 1880). Its university (1388–1797) was refounded 1919. Cologne suffered severely from aerial bombardment during World War II; 85% of the city and its three Rhine bridges were destroyed.

Colombes /kɒ'lɒmb/ suburb of Paris, France; population (1990) 79,100. It is the capital of Hauts-de-Seine *département*. Tyres, electronic equipment, and chemicals are manufactured.

Colombey-les-Deux-Eglises /ˌkɒlɒm'beɪ leɪ 'dɜːz eɪ'ɡliːz/ (French 'Colombey with the two churches') village in Haute-Marne, France, where General de Gaulle lived and was buried; population (1981) 700.

Colombia Republic of (*República de Colombia*)
area 1,141,748 sq km/440,715 sq mi
capital Bogotá
towns Medellín, Cali, Bucaramanga; ports Barranquilla, Cartagena, Buenaventura
physical the Andes mountains run N–S; flat coastland in W and plains (llanos) in E; Magdalena River runs N to Caribbean Sea; includes islands of Providencia, San Andrés, and Mapelo
features Zipaquira salt mine and underground cathedral; Lake Guatavita, source of the legend of 'El Dorado'
head of state and government Cesar Gaviria Trujillo from 1990
political system emergent democratic republic
political parties Liberal Party (PL), centrist; April 19 Movement (M-19); National Salvation Movement; Conservative Party, right-of-centre
exports emeralds (world's largest producer), coffee (world's second largest producer), cocaine (country's largest export), bananas, cotton, meat, sugar, oil, skins, hides, tobacco
currency peso
population (1990 est) 32,598,800 (mestizo 68%, white 20%, Amerindian 1%); growth rate 2.2% p.a.

life expectancy men 61, women 66; Indians 34
language Spanish
religion Roman Catholic 95%
literacy men 89%, women 87% (1987); Indians 40%
GDP $31.9 bn (1987); $1,074 per head
chronology
1886 Full independence achieved from Spain. Conservatives in power.
1930 Liberals in power.
1946 Conservatives in power.
1948 Left-wing mayor of Bogotá assassinated; widespread outcry.
1949 Start of civil war, 'La Violencia', during which over 250,000 people died.
1957 Hoping to halt the violence, Conservatives and Liberals agreed to form a National Front, sharing the presidency.
1970 National Popular Alliance (ANAPO) formed as a left-wing opposition to the National Front.
1974 National Front accord temporarily ended.
1975 Civil unrest because of disillusionment with the government.
1978 Liberals, under Julio Turbay, revived the accord and began an intensive fight against drug dealers.
1982 Liberals maintained their control of congress but lost the presidency. The Conservative president, Belisario Betancur, granted guerrillas an amnesty and freed political prisoners.
1984 Minister of justice assassinated by drug dealers; campaign against them stepped up.
1986 Virgilio Barco Vargas, Liberal, elected president by record margin.
1989 Drug cartel assassinated leading presidential candidate; Vargas declared antidrug war; bombing campaign by drug lords killed hundreds; police killed José Rodríguez Gacha, one of the most wanted cartel leaders.
1990 Cesar Gaviria Trujillo elected president. Liberals maintained control of congress.
1991 New constitution prohibited extradition of Colombians wanted for trial in other countries; several leading drug traffickers arrested. Oct: Liberal Party won general election.
1992 One of leading drug barons, Pablo Escobar, escaped from prison.

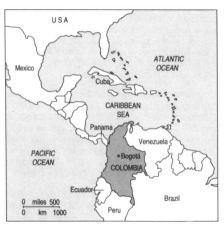

USA

Mexico

ATLANTIC OCEAN

Cuba

CARIBBEAN SEA

Panama

Venezuela

PACIFIC OCEAN

•Bogotá
COLOMBIA

Ecuador

Brazil

Peru

0 miles 500
0 km 1000

We occupy the centre of the universe and border upon every nation. Who else can claim as much?

On **Colombia** Simon Bolivar, letter of 1822

Colombo /kə'lʌmbəu/ capital and principal seaport of Sri Lanka, on the west coast near the mouth of the Kelani River; population (1990) 615,000, Greater Colombo about 1,000,000. It trades in tea, rubber, and cacao. It has iron- and steelworks and an oil refinery.

Colombo was mentioned as *Kalambu* about 1340, but the Portuguese renamed it in honour of the explorer Christopher Columbus. The Dutch seized it 1656 and surrendered it to Britain 1796. Since 1983 the chief government offices have

Colorado

been located at nearby Sri-Jayawardenapura, E of the city.

Colón /kɒ'lɒn/ second largest city in Panama, at the Caribbean end of the Panama Canal; population (1990) 140,900. It has a special economic zone (created 1948) used by foreign companies to avoid taxes on completed products in their home countries; $2 billion worth of goods passed through the zone in 1987, from dozens of countries and 600 companies. Unemployment in the city of Colón, outside the zone, was over 25% in 1991.

Founded 1850, and named *Aspinwall* 1852, it was renamed Colón 1890 in honour of the explorer Christopher Columbus.

Colón, Archipiélago de /kɒ'lɒn/ official name of the ♢Galápagos Islands.

Colorado /ˌkɒləˈrɑːdəʊ/ state of the W central USA; nicknamed Centennial State
area 269,700 sq km/104,104 sq mi
capital Denver
towns Colorado Springs, Aurora, Lakewood, Fort Collins, Greeley, Pueblo, Boulder
physical Great Plains in the E; the main ranges of the Rocky Mountains; high plateaux of the Colorado Basin in the W
features Rocky Mountain National Park; Pikes Peak; prehistoric cliff dwellings of the Mesa Verde National Park; Garden of the Gods (natural sandstone sculptures); Dinosaur and Great Sand Dunes national monuments; 'ghost' mining towns; ski resorts, including Aspen and Vail
products cereals, meat and dairy products, oil, coal, molybdenum, uranium, iron, steel, machinery
population (1990) 3,294,400
famous people Jack Dempsey, Douglas Fairbanks
history first visited by Spanish explorers in the 16th century; claimed for Spain 1706; east portion passed to the USA 1803 as part of the Louisiana Purchase, the rest 1845 and 1848 as a result of the Mexican War. It attracted fur traders, and Denver was founded following the discovery of gold 1858. Colorado became a state 1876.

Colorado /ˌkɒləˈrɑːdəʊ/ river in North America, rising in the Rocky Mountains and flowing 2,333 km/1,450 mi to the Gulf of California through Colorado, Utah, Arizona (including the Grand Canyon), and N Mexico. The many dams

along its course, including Hoover and Glen Canyon, provide power and irrigation water, but have destroyed wildlife and scenery, and very little water now reaches the sea. To the W of the river in SE California is the **Colorado Desert**, an arid area of 5,000 sq km/2,000 sq mi.

Colorado Springs city in Colorado, USA, 120 km/75 mi SE of Denver; population (1990) 281,100. At an altitude of about 1,800 m/6,000 ft, and surrounded by magnificent scenery, it was founded as a health resort 1871. A gold strike at nearby Cripple Creek 1892 aided its growth.

Colorado Springs is the home of the US Air Force Academy and the operations centre of the North American Air Defense Command.

Columbia /kəˈlʌmbiə/ river in W North America, over 1,950 km/1,218 mi long; it rises in British Columbia, Canada, and flows through Washington State, USA, to the Pacific below Astoria. It is harnessed for irrigation and power by the Grand Coulee and other great dams. It is rich in salmon.

Columbia /kəˈlʌmbiə/ capital of South Carolina, USA, on the Congaree River; population (1990) 98,100. Manufacturing includes textiles, plastics, electrical goods, fertilizers, and hosiery, but the chief product is fuel assemblies for nuclear reactors.

Columbia /kəˈlʌmbiə/ city in central Missouri, USA, NE of Jefferson City, seat of Boone County; population (1990) 69,100. It is the site of the University of Missouri 1853 and Stephens College 1833.

Columbia, District of /kəˈlʌmbiə/ seat of the federal government of the USA, coextensive with the city of Washington, situated on the Potomac River; area 178 sq km/69 sq mi. It was ceded by Maryland as the national capital site 1790.

Columbus /kəˈlʌmbəs/ capital of Ohio, USA, on the rivers Scioto and Olentangy; population (1990) 632,900. It has coalfield and natural gas resources nearby; its industries include the manufacture of cars, planes, missiles, and electrical goods.

Columbus /kəˈlʌmbəs/ city in W central Georgia, USA, SW of Macon, across the Chattahoochee River from Phenix City, Alabama; seat of Muscogee County; population (1990) 179,300. Industries include processed food, machinery, iron and steel, textiles, cotton, and peanuts. It is a distribution centre for surrounding farmlands, and lies just N of the US Army infantry base Fort Benning.

Colwyn Bay /ˈkɒlwɪn ˈbeɪ/ (Welsh *Bae Colwyn*) seaside town in Clwyd, N Wales, known as the 'garden resort of Wales'; population (1985 est) 26,000.

Communism Peak alternative form of Pik ♢Kommunizma, the highest mountain in the Pamirs.

Como /ˈkəʊməʊ/ city in Lombardy, Italy, on Lake Como at the foot of the Alps; population (1981)

95,500. Motorcycles, glass, silk, and furniture are produced here. The river Adda flows N–S through the lake, and the shores are extremely beautiful.

Como has a marble cathedral, built 1396–1732, and is a tourist resort.

Comodoro Rivadavia port in Patagonia, SE Argentina; population (1984) 120,000. Argentina's main oilfields and natural gas are nearby.

Comorin /ˈkɒmərɪn/ southernmost cape of the Indian subcontinent, in Tamil Nadu, where the Indian Ocean, Bay of Bengal, and Arabian Sea meet.

Comoros Federal Islamic Republic of (*Jumhūrīyat al-Qumur al-Itthādīyah al-Islāmīyah*)
area 1,862 sq km/719 sq mi
capital Moroni
towns Mutsamudu, Domoni, Fomboni
physical comprises the volcanic islands of Njazídja, Nzwani, and Mwali (formerly Grande Comore, Anjouan, Moheli); at N end of Mozambique Channel
features active volcano on Njazídja; poor tropical soil
head of state and government Said Mohammad Djohar (interim administration) from 1989
political system authoritarian nationalism
political parties Comoran Union for Progress (Udzima), nationalist Islamic; National Union for Congolese Democracy (UNDC), left of centre; Popular Democratic Movement (MDP), centrist
exports copra, vanilla, cocoa, sisal, coffee, cloves, essential oils
currency CFA franc
population (1990 est) 459,000; growth rate 3.1% p.a.
life expectancy men 48, women 52
languages Arabic (official), Comorian (Swahili and Arabic dialect), Makua, French
religions Muslim (official) 86%, Roman Catholic 14%
literacy 15%

GDP $198 million (1987); $468 per head
chronology
1975 Independence achieved from France, but Mayotte remained part of France. Ahmed Abdallah elected president. The Comoros joined the United Nations.
1976 Abdallah overthrown by Ali Soilih.
1978 Soilih killed by mercenaries working for Abdallah. Islamic republic proclaimed and Abdallah elected president.
1979 The Comoros became a one-party state; powers of the federal government increased.
1985 Constitution amended to make Abdallah head of government as well as head of state.
1989 Abdallah killed by French mercenaries who took control of government; under French and South African pressure, mercenaries left Comoros, turning authority over to French administration and interim president Said Mohammad Djohar.
1990 Antigovernment coup foiled.
1992 Third transitional government appointed. Antigovernment coup foiled.

Compiègne /ˌkɒmpiˈeɪn/ town in Oise *département*, France, on the river Oise near its confluence with the river Aisne; population (1990) 44,700. It has an enormous château, built by Louis XV. The armistices of 1918 and 1940 were signed (the latter by Hitler and Pétain) in a railway coach in the forest of Compiègne.

Conakry /ˌkɒnəˈkriː/ capital and chief port of the Republic of Guinea; population (1983) 705,300. It is on the island of Tumbo, linked with the mainland by a causeway and by rail with Kankan, 480 km/300 mi to the NE. Bauxite and iron ore are mined nearby.

Concepción /ˌkɒnsepsˈjɒn/ city in Chile, near the mouth of the river Bió-Bió; population (1990) 306,400. It is the capital of the province of Concepción. It is in a rich agricultural district and is also an industrial centre for coal, steel, paper, and textiles.

Concord /ˈkɒŋkəd/ capital of New Hampshire, USA, in the S central part of the state, on the Merrimack River, N of Manchester; population (1990) 36,000. Industries include granite, leather goods, electrical equipment, printed products, and wood products.

Concord /ˈkɒŋkəd/ town in Massachusetts, USA, now a suburb of Boston; population (1990) 17,300. Concord was settled 1635 and was the site of the first battle of the American Revolution, 19 April 1775. The writers Ralph Waldo Emerson, Henry Thoreau, Nathaniel Hawthorne, and Louisa May Alcott lived here.

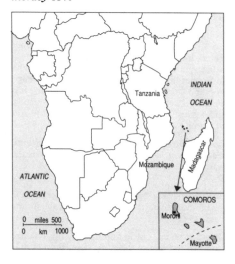

By the wide bridge that arched the flood,/ Their flag to freedom's breeze unfurled,/ Here once the embattled farmers stood,/ And fired the shot heard round the world.

On **Concord**, Massachusetts, Ralph Waldo Emerson 'Concord Hymn' 1836

Coney Island /'kəuni/ seaside resort on a peninsula in Brooklyn, in the SW of Long Island, New York, USA. It has been popular for ocean bathing and its amusement parks since the 1840s.

Congo Republic of (*République du Congo*)
area 342,000 sq km/132,012 sq mi
capital Brazzaville
towns chief port Pointe-Noire; N'Kayi, Loubomo
physical narrow coastal plain rises to central plateau, then falls into northern basin; Zaïre (Congo) River on the border with Zaire; half the country is rainforest
environment an estimated 93% of the rural population is without access to safe drinking water
features 70% of the population lives in Brazzaville, Pointe-Noire, or in towns along the railway linking these two places
head of state and government Pascal Lissouba from 1992
political system emergent democracy
political parties Pan-African Union for Social Democracy (UPADS), moderate left of centre; Congolese Labour Party (PCT), (Marxist-Leninist ideology abandoned 1990) left-wing; Union for Congolese Democracy (UDC), left of centre; National Union for Democracy and Progress (UNDP), left of centre
exports timber, petroleum, cocoa, sugar
currency CFA franc
population (1990 est) 2,305,000 (chiefly Bantu); growth rate 2.6% p.a.
life expectancy men 45, women 48
languages French (official); many African languages
religions animist 50%, Christian 48%, Muslim 2%
literacy men 79%, women 55% (1985 est)
GDP $2.1 bn (1983); $500 per head
chronology
1910 Became part of French Equatorial Africa.
1960 Achieved independence from France, with Abbé Youlou as the first president.
1963 Youlou forced to resign. New constitution approved, with Alphonse Massamba-Débat as president.
1964 The Congo became a one-party state.
1968 Military coup, led by Capt Marien Ngouabi, ousted Massamba-Débat.

1970 A Marxist state, the People's Republic of the Congo, was announced, with the PCT as the only legal party.
1977 Ngouabi assassinated. Col Yhombi-Opango became president.
1979 Yhombi-Opango handed over the presidency to the PCT, who chose Col Denis Sassou-Nguessou as his successor.
1984 Sassou-Nguessou elected for another five-year term.
1990 The PCT abandoned Marxist-Leninism and promised multiparty politics.
1991 1979 constitution suspended. Country renamed the Republic of Congo.
1992 New constitution approved and multiparty elections held, giving UPADS most assembly seats.

Congo /'kɒŋgəu/ former name (1960–71) of ◊Zaire.

Connacht /'kɒnɔːt/ province of the Republic of Ireland, comprising the counties of Galway, Leitrim, Mayo, Roscommon, and Sligo; area 17,130 sq km/6,612 sq mi; population (1991) 422,900. The chief towns are Galway, Roscommon, Castlebar, Sligo, and Carrick-on-Shannon. Mainly lowland, it is agricultural and stock-raising country, with poor land in the W.

The chief rivers are the Shannon, Moy, and Suck, and there are a number of lakes. The Connacht dialect is the national standard.

Connecticut /kə'netɪkət/ state in New England, USA; nicknamed Constitution State/Nutmeg State
area 13,000 sq km/5,018 sq mi
capital Hartford
towns Bridgeport, New Haven, Waterbury
physical highlands in the NW; Connecticut River
features Yale University; Mystic Seaport (reconstruction of 19th-century village, with restored ships)
products dairy, poultry, and market garden products; tobacco, watches, clocks, silverware, helicopters, jet engines, nuclear submarines

Connecticut

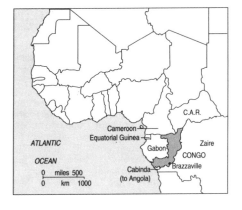

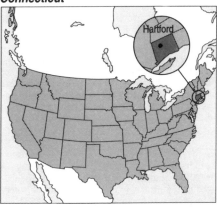

population (1990) 3,287,100
famous people Phineas T Barnum, George Bush, Katharine Hepburn, Harriet Beecher Stowe, Mark Twain
history settled by Puritan colonists from Massachusetts 1635, it was one of the Thirteen Colonies, and became a state 1788.

Connemara /ˌkɒnɪ'mɑːrə/ western part of County Galway, Republic of Ireland, an area of rocky coastline and mountainous scenery. There is fishing and tourism.

Constance /'kɒnstəns/ (German *Konstanz*) town in Baden-Württemberg, Germany, on the section of the river Rhine joining Lake Constance and Lake Untersee; population (1983) 69,100. Suburbs stretch across the frontier into Switzerland. Constance has clothing, machinery, and chemical factories, and printing works.

Constance, Lake /'kɒnstəns/ (German *Bodensee*) lake bounded by Germany, Austria, and Switzerland, through which the river Rhine flows; area 530 sq km/200 sq mi.

Constanţa /kɒn'stæntsə/ chief Romanian port on the Black Sea, capital of Constanţa region, and third largest city of Romania; population (1985) 323,000. It has refineries, shipbuilding yards, and food factories.

It is the exporting centre for the Romanian oilfields, to which it is connected by pipeline. It was founded as a Greek colony in the 7th century BC, and later named after the Roman emperor Constantine I (4th century AD). Ovid, the Roman poet, lived in exile here.

Constantine /ˌkɒnstən'tiːn/ city in NE Algeria; population (1983) 449,000. Products include carpets and leather goods. It was one of the chief towns of the Roman province of Numidia, but declined and was ruined, then restored 313 by Constantine the Great, whose name it bears. It was subsequently ruled by Vandals, Arabs, and Turks and was captured by the French 1837.

Contadora /ˌkɒntə'dɔːrə/ Panamanian island of the ▷Pearl Island group in the Gulf of Panama.

Conwy /'kɒnwɪ/ port in Wales on the river Conwy, Gwynedd; population (1981) 13,000. It was known until 1972 by the anglicized form *Conway*. Still surrounded by walls, Conwy has the ruins of a castle rebuilt by Edward I 1284.

Coober Pedy /'kuːbə 'piːdɪ/ (native Australian 'white man in a hole') town in the Great Central Desert, Australia, 700 km/437 mi NW of Adelaide; population (1976) 1,900. Opals were discovered 1915, and are mined amid a moonscape of diggings in temperatures up to 60°C/140°F.

Cooch Behar /'kuːtʃ bɪ'hɑː/ former princely state in India, merged into ▷West Bengal 1950.

Cookham-on-Thames /'kukəm ɒn 'temz/ village in Berkshire, England. The artist Stanley Spencer lived here for many years and a memorial gallery of his work was opened 1962.

Cook Islands /kuk/ group of six large and a number of smaller Polynesian islands 2,600 km/1,600 mi NE of Auckland, New Zealand; area 290 sq km/112 sq mi; population (1991) 19,000. Their main products include fruit, copra, and crafts. They became a self-governing overseas territory of New Zealand 1965.

The chief island, Rarotonga, is the site of Avarua, the seat of government. Niue, geographically part of the group, is separately administered. The Cook Islands were visited by Capt James Cook 1773, annexed by Britain 1888, and transferred to New Zealand 1901. They have common citizenship with New Zealand.

Cook, Mount /kuk/ highest point, 3,764 m/12,353 ft, of the Southern Alps, a range of mountains running through New Zealand.

Cook Strait /kuk/ strait dividing North Island and South Island, New Zealand. A submarine cable carries electricity from South to North Island.

Cooper's Creek /'kuːpə/ river, often dry, in Channel Country, SW Queensland, Australia.

Coorg /kuəg/ or *Kurg* mountainous district of the state of Karnataka in the Western Ghats of India. Formerly the princely state of Coorg, it was merged in Karnataka 1956.

Copán /kəu'pæn/ town in W Honduras; population (1983) 19,000. The nearby site of a Mayan city, including a temple and pyramids, was bought by John Stephens of the USA in the 1830s for $50.

Copenhagen /ˌkəupən'heɪgən/ (Danish *København*) capital of Denmark, on the islands of Zealand and Amager; population (1990) 1,337,100 (including suburbs).
features To the NE is the royal palace at Amalienborg; the 17th-century Charlottenburg Palace houses the Academy of Arts, and parliament meets in the Christiansborg Palace. The statue of Hans Christian Andersen's *Little Mermaid* (by Edvard Eriksen) is at the harbour entrance. The Tivoli amusement park is on the shore of the Öresund ('the Sound', between Copenhagen and S Sweden).
history Copenhagen was a fishing village until 1167, when the bishop of Roskilde built the castle on the site of the present Christiansborg palace. A settlement grew up, and it became the Danish capital 1443. The university was founded 1479. The city was under German occupation April 1940–May 1945.

Coral Sea /'kɒrəl/ or *Solomon Sea* part of the Pacific Ocean bounded by NE Australia, New Guinea, the Solomon Islands, Vanuatu, and New Caledonia. It contains numerous coral islands and reefs. The Coral Sea Islands are a territory of Australia; they comprise scattered reefs and islands over an area of about 1,000,000 sq km/386,000 sq mi. They are uninhabited except for a meteorological station on Willis Island. The Great Barrier Reef lies along its western edge, just off the east coast of Australia.

The naval battle of the Coral Sea 7–8 May 1942, which was fought between the USA and

Japan, mainly from aircraft carriers, checked the Japanese advance in the South Pacific.

Corby /'kɔːbi/ town in Northamptonshire, England; population (1988 est) 51,500. It developed in the 1930s from a village to a new town through the establishment of a steel-making industry. The steelworks closed 1979 and it is now an enterprise zone producing plastics.

Cordilleras, the /ˌkɔːdɪl'jeərəz/ mountainous western section of North America, with the Rocky Mountains and the coastal ranges parallel to the contact between the North American and the Pacific plates.

Córdoba /'kɔːdəbə/ city in central Argentina, on the Río Primero; population (1991) 1,179,000. It is the capital of Córdoba province. Main industries include cement, glass, textiles, and vehicles. Founded 1573, it has a university founded 1613, a military aviation college, an observatory, and a cathedral.

Córdoba /'kɔːdəbə/ capital of Córdoba province, Spain, on the river Guadalquivir; population (1991) 309,200. Paper, textiles, and copper products are manufactured here. It has many Moorish remains, including the mosque, now a cathedral, founded by 'Abd-ar-Rahman I 785, which is one of the largest Christian churches in the world. Córdoba was probably founded by the Carthaginians; it was held by the Moors 711–1236.

Corfe Castle /kɔːf/ village in the Isle of Purbeck, Dorset, S England, built around the ruins of a Norman castle destroyed in the Civil War. Industries include electronics and oil; tourism is important.

Corfu /kɔː'fuː/ (Greek **Kérkyra**) northernmost and second largest of the Ionian islands of Greece, off the coast of Epirus in the Ionian Sea; area 1,072 sq km/414 sq mi; population (1981) 96,500. Its businesses include tourism, fruit, olive oil, and textiles. Its largest town is the port of Corfu (Kérkyra), population (1981) 33,560. Corfu was colonized by the Corinthians about 700 BC. Venice held it 1386–1797, Britain 1815–64.

Corinth /'kɒrɪnθ/ (Greek **Kórinthos**) port in Greece, on the isthmus connecting the Peloponnese with the mainland; population (1981) 22,650. The rocky isthmus is bisected by the 6.5 km/4 mi Corinth Canal, opened 1893. The site of the ancient city-state of Corinth lies 7 km/4.5 mi SW of the port.
history Corinth was already a place of some commercial importance in the 9th century BC. At the end of the 6th century BC it joined the Peloponnesian League, and took a prominent part in the Persian and the Peloponnesian Wars. In 146 BC it was conquered by the Romans. The emperor Augustus (63 BC–AD 14) made it capital of the Roman province of Achaea. St Paul visited Corinth AD 51 and addressed two epistles to its churches. After many changes of ownership it became part of independent Greece 1822.

Corinth's ancient monuments include the ruined temple of Apollo (6th century BC).

Cork /kɔːk/ largest county of the Republic of Ireland, in the province of Munster; county town Cork; area 7,460 sq km/2,880 sq mi; population (1991) 409,800. It is agricultural, but there is also some copper and manganese mining, marble quarrying, and river and sea fishing. Natural gas and oil fields are found off the S coast at Kinsale.
It includes Bantry Bay and the village of Blarney. There is a series of ridges and vales running NE–SW across the county. The Nagles and Boggeraph mountains run across the centre, separating the two main rivers, the Blackwater and the Lee. Towns are Cobh, Bantry, Youghal, Fermoy, and Mallow.

Cork /kɔːk/ city and seaport of County Cork, on the river Lee, at the head of the long inlet of Cork harbour; population (1991) 127,000. Cork is the second port of the Republic of Ireland. The lower section of the harbour can berth liners, and the town has distilleries, shipyards, and iron foundries. St Finbarr's 7th-century monastery was the original foundation of Cork. It was eventually settled by Danes who were dispossessed by the English 1172.
University College (1845) became the University of Cork 1968. The city hall was opened 1937. There is a Protestant cathedral dedicated to the city's patron saint, St Finbarr, and a Roman Catholic cathedral of St Mary and St Finbarr.

Corniche /kɔː'niːʃ/ (French 'mountain ledge') *la Grande* (Great) *Corniche* is a road with superb alpine and coastal scenery, built between Nice and Menton, S France, by Napoleon; it rises to 520 m/1,700 ft. *La Moyenne* (Middle) and *la Petite* (Little) *Corniche* are supplementary parallel roads, the latter being nearest the coast.

Cornwall /'kɔːnwɔːl/ county in SW England including the Isles of Scilly (Scillies)

Cornwall

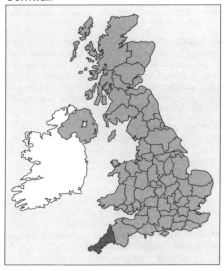

area (excluding Scillies) 3,550 sq km/1,370 sq mi
towns Truro (administrative headquarters), Camborne, Launceston; resorts of Bude, Falmouth, Newquay, Penzance, St Ives
physical Bodmin Moor (including Brown Willy 419 m/1,375 ft), Land's End peninsula, St Michael's Mount, rivers Tamar, Fowey, Fal, Camel
features Poldhu, site of first transatlantic radio signal 1901. The Stannary or Tinners' Parliament has six members from each of the four Stannary towns: Losthwithiel, Launceston, Helston, and Truro. The flag of St Piran, a white St George's cross on a black ground, is used by separatists
products electronics, spring flowers, tin (mined since Bronze Age, some workings renewed 1960s, though the industry has all but disappeared), kaolin (St Austell), fish
population (1991) 469,300
famous people John Betjeman, Humphry Davy, Daphne Du Maurier, William Golding
history the Stannary, established in the 11th century, ceased to meet 1752 but its powers were never rescinded at Westminster, and it was revived 1974 as a separatist movement.

There are more saints in Cornwall than there are in heaven.

On **Cornwall** Cornish proverb

Coromandel Coast /ˌkɒrə'mændl/ east coast of Tamil Nadu, India, between the Kistma river delta in the north and Point Calimere in the south.

Coromandel Peninsula /ˌkɒrə'mændl/ peninsula on North Island, New Zealand, E of Auckland.

Corpus Christi /ˌkɔːpəs 'krɪsti/ city and port in SE Texas, USA, on the Gulf of Mexico at the mouth of the river Nueces, SE of San Antonio; seat of Nueces County; population (1990) 257,500. Its main industries are oil refining and shipping, commercial fishing, and the processing and shipping of agricultural products. Corpus Christi was originally a small trading post.

Corregidor /kə'regɪdɔː/ island fortress off the Bataan Peninsula at the mouth of Manila Bay, Luzon, the Philippines. On 6 May 1942, Japanese forces captured Corregidor and its 10,000 US and Filipino defenders, completing their conquest of the Philippines. US forces recaptured Corregidor Feb 1945.

Corrèze /kɒ'reɪz/ river of central France flowing 89 km/55 mi from the Plateau des Millevaches, past Tulle, capital of Corrèze *département* (to which it gives its name), to join the Vézère River. It is used for generating electricity at Bar, 9.5 km/6 mi NW of Tulle.

Corrientes /ˌkɒri'entes/ city and river port of Argentina, on the Paraná River; population (1991) 267,700. Capital of Corrientes province, it is in a stock-raising district. Industries include tanning, sawmilling, and textiles.

Corse /kɔːs/ French name for ◊Corsica.

Corsica

Corsica /'kɔːsɪkə/ (French **Corse**) island region of France, in the Mediterranean off the west coast of Italy, N of Sardinia; it comprises the *départements* of Haute Corse and Corse du Sud
area 8,700 sq km/3,358 sq mi
capital Ajaccio (port)
physical mountainous; maquis vegetation
features Corsica's mountain bandits were eradicated 1931, but the tradition of the vendetta or blood feud lingers. The island is the main base of the Foreign Legion
government its special status involves a 61-member regional parliament with the power to scrutinize French National Assembly bills applicable to the island and propose amendments
products wine, olive oil
population (1986) 249,000, including just under 50% native Corsicans. There are about 400,000 *émigrés*, mostly in Mexico and Central America, who return to retire
languages French (official); the majority speak Corsican, an Italian dialect
famous people Napoleon
history the Phocaeans of Ionia founded Alalia about 570 BC, and were succeeded in turn by the Etruscans, the Carthaginians, the Romans, the Vandals, and the Arabs. In the 14th century Corsica fell to the Genoese, and in the second half of the 18th century a Corsican nationalist, Pasquale Paoli (1725–1807), led an independence movement. Genoa sold Corsica to France 1768. In World War II Corsica was occupied by Italy 1942–43. From 1962, French *pieds noirs* (refugees from Algeria, mainly vine growers, were settled in Corsica, and their prosperity helped to fan nationalist feeling, which demands an independent Corsica. This fuelled the National Liberation Front of Corsica (FNLC), banned 1983, which has engaged in some terrorist bombings (a truce began June 1988 but ended Jan 1991).

Cortona /kɔː'təʊnə/ town in Tuscany, N Italy, 22 km/13 mi SE of Arezzo; population (1981)

22,000. It is one of Europe's oldest cities, encircled by walls built by the Etruscans, and has a medieval castle and an 11th-century cathedral.

Corunna /kɒ'rʌnə/ (Spanish *La Coruña*) city in the extreme NW of Spain; population (1991) 251,300. It is the capital of Corunna province. Industry is centred on the fisheries; tobacco, sugar refining, and textiles are also important. The Armada sailed from Corunna 1588, and the town was sacked by Francis Drake 1589.

Cos /kɒs/ alternative spelling of ⟡Kos, a Greek island.

Cosenza /kəʊ'zentsə/ town in Calabria, S Italy, at the junction of the rivers Crati and Busento; population (1989) 105,300. It is the capital of Cosenza province and is an archiepiscopal see. Alaric, king of the Visigoths, is buried here.

Cossyra /kɒ'saɪrə/ ancient name for ⟡Pantelleria, Italian island in the Mediterranean.

Costa Blanca /ˌkɒstə 'blæŋkə/ (Spanish 'White Coast') Mediterranean coastline of Alicante and Murcia provinces, E Spain. The chief ports are Alicante and Cartagena; popular resorts include Benidorm and Torrevieja.

Costa Brava /ˌkɒstə 'brɑːvə/ (Spanish 'Wild Coast') Mediterranean coastline of NE Spain, stretching from Port-Bou on the French border southwards to Blanes, NE of Barcelona. It is noted for its irregular rocky coastline, small fishing villages, and resorts such as Puerto de la Selva, Palafrugell, Playa de Aro, and Lloret del Mar.

Costa del Sol /ˌkɒstə del 'sɒl/ (Spanish 'Coast of the Sun') Mediterranean coastline of Andalucia, S Spain, stretching for nearly 300 km/190 mi from Algeciras to Almeria. Málaga is the principal port and Marbella, Torremolinos, and Nerja are the chief tourist resorts.

Costa Rica Republic of (*República de Costa Rica*)
area 51,100 sq km/19,735 sq mi
capital San José
towns ports Limón, Puntarenas
physical high central plateau and tropical coasts; Costa Rica was once entirely forested, containing

an estimated 5% of the Earth's flora and fauna. By 1983 only 17% of the forest remained; half of the arable land had been cleared for cattle ranching, which led to landlessness, unemployment (except for 2,000 politically powerful families), and soil erosion; the massive environmental destruction also caused incalculable loss to the gene pool
environment one of the leading centres of conservation in Latin America, with more than 10% of the country protected by national parks, and tree replanting proceeding at a rate of 150 sq km/60 sq mi per year
features Poas Volcano; pre-Colombian ceremonial site of Guayabo
head of state and government Rafael Calderón from 1990
political system liberal democracy
political parties National Liberation Party (PLN), left of centre; Christian Socialist Unity Party (PUSC), centrist coalition; ten minor parties
exports coffee, bananas, cocoa, sugar, beef
currency colón
population (1990 est) 3,032,000 (including 1,200 Guaymi Indians); growth rate 2.6% p.a.
life expectancy men 71, women 76
language Spanish (official)
religion Roman Catholic 95%
literacy men 94%, women 93% (1985 est)
GDP $4.3 bn (1986); $1,550 per head
chronology
1821 Independence achieved from Spain.
1949 New constitution adopted. National army abolished. José Figueres, cofounder of the PLN, elected president; he embarked on ambitious socialist programme.
1958–73 Mainly conservative administrations.
1974 PLN regained the presidency and returned to socialist policies.
1978 Rodrigo Carazo, conservative, elected president. Sharp deterioration in the state of the economy.
1982 Luis Alberto Monge (PLN) elected president. Harsh austerity programme introduced to rebuild the economy. Pressure from the USA to abandon neutral stance and condemn Sandinista regime in Nicaragua.
1983 Policy of neutrality reaffirmed.
1985 Following border clashes with Sandinista forces, a US-trained antiguerrilla guard formed.
1986 Oscar Arias Sánchez won the presidency on a neutralist platform.
1987 Oscar Arias Sánchez won Nobel Prize for Peace for devising a Central American peace plan.
1990 Rafael Calderón (PUSC) elected president.

When a man has travelled all the way to Costa Rica, he does expect something strange.

On **Costa Rica** Anthony Trollope
The West Indies and the Spanish Main 1859

Côte d'Azur /ˈkəʊt dæ'zjʊə/ Mediterranean coast from Menton to St Tropez, France, renowned for its beaches; it is part of the region Provence-Alpes-Côte d'Azur.

Cotonou /,kɒtə'nuː/ chief port and largest city of Benin, on the Bight of Benin; population (1982) 487,000. Palm products and timber are exported. Although not the official capital, it is the seat of the president, and the main centre of commerce and politics.

Cotopaxi /,kɒtə'pæksi/ (Quechua 'shining peak') active volcano, situated to the S of Quito in Ecuador. It is 5,897 m/19,347 ft high and was first climbed 1872.

Cotswold Hills /'kɒtswəuld/ or **Cotswolds** range of hills in Avon and Gloucestershire, England, 80 km/50 mi long, between Bath and Chipping Camden. They rise to 333 m/1,086 ft at Cleeve Cloud, near Cheltenham, but average about 200 m/600 ft. The area is known for its picturesque villages, built with the local honey-coloured stone.

Cottbus /'kɒtbus/ industrial city (textiles, carpets, glassware) in the state of Brandenburg, Germany; population (1990) 128,000. It was formerly the capital of the East German district of Cottbus 1952–90.

Courtrai /kuə'treɪ/ (Flemish *Kortrijk*) town in Belgium on the river Lys, in West Flanders; population (1991) 76,100. It is connected by canal with the coast, and by river and canal with Antwerp and Brussels. It has a large textile industry, including damask, linens, and lace.

Covent Garden /'kɒvənt 'gɑːdn/ London square (named from the convent garden once on the site) laid out by Inigo Jones 1631. The buildings that formerly housed London's fruit and vegetable market (moved to Nine Elms, Wandsworth 1973) have been adapted for shops and restaurants. The Royal Opera House, also housing the Royal Ballet, is here; also the London Transport Museum.

The Theatre Museum, opened 1987, is in the Old Flower Market.

Coventry /'kɒvəntri/ industrial city in West Midlands, England; population (1984 est) 314,100. Manufacturing includes cars, electronic equipment, machine tools, and agricultural machinery. The poet Philip Larkin was born here.

features Coventry cathedral, opened 1962, was designed by Basil Spence and retains the ruins of the old cathedral, which was destroyed in an air raid Nov 1940; St Mary's Hall, built 1394–1414 as a guild centre; two gates of the old city walls (1356); Belgrade Theatre (1958); Coventry Art Gallery and Museum; Museum of British Road Transport; and Lanchester Polytechnic.

history The city originated when Leofric, Earl of Mercia and husband of Lady Godiva, founded a priory 1043. Industry began with bicycle manufacture 1870.

Cowes /kauz/ seaport and resort on the north coast of the Isle of Wight, England, on the Medina estuary, opposite Southampton Water; population (1981) 19,600. It is the headquarters of the Royal Yacht Squadron, which holds the annual Cowes Regatta, and has maritime industries. In East Cowes is Osborne House, once a residence of Queen Victoria, now used as a museum.

Cracow /'krækau/ alternative form of ◊Kraków, a Polish city.

Craigavon /,kreɪg'ævən/ town in Armagh, Northern Ireland; population (1981) 10,195. It was created from 1965 by the merging of Lurgan and Portadown and named after James Craigavon, the first prime minister of Northern Ireland (1921–40).

Craiova /kraɪ'əuvə/ town in S Romania, near the river Jiu; population (1985) 275,000. Industries include electrical engineering, food processing, textiles, fertilizers, and farm machinery.

Crater Lake /'kreɪtə 'leɪk/ lake in the crater which forms the remains of Mount Mazama in SW Oregon, USA.

Crawley /'krɔːli/ town in West Sussex, England, NE of Horsham; population (1981) 73,000. It was chartered by King John 1202 and developed as a 'new town' from 1946. Industries include plastics, engineering, and printing.

Cremona /krɪ'məunə/ city in Lombardy, Italy, on the river Po, 72 km/45 mi SE of Milan; population (1981) 81,000. It is the capital of Cremona province. Once a violin-making centre, it now produces food products and textiles. It has a 12th-century cathedral.

Crete /kriːt/ (Greek *Kríti*) largest Greek island in the E Mediterranean Sea, 100 km/62 mi SE of mainland Greece
area 8,378 sq km/3,234 sq mi
capital Khaniá (Canea)
towns Iráklion (Heraklion), Rethymnon, Aghios Nikolaos
products citrus fruit, olives, wine
population (1991) 536,900
language Cretan dialect of Greek
history it has remains of the Minoan civilization 3000–1400 BC and was successively under

Crete

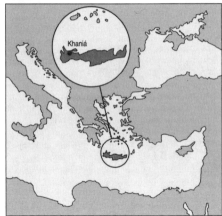

Roman, Byzantine, Venetian, and Turkish rule. The island was annexed by Greece 1913.

In 1941 it was captured by German forces from Allied troops who had retreated from the mainland and was retaken by the Allies 1944.

The people of Crete unfortunately make more history than they can consume locally.

On **Crete** Saki (H H Munro) 'The Jesting of Arlington Stringham' 1911

Creuse /krɜːz/ river in central France flowing 255 km/158 mi generally N from the Plateau des Millevaches to the Vienne River. It traverses Creuse *département*, to which it gives its name.

Crewe /kruː/ town in Cheshire, England; population (1981) 59,300. It owed its growth to its position as a railway junction; the chief construction workshops of British Rail are here. It is the centre of the dairy industry, providing cattle breeding, management, and animal health services. Other occupations include chemical works, clothing factories, and vehicle manufacture.

Crimea /kraɪˈmɪə/ northern peninsula on the Black Sea, an autonomous republic of Ukraine; formerly a region (1954–91)
area 27,000 sq km/10,425 sq mi
capital Simferopol
towns Sevastopol, Yalta
features mainly steppe, but southern coast is a holiday resort
products iron, oil
population 2.5 million (70% Russian, despite return of 150,000 Tatars since 1989)
history Crimea was under Turkish rule 1475–1774; a subsequent brief independence was ended by Russian annexation 1783. Crimea was the republic of Taurida 1917–20 and the Crimean Autonomous Soviet Republic from 1920 until occupied by Germany 1942–1944. It was then reduced to a region, its Tatar people being deported to Uzbekistan for collaboration. Although they were exonerated 1967 and some were allowed to return, others were forcibly re-exiled 1979. A drift back to their former homeland began 1987 and a federal ruling 1988 confirmed their right to residency. Since 1991 Crimea has sought to gain independence from the Ukraine; the latter has resisted all secessionist moves.

In a referendum organized by the regional soviet (council) 1991, citizens of the Crimean peninsula voted overwhelmingly in favour of restoring Crimea as an autonomous republic independent of the Ukraine. This referendum was opposed by representatives of the Tatars, who sought restriction of the voting solely to their community, and by the Ukrainian nationalist group Rukh. In Feb 1991, the Ukrainian Supreme Soviet voted to restore to the Crimea the status of an autonomous Soviet socialist republic within the Ukraine. In Sept 1991,

Crimea declared its independence but this was not recognized by Ukraine. In May 1992, the Crimean parliament declared the republic's sovereignty subject to a regional referendum planned for Aug. Ukrainian president Kravchuk responded by authorizing the use of 'all necessary means' to prevent Crimea's secession and demanded that the declaration be rescinded. The Crimean parliament acceded but declared its intention to go ahead with the referendum. Also in May, Russia voted to nullify the 1945 transfer of Crimea to the Ukraine and called for bilateral talks on the republic's status.

Whether Paradise is formed from the plans of the south coast of the Crimea, or vice versa, I don't know, but they must be from the same design.

On the **Crimea** John Foster Fraser *Round the World on a Wheel* 1899

Croagh Patrick /krəʊˈpætrɪk/ holy mountain rising to 765 m/2,510 ft in County Mayo, W Ireland, one of the three national places of pilgrimage in Ireland (with Lough Derg and Knock). An annual pilgrimage on the last Sunday of July commemorates St Patrick who fasted there for the 40 days of Lent AD 440.

Croatia Republic of
area 56,538 sq km/21,824 sq mi
capital Zagreb
towns chief port: Rijeka (Fiume); other ports: Zadar, Sibenik, Split, Dubrovnik
physical Adriatic coastline with large islands; very mountainous, with part of the Karst region and the Julian and Styrian Alps; some marshland
features popular sea resorts along the extensive Adriatic coastline
head of state Franjo Tudjman from 1990
head of government Nikica Valentic from 1993
political system emergent democracy
political parties Croatian Democratic Union (HDZ), right-wing, nationalist; Coalition of

National Agreement, centrist; Communist Party, reform-communist
products cereals, potatoes, tobacco, fruit, livestock, metal goods, textiles
currency Croatian dinar
population (1990) 4,760,000 including 75% Croats, 12% Serbs, and 1% Slovenes
language Croatian variant of Serbo-Croatian
religions Roman Catholic (Croats); Orthodox Christian (Serbs)
GNP $7.9 bn (1990); $1,660 per head
chronology
1918 Became part of the kingdom which united the Serbs, Croats, and Slovenes.
1929 The kingdom of Croatia, Serbia, and Slovenia became Yugoslavia. Croatia continued its campaign for autonomy.
1941 Became a Nazi puppet state following German invasion.
1945 Became constituent republic of Yugoslavia.
1970s Separatist demands resurfaced. Crackdown against anti-Serb separatist agitators.
1989 Formation of opposition parties permitted.
1990 April–May: Communists defeated by Tudjman-led Croatian Democratic Union (HDZ) in first free election since 1938. Sept: 'sovereignty' declared. Dec: new constitution adopted.
1991 Feb: assembly called for Croatia's secession. March: Serb-dominated Krajina announced secession from Croatia. June: Croatia declared independence; military conflict with Serbia; internal civil war ensued. July onwards: civil war intensified. Oct: Croatia formally seceded from Yugoslavia.
1992 Jan: United Nations peace accord reached in Sarajevo; Croatia's independence recognized by the European Community. March–April: UN peacekeeping forces drafted into Croatia. April: independence recognized by USA. May: became a member of the United Nations. Sept: Tudjman requested withdrawal of UN forces on expiry of mandate 1993.

Croydon /'krɔɪdn/ borough of S London, England; it includes the suburbs of Croydon, Purley, and Coulsdon
features 11th-century Lanfranc's palace, former residence of archbishops of Canterbury; Ashcroft Theatre, founded 1962; overspill office development from central London
industries engineering, electronics, foodstuffs, pharmaceuticals
population (1991) 290,600.

CT abbreviation for ◊*Connecticut*, a state of the USA.

Cuba Republic of (*República de Cuba*)
area 110,860 sq km/42,820 sq mi
capital Havana
towns Santiago de Cuba, Camagüey
physical comprises Cuba, the largest and westernmost of the West Indies, and smaller islands including Isle of Youth; low hills; Sierra Maestra mountains in SE
features 3,380 km/2,100 mi of coastline, with deep bays, sandy beaches, coral islands and reefs;

more than 1,600 islands surround the Cuban mainland
head of state and government Fidel Castro Ruz from 1959
political system communist republic
political party Communist Party of Cuba (PCC), Marxist-Leninist
exports sugar, tobacco, coffee, nickel, fish
currency Cuban peso
population (1990 est) 10,582,000; 37% are white of Spanish descent, 51% mulatto, and 11% are of African origin; growth rate 0.6% p.a.
life expectancy men 72, women 75
language Spanish
religions Roman Catholic 85%; also Episcopalians and Methodists
literacy men 96%, women 95% (1988)
disposable national income $15.8 bn (1983); $1,590 per head
chronology
1492 Christopher Columbus landed in Cuba and claimed it for Spain.
1898 USA defeated Spain in Spanish-American War; Spain gave up all claims to Cuba.
1901 Cuba achieved independence; Tomás Estrada Palma became first president of the Republic of Cuba.
1933 Fulgencia Batista seized power.
1944 Batista retired.
1952 Batista seized power again to begin an oppressive regime.
1953 Fidel Castro led an unsuccessful coup against Batista.
1956 Second unsuccessful coup by Castro.
1959 Batista overthrown by Castro. Constitution of 1940 replaced by a 'Fundamental Law', making Castro prime minister, his brother Raúl Castro his deputy, and Che Guevara his number three.
1960 All US businesses in Cuba appropriated without compensation; USA broke off diplomatic relations.
1961 USA sponsored an unsuccessful invasion at the Bay of Pigs. Castro announced that Cuba had become a communist state, with a Marxist-Leninist programme of economic development.
1962 Cuba expelled from the Organization of

American States. Soviet nuclear missiles installed but subsequently removed from Cuba at US insistence.
1965 Cuba's sole political party renamed Cuban Communist Party (PCC). With Soviet help, Cuba began to make considerable economic and social progress.
1972 Cuba became a full member of the Moscow-based Council for Mutual Economic Assistance.
1976 New socialist constitution approved; Castro elected president.
1976–81 Castro became involved in extensive international commitments, sending troops as Soviet surrogates, particularly to Africa.
1984 Castro tried to improve US-Cuban relations by discussing exchange of US prisoners in Cuba for Cuban 'undesirables' in the USA.
1988 Peace accord with South Africa signed, agreeing to withdrawal of Cuban troops from Angola.
1989 Reduction in Cuba's overseas military activities.
1991 Soviet troops withdrawn.
1992 Castro affirmed continuing support of communism.
1993 First direct parliamentary vote held. Communist seats were uncontested and all candidates won their seats by required majority.

Cubango /ku'bæŋgəʊ/ Portuguese name for the ◊Okavango Swamp in Africa.

Cúcuta /'ku:kətə/ capital of Norte de Santander department, NE Colombia; population (1985) 379,000. It is situated in a tax-free zone close to the Venezuelan border, and trades in coffee, tobacco, and cattle. It was a focal point of the independence movement and meeting place of the first Constituent Congress 1821.

Cuenca /'kweŋkə/ city in S Ecuador; population (1991) 193,000. It is the capital of Azuay province. Industries include chemicals, food processing, agricultural machinery, and textiles. It was founded by the Spanish 1557.

Cuenca /'kweŋkə/ city in Castilla–La Mancha, Spain, at the confluence of the rivers Júcar and Huécar; 135 km/84 mi SE of Madrid; population (1991) 45,800. It is the capital of Cuenca province. It has a 13th-century cathedral.

Cuiaba /ˌku:jə'ba:/ town in Brazil, on the river Cuiaba; population (1991) 389,000. It is the capital of Mato Grosso state. Gold and diamonds are worked nearby.

Culham /'kʌləm/ village near Oxford, England; site of a British nuclear research establishment.

Culiacán (Spanish *Culiacán Rosales*) capital of Sinaloa state, NW Mexico, on the Culiacán River; population (1980) 560,000. It trades in vegetables and textiles. It was founded 1599.

Cumberland /'kʌmbələnd/ former county of NW England, merged with Cumbria 1974.

After the Roman withdrawal, Cumberland became part of Strathclyde, a British kingdom. In 945 it passed to Scotland, in 1157 to England, and until the union of the English and Scottish crowns 1603 Cumberland was the scene of frequent battles between the two countries.

Cumberland /'kʌmbələnd/ city in NW Maryland, USA, in the Allegheny Mountains, on the Potomac River, directly S of Johnstown, Pennsylvania; seat of Allegheny County; population (1990) 23,700. Its industries include the mining and shipping of coal, sheet metal, iron products, and tyres. It was first an Indian village and then a trading post and fort before it was incorporated 1815.

Cumbernauld /ˌkʌmbə'nɔːld/ new town in Strathclyde, Scotland, 18 km/11 mi from Glasgow; population (1981) 48,000. It was founded 1955 to take in city overspill. In 1966 it won a prize as the world's best-designed community.

Cumbria /'kʌmbriə/ county in NW England
area 6,810 sq km/2,629 sq mi
towns Carlisle (administrative headquarters), Barrow, Kendal, Whitehaven, Workington, Penrith
physical Lake District National Park, including Scafell Pike 978 m/3,210 ft, the highest mountain in England; Helvellyn 950 m/3,118 ft; Lake Windermere, the largest lake in England, 17 km/10.5 mi long, 1.6 km/1 mi wide; other lakes (Derwentwater, Ullswater)
features Grizedale Forest sculpture project; Furness peninsula; atomic stations at Calder Hall and Sellafield (reprocessing plant), formerly Windscale
products the traditional coal, iron, and steel industries of the coast towns have been replaced by newer industries including chemicals, plastics, and electronics; in the N and E there is dairying, and West Cumberland Farmers is the country's largest agricultural cooperative

Cumbria

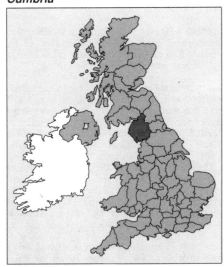

population (1991) 486,900
famous people birthplace of William Wordsworth at Cockermouth, and home at Grasmere; homes of Samuel Taylor Coleridge and Robert Southey at Keswick; John Ruskin's home, Brantwood, on Coniston Water; Thomas de Quincey; Beatrix Potter.

Cunene /kuː'neɪni/ or *Kunene* river rising near Nova Lisboa in W central Angola. It flows S to the frontier with Namibia, then W to the Atlantic Ocean; length 250 km/150 mi.

Curaçao /ˌkjuərə'səu/ island in the West Indies, one of the Netherlands Antilles; area 444 sq km/171 sq mi; population (1988) 148,500. The principal industry, dating from 1918, is the refining of Venezuelan petroleum.

Curaçao was colonized by Spain 1527, annexed by the Dutch West India Company 1634, and gave its name from 1924 to the group of islands renamed the Netherlands Antilles 1948. Its capital is the port of Willemstad.

Curitiba /ˌkuərɪ'tiːbə/ city in Brazil, on the Curitiba River; population (1991) 1,248,300. The capital of Paraná state, it dates from 1654 and makes paper, furniture, textiles, and chemicals. Coffee, timber, and maté (a beverage) are exported.

Curragh, the /'kʌrə/ horse-racing course on the Curragh plain in County Kildare, Republic of Ireland, where all five Irish Classic races are run. At one time used for hurdle races, it is now used for flat racing only. It is also the site of the national stud.

Racing has been held at the Curragh since the mid-1880s. The course is right-handed and in the shape of a horseshoe.

Cuttack /kʌ'tæk/ city and river port in E India, on the Mahanadi River delta; population (1981) 327,500. It was the capital of Orissa state until 1950. The old fort (Kataka) from which the town takes its name is in ruins.

Cuxhaven /'kukshɑːvən/ seaport in Germany on the southern side of the Elbe estuary, at its entrance into the North Sea; population (1983) 57,800. It acts as an outport for Hamburg.

Cuzco /'kuskəu/ city in S Peru, capital of Cuzco department, in the Andes Mountains, over 3,350 m/11,000 ft above sea level and 560 km/350 mi SE of Lima; population (1988) 255,000. It was founded in the 11th century as the ancient capital of the Inca empire and was captured by the Spanish conqueror Francisco Pizarro 1533.

The university was founded 1598. The city has a Renaissance cathedral and other relics of the early Spanish conquerors. There are many Inca remains and in the 1970s and 1980s the Inca irrigation canals and terracing nearby were being restored to increase cultivation.

Cwmbran /kum'brɑːn/ (Welsh 'Vale of the Crow') town in Wales, NW of Newport, on the Afon

Lywel, a tributary of the river Usk; population (1981) 45,000. It is the administrative headquarters of Gwent. It was established 1949 to provide a focus for new industrial growth in a depressed area, producing scientific instruments, car components, nylon, and biscuits.

Cyclades /'sɪklədiːz/ (Greek *Kikládhes*) group of about 200 Greek islands in the Aegean Sea, lying between mainland Greece and Turkey; area 2,579 sq km/996 sq mi; population (1981) 88,500. They include Andros, Melos, Paros, Naxos, and Siros, on which is the capital Hermoupolis.

Cymru /'kʌmri/ Welsh name for ◊Wales.

Cyprus *Greek Republic of Cyprus* (*Kypriakí Dimokratía*) in the south, and *Turkish Republic of Northern Cyprus* (*Kibris Cumhuriyeti*) in the north
area 9,251 sq km/3,571 sq mi, 37% in Turkish hands
capital Nicosia (divided between Greeks and Turks)
towns ports Limassol, Larnaca, Paphos (Greek); Morphou, and ports Kyrenia and Famagusta (Turkish)
physical central plain between two E–W mountain ranges
features archaeological and historic sites; Mount Olympus 1,953 m/6,406 ft (highest peak); beaches
heads of state and government Glafkos Clerides (Greek) from 1993, Rauf Denktaş (Turkish) from 1976
political system democratic divided republic
political parties Democratic Front (DIKO), centre-left; Progressive Party of the Working People (AKEL), socialist; Democratic Rally (DISY), centrist; Socialist Party (EDEK), socialist; *Turkish zone*: National Unity Party (NUP), Communal Liberation Party (CLP), Republican Turkish Party (RTP), New British Party (NBP)
exports citrus, grapes, raisins, Cyprus sherry, potatoes, clothing, footwear

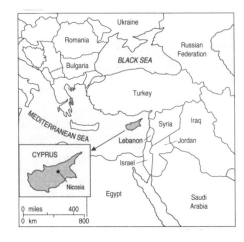

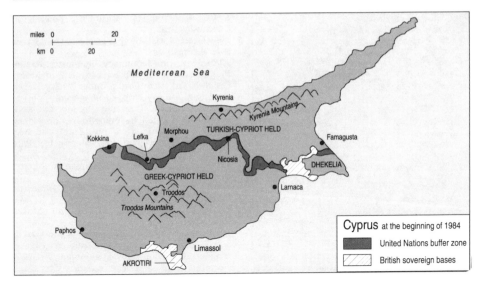

Cyprus at the beginning of 1984
- United Nations buffer zone
- British sovereign bases

currency Cyprus pound and Turkish lira
population (1990 est) 708,000 (Greek Cypriot 78%, Turkish Cypriot 18%); growth rate 1.2% p.a.
life expectancy men 72, women 76
languages Greek and Turkish (official), English
religions Greek Orthodox 78%, Sunni Muslim 18%
literacy 99% (1984)
GDP $3.7 bn (1987); $5,497 per head
chronology
1878 Came under British administration.
1955 Guerrilla campaign began against the British for enosis (union with Greece), led by Archbishop Makarios and General Grivas.
1956 Makarios and enosis leaders deported.
1959 Compromise agreed and Makarios returned to be elected president of an independent Greek-Turkish Cyprus.
1960 Independence achieved from Britain, with Britain retaining its military bases.
1963 Turks set up their own government in northern Cyprus. Fighting broke out between the two communities.
1964 United Nations peacekeeping force installed.
1971 Grivas returned to start a guerrilla war against the Makarios government.
1974 Grivas died. Military coup deposed Makarios, who fled to Britain. Nicos Sampson appointed president. Turkish army sent to northern Cyprus to confirm Turkish Cypriots' control; military regime in southern Cyprus collapsed; Makarios returned. Northern Cyprus declared itself the Turkish Federated State of Cyprus (TFSC), with Rauf Denktaş as president.
1977 Makarios died; succeeded by Spyros Kyprianou.
1983 An independent Turkish Republic of Northern Cyprus proclaimed but recognized only by Turkey.
1984 UN peace proposals rejected.
1985 Summit meeting between Kyprianou and Denktaş failed to reach agreement.

1988 Georgios Vassiliou elected president. Talks with Denktas began, under UN auspices.
1989 Peace talks abandoned.
1992 UN-sponsored peace talks collapsed.
1993 Democratic Rally leader Glafkos Clerides narrowly won presidential election.

Czechoslovakia former country in E central Europe, which came into existence as an independent republic 1918 after the break-up of the Austro–Hungarian empire at the end of World War I. It consisted originally of the Bohemian crownlands (Bohemia, Moravia, and part of Silesia) and Slovakia, the area of Hungary inhabited by Slavonic peoples. Czechoslovakia reverted to two independent states, the Czech Republic and Slovak Republic, Jan 1993.

Czech Republic (*Česká Republika*)
area 78,864 sq km/30,461 sq mi
capital Prague
towns Brno, Ostrava, Olomouc, Liberec, Plzeň, Ustí nad Labem, Hradec Králové
physical mountainous; rivers: Morava, Labe (Elbe), Vltava (Moldau)
environment considered in 1991 to be the most polluted country in E Europe. Pollution is worst in N Bohemia, which produced 70% of Czechoslovakia's coal and 45% of its coal-generated electricity. Up to 20 times the permissible level of sulphur dioxide is released over Prague, where 75% of the drinking water fails to meet the country's health standards
features summer and winter resort areas in Western Carpathian, Bohemian, and Sudetic mountain ranges
head of state Václav Havel from 1993
head of government Václav Klaus from 1993
political system emergent democracy
political parties Civic Democratic Party (CDP), right of centre; Civic Movement (CM), left of centre; Communist Party (CPCZ), left-wing; Czechoslovak People's Party, centrist nationalist

exports machinery, vehicles, coal, iron and steel, chemicals, glass, ceramics, clothing
currency new currency based on koruna
population (1991) 10,298,700 (with German and other minorities); growth rate 0.4% p.a.
life expectancy men 68, women 75
languages Czech (official)
religions Roman Catholic (75%), Protestant, Hussite, Orthodox
literacy 100%
GDP $26,600 million (1990); $2,562 per head
chronology
1526–1918 Under Habsburg domination.
1918 Independence achieved from Austro-Hungarian Empire; Czechs joined Slovaks in forming Czechoslovakia as independent nation.

1948 Communists assumed power in Czechoslovakia.
1969 Czech Socialist Republic created under new federal constitution.
1989 Nov: pro-democracy demonstrations in Prague; new political parties formed, including Czech-based Civic Forum under Václav Havel; Communist Party stripped of powers; political parties legalized. Dec: new 'grand coalition' government formed, including former dissidents; Havel appointed state president. Amnesty granted to 22,000 prisoners; calls for USSR to withdraw troops.
1990 Havel re-elected president in multiparty elections.
1991 Civic Forum split into Civic Democratic Party (CDP) and Civic Movement (CM); evidence of increasing Czech and Slovak separatism.
1992 June: Václav Klaus, leader of the Czech-based CDP, became prime minister; Havel resigned following Slovak gains in assembly elections. Aug: creation of separate Czech and Slovak states agreed.
1993 Jan: Czech Republic became sovereign state, with Klaus as prime minister. Havel elected president of the new republic. Admitted into United Nations, Conference on Security and Cooperation in Europe, and Council of Europe.

Częstochowa /ˌtʃenstə'xəuvə/ city in Poland, on the river Vistula, 193 km/120 mi SW of Warsaw; population (1990) 257,900. It produces iron goods, chemicals, paper, and cement. The basilica of Jasna Góra is a centre for Catholic pilgrims (it contains the painting known as the Black Madonna).

Dacca /'dækə/ alternative name for ◊Dhaka, the capital of Bangladesh.

Dacia /'deɪsɪə/ ancient region forming much of modern Romania. The various Dacian tribes were united around 60 BC, and for many years posed a threat to the Roman Empire; they were finally conquered by the Roman emperor Trajan AD 101–06, and the region became a province of the same name. It was abandoned to the invading Goths about 275.

Dadra and Nagar Haveli /də'drɑː 'nəgə'veli/ a Union Territory of W India since 1961; capital Silvassa; area 490 sq km/189 sq mi; population (1991) 138,500. It was formerly part of Portuguese Daman. It produces rice, wheat, millet, and timber.

Dagestan /ˌdægɪ'stɑːn/ autonomous republic of S Russia, situated E of the Caucasus, bordering the Caspian Sea; capital Makhachkala; area 50,300 sq km/19,421 sq mi; population (1982) 1,700,000. It is mountainous, with deep valleys, and its numerous ethnic groups speak a variety of distinct languages. Annexed 1723 from Iran, which strongly resisted Russian conquest, it became an autonomous republic 1921.

Dairen /ˌdaɪ'ren/ former name for the Chinese port of ◊Dalian, part of ◊Lüda.

Dakar /'dækɑː/ capital and chief port (with artificial harbour) of Senegal; population (1984) 1,000,000. It is an industrial centre, and there is a university, established 1957.
  Founded 1862, it was formerly the seat of government of French West Africa. In July 1940 an unsuccessful naval action was undertaken by British and Free French forces to seize Dakar as an Allied base.

Dakhla /'dɑːxlə/ port in Western Sahara; population (1982) 17,800. First established as a Spanish trading port 1476, it was known as *Villa Cisneros*.

Dakota /də'kəutə/ see ◊North Dakota and ◊South Dakota.

Dales or *Yorkshire Dales* series of river valleys in N England, running E from the Pennines in West Yorkshire. The principal valleys are Airedale, Nidderdale, Swaledale, Teesdale, Wensleydale, and Wharfedale. The Dales are highly scenic and popular with walkers; dry stone walls and barns are a regular feature of the landscape.

Dalian /ˌdɑːli'æn/ one of the two cities comprising the Chinese port of ◊Lüda.

Dallas /'dæləs/ commercial city in Texas, USA; population (1990) 1,006,900, metropolitan area (with Fort Worth) 3,885,400. Industries include banking, insurance, oil, aviation, aerospace, and electronics. Dallas–Fort Worth Regional Airport (opened 1973) is one of the world's largest. John F Kennedy was assassinated here 1963.
  Dallas is a cultural centre, with a symphony orchestra, opera, ballet, and theatre; there is an annual Texas State Fair. Southern Methodist University is here. Founded as a trading post 1844, it developed as the focus of a cotton area and then as a mineral and oil-producing centre, with banking and insurance operations. After World War II, growth increased rapidly.

Dalmatia /dæl'meɪʃə/ region divided between Croatia, Montenegro in Yugoslavia, and Bosnia-Herzegovina. The capital is Split. It lies along the eastern shore of the Adriatic Sea and includes a number of islands. The interior is mountainous. Important products are wine, olives, and fish. Notable towns in addition to the capital are Zadar, Sibenik, and Dubrovnik.
history Dalmatia became Austrian 1815 and by the treaty of Rapallo 1920 became part of the kingdom of the Serbs, Croats, and Slovenes (Yugoslavia from 1931), except for the town of Zadar (Zara) and the island of Lastovo (Lagosta), which, with neighbouring islets, were given to Italy until transferred to Yugoslavia 1947.

Daman part of the Union Territory of Daman and Diu, NW India.

Daman and Diu /də'mɑːn, 'diːuː/ Union Territory of W India; area 112 sq km/43 sq mi; capital Daman; population (1991) 101,400. *Daman* has an area of 72 sq km/28 sq mi. The port and capital, Daman, is on the west coast, 160 km/100 mi N of Bombay. The economy is based on tourism and fishing. *Diu* is an island off the Kathiawar peninsula with an area of 40 sq km/15 sq mi. The main town is also called Diu. The economy is based on tourism, coconuts, pearl millet, and salt.
history Daman was seized by Portugal 1531 and ceded to Portugal by the Shah of Gujarat 1539; Diu was captured by the Portuguese 1534. Both areas were annexed by India 1961 and were part of the Union Territory of Goa, Daman, and Diu until Goa became a separate state 1987.

Damaraland /də'mɑːrəlænd/ central region of Namibia, home of the nomadic Bantu-speaking Herero. Damaraland lies between the Namib and Kalahari deserts and is largely grassland.

Damascus /də'mæskəs/ (Arabic *Dimashq*) capital of Syria, on the river Barada, SE of Beirut; population (1981) 1,251,000. It produces silk, wood products, and brass and copper ware. Said to be the oldest continuously inhabited city in the world, Damascus was an ancient city even in Old Testament times. Most notable of the old buildings is the Great Mosque, completed as a Christian church in the 5th century.

The Assyrians destroyed Damascus about 733 BC. In 332 BC it fell to one of the generals of Alexander the Great; in 63 BC it came under Roman rule. In AD 635 it was taken by the Arabs, and has since been captured many times, by Egyptians, Mongolians, and Turks. In 1918, during World War I, it was taken from the Turks by the British with Arab aid and in 1920 became the capital of French-mandated Syria.

The 'street which is called straight' is associated with St Paul, who was converted while on the road to Damascus. The tomb of Saladin is here. The fortress dates from 1219.

Damietta /ˌdæmiˈetə/ English name for the Egyptian port of ◊Dumyat.

Damodar /ˈdæmədɑː/ Indian river flowing 560 km/ 350 mi from Chota Nagpur plateau in Bihar, through Bihar and West Bengal states to join the Hooghly River 40 km/25 mi SW of Calcutta. The Damodar Valley is an industrial centre with a hydroelectric project, combined with irrigation works.

Da Nang /ˈdɑːˈnæŋ/ port city (formerly Tourane) of southern Vietnam, 80 km/50 mi SE of Hué; population (1989) 370,600. Following the reunion of North and South Vietnam, the major part of the population was dispersed 1976 to rural areas.

Danbury /ˈdænbəri/ city in SW Connecticut, USA, NW of Bridgeport; population (1980) 60,470. Long a centre for the manufacturing of hats, Danbury's newer industries include electronics, publishing, chemicals, and furniture.

Danube /ˈdænjuːb/ (German **Donau**) second longest of European rivers, rising on the eastern slopes of the Black Forest, and flowing 2,858 km/ 1,776 mi across Europe to enter the Black Sea in Romania by a swampy delta.

The head of river navigation is Ulm, in Baden-Württemberg; Braila, Romania, is the limit for ocean-going ships. Cities on the Danube include Linz, Vienna, Bratislava, Budapest, Belgrade, Ruse, Braila, and Galati. A canal connects the Danube with the river Main, and thus with the Rhine river system. In 1992 the river was diverted in Slovakia to feed the controversial Gabcikovo Dam.

See how the wand'ring Danube flows,/ Realms and religions parting!

On the **Danube** attributed to Jonathan Swift

Danville /ˈdænvɪl/ city in S Virginia, USA, just above the North Carolina border, on the Dan River, SE of Roanoke; seat of Averett County; population (1990) 53,100. Danville is

Dardanelles

situated in a tobacco-growing area. Industries include tobacco processing and marketing, the manufacture of tools, textiles, and building materials.

Danzig /ˈdæntsɪg/ German name for the Polish port of ◊Gdańsk.

Dardanelles /ˌdɑːdəˈnelz/ (ancient name **Hellespont**, Turkish name **Çanakkale Boğazi**) Turkish strait connecting the Sea of Marmara with the Aegean Sea; its shores are formed by the Gallipoli peninsula on the NW and the mainland of Anatolia on the SE. It is 75 km/47 mi long and 5–6 km/3–4 mi wide.

The water of the Hellespont is the most important channel of water in the world ...

On the **Dardanelles** John Masefield
Gallipoli 1916

Dar el-Beida /ˈdɑːr el ˈbeɪdə/ Arabic name for the port of ◊Casablanca, Morocco.

Dar es Salaam /ˈdɑːr es səˈlɑːm/ (Arabic 'haven of peace') chief seaport in Tanzania, on the Indian Ocean, and capital of Tanzania until its replacement by Dodoma 1974; population (1985) 1,394,000.

It is the Indian Ocean terminus of the TanZam Railway, and a line also runs to the lake port of Kigoma; a road links it with Ndola in the Zambian copperbelt, and oil is carried to Zambia by pipeline from Dar es Salaam's refineries. University College (1963) became the University of Dar es Salaam 1970.

Dar es Salam lies like a dormant bush-baby curled round a blue bay fringed with coconuts and flamboyants.

On **Dar es Salaam** Elspeth Huxley
Forks and Hopes 1964

Darfur /dɑːˈfuə/ province in the W of the Republic of Sudan; area 196,555 sq km/75,920 sq mi; population (1983) 3,093,699. The capital is El Fasher (population 30,000). The area is a vast rolling plain. It produces gum arabic, and there is also some stock raising. Darfur was an independent sultanate until conquered by Egypt 1874.

Darien /ˈdeəriən/ former name for the Panama isthmus as a whole, and still the name of an eastern province of Panama; area 16,803 sq km/6,490 sq mi; population (1980) 26,500. The *Gulf of Darien*, part of the Caribbean Sea, lies between Panama and Colombia. The *Darien Gap* is the complex of swamp, jungle, and ravines, which long prevented the linking of the North and South American sections of the Pan-American Highway, stretching about 300 km/200 mi between Canitas, Panama, and Chigorodo, Colombia. At the Colombian end is the Great Atrato Swamp, 60 km/35 mi across and over 300 m/1,000 ft deep.

The *Darien Expedition* was a Scottish attempt to colonize the isthmus 1698–99, which failed disastrously owing to the climate and Spanish hostility. The British Trans-Americas Expedition, led by John Blashford-Snell, made the first motorized crossing 1972.

Darjeeling /dɑːˈdʒiːlɪŋ/ town and health resort in West Bengal, India, situated 2,150 m/7,000 ft above sea level, on the southern slopes of the Himalayas; population (1981) 57,600. It is connected by rail with Calcutta, 595 km/370 mi to the S. It is the centre of a tea-producing district.

Darkhan /dɑːˈkɑːn/ or *Darhan* industrial town in Outer Mongolia, near the border with Russia; population (1991) 90,000. Industries include the manufacture of cement and bricks, and to the S is Erdenet, where copper and molybdenum are mined.

Darling /ˈdɑːlɪŋ/ river in SE Australia, a tributary of the river Murray, which it joins at Wentworth. It is 3,075 km/1,910 mi long, and its waters are conserved in Menindee Lake (155 sq km/60 sq mi) and others nearby. The name comes from Sir Ralph Darling (1775–1858), governor of New South Wales 1825–31. The *Darling Range*, a ridge in W Australia, has a highest point of about 582 m/1,669 ft.

The *Darling Downs* in SE Queensland is an agricultural and stock-raising area.

Darlington /ˈdɑːlɪŋtən/ industrial town in Durham, England, on the river Skerne, near its junction with the river Tees; population (1991) 96,700. It has coal and ironstone mines, and produces iron and steel goods, and knitting wool. The world's first passenger railway was opened between Darlington and Stockton 27 Sept 1825.

Darmstadt /ˈdɑːmstæt/ town in the *Land* of Hessen, Germany, 29 km/18 mi S of Frankfurt-am-Main; population (1988) 134,000. Industries include iron founding and the manufacture of chemicals, plastics, and electronics. It is a centre of the European space industry. It has a ducal palace and a technical university.

Dartford /ˈdɑːtfəd/ industrial town in Kent, England, 27 km/17 mi SE of London; population (1991) 28,400. Cement, chemicals, and paper are manufactured. The *Dartford Tunnel* (1963) runs under the Thames to Purfleet, Essex. Congestion in the tunnel was relieved 1991 by the opening of the Queen Elizabeth II bridge.

Dartmoor /ˈdɑːtmuə/ plateau of SW Devon, England, over 1,000 sq km/400 sq mi in extent, of which half is some 300 m/1,000 ft above sea level. Most of Dartmoor is a National Park. The moor is noted for its wild aspect, and rugged blocks of granite, or 'tors', crown its higher points. The highest are *Yes Tor* 618 m/2,028 ft and *High Willhays* 621 m/2,039 ft. Devon's chief rivers have their sources on Dartmoor. There are numerous prehistoric remains. Near Hemerdon there are tungsten reserves.

Dartmoor Prison, opened 1809 originally to house French prisoners-of-war during the Napoleonic Wars, is at Princetown in the centre of the moor, 11 km/7 mi E of Tavistock. It is still used for category B prisoners.

Dartmouth /ˈdɑːtməθ/ English seaport at the mouth of the river Dart, 43 km/27 mi E of Plymouth, on the Devon coast; population (1981) 6,300. It is a centre for yachting and has an excellent harbour. The Britannia Royal Naval College dates from 1905.

Dartmouth /ˈdɑːtməθ/ port in Nova Scotia, Canada, on the NE of Halifax harbour; population (1986) 65,300. It is virtually part of the capital city itself. Industries include oil refining and shipbuilding.

Darwin /ˈdɑːwɪn/ capital and port in Northern Territory, Australia, in NW Arnhem Land; population (1986) 69,000. It serves the uranium mining site at Rum Jungle to the S. Destroyed 1974 by a cyclone, the city was rebuilt on the same site.

Darwin is the northern terminus of the rail line from Birdum; commercial fruit and vegetable growing is being developed in the area. Founded 1869, under the name of Palmerston, the city was renamed after Charles Darwin 1911.

Dasht-e-Kavir Desert /ˈdæʃti kæˈvɪə/ or *Dasht-i-Davir Desert* salt desert SE of Tehran, Iran; US forces landed here 1980 in an abortive mission to rescue hostages held at the American Embassy in Tehran.

Daugavpils /ˈdaugəfpɪlz/ (Russian *Dvinsk*) town in Latvia on the river Daugava (W Dvina); population (1985) 124,000. A fortress of the Livonian Knights 1278, it became the capital of Polish Livonia (former independent region until 1583, comprising most of present-day Latvia and Estonia). Industries include timber, textiles, engineering, and food products.

Daulaghiri /ˌdauləˈgɪəri/ mountain in the Himalayas, NW of Pokhara, Nepal; it rises to 8,172 m/2,681 ft.

Davao /ˈdɑːvau/ town in the Philippine Republic, at the mouth of the Davao River on the island of Mindanao; population (1990) 849,900. It is the capital of Davao province. It is the centre of a fertile district and trades in pearls, copra, rice, and corn.

Davenport /'dævnpɔːt/ city in SE Iowa, USA, S of Dubuque, directly across from Rock Island, Illinois, on the Mississippi River; seat of Scott County; population (1990) 95,300. It forms the 'Quad Cities' metropolitan area with the Illinois cities of Rock Island, Moline, and East Moline. Industries include aluminium, agriculture, and machinery parts.

Daventry /'dævəntri/ town in Northamptonshire, England, 19 km/12 mi W of Northampton; population (1989 est) 17,000. Because of its central position, it became the site of the BBC high-power radio transmitter 1925. Originally specializing in footwear manufacture, it received London and Birmingham overspill from the 1950s, and developed varied light industries.

Davos /dɑː'vəus/ town in an Alpine valley in Grisons canton, Switzerland 1,559 m/5,115 ft above sea level; population (1990) 10,400. It is recognized as a health resort and as a winter sports centre.

Dawson City /'dɔːsən/ town in Canada, capital until 1953 of Yukon Territory, at the junction of the Yukon and Klondike rivers; population (1986) 1,700. It was founded 1896, at the time of the Klondike gold rush, when its population was 25,000.

Dawson Creek /'dɔːsən/ town in British Columbia, Canada; population (1981) 11,500. It is the southeastern terminus of the Alaska Highway.

Dayton /'deɪtn/ city in Ohio, USA; population (1990) 182,000. It produces precision machinery, household appliances, and electrical equipment. It has an aeronautical research centre and a Roman Catholic university and was the home of aviators Wilbur and Orville Wright.

Daytona Beach /deɪ'təunə/ city on the Atlantic coast of Florida, USA; population (1990) 62,000. Economic activities include printing, commercial fishing, and manufacture of electronic equipment and metal products. It is also a resort. The Daytona International Speedway for motor racing is here.

DE abbreviation for ◊*Delaware*, a state of the USA.

Dead Sea large lake, partly in Israel and partly in Jordan, lying 394 m/1,293 ft below sea level; area 1,020 sq km/394 sq mi. The chief river entering it is the Jordan; it has no outlet and the water is very salty.
    Since both Israel and Jordan use the waters of the Jordan River, the Dead Sea has now dried up in the centre and is divided into two halves, but in 1980 Israel announced a plan to link it by canal with the Mediterranean. The Dead Sea Rift is part of the fault between the African and Arab plates.

The Dead Sea is but the Jordan's highway to heaven.

On the **Dead Sea** W M Thomson
The Hand and the Book 1859

Deal /diːl/ port and resort on the east coast of Kent, England; population (1989 est) 27,000. It was one of the Cinque Ports. Julius Caesar is said to have landed here 55 BC. The castle was built by Henry VIII and houses the town museum.

Dearborn /'dɪəbɔːn/ city in Michigan, USA, on the Rouge River 16 km/10 mi SW of Detroit; population (1990) 89,300. Settled 1795, it was the birthplace and home of Henry Ford, who built his first car factory here. Car manufacturing is still the main industry. Dearborn also makes aircraft parts, steel, and bricks.

Death Valley /deθ 'væli/ depression 225 km/140 mi long and 6–26 km/4–16 mi wide in SE California, USA. At 85 m/280 ft below sea level, it is the lowest point in North America. Bordering mountains rise to 3,000 m/10,000 ft. It is one of the world's hottest and driest places, with temperatures sometimes exceeding 51.7°C/125°F and an annual rainfall of less than 5 cm/2 in. Borax, iron ore, tungsten, gypsum, and salts are extracted.

Deauville /dəuvɪl/ holiday resort of Normandy in the Calvados *département*, France, on the English Channel and at the mouth of the Touques, opposite Trouville; population (1982) 4,800.

Debrecen /'debrətsen/ third largest city in Hungary, 193 km/120 mi E of Budapest, in the Great Plain (*Alföld*) region; population (1988) 217,000. It produces tobacco, agricultural machinery, and pharmaceuticals. Nationalist leader Lajos Kossuth declared Hungary independent of the Habsburgs here 1849. It is a commercial centre and has a university founded 1912.

Decatur /dɪ'keɪtə/ city in central Illinois, USA, on Lake Decatur; population (1990) 83,900. It has engineering, food processing, and plastics industries. It was founded 1829 and named after the US naval hero, Stephen Decatur.

Deccan /'dekən/ triangular tableland in E India, stretching between the Vindhya Hills in the N, and the Western and Eastern Ghats in the S.

Dee /diː/ river in Grampian Region, Scotland; length 139 km/87 mi. From its source in the Cairngorm Mountains, it flows E into the North Sea at Aberdeen (by an artificial channel). It is noted for salmon fishing.
    Also a river in Wales and England; length 112 km/70 mi. Rising in Bala Lake, Gwynedd, it flows into the Irish Sea W of Chester. There is another Scottish river Dee (61 km/38 mi) in Kirkcudbright.

Dehra Dun /'deərə 'duːn/ town in Uttar Pradesh, India; population (1981) 220,530. It is the capital

Delaware

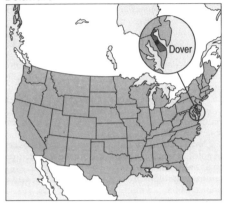

of Dehra Dun district. It has a military academy, a forest research institute, and a Sikh temple built 1699.

Delaware /ˈdeləweə/ state in northeastern USA; nickname First State/Diamond State
area 5,300 sq km/2,046 sq mi
capital Dover
towns Wilmington, Newark
physical divided into two physical areas, one hilly and wooded, the other gently undulating
features one of the most industrialized states; headquarters of the Dupont chemical firm; Rehoboboth Beach
products dairy, poultry, and market-garden produce; chemicals, motor vehicles, and textiles
population (1990) 666,200
famous people J P Marquand
history the first settlers were Dutch 1631 and Swedes 1638, but the area was captured by the British 1664. Delaware was made a separate colony 1704 and organized as a state 1776.
It was one of the original 13 states of the USA.

Delft /delft/ town in South Holland province, the Netherlands, on the Schie Canal, 14 km/9 mi NW of Rotterdam; population (1991) 89,300. It is known worldwide for its pottery and porcelain; other industries include electronic equipment and cables. There is a technical university (1863). The Dutch nationalist leader William the Silent was murdered here 1584. It is the birthplace of the artist Jan Vermeer.

That great parent of pottery.

On **Delft** William Beckford *Dreams Waking Thoughts and Incidents* 1783

Delhi /ˈdeli/ capital of India, comprising the walled city of *Old Delhi*, situated on the west bank of the river Jumna, and *New Delhi* to the S, largely designed by British architect Edwin Lutyens and chosen to replace Calcutta as the seat of government 1912 (completed 1929; officially inaugurated 1931). Delhi is the administrative

centre of the union territory of Delhi and India's largest commercial and communications centre; population (1981) 4,865,077.

Delhi /ˈdeli/ Union Territory of the Republic of India from 1956; capital Delhi; area 1,500 sq km/ 579 sq mi; population (1991) 9,370,400. It produces grain, sugar cane, fruit, and vegetables.

Delos /ˈdiːlɒs/ Greek island, smallest in the Cyclades group, in the SW Aegean Sea; area about 5 sq km/2 sq mi. The great temple of Apollo (4th century BC) is still standing.

Delray Beach /delˌreɪ ˈbiːtʃ/ city in SE Florida, USA, on the Atlantic Ocean, N of Fort Lauderdale; population (1990) 47,200. A tourist resort, it also relies economically on the cultivation of flowers.

Demerara /ˌdeməˈreərə/ river in Guyana, 215 mi/ 346 km long, rising in the centre of the country and entering the Atlantic Ocean at Georgetown. It gives its name to the country's chief sugarcane growing area, after which Demerara sugar is named.

Denbighshire /ˈdenbiʃə/ (Welsh *Sir Ddinbych*) former county of Wales, largely merged 1974, together with Flint and part of Merioneth, into Clwyd; a small area along the western border was included in Gwynedd. Denbigh, in the Clwyd Valley (population about 9,000), was the county town.

Den Haag /den ˈhɑːx/ Dutch form of The ◊Hague, a town in the Netherlands.

Den Helder /den ˈheldə/ port in North Holland province, the Netherlands, 65 km/40 mi N of Amsterdam, on the entrance to the North Holland Canal from the North Sea; population (1991) 61,400. It is a fishing port and naval base.

Denison /ˈdenɪsn/ city in NE Texas, USA, near the Red River and the Oklahoma border, N of Dallas; population (1990) 21,500. A distribution centre for grain and dairy products, its industries include textiles, wood products, and food processing.

Denmark Kingdom of (*Kongeriget Danmark*)
area 43,075 sq km/16,627 sq mi
capital Copenhagen

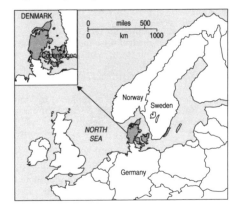

towns Aarhus, Odense, Aalborg, Esbjerg, all ports

physical comprises the Jutland peninsula and about 500 islands (100 inhabited) including Bornholm in the Baltic Sea; the land is flat and cultivated; sand dunes and lagoons on the W coast and long inlets (fjords) on the E; the main island is Sjælland (Zealand), where most of Copenhagen is located (the rest is on the island of Amager)

territories the dependencies of Faeroe Islands and Greenland

features Kronborg Castle in Helsingør (Elsinore); Tivoli Gardens (Copenhagen); Legoland Park in Sillund

head of state Queen Margrethe II from 1972

head of government Poul Nyrup Rasmussen from 1993

political system liberal democracy

political parties Social Democrats (SD), left of centre; Conservative People's Party (KF), moderate centre-right; Liberal Party (V), centre-left; Socialist People's Party (SF), moderate left-wing; Radical Liberals (RV), radical inter-nationalist, left of centre; Centre Democrats (CD), moderate centrist; Progress Party (FP), radical antibureaucratic; Christian People's Party (KrF), interdenominational, family values

exports bacon, dairy produce, eggs, fish, mink pelts, car and aircraft parts, electrical equipment, textiles, chemicals

currency kroner

population (1990 est) 5,134,000; growth rate 0% p.a.

life expectancy men 72, women 78

languages Danish (official); there is a German-speaking minority

religion Lutheran 97%

literacy 99% (1983)

GDP $142.1 bn (1992)

chronology

1940–45 Occupied by Germany.

1945 Iceland's independence recognized.

1947 Frederik IX succeeded Christian X.

1948 Home rule granted for Faeroe Islands.

1949 Became a founding member of NATO.

1960 Joined European Free Trade Association (EFTA).

1972 Margrethe II became Denmark's first queen in nearly 600 years.

1973 Left EFTA and joined European Economic Community (EEC).

1979 Home rule granted for Greenland.

1985 Strong non-nuclear movement in evidence.

1990 General election; another coalition government formed.

1992 Rejection of Maastricht Treaty in national referendum.

1993 Poul Schlüter resigned; replaced by Poul Nyrup Rasmussen at head of Social Democrat-led coalition government. Second referendum approved the Maastricht Treaty after modifications.

Denpasar /den'pɑːsɑː/ capital town of Bali in the Lesser Sunda Islands of Indonesia; population (1980) 88,100. Industries include food processing, machinery, papermaking and printing, and handicrafts. There is a university (1962) and, housed in the temple and palace, a museum of Balinese art.

The Dutch defeated the rajas of Bandung nearby 1906 during their conquest of the island.

Denver /'denvə/ city and capital of Colorado, USA, on the South Platte River, near the foothills of the Rocky Mountains; population (1990) 467,600, Denver–Boulder metropolitan area 1,848,300. It is a processing and distribution centre for a large agricultural area and for natural resources (minerals, oil, gas).

Denver was founded 1858 with the discovery of gold, becoming a mining-camp supply centre; coal is also mined nearby. There is a university, a mining school, many medical institutions, and a branch of the US mint.

> ... Denver is Olympian, impassive and inert. It is probably the most self-sufficient, isolated, self-contained and complacent city in the world.
>
> On **Denver** John Gunther *Inside USA* 1947

Deptford /'detfəd/ district in SE London, in the borough of Lewisham, mainly residential, with industries including engineering and chemicals. It was a major royal naval dockyard 1513-1869, on the south bank of the river Thames.

Derby /'dɑːbi/ industrial city in Derbyshire, England; population (1991) 214,000. Products include rail locomotives, Rolls-Royce cars and aero engines, chemicals, paper, and electrical, mining, and engineering equipment. The museum collections of Crown Derby china, the Rolls-Royce collection of aero engines, and the Derby Playhouse are here.

Derbyshire

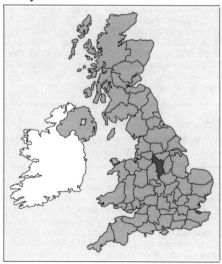

Derbyshire /'dɑːbiʃə/ county in N central England
area 2,630 sq km/1,015 sq mi
towns Matlock (administrative headquarters),
Derby, Chesterfield, Ilkeston
features Peak District National Park (including
Kinder Scout 636 m/2,088 ft); rivers: Derwent, Dove, Rother, Trent; Chatsworth House,
Bakewell (seat of the Duke of Devonshire);
Haddon Hall
products cereals; dairy and sheep farming; there
have been pit and factory closures, but the area
is being redeveloped, and there are large reserves
of fluorite
population (1991) 915,000
famous people Thomas Cook, Marquess
Curzon of Kedleston, Samuel Richardson.

Derg, Lough /dɜːg/ lake in County Donegal, NW
Ireland. The island (Station Island or St Patrick's
Purgatory) is the country's leading place of pilgrimage. Associated with St Patrick, a monastery
flourished here from early times.

Derry /'deri/ county of Northern Ireland
area 2,070 sq km/799 sq mi
towns Derry (county town, formerly
Londonderry), Coleraine, Portstewart
features rivers Foyle, Bann, and Roe; borders
Lough Neagh
products mainly agricultural, but farming is
hindered by the very heavy rainfall; flax, cattle,
sheep, food processing, textiles, light engineering
population (1981) 187,000
famous people Joyce Cary.

Derry /'deri/ (Gaelic *doire* 'a place of oaks') historic city and port on the river Foyle, County
Derry, Northern Ireland; population (1981)
89,100. Known as Londonderry until 1984,
Derry dates from the foundation of a monastery
by St Columba AD 546. James I of England
granted the borough and surrounding land to the
citizens of London and a large colony of imported
Protestants founded the present city which they
named Londonderry. Textiles and chemicals are
produced.

Derwent /'dɜːwənt/ river in North Yorkshire, NE
England; length 112 km/70 mi. Rising in the N
Yorkshire moors, it joins the river Ouse SE of
Selby.
  Other rivers of the same name in the UK are
found in Derbyshire (96 km/60 mi), Cumbria
(56 km/35 mi), and Northumberland (26 km/
16 mi).

Des Moines /dɪ 'mɔɪn/ capital and largest town in
Iowa, USA, on the Des Moines River, a tributary
of the Mississippi; population (1990) 193,200. It
is a major road, railway, and air centre. Industries
include printing, banking, insurance, and food
processing.

Dessau /'desau/ town in the state of Saxony-
Anhalt, Germany, on the river Mulde, 115 km/70
mi SW of Berlin; population (1990) 120,000. It
is the former capital of Anhalt duchy and state. It
manufactures chemicals, machinery, and chocolate and was the site of the Junkers aeroplane
works. The Bauhaus school of art was based in
Dessau 1925–33.

Detroit /dɪ'trɔɪt/ city in Michigan, USA, situated
on Detroit River; population (1990) 1,028,000,
metropolitan area 4,665,200. It is an industrial
centre with the headquarters of Ford, Chrysler,
and General Motors, hence its nickname,
Motown (from 'motor town'). Other manufactured products include metal products, machine
tools, chemicals, office machines, and pharmaceuticals. During the 1960s and 1970s Detroit
became associated with the 'Motown Sound' of
rock and soul music.
history Detroit was founded 1701 and is the
oldest US city of any size W of the original
colonies. It was captured from the French by the
British 1760 and passed to the USA 1796. In
1805 it was completely destroyed by fire but was
soon rebuilt. The opening of the Erie Canal 1825
aided development. Detroit grew rapidly after the
building of the first car factories, 1899–1903.
  There were significant race riots 1943 and
1967; a black mayor was elected 1973.

Detskoe Selo /'detskəjə sɪ'lou/ (Russian 'children's village') former name (1917–37) of
◊Pushkin, near St Petersburg, which was
renamed after the Russian poet 1937.

Deventer /'deivəntə/ town in Overijssel province,
the Netherlands, on the river IJssel, 45 km/
28 mi S of the IJsselmeer lake; population (1991)
67,400. It is an agricultural and transport centre
and produces carpets, precision equipment, and
packaging machinery.

Devil's Island /'devəlz 'aılənd/ (French *Ile du
Diable*) smallest of the Iles du Salut, off French
Guiana, 43 km/27 mi NW of Cayenne. The group
of islands was collectively and popularly known
by the name Devil's Island and formed a penal
colony notorious for its terrible conditions.
  Alfred Dreyfus was imprisoned here 1895–99.
Political prisoners were held on Devil's Island,
and dangerous criminals on St Joseph, where they
were subdued by solitary confinement in tiny cells
or subterranean cages. The largest island, Royale,
now has a tracking station for the French rocket
site at Kourou.

Devil's Marbles area of granite boulders, S
of Tennant Creek, off the Stuart Highway in
Northern Territory, Australia.

Devizes /dɪ'vaızız/ historic market town in
Wiltshire, England; population (1982) 13,000.
It was formerly known for its trade in cloth,
but is now a centre for brewing, engineering,
and food processing. Special features include
ancient earthworks and the shattered remains of
a Norman castle stormed by Oliver Cromwell
1645.

Devon /'devən/ or *Devonshire* county in SW
England
area 6,720 sq km/2,594 sq mi
towns Exeter (administrative headquarters),
Plymouth; resorts: Paignton, Torquay,
Teignmouth, and Ilfracombe
features rivers: Dart, Exe, Tamar; National
Parks: Dartmoor, Exmoor; Lundy bird sanctuary
and marine nature reserve in the Bristol Channel

Devon

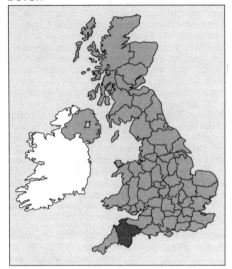

products mainly agricultural, with sheep and dairy farming; cider and clotted cream; kaolin in the S; Honiton lace; Dartington glass
population (1991) 1,008,300
famous people Francis Drake, John Hawkins, Charles Kingsley, Robert F Scott.

Dhaka /'dækə/ or **Dacca** capital of Bangladesh from 1971, in Dhaka region, W of the river Meghna; population (1984) 3,600,000. It trades in jute, oilseed, sugar, and tea and produces textiles, chemicals, glass, and metal products.
history A former French, Dutch, and English trading post, Dhaka became capital of East Pakistan 1947; it was handed over to Indian troops Dec 1971 to become capital of the new country of Bangladesh.

Dharan /ˌdɑː'rɑːn/ oil town in E Saudi Arabia, near the Gulf Coast; it was used as a military base during the Gulf War 1991.

Dhaulagiri /ˌdaʊlə'gɪəri/ mountain in the Himalayas of W central Nepal, rising to 8,172 m/26,811 ft.

Dhofar /'dəʊfɑː/ mountainous western province of Oman, on the border with Yemen; population (1982) 40,000. South Yemen supported guerrilla activity here in the 1970s, while Britain and Iran supported the government's military operations. The capital is Salalah, which has a port at Rasut.

Diego Garcia /di'eɪgəʊ gɑː'siːə/ island in the Chagos Archipelago, Indian Ocean, named after its Portuguese discoverer 1532. See ◊British Indian Ocean Territory.

Dieppe /diː'ep/ channel port at the mouth of the river Arques, Seine-Maritime *département*, N France; population (1990) 36,600. There are ferry services from its harbour to Newhaven and elsewhere; industries include fishing, shipbuilding, and pharmaceuticals.

Dijon /'diːʒɒn/city in Côte-d'Or *département*, E central France, capital of Burgundy region; population (1990) 151,600. As well as metallurgical, chemical, and other industries, it has a wine trade and is famed for its mustard.

Dinan town in Côtes-du-Nord *département*, N France, on the river Rance; population (1989 est) 14,200. The river is harnessed for tidal hydroelectric power.

Dinant /diː'nɒn/ ancient town in Namur province, Belgium, on the river Meuse; population (1991) 12,100. It is a tourist centre for the Ardennes.

Dinaric Alps /dɪ'nærɪk/ extension of the European Alps that runs parallel to the E Adriatic coast, stretching from Slovenia along the frontier between Croatia and Bosnia-Herzegovina into W Yugoslavia and N Albania. The highest peak is Durmitor at 2,522 m/8,274 ft.

Diomede /'daɪəmiːd/ two islands off the tip of the Seward peninsula, Alaska. *Little Diomede* (6.2 sq km/2.4 sq mi) belongs to the USA and is 3.9 km/2.4 mi from *Big Diomede* (29.3 sq km/11.3 sq mi), owned by Russia. They were first sighted by the Danish navigator Vitus Bering 1728.

District of Columbia /kə'lʌmbiə/ federal district of the USA, see ◊Washington.

Diu island off the Kathiawar peninsula, NW India, part of the Union Territory of ◊Daman and Diu.

Diyarbakir /dɪ'jɑːbəkɪə/ town in Asiatic Turkey, on the river Tigris; population (1990) 381,100. It has a trade in gold and silver filigree work, copper, wool, and mohair, and manufactures textiles and leather goods.

Djakarta variant spelling of ◊Jakarta, the capital of Indonesia.

Djibouti Republic of (*Jumhouriyya Djibouti*)
area 23,200 sq km/8,955 sq mi
capital (and chief port) Djibouti
towns Tadjoura, Obock, Dikhil
physical mountains divide an inland plateau from a coastal plain; hot and arid
features terminus of railway link with Ethiopia; Lac Assal salt lake is the second lowest point on Earth (-144 m/-471 ft)

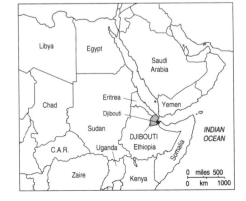

head of state and government Hassan Gouled Aptidon from 1977
political system authoritarian nationalism
political party People's Progress Assembly (RPP), nationalist
exports acts mainly as a transit port for Ethiopia
currency Djibouti franc
population (1990 est) 337,000 (Issa 47%, Afar 37%, European 8%, Arab 6%); growth rate 3.4% p.a.
life expectancy 50
languages French (official), Somali, Afar, Arabic
religion Sunni Muslim
literacy 20% (1988)
GDP $378 million (1987); $1,016 per head
chronology
1884 Annexed by France as part of French Somaliland.
1967 French Somaliland became the French Territory of the Afars and the Issas.
1977 Independence achieved from France; Hassan Gouled elected president.
1979 All political parties combined to form the People's Progress Assembly (RPP).
1981 New constitution made RPP the only legal party. Gouled re-elected. Treaties of friendship signed with Ethiopia, Somalia, Kenya, and Sudan.
1984 Policy of neutrality reaffirmed.
1987 Gouled re-elected for a final term.
1991 Amnesty International accused secret police of brutality.
1992 Djibouti elected member of UN Security Council 1993–95.

Jibouti is a strange remote place: situated on the equator, it yet has all the comforts and discomforts of a French provincial town.

On **Djibouti** Ladislas Farago *Abyssinia on the Eve* 1935

Djibouti /dʒɪˈbuːti/ chief port and capital of the Republic of Djibouti, on a peninsula 240 km/149 mi SW of Aden and 565 km/351 mi NE of Addis Ababa; population (1988) 290,000.
The city succeeded Obock as capital of French Somaliland 1896 and was the official port of Ethiopia 1897–1949.

Dneprodzerzhinsk /ˌnɪprədzəˈʒɪnsk/ port in Ukraine, on the river Dnieper, 48 km/30 mi NW of Dnepropetrovsk; population (1987) 279,000. It produces chemicals, iron, and steel.

Dnepropetrovsk /ˌnɪprəpɪˈtrɒfsk/ city in Ukraine, on the right bank of the river Dnieper; population (1987) 1,182,000. It is the centre of a major industrial region, with iron, steel, chemical, and engineering industries. It is linked with the Dnieper Dam, 60 km/37 mi downstream.

Dnieper /ˈniːpə/ or *Dnepr* river rising in the Smolensk region of Russia and flowing S through Belarus and Ukraine to enter the Black Sea E of Odessa; total length 2,250 km/1,400 mi.

Dodecanese

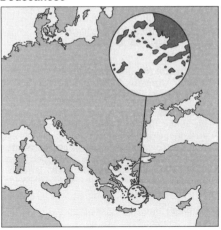

Dobruja /ˈdɒbrudʒə/ district in the Balkans, bounded N and W by the Danube and E by the Black Sea. It is low-lying, partly marshland, partly fertile steppe land. Constanta is the chief town. Dobruja was divided between Romania and Bulgaria 1878. In 1913, after the second Balkan War, Bulgaria ceded its part to Romania but received it back 1940, a cession confirmed by the peace treaty of 1947.

Docklands redevelopment area E of St Katherine's Dock, London, occupying the site of the former Wapping and Limehouse docks, the Isle of Dogs, and Royal Docks. The tallest building is the Canary Wharf tower. Designed to be a European business centre, Docklands is served by the London City airport (Stolport) and the Docklands Light Railway (DLR).

Dodecanese /ˌdəʊdekəˈniːz/ (Greek *Dhodhekánisos* 'twelve islands') group of islands in the Aegean Sea; area 1,028 sq m/2,663 sq km. Once Turkish, the islands were Italian 1912–47, when they were ceded to Greece. They include Rhodes and Kos. Chief products include fruit, olives, and sponges.

Dodge City /ˌdɒdʒ ˈsɪti/ city in SW Kansas, USA, on the Arkansas River; population (1990) 21,100. It was a noted frontier cattle town in the days of the Wild West.

Dodoma /ˈdəʊdəmə/ capital (replacing Dar es Salaam 1974) of Tanzania; 1,132 m/3,713 ft above sea level; population (1985) 85,000. It is a centre of communications, linked by rail with Dar es Salaam and Kigoma on Lake Tanganyika, and by road with Kenya to the N and Zambia and Malawi to the S.

Dogger Bank /ˈdɒgə/ submerged sandbank in the North Sea, about 115 km/70 mi off the coast of Yorkshire, England. In places the water is only 11 m/36 ft deep, but the general depth is 18–36 m/60–120 ft; it is a well-known fishing ground.

Dogs, Isle of /ˈdɒgz/ district of E London, England, part of the Greater London borough of ⟡Tower Hamlets.

Doha /'dəuhɑ:/ (Arabic *Ad Dawḥah*) capital and chief port of Qatar; population (1986) 217,000. Industries include oil refining, refrigeration plants, engineering, and food processing. It is the centre of vocational training for all the Persian Gulf states.

Doi Inthanon /ˌdɔɪ ɪn'θænən/ highest mountain in Thailand, rising to 2,595 m/8,513 ft SW of Chiang Mai in NW Thailand.

Dolgellau /dɒl'geɬi/ (formerly *Dolgelly*) market town at the foot of Cader Idris in Gwynedd, Wales, on the river Wnion; population (1989 est) 2,300. The town is also a tourist centre. Nearby are the Gwynfynydd ('White Mountain') and Clogau goldmines; a nugget from the latter has supplied gold for the wedding rings of royal brides since 1923.

Dominica Commonwealth of
area 751 sq km/290 sq mi
capital Roseau, with a deepwater port
towns Portsmouth, Marigot
physical second largest of the Windward Islands, mountainous central ridge with tropical rainforest
features of great beauty, it has mountains of volcanic origin rising to 1,620 m/5,317 ft; Boiling Lake (an effect produced by escaping subterranean gas)
head of state Clarence Seignoret from 1983
head of government Eugenia Charles from 1980
political system liberal democracy
political parties Dominica Freedom Party (DFP), centrist; Labour Party of Dominica (LPD), left-of-centre coalition
exports bananas, coconuts, citrus, lime, bay oil
currency E Caribbean dollar, pound sterling, French franc
population (1990 est) 94,200 (mainly black African in origin, but with a small Carib reserve of some 500); growth rate 1.3% p.a.
life expectancy men 57, women 59
language English (official), but the Dominican patois reflects earlier periods of French rule
media one independent weekly newspaper
religion Roman Catholic 80%
literacy 80%

GDP $91 million (1985); $1,090 per head
chronology
1763 Became British possession.
1978 Independence achieved from Britain. Patrick John, leader of Dominica Labour Party (DLP), elected prime minister.
1980 Dominica Freedom Party (DFP), led by Eugenia Charles, won convincing victory in general election.
1981 Patrick John implicated in plot to overthrow government.
1982 John tried and acquitted.
1985 John retried and found guilty. Regrouping of left-of-centre parties resulted in new Labour Party of Dominica (LPD). DFP, led by Eugenia Charles, re-elected.
1990 Charles elected to a third term.
1991 Integration into Windward Islands confederation proposed.

Dominican Republic (*República Dominicana*)
area 48,442 sq km/18,700 sq mi
capital Santo Domingo
towns Santiago de los Caballeros, San Pedro de Macoris
physical comprises eastern two-thirds of island of Hispaniola; central mountain range with fertile valleys
features Pico Duarte 3,174 m/10,417 ft, highest point in Caribbean islands; Santo Domingo is the oldest European city in the western hemisphere
head of state and government Joaquín Ricardo Balaguer from 1986
political system democratic republic
political parties Dominican Revolutionary Party (PRD), moderate, left of centre; Christian Social Reform Party (PRSC), independent socialist; Dominican Liberation Party (PLD), nationalist
exports sugar, gold, silver, tobacco, coffee, nickel
currency peso
population (1989 est) 7,307,000; growth rate 2.3% p.a.
life expectancy men 61, women 65
language Spanish (official)
religion Roman Catholic 95%
literacy men 78%, women 77% (1985 est)
GDP $4.9 bn (1987); $731 per head

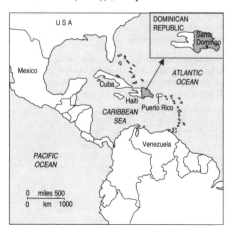

chronology
1492 Visited by Christopher Columbus.
1844 Dominican Republic established.
1930 Military coup established dictatorship of Rafael Trujillo.
1961 Trujillo assassinated.
1962 First democratic elections resulted in Juan Bosch, founder of the PRD, becoming president.
1963 Bosch overthrown in military coup.
1965 US Marines intervene to restore order and protect foreign nationals.
1966 New constitution adopted. Joaquín Balaguer, leader of PRSC, became president.
1978 PRD returned to power, with Silvestre Antonio Guzmán as president.
1982 PRD re-elected, with Jorge Blanco as president.
1985 Blanco forced by International Monetary Fund to adopt austerity measures to save the economy.
1986 PRSC returned to power, with Balaguer re-elected president.
1990 Balaguer re-elected by a small majority.

Don /dɒn/ river in Russia, rising to the S of Moscow and entering the northeast extremity of the Sea of Azov; length 1,900 km/1,180 mi. In its lower reaches the Don is 1.5 km/1 mi wide, and for about four months of the year it is closed by ice. Its upper course is linked with the river Volga by a canal.

Donau /'dəunau/ German name for the river ⟡Danube.

Donbas /ˌdɒn'bæs/ acronym for the ⟡Donets Basin, a coal-rich area in Ukraine.

Doncaster /'dɒŋkəstə/ town in South Yorkshire, England, on the river Don; population (1981) 81,600. It has a racecourse; famous races here are the St Leger (1776) in Sept and the Lincolnshire Handicap in March.
history Doncaster was originally a Roman station. Conisbrough, a ruined Norman castle to the SW, features in Walter Scott's novel *Ivanhoe* as Athelstan's stronghold. Coal, iron, and steel have been the dominant industries in this area for hundreds of years, though they have recently declined and are being replaced by other products, such as synthetic textiles.

Donegal /ˌdɒnɪ'gɔːl/ mountainous county in Ulster province in the NW of the Republic of Ireland, surrounded on three sides by the Atlantic Ocean; area 4,830 sq km/1,864 sq mi; population (1991) 127,900. The county town is Lifford; the market town and port of Donegal is at the head of Donegal Bay in the SW. Commercial activities include sheep and cattle raising, tweed and linen manufacture, and some deep-sea fishing. The river Erne hydroelectric project (1952) involved the building of large power stations at Ballyshannon.

Donets /dɒ'nets/ river rising in the Kursk region of Russia and flowing 1,080 km/670 mi through Ukraine to join the river Don 100 km/60 mi E of Rostov; see also ⟡Donets Basin.

Donets Basin /də'nets/ (abbreviated to *Donbas*)

area in Ukraine, situated in the bend formed by the rivers Don and Donets, which holds one of Europe's richest coalfields, together with salt, mercury, and lead.

Donetsk /dɒ'nets/ city in Ukraine; capital of Donetsk region, situated in the Donets Basin, a major coal-mining area, 600 km/372 mi SE of Kiev; population (1987) 1,090,000. It has blast furnaces, rolling mills, and other heavy industries.
It developed from 1871 when a Welshman, John Hughes, established a metallurgical factory, and the town was first called Yuzovka after him; it was renamed Stalino 1924, and Donetsk 1961.

Dongola /'dɒŋgələ/ town in the Northern Province of the Sudan, above the third cataract on the river Nile. It was founded about 1812 to replace *Old Dongola*, 120 km/75 mi upriver, which was destroyed by the Mamelukes. Old Dongola, a trading centre on a caravan route, was the capital of the Christian kingdom of ⟡Nubia between the 6th and 14th centuries.

Dongting /ˌduŋ'tɪŋ/ lake in Hunan province, China; area 10,000 sq km/4,000 sq mi.

Donnybrook /'dɒnibruk/ former village, now part of Dublin, Republic of Ireland, notorious until 1855 for riotous fairs.

Dorchester market town in Dorset, England, on the river Frome, N of Weymouth; population (1989 est) 15,000. It is the administrative centre for the county. The hill-fort Maiden Castle to the SW was occupied as a settlement from about 2000 BC. The novelist Thomas Hardy was born nearby.
In the 17th century the Puritans gained control of the town, using the money garnered from the town brewhouse monopoly to pay for poor relief from 1622. In 1642 Dorchester was a centre of Parliamentary revolt, but was captured by the Royalists 1643.

Dordogne /dɔː'dɔɪn/ river in SW France, rising in Puy-de-Dôme *département* and flowing 490 km/300 mi to join the river Garonne 23 km/14 mi N of Bordeaux. It gives its name to a *département* and is a major source of hydroelectric power.
The valley of the Dordogne is a popular tourist area, and the caves of the wooded valleys of its tributary, the Vézère, have signs of early human occupation. Famous sites include the Cro Magnon, Le Moustier, and Lascaux caves.

Dordrecht /'dɔːdrext/ or **Dort** river port on an island in the Maas, South Holland, the Netherlands, 19 km/12 mi SE of Rotterdam; population (1991) 110,400, metropolitan area of Dordrecht-Zwijndrecht 203,000. It has shipbuilding yards and makes heavy machinery, plastics, and chemicals.

Dorpat /'dɔːpæt/ German name for the Estonian city of ⟡Tartu.

Dorset /'dɔːsɪt/ county in SW England
area 2,650 sq km/1,023 sq mi
towns Dorchester (administrative headquarters), Poole, Shaftesbury, Sherborne; resorts: Bournemouth, Lyme Regis, Weymouth

Dorset

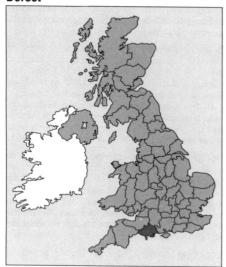

features Chesil Bank, a shingle bank along the coast 19 km/11 mi long; Isle of Purbeck, a peninsula where china clay and Purbeck 'marble' are quarried, and which includes Corfe Castle and the holiday resort of Swanage; Dorset Downs; Cranborne Chase; rivers Frome and Stour; Maiden Castle; Tank Museum at Royal Armoured Corps Centre, Bovington, where the cottage of the soldier and writer T E Lawrence is a museum
products Wytch Farm is the largest onshore oilfield in the UK
population (1988 est) 655,700
famous people Anthony Ashley Cooper, Thomas Hardy, Thomas Love Peacock.

Dort /dɔːt/ another name for ⏴Dordrecht, a port in the Netherlands.

Dortmund /ˈdɔːtmund/ industrial centre in the ⏴Ruhr, Germany, 58 km/36 mi NE of Düsseldorf; population (1988) 568,000. It is the largest mining town of the Westphalian coalfield and the southern terminus of the Dortmund–Ems Canal. The enlargement of the Wesel–Datteln Canal 1989, connecting Dortmund to the Rhine River, allows barges to travel between Dortmund and Rotterdam in the Netherlands. Industries include iron, steel, construction machinery, engineering, and brewing.

Dothan /ˈdəʊθn/ city in the southeast corner of Alabama, USA, SE of Montgomery; seat of Houston County; population (1990) 53,589. Its industries include fertilizer, clothing, furniture, vegetable oils, and hosiery. It is an agricultural and livestock marketing centre.

Douai /duːˈeɪ/ town in Nord *département*, France, on the river Scarpe; population (1990) 44,200. It has coal mines, iron foundries, and breweries. An English Roman Catholic college was founded there 1568 by English Catholics in exile. The Douai-Reims Bible, published 1582–1610,

influenced the translators of the King James Version.

Douala /duːˈɑːlə/ or *Duala* chief port and industrial centre (aluminium, chemicals, textiles, pulp) of Cameroon, on the Wouri river estuary; population (1981) 637,000. Known as Kamerunstadt until 1907, it was capital of German Cameroon 1885–1901.

Doubs /duː/ river in France and Switzerland, rising in the Jura Mountains and flowing 430 km/265 mi to join the river Saône. It gives its name to a *département*.

Douglas /ˈdʌɡləs/ capital of the Isle of Man in the Irish Sea; population (1986) 20,400. It is a holiday resort and terminus of shipping routes to and from Fleetwood and Liverpool; banking and financial services are important to the local economy.

Dounreay /ˈduːnreɪ/ experimental nuclear reactor site on the north coast of Scotland, 12 km/7 mi W of Thurso. Development started 1974 and continued until a decision was made 1988 to decommission the site by 1994.

Douro /ˈduərəu/ (Spanish *Duero*) river rising in N central Spain and flowing through N Portugal to the Atlantic at Porto; length 800 km/500 mi. Navigation at the river mouth is hindered by sand bars. There are hydroelectric installations. Vineyards (port and Mateus rosé) are irrigated with water from the river.

Dover /ˈdəʊvə/ market town and seaport on the southeast coast of Kent, England; population (1989 est) 36,000. It is Britain's nearest point to mainland Europe, being only 34 km/21 mi from Calais, France. Dover's development has been chiefly due to the cross-Channel traffic, which includes train, ferry, hovercraft, and other services. It was one of the Cinque Ports, part of England's defences against invasion after the Norman Conquest.
history Under Roman rule, Dover (*Portus Dubris*) was the terminus of Watling Street, and the beacon or 'lighthouse' in the grounds of the Norman castle dates from about AD 50, making it one of the oldest buildings in Britain.

Dover /ˈdəʊvə/ capital of Delaware, USA, located in the central part of the state, on the St Jones River, S of Wilmington; population (1990) 27,600. Industries include synthetic materials, adhesives, latex, resins, chemicals, food products, and space equipment.

Dover /ˈdəʊvə/ city in SE New Hampshire, USA, on the Cocheco River, NW of Portsmouth; seat of Strafford County; population (1990) 25,000. Industries include lumber, electronics, rubber, and aluminium products.

Dover, Strait of /ˈdəʊvə/ (French *Pas-de-Calais*) stretch of water separating England from France, and connecting the English Channel with the North Sea. It is about 35 km/22 mi long and 34 km/21 mi wide at its narrowest part. It is one of the world's busiest sea lanes.

By 1972 increasing traffic, collisions, and shipwrecks had become so frequent that traffic-routeing schemes were enforced.

Down /daun/ county in SE Northern Ireland, facing the Irish Sea on the E; area 2,470 sq km/ 953 sq mi; population (1981) 339,200. To the S are the Mourne Mountains, to the E Strangford sea lough. The county town is Downpatrick; the main industry is dairying.

Downs, North and South /daunz/ two lines of chalk hills in SE England; see ⟡ North Downs and ⟡ South Downs.

Downs, the /daunz/ roadstead (partly sheltered anchorage) off E Kent, England, between Deal and the Goodwin Sands. Several 17th-century naval battles took place here, including a defeat of Spain by the Dutch 1639.

Drakensberg /ˈdrɑːkənsbɜːg/ (Afrikaans 'dragon's mountain') mountain range in South Africa (Sesuto name **Quathlamba**), on the boundary of Lesotho and the Orange Free State with Natal. Its highest point is Thabana Ntlenyana, 3,350 m/ 10,988 ft, near which is Natal National Park.

Drammen town and port in SE Norway, on the Drammen Fjord, a branch of the Oslo Fjord; population (1990) 51,900. There are engineering, water power equipment, food processing, and brewing industries; paper and wood pulp are manufactured and shipped from here.

Drenthe /ˈdrentə/ low-lying northern province of the Netherlands
area 2,660 sq km/1,027 sq mi
capital Assen
towns Emmen, Hoogeveen
physical fenland and moors; well-drained clay and peat soils
products livestock, arable crops, horticulture, petroleum
population (1988) 437,000
history governed in the Middle Ages by provincial nobles and by bishops of Utrecht, Drenthe was eventually acquired by Charles V of Spain 1536. It developed following land drainage initiated in the mid-18th century and was established as a separate province of the Netherlands 1796.

Dresden /ˈdrezdən/ capital of the state of Saxony, Germany; population (1990) 520,000. Industries include chemicals, machinery, glassware, and musical instruments. It was one of the most beautiful German cities until its devastation by Allied fire-bombing 1945. Dresden county has an area of 6,740 sq km/2,602 sq mi and a population of 1,772,000.
history Under the elector Augustus II the Strong (1694–1733), it became a centre of art and culture. The manufacture of Dresden china, started in Dresden 1709, was transferred to Meissen 1710. The city was bombed by the Allies on the night 13–14 Feb 1945, 15.5 sq km/6 sq mi of the inner town being destroyed, and deaths being estimated at 35,000–135,000. Following the reunification of Germany 1990 Dresden once again became capital of Saxony.

Drogheda /ˈdrɔɪdə/ seaport near the mouth of the river Boyne, County Louth, Republic of Ireland. The port trades in cattle and textiles; chemicals and foodstuffs are produced. In 1649 the town was stormed by Oliver Cromwell, who massacred most of the garrison, and in 1690 it surrendered to William III after the Battle of the Boyne.

Drôme /drəum/ river in France, rising in Dauphiné Pre-Alps and flowing NW for 101 km/63 mi to join the river Rhône below Livron. It gives its name to a *département*.

Dubai /duːˈbaɪ/ one of the ⟡ United Arab Emirates.

Dublin /ˈdʌblɪn/ (Gaelic *Baile Atha Cliath*) capital and port on the east coast of the Republic of Ireland, at the mouth of the river Liffey, facing the Irish Sea; population (1986 est) 502,700, Greater Dublin (including Dún Laoghaire) 921,000. It is the site of one of the world's largest breweries (Guinness); other industries include textiles, pharmaceuticals, electrical goods, and machine tools.
features In the Georgian period many fine squares were laid out, and the Custom House (damaged in the 1921 uprising but later restored) survives. There is a Roman Catholic procathedral, St Mary's (1816); two Protestant cathedrals; and two universities, the University of Dublin and the National University of Ireland. Trinity College library contains the Book of Kells, a splendidly illuminated 8th-century gospel book produced at the monastery of Kells in County Meath, founded by St Columba. Other buildings are the City Hall (1779), the Four Courts (1796), the National Gallery, Dublin Municipal Gallery, National Museum, Leinster House (where the legislature, Dáil Eireann, sits), and the Abbey and Gate theatres.
history The city was founded 840 by the invading Danes, who were finally defeated 1014 at Clontarf, now a northern suburb of the city. Dublin was the centre of English rule from 1171 (exercised from Dublin Castle 1220) until 1922.

The most instantly talkative city in Europe.

On **Dublin** V S Pritchett in *Sunday Times Weekly Review* 2 July 1978

Dublin /ˈdʌblɪn/ county in Leinster province, Republic of Ireland, facing the Irish Sea; area 920 sq km/355 sq mi; population (1986) 1,021,000. It is mostly level and low-lying, but rises in the S to 753 m/2,471 ft in Kippure, part of the Wicklow Mountains. The river Liffey enters Dublin Bay. Dublin, the capital of the Republic of Ireland, and Dún Laoghaire are the two major towns.

Dubna /dubˈnɑː/ town in Russia, 40 km/25 mi W of Tula; population (1985) 61,000. It is a metal-working centre, and has the Volga Nuclear Physics Centre.

Dubrovnik /duːˈbrɒvnɪk/ (Italian *Ragusa*) city and port in Croatia on the Adriatic coast; population

(1985) 35,000. It manufactures cheese, liqueurs, silk, and leather.

Once a Roman station, Dubrovnik was for a long time an independent republic but passed to Austrian rule 1814–1919. During the 1991 civil war, Dubrovnik was placed under siege by Yugoslav federal forces (as part of its blockade of the Croatian coast) and subjected to frequent artillery barrages and naval shelling. The plight of the city and its residents attracted international concern during the siege; medieval buildings and works of art were destroyed.

Dubuque /dəˈbjuːk/ city in E central Iowa, USA, NE of Iowa City, just across the Mississippi River from the Wisconsin–Illinois border; population (1990) 57,546. An important port, it has shipbuilding and agricultural marketing facilities; industries include meatpacking, lumber, metals, and machinery.

Dudley /ˈdʌdli/ town NW of Birmingham, West Midlands, England; population (1981) 187,000. Industries include light engineering and clothing manufacture. There is a 13th-century castle.

Dufourspitze /duːˈfuəˌʃpɪtsə/ second highest of the alpine peaks, 4,634 m/15,203 ft high. It is the highest peak in the Monte Rosa group of the Pennine Alps on the Swiss-Italian frontier.

Duisburg river port and industrial city in North Rhine–Westphalia, Germany, at the confluence of the Rhine and Ruhr rivers; population (1987) 515,000. It is the largest inland river port in Europe. Heavy industries include oil refining and the production of steel, copper, zinc, plastics, and machinery.

Dukeries /ˈdjuːkəriz/ area of estates in Nottinghamshire, England, with magnificent noblemen's mansions, few now surviving. Thoresby Hall, said to be the largest house in England (about 365 rooms), was sold as a hotel 1989 and the contents dispersed.

Duluth /dəˈluːθ/ port in the USA on Lake Superior, by the mouth of the St Louis River, Minnesota; population (1990) 85,500. It manufactures steel, flour, timber, and dairy produce; it trades in iron ore and grain.

Dulwich /ˈdʌlɪdʒ/ suburb, part of the inner London borough of Southwark, England. It contains Dulwich College (founded 1619 by Edward Alleyn, an Elizabethan actor), the Horniman Museum (1901), with a fine ethnological collection, Dulwich Picture Gallery (1814), rebuilt 1953 after being bombed during World War II, Dulwich Park, and Dulwich Village.

Dumbarton /dʌmˈbɑːtn/ town in Strathclyde, Scotland; population (1981) 23,200. Industries include marine engineering, whisky distilling, and electronics.

Dumfries administrative headquarters of Dumfries and Galloway Region, Scotland; population (1981) 31,600. It is situated on the river Nith and has knitwear, plastics, and other industries.

Dumfries and Galloway /dʌmˈfriːs, ˈɡæləweɪ/ region of Scotland

Dumfries and Galloway

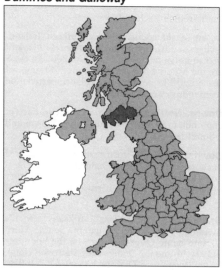

area 6,500 sq km/2,510 sq mi
towns Dumfries (administrative headquarters)
features Solway Firth; Galloway Hills, setting of John Buchan's *The Thirty-Nine Steps*; Glen Trool National Park; Ruthwell Cross, a runic cross dating from about 800 at the village of Ruthwell; Stranraer provides the shortest sea route to Ireland
products horses and cattle (for which the Galloway area was renowned), sheep, timber
famous people Robert I (Robert the Bruce), Robert Burns, Thomas Carlyle
population (1990 est) 147,000.

Dumfriesshire /dʌmˈfriːsʃə/ former county of S Scotland, merged 1975 into the region of Dumfries and Galloway.

Dumyat /dumˈjɑːt/ (English **Damietta**) town in Egypt at the mouth of the river Nile; population (1986) 121,200. Cotton goods are produced.

Duna /ˈduːnə/ Hungarian name for the river ◊Danube.

Dunarea /ˈduːnəriə/ Romanian name for the river ◊Danube.

Dunbar /dʌnˈbɑː/ port and resort in Lothian Region, Scotland; population (1981) 5,800. Torness nuclear power station is nearby. Oliver Cromwell defeated the Scots here 1650.

Dunbartonshire /dʌnˈbɑːtnʃə/ former county of Scotland, bordering the north bank of the Clyde estuary, on which stand Dunbarton (the former county town), Clydebank, and Helensburgh. It was merged 1975 into the region of Strathclyde.

Dundee /dʌnˈdiː/ city and fishing port, administrative headquarters of Tayside, Scotland, on the north side of the Firth of Tay; population (1988 est) 174,300. It is an important shipping and rail centre with marine engineering, watch and clock, and textile industries.

The city developed around the jute industry in the 19th century, and has benefited from the North Sea oil discoveries of the 1970s. There is a university (1967) derived from Queen's College (founded 1881), and other notable buildings include the Albert Institute (1867) and Caird Hall.

Dunedin /dʌnˈiːdn/ port on Otago Harbour, South Island, New Zealand; population (1991) 116,500. It is a road, rail, and air centre, with engineering and textile industries. The city was founded 1848 by members of the Free Church of Scotland. The university was established 1869.

The people are Scotch. They stopped here on their way from home to heaven—thinking they had arrived.

On **Dunedin** Mark Twain *More Tramps Abroad* 1897

Dunfermline /dʌnˈfɜːmlɪn/ industrial town near the Firth of Forth in Fife region, Scotland; population (1988 est) 42,700. It is the site of the naval base of Rosyth; industries include engineering, shipbuilding, electronics, and textiles. Many Scottish kings, including Robert the Bruce, are buried in Dunfermline Abbey. It is the birthplace of the industrialist Andrew Carnegie.

Dungeness /ˌdʌndʒəˈnes/ shingle headland on the south coast of Kent, England. It has nuclear power stations, a lighthouse, and a bird sanctuary.

Dunkirk /dʌnˈkɜːk/ (French *Dunkerque*) seaport on the north coast of France, in Nord *département*, on the Strait of Dover; population (1990) 71,100. Its harbour is one of the foremost in France, and it has widespread canal links with the rest of France and with Belgium; there is a ferry service to Ramsgate, England. Industries include oil refining, fishing, and the manufacture of textiles, machinery, and soap. Dunkirk was close to the front line during much of World War I, and in World War II, 337,131 Allied troops (including about 110,000 French) were evacuated from the beaches as German forces approached.

Dún Laoghaire /dʌnˈleərə/ (former name *Kingstown*) port and suburb of Dublin, Republic of Ireland; population (1986 est) 54,700. It is a terminal for ferries to Britain, and there are fishing industries.

Dunmow, Little /ˈdʌnməʊ/ village in Essex, England, scene every four years of the *Dunmow Flitch* trial (dating from 1111), in which a side of bacon is presented to any couple who 'will swear that they have not quarrelled nor repented of their marriage within a year and a day after its celebration'; they are judged by a jury whose members are all unmarried.

Dunstable /ˈdʌnstəbəl/ town in SW Bedfordshire, England, at the northern end of the Chiltern Hills; 48 km/30 mi NW of London; population (1984

Durham

est) 35,700. Whipsnade Zoo is nearby. Industries include printing and engineering.

Durazzo /duˈrætsəʊ/ Italian form of ◊Durrës, a port in Albania.

Durban /ˈdɜːbən/ principal port of Natal, South Africa, and second port of the republic; population (1985) 634,000, urban area 982,000. It exports coal, maize, and wool, imports heavy machinery and mining equipment, and is also a holiday resort.

Founded 1824 as Port Natal, it was renamed 1835 after General Benjamin d'Urban (1777–1849), lieutenant governor of the eastern district of Cape Colony 1834–37. Natal university (1949) is divided between Durban and Pietermaritzburg.

Durham /ˈdʌrəm/ county in NE England
area 2,440 sq km/942 sq mi
towns Durham (administrative headquarters), Darlington, Peterlee, Newton Aycliffe
features Beamish open-air industrial museum
products sheep and dairy produce; site of one of Britain's richest coalfields
population (1988 est) 596,800
famous people Elizabeth Barrett Browning, Anthony Eden.

Durham /ˈdʌrəm/ city and administrative headquarters of the county of Durham, England; population (1983) 88,600. Founded 995, it has a Norman cathedral dating from 1093, where the remains of the theologian and historian Bede were transferred 1370; the castle was built 1072 by William I and the university was founded 1832. Textiles, engineering, and coal mining are the chief industries.

Durham /ˈdɜːrəm/ city in N central North Carolina, USA, NW of Raleigh; seat of Durham County; population (1990) 136,600. Tobacco is the main industry, and other products include precision instruments, textiles, furniture, and lumber. Duke University is here.

Dyfed

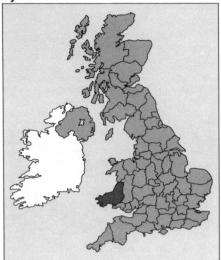

Durrës /'durəs/ chief port of Albania; population (1983) 72,000. It is a commercial and communications centre, with flour mills, soap and cigarette factories, distilleries, and an electronics plant. It was the capital of Albania 1912–21.

Dushanbe /ˌduːʃænˈbeɪ/ (formerly 1929–69 *Stalinabad*) capital of Tajikistan, 160 km/ 100 mi N of the Afghan frontier; population (1987) 582,000. It is a road, rail, and air centre. Industries include cotton mills, tanneries, meat-packing factories, and printing works. It is the seat of Tajik state university. A curfew was imposed Feb 1990–Jan 1991 in response to antigovernment rioting and pogroms; a state of emergency remained in force after Jan. In March–May 1992 antigovernment protests left more than 100 dead and in Aug protestors stormed the presidential palace demanding President Nabiyev's resignation. Nabiyev was seized while trying to flee the capital and resigned.

Düsseldorf /'dusəldɔːf/ industrial city of Germany, on the right bank of the river Rhine, 26 km/ 16 mi NW of Cologne, capital of North Rhine–Westphalia; population (1988) 561,000. It is a river port and the commercial and financial centre of the Ruhr area, with food processing, brewing, agricultural machinery, textile, and chemical industries.

Dutch East Indies former Dutch colony, which in 1945 became independent as ⟡Indonesia.

Dutch Guiana /gɪˈɑnə/ former Dutch colony, which in 1975 became independent as ⟡Surinam.

Dyfed /'dʌvɪd/ county in SW Wales
area 5,770 sq km/2,227 sq mi
towns Carmarthen (administrative headquarters), Llanelli, Haverfordwest, Aberystwyth, Cardigan, Lampeter
features Pembrokeshire Coast National Park; part of the Brecon Beacons National Park, including the Black Mountain range; part of the Cambrian Mountains, including Plynlimon Fawr, 752 m/2,468 ft; the village of Laugharne, at the mouth of the river Tywi, which was the home of the writer Dylan Thomas; the Museum of the Woollen Industry at Dre-fach Felindre, and the Museum of Welsh religious life at Tre'rddôl; anthracite mines produce about 50,000 tonnes a year
population (1988 est) 348,400
languages 46% Welsh, English
famous people Dafydd ap Gwilym, Giraldus Cambrensis.

Dzerzhinsk /dzə'ʒɪnsk/ city in Russia, on the Oka River, 32 km/20 mi W of Gorky; population (1987) 281,000. There are engineering, chemical, and timber industries.

Dzhambul /dʒæm'bul/ city in S Kazakhstan, in a fruit-growing area NE of Tashkent; population (1985) 303,000. Industries include fruit canning, sugar refining, and the manufacture of phosphate fertilizers.

Ealing /ˈiːlɪŋ/ borough of W Greater London
industries engineering, chemicals
population (1991) 283,600
history The first British sound-film studio was built here 1931, and 'Ealing comedies' became a noted genre in British film-making.

East Anglia /iːst ˈæŋgliə/ region of E England, formerly a Saxon kingdom, including Norfolk, Suffolk, and parts of Essex and Cambridgeshire. Norwich is the principal city of East Anglia. The University of East Anglia (UEA) was founded in Norwich 1962, and includes the Sainsbury Centre for Visual Arts, opened 1978, which has a collection of ethnographic art and sculpture. East Anglian ports such as Harwich and Felixstowe have greatly developed as trade with the rest of Europe increases.

Eastbourne /ˈiːstbɔːn/ English seaside resort in East Sussex, 103 km/64 mi SE of London; population (1989 est) 82,000. The old town was developed in the early 19th century as a model of town planning, largely owing to the 7th duke of Devonshire. The modern town extends along the coast for 5 km/3 mi.
To the E the South Downs terminate in Beachy Head.

Easter Island or *Rapa Nui* Chilean island in the S Pacific Ocean, part of the Polynesian group, about 3,500 km/2,200 mi W of Chile; area about 166 sq km/64 sq mi; population (1985) 2,000. It was first reached by Europeans on Easter Sunday 1722.
On it stand over 800 huge carved statues (moai) and the remains of boat-shaped stone houses, the work of neolithic peoples of unknown origin. The chief centre is Hanga-Roa.

East Germany /iːst ˈdʒɜːməni/ see ⟡Germany, East.

East Indies the Malay Archipelago; the Philippines are sometimes included. The term is also used to refer more generally to SE Asia.

East Kilbride /iːst kɪlˈbraɪd/ town in Strathclyde, Scotland; population (1985) 72,000. It was an old village developed as a new town from 1947 to take overspill from Glasgow, 11 km/6 mi to the NE. It is the site of the National Engineering Laboratory. There are various light industries and some engineering, including jet engines.

East London /iːst ˈlʌndən/ port and resort on the southeast coast of Cape Province, South Africa; population (1980) 160,582. Founded 1846 as *Port Rex*, its name was changed to East London 1848. It has a good harbour, is the terminus of a railway from the interior, and is a leading wool-exporting port.

East Lothian /iːst ˈləʊðiən/ former county of SE Scotland, merged with West Lothian and Midlothian 1975 in the new region of Lothian. Haddington was the county town.

East River /iːst/ tidal strait 26 km/16 mi long, between Manhattan and the Bronx, and Long Island, in New York, USA. It links Long Island Sound with New York Bay and is also connected, via the Harlem River, with the Hudson River. There are docks; most famous of its many bridges is the Brooklyn.

East Siberian Sea /iːst saɪˈbɪəriən/ part of the Arctic Ocean, off the north coast of Russia, between the New Siberian Islands and Chukchi Sea. The world's widest continental shelf, with an average width of nearly 650 km/404 mi, lies in the East Siberian Sea.

East St Louis /iːst seɪnt ˈluːɪs/ city in SW Illinois, USA, on the Mississippi River, across from St Louis, Missouri; population (1990) 40,900. A centre for the processing of livestock; its other industries include steel, paint materials, and machinery.

East Sussex /iːst ˈsʌsɪks/ county in SE England
area 1,800 sq km/695 sq mi
towns Lewes (administrative headquarters), Newhaven (cross-channel port), Brighton, Eastbourne, Hastings, Bexhill, Winchelsea, Rye
features Beachy Head, highest headland on the south coast at 180 m/590 ft, the east end of the ⟡South Downs; the Weald (including Ashdown Forest); Friston Forest; rivers: Ouse, Cuckmere, East Rother; Romney Marsh; the 'Long Man' chalk hill figure at Wilmington, near Eastbourne;

East Sussex

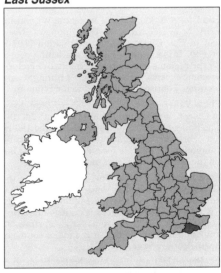

Herstmonceux, with a 15th-century castle (conference and exhibition centre) and adjacent modern buildings, site of the Greenwich Royal Observatory 1958–90; other castles at Hastings, Lewes, Pevensey, and Bodiam; Battle Abbey and the site of the Battle of Hastings; Michelham Priory; Sheffield Park garden; University of Sussex at Falmer, near Brighton, founded 1961 *products* electronics, gypsum, timber *population* (1988 est) 698,000 *famous people* former homes of Henry James at Rye, Rudyard Kipling at Burwash, Virginia Woolf at Rodmell.

East Timor /iːst 'tiːmɔː/ disputed territory on the island of Timor in the Malay Archipelago; prior to 1975, it was a Portuguese colony for almost 460 years *area* 14,874 sq km/5,706 sq mi *capital* Dili *products* coffee *population* (1980) 555,000 *history* Following Portugal's withdrawal 1975, East Timor was left with a literacy rate of under 10% and no infrastructure. Civil war broke out and the left-wing Revolutionary Front of Independent East Timor (Fretilin) occupied the capital, calling for independence. In opposition, troops from neighbouring Indonesia invaded the territory, declaring East Timor (*Loro Sae*) the 17th province of Indonesia July 1976. This claim is not recognized by the United Nations. (It has long been the aim of Indonesian military rulers to absorb the remaining colonial outposts in the East Indies.)

The war and its attendant famine are thought to have caused more than 100,000 deaths, but starvation had been alleviated by the mid-1980s, and the Indonesian government had built schools, roads, and hospitals. Fretilin guerrillas remained active, claiming to have the support of the local population.

In Nov 1991, at least 19 people were killed and 91 injured when Indonesian troops fired on pro-independence demonstrators.

Eau Claire /əu'kleə/ city in W central Wisconsin, USA, N of LaCrosse, on the Chippewa River; seat of Eau Claire County; population (1990) 56,850. It is a processing and marketing centre for the region's dairy farmers. Industries include machine parts, electronics, printing, and brewing. Tourism is important to the economy.

Ebbw Vale /ebuː 'veɪl/ (Welsh *Glyn Ebwy*) town in Gwent, Wales; population (1988 est) 24,100. The iron and steel industries ended in the 1970s, but tin-plate manufacture and engineering continued. To the E is Blaenavon, where the Big Pit (no longer working) is a tourist attraction.

Ebro /iːbrəu/ river in NE Spain, which rises in the Cantabrian Mountains and flows some 800 km/500 mi SE to meet the Mediterranean Sea SW of Barcelona. Zaragoza is on its course, and ocean-going ships can sail as far as Tortosa, 35 km/22 mi from its mouth. It is a major source of hydroelectric power.

Eccles /eklz/ town near Manchester, England, 8 km/5 mi W of Manchester, on the river Irwell and

Manchester Ship Canal; population (1981) 37,200. Industries include cotton textiles, machinery, and pharmaceuticals. Eccles cakes, rounded pastries with a dried fruit filling, originated here.

Ecuador Republic of (*República del Ecuador*) *area* 270,670 sq km/104,479 sq mi *capital* Quito *towns* Cuenca; chief port Guayaquil *physical* coastal plain rises sharply to Andes Mountains, which are divided into a series of cultivated valleys; flat, low-lying rainforest in E *environment* about 25,000 species became extinct 1965–90 as a result of environmental destruction *features* Ecuador is crossed by the equator, from which it derives its name; Galápagos Islands; Cotopaxi is world's highest active volcano; rich wildlife in rainforest of Amazon basin *head of state and government* Sixto Duran Ballen from 1992 *political system* emergent democracy *political parties* Social Christian Party (PSC), right-wing; United Republican Party (PUR), right-of-centre coalition; Ecuadorean Roldosist Party (PRE) *exports* bananas, cocoa, coffee, sugar, rice, fruit, balsa wood, fish, petroleum *currency* sucre *population* (1989 est) 10,490,000; (mestizo 55%, Indian 25%, European 10%, black African 10%); growth rate 2.9% p.a. *life expectancy* men 62, women 66 *languages* Spanish (official), Quechua, Jivaro, and other Indian languages *religion* Roman Catholic 95% *literacy* men 85%, women 80% (1985 est) *GDP* $10.6 bn (1987); $1,069 per head *chronology* *1830* Independence achieved from Spain. *1925–48* Great political instability; no president completed his term of office. *1948–55* Liberals in power.

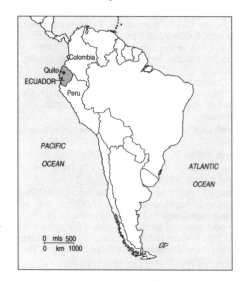

1956 First conservative president in 60 years.
1960 Liberals returned, with José Velasco as president.
1961 Velasco deposed and replaced by the vice president.
1962 Military junta installed.
1968 Velasco returned as president.
1972 A coup put the military back in power.
1978 New democratic constitution adopted.
1979 Liberals in power but opposed by right-and left-wing parties.
1982 Deteriorating economy provoked strikes, demonstrations, and a state of emergency.
1983 Austerity measures introduced.
1984–85 No party with a clear majority in the national congress; Febres Cordero narrowly won the presidency for the Conservatives.
1988 Rodrigo Borja Cevallos elected president for moderate left-wing coalition.
1989 Guerrilla left-wing group, *Alfaro Vive, Carajo* ('Alfaro lives, Dammit'), numbering about 1,000, laid down arms after nine years.
1992 PUR leader, Sixto Duran Ballen, elected president; PSC became largest party in congress.

Edam /ˈiːdæm/ town in the Netherlands on the river IJ, North Holland province; population (1987) 24,200. Founded as a customs post in the 13th century, Edam's prosperity in the 16th and 17th centuries was based upon its cheese trade; it is still famous today for its round cheeses covered in red wax.

Eddystone Rocks /ˈedɪstən/ rocks in the English Channel, 23 km/14 mi S of Plymouth. The lighthouse, built 1882, is the fourth on this exposed site.

Eden /ˈiːdn/ river in Cumbria, NW England; length 104 km/65 mi. From its source in the Pennines, it flows NW to enter the Solway Firth NW of Carlisle.

Edinburg /ˈednbɜːg/ city in SE Texas, USA, NW of Brownsville; seat of Hidalgo County; population (1990) 29,800. It is a shipping centre for the area's citrus fruit crops.

Edinburgh /ˈedɪnbərə/ capital of Scotland and administrative centre of the region of Lothian, near the southern shores of the Firth of Forth; population (1988 est) 433,500. A cultural centre, it holds a major annual festival of music and the arts. The university was established 1583. Industries include printing, publishing, banking, insurance, chemical manufactures, distilling, brewing, and some shipbuilding.
features Edinburgh Castle contains the 12th-century St Margaret's chapel, the oldest building in Edinburgh. The palace of Holyrood House was built in the 15th and 16th centuries on the site of a 12th-century abbey; it is the British sovereign's official Scottish residence. Rizzio, the Italian favourite of Mary Queen of Scots, was murdered here 1566 in her apartments. The Parliament House, begun 1632, is now the seat of the supreme courts. The Royal Scottish Academy and the National Gallery of Scotland (renovated 1989) in Classical style are by William Henry Playfair (1789–1857). The episcopal cathedral of St Mary, opened 1879, and St Giles parish church (mostly 15th-century) are the principal churches.

The Royal Observatory has been at Blackford Hill since 1896. The principal thoroughfares are Princes Street and the Royal Mile. The university has a famous medical school and the Koestler chair of parapsychology (instituted 1985), the only such professorship in the UK. The Heriot-Watt University (established 1885; university status 1966) is mainly a technical institution.
history In Roman times the site was occupied by Celtic peoples and about 617 was captured by Edwin of Northumbria, from whom the town took its name. The early settlement grew up around a castle on Castle Rock, while about a mile to the E another burgh, Canongate, developed around the abbey of Holyrood, founded 1128 by David I. It remained separate from Edinburgh until 1856. Robert Bruce made Edinburgh a burgh 1329, and established its port at Leith. In 1544 the town was destroyed by the English. After the union with England 1707, Edinburgh lost its political importance but remained culturally pre-eminent. During the 18th century, Edinburgh was known as the 'Athens of the North' because of its concentration of intellectual talent, for example, Adam Smith, David Hume, and Joseph Black.
Development of the area known as New Town was started 1767.

Pompous the boast, and yet a truth it speaks/A 'modern Athens', – fit for modern Greeks.

On **Edinburgh** James Hannay
Edinburgh Evening Courant 1860

Edirne /eˈdɪəneɪ/ town in European Turkey, on the river Maritsa, about 225 km/140 mi NW of Istanbul; population (1985) 86,700. Founded on the site of ancient Uscadama, it was formerly known as *Adrianople*, named after the Emperor Hadrian about AD 125.

Edmonton /ˈedməntən/ capital of Alberta, Canada, on the North Saskatchewan River; population (1986) 576,200. It is the centre of an oil and mining area to the N and also an agricultural and dairying region. Petroleum pipelines link Edmonton with Superior, Wisconsin, USA, and Vancouver, British Columbia.

Edmonton /ˈedməntən/ a locality, once a town, part of the London borough of Enfield. John Keats lived at Edmonton, and Charles Lamb lived and died here. The Bell Inn is referred to in William Cowper's poem 'John Gilpin'.

Edward, Lake /ˈedwəd/ lake in Uganda, area 2,150 sq km/830 sq mi, at about 900 m/3,000 ft above sea level in the Albertine rift valley. It was known as Lake Idi Amin Dada, after President Amin of Uganda, 1973–79.

Egmont, Mount /ˈegmɒnt/ (Maori *Taranaki*) symmetrical extinct volcano in North Island, New Zealand, situated S of New Plymouth; it is 2,517 m/8,260 ft high.

Egypt Arab Republic of (*Jumhuriyat Misr al-Arabiya*)
area 1,001,450 sq km/386,990 sq mi
capital Cairo
towns Gîza; ports Alexandria, Port Said, Suez, Damietta
physical mostly desert; hills in E; fertile land along Nile valley and delta; cultivated and settled area is about 35,500 sq km/13,700 sq mi
environment the building of the Aswan Dam (opened 1970) on the Nile has caused widespread salinization and an increase in waterborne diseases in villages close to Lake Nasser. A dramatic fall in the annual load of silt deposited downstream has reduced fertility of cropland and has led to coastal erosion and loss of sardine shoals
features Aswan High Dam and Lake Nasser; Sinai; remains of ancient Egypt (pyramids, Sphinx, Luxor, Karnak, Abu Simbel, El Faiyum)
head of state and government Hosni Mubarak from 1981
political system democratic republic
political parties National Democratic Party (NDP), moderate, left of centre; Socialist Labour Party, right of centre; Socialist Liberal Party, free-enterprise; New Wafd Party, nationalist
exports cotton and textiles, petroleum, fruit and vegetables
currency Egyptian pound
population (1989 est) 54,779,000; growth rate 2.4% p.a.
life expectancy men 57, women 60
languages Arabic (official); ancient Egyptian survives to some extent in Coptic
media there is no legal censorship, but the largest publishing houses, newspapers, and magazines are owned and controlled by the state, as is all television. Questioning of prevalent values, ideas, and social practices is discouraged
religions Sunni Muslim 95%, Coptic Christian 5%
literacy men 59%, women 30% (1985 est)
GDP $34.5 bn (1987); $679 per head
chronology
1914 Egypt became a British protectorate.
1936 Independence achieved from Britain. King Fuad succeeded by his son Farouk.
1946 Withdrawal of British troops except from Suez Canal Zone.
1952 Farouk overthrown by army in coup.

1953 Egypt declared a republic, with General Neguib as president.
1956 Neguib replaced by Col Gamal Nasser. Nasser announced nationalization of Suez Canal; Egypt attacked by Britain, France, and Israel. Cease-fire agreed because of US intervention.
1958 Short-lived merger of Egypt and Syria as United Arab Republic (UAR). Subsequent attempts to federate Egypt, Syria, and Iraq failed.
1967 Six-Day War with Israel ended in Egypt's defeat and Israeli occupation of Sinai and Gaza Strip.
1970 Nasser died suddenly, succeeded by Anwar Sadat.
1973 Attempt to regain territory lost to Israel led to fighting; cease-fire arranged by US secretary of state Henry Kissinger.
1977 Sadat's visit to Israel to address the Israeli parliament was criticized by Egypt's Arab neighbours.
1978–79 Camp David talks in the USA resulted in a treaty between Egypt and Israel. Egypt expelled from the Arab League.
1981 Sadat assassinated, succeeded by Hosni Mubarak.
1983 Improved relations between Egypt and the Arab world; only Libya and Syria maintained a trade boycott.
1984 Mubarak's party victorious in the people's assembly elections.
1987 Mubarak re-elected. Egypt readmitted to Arab League.
1988 Full diplomatic relations with Algeria restored.
1989 Improved relations with Libya; diplomatic relations with Syria restored. Mubarak proposed a peace plan.
1990 Gains for independents in general election.
1991 Participation in Gulf War on US-led side. Major force in convening Middle East peace conference in Spain.
1992 Outbreaks of violence between Muslims and Christians. Earthquake devastated Cairo.

Egypt is an acquired country, the gift of the river.

On **Egypt** Herodotus *History c.* 460 BC

Eiger /ˈaɪgə/ mountain peak in the ◊Swiss Alps.

Eilat /eɪˈlɑːt/ alternative spelling of ◊Elat, a port in Israel.

Eindhoven /ˈaɪndhəuvn/ town in North Brabant province, the Netherlands, on the river Dommel; population (1988) 381,000. Industries include electrical and electronic equipment; Philips Electronics have their headquarters here.

Eire /ˈeərə/ Gaelic name for the Republic of ◊Ireland.

Eisenach /ˈaɪzənæx/ industrial town (pottery, vehicles, machinery) in the state of Thuringia, Germany; population (1981) 50,700. Martin Luther made the first translation of the Bible into

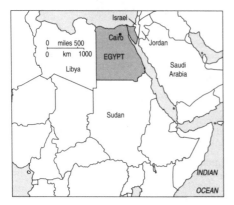

German in Wartburg Castle and the composer J S Bach was born here.

Eisenhower, Mount /ˈaɪzənˌhaʊə/ Rocky Mountain peak in Alberta, Canada, included in Banff National Park; it is 2,862 m/9,390 ft.

Ekaterinburg /eˌkætəriːnˈbɜːg/ (formerly 1924–90 *Sverdlovsk*) industrial town (copper, iron, platinum, engineering, and chemicals) in Russia in the eastern foothills of the Ural Mountains; population (1987) 1,331,000. Tsar Nicholas II and his family were murdered here 1918.

Ekaterinodar /eˌkætəriːnəʊˈdɑː/ pre-revolutionary name of ◊Krasnodar, an industrial town in Russia.

Ekaterinoslav /eˌkætəriːnəʊˈslɑːv/ pre-revolutionary name of ◊Dnepropetrovsk, the centre of an industrial region in Ukraine.

El Aaiún /ˌelaɪˈuːn/ Arabic name of ◊Laâyoune, the capital of Western Sahara.

Elat /eɪˈlɑːt/ or *Eilat* port at the head of the Gulf of Aqaba, Israel's only outlet to the Red Sea; population (1982) 19,500. Founded 1948, on the site of the Biblical Elath, it is linked by road with Beersheba. There are copper mines and granite quarries nearby, and a major geophysical observatory, opened 1968, is 16 km/10 mi to the N.

Elba /ˈelbə/ island in the Mediterranean Sea, 10 km/6 mi off the west coast of Italy; area 223 sq km/86 sq mi; population (1981) 35,000. Iron ore is exported from the capital, Portoferraio, to the Italian mainland, and there is a fishing industry. The small uninhabited island of *Monte Cristo*, 40 km/25 mi to the S, supplied the title of Alexandre Dumas's hero in his historical romance *The Count of Monte Cristo* 1844. Elba was Napoleon's place of exile 1814–15.

Lucky Napoleon!

On **Elba** Dylan Thomas, letter of 1947

Elbe /elb/ one of the principal rivers of Germany, 1,166 km/725 mi long, rising on the southern slopes of the Riesengebirge, Czech Republic, and flowing NW across the German plain to the North Sea.

Elberfeld /ˈelbəfelt/ German industrial town, merged with ◊Wuppertal 1929.

Elbing /ˈelbɪŋ/ German name for Elbląg, a Polish port.

Elbląg /ˈelblɒŋk/ Polish port 11 km/7 mi from the mouth of the river Elbląg, which flows into the Vistula Lagoon, an inlet of the Baltic Sea; population (1983) 115,900. It has shipyards, engineering works, and car and tractor factories.

Elbruz /elˈbruːs/ or *Elbrus* highest mountain (5,642 m/18,510 ft) on the continent of Europe, in the Caucasus, Georgia.

Elburz /elˈbʊəz/ volcanic mountain range in NW Iran, close to the southern shore of the Caspian

Sea; the highest point is Mount Damavand at 5,670 m/18,602 ft.

Elephanta /ˌeliˈfæntə/ island in Bombay harbour, Maharashtra, India, some 8 km/5 mi from Bombay. The Temple Caves (6th century), cut out of solid rock, have sculptures of many Hindu deities executed 450–740.

There was formerly a large stone elephant near the island's landing place.

El Faiyûm /el faɪˈjuːm/ city in N Egypt, 90 km/56 mi SW of Cairo; population (1985) 218,500. It was a centre of prehistoric culture; the crocodile god Sobek was worshipped nearby, and realistic mummy portraits dating from the 1st–4th centuries AD were found in the area.

El Ferrol /el feˈrɒl/ full name *El Ferrol del Caudillo* city and port in La Coruña province, on the northwest coast of Spain; population (1986) 88,000. It is a naval base and has a deep, sheltered harbour and shipbuilding industries. It is the birthplace of the Spanish dictator Francisco Franco.

Elgin /ˈelgɪn/ chief town of Moray District, Grampian Region, NE Scotland, on the river Lossie 8 km/5 mi S of its port of Lossiemouth on the southern shore of the Moray Firth; population (1988 est) 20,000. There are sawmills and whisky distilleries. Gordonstoun public school is nearby. Elgin Cathedral, founded 1224, was destroyed 1390.

Elgin /ˈeldʒɪn/ city in NE Illinois, USA, NW of Chicago, on the Fox River; population (1990) 77,000. Industries include electrical machinery and dairy products. The city was once the home of Elgin watches.

Elisabethville /ɪˈlɪzəbəθvɪl/ former name of ◊Lubumbashi, a town in Zaire.

Elizabeth /ɪˈlɪzəbəθ/ city in NE New Jersey, USA; population (1990) 110,000. Established 1664, it was the first English settlement in New Jersey. It has car, sewing-machine, and tool factories; oil refineries; and chemical works.

Elizavetpol /ɪˌlɪzəˈvetpɒl/ former name of ◊Kirovabad, an industrial town in Azerbaijan.

Elkhart /ˈelkɑːt/ city in N Indiana, USA, E of South Bend, where the Elkhart River meets the St Joseph River; population (1990) 43,600. Its factories produce mobile homes, firefighting apparatus, recreational vehicles, pharmaceuticals, and musical instruments.

Ellesmere /ˈelzmɪə/ second largest island of the Canadian Arctic archipelago, Northwest Territories; area 212,687 sq km/82,097 sq mi. It is for the most part barren or glacier-covered.

Ellesmere Port /ˈelzmɪə ˈpɔːt/ oil port and industrial town in Cheshire, England, on the river Mersey and the Manchester Ship Canal; population (1988 est) 79,700.

Formerly the biggest transshipment canal port in NW England, it now has the National Waterways Museum (opened 1976), with old narrow boats and a blacksmith's forge.

Ellice Islands /ˈelɪs/ former name of ◊Tuvalu, a group of islands in the W Pacific Ocean.

Ellis Island /'elɪs/ island in New York Harbor, USA; area 11 hectares/27 acres. A former reception centre for steerage-class immigrants during the immigration waves between 1892 and 1943 (12 million people passed through it 1892–1924), it was later used as a detention centre for non-residents without documentation, or for those who were being deported. It was declared a National Historic Site 1964 by President Lyndon Johnson.

The island was named after a Welshman, Samuel Ellis, who owned it in the late 18th century. In 1990 a museum of American immigration was established on the island.

Elmira /el'maɪrə/ city in S central New York, USA, on the Chemung River, W of Binghamton, below the Finger Lakes region; seat of Chemung County; population (1990) 33,700. It is the processing and marketing centre for the area's dairy and poultry farms. Other industries include business machinery, machine parts, aeroplanes, and fire engines. Elmira College (1853) is located here.

El Obeid /el əʊ'beɪd/ capital of Kordofan province, Sudan; population (1984) 140,025. Linked by rail with Khartoum, it is a market for cattle, gum arabic, and durra (Indian millet).

El Paso /el 'pæsəʊ/ city in Texas, USA, situated at the base of the Franklin Mountains, on the Rio Grande, opposite the Mexican city of Ciudad Juárez; population (1990) 515,300. It is the centre of an agricultural and cattle-raising area, and there are electronics, food processing, packing, and leather industries, as well as oil refineries and industries based on local iron and copper mines.

The city of the Four C's—climate, cotton, cattle, copper.

On **El Paso** John Gunther *Inside USA* 1947

El Salvador Republic of (*República de El Salvador*)
area 21,393 sq km/8,258 sq mi
capital San Salvador
towns Santa Ana, San Miguel
physical narrow coastal plain, rising to moun-

tains in N with central plateau
features smallest and most densely populated Central American country; Mayan archaeological remains
head of state and government Alfredo Cristiani from 1989
political system emergent democracy
political parties Christian Democrats (PDC), anti-imperialist; National Republican Alliance (ARENA), right-wing; National Conciliation Party (PCN), right-wing; Farabundo Marti Liberation Front (FMLN), left-wing
exports coffee, cotton, sugar
currency colón
population (1989 est) 5,900,000 (mainly of mixed Spanish and Indian ancestry; 10% Indian); growth rate 2.9% p.a.
life expectancy men 63, women 66
languages Spanish, Nahuatl
religion Roman Catholic 97%
literacy men 75%, women 69% (1985 est)
GDP $4.7 bn (1987); $790 per head
chronology
1821 Independence achieved from Spain.
1931 Peasant unrest followed by a military coup.
1961 Following a coup, PCN established and in power.
1969 'Soccer' war with Honduras.
1972 Allegations of human-rights violations and growth of left-wing guerrilla activities. General Carlos Romero elected president.
1979 A coup replaced Romero with a military-civilian junta.
1980 Archbishop Oscar Romero assassinated; country on verge of civil war. José Duarte became first civilian president since 1931.
1981 Mexico and France recognized the guerrillas as a legitimate political force but the USA actively assisted the government in its battle against them.
1982 Assembly elections boycotted by left-wing parties and held amid considerable violence.
1986 Duarte sought a negotiated settlement with the guerrillas.
1988 Duarte resigned.
1989 Alfredo Cristiani (ARENA) became president in rigged elections; rebel attacks intensified.
1991 United Nations-sponsored peace accord signed by representatives of the government and the socialist guerrilla group, the FMLN.
1992 Peace accord validated; FMLN became political party.

Elsinore /'elsɪnɔː/ another form of ◊Helsingør, a port on the northeast coast of Denmark.

Ely /'iːli/ city in Cambridgeshire, England, on the Great Ouse River 24 km/15 mi NE of Cambridge; population (1983) 11,000. It has sugar beet, paper, and engineering factories.
history It was the chief town of the former administrative district of the *Isle of Ely*, so called because the area was once cut off from the surrounding countryside by the fens. The Anglo-Saxon rebel Hereward the Wake had his stronghold here. The 11th-century cathedral is one of the largest in England. At the annual feast of St Etheldreda (Audrey), founder of a religious community at Ely in the 7th century, cheap, low-quality souvenirs were sold; the word 'tawdry',

a corruption of St Audrey, derives from this practice.

The Isle of Ely lying on the fens is like a starfish lying on a flat stone at low tide.

On **Ely** Hilaire Belloc *Hills and the Sea* 1906

Elyria /ɪˈlɪrɪə/ city in N Ohio, USA, on the Black River, W of Cleveland; seat of Lorain County; population (1990) 56,700. Industries include tools, electric motors, chromium hardware, chemicals, and automotive parts.

Emba /ˈembə/ river in Kazakhstan, 612 km/380 mi long, draining into the northern part of the Caspian Sea.

Emden /ˈemdən/ port in Lower Saxony, Germany, at the mouth of the river Ems; population (1984) 51,000. It is a fishing port and an export outlet for the river Ruhr, with which it is connected by the Dortmund–Ems Canal. There are oil refineries here.

Emi Koussi /ˌemɪˈkuːsi/ highest point of the Tibesti massif in N Chad, rising to 3,415 m/11,204 ft.

Emilia-Romagna /eˈmiːljə rəuˈmænjə/ region of N central Italy including much of the Po Valley; area 22,100 sq km/8,531 sq mi; population (1988) 3,924,000. The capital is Bologna; other towns include Reggio, Rimini, Parma, Ferrara, and Ravenna. Agricultural produce includes fruit, wine, sugar beet, beef, and dairy products; oil and natural-gas resources have been developed in the Po Valley.

Emmental /ˈeməntɑːl/ district in the valley of the Emme River, Berne, Switzerland, where a hard cheese of the same name has been made since the mid-15th century. The main town in Emmental is Langnau.

Enfield /ˈenfiːld/ borough of NE Greater London
features little remains of Edward VI's palace, but the royal hunting ground of Enfield Chase partly survives in the 'green belt'; the district of Edmonton, where John Keats and Charles and Mary Lamb once lived (the Lambs are buried there); the Bell Inn, referred to in William Cowper's poem 'John Gilpin';London's first regional park, Lea Valley, developed since the 1970s
industries engineering (the Royal Small Arms factory produced the Enfield rifle), textiles, furniture, and cement
population (1991) 249,100.

Engadine /ˈeŋɡədiːn/ upper valley of the river Inn in Switzerland, a winter sports resort.

England /ˈɪŋɡlənd/ largest division of the United Kingdom
area 130,357 sq km/50,318 sq mi
capital London
towns Birmingham, Cambridge, Coventry, Leeds, Leicester, Manchester, Newcastle-upon-Tyne, Nottingham, Oxford, Sheffield, York; ports Bristol, Dover, Felixstowe, Harwich, Liverpool, Portsmouth, Southampton

features variability of climate and diversity of scenery; among European countries, only the Netherlands is more densely populated
exports agricultural (cereals, rape, sugar beet, potatoes); meat and meat products; electronic (software) and telecommunications equipment (main centres Berkshire and Cambridge); scientific instruments; textiles and fashion goods; North Sea oil and gas, petrochemicals, pharmaceuticals, fertilizers; beer; china clay, pottery, porcelain, and glass; film and television

England: counties

county	administrative headquarters	area in sq km
Avon	Bristol	1,340
Bedfordshire	Bedford	1,240
Berkshire	Reading	1,260
Buckinghamshire	Aylesbury	1,880
Cambridgeshire	Cambridge	3,410
Cheshire	Chester	2,320
Cleveland	Middlesbrough	580
Cornwall	Truro	3,550
Cumbria	Carlisle	6,810
Derbyshire	Matlock	2,630
Devon	Exeter	6,720
Dorset	Dorchester	2,650
Durham	Durham	2,440
East Sussex	Lewes	1,800
Essex	Chelmsford	3,670
Gloucestershire	Gloucester	2,640
Hampshire	Winchester	3,770
Hereford & Worcester	Worcester	3,930
Hertfordshire	Hertford	1,630
Humberside	Kingston upon Hull	3,510
Isle of Wight	Newport	380
Kent	Maidstone	3,730
Lancashire	Preston	3,040
Leicestershire	Leicester	2,550
Lincolnshire	Lincoln	5,890
London, Greater		1,580
Manchester, Greater		1,290
Merseyside		650
Norfolk	Norwich	5,360
Northamptonshire	Northampton	2,370
Northumberland	Newcastle-upon-Tyne	5,030
North Yorkshire	Northallerton	8,320
Nottinghamshire	Nottingham	2,160
Oxfordshire	Oxford	2,610
Shropshire	Shrewsbury	3,490
Somerset	Taunton	3,460
South Yorkshire		1,560
Staffordshire	Stafford	2,720
Suffolk	Ipswich	3,800
Surrey	Kingston upon Thames	1,660
Tyne & Wear		540
Warwickshire	Warwick	1,980
West Midlands		900
West Sussex	Chichester	2,020
West Yorkshire		2,040
Wiltshire	Trowbridge	3,480

programmes, and sound recordings. Tourism is important. There are worldwide banking and insurance interests
currency pound sterling
population (1986) 47,255,000
languages English, with more than 100 minority languages
religions Christian, with the Church of England as the established church, 31,500,000; and various Protestant groups, of which the largest is the Methodist 1,400,000; Roman Catholic about 5,000,000; Muslim 900,000; Jewish 410,000; Sikh 175,000; Hindu 140,000.

There'll always be an England/While there's a country lane/ Wherever there's a cottage small/Beside a field of grain.

On **England** song by Ross Parker and Hugh Charles 1939

English Channel stretch of water between England and France, leading in the W to the Atlantic Ocean, and in the E via the Strait of Dover to the North Sea; it is also known as *La Manche* (French 'the sleeve') from its shape.

The English Channel is 450 km/280 mi long W–E; 27 km/17 mi wide at its narrowest (Cap Gris Nez–Dover) and 117 km/110 mi wide at its widest (Ushant–Land's End).

If the Almighty were to rebuild the world and asked me for advice I would have English Channels round every country.

On the **English Channel** Winston Churchill 1952

Enid /'i:nɪd/ city in N central Oklahoma, USA, N of Oklahoma City, seat of Garfield County; population (1990) 45,300. It is a processing and marketing centre for the surrounding region's poultry, cattle, dairy farms, and oil wells. Enid's founded 1893, when land in the Cherokee Strip region, which was originally put aside for Indian tribes, was opened to white settlement.

Eniwetok /ˌenɪ'wi:tɒk/ atoll in the Marshall Islands, in the central Pacific Ocean; population (1980) 453. It was taken from Japan by the USA 1944, which made the island a naval base; 43 atomic tests were conducted here from 1947. The inhabitants were resettled at Ujelang, but insisted on returning home 1980. Despite the clearance of nuclear debris and radioactive soil to the islet of Runit, high radiation levels persisted.

Ennis /'enɪs/ county town of County Clare, Republic of Ireland, on the river Fergus, 32 km/20 mi NW of Limerick; population (1981) 15,000. There are distilleries, flour mills, and furniture manufacturing.

Enniskillen /ˌenɪs'kɪlən/ county town of Fermanagh, Northern Ireland, between Upper and Lower Lough Erne; population (1989 est) 10,500. There is some light industry (engineer-

ing, food processing) and it has been designated for further industrial growth. A bomb exploded there at a Remembrance Day service Nov 1987, causing many casualties.

Enschede /'enskədeɪ/ textile manufacturing centre in Overijssel province, the Netherlands; population (1990) 146,000, urban area of *Enschede-Hengelo* 250,000.

Entebbe /en'tebi/ town in Uganda, on the northwest shore of Lake Victoria, 20 km/12 mi SW of Kampala, the capital; 1,136 m/3,728 ft above sea level; population (1983) 21,000. Founded 1893, it was the administrative centre of Uganda 1894–1962.

In 1976, a French aircraft was hijacked by a Palestinian liberation group. It was flown to Entebbe airport, where the hostages on board were rescued six days later by Israeli troops.

Enugu /e'nu:gu:/ town in Nigeria, capital of Anambra state; population (1983) 228,400. It is a coal-mining centre, with steel and cement works, and is linked by rail with Port Harcourt.

Epernay /ˌepeə'neɪ/ town in Marne *département*, Champagne-Ardenne region, France; population (1990) 27,700. Together with Reims, it is the centre of the champagne industry. Corks, casks, and bottles are manufactured alongside the production and storage of champagne.

Epinal /ˌeɪpɪ'næl/ capital of Vosges *département*, on the river Moselle, France; population (1990) 39,500. A cotton-textile centre, it dates from the 10th century.

Epirus /ɪ'paɪrəs/ (Greek *Ipiros*) region of NW Greece; area 9,200 sq km/3,551 sq mi; population (1981) 325,000. Its capital is Yannina, and it consists of the provinces (nomes) of Arta, Thesprotia, Yannina, and Preveza. There is livestock farming.

Epping Forest /'epɪŋ/ forest in Essex, SE England.

Epsom /'epsəm/ town in Surrey, England; population (with Ewell) (1988 est) 67,500. In the 17th century it was a spa producing Epsom salts. There is a racecourse, where the Derby and the Oaks horse races are held. The site of Henry VIII's palace of Nonsuch was excavated 1959.

Equatorial Guinea Republic of (*República de Guinea Ecuatorial*)
area 28,051 sq km/10,828 sq mi
capital Malabo (Bioko)
towns Bata, Mbini (Río Muni)
physical comprises mainland Río Muni, plus the small islands of Corisco, Elobey Grande and Elobey Chico, and Bioko (formerly Fernando Po) together with Annobón (formerly Pagalu)
features volcanic mountains on Bioko
head of state and government Teodoro Obiang Nguema Mbasogo from 1979
political system one-party military republic
political party Democratic Party of Equatorial Guinea (PDGE), militarily controlled
exports cocoa, coffee, timber
currency ekuele; CFA franc
population (1988 est) 336,000 (plus 110,000 estimated to live in exile abroad); growth rate 2.2% p.a.
life expectancy men 44, women 48

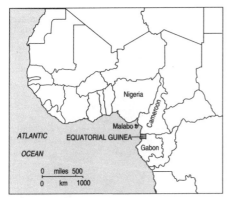

languages Spanish (official); pidgin English is widely spoken, and on Annobón (whose people were formerly slaves of the Portuguese) a Portuguese dialect; Fang and other African dialects spoken on Río Muni
religions nominally Christian, mainly Catholic, but in 1978 Roman Catholicism was banned
literacy 55% (1984)
GDP $90 million (1987); $220 per head
chronology
1778 Fernando Po (Bioko Island) ceded to Spain.
1885 Mainland territory came under Spanish rule; colony known as Spanish Guinea.
1968 Independence achieved from Spain. Francisco Macias Nguema became first president, soon assuming dictatorial powers.
1979 Macias overthrown and replaced by his nephew, Teodoro Obiang Nguema Mbasogo, who established a military regime. Macias tried and executed.
1982 Obiang elected president unopposed for another seven years. New constitution adopted.
1989 Obiang re-elected president.
1992 New constitution adopted; elections held, but president continued to nominate candidates for top government posts.

Erebus, Mount /'erɪbəs/ the world's southernmost active volcano, 3,794 m/12,452 ft high, on Ross Island, Antarctica. It contains a lake of molten lava, that scientists are investigating in the belief that it can provide a 'window' onto the magma beneath the Earth's crust.

Erfurt /'eəfuət/ city in Germany on the river Gera, capital of the state of Thuringia; population (1990) 217,000. It is in a rich horticultural area, and its industries include textiles, typewriters, and electrical goods.

Erie /'ɪəri/ city and port on the Pennsylvania bank of Lake Erie, USA; population (1990) 108,700. It has heavy industries and a trade in iron, grain, and freshwater fish. A French fort was built on the site 1753, and a permanent settlement was laid out 1795.

Erie, Lake /'ɪəri/ fourth largest of the Great Lakes of North America, connected to Lake Ontario by the Niagara River and bypassed by the Welland Canal; area 9,930 sq mi/25,720 sq km. It is linked to Lake Huron by Lake St Clair and the St

Clair and Detroit rivers and to the Hudson River by the New York State Barge Canal. It is an important component of the St Lawrence Seaway. Lake Erie ports include Cleveland and Toledo, Ohio; Erie, Pennsylvania; and Buffalo, New York. The shallowest of the Great Lakes, Lake Erie has become severely polluted from industrial and municipal waste.

Eritrea State of
area 125,000 sq km/48,250 sq mi
capital Asmara
towns Keren, Adigrat; ports: Asab, Massawa
physical coastline along the Red Sea 1,000 km/620 mi; narrow coastal plain that rises to an inland plateau
features Dahlak Islands
head of state and government Issaias Afewerki from 1993
political system emergent democracy
political parties Eritrean People's Liberation Front (EPLF), militant nationalist; Eritrean National Pact Alliance (ENPA), moderate, centrist
products coffee, salt, citrus fruits, grains, cotton
currency birr
population (1993) 4,000,000
languages Amharic (official), Tigrinya (official), Arabic, Afar, Bilen, Hidareb, Kunama, Nara, Rashaida, Saho, and Tigre
religions Muslim, Coptic Christian
chronology
1962 Annexed by Ethiopia; secessionist movement began.
1974 Ethiopian emperor Haile Selassie deposed by military; Eritrean People's Liberation Front (EPLF) continued struggle for independence.
1990 Port of Massawa captured by Eritrean forces.
1991 Ethiopian president Mengistu Haile Mariam overthrown. EPLF secured the whole of Eritrea. Ethiopian government acknowledged Eritrea's right to secede. Issaias Afewerki became secretary general of provisional government.
1992 Transitional ruling council established.
1993 National referendum supported independence; new state formally recognized by Ethiopia and UK. Transitional government established for

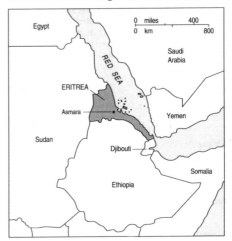

four-year period. Issaias Afwerki elected chairman of state council and president.

Erivan /ˌerɪ'væn/ alternative transliteration of ◊Yerevan, the capital of Armenia.

Erlangen /'eəlæŋən/ industrial town in Bavaria, Germany; population (1988) 100,000. Products include electrical equipment, textiles, glass, and beer.

Erzgebirge /'eətsgə,bɪəgə/ (German 'ore mountains') mountain range on the German-Czech frontier, where the rare metals uranium, cobalt, bismuth, arsenic, and antimony are mined. Some 145 km/90 mi long, its highest summit is Mount Klinovec (Keilberg), 1,244 m/4,080 ft, in the Czech Republic. In 1991, following the reunification of Germany, many uranium mines were closed and plans to clean up the heavily polluted region were formulated.

Erzurum /'eəzurum/ capital of Erzurum province, NE Turkey; population (1985) 253,000. It is a centre of agricultural trade and mining, and has a military base.

Esbjerg /'esbjɜːg/ port of Ribe county, Denmark, on the west coast of Jutland; population (1990) 81,500. It is the terminus of links with Sweden and the UK, and is a base for Danish North Sea oil exploration.

Eskilstuna /'eskɪlz,tuːnə/ town W of Stockholm, Sweden; population (1986) 88,400. It has iron foundries and steel and armament works; it was known in the 17th century for its swords and cutlery.

Eskişehir /es'kiːʃəhɪə/ city in Turkey, 200 km/125 mi W of Ankara; population (1985) 367,000. Products include meerschaum, chromium, magnesite, cotton goods, tiles, and aircraft.

Esquipulas /eskɪ'pʌləs/ pilgrimage town in Chiquimula department, SE Guatemala; seat of the 'Black Christ', which is a symbol of peace throughout Central America. In May 1986 five Central American presidents met here to discuss a plan for peace in the region.

Essen /'esən/ city in North Rhine–Westphalia, Germany; population (1988) 615,000. It is the administrative centre of the Ruhr region, situated between the rivers Emscher and Ruhr, and has textile, chemical, and electrical industries. Its 9th–14th-century minster is one of the oldest churches in Germany.

Essequibo /ˌesɪ'kwiːbəʊ/ longest river in Guyana, South America, rising in the Guiana Highlands of S Guyana; length 1,014 km/630 mi. Part of the district of Essequibo, which lies to the W of the river, is claimed by Venezuela.

Essex /'esɪks/ county in SE England
area 3,670 sq km/1,417 sq mi
towns Chelmsford (administrative headquarters), Colchester; ports: Harwich, Tilbury; resorts: Southend, Clacton
features former royal hunting ground of Epping Forest (controlled from 1882 by the City of London); the marshy coastal headland of the Naze; since 1111 at Great Dunmow the

Essex

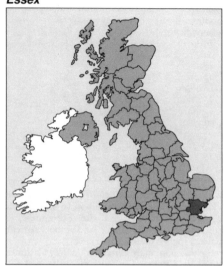

Dunmow flitch (side of cured pork) can be claimed every four years by any couple proving to a jury they have not regretted their marriage within the year (winners are few); Stansted, London's third airport
products dairying, cereals, fruit
population (1988 est) 1,529,500
famous people William Harvey.

Estonia Republic of
area 45,000 sq km/17,000 sq mi
capital Tallinn
towns Tartu, Narva, Kohtla-Järve, Pärnu
physical lakes and marshes in a partly forested plain; 774 km/481 mi of coastline; mild climate
features Lake Peipus and Narva River forming boundary with Russian Federation; Baltic islands, the largest of which is Saaremaa Island
head of state Lennart Meri from 1992
head of government Tiit Vahl from 1992
political system emergent democracy
political parties Estonian Popular Front (Rahvarinne), nationalist; Association for a Free Estonia, nationalist; Fatherland Group, right-wing; International Movement, ethnic Russian;

Estonian Green Party, environmentalist
products oil and gas (from shale), wood
products, flax, dairy and pig products
currency kroon
population (1989) 1,573,000 (Estonian 62%,
Russian 30%, Ukrainian 3%, Byelorussian 2%)
language Estonian, allied to Finnish
religion traditionally Lutheran
chronology
1918 Estonia declared its independence. March:
Soviet forces, who had tried to regain control from
occupying German forces during World War I,
were overthrown by German troops. Nov: Soviet
troops took control after German withdrawal.
1919 Soviet rule overthrown with help of British
navy; Estonia declared a democratic republic.
1934 Fascist coup replaced government.
1940 Estonia incorporated into USSR.
1941–44 German occupation during World War II.
1944 USSR regained control.
1980 Beginnings of nationalist dissent.
1988 Adopted own constitution, with power of
veto on all centralized Soviet legislation. Popular
Front (Rahvarinne) established to campaign for
democracy. Estonia's supreme soviet (state
assembly) voted to declare the republic
'sovereign' and autonomous in all matters except
military and foreign affairs; rejected by USSR as
unconstitutional.
1989 Estonian replaced Russian as main language.
1990 Feb: Communist Party monopoly of power
abolished; multiparty system established. March:
pro-independence candidates secured majority
after republic elections; coalition government
formed with Popular Front leader Edgar Savisaar
as prime minister; Arnold Rüütel became
president. May: prewar constitution partially
restored.
1991 March: independence plebiscite
overwhelmingly approved. Aug: full
independence declared after abortive anti-
Gorbachev coup; Communist Party outlawed.
Sept: independence recognized by Soviet
government and Western nations; admitted into
United Nations and Conference on Security and
Cooperation in Europe (CSCE).
1992 Jan: Savisaar resigned owing to his
government's inability to alleviate food and
energy shortages; new government formed by
Tiit Vahl. June: New constitution approved.
Sept: presidential election inconclusive; right-
wing Fatherland Group did well in general
election. Oct: Fatherland leader Lennart Meri
chosen by parliament to replace Rüütel.

Estoril /estɔrıl/ fashionable resort on the coast 20
km/13 mi W of Lisbon, Portugal; population
(1981) 16,000. There is a Grand Prix motor-
racing circuit.

Esztergom /'estəgɒm/ city on the Danube, NW
of Budapest, Hungary; population (1986)
31,000. It was the birthplace of St Stephen and
the former ecclesiastical capital of Hungary, with
a fine cathedral.

Etaples /eı'tɑːplə/ fishing port and seaside resort
on the Canche estuary, Pas de Calais *département*,
France; population (1985) 11,500. During World
War I it was a British base and hospital centre.

Ethiopia People's Democratic Republic of
(*Hebretesebawit Ityopia*, formerly also known as
Abyssinia)
area 1,096,900 sq km/423,403 sq mi
capital Addis Ababa
towns Dire Dawa, Harar
physical a high plateau with central mountain
range divided by Rift Valley; plains in E; source
of Blue Nile River
environment more than 90% of the forests of
the Ethiopian highlands have been destroyed
since 1900
features Danakil and Ogaden deserts; ancient
remains (in Aksum, Gondar, Lalibela, among
others); only African country to retain its
independence during the colonial period
head of state and government Meles Zenawi
from 1991
political system transition to democratic
socialist republic
political parties Ethiopian People's
Revolutionary Democratic Front (EPRDF),
nationalist, left of centre; Tigré People's
Liberation Front (TPLF); Ethiopian People's
Democratic Movement (EPDM); Oromo
People's Democratic Organization (OPDO)
exports coffee, pulses, oilseeds, hides, skins
currency birr
population (1992) 50,345,000 (Oromo 40%,
Amhara 25%, Tigré 12%, Sidamo 9%); growth
rate 2.5% p.a.
life expectancy 38
languages Amharic (official), Tigrinya,
Orominga, Arabic
religions Sunni Muslim 45%, Christian
(Ethiopian Orthodox Church, which has had its
own patriarch since 1976) 40%
literacy 35% (1988)
GDP $4.8 bn (1987); $104 per head.
chronology
1889 Abyssinia reunited by Menelik II.
1930 Haile Selassie became emperor.
1962 Eritrea annexed by Haile Selassie;
resistance movement began.
1974 Haile Selassie deposed and replaced by a
military government led by General Teferi Benti.
Ethiopia declared a socialist state.
1977 Teferi Benti killed and replaced by Col
Mengistu Haile Mariam.
1977–79 'Red Terror' period in which

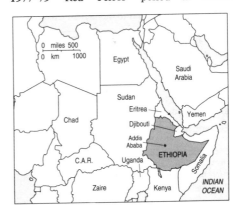

Mengistu's regime killed thousands of innocent people.
1981–85 Ethiopia spent at least $2 billion on arms.
1984 Workers' Party of Ethiopia (WPE) declared the only legal political party.
1985 Worst famine in more than a decade; Western aid sent and forcible internal resettlement programmes undertaken.
1987 New constitution adopted, Mengistu Mariam elected president. New famine; food aid hindered by guerrillas.
1988 Mengistu agreed to adjust his economic policies in order to secure IMF assistance. Influx of refugees from Sudan.
1989 Coup attempt against Mengistu foiled. Peace talks with Eritrean rebels mediated by former US president Carter reported some progress.
1990 Rebels captured port of Massawa.
1991 Mengistu overthrown; transitional government set up by EPRDF. EPLF secured Eritrea; Eritrea's right to secede recognized. Meles Zenawi elected Ethiopia's new head of state and government.
1993 Eritrean independence recognized after referendum.

I saw parch'd Abyssinia rise and sing.

On **Ethiopia** John Keats *Endymion* 1818

Etna volcano on the east coast of Sicily, 3,323 m/10,906 ft, the highest in Europe. About 90 eruptions have been recorded since 1800 BC, yet because of the rich soil, the cultivated zone on the lower slopes is densely populated, including the coastal town of Catania. The most recent eruption was in Dec 1985.

Eton /'iːtn/ town in Berkshire, England, on the north bank of the river Thames, opposite Windsor; population (1983 est) 3,500. *Eton College* is one of the UK's oldest, largest, and most prestigious public (private and fee-paying) schools. It was founded 1440.

The battle of Waterloo was won on the playing fields of Eton.

On **Eton** attributed to the Duke of Wellington

Euboea /juː'biːə/ (Greek *Evvoia*) mountainous island off the east coast of Greece, in the Aegean Sea; area 3,755 sq km/1,450 sq mi; about 177 km/110 mi long; population (1981) 188,410. Mount Delphi reaches 1,743 m/5,721 ft. The chief town, Chalcis, is connected by a bridge to the mainland.

Eugene /juː'dʒiːn/ city in W central Oregon, USA, S of Portland, on the Willamette River; population (1990) 112,700. It is a processing and shipping centre for agricultural products from the surrounding region; wood products are also manufactured.

Eupen-et-Malmédy /ɜː'pen eɪ mælme'diː/ region

of Belgium around the towns of Eupen and Malmédy. It was Prussian from 1814 until it became Belgian 1920 after a plebiscite; there was fierce fighting here in the German Ardennes offensive Dec 1944.

Euphrates /juː'freɪtiːz/ (Arabic *Furat*) river, rising in E Turkey, flowing through Syria and Iraq and joining the river Tigris above Basra to form the river Shatt-al-Arab, at the head of the Persian/Arabian Gulf; 3,600 km/2,240 mi in length. The ancient cities of Babylon, Eridu, and Ur were situated along its course.

There is no river to be compared to him.

On the **Euphrates** Gertrude Bell, letter of 1911

Eure /ɜː/ river rising in Orne *département*, France, and flowing SE, then N, to the river Seine; length 115 km/70 mi. Chartres is on its banks. It gives its name to two départements, Eure and Eure-et-Loire.

Europe /'juərəp/ second-smallest continent, occupying 8% of the Earth's surface
area 10,400,000 sq km/4,000,000 sq mi
largest cities (population over 1.5 million) Athens, Barcelona, Berlin, Birmingham, Bucharest, Budapest, Hamburg, Istanbul, Kharkov, Kiev, Lisbon, London, Madrid, Manchester, Milan, Moscow, Paris, Rome, St Petersburg, Vienna, Warsaw
features Mont Blanc (4,807 m/15,772 ft), the highest peak in the Alps; lakes (over 5,100 sq km/2,000 sq mi) include Ladoga, Onega, Vänern; rivers (over 800 km/500 mi) include the Volga, Danube, Dnieper Ural, Don, Pechora, Dniester, Rhine, Loire, Tagus, Ebro, Oder, Prut, Rhône
highest point Mount Elbruz (5,642 m/18,517 ft) in the Caucasus Mountains, the highest peak in Europe
geographical extremities of mainland Nordkynn in Norway in the N; Tarifa Point in Spain in the S; Cape Roca in Portugal in the W
physical conventionally occupying that part of Eurasia to the W of the Ural Mountains, N of the Caucasus Mountains and N of the Sea of Marmara; Europe lies entirely in the northern hemisphere between 36° N and the Arctic Ocean. About two-thirds of the continent is a great plain which covers the whole of European Russia and spreads westwards through Poland to the Low Countries and the Bay of Biscay. To the N lie the Scandinavian highlands rising to 2,472 m/8,110 ft at Glittertind in the Jotenheim range of Norway. To the S, a series of mountain ranges stretch E–W (Caucasus, Balkans, Carpathians, Apennines, Alps, Pyrenees, and Sierra Nevada).
A line from the Baltic to the Black Sea divides Europe between an eastern continental region and a western region characterized by a series of peninsulas that include Scandinavia (Norway, Sweden, and Finland), Jutland (Denmark and Germany), Iberia (Spain and Portugal), and Italy and the Balkans (Greece, Albania, Croatia, Slovenia, Bosnia-Herzegovina, Yugoslavia, Bulgaria, and European Turkey).

climate the greater part of Europe falls within the northern temperate zone which is modified by the Gulf Stream in the NW; Central Europe has warm summers and cold winters; the Mediterranean coast has comparatively mild winters and hot summers
products nearly 50% of the world's cars (Germany, France, Italy, Spain, Russia, Georgia, Ukraine, Latvia, Belarus, UK); 43% of its barley (Germany, Spain, France, UK); 41% of its rye (Poland, Germany); 31% of its oats (Poland, Germany, Sweden, France); 24% of its wheat (France, Germany, UK, Romania); 70% of its olive oil (Italy, Spain, Greece). The rate of fertilizer consumption on agricultural land is four times greater than that in any other continent
population (1985) 496 million (excluding Turkey and the ex-Soviet republics); annual growth rate 0.3%, projected population of 512 million by the year 2000
languages mostly Indo-European, with a few exceptions, including Finno-Ugrian (Finnish and Hungarian), Basque and Altaic (Turkish); apart from a fringe of Celtic, the NW is Germanic; Letto-Lithuanian languages separate the Germanic from the Slavonic tongues of E Europe; Romance languages spread E–W from Romania through Italy and France to Spain and Portugal
religions Christianity (Protestant, Roman Catholic, Eastern Orthodox), Muslim (Turkey, Albania, Yugoslavia, Bulgaria), Judaism.

Euskal Herria / juːskælə'riːə/ Basque name for the ▷Basque Country.

Evansville /'evənzvɪl/ industrial city in SW Indiana, USA, on the Ohio River; population (1990) 126,300. Industries include pharmaceuticals and plastics. The University of Evansville is here.
During the Civil War, Evansville was an important station on the Underground Railroad. Abraham Lincoln spent his boyhood in nearby Spencer County.

Everest, Mount /'evərɪst/ (Chinese *Qomolungma* 'goddess mother of the snows/ world'; Nepalese *Sagarmatha* 'head of the earth') the world's highest mountain above sea level, in the Himalayas, on the China–Nepal frontier; height 8,872 m/29,118 ft (recently measured by satellite to this new height from the former official height of 8,848 m/29,028 ft). It was first climbed by New Zealand mountaineer Edmund Hillary and Sherpa Tenzing Norgay 1953. More than 360 climbers have reached the summit; over 100 have died during the ascent.
The English name comes from George Everest (1790–1866), surveyor general of India. In 1987 a US expedition obtained measurements of K2 that disputed Everest's 'highest mountain' status, but the recent satellite measurements have established Mount Everest as the highest.

Everglades /'evəgleɪdz/ area of swamps, marsh, and lakes in S Florida; area 5,000 sq mi/ 12,950 sq km. A national park covers the southern tip.
Formed by overflow of Lake Okeechobee after heavy rains, it is one of the wildest areas

in the USA, noted for its distinctive plant and animal life. The only human residents are several hundred Seminole, a North American Indian people. Large drainage programmes have reduced the flow of water from the lake southwards, threatening the region's ecological balance.

Evesham /'iːvʃəm/ town in Hereford and Worcester, England, on the river Avon SE of Worcester; population (1990 est) 18,000. Fruit and vegetables are grown in the fertile *Vale of Evesham*. In the Battle of Evesham, 4 Aug 1265, during the Barons' Wars, Edward, Prince of Wales, defeated Simon de Montfort, who was killed.

Evreux /ev'rɜː/ capital of Eure *département*, situated in the Iton River valley, in NW France; population (1990) 51,500. It produces pharmaceuticals, rubber, and textiles; there are also printing and electronic companies.
Evreux dates from Gallo-Roman times. It has a 12th–16th-century cathedral and a 15th-century bishop's palace. The town was badly bombed during World War II.

Evvoia /'eviə/ Greek name for the island of ▷Euboea.

Exeter /'eksɪtə/ city, administrative headquarters of Devon, England, on the river Exe; population (1983 est) 101,800. It has medieval, Georgian, and Regency architecture, including a cathedral (1280–1369), a market centre, and a university (1955). It manufactures agricultural machinery, pharmaceuticals, and textiles.

Exmoor /'eksmuə/ moorland in Devon and Somerset, England, forming (with the coast from Minehead to Combe Martin) a National Park since 1954. It includes Dunkery Beacon, 520 m/ 1,707 ft, and the Doone Valley.

Exmouth resort town (sailing centre) and port in Devon, SW England, at the mouth of the river Exe; population (1982) 28,800. There are engineering industries. The port was permanently closed to commercial vessels 1989.

Extremadura /ˌestrəmə'duərə/ autonomous region of W Spain including the provinces of Badajoz and Cáceres; area 41,600 sq km/ 16,058 sq mi; population (1986) 1,089,000. Irrigated land is used for growing wheat; the remainder is either oak forest or used for pig or sheep grazing.

Eyre, Lake /eə/ Australia's largest lake, in central South Australia, which frequently runs dry, becoming a salt marsh in dry seasons; area up to 9,000 sq km/3,500 sq mi. It is the continent's lowest point, 12 m/39 ft below sea level.

Eyre Peninsula /eə/ peninsula in South Australia, which includes the iron and steel city of Whyalla. Over 50% of the iron used in Australia's steel industry is mined at Iron Knob; the only seal colony on mainland Australia is at Point Labatt.

Faenza /faɪ'entsə/ city on the river Lamone in Ravenna province, Emilia-Romagna, Italy; population (1985) 54,900. It has many medieval remains, including the 15th-century walls. It gave its name to 'faience' pottery, a type of tin-glazed earthenware first produced here.

Faeroe Islands /'feərəu/ or **Faeroes** alternative spelling of the ♢Faroe Islands, in the N Atlantic.

Fagatogo /'faːgə'təugəu/ capital of American Samoa, situated on Pago Pago Harbour, Tutuila Island; population (1980) 30,124.

Fairbanks /'feəbæŋks/ town in central Alaska, situated on the Chena Slough, a tributary of the Tanana River; population (1990) 30,800. Founded 1902, it became a gold-mining and fur-trading centre and the terminus of the Alaska Railroad and the Pan-American Highway.

Fairfield /'feəfiːld/ city in W central California, USA, SW of Sacramento; seat of Solano County; population (1990) 77,200. It is a trading centre for the area's fruits, grains, and dairy products.

Faisalabad /'faɪsələbæd/ city in Punjab province, Pakistan, 120 km/75 mi W of Lahore; population (1981) 1,092,000. It trades in grain, cotton, and textiles.

Faizabad /,faɪzə'bæd/ town in Uttar Pradesh, N India; population (1981) 143,167. It lies at the head of navigation of the river Ghaghara and has sugar refineries and an agricultural trade.

Falaise /fə'leɪz/ town 32 km/20 mi SE of Caen, in Normandy, France; population (1982) 8,820. It is a market centre, manufacturing cotton and leather goods and trading in livestock and cheese. The 12th–13th-century castle was that of the first dukes of Normandy, and William the Conqueror was born here 1027.

Falkirk /'fɔːlkɜːk/ town in Central Region, Scotland, 37 km/23 mi W of Edinburgh; population (1981) 37,700. An iron-founding centre, Falkirk has brewing, distilling, tanning, and chemical industries.

Falkland Islands /'fɔːklənd/ (Argentine *Islas Malvinas*) British crown colony in the S Atlantic **area** 12,173 sq km/4,700 sq mi, made up of two main islands: East Falkland 6,760 sq km/

2,610 sq mi, and West Falkland 5,413 sq km/ 2,090 sq mi
capital Stanley; new port facilities opened 1984, Mount Pleasant airport 1985
features in addition to the two main islands, there are about 200 small islands, all with wild scenery and rich bird life
products wool, alginates (used as dyes and as a food additive) from seaweed beds
population (1986) 1,916
government a governor is advised by an executive council, and a mainly elected legislative council. Administered with the Falklands, but separate dependencies of the UK, are South Georgia and the South Sandwich Islands; see also ♢British Antarctic Territory.
history the first European to visit the islands was Englishman John Davis 1592, and at the end of the 17th century they were named after Lord Falkland, treasurer of the British navy. West Falkland was settled by the French 1764. The first British settlers arrived 1765; Spain bought out a French settlement 1766, and the British were ejected 1770–71, but British sovereignty was never ceded, and from 1833, when a few Argentines were expelled, British settlement was continuous. Argentina asserts its succession to the Spanish claim to the 'Islas Malvinas', but the inhabitants oppose cession. Occupied by Argentina April 1982, the islands were recaptured by British military forces in May–June of the same year. In April 1990 Argentina's congress declared the Falkland Islands and other British-held South Atlantic islands part of the new Argentine province of Tierra del Fuego.

A war declared for the empty sound of an ancient title to a Magellanick rock, would raise the indignation of the earth against us.

On the **Falkland Islands** Samuel Johnson
Thoughts on the late Transactions Respecting Falklands Islands 1771

Falkland Islands

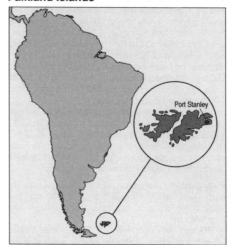

Fall River /ˌfɔːl ˈrɪvə/ city and port in Massachusetts; population (1990) 92,700. It stands at the mouth of the Taunton River, over the Little Fall River, which gave it its name. Textiles and clothing, rubber, paper, and plastics are among the goods produced.

Falmouth /ˈfælməθ/ port on the south coast of Cornwall, England, on the estuary of the river Fal; population (1981) 18,500. There are ship-repairing and marine engineering industries.

Famagusta /ˌfæməˈgustə/ seaport on the east coast of Cyprus, in the Turkish Republic of Northern Cyprus; population (1985) 19,500. It was the chief port of the island until the Turkish invasion 1974.

It has many ruined Gothic churches, the result of its having been taken from the Venetians by the Turks 1571. It is also famous for Othello's Tower, on whose story Shakespeare based his play.

Fao /fau/ or **Faw** oil port on a peninsula at the mouth of the Shatt-al-Arab River in Iraq. Iran launched a major offensive against Iraq 1986, capturing Fao for two years.

Far East geographical term for all Asia east of the Indian subcontinent.

Fareham /ˈfeərəm/ town in Hampshire, England, 10 km/6 mi NW of Portsmouth; population (1988 est) 100,000. Bricks, ceramics, and rope are made and there is engineering and boat-building as well as varied light industries.

Fargo /ˈfɑːgəu/ city in SE North Dakota, USA, across the Red River from Moorhead, Minnesota; seat of Cass County; population (1990) 74,100. The largest city in the state, it is a centre for processing and distributing agricultural products and farm machinery. Chemicals and building materials are also manufactured.

Farnborough /ˈfɑːnbərə/ town in Hampshire, England, N of Aldershot; population (1987 est) 42,800. Experimental work is carried out at the Royal Aircraft Establishment. Aeronautical displays are given at the biennial air show. The mansion of Farnborough Hill was occupied by Napoleon III and the Empress Eugénie, and she, her husband, and her son, are buried in a mausoleum at the Roman Catholic church she built.

Farne Islands /fɑːn/ rocky island group in the North Sea, off Northumberland, England. The islands are a sanctuary for birds and grey seals.

A chapel stands on the site of the hermitage at St Cuthbert on Inner Farne. There are two lighthouses; the Longstone lighthouse was the scene of the rescue of shipwrecked sailors by Grace Darling.

Farnham /ˈfɑːnəm/ town in Surrey, England, on the river Wey; population (1985 est) 20,900. The parish church was once part of Waverley Abbey (1128), the first Cistercian house in England; Walter Scott named his first novel after the abbey.

Faroe Islands /ˈfeərəu/ or **Faeroe Islands** or **Faeroes** (Danish **Faerøerne** 'Sheep Islands') island group (18 out of 22 inhabited) in the N Atlantic, between the Shetland Islands and Iceland, forming an outlying part of Denmark
area 1,399 sq km/540 sq mi; largest islands are Strømø, Østerø, Vagø, Suderø, Sandø, and Bordø
capital Thorshavn on Strømø, population (1986) 15,287
products fish, crafted goods
currency Danish krone
population (1986) 46,000
languages Faeroese, Danish
government since 1948 the islands have had full self-government; they do not belong to the EC
history first settled by Norsemen in the 9th century, the Faroes were a Norwegian province 1380–1709. Their parliament was restored 1852. They withdrew from the European Free Trade Association (EFTA) 1972.

Fars /fɑːs/ province of SW Iran, comprising fertile valleys among mountain ranges running NW–SE; population (1982) 2,035,600; area 133,300 sq km/51,487 sq mi. The capital is Shiraz, and there are imposing ruins of Cyrus the Great's city of Parargardae and of Persepolis.

Fayetteville /ˈfeɪətvɪl/ city in S central North Carolina, USA, on the Cape Fear river, S of Durham and Chapel Hill and SW of Raleigh; population (1990) 75,700. Its industries include processing of the area's agricultural products, tools, textiles, and lumber. It was named after the Marquis de Lafayette.

Fayetteville /ˈfeɪətvɪl/ city in the northwest corner of Arkansas, USA, SE of Fort Smith; seat of Washington County; population (1990) 42,100. It is an agricultural trading centre. Its main industry is poultry processing; others include lumber, clothing, and tools.

Fécamp /ˈfeɪkɒm/ seaport and resort of France, NE of Le Havre in the *département* of Seine Maritime; industries include shipbuilding and fishing; population (1990) 21,100. Benedictine liqueur was first produced here in the early 16th century.

Felixstowe /ˈfiːlɪkstəu/ port and resort opposite Harwich in Suffolk, England, between the Orwell and Deben estuaries; population (1981) 21,000. It is Britain's busiest container port, and also has ferry services to Gothenburg, Rotterdam, and Zeebrugge.

Fens, the /fenz/ level, low-lying tracts of land in E England, W and S of the Wash, about 115 km/70 mi N–S and 55 km/34 mi E–W. They fall within the counties of Lincolnshire, Cambridgeshire, and Norfolk, consisting of a huge area, formerly a bay of the North Sea, but now crossed by numerous drainage canals and forming some of the most productive agricultural land in Britain. The peat portion of the Fens is known as the **Bedford Level**.

The first drainage attempts were made by the Romans. After the Norman conquest an earthwork 100 km/60 mi long was constructed as a barrage against the sea. In 1634 the 4th Earl of Bedford brought over the Dutch water-engineer Cornelius Vermuyden (c. 1596–1683) who introduced Dutch methods. Burwell Fen and Wicken Fen, NE of Cambridge, have been preserved undrained as nature reserves.

Ferghana /fə'gɑːnə/ town in Uzbekistan, in the fertile Ferghana Valley; population (1987) 203,000. It is the capital of the major cotton- and fruit-growing Ferghana region; nearby are petroleum fields.

The *Ferghana Valley* is divided between the republics of Uzbekistan, Kyrgyzstan, and Tajikistan and has been the scene of interethnic violence among Uzbek, Meskhetian, and Kyrgyz communities.

Fermanagh /fə'mænə/ county in the southern part of Northern Ireland
area 1,680 sq km/648 sq mi
towns Enniskillen (county town), Lisnaskea, Irvinestown
physical in the centre is a broad trough of low-lying land, in which lie Upper and Lower Lough Erne
products mainly agricultural; livestock, tweeds, clothing
population (1989 est) 52,000.

Fernando Po /fə'nændəu 'pəu/ former name (until 1973) of the island of ◊Bioko, part of Equatorial Guinea.

Ferrara /fə'rɑːrə/ industrial city and archbishopric in Emilia-Romagna region, N Italy, on a branch of the Po delta 52 km/32 mi W of the Adriatic Sea; population (1988) 143,000. There are chemical industries and textile manufacturers.

It has the Gothic castle of its medieval rulers, the House of Este, palaces, museums, and a cathedral, consecrated 1135. The university was founded 1391. Italian religious reformer Girolamo Savonarola was born here, and the poet Torquato Tasso was confined in the asylum 1579-86.

Ferrol /fe'rəul/ alternative name for ◊El Ferrol, a city and port in Spain.

Fertö tó /'feətəutəu/ Hungarian name for the ◊Neusiedler See, a lake in Austria and Hungary.

Fès /fez/ or *Fez* former capital of Morocco 808–1062, 1296–1548, and 1662–1912, in a valley N of the Great Atlas Mountains, 160 km/100 mi E of Rabat; population (1982) 563,000. Textiles, carpets, and leather are manufactured, and the *fez*, a brimless hat worn in S and E Mediterranean countries, is traditionally said to have originated here. Kairwan Islamic University dates from 859; a second university was founded 1961.

Fez,/Where all is Eden, or a wilderness.

On **Fès** Lord Byron
Don Juan 1819–24

Fez /fez/ alternative spelling of Fès, a city in Morocco.

Fezzan /fe'zɑːn/ former province of Libya, a desert region with many oases, and with rock

Fife

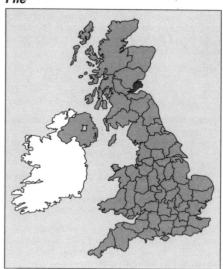

paintings from about 3000 BC. It was captured from Italy 1942, and placed under French control until 1951 when it became a province of the newly independent United Kingdom of Libya. It was split into smaller divisions 1963.

Fichtelgebirge /'fɪxtəlgə,bɪəgə/ chain of mountains in Bavaria, Germany, on the Czech border. The highest peak is the *Schneeberg*, 1,051 m/3,448 ft. There are granite quarries, uranium mining, china and glass industries, and forestry.

Fiesole /fjeɪzəuleɪ/ resort town 6 km/4 mi NE of Florence, Italy, with many Etruscan and Roman relics; population (1971) 14,400. The Romanesque cathedral was completed 1028.

Fife /faɪf/ region of E Scotland (formerly the county of Fife), facing the North Sea and Firth of Forth
area 1,300 sq km/502 sq mi
towns administrative headquarters Glenrothes; Dunfermline, St Andrews, Kirkcaldy, Cupar
physical the only high land is the Lomond Hills, in the NW; chief rivers Eden and Leven
features Rosyth naval base and dockyard (used for nuclear submarine refits) on northern shore of the Firth of Forth; Tentsmuir, possibly the earliest settled site in Scotland. The ancient palace of the Stuarts was at Falkland, and eight Scottish kings are buried at Dunfermline
products potatoes, cereals, electronics, petrochemicals (Mossmorran), light engineering
population (1991) 339,200.

Figueras town in Catalonia, NE Spain; population (1991) 33,450. It was the home of the Surrealist painter Salvador Dalí. The Salvador Dalí museum, opened 1974, was designed by the artist.

Fiji Republic of
area 18,333 sq km/7,078 sq mi
capital Suva

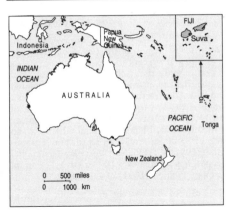

towns ports Lautoka and Levuka
physical comprises 844 Melanesian and Polynesian islands and islets (about 110 inhabited), the largest being Viti Levu (10,429 sq km/4,028 sq mi) and Vanua Levu (5,550 sq km/2,146 sq mi); mountainous, volcanic, with tropical rainforest and grasslands
features almost all islands surrounded by coral reefs; high volcanic peaks; crossroads of air and sea services between N America and Australia
head of state Ratu Sir Penaia Ganilau from 1987
head of government Col Sitiveni Rabuka from 1992
political system democratic republic
political parties Alliance Party (AP), moderate centrist Fijian; National Federation Party (NFP), moderate left-of-centre Indian; Fijian Labour Party (FLP), left-of-centre Indian; United Front, Fijian
exports sugar, coconut oil, ginger, timber, canned fish, gold; tourism is important
currency Fiji dollar
population (1989 est) 758,000 (46% Fijian, holding 80% of the land communally, and 49% Indian, introduced in the 19th century to work the sugar crop); growth rate 2.1% p.a.
life expectancy men 67, women 71
languages English (official), Fijian, Hindi
religions Hindu 50%, Methodist 44%
literacy men 88%, women 77% (1980 est)
GDP $1.2 bn (1987); $1,604 per head
chronology
1874 Fiji became a British crown colony.
1970 Independence achieved from Britain; Ratu Sir Kamisese Mara elected as first prime minister.
1987 April: general election brought to power an Indian-dominated coalition led by Dr Timoci Bavadra. May: military coup by Col Sitiveni Rabuka removed new government at gunpoint; Governor General Ratu Sir Penaia Ganilau regained control within weeks. Sept: second military coup by Rabuka proclaimed Fiji a republic and suspended the constitution. Oct: Fiji ceased to be a member of the Commonwealth. Dec: civilian government restored with Rabuka retaining control of security as minister for home affairs.
1990 New constitution, favouring indigenous Fijians, introduced.
1992 General election produced coalition government; Rabuka named as president.

Tahiti was put on the map by Gauguin, Fiji by Qantas and Pan Am(erican Airways).
On **Fiji** George Mikes *Boomerang—Australia Rediscovered* 1968

Fingal's Cave /'fɪŋgəlz/ cave on the island of Staffa, Inner Hebrides, Scotland. It is lined with natural basalt columns, and is 60 m/200 ft long and 20 m/65 ft high. Fingal, based on the Irish hero Finn Mac Cumhaill, was the leading character in Macpherson's Ossianic forgeries. Visited by the German Romantic composer Felix Mendelssohn 1829, the cave was the inspiration of his *Hebrides* overture, otherwise known as *Fingal's Cave*.

Finistère /fɪnɪs'teə/ (Latin *finis terrae* 'land's end') *département* of Brittany, NW France, occupying a headland with the English Channel to the N and the Bay of Biscay to the S; area 7,030 sq km/2,714 mi; population (1990) 838,700. The administrative centre is Quimper; Brest and Douarnenez are the chief towns. Inland there is heathland and fertile valleys; the Arrée and Black mountains run E–W. Horses and cattle are raised, and there is agriculture and seasonal fishing.
history The area was settled by Celts in the 5th century BC, conquered by the Romans 56 BC, and invaded by Anglo-Saxons in the 5th century AD.

Finisterre Cape /fɪnɪs'teə/ promontory in the extreme NW of Spain.

Finland Republic of (*Suomen Tasavalta*)
area 338,145 sq km/130,608 sq mi
capital Helsinki
towns Tampere, Rovaniemi, Lahti; ports Turku, Oulu
physical most of the country is forest, with low hills and about 60,000 lakes; one-third is within the Arctic Circle; archipelago in S; includes Åland Islands
features Helsinki is the most northerly national capital on the European continent; at the 70th parallel there is constant daylight for 73 days in summer and 51 days of uninterrupted night in winter

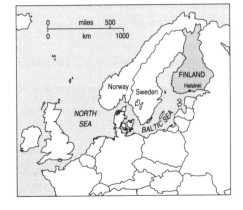

head of state Mauno Koivisto from 1982
head of government Esko Aho from 1991
political system democratic republic
political parties Social Democratic Party
(SDP), moderate, left of centre; National
Coalition Party (KOK), moderate, right of
centre; Centre Party (KP), centrist, rural-
oriented; Finnish People's Democratic League
(SKDL), left-wing; Swedish People's Party
(SFP), independent Swedish-oriented; Finnish
Rural Party (SMP), farmers and small businesses;
Democratic Alternative, left-wing; Green Party
exports metal, chemical, and engineering products (icebreakers and oil rigs), paper, sawn
wood, clothing, fine ceramics, glass, furniture
currency markka
population (1989 est) 4,990,000; growth rate
0.5% p.a.
life expectancy men 70, women 78
languages Finnish 93%, Swedish 6% (both
official), small Saami and Russian-speaking
minorities
religions Lutheran 97%, Eastern Orthodox
1.2%
literacy 99%
GDP $77.9 bn (1987); $15,795 per head
chronology
1809 Finland annexed by Russia.
1917 Independence declared from Russia.
1920 Soviet regime acknowledged independence.
1939 Defeated by USSR in Winter War.
1941 Allowed Germany to station troops in
Finland to attack USSR; USSR bombed Finland.
1944 Concluded separate armistice with USSR.
1948 Finno-Soviet Pact of Friendship, Cooperation, and Mutual Assistance signed.
1955 Finland joined the United Nations and the
Nordic Council.
1956 Urho Kekkonen elected president; re-
elected 1962, 1968, 1978.
1973 Trade treaty with European Economic
Community signed.
1977 Trade agreement with USSR signed.
1982 Koivisto elected president; re-elected 1988.
1989 Finland joined Council of Europe.
1991 Big swing to the centre in general election.
New coalition government formed.
1992 Formal application for European Community (EC) membership.

Finland, Gulf of /'fɪnlənd/ eastern arm of the
◊Baltic Sea, separating Finland from Estonia.

Finsteraarhorn /ˌfɪnstərˈɑːhɔːn/ highest mountain,
4,274 m/14,020 ft, in the Bernese Alps,
Switzerland.

Firenze /fɪˈrentseɪ/ Italian form of ◊Florence, a
city in Italy.

Fishguard /'fɪʃɡɑːd/ (Welsh *Abergwaun*) seaport
on an inlet on the south side of Fishguard Bay,
Dyfed, SW Wales; population about 5,000. There
is a ferry service to Rosslare in the Republic of
Ireland.

Fitchburg /'fɪtʃbɜːɡ/ city in N Massachusetts,
USA, on the Nashua River, N of Worcester;
population (1990) 41,200. Industries include
paper, textiles, furniture, clothing, and foundry
products.

FL abbreviation for ◊*Florida*, a state of the USA.

Flanders /'flɑːndəz/ (French *Flandre*, Flemish
Vlaanderen) region of the Low Countries that
in the 8th and 9th centuries extended from Calais
in France to the river Scheldt and is now covered
by the Belgian provinces of Oost Vlaanderen
and West Vlaanderen (East and West Flanders),
the French *département* of Nord, and part of
the Dutch province of Zeeland. The language
is Flemish. East Flanders, capital Ghent, has an
area of 3,000 sq km/1,158 sq mi and a population (1991) of 1,335,700. West Flanders, capital
Bruges, has an area of 3,100 sq km/1,197 sq mi
and a population (1991) of 1,106,800.

Fleet Street /fliːt/ street in London, England
(named after the subterranean river Fleet), traditionally the centre of British journalism. It runs
from Temple Bar eastwards to Ludgate Circus.
With adjoining streets it contained the offices
and printing works of many leading British newspapers until the mid-1980s, when most moved to
sites farther from the centre of London.

Fleetwood /'fliːtwud/ port and seaside resort in
Lancashire, England, at the mouth of the river
Wyre; population (1981) 24,500. The fishing
industry has declined, but the port still handles
timber, petroleum, and chemicals. Ferry services
operate to the Isle of Man and Belfast.

Flensburg /'flensbɜːɡ/ port on the east coast of
Schleswig-Holstein, Germany, with shipyards
and breweries; population (1984) 86,700. Rum
and smoked eels are specialities.

Flevoland /'fleɪvəulænd/ (formerly
IJsselmeerpolders) low-lying province of the
Netherlands established 1986
area 1,410 sq km/544 sq mi
capital Lelystad
towns Dronten, Almere
population (1988) 194,000
history created 1986 out of land reclaimed from
IJsselmeer 1950–68.

Flint /flɪnt/ city in Michigan, USA, on the Flint
River, 90 km/56 mi NW of Detroit; population
(1990) 140,800. Car manufacturing is the chief
industry but it is declining.

Flintshire /'flɪntʃə/ (Welsh *Sir y Fflint*) former
county of Wales, and smallest of the Welsh counties. It was merged 1974, with Denbigh and part
of Merioneth, into the new county of Clwyd; the
county town of Mold became the administrative
headquarters of the new region.

Florence /'flɒrəns/ (Italian *Firenze*) capital of
Tuscany, N Italy, 88 km/55 mi from the mouth
of the river Arno; population (1988) 421,000. It
has printing, engineering, and optical industries;
many crafts, including leather, gold and silver
work, and embroidery; and its art and architecture
attract large numbers of tourists. Notable medieval and Renaissance citizens included the writers
Dante and Boccaccio, and the artists Giotto,
Leonardo da Vinci, and Michelangelo.
features Florence's architectural treasures
include the Ponte Vecchio (1345); the Pitti and

Vecchio palaces; the churches of Santa Croce and Santa Maria Novella; the cathedral of Santa Maria del Fiore (1314); and the Uffizi Gallery, which has one of Europe's finest art collections, based on that of the Medicis.

history The Roman town of Florentia was founded in the 1st century BC on the site of the Etruscan town of Faesulae. It was besieged by the Goths AD 405 and visited by Charlemagne 786. In 1052, Florence passed to Countess Matilda of Tuscany (1046–1115), and from the 11th century onwards gained increasing autonomy. In 1198 it became an independent republic, with new city walls, and governed by a body of 12 citizens. In the 13th–14th centuries, the city was the centre of the struggle between the Guelphs (papal supporters) and Ghibellines (supporters of the Holy Roman emperor). Despite this, Florence became immensely prosperous and went on to reach its cultural peak during the 14th–16th centuries.

From the 15th to the 18th century, the Medici family, originally bankers, were the predominant power, in spite of their having been twice expelled by revolutions. In the first of these, in 1493, a year after Lorenzo de' Medici's death, a republic was proclaimed (with Machiavelli as secretary) that lasted until 1512. From 1494 to 1498, the city was under the control of religious reformer Savonarola. In 1527, the Medicis again proclaimed a republic, which lasted through many years of gradual decline until 1737, when the city passed to Maria Theresa of Austria. From 1737 the city was ruled by the Habsburg imperial dynasty. The city was badly damaged in World War II and by floods 1966.

It still has the sparks of the old Renaissance genius but no longer the material around them to burst into flames.

On **Florence** Peter Nichols *Italia, Italia* 1973

Florence /'flɒrəns/ city in NW Alabama, USA, on the Tennessee River near the Tennessee Valley Authority's Wilson Dam, NW of Birmingham; seat of Lauderdale County; population (1990) 36,400. Industries include agricultural and poultry products, building materials, lumber, and fertilizers.

Florence /'flɒrəns/ city in NE South Carolina, USA, NW of Myrtle Beach; seat of Florence County; population (1990) 29,800. It is a centre of the trucking industry, serving as a terminus for many companies. Other industries include dairy products, fertilizers, film, furniture, machined goods, and clothing.

Florianópolis /ˌflɒriə'nɒpəlis/ seaport and resort on Santa Caterina Island, Brazil; population (1980) 153,500. It is linked to the mainland by two bridges, one of which is the largest expansion bridge in Brazil.

Florida /'flɒrɪdə/ southeasternmost state of the USA; mainly a peninsula jutting into the Atlantic, which it separates from the Gulf of Mexico; nickname Sunshine State

Florida

area 152,000 sq km/58,672 sq mi

capital Tallahassee

towns Miami, Tampa, Jacksonville

population (1990) 12,937,900, one of the fastest-growing of the states; including 15% nonwhite; 10% Hispanic, (especially Cuban)

physical 50% forested; lakes (including Okeechobee 1,000 sq km/695 sq mi/; Everglades National Park (5,000 sq km/1,930 sq mi), with birdlife, cypresses, alligators

features Palm Beach island resort, between the lagoon of Lake Worth and the Atlantic; Florida Keys; John F Kennedy Space Center at Cape Canaveral; Disney World theme park; beach resorts on Gulf and on Atlantic; Daytona International Speedway

products citrus fruits, melons, vegetables, fish, shellfish, phosphates, chemicals, electrical and electronic equipment, aircraft, fabricated metals

famous people Chris Evert, Henry Flagler, James Weldon Johnson, Sidney Poitier, Philip Randolph, Joseph Stilwell

history discovered by Ponce de Leon and under Spanish rule from 1513 until its cession to England 1763; returned to Spain 1783 and purchased by the USA 1819, becoming a state 1845.

More recently, Florida has become a banking centre, a development often partially attributed to the sizable inflow of cash derived from the traffic in illegal drugs from Latin America. Florida State University is the US Federal Centre for magnet research.

Florida Keys /'flɒrɪdə/ series of small coral islands that curve over 240 km/150 mi SW from the southern tip of Florida. The most important are Key Largo and Key West (with a US naval and air station); they depend on fishing and tourism.

Flow Country /'fləʊ ˌkʌntri/ wilderness and wildlife habitat in Caithness and Sunderland districts, Highland Region, N Scotland. Under threat from commercial afforestation, it consists of treeless peat bog and various lochs, which are home to many rare plants and wading birds.

Flushing /'flʌʃɪŋ/ (Dutch *Vlissingen*) port on Walcheren Island, Zeeland, the Netherlands; population (1987) 44,900. It stands at the entrance to the Scheldt estuary, one of the principal sea routes into Europe. Industries include fishing, shipbuilding, and petrochemicals, and there is a ferry service to Harwich.

Foggia /'fɒdʒə/ city of Puglia region, S Italy; population (1988) 159,000. The cathedral, dating from about 1170, was rebuilt after an earthquake 1731. Natural gas is found nearby.

Folkestone /'fəukstən/ port and holiday resort on the southeast coast of Kent, England, 10 km/6 mi SW of Dover; population (1983) 44,200. There are ferry and hovercraft services to and from Boulogne and Zeebrugge. It is the birthplace of the physician William Harvey.

Fontainebleau /'fɒntɪnbləu/ town to the SE of Paris, in Seine-et-Marne *département*; population (1982) 18,753. The château was built by François I in the 16th century. Mme de Montespan lived there in the reign of Louis XIV, and Mme du Barry in that of Louis XV. Napoleon signed his abdication there 1814. Nearby is the village of Barbizon, the haunt of several 19th-century painters (known as the Barbizon School).

Foochow /,fuː'tʃau/ alternative transcription of ▷Fuzhou, a port and capital of Fujian province, SE China.

Foreland, North and South /'fɔːlənd/ headlands on the Kent coast, England. *North Foreland*, with one lighthouse, lies 4 km/2.5 mi E of Margate; *South Foreland*, with two, lies 4.8 km/3 mi NE of Dover.

Forfarshire /'fɔːfəʃə/ former name of Angus, which was absorbed in Tayside, Scotland, 1975.

Forli /fɔː'liː/ city and market centre in Emilia-Romagna region, NE Italy, S of Ravenna; population (1988) 110,000. Felt, majolica, and paper are manufactured.

Formentera /,fɔːmən'teərə/ smallest inhabited island in the Spanish Balearic Islands, lying S of Ibiza; area 93 sq km/36 sq mi; population (1981) 3,500. The chief town is San Francisco Javier and the main port is La Sabina. The main industry is tourism.

Formentor, Cape /,fɔːmen'tɔː/ northern extremity of ▷Majorca, in the Balearic Islands of the W Mediterranean Sea.

Formosa /fɔː'məusə/ alternative name for ▷Taiwan.

Fortaleza (also called *Ceará*) industrial port in NE Brazil; population (1991) 1,708,700. It has textile, flour-milling, and sugar-refining industries.

Fort Collins /fɔːt 'kɒlɪnz/ city in N Colorado, USA, on the Cache de la Poudre river, NE of Boulder; seat of Larimer County; population (1990) 87,800. It is the processing and marketing centre for the surrounding agricultural area. Industries include engines, cement, plastics, film, and prefabricated metal buildings. The city was established as a fort 1864 for the protection of travellers on the Overland Trail.

Fort-de-France /fɔː də 'frɒns/ capital, chief commercial centre of Martinique, West Indies, at the mouth of the Madame River; population (1990) 101,500. It trades in sugar, rum, and cacao.

The Empress Josephine was born at Trois-Ilets on the south side of Fort-de-France Bay.

Forth /fɔːθ/ river in SE Scotland, with its headstreams rising on the northeast slopes of Ben Lomond. It flows approximately 72 km/45 mi to Kincardine where the *Firth of Forth* begins. The Firth is approximately 80 km/50 mi long, and is 26 km/16 mi wide where it joins the North Sea.

At Queensferry near Edinburgh are the Forth rail (1890) and road (1964) bridges. The *Forth and Clyde Canal* (1768–90) across the lowlands of Scotland links the Firth with the river Clyde, Grangemouth to Bowling (53 km/33 mi). A coalfield was located beneath the Firth of Forth 1976.

Fort Lamy /læ'miː/ former name of ▷N'djamena, the capital of Chad.

Fort Lauderdale /fɔːt 'lɔːdədeɪl/ city in SE coastal Florida, USA, just N of Miami; seat of Broward County; population (1990) 149,400. The city's main industry is tourism. Channels for boating cross the city, Atlantic Ocean beaches line it on the E, and deep-water Port Everglades to the S allows ocean-going vessels to dock. A fort was built here 1837 during the Seminole War.

Fort Myers /fɔːt 'maɪəz/ city in SW Florida, USA, on the Caloosahatchee River, SE of St Petersburg; seat of Lee County; population (1990) 45,200. It is a shipping centre for its fish, fruit, and vegetable products. Tourism is also an important industry. The surrender of Holatto-Micco, the last Seminole chief, took place here 1858.

Fort Pierce /fɔːt 'pɪəs/ city in E central Florida, USA, where the Indian River flows into the Atlantic Ocean; seat of St Lucie County; population (1990) 36,800. It is an important transportation centre for the fruit and vegetable crops of the surrounding area. Fishing and tourism are significant industries.

Fort Smith /fɔːt 'smɪθ/ city in W Arkansas, USA, on the Arkansas River where it crosses the Oklahoma–Arkansas border, SW of Fayetteville; population (1990) 72,800. The site of coal and natural-gas mines, the city's industries include furniture, cars, paper, plastics, and metals. Fort Smith national historic site is here.

Fort Walton Beach /fɔːt ,wɔːltən 'biːt/ city in the northwestern panhandle of Florida, USA, on the Gulf of Mexico, E of Pensacola; population (1990) 21,500. It is a tourist and fishing centre.

Fort Wayne /weɪn/ town in NE Indiana, USA; population (1990) 173,100. Industries include electrical goods, electronics, and farm machinery. A fort was built here against the North American Indians 1794 by General Anthony Wayne (1745–1796), hero of a surprise attack on a British force at Stony Point, New York, 1779, which earned him the nickname 'Mad Anthony'.

Fort Worth /wɜːθ/ city in NE Texas, USA; population (1990) 447,600. Formerly an important cattle area, it is now a grain, petroleum, aerospace, and railway centre serving the southern USA.

Fos-sur-Mer /ˈfɒs sjuə ˈmeə/ harbour and medieval township near Marseille, France, forming the southern focus of a direct Rhône–Rhine route to the North Sea.

Foveaux Strait /ˈfɒvəʊ/ stretch of water between the extreme south of South Island, New Zealand, and Stewart Island, New Zealand. It is a fishing area, producing a good oyster catch.

Fowey /ˈfɔɪ/ port and resort in Cornwall, England, near the mouth of the Fowey estuary; population, with St Austell (1988 est) 2,600. It is an outlet for the Cornish china clay mining industry.

Foyle /fɔɪl/ sea lough on the north coast of Ireland, traversed by the frontier of Northern Ireland and the Irish Republic.

France French Republic (*République Française*)
area (including Corsica) 543,965 sq km/209,970 sq mi
capital Paris
towns Lyons, Lille, Bordeaux, Toulouse, Nantes, Strasbourg; ports Marseille, Nice, Le Havre
physical rivers Seine, Loire, Garonne, Rhône, Rhine; mountain ranges Alps, Massif Central, Pyrenees, Jura, Vosges, Cévennes; the island of Corsica
territories Guadeloupe, French Guiana, Martinique, Réunion, St Pierre and Miquelon, Southern and Antarctic Territories, New Caledonia, French Polynesia, Wallis and Futuna
features Ardennes forest, Auvergne mountain region, Riviera, Mont Blanc (4,810 m/15,781 ft), caves of Dordogne with relics of early humans
head of state François Mitterrand from 1981
head of government Edouard Balladur from 1993
political system liberal democracy
political parties Socialist Party (PS), left of centre; Rally for the Republic (RPR), neo-Gaullist conservative; Union for French Democracy (UDF), centre-right; Republican

Party (RP), centre-right; French Communist Party (PCF), Marxist-Leninist; National Front, far right; Greens, environmentalist
exports fruit (especially apples), wine, cheese, wheat, cars, aircraft, iron and steel, petroleum products, chemicals, jewellery, silk, lace; tourism
currency franc
population (1990 est) 56,184,000 (including 4,500,000 immigrants, chiefly from Portugal, Algeria, Morocco, and Tunisia); growth rate 0.3% p.a.
life expectancy men 71, women 79
language French (regional languages include Basque, Breton, Catalan, and the Provençal dialect)
religions Roman Catholic 90%, Protestant 2%, Muslim 1%
literacy 99% (1984)
GNP $1,324.9 bn (1992)
chronology
1944–46 Provisional government headed by General Charles de Gaulle; start of Fourth Republic.
1954 Indochina achieved independence.
1956 Morocco and Tunisia achieved independence.
1957 Entry into European Economic Community.
1958 Recall of de Gaulle after Algerian crisis; start of Fifth Republic.
1959 De Gaulle became president.
1962 Algeria achieved independence.
1966 France withdrew from military wing of NATO.
1968 'May events' uprising of students and workers.
1969 De Gaulle resigned after referendum defeat; Georges Pompidou became president.
1974 Giscard d'Estaing elected president.
1981 François Mitterrand elected Fifth Republic's first socialist president.
1986 'Cohabitation' experiment, with the conservative Jacques Chirac as prime minister.
1988 Mitterrand re-elected. Moderate socialist Michel Rocard became prime minister. Matignon Accord on future of New Caledonia approved by referendum.
1991 French forces were part of the US-led coalition in the Gulf War. Edith Cresson became prime minister; Mitterrand's popularity rating fell rapidly.
1992 March: Socialist Party humiliated in regional and local elections; Greens and National Front polled strongly. April: Cresson replaced by Pierre Bérégovoy. Sept: referendum narrowly endorsed Maastricht Treaty.
1993 Socialist Party suffered heavy defeat in National Assembly elections. Edouard Balladur appointed prime minister; 'cohabitation' government re-established.

France has no friends, only interests.

On **France** Charles de Gaulle

Franche-Comté /frɒnʃ kɒnˈteɪ/ region of E France; area 16,200 sq km/6,253 sq mi; population

France: regions and départements

region and département	capital	area in sq km
Alsace		8,300
Bas-Rhin	Strasbourg	
Haut-Rhin	Colmar	
Aquitaine		41,300
Dordogne	Périgueux	
Gironde	Bordeaux	
Landes	Mont-de-Marsan	
Lot-et-Garonne	Agen	
Pyrénées-Atlantiques	Pau	
Auvergne		26,000
Allier	Moulins	
Cantal	Aurillac	
Haute-Loire	Le Puy	
Puy-de-Dôme	Clermont-Ferrand	
Basse-Normandie		17,600
Calvados	Caen	
Manche	Saint-Lô	
Orne	Alençon	
Brittany (Bretagne)		27,200
Côtes-du-Nord	St Brieuc	
Finistère	Quimper	
Ille-et-Vilaine	Rennes	
Morbihan	Vannes	
Burgundy (Bourgogne)		31,600
Côte-d'Or	Dijon	
Nièvre	Nevers	
Saône-et-Loire	Mâcon	
Yonne	Auxerre	
Centre		39,200
Cher	Bourges	
Eure-et-Loir	Chartres	
Indre	Châteauroux	
Indre-et-Loire	Tours	
Loir-et-Cher	Blois	
Loiret	Orléans	
Champagne-Ardenne		25,600
Ardennes	Charleville-Mézières	
Aube	Troyes	
Marne	Châlons-sur-Marne	
Haute-Marne	Chaumont	
Corsica		8,700
Haute-Corse	Bastia	
Corse du Sud	Ajaccio	
Franche-Comté		16,200
Doubs	Besançon	
Jura	Lons-le-Saunier	
Haute-Saône	Vesoul	
Terr. de Belfort	Belfort	
Haute-Normandie		12,300
Eure	Evreux	
Seine-Maritime	Rouen	
Ile de France		12,000
Essonne	Evry	
Val-de-Marne	Créteil	
Val-d'Oise	Cergy-Pontoise	
Ville de Paris		
Seine-et-Marne	Melun	
Hauts-de-Seine	Nanterre	
Seine-Saint-Denis	Bobigny	
Yvelines	Versailles	
Languedoc-Roussillon		27,400
Aude	Carcassonne	
Gard	Nîmes	
Hérault	Montpellier	
Lozère	Mende	
Pyrénées-Orientales	Perpignan	
Limousin		16,900
Corrèze	Tulle	
Creuse	Guéret	
Haute-Vienne	Limoges	
Lorraine		23,600
Meurthe-et-Mselle	Nancy	
Meuse	Bar-le-Duc	
Moselle	Metz	
Vosges	Epinal	
Midi-Pyrénées		45,300
Ariège	Foix	
Aveyron	Rodez	
Haute-Garonne	Toulouse	
Gers	Auch	
Lot	Cahors	
Hautes-Pyrénées	Tarbes	
Tarn	Albi	
Tarn-et-Garonne	Montauban	
Nord-Pas-de-Calais		12,400
Nord	Lille	
Pas-de-Calais	Arras	
Pays de la Loire		32,100
Loire-Atlantique	Nantes	
Maine-et-Loire	Angers	
Mayenne	Laval	
Sarthe	Le Mans	
Vendée	La Roche-sur-Yon	
Picardie		19,400
Aisne	Laon	
Oise	Beauvais	
Somme	Amiens	
Poitou-Charentes		25,800
Charente	Angoulême	
Charente-Maritime	La Rochelle	
Deux-Sèvres	Niort	
Vienne	Poitiers	
Provence-Alpes-Côte d'Azur		31,400
Alpes-de-Haute-Provence	Digne	
Hautes-Alpes	Gap	
Alpes-Maritimes	Nice	
Bouches-du-Rhône	Marseille	
Var	Draguignan	
Vaucluse	Avignon	
Rhône-Alpes		43,700
Ain	Bourg-en-Bresse	
Ardèche	Privas	
Drôme	Valence	
Isère	Grenoble	
Loire	St Étienne	
Rhône	Lyon	
Savoie	Chambéry	
Haute-Savoie	Annecy	

(1987) 1,086,000. Its capital is Besançon, and it includes the *départements* of Doubs, Jura, Haute Saône, and Territoire de Belfort. In the mountainous Jura, there is farming and forestry, and elsewhere there are engineering and plastics industries.

Once independent and ruled by its own count, it was disputed by France, Burgundy, Austria, and Spain from the 9th century until it became a French province under the Treaty of Nijmegen 1678.

Frankfort /'fræŋkfət/ capital of Kentucky, USA, located in the N central part of the state, on the Kentucky River, E of Louisville; population (1990) 26,000. Industries include bourbon (whisky), electronic equipment, furniture, and footwear. Frankfort became the capital of Kentucky 1786.

Frankfurt-am-Main /'fræŋkfɜːt æm 'maɪn/ city in Hessen, Germany, 72 km/45 mi NE of Mannheim; population (1988) 592,000. It is a commercial and banking centre, with electrical and machine industries, and an inland port on the river Main. An international book fair is held here annually.

history Frankfurt was a free imperial city 1372–1806, when it was incorporated into Prussia. It is the birthplace of the poet Goethe. It was the headquarters of the US zone of occupation in World War II and of the Anglo-US zone 1947–49.

Skyscrapers (most of them banking headquarters) sprouted all over Frankfurt; the town was dubbed 'Mainhattan' and 'Bankfurt'.

On **Frankfurt-am-Main** Gale Wiley in *International Herald Tribune* 6 Feb 1979

Frankfurt-an-der-Oder /'fræŋkfɜːt æn deə 'əudə/ industrial town in the state of Brandenburg, Germany, 80 km/50 mi SE of Berlin; population (1990) 87,000. It was the former capital of the East German district of Frankfurt 1952–90. It is linked by the river Oder and its canals to the rivers Vistula and Elbe. Industries include semiconductors, chemicals, engineering, paper, and leather.

Franklin /'fræŋklɪn/ district of Northwest Territories, Canada; area 1,422,550 sq km/549,250 sq mi. Fur trapping is important to the economy.

Franz Josef Land /'frænts 'jəuzef/ (Russian *Zemlya Frantsa Iosifa*) archipelago of over 85 islands in the Arctic Ocean, E of Spitsbergen and NW of Novaya Zemlya, Russia; area 20,720 sq km/8,000 sq mi. There are scientific stations on the islands.

Fraser /'freɪzə/ river in British Columbia, Canada. It rises in the Yellowhead Pass of the Rockies and flows NW, then S, then W to the Strait of Georgia. It is 1,370 km/850 mi long and rich in salmon.

Fray Bentos /'fraɪ 'bentɒs/ river port in Uruguay;

population (1985) 20,000. Linked by a bridge over the Uruguay with Puerto Unzué in Argentina (1976), it is known for meat-packing, particularly of corned beef.

Fredericton /'fredrɪktən/ capital of New Brunswick, Canada, on the St John River; population (1986) 44,000. It was known as *St Anne's Point* until 1785 when it was named after Prince Frederick, second son of George III.

Fredrikstad /'fredrɪkstæd/ Norwegian port at the mouth of the river Glomma, on the Oslo Fjord; population (1991) 26,500. It is a centre of the timber trade and has shipyards; there are also canning, paint, porcelain, and plastic industries.

Fredrikstad was founded 1567 by Frederick II, king of Denmark and Norway, as a fortress town and remains of the fortifications can be seen in Gamlebyen (the old town).

Freetown /'friːtaun/ capital of Sierra Leone, W Africa; population (1988) 470,000. It has a naval station and a harbour. Industries include cement, plastics, footwear, and oil refining. Platinum, chromite, diamonds, and gold are traded. It was founded as a settlement for freed slaves in the 1790s.

Freiburg-im-Breisgau /'fraɪbuəg ɪm 'braɪsgau/ industrial city (pharmaceuticals, precision instruments) in Baden-Württemberg, Germany; population (1988) 186,000. It has a university and a 12th-century cathedral.

Fremantle /'friːmæntl/ chief port of Western Australia, at the mouth of the Swan River, SW of Perth; population (1981) 23,780. It has shipbuilding yards, sawmills, and iron foundries and exports wheat, fruit, wool, and timber. It was founded as a penal settlement 1829.

French Antarctica /'frentʃ ænt'ɑːktɪkə/ *French Southern and Antarctic Territories* territory created 1955; area 10,100 sq km/3,900 sq mi; population about 200 research scientists. It includes Adélie Land on the Antarctic continent, the Kerguelen and Crozet archipelagos, and St Paul and Nouvelle Amsterdam islands in the southern seas. It is administered from Paris.

Port-aux-Français on Kerguelen is the chief centre, with several research stations. There are also research stations on Nouvelle Amsterdam and in Adélie Land and a meteorological station on Possession Island in the Crozet archipelago. St Paul is uninhabited. In 1988 French workers, who were illegally building an airstrip, thus violating a United Nations treaty on Antarctica, attacked Greenpeace workers.

French Guiana /giː'ɑːnə/ (French *Guyane Française*) French overseas *département* from 1946, and administrative region from 1974, on the north coast of South America, bounded W by Surinam and E and S by Brazil
area 83,500 sq km/32,230 sq mi
capital Cayenne
towns St Laurent
features Eurospace rocket launch pad at Kourou; Iles du Salut, which include Devil's Island

French Guiana

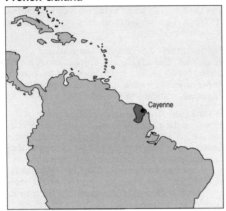

Cayenne

products timber, shrimps, gold
currency franc
population (1987) 89,000
languages 90% Creole, French, Amerindian
famous people Alfred Dreyfus
history first settled by France 1604, the territory became a French possession 1817; penal colonies, including Devil's Island, were established from 1852; by 1945 the shipments of convicts from France ceased.

The status changed to an overseas *départment* 1946, and an administrative region 1974.

French Polynesia /ˌpɒlɪ'niːziə/ French Overseas Territory in the S Pacific, consisting of five archipelagos: Windward Islands, Leeward Islands (the two island groups comprising the Society Islands), Tuamotu Archipelago (including Gambier Islands), Tubuai Islands, and Marquesas Islands
total area 3,940 sq km/1,521 sq mi
capital Papeete on Tahiti
products cultivated pearls, coconut oil, vanilla; tourism is important
population (1990) 199,100
languages Tahitian (official), French
government a high commissioner (Alain Ohrel) and Council of Government; two deputies are returned to the National Assembly in France
history first visited by Europeans 1595; French Protectorate 1843; annexed to France 1880–82; became an Overseas Territory, changing its name from French Oceania 1958; self-governing 1977. Following demands for independence in New Caledonia 1984–85, agitation increased also in Polynesia.

French Somaliland /sə'mɑːlilænd/ former name (until 1967) of ◊Djibouti, in E Africa.

French Sudan /suː'dɑːn/ former name (1898–1959) of ◊Mali.

French West Africa group of French colonies administered from Dakar 1895–1958. They are now Senegal, Mauritania, Sudan, Burkina Faso, Guinea, Niger, Ivory Coast, and Benin.

Fresno /'freznəʊ/ city in central California, USA, SE of San Jose, seat of Fresno County; popu-

lation (1990) 354,200. It is the processing and marketing centre for the fruits and vegetables of the San Joaquin Valley. Industries include glass, machinery, fertilizers, and vending machines.

Fribourg /'friːbuə/ (German *Freiburg*) city in W Switzerland, on the river Sarine, capital of the canton of Fribourg; population (1990) 33,900. It is renowned for its food products, such as the cheese of the Gruyère district.

Friendly Islands another name for ◊Tonga, a country in the Pacific.

Friesland /'friːzlənd/ maritime province of the N Netherlands, which includes the Frisian Islands and land that is still being reclaimed from the former Zuyder Zee; the inhabitants of the province are called Frisians
area 3,400 sq km/1,312 sq mi
capital Leeuwarden
towns Drachten, Harlingen, Sneek, Heerenveen
features sailing is popular; the *Elfstedentocht* (skating race on canals through 11 towns) is held in very cold winters
products livestock (Friesian cattle originated here; black Friesian horses), dairy products, small boats
population (1990) 600,000
history Ruled as a county of the Holy Roman Empire during the Middle Ages, Friesland passed to Saxony 1498 and, after a revolt, to Charles V of Spain. In 1579 it subscribed to the Treaty of Utrecht, opposing Spanish rule. In 1748 its stadholder, Prince William IV of Orange, became stadholder of all the United Provinces of the Netherlands.

Frisian Islands /'friːziən/ chain of low-lying islands 5–32 km/3–20 mi off the northwest coasts of the Netherlands and Germany, with a northerly extension off the west coast of Denmark. They were formed by the sinking of the intervening land. *Texel* is the largest and westernmost island, at the southern end of the chain.

Friuli-Venezia Giulia /fri'uːli vɪ'netsiə 'dʒuːliə/ autonomous agricultural and wine-growing region of NE Italy, bordered to the E by Slovenia; area 7,800 sq km/3,011 sq mi; population (1990) 1,201,000. Cities include Udine (the capital), Gorizia, Pordenone, and Trieste.

Formed 1947 from the province of Venetian Friuli and part of Eastern Friuli, to which Trieste was added after its cession to Italy 1954, it was granted autonomy 1963. A Slav minority numbers about 100,000, and in Friuli there is a movement for complete independence.

Frunze /'fruːnzi/ (formerly *Pishpek*) former name (1926–92) of ◊Bishkek in Kyrgyzstan.

Fujairah /fu'dʒaɪərə/ or *Fujayrah* one of the seven constituent member states of the United Arab Emirates; area 1,150 sq km/450 sq mi; population (1985) 54,000.

Fujian /ˌfuːdʒi'æn/ or *Fukien* province of SE China, bordering Taiwan Strait, opposite Taiwan
area 123,100 sq km/47,517 sq mi
capital Fuzhou

physical dramatic mountainous coastline
features being developed for tourists; designated
as a pace-setting province for modernization 1980
products sugar, rice, special aromatic teas,
tobacco, timber, fruit
population (1990) 30,048,000.

Fujiyama /ˌfuːdʒiˈjɑːmə/ or *Mount Fuji* Japanese
volcano and highest peak, on Honshu Island,
near Tokyo; height 3,778 m/12,400 ft. Extinct
since 1707, it has a Shinto shrine and a weather
station on its summit. Fuji has long been revered
for its picturesque cone-shaped crater peak, and
figures prominently in Japanese art, literature,
and religion.

---

*Fujiyama ... gathers in its skirts from
the surrounding plain and rises, plumed
with white across the sky – faultless, utterly
symmetrical.*

On **Fujiyama** Peter Quennell *A Superficial
Journal through Tokyo and Peking* 1932

Fukien /ˌfuːˈkjen/ alternative transcription of
◇Fujian, a province of SE China.

Fukuoka /ˌfuːkuːˈəukə/ formerly *Najime* Japanese
industrial port on the northwest coast of Kyushu
Island; population (1990) 1,237,100. It produces
chemicals, textiles, paper, and metal goods.

Fukushima /ˌfuːkuːˈʃiːmə/ city in the N of Honshu
Island, Japan; population (1990) 277,500. It has
a silk industry.

Fukuyama /ˌfuːkuːˈjɑːmə/ port in the SW of
Honshu Island, Japan, at the mouth of the Ashida
River; population (1990) 365,600. Exports
include cotton and rubber.

Funabashi /ˌfuːnəˈbæʃi/ city in Kanto region,
Honshu Island, E of Tokyo, Japan; population
(1990) 533,300.

Funafuti /ˌfuːnəˈfuːti/ atoll consisting of 30 islets in
the W Pacific and capital of the state of Tuvalu;
area 2.8 sq km/1.1 sq mi; population 900.

Funchal /funˈʃɑːl/ capital and chief port of the
Portuguese island of Madeira, on the south
coast; population (1980) 100,000. Tourism and
Madeira wine are the main industries.

Founded 1421 by the Portuguese navigator
João Gonçalves Zarco, it was under Spanish rule
1580–1640 and under Britain 1801 and 1807–14.

Fundy, Bay of /ˈfʌndi/ Canadian Atlantic inlet
between New Brunswick and Nova Scotia, with a
rapid tidal rise and fall of 18 m/60 ft (harnessed for
electricity since 1984). In summer, fog increases
the dangers to shipping.

Fünen /ˈfjuːnən/ German form of ◇Fyn, an island
forming part of Denmark.

Fünfkirchen /ˈfunfkɪəkən/ (German 'Five
Churches') German name for ◇Pécs, a town in
SW Hungary.

Furness /ˈfɜːnɪs/ peninsula in England, formerly a
detached northern portion of Lancashire, separated from the main part by Morecambe Bay.
In 1974 it was included in the new county
of Cumbria. Barrow is its ship-building and
industrial centre.

Fürth /ˈfjuət/ town in Bavaria, Germany, adjoining
Nuremberg; population (1984) 98,500. It has
electrical, chemical, textile, and toy industries.

Fushun /ˈfuːˈʃʌn/ coalmining and oil-refining centre in Liaoning province, China, 40 km/25 mi E
of Shenyang; population (1984) 636,000. It has
aluminium, steel, and chemical works.

Fuzhou /ˈfuːˈdʒəu/ or *Foochow* industrial port and
capital of Fujian province, SE China; population
(1989) 1,270,000. It is a centre for shipbuilding
and steel production; rice, sugar, tea, and fruit
pass through the port. There are joint foreign and
Chinese factories.

The Mazu (Matsu) island group, occupied by
the Nationalist Chinese, is offshore.

Fyn /fjuːn/ (German *Fünen*) island forming part
of Denmark and lying between the mainland
and Zealand; capital Odense; area 2,976 sq km/
1,149 sq mi; population (1984) 454,000.

GA abbreviation for ◊ *Georgia*, a state of the USA.

Gabcikovo Dam hydroelectric dam on the river Danube, at the point where it crosses the frontier between Hungary and Slovakia in Czechoslovakia. A treaty agreeing to its construction was signed by the two countries 1977, but work was suspended 1989 after Hungary withdrew its support for a scheme to divert water from the river. Czechoslovakia resumed work 1991, despite warnings from scientists and environmentalists that the scheme would destroy valuable wetlands in the Danube valley.
A dramatic reduction in river flow resulted from Czechoslovakia's first attempts to divert water Nov 1992, prompting the setting up of an investigative committee, under the auspices of the European Community (EC) and involving both parties concerned, to reassess the project.

Gabès /ˈɡɑːbes/ port in E Tunisia; population (1984) 92,300. Fertilizers and dates are exported. The town stands on the site of the Roman town of *Tacapae*.

Gabon Gabonese Republic (*République Gabonaise*)
area 267,667 sq km/103,319 sq mi
capital Libreville
towns Port-Gentil and Owendo (ports); Masuku (Franceville)

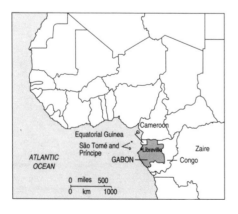

physical virtually the whole country is tropical rainforest; narrow coastal plain rising to hilly interior with savanna in E and S; Ogooué River flows N–W
features Schweitzer hospital at Lambaréné; Trans-Gabonais railway
head of state and government Omar Bongo from 1967
political system emergent democracy
political parties Gabonese Democratic Party (PDG), nationalist; Morena Movement of National Recovery, left of centre
exports petroleum, manganese, uranium, timber
currency CFA franc
population (1988) 1,226,000 including 40 Bantu groups; growth rate 1.6% p.a.
life expectancy men 47, women 51
languages French (official), Bantu
religions 96% Christian (Roman Catholic 65%), small Muslim minority (1%), animist 3%
literacy men 70%, women 53% (1985 est)
GDP $3.5 bn (1987); $3,308 per head
chronology
1889 Gabon became part of the French Congo.
1960 Independence from France achieved; Léon M'ba became the first president.
1964 Attempted coup by rival party foiled with French help. M'ba died; he was succeeded by his protégé Albert-Bernard Bongo.
1968 One-party state established.
1973 Bongo re-elected; converted to Islam, he changed his first name to Omar.
1986 Bongo re-elected.
1989 Coup attempt against Bongo defeated.
1990 PDG won first multiparty elections since 1964 amidst allegations of ballot-rigging.

Gaborone /ˌɡæbəˈrəuni/ capital of Botswana, mainly an administrative and government-service centre; population (1990) 341,100. The University of Botswana and Swaziland (1976) is here. The town developed after it replaced Mafikeng as the country's capital 1965.
Light industries include textiles, brewing, printing and publishing, and construction.

Gadsden /ˈɡædzdən/ city in NE Alabama, USA, on the Coosa River, SE of Huntsville; seat of Etowah County; population (1990) 42,500. It is a distribution centre for the area's livestock, poultry, and dairy products. Industries include manganese, bauxite, coal, timber, steel, rubber products, electrical machinery parts, and farm equipment.

Gafsa /ˈɡæfsə/ oasis town in central Tunisia, centre of a phosphate-mining area; population (1984) 60,900.

Gainesville /ˈɡeɪnzvɪl/ city in N Florida, USA, SW of Jacksonville; seat of Alachua County; population (1990) 84,700. Its industries include electronic parts, concrete, and wooden products. The University of Florida 1853 is here.

Gainsborough /ˈɡeɪnzbərə/ market town in Lincolnshire, England; population (1985) 18,700. It is an agricultural marketing centre with flour mills and the manufacture of agricultural machinery. It stands on the river Trent, which periodically rises in a tidal wave, the 'eagre'.

Galápagos Islands /gəˈlæpəgəs/ (official name *Archipiélago de Colón*) group of 15 islands in the Pacific, belonging to Ecuador; area 7,800 sq km/3,000 sq mi; population (1982) 6,120. The capital is San Cristóbal on the island of the same name. The islands are a nature reserve. Their unique fauna (including giant tortoises, iguanas, penguins, flightless cormorants, and Darwin's finches), which inspired Charles Darwin to formulate the principle of evolution by natural selection, is under threat from introduced species.

Galaţi /gæˈlæts/ (German *Galatz*) port on the river Danube in Romania; population (1985) 293,000. Industries include shipbuilding, iron, steel, textiles, food processing, and cosmetics.

Galdhøpiggen mountain in the Jotunheim range, Norway, 2,469 m/8,100 ft. It is counted as second to Glittertind, 2,472 m/8,110 ft (with glacier ice), although without ice and snow Galdhøpiggen is the higher of the two.

Galicia /gəˈlɪsɪə/ region of central Europe, extending from the northern slopes of the Carpathian Mountains to the Czechoslovak-Romanian border. Once part of the Austrian Empire, it was included in Poland after World War I and divided 1945 between Poland and the USSR.

Galicia /gəˈlɪsɪə/ mountainous but fertile autonomous region of NW Spain, formerly an independent kingdom; area 29,400 sq km/11,348 sq mi; population (1986) 2,785,000. It includes La Coruña, Lugo, Orense, and Pontevedra. Industries include fishing and the mining of tungsten and tin. The language is similar to Portuguese.

Gällivare /ˈjelɪvɑːrə/ iron-mining town above the Arctic Circle in Norrbotten county, N Sweden; population (1991) 22,400.

Galloway /ˈgæləweɪ/ ancient area of SW Scotland, now part of the region of Dumfries and Galloway.

Galveston /ˈgælvəstən/ Gulf of Mexico port on Galveston Island in Texas, USA; population (1990) 59,000. It exports cotton, petroleum, wheat, and timber and has chemical works and petroleum refineries. In 1900, 8,000 people died in one of the hurricanes that periodically hit the region.

Galway /ˈgɔːlweɪ/ county on the west coast of the Republic of Ireland, in the province of Connacht; area 5,940 sq km/2,293 sq mi; population (1991) 180,300. Towns include Galway (county town), Ballinasloe, Tuam, Clifden, and Loughrea (near which deposits of lead, zinc, and copper were found 1959).
The E is low-lying. In the S are the Slieve Aughty Mountains and Galway Bay, with the Aran Islands. To the W of Lough Corrib is Connemara, a wild area of moors, hills, lakes, and bogs. The Shannon is the principal river.

Galway /ˈgɔːlweɪ/ fishing port and county town of County Galway, Republic of Ireland; population (1991) 50,800. It produces textiles and chemi-

cals. University College is part of the national university, and Galway Theatre stages Irish Gaelic plays.

Gambia /ˈgæmbɪə/ river in W Africa, which gives its name to The Gambia; 1,000 km/620 mi long. It rises in Guinea and flows W through Gambia to the Atlantic Ocean.

Gambia Republic of The
area 10,402 sq km/4,018 sq mi
capital Banjul
towns Serekunda, Bakau, Georgetown
physical banks of the river Gambia flanked by low hills
features smallest state in black Africa; stone circles; Karantaba obelisk marking spot where Mungo Park began his journey to the Niger River 1796
head of state and government Dawda Jawara from 1970
political system liberal democracy
political parties Progressive People's Party (PPP), moderate centrist; National Convention Party (NCP), left of centre
exports groundnuts, palm oil, fish
currency dalasi
population (1990 est) 820,000; growth rate 1.9% p.a.
life expectancy 42 (1988 est)
languages English (official), Mandinka, Fula and other native tongues
media no daily newspaper; two official weeklies sell about 2,000 copies combined and none of the other independents more than 700
religions Muslim 90%, with animist and Christian minorities
literacy men 36%, women 15% (1985 est)
GDP $189 million (1987); $236 per head
chronology
1843 The Gambia became a crown colony.
1965 Independence achieved from Britain as a constitutional monarchy within the Commonwealth, with Dawda K Jawara as prime minister.
1970 Declared itself a republic, with Jawara as president.
1972 Jawara re-elected.
1981 Attempted coup foiled with the help of Senegal.
1982 Formed with Senegal the Confederation of Senegambia; Jawara re-elected.

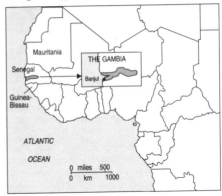

1987 Jawara re-elected.
1989 Confederation of Senegambia dissolved.
1990 Gambian troops contributed to the stabilizing force in Liberia.

The Gambia is one of the most oddly shaped countries in the world; it looks like an earthworm, and fits around the Gambia River like a long, tight wrinkled sleeve.

On **The Gambia** John Gunther
Inside Africa 1955

Gambier Islands /ˈɡæmbiə/ island group, part of French Polynesia, administered with the Tuamotu Archipelago; area 36 sq km/14 sq mi; population (1983) 582. It includes four coral islands and many small islets. The main island is Mangareva, with its town Rikitea.

Ganges /ˈɡændʒiːz/ (Hindi *Ganga*) major river of India and Bangladesh; length 2,510 km/1,560 mi. It is the most sacred river for Hindus.

Its chief tributary is the Jumna (*Yamuna*, length 1,385 km/860 mi), which joins the Ganges near Allahabad, where there is a sacred bathing place. The Ganges is joined in its delta in Bangladesh by the river Brahmaputra, and its most commercially important and westernmost channel to the Bay of Bengal is the *Hooghly*. It receives untreated sewage and chemical waste from more than 100 cities. The political leaders M K Gandhi, Nehru, and Indira Gandhi were all cremated on the banks of the Jumna at Delhi.

Deforestation of the Ganges watershed has decreased the river's dry-season flow by 20%; the area regularly flooded in the wet season has increased from 24.3 to 40.47 million hectares/60 to 100 million acres; the annual cost of flood damage has risen from $120 million to $1 billion.

Gannet Peak /ˈɡænɪt/ the highest peak in Wyoming, USA, rising to 4,207 m/13,804 ft. It is in the Rocky Mountains.

Gansu /ˌɡænˈsuː/ or *Kansu* province of NW China
area 530,000 sq km/204,580 sq mi
capital Lanzhou
features subject to earthquakes; the 'Silk Road' (now a motor road) passed through it in the Middle Ages, carrying trade to central Asia
products coal, oil, hydroelectric power from the Huang He (Yellow) River
population (1990) 22,371,000, including many Muslims.

Gaoxiong /ˈɡaʊʃiˈɒŋ/ mainland Chinese form of ▷Kaohsiung, a port in Taiwan.

Garching /ˈɡɑːkɪŋ/ town N of Munich, Germany, site of a nuclear research centre.

Gard /ɡɑː/ French river, 133 km/83 mi long, a tributary of the Rhône, which it joins above Beaucaire. It gives its name to Gard *département* in Languedoc-Roussillon region.

Garda, Lake /ˈɡɑːdə/ largest lake in Italy; situated

on the border between the regions of Lombardia and Veneto; area 370 sq km/143 sq mi.

Garonne /ɡæˈrɒn/ river in SW France, rising on the Spanish side of the Pyrenees and flowing to the Gironde estuary; length 560 km/350 mi.

Gary /ˈɡæri/ city in NW Indiana, USA; population (1990) 116,600. It contains the steel and cement works of the US Steel Corporation and was named after E H Gary (1846–1927), its chairman. Cutbacks in steel production have left the city economically depressed.

Gaspé Peninsula /ɡæˈspeɪ/ mountainous peninsula in SE Québec, Canada; area 29,500 sq km/11,390 sq mi. It has fishing and lumbering industries.

Gastonia /ɡæˈstəʊniə/ city in SW North Carolina, USA, directly W of Charlotte; seat of Gaston County; population (1990) 54,700. Its most important industry is textiles. It was the site of a violent labour strike 1929.

Gateshead /ˈɡeɪtshed/ port in Tyne and Wear, NE England; population (1981) 81,000. It is situated on the south bank of the river Tyne, opposite Newcastle-upon-Tyne. Formerly a port for the Tyne coalfields and a railway workshop centre, present-day industries include engineering, chemicals, and glass.

Gatwick /ˈɡætwɪk/ site of Gatwick Airport, West Sussex, England, constructed 1956–58. One of London's three international airports, it is situated 42 km/26 mi S from London's city centre. In 1989 its two terminals handled 21,293,200 passengers.

Gatwick Airport is connected by rail to Victoria Station, London.

Gaya /ˈɡaɪə/ ancient city in Bihar state, NE India; population (1986) 200,000. It is a centre of pilgrimage for Buddhists and Hindus with many temples and shrines. A bo tree at nearby Buddh Gaya is said to be a direct descendant of the original tree under which the Buddha sat.

Gaza /ˈɡɑːzə/ capital of the Gaza Strip, once a Philistine city; population (1980 est) 120,000. It was the scene of three World War I battles.

Gazankulu /ˌɡɑːzənˈkuːluː/ Black National State in Transvaal province, South Africa, with self-governing status from 1971; population (1985) 497,200.

Gaza Strip /ˈɡɑːzə/ strip of land on the Mediterranean sea, under Israeli administration; capital Gaza; area 363 sq km/140 sq mi; population (1989) 645,000 of which 446,000 are refugees.

Part of the British mandate of Palestine until 1948, it was then occupied by Egypt. It was invaded by Israel 1956, reoccupied 1967, and retained 1973. Clashes between the Israeli authorities and the Arab Palestinian inhabitants escalated to Intifada (uprising) 1988. In April 1992 the UN Security Council issued a statement condemning Israel for allowing 'the continued deterioration of the situation in the Gaza Strip', after clashes between Israeli troops and

Palestinian demonstrators left five Palestinians dead and more than 60 wounded. In 1993 a preliminary accord was signed with Israel, outlining principles for interim Palestinian self-rule in the Gaza Strip.

Gaziantep /ˌgæziæn'tep/ Turkish city 185 km/115 mi NE of Adana; population (1990) 603,400. It has textile and tanning industries. Until 1922 it was known as *Antep* or *Aintab*.

Gdańsk /gdænsk/ (German *Danzig*) Polish port; population (1990) 465,100. Oil is refined, and textiles, televisions, and fertilizers are produced. During the 1980s there were repeated anti-government strikes at the Lenin shipyards.

Formerly a member of the Hanseatic League, it was in almost continuous Prussian possession 1793–1919, when it again became a free city under the protection of the League of Nations. The annexation of the city by Germany marked the beginning of World War II. It reverted to Poland 1945, when the churches and old merchant houses were restored. The Lenin shipyards were the birthplace of Solidarity, the Polish resistance movement to pro-Soviet communism, 1981.

Gdynia /'gdɪnjə/ port in N Poland; population (1990) 251,500. It was established 1920 to give newly constituted Poland a sea outlet to replace lost Gdańsk. It has a naval base and shipyards and is now part of the 'Tri-city', which includes Sopot and Gdańsk.

Geelong /dʒɪ'lɒŋ/ industrial port in S Victoria, Australia; population (1986) 148,300. In addition to oil refining and trade in grain, it produces aluminium, motor vehicles, textiles, glass, and fertilizers.

Gelderland /'geldəlænd/ (English *Guelders*) province of the E Netherlands
area 5,020 sq km/1,938 sq mi
capital Arnhem
towns Apeldoorn, Nijmegen, Ede
products livestock, textiles, electrical goods
population (1991) 1,817,000
history in the Middle Ages Gelderland was divided into Upper Gelderland (Roermond in N Limburg) and Lower Gelderland (Nijmegen, Arnhem, Zutphen). These territories were inherited by Charles V of Spain, but when the revolt against Spanish rule reached a climax 1579, Lower Gelderland joined the United Provinces of the Netherlands.

Gelsenkirchen /ˌgelzən'kɪəkən/ industrial city in the Ruhr, Germany, 25 km/15 mi W of Dortmund; population (1988) 284,000. It has iron, steel, chemical, and glass industries.

Geneva /dʒɪ'niːvə/ (French *Genève*) Swiss city, capital of Geneva canton, on the shore of Lake Geneva; population (1990) city 167,200; canton 376,000. It is a point of convergence of natural routes and is a cultural and commercial centre. Industries include the manufacture of watches, scientific and optical instruments, foodstuffs, jewellery, and musical boxes.

The site on which Geneva now stands was the chief settlement of the Allobroges, a central European tribe who were annexed to Rome 121 BC; Julius Caesar built an entrenched camp here. In the Middle Ages, Geneva was controlled by the prince-bishops of Geneva and the rulers of Savoy. Under the Protestant theologian John Calvin, it became a centre of the Reformation 1536–64; the Academy, which he founded 1559, became a university 1892. Geneva was annexed by France 1798; it was freed 1814 and entered the Swiss Confederation 1815. In 1864 the International Red Cross Society was established in Geneva. It was the headquarters of the League of Nations, whose properties in Geneva passed 1946 into the possession of the United Nations.

This city of wealth by stealth.

On **Geneva** Robert Morley *More Morley* 1978

Geneva, Lake /dʒɪ'niːvə/ (French *Lac Léman*) largest of the central European lakes, between Switzerland and France; area 580 sq km/225 sq mi.

Genf /genf/ German form of ◊Geneva, Switzerland.

Gennesaret, Lake of /gɪ'nezərɪt/ another name for Lake ◊Tiberias (Sea of Galilee) in N Israel.

Genoa /'dʒenəuə/ (Italian *Genova*) historic city in NW Italy, capital of Liguria; population (1989) 706,700. It is Italy's largest port; industries include oil-refining, chemicals, engineering, and textiles.

history Decline followed its conquest by the Lombards 640, but from the 10th century it established a commercial empire in the W Mediterranean, pushing back the Muslims, and founding trading posts in Corsica, Sardinia, and N Africa; during the period of the Crusades, further colonies were founded in the kingdom of Jerusalem and on the Black Sea, where Genoese merchants enjoyed the protection of the Byzantine empire. At its peak about 1300, the city had a virtual monopoly of European trade with the East. Strife between lower-class Genoese and the ruling mercantile-aristocratic oligarchy led to weakness and domination by a succession of foreign powers, including Pope John XXII (1249–1334), Robert of Anjou, king of Naples (1318–43), and Charles VI of France (1368–1422). During the 15th century, most of its trade and colonies were taken over by Venice or the Ottomans.

Rebuilt after World War II, it became the busiest port on the Mediterranean, and the first to build modern container facilities. The nationalist Giuseppe Mazzini and the explorer Christopher Columbus were born here.

The dock-front of Genoa is marvellous. Such heat and colours and dirt and noise and loud wicked alleys with all the washing of the world hanging from the high windows.

On **Genoa** Dylan Thomas, letter of 1947

Genova /'dʒenəuə/ Italian form of ⟡Genoa, a city in Italy.

Georgetown /'dʒɔːdʒtaun/ capital and port of Guyana, situated at the mouth of the Demerara River on the Caribbean coast; population (1983) 188,000. There is food processing and shrimp fishing.

Founded 1781 by the British, it was held 1784–1812 by the Dutch (who renamed it Stabroek), and ceded to Britain 1814. The town's old buildings are wooden and built on stilts. St George's cathedral (1892) is one of the world's tallest wooden buildings at 43.5 m/143 ft. There is a university (1963).

Georgetown /'dʒɔːdʒtaun/ or **Penang** chief port of the Federation of Malaysia, and capital of Penang, on the island of Penang; population (1980) 250,600. It produces textiles and toys.

It is named after King George III of Great Britain.

Georgia, Republic of
area 69,700 sq km/26,911 sq mi
capital Tbilisi
towns Kutaisi, Rustavi, Batumi, Sukhumi
physical largely mountainous with a variety of landscape from the subtropical Black Sea shores to the ice and snow of the crest line of the Caucasus; chief rivers are Kura and Rioni
features holiday resorts and spas on the Black Sea; good climate; two autonomous republics, Abkhazia and Adzharia; one autonomous region, South Ossetia
interim head of state Eduard Shevardnadze from 1992
head of government Tengiz Sigua from 1992
political system transitional
political parties over 100 including: Georgian Popular Front, moderate nationalist; National Democratic Party of Georgia, radical nationalist; the National Independence Party; the Monarchist Party
products tea, citrus and orchard fruits, tung oil, tobacco, vines, silk, hydroelectricity
population (1990) 5,500,000 (70% Georgian, 8% Armenian, 8% Russian, 6% Azeri, 3% Ossetian, 2% Abkhazian)
language Georgian

religion Georgian Church, independent of the Russian Orthodox Church since 1917
chronology
1918–21 Independent republic.
1921 Uprising quelled by Red Army and Soviet republic established.
1922–36 Linked with Armenia and Azerbaijan as the Transcaucasian Republic.
1936 Became separate republic within the Soviet Union.
1981–88 Increasing demands for autonomy, spearheaded from 1988 by the Georgian Popular Front.
1989 March–April: Abkhazis demanded secession from Georgia, provoking inter-ethnic clashes. April: Georgian Communist Party (GCP) leadership purged. July: state of emergency imposed in Abkhazia; inter-ethnic clashes in South Ossetia. Nov: economic and political sovereignty declared.
1990 March: GCP monopoly ended. Oct: nationalist coalition triumphed in supreme soviet elections. Nov: Zviad Gamsakhurdia became president. Dec: GCP seceded from Communist Party of USSR; calls for Georgian independence.
1991 April: declared independence. May: Gamsakhurdia popularly elected president. Aug: GCP outlawed and all relations with USSR severed. Sept: anti-Gamsakhurdia demonstrations; state of emergency declared. Dec: Georgia failed to join new Commonwealth of Independent States (CIS).
1992 Jan: Gamsakhurdia fled to Armenia; Sigua appointed prime minister; Georgia admitted into Conference on Security and Cooperation in Europe (CSCE). Eduard Shevardnadze appointed interim president. July: admitted into United Nations (UN). Oct: Shevardnadze elected chair of new parliament. Clashes in South Ossetia and Abkhazia continued.

Georgia /'dʒɔːdʒə/ state in southeastern USA; nicknames Empire State of the South/Peach State
area 58,904 sq mi/152,600 sq km
capital Atlanta
towns Columbus, Savannah, Macon
features Okefenokee National Wildlife Refuge (656 sq mi/1,700 sq km), Sea Islands, historic Savannah
products poultry, livestock, tobacco, maize, peanuts, cotton, soya beans, china clay, crushed granite, textiles, carpets, aircraft, paper products

Georgia

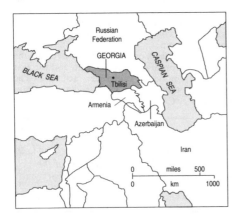

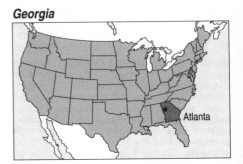

population (1990) 6,478,200
famous people Jim Bowie; Erskine Caldwell; Jimmy Carter; Ray Charles; Ty Cobb; Bobby Jones; Martin Luther King, Jr; Margaret Mitchell; James Oglethorpe; Jackie Robinson
history explored 1540 by Hernando de Soto; claimed by the British and named after George II of England; founded 1733 as a colony for the industrious poor by James Oglethorpe, a philanthropist; one of the original 13 states of the USA.

Gera /ˈgeərə/ industrial city (textiles, electronics) in the state of Thuringia, Germany, on the White Elster River; population (1990) 130,000.
It was the former capital of the East German district of Gera 1952–90.

German Ocean German name for the ◊North Sea.

Germany Federal Republic of (*Bundesrepublik Deutschland*)
area 357,041 sq km/137,853 sq mi
capital Berlin
towns Cologne, Munich, Essen, Frankfurt-am-Main, Dortmund, Stuttgart, Düsseldorf, Leipzig, Dresden, Chemnitz, Magdeburg; ports Hamburg, Kiel, Cuxhaven, Bremerhaven, Rostock
physical flat in N, mountainous in S with Alps; rivers Rhine, Weser, Elbe flow N, Danube flows SE, Oder, Neisse flow N along Polish frontier; many lakes, including Müritz
environment acid rain causing *Waldsterben* (tree death) affects more than half the country's forests; industrial E Germany has the highest sulphur-dioxide emissions in the world per head of population
features Black Forest, Harz Mountains, Erzgebirge (Ore Mountains), Bavarian Alps, Fichtelgebirge, Thüringer Forest
head of state Richard von Weizsäcker from 1984
head of government Helmut Kohl from 1982
political system liberal democratic federal republic
political parties Christian Democratic Union (CDU), right of centre; Christian Social Union (CSU), right of centre; Social Democratic Party (SPD), left of centre; Free Democratic Party (FDP), liberal; Greens, environmentalist; Republicans, far right; Party of Democratic Socialism (PDS), reform-communist (formerly Socialist Unity Party: SED)
exports machine tools (world's leading exporter), cars, commercial vehicles, electronics, industrial goods, textiles, chemicals, iron, steel, wine, lignite (world's largest producer), uranium, coal, fertilizers, plastics
currency Deutschmark
population (1990) 78,420,000 (including nearly 5,000,000 'guest workers', *Gastarbeiter*, of whom 1,600,000 are Turks; the rest are Yugoslav, Italian, Greek, Spanish, and Portuguese); growth rate –0.7% p.a.
life expectancy men 68, women 74
languages German, Sorbian
religions Protestant 42%, Roman Catholic 35%
literacy 99% (1985)
GNP $1,775.1 bn (1992)

chronology
1945 Germany surrendered; country divided into four occupation zones (US, French, British, Soviet).
1948 Blockade of West Berlin.
1949 Establishment of Federal Republic under the 'Basic Law' Constitution with Konrad Adenauer as chancellor; establishment of the German Democratic Republic as an independent state.
1953 Uprising in East Berlin suppressed by Soviet troops.
1954 Grant of full sovereignty to both West Germany and East Germany.
1957 West Germany was a founder-member of the European Economic Community; recovery of Saarland from France.
1961 Construction of Berlin Wall.
1963 Retirement of Chancellor Adenauer.
1964 Treaty of Friendship and Mutual Assistance signed between East Germany and USSR.
1969 Willy Brandt became chancellor of West Germany.
1971 Erich Honecker elected SED leader in East Germany.
1972 Basic Treaty between West Germany and East Germany; treaty ratified 1973, normalizing relations between the two.
1974 Resignation of Brandt; Helmut Schmidt became chancellor.
1975 East German friendship treaty with USSR renewed for 25 years.
1982 Helmut Kohl became chancellor.
1987 Official visit of Honecker to the Federal Republic.
1988 Death of Franz-Josef Strauss, leader of the West German Bavarian CSU.
1989 West Germany: rising support for far right in local and European elections, declining support for Kohl. East Germany: mass exodus to West Germany began. Honecker replaced by Egon Krenz. National borders opened in Nov, including Berlin Wall. Reformist Hans Modrow appointed prime minister. Krenz replaced.
1990 March: East German multiparty elections

Germany: states

state	capital	area in sq km
Baden-Württemberg	Stuttgart	35,800
Bavaria	Munich	70,600
Berlin	Berlin	880
Brandenburg	Potsdam	25,900
Bremen	Bremen	400
Hamburg	Hamburg	760
Hessen	Wiesbaden	21,100
Lower Saxony	Hanover	47,400
Mecklenburg–West Pomerania	Schwerin	22,900
North Rhine–Westphalia	Düsseldorf	34,100
Rhineland-Palatinate	Mainz	19,800
Saarland	Saarbrücken	2,570
Saxony	Dresden	17,050
Saxony-Anhalt	Magdeburg	25,900
Schleswig-Holstein	Kiel	15,700
Thuringia	Erfurt	15,500

won by a coalition led by the right-wing CDU. 3 Oct: official reunification of East and West Germany. 2 Dec: first all-German elections since 1932, resulting in a victory for Kohl.

1991 Kohl's popularity declined after tax increase. The CDU lost its Bundesrat majority to the SPD. Racism continued with violent attacks on foreigners.

1992 Neo-Nazi riots against immigrants continued.

The most civilized nations of modern Europe issued from the woods of Germany, and in the rude institutions of those barbarians we may still distinguish the original principles of our present laws and manners.

On **Germany** Edward Gibbon *The Decline and Fall of the Roman Empire* 1776–88

Germany, East /dʒɜːməni/ (German Democratic Republic, GDR) country 1949–90, formed from the Soviet zone of occupation in the partition of Germany following World War II. East Germany became a sovereign state 1954, and was reunified with West Germany Oct 1990.

Germany, West /dʒɜːməni/ (Federal Republic of Germany) country 1949–90, formed from the British, US, and French occupation zones in the partition of Germany following World War II; reunified with East Germany Oct 1990.

Germiston /dʒɜːmɪstən/ city in the Transvaal, South Africa; population (1980) 155,435. Industries include gold refining, chemicals, steel, and textiles.

Gerona /xeˈrəunə/ town in Catalonia, NE Spain, capital of Gerona province; population (1991) 66,900. Industries include textiles and chemicals. There are ferry links with Ibiza, Barcelona, and Málaga.

Gers /ʒeə/ river in France, 178 km/110 mi in length; it rises in the Lannemezan Plateau and flows N to join the river Garonne 8 km/5 mi above Agen. It gives its name to a *département* in Midi-Pyrénées region.

Gethsemane /geθˈsemənɪ/ site of the garden where Judas Iscariot, according to the New Testament, betrayed Jesus. It is on the Mount of Olives, E of Jerusalem. When Jerusalem was divided between Israel and Jordan 1948, Gethsemane fell within Jordanian territory.

Gezira, El /gɪˈzɪərə/ plain in the Republic of Sudan, between the Blue and White Nile rivers. The cultivation of cotton, sorghum, wheat, and groundnuts is made possible by irrigation.

Ghaghara /ˈgɑːgərə/ or **Gogra** river in N India, a tributary of the Ganges. It rises in Tibet and flows through Nepal and the state of Uttar Pradesh; length 1,000 km/620 mi.

Ghana /ˈgɑːnə/ Republic of
area 238,305 sq km/91,986 sq mi
capital Accra
towns Kumasi, and ports Sekondi-Takoradi, Tema
physical mostly tropical lowland plains; bisected by river Volta
environment forested areas have shrunk from

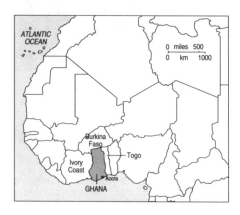

8.2 million sq km/3.17 million sq mi at the beginning of the 20th century to 1.9 million sq km/ 730,000 sq mi by 1990
features world's largest artificial lake, Lake Volta; relics of traditional kingdom of Ashanti: 32,000 chiefs and kings
head of state and government Jerry Rawlings from 1981
political system military republic
exports cocoa, coffee, timber, gold, diamonds, manganese, bauxite
currency cedi
population (1990 est) 15,310,000; growth rate 3.2% p.a.
life expectancy men 50, women 54
languages English (official) and African languages
media all media are government-controlled and the two daily newspapers are government-owned
religion animist 38%, Muslim 30%, Christian 24%
literacy men 64%, women 43% (1985 est)
GNP $3.9 bn (1983); $420 per head
chronology
1957 Independence achieved from Britain, within the Commonwealth, with Kwame Nkrumah as president.
1960 Ghana became a republic.
1964 Ghana became a one-party state.
1966 Nkrumah deposed and replaced by General Joseph Ankrah.
1969 Ankrah replaced by General Akwasi Afrifa, who initiated a return to civilian government.
1970 Edward Akufo-Addo elected president.
1972 Another coup placed Col Acheampong at the head of a military government.
1978 Acheampong deposed in a bloodless coup led by Frederick Akuffo; another coup put Flight-Lt Jerry Rawlings in power.
1979 Return to civilian rule under Hilla Limann.
1981 Rawlings seized power again, citing the incompetence of previous governments. All political parties banned.
1989 Coup attempt against Rawlings foiled.
1992 New multiparty constitution approved. Partial lifting of ban on political parties. Nov: Rawlings won presidency in national elections.

Ghats, Eastern and Western /gɔːts/ twin mountain ranges in S India, E and W of the central plateau; a few peaks reach about 3,000 m/9,800 ft. The name is a European misnomer, the Indian word *ghat* meaning 'pass', not 'mountain'.
They are connected by the Nilgiri Hills.

Ghent /gent/ (Flemish *Gent*, French *Gand*) city and port in East Flanders, NW Belgium; population (1991) 230,200. Industries include textiles, chemicals, electronics, and metallurgy. The cathedral of St Bavon (12th–14th centuries) has paintings by van Eyck and Rubens.

Giant's Causeway stretch of basalt columns forming a headland on the north coast of Antrim, Northern Ireland. It was formed by an outflow of lava in Tertiary times that has solidified in polygonal columns.

Gibraltar /dʒɪˈbrɔːltə/ British dependency, situated on a narrow rocky promontory in S Spain

Gibraltar

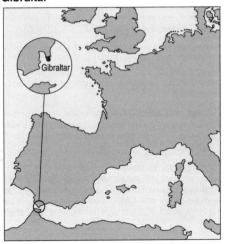

area 6.5 sq km/2.5 sq mi
features strategic naval and air base, with NATO underground headquarters and communications centre; colony of Barbary apes; the frontier zone is adjoined by the Spanish port of La Línea
exports mainly a trading centre for the import and re-export of goods
population (1988) 30,000
history captured from Spain 1704 by English admiral George Rooke (1650–1709), it was ceded to Britain under the Treaty of Utrecht 1713. A referendum 1967 confirmed the wish of the people to remain in association with the UK, but Spain continues to claim sovereignty and closed the border 1969–85. In 1989, the UK government announced it would reduce the military garrison by half. Ground troop withdrawals began March 1991, but navy and airforce units remained.
currency Gibraltar government notes and UK coinage
language English
religion mainly Roman Catholic
government the governor has executive authority, with the advice of the Gibraltar council, and there is an elected house of assembly (chief minister Joe Bossano from 1988).

The greatest drawback to the charms of Gibraltar has seemed to us to be the difficulty in leaving it. It is a beautiful prison.

On **Gibraltar** Augustus Hare *Wanderings in Spain* 1885

Gibraltar, Strait of /dʒɪˈbrɔːltə/ strait between North Africa and Spain, with the Rock of Gibraltar on the north side and Jebel Musa on the south, the so-called Pillars of Hercules.

Gibson Desert /ˈgɪbsən/ desert in central Western Australia, between the Great Sandy Desert to the N and the Great Victoria Desert in the S; area 220,000 sq km/85,000 sq mi.

Giessen /ˈgiːsən/ manufacturing town (machine tools, rubber, leather, tobacco) on the river Lahn, Hessen, Germany; population (1984) 71,800. Its university was established 1605.

Gijón /xiːˈxɒn/ port on the Bay of Biscay, Oviedo province, N Spain; population (1991) 260,200. It produces iron, steel, chemicals, and oil; it is an outlet for the coalmines of Asturias and is a major fishing and shipbuilding centre.

Gilbert and Ellice Islands /ˈgɪlbət, ˈelɪs/ former British colony in the Pacific, known since independence 1978 as the countries of ⋄Tuvalu and ⋄Kiribati.

Gilgit /ˈgɪlgɪt/ mountainous region on the northwest frontier of Kashmir, under the rule of Pakistan; area 38,021 sq km/14,676 sq mi. It is drained by the Gilgit and Indus rivers. The region's town, Gilgit, was formerly a Buddhist centre.

Gippsland Lakes /ˈgɪpslænd/ series of shallow lagoons on the coast of Victoria, Australia. The main ones are Wellington, Victoria and King (broadly interconnected), and Reeve.

Girgenti /dʒɪəˈdʒenti/ former name (until 1927) of ⋄Agrigento, a town in Italy.

Gironde /ʒɪˈrɒnd/ navigable estuary 80 km/50 mi long, formed by the mouths of the Garonne, length 580 km/360 mi, and Dordogne rivers, in SW France. The Lot, length 480 km/300 mi, is a tributary of the Garonne.

Gisborne /ˈgɪzbən/ port on the east coast of North Island, New Zealand, exporting dairy products, wool and meat; population (1991) 31,400.

Gîza, El /ˈgiːzə/ or **al-Jizah** site of the Great Pyramids and Sphinx; a suburb of Cairo, Egypt; population (1983) 1,500,000. It has textile and film industries.

Glace Bay /gleɪs/ port on Cape Breton Island, Nova Scotia, Canada, centre of a coal-mining area; population (1986) 20,500.

Glamorgan /gləˈmɔːgən/ (Welsh **Morgannwg**) three counties of S Wales—Mid Glamorgan, South Glamorgan, and West Glamorgan—created 1974 from the former county of Glamorganshire.

Glasgow /ˈglæzgəʊ/ city and administrative headquarters of Strathclyde, Scotland; population (1991) 654,500. Industries include engineering, chemicals, printing, and distilling.

Notable buildings include the 12th-century cathedral of St Mungo; the Cross Steeple (part of the historic Tolbooth); the universities of Glasgow, established 1451 (present buildings constructed 1868–70 to designs by George Gilbert Scott) and Strathclyde, established 1964; the Royal Exchange; the Stock Exchange; Kelvingrove Art Gallery (Impressionist collection); the Glasgow School of Art, designed by C R Mackintosh; the Burrell Collection at Pollock Park, bequeathed by shipping magnate William Burrell (1861–1958); and Mitchell Library.

Direct transatlantic flights from Glasgow International Airport were introduced 1990.

Glastonbury /ˈglæstənbəri/ market town in Somerset, England, on the river Brue; population (1981) 6,800. There is light industry and tourism. Nearby are two excavated lake villages thought to have been occupied for about 150 years before the Romans came to Britain. **Glastonbury Tor**, a hill with a ruined church tower, rises to 159 m/522 ft.

The first church on the site was traditionally founded in the 1st century by Joseph of Arimathea. Legend has it that he brought the Holy Grail to Glastonbury. The ruins of the Benedictine abbey built in the 10th and 11th centuries by Dunstan and his followers were excavated 1963 and the site of the grave of King Arthur and Queen Guinevere was thought to have been identified. One of Europe's largest pop festivals is held outside Glastonbury most years in June.

Glencoe /glenˈkaʊ/ glen in Strathclyde Region, Scotland, where members of the Macdonald clan were massacred 1692. John Campbell, Earl of Breadalbane, was the chief instigator. It is now a winter sports area.

Gleneagles /glenˈiːgəlz/ glen in Tayside, Scotland, famous for its golf course and for the **Gleneagles Agreement**, formulated 1977 at the Gleneagles Hotel by Commonwealth heads of government, that 'every practical step (should be taken) to discourage contact or competition by their nationals' with South Africa, in opposition to apartheid.

Glenrothes /glenˈrɒθɪs/ town and administrative headquarters of Fife, Scotland, 10 km/6 mi N of Kirkcaldy, developed as a new town from 1948; population (1989 est) 38,100. Industries include electronics, plastics, and paper.

Glens Falls /ˌglenz ˈfɔːlz/ city in E central New York, USA, N of Albany and S of lakes George and Champlain; population (1990) 15,000. Situated in Warren County by a waterfall in the Hudson River, its industries include clothing, paper, machinery parts, insurance, and tourism.

Glittertind /ˈglɪtətɪn/ highest mountain in Norway, rising to 2,472 m/8,110 ft (including glacier ice) in the Jotunheim range. Glittertind is measured as the highest mountain, but if the measurement was exclusive of ice and snow Galdhøpiggen would be the highest, at 2,469 m/8,100 ft.

Gliwice /glɪˈviːtseɪ/ city in Katowice region, S Poland, formerly in German Silesia; population (1990) 214,200. It has coal-mining, iron, steel, and electrical industries. It is connected to the river Oder by the Gliwice Canal.

Glomma /ˈglɒmə/ river in Norway, 570 km/350 mi long. The largest river in Scandinavia, it flows into the Skagerrak (an arm of the North Sea) at Fredrikstad.

Glos abbreviation for ⋄**Gloucestershire**, an English county.

Gloucestershire

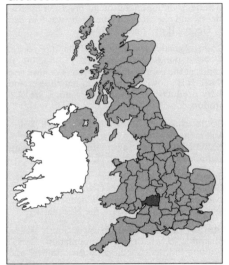

Gloucester /ˈglɒstə/ city, port, and administrative headquarters of Gloucestershire, England; population (1991) 91,800. Industries include the manufacture of aircraft and agricultural machinery. Its 11th–14th-century cathedral has a Norman nucleus and additions in every style of Gothic.

The Museum of Advertising and Packaging was established 1984 by Robert Opie.

Gloucester /ˈglɒstə/ city in NE Massachusetts, USA, on Cape Ann, NE of Boston; population (1990) 28,700. A famous fishing port, its industries include tourism and fish processing, especially lobster, whiting, and cod.

Gloucestershire /ˈglɒstəʃə/ county in SW England
area 2,640 sq km/1,019 sq mi
towns Gloucester (administrative headquarters), Stroud, Cheltenham, Tewkesbury, Cirencester
features Cotswold Hills; river Severn and tributaries; Berkeley Castle, where Edward II was murdered; Prinknash Abbey, where pottery is made; Cotswold Farm Park, near Stow-on-the-Wold, which has rare and ancient breeds of farm animals
products cereals, fruit, dairy products; engineering, coal in the Forest of Dean
population (1991) 520,600
famous people Edward Jenner, John Keble, Gustav Holst.

Goa /ˈgəuə/ state of India
area 3,700 sq km/1,428 sq mi
capital Panaji
population (1991) 1,168,600
history captured by the Portuguese 1510; the inland area was added in the 18th century. Goa was incorporated into India as a Union Territory with Daman and Diu 1961 and became a state 1987.

Gobi Desert /ˈgəubi/ Asian desert divided between the Mongolian People's Republic and Inner Mongolia, China; 800 km/500 mi N–S, and 1,600 km/1,000 mi E–W. It is rich in fossil remains of extinct species.

The Great Hungry Desert.
Chinese name for the **Gobi Desert**

Godalming /ˈgɒdlmɪŋ/ town in Surrey, SE England; population (1981) 18,200. Industries include light engineering and textiles. The writer Aldous Huxley lived here.
Godalming was once the centre of the Surrey wool trade. Charterhouse School is to the N; it moved here from London 1872.

Godavari /gəuˈdɑːvəri/ river in central India, flowing from the Western Ghats to the Bay of Bengal; length 1,450 km/900 mi. It is sacred to Hindus.

Godthaab /ˈgɒdhɔːb/ (Greenlandic *Nuuk*) capital and largest town of Greenland; population (1982) 9,700. It is a storage centre for oil and gas, and the chief industry is fish processing.

Gogra /ˈgɒgrə/ alternative transcription of the river ▷Ghaghara in India.

Golan Heights /ˈgəulæn/ (Arabic *Jawlan*) plateau on the Syrian border with Israel, bitterly contested in the Arab-Israeli Wars and annexed by Israel 14 Dec 1981.

Gold Coast /ˈgəuld kəust/ former name for ▷Ghana, but historically the west coast of Africa from Cape Three Points to the Volta River, where alluvial gold is washed down. Portuguese and French navigators visited this coast in the 14th century, and a British trading settlement developed into the colony of the Gold Coast 1618. With its dependencies of Ashanti and Northern Territories plus the trusteeship territory of Togoland, it became Ghana 1957. The name is also used for many coastal resort areas—for example, in E Australia and Florida, USA.

Gold Coast /ˈgəuld kəust/ resort region on the east coast of Australia, stretching 32 km/20 mi along the coast of Queensland and New South Wales S of Brisbane; population (1986) 219,000.

Golden Gate /ˈgəuldən ˈgeɪt/ strait in California, USA, linking San Francisco Bay with the Pacific, spanned by a suspension bridge that was completed 1937. The longest span is 1,280 m/4,200 ft.

Gondar /ˈgɒndɑː/ town in Ethiopia about 2,300 m/7,500 ft above sea level and 40 km/25 mi N of Lake Tana; population (1984) 69,000. It is the capital of a region by the same name. Cattle, grain, and seed are traded. It was the capital of Ethiopia 1632–1855.

Goodwin Sands /ˈgudwɪn/ sandbanks off the coast of Kent, England, exposed at low tide, and famous for wrecks. According to legend, they are the remains of the island of Lomea, owned by Earl Godwin in the 11th century.

Goose Bay /ˈguːs ˈbeɪ/ settlement at the head of Lake Melville on the Labrador coast of

Newfoundland, Canada. In World War II it was used as a staging post by US and Canadian troops on their way to Europe. Until 1975 it was used by the US Air Force as a low-level-flying base.

Gorakhpur /'gɔːrəkpuə/ city in Uttar Pradesh, N India, situated on the Rapti River, at the centre of an agricultural region producing cotton, rice, and grain; population (1981) 306,000.

Gorgonzola /ˌgɔːgən'zəulə/ town NE of Milan in Lombardy, Italy; population (1971 est) 13,000. It is famous for its blue-veined cheese. Bel Paese is also made here.

Gorizia /gɒ'rɪtsiə/ town in Friuli-Venezia-Giulia region, N Italy, on the Isonzo, SE of Udine; population (1981) 41,500. Industries include textiles, furniture, and paper. It has a 16th-century castle, and was a cultural centre during Habsburg rule.

Gorky /'gɔːki/ (Russian *Gor'kiy*) name 1932–90 of ⋄Nizhny Novgorod, a city in central Russia.

Görlitz /'gɜːlɪts/ manufacturing town (rolling stock) in the state of Saxony, Germany; population (1981) 81,000.

Gorlovka /gɔ:'ləvkə/ industrial town (coalmining, chemicals, engineering) on the Donets Basin coalfield, Ukraine; population (1987) 345,000.

Goshen /'gəuʃn/ city in N Indiana, USA, on the Elkhart River, SE of South Bend; seat of Elkhart County; population (1990) 23,700. It is situated in an agricultural area and serves as a market town. Industries include steel, rubber, electrical, and building products.

Gosport /'gɒspɔːt/ naval port opposite Portsmouth, Hampshire, England; population (1985 est) 82,000. Industries include chemicals and engineering.

Göteborg /ˌjɜːtə'bɔːrj/ (German *Gothenburg*) port and industrial city (ships, vehicles, chemicals) on the west coast of Sweden, at the mouth of the Göta River; population (1990) 433,000. It is Sweden's second largest city and is linked with Stockholm by the Göta Canal (built 1832).
    The city was founded 1619 by King Gustavus Adolphus. There is a cathedral (1633) and a technical university (1829).

Gotha /'gəutə/ town in Thuringia, Germany, former capital of the duchy of Saxe-Coburg-Gotha; population (1981) 57,600. It has a castle and two observatories; pottery, soap, textiles, precision instruments, and aircraft are manufactured here.

Gothenburg /'gɒθənbɜːg/ German form of ⋄Göteborg, a city in Sweden.

Gotland /'gɒtlənd/ Swedish island in the Baltic Sea; area 3,140 sq km/1,212 sq mi; population (1990) 57,100. The capital is Visby. Its products are mainly agricultural (sheep and cattle), and there is tourism. It was an area of dispute between Sweden and Denmark but became part of Sweden 1645.

Göttingen /'gɜːtɪŋən/ town in Lower Saxony, Ger-

many; population (1988) 134,000. Industries include printing, publishing, precision instruments, and chemicals. Its university was founded by George II of England 1734.

Gouda /'gaudə/ town in South Holland province, W Netherlands; population (1991) 65,900. It stands at the confluence of the rivers Gouwe and IJssel. It is known for its round, flat cheeses and dairy products; stoneware and candles are also produced.

Goulburn /'gəulbɜːn/ town in New South Wales, Australia, SW of Sydney; population (1983) 22,500. It is an agricultural centre, and manufactures bricks, tiles, and pottery.

Gower Peninsula /'gauə/ (Welsh *Penrhyn Gŵyr*) peninsula in West Glamorgan, S Wales, extending into the Bristol Channel. There is tourism on the south coast; the N is marshy.

Grafton /'grɑːftən/ town in New South Wales, Australia, S of Brisbane; population (1985) 17,600. Industries include sugar, timber, and dairy products.

Graham Land /'greɪəm/ mountainous peninsula in Antarctica, formerly a dependency of the Falkland Islands, and from 1962 part of the British Antarctic Territory. It was discovered by John Biscoe 1832 and until 1934 was thought to be an archipelago.

Grahamstown /'greɪəmztaun/ town in SE Cape Province, South Africa; population (1985) 75,000. It is the seat of Rhodes University, established 1951, founded 1904 as Rhodes University College.

Grampian /'græmpiən/ region of Scotland
area 8,600 sq km/3,320 sq mi
towns Aberdeen (administrative headquarters)
features part of the Grampian Mountains (the Cairngorm Mountains); valley of the river Spey,

Grampian

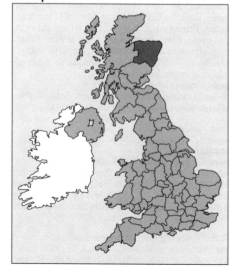

with its whisky distilleries; Balmoral Castle (royal residence on the river Dee near Braemar, bought by Prince Albert 1852, and rebuilt in Scottish baronial style); Braemar Highland Games in Aug *products* beef cattle (Aberdeen Angus and Beef Shorthorn), fishing, North Sea oil service industries, tourism (winter skiing) *population* (1991) 493,200 *famous people* John Barbour, James Ramsay MacDonald, Alexander Cruden.

Grampian Mountains /ˈɡræmpiən/ range that separates the Highlands from the Lowlands of Scotland, running NE from Strathclyde. It takes in the S Highland region (which includes **Ben Nevis**, the highest mountain in the British Isles at 1,340 m/4,406 ft), northern Tayside, and the southern border of Grampian Region itself (the Cairngorm Mountains, which include **Ben Macdhui** 1,309 m/4,296 ft). The region includes Aviemore, a winter holiday and sports centre.

Grampians /ˈɡræmpiənz/ western end of Australia's eastern highlands, in Victoria; the highest peak is Mount William, 1,167 m/3,829 ft.

Granada /ɡrəˈnɑːdə/ city in the Sierra Nevada in Andalusia, S Spain; population (1986) 281,000. It produces textiles, soap, and paper. The *Alhambra*, a fortified hilltop palace, was built in the 13th and 14th centuries by the Moorish kings. *history* Founded by the Moors in the 8th century, it became the capital of an independent kingdom 1236–1492, when it was the last Moorish stronghold to surrender to the Spaniards. Ferdinand and Isabella, the first sovereigns of a united Spain, are buried in the cathedral (built 1529–1703).

The Moors' once affluent and beautiful capital, the Damascus of the West.

On **Granada** Josiah Conder *The Modern Traveller* 1830

Granada /ɡrəˈnɑːdə/ Nicaraguan city on the northwestern shore of Lake Nicaragua; population (1985) 89,000. It has shipyards and manufactures sugar, soap, clothing, and furniture. Founded 1523, it is the oldest city in Nicaragua.

Gran Chaco /ɡræn ˈtʃɑːkəʊ/ large lowland plain in N Argentina, W Paraguay, and SE Bolivia; area 650,000 sq km/251,000 sq mi. It consists of swamps, forests (a source of quebracho timber), and grasslands, and there is cattle-raising.

Grand Canal (Chinese *Da Yune*) the world's longest canal. It is 1,600 km/1,000 mi long and runs N from Hangzhou to Tianjin, China; it is 30–61 m/100–200 ft wide, and reaches depths of over 1.5 km/1 mi. The earliest section was completed 486 BC, and the northern section was built AD 1282–92, during the reign of Kublai Khan.

Grand Canyon /ɡrænd ˈkænjən/ gorge of multi-coloured rock strata cut by and containing the Colorado River, N Arizona, USA. It is 350 km/

217 mi long, 6–29 km/4–18 mi wide, and reaches depths of over 1.7 km/1.1 mi. It was made a national park 1919. Millions of tourists visit the canyon each year.

... the one great sight which every American should see.

On the **Grand Canyon** attributed to Theodore Roosevelt 1903

Grand Falls /ɡrænd ˈfɔːlz/ town in Newfoundland, Canada; population (1986) 9,100. It is the site of large paper and pulp mills.

Grand Forks /ɡrænd ˈfɔːks/ city in E central North Dakota, USA, on the Minnesota border, on the Red River, N of Fargo; seat of Grand Forks County; population (1990) 49,400. It serves the surrounding agricultural area; most of its industries, such as food-processing mills and fertilizer plants, are associated with agriculture.

Grand Rapids /ɡrænd ˈræpɪdz/ city in W Michigan, USA, on the Grand River; population (1990) 189,100. It produces furniture, motor bodies, plumbing fixtures, and electrical goods. A fur-trading post was founded here 1826, and the furniture industry developed in the 1840s.

Grand Teton /ɡrænd ˈtiːtn/ highest point of the spectacular Teton mountain range, NW Wyoming, USA, rising to 4,197 m/13,770 ft. Grand Teton National Park was established 1929.

Granite City /ˈɡrænɪt ˈsɪti/ city in SW Illinois, USA, across the Mississippi River from St Louis, Missouri; population (1990) 32,800. An industrial city with its own port on the Chain of Rocks Canal, it manufactures steel products, car frames, and building materials.

Grantham /ˈɡrænθəm/ market town in SE Lincolnshire, England; population (1988 est) 35,200. It is an agricultural centre, dating from Saxon times. The former prime minister Margaret Thatcher was born here.

Grasmere /ˈɡrɑːsmɪə/ English lake and village in the Lake District, Cumbria, associated with many writers. William Wordsworth and his sister Dorothy lived at Dove Cottage (now a museum) 1799–1808; Thomas de Quincey later made his home in the same house. Both Samuel Coleridge and Wordsworth are buried in the churchyard of St Oswald's.

Grasse /ɡrɑːs/ town near Cannes, SE France; population (1990) 42,100. It is the centre of a perfume-manufacturing region, and flowers are grown on a large scale for this purpose.

Graubünden /ɡraʊˈbʊndən/ (French *Grisons*) Swiss canton, the largest in Switzerland; area 7,106 sq km/2,743 sq mi; population (1990) 170,400. The inner valleys are the highest in Europe, and the main sources of the river Rhine rise here. It also includes the resort of Davos and, in the Upper Engadine, St Moritz. The capital is Chur. Romansch is still widely spoken. Graubünden entered the Swiss Confederation 1803.

Gravesend /ˌɡreɪvz'end/ town on the river Thames, Kent, SE England, linked by ferry with Tilbury opposite; population (1981) 56,200. Industries include electrical goods, engineering, and printing.

Graz /ɡrɑːts/ capital of Styria province, and second largest city in Austria; population (1981) 243,400. Industries include engineering, chemicals, iron, and steel. It has a 15th-century cathedral and a university founded 1573. Lippizaner horses are bred near here.

Great Australian Bight broad bay of the Indian Ocean in S Australia, notorious for storms. It was discovered by Dutch navigator Captain Thyssen 1627. The coast was charted by the English explorer Captain Matthew Flinders 1802.

Great Barrier Reef chain of coral reefs and islands about 2,000 km/1,250 mi long, off the east coast of Queensland, Australia, at a distance of 15–45 km/10–30 mi. It is believed to be the world's largest living organism and forms an immense natural breakwater, the coral rock forming a structure larger than all human-made structures on Earth combined. The reef is in danger from large numbers of starfish, which are reported to have infested 35% of the reef. Some scientists fear the entire reef will disappear within 50 years.

Annually, a few nights after the full Moon in Nov, 135 species of hard coral release their eggs and sperm for fertilization and the sea turns pink. The phenomenon, one of the wonders of the natural world, was discovered 1983, and is triggered by a mechanism dependent on the Moon, the tides, and water temperatures.

Great Bear Lake lake on the Arctic Circle, in the Northwest Territories, Canada; area 31,800 sq km/12,275 sq mi.

Great Britain /ɡreɪt 'brɪtn/ official name for ⟁England, ⟁Scotland, and ⟁Wales, and the adjacent islands (except the Channel Islands and the Isle of Man) from 1603, when the English and Scottish crowns were united under James I of England (James VI of Scotland). With Northern Ireland, it forms the United Kingdom.

Great Dividing Range E Australian mountain range, extending 3,700 km/2,300 mi N–S from Cape York Peninsula, Queensland, to Victoria. It includes the Carnarvon Range, Queensland, which has many Aboriginal cave paintings, the Blue Mountains in New South Wales, and the Australian Alps.

Great Falls /ɡreɪt 'fɔːlz/ city in central Montana, USA, on the Missouri River, NE of Helena; seat of Cascade County; population (1990) 55,000. Its main industries are involved with the processing of copper and zinc from nearby mines. The processing of agricultural products and oil

refining is also important. The city is named after nearby waterfalls, first discovered 1805 by Meriwether Lewis and William Clark.

Great Lake Australia's largest freshwater lake, 1,030 m/3,380 ft above sea level, in Tasmania; area 114 sq km/44 sq mi. It is used for hydroelectric power and is a tourist attraction.

Great Lakes series of five freshwater lakes along the US–Canadian border: Lakes Superior, Michigan, Huron, Erie, and Ontario; total area 245,000 sq km/94,600 sq mi. Interconnecting canals make them navigable by large ships, and they are drained by the St Lawrence River. The whole forms the St Lawrence Seaway. They are said to contain 20% of the world's surface fresh water.

These lakes, called 'Great' with reason, are the fresh-water Mediterranean of the Western Hemisphere .... They are the Middle West's equivalent of a coastline.

On the **Great Lakes** John Gunther
Inside USA 1947

Great Plains semi-arid region to the E of the Rocky Mountains, USA, stretching as far as the 100th meridian of longitude through Oklahoma, Kansas, Nebraska, and the Dakotas. The plains, which cover one-fifth of the USA, extend from Texas in the S over 2,400 km/1,500 mi N to Canada. Ranching and wheat farming have resulted in over-use of water resources to such an extent that available farmland has been reduced by erosion.

Great Rift Valley longest 'split' in the Earth's surface; see ⟁Rift Valley, Great.

Great Sandy Desert desert in N Western Australia; 415,000 sq km/160,000 sq mi. It is also the name of an arid region in S Oregon, USA.

Great Slave Lake lake in the Northwest Territories, Canada; area 28,450 sq km/10,980 sq mi. It is the deepest lake (615 m/2,020 ft) in North America.

Great Yarmouth /ɡreɪt 'jɑːməθ/ alternative name for the resort and port of ⟁Yarmouth in Norfolk, England.

Greece Hellenic Republic (*Elliniki Dimokratia*)
area 131,957 sq km/50,935 sq mi
capital Athens
towns Larisa; ports Piraeus, Thessaloníki, Patras, Iráklion
physical mountainous; a large number of islands, notably Crete, Corfu, and Rhodes
environment acid rain and other airborne pollutants are destroying the Classical buildings and ancient monuments of Athens
features Corinth canal; Mount Olympus; the Acropolis; many classical archaeological sites; the Aegean and Ionian Islands
head of state Constantine Karamanlis from 1990

head of government Andreas Papandreou from 1993
political system democratic republic
political parties Panhellenic Socialist Movement (PASOK), democratic socialist; New Democracy Party (ND), centre-right; Democratic Renewal (DR); Communist Party; Greek Left Party
exports tobacco, fruit, vegetables, olives, olive oil, textiles, aluminium, iron and steel
currency drachma
population (1992) 10,288,000; growth rate 0.3% p.a.
life expectancy men 72, women 76
language Greek
religion Greek Orthodox 97%
literacy men 96%, women 89% (1985)
GDP $79.2 bn (1992)
chronology
1829 Independence achieved from Turkish rule.
1912–13 Balkan Wars; Greece gained much land.
1941–44 German occupation of Greece.
1946 Civil war between royalists and communists; communists defeated.
1949 Monarchy re-established with Paul as king.
1964 King Paul succeeded by his son Constantine.
1967 Army coup removed the king; Col George Papadopoulos became prime minister. Martial law imposed, all political activity banned.
1973 Republic proclaimed, with Papadopoulos as president.
1974 Former premier Constantine Karamanlis recalled from exile to lead government. Martial law and ban on political parties lifted; restoration of the monarchy rejected by a referendum.
1975 New constitution adopted, making Greece a democratic republic.
1980 Karamanlis resigned as prime minister and was elected president.
1981 Greece became full member of European Economic Community. Andreas Papandreou elected Greece's first socialist prime minister.
1983 Five-year military and economic cooperation agreement signed with USA; ten-year economic cooperation agreement signed with USSR.
1985 Papandreou re-elected.

1988 Relations with Turkey improved. Mounting criticism of Papandreou.
1989 Papandreou defeated. Tzannis Tzannetakis became prime minister; his all-party government collapsed. Xenophon Zolotas formed new unity government. Papandreou charged with corruption.
1990 New Democracy Party (ND) won half of parliamentary seats in general election but no outright majority; Constantine Mitsotakis became premier; formed new all-party government. Karamanlis re-elected president.
1992 Papandreou acquitted. Greece opposed recognition of the Yugoslav breakaway republic of Macedonia.
1993 Parliament ratified the Maastricht Treaty. PASOK won general election and Papandreou returned as prime minister.

Marvellous things happen in Greece ... Greece is still a sacred precinct—and my belief is it will remain so until the end of time.

On **Greece** Henry Miller *The Colossus of Maroussi* 1942

Greeley /'griːli/ city in N Colorado, USA, at the point where the Cache de Poudre River flows into the South Platte River, NE of Boulder; population (1990) 60,500. A distribution centre for the surrounding agricultural area, its main industry is the processing of sugar beet.

Green Bay /ˌgriːn 'beɪ/ city in NE Wisconsin, USA, where the Little Fox River flows into Green Bay on Lake Michigan; seat of Brown County; population (1990) 96,400. It is a port of entry to the USA through the St Lawrence Seaway and serves as a distribution centre. Industries include paper and food products.

Greenland /'griːnlənd/ (Greenlandic **Kalaalit Nunaat**) world's largest island, lying between the North Atlantic and Arctic Oceans east of North America
area 2,175,600 sq km/840,000 sq mi
capital Godthaab (Greenlandic **Nuuk**)
features the whole of the interior is covered by a vast ice sheet (the remnant of the last glaciation, part of the N Polar icecap); the island has an important role strategically and in civil aviation, and shares military responsibilities with the USA; there are lead and cryolite deposits, and offshore oil is being explored
economy fishing and fish-processing
population (1990) 55,500; Inuit (Ammassalik Eskimoan), Danish, and other European
language Greenlandic (Ammassalik Eskimoan)
history Greenland was discovered about 982 by Eric the Red, who founded colonies on the W coast soon after Eskimos from the North American Arctic had made their way to Greenland. Christianity was introduced to the Vikings about 1000. In 1261 the Viking colonies accepted Norwegian sovereignty, but early in the 15th

century all communication with Europe ceased, and by the 16th century the colonies had died out, but the Eskimos had moved on to the E coast. It became a Danish colony in the 18th century, and following a referendum 1979 was granted full internal self-government 1981.

Greenland Sea /ˈɡriːnlənd/ area of the ◊Arctic Ocean between Spitsbergen and Greenland, and N of the Norwegian Sea.

Greenock /ˈɡriːnək/ port on the S shore of the Firth of Clyde, Strathclyde, Scotland; population (1981) 59,000. Industries include shipbuilding, engineering, and electronics. It is the birthplace of the engineer and inventor James Watt.

Greensboro /ˈɡriːnzbʌrə/ city in N central North Carolina, USA, W of Durham; seat of Guilford County; population (1990) 183,500. It is noted for its textile, chemical, and tobacco industries. Many schools are located here, including Guilford College (1834) and the University of North Carolina at Greensboro (1891). The Battle of Guilford Courthouse (1781) was fought nearby.

Greenville /ˈɡriːnvɪl/ city in NW South Carolina, USA, on the Reedy River, near the foothills of the Blue Ridge Mountains, SW of Spartanburg; seat of Greenville County; population (1990) 44,900. It is known as a major textile manufacturing centre. Other industries include lumber and chemicals.

Greenwich /ˈɡrɛnɪdʒ/ inner borough of Greater London, England
features the *Queen's House* 1637, designed by Inigo Jones, the first Palladian-style building in England, since 1937 housing the National Maritime Museum; the *Royal Naval College*, designed by Christopher Wren 1694 as a naval hospital to replace a palace previously on this site (the birthplace of Henry VIII, Mary, Elizabeth I), and used as a college since 1873; the *Royal Observatory* (founded here 1675). The source of Greenwich Mean Time was moved to Herstmonceux 1958, and then to Cambridge 1990, but the Greenwich meridian (0°) remains unchanged. Part of the buildings have been taken over by the National Maritime Museum, and named Flamsteed House after the first Astronomer Royal. The *Cutty Sark*, one of the great tea clippers, is preserved as a museum of sail and Francis Chichester's *Gipsy Moth IV* is also here. The borough also includes **Woolwich**, with the Royal Arsenal, and Eltham Palace 1300
population (1991) 200,800.

The most delightful spot of ground in Great Britain ... the best air, the best prospect, and the best conversation in England.

On **Greenwich** Daniel Defoe *A Tour through the Whole Island of Great Britain* 1724–27

Grenada
area (including the Grenadines, notably Carriacou) 340 sq km/131 sq mi

capital St George's
towns Grenville, Hillsborough (Carriacou)
physical southernmost of the Windward Islands; mountainous
features Grand-Anse beach; Annandale Falls; the Great Pool volcanic crater
head of state Elizabeth II from 1974 represented by governor general
head of government Nicholas Braithwaite from 1990
political system emergent democracy
political parties New National Party (NNP), centrist; Grenada United Labour Party (GULP), nationalist, left of centre; National Democratic Congress (NDC), centrist
exports cocoa, nutmeg, bananas, mace
currency Eastern Caribbean dollar
population (1990 est) 84,000, 84% of black African descent; growth rate –0.2% p.a.
life expectancy 69
language English (official); some French patois spoken
religion Roman Catholic 60%
literacy 85% (1985)
GDP $139 million (1987); $1,391 per head
chronology
1974 Independence achieved from Britain; Eric Gairy elected prime minister.
1979 Gairy removed in bloodless coup led by Maurice Bishop; constitution suspended and a People's Revolutionary Government established.
1982 Relations with the USA and Britain deteriorated as ties with Cuba and the USSR strengthened.
1983 After Bishop's attempt to improve relations with the USA, he was overthrown by left-wing opponents. A coup established the Revolutionary Military Council (RMC), and Bishop and three colleagues were executed. The USA invaded Grenada, accompanied by troops from other E Caribbean countries; RMC overthrown, 1974 constitution reinstated.
1984 The newly formed NNP won 14 of the 15 seats in the house of representatives and its leader, Herbert Blaize, became prime minister.
1989 Herbert Blaize lost leadership of NNP remaining as head of government; he died and was succeeded by Ben Jones.

1990 Nicholas Braithwaite of the NDC became prime minister.
1991 Integration into Windward Islands confederation proposed.

Grenadines /ˈɡrenədiːnz/ chain of about 600 small islands in the Caribbean Sea, part of the group known as the Windward Islands. They are divided between ▷St Vincent and ▷Grenada.

Grenoble /ɡrəˈnəʊbəl/ alpine town in the Isère *département*, Rhône-Alpes region, SE France; population (1990) 154,000. Industries include engineering, nuclear research, hydroelectric power, computers, technology, chemicals, plastics, and gloves. It was the birthplace of the novelist Stendhal, commemorated by a museum, and the Beaux Arts gallery has a modern collection. There is a 12th–13th-century cathedral, a university (1339), and the Institut Laue-Langevin for nuclear research. The 1968 Winter Olympics were held here.

Grimsby /ˈɡrɪmzbi/ fishing port in Humberside, England; population (1991) 88,900.
It declined in the 1970s when Icelandic waters were closed to British fishing fleets.

Grisons /ɡriːˈsɒn/ French name for the Swiss canton of ▷Graubünden.

Grodno /ˈɡrɒdnəʊ/ industrial town in the Republic of Belarus, on the Sozh River; population (1987) 263,000. Part of Lithuania from 1376, it passed to Poland 1596, Russia 1795, Poland 1920, and the USSR 1939.

Groningen /ˈɡrəʊnɪŋən/ most northerly province of the Netherlands
area 2,350 sq km/907 sq mi
capital Groningen
towns Hoogezand-Sappemeer, Stadskanaal, Veendam, Delfzijl, Winschoten
physical Ems estuary, innermost W Friesian Islands
products natural gas, arable crops, dairy produce, sheep, horses
population (1991) 554,600
history under the power of the bishops of Utrecht from 1040, Groningen became a member of the Hanseatic League 1284. Taken by Spain 1580, it was recaptured by Maurice of Nassau 1594.

Grossglockner /ɡrəʊsˈɡlɒknə/ highest mountain in Austria, rising to 3,797 m/12,457 ft in the Hohe Tauern range of the Tirol Alps.

Grozny /ˈɡrɒzni/ capital of the Checheno-Ingush autonomous republic, Russia; population (1987) 404,000. It is an oil-producing centre.

Gruyère /ɡruːˈjeə/ district in Fribourg canton, W Switzerland, renowned for its pale yellow cheese with large holes. Gruyère centres on the Saane Valley. Bulle is the present-day capital, Gruyères the historic capital. In addition to cheese production, cattle are raised and chocolate is made here.

Guadalajara /ˌɡwɑːdələˈhɑːrə/ industrial city (textiles, glass, soap, pottery), capital of Jalisco state, W Mexico; population (1990) 2,847,000.

It is a key communications centre. It has a 16th–17th-century cathedral, the Governor's Palace, and an orphanage with murals by the Mexican painter José Orozco (1883–1949).

Guadalcanal /ˌɡwɑːdlkəˈnæl/ largest of the Solomon Islands; area 6,500 sq km/2,510 sq mi; population (1987) 71,000. Gold, copra, and rubber are produced. During World War II it was the scene of a battle that was won by US forces after six months of fighting.

Guadeloupe /ˌɡwɑːdəˈluːp/ island group in the Leeward Islands, West Indies, an overseas *département* of France; area 1,705 sq km/658 sq mi; population (1990) 387,000. The main islands are Basse-Terre, on which is the chief town of the same name, and Grande-Terre. Sugar refining and rum distilling are the main industries.

Guam /ɡwɑːm/ largest of the Mariana Islands in the W Pacific, an unincorporated territory of the USA
area 540 sq km/208 sq mi
capital Agaña
towns Apra (port), Tamuning
features major US air and naval base, much used in the Vietnam War; tropical, with much rain
products sweet potatoes, fish; tourism is important
currency US dollar
population (1990) 132,800
languages English, Chamorro (basically Malay-Polynesian)
religion 96% Roman Catholic
government popularly elected governor (Ricardo Bordallo from 1985) and single-chamber legislature
recent history ceded by Spain to the USA 1898; occupied by Japan 1941–44. Guam achieved full US citizenship and self-government from 1950. A referendum 1982 favoured the status of a commonwealth, in association with the USA.

Guanch Republic /ɡwɑːntʃ/ proposed name for an independent state in the ▷Canary Islands.

Guangdong /ˌɡwæŋˈdʊŋ/ or *Kwantung* province of S China
area 231,400 sq km/89,320 sq mi
capital Guangzhou
features tropical climate; Hainan, Leizhou peninsula, and the foreign enclaves of Hong Kong and Macao in the Pearl river delta
products rice, sugar, tobacco, minerals, fish
population (1990) 62,829,000.

Guangxi /ˈɡwæŋˈʃiː/ or *Kwangsi Chuang* autonomous region in S China
area 220,400 sq km/85,074 sq mi
capital Nanning
products rice, sugar, fruit
population (1990) 42,246,000, including the Zhuang people, allied to the Thai, who form China's largest ethnic minority.

Guangzhou /ˌɡwæŋˈdʒəʊ/ or *Kwangchow* or *Canton* capital of Guangdong province, S China; population (1989) 3,490,000. Industries include shipbuilding, engineering, chemicals, and textiles.

Sun Yat-sen Memorial Hall, a theatre, commemorates the politician, who was born nearby and founded the university. There is a rail link with Beijing, and one is planned with Liuzhai. *history* It was the first Chinese port opened to foreign trade, the Portuguese visiting it 1516, and was a treaty port from 1842 until its occupation by Japan 1938.

Guantánamo /gwæn'tɑːnəmɔu/ capital of a province of the same name in SE Cuba; population (1989) 200,400. It is a trading centre in a fertile agricultural region producing sugar. Iron, copper, chromium, and manganese are mined nearby. There is a US naval base, for which the Cuban government has refused to accept rent since 1959.

Guatemala Republic of (*República de Guatemala*)
area 108,889 sq km/42,031 sq mi
capital Guatemala City
towns Quezaltenango, Puerto Barrios (naval base)
physical mountainous; narrow coastal plains; limestone tropical plateau in N; frequent earthquakes
environment between 1960 and 1980 nearly 57% of the country's forest was cleared for farming
features Mayan archaeological remains, including site at Tikal
head of state and government Jorge Serrano Elías from 1991
political system democratic republic
political parties Guatemalan Christian Democratic Party (PDCG), Christian centre-left; Centre Party (UCN), centrist; National Democratic Cooperation Party (PDCN), centre-right; Revolutionary Party (PR), radical; Movement of National Liberation (MLN), extreme right-wing; Democratic Institutional Party (PID), moderate conservative; Solidarity Action Movement (MAS), right-wing
exports coffee, bananas, cotton, sugar, beef
currency quetzal
population (1990 est) 9,340,000 (Mayaquiche Indians 54%, mestizos (mixed race) 42%); growth rate 2.8% p.a. (87% of under-fives suffer from malnutrition)
life expectancy men 57, women 61

languages Spanish (official); 40% speak 18 Indian languages
religion Roman Catholic 80%, Protestant 20%
literacy men 63%, women 47% (1985 est)
GDP $7 bn (1987); $834 per head
chronology
1839 Independence achieved from Spain.
1954 Col Carlos Castillo became president in US-backed coup, halting land reform.
1963 Military coup made Col Enrique Peralta president.
1966 Cesar Méndez elected president.
1970 Carlos Araña elected president.
1974 General Kjell Laugerud became president. Widespread political violence precipitated by the discovery of falsified election returns in March.
1978 General Fernando Romeo became president.
1981 Growth of antigovernment guerrilla movement.
1982 General Angel Anibal became president. Army coup installed General Ríos Montt as head of junta and then as president; political violence continued.
1983 Montt removed in coup led by General Mejía Victores, who declared amnesty for the guerrillas.
1985 New constitution adopted; PDCG won congressional elections; Vinicio Cerezo elected president.
1989 Coup attempt against Cerezo foiled. Over 100,000 people killed, and 40,000 reported missing since 1980.
1991 Jorge Serrano Elías (MAS) elected president. Diplomatic relations with Belize established.

Guatemala City /ˌgwɑːtə'mɑːlə/ capital of Guatemala; population (1983) 1,300,000. It produces textiles, tyres, footwear, and cement. It was founded 1776 when its predecessor (Antigua) was destroyed in an earthquake. It was severely damaged by another earthquake 1976.

Guavio Dam hydroelectric rockfill dam in central Colombia, opened 1990. It is the highest in South America, 245 m/804 ft, and has a generating capacity of 1,600 megawatts.

Guayaquil /ˌgwaɪə'kiːl/ largest city and chief port of Ecuador; population (1986) 1,509,100. The economic centre of Ecuador, Guayaquil manufactures machinery and consumer goods, processes food, and refines petroleum. It was founded 1537 by the Spanish explorer Francisco de Orellana.

Guelders /'geldəz/ another name for ◊Gelderland, a province of the Netherlands.

Guelph /gwelf/ industrial town and agricultural centre in SE Ontario, Canada, on the Speed River; population (1981) 71,250. Industries include food processing, electrical goods, and pharmaceuticals.

Guernica /geə'niːkə/ town in the Basque province of Vizcaya, N Spain; population (1981) 18,000. Here the Castilian kings formerly swore to respect the rights of the Basques. It was almost completely

destroyed 1937 by German bombers aiding General Franco in the Spanish Civil War and rebuilt 1946. The bombing inspired a painting by Pablo Picasso and a play by dramatist Fernando Arrabal (1932–).

Guernsey /'gɜːnzi/ second largest of the Channel Islands; area 63 sq km/24.3 sq mi; population (1986) 55,500. The capital is St Peter Port. Products include electronics, tomatoes, flowers, and, more recently, butterflies; from 1975 it has been a major financial centre. Guernsey cattle, which are a distinctive pale fawn colour and give rich creamy milk, originated here.

Guernsey has belonged to the English crown since 1066, but was occupied by German forces 1940–45.

The island has no jury system; instead, it has a Royal Court with 12 jurats (full-time unpaid jurors appointed by an electoral college) with no legal training. This system dates from Norman times. Jurats cannot be challenged or replaced.

Guiana /gi'ɑːnə/ northeastern part of South America that includes ▷French Guiana, ▷Guyana, and ▷Surinam.

Guildford /'gɪlfəd/ city in Surrey, S England, on the river Wey; population (1981) 57,000. It has a ruined Norman castle, a cathedral (1936–61), and the University of Surrey (1966). There is a cattle market, and industries include flour-milling, plastics, and engineering.

Guilin /ˌgweɪ'lɪn/ or **Kweilin** principal tourist city of S China, on the Li River, Guangxi province; population (1984 est) 446,900. The dramatic limestone mountains are a tourist attraction.

Guinea Republic of (*République de Guinée*)
area 245,857 sq km/94,901 sq mi
capital Conakry
towns Labé, Nzérékoré, Kankan
physical flat coastal plain with mountainous interior; sources of rivers Niger, Gambia, and Senegal; forest in SE
environment large amounts of toxic waste from industrialized countries have been dumped in Guinea
features Fouta Djallon, area of sandstone plateaus, cut by deep valleys
head of state and government Lansana Conté from 1984

political system military republic
political parties none since 1984
exports coffee, rice, palm kernels, alumina, bauxite, diamonds
currency syli or franc
population (1990 est) 7,269,000 (chief peoples are Fulani, Malinke, Susu); growth rate 2.3% p.a.
life expectancy men 39, women 42
languages French (official), African languages
media state-owned, but some criticism of the government tolerated; no daily newspaper
religions Muslim 85%, Christian 10%, local 5%
literacy men 40%, women 17% (1985 est)
GNP $1.9 bn (1987); $369 per head
chronology
1958 Full independence achieved from France; Sékou Touré elected president.
1977 Strong opposition to Touré's rigid Marxist policies forced him to accept return to mixed economy.
1980 Touré returned unopposed for fourth seven-year term.
1984 Touré died. Bloodless coup established a military committee for national recovery, led by Col Lansana Conté.
1985 Attempted coup against Conté while he was out of the country was foiled by loyal troops.
1990 Sent troops to join the multinational force that attempted to stabilize Liberia.
1991 Antigovernment general strike by National Confederation of Guinea Workers (CNTG).

Guinea-Bissau Republic of (*República da Guiné-Bissau*)
area 36,125 sq km/13,944 sq mi
capital Bissau
towns Mansôa, São Domingos
physical flat coastal plain rising to savanna in E
features the archipelago of Bijagós
head of state and government João Bernardo Vieira from 1980
political system socialist pluralist republic
political party African Party for the Independence of Portuguese Guinea and Cape Verde (PAIGC), nationalist socialist
exports rice, coconuts, peanuts, fish, timber
currency peso
population (1989 est) 929,000; growth rate 2.4% p.a.

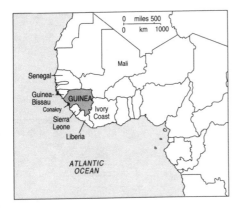

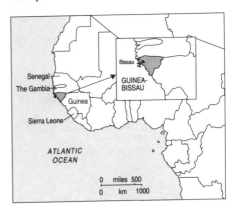

life expectancy 42; 1990 infant mortality rate was 14.8%
languages Portuguese (official), Crioulo (Cape Verdean dialect of Portuguese), African languages
religions animism 54%, Muslim 38%, Christian 8%
literacy men 46%, women 17% (1985 est)
GDP $135 million (1987); $146 per head
chronology
1956 PAIGC formed to secure independence from Portugal.
1973 Two-thirds of the country declared independent, with Luiz Cabral as president of a state council.
1974 Independence achieved from Portugal.
1980 Cape Verde decided not to join a unified state. Cabral deposed, and João Vieira became chair of a council of revolution.
1981 PAIGC confirmed as the only legal party, with Vieira as its secretary general.
1982 Normal relations with Cape Verde restored.
1984 New constitution adopted, making Vieira head of government as well as head of state.
1989 Vieira re-elected.
1991 Other parties legalized.
1992 Multiparty electoral commission established.

Guinea Coast /ˈgɪni/ or *Gulf of Guinea* coast of W Africa between Cape Palmas, Liberia, and Cape Lopez, Gabon. The coastline features the Bight of Benin and the Bight of Bonny, and the rivers Volta, Niger, and Ogowé reach the sea here.

Guiyang /ˌgweɪˈjæŋ/ or *Kweiyang* capital and industrial city of Guizhou province, S China; population (1989) 1,490,000. Industries include metals and machinery.

Guizhou /ˌgweɪˈdʒəʊ/ or *Kweichow* province of S China
area 174,000 sq km/67,164 sq mi
capital Guiyang
products rice, maize, nonferrous minerals
population (1990) 32,392,000.

Gujarat /ˌgʊdʒəˈrɑːt/ or *Gujerat* state of W India
area 196,000 sq km/75,656 sq mi
capital Ahmedabad
features heavily industrialized; includes most of the Rann of Kutch; the Gir Forest (the last home of the wild Asian lion)
products cotton, petrochemicals, oil, gas, rice, textiles
languages Gujarati (Gujerati), Hindi
population (1991) 41,174,000.

Gujranwala /ˌgʊdʒrənˈwɑːlə/ city in Punjab province, Pakistan; population (1981) 597,000. It is a centre of grain trading. The city is a former Sikh capital and the birthplace of Sikh leader Ranjit Singh (1780–1839).

Gujrat /ˌgʊdʒˈrɑːt/ city in Punjab province, E Pakistan, N of Lahore; products include cotton, pottery, brassware, and furniture; population (1981) 154,000. It occupies the site of a fort built 1580 by the Mogul ruler Akbar. Gujrat was the

Gujarat

scene of the final battle between the British and the Sikhs in the Sikh Wars 1845–49; the British subsequently annexed the Punjab.

Gulfport /ˈgʌlfpɔːt/ city in SE Mississippi, USA, on the Gulf of Mexico, W of Biloxi and E of New Orleans, Louisiana; seat of Harrison County and a port of entry to the USA; population (1990) 40,700. It is a major shipping point for lumber, cotton, and food products.

Gulf States oil-rich countries sharing the coastline of the Persian Gulf (Bahrain, Iran, Iraq, Kuwait, Oman, Qatar, Saudi Arabia, and the United Arab Emirates). In the USA, the term refers to those states bordering the Gulf of Mexico (Alabama, Florida, Louisiana, Mississippi, and Texas).

Except for Iran and Iraq, the Persian Gulf States formed a Gulf Co-operation Council (GCC) 1981.

Guyana Cooperative Republic of
area 214,969 sq km/82,978 sq mi
capital and port Georgetown
towns New Amsterdam, Mabaruma

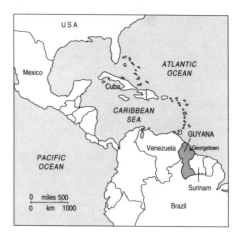

physical coastal plain rises into rolling highlands with savanna in S; mostly tropical rainforest
features Mount Roraima; Kaietur National Park, including Kaietur Fall on the Potaro (tributary of Essequibo) 250 m/821 ft
head of state Desmond Hoyte from 1985
head of government Cheddi Jagan from 1992
political system democratic republic
political parties People's National Congress (PNC), Afro-Guyanan nationalist socialist; People's Progressive Party (PPP), Indian Marxist-Leninist
exports sugar, rice, rum, timber, diamonds, bauxite, shrimps, molasses
currency Guyanese dollar
population (1989 est) 846,000 (51% descendants of workers introduced from India to work the sugar plantations after the abolition of slavery, 30% black, 5% Amerindian); growth rate 2% p.a.
life expectancy men 66, women 71
languages English (official), Hindi, Amerindian
media one government-owned daily newspaper; one independent paper published three times a week, on which the government puts pressure by withholding foreign exchange for newsprint; one weekly independent in the same position. There is also legislation that restricts exchange of information between public officials, government, and the press
religions Christian 57%, Hindu 33%, Sunni Muslim 9%
literacy men 97%, women 95% (1985 est)
GNP $359 million (1987); $445 per head
chronology
1831 Became British colony under name of British Guiana.
1953 Assembly elections won by left-wing PPP; Britain suspended constitution and installed interim administration, fearing communist takeover.
1961 Internal self-government granted; Cheddi Jagan became prime minister.
1964 PNC leader Forbes Burnham led PPP-PNC coalition.
1966 Independence achieved from Britain.
1970 Guyana became a republic within the Commonwealth.
1981 Forbes Burnham became first executive president under new constitution.
1985 Burnham died; succeeded by Desmond Hoyte.
1992 PPP had decisive victory in assembly elections; Jagan returned as prime minister.

The form of government is a mild despotism, tempered by sugar.

On **Guyana** Anthony Trollope *The West Indies and the Spanish Main* 1859

Gwalior /ˈgwɑːliɔː/ city in Madhya Pradesh, India; population (1981) 543,862. It was formerly a small princely state and has Jain and Hindu monuments.

Gwent

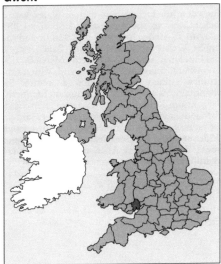

Gwent /gwent/ county in S Wales
area 1,380 sq km/533 sq mi
towns Cwmbran (administrative headquarters), Abergavenny, Newport, Tredegar
features Wye Valley; Tintern Abbey; Legionary Museum at Caerleon, and Roman amphitheatre; Chepstow and Raglan castles
products salmon and trout from the Wye and Usk rivers; iron and steel at Llanwern
population (1991) 432,300
languages 2.5% Welsh, English
famous people Aneurin Bevan and Neil Kinnock, both born in Tredegar; Alfred Russel Wallace.

Gwynedd /ˈgwɪnəð/ county in NW Wales
area 3,870 sq km/1,494 sq mi
towns Caernarvon (administrative headquarters), Bangor

Gwynedd

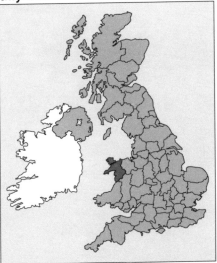

features Snowdonia National Park including Snowdon (the highest mountain in Wales, with a rack railway to the top from Llanberis) 1,085 m/3,561 ft, Cader Idris 892 m/2,928 ft, and the largest Welsh lake, Llyn Tegid (Bala Lake) 6 km/4 mi long; Anglesey, across the Menai Straits; Lleyn Peninsula and Bardsey Island, with a 6th-century ruined abbey, once a centre for pilgrimage; Welsh Slate Museum at Llanberis; Sergontium Roman Fort Museum; Caernarvon, Criccieth, and Harlech castles; Bodnant Garden; the fantasy resort of Portmeirion, built by Clough Williams-Ellis. In 1990 features of the manor house Pen y Bryn at Aber, near Bangor, were identified as surviving from the royal palace of Llewellyn I and II. The Clogau mine at Bontddu supplies the gold for royal wedding rings

products cattle, sheep, gold (at Dolgellau), textiles, electronics, slate
population (1991) 235,450
languages 61% Welsh, English
famous people Edward II, T E Lawrence.

Gyandzha /gjæn'dʒɑː/ city in Azerbaijan; industries include cottons, woollens, and processed foods; population (1987) 270,000. It was known as *Elizavetpol* 1804–1918, *Gandzha* before 1804 and again 1918–35, and *Kirovabad* 1935–89.

Györ /djɜː/ industrial city (steel, vehicles, textiles, foodstuffs) in NW Hungary, near the frontier with the Slovak Republic; population (1988) 131,000. It is linked to lake Neusiedler See by canal.

Haarlem /'hɑːləm/ industrial city and capital of the province of North Holland, the Netherlands, 20 km/12 mi W of Amsterdam; population (1991) 149,500. At Velsea to the N a road-rail tunnel runs under the North Sea Canal, linking North and South Holland. Industries include chemicals, pharmaceuticals, textiles, and printing. Haarlem is renowned for flowering bulbs and has a 15th–16th-century cathedral and a Frans Hals museum.

Hackney /'hækni/ inner borough of N central Greater London
features Hackney Downs and Hackney Marsh, formerly the haunt of highwaymen, now a leisure area; includes *Shoreditch*, site of England's first theatre (The Theatre) 1576; *Hoxton*, with the Geffrye Museum of the domestic arts; *Stoke Newington*, where the writer Daniel Defoe once lived. The horse-drawn *hackney carriage* is so named because horses were bred in Hackney in the 14th century
population (1991) 164,200.

Haddington /'hædɪŋtən/ agricultural market town in Lothian, Scotland, on the river Tyne, 16 km/10 mi SW of Dunbar; population (1989 est) 8,200. The Protestant reformer John Knox was born here.

Hadhramaut /,hɑːdrə'maut/ district of Yemen, which was formerly ruled by Arab chiefs in protective relations with Britain. A remote plateau region at 1,400 m/4,500 ft, it was for a long time unknown to westerners and later attracted such travellers as Harry St John Philby and Freya Stark. Cereals, tobacco, and dates are grown by settled farmers, and there are nomadic Bedouin. The chief town is Mukalla.

Hagen /'hɑːgən/ industrial city in the Ruhr, North Rhine–Westphalia, Germany, at the confluence of the Ennepe and Volme rivers; population (1988) 206,000. It produces iron, steel, textiles, paper, and food products.
It was an agricultural village until mining became important in the late Middle Ages.

Hagerstown /'heɪgəztaun/ city in NW Maryland, USA, on Antietam Creek, NW of Baltimore and just S of the Pennsylvania border; population

(1980) 34,132. Industries include engine and missile parts and furniture.

Hague, The /heɪg/ (Dutch *'s-Gravenhage* or *Den Haag*) capital of the province of South Holland and seat of the Netherlands government, linked by canal with Rotterdam and Amsterdam; population (1991) 444,200.
features The Gothic Hall of Knights (1280) in the Binnenhof, where the annual opening of parliament takes place, the 16th-century town hall, and the Peace Palace (1913), which houses the United Nations International Court of Justice. The seaside resort of *Scheveningen* (patronized by Wilhelm II and Winston Churchill), with its Kurhaus, is virtually incorporated.
history The city grew up around a hunting lodge after a castle was built nearby 1248. The States-General (parliament) established itself here 1585, and from the 17th century the city was a centre of European diplomacy.

Haifa /'haɪfə/ port in NE Israel, at the foot of Mount Carmel; population (1988) 222,600. Industries include oil refining and chemicals. It is the capital of a district of the same name.
history Haifa was taken by the crusaders 1100, Napoleon 1799, and Egypt 1839. It was surrendered by Egypt to Turkey 1840. Occupied by the British 1918, it became part of mandated Palestine 1922. The Arabs surrendered to Israeli rule during the Arab–Israeli War 1948–49.

Hainan /,haɪ'næn/ island in the South China Sea; area 34,000 sq km/13,124 sq mi; population (1990) 6,557,000. The capital is Haikou. In 1987 Hainan was designated a Special Economic Zone; in 1988 it was separated from Guangdong and made a new province.

Hainaut /eɪ'nəu/ industrial province of SW Belgium; capital Mons; area 3,800 sq km/1,467 sq mi; population (1991) 1,278,800. It produces coal, iron, and steel.

Haiphong /,haɪ'foŋ/ industrial port in N Vietnam; population (1989) 456,000. Among its industries are shipbuilding and the making of cement, plastics, phosphates, and textiles.

Haiti Republic of (*République d'Haïti*)

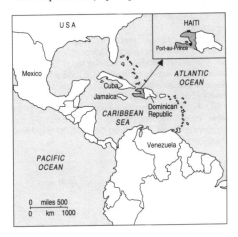

area 27,750 sq km/10,712 sq mi
capital Port-au-Prince
towns Cap-Haïtien, Gonaïves, Les Cayes
physical mainly mountainous and tropical; occupies W third of Hispaniola Island in Caribbean Sea; seriously deforested
features oldest black republic in the world; only French-speaking republic in the Americas; island of La Tortuga off N coast was formerly a pirate lair
interim head of state Joseph Nerette from 1991
head of government Marc Bazin from 1992
political system transitional
political party National Progressive Party (PNP), right-wing military
exports coffee, sugar, sisal, cotton, cocoa, bauxite
currency gourde
population (1990 est) 6,409,000; growth rate 1.7% p.a.; one of highest population densities in the world; about 1.5 million Haitians live outside Haiti (in USA and Canada); about 400,000 live in virtual slavery in the Dominican Republic, where they went or were sent to cut sugar cane
life expectancy men 51, women 54
languages French (official, spoken by literate 10% minority), Creole (spoken by 90% black majority)
religion Christian 95% of which 80% Roman Catholic, voodoo 4%
literacy men 40%, women 35% (1985 est)
GDP $2.2 bn (1987); $414 per head
chronology
1804 Independence achieved from France.
1915 Haiti invaded by USA; remained under US control until 1934.
1957 Dr François Duvalier (Papa Doc) elected president.
1964 Duvalier pronounced himself president for life.
1971 Duvalier died, succeeded by his son, Jean-Claude (Baby Doc); thousands murdered during Duvalier era.
1986 Duvalier deposed; replaced by Lt-Gen Henri Namphy as head of a governing council.
1988 Feb: Leslie Manigat became president. Namphy staged a military coup in June, but another coup in Sept led by Brig-Gen Prosper Avril replaced him with a civilian government under military control.
1989 Coup attempt against Avril foiled; US aid resumed.
1990 Opposition elements expelled; Ertha Pascal-Trouillot acting president.
1991 Jean-Bertrand Aristide elected president but later overthrown in military coup led by Brig-Gen Raoul Cedras. Efforts to reinstate Aristide failed. Joseph Nerette became interim head of state.
1992 Economic sanctions imposed since 1991 were eased by the USA but increased by the Organization of American States (OAS). Marc Bazin appointed premier.

It is the destiny of the people of Haiti to suffer.

On **Haiti** President Duvalier

Hakodate /,hækəu'dɑːteɪ/ port in Hokkaido, Japan; population (1990) 307,300. It was the earliest port opened to the West after the period of isolation, 1854.

It is the realisation of all a sailor's dreams of a harbour.

On **Hakodate** Rutherford Alcock
The Capital of the Tycoon 1863

Halab /həˈlæb/ Arabic name of ♢Aleppo, a city in Syria.

Halabja /həˈlæbdʒə/ Kurdish town near the Iran border in Sulaymaniyah province, NE Iraq. In Aug 1988 international attention was focused on the town when Iraqi planes dropped poison gas, killing 5,000 of its inhabitants.

Halifax /ˈhælɪfæks/ capital of Nova Scotia, E Canada's main port; population (1986) 296,000. Its industries include oil refining and food processing.
There are six military bases in Halifax and it is a major centre of oceanography. It was founded by British settlers 1749.

Halifax /ˈhælɪfæks/ woollen textile town in West Yorkshire, England, on the river Calder; population (1981) 87,500. The cloth trade dates from the 15th century. Present-day industries include textiles, carpets, clothing, and engineering.
features St John's parish church is Perpendicular Gothic; All Souls' is by Gilbert Scott (built for a mill owner named Ackroyd, whose home, Bankfield, is now a museum); the Town Hall is by Charles Barry; the Piece Hall of 1779 (former cloth market) has been adapted to modern use; the surviving gibbet (predecessor of the guillotine) was used to behead cloth stealers 1541–1650.

Halle /ˈhælə/ industrial city (salt, chemicals, lignite) on the river Saale, in the state of Saxony-Anhalt, Germany; population (1990) 240,000. It was the capital of the East German district of Halle 1952–90.

Hamadán /,hæməˈdɑːn/ city in NW Iran on the site of Ecbatana, capital of the Medes (an ancient Indo-European people); population (1986) 274,300. Carpets and rugs are traded. The tomb of the Arab philosopher and physician Avicenna is here.

Hamamatsu /,hæməˈmætsuː/ industrial city (textiles, chemicals, motorcycles) in Chubu region, central Honshu Island, Japan; population (1990) 534,600.

Hambledon /ˈhæmbəldən/ village in SE Hampshire, England. The first cricket club was founded here 1750.

Hamburg /ˈhæmbɜːg/ largest inland port of Europe, in Germany, on the river Elbe; population (1988) 1,571,000. Industries include oil, chemicals, electronics, and cosmetics.

It is capital of the *Land* of Hamburg, and an archbishopric from 834. In alliance with Lübeck, it founded the Hanseatic League.

Hamburg /'hæmbɜːg/ administrative region (German *Land*) of Germany
area 760 sq km/293 sq mi
capital Hamburg
features comprises the city and surrounding districts; there is a university, established 1919, and the Hamburg Schauspielhaus is one of Germany's leading theatres
products refined oil, chemicals, electrical goods, ships, processed food
population (1990 est) 1,626,000
religion 74% Protestant, 8% Roman Catholic
history in 1510 the emperor Maximilian I made Hamburg a free imperial city, and in 1871 it became a state of the German Empire.

Hameln /'hæməln/ (English form **Hamelin**) town in Lower Saxony, Germany; population (1984) 56,300. Old buildings include the *Rattenhaus* (rat-catcher's house). Hameln is the setting for the Pied Piper legend.

Hamelin town's in Brunswick/By famous Hanover city;/The river Weser, deep and wide,/Washes its wall on the southern side;/A pleasanter spot you never espied.

On **Hameln** (Hamelin) Robert Browning
The Pied Piper of Hamelin 1842

Hamersley Range /'hæməzli/ range of hills above the Hamersley Plateau, Western Australia, with coloured rocks and river gorges, as well as iron reserves.

Hamilton /'hæmɔltən/ port in Ontario, Canada; population (1986) 557,000. Linked with Lake Ontario by the Burlington Canal, it has a hydroelectric plant and steel, heavy machinery, electrical, chemical, and textile industries.

Hamilton /'hæmɔltən/ industrial and university town on North Island, New Zealand, on Waikato River; population (1991) 101,300. It trades in forestry, horticulture, and dairy-farming products. Waikato University was established here 1964.

Hamilton /'hæmltən/ city in the southwest corner of Ohio, USA, on the Great Miami River, NW of Cincinnati; seat of Butler County; population (1990) 61,400. Its industries include livestock processing, metal products, paper, and building materials.

Hamilton /'hæmɔltən/ town in Strathclyde, Scotland; population (1988 est) 51,300. Industries include textiles, electronics, and engineering.

Hamilton /'hæmɔltən/ capital (since 1815) of Bermuda, on Bermuda Island; population about

(1980) 1,617. It has a deep-sea harbour. Hamilton was founded 1612.

Hamm /hæm/ industrial town in North Rhine–Westphalia, Germany; population (1988) 166,000. There are coal mines and chemical and engineering industries.

Hammerfest /'hæmɔfest/ fishing port in NW Norway, the northernmost town in Europe; population (1991) 6,900.

Hammersmith and Fulham inner borough of W central Greater London, N of the Thames
features Parish Church of St Paul (1631); Lyric Theatre (1890); Olympia exhibition centre; Fulham Palace, residence of the Bishop of London; 18th-century Hurlingham Club; White City Stadium
population (1991) 136,500.

Hammond /'hæmɔnd/ city in the northwest corner of Indiana, USA, on the Calumet River, just S of Chicago, Illinois; population (1990) 84,200. It is a major transportation centre, connecting to Lake Michigan via the Calumet Canal. Industries include soap, cereal products, publishing, railroad equipment, and transportation facilities for the city's surrounding steel plants and oil refineries.

Hampshire /'hæmpʃɔ/ county of S England
area 3,770 sq km/1,455 sq mi
towns Winchester (administrative headquarters), Southampton, Portsmouth, Gosport
features New Forest, area 373 sq km/144 sq mi, a Saxon royal hunting ground; the river Test which is renowned for its trout fishing; Hampshire Basin, where Britain has onshore and offshore oil; Danebury, 2,500-year-old Celtic hillfort; Beaulieu (including National Motor Museum); Broadlands (home of Lord Mountbatten); Highclere (home of the Earl of Carnarvon,

Hampshire

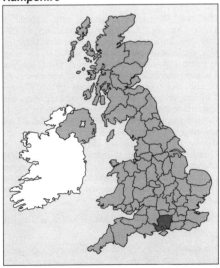

with gardens by Capability Brown); Hambledon, where the first cricket club was founded 1750; site of the Roman town of Silchester, the only one in Britain known in such detail; Jane Austen's cottage 1809–17 is a museum
products agricultural including watercress growing; oil from refineries at Fawley; chemicals, pharmaceuticals, electronics
population (1991) 1,511,900
famous people Jane Austen, Charles Dickens, Gilbert White.

Hampstead /'hæmpstɪd/ district of N London, part of the borough of Camden.

Hangchow /ˌhæŋ'tʃau/ alternative transcription of ♢Hangzhou, a port in Zhejiang province, China.

Hangzhou /ˌhæŋ'dʒəu/ or *Hangchow* port and capital of Zhejiang province, China; population (1989) 1,330,000. It has jute, steel, chemical, tea, and silk industries.
Hangzhou has fine landscaped gardens and was the capital of China 1127–1278 under the Sung dynasty.

Hanley /'hænli/ one of the old Staffordshire pottery towns in England, now part of Stoke-on-Trent.

Hannibal /'hænɪbəl/ town in Missouri, USA, population (1990) 18,000. It is a centre for railway and river traffic and trades in agricultural and dairy products, cement, steel, and metal goods. Mark Twain lived here as a boy and made it the setting for *The Adventures of Huckleberry Finn*.

Hanoi /hæ'nɔɪ/ capital of Vietnam, on the Red River; population (1989) 1,088,900. Central Hanoi has one of the highest population densities in the world: 1,300 people per hectare/3,250 per acre. Industries include textiles, paper, and engineering.
Captured by the French 1873, it was the capital of French Indochina 1887–1946. It was the capital of North Vietnam 1954–76. Hanoi University was founded 1918.

Hanover /'hænəuvə/ industrial city, capital of Lower Saxony, Germany; population (1988) 506,000. Industries include machinery, vehicles, electrical goods, rubber, textiles, and oil refining.
From 1386, it was a member of the Hanseatic League, and from 1692 capital of the electorate of Hanover (created a kingdom 1815). George I of England was also Elector of Hanover, and the two countries shared the same monarch until the accession of Victoria 1837. Since Salic Law meant a woman could not rule in Hanover, the throne passed to her uncle, Ernest, Duke of Cumberland. His son was forced by Bismarck to abdicate 1866, and Hanover became a Prussian province. In 1946, Hanover was merged with Brunswick and Oldenburg to form the *Land* of Lower Saxony.

Hants abbreviation for ♢*Hampshire*, an English county.

Hanyang /ˌhæn'jæŋ/ former Chinese city, now merged in ♢Wuhan, in Hubei province.

Harare /hə'rɑːri/ capital of Zimbabwe, on the Mashonaland plateau, about 1,525 m/5,000 ft above sea level; population (1982) 656,000. It is the centre of a rich farming area (tobacco and maize), with metallurgical and food processing industries.
The British occupied the site 1890 and named it Fort Salisbury in honour of Lord Salisbury, then prime minister of the UK. It was capital of the Federation of Rhodesia and Nyasaland 1953–63.

Harbin /ˌhɑː'bɪn/ or **Haerhpin** or **Pinkiang** port on the Songhua River, NE China, capital of Heilongjiang province; population (1989) 2,800,000. Industries include metallurgy, machinery, paper, food processing, and sugar refining, and it is a major rail junction. Harbin was developed by Russian settlers after Russia was granted trading rights here 1896, and more Russians arrived as refugees after the October Revolution 1917.

Hardwar /hə'dwɑː/ town in Uttar Pradesh, India, on the right bank of the river Ganges; population (1981) 115,513. The name means 'door of Hari' (or Vishnu). It is one of the holy places of the Hindu religion and a pilgrimage centre. The *Kumbhmela* festival, held every 12th year in honour of the god Siva, attracts about 1 million pilgrims.

Harfleur /ɑː'flɜː/ port in NW France; population (1985) 9,700. It was important in medieval times, but was later superseded by Le Havre.

Hargeisa /hɑː'geɪsə/ trading centre (meat, livestock, skins) in NW Somalia; population (1988) 400,000. The town has been severely damaged by the civil war and much of the population has fled.
Hargeisa became the capital of British Somaliland 1941.

Haringey /'hærɪŋgeɪ/ borough of N Greater London. It includes the suburbs of Wood Green, Tottenham, and Hornsey
features Alexandra Palace, with a park; Finsbury Park (once part of Hornsey Wood); Tottenham with Bruce Castle, originally built in the 16th century on land that once belonged to the Scottish king Robert the Bruce's father; Rowland Hill, inventor of the postage stamp, once ran a school here
population (1991) 187,300.

Harlech /'hɑːlex/ town in Gwynedd, N Wales; population (1989 est) 1,300. The castle, now in ruins, was built by the English king Edward I 1283–89. It was captured by the Welsh chieftain Owen Glendower 1404–08 and by the Yorkists in the Wars of the Roses 1468. Harlech is now a centre for visiting Snowdonia National Park.
The song 'March of the Men of Harlech' originated in the siege when the town was captured 1468 by the Yorkists in the Wars of the Roses.

Harlem /'hɑːləm/ commercial and residential district of NE Manhattan, New York City, USA. Originally a Dutch settlement 1658, it developed as a black ghetto after World War I. Harlem's heyday was in the 1920s, when it established its reputation as the intellectual, cultural, and entertainment centre of black America. It is known as a centre for music, particularly jazz.

Harlingen /'hɑːlɪndʒən/ city in the southeast corner of Texas, USA, S of Corpus Christi and just N of the Mexican border; population (1990) 48,735. Connected to the Rio Grande by an intracoastal waterway, it serves as the processing and marketing area for the lower Rio Grande Valley. Industries include citrus-fruit processing and cotton products.

Harper's Ferry /'hɑːpəz 'feri/ village in W Virginia, USA, where the Potomac and Shenandoah rivers meet. In 1859 antislavery leader John Brown seized the federal government's arsenal here, an action that helped precipitate the Civil War.

Harris /'hærɪs/ southern part of Lewis with Harris, in the Outer Hebrides Islands off Scotland; area 500 sq km/193 sq mi; population (1971) 2,900. It is joined to Lewis by a narrow isthmus. Harris tweeds are produced here.

Harrisburg /'hærɪsbɜːg/ capital city of Pennsylvania, USA, located in the S central part of the state, on the Susquehanna River; seat of Dauphin County; population (1990) 52,400. Industries include steel, railroad equipment, food processing, printing and publishing, and clothing.

Harrogate /'hærəgət/ resort and spa in North Yorkshire, England; population (1987 est) 69,300. There is a US communications station at Menwith Hill.

Harrow borough of NW Greater London
features Harrow School (1571)
population (1991) 194,300.

Hartford /'hɑːtfəd/ capital city of Connecticut, USA, located in the N central part of the state, on the Connecticut River, NE of Waterbury; population (1990) 139,700. Industries include insurance, firearms, business office equipment, and tools. The Fundamental Orders of Connecticut, the first constitution that created a democratic government, were signed here 1639.

Hartz Mountains /'hɑːts/ range running N–S in Tasmania, Australia, with two remarkable peaks: Hartz Mountain (1,254 m/4,113 ft) and Adamsons Peak (1,224 m/4,017 ft).

Harwich /'hærɪdʒ/ seaport in Essex, England; with ferry services to Scandinavia and NW Europe; population (1988 est) 15,500. Reclamation of Bathside Bay mudflats is making it a rival, as a port, to Felixstowe.

Haryana /ˌhæri'ɑːnə/ state of NW India
area 44,200 sq km/17,061 sq mi
capital Chandigarh
features part of the Ganges plain; a centre of Hinduism
products sugar, cotton, oilseed, textiles, cement, iron ore

population (1991) 16,317,700
language Hindi.

Hastings /'heɪstɪŋz/ resort in East Sussex, England; population (1981) 74,800. The chief of the Cinque Ports, it has ruins of a Norman castle. It is adjoined by *St Leonard's*, developed in the 19th century. The wreck of the Dutch East Indiaman *Amsterdam* 1749 is under excavation.

Hatfield /'hætfiːld/ town in Hertfordshire, England, 8 km/5 mi E of St Albans; population (1981) 29,000. Designated a new town 1948, it has engineering industries. It was the site of the 12th-century palace of the Bishops of Ely, which was seized by Henry VIII and inhabited by Edward VI and Elizabeth I prior to their accession. James I gave the palace in part exchange to Robert Cecil, 1st Earl of Salisbury, who replaced it 1611 with the existing Jacobean mansion, Hatfield House.

Hatteras /'hætərəs/ cape on the coast of N Carolina, USA. It is a noted spot for shipwrecks, and is nicknamed 'the Graveyard of the Atlantic'.

Haute-Normandie /'əʊt ˌnɔːmən'diː/ or *Upper Normandy* coastal region of NW France lying between Basse-Normandie and Picardy and bisected by the river Seine; area 12,300 sq km/4,757 sq mi; population (1986) 1,693,000. It comprises the *départements* of Eure and Seine-Maritime; its capital is Rouen. Major ports include Dieppe and Fécamp. The area has many beech forests.

Havana /hə'vænə/ capital and port of Cuba, on the northwest coast of the island; population (1989) 2,096,100. Products include cigars and tobacco, sugar, coffee, and fruit. The palace of the Spanish governors and the stronghold of La Fuerza (1583) survive. Tourism ended when Fidel Castro came to power 1959.
history Founded on the south coast as *San Cristolbál de la Habana* by Spanish explorer Diego Velásquez 1515, it was moved to its present site on a natural harbour 1519. It

Haryana

became the capital of Cuba in the late 16th century. Taken by Anglo-American forces 1762, it was returned to Spain 1763 until independence 1898. The blowing up of the US battleship *Maine* in the harbour that year began the Spanish-American War.

... is the chiefest port that the king of Spaine hath in all the countreys of the Indies, and of great importance ...

On **Havana** John Chilton 1570

Haverhill /'heɪvənl/ city in NE Massachusetts, USA, on the Merrimac River, N of Boston; population (1990) 51,418. Products include paints, chemicals, machine tools, and shoes.

Havering borough of NE Greater London *population* (1991) 224,400.

Havre, Le /ɑːvrə/ see ◊ Le Havre, a port in France.

Hawaii /hə'waiiː/ Pacific state of the USA; nickname Aloha State
area 16,800 sq km/6,485 sq mi
capital Honolulu on Oahu
towns Hilo
physical Hawaii consists of a chain of some 20 volcanic islands, of which the chief are (1) *Hawaii*, noted for Mauna Kea (4,201 m/ 13,788 ft), the world's highest island mountain (site of a UK infrared telescope) and Mauna Loa (4,170 m/13,686 ft), the world's largest active volcanic crater; (2) *Maui*, the second largest of the islands; (3) *Oahu*, the third largest, with greatest concentration of population and tourist attractions—for example, Waikiki beach and the Pearl Harbor naval base; (4) *Kauai*; and (5) *Molokai*, site of a historic leper colony
products sugar, coffee, pineapples, flowers, women's clothing
population (1990) 1,108,200 (34% European, 25% Japanese, 14% Filipino, 12% Hawaiian, 6% Chinese)

language English
religions Christianity; Buddhist minority
famous people Father Joseph Damien, Kamehameha I
history a Polynesian kingdom from the 6th century until 1893; Hawaii became a republic 1894; ceded itself to the USA 1898, and became a US territory 1900. Japan's air attack on Pearl Harbor 7 Dec 1941 crippled the US Pacific fleet and turned the territory into an armed camp, under martial law, for the remainder of the war. Hawaii became a state 1959. Tourism is the chief source of income. Capt James Cook, who called Hawaii the Sandwich Islands, was the first known European visitor 1778.

O, how my spirit languishes/ To step ashore in the Sanguishes ...

On **Hawaii** (the Sandwich Islands)
Robert Louis Stevenson, letter of 1889

Hawarden /'hɑːdn/ (Welsh *Penarlâg*) town in Clwyd, N Wales; population 8,500. The Liberal politician William Gladstone lived at Hawarden Castle for many years, and founded St Deiniol's theological library in Hawarden.

Hawkesbury /'hɔːksbəri/ river in New South Wales, Australia; length 480 km/300 mi. It rises in the Great Dividing Range and reaches the Tasman Sea at Broken Bay. It is a major source of Sydney's water.

Haworth /'hauəθ/ village in West Yorkshire, home of the Brontë family of writers (their house is now a museum). Haworth is now part of the town of Keighley.

Hay-on-Wye /'heɪ ɒn 'waɪ/ (Welsh *Y Gelli*) town in Powys, Wales, on the south bank of the river Wye, known as the 'town of books' because of the huge secondhand bookshop started there 1961 by Richard Booth; it was followed by others.

Heard Island and McDonald Islands /hɜːd/ group of islands forming an Australian external territory in the S Indian Ocean, about 4,000 km/ 2,500 mi SW of Fremantle; area 410 km/ 158 sq mi. They were discovered 1833, annexed by Britain 1910, and transferred to Australia 1947. *Heard Island*, 42 km/26 mi by 19 km/ 12 mi, is glacier-covered, although the volcanic mountain *Big Ben* (2,742 m/9,000 ft) is still active. A weather station was built 1947. *Shag Island* is 8 km/5 mi to the N and the craggy McDonalds are 42 km/26 mi to the W.

Heathrow /,hiːθ'rəu/ major international airport to the W of London, approximately 24 km/14 mi from the city centre. It is one of the world's busiest, with four terminals. It was linked with the London underground system 1977.

Hebei /,hʌ'beɪ/ or *Hopei* or *Hupei* province of N China
area 202,700 sq km/78,242 sq mi
capital Shijiazhuang

Hawaii

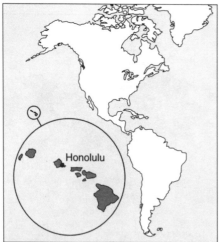

features includes special municipalities of Beijing and Tianjin
products cereals, textiles, iron, steel
population (1990) 61,082,000.

Hebrides /'hebrıdi:z/ group of more than 500 islands (fewer than 100 inhabited) off W Scotland; total area 2,900 sq km/1,120 sq mi. The Hebrides were settled by Scandinavians during the 6th to 9th centuries and passed under Norwegian rule from about 890 to 1266.
   The *Inner Hebrides* are divided between Highland and Strathclyde regions, and include Skye, Mull, Jura, Islay, Iona, Rum, Raasay, Coll, Tiree, Colonsay, Muck, and uninhabited Staffa. The *Outer Hebrides* form the islands area of the Western Isles administrative area, separated from the Inner Hebrides by the Little Minch. They include Lewis with Harris, North Uist, South Uist, Barra, and St Kilda.

Hebron /'hebrən/ (Arabic *El Khalil*) town on the West Bank of the Jordan, occupied by Israel 1967; population (1967) 43,000, including 4,000 Jews. It is a front-line position in the confrontation between Israelis and Arabs in the Intifada. Within the mosque is the traditional site of the tombs of Abraham, Isaac, and Jacob.

Hefei /ˌhʌ'feɪ/ or *Hofei* capital of Anhui province, China; population (1989) 980,000. Products include textiles, chemicals, and steel.

Heidelberg /'haɪdlbɜ:g/ town on the south bank of the river Neckar, 19 km/12 mi SE of Mannheim, in Baden-Württemberg, Germany; population (1988) 136,000. Heidelberg University, the oldest in Germany, was established 1386. The town is overlooked by the ruins of its 13th–17th-century castle, 100 m/330 ft above the river.

Heidelberg /'haɪdlbɜ:g/ village near Melbourne, Australia, that gave its name to the *Heidelberg School*—a group of Impressionist artists (including Tom Roberts, Arthur Streeton, and Charles Conder) working in teaching camps in the neighbourhood.
   Flourishing 1888–90, the school had its most famous exhibition 1889, the '9 by 5', from the size of the cigar-box lids used.

Heilbronn /'haɪlbrɒn/ river port in Baden-Württemberg, Germany, on the river Neckar, N of Stuttgart; population (1988) 112,000. It trades extensively in wine.

Heilongjiang /ˌheɪˌluŋdʒi'æŋ/ or *Heilungkiang* province of NE China, in Manchuria
area 463,600 sq km/178,950 sq mi
capital Harbin
features China's largest oilfield, near Anda
products cereals, gold, coal, copper, zinc, lead, cobalt
population (1990) 35,215,000.

Heilungkiang /ˌheɪˌluŋki'æŋ/ former name of ¢Heilongjiang, a Chinese province.

Hejaz /hi:'dʒæz/ former independent kingdom, merged 1932 with Nejd to form Saudi Arabia;

population (1970) 2,000,000. The capital is Mecca.

Helena /'helənə/ capital of Montana, USA, located in the W central part of the state, near the Big Belt Mountains, S of the Missouri River; population (1990) 24,600. It was settled after gold was discovered 1864. Industries include agricultural products, machine parts, ceramics, paints, sheet metal, and chemicals.

Heligoland /'helɪgəulænd/ island in the North Sea, one of the North Frisian Islands; area 0.6 sq km/0.2 sq mi. It is administered by the state of Schleswig-Holstein, Germany, having been ceded to Germany by Britain 1890 in exchange for Zanzibar. It was used as a naval base in both world wars.

Hellespont /'helɪspɒnt/ former name of the ¢Dardanelles, the strait that separates Europe from Asia.

Helmand /'helmənd/ longest river in Afghanistan. Rising in the Hindu Kush, W of Kabul, it flows SW for 1,125 km/703 mi before entering the marshland surrounding Lake Saberi on the Iranian frontier.

Helsingborg /'helsɪŋbɔ:g/ (Swedish *Hälsingborg*) port in SW Sweden, linked by ferry with Helsingør across Øre Sound; population (1990) 109,300. Industries include copper smelting, rubber and chemical manufacture, and sugar refining.

Helsingfors /ˌhelsɪŋ'fɔ:ʃ/ Swedish name for ¢Helsinki, the capital of Finland.

Helsingør /ˌhelsɪŋ'ɜ:/ (English *Elsinore*) port in NE Denmark; population (1990) 56,800. It is linked by ferry with Helsingborg across Øre Sound; Shakespeare made it the scene of *Hamlet*.

Helsinki /'helsɪŋki/ (Swedish *Helsingfors*) capital and port of Finland; population (1990) 492,400, metropolitan area 978,000. Industries include shipbuilding, engineering, and textiles. The homes of the architect Eliel Saarinen and the composer Jean Sibelius outside the town are museums.
   Helsinki was founded 1550 by King Gustav Vasa of Sweden, N of its present location. It was ceded to Russia 1809 and regained independence 1917. The city contains the parliament house, an 18th-century cathedral, and two universities. It is rich in fin-de-siècle buildings (many by the German-born architect Carl Ludwig Engel), rivalling even Barcelona and Vienna. The pale colour of the local building material has given it the name of the 'white city of Europe'. The port is kept open by ice-breakers in winter.

Helvellyn /hel'velɪn/ peak of the English Lake District in Cumbria, 950 m/3,118 ft high.

*Helvellyn slopes where Coleridge crept/
And Southey sprang/ Where Wordsworth
hoisted up with ropes/Took out his fountain
pen and sang.*

On **Helvellyn** anonymous

Helvetia /hel'viːʃə/ region, corresponding to W Switzerland, occupied by the Celtic Helvetii 1st century BC–5th century AD. In 58 BC Caesar repulsed their invasion of southern Gaul at Bibracte (near Autun) and Helvetia became subject to Rome.

Hemel Hempstead /'heməl 'hempstɪd/ new town in Hertfordshire, England; population (1986 est) 81,000. Industries include manufacture of paper, electrical goods, and office equipment.

Henan /ˌhʌ'næn/ or **Honan** province of E central China
area 167,000 sq km/64,462 sq mi
capital Zhengzhou
features river plains of the Huang He (Yellow River); ruins of Xibo, the 16th-century BC capital of the Shang dynasty, were discovered here in the 1980s
products cereals, cotton
population (1990) 85,510,000.

Hendon /'hendən/ residential district in the borough of Barnet, Greater London, England. The Metropolitan Police Detective Training and Motor Driving schools are here, and the RAF Museum (1972) includes the Battle of Britain Museum (1980).

Henley-on-Thames /'henli ɒn 'temz/ town in Oxfordshire, England; population (1984) 11,000. The rowing regatta, held here annually since 1839, is in July; Henley Management College, established 1946, was the first in Europe.

Henzada /hen'zɑːdə/ city in S central Myanmar (Burma), on the Irrawaddy River; population (1980 est) 284,000. Products include rice and potatoes.

Heraklion /hɪ'ræklɪən/ alternative name for ◊Iráklion, a Greek port.

Herat /he'ræt/ capital of Herat province, and the largest city in W Afghanistan, on the north banks of the Hari Rud River; population (1980) 160,000. A principal road junction, it was a great city in ancient and medieval times.

Hérault /e'rəu/ river in S France, 160 km/100 mi long, rising in the Cévennes Mountains and flowing into the Gulf of Lyons near Agde. It gives its name to a *département*.

Herculaneum /ˌhɜːkjuˈleɪnɪəm/ ancient city of Italy between Naples and Pompeii. Along with Pompeii, it was buried when the volcano Vesuvius erupted AD 79. It was excavated from the 18th century onwards.

Hercules, Pillars of /'hɜːkjuliːz/ rocks (at Gibraltar and Ceuta) which guard the western entrance to the Mediterranean Sea.

Hereford /'herɪfəd/ town in the county of Hereford and Worcester, on the river Wye, England; population (1991) 49,000. Products include cider, beer, and metal goods. The cathedral, which was begun 1079, contains a *Mappa Mundi*, a medieval map of the world. An appeal was

Hereford and Worcester

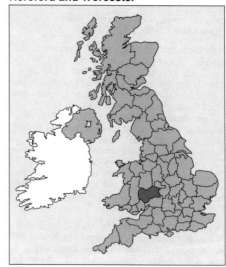

launched 1988 and the money was raised to restore and preserve the map.

Hereford and Worcester /'herɪfəd, 'wʊstə/ county in W central England
area 3,930 sq km/1,517 sq mi
towns Worcester (administrative headquarters), Hereford, Kidderminster, Evesham, Ross-on-Wye, Ledbury
features rivers: Wye, Severn; Malvern Hills (high point Worcester Beacon, 425 m/1,395 ft) and Black Mountains; fertile Vale of Evesham; Droitwich, once a Victorian spa, reopened its baths 1985 (the town lies over a subterranean brine reservoir with waters buoyant enough to support a laden tea tray)
products mainly agricultural: apples, pears, cider; hops, vegetables, Hereford cattle; carpets; porcelain; some chemicals and engineering
population (1991) 667,800
famous people Edward Elgar, A E Housman, William Langland, John Masefield.

Hermon /'hɜːmən/ (Arabic *Jebel esh-Sheikh*) snow-topped mountain, 2,814 m/9,232 ft high, on the Syria–Lebanon border. According to tradition, Jesus was transfigured here.

Herne /'heənə/ industrial city in North Rhine–Westphalia, Germany; population (1988) 171,000. Industries include civil engineering, coal mining, construction, and chemicals.

Herne Bay /'hɜːn/ seaside resort in Kent, SE England, 11 km/7 mi N of Canterbury; population (1985 est) 27,800.

Herstmonceux /ˌhɜːstmən'suː/ village 11 km/7 mi N of Eastbourne, East Sussex, England. From 1958 the buildings of the Royal Greenwich Observatory were here, alongside the 15th-century castle. The Observatory moved from Herstmonceux to Cambridge 1990, but its fine antique telescopes remained as the centrepiece of a proposed scientific theme park.

Hertfordshire

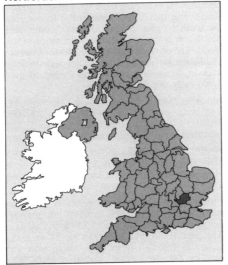

Hertford /'hɑːfəd/ administrative headquarters of Hertfordshire, SE England, on the river Lea; population (1989 est) 25,000. There are brewing, engineering, and brick industries.

Hertfordshire /'hɑːtfədʃə/ county in SE England
area 1,630 sq km/629 sq mi
towns Hertford (administrative headquarters), St Albans, Watford, Hatfield, Hemel Hempstead, Bishop's Stortford, Letchworth (the first garden city, followed by Welwyn 1919 and Stevenage 1947)
features rivers: Lea, Stort, Colne; part of the Chiltern Hills; Hatfield House; Knebworth House (home of Lord Lytton); Brocket Hall (home of Palmerston and Melbourne); home of G B Shaw at Ayot St Lawrence; Berkhamsted Castle (Norman); Rothamsted agricultural experimental station
products engineering, aircraft, electrical goods, paper and printing; general agricultural goods
population (1991) 951,500
famous people Henry Bessemer, Cecil Rhodes, Graham Greene.

Hertogenbosch see ◊'s-Hertogenbosch, a city in the Netherlands.

Herts abbreviation for ◊*Hertfordshire*, an English county.

Herzegovina /ˌheətsə'gɒvɪnə/ or *Hercegovina* part of ◊Bosnia-Herzegovina (which was formerly, until 1992, a republic of Yugoslavia).

Hessen /'hesən/ administrative region (German *Land*) of Germany
area 21,100 sq km/8,145 sq mi
capital Wiesbaden
towns Frankfurt-am-Main, Kassel, Darmstadt, Offenbach-am-Main
features valleys of the rivers Rhine and Main; Taunus Mountains, rich in mineral springs, as at Homburg and Wiesbaden
products wine, timber, chemicals, cars, electrical engineering, optical instruments

population (1988) 5,550,000
religion Protestant 61%, Roman Catholic 33%
history until 1945, Hessen was divided in two by a strip of Prussian territory, the southern portion consisting of the valleys of the rivers Rhine and the Main, the northern being dominated by the Vogelsberg Mountains (744 m/2,442 ft). Its capital was Darmstadt.

HI abbreviation for ◊*Hawaii*, a state of the USA.

Hialeah /ˌhaɪə'liːə/ city in SE Florida, USA, just NW of Miami; population (1990) 188,000. Industries include clothing, furniture, plastics, and chemicals. Hialeah Park racetrack is a centre for horse racing.

Hickory /'hɪkəri/ city in W central North Carolina, USA, in the foothills of the Blue Ridge Mountains, NW of Charlotte; population (1990) 28,300.
   The city's main industry is hosiery manufacture. Other products include rope, cotton, and wagons.

Higashi-Osaka /hɪˌgæʃiəu'sɑːkə/ industrial city (textiles, chemicals, engineering), an eastern suburb of Osaka, Kinki region, Honshu Island, Japan; population (1990) 518,300.

High Country in New Zealand, the generally mountainous land, many peaks rising to heights between 8,000–13,000 ft/2,400–3,800 m, most of which are on South Island. The lakes, fed by melting snow, are used for hydroelectric power, and it is a skiing, mountaineering, and tourist area.

Highland /'haɪlənd/ region of Scotland
area 26,100 sq km/10,077 sq mi
towns Inverness (administrative headquarters), Thurso, Wick
features comprises almost half the country; Grampian Mountains; Ben Nevis (highest peak in the UK); Loch Ness, Caledonian Canal; Inner Hebrides; the Queen Mother's castle of Mey

Highland

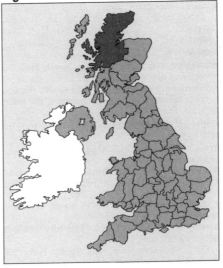

at Caithness; John O'Groats' House; Dounreay (with Atomic Energy Authority's prototype fast reactor and a nuclear processing plant)
products oil services, winter sports, timber, livestock, grouse and deer hunting, salmon fishing
population (1991) 209,400
famous people Alexander Mackenzie, William Smith.

Highlands /'haɪləndz/ one of the three geographical divisions of Scotland, lying to the N of a geological fault line that stretches from Stonehaven in the North Sea to Dumbarton on the Clyde. It is a mountainous region of hard rocks, shallow infertile soils, and high rainfall.

Speak weel o' the Hielands, but dwell in the Laigh.

On the **Highlands** old Scottish saying

High Point /'haɪ pɔɪnt/ city in N central North Carolina, USA, SW of Greensboro; population (1990) 69,500. The furniture industry is very important to the economy; the Southern Furniture Market is held four times a year. Hosiery is also manufactured.

High Wycombe /'wɪkəm/ market town in Buckinghamshire, on the river Wye, England; population (1981) 60,500. RAF Strike Command has its underground headquarters (built 1984) beneath the Chiltern Hills nearby, a four-storey office block used as Joint Headquarters (army, navy, and air force) in the Gulf War 1991. Products include furniture.

Hildesheim /'hɪldəshaɪm/ industrial town in Lower Saxony, Germany, linked to the Mittelland Canal; population (1988) 101,000. Products include electronics and hardware. A bishopric from the 9th century, Hildesheim became a free city of the Holy Roman Empire in the 13th century. It was under Prussian rule 1866–1945.

Hillingdon /'hɪlɪŋdən/ outer borough of W London
features Heathrow airport (built on the site of a Neolithic settlement); Jacobean mansion (Swakeleys) at Ickenham; Brunel University 1966; Grand Union Canal; includes Uxbridge
population (1991) 225,800.

Hilversum /'hɪlvəsum/ town in North Holland province of the Netherlands, 27 km/17 mi SE of Amsterdam; population (1991) 84,600. Besides being a summer resort, Hilversum is the main centre of Dutch broadcasting.

Himachal Pradesh /hɪ'mɑːtʃəl prə'deʃ/ state of NW India
area 55,700 sq km/21,500 sq mi
capital Simla
features mainly agricultural state, one-third forested, with softwood timber industry
products timber, grain, rice, fruit
population (1991) 5,111,000; mainly Hindu
language Pahari
history created as a Union Territory 1948, it became a full state 1971.
Certain hill areas were transferred to Himachal Pradesh from the Punjab 1966.

Himalayas /ˌhɪmə'leɪəz/ vast mountain system of central Asia, extending from the Indian states of Kashmir in the W to Assam in the E, covering the southern part of Tibet, Nepal, Sikkim, and Bhutan. It is the highest mountain range in the world. The two highest peaks are *Mount Everest* and *Kangchenjunga*. Other major peaks include Makalu, Annapurna, and Nanga Parbat, all over 8,000 m/26,000 ft.

Hinckley /'hɪŋkli/ market town in Leicestershire, England; population (1987 est) 25,700. Industries include engineering and the manufacture of footwear and hosiery.

Hindenburg /'hɪndənbɜːg/ German name 1915–45 of the Polish city of ◊ Zabrze, in honour of General Hindenburg, German field marshal and president 1925–34.

Hindu Kush /'hɪndu: 'kuʃ/ mountain range in central Asia, length 800 km/500 mi, greatest height Tirich Mir, 7,690 m/25,239 ft, in Pakistan. The narrow *Khyber Pass* (53 km/33 mi long) separates Pakistan from Afghanistan and was used by Zahir and other invaders of India. The present road was built by the British in the Afghan Wars.

Hindustan /ˌhɪndu:'stɑːn/ ('land of the Hindus') the whole of India, but more specifically the plain of the Ganges and Jumna rivers, or that part of India N of the Deccan.

Hiroshima /hɪ'rɒʃɪmə/ industrial city and port on the south coast of Honshu Island, Japan, destroyed by the first wartime use of an atomic bomb 6 Aug 1945. The city has largely been rebuilt since the war; population (1990) 1,085,700.
Towards the end of World War II the city was utterly devastated by the US atomic bomb. More than 10 sq km/4 sq mi were obliterated, with very heavy damage outside that area. Casualties totalled at least 137,000 out of a population of 343,000: 78,150 were found dead, others died later.

Hispaniola /ˌhɪspæni'əulə/ (Spanish 'little Spain') West Indian island, first landing place of

Himachal Pradesh

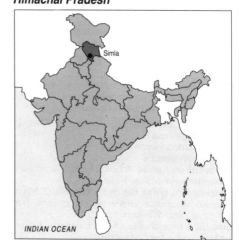

Columbus in the New World, 6 Dec 1492; it is now divided into Haiti and the Dominican Republic.

Hitachi /hɪˈtɑːtʃi/ city on Honshu Island, Japan; population (1990) 202,100. The chief industry is the manufacture of electrical and electronic goods.

Hitchin /ˈhɪtʃɪn/ market town in Hertfordshire, England, 48 km/30 mi NW of London; population (1985) 30,000. Industries include engineering and flour milling, and parchment is manufactured. The cultivation and distillation of lavender, introduced from Naples in the 16th century, still continues.

Hobart /ˈhəʊbɑːt/ capital and port of Tasmania, Australia; population (1986) 180,000. Products include zinc, textiles, and paper. Founded 1804 as a penal colony, it was named after Lord Hobart, then secretary of state for the colonies.
The University of Tasmania, established 1890, is at Hobart.

Hoboken /ˈhəʊbəʊkən/ city and port in NE New Jersey, USA, on the Hudson River; population (1990) 33,397. Industries include electronics, electrical equipment, and chemicals.

Ho Chi Minh City /ˌhəʊ tʃiː ˈmɪn/ (until 1976 *Saigon*) chief port and industrial city of S Vietnam; population (1989) 3,169,100. Industries include shipbuilding, textiles, rubber, and food products. Saigon was the capital of the Republic of Vietnam (South Vietnam) from 1954 to 1976, when it was renamed.

… a French city flowering alone out of a tropical swamp in the farthest corner of Asia.

On **Ho Chi Minh City** Osbert Sitwell
Escape with Me 1939

Hodeida /hɒˈdeɪdə/ or *Al Hudaydah* Red Sea port of Yemen; population (1986) 155,000. It trades in coffee and spices.

Hofei /ˌhəʊˈfeɪ/ alternative transcription of ◊*Hefei*, a city in China.

Hoggar /ˈhɒgə/ alternative spelling of ◊*Ahaggar*, a plateau in the Sahara Desert.

Hohhot /ˌhɒˈhɒt/ or *Huhehot* city and capital of Inner Mongolia (Nei Mongol) autonomous region, China; population (1989) 870,000. Industries include textiles, electronics, and dairy products. There are Lamaist monasteries and temples here.

Hokkaido /hɒˈkaɪdəʊ/ (formerly until 1868 *Yezo* or *Ezo*) northernmost of the four main islands of Japan, separated from Honshu to the S by Tsugaru Strait and from Sakhalin to the N by Soya Strait; area 83,500 sq km/32,231 sq mi; population (1986) 5,678,000, including 16,000 Ainus. The capital is Sapporo. Natural resources include coal, mercury, manganese, oil and natural gas, timber, and fisheries. Coal mining and agriculture are the main industries.

Snow-covered for half the year, Hokkaido was little developed until the Meiji Restoration 1868 when disbanded samurai were settled here. Intensive exploitation followed World War II, including heavy and chemical industrial plants, development of electric power, and dairy farming. An artificial harbour has been constructed at Tomakomai, and an undersea rail tunnel links Hakodate with Aomori (Honshu) but remains as yet closed to public transport.

Holland /ˈhɒlənd/ popular name for the ◊Netherlands; also two provinces of the Netherlands, see ◊North Holland and ◊South Holland.

Holland, parts of /ˈhɒlənd/ former separate administrative county of SE Lincolnshire, England.

Hollywood /ˈhɒliwʊd/ city in SE Florida, USA, on the Atlantic Ocean, S of Fort Lauderdale and N of Miami; population (1990) 121,700. Famous as the centre of the film industry, its principal industry is tourism.

Holyhead /ˌhɒliˈhed/ (Welsh *Caergybi*) seaport on the north coast of Holyhead Island, off Anglesey, Gwynedd, N Wales; population (1988 est) 13,000. Holyhead Island is linked by road and railway bridges with Anglesey, and there are regular sailings between Holyhead and Dublin.

Holy Island /ˈhəʊli/ or *Lindisfarne* island in the North Sea, area 10 sq km/4 sq mi, 3 km/2 mi off Northumberland, England, with which it is connected by a causeway. St Aidan founded a monastery here 635.

Holy Loch /ˈhəʊli ˈlɒx/ western inlet of the Firth of Clyde, W Scotland, with a US nuclear submarine base.

Homburg /ˈhɒmbɜːg/ or *Bad Homburg* town and spa at the foot of the Taunus Mountains, Germany; population (1984) 41,800. It has given its name to a soft felt hat for men, made fashionable by Edward VII of England.

Home Counties the counties in close proximity to London, England: Hertfordshire, Essex, Kent, Surrey, and formerly Middlesex.

Homs /hɒmz/ or *Hums* city, capital of Homs district, W Syria, near the Orontes River; population (1981) 355,000. Silk, cereals, and fruit are produced in the area, and industries include silk textiles, oil refining, and jewellery. Zenobia, Queen of Palmyra, was defeated at Homs by the Roman emperor Aurelian 272.

Honan /ˌhəʊˈnæn/ alternative name of ◊*Henan*, a province of China.

Hondo /ˈhɒndəʊ/ another name for ◊*Honshu*, an island of Japan.

Honduras Republic of (*República de Honduras*)
area 112,100 sq km/43,282 sq mi
capital Tegucigalpa
towns San Pedro Sula; ports La Ceiba, Puerto Cortés
physical narrow tropical coastal plain with mountainous interior, Bay Islands
features archaeological sites; Mayan ruins at Copán

head of state and government Rafael Leonardo Callejas from 1990
political system democratic republic
political parties Liberal Party of Honduras (PLH), centre-left; National Party (PN), right-wing
exports coffee, bananas, meat, sugar, timber (including mahogany, rosewood)
currency lempira
population (1989 est) 5,106,000 (90% mestizo (mixed), 10% Indians and Europeans); growth rate 3.1% p.a.
life expectancy men 58, women 62
languages Spanish (official), English, Indian languages
religion Roman Catholic 97%
literacy men 61%, women 58% (1985 est)
GDP $3.5 bn (1987); $758 per head
chronology
1838 Independence achieved from Spain.
1980 After more than a century of mostly military rule, a civilian government was elected, with Dr Roberto Suazo as president; the commander in chief of the army, General Gustavo Alvarez, retained considerable power.
1983 Close involvement with the USA in providing naval and air bases and allowing Nicaraguan counter-revolutionaries ('Contras') to operate from Honduras.
1984 Alvarez ousted in coup led by junior officers, resulting in policy review towards USA and Nicaragua.
1985 José Azcona elected president after electoral law changed, making Suazo ineligible for presidency.
1989 Government and opposition declared support for Central American peace plan to demobilize Nicaraguan Contras based in Honduras; Contras and their dependents in Honduras in 1989 thought to number about 55,000.
1990 Rafael Callejas (PN) inaugurated as president.
1992 Border dispute with El Salvador dating from 1861 finally resolved.

Hong Kong /ˈhɒŋ ˈkɒŋ/ British crown colony SE of China, in the South China Sea, comprising Hong

Kong Island; the Kowloon Peninsula; many other islands, of which the largest is Lantau; and the mainland New Territories. It is due to revert to Chinese control 1997.
area 1,070 sq km/413 sq mi
capital Victoria (Hong Kong City)
towns Kowloon, Tsuen Wan (in the New Territories)
features an enclave of Kwantung province, China, it has one of the world's finest natural harbours; Hong Kong Island is connected with Kowloon by undersea railway and ferries; a world financial centre, its stock market has four exchanges; across the border of the New Territories in China itself is the Shenzhen special economic zone
exports textiles, clothing, electronic goods, clocks, watches, cameras, plastic products; a large proportion of the exports and imports of S China are transshipped here; tourism is important
currency Hong Kong dollar
population (1986) 5,431,000; 57% Hong Kong Chinese, most of the remainder refugees from the mainland
languages English, Chinese
media Hong Kong has the most free press in Asia but its freedoms are not enshrined in law
religions Confucianist, Buddhist, Taoist, with Muslim and Christian minorities
government Hong Kong is a British dependency administered by a crown-appointed governor (Chris Patten from 1992) who presides over an unelected executive council, composed of 4 ex-officio and 11 nominated members, and a legislative council composed of 3 ex-officio members, 29 appointees, and 24 indirectly elected members
history formerly part of China, Hong Kong Island was occupied by Britain 1841, during the first of the Opium Wars, and ceded by China under the 1842 Treaty of Nanking. The Kowloon Peninsula was acquired under the 1860 Beijing (Peking) Convention and the New Territories secured on a 99-year lease from 1898. The colony, which developed into a major centre for Sino-British trade during the late 19th and early 20th centuries, was occupied by Japan 1941–45. The restored British administration promised,

Hong Kong

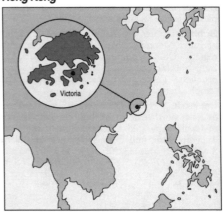

after 1946, to increase self-government. These plans were shelved, however, after the 1949 Communist revolution in China. During the 1950s almost 1 million Chinese (predominantly Cantonese) refugees fled to Hong Kong. Immigration continued during the 1960s and 1970s, raising the colony's population from 1 million in 1946 to 5 million in 1980, leading to the imposition of strict border controls during the 1980s. Since 1975, 160,000 Vietnamese boat people have fled to Hong Kong; in 1991 some 61,000 remained. The UK government began forced repatriation 1989.

Hong Kong's economy expanded rapidly during the corresponding period and the colony became one of Asia's major commercial, financial, and industrial centres, boasting the world's busiest container port from 1987. As the date (1997) for the termination of the New Territories' lease approached, negotiations on Hong Kong's future were opened between Britain and China 1982. These culminated in a unique agreement, signed in Beijing 1984, in which Britain agreed to transfer full sovereignty of the islands and New Territories to China 1997 in return for Chinese assurance that Hong Kong's social and economic freedom and capitalist lifestyle would be preserved for at least 50 years.

Under this 'one country, two systems' agreement, in 1997 Hong Kong would become a special administrative region within China, with its own laws, currency, budget, and tax system, and would retain its free-port status and authority to negotiate separate international trade agreements. In preparation for its withdrawal from the colony, the British government introduced indirect elections to select a portion of the new legislative council 1984, and direct elections for seats on lower-tier local councils 1985. A Sino-British joint liaison group was also established to monitor the functioning of the new agreement, and a 59-member committee (including 25 representatives from Hong Kong) formed in Beijing 1985 to draft a new constitution. In Dec 1989 the UK government granted British citizenship to 225,000 Hong Kong residents, beginning 1997. In March 1990 the 59-member committee agreed to a 'Basic Law' with 18 directly-elected members of the legislative council from 1991, rising to 30 in 2003 (out of a total of 60). In Sept 1991, the liberal United Democrats won 16 of the 18 seats in the territory's legislature; they were led by lawyer Martin Lee and teacher Szeto Wah (who had spearheaded Hong Kong's 1989 pro-democracy demonstrations in the wake of the Tiananmen Square massacre in Beijing). Earlier in the year, the UK and China agreed on the construction of a new airport and seaport.

A borrowed place living on borrowed time.

On **Hong Kong** anonymous, quoted in
The Times 5 March 1981

Honiara /ˌhɒniˈɑːrə/ port and capital of the Solomon Islands, on the northwest coast of Guadalcanal Island, on the river Mataniko; population (1985) 26,000.

It was claimed for Spain 1568. After World War II, it replaced Tulagi as capital.

Honiton /ˈhʌnɪtən/ market town in Devon, SW England, on the river Otter; population (1981) 6,600. Its handmade pillow-lace industry is undergoing a revival.

Honolulu /ˌhɒnəˈluːluː/ (Hawaiian 'sheltered bay') capital city and port of Hawaii, on the south coast of Oahu; population (1990) 365,300. It is a holiday resort, noted for its beauty and tropical vegetation, with some industry.

Pearl Harbor and Hickam Air Force Base are 11 km/7 mi to the NW. William Brown, a British sea captain, was the first European to see Honolulu, 1794. It became the Hawaiian royal capital in the 19th century.

Honshu /ˈhɒnʃuː/ principal island of Japan. It lies between Hokkaido to the NE and Kyushu to the SW; area 231,100 sq km/89,205 sq mi, including 382 smaller islands; population (1986) 97,283,000. A chain of volcanic mountains runs along the island, which is subject to frequent earthquakes. The main cities are Tokyo, Yokohama, Osaka, Kobe, Nagoya, and Hiroshima.

Hooghly /ˈhuːɡli/ or *Hugli* river and town in West Bengal, India; population (1981) 125,193. The river is the western stream of the Ganges delta. The town is on the site of a factory set up by the East India Company 1640, which was moved to Calcutta, 40 km/25 mi downstream, 1686–90.

Hook of Holland /ˈhuk əv ˈhɒlənd/ (Dutch *Hoek van Holland* 'corner of Holland') small peninsula and village in South Holland, the Netherlands; the terminus for ferry services with Harwich (Parkeston Quay), England.

Hopei /ˌhəuˈpeɪ/ alternative transcription of ◊*Hebei*, a province of China.

Hopkinsville /ˈhɒpkɪnzvɪl/ city in SW Kentucky, USA, SW of Louisville; population (1990) 29,800. It is a marketplace for tobacco and livestock.

Hormuz /hɔːˈmuːz/ or *Ormuz* small island, area 41 sq km/16 sq mi, in the Strait of Hormuz, belonging to Iran. It is strategically important because oil tankers leaving the Persian Gulf for Japan and the West have to pass through the strait to reach the Arabian Sea.

It was occupied by the Portuguese 1515–1622.

Horsham /ˈhɔːʃəm/ town and market centre on the river Arun, in West Sussex, England, 26 km/16 mi SE of Guildford; population (1988 est) 40,000. The public school Christ's Hospital is about 3 km/2 mi to the SW. The poet Percy Bysshe Shelley was born here.

Houma /ˈhəumə/ city in S Louisiana, USA, on the gulf intracoastal waterway, SW of New Orleans; population (1990) 30,500. The seat of Terrebonne parish, it is a supply centre for offshore oil rigs in the Gulf of Mexico. Industries include shellfish processing and sugar refining.

Hounslow /'haunzləu/ borough of W Greater London
features Hounslow Heath, formerly the haunt of highwaymen; *Chiswick*, with the Palladian villa by Burlington, and the artist William Hogarth's home (now a museum); *Heston*, site of London's first civil airport established 1919; *Brentford*, reputed site of Caesar's crossing of the Thames 54 BC, and the duke of Northumberland's seat at Syon House; and *Isleworth*, Osterley, home of the economist Thomas Gresham
population (1991) 194,100.

Houston /'hju:stən/ port in Texas, USA; linked by canal to the Gulf of Mexico; population (1990) 1,630,600. It is a major centre of the petroleum industry and of finance and commerce. It is also one of the busiest US ports.
   Houston was first settled 1826. Its modern growth dates from the discovery of oil nearby 1901 and the completion of the Houston Ship Channel 1914.

Hove /həuv/ residential town and seaside resort in East Sussex, England, adjoining Brighton to the W; population (1987 est) 90,400.

Howrah /'haurə/ or *Haora* city of West Bengal, India, on the right bank of the river Hooghly, opposite Calcutta; population (1981) 742,298. The capital of Howrah district, it has jute and cotton factories; rice, flour, and saw mills; chemical factories; and engineering works. Howrah suspension bridge, opened 1943, spans the river.

Huallaga River /waɪ'a:gə/ tributary of the Marayon River in NE Peru. The upper reaches of the river valley are used for growing coca, a major source of the drug cocaine.

Huambo /'wa:mbəu/ town in central Angola; population (1970) 61,885. Founded 1912, it was known as *Nova Lisboa* ('New Lisbon') 1928–78. It is an agricultural centre.

Huang He /'hwæŋ 'həu/ or *Hwang Ho* river in China; length 5,464 km/3,395 mi. It takes its name (meaning 'yellow river') from its muddy waters. Formerly known as 'China's sorrow' because of disastrous floods, it is now largely controlled through hydroelectric works and flood barriers.
   The flood barriers, however, are ceasing to work because the silt is continually raising the river bed.

Huangshan Mountains /,hwæŋ'ʃa:n/ mountain range in S Anhui province, China; the highest peak is Lotus Flower, 1,873 m/5,106 ft.

Huáscaran /,wa:skə'ra:n/ extinct volcano in the Andes; the highest mountain in Peru, 6,768 m/22,205 ft.

Hubei /,hu:'beɪ/ or *Hupei* province of central China, through which flow the river Chang Jiang and its tributary the Han Shui
area 187,500 sq km/72,375 sq mi
capital Wuhan

features high land in the W, the river Chang breaking through from Sichuan in gorges; elsewhere low-lying, fertile land; many lakes
products beans, cereals, cotton, rice, vegetables, copper, gypsum, iron ore, phosphorous, salt
population (1990) 53,969,000.

Huddersfield /'hʌdəz,fi:ld/ industrial town in West Yorkshire, on the river Colne, linked by canal with Manchester and other N England centres; population (1988 est) 121,800. A village in Anglo-Saxon times, it was a thriving centre of woollen manufacture by the end of the 18th century; industries now include dyestuffs, chemicals, and electrical and mechanical engineering.

Hudson /'hʌdsən/ river of the northeastern USA; length 485 km/300 mi. It rises in the Adirondack Mountains and flows S, emptying into a bay of the Atlantic Ocean at New York City.
   First sighted at its mouth by the Italian navigator Giovanni da Verrazano 1524, the river was explored upstream as far as modern Albany 1609 by English navigator Henry Hudson, after whom it is named.

Hudson Bay /'hʌdsən/ inland sea of NE Canada, linked with the Atlantic Ocean by *Hudson Strait* and with the Arctic Ocean by Foxe Channel; area 1,233,000 sq km/476,000 sq mi. It is named after Henry Hudson, who reached it 1610.

A gigantic natural refrigerator.

On **Hudson Bay** Alistair Horne *Canada and the Canadians* 1961

Hué /hu:'eɪ/ town in central Vietnam, formerly capital of Annam, 13 km/8 mi from the China Sea; population (1989) 211,100. The Citadel, within which is the Imperial City enclosing the palace of the former emperor, lies to the W of the Old City on the north bank of the Huong (Perfume) River; the New City is on the south bank.
   Hué was once an architecturally beautiful cultural and religious centre, but large areas were devastated, with many casualties, during the Battle of Hué 31 Jan–24 Feb 1968, when US and South Vietnamese forces retook the city after Vietcong occupation.

... a pleasant little town with something of the leisurely air of a cathedral city in the West of England.

On **Hué** William Somerset Maugham 'The Gentleman in the Parlour' 1930

Huelva /'welvə/ port and capital of Huelva province, Andalusia, SW Spain, near the mouth of the river Odiel; population (1991) 143,600. Industries include shipbuilding, oil refining, fisheries, and trade in ores from Río Tinto. Columbus began and ended his voyage to America at nearby Palos de la Frontera.

Huesca /'weskə/ capital of Huesca province in Aragon, N Spain; population (1991) 50,000. Industries include engineering and food processing. Among its buldings are a fine 13th-century cathedral and the former palace of the kings of Aragon.

Huhehot /ˌhuːhɜːˈhəʊt/ former name of ◊Hohhot, a city in Inner Mongolia.

Hull /hʌl/ officially *Kingston upon Hull* city and port on the north bank of the Humber estuary, England, where the river Hull flows into it; population (1991) 252,200. It is linked with the south bank of the estuary by the Humber Bridge. Industries include fish processing, vegetable oils, flour milling, electrical goods, textiles, paint, pharmaceuticals, chemicals, caravans, and aircraft.

Notable buildings include 13th-century Holy Trinity Church, Guildhall, Ferens Art Gallery 1927, and the university 1954.

There are ferries to Rotterdam and Zeebrugge. Since the building of the Queen Elizabeth Dock 1971, the port's roll-on/roll-off freight traffic expanded rapidly in the 1980s.

Humber estuary in NE England formed by the Ouse and Trent rivers, which meet E of Goole and flow 60 km/38 mi to enter the North Sea below Spurn Head. The main ports are Kingston-upon-Hull on the north side, and Grimsby on the south side. The *Humber Bridge* (1981) is the longest single-span suspension bridge in the world.

Humberside /'hʌmbəsaɪd/ county of NE England; recommended 1991 for break-up, returning Cleethorpes, Glanford, Great Grimsby, and Scunthorpe to Lincolnshire
area 3,510 sq km/1,355 sq mi
towns Hull (administrative headquarters), Grimsby, Scunthorpe, Goole, Cleethorpes
features Humber Bridge; fertile Holderness peninsula; Isle of Axholme; bounded by rivers

Humberside

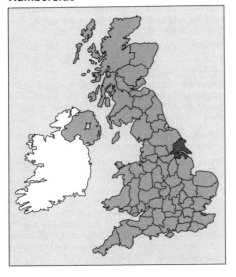

Trent, Don, Idle, and Torne, where medieval open-field strip farming is still practised
products petrochemicals, refined oil, processed fish, cereals, root crops, cattle
population (1991) 845,200
famous people Andrew Marvell, John Wesley, Amy Johnson.

Hunan /ˌhuːˈnæn/ province of S central China
area 210,500 sq km/81,253 sq mi
capital Changsha
features Dongting Lake; farmhouse in Shaoshan village where Mao Zedong was born
products rice, tea, tobacco, cotton; nonferrous minerals
population (1990) 60,660,000.

Hungary Republic of (*Magyar Köztársaság*)
area 93,032 sq km/35,910 sq mi
capital Budapest
towns Miskolc, Debrecen, Szeged, Pécs
physical Great Hungarian Plain covers E half of country; Bakony Forest, Lake Balaton, and Transdanubian Highlands in the W; rivers Danube, Tisza, and Raba
environment an estimated 35%–40% of the population live in areas with officially 'inadmissible' air and water pollution. In Budapest lead levels have reached 30 times the maximum international standards
features more than 500 thermal springs; Hortobágy National Park; Tokay wine area
head of state Arpád Göncz from 1990
head of government József Antall from 1990
political system emergent democratic republic
political parties over 50, including Hungarian Socialist Party (HSP), left of centre; Hungarian Democratic Forum (MDF), umbrella prodemocracy grouping; Alliance of Free Democrats (SzDSz), radical free-market opposition group heading coalition with Alliance of Young Democrats, Social Democrats, and Smallholders Party, right-wing
exports machinery, vehicles, iron and steel, chemicals, fruit and vegetables
currency forint
population (1990 est) 10,546,000 (Magyar

92%, Romany 3%, German 2.5%; Hungarian minority in Romania has caused some friction between the two countries); growth rate 0.2% p.a.
life expectancy men 67, women 74
language Hungarian (or Magyar), one of the few languages of Europe with non-Indo-European origins; it is grouped with Finnish, Estonian, and others in the Finno-Ugric family
religions Roman Catholic 67%, other Christian denominations 25%
literacy men 99.3%, women 98.5% (1980)
GDP $26.1 bn (1987); $2,455 per head
chronology
1918 Independence achieved from Austro-Hungarian empire.
1919 A communist state formed for 133 days.
1920–44 Regency formed under Admiral Horthy, who joined Hitler's attack on the USSR.
1945 Liberated by USSR.
1946 Republic proclaimed; Stalinist regime imposed.
1949 Soviet-style constitution adopted.
1956 Hungarian national uprising; workers' demonstrations in Budapest; democratization reforms by Imre Nagy overturned by Soviet tanks, Kádár installed as party leader.
1968 Economic decentralization reforms.
1983 Competition introduced into elections.
1987 VAT and income tax introduced.
1988 Kádár replaced by Károly Grosz. First free trade union recognized; rival political parties legalized.
1989 May: border with Austria opened. July: new four-person collective leadership of HSWP. Oct: new 'transitional constitution' adopted, founded on multiparty democracy and new presidentialist executive. HSWP changed name to Hungarian Socialist Party, with Nyers as new leader. Kádár 'retired'.
1990 HSP reputation damaged by 'Danubegate' bugging scandal. March–April: elections won by right-of-centre coalition, headed by Hungarian Democratic Forum (MDF). May: József Antall, leader of the MDF, appointed premier. Aug: Árpád Göncz elected president.
1991 Jan: devaluation of currency. June: legislation approved to compensate owners of land and property expropriated under communist government. Last Soviet troops departed. Dec: European Community (EC) association pact signed.
1992 March: EC pact came into effect.

Hunter /ˈhʌntə/ river in New South Wales, Australia, which rises in the Mount Royal Range and flows into the Pacific Ocean near Newcastle, after a course of about 465 km/290 mi. Although the river is liable to flooding, the Hunter Valley has dairying and market gardening, and produces wines.

Huntingdon /ˈhʌntɪŋdən/ town in Cambridgeshire, E England, on the river Ouse, 26 km/16 mi NW of Cambridge; population (1988 est) 15,100. It is a market town with a number of light industries. A bridge built 1332 connects Huntingdon with Godmanchester on the south bank of the river, and the two towns were united 1961. Samuel Pepys and Oliver Cromwell attended the grammar school founded 1565 in a 12th-century building, formerly part of the medieval hospital; it was opened 1962 as a Cromwell museum. The Environmental Information Centre was opened 1989.

Huntingdonshire /ˈhʌntɪŋdənʃə/ former English county, merged 1974 into a much enlarged Cambridgeshire.

Huntington /ˈhʌntɪŋtən/ city in W West Virginia, USA, across the Ohio River from Ohio, NW of Charleston; seat of Cabell County; population (1990) 54,800. It is an important transportation centre for coal mined to the S. Other industries include chemicals; metal, wood, and glass products; tobacco; and fruit processing.

Huntsville /ˈhʌntsvɪl/ town in NE Alabama, USA; population (1990) 159,800. It is the site of an aerospace research centre called the Marshall Space Flight Center.

Hunza /ˈhunzə/ former princely state in NW Kashmir; area 10,101 sq km/3,900 sq mi; capital Baltit (also known as Hunza). It recognized the sovereignty of the Maharajah of Jammu and Kashmir 1869, combined with the state of Nagar 1888, and became part of the Gilgit Agency 1889. Since 1948 it has been administered by Pakistan.

Hupei /ˌhuːˈpeɪ/ alternative transcription of ▷Hebei, a province of China.

Huron /ˈhjuərən/ second largest of the Great Lakes of North America, on the US–Canadian border; area 23,160 sq mi/60,000 sq km. It includes Georgian Bay, Saginaw Bay, and Manitoulin Island.
It receives Lake Superior's waters through the Sault Ste Marie River, and Lake Michigan's through the Straits of Mackinac. It drains south into Lake Erie through the St Clair River–Lake St Clair–Detroit River system. There are a number of small Michigan and Ontario ports on its shores. Jesuit missionaries established the first European settlement, on Georgian Bay, 1638.

Hurstmonceux /ˌhɜːstmənˈsjuː/ alternative spelling of ▷Herstmonceux, a village in East Sussex, England.

Hvannadalshnjukur /ˈvænədælsˌnuːkə/ highest peak in Iceland, rising to 2,119 m/6,952 ft in the southeast of the country.

Hwange /ˈhwæŋeɪ/ (formerly until 1982 *Wankie*) coalmining town in Zimbabwe; population (1982) 39,200. Hwange National Park is nearby.

Hwang Ho /ˌhwæŋˈhəu/ alternative transcription of ▷Huang He, a river in China.

Hydaspes /haɪˈdæspiːz/ classical name of the river ▷Jhelum, in Pakistan and Kashmir.

Hyde Park /haɪd/ one of the largest open spaces in London, England. It occupies about 146 ha/350 acres in Westminster. It adjoins Kensington Gardens, and includes the Serpentine, a boating lake with a 'lido' for swimming. Rotten Row is a famous riding track in the park. In 1851 the Great Exhibition was held here.

Hyderabad /ˈhaɪdərəbæd/ capital city of the S central Indian state of Andhra Pradesh, on the river Musi; population (1981) 2,528,000. Products include carpets, silks, and metal inlay work. It was formerly the capital of the state of Hyderabad. Buildings include the Jama Masjid mosque and Golconda fort.

Hyderabad /ˈhaɪdərəbæd/ city in Sind province, SE Pakistan; population (1981) 795,000. It produces gold, pottery, glass, and furniture. The third largest city of Pakistan, it was founded 1768.

Hyères /iːˈeə/ town on the Côte d'Azur in the *département* of Var, S France; population (1990) 50,100. It has a mild climate, and is a winter health resort. Industries include olive-oil pressing and the export of violets, strawberries, and vegetables.

Hyphasis /ˈhɪfəsɪs/ classical name of the river ▷ Beas, in India.

Hythe /haɪð/ seaside resort (former Cinque Port) in the Romney Marsh area of Kent, SE England; population (1981) 12,700. There is horticulture and plastic industry.

IA abbreviation for ⟡*Iowa*, a state of the USA.

Iaşi /ˈjæʃi/ (German *Jassy*) city in NE Romania; population (1985) 314,000. It has chemical, machinery, electronic, and textile industries. It was the capital of the principality of Moldavia 1568–89.

Ibadan /ɪˈbædən/ city in SW Nigeria and capital of Oyo state; population (1981) 2,100,000. Industries include chemicals, electronics, plastics, and vehicles.

Ibagué /ˌiːbæˈgeɪ/ capital of Tolima department, W central Colombia; population (1985) 293,000. Leather goods and a local drink, Mistela, are produced here.

Iberia /aɪˈbɪəriə/ name given by ancient Greek navigators to the Spanish peninsula, derived from the river Iberus (Ebro). Anthropologists have given the name '*Iberian*' to a Neolithic people, traces of whom are found in the Spanish peninsula, southern France, the Canary Isles, Corsica, and part of North Africa.

Ibiza /ɪˈbiːθə/ one of the Balearic Islands, a popular tourist resort; area 596 sq km/230 sq mi; population (1986) 45,000. The capital and port, also called Ibiza, has a cathedral.

Içel /iːˈtʃel/ another name for ⟡Mersin, a city in Turkey.

Iceland Republic of (*Lýdveldid Ísland*)

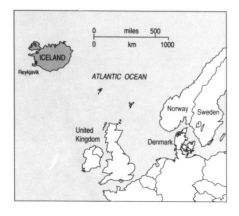

area 103,000 sq km/39,758 sq mi
capital Reykjavík
towns Akureyri, Akranes
physical warmed by the Gulf Stream; glaciers and lava fields cover 75% of the country; active volcanoes (Hekla was once thought the gateway to Hell), geysers, hot springs, and new islands created offshore (Surtsey in 1963); subterranean hot water heats 85% of Iceland's homes
features Thingvellir, where the oldest parliament in the world first met AD 930; shallow lake Mývatn (38 sq km/15 sq mi) in N
head of state Vigdís Finnbogadóttir from 1980
head of government Davíd Oddsson from 1991
political system democratic republic
political parties Independence Party (IP), right of centre; Progressive Party (PP), radical socialist; People's Alliance (PA), socialist; Social Democratic Party (SDP), moderate, left of centre; Citizens' Party, centrist; Women's Alliance, women-and family-oriented
exports cod and other fish products, aluminium, diatomite
currency krona
population (1990 est) 251,000; growth rate 0.8% p.a.
life expectancy men 74, women 80
language Icelandic, the most archaic Scandinavian language, in which some of the finest sagas were written
religion Evangelical Lutheran 95%
literacy 99.9% (1984)
GDP $3.9 bn (1986); $16,200 per head
chronology
1944 Independence achieved from Denmark.
1949 Joined NATO and Council of Europe.
1953 Joined Nordic Council.
1976 'Cod War' with UK.
1979 Iceland announced 320-km/200-mi exclusive fishing zone.
1983 Steingrímur Hermannsson appointed to lead a coalition government.
1985 Iceland declared itself a nuclear-free zone.
1987 New coalition government formed by Thorsteinn Pálsson after general election.
1988 Vigdís Finnbogadóttir re-elected president for a third term; Hermannsson led new coalition.
1991 Davíd Oddsson led new IP–SDP (Independence Party and Social Democratic Party) centre-right coalition, becoming prime minister in the general election.
1992 Iceland defied world ban by resuming whaling industry.

Fortunate island, Where all men are equal But not vulgar—not yet.

On **Iceland** W H Auden *Iceland Revisited* 1964

Ichang /ˌiːˈtʃæŋ/ alternative form of ⟡Yichang, a port in China.

Iconium /aɪˈkəʊniəm/ city of ancient Turkey; see ⟡Konya.

ID abbreviation for ⟡*Idaho*, a state of the USA.

Idaho

Idaho /'aɪdəhəu/ state of northwestern USA; nickname Gem State
area 216,500 sq km/83,569 sq mi
capital Boise
towns Pocatello, Idaho Falls
features Rocky Mountains; Snake River, which runs through Hell's Canyon (2,330 m/7,647 ft), the deepest in North America, and has the National Reactor Testing Station on the plains of its upper reaches; Sun Valley ski and summer resort; Craters of the Moon National Monument; Nez Percé National Historic Park
products potatoes, wheat, livestock, timber, silver, lead, zinc, antimony
population (1990) 1,006,700
history part of the Louisiana Purchase 1803; first permanently settled 1860 after the discovery of gold, Idaho became a state 1890.

Idi Amin Dada, Lake /'ɪdi ɑ'miːn 'dɑːdɑː/ former name (1973–79) of Lake ◊Edward in Uganda/ Zaire.

If /iːf/ small French island in the Mediterranean about 3 km/2 mi off Marseille, with a castle, Château d'If, built about 1529. This was used as a state prison and is the scene of the imprisonment of Dante in Alexandre Dumas's novel *The Count of Monte Cristo.*

Ifni /'ɪfni/ former Spanish overseas province in SW Morocco 1860–1969; area 1,920 sq km/ 740 sq mi. The chief town is Sidi Ifni.

Igls /'iːgls/ winter sports resort in the Austrian Tyrol, near Innsbruck; it was the venue for the 1964 Winter Olympics.

Iguaçú Falls /,iːgwæ'suː/ or *Iguassú Falls* waterfall in South America, on the border between Brazil and Argentina. The falls lie 19 km/12 mi above the junction of the river Iguaçú with the Paraná. The falls are divided by forested rocky islands and form a spectacular tourist attraction. The water plunges in 275 falls, many of which have separate names. They have a height of 82 m/ 269 ft and a width of about 4 km/2.5 mi.

IJsselmeer /'aɪsəlmɪə/ lake in the Netherlands, area 1,217 sq km/470 sq mi. It was formed 1932 after the Zuider Zee was cut off from the North Sea by a dyke 32 km/20 mi long (the *Afsluitdijk*); it has been freshwater since 1944. The rivers Vecht, IJssel, and Zwatewater empty into the lake. Four polders have been reclaimed,

primarily for agriculture: Wieringermeer Polder 193 sq km/75 sq mi (1930); Northeast Polder 469 sq km/181 sq mi (1942); East Flevoland Polder 528 sq km/204 sq mi (1957); and South Flevoland Polder 430 sq km/166 sq mi (1968).

IL abbreviation for ◊ *Illinois*, a state of the USA.

Ile-de-France /'iːl də 'frɒns/ region of N France; area 12,000 sq km/4,632 sq mi; population (1986) 10,251,000. It includes the French capital, Paris, and the towns of Versailles, Sèvres, and St-Cloud and comprises the *départements* of Essonne, Val-de-Marne, Val d'Oise, Ville de Paris, Seine-et-Marne, Hauts-de-Seine, Seine-Saint-Denis, and Yvelines. From here the early French kings extended their authority over the whole country.

Ilfracombe /'ɪlfrəkuːm/ resort on the north coast of Devon, SW England; population (1981) 10,500. There is a 14th-century chapel.

Ilkeston /'ɪlkɪstən/ town in SE Derbyshire, England; population (1989 est) 35,000. Products include clothing and plastics.

Ilkley /'ɪlkli/ town in West Yorkshire, England, noted for nearby *Ilkley Moor*; population (1981) 13,000.

Ille /iːl/ French river 45 km/28 mi long, which rises in Lake Boulet and enters the Vilaine at Rennes. It gives its name to the *département* of Ille-et-Vilaine in Brittany.

Illimani /,iːljɪ'mɑːni/ highest peak in the Bolivian Andes, rising to 6,402 m/21,004 ft E of the city of La Paz.

Illinois /,ɪlə'nɔɪ/ midwest state of the USA; nickname Land of Lincoln/Prairie State
area 146,100 sq km/56,395 sq mi
capital Springfield
towns Chicago, Rockford, Peoria, Decatur, Aurora
features Lake Michigan; rivers: Mississippi, Illinois, Ohio, Rock; Cahokia Mounds, the largest group of prehistoric earthworks in the USA; the Lincoln Home National Historic Site, Springfield; the University of Chicago; Mormon leader Joseph Smith's home, Nauvoo; the Art Institute and Field Museum, Chicago
products soya beans, cereals, meat and dairy products, machinery, electrical and electronic equipment
population (1990) 11,430,600
famous people Jane Addams, Saul Bellow,

Illinois

India: states and union territories

state	capital	area in sq km
Andhra Pradesh	Hyderabad	276,800
Arunachal Pradesh	Itanagar	83,600
Assam	Dispur	78,400
Bihar	Patna	173,900
Goa	Panaji	3,700
Gujarat	Gandhinagar	196,000
Haryana	Chandigarh	44,200
Himachal Pradesh	Simla	55,700
Jammu and Kashmir	Srinagar	101,300
Karnataka	Bangalore	191,800
Kerala	Trivandrum	38,900
Madhya Pradesh	Bhopal	442,800
Maharashtra	Bombay	307,800
Manipur	Imphal	22,400
Meghalaya	Shillong	22,500
Mizoram	Aizawl	21,100
Nagaland	Kohima	16,500
Orissa	Bhubaneswar	155,800
Punjab	Chandigarh	50,400
Rajasthan	Jaipur	342,200
Sikkim	Gangtok	7,300
Tamil Nadu	Madras	130,100
Tripura	Agartala	10,500
Uttar Pradesh	Lucknow	294,400
West Bengal	Calcutta	87,900

union territory		
Andaman and Nicobar Islands	Port Blair	8,200
Chandigarh	Chandigarh	114
Dadra and Nagar Haveli	Silvassa	490
Daman and Diu	Daman	110
Delhi	Delhi	1,500
Lakshadweep	Kavaratti	32
Pondicherry	Pondicherry	492

Mother Cabrini, Clarence Darrow, Enrico Fermi, Ernest Hemingway, Jesse Jackson, Edgar Lee Masters, Ronald Reagan, Louis Sullivan, Frank Lloyd Wright
history originally explored by the French in the 17th century, and ceded to Britain by the French 1763, Illinois passed to American control 1783 and became a state 1818.

Ilorin /ɪˈlɔːrɪn/ capital of Kwara state, Nigeria; population (1983) 344,000. It trades in tobacco and wood products.

Imbros /ˈɪmbrɒs/ (Turkish *Imroz*) island in the Aegean Sea; area 280 sq km/108 sq mi; population (1970) 6,786. Occupied by Greece in World War I, it became Turkish under the Treaty of Lausanne 1923.

Immingham /ˈɪmɪŋəm/ town on the river Humber, Humberside, NE England; population (1981) 11,500. It is a bulk cargo handling port, with petrochemical works and oil refineries.

Imphal /ɪmˈfɑːl/ capital of Manipur state on the

Manipur River, India; population (1981) 156,622; a communications and trade centre (tobacco, sugar, fruit). It was besieged March–June 1944, when Japan invaded Assam, but held out with the help of supplies dropped by air.

Imroz /ˈɪmrɒz/ Turkish form of ◊Imbros, an island in the Aegean Sea.

IN abbreviation for ◊*Indiana*, a state of the USA.

Inchon /ˈɪnˈtʃɒn/ (formerly *Chemulpo*) chief port of Seoul, South Korea; population (1990) 1,818,300. It produces steel and textiles.

Independence /ˌɪndɪˈpendəns/ city in W Missouri, USA; population (1990) 112,300. Industries include steel, Portland cement, petroleum refining, and flour milling. President Harry S Truman was raised here, and it is the site of the Truman Library and Museum.

India /ˈɪndiˈænə/ Republic of (Hindi *Bharat*)
area 3,166,829 sq km/1,222,396 sq mi
capital Delhi
towns Bangalore, Hyderabad, Ahmedabad, Kanpur, Pune, Nagpur; ports Calcutta, Bombay, Madras
physical Himalaya mountains on N border; plains around rivers Ganges, Indus, Brahmaputra; Deccan peninsula S of the Narmada River forms plateau between Western and Eastern Ghats mountain ranges; desert in W; Andaman and Nicobar Islands, Lakshadweep (Laccadive Islands)
environment the controversial Narmada Valley Project is the world's largest combined hydroelectric irrigation scheme. In addition to displacing a million people, the damming of the holy Narmada River will submerge large areas of forest and farmland and create problems of waterlogging and salinization
features Taj Mahal monument; Golden Temple, Amritsar; archaeological sites and cave paintings (Ajanta); world's second most populous country
head of state Shankar Dayal Sharma from 1992
head of government P V Narasimha Rao from 1991
political system liberal democratic federal republic
political parties All India Congress Committee (I), or Congress (I), cross-caste and cross-religion, left of centre; Janata Dal, left of centre; Bharatiya Janata Party (BJP), conservative Hindu-chauvinist; Communist Party of India (CPI), Marxist-Leninist; Communist Party of India–Marxist (CPI–M), West Bengal–based moderate socialist
exports tea (world's largest producer), coffee, fish, iron and steel, leather, textiles, clothing, polished diamonds
currency rupee
population (1991 est) 844,000,000 (920 women to every 1,000 men); growth rate 2.0% p.a.
life expectancy men 56, women 55
languages Hindi (widely spoken in N India), English, and 14 other official languages: Assamese, Bengali, Gujarati, Kannada,

over 1,200 people, mainly Muslims, following destruction of a mosque in Ayodhya, N India, by Hindu extremists.
1993 Sectarian violence in Bombay left 500 dead.

Well, India is a country of nonsense.

On **India** M K Gandhi

Indiana state of the midwest USA; nickname Hoosier State
area 93,700 sq km/36,168 sq mi
capital Indianapolis
towns Fort Wayne, Gary, Evansville, South Bend
features Wabash River; Wyandotte Cavern; Indiana Dunes National Lakeshore; Indianapolis Motor Speedway and Museum; George Rogers National Historic Park, Vincennes; Robert Owen's utopian commune, New Harmony; Lincoln Boyhood National Memorial
products maize, pigs, soya beans, limestone, machinery, electrical goods, coal, steel, iron, chemicals
population (1990) 5,544,200
famous people Hoagy Carmichael, Eugene V Debs, Theodore Dreiser, Michael Jackson, Cole Porter, Wilbur Wright
history first white settlements established 1731–35 by French traders; ceded to Britain by the French 1763; passed to American control 1783; became a state 1816.

Indianapolis /ˌɪndɪəˈnæpəlɪs/ capital and largest city of Indiana, on the White River; population (1990) 742,000. It is an industrial centre and venue of the 'Indianapolis 500' car race.

Indian Ocean ocean between Africa and Australia, with India to the N, and the southern boundary being an arbitrary line from Cape Agulhas to S Tasmania; area 73,500,000 sq km/28,371,000 sq mi; average depth 3,872 m/12,708 ft. The greatest depth is the Java Trench 7,725 m/25,353 ft.

Kashmiri, Malayalam, Marathi, Oriya, Punjabi, Sanskrit, Sindhi, Tamil, Telugu, Urdu
media free press; government-owned broadcasting
religions Hindu 80%, Sunni Muslim 10%, Christian 2.5%, Sikh 2%
literacy men 57%, women 29% (1985 est)
GDP $220.8 bn (1987); $283 per head
chronology
1947 Independence achieved from Britain.
1950 Federal republic proclaimed.
1964 Death of Prime Minister Nehru. Border war with Pakistan over Kashmir.
1966 Indira Gandhi became prime minister.
1971 War with Pakistan leading to creation of Bangladesh.
1975–77 State of emergency proclaimed.
1977–79 Janata Party government in power.
1980 Indira Gandhi returned in landslide victory.
1984 Indira Gandhi assassinated; Rajiv Gandhi elected with record majority.
1987 Signing of 'Tamil' Colombo peace accord with Sri Lanka; Indian Peacekeeping Force (IPKF) sent there. Public revelation of Bofors corruption scandal.
1988 New opposition party, Janata Dal, established by former finance minister V P Singh.
1989 Congress (I) lost majority in general election, after Gandhi associates implicated in financial misconduct; Janata Dal minority government formed, with V P Singh as prime minister.
1990 Central rule imposed in Jammu and Kashmir. V P Singh resigned; new minority Janata Dal government formed by Chandra Shekhar. Interethnic and religious violence in Punjab and elsewhere.
1991 Central rule imposed in Tamil Nadu. Shekhar resigned; elections called for May. May: Rajiv Gandhi assassinated. June: elections resumed, resulting in a Congress (I) minority government led by P V Narasimha Rao. Separatist violence continued.
1992 Congress (I) won control of state assembly and a majority in parliament in Punjab state elections. Widespread communal violence killed

The Injian Ocean sets and smiles / So sof',
so bright, so bloomin' blue.

On the **Indian Ocean** Rudyard Kipling
'For to Admire' 1894

Indiana

Indochina /ˌɪndəʊˈtʃaɪnə/ former collective name for ⬧Cambodia, ⬧Laos, and ⬧Vietnam, which became independent after World War II.

Indonesia Republic of (*Republik Indonesia*)
area 1,919,443 sq km/740,905 sq mi
capital Jakarta
towns Bandung; ports Surabaya, Semarang, Tandjungpriok
physical comprises 13,677 tropical islands, of the Greater Sunda group (including Java and Madura, part of Borneo (Kalimantan), Sumatra, Sulawesi and Belitung), and the Lesser Sundas/Nusa Tenggara (including Bali, Lombok, Sumbawa, Sumba, Flores, and Timor), as well as Malaku/Moluccas and part of New Guinea (Irian Jaya)
environment comparison of primary forest and 30-year-old secondary forest has shown that logging in Kalimantan has led to a 20% decline in tree species
head of state and government T N J Suharto from 1967
political system authoritarian nationalist republic
political parties Golkar, ruling military-bureaucrat, farmers' party; United Development Party (PPP), moderate Islamic; Indonesian Democratic Party (PDI), nationalist Christian
exports coffee, rubber, timber, palm oil, coconuts, tin, tea, tobacco, oil, liquid natural gas
currency rupiah
population (1989 est) 187,726,000 (including 300 ethnic groups); growth rate 2% p.a.; Indonesia has the world's largest Muslim population; Java is one of the world's most densely populated areas
life expectancy men 52, women 55
languages Indonesian (official), closely allied to Malay; Javanese is the most widely spoken local dialect
religions Muslim 88%, Christian 10%, Buddhist and Hindu 2%
literacy men 83%, women 65% (1985 est)
GDP $69.7 bn (1987); $409 per head
chronology
17th century Dutch colonial rule established.
1942 Occupied by Japan; nationalist government established.

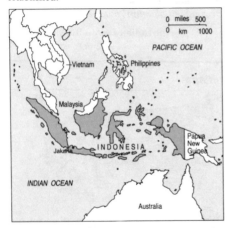

1945 Japanese surrender; nationalists declared independence under Sukarno.
1949 Formal transfer of Dutch sovereignty.
1950 Unitary constitution established.
1963 Western New Guinea (Irian Jaya) ceded by the Netherlands.
1965–66 Attempted communist coup; General Suharto imposed emergency administration, carried out massacre of hundreds of thousands.
1967 Sukarno replaced as president by Suharto.
1975 Guerrillas seeking independence for S Moluccas seized train and Indonesian consulate in the Netherlands, held Western hostages.
1976 Forced annexation of former Portuguese colony of East Timor.
1986 Institution of 'transmigration programme' to settle large numbers of Javanese on sparsely populated outer islands, particularly Irian Jaya.
1988 Partial easing of travel restrictions to East Timor. Suharto re-elected for fifth term.
1989 Foreign debt reaches $50 billion; Western creditors offer aid on condition that concessions are made to foreign companies and that austerity measures are introduced.
1991 Democracy forums launched to promote political dialogue. Massacre in East Timor.
1992 The ruling Golkar party won the assembly elections.

Indore /ɪnˈdɔː/ city in Madhya Pradesh, India; population (1981) 829,327. A former capital of the princely state of Indore, it now produces cotton, chemicals, and furniture.

Indre /ˈændrə/ river rising in the Auvergne Mountains, France, and flowing NW for 260 km/165 mi to join the Loire below Tours. It gives its name to the *départements* of Indre and Indre-et-Loire.

Indus /ˈɪndəs/ river in Asia, rising in Tibet and flowing 3,180 km/1,975 mi to the Arabian Sea. In 1960 the use of its waters, including those of its five tributaries, was divided between India (rivers Ravi, Beas, Sutlej) and Pakistan (rivers Indus, Jhelum, Chenab).

Inhambane /ˌɪnjəmˈbɑːnə/ seaport on the southeast coast of Mozambique, 370 km/231 mi NE of Maputo; population (1980) 56,000.

Inland Sea (Japanese *Seto Naikai*) arm of the Pacific Ocean, 390 km/240 mi long, almost enclosed by the Japanese islands of Honshu, Kyushu, and Shikoku. It has about 300 small islands.

Inn /ɪn/ river in S central Europe, tributary of the Danube. Rising in the Swiss Alps, it flows 507 km/317 mi NE through Austria and into Bavaria, Germany, where it meets the Danube at Passau.

Innsbruck /ˈɪnzbruk/ capital of Tirol state, W Austria; population (1981) 117,000. It is a tourist and winter sports centre and a route junction for the Brenner Pass. The 1964 and 1976 Winter Olympics were held here.

Interlaken /ˈɪntəˌlɑːkən/ chief town of the Bernese Oberland, on the river Aar between lakes Brienz and Thun, Switzerland; population (1985) 13,000. The site was first occupied 1130 by a monastery, suppressed 1528.

Invercargill /ˌɪnvəˈkɑːgəl/ city on the south coast of South Island, New Zealand; population (1991) 56,100. It has saw mills and meatpacking and aluminium-smelting plants.

The Last Lamp-post in the World.

On **Invercargill** Rudyard Kipling *Something of Myself* 1937

Inverness /ˌɪnvəˈnes/ town in Highland Region, Scotland, lying in a sheltered site at the mouth of the river Ness; population (1989 est) 61,000. It is a tourist centre with tweed, tanning, engineering, and distilling industries.

Inverness-shire /ˌɪnvəˈnesʃə/ largest of the former Scottish counties, it was merged into Highland Region 1975.

Inyangani /ˌɪnjæŋˈgɑːni/ highest peak in Zimbabwe, rising to 2,593 m/8,507 ft near the Mozambique frontier in NE Zimbabwe.

Inyokern /ɪnjəʊˈkɜːn/ village in the Mojave Desert, California, USA, 72 km/45 mi NW of Mojave. It is the site of a US naval ordnance test station, founded 1944, carrying out research in rocket flight and propulsion.

IOM abbreviation for *Isle of ◊Man*, an island in the Irish Sea.

Iona /aɪˈəʊnə/ island in the Inner Hebrides; area 850 hectares/2,100 acres. A centre of early Christianity, it is the site of a monastery founded 563 by St Columba. It later became a burial ground for Irish, Scottish, and Norwegian kings. It has a 13th-century abbey.

Ionian Islands /aɪˈəʊniən/ (Greek *Ionioi Nisoi*) island group off the west coast of Greece; area 860 sq km/332 sq mi. A British protectorate from 1815 until their cession to Greece 1864, they include *Cephalonia* (Greek *Kefallínia*); *Corfu* (*Kérkyra*), a Venetian possession 1386–1797; *Cythera* (*Kíthira*); *Ithaca* (*Itháki*), the traditional home of Odysseus; *Leukas* (*Levkás*); *Paxos* (*Paxoí*); and *Zante* (*Zákynthos*).

Ionian Sea /aɪˈəʊniən/ part of the Mediterranean Sea that lies between Italy and Greece, to the S of the Adriatic Sea, and containing the Ionian Islands.

IOW abbreviation for ◊*Isle of Wight*, an island and county off the coast of S England.

Iowa /ˈaɪəʊə/ state of the midwest USA; nickname Hawkeye State
area 145,800 sq km/56,279 sq mi
capital Des Moines
towns Cedar Rapids, Davenport, Sioux City
features Grant Wood Gallery, Davenport; Herbert Hoover birthplace, and museum near West Branch; 'Little Switzerland' region in the NE, overlooking the Mississippi River; Effigy Mounds National Monument, near Marquette, a prehistoric Indian burial site
products cereals, soya beans, pigs and cattle, chemicals, farm machinery, electrical goods, hardwood lumber, minerals

Iowa

population (1990) 2,776,800
famous people 'Bix' Beiderbecke, Buffalo Bill, Herbert Hoover, Glenn Miller, Lillian Russell, Grant Wood
history part of the Louisiana Purchase 1803, Iowa remains an area of small farms. It became a state 1846.

Iowa City /ˈaɪəwə ˈsɪti/ city in E Iowa, USA, on the Iowa River, S of Cedar Rapids, seat of Johnson County; population (1990) 59,700. It is a distribution centre for the area's agricultural products. Other industries include printed matter and building materials. The University of Iowa (1847) is here.

Ipoh /ˈiːpəʊ/ capital of Perak state, Peninsular Malaysia; population (1980) 301,000. The economy is based on tin mining.

Ipswich /ˈɪpswɪtʃ/ river port on the Orwell estuary, administrative headquarters of Suffolk, England; population (1981) 120,500. Industries include engineering and the manufacture of textiles, plastics, and electrical goods. Ipswich was an important wool port in the 16th century.
It was the birthplace of Cardinal Wolsey and home of the painter Thomas Gainsborough.

Iquique /ɪˈkiːkeɪ/ city and seaport in N Chile, capital of the province of Tarapaca; population (1985) 120,700. It exports sodium nitrate from its desert region.

Iquitos /ɪˈkiːtɒs/ river port on the Amazon, in Peru, also a tourist centre for the rainforest; population (1988) 248,000.

Iráklion /ɪˈræklɪən/ or *Heraklion* chief commercial port and largest city of Crete, Greece; population (1981) 102,000. There is a ferry link to Piraeus on the mainland. The archaeological museum contains a fine collection of antiquities from the island.

Iran Islamic Republic of (*Jomhori-e-Islami-e-Irân*; until 1935 *Persia*)
area 1,648,000 sq km/636,128 sq mi
capital Tehran
towns Isfahan, Mashhad, Tabriz, Shiraz, Ahvaz; chief port Abadan
physical plateau surrounded by mountains, including Elburz and Zagros; Lake Rezayeh; Dasht-Ekavir Desert; occupies islands of Abu Musa, Greater Tunb and Lesser Tunb in the Gulf
features ruins of Persepolis; Mount Demavend 5,670 m/18,603 ft

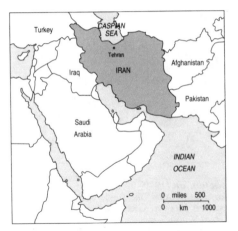

Leader of the Islamic Revolution Seyed Ali Khamenei from 1989
head of government Ali Akbar Hoshemi Rafsanjani from 1989
political system authoritarian Islamic republic
political party Islamic Republican Party (IRP), fundamentalist Islamic
exports carpets, cotton textiles, metalwork, leather goods, oil, petrochemicals, fruit
currency rial
population (1989 est) 51,005,000 (including minorities in Azerbaijan, Baluchistan, Khuzestan/Arabistan, and Kurdistan); growth rate 3.2% p.a.
life expectancy men 57, women 57
languages Farsi (official), Kurdish, Turkish, Arabic, English, French
religion Shi'ite Muslim (official) 92%, Sunni Muslim 5%, Zoroastrian 2%, Jewish, Baha'i, and Christian 1%
literacy men 62%, women 39% (1985 est)
GDP $86.4 bn (1987); $1,756 per head
chronology
1946 British, US, and Soviet forces left Iran.
1951 Oilfields nationalized by Prime Minister Muhammad Mossadeq.
1953 Mossadeq deposed and the US-backed shah took full control of the government.
1975 The shah introduced single-party system.
1978 Opposition to the shah organized from France by Ayatollah Khomeini.
1979 Shah left the country; Khomeini returned to create Islamic state. Revolutionaries seized US hostages at embassy in Tehran; US economic boycott.
1980 Start of Iran–Iraq War.
1981 US hostages released.
1984 Egyptian peace proposals rejected.
1985 Fighting intensified in Iran–Iraq War.
1988 Cease-fire; talks with Iraq began.
1989 Khomeini called for the death of British writer Salman Rushdie. June: Khomeini died; Ali Khamenei elected interim Leader of the Revolution; speaker of Iranian parliament Hoshemi Rafsanjani elected president. Secret oil deal with Israel revealed.
1990 Generous peace terms with Iraq accepted. Normal relations with UK restored.
1991 Imprisoned British business executive

released. Nearly one million Kurds arrived in Iran from Iraq, fleeing persecution by Saddam Hussein after the Gulf War.
1992 Pro-Rafsanjani moderates won assembly elections.

Persia consists of two parts: a desert with salt, and a desert without salt.

On **Iran** old saying

Iraq Republic of (*al Jumhouriya al 'Iraqia*)
area 434,924 sq km/167,881 sq mi
capital Baghdad
towns Mosul and port of Basra
physical mountains in N, desert in W; wide valley of rivers Tigris and Euphrates NW–SE
environment a chemical-weapons plant covering an area of 65 sq km/25 sq mi, situated 80 km/50 mi NW of Baghdad, has been described by the UN as the largest toxic waste dump in the world
features reed architecture of the marsh Arabs; ancient sites of Eridu, Babylon, Nineveh, Ur, Ctesiphon
head of state and government Saddam Hussein al-Tikriti from 1979
political system one-party socialist republic
political party Arab Ba'ath Socialist Party, nationalist socialist
exports oil (prior to UN sanctions), wool, dates (80% of world supply)
currency Iraqi dinar
population (1989 est) 17,610,000 (Arabs 77%, Kurds 19%, Turks 2%); growth rate 3.6% p.a.
life expectancy men 62, women 63
languages Arabic (official); Kurdish, Assyrian, Armenian
religions Shi'ite Muslim 60%, Sunni Muslim 37%, Christian 3%
literacy men 68%, women 32% (1980 est)
GDP $42.3 bn (1987); $3,000 per head
chronology
1920 Iraq became a British League of Nations protectorate.
1921 Hashemite dynasty established, with Faisal I installed by Britain as king.
1932 Independence achieved from British protectorate status.

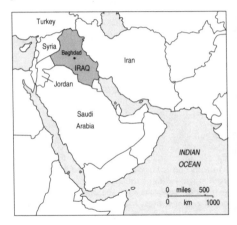

1958 Monarchy overthrown; Iraq became a republic.
1963 Joint Ba'athist-military coup headed by Col Salem Aref.
1968 Military coup put Maj-Gen al-Bakr in power.
1979 Al-Bakr replaced by Saddam Hussein.
1980 War between Iraq and Iran broke out.
1985 Fighting intensified.
1988 Cease-fire; talks began with Iran. Iraq used chemical weapons against Kurdish rebels seeking greater autonomy.
1989 Unsuccessful coup against President Hussein; Iraq launched ballistic missile in successful test.
1990 Peace treaty favouring Iran agreed. Aug: Iraq invaded and annexed Kuwait, precipitating another Gulf crisis. US forces massed in Saudi Arabia at request of King Fahd. United Nations resolutions ordered Iraqi withdrawal from Kuwait and imposed total trade ban on Iraq; UN resolution sanctioning force approved. All foreign hostages released.
1991 16 Jan: US-led forces began aerial assault on Iraq; Iraq's infrastructure destroyed by bombing. 23–28 Feb: land–sea–air offensive to free Kuwait successful. Uprisings of Kurds and Shi'ites brutally suppressed by surviving Iraqi troops. Allied troops withdrew after establishing 'safe havens' for Kurds in the north, leaving a rapid-reaction force near the Turkish border. Allies threatened to bomb strategic targets in Iraq if full information about nuclear facilities denied to UN.
1992 UN imposed a 'no-fly zone' over S Iraq to protect Shi'ites.
1993 Jan: Iraqi incursions into 'no-fly zone' prompted US-led alliance aircraft to bomb 'strategic' targets in Iraq. USA also bombed a target in Baghdad in retaliation for an alleged plot against former president Bush, claiming 'self-defence'. Continued persecution of Shi'ites in S.

Ireland /ˈaɪələnd/ one of the British Isles, lying to the W of Great Britain, from which it is separated

by the Irish Sea. It comprises the provinces of Ulster, Leinster, Munster, and Connacht, and is divided into the Republic of Ireland (which occupies the S, centre, and NW of the island) and Northern Ireland (which occupies the NE corner and forms part of the United Kingdom).

The centre of Ireland is a lowland, about 60–120 m/200–400 ft above sea level; hills are mainly around the coasts, although there are a few peaks over 1,000 m/3,000 ft high, the highest being Carrantuohill ('the inverted reaping hook'), 1,040 m/3,415 ft, in Macgillicuddy's Reeks, County Kerry. The entire western coastline is an intricate alternation of bays and estuaries. Several of the rivers flow in sluggish courses through the central lowland and then cut through fjordlike valleys to the sea. The Shannon in particular falls 30 m/100 ft in its last 26 km/16 mi above Limerick, and is used to produce hydroelectric power.

The lowland bogs that cover parts of central Ireland are intermingled with fertile limestone country where dairy farming is the chief occupation. The bogs are an important source of fuel in the form of peat, Ireland being poorly supplied with coal.

The climate is mild, moist, and changeable. The annual rainfall on the lowlands varies from 76 cm/30 in in the E to 203 cm/80 in in some western districts, but much higher falls are recorded in the hills.

Ireland, Northern /ˈaɪələnd/ constituent part of the United Kingdom
area 13,460 sq km/5,196 sq mi
capital Belfast
towns Londonderry, Enniskillen, Omagh, Newry, Armagh, Coleraine
features Mourne Mountains, Belfast Lough and Lough Neagh; Giant's Causeway; comprises the six counties (Antrim, Armagh, Down, Fermanagh, Derry, and Tyrone) that form part of Ireland's northernmost province of Ulster
exports engineering, especially shipbuilding, textile machinery, aircraft components; linen and synthetic textiles; processed foods, especially dairy and poultry products
currency pound sterling
population (1988 est) 1,578,100
language English
religion Protestant 54%, Roman Catholic 31%
famous people Viscount Montgomery, Lord Alanbrooke

Ireland, Northern: counties

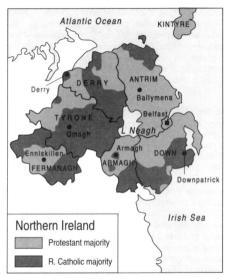

Northern Ireland
▢ Protestant majority
■ R. Catholic majority

county	administrative headquarters	area sq km
Ulster province		
Antrim	Belfast	2,830
Armagh	Armagh	1,250
Down	Downpatrick	2,470
Fermanagh	Enniskillen	1,680
Derry	Derry	2,070
Tyrone	Omagh	3,160
	Total area	13,460

government direct rule from the UK since 1972. Northern Ireland is entitled to send 12 members to the Westminster Parliament. The province costs the UK government £3 billion annually
history the creation of Northern Ireland dates from 1921 when the mainly Protestant counties of Ulster withdrew from the newly established Irish Free State. Spasmodic outbreaks of violence by the IRA continued, but only in 1968–69 were there serious disturbances arising from Protestant political dominance and discrimination against the Roman Catholic minority in employment and housing. British troops were sent to restore peace and protect Catholics, but disturbances continued and in 1972 the parliament at Stormont was prorogued and superseded by direct rule from Westminster. Under the Anglo-Irish Agreement 1985, the Republic of Ireland was given a consultative role (via an Anglo-Irish conference) in the government of Northern Ireland, but agreed that there should be no change in its status except by majority consent. The agreement was approved by Parliament, but all 12 Ulster members gave up their seats, so that by-elections could be fought as a form of 'referendum' on the views of the province itself. A similar boycotting of the Northern Ireland Assembly led to its dissolution 1986 by the UK government. Job discrimination was outlawed under the Fair Employment Act 1975, but in 1993 75% of the unemployed were Catholic; residential integration was still sparse, with 650,000 people living in areas that were 90% Catholic or Protestant. Between 1969 and 1991 violence had claimed 2,872 lives in Northern Ireland; another 94 people were killed 1991. At the end of 1991 there were 11,000 regular soldiers and 6,000 Ulster Defence Regiment members in Northern Ireland.

The question of Northern Ireland's political future was debated in talks held in Belfast April–Sept 1991—the first direct negotiations between the political parties for 16 years. Follow-up talks between the British government and the main Northern Ireland parties Sept–Nov 1992 made little progress. In Sept 1993 a peace initiative put forward by the Social Democratic Labour Party and Sinn Fein (political wing of the outlawed Irish Liberation Army [IRA]), both Catholic nationalist parties, foundered in the wake of renewed violence. It was followed by intensive bilateral talks between the British government and the main Northern Ireland parties, and in Dec 1993 London and Dublin issued a joint peace proposal, the Downing Street Declaration, for consideration by all parties.

Ireland, Republic of: counties

county	administrative headquarters	area in in sq km
Ulster province		
Cavan	Cavan	1,890
Donegal	Lifford	4,830
Monaghan	Monaghan	1,290
Munster province		
Clare	Ennis	3,190
Cork	Cork	7,460
Kerry	Tralee	4,700
Limerick	Limerick	2,690
Tipperary (N)	Nenagh	2,000
Tipperary (S)	Clonmel	2,260
Waterford	Waterford	1,840
Leinster province		
Carlow	Carlow	900
Dublin	Dublin	920
Kildare	Naas	1,690
Kilkenny	Kilkenny	2,060
Laois	Port Laoise	1,720
Longford	Longford	1,040
Louth	Dundalk	820
Meath	Trim	2,340
Offaly	Tullamore	2,000
Westmeath	Mullingar	1,760
Wexford	Wexford	2,350
Wicklow	Wicklow	2,030
Connacht province		
Galway	Galway	5,940
Leitrim	Carrick-on-Shannon	1,530
Mayo	Castlebar	5,400
Roscommon	Roscommon	2,460
Sligo	Sligo	1,800
Total area		68,910

Ireland, Republic of (*Eire*)
area 70,282 sq km/27,146 sq mi
capital Dublin
towns ports Cork, Dun Laoghaire, Limerick, Waterford
physical central plateau surrounded by hills; rivers Shannon, Liffey, Boyne
features Bog of Allen, source of domestic and national power; Macgillicuddy's Reeks, Wicklow Mountains; Lough Corrib, lakes of Killarney; Galway Bay and Aran Islands
head of state Mary Robinson from 1990
head of government Albert Reynolds from 1992
political system democratic republic
political parties Fianna Fáil (Soldiers of

Destiny), moderate centre-right; Fine Gael (Irish Tribe), moderate centre-left; Labour Party, moderate, left of centre; Progressive Democrats, radical free-enterprise
exports livestock, dairy products, Irish whiskey, microelectronic components, mining and engineering products, chemicals, clothing
currency punt
population (1989 est) 3,734,000; growth rate 0.1% p.a.
life expectancy men 70, women 76
languages Irish Gaelic and English (both official)
religion Roman Catholic 94%
literacy 99% (1984)
GDP $48.8 (1992)
chronology
1916 Easter Rising: nationalists against British rule seized the Dublin general post office and proclaimed a republic; the revolt was suppressed by the British army and most of the leaders were executed.
1918–21 Guerrilla warfare against British army led to split in rebel forces.
1921 Anglo-Irish Treaty resulted in creation of the Irish Free State (Southern Ireland).
1937 Independence achieved from Britain.
1949 Eire left the Commonwealth and became the Republic of Ireland.
1973 Fianna Fáil defeated after 40 years in office; Liam Cosgrave formed a coalition government.
1977 Fianna Fáil returned to power, with Jack Lynch as prime minister.
1979 Lynch resigned, succeeded by Charles Haughey.
1981 Garret FitzGerald formed a coalition.
1983 New Ireland Forum formed, but rejected by the British government.
1985 Anglo-Irish Agreement signed.
1986 Protests by Ulster Unionists against the agreement.
1987 General election won by Charles Haughey.
1988 Relations with UK at low ebb because of disagreement over extradition decisions.
1989 Haughey failed to win majority in general election. Progressive Democrats given cabinet positions in coalition government.
1990 Mary Robinson elected president; John Bruton became Fine Gael leader.
1992 Jan: Haughey resigned after losing parliamentary majority. Feb: Albert Reynolds became Fianna Fáil leader and prime minister. June: National referendum approved ratification of Maastricht Treaty. Nov: Reynolds lost confidence vote; election result inconclusive.
1993 Fianna Fáil–Labour coalition formed. Joint Anglo–Irish peace proposal for Northern Ireland issued, the Downing Street Declaration.

Love is never defeated, and I could add, the history of Ireland proves that.

On **Ireland** Pope John Paul II, speech in Galway 1979

Irian Jaya /ˈɪriən ˈdʒaɪə/ W portion of the island of New Guinea, part of Indonesia
area 420,000 sq km/162,000 sq mi
capital Jayapura
population (1989) 1,555,700
history part of the Dutch East Indies 1828 as Western New Guinea; retained by the Netherlands after Indonesian independence 1949 but ceded to Indonesia 1963 by the United Nations and remained part of Indonesia by an 'Act of Free Choice' 1969. In the 1980s, 283,500 hectares/700,000 acres were given over to Indonesia's controversial transmigration programme for the resettlement of farming families from overcrowded Java, causing destruction of rainforests and displacing indigenous people. In 1989 Indonesia began construction of a space launching pad on the island of Biak, near the equator where the Earth's atmosphere is least thick.

Irkutsk /ɪəˈkutsk/ city in S Siberian Russia; population (1987) 609,000. It produces coal, iron, steel, and machine tools. Founded 1652, it began to grow after the Trans-Siberian railway reached it 1898.

Ironbridge Gorge /ˈaɪənbrɪdʒ/ site, near Telford New Town, Shropshire, England, of the Iron Bridge (1779), one of the first and most striking products of the Industrial Revolution in Britain: it is now part of an open-air museum of industrial archaeology.

Iron Gate (Romanian *Porţile de Fier*) narrow gorge, interrupted by rapids, in Romania. A hydroelectric scheme undertaken 1964–70 by Romania and Yugoslavia transformed this section of the river Danube into a lake 145 km/90 mi long and eliminated the rapids as a navigation hazard. Before flooding, in 1965, an archaeological survey revealed Europe's oldest urban settlement, Lepenski Vir.

Irrawaddy /ˌɪrəˈwɒdi/ (Myanmar *Ayeryarwady*) chief river of Myanmar (Burma), flowing roughly N–S for 2,090 km/1,300 mi across the centre of the country into the Bay of Bengal. Its sources are the Mali and N'mai rivers; its chief tributaries are the Chindwin and Shweli.

Irvine /ˈɜːvɪn/ new town in Strathclyde, W Scotland; population (1989 est) 57,000. It overlooks the Isle of Arran, and is a holiday resort.

Ischia /ˈɪskiə/ volcanic island about 26 km/16 mi SW of Naples, Italy, in the Tyrrhenian Sea; population (1985) 26,000. It has mineral springs (known to the Romans) and beautiful scenery, and is a holiday resort.

Ise /iːˈseɪ/ city SE of Kyoto, on Honshu Island, Japan; population (1990) 104,200. It is the site of the most sacred Shinto shrine, dedicated to the sun-goddess Amaterasu, rebuilt every 20 years in the form of a perfect thatched house of the 7th century BC and containing the octagonal mirror of the goddess.

Isère /ɪ'zeə/river in SE France, 290 km/180 mi long, a tributary of the Rhône. It gives its name to the *département* of Isère.

Isfahan /ˌɪsfə'hɑːn/ or *Esfahan* industrial city (steel, textiles, carpets) in central Iran; population (1986) 1,001,000. It was the ancient capital (1598–1722) of Abbas I, and its features include the Great Square, Grand Mosque, and Hall of Forty Pillars.

He who has not been to Isfahan has not seen half the world.

On **Isfahan** Persian saying

Isis /'aɪsɪs/ the upper stretches of the river Thames, England, above Oxford.

Iskandariya /ˌɪˌskændə'riːə/ Arabic name for the Egyptian port of ◊Alexandria.

Iskenderun /ɪs'kendəruːn/ port, naval base, and steel-manufacturing town in Turkey; population (1980) 125,000. It was founded by Alexander the Great 333 BC and called *Alexandretta* until 1939.

Islamabad /ɪz'læməbæd/ capital of Pakistan from 1967, in the Potwar district, at the foot of the Margala Hills and immediately NW of Rawalpindi; population (1981) 201,000. The city was designed by Constantinos Doxiadis in the 1960s. The Federal Capital Territory of Islamabad has an area of 907 sq km/350 sq mi and a population (1985) of 379,000.

Islay /'aɪleɪ/ southernmost island of the Inner Hebrides, Scotland, in Strathclyde region, separated from Jura by the Sound of Islay; area 610 sq km/235 sq mi; population (1981) 3,800. The principal towns are Bowmore and Port Ellen. It produces malt whisky, and its wildlife includes eagles and rare wintering geese.

Isle of Ely /'iːli/ former county of England, in East Anglia. It was merged with Cambridgeshire 1965.

Isle of Man see ◊Man, Isle of.

Isle of Wight see ◊Wight, Isle of.

Islington /'ɪzlɪŋtən/ inner borough of N Greater London, including the suburbs of Islington and Finsbury
features 19th-century squares and terraces in Highbury, Barnsbury, Canonbury; Wesley Museum in City Road
population (1991) 169,200
history Mineral springs (Sadler's Wells) in Clerkenwell were exploited in conjunction with a music-hall in the 17th century, and Lilian Baylis developed a later theatre as an 'Old Vic' annexe.

Ismailia /ˌɪzmaɪ'liːə/ city in NE Egypt; population (1985) 191,700. It was founded 1863 as the headquarters for construction of the Suez Canal and was named after the Khedive Ismail.

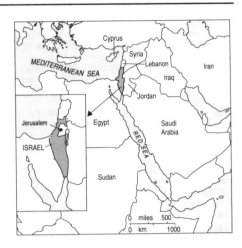

Israel State of (*Medinat Israel*)
area 20,800 sq km/8,029 sq mi (as at 1949 armistice)
capital Jerusalem (not recognized by the United Nations)
towns ports Tel Aviv/Jaffa, Haifa, Acre, Eilat; Bat-Yam, Holon, Ramat Gan, Petach Tikva, Beersheba
physical coastal plain of Sharon between Haifa and Tel Aviv noted since ancient times for fertility; central mountains of Galilee, Samariq, and Judea; Dead Sea, Lake Tiberias, and river Jordan Rift Valley along the E are below sea level; Negev Desert in the S; Israel occupies Golan Heights, West Bank, and Gaza
features historic sites: Jerusalem, Bethlehem, Nazareth, Masada, Megiddo, Jericho; caves of the Dead Sea scrolls
head of state Ezer Weizman from 1993
head of government Yitzhak Rabin from 1992
political system democratic republic
political parties Israel Labour Party, moderate, left of centre; Consolidation Party (Likud), right of centre
exports citrus and other fruit, avocados, chinese leaves, fertilizers, diamonds, plastics, petrochemicals, textiles, electronics (military, medical, scientific, industrial), electro-optics, precision instruments, aircraft and missiles
currency shekel
population (1989 est) 4,477,000 (including 750,000 Arab Israeli citizens and over 1 million Arabs in the occupied territories); under the Law of Return 1950, 'every Jew shall be entitled to come to Israel as an immigrant'; those from East and E Europe are Ashkenazim, and those from Mediterranean Europe (Spain, Portugal, Italy, France, Greece) and Arab N Africa are Sephardim (over 50% of the population is now of Sephardic descent). Between Jan 1990 and April 1991, 250,000 Soviet Jews emigrated to Israel. An Israeli-born Jew is a Sabra. About 500,000 Israeli Jews are resident in the USA. Growth rate 1.8% p.a.
life expectancy men 73, women 76
languages Hebrew and Arabic (official); Yiddish, European and W Asian languages

religions Israel is a secular state, but the predominant faith is Judaism 83%; also Sunni Muslim, Christian, and Druse
literacy Jewish 88%, Arab 70%
GDP $35 bn (1987); $8,011 per head
chronology
1948 Independent State of Israel proclaimed with David Ben-Gurion as prime minister; attacked by Arab nations, Israel won the War of Independence. Many displaced Arabs settled in refugee camps in the Gaza Strip and West Bank.
1952 Col Gamal Nasser of Egypt stepped up blockade of Israeli ports and support of Arab guerrillas in Gaza.
1956 Israel invaded Gaza and Sinai.
1959 Egypt renewed blockade of Israeli trade through Suez Canal.
1963 Ben-Gurion resigned, succeeded by Levi Eshkol.
1964 Palestine Liberation Organization (PLO) founded with the aim of overthrowing the state of Israel.
1967 Israel victorious in the Six-Day War. Gaza, West Bank, E Jerusalem, Sinai, and Golan Heights captured.
1968 Israel Labour Party formed, led by Golda Meir.
1969 Golda Meir became prime minister.
1973 Yom Kippur War: Israel attacked by Egypt and Syria.
1974 Golda Meir succeeded by Yitzhak Rabin.
1975 Suez Canal reopened.
1977 Menachem Begin elected prime minister. Egyptian president addressed the Knesset.
1978 Camp David talks.
1979 Egyptian–Israeli agreement signed. Israel agreed to withdraw from Sinai.
1980 Jerusalem declared capital of Israel.
1981 Golan Heights formally annexed.
1982 Israel pursued PLO fighters into Lebanon.
1983 Peace treaty between Israel and Lebanon signed but not ratified.
1985 Formation of government of national unity with Labour and Likud ministers.
1986 Yitzhak Shamir took over from Peres under power-sharing agreement.
1987 Outbreak of Palestinian uprising (Intifada) in West Bank and Gaza.
1988 Criticism of Israel's handling of Palestinian uprising in occupied territories; PLO acknowledged Israel's right to exist.
1989 New Likud–Labour coalition government formed under Shamir. Limited progress achieved on proposals for negotiations leading to elections in occupied territories.
1990 Coalition collapsed due to differences over peace process; international condemnation of Temple Mount killings. New Shamir right-wing coalition formed.
1991 Shamir gave cautious response to Middle East peace proposals. Some Palestinian prisoners released. Peace talks began in Madrid.
1992 Jan: Shamir lost majority in Knesset when ultra-orthodox party withdrew from coalition. June: Labour Party, led by Yitzhak Rabin, won elections; coalition formed under Rabin. Aug: US–Israeli loan agreement signed. Dec: 400 Palestinians summarily expelled, in the face of

international criticism.
1993 Jan: UN condemned expulsion of Palestinians. Ban on contacts with PLO formally lifted. Feb: 100 of the expelled Palestinians allowed to return. March: Ezer Weizman elected president; Binyamin 'Bibi' Netanyahu elected leader of Likud party. Middle East peace talks restarted. Sept: Israeli–PLO preliminary accord signed outlining principles for Palestinian self-rule in Gaza Strip and West Bank town of Jericho.

In Israel, in order to be a realist, you must believe in miracles.

On **Israel** David Ben-Gurion, comment on CBS-TV 5 Oct 1956

Istanbul /ˌɪstænˈbʊl/ city and chief seaport of Turkey; population (1990) 6,620,200. It produces textiles, tobacco, cement, glass, and leather. Founded as *Byzantium* about 660 BC, it was renamed *Constantinople* AD 330 and was the capital of the Byzantine Empire until captured by the Turks 1453. As *Istamboul* it was capital of the Ottoman Empire until 1922.
features the harbour of the Golden Horn; Hagia Sophia (Emperor Justinian's church of the Holy Wisdom, 537, now a mosque); Sultan Ahmet Mosque, known as the Blue Mosque, from its tiles; Topkapi Palace of the Sultans (with a harem of 400 rooms), now a museum. The Selimye Barracks in the suburb of *Usküdar* (Scutari) was used as a hospital in the Crimean War; the rooms used by Florence Nightingale, with her personal possessions, are preserved as a museum.

It is the real keystone of power; for he who possesses it may rule the world.

On **Istanbul** Napoleon I 1816

Itaipu /iːˈtaɪpuː/ world's largest hydroelectric plant, situated on the Paraná River, SW Brazil. A joint Brazilian-Paraguayan venture, it came into operation 1984; it supplies hydroelectricity to a wide area.

Italy Republic of (*Repubblica Italiana*)
area 301,300 sq km/116,332 sq mi
capital Rome
towns Milan, Turin; ports Naples, Genoa, Palermo, Bari, Catania, Trieste
physical mountainous (Maritime Alps, Dolomites, Apennines) with narrow coastal lowlands; rivers Po, Adige, Arno, Tiber, Rubicon; islands of Sicily, Sardinia, Elba, Capri, Ischia, Lipari, Pantelleria; lakes Como, Maggiore, Garda
environment Milan has the highest recorded level of sulphur-dioxide pollution of any city in the world. The Po River, with pollution ten times higher than officially recommended levels, is estimated to discharge around 250 tonnes of arsenic into the sea each year

features continental Europe's only active volcanoes: Vesuvius, Etna, Stromboli; historic towns include Venice, Florence, Siena, Rome; Greek, Roman, Etruscan archaeological sites **head of state** Oscar Luigi Scalfaro from 1992 **head of government** Carlo Azeglio Ciampi from 1993 **political system** democratic republic **political parties** Christian Democratic Party (DC), Christian, centrist; Democratic Party of the Left (PDS), pro-European socialist; Italian Socialist Party (PSI), moderate socialist; Italian Social Movement–National Right (MSI–DN), neofascist; Italian Republican Party (PRI), social democratic, left of centre; Italian Social Democratic Party (PSDI), moderate, left of centre; Liberals (PLI), right of centre

Italy: regions

region	capital	area in sq km
Abruzzi	Aquila	10,800
Basilicata	Potenza	10,000
Calabria	Catanzaro	15,100
Campania	Naples	13,600
Emilia-Romagna	Bologna	22,100
Friuli-Venezia Giulia*	Udine	7,800
Lazio	Rome	17,200
Liguria	Genoa	5,400
Lombardy	Milan	23,900
Marche	Ancona	9,700
Molise	Campobasso	4,400
Piedmont	Turin	25,400
Puglia	Bari	19,300
Sardinia*	Cagliari	24,100
Sicily*	Palermo	25,700
Trentino-Alto Adige*	Trento**	13,600
Tuscany	Florence	23,000
Umbria	Perugia	8,500
Valle d'Aosta*	Aosta	3,300
Veneto	Venice	18,400
		301,300

* special autonomous regions
** also Bolzano-Bozen

exports wine (world's largest producer), fruit, vegetables, textiles (Europe's largest silk producer), clothing, leather goods, motor vehicles, electrical goods, chemicals, marble (Carrara), sulphur, mercury, iron, steel **currency** lira **population** (1990 est) 57,657,000; growth rate 0.1% p.a. **life expectancy** men 73, women 80 (1989) **language** Italian; German, French, Slovene, and Albanian minorities **religion** Roman Catholic 100% (state religion) **literacy** 97% (1989) **GDP** $1,223.6 bn (1992) **chronology** **1946** Monarchy replaced by a republic. **1948** New constitution adopted. **1954** Trieste (claimed by Yugoslavia after World War II) was divided between Italy and Yugoslavia. **1976** Communists proposed establishment of broad-based, left–right government, the 'historic compromise'; rejected by Christian Democrats. **1978** Christian Democrat Aldo Moro, architect of the historic compromise, kidnapped and murdered by Red Brigade guerrillas infiltrated by Western intelligence agents. **1983** Bettino Craxi, a Socialist, became leader of broad coalition government. **1987** Craxi resigned; succeeding coalition fell within months. **1988** Christian Democrats' leader Ciriaco de Mita established a five-party coalition including the Socialists. **1989** De Mita resigned after disagreements within his coalition government; succeeded by Giulio Andreotti. De Mita lost leadership of Christian Democrats; Communists formed 'shadow government'. **1991** Referendum approved electoral reform. **1992** April: ruling coalition lost its majority in general election; President Cossiga resigned, replaced by Oscar Luigi Scalfaro in May. Giuliano Amato, deputy leader of PDS, accepted premiership. Sept: lira devalued and Italy withdrew from the Exchange Rate Mechanism. **1993** Investigation of corruption network exposed Mafia links with several notable politicians, including Craxi and Andreotti. Craxi resigned Socialist Party leadership; replaced by Giorgio Benvenuto and then Ottaviano del Turro. Referendum supported end of proportional-representation electoral system. Amato resigned premiership; Carlo Ciampi, with no party allegiance, named as his successor.

Ithaca /ˈiθəkə/ (Greek **Itháki**) Greek island in the Ionian Sea, area 93 sq km/36 sq mi. Important in pre-Classical Greece, Ithaca was (in Homer's poem) the birthplace of Odysseus, though this is sometimes identified with the island of Leukas (some archaeologists have equated ancient Ithaca with Leukas rather than modern Ithaca).

Ivanovo /iˈvɑːnəvəu/ capital of Ivanovo region, Russia, 240 km/150 mi NE of Moscow; population (1987) 479,000. Industries include textiles, chemicals, and engineering.

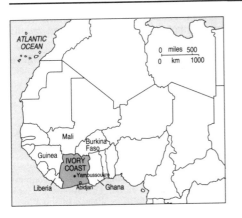

Iviza alternative spelling of ◊Ibiza, one of the Balearic Islands.

Ivory Coast Republic of (*République de la Côte d'Ivoire*)
area 322,463 sq km/124,471 sq mi
capital Yamoussoukro
towns Bouaké, Daloa, Man; ports Abidjan, San-Pédro
physical tropical rainforest (diminishing as exploited) in S; savannah and low mountains in N
environment an estimated 85% of the country's forest has been destroyed by humans
features Vridi canal, Kossou dam, Monts du Toura
head of state and government Félix Houphouët-Boigny from 1960
political system emergent democratic republic
political party Democratic Party of the Ivory Coast (PDCI), nationalist, free-enterprise
exports coffee, cocoa, timber, petroleum products
currency franc CFA
population (1990 est) 12,070,000; growth rate 3.3% p.a.
life expectancy men 52, women 55 (1989)
languages French (official), over 60 native dialects

media the government has full control of the media
religions animist 65%, Muslim 24%, Christian 11%
literacy 35% (1988)
GDP $7.6 bn (1987); $687 per head
chronology
1904 Became part of French West Africa.
1958 Achieved internal self-government.
1960 Independence achieved from France, with Félix Houphouët-Boigny as president of a one-party state.
1985 Houphouët-Boigny re-elected, unopposed.
1986 Name changed officially from Ivory Coast to Côte d'Ivoire.
1990 Houphouët-Boigny and PDCI re-elected.

Iwo Jima /,i:wəu'dʒi:mə/ largest of the Japanese Volcano Islands in the W Pacific Ocean, 1,222 km/760 mi S of Tokyo; area 21 sq km/8 sq mi. Annexed by Japan 1891, it was captured by the USA 1945 after fierce fighting. It was returned to Japan 1968.

Izhevsk /i:'ʒefsk/ industrial city in central Russia, capital of Udmurt Autonomous Republic; population (1987) 631,000. Industries include steel, agricultural machinery, machine tools, and armaments. It was founded 1760.

Izmir /ɪz'mɪə/ (formerly *Smyrna*) port and naval base in Turkey; population (1990) 1,757,400. Products include steel, electronics, and plastics. The largest annual trade fair in the Middle East is held here. It is the headquarters of North Atlantic Treaty Organization SE Command.
history Originally Greek (founded about 1000 BC), it was of considerable importance in ancient times, vying with Ephesus and Pergamum as the first city of Asia. It was destroyed by Tamerlane 1402 and became Turkish 1424. It was developed in the 16th century as an international trading centre, much of its trade gained at the expense of Venice. It was occupied by the Greeks 1919 but retaken by the Turks 1922; in the same year it was largely destroyed by fire.

J

Jabalpur /ˌdʒʌbəl'puə/ industrial city on the Narbarda River in Madhya Pradesh, India; population (1981) 758,000. Products include textiles, oil, bauxite, and armaments.

Jablonec /'jæblɒnets/ town in the Czech Republic, on the river Neisse, NE of Prague; population (1991) 45,900. It has had a glass industry since the 14th century.

Jackson /'dʒæksən/ largest city and capital of Mississippi, on the Pearl River; population (1990) 196,600. It produces furniture, cottonseed oil, and iron and steel castings, and owes its prosperity to the discovery of gas fields to the S in the 1930s. Named after Andrew Jackson, later president, it dates from 1821 and was virtually destroyed by Union troops 1863, during the American Civil War.

Jackson /'dʒækson/ city in S Michigan, USA, on the Grand River, S of Lansing; seat of Jackson County; population (1990) 37,400. Its industries include motor-vehicle and aircraft parts, tools, plastics, and air-conditioning equipment. The Republican Party was formed here 1854.

Jacksonville /'dʒæksənvɪl/ port, resort, and commercial centre in NE Florida, USA; population (1990) 673,000. The port has naval installations and ship-repair yards. To the N the Cross-Florida Barge Canal links the Atlantic with the Gulf of Mexico. Manufactured goods include wood and paper products, chemicals, and processed food.

Jacobabad /ˌdʒeɪkəbə'bæd/ city in Sind province, SE Pakistan, 400 km/250 mi NE of Karachi; population (1981) 80,000. Founded by General John Jacob as a frontier post, the city now trades in wheat, rice, and millet.

It has a low annual rainfall (about 5 cm/2 in) and temperatures are among the highest in the Indian subcontinent—up to 53°C/127°F.

Jaén /xɑː'en/ capital of Jaén province, S Spain, on the river Guadalbullon; population (1991) 105,500. It has remains of its Moorish walls and citadel.

Jaffa /'dʒæfə/ (biblical name *Joppa*) port in W Israel, part of Tel Aviv from 1950. It was captured during the Crusades in the 12th century, by the French emperor Napoleon 1799, and by the British field marshal Allenby 1917.

Jaffna /'dʒæfnə/ capital of Jaffna district, Northern Province, Sri Lanka; population (1990) 129,000. It was the focal point of Hindu Tamil nationalism and the scene of recurring riots during the 1980s.

Jaipur /ˌdʒaɪ'puə/ capital of Rajasthan, India; population (1981) 1,005,000. It was formerly the capital of the state of Jaipur, which was merged with Rajasthan 1949. Products include textiles and metal products.

Features include the Jantar Mantar observatory, built by Maharajah Jai Singh II 1728; it contains the world's largest sundial. The five-storey, pink sandstone Hawa Mahal or Palace of the Winds (1799) is a notable landmark. The city is also noted for its blue pottery.

Jakarta /dʒə'kɑːtə/ or *Djakarta* (former name until 1949 *Batavia*) capital of Indonesia on the NW coast of Java; population (1980) 6,504,000. Industries include textiles, chemicals, and plastics; a canal links it with its port of Tanjung Priok where rubber, oil, tin, coffee, tea, and palm oil are among its exports; also a tourist centre. It was founded by Dutch traders 1619.

Jalalabad /dʒə'lɑːləbɑːd/ capital of Nangarhar province, E Afghanistan, on the road from Kabul to Peshawar in Pakistan; population (1984 est) 61,900 (numbers swelled to over 1 million during the civil war).

Jalgava /'jælgəvə/ or *Jelgava*, formerly (until 1917) *Mitau* town in W Latvia, 48 km/30 mi S of Riga; population (1985) 70,000. Industries include textiles and sugar-refining. The town was founded 1265 by Teutonic knights.

Jamaica
area 10,957 sq km/4,230 sq mi
capital Kingston
towns Montego Bay, Spanish Town, St Andrew
physical mountainous tropical island
features Blue Mountains (so called because of the haze over them) renowned for their coffee; partly undersea ruins of pirate city of Port Royal, destroyed by an earthquake 1692
head of state Elizabeth II from 1962, represented by governor general
head of government P J Patterson from 1992
political system constitutional monarchy
political parties Jamaica Labour Party (JLP), moderate, centrist; People's National Party (PNP), left of centre
exports sugar, bananas, bauxite, rum, cocoa, coconuts, liqueurs, cigars, citrus
currency Jamaican dollar
population (1990 est) 2,513,000 (African 76%, mixed 15%, Chinese, Caucasian, East Indian); growth rate 2.2% p.a.
life expectancy men 75, women 78 (1989)
languages English, Jamaican creole
media one daily newspaper 1834–1988 (except 1973–82), privately owned; sensational evening and weekly papers
religions Protestant 70%, Rastafarian
literacy 82% (1988)

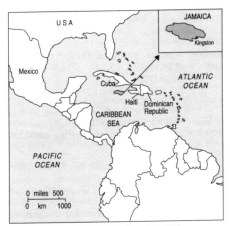

GDP $2.9 bn; $1,187 per head (1989)
chronology
1494 Columbus reached Jamaica.
1509–1655 Occupied by Spanish.
1655 Captured by British.
1944 Internal self-government introduced.
1962 Independence achieved from Britain, with Alexander Bustamante of the JLP as prime minister.
1967 JLP re-elected under Hugh Shearer.
1972 Michael Manley of the PNP became prime minister.
1980 JLP elected, with Edward Seaga as prime minister.
1983 JLP re-elected, winning all 60 seats.
1988 Island badly damaged by Hurricane Gilbert.
1989 PNP won a decisive victory with Michael Manley returning as prime minister.
1992 Manley resigned, succeeded by P J Patterson.

Jammu /ˈdʒʌmuː/ winter capital of the state of Jammu and Kashmir, India; population (1981) 206,100. It stands on the river Tavi and was linked to India's rail system 1972.

Jammu and Kashmir /ˈdʒʌmuː, ˌkæʃˈmɪə/ state of N India

Jammu and Kashmir

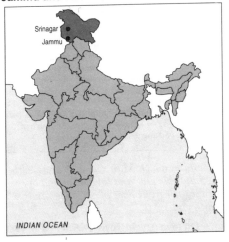

area 101,300 sq km/39,102 sq mi
capital Jammu (winter); Srinagar (summer)
towns Leh
products timber, grain, rice, fruit, silk, carpets
population (1991) 7,718,700 (Indian-occupied territory)
history part of the Mogul Empire from 1586, Jammu came under the control of Gulab Singh 1820. In 1947 Jammu was attacked by Pakistan and chose to become part of the new state of India. Dispute over the area (see ⟡ Kashmir) caused further hostilities 1971 between India and Pakistan (ended by the Simla agreement 1972). Since then, separatist agitation has developed, complicating the territorial dispute between India and Pakistan.

Jamnagar /dʒæmˈnʌɡə/ city in Gujarat, India, on the Gulf of Kutch, SW of Ahmedabad; population (1981) 317,000. Its port is at Bedi.

Jamshedpur /ˌdʒʌmʃedˈpuə/ city in Bihar, India; population (1981) 439,000. It was built 1909 and takes its name from the industrialist Jamsheedji Tata who founded the Tata iron and steel works here and in Bombay.

Janesville /ˈdʒeɪnzvɪl/ city in S Wisconsin, USA, on the Rock River, SE of Madison; seat of Rock County; population (1990) 52,100. Industries include cars and car parts, building materials, and electronic equipment. It is a processing and marketing centre for the area's agricultural products.

Jan Mayen /ˈjæn ˈmaɪən/ Norwegian volcanic island in the Arctic Ocean, between Greenland and Norway; area 380 sq km/147 sq mi. It is named after a Dutchman who visited it about 1610, and was annexed by Norway 1929.

Japan (*Nippon*)
area 377,535 sq km/145,822 sq mi
capital Tokyo
towns Fukuoka, Kitakyushu, Kyoto, Sapporo; ports Osaka, Nagoya, Yokohama, Kobe, Kawasaki
physical mountainous, volcanic; comprises over 1,000 islands, the largest of which are Hokkaido, Honshu, Kyushu, and Shikoku
features Mount Fuji, Mount Aso (volcanic), Japan Alps, Inland Sea archipelago

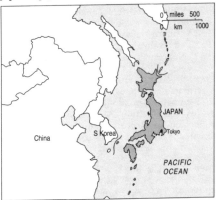

head of state (figurehead) Emperor Akihito (Heisei) from 1989
head of government Morihiro Hosokawa from 1993
political system liberal democracy
political parties Liberal Democratic Party (LDP), right of centre; Social Democratic Party of Japan (SDJP), former Socialist Party, left of centre but moving towards centre; Komeito (Clean Government Party), Buddhist, centrist; Democratic Socialist Party, centrist; Japanese Communist Party (JCP), socialist; Shinseito (Japan Renewal Party), right-wing reformist
exports televisions, cassette and video recorders, radios, cameras, computers, robots, other electronic and electrical equipment, motor vehicles, ships, iron, steel, chemicals, textiles
currency yen
population (1990 est) 123,778,000; growth rate 0.5% p.a.
life expectancy men 76, women 82 (1989)
language Japanese
religions Shinto, Buddhist (often combined), Christian; 30% claim a personal religious faith
literacy 99% (1989)
GDP $3,674.5 bn (1992)
chronology
1867 End of shogun rule; executive power passed to emperor. Start of modernization of Japan.
1894–95 War with China; Formosa (Taiwan) and S Manchuria gained.
1902 Formed alliance with Britain.
1904–05 War with Russia; Russia ceded southern half of Sakhalin.
1910 Japan annexed Korea.
1914 Joined Allies in World War I.
1918 Received German Pacific islands as mandates.
1931–32 War with China; renewed 1937.
1941 Japan attacked US fleet at Pearl Harbor 7 Dec.
1945 World War II ended with Japanese surrender. Allied control commission took power. Formosa and Manchuria returned to China.
1946 Framing of 'peace constitution'. Emperor Hirohito became figurehead ruler.
1952 Full sovereignty regained.
1958 Joined United Nations.
1968 Bonin and Volcano Islands regained.
1972 Ryukyu Islands regained.
1974 Prime Minister Tanaka resigned over Lockheed bribes scandal.
1982 Yasuhiro Nakasone elected prime minister.
1985 Yen revalued.
1987 Noboru Takeshita chosen to succeed Nakasone.
1988 Recruit scandal cast shadow over government and opposition parties.
1989 Emperor Hirohito (Shōwa) died; succeeded by his son Akihito. Many cabinet ministers implicated in Recruit scandal and Takeshita resigned; succeeded by Sosuke Uno. Aug: Uno resigned after sex scandal; succeeded by Toshiki Kaifu.
1990 Feb: new house of councillors' elections won by LDP. Public-works budget increased by 50% to encourage imports.
1991 Japan contributed billions of dollars to the

Gulf War and its aftermath. Kaifu succeeded by Kiichi Miyazawa.
1992 Over 100 politicians implicated in new financial scandal. Emperor Akihito made first Japanese imperial visit to China. Trade surpluses reached record levels.
1993 Worst recession of postwar era; trade surpluses, however, again reached record levels. LDP lost its majority in general election, ending 38 years in power; Miyazawa resigned. Morihiro Hosokawa chosen as premier, heading non-LDP seven-party coalition.

Japan, Sea of /dʒə'pæn/ sea separating Japan from the mainland of Asia.

Jarrow /'dʒærəʊ/ town in Tyne and Wear, NE England, on the south bank of the Tyne, 10 km/6 mi E of Newcastle and connected with the north bank by the Tyne Tunnel (1967); population (1988 est) 27,000. The closure of Palmer's shipyard in Jarrow 1933 prompted the unemployed to march to London, a landmark event of the Depression.

Jassy /'jæsi/ German name for the Romanian city of ◊Iaşi.

Java /'dʒɑːvə/ or *Jawa* most important island of Indonesia, situated between Sumatra and Bali
area (with the island of Madura) 132,000 sq km/51,000 sq mi
capital Jakarta (also capital of Indonesia)
towns ports include Surabaya and Semarang
physical about half the island is under cultivation, the rest being thickly forested. Mountains and sea breezes keep temperatures down, but humidity is high, with heavy rainfall from Dec to March
features a chain of mountains, some of which are volcanic, runs along the centre, rising to 2,750 m/9,000 ft. The highest mountain, Semeru (3,676 m/12,060 ft), is in the E
products rice, coffee, cocoa, tea, sugar, rubber, quinine, teak, petroleum
population (with Madura; 1989) 107,513,800, including people of Javanese, Sundanese, and Madurese origin, with differing languages
religion predominantly Muslim
history fossilized early human remains (*Homo erectus*) were discovered 1891–92. In central Java there are ruins of magnificent Buddhist monuments and of the Sivaite temple in Prambanan. The island's last Hindu kingdom, Majapahit, was destroyed about 1520 and followed by a number of short-lived Javanese kingdoms. The Dutch East India company founded a factory 1610. Britain took over during the Napoleonic period, 1811–16, and Java then reverted to Dutch control. Occupied by Japan 1942–45, Java then became part of the republic of Indonesia.

A world in which Man coaxes Nature with inexhaustible love and labour instead of coercing her as he does in Australia and North America.

On **Java** Arnold Toynbee *East to West* 1958

Jedburgh /'dʒedbərə/ small town in the Borders Region, SE Scotland, on Jed Water; population (1981) 4,200. It has the remains of a 12th-century abbey.

Jedda /'dʒedə/ alternative spelling for the Saudi Arabian port ◊Jiddah.

Jefferson City /ˌdʒefəsn 'sɪti/ capital of Missouri, USA, located in the central part of the state, W of St Louis, on the Mississippi River; population (1990) 35,500. Industries include agricultural products, shoes, electrical appliances, and cosmetics.

Jehol /ˌdʒʌ'hɒl/ former name for the city of ◊Chengde in NE Hebei province, N China.

Jena /'jeɪnə/ town SE of Weimar, in the state of Thuringia, Germany, population (1990) 110,000. Industries include the Zeiss firm of optical-instrument makers, founded 1846. Here in 1806 Napoleon defeated the Prussians, and Schiller and Hegel taught at the university, which dates from 1558.

Jerez de la Frontera /xe'reθ deɪ lɑ: frɒn'teərə/ city in Andalusia, SW Spain; population (1991) 184,000. It is famed for sherry, the fortified wine to which it gave its name.

Jersey /'dʒɜːzi/ largest of the Channel Islands; capital St Helier; area 117 sq km/45 sq mi; population (1990) 57,500. It is governed by a lieutenant-governor representing the English crown and an assembly. Jersey cattle were originally bred here. Jersey gave its name to a woollen garment.
    The island was occupied 1940–45 by German forces. Jersey zoo (founded 1959 by Gerald Durrell) is engaged in breeding some of the world's endangered species.

Jersey City /'dʒɜːzi/ city of NE New Jersey, USA; population (1990) 228,500. It faces Manhattan Island, to which it is connected by tunnels. A former port, it is now an industrial centre.

Jerusalem /dʒə'ruːsələm/ ancient city of Palestine, divided 1948 between Jordan and the new republic of Israel; area (pre-1967) 37.5 sq km/ 14.5 sq mi, (post-1967) 108 sq km/42 sq mi, including areas of the West Bank; population (1989) 500,000, about 350,000 Israelis and 150,000 Palestinians. In 1950 the western New City was proclaimed as the Israeli capital, and, having captured from Jordan the eastern Old City 1967, Israel affirmed 1980 that the united city was the country's capital; the United Nations does not recognize the claim.
features Seven gates into the Old City through the walls built by Selim I (1467–1520); buildings: the Church of the Holy Sepulchre (built by Emperor Constantine 335) and the mosque of the Dome of the Rock. The latter stands on the site of the Temple built by King Solomon in the 10th century BC, and the Western ('wailing') Wall, held sacred by Jews, is part of the walled platform on which the Temple once stood. The Hebrew University of Jerusalem opened 1925.
religions Christianity, Judaism, and Muslim,

with Roman Catholic, Anglican, Eastern Orthodox, and Coptic bishoprics. In 1967 Israel guaranteed freedom of access of all faiths to their holy places.

Jervis Bay /'dʒɑːvɪs/ deep bay on the coast of New South Wales, Australia, 145 km/90 mi SW of Sydney. The federal government in 1915 acquired 73 sq km/28 sq mi here to create a port for Canberra. It forms part of the Australian Capital Territory and is the site of the Royal Australian Naval College.

Jewish Autonomous Region part of the Khabarovsk Territory, E Siberian Russia, on the river Amur; capital Birobidzhan; area 36,000 sq km/13,900 sq mi; population (1986) 211,000. Industries include textiles, leather, metallurgy, light engineering, agriculture, and timber. It was established as a Jewish National District 1928 and became an Autonomous Region 1934 but became only nominally Jewish after the Stalinist purges 1936–47 and 1948–49.

Jhansi /'dʒɑːnsi/ city in Uttar Pradesh, NE India, 286 km/178 mi SW of Lucknow; population (1981) 281,000. It is a railway and road junction and a market centre. It was founded 1613, and was the scene of a massacre of British civilians 1857.

Jhelum /'dʒiːləm/ river rising in Kashmir and flowing into Pakistan; length about 720 km/450 mi. The Mangla Dam 1967, one of the world's largest earth-filled dams, stores flood waters for irrigation and hydroelectricity. The Jhelum is one of the five rivers that give Punjab its name and was known in the ancient world as the *Hydaspes*, on whose banks Alexander the Great won a battle 326 BC.

Jiangsu /dʒi,æŋ'suː/ or *Kiangsu* province on the coast of E China
area 102,200 sq km/39,449 sq mi
capital Nanjing
features the swampy mouth of the river Chang Jiang; the special municipality of Shanghai
products cereals, rice, tea, cotton, soya beans, fish, silk, ceramics, textiles, coal, iron, copper, cement
population (1990) 67,057,000
history Jiangsu was originally part of the Wu kingdom, and Wu is still a traditional local name for the province. Jiangsu's capture by Japan in 1937 was an important step in that country's attempt to conquer China.

Jiangxi /dʒi,æŋ'ʃiː/ or *Kiangsi* province of SE China
area 164,800 sq km/63,613 sq mi
capital Nanchang
products rice, tea, cotton, tobacco, porcelain, coal, tungsten, uranium
population (1990) 37,710,000
history the province was Mao Zedong's original base in the first phase of the Communist struggle against the Nationalists.

Jibuti /dʒɪ'buti/ variant spelling of ◊Djibouti, a republic of NE Africa.

Jiddah /'dʒɪdə/ or **Jedda** port in Hejaz, Saudi Arabia, on the eastern shore of the Red Sea; population (1986) 1,000,000. Industries include cement, steel, and oil refining. Pilgrims pass through here on their way to Mecca.

Jilin /,dʒi:'lɪn/ or **Kirin** province of NE China in central Manchuria
area 187,000 sq km/72,182 sq mi
capital Changchun
population (1990) 24,659,000.

Jinan /,dʒi:'næn/ or **Tsinan** city and capital of Shandong province, China; population (1989) 2,290,000. It has food-processing and textile industries.

Jingdezhen /,dʒɪŋdə'dʒen/ or **Chingtechen** or **Fou-liang** town in Jiangxi, China. Ming blue-and-white china was produced here, the name of the clay (kaolin) coming from Kaoling, a hill east of Jingdezhen; some of the best Chinese porcelain is still made here.

Jinja /'dʒɪndʒə/ town in Busoga Province, Uganda, on the Victoria Nile E of Kampala; population (1983) 45,000. Nearby is the Owen Falls Dam (1954).

Jinsha Jiang /,dʒɪn'ʃɑː dʒi:'æŋ/ river that rises in SW China and forms the Chang Jiang (Yangtze) at Yibin.

Jodhpur /,dʒɒd'puə/ city in Rajasthan, India, formerly capital of Jodhpur princely state, founded 1459 by Rao Jodha; population (1981) 493,600. It is a market centre and has the training college of the Indian air force, an 18th-century Mogul palace, and a red sandstone fort. A style of riding breeches is named after it.

Jogjakarta /,jɒgjə'jɑːtə/ alternative spelling of ◊Yogyakarta, a city in Indonesia.

Johannesburg /dʒəu'hænɪsbɜːg/ largest city of South Africa, situated on the Witwatersrand River in Transvaal; population (1985) 1,609,000. It is the centre of a large gold-mining industry; other industries include engineering works, meat-chilling plants, and clothing factories.

Notable buildings include the law courts, Escom House (Electricity Supply Commission), the South African Railways Administration Building, the City Hall, Chamber of Mines and Stock Exchange, the Witwatersrand (1921) and Rand Afrikaans (1966) universities, and the Union Observatory. Johannesburg was founded after the discovery of gold 1886 and was probably named after Jan (Johannes) Meyer, the first mining commissioner.

John o' Groats /ə'grəuts/ village in NE Highland Region, Scotland, about 3 km/2 mi W of Duncansby Head, proverbially Britain's northernmost point. It is named after the Dutchman John de Groot, who built a house there in the 16th century.

Johnson City /,dʒɒnsn 'sɪti/ city in NE Tennessee, USA, just below the Virginia border, in the Appalachian Mountains NE of Knoxville; population (1990) 49,400. Industries include tobacco, furniture, building materials, metals, textiles, and food processing.

Johnston Atoll /'dʒɒnstən/ coral island in the mid-Pacific, lying between the Marshall Islands and Hawaii; area 2.8 sq km/1.1 sq mi. The island is only 2.4 m/8 ft above sea level and subject to hurricanes and tidal waves. It has the status of a National Wildlife Refuge but was contaminated by fallout from nuclear-weapons testing 1962, and has since 1971 been used as a repository for chemical weapons left over from the Korean and Vietnam wars. An unincorporated territory of the USA, it is administered by the US Defense Nuclear Agency (DNA).

Johnstown /'dʒɒnztaun/ city in SW Pennsylvania, USA, on the Conemaugh River, E of Pittsburgh; population (1990) 28,100. Industries include steel, coal and coal by-products, chemicals, building materials, and clothing. Johnstown was the victim of disastrous floods 1889.

Johor /dʒəu'hɔː/ state in S Peninsular Malaysia; capital Johor Baharu; area 19,000 sq km/7,334 sq mi; population (1990) 2,106,500. The southernmost point of mainland Asia, it is joined to Singapore by a causeway. It is mainly forested, with swamps. There is bauxite and iron.

Joliet /,dʒəuli'et/ city in NE Illinois, USA, on the Des Plaines River, SW of Chicago; seat of Will County; population (1990) 76,800. It is a centre for barge traffic. Industries include building materials, chemicals, oil refining, heavy construction machinery, and paper. The city is named after the explorer Louis Joliet.

Jonglei Canal N African civil-engineering project to divert the White Nile in order to improve the water supply to Egypt from Sudan. The project was conceived in the 1940s but was not begun until 1982, and work was halted by the outbreak 1983 of civil war in Sudan.

Jönköping /jɜːnt'ʃɜːpɪŋ/ town at the south end of Lake Vättern, Sweden; population (1990) 111,500. It is an industrial centre in an agricultural and forestry region.

Joplin /'dʒɒplɪn/ city in SW Missouri, USA, W of Springfield; population (1990) 41,000. Industries include zinc and lead smelting, leather goods, and furniture.

Joppa /'dʒɒpə/ ancient name of ◊Jaffa, a port in W Israel.

Jordan Hashemite Kingdom of (*Al Mamlaka al Urduniya al Hashemiyah*)
area 89,206 sq km/34,434 sq mi (West Bank 5,879 sq km/2,269 sq mi)
capital Amman
towns Zarqa, Irbid, Aqaba (the only port)
physical desert plateau in E; rift valley separates E and W banks of the river Jordan
features lowest point on Earth below sea level in the Dead Sea (−396 m/−1,299 ft); archaeological sites at Jerash and Petra
head of state King Hussein ibn Talai from 1952
head of government Mudar Badran from 1989
political system constitutional monarchy
political parties none

exports potash, phosphates, citrus, vegetables
currency Jordanian dinar
population (1990 est) 3,065,000 (including Palestinian refugees); West Bank (1988) 866,000; growth rate 3.6% p.a.
life expectancy men 67, women 71
languages Arabic (official), English
religions Sunni Muslim 92%, Christian 8%
literacy 71% (1988)
GDP $4.3 bn (1987); $1,127 per head (1988)
chronology
1946 Independence achieved from Britain as Transjordan.
1949 New state of Jordan declared.
1950 Jordan annexed West Bank.
1953 Hussein ibn Talai officially became king of Jordan.
1958 Jordan and Iraq formed Arab Federation that ended when the Iraqi monarchy was deposed.
1967 Israel captured and occupied West Bank. Martial law imposed.
1976 Lower house dissolved, political parties banned, elections postponed until further notice.
1982 Hussein tried to mediate in Arab-Israeli conflict.
1984 Women voted for the first time.
1985 Hussein and Yassir Arafat put forward framework for Middle East peace settlement. Secret meeting between Hussein and Israeli prime minister.
1988 Hussein announced decision to cease administering the West Bank as part of Jordan, passing responsibility to Palestine Liberation Organization, and the suspension of parliament.
1989 Prime Minister Zaid al-Rifai resigned; Hussein promised new parliamentary elections following criticism of economic policies. Riots over price increases up to 50% following fall in oil revenues. First parliamentary elections for 22 years; Muslim Brotherhood won 25 of 80 seats but exiled from government; martial law lifted.
1990 Hussein unsuccessfully tried to mediate after Iraq's invasion of Kuwait. Massive refugee problems as thousands fled to Jordan from Kuwait and Iraq.
1991 24 years of martial law ended; ban on political parties lifted.
1992 Political parties allowed to register.

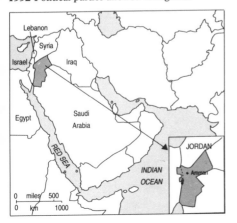

Nowhere more than in Jordan have I felt a sense that this is where the world began.

On **Jordan** Edward Heath *Travels, People and Places in My Life* 1977

Jordan /'dʒɔːdn/ river rising on Mount Hermon, Syria, at 550 m/1,800 ft above sea level and flowing S for about 320 km/200 mi via the Sea of Galilee to the Dead Sea, 390 m/1,290 ft below sea level. It occupies the northern part of the Great Rift Valley; its upper course forms the boundary of Israel with Syria and the kingdom of Jordan; its lower course runs through Jordan; the West Bank has been occupied by Israel since 1967.

Jotunheim /'jəutunhaɪm/ mountainous region of S Norway, containing the highest mountains in Scandinavia, Glittertind (2,472 m/8,110 ft) and Galdhøpiggen (2,469 m/8,100 ft). In Norse mythology it is the home of the giants.

Jounieh /'dʒuːniə/ port on the Mediterranean coast of Lebanon, 15 km/9 mi N of Beirut. It was the centre of an anti-Syrian enclave during the civil war.

Juan Fernández Islands /'dʒuːən fə'nændez, Spanish 'xwæn feə'nændeθ/ three small volcanic Pacific islands belonging to Chile; almost uninhabited. The largest is Más-a-Tierra (also sometimes called Juan Fernández Island), where Alexander Selkirk was marooned 1704–09. The islands were named after the Spanish navigator who reached them 1563.

Juba /'dʒuːbə/ river in E Africa, formed at Dolo, Ethiopia, by the junction of the Ganale Dorya and Dawa rivers. It flows S for about 885 km/550 mi through the Somali Republic (of which its valley is the most productive area) into the Indian Ocean.

Juba /'dʒuːbə/ capital of Southern Region, Sudan Republic; situated on the left bank of the White Nile, at the head of navigation above Khartoum, 1,200 km/750 mi to the N; population (1973) 56,700.

Jubbulpore /ˌdʒʌbəl'puə/ alternative name for the city of ♢Jabalpur in India.

Jugoslavia /ˌjuːgəu'slaːviə/ alternative spelling of ♢Yugoslavia.

Jumna /'dʒʌmnə/ or *Yamuna* river in India, 1,385 km/860 mi in length, rising in the Himalayas, in Uttar Pradesh, and joining the river Ganges near Allahabad, where it forms a sacred bathing place. Agra and Delhi are also on its course.

Juneau /'dʒuːnəu/ ice-free port and state capital of Alaska, USA, on Gastineau Channel in the S Alaska panhandle; population (1980) 19,528. Juneau is the commercial and distribution centre for the fur-trading and mining of the Panhandle region; also important are salmon fishing, fish processing, and lumbering.

Jungfrau /'juŋfrau/ (German 'maiden') mountain in the Bernese Oberland, Switzerland; 4,166 m/ 13,669 ft high. A railway ascends to the plateau of the Jungfraujoch, 3,456 m/11,340 ft, where there is a winter-sports centre.

Jura /'dʒuərə/ island of the Inner Hebrides; area 380 sq km/147 sq mi; population (with Colonsay, 1971) 343. It is separated from Scotland by the Sound of Jura. The whirlpool Corryvreckan (Gaelic 'Brecan's cauldron') is off the north coast.

Jura Mountains /'dʒuərə/ series of parallel mountain ranges running SW–NE along the French-Swiss frontier between the rivers Rhône and Rhine, a distance of 250 km/156 mi. The highest peak is *Crête de la Neige*, 1,723 m/5,650 ft.

The mountains give their name to a *département* of France, and in 1979 a Jura canton was established in Switzerland, formed from the French-speaking areas of Berne.

Jutland /'dʒʌtlənd/ (Danish *Jylland*) peninsula of N Europe; area 29,500 sq km/11,400 sq mi. It is separated from Norway by the Skagerrak and from Sweden by the Kattegat, with the North Sea to the W. The larger northern part belongs to Denmark, the southern part to Germany.

Jylland /'juːlæn/ Danish name for the mainland of Denmark, the northern section of the Jutland peninsula. The chief towns are Aalborg, Aarhus, Esbjerg, Fredericia, Horsens, Kolding, Randers, and Vejle.

K2 or *Chogori* second highest mountain above sea level, 8,611 m/28,261 ft, in the Karakoram range, Kashmir, N India. It was first climbed 1954 by an Italian expedition.

The peak was designated 'K2' by Lt-Col Henry Godwin-Austen in his surveying log since it was the second peak in the Karakorams to be surveyed and mapped. It was later unofficially called Mount Godwin-Austen in his honour, although this alternative name has largely gone out of use.

Kabardino-Balkar /ˌkæbəˌdiːnəʊˈbælkə/ autonomous republic of Russia, in the N Caucasus Mountains; capital Nalchik; area 12,500 sq km/ 4,825 sq mi; population (1989) 760,000. Under Russian control from 1557, it was annexed 1827; it was an autonomous republic of the USSR 1936–91.

Kabinda /kəˈbɪndə/ part of Angola; see ◊ Cabinda.

Kabul /ˈkɑːbul/ capital of Afghanistan, 2,100 m/ 6,900 ft above sea level, on the river Kabul; population (1984) 1,179,300. Products include textiles, plastics, leather, and glass. It commands the strategic routes to Pakistan via the Khyber Pass.

Kabul has been in existence for over 3,000 years. It became the capital 1776, was captured by the British 1839 and 1879, and was under Soviet control 1979–89. In 1992 the city saw fierce fighting during the mujaheddin takeover and ousting of the Soviet-backed Najibullah regime. There is a university (1931), the tomb of Zahir (Babur), founder of the Mogul empire, and the Dar ol-Aman palace, which houses the parliament and government departments.

Kabwe /ˈkɑːbweɪ/ town in central Zambia (formerly *Broken Hill*); population (1988) 200,300. It is a mining centre (copper, cadmium, lead, and zinc).

Kaduna /kəˈduːnə/ town in N Nigeria, on the Kaduna River; population (1983) 247,000. It is a market centre for grain and cotton; industries include textiles, cars, timber, pottery, and oil refining.

Kafue /kəˈfuːeɪ/ river in central Zambia, a tributary of the Zambezi, 965 km/600 mi long. The upper reaches of the river form part of the Kafue national park (1951). Kafue town, 44 km/27 mi S of Lusaka, is the centre of Zambia's heavy industry; population (1980) 35,000.

Kagoshima /ˌkægəˈʃiːmə/ industrial city (Satsumayaki porcelain) and port on Kyushu Island, SW Japan; population (1990) 536,700.

Kaieteur /ˌkaɪəˈtuə/ waterfall on the river Potaro, a tributary of the Essequibo, Guyana. At 250 m/ 822 ft, it is five times as high as Niagara Falls.

Kaifeng /ˌkaɪˈfʌŋ/ former capital of China, 907–1127, and of Honan province; population (1984) 619,200. It has lost its importance because of the silting-up of the nearby Huang He River.

Kaikouras /kaɪˈkuərəz/ double range of mountains in the NE of South Island, New Zealand, separated by the Clarence River, and reaching 2,885 m/9,465 ft.

Kaingaroa /ˌkaɪŋəˈrəʊə/ forest NE of Lake Taupo in North Island, New Zealand, one of the world's largest planted forests.

Kairouan /ˌkaɪəˈwɑːn/ Muslim holy city in Tunisia, N Africa, S of Tunis; population (1984) 72,200. It is a centre of carpet production. The city, said to have been founded AD 617, ranks after Mecca and Medina as a place of pilgrimage.

Kaiserslautern /ˌkaɪzəzˈlaʊtən/ industrial town (textiles, cars) in Germany, in the Rhineland-Palatinate, 48 km/30 mi W of Mannheim; population (1983) 98,700. It dates from 882; the castle from which it takes its name was built by Frederick Barbarossa 1152 and destroyed by the French 1703.

Kakadu /ˌkækəˈduː/ national park E of Darwin in the Alligator Rivers Region of Arnhem Land, Northern Territory, Australia. Established 1979, it overlies one of the richest uranium deposits in the world. As a result of this, the park has become the focal point of controversy between conservationists and mining interests.

Kalahari Desert /ˌkæləˈhɑːri/ semi-desert area forming most of Botswana and extending into Namibia, Zimbabwe, and South Africa; area about 900,000 sq km/347,400 sq mi. The only permanent river, the Okavango, flows into a delta in the NW forming marshes rich in wildlife. Its inhabitants are the nomadic Kung.

Kalamazoo /ˌkæləməˈzuː/ (American Indian 'boiling pot') city in SW Michigan, on the Kalamazoo River, SW of Lansing; seat of Kalamazoo County; population (1990) 80,300. Its industries include the processing of the area's agricultural products, car and transportation machinery parts, chemicals, and metal and paper products.

Kalgan /ˌkɑːlˈgɑːn/ city in NE China, now known as ◊ Zhangjiakou.

Kalgoorlie /kælˈguəli/ town in Western Australia, 545 km/340 mi NE of Perth, amalgamated with Boulder 1966; population (1989 est) 26,800. Gold has been mined here since 1893.

Kalimantan /ˌkælɪ'mæntən/ province of the republic of Indonesia occupying part of the island of Borneo
area 543,900 sq km/210,000 sq mi
towns Banjermasin and Balikpapan
physical mostly low-lying, with mountains in the N
products petroleum, rubber, coffee, copra, pepper, timber
population (1989 est) 8,677,500.

Kalinin /kə'liːnɪn/ former name (1932–91) of ◊Tver, a city in Russia.

Kaliningrad /kə'liːnɪngræd/ formerly *Königsberg* Baltic naval base in W Russia; population (1987) 394,000. Industries include engineering and paper. It was the capital of East Prussia until the latter was divided between the USSR and Poland 1945 under the Potsdam Agreement, when it was renamed in honour of President Kalinin.

Kalmar /'kælmɑː/ port on the southeast coast of Sweden; population (1990) 56,200. Industries include paper, matches, and the Orrefors glassworks.

Kalmyk /'kælmək/ or *Kalmuck* autonomous republic in central Russia, on the Caspian Sea; area 75,900 sq km/29,300 sq mi; population (1989) 322,000; capital Elista. Industry is mainly agricultural. It was settled by migrants from China in the 17th century. The autonomous Soviet republic was abolished 1943–57 because of alleged collaboration of the people with the Germans during the siege of Stalingrad, but restored 1958.

Kaluga /kə'luːgə/ town in central Russia, on the river Oka, 160 km/100 mi SW of Moscow, capital of Kaluga region; population (1987) 307,000. Industries include hydroelectric installations and engineering works, telephone equipment, chemicals, and measuring devices.

Kamakura /ˌkæmə'kuərə/ city on Honshu Island, Japan, near Tokyo; population (1990) 174,300. It was the seat of the first shogunate 1192–1333, which established the rule of the samurai class, and the Hachimangu Shrine is dedicated to the gods of war; the 13th-century statue of Buddha (Daibutsu) is 13 m/43 ft high. From the 19th century, artists and writers (for example, the novelist Kawabata) settled here.

Kamara'n /ˌkæmə'rɑːn/ island in the Red Sea, formerly belonging to South Yemen, and occupied by North Yemen 1972. It was included in the territory of the Yemen Republic formed 1990; area 180 sq km/70 sq mi.

Kamchatka /kæm'tʃætkə/ mountainous peninsula separating the Bering Sea and Sea of Okhotsk, forming (together with the Chukchi and Koryak national districts) a region of E Siberian Russia. Its capital, Petropavlovsk, is the only town; agriculture is possible only in the S. Most of the inhabitants are fishers and hunters.

Kamet /'kʌmeɪt/ Himalayan mountain 7,756 m/ 25,447 ft high on the Tibet–India border. The

Britons Francis Smythe (1900–1949) and Eric Shipton (1907–1977) were in the group that made the first ascent 1931.

Kampala /kæm'pɑːlə/ capital of Uganda, on Lake Victoria; population (1983) 455,000. It is linked by rail with Mombasa. Products include tea, coffee, textiles, fruit, and vegetables.
Built on six hills at an altitude of about 1,220 m/ 4,000 ft, it was the capital of the kingdom of Buganda in the 19th century, became the headquarters of the Imperial British East Africa Company 1890, and was made capital of Uganda 1962.

Kampuchea /ˌkæmpu'tʃiːə/ former name (1975–89) of ◊Cambodia.

Kananga /kə'næŋgə/ chief city of Kasai Occidental region, W central Zaire, on the Lulua River; population (1984) 291,000. It was known as *Luluabourg* until 1966.

Kanazawa /ˌkænə'zɑːwə/ industrial city (textiles and porcelain) on Honshu Island, in Chubu region, Japan, 160 km/100 mi NNW of Nagoya; population (1990) 442,900. Kanazawa was a feudal castle town from the 16th century and has a number of old samurai residences.

Kanchenjunga /ˌkæntʃən'dʒʊŋgə/ variant spelling of ◊Kangchenjunga, a Himalayan mountain.

Kandahar /ˌkændə'hɑː/ city in Afghanistan, 450 km/ 280 mi SW of Kabul, capital of Kandahar province and a trading centre, with wool and cotton factories; population (1984) 203,200. It is surrounded by a mud wall 8 m/25 ft high. When Afghanistan became independent 1747, Kandahar was its first capital.

Kandy /'kændi/ city in central Sri Lanka, on the Mahaweli River; capital of a district of the same name; population (1990) 104,000. Products include tea. One of the most sacred Buddhist shrines, the Dalada Maligawa, is situated in Kandy; it contains an alleged tooth of the Buddha.
The city was capital of the kingdom of Kandy 1480–1815, when it was captured by the British. The chief campus of the University of Sri Lanka (1942) is at Peradenia, 5 km/3 mi away.

Kangchenjunga /ˌkæntʃən'dʒʊŋgə/ Himalayan mountain on the Nepal–Sikkim border, 8,586 m/ 28,170 ft high, 120 km/75 mi SE of Mount Everest. The name means 'five treasure houses of the great snows'. Kangchenjunga was first climbed by a British expedition 1955.

Ka Ngwane /kæŋ'gwɑːneɪ/ black homeland in Natal province, South Africa; population (1985) 392,800. It achieved self-governing status 1971.

Kankakee /ˌkæŋkə'kiː/ city in NE Illinois, on the Kankakee River, S of Chicago; seat of Kankakee County; population (1990) 27,600. It is a distribution centre for corn. Industries also include building materials, furniture, pharmaceuticals, and farm machinery.

Kano /'kɑːnəʊ/ capital of Kano state in N Nigeria, trade centre of an irrigated area; population

(1983) 487,100. Products include bicycles, glass, furniture, textiles, and chemicals. Founded about 1000 BC, Kano is a walled city, with New Kano extending beyond the walls.

Kanpur /ˌkɑːnˈpuə/ formerly **Cawnpore** capital of Kanpur district, Uttar Pradesh, India, SW of Lucknow, on the river Ganges; a commercial and industrial centre (cotton, wool, jute, chemicals, plastics, iron, steel); population (1981) 1,688,000.

Kansas /ˈkænzəs/ state in central USA; nickname Sunflower State
area 213,200 sq km/82,296 sq mi
capital Topeka
towns Kansas City, Wichita, Overland Park
features Dodge City, once 'cowboy capital of the world'; Eisenhower Center, Abilene; Fort Larned and Fort Scott; Pony Express station, Hanover; Wichita Cowtown, a frontier-era reproduction
products wheat, cattle, coal, petroleum, natural gas, aircraft, minerals
population (1990) 2,477,600
famous people Amelia Earhart; Dwight D Eisenhower; William Inge; Buster Keaton; Carry Nation; Charlie Parker
history explored by Francisco de Coronado for Spain 1541 and La Salle for France 1682; ceded to the USA 1803 as part of the Louisiana Purchase.

A kind of gravity point for American democracy.

On **Kansas** John Gunther *Inside USA* 1947

Kansas City /ˈkænzəs/ twin city in the USA at the confluence of the Missouri and Kansas rivers, partly in Kansas and partly in Missouri; population (1990) of Kansas City (Kansas) 149,800, Kansas City (Missouri) 435,100. It is a market and agricultural distribution centre and one of the chief livestock centres of the USA. Kansas City, Missouri, has car-assembly plants and Kansas City, Kansas, has the majority of offices.
history The city was founded as a trading post by French fur trappers about 1826. In the 1920s and 1930s Kansas City was run by boss Tom Pendergast, of the Ready-Mix Concrete Company, and in the nightclubs on Twelfth Street

under his 'protection' jazz musicians such as Lester Young, Count Basie, and Charlie Parker performed.

Kansu /ˌkænˈsuː/ alternative spelling for the Chinese province ⟡Gansu.

Kanto /ˈkæntəu/ flat, densely populated region of E Honshu Island, Japan; area 32,377 sq km/ 12,505 sq mi; population (1988) 37,867,000. The chief city is Tokyo.

Kanton and Enderbury /ˈkæntən, ˈendəbəri/ two atolls in the Phoenix group, which forms part of the Republic of Kiribati. They were a UK–US condominium (joint rule) 1939–80. There are US aviation, radar, and tracking stations here.

Kaohsiung /ˌkau ʃiˈuŋ/ city and port on the west coast of Taiwan; population (1990) 1,393,200. Industries include aluminium ware, fertilizers, cement, oil refineries, iron and steel works, shipyards, and food processing. Kaohsiung began to develop as a commercial port after 1858; its industrial development came about while it was occupied by Japan 1895–1945.

Kara Bogaz Gol /kəˈrɑː bəˈgæz ˈgɒl/ shallow gulf of the Caspian Sea, Turkmenistan; area 20,000 sq km/8,000 sq mi. Rich deposits of sodium chloride, sulphates, and other salts have formed by evaporation.

Karachi /kəˈrɑːtʃi/ largest city and chief seaport of Pakistan, and capital of Sind province, NW of the Indus delta; population (1981) 5,208,000. Industries include engineering, chemicals, plastics, and textiles. It was the capital of Pakistan 1947–59.

Karafuto /ˌkɑːrəˈfuːtəu/ Japanese name for ⟡Sakhalin Island.

Karaganda /ˌkærəgənˈdɑː/ industrial town (coal, copper, tungsten, manganese) in Kazakhstan, linked by canal with the Irtysh River; capital of Karaganda region; population (1989) 614,000.

Karaikal /ˌkærɪˈkɑːl/ small port in India, 250 km/ 155 mi S of Madras, at the mouth of the right branch of the Cauvery delta. On a tract of land acquired by the French 1739, it was transferred to India 1954, confirmed by treaty 1956. See also ⟡Pondicherry.

Kara-Kalpak /kəˈrɑː kælˈpɑːk/ autonomous republic of Uzbekistan
area 158,000 sq km/61,000 sq mi
capital Nukus
towns Munyak
products cotton, rice, wheat, fish
population (1989) 1,214,000
history named after the Kara-Kalpak ('black hood') people who live S of the Sea of Aral and were conquered by Russia 1867. An autonomous Kara-Kalpak region was formed 1926 within Kazakhstan, transferred to the Soviet republic 1930, made a republic 1932, and attached to Uzbekistan 1936.

Karakoram /ˌkærəˈkɔːrəm/ mountain range in central Asia, divided among China, Pakistan,

Kansas

Topeka

and India. Peaks include K2, Masharbrum, Gasharbrum, and Mustagh Tower. **Ladakh** subsidiary range is in NE Kashmir on the Tibetan border.

Karakoram Highway /ˌkærəˈkɔːrəm/ road constructed by China and Pakistan and completed 1978; it runs 800 km/500 mi from Havelian (NW of Rawalpindi), via Gilgit in Kashmir and the Khunjerab Pass (4,800 m/16,000 ft) to Kashi in China.

Karakorum /ˌkærəˈkɔːrəm/ ruined capital of Mongol ruler Genghis Khan, SW of Ulaanbaatar in Mongolia.

Kara-Kum /kəˈrɑː ˈkuːm/ sandy desert occupying most of Turkmenistan; area about 310,800 sq km/120,000 sq mi. It is crossed by the Caspian railway.

Kara Sea /ˈkɑːrə/ (Russian **Kavaskoye More**) part of the Arctic Ocean off the north coast of Russia, bounded to the NW by the island of Novaya Zemlya and to the NE by Severnaya Zemlya. Novy Port on the Gulf of Ob is the chief port, and the Yenisei River also flows into it.

Karbala /ˈkɑːbələ/ alternative spelling for ◊Kerbela, a holy city in Iraq.

Karelia /kəˈriːliə/ autonomous republic of NW Russia
area 172,400 sq km/66,550 sq mi
capital Petrozavodsk
towns Vyborg
physical mainly forested
features Lake Ladoga
products fishing, timber, chemicals, coal
population (1989) 792,000
history Karelia was annexed to Russia by Peter the Great 1721 as part of the grand duchy of Finland. In 1917 part of Karelia was retained by Finland when it gained its independence from Russia. The remainder became an autonomous region 1920 and an autonomous republic 1923 of the USSR. Following the wars of 1939-40 and 1941-44, Finland ceded 46,000 sq km/18,000 sq mi of Karelia to the USSR. Part of this territory was incorporated in the Russian Soviet Republic and part in the Karelian autonomous republic. A movement for the reunification of Russian and Finnish Karelia emerged in the late 1980s.

Karelian Isthmus /kəˈriːliən ˈɪsməs/ strip of land between Lake Ladoga and the Gulf of Finland, Russia, with St Petersburg at the southern extremity and Vyborg at the northern. Finland ceded it to the USSR 1940–41 and from 1947.

Kargopol /ˈkɑːɡəpɒl/ town on the left bank of the Onega River, NW Russia; population (1980) 1,200. It developed with the salt trade from about the 14th century and has several outstanding 17th-century churches, notably the Church of St Vladimir (1653) and the Blagoveshchenie Church (1682–92). The town gives its name to the hunting and fishing Kargopol culture, which flourished in the area during the late Mesolithic and Neolithic periods.

Kariba Dam /kəˈriːbə/ concrete dam on the

Zambezi River, on the Zambia–Zimbabwe border, about 386 km/240 mi downstream from the Victoria Falls, constructed 1955–60 to supply power to both countries.

The dam crosses Kariba Gorge, and the reservoir, Lake Kariba, has important fisheries.

Karl-Marx-Stadt /ˈkɑːl ˈmɑːks ʃtæt/ former name (1953–90) of ◊Chemnitz, a city in Germany.

Karlovy Vary /ˈkɑːləvi ˈvɑːri/ (German **Karlsbad**) spa in the Bohemian Forest, Czech Republic, celebrated from the 14th century for its alkaline thermal springs; population (1991) 56,300.

Karlsbad /ˈkɑːlzbæd/ German name of ◊Karlovy Vary, a town in the Czech Republic.

Karlsruhe /ˈkɑːlzruːə/ industrial town (nuclear research, oil refining) in Baden-Württemberg, Germany; population (1988) 263,100.

Karnak /ˈkɑːnæk/ village of modern Egypt, on the east bank of the river Nile, that gives its name to the temple of Ammon (constructed by Seti I and Ramses I) around which the major part of the ancient city of Thebes was built. An avenue of rams leads to Luxor.

Karnataka /kəˈnɑːtəkə/ formerly (until 1973) **Mysore** state in SW India
area 191,800 sq km/74,035 sq mi
capital Bangalore
products mainly agricultural; minerals include manganese, chromite, and India's only sources of gold and silver
population (1991) 44,817,400
language Kannada
famous people Hyder Ali, Tippu Sultan.

Kärnten /ˈkeəntən/ German name for ◊Carinthia, a province of Austria.

Karroo /kəˈruː/ two areas of semi-desert in Cape Province, South Africa, divided into the **Great Karroo** and **Little Karroo** by the Swartberg Mountains. The two Karroos together have an area of about 260,000 sq km/100,000 sq mi.

Karnataka

Bangalore

INDIAN OCEAN

Kasai /kɑːˈsaɪ/ river that rises in Angola and forms the frontier with Zaire before entering Zaire and joining the Zaïre River, of which it is the chief tributary. It is 2,100 km/1,300 mi long and is rich in alluvial diamonds.

Kashgar /ˌkæʃˈgɑː/ former name of Kashi, a town in China.

Kashi /ˌkɑːˈʃiː/ formerly **Kashgar** oasis town in Xinjiang Uyghur autonomous region, China, on the river Kaxgar He. It is the capital of Kashi district, which adjoins the Kirghiz and Tadzic republics, Afghanistan, and Jammu and Kashmir; population (1973) 180,000. It is a trading centre, the Chinese terminus of the Karakoram Highway, and a focus of Muslim culture.

Kashmir /ˌkæʃˈmɪə/ Pakistan-occupied area, 30,445 sq mi/78,900 sq km, in the NW of the former state of Kashmir, now Jammu and Kashmir. Azad ('free') Kashmir in the W has its own legislative assembly based in Muzaffarabad while Gilgit and Baltistan regions to the N and E are governed directly by Pakistan. The Northern Areas are claimed by India and Pakistan

towns Gilgit, Skardu
features W Himalayan peak Nanga Parbat 8,182 m/26,660 ft, Karakoram Pass, Indus River, Baltoro Glacier
population 1,500,000.

Kassel /ˈkæsəl/ industrial town (engineering, chemicals, electronics) in Hessen, Germany, on the river Fulda; population (1988) 185,000. There is the spectacular Wilhelmshöhe mountain park, and the Grimm Museum commemorates the compilers of fairy tales who lived here.

Katanga /kəˈtæŋɡə/ former name of the ⟩Shaba region in Zaire.

Kathiawar /ˌkætiəˈwɑː/ peninsula on the west coast of India. Formerly occupied by a number of princely states, all Kathiawar (60,723 sq km/ 23,445 sq mi) had been included in Bombay state by 1956 but was transferred to Gujarat 1960. Mahatma Gandhi was born in Kathiawar at Porbandar.

Katmai /ˈkætmaɪ/ active volcano in Alaska, USA, 2,046 m/6,715 ft. Its major eruption 1912 created the 'Valley of Ten Thousand Smokes'. Katmai National Park, area 6,922 sq mi/17,928 sq km, was designated 1980.

Katmandu /ˌkætmənˈduː/ or **Kathmandu** capital of Nepal; population (1981) 235,000. Founded in the 8th century on an ancient pilgrim and trade route from India to Tibet and China, it has a royal palace, Buddhist temples, and monasteries.

Katowice /ˌkætəʊˈviːtseɪ/ industrial city (anthracite, iron and coal mining, iron foundries, smelting works, machine shops) in Upper Silesia, S Poland; population (1990) 366,800.

Kattegat /ˈkætɪɡæt/ sea passage between Denmark and Sweden. It is about 240 km/150 mi long and 135 km/85 mi wide at its broadest point.

Kaunas /ˈkaunəs/ (formerly until 1917 **Kovno**) industrial river port (textiles, chemicals, agricul-

tural machinery) in Lithuania, on the Niemen River; population (1987) 417,000. It was the capital of Lithuania 1910–40.

Kawasaki /ˌkauəˈsɑːki/ industrial city (iron, steel, shipbuilding, chemicals, textiles) on Honshu Island, Japan; population (1990) 1,173,600.

Kayah State /ˈkaɪə/ division of Myanmar (formerly Burma), area 11,900 sq km/4,600 sq mi, formed 1954 from the Karenni states (Kantarrawaddy, Bawlake, and Kyebogyi) and inhabited mainly by the Karen people. Kayah State has a measure of autonomy.

Kayseri /ˈkaɪsəri/ (ancient name **Caesarea Mazaca**) capital of Kayseri province, central Turkey; population (1990) 421,400. It produces textiles, carpets, and tiles. In Roman times it was capital of the province of Cappadocia.

Kazakhstan Republic of
area 2,717,300 sq km/1,049,150 sq mi
capital Alma-Ata
towns Karaganda, Semipalatinsk, Petropavlovsk
physical Caspian and Aral seas, Lake Balkhash; Steppe region
features Baikonur Cosmodrome (space launch site at Tyuratam, near Baikonur)
head of state Nursultan Nazarbayev from 1990
head of government Sergey Tereshchenko from 1991
political system emergent democracy
political party Independent Socialist Party of Kazakhstan (SPK)
products grain, copper, lead, zinc, manganese, coal, oil
population (1990) 16,700,000 (40% Kazakh, 38% Russian, 6% Germans, 5% Ukrainians)
language Russian; Kazakh, related to Turkish
religion Sunni Muslim
chronology
1920 Autonomous republic in USSR.
1936 Joined the USSR and became a full union republic.
1950s Site of Nikita Khrushchev's ambitious 'Virgin Lands' agricultural extension programme.

1960s A large influx of Russian settlers turned the Kazakhs into a minority in their own republic.
1986 Riots in Alma-Alta after Gorbachev ousted local communist leader.
1989 June: Nazarbayev became leader of the Kazakh Communist Party (KCP) and instituted economic and cultural reform programmes.
1990 Feb: Nazarbayev became head of state.
1991 March: support pledged for continued union with USSR; Aug: Nazarbayev condemned attempted anti-Gorbachev coup; CP abolished and replaced by Independent Socialist Party of Kazakhstan (SPK). Dec: joined new Commonwealth of Independent States (CIS); independence recognized by USA.
1992 Jan: admitted into Conference on Security and Cooperation in Europe (CSCE). March: became a member of the United Nations (UN). May: trade agreement with USA.

Kazan /kə'zæn/ capital of Tatarstan, central Russia, on the river Volga; population (1989) 1,094,000. It is a transport, commercial, and industrial centre (engineering, oil refining, petrochemicals, textiles, large fur trade). Formerly the capital of a Tatar khanate, Kazan was captured by Ivan IV 'the Terrible' 1552.
   The 'Black Virgin of Kazan', an icon so called because blackened with age, was removed to Moscow (1612–1917), where the great Kazan Cathedral was built to house it 1631; it is now in the USA. Among miracles attributed to its presence were the defeat of Poland 1612 and of Napoleon at Moscow 1812.

Kebnekaise /'kebnəkaısə/ highest peak in Sweden, rising to 2,111 m/6,926 ft in the Kolen range, W of the town of Kiruna.

Kecskemét /'ketʃkımeıt/ town in Hungary, situated on the Hungarian plain SE of Budapest; population (1989) 106,000. It is a trading centre of an agricultural region.

Kedah /'kedə/ state in NW Peninsular Malaysia; capital Alor Setar; area 9,400 sq km/3,628 sq mi; population (1990) 1,412,800. Products include rice, rubber, tapioca, tin, and tungsten. Kedah was transferred by Thailand to Britain 1909, and was one of the Unfederated Malay States until 1948.

Keeling Islands /'ki:lıŋ/ another name for the ◊Cocos Islands, an Australian territory.

Keelung /,ki:'luŋ/ or **Chi-lung** industrial port (shipbuilding, chemicals, fertilizer) on the north coast of Taiwan, 24 km/15 mi NE of Taipei; population (1990) 352,900.

Keewatin /ki:'weıtın/ eastern district of Northwest Territories, Canada, including the islands in Hudson and James bays
area 590,935 sq km/228,160 sq mi
towns (trading posts) Chesterfield Inlet, Eskimo Point, and Coral Harbour (site of an air base set up during World War II)
physical upland plateau in the N, the S low and level, covering the greater part of the Arctic prairies of Canada; numerous lakes

products furs (trapping is main occupation)
history Keewatin District formed 1876, under the administration of Manitoba; it was transferred to Northwest Territories 1905, and in 1912 lost land S of 60° N to Manitoba and Ontario.

Kefallinia /,kefəlı'ni:ə/ (English **Cephalonia**) largest of the Ionian Islands off the west coast of Greece; area 935 sq km/360 sq mi; population (1981) 31,300. It was devastated by an earthquake 1953 that destroyed the capital Argostolion.

Keflavik /'kepləvık/ fishing port in Iceland, 35 km/22 mi SW of Reykjavik; population (1986) 7,500. Its international airport was built during World War II by US forces (who called it Meeks Field). Keflavik became a NATO base 1951.

Keighley /'ki:θli/ industrial town (wool, engineering) on the river Aire, NW of Bradford in West Yorkshire, England; population (1981) 57,800.
   Haworth, home of the Brontë family of writers, is now part of Keighley.

Kelantan /ke'læntən/ state in NE Peninsular Malaysia; capital Kota Baharu; area 14,900 sq km/5,751 sq mi; population (1990) 1,220,100. It produces rice, rubber, copra, tin, manganese, and gold. Kelantan was transferred by Siam to Britain 1909 and until 1948 was one of the Unfederated Malay States.

Kemerovo /'kemırəuvəu/ coal-mining town in W Siberian Russia, centre of Kuznetsk coal basin; population (1987) 520,000. It has chemical and metallurgical industries. The town, which was formed out of the villages of Kemerovo and Shcheglovisk, was known as Shcheglovisk 1918–32.

Kendal /'kendl/ town in Cumbria, England, on the river Kent; population (1981) 23,500. It is an industrial centre (light industry; agricultural machinery and, since the 14th century, wool) and tourist centre for visitors to the Lake District.

Kenilworth /'kenlwɜ:θ/ town in Warwickshire, England; population (1986 est) 21,400. The Norman castle, celebrated in Walter Scott's novel *Kenilworth*, became a royal residence (Edward II relinquished his crown here 1327). It was enlarged by John of Gaunt and later by the Earl of Leicester, who entertained Elizabeth I here 1575, but was dismantled after the Civil War.
   The ruins were given to the British nation by the 1st Lord Kenilworth 1937.

Kennewick /'kenəwık/ city in SE Washington, USA, on the Columbia River, SE of Seattle; population (1990) 42,200. Dams built on the Columbia and Snake rivers provide irrigation for the area's grape, sugar beet, alfalfa, and corn crops.

Kenosha /kə'nəuʃə/ (Indian 'pike' or 'pickerel') city in the southeast corner of Wisconsin, USA, on Lake Michigan, SE of Milwaukee, seat of Kenosha County; population (1980) 77,685. Its industries include food-processing equipment, fertilizers, motor vehicles, textiles and clothing, and food products.

Kensington and Chelsea /'kenzɪŋtən, 'tʃelsi/ inner borough of Greater London, England, N of the river Thames
features Kensington Gardens; museums—Victoria and Albert, Natural History, Science; Imperial College of Science and Technology 1907; Commonwealth Institute; Kensington Palace; Holland House (damaged in World War II, and partly rebuilt as a youth hostel); Leighton House. The annual Notting Hill Carnival, held each August from 1966, is the largest street carnival in Europe
population (1991) 127,600.

Kent /kent/ county in SE England, nicknamed the 'garden of England'
area 3,730 sq km/1,440 sq mi
towns Maidstone (administrative headquarters), Canterbury, Chatham, Rochester, Sheerness, Tunbridge Wells; resorts: Folkestone, Margate, Ramsgate
features traditionally, a 'man of Kent' comes from E of the Medway and a 'Kentish man' from W Kent; New Ash Green, a new town; Romney Marsh; the Isles of Grain, Sheppey (on which is the resort of Sheerness, formerly a royal dockyard) and Thanet; Weald (agricultural area); rivers: Darent, Medway, Stour; Leeds Castle (converted to a palace by Henry VIII); Hever Castle (where Henry VIII courted Anne Boleyn); Chartwell (Churchill's country home), Knole, Sissinghurst Castle and gardens; the Brogdale Experimental Horticulture Station at Faversham has the world's finest collection of apple and other fruit trees; the former RAF Manston became Kent International Airport 1989
products hops, apples, soft fruit, coal, cement, paper
population (1991) 1,485,600

Kent

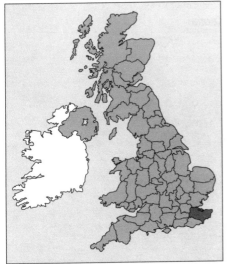

Kentucky

famous people Edward Heath, Christopher Marlowe.

Kent, Sir, —everybody knows Kent—Apples, cherries, hops and women.

On **Kent** Charles Dickens *Pickwick Papers* 1837 (Jingle)

Kentucky /ken'tʌki/ state in S central USA; nickname Bluegrass State
area 104,700 sq km/40,414 sq mi
capital Frankfort
towns Louisville, Lexington, Owensboro, Covington, Bowling Green
features bluegrass country; horse racing at Louisville (Kentucky Derby); Mammoth Cave National Park (main cave 6.5 km/4 mi long, up to 38 m/125 ft high, where Indian councils were once held); Abraham Lincoln's birthplace at Hodgenville; Fort Knox, US gold-bullion depository
products tobacco, cereals, textiles, coal, whisky, horses, transport vehicles
population (1990) 3,365,300
famous people Muhammad Ali, Daniel Boone, Louis D Brandeis, Kit Carson, Henry Clay, D W Griffith, Thomas Hunt Morgan, Harland 'Colonel' Sanders, Robert Penn Warren
history Kentucky was the first region W of the Alleghenies settled by American pioneers. James Harrod founded Harrodsburg 1774; in 1775 Daniel Boone, who blazed his Wilderness Trail 1767, founded Boonesboro. Originally part of Virginia, Kentucky became a state 1792. Badly divided over the slavery question, the state was racked by guerrilla warfare and partisan feuds during the Civil War.

The sun shines bright in the old Kentucky home.

On **Kentucky** Stephen Foster 'My Old Kentucky Home' 1852

Kenya Republic of (*Jamhuri ya Kenya*)
area 582,600 sq km/224,884 sq mi
capital Nairobi
towns Kisumu, port Mombasa

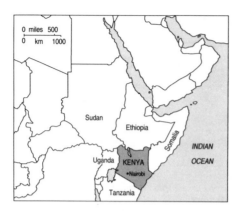

physical mountains and highlands in W and centre; coastal plain in S; arid interior and tropical coast
environment the elephant faces extinction as a result of poaching
features Great Rift Valley, Mount Kenya, Lake Nakuru (salt lake with world's largest colony of flamingos), Lake Turkana (Rudolf), national parks with wildlife, Malindini Marine Reserve, Olduvai Gorge
head of state and government Daniel arap Moi from 1978
political system authoritarian nationalism
political parties Kenya African National Union (KANU), nationalist, centrist; National Democratic Party (NDP), centrist (launched 1991, not accepted by government)
exports coffee, tea, pineapples, petroleum products
currency Kenya shilling
population (1990 est) 25,393,000 (Kikuyu 21%, Luo 13%, Luhya 14%, Kelenjin 11%; Asian, Arab, European); growth rate 4.2% p.a.
life expectancy men 59, women 63 (1989)
languages Kiswahili (official), English; there are many local dialects
religions Protestant 38%, Roman Catholic 28%, indigenous beliefs 26%, Muslim 6%
literacy 50% (1988)
GDP $6.9 bn (1987); $302 per head (1988)
chronology
1895 British East African protectorate established.
1920 Kenya became a British colony.
1944 African participation in politics began.
1950 Mau campaign began.
1953 Nationalist leader Jomo Kenyatta imprisoned by British authorities.
1956 Mau Mau campaign defeated, Kenyatta released.
1963 Achieved internal self-government, with Kenyatta as prime minister.
1964 Independence achieved from Britain as a republic within the Commonwealth, with Kenyatta as president.
1967 East African Community (EAC) formed with Tanzania and Uganda.
1977 Collapse of EAC.
1978 Death of Kenyatta. Succeeded by Daniel arap Moi.

1982 Attempted coup against Moi foiled.
1983 Moi re-elected unopposed.
1984 Over 2,000 people massacred by government forces at Wajir.
1985–86 Thousands of forest villagers evicted and their homes destroyed to make way for cash crops.
1988 Moi re-elected. 150,000 evicted from state-owned forests.
1989 Moi announced release of all known political prisoners. Confiscated ivory burned in attempt to stop elephant poaching.
1990 Despite antigovernment riots, Moi refused multiparty politics.
1991 Increasing demands for political reform; Moi promised multiparty politics.
1992 Constitutional amendment passed.

Kenya, Mount /ˈkenjə/ or *Kirinyaga* extinct volcano from which Kenya takes its name, 5,199 m/ 17,057 ft; the first European to climb it was Halford Mackinder 1899.

Kerala /ˈkerələ/ state of SW India, formed 1956 from the former princely states of Travancore and Cochin
area 38,900 sq km/15,015 sq mi
capital Trivandrum
features most densely populated, and most literate (60%), state of India; strong religious and caste divisions make it politically unstable
products tea, coffee, rice, oilseed, rubber, textiles, chemicals, electrical goods
population (1991) 29,011,200
languages Kannada, Malayalam, Tamil.

Kerbela /ˈkɜːbələ/ or *Karbala* holy city of the Shi'ite Muslims, 96 km/60 mi SW of Baghdad, Iraq; population (1985) 184,600. Kerbela is built on the site of the battlefield where Husein, son of Ali and Fatima, was killed 680 while defending his succession to the khalifate; his tomb in the city is visited every year by many pilgrims.

Kerch /keətʃ/ port in the Crimea, Ukraine, at the eastern end of Kerch peninsula, an iron-producing area; population (1987) 173,000. Built on the site of an ancient Greek settlement, Kerch belonged to Russia from 1783.

Kerala

Kerguelen Islands /'kɜːgəlɪn/ or *Desolation Islands* volcanic archipelago in the Indian Ocean, part of the French Southern and Antarctic Territories; area 7,215 km/2,787 sq mi.

It was discovered 1772 by the Breton navigator Yves de Kerguelen and annexed by France 1949. Uninhabited except for scientists (centre for joint study of geomagnetism with Russia), the islands support a unique wild cabbage containing a pungent oil.

Kérkyra /'keəkɪrə/ Greek form of ⍉Corfu, an island in the Ionian Sea.

Kermadec Islands /kɜːmədek, kə'mædek/ volcanic group, a dependency of New Zealand since 1887; area 30 sq km/12 sq mi. They are uninhabited except for a meteorological station on the largest island, Raoul.

Kerman /kə'mɑːn/ town in Kerman province SE Iran; population (1986) 254,800. It is a centre for the mining of copper and precious metals.

Kermanshah /,kɜːmæn'ʃɑː/ former name (until 1980) of the town of ⍉Bakhtaran in NW Iran.

Kernow /'kɜːnəu/ Celtic name for ⍉Cornwall, an English county.

Kerry /'keri/ county of Munster province, Republic of Ireland, E of Cork
area 4,700 sq km/1,814 sq mi
county town Tralee
physical western coastline deeply indented; northern part low-lying, but in the S are the highest mountains in Ireland, including Carrantuohill 1,041 m/3,417 ft, the highest peak in Ireland; many rivers and lakes
features Macgillycuddy's Reeks, Lakes of Killarney
products engineering, woollens, shoes, cutlery; tourism is important
population (1991) 121,700.

Kesteven, Parts of /kes'tiːvən/ area of SW Lincolnshire, England, formerly an administrative unit with county offices at Sleaford 1888–1974.

Key West /'kiː 'west/ town at the tip of the Florida peninsula, USA; population (1980) 24,400. As a tourist resort, it was popularized by the novelist Ernest Hemingway.

Khabarovsk /,kæbə'rɒfsk, Russian xə'bærəfsk/ industrial city (oil refining, saw milling, meat packing) in SE Siberian Russia; population (1987) 591,000.

Khabarovsk /,kæbə'rɒfsk/ territory of SE Siberian Russia, bordering the Sea of Okhotsk and drained by the Amur River; area 824,600 sq km/ 318,501 sq mi; population (1985) 1,728,000. The capital is Khabarovsk. Mineral resources include gold, coal, and iron ore.

Khajurāho /,kædʒu'rɑːhəu/ town in Madhya Pradesh, central India, the former capital of the Candella monarchs. It has 35 sandstone temples —Jain, Buddhist, and Hindu—built in the 10th and 11th centuries. The temples are covered inside and out with erotic sculpture symbolizing mystic union with the deity.

Khardungla Pass /'kɑːlsə/ road linking the Indian town of Leh with the high-altitude military outpost on the Siachen Glacier at an altitude of 5,662 m/1,744 ft in the Karakoram range, Kashmir. It is thought to be the highest road in the world.

Kharga /'kɑːgə/ or *Kharijah* oasis in the Western Desert of Egypt, known to the Romans, and from 1960 headquarters of the New Valley irrigation project. An area twice the size of Italy is watered from natural underground reservoirs.

Kharg Island /,kɑː'duŋ 'lɑː/ small island in the Persian Gulf used by Iran as a deep-water oil terminal. Between 1982 and 1988 Kharg Island came under frequent attack during the Iran–Iraq War.

Kharkov /'kɑːkɒf/ capital of the Kharkov region, E Ukraine, 400 km/250 mi E of Kiev; population (1987) 1,587,000. It is a railway junction and industrial city (engineering, tractors), close to the Donets Basin coalfield and Krivoy Rog iron mines. Kharkov was founded 1654 as a fortress town.

Khartoum /kɑː'tuːm/ capital and trading centre of Sudan, at the junction of the Blue and White Nile; population (1983) 476,000, and of Khartoum North, across the Blue Nile, 341,000. Omdurman is also a suburb of Khartoum, giving the urban area a population of over 1.3 million.

It was founded 1830 by Mehemet Ali. General Gordon was killed at Khartoum by the Mahdist rebels 1885. A new city was built after the site was recaptured by British troops under Kitchener 1898.

Kherson /kɜː'sɒn/ port in S Ukraine, on the river Dnieper, capital of Kherson region; population (1987) 358,000. Industries include shipbuilding, soap, and tobacco manufacture.

Kherson was founded 1778 by army commander Potemkin as the first Rirst Russian naval base on the Black Sea.

Khmer Republic /kmeə/ former name (1970–76) of ⍉Cambodia.

Khorramshahr /,kɔːrəm'ʃɑː/ former port and oil-refining centre in Iran, on the Shatt-al-Arab River and linked by bridge to the island of Abadan. It was completely destroyed in the 1980s by enemy action in the Iran–Iraq War.

Khulna /'kulnə/ capital of Khulna region, SW Bangladesh, situated close to the Ganges delta; population (1981) 646,000. Industry includes shipbuilding and textiles; it trades in jute, rice, salt, sugar, and oilseed.

Khuzestan /'kuːzɪstɑːn/ province of SW Iran, which includes the chief Iranian oil resources; population (1986) 2,702,533. Towns include

Ahvaz (capital) and the ports of Abadan and Khuninshahr. There have been calls for Sunni Muslim autonomy, under the name Arabistan.

Khyber Pass /'kaɪbə/ pass 53 km/33 mi long through the mountain range that separates Pakistan from Afghanistan. The Khyber Pass was used by invaders of India. The present road was constructed by the British during the Afghan Wars.

Kiangsi alternative spelling of ▷Jiangxi, a province of China.

Kiangsu alternative spelling of ▷Jiangsu, a province of China.

Kidderminster /'kɪdə,mɪnstə/ market town in the West Midlands of England, on the river Stour; population (1988 est) 54,200. It has had a carpet industry from about 1735.

Kiel /kiːl/ Baltic port (fishing, shipbuilding, electronics, engineering) in Germany; capital of Schleswig-Holstein; population (1988) 244,000. Kiel Week in June is a yachting meeting.

Kiel Canal /kiːl/ (formerly *Kaiser Wilhelm Canal*) waterway 98.7 km/61 mi long that connects the Baltic with the North Sea. Built by Germany in the years before World War I, the canal allowed the German navy to move from Baltic bases to the open sea without travelling through international waters.

Kielce /ki'eltseɪ/ city in central Poland, NE of Kraków; population (1990) 214,200. It is an industrial rail junction (chemicals, metals).

Kiev /'kiːef/ capital of Ukraine, industrial centre (chemicals, clothing, leatherwork), on the confluence of the Desna and Dnieper rivers; population (1987) 2,554,000. It was the capital of Russia in the Middle Ages.
features St Sophia cathedral (11th century) and Kiev-Pechersky Monastery (both now museums) survive, and also remains of the Golden Gate, an arched entrance to the old walled city, built 1037, the gate is surmounted by the small Church of the Annunciation. The Kiev ballet and opera are renowned.
history Kiev was founded in the 5th century by Vikings. The Slav domination of Russia began with the rise of Kiev, the 'mother of Russian cities'; Kiev replaced Novgorod as the capital of Russia 882 and was the original centre of the Orthodox Christian faith 988. The city was occupied by Germany 1941, and was the third largest city of the USSR.

Your tongue will lead you to Kiev.

On **Kiev** proverb

Kigali /kɪ'gɑːli/ capital of Rwanda, central Africa, 80 km/50 mi E of Lake Kivu; population (1981) 157,000. Products include coffee, hides, shoes, paints, and varnishes; there is tin mining.

Kigali was under German colonial administration from 1895, and under Belgium 1919–62, when it became capital of independent Rwanda.

Kigoma /kɪ'gəʊmə/ town and port on the eastern shore of Lake Tanganyika, Tanzania, at the western terminal of the railway from Dar es Salaam; population (1978) 50,044. It trades in timber, cotton, and tobacco.

Kildare /kɪl'deə/ county of Leinster province, Republic of Ireland, S of Meath
area 1,690 sq km/652 sq mi
county town Naas
physical wet and boggy in the N
features part of the Bog of Allen; the village of Maynooth, with a training college for Roman Catholic priests; the Curragh, a plain that is the site of the national stud and headquarters of Irish horse racing
products oats, barley, potatoes, cattle
population (1991) 122,516.

Kilimanjaro /,kɪlɪmæn'dʒɑːrəʊ/ volcano in Tanzania, the highest mountain in Africa, 5,900 m/19,364 ft.

Kilkenny /kɪl'keni/ county of Leinster province, Republic of Ireland, E of Tipperary
area 2,060 sq km/795 sq mi
county town Kilkenny
features river Nore
products agricultural, coal
population (1991) 73,600.

Killarney /kɪ'lɑːni/ market town in County Kerry, Republic of Ireland; population (1991) 7,300. A famous beauty spot in Ireland, it has Macgillycuddy's Reeks (a range of mountains) and the Lakes of Killarney to the SW.

Killeen /kɪ'liːn/ city in central Texas, S of Fort Worth and N of Austin; population (1990) 63,500. Although some concrete is produced here, the economy relies heavily on the nearby army base, Fort Hood.

Kilmarnock /kɪl'mɑːnək/ town in Strathclyde region, Scotland, 32 km/20 mi SW of Glasgow; population (1990 est) 48,000. Products include carpets, agricultural machinery, and whisky; Robert Burns's first book of poems was published here 1786.

Kimberley /'kɪmbəli/ diamond-mining town in Cape Province, South Africa, 153 km/95 mi NW of Bloemfontein; population (1980) 144,923. Its mines have been controlled by De Beers Consolidated Mines since 1887.

Kimberley /'kɪmbəli/ diamond site in Western Australia, found 1978–79, estimated to have 5% of the world's known gem-quality stones and 50% of its industrial diamonds.

Kincardineshire /kɪn'kɑːdɪnʃə/ former county of E Scotland, merged 1975 into Grampian Region. The county town was Stonehaven.

King's County older name of ▷Offaly, an Irish county.

King's Lynn /kɪŋz ˈlɪn/ port and market town at the mouth of the Great Ouse River, Norfolk, E England; population (1987 est) 35,500. A thriving port in medieval times, it was called Lynn until its name was changed by Henry VIII.

Kingsport /ˈkɪŋzpɔːt/ city in NE Tennessee, USA, on the Holston River, NE of Knoxville, near the Virginia border; population (1990) 36,400. Products include plastics, chemicals, textiles, paper and printing, and cement. Fort Patrick Henry, built here 1776, protected the Wilderness Road.

Kingston /ˈkɪŋstən/ capital and principal port of Jamaica, West Indies, the cultural and commercial centre of the island; population (1983) 101,000, metropolitan area 525,000. Founded 1693, Kingston became the capital of Jamaica 1872.

Kingston /ˈkɪŋstən/ town in E Ontario, Canada, on Lake Ontario; population (1981) 60,300. Industries include shipbuilding yards, engineering works, and grain elevators. It grew from 1782 around the French Fort Frontenac, was captured by the English 1748, and renamed in honour of George III.

Kingston upon Hull /ˈkɪŋstən əpɒn ˈhʌl/ official name of ◊Hull, a city in NE England.

Kingston upon Thames /ˈkɪŋstən əpɒn ˈtemz/ borough of SW Greater London, England, on the south bank of the river Thames; administrative headquarters of Surrey
features the coronation stone of the Saxon kings is still preserved here
industries metalworking, plastics, paint
population (1991) 130,300.

Kingstown /ˈkɪŋstaun/ former name for ◊Dún Laoghaire, a port near Dublin, Ireland.

Kingstown /ˈkɪŋztaun/ capital and principal port of St Vincent and the Grenadines, West Indies, in the SW of the island of St Vincent; population (1989) 29,400.

King-Te-Chen alternative spelling of ◊Jingdezhen, a town in China.

Kinki /ˈkɪŋki/ region of S Honshu Island, Japan; population (1988) 22,105,000; area 33,070 sq km/ 12,773 sq mi. The chief city is Osaka.

Kinshasa /kɪnˈʃɑːsə/ (formerly *Léopoldville*) capital of Zaire on the river Zaïre, 400 km/ 250 mi inland from the port of Matadi; population (1984) 2,654,000. Industries include chemicals, textiles, engineering, food processing, and furniture. It was founded by the explorer Henry Stanley 1887.

There is no nonsense about aesthetics. Léopoldville puts forth a note of practicality, commerce and good will.

On **Kinshasa** (formerly Léopoldville) John Gunther *Inside Africa* 1955

Kirghizia alternative form of ◊Kyrgyzstan, a country in central Asia.

Kiribati Republic of
area 717 sq km/277 sq mi
capital and port Bairiki (on Tarawa Atoll)
physical comprises 33 Pacific coral islands: the Kiribati (Gilbert), Rawaki (Phoenix), Banaba (Ocean Island), three of the Line Islands including Kiritimati (Christmas Island)
environment the islands are threatened by the possibility of a rise in sea level caused by global warming. A rise of approximately 30 cm/1 ft by the year 2040 will make existing fresh water brackish and undrinkable
features island groups crossed by equator and International Date Line
head of state and government Teatao Teannaki from 1991
political system liberal democracy
political parties National Progressive Party, governing faction; opposition parties: Christian Democratic Party and the Kiribati United Party
exports copra, fish
currency Australian dollar
population (1990 est) 65,600 (Micronesian); growth rate 1.7% p.a.
languages English (official), Gilbertese
religions Roman Catholic 48%, Protestant 45%
literacy 90% (1985)
GDP $26 million (1987); $430 per head (1988)
chronology
1892 Gilbert and Ellice Islands proclaimed a British protectorate.
1937 Phoenix Islands added to colony.
1950s UK tested nuclear weapons on Kiritimati (formerly Christmas Island).
1962 USA tested nuclear weapons on Kiritimati.
1975 Ellice Islands separated to become Tuvalu.
1977 Gilbert Islands granted internal self-government.
1979 Independence achieved from Britain, within the Commonwealth, as the Republic of Kiribati, with Ieremia Tabai as president.
1982 and 1983 Tabai re-elected.
1985 Fishing agreement with Soviet state-owned company negotiated, prompting formation of Kiribati's first political party, the opposition Christian Democrats.
1987 Tabai re-elected.

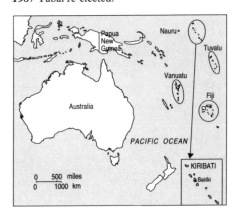

1991 Tabai re-elected but not allowed under constitution to serve further term; Teatao Teannaki elected president.

Kirin /ˌkiːˈrɪn/ alternative name for ▷ Jilin, a Chinese province.

Kiritimati (formerly **Christmas Island**) island in Kiribati, in the central Pacific; one of the Line Islands.

Kirkcaldy /kərˈkɒdɪ/ seaport on the Firth of Forth, Fife region, Scotland; population (1987 est) 49,200. Products include floor coverings and paper. It is the birthplace of the economist Adam Smith and the architect Robert Adam.

Kirkcudbright /kəˈkuːbri/ former county of S Scotland, merged 1975 into Dumfries and Galloway Region. The county town was Kirkcudbright.

Kirkuk /kɜːˈkuk/ town in NE Iraq; population (1985) 208,000. It is the centre of a major oil-field. Formerly it was served by several pipelines providing outlets to Lebanon, Syria, and other countries, but closures caused by the Iran–Iraq War left only the pipeline to Turkey operational.

Kirkwall /ˈkɜːkwɔːl/ administrative headquarters and port of the Orkneys, Scotland, on the north coast of the largest island, Mainland; population (1990 est) 6,700. The Norse cathedral of St Magnus dates from 1137.

Kirov /ˈkɪərɒf/ (formerly until 1934 **Vyatka**) town NE of Nizhny Novgorod, on the river Vyatka, central Russia; population (1987) 421,000. It is a rail and industrial centre for rolling stock, tyres, clothing, toys, and machine tools.

Kirovabad /ˌkɪrəvəˈbæd/ city in Azerbaijan; population (1987) 270,000. Industries include cottons, woollens, and processed foods. The city was known as **Gandzha** until 1804 and again 1918–35, and as **Elizavetpol** 1804–1918.

Kirovograd /ˌkɪrəvəˈgræd/ city in central Ukraine; population (1987) 269,000. Manufacturing includes agricultural machinery and food processing. The city is on a lignite field. It was known as **Yelizavetgrad** until 1924 and **Zinovyevsk** 1924–36.

Kirriemuir /ˌkɪrɪˈmjuə/ market town of Tayside, Scotland, called 'Thrums' in James Barrie's novels; it is his birthplace.

Kisangani /ˌkɪsænˈɡɑːni/ (formerly until 1966 **Stanleyville**) town in NE Zaire, on the upper Zaïre River, below Stanley Falls; population (1984) 283,000. It is a communications centre.

Kishinev Russian name for ▷ Chişinău, the capital of Moldova.

Kitakyushu /ˌkiːtəˈkjuːʃuː/ industrial port city (coal, steel, chemicals, cotton thread, plate glass, alcohol) in Japan, on the Hibiki Sea, N Kyushu Island, formed 1963 by the amalgamation of Moji, Kokura, Tobata, Yawata, and Wakamatsu; population (1990) 1,026,500. A tunnel (1942) links it with Honshu.

Kitchener /ˈkɪtʃɪnə/ city in SW Ontario, Canada; population (1986) 151,000, metropolitan area (with Waterloo) 311,000. Manufacturing includes agricultural machinery and tyres. Settled by Germans from Pennsylvania in the 1800s, it was known as Berlin until 1916.

Kitimat /ˈkɪtɪmæt/ port near Prince Rupert, British Columbia, Canada; population (1981) 4,300. Founded 1955, it has one of the world's largest aluminium smelters, powered by the Kemano hydroelectric scheme.

Kitwe /ˈkɪtweɪ/ commercial centre for the Zambian copperbelt; population (1988) 472,300. To the S are Zambia's emerald mines.

Kitzbühel /ˈkɪtsbjuːəl/ winter-sports resort in the Austrian Tirol, NE of Innsbruck; population (1985) 9,000.

Kivu /ˈkiːvuː/ lake in the Great Rift Valley between Zaire and Rwanda, about 105 km/65 mi long. The chief port is Bukavu.

Klaipeda /ˈklaɪpɪdə/ (formerly **Memel**) port in Lithuania, on the Baltic coast at the mouth of the river Dange; population (1987) 201,000. Industries include shipbuilding and iron foundries; it trades in timber, grain, and fish. It was founded on the site of a local fortress 1252 as the castle of Memelburg by the Teutonic Knights, joined the Hanseatic League soon after, and has changed hands among Sweden, Russia, and Germany. Lithuania annexed Klaipeda 1923, and after German occupation 1939–45 it was restored to Soviet Lithuania 1945–91.

Klondike /ˈklɒndaɪk/ former gold-mining area in Yukon, Canada, named after the river valley where gold was found 1896. About 30,000 people moved there during the following 15 years. Silver is still mined there.

Klosters /ˈkləʊstəz/ fashionable alpine skiing resort (altitude 1,191 m/3,908 ft) in Grisons canton, E Switzerland, on the river Landquart, 10 km/6 mi NE of Davos. To the E, Kloster Pass leads to Austria.

Knaresborough /ˈneəzbərə/ market town in North Yorkshire, England, 6 km/4 mi NE of Harrogate; population (1987 est) 14,200. It has a castle dating from about 1070.

Knock /nɒk/ village in County Mayo, W Ireland, known for its church shrine (the **Basilica of Our Lady, Queen of Ireland**), one of three national places of pilgrimage (with Lough Derg and Croagh Patrick). On 21 Aug 1879 it was the scene of an alleged apparition of the Virgin Mary, St Joseph, and St John to some 14 people. Pope John Paul II (the first pope to set foot on Irish soil) celebrated Mass for the Sick here 100 years later, 1979. Horan International Airport (1986) receives transatlantic flights.

Knokke-Heist /ˈnɒkəhaɪst/ fashionable resort town on the coast of West Flanders, Belgium, NE of Blankenberge; population (1991) 31,700.

The fishing village of Knokke merged with the residential suburb of Het Zoute 1880 and, as the town developed, other villages were absorbed. It has a casino and one of the largest golf courses in Europe.

Knoxville /nɒksvɪl/ city in E Tennessee, USA; population (1990) 165,100. It is the centre of a mining and agricultural region, and the administrative headquarters of the Tennessee Valley Authority.

The University of Tennessee, founded 1794, is here, and Oak Ridge National Laboratory, one of the world's largest nuclear research facilities, is nearby.

Knutsford /nʌtsfəd/ town in Cheshire, England; population (1990 est) 14,000. There is engineering, and chemicals and scientific instruments are produced. The novelist Elizabeth Gaskell, who lived in Knutsford for 22 years and is buried there, wrote a social study of it under the name Cranford.

Kobarid /kəʊbərɪd/ formerly *Caporetto* village on the river Isonzo in NW Slovenia. Originally in Hungary, it was in Italy from 1918, and in 1947 became Kobarid. During World War I, German-Austrian troops defeated Italian forces there 1917.

Kobe /kəʊbeɪ/ deep-water port in S Honshu, Japan; population (1990) 1,477,400. *Port Island*, created 1960–68 from the rock of nearby mountains, area 5 sq km/2 sq mi, is one of the world's largest construction projects.

København /kɜːbən'haʊn/ Danish name for ◊Copenhagen, the capital of Denmark.

Koblenz /kəʊblents/ city in the Rhineland-Palatinate, Germany, at the junction of the rivers Rhine and Mosel; population (1988) 110,000. The city dates from Roman times. It is a centre of communications and the wine trade, with industries (shoes, cigars, paper).

Kodiak /kəʊdiæk/ island off the south coast of Alaska, site of a US naval base; area 9,505 sq km/3,670 sq mi. It is the home of the Kodiak bear, the world's largest bear. The town of Kodiak is one of the largest US fishing ports (mainly salmon).

Kokand /kəˈkænd/ oasis town in Uzbekistan; population (1981) 156,000. It was the capital of Kokand khanate when annexed by Russia 1876. Industries include fertilizers, cotton, and silk.

Kokomo /kəʊkəməʊ/ city in N central Indiana, USA, on Wildcat Creek, N of Indianapolis and SW of Fort Wayne; seat of Howard County; population (1990) 45,000.

The city's industries produce car, radio, and plumbing parts; steel and wire; and electrical machinery. The first car to use petrol was invented and tested here 1893.

Koko Nor /kəʊkəʊ 'nɔː/ Mongolian form of ◊Qinghai, a province of China.

Kola Peninsula /kəʊlə/ (Russian *Kol'skiy Poluostrov*) peninsula in N Russia, bounded S

and E by the White Sea and N by the Barents Sea; area 129,500 sq km/50,000 sq mi; population 1.3 million (of whom 2,000 are Saami). Kola is coterminous with Murmansk region. Apatite and other minerals are exported.

Heavy pollution from nickel smelting has damaged a large area of forest. The acid (SO2) emissions from nickel smelting have caused *Waldsterben* (forest die-back) as far away as neighbouring Finland. To the NW the low-lying granite plateau adjoins Norway's thinly populated county of Finnmark.

There is a heavy concentration of army, naval, and air bases in the Kola Peninsula.

Kolchugino /kɒlˈtʃuːɡɪnəʊ/ former name (to 1925) of ◊Leninsk-Kuznetsky, a town in Russia.

Kolhapur /ˌkəʊləˈpʊə/ industrial city and film production centre in Maharashtra, India; population (1981) 346,000. It is also an ancient Buddhist site and a centre of pilgrimage.

Köln /kɜːln/ German form of ◊Cologne, a city in Germany.

Kolwezi /kɒlˈweɪzi/ mining town (copper and cobalt) in Shaba province, SE Zaire; population (1985) 82,000. In 1978 former police of the province invaded from Angola and massacred some 650 of the inhabitants.

Komi /kəʊmi/ autonomous republic in N central Russia; area 415,900 sq km/160,580 sq mi; population (1986) 1,200,000. Its capital is Syktyvkar.

Kommunizma, Pik /ˌkɒmuˈnɪzmə/ or *Communism Peak* highest mountain in the Pamirs, a mountain range in Tajikistan; 7,495 m/24,599 ft. As part of the former USSR, it was known as *Mount Garmo* until 1933 and *Mount Stalin* 1933–62.

Kongur Shan /ˌkəʊʊˈnʊə/ mountain peak in China, 7,719 m/25,325 ft high, part of the Pamir range. The 1981 expedition that first reached the summit was led by British climber Chris Bonington.

Königsberg /kɜːnɪɡzbeəɡ/ former name of ◊Kaliningrad, a port in Russia.

Konstanz /kɒnstænts/ German form of ◊Constance, a town in Germany.

Konya /kɒnjə/ (Roman *Iconium*) city in SW central Turkey; population (1990) 513,300. Carpets and silks are made here, and the city contains the monastery of the dancing dervishes.

Kordofan /ˌkɔːdəˈfɑːn/ province of central Sudan, known as the 'White Land'; area 146,990 sq km/ 56,752 sq mi; population (1983) 3,093,300. Although it has never been an independent state, it has a character of its own. It is mainly undulating plain, with acacia scrub producing gum arabic, marketed in the chief town El Obeid. Formerly a rich agricultural region, it has been overtaken by desertification.

Korea /kə'rɪə/ peninsula in E Asia, divided into north and south; see ⏵Korea, North, and ⏵Korea, South.

Korea, North Democratic People's Republic of (*Chosun Minchu-chui Inmin Konghwa-guk*)
area 120,538 sq km/46,528 sq mi
capital Pyongyang
towns Chongjin, Nampo, Wonsan
physical wide coastal plain in W rising to mountains cut by deep valleys in interior
environment the building of a hydroelectric dam at Kumgangsan on a tributary of the Han River has been opposed by South Korea as a potential flooding threat to central Korea
features separated from South Korea by a military demarcation line; the richer of the two Koreas in mineral resources (copper, iron ore, graphite, tungsten, zinc, lead, magnesite, gold, phosphor, phosphates)
head of state Kim Il Sung from 1972 (also head of Korean Workers' Party)
head of government Kang Song San from 1992
political system communism
political parties Korean Workers' Party (KWP), Marxist-Leninist-Kim Il Sungist (leads Democratic Front for the Reunification of the Fatherland, including North Korean Democratic Party and Religious Chungwoo Party)
exports coal, iron, copper, textiles, chemicals
currency won
population (1990 est) 23,059,000; growth rate 2.5% p.a.
life expectancy men 67, women 73 (1989)
language Korean
religions traditionally Buddhist, Confucian, but religious activity curtailed by the state
literacy 99% (1989)
GNP $20 bn; $3,450 per head (1988)
chronology
1910 Korea formally annexed by Japan.
1945 Russian and US troops entered Korea, forced surrender of Japanese, and divided the country in two. Soviet troops occupied North Korea.
1948 Democratic People's Republic of Korea declared.
1950 North Korea invaded South Korea to unite the nation, beginning the Korean War.
1953 Armistice agreed to end Korean War.

1961 Friendship and mutual assistance treaty signed with China.
1972 New constitution, with executive president, adopted. Talks took place with South Korea about possible reunification.
1980 Reunification talks broke down.
1983 Four South Korean cabinet ministers assassinated in Rangoon, Burma (Myanmar), by North Korean army officers.
1985 Increased relations with the USSR.
1989 Increasing evidence shown of nuclear-weapons development.
1990 Diplomatic contacts with South Korea and Japan suggested the beginning of a thaw in North Korea's relations with the rest of the world.
1991 Became a member of the United Nations. Signed nonaggression agreement with South Korea; agreed to ban nuclear weapons.
1992 Signed Nuclear Safeguards Agreement, allowing international inspection of its nuclear facilities. Also signed a pact with South Korea for mutual inspection of nuclear facilities. Passed legislation making foreign investment in the country attractive.
1993 Threatened to withdraw from Nuclear Non-Proliferation Treaty.

Korea, South Republic of Korea (*Daehan Minguk*)
area 98,799 sq km/38,161 sq mi
capital Seoul
towns Taegu, ports Pusan, Inchon
physical southern end of a mountainous peninsula separating the Sea of Japan from the Yellow Sea
features Chomsongdae (world's earliest observatory); giant Popchusa Buddha; granite peaks of Soraksan National Park
head of state Kim Young Sam from 1992
head of government Hwang In Sung from 1993
political system emergent democracy
political parties Democratic Liberal Party (DLP), right of centre; Democratic Party, left of centre; Unification National Party (UNP), right of centre
exports steel, ships, chemicals, electronics, textiles and clothing, plywood, fish
currency won
population (1990 est) 43,919,000; growth rate 1.4% p.a.

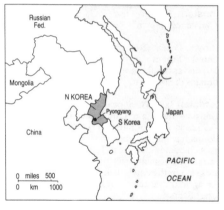

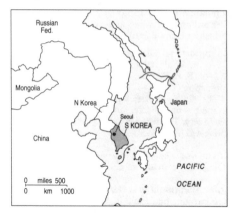

life expectancy men 66, women 73 (1989)
language Korean
media freedom of the press achieved 1987; large numbers of newspapers with large circulations. It is prohibited to say anything favourable about North Korea
religions traditionally Buddhist, Confucian, and Chondokyo; Christian 28%
literacy 92% (1989)
GNP $171bn (1988); $2,180 per head (1986)
chronology
1910 Korea formally annexed by Japan.
1945 Russian and US troops entered Korea, forced surrender of Japanese, and divided the country in two. US military government took control of South Korea.
1948 Republic proclaimed.
1950–53 War with North Korea.
1960 President Syngman Rhee resigned amid unrest.
1961 Military coup by General Park Chung-Hee. Industrial growth programme.
1979 Assassination of President Park.
1980 Military takeover by General Chun Doo Hwan.
1987 Adoption of more democratic constitution after student unrest. Roh Tae Woo elected president.
1988 Former president Chun, accused of corruption, publicly apologized and agreed to hand over his financial assets to the state. Seoul hosted Summer Olympic Games.
1989 Roh reshuffled cabinet, threatened crackdown on protesters.
1990 Two minor opposition parties united with Democratic Justice Party to form ruling Democratic Liberal Party. Diplomatic relations established with the USSR.
1991 Violent mass demonstrations against the government. New opposition grouping, the Democratic Party, formed. Entered United Nations. Nonaggression and nuclear pacts signed with North Korea.
1992 DLP lost absolute majority in March general election; substantial gains made by Democratic Party and newly formed UNP, led by Chung Ju Wong. Diplomatic relations with China established. Dec: Kim Young Sam, DLP candidate, won the presidential election.

Korinthos /'kɒrɪnθɒs/ Greek form of ◊Corinth, a port in Greece.

Kortrijk /'kɔːtraɪk/ Flemish form of ◊Courtrai, a town in Belgium.

Kos /kɒs/ or **Cos** fertile Greek island, one of the Dodecanese, in the Aegean Sea; area 287 sq km/111 sq mi. It gives its name to the Cos lettuce.

Kościusko /ˌkɒsɪ'ʌskəʊ/ highest mountain in Australia (2,229 m/7,316 ft), in New South Wales.
   The mineralogist Paul Strzelecki, who was born in Prussian Poland, climbed the mountain 1840. He thought one of the summit tops resembled the grave of the Polish revolutionary hero Tadeusz Kościuszko, and named it after him.

Košice /'kɒʃɪtseɪ/ town in the Slovak Republic; population (1991) 234,800 (92% Magyar-speaking). It has a textile industry and is a road centre. Košice was in Hungary until 1920 and 1938–45.

Kosovo /'kɒsəvəʊ/ autonomous region (1974–90) in S Serbia, Yugoslavia; capital Priština; area 10,900 sq km/4,207 sq mi; population (1986) 1,900,000, consisting of about 200,000 Serbs and about 1.7 million Albanians. Products include wine, nickel, lead, and zinc. Since it is largely inhabited by Albanians and bordering on Albania, there have been demands for unification with that country, while in the late 1980s Serbians agitated for Kosovo to be merged with the rest of Serbia. A state of emergency was declared Feb 1990 after fighting broke out between ethnic Albanians, police, and the Slavonic minority. The parliament and government were dissolved July 1990 and the Serbian parliament formally annexed Kosovo Sept 1990.
   The Serbian invasion brought Kosovo to the brink of civil war. Albanian institutions and media were supressed, and 'emergency legislation' used to rid industry of Albanian employees at all levels. In 1991 the Kosovo assembly, though still technically dissolved, organized a referendum on sovereignty which received 99% suppott. It elected a provisional government, headed by Bujar Bukoshi, which was recognized by Albania Oct 1991.
   In May 1992 the Albanian majority held unsanctioned elections, chooosing Ibrahim Rugova as president and selecting a 130-member parliament. Serbia regarded the elections as illegal but allowed them to proceed.

Kota Bharu /'kəʊtə 'bɑːruː/ capital of Kelantan, Malaysia; population (1980) 170,600. The local economy centres on fishing; there is batik craft work also.

Kota Kinabalu /'kəʊtə ˌkɪnəbə'luː/ (formerly until 1968 *Jesselton*) capital and port in Sabah, Malaysia; population (1980) 59,500. Exports include rubber and timber. Originally named after Sir Charles Jessel of the Chartered Company which rebuilt the town, it is now named after Mount Kinabulu.

Kourou /ku'ruː/ river and second-largest town of French Guiana, NW of Cayenne, site of the Guiana Space Centre of the European Space Agency.
   Situated near the equator, it is an ideal site for launches of satellites into geostationary orbit.

Kowloon /ˌkau'luːn/ peninsula on the Chinese coast forming part of the British crown colony of Hong Kong; the town of Kowloon is a residential area.
   The walled city of Kowloon was created 1898, when Britain leased the New Territories from China: China reserved jurisdiction over a disused fort, a site of 2.6 hectares/6.5 acres, and although the UK did not recognize this, some legal anomalies remained (for example, criminals captured in the Walled City were until 1960 deported to China). Although it lacked running water and other amenities, the Walled City became

the world's most densely populated area, with 33,000 residents in 1987 and 1,000 industrial and commercial enterprises. The wall was dismantled 1943. Its 500 buildings of up to 14 floors are less than a metre apart. The Walled City was being cleared 1991, to be torn down.

Kragujevac /'krægu:jeɪvæts/ garrison town and former capital (1818–39) of Serbia, Yugoslavia; population (1981) 165,000.

Krajina /'kraɪnə/ region on the frontier between Croatia and Bosnia-Herzegovina; the chief town is Knin. Dominated by Serbs, the region proclaimed itself an autonomous Serbian province after Croatia declared its independence from Yugoslavia 1991. It was the scene of intense inter-ethnic fighting during the civil war in Croatia 1991–92 and, following the cease-fire Jan 1992, 10,000 UN troops were deployed here and in E and W Slavonia.

Krakatoa /ˌkrækə'təuə/ (Indonesian *Krakatau*) volcanic island in Sunda strait, Indonesia, that erupted 1883, causing 36,000 deaths on Java and Sumatra by the tidal waves that followed. The island is now uninhabited.

Kraków /'krækau/ or **Cracow** city in Poland, on the river Vistula; population (1990) 750,500. It is an industrial centre producing railway wagons, paper, chemicals, and tobacco. It was capital of Poland about 1300–1595.

Founded about 1400, its university, at which the astronomer Copernicus was a student, is one of the oldest in central Europe. There is a 14th-century Gothic cathedral.

Kramatorsk /ˌkræmə'tɔ:sk/ industrial town in E Ukraine, in the Donets Basin, N of Donetsk; population (1987) 198,000. Industries include coal-mining machinery, steel, ceramics, and railway repairs.

Krasnodar /ˌkræsnəu'dɑ:/ (formerly until 1920 *Ekaterinodar*) industrial town at the head of navigation of the river Kuban, in SW Russia; population (1987) 623,000. It is linked by pipeline with the Caspian oilfields.

Krasnodar /ˌkræsnəu'dɑ:/ territory of SW Russia, in the N Caucasus Mountains, adjacent to the Black Sea; area 83,600 sq km/32,290 sq mi; population (1985) 4,992,000. The capital is Krasnodar. In addition to stock rearing and the production of grain, rice, fruit, and tobacco, oil is refined.

Krasnoyarsk /ˌkræsnəu'jɑ:sk/ industrial city (locomotives, paper, timber, cement, gold refining, and a large hydroelectric works) in central Siberian Russia; population (1987) 899,000. There is an early-warning and space-tracking radar phased array at nearby Abalakova.

Krasnoyarsk /ˌkræsnəu'jɑ:sk/ territory of Russia in central Siberia stretching N to the Arctic Ocean; area 2,401,600 sq km/927,617 sq mi; population (1985) 3,430,000. The capital is Krasnoyarsk. It is drained by the Yenisei River. Mineral

resources include gold, graphite, coal, iron ore, and uranium.

Krefeld /'kreɪfelt/ industrial town near the river Rhine; 52 km/32 mi NW of Cologne, Germany; population (1988) 217,000. Industries include chemicals, textiles, and machinery. It is situated on the Westphalian coalfield.

Kremenchug /ˌkreɪmen'tʃu:g/ industrial town on the river Dnieper, in central Ukraine; population (1987) 230,000. Manufacturing includes road-building machinery, railway wagons, and processed food.

Krivoi Rog /krɪ'vɔɪ 'rɒg/ (Russian 'crooked horn') town in central Ukraine, 130 km/80 mi SW of Dnepropetrovsk; population (1987) 698,000. The surrounding district is rich in iron ore, and there is a metallurgical industry.

Kruger National Park /'kru:gə/ game reserve in NE Transvaal, South Africa, between the Limpopo and Crocodile rivers; it is the largest in the world (about 20,720 sq km/8,000 sq mi). The Sabie Game Reserve was established 1898 by President Kruger, and the park declared 1926.

Krugersdorp /'kru:gəzdɔ:p/ mining town in the Witwatersrand district, Transvaal, South Africa; population (1980) 103,000. Manganese, uranium, and gold are mined.

KS abbreviation for ◊ *Kansas*, a state of the USA.

Kuala Lumpur /'kwɑ:lə 'lumpuə/ capital of the Federation of Malaysia; area 240 sq km/93 sq mi; population (1990) 1,237,900. The city developed after 1873 with the expansion of tin and rubber trading; these are now its main industries. Formerly within the state of Selangor, of which it was also the capital, it was created a federal territory 1974.

Kuban /ku:'bɑ:n/ river in E Europe, rising in Georgia and flowing through Russia to the Sea of Azov; length 906 km/563 mi.

Kuching /'ku:tʃɪŋ/ capital and port of Sarawak state, E Malaysia, on the Sarawak River; population (1980) 74,200.

Kufra /'ku:frə/ group of oases in the Libyan Desert, N Africa, SE of Tripoli. By the 1970s the vast underground reservoirs were being used for irrigation.

Kumamoto /ˌku:mə'məutəu/ city on Kyushu Island, Japan, 80 km/50 mi E of Nagasaki; population (1990) 579,300. A military stronghold until the 19th century, the city is now a centre for fishing, food processing, and textile industries.

Kumasi /ku:'mɑ:si/ second largest city in Ghana, W Africa, capital of Ashanti region, with trade in cocoa, rubber, and cattle; population (1984) 376,200.

history From the late 17th century until 1901, when it was absorbed into the British Gold Coast Colony, Kumasi was capital of the Ashanti confederation.

In 1874 the Ashanti king's palace was destroyed by a British military force under General Wolseley and in 1896 the city was occupied by the British for a second time. During an Ashanti revolt 1900, Sir Frederic Hodgson, governor of the Gold Coast Colony, and a small garrison were besieged in the fort at Kumasi from March to June.

Kumayri /ˌkumaɪˈriː/ (formerly until 1990 *Leninakan*) town in Armenia, 40 km/25 m NW of Yerevan; population (1987) 228,000. Industries include textiles and engineering. It was founded 1837 as a fortress called Alexandropol. The city was virtually destroyed by an earthquake 1926 and again 1988.

Kunlunshan /ˈkunlun ˈʃɑːn/ mountain range on the edge of the great Tibetan plateau, China; 4,000 km/2,500 mi E–W; highest peak Muztag (7,282 m/23,900 ft).

Kunming /ˌkunˈmɪŋ/ (formerly *Yunnan*) capital of Yunnan province, China, on Lake Dian Chi, about 2,000 m/6,500 ft above sea level; population (1989) 1,500,000. Industries include chemicals, textiles, and copper smelted with nearby hydroelectric power.

Kurdistan /ˌkɜːdɪˈstɑːn/ or *Kordestan* hilly region in SW Asia near Mount Ararat, where the borders of Iran, Iraq, Syria, Turkey, Armenia, and Azerbaijan meet; area 193,000 sq km/ 74,600 sq mi; total population around 18 million.

Kure /ˈkuəreɪ/ naval base and port 32 km/20 mi SE of Hiroshima, on the south coast of Honshu Island, Japan; population (1990) 216,700. Industries include shipyards and engineering works.

Kuria Muria /ˈkuərɪə ˈmuərɪə/ group of five islands in the Arabian Sea, off the south coast of Oman; area 72 sq km/28 sq mi.

Kuril Islands /kuˈriːlz/ or *Kuriles* chain of about 50 small islands stretching from the NE of Hokkaido, Japan, to the S of Kamchatka, Russia; area 14,765 sq km/5,700 sq mi; population (1990) 25,000. Some of them are of volcanic origin. Two of the Kurils (Etorofu and Kunashiri) are claimed by Japan and Russia.

The Kurils were discovered 1634 by a Russian navigator and were settled by Russians. Japan seized them 1875 and held them until 1945, when under the Yalta agreement they were returned to the USSR. Japan still claims the southernmost two (Etorofu and Kunashiri) and also the nearby small islands of Habomai and Shikotan (not part of the Kurils). The question of the S Kurils prevents signature of a Japanese–Russian peace treaty.

Kursk /kuəsk/ capital city of Kursk region of W Russia; population (1987) 434,000. Industries include chemicals, machinery, alcohol, and tobacco. It dates from the 9th century.

Kūt-al-Imāra /ˈkuːt æl ɪˈmɑːrə/ or *al Kūt* city in Iraq, on the river Tigris; population (1985) 58,600. It is a grain market and carpet-manufacturing centre. In World War I it was under siege by Turkish forces from Dec 1915 to April 1916, when the British garrison surrendered.

Kutch, Rann of /kʌtʃ/ salt-marsh area in Gujarat state, India, that forms two shallow lakes (the *Great Rann* and the *Little Rann*) in the wet season and is a salt-covered desert in the dry. It takes its name from the former princely state of Kutch, which it adjoined. An international tribunal 1968 awarded 90% of the Rann of

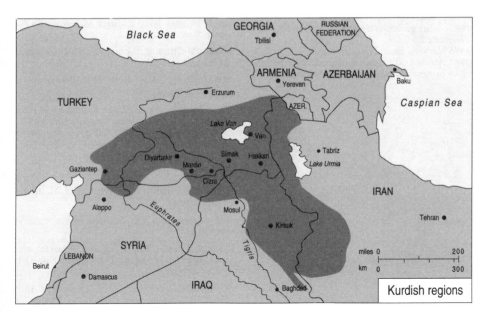
Kurdish regions

Kutch to India and 10% (about 800 sq km/ 300 sq mi) to Pakistan, the latter comprising almost all the elevated area above water the year round.

Kuwait State of (*Dowlat al Kuwait*)
area 17,819 sq km/6,878 sq mi
capital Kuwait (also chief port)
towns Jahra, Ahmadi, Fahaheel
physical hot desert; islands of Failaka, Bubiyan, and Warba at NE corner of Arabian Peninsula
environment during the Gulf War 1990–91, 650 oil wells were set alight and about 300,000 tonnes of oil were released into the waters of the Gulf leading to pollution haze, photochemical smog, acid rain, soil contamination, and water pollution
features there are no rivers and rain is light; the world's largest desalination plants, built in the 1950s
head of state and government Jabir al-Ahmad al-Jabir al-Sabah from 1977
political system absolute monarchy
political parties none
exports oil
currency Kuwaiti dinar
population (1990 est) 2,080,000 (Kuwaitis 40%, Palestinians 30%); growth rate 5.5% p.a.
life expectancy men 72, women 76 (1989)
languages Arabic 78%, Kurdish 10%, Farsi 4%
religion Sunni Muslim 45%, Shi'ite minority 30%
literacy 71% (1988)
GNP $19.1 bn; $10,410 per head (1988)
chronology
1914 Britain recognized Kuwait as an independent sovereign state.
1961 Full independence achieved from Britain, with Sheik Abdullah al-Salem al-Sabah as emir.
1965 Sheik Abdullah died; succeeded by his brother, Sheik Sabah.
1977 Sheik Sabah died; succeeded by Crown Prince Jabir.
1983 Shi'ite guerrillas bombed targets in Kuwait; 17 arrested.
1986 National assembly suspended.
1987 Kuwaiti oil tankers reflagged, received US Navy protection; missile attacks by Iran.

1988 Aircraft hijacked by pro-Iranian Shi'ites demanding release of convicted guerrillas; Kuwait refused.
1989 Two of the convicted guerrillas released.
1990 Prodemocracy demonstrations suppressed. Kuwait annexed by Iraq. Emir set up government in exile in Saudi Arabia.
1991 Feb: Kuwait liberated by US-led coalition forces; extensive damage to property and environment. New government omitted any opposition representatives. Trials of alleged Iraqi collaborators criticized.
1992 Reconstituted national assembly elected.

Kuwait City / kuː'weɪt/ (Arabic *Al Kuwayt*) (formerly *Qurein*) chief port and capital of the state of Kuwait, on the southern shore of Kuwait Bay; population (1985) 44,300, plus the suburbs of: Hawalli, population (1985) 145,100; Jahra, population (1985) 111,200; and as-Salimiya, population (1985) 153,400. Kuwait is a banking and investment centre.

Kuzbas /ˌkuz'bæs/ acronym for ◊Kuznetsk Basin.

Kuznetsk Basin /kuz'netsk/ (abbreviated to *Kuzbas*) industrial area in Kemorovo region, E Siberian Russia, lying on the Tom River N of the Altai Mountains; development began in the 1930s. It takes its name from the old town of Kuznetsk.

Kwa Ndebele /ˌkwɑːndə'beɪli/ black homeland in Transvaal province, South Africa; population (1985) 235,800. It achieved self-governing status 1981.

Kwangchow alternative transliteration of ◊Guangzhou, a city in China.

Kwangchu /ˌkwæŋ'dʒuː/ or *Kwangju* capital of South Cholla province, SW South Korea; population (1990) 1,144,700. It is at the centre of a rice-growing region. A museum in the city houses a large collection of Chinese porcelain dredged up 1976 after lying for over 600 years on the ocean floor.

Kwangsi-Chuang alternative transliteration of ◊Guangxi, a region of China.

Kwangtung alternative transliteration of ◊Guangdong, a province of China.

Kwa Zulu /kwɑː'zuːluː/ black homeland in Natal province, South Africa; population (1985) 3,747,000. It achieved self-governing status 1971.

Kweichow alternative transliteration of ◊Guizhou, a province of China.

Kweilin alternative transliteration of ◊Guilin, a city in China.

KY abbreviation for ◊*Kentucky*, a state of the USA.

Kyoga /ki'əʊɡə/ lake in central Uganda; area 4,425 sq km/1,709 sq mi. The Victoria Nile River passes through it.

Kyoto /ki'əʊtəʊ/ former capital of Japan 794–1868 (when the capital was changed to Tokyo) on

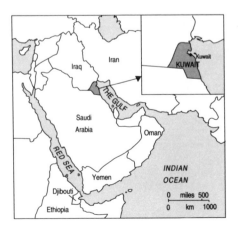

Honshu Island, linked by canal with Biwa Lake; population (1989) 1,407,300. Industries include electrical, chemical, and machinery plants; silk weaving; and the manufacture of porcelain, bronze, and lacquerware.

features The city's more than 2,000 temples and shrines include Tō-ji (1380), Kiyomizu-dera (1633), Ryōan-ji with its 15th-century Zen rock and sand garden, Sanjusangendo (1266), and the former Ashikaga shoguns' villas Kinkaku-ji and Ginkaku-ji (the 'gold and silver pavilions'). Other features are the Gion teahouse district with traditional geishas, the silk-weavers' district of Nishijin, 17th-century sake warehouses in Fushimi, Momoyama castle, and Japan's oldest theatre, the Minamiza kabuki theatre (early 1600s).

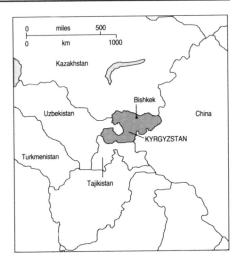

In Kioto the streets are almost as clean as the floors.

On **Kyoto** Lilian Leland *Travelling Alone, A Woman's Journey Round the World* 1890

Kyrenia /kar'ri:niə/ port in Turkish-occupied Cyprus, about 20 km/12 mi N of Nicosia; population (1985) 7,000. The Turkish army landed at Kyrenia during the 1976 invasion, and the town was temporarily evacuated.

Kyrgyzstan Republic of
area 198,500 sq km/76,641 sq mi
capital Bishkek (formerly Frunze)
towns Osh, Przhevalsk, Kyzyl-Kiya, Tormak
physical mountainous, an extension of the Tian Shan range
head of state Askar Akayev from 1990
head of government Tursunbek Chyngyshev from 1991
political system emergent democracy
political party Democratic Kyrgyzstan, nationalist reformist; Asaba (Banner) Party and Free Kyrgyzstan Party, both opposition groupings
products cereals, sugar, cotton, coal, oil, sheep, yaks, horses
population (1990) 4,400,000 (52% Kyrgyz, 22% Russian, 13% Uzbek, 3% Ukrainian, 2% German)
language Kyrgyz, a Turkic language
religion Sunni Muslim
chronology
1917–1924 Part of independent Turkestan.
1924 Became autonomous republic within USSR.
1936 Became full union republic within USSR.
1990 June: ethnic clashes resulted in state of emergency being imposed in Bishkek. Nov: Askar Akayev chosen as state president.
1991 March: Kyrgyz voters endorsed maintenance of Union in USSR referendum. Aug: President Akayev condemned anti-Gorbachev attempted coup in Moscow; Kyrgyz Communist Party, which supported the coup, suspended. Oct: Akayev directly elected president. Dec: joined Commonwealth of Independent States (CIS) and independence recognized by USA.
1992 Jan: admitted into Conference on Security and Cooperation in Europe (CSCE); March: became a member of the United Nations (UN).

Kyushu /'kju:u:/ southernmost of the main islands of Japan, separated from Shikoku and Honshu islands by Bungo Channel and Suo Bay, but connected to Honshu by bridge and rail tunnel
area 42,150 sq km/16,270 sq mi, including about 370 small islands
capital Nagasaki
cities Fukuoka, Kumamoto, Kagoshima
physical mountainous, volcanic, with subtropical climate
features the active volcano Aso-take (1,592 m/5,225 ft), with the world's largest crater
products coal, gold, silver, iron, tin, rice, tea, timber
population (1986) 13,295,000.

Kyustendil /'kju:stəndɪl/ town with hot springs in SW Bulgaria, SW of Sofia; population about 25,000.

Kyzyl-Kum /kɪ'zi:l 'ku:m/ desert in Kazakhstan and Uzbekistan, between the Sur-Darya and Amu-Darya rivers; area about 300,000 sq km/116,000 sq mi. It is being reclaimed for cultivation by irrigation and protective tree planting.

LA abbreviation for ▷ *Louisiana*, a state of the USA; ▷ *Los Angeles*, a city in California, USA.

Laâyoune /lɑː ˈjuːn/ (Arabic *El Aaiún*) capital of Western Sahara; population (1982) 97,000. It has expanded from a population of 25,000 in 1970 as a result of Moroccan investment (Morocco lays claim to Western Sahara).

Labrador /ˈlæbrədɔː/ area of NE Canada, part of the province of Newfoundland, lying between Ungava Bay on the NW, the Atlantic Ocean on the E, and the Strait of Belle Isle on the SE; area 266,060 sq km/102,699 sq mi; population (1986) 28,741. It consists primarily of a gently sloping plateau with an irregular coastline of numerous bays, fjords, inlets, and cliffs (60–120 m/200–400 ft high). Industries include fisheries, timber and pulp, and many minerals. Hydroelectric resources include Churchill Falls on Churchill River, where one of the world's largest underground power houses is situated.

The Canadian Air Force base in Goose Bay is on land claimed by the Innu (or Montagnais-Naskapi) Indian people, who call themselves a sovereign nation (in 1989 they numbered 9,500).

... I believe that this was the land that God allotted to Cain.

On **Labrador** Jacques Cartier 1534

Labuan /ləˈbuːən/ flat, wooded island off NW Borneo, a Federal Territory of East Malaysia; area 100 sq km/39 sq mi; population (1980) 12,000. Its chief town and port is Victoria, population 3,200. Labuan was ceded to Great Britain 1846, and from 1963 it was included in Sabah, a state of the Federation of Malaysia.

Laccadive, Minicoy, and Amindivi Islands /ˈlækədɪv, ˈmɪnɪkɔɪ, ˌæmɪnˈdiːvi/ former name of the Indian island group ▷ Lakshadweep.

La Ceiba /lɑː ˈseɪbə/ chief Atlantic port of Honduras; population (1989) 71,600. It exports fruit, especially bananas and pineapples.

Lachlan /ˈlæklən/ river that rises in the Blue Mountains, Australia; a tributary of the Murrumbidgee; length 1,485 km/920 mi.

La Condamine /lɑː ˌkɒndəˈmiːn/ commune of Monaco, SW of Monte Carlo. It is a seaside resort.

La Crosse /ləˈkrɒs/ city in SW Wisconsin, USA, at the confluence of the Black, La Crosse, and Mississippi rivers, NW of Madison; seat of La Crosse County; population (1990) 51,000. The processing and marketing centre for the area's agricultural products, it also manufactures plastics, rubber products, and electrical machinery. The city began as a French trading post and grew as a lumber town.

Ladakh /ləˈdɑːk/ subsidiary range of the Karakoram Mountains and district of NE Kashmir, India, on the border of Tibet; chief town Leh. After China occupied Tibet 1951, it made claims on the area.

Climate arid. The atmosphere like dry ice ... To think about Ladakh at all is to experience a sort of disorientation.

On **Ladakh** Ved Mehta *Portrait of India* 1970

Ladoga /ˈlædəgə/ (Russian *Ladozhskoye*) largest lake on the continent of Europe, in Russia, just NE of St Petersburg; area 18,400 sq km/7,100 sq mi. It receives the waters of several rivers, including the Svir, which drains Lake Onega and runs to the Gulf of Finland by the river Neva.

Ladrones /ləˈdrəʊnɪz/ Spanish name (meaning 'thieves') of the ▷ Mariana Islands.

Ladysmith /ˈleɪdɪsmɪθ/ town in Natal, South Africa, 185 km/115 mi NW of Durban, on the Klip. It was besieged by the Boers, 2 Nov 1899–28 Feb 1900, during the South African War. Ladysmith was named in honour of the wife of Henry Smith, a British soldier and colonial administrator.

Lafayette /ˌlæfeɪˈet/ city in S Louisiana, USA, on the Vermilion River, W of New Orleans and SW of Baton Rouge; seat of Lafayette parish; population (1990) 94,440. Its economy centres around the area's oil industry. Settled by Acadians from Nova Scotia in the late 1700s, Lafayette is in the heart of the area of Louisiana that is associated with French-speaking Cajuns (descendents of the settlers).

Lafayette /ˌlæfeɪˈet/ city in W central Indiana, USA, on the Wabash River, NW of Indianapolis, seat of Tippecanoe County; population (1990) 43,760. A distribution centre for the area's agricultural products, its industries also include building materials, chemicals, wire, pharmaceuticals, and automobile parts.

Lagos /ˈleɪgɒs/ chief port and former capital of Nigeria, located at the western end of an island in a lagoon and linked by bridges with the mainland via Iddo Island; population (1983) 1,097,000. Industries include chemicals, metal products, and fish. One of the most important slaving ports, Lagos was bombarded and occupied by the British 1851, becoming the colony of Lagos 1862. Abuja was established as the new capital 1982 and received official recognition 1992.

Lahore /ləˈhɔː/ capital of the province of Punjab and second city of Pakistan; population (1981) 2,920,000. Industries include engineering, textiles, carpets, and chemicals. It is associated with the Mogul rulers Akbar, Jahangir, and Aurangzeb, whose capital it was in the 16th and 17th centuries.

Laibach /ˈlaɪbæx/ German name of ⟡Ljubljana, a city in Slovenia.

Lake Charles /ˌleɪk ˈtʃɑːlz/ city in SW Louisiana, USA, on the Calcasieu River, SW of Baton Rouge, seat of Calcasieu parish; population (1990) 70,580. It is a port of entry on the Gulf of Mexico via a deep-water channel in the Calcasieu River. Most of the city's industries are related to the area's oil and gas resources and major crop, rice.

Lake District region in Cumbria, England; area 1,800 sq km/700 sq mi. It contains the principal English lakes, which are separated by wild uplands rising to many peaks, including Scafell Pike (978 m/3,210 ft).
    Windermere, in the SE, is connected with Rydal Water and Grasmere. The westerly Scafell range extends S to the Old Man of Coniston overlooking Coniston Water, and N to Wastwater. Ullswater lies in the NE of the district, with Hawes Water and Thirlmere nearby. The river Derwent flows N through Borrowdale forming Derwentwater and Bassenthwaite. W of Borrowdale lie Buttermere, Crummock Water, and, beyond, Ennerdale Water.
    The Lake District has associations with the writers Wordsworth, Coleridge, Southey, De Quincey, Ruskin, and Beatrix Potter and was made a national park 1951.

Lake Havasu City /ˈhævəsuː/ town in Arizona, USA, developed as a tourist resort. Old London Bridge was transported and reconstructed here 1971.

Lakeland /ˈleɪklənd/ city in W central Florida, USA, NE of Tampa and SW of Orlando, in the lake region and citrus belt; population (1990) 70,576. It serves as a centre for the area's citrus products, but its economy depends mainly on its reputation as a winter resort.

Lakshadweep /læk'ʃædwiːp/ group of 36 coral islands, 10 inhabited, in the Indian Ocean, 320 km/200 mi off the Malabar coast; area 32 sq km/12 sq mi; population (1991) 51,700. The administrative headquarters are on Kavaratti Island. Products include coir, copra, and fish. The religion is Muslim. The first Western visitor was Vasco da Gama 1499. The islands were British from 1877 until Indian independence

and were created a Union Territory of the Republic of India 1956. Formerly known as the Laccadive, Minicoy, and Amindivi Islands, they were renamed Lakshadweep 1973.

La Mancha /læ ˈmæntʃə/ (Arabic *al mansha* the 'dry land') former province of Spain now part of the autonomous region of Castilla-La Mancha. The fictional travels of Cervantes's *Don Quixote de la Mancha* 1605 begin here.

Lambeth /ˈlæmbəθ/ inner borough of S central Greater London
features Lambeth Palace (chief residence of the archbishop of Canterbury since 1197); Tradescant Museum of gardening history; the South Bank (including Royal Festival Hall, National Theatre); the Oval (headquarters of Surrey County Cricket Club from 1846) at Kennington, where the first England–Australia test match was played 1880; Brixton Prison
population (1991) 220,100.

Lammermuir Hills /ˈlæməmjuə/ range of hills dividing Lothian and Borders regions, Scotland, running from Gala Water to St Abb's Head. The highest point is Meikle Says Law (533 m/1,750 ft).

Lamu /ˈlɑːmuː/ island off the east coast of Kenya, 200 km/124 mi NE of Mombasa; population (1991 est) 12,000. The chief centre, Lamu Town, was formerly the focal point of a coastal city-state, trading in ivory, cowries, spices, tortoiseshell, oil seed, and grain. The island's economy declined following the abolition of slavery 1907, but in recent years tourism has developed alongside traditional crafts such as dhow building.

Lanark /ˈlænək/ former county town of Lanarkshire, Scotland; now capital of Clydesdale district, Strathclyde region; population (1981) 9,800. William Wallace once lived here, and later returned to burn the town and kill the English sheriff. *New Lanark* to the S, founded 1785 by Robert Owen, was a socialist 'ideal village' experiment.

Lanarkshire /ˈlænəkʃə/ former county of Scotland, merged 1975 into the region of Strathclyde. The county town was Lanark.

Lancashire /ˈlæŋkəʃə/ county in NW England
area 3,040 sq km/1,173 sq mi
towns Preston (administrative headquarters), which forms part of Central Lancashire New Town (together with Fulwood, Bamber Bridge, Leyland, and Chorley); Lancaster, Accrington, Blackburn, Burnley; ports Fleetwood and Heysham; seaside resorts Blackpool, Morecambe, and Southport
features the river Ribble; the Pennines; the Forest of Bowland (moors and farming valleys); Pendle Hill
products formerly a world centre of cotton manufacture, now replaced with high-technology aerospace and electronics industries
population (1991) 1,365,100
famous people Kathleen Ferrier, Gracie Fields, George Formby, Rex Harrison.

Lancashire

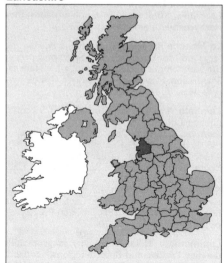

Lancaster /'læŋkəstə/ city in Lancashire, England, on the river Lune; population (1991) 125,600. It was the former county town of Lancashire (now Preston). Industries include paper, furniture, plastics, and chemicals. A castle here, which incorporates Roman work, was captured by Cromwell during the Civil War. The university was founded 1964.

Lancaster /'læŋkəstə/ city in Pennsylvania, USA, 115 km/70 mi W of Philadelphia; population (1990) 55,550. It produces textiles and electrical goods. It was capital of the USA briefly 1777, and was the state capital 1799–1812.

Lanchow /,læn'tʃau/ alternative transcription of ⟡Lanzhou, a city in China.

Lancs abbreviation for ⟡*Lancashire*, an English county.

Landes /lɒnd/ sandy, low-lying area in SW France, along the Bay of Biscay, about 12,950 sq km/ 5,000 sq mi in extent. Formerly covered with furze and heath, it has in many parts been planted with pine and oak forests. It gives its name to a *département* and extends into the *départements* of Gironde and Lot-et-Garonne. There is a testing range for rockets and missiles at Biscarosse, 72 km/45 mi SW of Bordeaux. There is an oilfield at Parentis-en-Born.

Land's End /'lændz 'end/ promontory of W Cornwall, 15 km/9 mi WSW of Penzance, the westernmost point of England.

A group of dangerous rocks, the Longships, extend a mile out beyond Land's End; they are marked by a lighthouse.

The wildest most impressive place I ever saw on the coasts of Britain.

On **Land's End** Thomas Carlyle 1882

Landskrona /læendz'kruːnə/ town and port in Sweden, on the Sound, 32 km/20 mi N of Malmö; population (1990) 36,300. Industries include shipyards, machinery, chemicals, and sugar refining. Carl XI defeated the Danes off Landskrona 1677.

Languedoc /,lɑːŋgə'dɒk/ former province of S France, bounded by the river Rhône, the Mediterranean Sea, and the regions of Guienne and Gascony.

It took its name from the Romance Provençal language widely spoken in S France in the Middle Ages and known as *langue d'oc* (*oc* meaning 'yes'). The French spokeñ N of the Loire, with which it was in competition, was known as *langue d'oïl* (*oïl* meaning 'yes').

Languedoc-Roussillon /,lɑːŋgə'dɒk ,ruːsiː'jɒn/ region of S France, comprising the *départements* of Aude, Gard, Hérault, Lozère, and Pyrénées-Orientales; area 27,400 sq km/10,576 sq mi; population (1986) 2,012,000. Its capital is Montpellier, and products include fruit, vegetables, wine, and cheese.

Lansing /'lænsɪŋ/ capital of Michigan, USA, at the confluence of the Grand and Red Cedar rivers; population (1990) 127,300. Manufacturing includes motor vehicles, diesel engines, and pumps.

Lanzarote /,læenzə'rɒti/ most easterly of the Spanish Canary Islands; area 795 sq km/307 sq mi; capital Arrecife. The desertlike volcanic landscape is dominated by the Montañas de Fuego ('Mountains of Fire') with more than 300 volcanic cones.

Lanzhou /,læn'dʒəu/ or *Lanchow* capital of Gansu province, China, on the river Huang He, 190 km/120 mi S of the Great Wall; population (1989) 1,480,000. Industries include oil refining, chemicals, fertilizers, and synthetic rubber.

Laois /liːʃ/ or *Laoighis* county in Leinster province, Republic of Ireland
area 1,720 sq km/664 sq mi
county town Port Laoise
physical flat except for the Slieve Bloom Mountains in the NW
products sugar beet, dairy products, woollens, agricultural machinery
population (1991) 52,300.

Laon /lɒŋ/ capital of Aisne *département*, Picardie, N France; 120 km/75 mi NE of Paris; population (1990) 28,700. It was the capital of France and a royal residence until the 10th century. It has a 12th-century cathedral.

Laos Lao People's Democratic Republic (*Saathiaranagroat Prachhathippatay Prachhachhon Lao*)
area 236,790 sq km/91,400 sq mi
capital Vientiane
towns Luang Prabang (the former royal capital), Pakse, Savannakhet
physical landlocked state with high mountains in E; Mekong River in W; jungle covers nearly 60% of land

features Plain of Jars, where prehistoric people carved stone jars large enough to hold a person
head of state Nouhak Phoumsavan from 1992
head of government General Khamtay Siphandon from 1991
political system communism, one-party state
political party Lao People's Revolutionary Party (only legal party)
exports hydroelectric power from the Mekong is exported to Thailand, timber, teak, coffee, electricity
currency new kip
population (1990 est) 4,024,000 (Lao 48%, Thai 14%, Khmer 25%, Chinese 13%); growth rate 2.2% p.a.
life expectancy men 48, women 51 (1989)
languages Lao (official), French
religions Theravāda Buddhist 85%, animist beliefs among mountain dwellers
literacy 45% (1991)
GNP $500 million (1987); $180 per head (1988)
chronology
1893–1945 Laos was a French protectorate.
1945 Temporarily occupied by Japan.
1946 Retaken by France.
1950 Granted semi-autonomy in French Union.
1954 Independence achieved from France.
1960 Right-wing government seized power.
1962 Coalition government established; civil war continued.
1973 Vientiane cease-fire agreement. Withdrawal of US, Thai, and North Vietnamese forces.
1975 Communist-dominated republic proclaimed with Prince Souphanouvong as head of state.
1986 Phoumi Vongvichit became acting president.
1988 Plans announced to withdraw 40% of Vietnamese forces stationed in the country.
1989 First assembly elections since communist takeover.
1991 Constitution approved. Kaysone Phomvihane elected president. General Khamtay Siphandon named as new premier.
1992 Phomvihane died; replaced by Nouhak Phoumsavan.

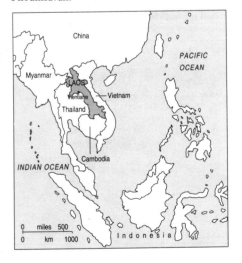

La Palma one of the Spanish Canary Islands; see ⏵Palma, La.

La Pampa /læ 'pæmpə/ province in Argentina, characterized by the plains of the eastern pampas.

La Paz /læ 'pæz/ capital city of Bolivia, in Murillo province, 3,800 m/12,400 ft above sea level; population (1988) 1,049,800. Products include textiles and copper. Founded by the Spanish 1548 as Pueblo Nuevo de Nuestra Señōra de la Paz, it has been the seat of government since 1898.

Lapland /'læplænd/ region of Europe within the Arctic Circle in Norway, Sweden, Finland, and the Kola Peninsula of NW Russia, without political definition. Its chief resources are chromium, copper, iron, timber, hydroelectric power, and tourism. The indigenous population are the Saami (formerly known as Lapps), a semi-nomadic herding people. Lapland has low temperatures, with three months' continuous daylight in summer and three months' continuous darkness in winter. There is summer agriculture.

La Plata /læ 'plɑːtə/ capital of Buenos Aires province, Argentina; population (1980) 560,300. Industries include meat packing and petroleum refining. It was founded 1882.

la Plata, Río de an estuary in South America; see ⏵Plata, Río de la.

Laptev Sea /'læptev/ part of the Arctic Ocean off the north coast of Russia between Taimyr Peninsula and New Siberian Island.

Laramie /'lærəmi/ town in Wyoming, USA, on the Laramie Plains, a plateau 2,300 m/7,500 ft above sea level, bounded N and E by the Laramie Mountains; population (1990) 26,680. The Laramie River, on which it stands, is linked with the Missouri via the Platte.
    A staging post on the overland trail and Pony Express route, Laramie features in Western legend.

Larderello /ˌlɑːdə'reləu/ site in the Tuscan hills, NE Italy, where the sulphur springs were used by the Romans for baths and exploited for boric acid in the 18th–19th centuries. Since 1904 they have been used to generate electricity; the water reaches 220°C/396°F.

Laredo /lə'reɪdəu/ city on the Rio Grande, Texas, USA; population (1990) 122,900.
    It was founded 1755. Industries include oil refining and metal processing. Laredo State University is here. *Nuevo Laredo*, Mexico, on the opposite bank, is a textile centre; population (1980) 203,300. It was considered part of Laredo until the international border was established 1848. There is much cross-border trade.

La Rioja /læ ri'ɒxə/ region of N Spain; area 5,000 sq km/1,930 sq mi; population (1986) 263,000. The river Ebro passes through the region, but it is a tributary, the Río Oja, which gives its name to the region. The capital is Logroño. La Rioja is known for its woody red and white wines.

Larisa /ləˈrɪsə/ town in Thessaly, Greece, S of Mount Olympus; population (1981) 102,000. Industries include textiles and agricultural produce.

Larne /lɑːn/ seaport of County Antrim, Northern Ireland, on Lough Larne, terminus of sea routes to Stranraer, Liverpool, Dublin, and other places; population (1981) 18,200.

La Rochelle /ˌlæ rɒˈʃel/ fishing port in W France; population (1990) 73,700. It is the capital of Charente-Maritime *département*. Industries include shipbuilding, chemicals, and motor vehicles. A Huguenot stronghold, it was taken by Cardinal Richelieu in the siege of 1627–28.

Las Cruces city in S New Mexico, on the Rio Grande, N of the Mexican border; population (1990) 62,120. It is a processing centre for the area's crops, such as pecans, cotton, and vegetables. White Sands Missile Range is nearby.

Lashio /ˈlæʃiəu/ town in N Myanmar (Burma), about 200 km/125 mi NE of Mandalay. The Burma Road to Kunming in China, constructed 1938, starts here.

Las Palmas /læs ˈpælməs/ or **Las Palmas de Gran Canaria** tourist resort on the northeast coast of Gran Canaria, Canary Islands; population (1991) 347,700. Products include sugar and bananas.

La Spezia /læ ˈspetsiə/ port in NW Italy, chief Italian naval base; population (1988) 107,000. Industries include shipbuilding, engineering, electrical goods, and textiles. The English poet Percy Bysshe Shelley drowned in the Gulf of Spezia.

Las Vegas /læs ˈveɪgæs/ city in Nevada, USA, known for its nightclubs and gambling casinos; population (1990) 258,300. Las Vegas entertains millions of visitors each year and is an important convention centre. Founded 1855 in a ranching area, the modern community developed with the coming of the railroad 1905. The first casino-hotel opened 1947.

If you aim to leave Las Vegas with a small fortune, go there with a large one.

On **Las Vegas** anonymous saying

Latakia /ˌlætəˈkiːə/ port with tobacco industries in NW Syria; population (1981) 197,000.

Latin America countries of South and Central America (also including Mexico) in which Spanish, Portuguese, and French are spoken.

Lattakia /ˌlætəˈkiːə/ alternative form of ▷Latakia, a port in Syria.

Latvia Republic of
area 63,700 sq km/24,595 sq mi
capital Riga
towns Daugavpils, Liepāja, Jurmala, Jelgava, Ventspils

physical wooded lowland (highest point 312 m/1,024 ft), marshes, lakes; 472 km/293 mi of coastline; mild climate
features Western Dvina River; Riga is largest port on the Baltic after Leningrad
head of state Anatolijs Gorbunov from 1988
head of government Ivars Godmanis from 1990
political system emergent democratic republic
political parties Latvian Popular Front, nationalist; Latvian Social-Democratic Workers' Party
products electronic and communications equipment, electric railway carriages, motorcycles, consumer durables, timber, paper and woollen goods, meat and dairy products
currency Latvian rouble
population (1990) 2,700,000 (52% Latvian, 34% Russian, 5% Byelorussian, 3% Ukrainian)
language Latvian
religions mostly Lutheran Protestant, with a Roman Catholic minority
chronology
1917 Soviets and Germans contested for control of Latvia.
1918 Feb: Soviet forces overthrown by Germany. Nov: Latvia declared independence. Dec: Soviet rule restored after German withdrawal.
1919 Soviet rule overthrown by British naval and German forces May–Dec; democracy established.
1934 Coup replaced established government.
1939 German-Soviet secret agreement placed Latvia under Russian influence.
1940 Incorporated into USSR as constituent republic.
1941–44 Occupied by Germany.
1944 USSR regained control.
1980 Nationalist dissent began to grow.
1988 Latvian Popular Front established to campaign for independence. Prewar flag readopted; official status given to Latvian language.
1989 Popular Front swept local elections.
1990 Jan: Communist Party's monopoly of power abolished. March–April: Popular Front secured majority in elections. April: Latvian Communist Party split into pro-independence and pro-Moscow wings. May: unilateral declaration of independence from USSR, subject to transitional period for negotiation.

1991 Jan: Soviet troops briefly seized key installations in Riga. March: overwhelming vote for independence in referendum. Aug: full independence declared at time of anti-Gorbachev coup; Communist Party outlawed. Sept: independence recognized by Soviet government and Western nations; United Nations (UN) membership granted; admitted into Conference on Security and Cooperation in Europe (CSCE). **1992** US reopened its embassy in Latvia. Russia began pullout of ex-Soviet troops, to be completed 1994. July: curbing of rights of non-citizens in Latvia prompted Russia to request minority protection by UN.

Laugharne /lɑːn/ (Welsh **Talacharn**) village at the mouth of the river Towey, Dyfed, Wales. The home of the poet Dylan Thomas, it features in his work 'Under Milk Wood'.

Launceston /'lɔːnsəstən/ port in NE Tasmania, Australia, on the Tamar River; population (1986) 88,500. Founded 1805, its industries include woollen blanket weaving, saw milling, engineering, furniture and pottery making, and railway workshops.

Lausanne /ləu'zæn/ resort and capital of Vaud canton, W Switzerland, above the north shore of Lake Geneva; population (1990) 123,200. Industries include chocolate, scientific instruments, and publishing.

La Vendée French river; see ⇨Vendée, La.

Lawrence /'lɒrəns/ city in NE Kansas, USA, on the Kansas River between Topeka to the W and Kansas City to the E; seat of Douglas County; population (1990) 65,600. Its main industries are food processing and chemicals.

Lawrence /'lɒrəns/ town in Massachusetts, USA; population (1990) 70,200. Industries include textiles, clothing, paper, and radio equipment. The town was established 1845 to utilize power from the Merrimack Rapids on a site first settled 1655.

Lawton /'lɔːtn/ city in SW Oklahoma, USA, on Cache Creek, SW of Oklahoma City, seat of Comanche County; population (1980) 80,054. Processing the area's agricultural products is the city's main industry. Fort Sill, an army base, is to the N.

Lazio /'lætsiəu/ (Roman **Latium**) region of W central Italy; area 17,200 sq km/6,639 sq mi; capital Rome; population (1990) 5,191,500. Products include olives, wine, chemicals, pharmaceuticals, and textiles. Home of the Latins from the 10th century BC, it was dominated by Romans from the 4th century BC.

Lea /liː/ river that rises in Bedfordshire, England, and joins the river Thames at Blackwall.

Leamington /'lemɪŋtən/ officially **Royal Leamington Spa** town and health resort in the West Midlands, England, on the river Leam, adjoining Warwick; population (1985) 56,500. The Royal Pump Room offers spa treatment.

Leatherhead /'leðəhed/ town in Surrey, England, SW of London, on the river Mole at the foot of

the North Downs; population (1985) 40,300. It has industrial research stations, the Thorndike Theatre (1968), and the Royal School for the Blind (1799).

Lebanon Republic of (*al-Jumhouria al-Lubnaniya*)
area 10,452 sq km/4,034 sq mi
capital and port Beirut
towns ports Tripoli, Tyre, Sidon
physical narrow coastal plain; Bekka valley N–S between Lebanon and Anti-Lebanon mountain ranges
features Mount Hermon; Chouf Mountains; archaeological sites at Baalbeck, Byblos, Tyre; until the civil war, the financial centre of the Middle East
head of state Elias Hrawi from 1989
head of government Rafik al-Hariri from 1992
political system emergent democratic republic
political parties Phalangist Party, Christian, radical, right-wing; Progressive Socialist Party (PSP), Druse, moderate, socialist; National Liberal Party (NLP), Maronite, centre-left; Parliamentary Democratic Front, Sunni Muslim, centrist; Lebanese Communist Party (PCL), nationalist, communist
exports citrus and other fruit, vegetables; industrial products to Arab neighbours
currency Lebanese pound
population (1990 est) 3,340,000 (Lebanese 82%, Palestinian 9%, Armenian 5%); growth rate –0.1% p.a.
life expectancy men 65, women 70 (1989)
languages Arabic, French (both official), Armenian, English
religions Muslim 57% (Shiite 33%, Sunni 24%), Christian (Maronite and Orthodox) 40%, Druse 3%
literacy 75% (1989)
GNP $1.8 bn; $690 per head (1986)
chronology
1920–41 Administered under French mandate.
1944 Independence achieved.
1948–49 Lebanon joined first Arab war against Israel. Palestinian refugees settled in the south.
1964 Palestine Liberation Organization (PLO) founded in Beirut.

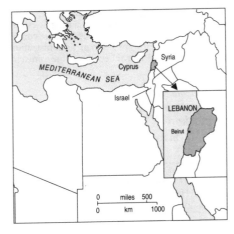

1967 More Palestinian refugees settled in Lebanon.
1971 PLO expelled from Jordan; established headquarters in Lebanon.
1975 Outbreak of civil war between Christians and Muslims.
1976 Cease-fire agreed; Syrian-dominated Arab deterrent force formed to keep the peace but considered by Christians as a occupying force.
1978 Israel invaded S Lebanon in search of PLO fighters. International peacekeeping force established. Fighting broke out again.
1979 Part of S Lebanon declared an 'independent free Lebanon'.
1982 Bachir Gemayel became president but was assassinated before he could assume office; succeeded by his brother Amin Gemayel. Israel again invaded Lebanon. Palestinians withdrew from Beirut under supervision of international peacekeeping force. PLO moved its headquarters to Tunis.
1983 Agreement reached for the withdrawal of Syrian and Israeli troops but abrogated under Syrian pressure.
1984 Most of international peacekeeping force withdrawn. Muslim militia took control of W Beirut.
1985 Lebanon in chaos; many foreigners taken hostage.
1987 Syrian troops sent into Beirut.
1988 Agreement on a Christian successor to Gemayel failed; he established a military government; Selim al-Hoss set up rival government; threat of partition hung over the country.
1989 Christian leader General Michel Aoun declared 'war of liberation' against Syrian occupation; Saudi Arabia and Arab League sponsored talks that resulted in new constitution recognizing Muslim majority; René Muhawad named president, assassinated after 17 days in office; Elias Hrawi named successor; Aoun occupied presidential palace, rejected constitution.
1990 Release of Western hostages began. General Aoun surrendered and legitimate government restored, with Umar Karami as prime minister.
1991 Government extended control to the whole country. Treaty of cooperation with Syria signed. More Western hostages released. General Aoun pardoned.
1992 Karami resigned as prime minister; succeeded by Rashid al-Solh. Remaining Western hostages released. General election boycotted by many Christians; pro-Syrian administration reelected; Rafik al-Hariri became prime minister.

Why should we kid ourselves? This is a parvenu nation. We never fought for our independence. We never had a state.

On **Lebanon** Edouard Saab quoted in
New York Times 10 April 1976

Lebanon /'lebənən/ city in SE Pennsylvania, USA, NE of Harrisburg, seat of Lebanon county; population (1990) 24,800. Industries include iron and steel products, textiles, clothing, and chemicals.

Lebda /'lebdə/ former name of ▷Homs, a city in Syria.

Lebowa /lə'bəuə/ black homeland in Transvaal province, South Africa; population (1985) 1,836,000. It achieved self-governing status 1972.

Leeds /li:dz/ city in West Yorkshire, England, on the river Aire; population (1991 est) 674,400. Industries include engineering, printing, chemicals, glass, and woollens. Notable buildings include the Town Hall designed by Cuthbert Brodrick, Leeds University (1904), the Art Gallery (1844), Temple Newsam (birthplace of Henry Darnley 1545, now a museum), and the Cistercian Abbey of Kirkstall (1147). It is a centre of communications where road, rail, and canal (to Liverpool and Goole) meet.

Opera North is based here, and the Leeds International Pianoforte Competition is held here every three years. A new museum, due to open 1996, will share the Royal Armouries collection, exchanging items with the Tower of London.

Leeuwarden /'leɪwɑːdn/ city in the Netherlands, on the Ee River; population (1991) 85,700. It is the capital of Friesland province. A marketing centre, it also makes gold and silver ware. After the draining of the Middelzee fenlands, the town changed from a port to an agricultural market town. Notable buildings include the palace of the stadholders of Friesland and the church of St Jacob.

Leeward Islands /'li:wəd/ (1) group of islands, part of the Society Islands, in French Polynesia, S Pacific; (2) general term for the northern half of the Lesser Antilles in the West Indies; (3) former British colony in the West Indies (1871–1956) comprising Antigua, Montserrat, St Christopher/ St Kitts–Nevis, Anguilla, and the Virgin Islands.

Leghorn /'legho:n/ former English name for the Italian port ▷Livorno.

Leh /leɪ/ capital of Ladakh region, E Kashmir, India, situated E of the river Indus, 240 km/ 150 mi E of Srinagar. Leh is the nearest supply base to the Indian army outpost on the Siachen Glacier.

Le Havre /lə 'hɑːvrə/ industrial port (engineering, chemicals, oil refining) in Normandy, NW France, on the river Seine; population (1990) 197,200. It is the largest port in Europe, and has transatlantic passenger links.

Leicester /'lestə/ industrial city (food processing, hosiery, footwear, engineering, electronics, printing, plastics) and administrative headquarters of Leicestershire, England, on the river Soar; population (1991) 270,600. Founded AD 50 as the Roman ***Ratae Coritanorum***, Leicester is one of the oldest towns in England. The guildhall dates from the 14th century and ruined Bradgate House was the home of Lady Jane Grey. Leicester University was established 1957.

Leicestershire

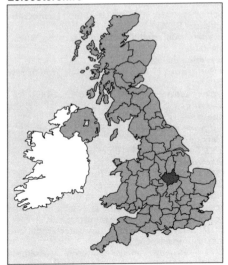

Leicestershire /ˈlestəʃə/ county in central England
area 2,550 sq km/984 sq mi
towns Leicester (administrative headquarters), Loughborough, Melton Mowbray
features Rutland district (formerly England's smallest county, with Oakham as its county town); Rutland Water, one of Europe's largest reservoirs; Charnwood Forest; Vale of Belvoir
products horses, cattle, sheep, dairy products, coal
population (1991) 860,500
famous people C P Snow, Thomas Babington Macaulay, Titus Oates.

Leics abbreviation for ▷*Leicestershire*, an English county.

Leiden /ˈlaɪdn/ or *Leyden* city in South Holland province, the Netherlands; population (1991) 111,900. Industries include textiles and cigars. It has been a printing centre since 1580, with a university established 1575. It is linked by canal to Haarlem, Amsterdam, and Rotterdam. The painters Rembrandt and Jan Steen were born here.

Leinster /ˈlenstə/ southeastern province of the Republic of Ireland, comprising the counties of Carlow, Dublin, Kildare, Kilkenny, Laois, Longford, Louth, Meath, Offaly, Westmeath, Wexford, and Wicklow; area 19,630 sq km/ 7,577 sq mi; capital Dublin; population (1991) 1,860,000.

Leipzig /ˈlaɪpzɪg/ city in W Saxony, Germany, 145 km/90 mi SW of Berlin; population (1986) 552,000. Products include furs, leather goods, cloth, glass, cars, and musical instruments.

Leith /liːθ/ port in Scotland, S of the Firth of Forth, incorporated in Edinburgh 1920. Leith was granted to Edinburgh as its port by Robert Bruce 1329.

Leitrim /ˈliːtrɪm/ county in Connacht province, Republic of Ireland, bounded NW by Donegal Bay
area 1,530 sq km/591 sq mi
county town Carrick-on-Shannon
features rivers: Shannon, Bonet, Drowes, and Duff
products potatoes, cattle, linen, woollens, pottery, coal, iron, lead
population (1991) 25,300.

Léman, Lac /ˌlæk ləˈmɒn/ French name for Lake ▷Geneva, in Switzerland.

Le Mans /lə ˈmɒn/ industrial town in Sarthe *département*, W France; population (1990) 148,500, conurbation 191,000. It has a motor-racing circuit where the annual endurance 24-hour race (established 1923) for sports cars and their prototypes is held.

Lemberg /ˈlembeək/ German name of ▷Lviv, a city in Ukraine.

Lemnos /ˈlemnɒs/ (Greek *Límnos)* Greek island in the north of the Aegean Sea
area 476 sq km/184 sq mi
towns Kastron, Mudros
physical of volcanic origin, rising to 430 m/ 1,411 ft
products mulberries and other fruit, tobacco, sheep
population (1981) 15,700.

Lena /ˈliːnə/, Russian /ˈljenə/ longest river in Asiatic Russia, 4,400 km/2,730 mi, with numerous tributaries. Its source is near Lake Baikal, and it empties into the Arctic Ocean through a delta 400 km/240 mi wide. It is ice-covered for half the year.

Leninakan /ˌlenɪnəˈkɑːn/ former name (to 1990) of ▷Kumayri, a town in Armenia.

Leningrad /ˈlenɪngræd/ former name (1924–91) of the Russian city ▷St Petersburg.

Leninsk-Kuznetsky /ˈlenɪnsk kuzˈnetski/ town in Kemerovo region, Siberian Russia, on the Inya River, 320 km/200 mi SSE of Tomsk; population (1985) 110,000. It is a mining centre. Formerly *Kolchugino*, the town was renamed Leninsk-Kuznetsky 1925.

León /leɪˈɒn/ city in W Nicaragua; population (1985) 101,000. Industries include textiles and food processing. Founded 1524, it was the capital of Nicaragua until 1855.

It had the appearance of old and aristocratic respectability, which no other city in Central America possessed.

On **León** John Lloyd Stephens *Incidents of Travel in Central America, Chiapas and Yucatan* 1841–43

León /leɪˈɒn/ city in Castilla-León, Spain; population (1991) 146,300. It was the capital of the kingdom of Léon from the 10th century until 1230, when it was merged with Castile.

Léon de los Aldamas /leɪˈɒn/ industrial city (leather goods, footwear) in central Mexico; population (1986) 947,000.

Léopoldville /liːəpəʊldvɪl/ former name (until 1966) of ◊Kinshasa, a city in Zaire.

Le Puy /ləˈpwiː/ capital of Haute-Loire *département*, Auvergne, SE France; population (1990) 23,400. It is dramatically situated on a rocky plateau, and has a 12th-century cathedral.

Lérida /ˈlerɪdə/ (Catalan *Lleida*) capital of Lérida province, N Spain, on the river Segre; 132 km/82 mi W of Barcelona; population (1991) 119,200. Industries include leather, paper, glass, and cloth. Lérida was captured by the Roman general Julius Caesar 49 BC. It has a palace of the kings of Aragon.

Lerwick /ˈlɜːwɪk/ port in Shetland, Scotland; population (1987 est) 7,500. It is the administrative headquarters of Shetland. Main occupations include fishing and oil. Hand-knitted shawls are a speciality. A Viking tradition survives in the Jan festival of Up-Helly-Aa when a replica of a longship is burned.

Lesbos /ˈlezbɒs/ alternative spelling of ◊Lesvos, an island in the Aegean Sea.

Lesotho Kingdom of
area 30,355 sq km/11,717 sq mi
capital Maseru
towns Teyateyaneng, Mafeteng, Roma, Quthing
physical mountainous with plateaus, forming part of South Africa's chief watershed
features Lesotho is an enclave within South Africa
head of state King Letsie III from 1990
head of government Ntsu Mokhehle from 1993
political system constitutional monarchy
political parties Basotho National Party

(BNP), traditionalist, nationalist; Basutoland Congress Party (BCP); Basotho Democratic Alliance (BDA)
exports wool, mohair, diamonds, cattle, wheat, vegetables
currency maluti
population (1990 est) 1,757,000; growth rate 2.7% p.a.
life expectancy men 59, women 62 (1989)
languages Sesotho, English (official), Zulu, Xhosa
religions Protestant 42%, Roman Catholic 38%
literacy 59% (1988)
GNP $408 million; $410 per head (1988)
chronology
1868 Basutoland became a British protectorate.
1966 Independence achieved from Britain, within the Commonwealth, as the Kingdom of Lesotho, with Moshoeshoe II as king and Chief Leabua Jonathan as prime minister.
1970 State of emergency declared and constitution suspended.
1973 Progovernment interim assembly established; BNP won majority of seats.
1975 Members of the ruling party attacked by guerrillas backed by South Africa.
1985 Elections cancelled because no candidates opposed BNP.
1986 South Africa imposed border blockade, forcing deportation of 60 African National Congress members. General Lekhanya ousted Chief Jonathan in coup. National assembly abolished. Highlands Water Project agreement signed with South Africa.
1990 Moshoeshoe II dethroned by military council; replaced by his son Mohato as King Letsie III.
1991 Lekhanya ousted in military coup led by Col Elias Tutsoane Ramaema. Political parties permitted to operate.
1992 Ex-king Moshoeshoe returned from exile.
1993 Free elections ended military rule; Ntsu Mokhehle of BCP became prime minister.

Lesvos /ˈlezvɒs/ Greek island in the Aegean Sea, near the coast of Turkey
area 2,154 sq km/831 sq mi
capital Mytilene
products olives, wine, grain
population (1981) 104,620
history ancient name Lesbos; an Aeolian settlement, the home of the poets Alcaeus and Sappho; conquered by the Turks from Genoa 1462; annexed to Greece 1913.

Letchworth /ˈletʃwəθ/ town in Hertfordshire, England, 56 km/35 mi NNW of London; population (1989 est) 32,700. Industries include clothing, furniture, scientific instruments, light metal goods, and printing. It was founded 1903 as the first English garden city.

Le Touquet /lə ˈtuːkeɪ/ resort in N France, at the mouth of the river Canche; it was fashionable in the 1920s–1930s.

Levant /lɪˈvænt/ the E Mediterranean region, or more specifically, the Mediterranean coastal regions of Turkey, Syria, Lebanon, and Israel.

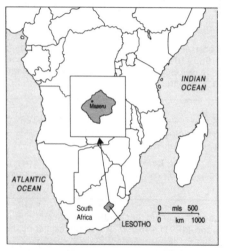

Leven /'liːvən/ town in Fife Region, Scotland; at the mouth of the river Leven, where it meets the Firth of Forth; population (1987 est) 9,200. It has timber, paper, and engineering industries.

Leven, Loch /'liːvən/ lake in Tayside Region, Scotland; area 16 sq km/6 sq mi. It is drained by the river Leven, and has seven islands; Mary Queen of Scots was imprisoned 1567–68 on Castle Island. It has been a national nature reserve since 1964. Leven is also the name of a sea loch in Strathclyde, Scotland.

Leverkusen /'leɪvəkuːzən/ river port in North Rhine–Westphalia, Germany, 8 km/5 mi N of Cologne; population (1988) 155,000. It has iron, steel, and chemical industries.

Lewes /'luːɪs/ market town (administrative headquarters) in E Sussex, England, on the river Ouse; population (1989 est) 15,000. There is light engineering, brewing, and printing. The Glyndebourne music festival is held nearby. Simon de Montfort defeated Henry III here 1264; there is a house which once belonged to Anne of Cleves, and a castle. The town is known for its 5th Nov celebrations.

Lewisham /'luːɪʃəm/ inner borough of SE Greater London
features Deptford shipbuilding yard (1512–1869), the explorer Francis Drake was knighted and Peter the Great, Tsar of Russia, worked here; Crystal Palace (re-erected at Sydenham 1854) site now partly occupied by the National Sports Centre; the poet James Elroy Flecker was born here
population (1991) 215,300.

Lewiston /'luːɪstən/ city in SW Maine, USA, across the Androscoggin River from Auburn; population (1990) 39,750. It has textile, shoe, and clothing industries. Bates College 1855 is here.

Lewis with Harris /'luːɪs/ largest island in the Outer Hebrides; area 2,220 sq km/857 sq mi; population (1981) 23,400. Its main town is Stornoway. It is separated from NW Scotland by the Minch. There are many lakes and peat moors. Harris is famous for its tweeds.

Lexington /'leksɪŋtən/ or *Lexington–Fayette* town in Kentucky, USA, centre of the bluegrass country; population (1990) 225,400. Racehorses are bred in the area, and races and shows are held.
There is a tobacco market and the University of Kentucky (1865).

Lexington /'leksɪŋtən/ town in Massachusetts, USA; population (1990) 28,970. Industries include printing and publishing. The Battle of Lexington and Concord 19 April 1775 opened the American War of Independence.

Leyden alternative form of ♢Leiden, a city in the Netherlands.

Leyland /'leɪlənd/ industrial town in Lancashire, England; population (1988 est) 100,200. Industries include motor vehicles, paint, and rubber.

The Rover Group (previously British Leyland), the largest of British firms producing cars, buses, and lorries, has its headquarters here.

Lhasa /'lɑːsə/ ('the Forbidden City') capital of the autonomous region of Tibet, China, at 5,000 m/16,400 ft; population (1982) 105,000. Products include handicrafts and light industry. The holy city of Lamaism, Lhasa was closed to Westerners until 1904, when members of a British expedition led by Col Francis E Younghusband visited the city. It was annexed with the rest of Tibet 1950–51 by China, and the spiritual and temporal head of state, the Dalai Lama, fled in 1959 after a popular uprising against Chinese rule. Monasteries have been destroyed and monks killed, and an influx of Chinese settlers has generated resentment. In 1988 and 1989 nationalist demonstrators were shot by Chinese soldiers.

Liao /li'au/ river in NE China, frozen Dec–March; the main headstream rises in the mountains of Inner Mongolia and flows E, then S to the Gulf of Liaodong; length 1,450 km/900 mi.

Liaoning /li,au'nɪŋ/ province of NE China
area 151,000 sq km/58,300 sq mi
capital Shenyang
towns Anshan, Fushun, Liaoyang
features one of China's most heavily industrialized areas
products cereals, coal, iron, salt, oil
population (1990) 39,460,000
history developed by Japan 1905–45, including the *Liaodong Peninsula*, whose ports had been conquered from the Russians.

Liaoyang /li'au 'jæŋ/ industrial city (engineering, textiles) in Liaoning province, China; population (1970) 250,000. In 1904 Russia was defeated by Japan here.

Libau /'liːbau/ German name of the Latvian port ♢Liepāja.

Liberia Republic of
area 111,370 sq km/42,989 sq mi
capital and port Monrovia
towns ports Buchanan, Greenville
physical forested highlands; swampy tropical coast where six rivers enter the sea

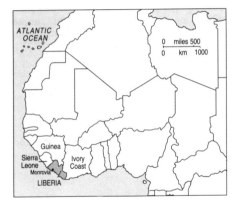

features nominally the world's largest merchant navy as minimal registration controls make Liberia's a flag of convenience; the world's largest rubber plantations
head of state and government Bismark Kuyon heading collective presidency
political system emergent democratic republic
political parties National Democratic Party of Liberia (NDLP), nationalist; Liberian Action Party (LAP); Liberian Unity Party (LUP); United People's Party (UPP); Unity Party (UP)
exports iron ore, rubber (Africa's largest producer), timber, diamonds, coffee, cocoa, palm oil
currency Liberian dollar
population (1990 est) 2,644,000 (95% indigenous); growth rate 3% p.a.
life expectancy men 53, women 56 (1989)
languages English (official), over 20 Niger-Congo languages
media two daily newspapers, one published under government auspices, the other independent and with the largest circulation
religions animist 65%, Muslim 20%, Christian 15%
literacy men 47%, women 23% (1985 est)
GNP $973 million; $450 per head (1988)
chronology
1847 Founded as an independent republic.
1944 William Tubman elected president.
1971 Tubman died; succeeded by William Tolbert.
1980 Tolbert assassinated in coup led by Samuel Doe, who suspended the constitution and ruled through a People's Redemption Council.
1984 New constitution approved. NDPL founded by Doe.
1985 NDPL won decisive victory in allegedly rigged general election. Unsuccessful coup against Doe.
1990 Rebels under former government minister Charles Taylor controlled nearly entire country by July. Doe killed during a bloody civil war between rival rebel factions. Amos Sawyer became interim head of government.
1991 Amos Sawyer re-elected president. Rebel leader Charles Taylor agreed to work with Sawyer. Peace agreement failed but later revived; UN peacekeeping force drafted into republic.
1992 Monrovia under siege by Taylor's rebel forces.
1993 Peace agreement between opposing groups signed in Benin, under OAU/UN auspices. Interim collective presidency established prior to elections, planned for 1994.

Libreville /ˈliːbrəviːl/ (French 'free town') capital of Gabon, on the estuary of the river Gabon; population (1988) 352,000. Products include timber, oil, and minerals. It was founded 1849 as a refuge for slaves freed by the French. Since the 1970s the city has developed rapidly due to the oil trade.
 Libreville was capital of French Equatorial Guinea 1888–1904.

Libya Great Socialist People's Libyan Arab Jamahiriya (*al-Jamahiriya al-Arabiya al-Libya al-Shabiya al-Ishtirakiya al-Uzma*)
area 1,759,540 sq km/679,182 sq mi

capital Tripoli
towns ports Benghazi, Misurata, Tobruk
physical flat to undulating plains with plateaus and depressions stretch S from the Mediterranean coast to an extremely dry desert interior
environment plan to pump water from below the Sahara to the coast risks rapid exhaustion of nonrenewable supply
features Gulf of Sirte; rock paintings of about 3000 BC in the Fezzan; Roman city sites include Leptis Magna, Sabratha
head of state and government Moamer al-Khaddhafi from 1969
political system one-party socialist state
political party Arab Socialist Union (ASU), radical, left-wing
exports oil, natural gas
currency Libyan dinar
population (1990 est) 4,280,000 (including 500,000 foreign workers); growth rate 3.1% p.a.
life expectancy men 64, women 69 (1989)
language Arabic
religion Sunni Muslim 97%
literacy 60% (1989)
GNP $20 bn; $5,410 per head (1988)
chronology
1911 Conquered by Italy.
1934 Colony named Libya.
1942 Divided into three provinces: Fezzan (under French control), Cyrenaica, Tripolitania (under British control).
1951 Achieved independence as the United Kingdom of Libya, under King Idris.
1969 King deposed in a coup led by Col Moamer al-Khaddhafi. Revolution Command Council set up and the Arab Socialist Union (ASU) proclaimed the only legal party.
1972 Proposed federation of Libya, Syria, and Egypt abandoned.
1980 Proposed merger with Syria abandoned. Libyan troops began fighting in Chad.
1981 Proposed merger with Chad abandoned.
1986 US bombing of Khaddhafi's headquarters, following allegations of his complicity in terrorist activities.
1988 Diplomatic relations with Chad restored.
1989 USA accused Libya of building a chemical-weapons factory and shot down two Libyan planes; reconciliation with Egypt.

1992 Khaddhafi under international pressure to extradite suspected Lockerbie and UTA (Union de Transports Aerians) bombers for trial outside Libya; sanctions imposed.

Lichfield /'lɪtʃfiːld/ town in the Trent Valley, Staffordshire, England; population (1987 est) 28,300. The cathedral, 13th–14th century, has three spires. The writer Samuel Johnson was born here.

Liechtenstein Principality of (*Fürstentum Liechtenstein*)
area 160 sq km/62 sq mi
capital Vaduz
towns Balzers, Schaan, Ruggell
physical landlocked Alpine; includes part of Rhine Valley in W
features no airport or railway station; easy tax laws make it an international haven for foreign companies and banks (some 50,000 companies are registered)
head of state Prince Hans Adam II from 1989
head of government Hans Brunhart from 1978
political system constitutional monarchy
political parties Fatherland Union (VU); Progressive Citizens' Party (FBP)
exports microchips, dental products, small machinery, processed foods, postage stamps
currency Swiss franc
population (1990 est) 30,000 (33% foreign); growth rate 1.4% p.a.
life expectancy men 78, women 83 (1989)
languages German (official); an Alemannic dialect is also spoken
religions Roman Catholic 87%, Protestant 8%
literacy 100% (1989)
GNP $450 million (1986)
GDP $1 bn (1987); $32,000 per head
chronology
1342 Became a sovereign state.
1434 Present boundaries established.
1719 Former counties of Schellenberg and Vaduz constituted as the Principality of Liechtenstein.
1921 Adopted Swiss currency.
1923 United with Switzerland in a customs union.
1938 Prince Franz Josef II came to power.

1984 Prince Franz Joseph II handed over power to Crown Prince Hans Adam. Vote extended to women in national elections.
1989 Prince Franz Joseph II died; Hans Adam II succeeded him.
1990 Became a member of the United Nations (UN).
1991 Became seventh member of European Free Trade Association (EFTA).

Liège /li'eɪʒ/ (German *Luik*) industrial city (weapons, textiles, paper, chemicals), capital of Liège province in Belgium, SE of Brussels, on the river Meuse; population (1991) 194,600. The province of Liège has an area of 3,900 sq km/ 1,505 sq mi and a population (1991) of 999,600.

Liepāja /lɪ'paɪə/ (German *Libau*) naval and industrial port in Latvia; population (1985) 112,000. The Knights of Livonia founded Liepāja in the 13th century. Industries include steel, engineering, textiles, and chemicals.

Liffey /'lɪfi/ river in the eastern Republic of Ireland, flowing from the Wicklow Mountains to Dublin Bay; length 80 km/50 mi.

Liguria /lɪ'gjʊərɪə/ coastal region of NW Italy, which includes the resorts of the Italian Riviera, lying between the western Alps and the Mediterranean Gulf of Genoa. The region comprises the provinces of Genova, La Spezia, Imperia, and Savona, with an area of 5,418 sq km/2,093 sq mi and a population (1990) of 1,719,200. Genoa is the chief town and port.

Lille /liːl/ (Flemish *Ryssel*) industrial city (textiles, chemicals, engineering, distilling), capital of Nord-Pas-de-Calais, France, on the river Deûle; population (1990) 178,300, metropolitan area 936,000. The world's first entirely automatic underground system was opened here 1982.
    Originally a village on an island (*l'île* island), it was fortified in the 11th century. During the Middle Ages it was capital of Flanders. It was claimed by Louis XIV 1667, captured by the duke of Marlborough 1708, and ceded to France by the Treaty of Utrecht 1713. There is a 17th-century fortress (built by Sébastien Le Prestre de Vauban), a Pasteur Institute, and a university (1887).

Lilongwe /lɪ'lɒŋweɪ/ capital of Malawi since 1975, on the Lilongwe River; population (1987) 234,000. Products include tobacco and textiles. Capital Hill, 5 km/3 mi from the old city, is the site of government buildings and offices.

Lima /'liːmə/ capital of Peru, an industrial city (textiles, chemicals, glass, cement), with its port at Callao; population (1988) 418,000, metropolitan area 4,605,000. Founded by the Spanish conquistador Francisco Pizarro 1535, it was rebuilt after destruction by an earthquake 1746.
    Survivals of the colonial period are the university (1551), cathedral (1746), government palace (the rebuilt palace of the viceroys), and the senate house (once the headquarters of the Inquisition).

Lima /'laɪmə/ city in NW Ohio, USA, on the Ottawa River, N of Dayton; seat of Allen County; population (1990) 45,550. Industries include motor-vehicle and aircraft parts, heavy machinery, electrical products, and oil processing.

Limassol /'lɪməsɒl/ port in S Cyprus in Akrotiri Bay; population (1985) 120,000. Products include cigarettes and wine. Richard I of England married Berengaria of Navarre here 1191. The town's population increased rapidly with the influx of Greek Cypriot refugees after the Turkish invasion 1974.

Limbourg /læm'buə/ province of N Belgium, bounded N and E by the Netherlands
area 2,422 sq km/935 sq mi
population (1991) 750,400
capital Hasselt
towns Genk, Tongeren
products sugar refining, food processing
physical river Demer; Kempen heathland in the N; rich coalfields; agriculture in the S
history formerly part of the feudal duchy of Limburg (which was divided 1839 into today's Belgian and Dutch provinces).

Limburg /'lɪmbɜːg/ southernmost province of the Netherlands
area 2,170 sq km/838 sq mi
capital Maastricht
towns Heerlen, Roermond, Weert
physical river Maas (Meuse); sandy soils in river plain, marl soils in S; becomes hilly towards S
features a monument marks the *Drielandenpunt*, where the Dutch, German, and Belgian borders meet
products chemicals, cement, fertilizer; mixed arable farming and horticulture are also important. The former coal industry is still remembered at Kerkrade, alleged site of the first European coal mine
population (1991) 1,109,900
history formerly part of the duchy of Limburg (which was divided 1839 into today's Dutch and Belgian provinces).

Limehouse /'laɪmhaus/ district in E London; part of Tower Hamlets.

Limerick /'lɪmərɪk/ county in the SW Republic of Ireland, in Munster province
area 2,690 sq km/1,038 sq mi
county town Limerick
physical fertile, with hills in the S
products dairy products
population (1991) 161,900.

Limerick /'lɪmərɪk/ county town of Limerick, Republic of Ireland, the main port of W Ireland, on the Shannon estuary; population (1991) 52,000. It was founded in the 12th century.

Limoges /lɪ'məuʒ/ city and capital of Limousin, France; population (1990) 136,400. Fine enamels were made here in the medieval period, and it is the centre of the modern French porcelain industry. Other industries include textiles, electrical equipment, and metal goods. The city was sacked by the Black Prince, the eldest son of Edward III of England, 1370.

Limousin /ˌlɪmu:'zæn/ former province and modern region of central France; area 16,900 sq km/

6,544 sq mi; population (1986) 736,000. It consists of the *départements* of Corréze, Creuse, and Haute-Vienne. The chief town is Limoges. A thinly populated and largely unfertile region, it is crossed by the mountains of the Massif Central. Fruit and vegetables are produced in the more fertile lowlands. Kaolin is mined.

... the country that has bred more popes and fewer lovers than any other in the world.

On **Limousin** Jean Giraudoux 1953

Limpopo /lɪm'pəupəu/ river in SE Africa, rising in the Transvaal and reaching the Indian Ocean in Mozambique; length 1,600 km/1,000 mi.

The great greygreen, greasy Limpopo River, all set about with fever trees.

On the **Limpopo** Rudyard Kipling
Just So Stories 1902

Lincoln /'lɪŋkən/ industrial city in Lincolnshire, England; population (1988 est) 80,600. Manufacturing includes excavators, cranes, gas turbines, power units for oil platforms, and cosmetics. Under the Romans it was the flourishing colony of *Lindum*, and in the Middle Ages it was a centre for the wool trade. Paulinus built a church here in the 7th century, and the 11th–15th-century cathedral has the earliest Gothic work in Britain. The 12th-century High Bridge is the oldest in Britain still to have buildings on it.

Lincoln /'lɪŋkən/ industrial city and capital of Nebraska, USA; population (1990) 192,000. Industries include engineering, pharmaceuticals, electronic and electrical equipment, and food processing. It was known as *Lancaster* until 1867, when it was renamed after Abraham Lincoln and designated the state capital.

Lincolnshire /'lɪŋkənʃə/ county in E England
area 5,890 sq km/2,274 sq mi
towns Lincoln (administrative headquarters), Skegness
physical Lincoln Wolds; marshy coastline; the Fens in the SE; rivers: Witham, Welland
features 16th-century Burghley House; Belton House, a Restoration mansion
products cattle, sheep, horses, cereals, flower bulbs, oil
population (1991) 573,900
famous people Isaac Newton, Alfred Tennyson, Margaret Thatcher, John Wesley.

Lincs abbreviation for *Lincolnshire*, an English county.

Lindisfarne /'lɪndɪsfɑːn/ site of a monastery off the coast of Northumberland, England; see ✧Holy Island.

Line Islands /laɪn/ group of coral islands in the Pacific Ocean; population (1990) 4,800. Products include coconut and guano. Eight of the islands belong to Kiribati, and two (Palmyra and Jarvis) are administered by the USA.

Linköping /'lɪntʃɜːpɪŋ/ industrial town in SE Sweden; 172 km/107 mi SW of Stockholm; population (1990) 122,300. Industries include hosiery, aircraft and engines, and tobacco. It has a 12th-century cathedral.

Linlithgow /lɪn'lɪθgəu/ tourist centre in Lothian Region, Scotland; population (1987 est) 11,300. Linlithgow Palace, now in ruins, was once a royal residence, and Mary Queen of Scots was born there.

Linlithgowshire /lɪn'lɪθgəuʃə/ former name of West Lothian, now included in Lothian Region, Scotland.

Linz /lɪnts/ industrial port (iron, steel, metalworking) on the river Danube in N Austria; population (1981) 199,900. It is associated with the composers Mozart and Bruckner.

Lipari Islands /'lɪpəri/ or *Aeolian Islands* volcanic group of seven islands off NE Sicily, including *Lipari* (on which is the capital of the same name), *Stromboli* (active volcano 926 m/3,038 ft high), and *Vulcano* (also with an active volcano); area 114 sq km/44 sq mi. In Greek mythology, the god Aeolus kept the winds imprisoned in a cave on the Lipari Islands.

Lippe /'lɪpə/ river of N Germany flowing into the river Rhine; length 230 km/147 mi; also a former German state, now part of North Rhine–Westphalia.

Lisboa /liːʒ'bɔːə/ Portuguese form of ◊ Lisbon, the capital of Portugal.

Lisbon /'lɪzbən/ (Portuguese *Lisboa*) city and capital of Portugal, in the SW of the country,

on the tidal lake and estuary formed by the river Tagus; population (1984) 808,000. Industries include steel, textiles, chemicals, pottery, shipbuilding, and fishing. It has been the capital since 1260 and reached its peak of prosperity in the period of Portugal's empire during the 16th century. In 1755 an earthquake killed 60,000 people and destroyed much of the city.

Lisburn /'lɪzbɜːn/ cathedral city and market town in Antrim, Northern Ireland, on the river Lagan; population (1989 est) 95,500. It produces linen and furniture.

Lisieux /liːz'jɜː/ town in Calvados *département*, France, SE of Caen; population (1990) 24,500. St Thérèse of Lisieux spent her religious life in the Carmelite convent here, and pilgrims visit her tomb.

Litani /lɪ'tɑːni/ river rising near Baalbek in the Anti-Lebanon Mountains of E Lebanon. It flows NE–SW through the Bekka Valley, then E to the Mediterranean 8 km/5 mi N of Tyre. Israeli forces invaded Lebanon as far as the Litani River 1978.

Lithuania Republic of
area 65,200 sq km/25,174 sq mi
capital Vilnius
towns Kaunas, Klaipeda, Siauliai, Panevezys
physical central lowlands with gentle hills in W and higher terrain in SE; 25% forested; some 3,000 small lakes, marshes, and complex sandy coastline
features river Nemen; white sand dunes on Kursiu Marios lagoon
head of state Vytautas Landsbergis from 1990
head of government Aleksandras Abisala from 1992
political system emergent democracy
political parties Sajudis (Lithuanian Restructuring Movement), nationalist; Democratic Party, centrist; Humanism and Progress Party, reformist; Social Democratic Party, left of centre; Green Party, ecological; Christian Democratic Party, right of centre; Democratic Labour Party, 'reform communist'
products heavy engineering, electrical goods, shipbuilding, cement, food processing, bacon,

Lincolnshire

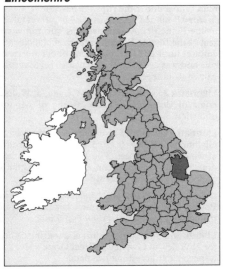

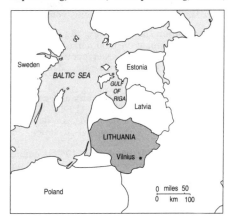

dairy products, cereals, potatoes
currency Lithuanian rouble
population (1990) 3,700,000 (Lithuanian 80%, Russian 9%, Polish 7%, Byelorussian 2%)
language Lithuanian
religion predominantly Roman Catholic
chronology
1918 Independence declared following withdrawal of German occupying troops at end of World War I; USSR attempted to regain power.
1919 Soviet forces overthrown by Germans, Poles, and nationalist Lithuanians; democratic republic established.
1920–39 Province and city of Vilnius occupied by Poles.
1926 Coup overthrew established government; Antanas Smetona became president.
1939 Secret German-Soviet agreement brought most of Lithuania under Soviet influence.
1940 Incorporated into USSR as constituent republic.
1941 Lithuania revolted against USSR and established own government. During World War II Germany again occupied the country.
1944 USSR resumed rule.
1944–52 Lithuanian guerrillas fought USSR.
1972 Demonstrations against Soviet government.
1980 Growth in nationalist dissent, influenced by Polish example.
1988 Popular front formed, the Sajudis, to campaign for increased autonomy.
1989 Lithuanian declared the state language; flag of independent interwar republic readopted. Communist Party (CP) split into pro-Moscow and nationalist wings. Communist local monopoly of power abolished.
1990 Feb: nationalist Sajudis won elections. March: Vytautas Landsbergis became president; unilateral declaration of independence resulted in temporary Soviet blockade.
1991 Jan: Albertas Shiminas became prime minister. Soviet paratroopers briefly occupied key buildings in Vilnius. Sept: independence recognized by Soviet government and Western nations; Gediminas Vagnorius elected prime minister; CP outlawed; admitted into United Nations (UN) and Conference on Security and Cooperation in Europe (CSCE).
1992 July: Aleksandras Abisala became prime minister. Nov: Democratic Labour Party, led by Algirdas Brazauskas, won majority vote.

Littlehampton /ˌlɪtl'hæmptən/ seaside resort in W Sussex, England, at the mouth of the river Arun, 16 km/10 mi SE of Chichester; population (1981) 22,000. There is light engineering and production of foodstuffs.

Little Rock /'lɪtl rɒk/ largest city and capital of Arkansas, USA; population (1990) 175,800. Products include metal goods, oil-field and electronic equipment, chemicals, clothing, and processed food.

Federal troops were sent here 1957 to enforce the integration of all-white Central High School.

Liverpool /'lɪvəpuːl/ city, seaport, and administrative headquarters of Merseyside, NW England; population (1991 est) 448,300. In the 19th and early 20th centuries it exported the textiles of Lancashire and Yorkshire. Liverpool is the UK's chief Atlantic port with miles of specialized, mechanized quays on the river Mersey.

The Liverpool Institute for the Performing Arts occupies the old Liverpool Institute where former Beatles Paul McCartney and George Harrison went to school. It is planned as an educational centre for people involved in graphic design, performance, production, and video.

Livingston /'lɪvɪŋstən/ industrial new town (electronics, engineering) in W Lothian, Scotland, established 1962; population (1985) 40,000.

Livingstone /'lɪvɪŋstən/ (formerly **Maramba**) town in Zambia; population (1988) 98,500. Founded 1905, it was named after the Scottish explorer David Livingstone, and was capital of N Rhodesia 1907–35. The Victoria Falls are nearby.

Livorno /lɪ'vɔːnəʊ/ (English **Leghorn**) industrial port in W Italy; population (1988) 173,000. Industries include shipbuilding, distilling, and motor vehicles. A fortress town since the 12th century, it was developed by the Medici family. It has a naval academy and is also a resort.

A thriving, business-like, matter-of-fact place, where idleness is shouldered out of the way by commerce.

On **Livorno** Charles Dickens *Pictures from Italy* 1846

Lizard Point /'lɪzəd/ southernmost point of England in Cornwall. The coast is broken into small bays overlooked by two cliff lighthouses.

Ljubljana /luː'bljɑːnə/ (German **Laibach**) capital and industrial city (textiles, chemicals, paper, leather goods) of Slovenia, near the confluence of the rivers Ljubljanica and Sava; population (1981) 305,200. It has a nuclear research centre and is linked with S Austria by the Karawanken road tunnel under the Alps (1979–83).
history Founded by Augustus 34 BC, it was taken by the French 1809 and made the governmental seat of the Illyrian Provinces (part of Napoleon's empire) until 1813. It then came under Austrian rule and was ceded to Yugoslavia 1918; Slovenia gained its independence 1992.

Llanberis /ɬæn'berɪs/ village in Gwynedd, Wales, point of departure for the ascent of Mount Snowdon.

Llandaff /'ɬændəf/ (Welsh **Llandaf**) town in S Glamorgan, Wales, 5 km/3 mi NW of Cardiff, of which it forms part. The 12th-century cathedral, heavily restored, contains Jacob Epstein's sculpture 'Christ in Majesty'.

Llandrindod Wells /ɬæn'drɪndɒd 'welz/ spa in Powys, E Wales, administrative headquarters of the county; population (1989 est) 5,000.

Llandudno /ɬæn'dɪdnəʊ/ resort and touring centre for N Wales, in Gwynedd. Great Orme's Head is a spectacular limestone headland.

Llanelli /ɬæn'eɪi/ (formerly **Llanelly**) industrial port in Dyfed, Wales; population (1989 est) 41,400. Industries include tinplate and copper smelting.

Llanfair P G /ɬænvaɪə/ village in Anglesey, Wales; full name *Llanfairpwllgwyngyllgogery-chwyrndrobwllllantysiliogogogoch* (St Mary's church in the hollow of the white hazel near the rapid whirlpool of St Tysillio's church, by the red cave), the longest place name in the UK.

Lleyn /ɬiːn/ (Welsh *Llŷn*) peninsula in Gwynedd, N Wales, between Cardigan Bay and Caernarvon Bay. It includes the resort of Pwllheli. Bardsey Island at its tip is the traditional burial place of 20,000 saints.

Lobito /lə'biːtəu/ port in Angola; population (1970) 60,000. It is linked by rail with Beira in Mozambique, via the Zaire and Zambia copperbelt.

Locarno /lə'kɑːnəu/ health resort in the Ticino canton of Switzerland on the north shore of Lago Maggiore, W of Bellinzona; population (1990) 14,200.
    Formerly in the duchy of Milan, it was captured by the Swiss 1803.

Lochaber /lɒ'xɑːbə/ wild mountainous district of Highland Region, Scotland, including Ben Nevis, the highest mountain in the British Isles. Fort William is the chief town of the area. It is the site of large hydroelectric installations.

Loch Ness /lɒx 'nes/ see ⟡Ness, Loch.

Lodi /'ləudi/ town in Italy, 30 km/18 mi SE of Milan; population (1980) 46,000. It is a market centre for agricultural produce; fertilizers, agricultural machinery, and textiles are produced. Napoleon's defeat of Austria at the battle of Lodi 1796 gave him control of Lombardy.

Łódź /lɒdz, Polish wuːtʃ/ industrial town (textiles, machinery, dyes) in central Poland, 120 km/75 mi SW of Warsaw; population (1990) 848,300.

Lofoten and Vesterålen /'ləufəutn, 'vestərɑːlən/ island group off NW Norway; area 4,530 sq km/ 1,750 sq mi. Hinnøy, in the Vesterålens, is the largest island of Norway. The surrounding waters are rich in cod and herring. The **Maelström**, a large whirlpool hazardous to ships, which gives its name to similar features elsewhere, occurs in one of the island channels.

Logroño /lə'grəunjəu/ market town and capital of La Rioja, N Spain, on the river Ebro; population (1991) 126,800. It is the centre of a wine-producing region.

Loir /lwɑː/ French river, rising N of Illiers in the *département* of Eure-et-Loir and flowing SE, then SW to join the Sarthe near Angers; 310 km/ 500 mi. It gives its name to the *départements* of Loir-et-Cher and Eure-et-Loir.

Loire /lwɑː/ longest river in France, rising in the Cévennes Mountains, at 1,350 m/4,430 ft and flowing for 1,050 km/650 mi first N then W

until it reaches the Bay of Biscay at St Nazaire, passing Nevers, Orléans, Tours, and Nantes. It gives its name to the *départements* of Loire, Haute-Loire, Loire-Atlantique, Indre-et-Loire, Maine-et-Loire, and Saône-et-Loire. There are many châteaux and vineyards along its banks.

One of the most wonderful rivers in the world, mirroring from sea to source a hundred cities and five hundred towers.

On the **Loire** Oscar Wilde, letter of 1880

Loiret /lwɑː'reɪ/ French river, 11 km/7 mi long, rising near Olivet and joining the Loire 8 km/ 5 mi below Orléans. It gives its name to Loiret *département*.

Lombardy /'lɒmbədi/ (Italian *Lombardia*) region of N Italy, including Lake Como; capital Milan; area 23,900 sq km/9,225 sq mi; population (1990) 8,939,400. It is the country's chief industrial area (chemicals, pharmaceuticals, engineering, textiles).

Lombok /'lɒmbɒk/ (Javanese 'chili pepper') island of Indonesia, E of Java, one of the Sunda Islands; area 4,730 sq km/1,826 sq mi; population (1980) 1,957,000. The chief town is Mataram. It has a fertile plain between northern and southern mountain ranges.

Lomé /'ləumeɪ/ capital and port of Togo; population (1983) 366,000. It is a centre for gold, silver, and marble crafts; industries include steel production and oil refining.

Lomond, Loch /'ləumənd/ largest freshwater Scottish lake, 37 km/21 mi long, area 70 sq km/ 27 sq mi, divided between Strathclyde and Central regions. It is overlooked by the mountain **Ben Lomond** (973 m/3,192 ft) and is linked to the Clyde estuary.

Lompoc /'lɒmpəuk/ city in SW California, USA, near the Pacific Ocean, W of Santa Barbara; population (1990) 37,650. Industries include the processing of oil from the city's oil wells.

London /'lʌndən/ capital of England and the United Kingdom, on the river Thames; area 1,580 sq km/610 sq mi; population (1991) 6,378,600, larger metropolitan area about 9 million. The *City of London*, known as the 'square mile', area 274 hectares/677 acres, is the financial and commercial centre of the UK. *Greater London* (see ⟡London, Greater) from 1965 comprises the City of London and 32 boroughs. Popular tourist attractions include the Tower of London, St Paul's Cathedral, Buckingham Palace, and Westminster Abbey.
    Roman *Londinium* was established soon after the Roman invasion AD 43; in the 2nd century London became a walled city; by the 11th century, it was the main city of England and gradually extended beyond the walls to link with the originally separate Westminster. Throughout the 19th century London was the largest city in the world (in population).

features The Tower of London, built by William the Conqueror on a Roman site, houses the crown jewels and the royal armouries; 15th-century Guildhall; the Monument (a column designed by Christopher Wren) marks the site in Pudding Lane where the Great Fire of 1666 began; Mansion House (residence of the lord mayor); Barbican arts and conference centre; Central Criminal Court (Old Bailey) and the Inner and Middle Temples; Covent Garden, once a vegetable market, is now a tourist shopping and entertainment area.

architecture London contains buildings in all styles of English architecture since the 11th century. *Norman*: the White Tower, Tower of London; St Bartholomew's, Smithfield; the Temple Church. *Gothic*: Westminster Abbey; Westminster Hall; Lambeth Palace; Southwark Cathedral. *Tudor*: St James's Palace; Staple Inn. *17th century*: Banqueting Hall, Whitehall (Inigo Jones); St Paul's, Kensington Palace; many City churches (Wren). *18th century*: Somerset House (Chambers); St Martin-in-the-Fields; Buckingham Palace. *19th century*: British Museum (Neo-Classical); Houses of Parliament; Law Courts (Neo-Gothic); Westminster Cathedral (Byzantine style). *20th century*: Lloyd's of London.

government There has since 1986 been no central authority for Greater London; responsibility is divided between individual boroughs and central government.

The City of London has been governed by a corporation from the 12th century. Its structure and the electoral procedures for its common councillors and aldermen are medievally complex, and it is headed by the lord mayor (who is, broadly speaking, nominated by the former and elected annually by the latter). After being sworn in at the Guildhall, he or she is presented the next day to the lord chief justice at the Royal Courts of Justice in Westminster, and the *Lord Mayor's Show* is a ceremonial procession there in November.

commerce and industry From Saxon times the Port of London dominated the Thames from Tower Bridge to Tilbury; its activity is now centred outside the metropolitan area, and

London: population figures

year	population
1563	+90,000
1583	c. 120,000
1600	-200,000
1666	c. 350,000
c. 1700	650,000
1800	865,000
1821	+1,000,000
1830	1,500,000
1888	-5,000,000
1900	4,500,000*
1951	8,346,000
1961	7,992,600
1971	7,452,500
1981	6,696,200
1991	6,378,600

* county of London

Greater London

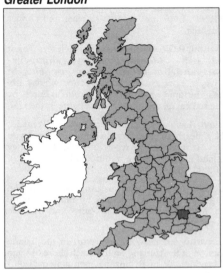

downstream Tilbury has been extended to cope with container traffic. The prime economic importance of modern London is as a financial centre. There are various industries, mainly on the outskirts. There are also recording, broadcasting, television, and film studios; publishing companies; and the works and offices of the national press. Tourism is important. Some of the docks in the East End of London, once the busiest in the world, have been sold to the Docklands Development Corporation, which has built offices, houses, factories, and a railway. Work on *Canary Wharf*, the world's largest office development project, was temporarily halted 1992 after its developers went into receivership.

education and entertainment Museums: British, Victoria and Albert, Natural History, Science museums; galleries: National and Tate. London University is the largest in Britain, while the Inns of Court have been the training school for lawyers since the 13th century. London has been the centre of English drama since its first theatre was built by James Burbage 1576.

The Greater London Boroughs are: Barking and Dagenham, Barnet, Bexley, Brent, Bromley, Camden, Croydon, Ealing, Enfield, Greenwich, Hackney, Hammersmith and Fulham, Haringey, Harrow, Havering, Hillingdon, Hounslow, Islington, Kensington and Chelsea, Kingston upon Thames, Lambeth, Lewisham, Merton, Newham, Redbridge, Richmond-upon-Thames, Southwark, Sutton, Tower Hamlets, Waltham Forest, Wandsworth, and Westminster.

Crowds without company, and dissipation without pleasure.

On **London** Edward Gibbon *Memoirs* 1796

London /ˈlʌndən/ city in SW Ontario, Canada, on the river Thames, 160 km/100 mi SW of

Toronto; population (1986) 342,000. The centre of a farming district, it has tanneries, breweries, and factories making hosiery, radio and electrical equipment, leather, and shoes. It dates from 1826 and is the seat of the University of Western Ontario.

A Shakespeare festival is held in Stratford, about 30 km/18 mi to the NW.

Londonderry /ˌlʌndən'deri/ former name (until 1984) of the county and city of ♢Derry in Northern Ireland.

London, Greater /'lʌndən/ metropolitan area of London, England, comprising the City of London, which forms a self-governing enclave, and 32 surrounding boroughs; area 1,580 sq km/ 610 sq mi; population (1991) 6,378,600. Certain powers were exercised over this whole area by the Greater London Council (GLC) until its abolition 1986.

Long Beach /'lɒŋ biːtʃ/ city in SW California, USA; population (1990) 429,400. A port and industrial city, it also has oil wells and a naval shipyard. Manufactured goods include aircraft, ships, petroleum products, chemicals, fabricated metals, electronic equipment, and processed food. It is also a convention centre. Long Beach forms part of Greater Los Angeles and adjoins the San Pedro harbour of Los Angeles.

Longchamp /'lɒŋʃɒm/ pleasure resort and racecourse in Paris, France, in the Bois de Boulogne. It is on the site of a former nunnery founded 1260, suppressed 1790. Many races in France are run at Longchamp including the most prestigous open-age group race in Europe, the *Prix de L'Arc de Triomphe*, which attracts a top-quality field every Oct.

Longford /'lɒŋfəd/ county of Leinster province, Republic of Ireland
area 1,040 sq km/401 sq mi
county town Longford
features rivers: Camlin, Inny, Shannon (the western boundary); several lakes
population (1991) 30,300.

Long Island /ˌlɒŋ 'aɪlənd/ island E of Manhattan and SE of Connecticut, USA, separated from the mainland by Long Island Sound and the East River; 120 mi/193 km long by about 30 mi/48 km wide; area 1,400 sq mi/3,627 sq km; population (1984) 6,818,480.

It includes two boroughs of of New York City (Queens and Brooklyn), John F Kennedy airport, suburbs, and resorts.

It also has Brookhaven National Laboratory for atomic research, the world's largest automotive museum, the New York Aquarium, and a whaling museum. The popular resort of Coney Island is actually a peninsula in the SW, with a boardwalk 3 km/2 mi long.

Longmont /'lɒŋmɒnt/ city in N central Colorado, USA, in the Rocky Mountain foothills, S of Fort Collins and NE of Boulder; population (1990) 51,550. Industries include business machinery, sugar-beet refining, and recreational vehicles.

Longview /'lɒŋvjuː/ city in E Texas, USA, E of Dallas; seat of Gregg County; population (1990) 70,310. In the heart of the oil fields of E Texas, Longview's industries are mainly oil and natural-gas processing.

Lop Nor /ˌlɒp 'nɔː/ series of shallow salt lakes with shifting boundaries in the Taklimakan Shamo (desert) in Xinjiang Uyghur, NW China. Marco Polo, the Venetian traveller, visited Lop Nor, then a single lake of considerable extent, about 1273. The area is used for atomic tests.

Lorain /lə'reɪn/ city in N central Ohio, USA, on Lake Erie, NW of Akron and SW of Cleveland; population (1990) 71,240. An important Great Lakes port, it has shipbuilding yards and manufactures cars and heavy construction equipment.

Lord Howe Island /ˌlɔːd 'hau/ volcanic island and dependency of New South Wales, Australia, 700 km/435 mi NE of Sydney; area 15 sq km/ 6 sq mi; population (1984) 300. It is a tourist resort and heritage area because of its scenery and wildlife. The woodhen is a bird found only here.

Lorestan alternative form of ♢Luristan, a province of Iran.

Loreto /lə'retəu/ town in the Marche region of central Italy; population (1981) 10,600. The town allegedly holds the Virgin Mary's house, carried there by angels from Nazareth; hence Our Lady of Loreto is the patron saint of aviators.

Lorient /ˌlɔːri'ɒn/ commercial and naval port in Brittany, NW France; population (1990) 61,600. Industries include fishing and shipbuilding.

Lorraine /lɒ'reɪn/ region of NE France in the upper reaches of the Meuse and Moselle rivers; bounded N by Belgium, Luxembourg, and Germany and E by Alsace; area 23,600 sq km/9,095 sq mi; population (1986) 2,313,000. It comprises the *départements* of Meurthe-et-Moselle, Meuse, Moselle, and Vosges, and its capital is Nancy. There are deposits of coal, iron ore, and salt; grain, fruit, and livestock are farmed. In 1871 the region was ceded to Germany as part of Alsace-Lorraine.

Los Angeles /lɒs 'ændʒəliːz/ city and port in SW California, USA; population (1990) 3,485,400, metropolitan area of Los Angeles–Long Beach 14,531,530. Industries include aerospace, electronics, motor vehicles, chemicals, clothing, printing, and food processing.
features Hollywood, centre of the film industry since 1911; the Hollywood Bowl concert arena; observatories at Mount Wilson and Mount Palomar; Disneyland; the Huntingdon Art Gallery and Library; and the John Paul Getty museum of art, (a re-creation of a Roman villa in Herculaneum) which from 1995 will house only the Classical collection, the rest of its exhibits being housed in the enlarged John Paul Getty Centre across the town in Brentwood.
history Los Angeles was established as a Spanish settlement 1781, but it was a farming region with orange groves until the early 20th century, when it annexed neighbouring communities and

acquired distant water supplies, a deepwater port, and the film industry. In the 1920s large petroleum deposits were found in the area. The aircraft industry, with its need for year-round flying weather, developed here soon after and grew rapidly with the advent of World War II.

In 1992 five days of racial disturbances, resulting in more than 50 deaths and extensive damage, followed a judge's acquittal of four white police officers charged with the beating of a black motorist. The nationwide outrage caused by the videotape of the beating led to a second trial, in which the four officers were indicted.

Greater Los Angeles comprises 86 towns, including Long Beach, Redondo Beach, Venice, Santa Monica, Burbank, Compton, Beverly Hills, Glendale, Pasadena, and Pomona. It covers 10,000 sq km/4,000 sq mi.

Nineteen suburbs in search of a metropolis.

On **Los Angeles** H L Mencken *Americana* 1928

Lossiemouth /ˌlɒsiˈmauθ/ fishing port and resort in Grampian Region, Scotland; population (1988 est) 7,300. The politician Ramsay MacDonald was born and buried here.

Lot /lɒu/ French river; see ◊Gironde.

Lothian /ˈləuðiən/ region of Scotland
area 1,800 sq km/695 sq mi
towns Edinburgh (administrative headquarters), Livingston
features hills: Lammermuir, Moorfoot, Pentland; Bass Rock in the Firth of Forth, noted for seabirds
products bacon, vegetables, coal, whisky, engineering, electronics
population (1991) 723,700
famous people Alexander Graham Bell, Arthur Conan Doyle, R L Stevenson.

Lothian

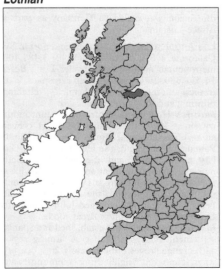

Louisiana

Loughborough /ˈlʌfbərə/ industrial town in Leicestershire, England; population (1988 est) 52,000. Occupations include engineering, bellfounding, electrical goods, and knitwear. The university of technology was established 1966.

Louisiana /luˌiːziˈænə/ state in southern USA; nickname Pelican State
area 135,900 sq km/52,457 sq mi
capital Baton Rouge
towns New Orleans, Shreveport, Lafayette, Lake Charles
features New Orleans French Quarter: jazz, restaurants, Mardi Gras; Cajun country and the Mississippi River delta; Jean Lafitte National Park and Chalmette National Historical Park; plantation homes near Natchitoches
products rice, cotton, sugar, oil, natural gas, chemicals, sulphur, fish and shellfish, salt, processed foods, petroleum products, timber, paper
population (1990) 4,220,000; including Cajuns, descendants of 18th-century religious exiles from Canada, who speak a French dialect
famous people Louis Armstrong, P G T Beauregard, Huey Long
history explored by the Spanish Piñeda 1519, Cabeza de Vaca 1528, and De Soto 1541 and by the French explorer La Salle 1862, who named it after Louis XIV and claimed it for France. It became Spanish 1762–1800, then French, then passed to the USA 1803 under the Louisiana Purchase; admitted to the Union as a state 1812.

Louisville /ˈluːiːvɪl/ industrial city and river port on the Ohio River, Kentucky, USA; population (1990) 269,000. Products include electrical goods, agricultural machinery, motor vehicles, tobacco, and whisky. It is the home of the Kentucky Fair and Exposition Center, and the Kentucky Derby.

Lourdes /luəd/ town in Midi-Pyrénées region, SW France, on the Gave de Pau River; population (1982) 18,000. Its Christian shrine to St Bernadette has a reputation for miraculous cures and Lourdes is an important Catholic pilgrimage centre. The young peasant girl Bernadette Soubirous was shown the healing springs of the Grotte de Massabielle by a vision of the Virgin Mary 1858.

Lourenço Marques /ləˈrensəu ˈmaːks/ former name of ◊Maputo, the capital of Mozambique.

Louth /lauð/ smallest county of the Republic

of Ireland, in Leinster province; county town Dundalk; area 820 sq km/317 sq mi; population (1991) 90,700.

Louvain /luːˈvæn/ (Flemish *Leuven*) industrial town in Brabant province, central Belgium; population (1991) 85,000. Manufacturing includes fertilizers and food processing. Its university dates from 1425 and there is a Science City.

Loveland /ˈlʌvlənd/ city in N central Colorado, USA, S of Fort Collins; population (1990) 37,350. It is a processing and marketing centre for the agricultural products of the area. Tourism is important to the economy, and Rocky Mountain National Park is to the W.

Low Countries region of Europe that consists of Belgium and the Netherlands, and usually includes Luxembourg.

Lowell /ˈləuəl/ city in Massachusetts, USA; population (1990) 103,400. Industries include electronics, plastics, and chemicals. Lowell was a textile centre in the 19th century; a substantial part of the old city was designated a national park 1978 as a birthplace of the US industrial revolution. Wang Laboratories moved its headquarters here 1978.

Lower Austria (German *Niederösterreich*) largest federal state of Austria; drained by the river Danube; area 19,200 sq km/7,411 sq mi; population (1987) 1,426,000. Its capital is St Pölten. In addition to wine, sugar beet, and grain, there are reserves of oil. Manufactured products include textiles, chemicals, and metal goods.

Lower California English name for ▷Baja California, Mexico.

Lower Saxony /ˈsæksəni/ (German *Niedersachsen*) administrative region (German *Land*) of N Germany
area 47,400 sq km/18,296 sq mi
capital Hanover
towns Brunswick, Osnabrück, Oldenburg, Göttingen, Wolfsburg, Salzgitter, Hildesheim
features Lüneburg Heath
products cereals, cars, machinery, electrical engineering
population (1988) 7,190,000
religion 75% Protestant, 20% Roman Catholic
history formed 1946 from Hanover, Oldenburg, Brunswick, and Schaumburg-Lippe.

Lowestoft /ˈləustɒft/ resort town and most easterly port in Britain, in Suffolk, England, 62 km/38 mi NE of Ipswich; population (1990 est) 58,000. There is fishing and fish processing, and radar and electrical equipment is produced. *Lowestoft Ness* is the most easterly point in England. The composer Benjamin Britten was born here.

Lozère /ləuˈzeə/ *département* in Languedoc-Roussillon region, S France. Occupying a section of the Cévennes Mountains, it rises in Finiels to 1,702 m/5,584 ft.

Lualaba /ˌluːəˈlɑːbə/ another name for the upper reaches of the river Zaïre in Africa, as it flows N through Zaire from near the Zambia border.

Luanda /luːˈændə/ (formerly *Loanda*) capital and industrial port (cotton, sugar, tobacco, timber, textiles, paper, oil) of Angola; population (1988) 1,200,000. Founded 1575, it became a Portuguese colonial administrative centre as well as an outlet for slaves transported to Brazil.

Luang Prabang /luːˈæŋ prɑːˈbæŋ/ or *Louangphrabang* Buddhist religious centre in Laos, on the Mekong River at the head of river navigation; population (1984) 44,244. It was the capital of the kingdom of Luang Prabang, incorporated in Laos 1946, and the royal capital of Laos 1946–75.

Lubbock /ˈlʌbək/ city in NW Texas, USA, S of Amarillo; seat of Lubbock County; population (1990) 186,200. Industries include heavy farm and construction machinery, cotton, sorghum, and mobile homes. The rock-and-roll singer Buddy Holly was born here.

Lübeck /ˈluːbek/ seaport of Schleswig-Holstein, Germany, on the Baltic Sea, 60 km/37 mi NE of Hamburg; population (1988) 209,000. Industries include machinery, aeronautical and space equipment, steel, ironwork, ship-building and servicing, fish canning; Lübeck is known for its wine trade and its marzipan.

Founded 1143, it has five Gothic churches and a cathedral dating from 1173. The Holstentor (1477) with its twin towers is the emblem of the city. Once head of the powerful Hanseatic League, it later lost much of its trade to Hamburg and Bremen, but improved canal and port facilities helped it to retain its position as a centre of Baltic trade. Lübeck was a free state of both the empire and the Weimar Republic.

The name Lübeck is of Wendish (the language of the Wends, a NW Slavonic people) origin and means 'lovely one'.

Lublin /ˈlublɪn/ city in Poland, on the Bystrzyca River, 150 km/95 mi SE of Warsaw; population (1990) 351,400. Industries include textiles, engineering, aircraft, and electrical goods. A trading centre from the 10th century, it has an ancient citadel, a 16th-century cathedral, and a university (1918). A council of workers and peasants proclaimed Poland's independence at Lublin 1918, and a Russian-sponsored committee of national liberation, which proclaimed itself the provincial government of Poland at Lublin 31 Dec 1944, was recognized by Russia five days later.

Lubumbashi /ˌluːbumˈbæʃi/ (formerly until 1986 *Elisabethville*) town in Zaire, on the Lualaba River; population (1984) 543,000. It is the chief commercial centre of the Shaba copper-mining region.

Lucca /ˈlukə/ city in NW Italy; population (1981) 91,246. It was an independent republic from 1160 until its absorption into Tuscany 1847. The composer Giacomo Puccini was born here.

Lucerne /luːˈsɜːn/ (German *Luzern*) capital and tourist centre of Lucerne canton, Switzerland, on the river Reuss where it flows out of Lake Lucerne; population (1990) city 59,400, canton

319,500. It developed around the Benedictine monastery, established about 750, and owes its prosperity to its position on the St Gotthard road and railway.

Lucerne, Lake /luːˈsɜːn/ (German *Vierwaldstättersee*) scenic lake in central Switzerland; area 114 sq km/44 sq mi.

Lucknow /ˈlʌknau/ capital and industrial city (engineering, chemicals, textiles, many handicrafts) of the state of Uttar Pradesh, India; population (1981) 1,007,000. During the Indian Mutiny against British rule, it was besieged 2 July–16 Nov 1857.

Lüda /ˌluːˈdɑː/ or *Lü-ta* or *Hüta* industrial port (engineering, chemicals, textiles, oil refining, shipbuilding, food processing) in Liaoning, China, on Liaodong Peninsula, facing the Yellow Sea; population (1986) 4,500,000. It comprises the naval base of Lüshun (known under 19th-century Russian occupation as Port Arthur) and the commercial port of Dalien (formerly Talien/Dairen).

Both were leased to Russia (which needed an ice-free naval base) 1898, but were ceded to Japan after the Russo-Japanese War; Lüshun was under Japanese siege June 1904–Jan 1905. After World War II, Lüshun was occupied by Russian airborne troops (it was returned to China 1955) and Russia was granted shared facilities at Dalien (ended on the deterioration of Sino-Russian relations 1955).

Lüderitz /ˈluːdərɪts/ port on Lüderitz Bay, Namibia; population (1970) 6,500. It is a centre for diamond-mining. The town, formerly a German possession, was named after a German merchant who acquired land here 1883.

Ludlow /ˈlʌdləu/ market town in Shropshire, England, on the river Teme, 42 km/26 mi S of Shrewsbury; population (1990 est) 8,000. Products include agricultural machinery, precision engineering, and clothing. Milton's masque *Comus* was first presented at Ludlow Castle 1634.

Ludwigshafen /ˈluːdvɪgzˌhɑːfən/ city and Rhine river port, Rhineland-Palatinate, Germany; population (1988) 152,000. Industries include chemicals, dyes, fertilizers, plastics, and textiles.

Lugano /luːˈgɑːnəu/ resort town on Lake Lugano, Switzerland; population (1990) 26,000. Industries include engineering and clothing.

Lugano, Lake /luːˈgɑːnəu/ lake partly in Italy, between lakes Maggiore and Como, and partly in Switzerland; area 49 sq km/19 sq mi.

Lugansk /luːˈgænsk/ (formerly 1935–58 and 1970–89 *Voroshilovgrad*) industrial city (locomotives, textiles, mining machinery) in Ukraine; population (1987) 509,000.

Luik /laɪk/ Flemish name of ▷Liège, a town in Belgium.

Luleå /ˈluːliɔː/ port in N Sweden, on the Gulf of Bothnia at the mouth of the river Luleå; population (1990) 68,400. It is the capital of Norrbotten county. Exports include iron ore and timber in ice-free months.

Lund /lund/ city in Malmöhus county, SW Sweden; 16 km/10 mi NE of Malmö; population (1990) 87,700. It has an 11th-century Romanesque cathedral and a university established 1666. The treaty of Lund was signed 1676 after Carl XI had defeated the Danes.

..when our Saviour was born, Lund was in its glory.

On **Lund** old Swedish saying

Lundy Island /ˈlʌndi/ rocky island at the entrance to the Bristol Channel, 19 km/12 mi NW of Hartland Point, Devon, England; area 9.6 sq km/3.7 sq mi; population (1975) 40. Formerly used by pirates and privateers as a lair, it is now a National Trust bird sanctuary and the first British marine nature reserve (1987). It has Bronze and Iron Age field systems, which can be traced by their boundaries which stand up above the surface.

Lüneburg /ˈluːnəbɜːg/ town in Lower Saxony, Germany, on the river Ilmenau; population (1985) 61,000. Industries include chemicals, paper, and ironworks. It is a health resort.

Lüneburg Heath /ˈluːnəbɜːg/ (German *Lüneburger Heide*) area in Lower Saxony, Germany, between the Elbe and Aller rivers. It was here that more than a million German soldiers surrendered to British general Montgomery 4 May 1945.

Luoyang /ˌluːəuˈjæŋ/ or *Loyang* industrial city (machinery and tractors) in Henan province, China, S of the river Huang He; population 1,114,000. It was formerly the capital of China and an important Buddhist centre in the 5th and 6th centuries.

The White Horse temple (Baimasi), 8 km/5 mi NE of the city, dates from the Ming dynasty and was restored in the 1950s; the original monastery, built AD 75, was one of the earliest Buddhist buildings in China. There are cave temples dating from the Northern Wei dynasty (386–534) to the S.

Lurgan /ˈlɜːgən/ see ▷Craigavon, a town in Northern Ireland.

Luristan /ˌluərɪˈstɑːn/ or *Lorestan* mountainous province in W Iran; area 28,800 sq km/11,117 sq mi; population (1986) 1,367,000. The capital is Khorramabad. The province is inhabited by Lur tribes who live by their sheep and cattle. Excavation in the area has revealed a culture of the 8th–7th century BC with bronzes decorated with animal forms; its origins are uncertain.

Lusaka /luːˈsɑːkə/ capital of Zambia from 1964 (of Northern Rhodesia 1935–64), 370 km/230 mi NE of Livingstone; it is a commercial and agricultural centre (flour mills, tobacco factories, vehicle assembly, plastics, printing); population (1988) 870,000.

Lüshun-Dalien /ˌluːˈʃuːn ˌdɑːˈlɪən/ see ⏵Lüda, a port in China.

Lusitania /ˌluːsɪˈteɪnɪən/ ancient area of the Iberian peninsula, roughly equivalent to Portugal. Conquered by Rome in 139 BC, the province of Lusitania rebelled periodically until it was finally conquered by Pompey 73–72 BC.

Lü-ta /ˌluːˈdɑː/ alternative transcription of ⏵Lüda, a port in China.

Luton /ˈluːtn/ industrial town in Bedfordshire, England, 53 km/33 mi SW of Cambridge; population (1991) 167,300. Luton airport is a secondary airport for London. Manufacturing includes cars, chemicals, electrical goods, ballbearings, as well as, traditionally, hats.

Luton Hoo, a Robert Adam mansion, was built 1762.

Lützen /ˈlutsən/ town in Saxony-Anhalt, Germany, SW of Leipzig, where in 1632 Gustavus Adolphus, king of Sweden, defeated the German commander Wallenstein in the Thirty Years' War; Gustavus was killed in the battle. The French emperor Napoleon Bonaparte overcame the Russians and Prussians here 1813.

Luxembourg Grand Duchy of (*Grand-Duché de Luxembourg*)
area 2,586 sq km/998 sq mi
capital Luxembourg
towns Esch-sur-Alzette, Dudelange
physical on the river Moselle; part of the Ardennes (Oesling) forest in N
features seat of the European Court of Justice, Secretariat of the European Parliament, international banking centre; economically linked with Belgium
head of state Grand Duke Jean from 1964
head of government Jacques Santer from 1984
political system liberal democracy
political parties Christian Social Party (PCS), moderate, left of centre; Luxembourg Socialist Workers' Party (POSL), moderate, socialist; Democratic Party (PD), centre-left; Communist Party of Luxembourg, pro-European left-wing

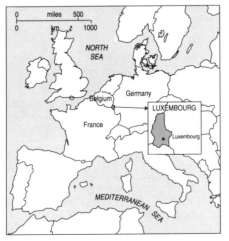

exports pharmaceuticals, synthetic textiles, steel
currency Luxembourg franc
population (1990 est) 369,000; growth rate 0% p.a.
life expectancy men 71, women 78 (1989)
languages French (official), local Letzeburgesch, German
religion Roman Catholic 97%
literacy 100% (1989)
GNP $4.9 bn; $13,380 per head (1988)
chronology
1354 Became a duchy.
1482 Under Habsburg control.
1797 Ceded, with Belgium, to France.
1815 Treaty of Vienna created Luxembourg a grand duchy, ruled by the king of the Netherlands.
1830 With Belgium, revolted against Dutch rule.
1890 Link with Netherlands ended with accession of Grand Duke Adolphe of Nassau-Weilburg.
1948 With Belgium and the Netherlands, formed the Benelux customs union.
1960 Benelux became fully effective economic union.
1961 Prince Jean became acting head of state on behalf of his mother, Grand Duchess Charlotte.
1964 Grand Duchess Charlotte abdicated; Prince Jean became grand duke.
1974 Dominance of PSC challenged by Socialists.
1979 PSC regained pre-eminence.
1991 Pact agreeing European free-trade area signed in Luxembourg.
1992 Vote in favour of ratification of Maastricht Treaty on European union.

Luxembourg /ˈlʌksəmbɜːɡ, French ˌluksæmˈbuəɡ/ capital of the country of Luxembourg, on the Alzette and Petrusse rivers; population (1985) 76,000. The 16th-century Grand Ducal Palace, European Court of Justice, and European Parliament secretariat are situated here, but plenary sessions of the parliament are now held only in Strasbourg, France. Products include steel, chemicals, textiles, and processed food.

A Roman fortress was built on the rocky plateau to control movement between France and Germany. The old town, including the 17th-century cathedral and town hall (1830–38), stands on the plateau; more recent development has spread across lower areas linked by bridges.

Luxembourg /ˈlʌksəmbɜːɡ, French ˌluksæmˈbuəɡ/ province of SE Belgium
area 4,400 sq km/1,698 sq mi
capital Arlon
towns Bastogne, St Hubert, Bouillon
products dairy products, iron and steel, tobacco
physical situated in the SE Ardennes and widely forested; rivers Ourthe, Semois, and Lesse
population (1991) 232,800
history formerly part of the Grand Duchy of Luxembourg, it became a Belgian province 1831.

Luxor /ˈlʌksɔː/ (Arabic *al-Uqsur*) small town in Egypt on the east bank of the river Nile. The ancient city of Thebes is on the west bank, with the temple of Luxor built by Amenhotep III (c. 1411–1375 BC) and the tombs of the pharaohs in the Valley of the Kings.

Luzern /luːtˈseən/ German name of ⏵Lucerne, a town in Switzerland.

Luzon /lu:'zɒn/ largest island of the Philippines; area 108,130 sq km/41,750 sq mi; capital Quezon City; population (1970) 18,001,270. The chief city is Manila, capital of the Philippines. Products include rice, timber, and minerals. It has US military bases.

In 1991 the volcanic Mount Pinatubo, 88 km/ 55 mi N of Manila, erupted after lying dormant for 600 years. Volcanic ash covered an area of 2,600 sq km/1,000 sq mi and brought economic catastrophe. The USA abandoned Clark Air Base, although Subic Bay naval base, their chief station in the W Pacific, was to be restored, and was leased for a further 10 years.

Lviv /lvɒf/ (Russian **Lvov**, Polish **Lwów**, German **Lemberg**) capital and industrial city of Lviv region, Ukraine; population (1987) 767,000. Industries include textiles, metals, and engineering. The university was founded 1661. Lviv was formerly a trade centre on the Black Sea– Baltic route. Founded in the 13th century by a Galician prince (the name means 'city of Leo' or 'city of Lev'), it was Polish 1340–1772, Austrian 1772–1919, Polish 1919–39, and annexed by the USSR 1945. It was the site of violent nationalist demonstrations Oct 1989; in 1991, Ukraine became independent.

Lvov /ljvɔf/ Russian form of ◊Lviv, a city in Ukraine.

Lwów /lvu:f/ Polish form of ◊Lviv, a city in Ukraine.

Lyme Regis /laɪm 'ri:dʒɪs/ seaport and resort in Dorset, S England; population (1981) 3,500. The rebel duke of Monmouth, claimant to the English crown, landed here 1685. The Cobb (a massive stone pier) features in Jane Austen's *Persuasion* 1818 and John Fowles's *The French Lieutenant's Woman* 1969.

Lymington /'lɪmɪŋtən/ port and yachting centre in Hampshire, S England; 8 km/5 mi SW of Southampton; population (1990 est) 15,000. It has a ferry link with the Isle of Wight.

Lynchburg /'lɪntʃbɜːg/ city in S central Virginia, USA, on the James River, NE of Roanoke; population (1990) 66,050. Industries include clothing, paper and rubber products, and machine parts.

Lynn /lɪn/ industrial city in Massachusetts, USA, on Massachusetts Bay; population (1990) 81,240. Founded as **Saugus** 1629, it was renamed 1637 after King's Lynn, England.

Lynton and Lynmouth /'lɪntən, 'lɪnməθ/ twin resort towns on the north coast of Devon, SW England. The fishing village of Lynmouth is linked by an alpine road and cliff railway (1890) to Lynton, which lies 152 m/500 ft above at the top of a cliff. In Aug 1952, 22.5 cm/9 in of rainfall within 24 hours on nearby Exmoor caused disastrous flooding at Lynmouth, leaving 31 people dead and the harbour and over 100 buildings severely damaged. The harbour, a Rhenish tower, and much of the town were later rebuilt.

Lyon /'li:ɒn/ (English **Lyons**) industrial city (textiles, chemicals, machinery, printing) and capital of Rhône *département*, Rhône-Alpes region, and third largest city of France, at the confluence of the rivers Rhône and Saône, 275 km/170 mi NNW of Marseille; population (1990) 422,400, conurbation 1,221,000. Formerly a chief fortress of France, it was the ancient **Lugdunum**, taken by the Romans 43 BC.

Lyons /'li:ɒn/ English form of ◊Lyon, a city in France.

Lytham St Annes /'lɪðəm sənt 'ænz/ resort in Lancashire, England, on the river Ribble; 10 km/ 6 mi SE of Blackpool; population (1982) 39,600. It has a championship golf course.

Maas /mɑːs/ Dutch or Flemish name for the river ⏵Meuse.

Maastricht /mɑːˈstrɪxt/ industrial city (metallurgy, textiles, pottery) and capital of the province of Limburg, the Netherlands, on the river Maas, near the Dutch-Belgian frontier; population (1991) 117,400. Maastricht dates from Roman times. It was the site of the Maastricht summit Dec 1991.

McAllen /məˈkælən/ city in S Texas, USA, just N of the Mexican border formed by the Rio Grande, SE of Laredo; population (1990) 84,000. Industries include oil refining and the processing of agricultural products from the Rio Grande Valley. It is a US port of entry for Mexicans, and many of the city's inhabitants are Spanish-speaking.

Macao /məˈkau/ Portuguese possession on the south coast of China, about 65 km/40 mi W of Hong Kong, from which it is separated by the estuary of the Canton River; it consists of a peninsula and the islands of Taipa and Colôane
area 17 sq km/7 sq mi
capital Macao, on the peninsula
features the peninsula is linked to Taipa by a bridge and to Colôane by a causeway, both 2 km/1 mi long

Macao

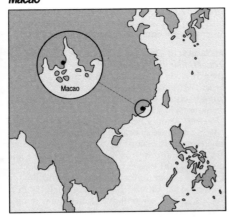

currency pataca
population (1986) 426,000
languages Cantonese; Portuguese (official)
religions Buddhist, with 6% Catholic minority
government Under the constitution ('organic statute') of 1990, Macao enjoys political autonomy. Executive power is held by the governor. The governor works with a cabinet of five appointed secretaries and confers with a 10-member consultative council and a 23-member legislative council, comprising seven government appointees, eight indirectly elected by business associations and eight directly elected by universal suffrage. The legislative council frames internal legislation, but any bills passed by less than a two-thirds majority can be vetoed by the governor. A number of 'civic associations' and interest groups function, sending representatives to the legislative council.
history Macao was first established as a Portuguese trading and missionary post in the Far East 1537, and was leased from China 1557. It was annexed 1849 and recognized as a Portuguese colony by the Chinese government in a treaty 1887. The port declined in prosperity during the late 19th and early 20th centuries, as its harbour silted up and international trade was diverted to Hong Kong and the new treaty ports. The colony thus concentrated instead on local 'country trade' and became a centre for gambling and, later, tourism.

In 1951 Macao became an overseas province of Portugal, sending an elected representative to the Lisbon parliament. After the Portuguese revolution 1974, it became a 'special territory' and was granted considerable autonomy under a governor appointed by the Portuguese president.

In 1986 negotiations opened between the Portuguese and the Chinese governments over the question of the return of Macao's sovereignty under 'one country, two systems' terms similar to those agreed by China and the UK for Hong Kong. These negotiations were concluded April 1987 by the signing of the Macao Pact, under which Portugal agreed to hand over sovereignty to the People's Republic Dec 1999, and China agreed in return to guarantee to maintain the port's capitalist economic and social system for at least 50 years.

In May 1990 administrative, economic, and financial autonomy was secured from Portugal; this followed the approval of a new 'organic statute' for the territory by both Portugal's parliament and Macao's legislative assembly. In Jan 1991 Macao acceded to GATT (the General Agreement on Tariffs and Trade).

—

... a bit of medieval Portugal transported half the world away.

On **Macao** Crosbie Garstin *The Dragon and the Lotus* 1923

—

Macassar /məˈkæsə/ another name for ⏵Ujung Pandang, a port in the Celebes (Sulawesi), Indonesia.

Macdonnell Ranges /ˌmækdə'nel/ mountain range in central Australia, Northern Territory, with the town of Alice Springs; highest peak Mount Zeil 1,510 m/4,955 ft.

Macedonia /ˌmæsɪ'dəʊnɪə/ (Greek *Makedhonia*) mountainous region of N Greece, part of the ancient country of Macedonia which was divided between Serbia, Bulgaria, and Greece after the Balkan Wars of 1912–13. Greek Macedonia is bounded W and N by Albania and the Former Yugoslav Republic of Macedonia; area 34,177 sq km/13,200 sq mi; population (1991) 2,263,000. The chief city is Thessaloniki. Fertile valleys produce grain, olives, grapes, tobacco, and livestock. Mount Olympus rises to 2,918 m/9,570 ft on the border with Thessaly.

Macedonia Former Yugoslav Republic of
area 25,700 sq km/9,920 sq mi
capital Skopje
physical mountainous; rivers: Struma, Vardar; Mediterranean climate with hot summers
head of state Kiro Gligorov from 1990
head of government Branko Crvenkovski from 1992
political system emergent democracy
political parties Internal Macedonian Revolutionary Organization–Democratic Party for Macedonian National Unity (VMRO–DMPNE), centrist; League of Communists, left-wing
population (1990) 1,920,000
language Macedonian, closely allied to Bulgarian and written in Cyrillic
religion Macedonian Orthodox Christian
chronology
1913 Ancient country of Macedonia divided between Serbia, Bulgaria, and Greece.
1918 Serbian part included in what was to become Yugoslavia.
1941–44 Occupied by Bulgaria.
1945 Created a republic within Yugoslav Socialist Federation.
1980 Rise of nationalism after death of Yugoslav leader Tito.
1991 'Socialist' dropped from republic's name. Referendum supported independence.
1992 Independence declared, but international recognition withheld because of objections to name by Greece.
1993 Sovereignty recognized by UK and Albania; won United Nations membership, with Greece's approval, under provisional name of Former Yugoslav Republic of Macedonia.

Maceió /ˌmæseɪ'əʊ/ industrial town (sugar, tobacco, textile, timber) in NE Brazil, capital of Alagaos state with its port at Jaraguá; population (1991) 699,800.

Macgillycuddy's Reeks /mə'gɪlɪˌkʌdiz 'riːks/ range of mountains in SW Ireland lying W of Killarney, in County Kerry; Carrauntoohill 1,041 m/3,414 ft is the highest peak in Ireland.

Machu Picchu /'mɑːtʃuː 'piːktʃuː/ ruined Inca city in Peru, built about AD 1500, NW of Cuzco, discovered 1911 by Hiram Bingham. It stands at the top of cliffs 300 m/1,000 ft high and contains the well-preserved remains of houses and temples.

Macias Nguema /mə'siːəs əŋ'gweɪmə/ former name (until 1979) of the island of ◊Bioko, part of Equatorial Guinea.

Mackenzie River /mə'kenzi/ river in the Northwest Territories, Canada, flowing NW from Great Slave Lake to the Arctic Ocean; about 1,800 km/1,120 mi long.

McKinley, Mount /mə'kɪnli/ or *Denali* peak in Alaska, USA, the highest in North America, 6,194 m/20,320 ft; named after US president William McKinley.

Mâcon /'mɑːkɒŋ/ capital of the French *département* of Saône-et-Loire, on the river Saône, 72 km/45 mi N of Lyon; population (1990) 38,500. It produces wine.

Macon /'meɪkən/ city in central Georgia, USA, on the Ocmulgee River, NE of Columbus; seat of Bibb County; population (1990) 106,600. An industrial city, Macon produces textiles, building materials, farm machinery, and chemicals; it processes fruits, pecans, and the special kaolin clay that is found nearby.

Macquarie Island /mə'kwɒri/ outlying Australian territorial possession, a Tasmanian dependency, some 1,370 km/850 mi SE of Hobart; area 170 sq km/65 sq mi; it is uninhabited except for an Australian government research station.

Madagascar Democratic Republic of (*Repoblika Demokratika n`i Madagaskar*)
area 587,041 sq km/226,598 sq mi
capital Antananarivo
towns chief port Toamasina, Antseranana, Fianarantsoa, Toliary
physical temperate central highlands; humid valleys and tropical coastal plains; arid in S
environment according to 1990 UN figures, 93% of the forest area has been destroyed and about 100,000 species have been made extinct
features one of the last places to be inhabited, it evolved in isolation with unique animals (such as the lemur, now under threat from deforestation)
head of state Didier Ratsiraka from 1975

head of government Guy Razanamasy from 1991
political system emergent democratic republic
political parties National Front for the Defence of the Malagasy Socialist Revolution (FNDR); AKFM-Congress and AKFM-Renewal, both left-of-centre; Social Democratic Party (PSD), centre-left
exports coffee, cloves, vanilla, sugar, chromite, shrimps
currency Malagasy franc
population (1990 est) 11,802,000, mostly of Malayo-Indonesian origin; growth rate 3.2% p.a.
life expectancy men 50, women 53 (1989)
languages Malagasy (official), French, English
religions animist 50%, Christian 40%, Muslim 10%
literacy 53% (1988)
GNP $2.1 bn (1987); $280 per head (1988)
chronology
1885 Became a French protectorate.
1896 Became a French colony.
1960 Independence achieved from France, with Philibert Tsiranana as president.
1972 Army took control of the government.
1975 Martial law imposed under a national military directorate. New Marxist constitution proclaimed the Democratic Republic of Madagascar, with Didier Ratsiraka as president.
1976 Front-Line Revolutionary Organization (AREMA) formed.
1977 National Front for the Defence of the Malagasy Socialist Revolution (FNDR) became the sole legal political organization.
1980 Ratsiraka abandoned Marxist experiment.
1983 Ratsiraka re-elected, despite strong opposition from radical socialist National Movement for the Independence of Madagascar (MONIMA) under Monja Jaona.
1989 Ratsiraka re-elected for third term after restricting opposition parties.
1990 Political opposition legalized; 36 new parties created.
1991 Antigovernment demonstrations; opposition to Ratsiraka led to general strike. Nov: Ratsiraka formed new unity government.
1992 Constitutional reform approved.

Madeira

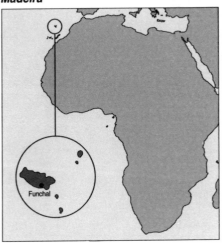

... one of the greatest and richest isles of the World ...

On **Madagascar** Marco Polo 1320

Madeira /mə'dɪərə/ group of islands forming an autonomous region of Portugal off the northwest coast of Africa, about 420 km/260 mi N of the Canary Islands. Madeira, the largest, and Porto Santo are the only inhabited islands. The Desertas and Selvagens are uninhabited islets. Their mild climate makes them a year-round resort
area 796 sq km/308 sq mi
capital Funchal, on Madeira
physical Pico Ruivo, on Madeira, is the highest mountain at 1,861 m/6,106 ft
products Madeira (a fortified wine), sugar cane, fruit, fish, handicrafts
population (1986) 269,500
history Portuguese from the 15th century; occupied by Britain 1801 and 1807–14. In 1980 Madeira gained partial autonomy but remained a Portuguese overseas territory.

Madeira River /mə'dɪərə/ river of W Brazil; length 3,250 km/2,020 mi. It is formed by the rivers Beni and Mamoré, and flows NE to join the Amazon.

Madhya Bharat /'mʌdjə 'baːrət/ state of India 1950–56. It was a union of 24 states of which Gwalior and Indore were the most important. In 1956 Madhya Bharat was absorbed in Madhya Pradesh.

Madhya Pradesh /'mʌdjə prə'deʃ/ state of central India; the largest of the Indian states
area 442,700 sq km/170,921 sq mi
capital Bhopal
towns Indore, Jabalpur, Gwalior, Durg-Bhilainagar, Raipur, Ujjain
products cotton, oilseed, sugar, textiles, engineering, paper, aluminium

Madhya Pradesh

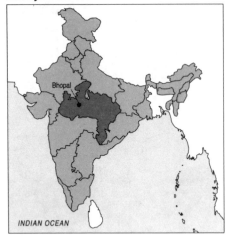

population (1991) 66,135,400
language Hindi
history formed 1950 from the former British province of Central Provinces and Berar and the princely states of Makrai and Chattisgarh; lost some southwestern districts 1956, including Nagpur, and absorbed Bhopal, Madhya Bharat, and Vindhya Pradesh. In 1984 some 2,600 people died in Bhopal from an escape of poisonous gas.

Madison /'mædɪsən/ capital of Wisconsin, USA, 193 km/120 mi NW of Chicago, between lakes Mendota and Monona; population (1990) 191,300. Products include agricultural machinery and medical equipment.

Madras /mə'drɑːs/ industrial port (cotton, cement, chemicals, iron, and steel) and capital of Tamil Nadu, India, on the Bay of Bengal; population (1981) 4,277,000. Fort St George 1639 remains from the East India Company when Madras was the chief port on the east coast. Madras was occupied by the French 1746–48 and shelled by the German ship *Emden* 1914, the only place in India attacked in World War I.

Madras /mə'drɑːs/ former name of ◊Tamil Nadu, a state of India.

Madrid /mə'drɪd/ industrial city (leather, chemicals, furniture, tobacco, paper) and capital of Spain and of Madrid province; population (1991) 2,984,600. Built on an elevated plateau in the centre of the country, at 655 m/2,183 ft it is the highest capital city in Europe and has excesses of heat and cold. Madrid province has an area of 8,000 sq km/3,088 sq mi and a population of 4,855,000. Madrid began as a Moorish citadel captured by Castile 1083, became important in the times of Charles V and Philip II, and was designated capital 1561.

Features include the Real Academia de Bellas Artes (1752), the Prado Museum (1785), and the royal palace (1764). During the Spanish Civil War, Madrid was besieged by the Nationalists 7 Nov 1936–28 March 1939.

In winter especially she seems a capital half-frozen in the attitudes of a past generation ... Upon Madrid herself change has fallen like a pile of concrete.

On **Madrid** James Morris in *Encounter* Oct 1964

Madura /mə'duərə/ island in Indonesia, off Surabaya, Java; one of the Sunda Islands
area 4,564 sq km/1,762 sq mi; with offshore islands, more than 5,000 sq km/2,000 sq mi
capital Pamekasan
features central hills rising to 480 m/1,545 ft; forested
products rice, tobacco, salt, cattle, fish
population (1970) 2,447,000
history see ◊Java.

Madurai /'mædjuraɪ/ city in Tamil Nadu, India; population (1981) 904,000. Founded in the 5th century BC, it is the site of the 16th–17th-century temple of Meenakshi. Industries include textiles (cotton) and brassware manufacturing.

Maeander /mi'ændə/ anglicized form of the ancient Greek name of the river ◊Menderes in Turkey.

Maestricht alternative form of ◊Maastricht, a city in the Netherlands.

Mafeking /'mæfɪkɪŋ/ former name of ◊Mafikeng, a town in South Africa.

Mafikeng /'mæfɪkeŋ/ (formerly until 1980 *Mafeking*) town in Bophuthatswana, South Africa. It was the capital of Bechuanaland, and the British officer Robert Baden-Powell held it under Boer siege 12 Oct 1899–17 May 1900.

Magadan /ˌmægə'dɑːn/ port for the gold mines in E Siberian Russia, off the northern shore of the Sea of Okhotsk; population (1985) 142,000.

Magdeburg /'mægdəbɜːg/ industrial city (vehicles, paper, textiles, machinery) and capital of Saxony-Anhalt, Germany, on the river Elbe; population (1990) 290,000. A former capital of Saxony, Magdeburg became capital of Saxony-Anhalt on German reunification 1990. In 1938 the city was linked by canal with the Rhine and Ruhr rivers.

Magdeburg was a member of the Hanseatic League, and has a 13th-century Gothic cathedral. Magdeburg county has an area of 11,530 sq km/4,451 sq mi, and a population of 1,250,000.

Magellan, Strait of /mə'gelən/ channel between South America and Tierra del Fuego, named after the Portuguese navigator Ferdinand Magellan. It is 595 km/370 mi long, and joins the Atlantic and Pacific oceans.

Magenta /mə'dʒentə/ town in Lombardy, Italy, 24 km/15 mi W of Milan, where France and Sardinia defeated Austria 1859 during the struggle for Italian independence. Magenta dye was named in honour of the victory.

Maggiore, Lago /mə'dʒɔːreɪ/ lake partly in Italy,

partly in the Swiss canton of Ticino, with Locarno on its northern shore, 63 km/39 mi long and up to 9 km/5.5 mi wide (area 212 sq km/ 82 sq mi), with fine scenery.

Maghreb /'mʌgrəb/ name for NW Africa (Arabic 'far west', 'sunset'). The Maghreb powers —Algeria, Libya, Morocco, Tunisia, and Western Sahara—agreed on economic coordination 1964–65, with Mauritania cooperating from 1970. In 1989 these countries formed an economic union known as the Arab Maghreb Union. Chad and Mali are sometimes included. Compare ⊅ Mashraq, the Arab countries of the E Mediterranean.

Magnitogorsk /mæg'niːtəugɔːsk/ industrial town (steel, motor vehicles, tractors, railway rolling stock) in Chelyabinsk region, Russia, on the eastern slopes of the Ural Mountains; population (1987) 430,000. It was developed in the 1930s to work iron, manganese, bauxite, and other metals in the district.

Mahabad /'mʌhəbʌd/ Kurdish town in Azerbaijan, W Iran, population (1983) 63,000. Occupied by Russian troops 1941, it formed the centre of a short-lived republic (1945–46) before being reoccupied by the Iranians. In the 1980s Mahabad was the focal point of resistance by Iranian Kurds against the Islamic republic.

Maharashtra /ˌmɑːhəˈræʃtrə/ state in W central India
area 307,800 sq km/118,811 sq mi
capital Bombay
towns Pune, Nagpur, Ulhasnagar, Sholapur, Nasik, Thana, Kolhapur, Aurangabad, Sangli, Amravati
features cave temples of Ajanta, containing 200 BC–7th century AD Buddhist murals and sculptures; Ellora cave temples 6th–9th century with Buddhist, Hindu, and Jain sculptures
products cotton, rice, groundnuts, sugar, minerals
population (1991) 78,706,700
language Marathi 50%

Maharashtra

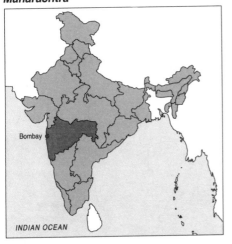

religions Hindu 80%, Parsee, Jain, and Sikh minorities
history formed 1960 from the southern part of the former Bombay state.

Mahón /mɑːˈɒn/ or *Port Mahon* capital and port of the Spanish island of Minorca; population (1981) 21,900. Probably founded by the Carthaginians, it was under British occupation 1708–56 and 1762–82.

Maidenhead /'meɪdnhed/ town in Berkshire, S England, 40 km/25 mi W of London, on the river Thames; population (1983) 48,500. Industries include computer software, plastics, pharmaceuticals, and printing. It is a boating centre.

Maidstone /'meɪdstəun/ town in Kent, SE England, on the river Medway, administrative headquarters of the county; population (1986 est) 72,000. Industries include agricultural machinery and paper.
Maidstone has the ruins of All Saints' College 1260. The Elizabethan Chillington Manor is an art gallery and museum.

Maikop /maɪˈkɒp/ capital of Adyge autonomous region of Russia on the river Bielaia, with timber mills, distilleries, tanneries, and tobacco and furniture factories; population (1985) 140,000. Oilfields, discovered 1900, are linked by pipeline with Tuapse on the Black Sea.

Main /maɪn/ river in central western Germany, 515 km/320 mi long, flowing through Frankfurt to join the river Rhine at Mainz. A canal links it with the Danube.

Maine /meɪn/ French river, 11 km/7 mi long, formed by the junction of the Mayenne and Sarthe; it enters the Loire below Angers, and gives its name to Maine-et-Loire *département*.

Maine /meɪn/ northeasternmost state of the USA, largest of the New England states; nickname Pine Tree State
area 86,200 sq km/33,273 sq mi
capital Augusta
towns Portland, Lewiston, Bangor
physical Appalachian Mountains; 80% of the state is forested
features Acadia National Park, including Bar Harbor and most of Mount Desert Island; Baxter State Park, including Mount Katahdin; Roosevelt's Campobello International Park;

Maine

Augusta

canoeing along the Allagush Wilderness Water-way
products dairy and market garden produce, paper, pulp, timber, footwear, textiles, fish, lobster; tourism is important
population (1990) 1,228,000
famous people Henry Wadsworth Longfellow, Kate Douglas Wiggin, Edward Arlington Robinson, Edna St Vincent Millay
history permanently settled by the British from 1623; absorbed by Massachusetts 1691; became a state 1820.

Mainz /maɪnts/ (French **Mayence**) capital of Rhineland-Palatinate, Germany, on the Rhine, 37 km/23 mi WSW of Frankfurt-am-Main; population (1988) 189,000. In Roman times it was a fortified camp and became the capital of Germania Superior. Printing was possibly invented here about 1448 by Johann Gutenberg.

Majorca /məˈjɔːkə/ (Spanish **Mallorca**) largest of the Balearic Islands, belonging to Spain, in the W Mediterranean
area 3,640 sq km/1,405 sq mi
capital Palma
features the highest mountain is Puig Mayor, 1,445 m/4,741 ft
products olives, figs, oranges, wine, brandy, timber, sheep; tourism is the mainstay of the economy
population (1981) 561,215
history captured 797 by the Moors, it became the kingdom of Majorca 1276, and was united with Aragon 1343.

Makeyevka /məˈkeɪəfkə/ (formerly until 1931 **Dmitrievsk**) industrial city (coal, iron, steel, and chemicals) in the Donets Basin, SE Ukraine; population (1987) 455,000.

Makhachkala /məˌkætʃkəˈlɑː/ (formerly until 1922 **Port Petrovsk**) capital of Dagestan, Russia, on the Caspian Sea, SE of Grozny, from which pipelines bring petroleum to Makhachkala's refineries; other industries include shipbuilding, meat packing, chemicals, matches, and cotton textiles; population (1987) 320,000.

Malabar Coast /ˈmæləbɑː ˈkəʊst/ coastal area of Karnataka and Kerala states, India, lying between the Arabian Sea and the Western Ghats; about 65 km/40 mi W–E, 725 km/450 mi N–S. A fertile area with heavy rains, it produces food grains, coconuts, rubber, spices; also teak, ebony, and other woods. Lagoons fringe the shore. A district of Tamil Nadu transferred 1956 to Kerala was formerly called the Malabar Coast.

Malabo /məˈlɑːbəʊ/ port and capital of Equatorial Guinea, on the island of Bioko; population (1983) 15,253. It was founded in the 1820s by the British as **Port Clarence**. Under Spanish rule it was known as **Santa Isabel** until 1973.

Malacca /məˈlækə/ or **Melaka** state of W Peninsular Malaysia; capital Malacca; area 1,700 sq km/656 sq mi; population (1980) 465,000 (about 70% Chinese). Products include rubber, tin, and wire. The town originated in the

13th century as a fishing village frequented by pirates, and later developed into a trading port. Portuguese from 1511, then Dutch from 1641, it was ceded to Britain 1824, becoming part of the Straits Settlements.

Whoever is Lord of Malacca has his hand on the throat of Venice.

On **Malacca** Tom Pires about 1515

Malacca, Strait of /məˈlækə/ channel between Sumatra and the Malay Peninsula; length 965 km/600 mi; it narrows to less than 38 km/24 mi wide. It carries all shipping between the Indian Ocean and the South China Sea.

Málaga /ˈmæləgə/ industrial seaport (sugar refining, distilling, brewing, olive-oil pressing, shipbuilding) and holiday resort in Andalusia, Spain; capital of Málaga province on the Mediterranean; population (1991) 524,800. Founded by the Phoenicians and taken by the Moors 711, Málaga was capital of the Moorish kingdom of Malaga from the 13th century until captured 1487 by the Catholic monarchs Ferdinand and Isabella.

Malagasy Republic /ˌmæləˈgæsi/ former name (1958–75) of ⟡Madagascar.

Malatya /ˌmælətˈjɑː/ capital of a province of the same name in E central Turkey, lying W of the river Euphrates; population (1990) 281,800.

Malawi Republic of (**Malaŵi**)
area 118,000 sq km/45,560 sq mi
capital Lilongwe
towns Blantyre (largest city and commercial centre), Mzuzu, Zomba
physical landlocked narrow plateau with rolling plains; mountainous W of Lake Malawi
features one-third is water, including lakes Malawi, Chilara, and Malombe; Great Rift Valley; Nyika, Kasungu, and Lengare national parks; Mulanje Massif; Shire River

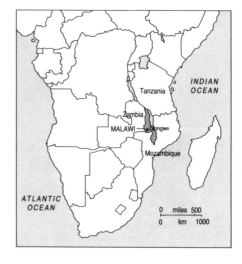

head of state and government Hastings
Kamusu Banda from 1966 for life
political system one-party republic
political party Malawi Congress Party (MCP),
multiracial, right-wing
exports tea, tobacco, cotton, peanuts, sugar
currency kwacha
population (1990 est) 9,080,000 (nearly 1 million refugees from Mozambique); growth rate
3.3% p.a.
life expectancy men 46, women 50 (1989)
languages English, Chichewa (both official)
religions Christian 75%, Muslim 20%
literacy 25% (1989)
GNP $1.2 bn (1987); $160 per head (1988)
chronology
1891 Became the British protectorate Nyasaland.
1964 Independence achieved from Britain, within
the Commonwealth, as Malawi.
1966 Became a one-party republic, with Hastings
Banda as president.
1971 Banda was made president for life.
1977 Banda released some political detainees and
allowed greater freedom of the press.
1986–89 Influx of nearly 1 million refugees from
Mozambique.
1992 Calls for multiparty politics. Countrywide
industrial riots caused many fatalities. Western
aid suspended over human-rights violations.

Malawi, Lake /məˈlɑːwi/ or **Lake Nyasa** African
lake, bordered by Malawi, Tanzania, and
Mozambique, formed in a section of the Great
Rift Valley. It is about 500 m/1,650 ft above sea
level and 560 km/350 mi long, with an area of
37,000 sq km/14,280 sq mi. It is intermittently
drained to the S by the river Shiré into the
Zambezi.

Malay Peninsula /məˈleɪ/ southern projection of
the continent of Asia, lying between the Strait
of Malacca, which divides it from Sumatra, and
the China Sea.
   The northern portion is partly in
Myanmar (formerly Burma), partly in Thailand;
the south forms part of Malaysia. The island of
Singapore lies off its southern extremity.

Malaysia
area 329,759 sq km/127,287 sq mi
capital Kuala Lumpur
towns Johor Baharu, Ipoh, Georgetown
(Penang), Kuching in Sarawak, Kota Kinabalu in
Sabah
physical comprises Peninsular Malaysia (the
nine Malay states—Johore, Kedah, Kelantan,
Negri Sembilan, Pahang, Perak, Perlis, Selangor,
Trengganu—plus Malacca and Penang); and E
Malaysia (Sabah and Sarawak); 75% tropical
jungle; central mountain range; swamps in E
features Mount Kinabalu (highest peak in SE
Asia); Niah caves (Sarawak)
head of state Rajah Azlan Muhibuddin Shah
(sultan of Perak) from 1989
head of government Mahathir bin Mohamad
from 1981
political system liberal democracy
political parties New United Malays' National
Organization (UMNO Baru), Malay-oriented

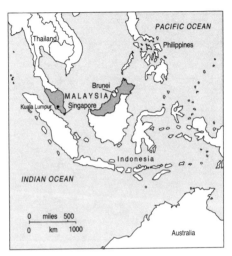

nationalist; Malaysian Chinese Association
(MCA), Chinese-oriented conservative; Gerakan
Party, Chinese-oriented, left of centre; Malaysian
Indian Congress (MIC), Indian-oriented; Democratic Action Party (DAP), left of centre, multiracial but Chinese-dominated; Pan-Malayan
Islamic Party (PAS), Islamic; Semangat '46
(Spirit of 1946), moderate, multiracial
exports pineapples, palm oil, rubber, timber,
petroleum (Sarawak), bauxite
currency ringgit
population (1990 est) 17,053,000 (Malaysian
47%, Chinese 32%, Indian 8%, others 13%);
growth rate 2% p.a.
life expectancy men 65, women 70 (1989)
languages Malay (official), English, Chinese,
Indian, and local languages
religions Muslim (official), Buddhist, Hindu,
local beliefs
literacy 80% (1989)
GNP $34.3 bn; $1,870 per head (1988)
chronology
1786 Britain established control.
1826 Became a British colony.
1963 Federation of Malaysia formed, including Malaya, Singapore, Sabah (N Borneo), and
Sarawak (NW Borneo).
1965 Secession of Singapore from federation.
1969 Anti-Chinese riots in Kuala Lumpur.
1971 Launch of *bumiputra* ethnic-Malay-oriented
economic policy.
1981 Election of Dr Mahathir bin Mohamad as
prime minister.
1982 Mahathir bin Mohamad re-elected.
1986 Mahathir bin Mohamad re-elected.
1987 Arrest of over 100 opposition activists, including DAP leader, as Malay-Chinese
relations deteriorated.
1988 Split in ruling UMNO party over Mahathir's
leadership style; new UMNO formed.
1989 Semangat '46 set up by former members
of UMNO including ex-premier Tunku Abdul
Rahman.
1990 Mahathir bin Mohamad re-elected.
1991 New economic growth programme
launched.

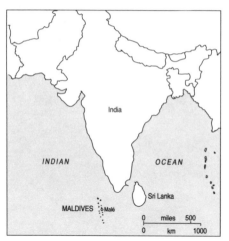

Maldives Republic of (*Divehi Jumhuriya*)
area 298 sq km/115 sq mi
capital Malé
towns Seenu
physical comprises 1,196 coral islands, grouped
into 12 clusters of atolls, largely flat, none bigger
than 13 sq km/5 sq mi, average elevation 1.8 m/
6 ft; 203 are inhabited
environment the threat of rising sea level has
been heightened by the frequency of flooding in
recent years
features tourism developed since 1972
head of state and government Maumoon
Abdul Gayoom from 1978
political system authoritarian nationalism
political parties none; candidates elected on the
basis of personal influence and clan loyalties
exports coconuts, copra, bonito (fish related to
tuna), garments
currency Rufiya
population (1990 est) 219,000; growth rate
3.7% p.a.
life expectancy men 60, women 63 (1989)
languages Divehi (Sinhalese dialect), English
religion Sunni Muslim
literacy 36% (1989)
GNP $69 million (1987); $410 per head (1988)
chronology
1887 Became a British protectorate.
1953 Long a sultanate, the Maldive Islands
became a republic within the Commonwealth.
1954 Sultan restored.
1965 Achieved full independence outside the
Commonwealth.
1968 Sultan deposed; republic reinstated with
Ibrahim Nasir as president.
1978 Nasir retired; replaced by Maumoon Abdul
Gayoom.
1982 Rejoined the Commonwealth.
1983 Gayoom re-elected.
1985 Became a founder member of South
Asian Association for Regional Cooperation
(SAARC).
1988 Gayoom re-elected. Coup attempt by
mercenaries, thought to have the backing of
former president Nasir, was foiled by Indian
paratroops.

Maldon /'mɔːldən/ English market town in Essex,
at the mouth of the river Chelmer; population
(1985 est) 15,500. It was the scene of a battle in
which the East Saxons were defeated by the Danes
991, commemorated in the Anglo-Saxon poem
The Battle of Maldon.

Malé /'mɑːleɪ/ capital and chief atoll of the Maldives in the Indian Ocean; population (1990)
55,100. It trades in copra, breadfruit, fish, and
palm products; it is also a growing tourist centre.

Mali Republic of (*République du Mali*)
area 1,240,142 sq km/478,695 sq mi
capital Bamako
towns Mopti, Kayes, Ségou, Timbuktu
physical landlocked state with river Niger and
savanna in S; part of the Sahara in N; hills in NE;
Senegal River and its branches irrigate the SW
environment a rising population coupled with
recent droughts has affected marginal agriculture.
head of state and government Alpha Oumar
Konare from 1992
political system emergent democratic republic
political parties Alliance for Democracy in
Mali (ADEMA), centrist; National Committee
for Democratic Initiative (CNID), centre-left;
Sudanese Union–African Democratic Rally (US–
RDA), Sudanese nationalist
exports cotton, peanuts, livestock, fish
currency franc CFA
population (1990 est) 9,182,000; growth rate
2.9% p.a.
life expectancy men 44, women 47 (1989)
languages French (official), Bambara
religion Sunni Muslim 90%, animist 9%,
Christian 1%
literacy 10% (1989)
GNP $1.6 bn (1987); $230 per head (1988)
chronology
1895 Came under French rule.
1959 With Senegal, formed the Federation
of Mali.
1960 Became the independent Republic of Mali,
with Modibo Keita as president.
1968 Keita replaced in an army coup by Moussa
Traoré.
1974 New constitution made Mali a one-party
state.
1976 New national party, the Malian
People's Democratic Union, announced.

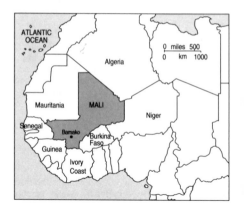

1983 Agreement between Mali and Guinea for eventual political and economic integration signed.
1985 Conflict with Burkina Faso lasted five days; mediated by International Court of Justice.
1991 Demonstrations against one-party rule. Moussa Traoré ousted in a coup led by Lt-Col Amadou Toumani Toure. New multiparty constitution agreed, subject to referendum.
1992 Referendum endorsed new democratic constitution. Alliance for Democracy in Mali (ADEMA) won multiparty elections; Alpha Oumar Konare elected president.

Malines /mæ'liːn/ French name for ⟡Mechelen, a city in Belgium.

Mallorca Spanish form of ⟡Majorca, an island in the Mediterranean.

Malmédy /'mælmədi/ town in Liège, E Belgium, 40 km/25 mi S of Aachen, in the region of Eupen et Malmédy; population (1991) 10,300.

Malmö /'mælməu/ industrial port (shipbuilding, engineering, textiles) in SW Sweden, situated across the Öresund from Copenhagen, Denmark; population (1990) 233,900. Founded in the 12th century, Malmö is Sweden's third largest city.

Malta Republic of (*Repubblika Ta'Malta*)
area 320 sq km/124 sq mi
capital and port Valletta
towns Rabat; port of Marsaxlokk
physical includes islands of Gozo 67 sq km/26 sq mi and Comino 2.5 sq km/1 sq mi
features occupies strategic location in central Mediterranean; large commercial dock facilities
head of state Vincent Tabone from 1989
head of government Edward Fenech Adami from 1987
political system liberal democracy
political parties Malta Labour Party (MLP), moderate, left-of-centre; Nationalist Party, Christian, centrist, pro-European
exports vegetables, knitwear, handmade lace, plastics, electronic equipment
currency Maltese lira
population (1990 est) 373,000; growth rate 0.7% p.a.
life expectancy men 72, women 77 (1987)

languages Maltese, English
religion Roman Catholic 98%
literacy 90% (1988)
GNP $1.6 bn; $4,750 per head (1988)
chronology
1814 Annexed to Britain by the Treaty of Paris.
1947 Achieved self-government.
1955 Dom Mintoff of the Malta Labour Party (MLP) became prime minister.
1956 Referendum approved MLP's proposal for integration with the UK. Proposal opposed by the Nationalist Party.
1958 MLP rejected the British integration proposal.
1962 Nationalists elected, with Borg Olivier as prime minister.
1964 Independence achieved from Britain, within the Commonwealth. Ten-year defence and economic aid treaty with UK signed.
1971 Mintoff re-elected. 1964 treaty declared invalid and negotiations began for leasing the NATO base in Malta.
1972 Seven-year NATO agreement signed.
1974 Became a republic.
1979 British military base closed.
1984 Mintoff retired and was replaced by Mifsud Bonnici as prime minister and MLP leader.
1987 Edward Fenech Adami (Nationalist) elected prime minister.
1989 Vincent Tabone elected president. USA–USSR summit held offshore.
1990 Formal application for EC membership.
1992 Nationalist Party returned to power in general election.

It is a whole Rock cover'd with very little Earth.

On **Malta** Lady Mary Wortley Montagu, letter of 1718

Maluku /mə'luːkuː/ or *Moluccas* group of Indonesian islands

Maluku (Moluccas)

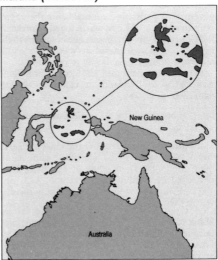

area 74,500 sq km/28,764 sq mi
capital Ambon, on Amboina
population (1989 est) 1,814,000
history as the Spice Islands, they were formerly part of the Netherlands East Indies; the S Moluccas attempted secession from the newly created Indonesian republic from 1949; exiles continued agitation in the Netherlands.

Malvern /ˈmɔːlvən/ English spa town in Hereford and Worcester, on the east side of the *Malvern Hills*, which extend for about 16 km/10 mi and have their high point in Worcester Beacon 425 m/1,395 ft; population (1981) 32,000. The *Malvern Festival* 1929–39, associated with the playwright G B Shaw and the composer Edward Elgar, was revived 1977.

Malvinas /mælˈviːnəs/ or *Islas Malvinas* Argentine name for the ◊Falkland Islands.

Mammoth Cave vast limestone cavern in Mammoth Cave National Park 1936, Kentucky, USA. The main cave is 6.5 km/4 mi long, and rises to a height of 38 m/125 ft; it is known for its stalactites and stalagmites. Indian councils were once held here.

Man, Isle of /mæn/ island in the Irish Sea, a dependency of the British crown, but not part of the UK
area 570 sq km/220 sq mi
capital Douglas
towns Ramsey, Peel, Castletown
features Snaefell 620 m/2,035 ft; annual TT (Tourist Trophy) motorcycle races, gambling casinos, Britain's first free port, tax haven; tailless Manx cat
products light engineering products; tourism, banking, and insurance are important
currency the island produces its own coins and notes in UK currency denominations
population (1986) 64,300
language English (Manx, nearer to Scottish than Irish Gaelic, has been almost extinct since the 1970s)
government crown-appointed lieutenant-governor, a legislative council, and the representative House of Keys, which together make up the Court of Tynwald, passing laws subject to the royal assent. Laws passed at Westminster only affect the island if specifically so provided
history Norwegian until 1266, when the island was ceded to Scotland; it came under UK administration 1765.

Man. abbreviation for ◊*Manitoba*, a Canadian province.

Managua /məˈnɑːɡwə/ capital and chief industrial city of Nicaragua, on the lake of the same name; population (1985) 682,000. It has twice been destroyed by earthquake and rebuilt, 1931 and 1972; it was also badly damaged during the civil war in the late 1970s.

Manama /məˈnɑːmə/ (Arabic *Al Manamah*) capital and free trade port of Bahrain, on Bahrain Island; population (1988) 152,000. It handles oil and entrepôt trade.

Manaus /məˈnaus/ capital of Amazonas, Brazil, on the Rio Negro, near its confluence with the Amazon; population (1991) 996,700. It can be reached by sea-going vessels, although it is 1,600 km/1,000 mi from the Atlantic. Formerly a centre of the rubber trade, it developed as a tourist centre in the 1970s.

Manawatu /ˌmænəˈwɑːtuː/ river in North Island, New Zealand, rising in the Ruahine Range. *Manawatu Plain* is a rich farming area, specializing in dairying and fat lamb production.

Mancha /ˈmæntʃə/ see ◊La Mancha, a former province of Spain.

Manche, La /mɒnʃ/ French name for the ◊English Channel. It gives its name to a French *département*.

Manchester /ˈmæntʃɪstə/ city in NW England, on the river Irwell, 50 km/31 mi E of Liverpool. It is a manufacturing (textile machinery, chemicals, rubber, processed foods) and financial centre; population (1991) 397,400. It is linked by the Manchester Ship Canal, built 1894, to the river Mersey and the sea. In 1992 the UK government pledged 55 million towards construction costs in a bid to host the Olympic Games in the year 2000.
features It is the home of the Hallé Orchestra, the Northern College of Music, the Royal Exchange (built 1869, now a theatre), a town hall (by Alfred Waterhouse), and a Cotton Exchange (now a leisure centre). The Castlefield Urban Heritage Park includes the Granada television studios, including the set of the soap opera *Coronation Street*, open to visitors, and also the Greater Manchester Museum of Science and Industry.
history Originally a Roman camp, Manchester is mentioned in the Domesday Book, and by the 13th century was already a centre for the wool trade. Its damp climate made it ideal for cotton, introduced in the 16th century, and in the 19th century the Manchester area was a world centre of manufacture, using cotton imported from North America and India. After 1945 there was a sharp decline, and many disused mills were refurbished to provide alternative industrial uses.
Long a hub of Radical thought, Manchester has always been a cultural and intellectual centre; it was the original home of the *Guardian* newspaper (founded as the *Manchester Guardian* 1821). Its pop-music scene flourished in the 1980s.

Manchester, the curse of the Ministry of Health, the despair of the architect, the salvation of the umbrella trade.

On **Manchester** *Manchester Guardian* 1941

Manchester, Greater /ˈmæntʃɪstə/ metropolitan county (1974–86) of NW England, replaced by a residuary body 1986 that covers some of its former functions
area 1,290 sq km/498 sq mi
towns Manchester, Bolton, Oldham, Rochdale, Salford, Stockport, and Wigan

Greater Manchester

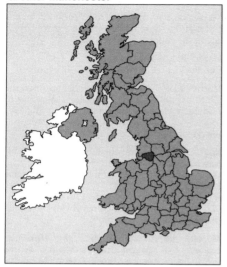

Manipur

features Manchester Ship Canal links it with the river Mersey and the sea; Old Trafford cricket ground at Stretford, and the football ground of Manchester United
products industrial
population (1991) 2,455,200
famous people John Dalton, James Joule, Emmeline Pankhurst, Gracie Fields, Anthony Burgess.

Manchuria /mæn'tʃuəriə/ European name for the northeast region of China, comprising the provinces of Heilongjiang, Jilin, and Liaoning. It was united with China by the Manchu dynasty 1644, but as the Chinese Empire declined, Japan and Russia were rivals for its control. The Russians were expelled after the Russo-Japanese War 1904–05, and in 1932 Japan consolidated its position by creating a puppet state, *Manchukuo*, which disintegrated on the defeat of Japan in World War II.

Mandalay /,mændə'leɪ/ chief town of the Mandalay division of Myanmar (formerly Burma), on the river Irrawaddy, about 495 km/370 mi N of Yangon (Rangoon); population (1983) 533,000.
Founded by King Mindon Min 1857, it was capital of Burma 1857–85, and has many pagodas, temples, and monasteries.

Mangalore /,mæŋgə'lɔː/ industrial port (textiles, timber, food-processing) at the mouth of the Netravati River in Karnataka, S India; population (1981) 306,000.

Manhattan /mæn'hætn/ island 20 km/12.5 mi long and 4 km/2.5 mi wide, lying between the Hudson and East rivers and forming a borough of the city of New York, USA. It includes the Wall Street business centre, Broadway and its theatres, Carnegie Hall (1891), the World Trade Centre (1973), the Empire State Building (1931), the United Nations headquarters (1952), Madison Square Garden, and Central Park.

First settled by the Dutch who bought the island from Algonquin Indians 1626, it was ceded to the British 1674. George Washington was sworn in as first president of the USA at Federal Hall.

Manila /mə'nɪlə/ industrial port (textiles, tobacco, distilling, chemicals, shipbuilding) and capital of the Philippines, on the island of Luzon; population (1990) 1,598,900, metropolitan area (including Quezon City) 5,926,000.
history Manila was founded by Spain 1571, captured by the USA 1898; in 1945 during World War II the old city to the S of the river Pasig was reduced to rubble in fighting between US and Japanese troops. It was replaced as capital by Quezon City 1948–76.

Manipur /,mʌnɪ'puə/ state of NE India
area 22,400 sq km/8,646 sq mi
capital Imphal
features Loktak Lake; original Indian home of polo
products grain, fruit, vegetables, sugar, textiles, cement
population (1991) 1,826,700
language Hindi
religion Hindu 70%
history administered from the state of Assam until 1947 when it became a Union Territory. It became a state 1972.

Manitoba /,mænɪ'təubə/ prairie province of Canada
area 650,000 sq km/250,900 sq mi
capital Winnipeg
features lakes Winnipeg, Winnipegosis, and Manitoba (area 4,700 sq km/1,814 sq mi); 50% forested
exports grain, manufactured foods, beverages, machinery, furs, fish, nickel, zinc, copper, and the world's largest caesium deposits
population (1991) 1,092,600
history trading posts and forts were built here by fur traders in the 18th century. What came to be known as the Red River settlement was first colonized 1811 by dispossessed Scottish Highlanders.

Manitoba

Winnipeg

The colony became the Canadian province of Manitoba 1870 after the Riel Rebellion 1869 ended. The area of the province was extended 1881 and 1912.

Manitoba, Lake /,mænɪ'təubə/ lake in Manitoba province, Canada, which drains into Lake Winnipeg to the NE through the river Dauphin; area 4,700 sq km/1,800 sq mi.

Manizales /,mænɪ'saːles/ city in the Central Cordillera in W Colombia 2,150 m/7,000 ft above sea level, centre of a coffee-growing area; population (1985) 328,000. It is linked with Mariquita by the world's longest overhead cable transport system, 72 km/45 mi long.

Mannheim /'mænhaɪm/ industrial city (heavy machinery, glass, earthenware, chemicals) on the Rhine in Baden-Württemberg, Germany; population (1988) 295,000. The modern symphony orchestra, with its balance of instruments and the vital role of the conductor, originated at Mannheim in the 18th century when the ruler of the Rhine Palatinate, Carl Theodor, assembled the finest players of his day.

Mansfield /'mænsfiːld/ industrial city (car parts, steel and rubber products) in N central Ohio, USA, NE of Columbus, seat of Richland County; population (1990) 50,600.

Mansûra /mæn'suərə/ industrial town (cotton) and capital of Dakahlia province, NE Egypt, on the Damietta branch of the river Nile; population (1986 est) 357,800. Mansûra was founded about 1220.

Mantua (Italian *Mantova*) capital of Mantua province, Lombardy, Italy, on an island of a lagoon of the river Mincio, SW of Verona; industry (chemicals, brewing, printing); population (1981) 60,866. The poet Virgil was born near Mantua, which dates from Roman times; it has Gothic palaces and a cathedral founded in the 12th century.

Maputo /mə'puːtəu/ (formerly until 1975 *Lourenço Marques*) capital of Mozambique, and Africa's second largest port, on Delagoa Bay; population (1987) 1,006,800. Linked by rail with Zimbabwe and South Africa, it is a major outlet for minerals, steel, textiles, processed foods, and furniture.

Maracaibo /,mærə'kaɪbəu/ oil-exporting port in Venezuela, on the channel connecting Lake Maracaibo with the Gulf of Venezuela; population (1989) 1,365,308. It is the second largest city in the country.

Maracaibo, Lake /,mærə'kaɪbəu/ lake in NW Venezuela; area 14,000 sq km/5,400 sq mi.

Marbella /maː'beɪə/ port and tourist resort on the Costa del Sol between Málaga and Algeciras in Andalucia, S Spain; population (1991) 80,645. There are three bullrings, a Moorish castle, and the remains of a medieval defensive wall.

Marburg /'maːbɜːg/ manufacturing town (chemicals, machinery, pottery) in Hessen, Germany, on the river Lahn, 80 km/50 mi N of Frankfurt-am-Main; population (1984) 77,300. The university was founded 1527 as a centre of Protestant teaching. Martin Luther and Ulrich Zwingli disputed on religion at Marburg 1529.

Marche, Le /'maːkeɪ/ (English *the Marches*) region of E central Italy consisting of the provinces of Ancona, Ascoli Piceno, Macerata, and Pesaro e Urbino; capital Ancona; area 9,700 sq km/3,744 sq mi; population (1990) 1,435,600.

Marches /'maːtʃɪz/ boundary areas of England with Wales, and England with Scotland. In the Middle Ages these troubled frontier regions were held by lords of the Marches.

Margate /'maːgeɪt/ town and seaside resort on the north coast of Kent, SE England; population (1981) 53,300. Industries include textiles and scientific instruments. It has a fine promenade and beach.

Mari /'maːri/ autonomous republic of Russia, E of Nizhny Novgorod and W of the Ural Mountains
area 23,200 sq km/8,900 sq mi
capital Yoshkar-Ola
features the Volga flows through the SW; 60% is forested
products timber, paper, grain, flax, potatoes, fruit
population (1989) 750,000; about 43% are ethnic Mari
history the Mari were conquered by Russia 1552. Mari was made an autonomous region 1920 and became an autonomous republic 1936.

Mariana Islands /,mæri'aːnəz/ or *Marianas* archipelago in the NW Pacific E of the Philippines, divided politically into *Guam* (an unincorporated territory of the USA) and *Northern Marianas* (a commonwealth of the USA with its own internal government, of 16 mountainous islands, extending 560 km/350 mi N from Guam)
area 480 sq km/185 sq mi
capital Garapan on Saipan
government own constitutionally elected government
products sugar, coconuts, coffee
currency US dollar
population (1988) 21,000, mainly Micronesian
languages Chamorro 55%, English
religion mainly Roman Catholic

Mariana Islands

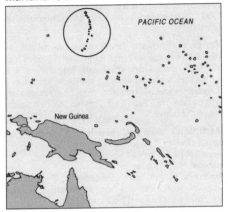

history sold to Germany by Spain 1899. The islands were mandated by the League of Nations to Japan 1918, and taken by US Marines 1944–45 in World War II. The islands were part of the US Trust Territory of the Pacific 1947–78. Since 1978 they have been a commonwealth of the USA.

Mariánské Lázně /ˈmæriaːnskeɪ ˈlaːzniei/ (German *Marienbad*) spa town in the Czech Republic; population (1981) 17,950. An international reputation for its healthy waters was established before World War II. The water of its springs, which contains Glauber's salts, has been used medicinally since the 16th century.

Maribor /ˈmærɪbɔː/ (German *Marburg*) town and resort on the river Drave in Slovenia, with a 12th-century cathedral and some industry (boots and shoes, railway rolling stock); population (1981) 185,500. Maribor dates from Roman times.

Marienbad /ˈmæriənbæd, German məˈriːənbaːt/ German name of ◊Mariánské Lázně, a spa town in the Czech Republic.

Marietta /ˌmæriˈetə/ industrial town (plastics, metal products, chemicals, and office equipment) in SE Ohio, USA; seat of Washington County; population (1990) 15,000. It lies where the Muskingum River flows into the Ohio River, SE of Columbus, and was the first permanent settlement (1788) in Ohio.

Maritsa /məˈrɪtsə/ (Greek *Hevros*; Turkish *Meric*) river rising in the Rhodope Mountains, Bulgaria, which forms the Greco-Turkish frontier before entering the Aegean Sea near Enez; length 440 km/275 mi.

Mariupol /ˌmæriˈuːpəl/ industrial port (iron, steel) in E Ukraine, on the Sea of Azov; population (1987) 529,000. It was named *Zhdanov* 1948, in honour of Andrei Zhdanov (1896–1948), but reverted to its former name 1989 following the Communist Party's condemnation of Zhdanov as having been one of the chief organizers of the Stalinist mass repressions of the 1930s and 1940s.

Marlborough /ˈmɔːlbrə/ market town in

Wiltshire, England, on the river Kennet, 122 km/76 mi W of London; population (1985 est) 6,900. There is engineering, tanning, and tourism. It is the site of Marlborough College (1843), a public school.

Marmara, Sea of /ˈmɑːmərə/ small inland sea separating Turkey in Europe from Turkey in Asia, connected through the Bosporus with the Black Sea, and through the Dardanelles with the Aegean; length 275 km/170 mi, breadth up to 80 km/50 mi.

Marne /mɑːn/ river in France which rises in the plateau of Langres and joins the Seine at Charenton near Paris; length 5,251 km/928 mi. It gives its name to the *départements* of Marne, Haute Marne, Seine-et-Marne, and Val de Marne; and to two battles of World War I.

Marquesas Islands /mɑːˈkeɪzəz/ (French *Iles Marquises*) island group in French Polynesia, lying N of the Tuamotu Archipelago; area 1,270 sq km/490 sq mi; population (1988) 7,500. The administrative headquarters is Atuona on Hiva Oa. The islands were annexed by France 1842.

Marquises, Iles /mɑːˈkiːz/ French form of ◊Marquesas Islands, part of French Polynesia.

Marrakesh /ˌmærəˈkeʃ/ historic town in Morocco in the foothills of the Atlas Mountains, about 210 km/130 mi S of Casablanca; population (1982) 549,000. It is a tourist centre, and has textile, leather, and food processing industries. Founded 1062, it has a medieval palace and mosques, and was formerly the capital of Morocco.

Fez is Europe but closed; Marrakesh is Africa, but open. Fez is black, white and grey; Marrakesh is red.

On **Marrakesh** John Gunther
Inside Africa 1955

Marsala /mɑːˈsaːlə/ port in W Sicily, Italy, notable for the sweet wine of the same name; population (1980) 85,000. The nationalist leader Giuseppe Garibaldi landed here 1860 at the start of his campaign to capture Sicily for Italy.

Marseille /mɑːˈseɪ/ (English *Marseilles*) chief seaport of France, industrial centre (chemicals, oil refining, metallurgy, shipbuilding, food processing), and capital of the *département* of Bouches-du-Rhône, on the Golfe du Lion, Mediterranean Sea; population (1990) 807,700.

It is surrounded by hills and connected with the river Rhône by a canal, and there are several offshore islands including If. Its university was founded 1409. In 1991 a grotto was discovered near Marseille, accessible only by an underwater passage. It contains prehistoric wall paintings showing people and animals, which may date from 20,000–12,000 BC; at that time the cave would have been accessible by land as the sea-level was much lower.

history Marseille was founded by mariners of Phocaea in Asia Minor 600 BC. Under the Romans it was a free city, and then, after suffering successive waves of invaders, became in the 13th century an independent republic, until included in France 1481. Much of the old quarter was destroyed by Germany 1943.

Marshall city in NE Texas, USA, SE of Dallas and across the border from Shreveport, Louisiana; seat of Harrison County; population (1990) 23,700. Industries include food processing, cotton, clothing, building materials, vehicle parts, and chemicals.

Marshall Islands
area 180 sq km/69 sq mi
capital Dalap-Uliga-Darrit (on Majuro atoll)
physical comprises the Radak (13 islands) and Ralik (11 islands) chains in the W Pacific
features include two atolls used for US atom-bomb tests 1946–63, Eniwetok and Bikini, where radioactivity will last for 100 years; and Kwajalein atoll (the largest) which has a US intercontinental missile range
head of state and government Amata Kabua from 1991
political system liberal democracy
political parties no organized party system
products copra, phosphates, fish; tourism is important
currency US dollar
population (1990) 31,600
language English (official)
religions Christian, mainly Roman Catholic, and local faiths
GNP $512 million; $16,516 per head (1990)
chronology
1855 Occupied by Germany.
1914 Occupied by Japan.
1920–45 Administered by Japan under United Nations mandate.
1946–63 Eniwetok and Bikini atolls used for US atom-bomb tests; islanders later demanded rehabilitation and compensation for the damage.
1947 Became part of the UN Pacific Islands Trust Territory, administered by the USA.
1986 Compact of free association with USA granted islands self-government, with USA

retaining military control and taking tribute.
1990 UN trust status terminated.
1991 Independence agreed; UN membership granted.

Martha's Vineyard /ˈmɑːθəz ˈvɪnjəd/ island 32 km/20 mi long off the coast of Cape Cod, Massachusetts, USA; chief town Edgartown. It is the former home of whaling captains, and now a summer resort. When the first English settlers arrived here they found wild grapes in abundance; hence the name.

Martinique /ˌmɑːtɪˈniːk/ French island in the West Indies (Lesser Antilles)
area 1,079 sq km/417 sq mi
capital Fort-de-France
features several active volcanoes; Napoleon's empress Josephine was born in Martinique, and her childhood home is now a museum
products sugar, cocoa, rum, bananas, pineapples
population (1990) 359,600
history Martinique was reached by Spanish navigators 1493, and became a French colony 1635; since 1972 it has been a French overseas region.

Martinique is France. Arriving from Trinidad you feel you have crossed not the Caribbean, but the English Channel.

On **Martinique** V S Naipul *The Middle Passage* 1962

Martin's Hundred /ˈmɑːtɪnz/ plantation town established in Virginia, USA, 1619 and eliminated by an Indian massacre three years later. Its remains, the earliest extensive trace of British colonization in America, were discovered 1970.

Mary /ˈmɑːriː/ town in SE Turkmenistan, on the Murghab River; population (1985) 85,000. It is situated in a cotton-growing oasis in the Kara Kum Desert, near where Alexander the Great founded a city.

Maryborough /ˈmɛərɪbərə/ Australian coastal and market town (grain, livestock) in SE Queensland;

Martinique

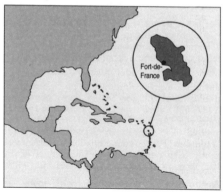

Maryland

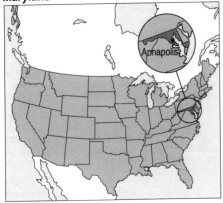

population (1988 est) 23,000. Industries include coal and gold-mining, iron, and steel.

Maryborough /'meəribərə/ former name of ◊Port Laoise, the county town of County Laois in the Republic of Ireland. The name gradually went out of use during the 1950s.

Maryland /'meərilænd/ state of eastern USA; nickname Old Line State/Free State
area 31,600 sq km/12,198 sq mi
capital Annapolis
cities Baltimore, Silver Spring, Dundalk, Bethesda
features Chesapeake Bay, an inlet of the Atlantic Ocean; horse racing (the Preakness Stakes at Baltimore); yacht racing at the US Naval Academy at Annapolis; historic Fort McHenry; Fort Meade, a government electronic-listening centre; Baltimore harbour
products poultry, dairy products, machinery, steel, cars and parts, electric and electronic equipment, chemicals, fish and shellfish
population (1990) 4,781,500
famous people Stephen Decatur, Francis Scott Key, Edgar Allan Poe, Frederick Douglass, Harriet Tubman, Upton Sinclair, H L Mencken, Babe Ruth, Billie Holiday
history one of the original 13 states, first settled 1634; it became a state 1788.

Mascara /mæs'kɑːrə/ town and wine-trade centre, 96 km/60 mi SE of Oran, Algeria; the headquarters of Abd-el-Kader (c. 1807–83) who fought the French invasion of Algeria 1830–47, Mascara being captured 1841.

Maseru /mə'seəruː/ capital of Lesotho, southern Africa, on the Caledon River; population (1986) 289,000. Founded 1869, it is a centre for trade and diamond processing.

Mashhad /mæʃ'hæd/ or *Meshed* holy city of the Shi'ites and industrial centre (carpets, textiles, leather goods) in NE Iran; population (1986) 1,464,000. It is the second largest city in Iran.

Mashonaland /mə'ʃɒnəlænd/ eastern Zimbabwe, the land of the Shona people, now divided into three administrative regions (Mashonaland East, Mashonaland Central, and Mashonaland West). Granted to the British South Africa Company 1889, it was included in Southern Rhodesia 1923. The Zimbabwe ruins are here. Prime Minister Robert Mugabe is a Shona.

Mashraq /mæʃ'rɑːk/ (Arabic 'east') the Arab countries of the E Mediterranean: Egypt, Sudan, Jordan, Syria, and Lebanon. The term is contrasted with Maghreb, comprising the Arab countries of NW Africa.

Masirah Island /mə'sɪərə/ island in the Arabian Sea, part of the sultanate of Oman, formerly used as an air staging post by British forces on their way to and from the Far East.

Mason–Dixon Line /'meɪsən 'dɪksən/ in the USA, the boundary line between Maryland and Pennsylvania (latitude 39° 43' 26.3" N), named after Charles Mason (1730–1787) and Jeremiah Dixon (died 1777), English astronomers and surveyors who surveyed it 1763–67. It was popularly seen as dividing the North from the South.

Massachusetts /ˌmæsə'tʃuːsɪts/ state of northeastern USA; nickname Bay State/Old Colony State
area 21,500 sq km/8,299 sq mi
capital Boston
towns Worcester, Springfield, New Bedford, Brockton, Cambridge
population (1990) 6,016,400
features Boston landmarks; Harvard University and the Massachusetts Institute of Technology, Cambridge; Cape Cod National Seashore; New Bedford and the islands of Nantucket and Martha's Vineyard, former whaling ports; Berkshire Hills with Tanglewood and other performing-arts centres; the battlefields of Lexington and Concord near Minute Man National Historical Park; Salem, site of witch trials; Plymouth Rock
products electronic, communications, and optical equipment; precision instruments; non-electrical machinery; fish; cranberries; dairy products

Massachusetts

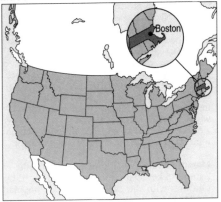

famous people Samuel Adams, Louis Brandeis, Emily Dickinson, Ralph Waldo Emerson, Robert Goddard, Nathaniel Hawthorne, Oliver Wendell Holmes, Winslow Homer, William James, John F Kennedy, Robert Lowell, Paul Revere, Henry Thoreau, Daniel Webster
history one of the original 13 states, it was first settled 1620 by the Pilgrims at Plymouth. After the Boston Tea Party 1773, the American Revolution began at Lexington and Concord 19 April 1775, and the British evacuated Boston the following year. Massachusetts became a state 1788.

Massawa /mə'saːwə/ chief port of Eritrea, on the Red Sea, with salt production and pearl fishing; population (1980) 33,000. It is one of the hottest inhabited places in the world, the temperature reaching 46°C/115°F in May. Massawa was an Italian possession 1885–1941.

Massif Central /mæ'siːf sɒn'trɑːl/ mountainous plateau region of S central France; area 93,000 sq km/36,000 sq mi, highest peak Puy de Sancy, 1,886 m/6,188 ft. The region is a source of hydro-electricity.

Masulipatnam /mə,suːlɪpət'næm/ or **Manchilipatnam**, also **Bandar** Indian seaport in Andhra Pradesh, at the mouth of the northern branch of the river Kistna; population (1981) 138,500.

Masurian Lakes /mə'suərɪən/ lakes in Poland (former East Prussia) which in 1914–15 were the scene of battles in which the Germans defeated the Russians.

Matabeleland /,mætə'biːlilænd/ western portion of Zimbabwe between the Zambezi and Limpopo rivers, inhabited by the Ndebele people
area 181,605 sq km/70,118 sq mi
towns Bulawayo
features rich plains watered by tributaries of the Zambezi and Limpopo, with mineral resources
language Matabele
famous people Joshua Nkomo
history Matabeleland was granted to the British South Africa Company 1889 and occupied 1893 after attacks on white settlements in Mashonaland; in 1923 it was included in Southern Rhodesia. It is now divided into two administrative regions (Matabeleland North and Matabeleland South). Joshua Nkomo was accused of plotting to overthrow the post-independence government of Zimbabwe and then expelled from the cabinet 1981. Zimbabwe African People's Union (ZAPU) supporters, mostly drawn from the Ndebele people, began a loosely organized armed rebellion against the Zimbabwe African National Union (ZANU) government of Robert Mugabe. The insurgency was brought to an end April 1988, when a unity agreement was reached between ZANU and ZAPU and Nkomo was appointed minister of state in the office of vice president.

Matadi /mə'taːdi/ chief port of Zaire on the river Zaïre, 115 km/70 mi from its mouth, linked by oil pipelines with the capital Kinshasa; population (1984) 144,700.

Matanzas /mə'tænsəs/ industrial port (tanning, textiles, sugar) in NW Cuba; population (1986) 105,400. Founded 1693, it became a major centre of coffee, tobacco, and sugar production.

Matlock /'mætlɒk/ spa town with warm springs, administrative headquarters of Derbyshire, England; population (1990 est) 15,000. Formerly known as a centre for hydropathic treatment, it manufactures textiles and high-tech products and caters for tourists.

Mato Grosso /'mætəu 'grɒsəu/ (Portuguese 'dense forest') area of SW Brazil, now forming two states, with their capitals at Cuiaba and Campo Grande. The forests, now depleted, supplied rubber and rare timbers; diamonds and silver are mined.

Matsue /'mɑːtsue/ city NW of Osaka on Honshu Island, Japan; population (1990) 142,900. It has remains of a castle, fine old tea houses, and the Izumo Grand Shrine (dating in its present form from 1744).

Matsuyama /,mɑːtsu'jama/ largest city on Shikoku Island, Japan; population (1990) 443,300. Industries include agricultural machinery, textiles, and chemicals. There is a feudal fortress (1634).

Matterhorn /'mætəhɔːn/ (French *le Cervin*, Italian *il Cervino*) mountain peak in the Alps on the Swiss-Italian border; 4,478 m/14,690 ft.
It was first climbed 1865 by English mountaineer Edward Whymper (1840–1911); four members of his party of seven were killed when a rope broke during their descent.

… it was considered to be the most inaccessible of all mountains, even by those who ought to have known better.

On the **Matterhorn** Edward Whymper
The Ascent of the Matterhorn 1880

Mauna Loa /,maunə'ləuə/ active volcano rising to a height of 4,169 m/13,678 ft on the Pacific island of Hawaii; it has numerous craters, including the second largest active crater in the world.

Mauritania Islamic Republic of (*République Islamique de Mauritanie*)
area 1,030,700 sq km/397,850 sq mi
capital Nouakchott
towns port of Nouadhibou, Kaédi, Zouérate
physical valley of river Senegal in S; remainder arid and flat
features part of the Sahara Desert; dusty sirocco wind blows in March
head of state and government Maaouia Ould Sid Ahmed Taya from 1984
political system emergent democratic republic
political parties Democratic and Social Republican Party (PRDS), centre-left, militarist; Union of Democratic Forces (UFD), centre-left; Rally for Democracy and National Unity (RDUN), centrist; Mauritian Renewal Party (PMR), centrist; Umma, Islamic fundamentalist; Socialist

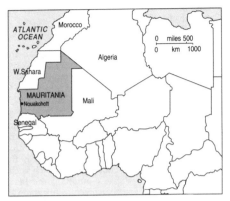

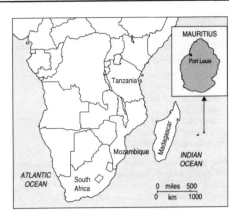

and Democratic Popular Front Union (UDSP),
left-of-centre
exports iron ore, fish, gypsum
currency ouguiya
population (1990 est) 2,038,000 (30% Arab-
Berber, 30% black Africans, 30% Haratine—
descendants of black slaves, who remained slaves
until 1980); growth rate 3% p.a.
life expectancy men 43, women 48 (1989)
languages French (official), Hasaniya Arabic,
black African languages
religion Sunni Muslim 99%
literacy 17% (1987)
GNP $843 million; $480 per head (1988)
chronology
1903 Became a French protectorate.
1960 Independence achieved from France, with
Moktar Ould Daddah as president.
1975 Western Sahara ceded by Spain. Mauritania
occupied the southern area and Morocco the
north. Polisario Front formed in Sahara to resist
the occupation by Mauritania and Morocco.
1978 Daddah deposed in bloodless coup;
replaced by Mohamed Khouna Ould Haidalla.
Peace agreed with Polisario Front.
1981 Diplomatic relations with Morocco broken.
1984 Haidalla overthrown by Maaouia Ould
Sid Ahmed Taya. Polisario regime formally
recognized.
1985 Relations with Morocco restored.
1989 Violent clashes between Mauritanians
and Senegalese. Arab-dominated government
expelled thousands of Africans into N Senegal;
governments had earlier agreed to repatriate each
other's citizens (about 250,000).
1991 Amnesty for political prisoners. Multiparty
elections promised. Calls for resignation of
President Taya.
1992 First multiparty elections won by ruling PRDS. Diplomatic relations with Senegal
resumed.

Mauritius Republic of
area 1,865 sq km/720 sq mi; the island of
Rodrigues is part of Mauritius; there are several
small island dependencies
capital Port Louis
towns Beau Bassin-Rose Hill, Curepipe, Quatre
Bornes
physical mountainous, volcanic island surrounded by coral reefs

features unusual wildlife includes flying fox and
ostrich; it was the home of the dodo (extinct from
about 1680)
interim head of state Veerasamy Ringadoo
from 1992
head of government Aneerood Jugnauth
from 1982
political system liberal democratic republic
political parties Mauritius Socialist Movement (MSM), moderate socialist-republican;
Mauritius Labour Party (MLP), centrist, Hindu-
oriented; Mauritius Social Democratic Party
(PMSD), conservative, Francophile; Mauritius
Militant Movement (MMM), Marxist-
republican; Rodriguais People's Organization
(OPR), left-of-centre
exports sugar, knitted goods, tea
currency Mauritius rupee
population (1990 est) 1,141,900, 68% of Indian
origin; growth rate 1.5% p.a.
life expectancy men 64, women 71 (1989)
languages English (official), French, creole,
Indian languages
religions Hindu 51%, Christian 30%, Muslim
17%
literacy 94% (1989)
GNP $1.4 bn (1987); $1,810 per head (1988)
chronology
1814 Annexed to Britain by the Treaty of Paris.
1968 Independence achieved from Britain
within the Commonwealth, with Seewoosagur
Ramgoolam as prime minister.
1982 Aneerood Jugnauth became prime minister.
1983 Jugnauth formed a new party, the Mauritius
Socialist Movement. Ramgoolam appointed governor general. Jugnauth formed a new coalition
government.
1985 Ramgoolam died, succeeded by Veersamy
Ringadoo.
1987 Jugnauth's coalition re-elected.
1990 Attempt to create a republic failed.
1991 Jugnauth's ruling MSM–MMM–OPR coalition won general election; pledge to secure
republican status by 1992.
1992 Mauritius became a republic while remaining a member of the Commonwealth. Ringadoo
became interim president.

Mayagüez /ˈmaɪəgwez/ port in W Puerto Rico,
Greater Antilles, with needlework industry and a

US agricultural experimental station; population (1986 est) 98,900.

Mayence /mar'pns/ French name for the German city of ◊Mainz.

Mayenne /mar'en/ *département* of W France in Pays-de-Loire region
area 5,212 sq km/2,033 sq mi
capital Laval
features river Mayenne
products iron, slate; paper
population (1990) 278,000.

Mayenne /mar'en/ river in W France which gives its name to the *département* of Mayenne; length 200 km/125 mi. It rises in Orne, flows in a generally southerly direction through Mayenne and Maine-et-Loire, and joins the river Sarthe just above Angers to form the Maine.

Mayfair /'merfeə/ district of Westminster in London, England, vaguely defined as lying between Piccadilly and Oxford Street, and including Park Lane; formerly a fashionable residential district, but increasingly taken up by offices.

Maynooth /mer'nu:θ/ village in Kildare, Republic of Ireland, with a Roman Catholic training college for priests; population (1981) 3,388.

Mayo /'merəu/ county in Connacht province, Republic of Ireland
area 5,400 sq km/2,084 sq mi
towns Castlebar (administrative town)
features Lough Conn; wild Atlantic coast scenery; Achill Island; the village of Knock, where two women claimed a vision of the Virgin with two saints 1879, now a site of pilgrimage; Croagh Patrick 765 m/2,510 ft, the mountain where St Patrick spent the 40 days of Lent in 441; pilgrims climb the mountain on the last Sunday of July each year
products sheep and cattle farming; fishing
population (1991) 110,700.

Mayotte /mar'pt/ or *Mahore* island group of the Comoros, off the east coast of Africa, a *collectivité territoriale* of France by its own wish. The two main islands are Grande Terre and Petite Terre
area 374 sq km/144 sq mi
capital Dzaoudzi
products coffee, copra, vanilla, fishing
languages French, Swahili
population (1987) 73,000
history a French colony 1843–1914, and later, with the Comoros, an overseas territory of France. In 1974, Mayotte voted to remain a French dependency.

Mbabane /əmbɑː'bɑːneɪ/ capital (since 1902) of Swaziland, 160 km/100 mi W of Maputo, in the Dalgeni Hills; population (1986) 38,000. Mining and tourism are important.

Mboma another spelling of ◊Boma, a Zairean port.

MD abbreviation for ◊*Maryland*, a state of the USA.

ME abbreviation for ◊*Maine*, a state of the USA.

Meath /mi:ð/ county in the province of Leinster, Republic of Ireland
area 2,340 sq km/903 sq mi
county town Trim
features Tara Hill, 155 m/509 ft high, was the site of a palace and coronation place of many kings of Ireland (abandoned in the 6th century) and St Patrick preached here
products sheep, cattle
population (1991) 105,600.

Mecca /'mekə/ (Arabic *Makkah*) city in Saudi Arabia and, as birthplace of Muhammad, the holiest city of the Islamic world; population (1974) 367,000. In the centre of Mecca is the Great Mosque, in the courtyard of which is the Kaaba, the sacred shrine containing the black stone believed to have been given to Abraham by the angel Gabriel.

It also contains the well Zam-Zam, associated by tradition with the biblical characters Hagar and Ishmael. Most pilgrims come via the port of Jiddah.

Our notions of Mecca must be drawn from the Arabians; as no unbeliever is permitted to enter the city, our travellers are silent.

On **Mecca** Edward Gibbon *The Decline and Fall of the Roman Empire* 1776–88

Mechelen /'mexələ/ (French *Malines*) industrial city (furniture, carpets, textiles) and market-gardening centre in Antwerp province, N Belgium, which gave its name to Mechlin lace; population (1991) 75,300.

Mecklenburg–West Pomerania /'meklənbɜːg/ (German *Mecklenburg-Vorpommern*) administrative *Land* (state) of Germany
area 22,887 sq km/8,840 sq mi
capital Schwerin
towns Rostock, Wismar, Stralsund, Neu-brandenburg
products fish, ships, diesel engines, electronics, plastics, chalk
population (1990) 2,100,000
history the state was formerly the two grand duchies of Mecklenburg-Schwerin and Mecklenburg-Strelitz, which became free states of the Weimar Republic 1918–34, and were joined 1946 with part of Pomerania to form a region of East Germany. In 1952 it was split into the districts of Rostock, Schwerin, and Neubrandenburg. Following German reunification 1990, the districts were abolished and Mecklenburg–West Pomerania was reconstructed as one of the five new states of the Federal Republic.

Medan /mə'dɑːn/ seaport and economic centre of the island of Sumatra, Indonesia; population (1980) 1,379,000. It trades in rubber, tobacco, and palm oil.

Medellín /ˌmeðer'iːn/ industrial town (textiles, chemicals, engineering, coffee) in the Central Cordillera, Colombia, 1,538 m/5,048 ft above

Mediterranean Sea

sea level; population (1985) 2,069,000. It is the second city of Colombia, and its drug capital, with 7,000 violent deaths in 1990.

There is a museum dedicated to local artist Fernando Botero.

Medford /'medfəd/ city in SW Oregon, USA, S of Eugene; seat of Jackson County; population (1990) 47,000. It is a summer resort, and tourism is important to the economy. Other industries include processing of the area's agricultural crops and dairy products.

Medina /me'di:nə/ (Arabic *Madinah*) Saudi Arabian city, about 355 km/220 mi N of Mecca; population (1974) 198,000. It is the second holiest city in the Islamic world, and is believed to contain the tomb of Muhammad. It produces grain and fruit.

It also contains the tombs of the caliphs or Muslim leaders Abu Bakr, Omar, and Fatima, Muhammad's daughter.

Mediterranean Sea /,medɪtə'reɪnɪən/ inland sea separating Europe from N Africa, with Asia to the E; extreme length 3,700 km/2,300 mi; area 2,966,000 sq km/1,145,000 sq mi. It is linked to the Atlantic Ocean (at the Strait of Gibraltar), Red Sea, and Indian Ocean (by the Suez Canal), Black Sea (at the Dardanelles and Sea of Marmara). The main subdivisions are the Adriatic, Aegean, Ionian, and Tyrrhenian seas. It is highly polluted.

The Mediterranean is almost tideless, and is saltier and warmer than the Atlantic; shallows from Sicily to Cape Bon (Africa) divide it into an east and a west basin. Dense salt water forms a permanent deep current out into the Atlantic.

The Mediterranean is severely endangered by human and industrial waste pollution; 100 million people live along the coast, 85% of sewage near the coast is discharged directly into the water, and it is regularly crossed by oil tankers. The Barcelona Convention 1976 to clean up the Mediterranean was signed by 17 countries and led to a ban on dumping of mercury, cadmium, persistent plastics, DDT, crude oil, and hydrocarbons.

Médoc /meɪ'dɒk/ French district bordering the Gironde in Aquitaine region, N of Bordeaux. It is famed for its claret wines, Margaux and St

Julien being two well-known names. Lesparre and Pauillac are the chief towns.

Medway /'medweɪ/ river of SE England, rising in Sussex and flowing through Kent and the *Medway towns* (Chatham, Gillingham, Rochester) to Sheerness, where it enters the Thames; it is about 96 km/60 mi long. In local tradition it divides the 'Men of Kent', who live to the E, from the 'Kentish Men', who live to the W. It is polluted by industrial waste.

Meerut /'mɪərət/ industrial city (chemicals, soap, food processing) in Uttar Pradesh, N India; population (1981) 538,000. The Indian Mutiny began here 1857.

Meghalaya /,megə'leɪə/ state of NE India
area 22,500 sq km/8,685 sq mi
capital Shillong
features mainly agricultural and comprises tribal hill districts
products potatoes, cotton, jute, fruit
minerals coal, limestone, white clay, corundum, sillimanite
population (1991) 1,760,600, mainly Khasi, Jaintia, and Garo
religion Hindu 70%
languages various.

Mekele /'meɪkəleɪ/ capital of Tigray region, N Ethiopia; population (1984) 62,000. It trades in salt, incense, and resin.

Meknès /mek'nes/ (Spanish *Mequinez*) city in N Morocco, known for wine and carpetmaking; population (1981) 487,000. One of Morocco's four imperial cities, it was the capital until 1728, and is the site of the tomb of Sultan Moulay Ismail.

Mekong /,mi:'kɒŋ/ river rising as the Za Qu in Tibet and flowing to the South China Sea, through a vast delta (about 200,000 sq km/77,000 sq mi); length 4,425 km/2,750 mi. It is being developed for irrigation and hydroelectricity by Cambodia, Laos, Thailand, and Vietnam.

Meghalaya

Shillong

INDIAN OCEAN

Melanesia

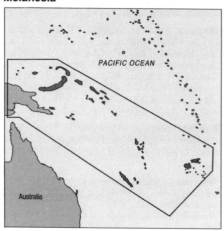

Melaka /məˈlækə/ Malaysian form of ⟡Malacca, a state of Peninsular Malaysia.

Melanesia /ˌmeləˈniːziə/ islands in the SW Pacific between Micronesia to the N and Polynesia to the E, embracing all the islands from the New Britain archipelago to Fiji.

Melbourne /ˈmelbən/ capital of Victoria, Australia, near the mouth of the river Yarra; population (1990) 3,080,000. Industries include engineering, shipbuilding, electronics, chemicals, food processing, clothing, and textiles.

Founded 1835, it was named after the British prime minister Lord Melbourne 1837, grew in the wake of the gold rushes, and was the seat of the Commonwealth government 1901–27. It is the country's second largest city, with three universities, and was the site of the 1956 Olympics.

Imagine a huge chessboard flung on to the earth, and you have what is the true and characteristic Melbourne.

On **Melbourne** Francis Adams
The Australians 1893

Melbourne /ˈmelbən/ industrial city (food processing and electronic and aviation equipment) on the east coast of Florida, USA, on the Indian River, SE of Orlando; population (1990) 59,600. Tourism is also important to the economy.

Melilla /meˈlɪljə/ port and military base on the northeast coast of Morocco; area 14 sq km/5 sq mi; population (1986) 56,000. It was captured by Spain 1496 and is still under Spanish rule. Also administered from Melilla are three other Spanish possessions: Peña ('rock') de Velez de la Gomera, Peña d'Alhucemas, and the Chaffarine Islands.

Melos /ˈmiːlɒs/ (modern Greek *Mílos*) Greek island in the Aegean, one of the Cyclades; area

155 sq km/60 sq mi. The sculpture of *Venus de Milo* was discovered here 1820 (now in the Louvre). The capital is Plaka.

Melrose /ˈmelrəuz/ town in Borders Region, Scotland. The heart of King Robert the Bruce is buried here and the ruins of Melrose Abbey 1136 are commemorated in verse by Sir Walter Scott.

Melton Mowbray /ˈmeltən ˈməubreɪ/ market town in Leicestershire, England, on the river Eye; population (1981) 24,500. A hunting and horse-breeding centre, it is also known for pork pies and Stilton cheeses.

Memel /ˈmeɪməl/ German name for ⟡Klaipeda, a port in Lithuania.

Memphis /ˈmemfɪs/ industrial port city (pharmaceuticals, food processing, cotton, timber, tobacco) on the Mississippi River, in Tennessee, USA; population (1990) 610,300. The French built a fort here 1739, but Memphis was not founded until 1819. Its musical history includes Beale Street, home of the blues composer W C Handy, and Graceland, the home of Elvis Presley; its recording studios and record companies (Sun 1953–68, Stax 1960–75) made it a focus of the music industry.

During the Civil War the city was captured 1862 by Union forces after a river battle. Martin Luther King, Jr, was assassinated here 4 April 1968.

Menai Strait /ˈmenaɪ/ (Welsh *Afon Menai*) channel of the Irish Sea, dividing Anglesey from the Welsh mainland; about 22 km/14 mi long, up to 3 km/2 mi wide. It is crossed by Thomas Telford's suspension bridge 1826 (reconstructed 1940) and Robert Stephenson's tubular rail bridge 1850.

Menam /miːˈnæm/ another name for the river ⟡Chao Phraya, in Thailand.

Menderes /ˌmendəˈres/ (Turkish *Büyük Menderes*) river in European Turkey, about 400 km/250 mi long, rising near Afyonkarahisar and flowing along a winding course into the Aegean Sea. The word 'meander' is derived from the ancient Greek name for the river.

Mendip Hills or *Mendips* range of limestone hills in S England, stretching nearly 40 km/25 mi SE–NW from Wells in Somerset toward the Bristol Channel. There are many cliffs, scars, and caverns, notably *Cheddar Gorge*. The highest peak is *Blackdown* (326 m/1,068 ft).

Mendoza /menˈdəusə/ capital of the Argentine province of the same name; population (1991) 121,700. Founded 1561, it developed because of its position on the Trans-Andean railway; it lies at the centre of a wine-producing area.

Menindee /məˈnɪndi/ village and sheep centre on the Darling River in New South Wales, Australia. It is the centre of a scheme for conserving the waters of the Darling in *Menindee Lake* (155 sq km/60 sq mi) and other lakes nearby.

Menorca /meˈnɔːkə/ Spanish form of ⟡Minorca, one of the Balearic Islands.

Menton /mɒnˈtɒn/ (Italian **Mentone**) resort on the French Riviera, close to the Italian frontier; population (1990) 29,500. It belonged to the princes of Monaco from the 14th century until briefly independent 1848–60, when the citizens voted to merge with France.

Mequinez /ˌmekɪˈneθ/ Spanish name for ▷Meknés, a town in Morocco.

Mérida capital of Yucatán state, Mexico, a centre of the sisal industry; population (1986) 580,000. It was founded 1542, and has a cathedral 1598. Its port on the Gulf of Mexico is Progreso.

Meriden /ˈmerɪdən/ industrial city (plastics, silver, and electronics) in S central Connecticut, USA, E of Waterbury; population (1990) 59,500.

Merionethshire /ˌmeriˈɒnəθʃə/ (Welsh **Sir Feirionnydd**) former county of N Wales, included in the new county of Gwynedd 1974. Dolgellau was the administrative town.

Mersey /ˈmɜːzi/ river in NW England; length 112 km/70 mi. Formed by the confluence of the Goyt and Etherow rivers, it flows W to join the Irish Sea at Liverpool Bay. It is linked to the Manchester Ship Canal. It is polluted by industrial waste, sewage, and chemicals.

Merseyside /ˈmɜːzisaɪd/ metropolitan county (1974–86) of NW England, replaced 1986 by a residuary body which covers some of its former functions
area 650 sq km/251 sq mi
towns Liverpool; Bootle, Birkenhead, St Helens, Wallasey, Southport
features river Mersey; Merseyside Innovation Centre (MIC), linked with Liverpool University and Polytechnic; Prescot Museum of clock- and watch-making; Speke Hall (Tudor), and Croxteth Hall and Country Park (a working country estate open to the public)

Merseyside

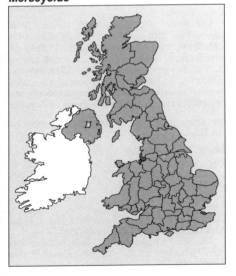

products chemicals, electrical goods, vehicles
population (1991) 1,376,800
famous people the Beatles, William Ewart Gladstone, George Stubbs.

Mersin /meəˈsiːn/ or **İçel** Turkish industrial free port (chrome, copper, textiles, oil refining); population (1990) 422,400.

Merthyr Tydfil /ˈmɜːθə ˈtɪdvɪl/ industrial town (light engineering, electrical goods) in Mid Glamorgan, Wales, UK; population (1982) 60,000. It was formerly a centre of the Welsh coal and steel industries.

Merton /ˈmɜːtn/ borough of SW Greater London, including the districts of Wimbledon, Merton, Mitcham, and Morden
features part of Wimbledon Common (includes Caesar's Camp – an Iron Age fort); All England Tennis Club 1877
population (1991) 161,800.

Mesa Verde /ˈmeɪsə ˈvɜːdi/ (Spanish 'green table') wooded clifftop in Colorado, USA, with Pueblo dwellings, called the Cliff Palace, built into its side. Dating from about 1000 BC, with 200 rooms and 23 circular ceremonial chambers (kivas), it had an estimated population of about 400 people and was probably a regional centre.

Meshed /meˈʃed/ variant spelling of ▷Mashhad, a town in Iran.

Mesopotamia /ˌmesəpəˈteɪmiə/ land between the Tigris and Euphrates rivers, now part of Iraq. Here the civilizations of Sumer and Babylon flourished. Sumer (3500 BC) may have been the earliest civilization.

Messina /meˈsiːnə/ city and port in NE Sicily; population (1988) 271,000. It produces soap, olive oil, wine, and pasta. Originally an ancient Greek settlement (Zancle), it was taken first by Carthage and then by Rome. It was rebuilt after an earthquake 1908.

Messina, Strait of /meˈsiːnə/ channel in the central Mediterranean separating Sicily from mainland Italy; in Greek legend a monster (Charybdis), who devoured ships, lived in the whirlpool on the Sicilian side, and another (Scylla), who devoured sailors, in the rock on the Italian side. The classical hero Odysseus passed safely between them.

Metz /mets, French mes/ industrial city (shoes, metal goods, tobacco) in Lorraine region, NE France, on the Moselle River; population (1990) 123,900. Part of the Holy Roman Empire 870–1552, it became one of the great frontier fortresses of France, and was in German hands 1871–1918.

Meurthe /mɜːt/ river rising in the Vosges Mountains in NE France and flowing NW to join the Moselle at Frouard, near Nancy; length 163 km/ 102 mi. It gives its name to the *département* of Meurthe-et-Moselle.

Meuse /mɜːz/ (Dutch **Maas**) river flowing through France, Belgium, and the Netherlands;

length 900 km/560 mi. It was a line of battle in both world wars. It gives its name to a French *département*.

Mewar /me'wɑː/ another name for ◊Udaipur, a city in India.

Mexicali /ˌmeksɪ'kæli/ city in NW Mexico; population (1984) 500,000. It produces soap and cottonseed oil. The availability of cheap labour attracts many US companies (Hughes Aerospace, Rockwell International, and others).

Mexico United States of (*Estados Unidos Mexicanos*)
area 1,958,201 sq km/756,198 sq mi
capital Mexico City
towns Guadalajara, Monterrey; port Veracruz
physical partly arid central highlands; Sierra Madre mountain ranges E and W; tropical coastal plains
environment during the 1980s, smog levels in Mexico City exceeded World Health Organization standards on more than 300 days of the year. Air is polluted by 130,000 factories and 2.5 million vehicles
features Rio Grande; 3,218 km/2,000 mi frontier with USA; resorts Acapulco, Cancun, Mexicali, Tijuana; Baja California, Yucatán peninsula; volcanoes, including Popocatepetl; pre-Columbian archaeological sites
head of state and government Carlos Salinas de Gortari from 1988
political system federal democratic republic
political parties Institutional Revolutionary Party (PRI), moderate, left-wing; National Action Party (PAN), moderate Christian socialist
exports silver, gold, lead, uranium, oil, natural gas, handicrafts, fish, shellfish, fruits and vegetables, cotton, machinery
currency peso
population (1990 est) 88,335,000 (60% mixed descent, 30% Indian, 10% Spanish descent); 50% under 20 years of age; growth rate 2.6% p.a.
life expectancy men 67, women 73
languages Spanish (official) 92%, Nahuatl, Maya, Mixtec
religion Roman Catholic 97%
literacy men 92%, women 88% (1989)

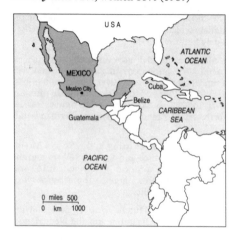

GNP $126 bn (1987); $2,082 per head
chronology
1821 Independence achieved from Spain.
1846–48 Mexico at war with USA; loss of territory.
1848 Maya Indian revolt suppressed.
1864–67 Maximilian of Austria was emperor of Mexico.
1917 New constitution introduced, designed to establish permanent democracy.
1983–84 Financial crisis.
1985 Institutional Revolutionary Party (PRI) returned to power. Earthquake in Mexico City.
1986 IMF loan agreement signed to keep the country solvent until at least 1988.
1988 PRI candidate Carlos Salinas Gotari elected president. Debt reduction accords negotiated with USA.
1991 PRI won general election. President Salinas promised constitutional reforms.
1992 Public outrage following Guadalajara gas-explosion disaster.

Poor Mexico, so far from God, and so close to the United States!

On **Mexico** attributed to President Porfirio Díaz

Mexico City /'meksɪkəu/ (Spanish *Ciudad de México*) capital, industrial (iron, steel, chemicals, textiles), and cultural centre of Mexico, 2,255 m/7,400 ft above sea level on the southern edge of the central plateau; population (1986) 18,748,000. It is thought to be one of the world's most polluted cities because of its position in a volcanic basin 2,000 m/7,400 ft above sea level. Pollutants gather in the basin causing a smog cloud.

Notable buildings include the 16th-century cathedral, the national palace, national library, Palace of Justice, and national university; the Ministry of Education has murals 1923–27 by Diego Rivera.

The city dates from about 1325, when the Aztec capital Tenochtitlán was founded on an island in Lake Texcoco. This city was levelled 1521 by the Spaniards, who in 1522 founded a new city on the site. It was the location of the 1968 Summer Olympics. In 1984, the explosion of a liquefied gas tank caused the deaths of over 450 people, and in 1985, over 2,000 were killed by an earthquake.

MI abbreviation for ◊*Michigan*, a state of the USA.

Miami /maɪ'æmi/ industrial city (food processing, transportation and electronic equipment, clothing, and machinery) and port in Florida, USA; population (1990) 358,500. It is the hub of finance, trade, and air transport for the USA, Latin America, and the Caribbean. There has been an influx of immigrants from Cuba, Haiti, Mexico, and South America since 1959.

The first permanent non-Indian settlement dates from the 1870s. In 1896 a railway was extended to Miami, and the city was subsequently

Michigan

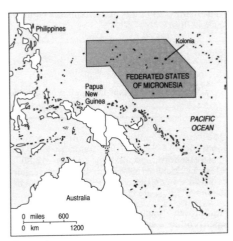

promoted as a tourist resort for its beaches. It is also a centre for oceanographic research.

Michigan /'mɪʃɪgən/ state in N central USA; nickname Wolverine State/Great Lake State
area 151,600 sq km/58,518 sq mi
capital Lansing
cities Detroit, Grand Rapids, Flint
features Great Lakes: Superior, Michigan, Huron, Erie; Porcupine Mountains; Muskegon, Grand, St Joseph, and Kalamazoo rivers; over 50% forested; Isle Royale National Park; Pictured Rocks and Sleeping Bear national seashores; Henry Ford Museum and Greenfield Village, Dearborn
products motor vehicles and equipment; nonelectrical machinery; iron and steel; chemicals; pharmaceuticals; dairy products
population (1990) 9,295,300
famous people Edna Ferber, Gerald Ford, Henry Ford, Jimmy Hoffa, Iggy Pop, Diana Ross
history first settled 1668 at Sault Sainte Marie; present-day Detroit settled 1701; passed to the British 1763 and to the USA 1796; statehood achieved 1837.

Michigan, Lake /'mɪʃɪgən/ lake in N central USA, one of the Great Lakes; area 58,000 sq km/22,390 sq mi. Chicago and Milwaukee are its main ports.

Micronesia /ˌmaɪkrəʊ'niːziə/ group of islands in the Pacific Ocean lying N of Melanesia, including the Federated States of Micronesia, Belau, Kiribati, the Mariana and Marshall Islands, Nauru, and Tuvalu.

Micronesia Federated States of (FSM)
area 700 sq km/270 sq mi
capital Kolonia in Pohnpei state
towns Moen, in Chuuk state; Lelu, in Kosrae state; Colonia, in Yap state
physical an archipelago in the W Pacific
features equatorial, volcanic island chain, with extensive coral, limestone, and lava shores
head of state and government Bailey Olter from 1991
political system democratic federal state
political parties no formally organized political parties
products copra, fish products, tourism
currency US dollar
population (1985) 91,440
languages English (official) and local languages
religion Christianity

chronology
16th century Colonized by Spain.
1885 Purchased from Spain by Germany.
1914 Occupied by Japan.
1920 Administered by Japan under League of Nations mandate.
1944 Occupied by USA.
1947 Became part of the UN Pacific Islands Trust Territory, administered by the USA.
1982 Compact of Free Association signed with USA.
1990 UN trust status terminated. Independent state established, with USA responsible for defence and foreign affairs.
1991 First independent president elected. Entered into UN membership.

Middelburg /'mɪdlbɜːg/ industrial town (engineering, tobacco, furniture) in the SW Netherlands, capital of Zeeland and a former Hanseatic town; population (1991) 39,600.

Middle East indeterminate area now usually taken to include the Balkan States, Egypt, and SW Asia. Until the 1940s, this area was generally called the Near East, and the term Middle East referred to the area from Iran to Burma (now Myanmar).

Middle Range or *Middleback Range* mountain range in the NE of Eyre Peninsula, South Australia, about 65 km/40 mi long, parallel with the west coast of Spencer Gulf.

Middlesbrough /'mɪdlzbrə/ industrial town and port on the Tees, Cleveland, England, commercial and cultural centre of the urban area formed by Stockton-on-Tees, Redcar, Billingham, Thornaby, and Eston; population (1991) 141,100. Formerly a centre of heavy industry, it diversified its products in the 1960s.

Middlesex /'mɪdlseks/ former English county, absorbed by Greater London 1965. It was settled in the 6th century by Saxons, and its name comes from its position between the kingdoms of the East and West Saxons. Contained within the Thames basin, it provided good agricultural land before it was built over.

The name is still used, as in Middlesex County Cricket Club.

An acre in Middlesex is better than a principality in Utopia.

On **Middlesex** Thomas Babington Macaulay
Lord Bacon 1837

Middletown /'mɪdltaun/ city in S central Connecticut, USA, on the Connecticut River, S of Hartford; population (1990) 42,800. Industries include insurance, banking, vehicle parts, electronics, hardware, and paper products. Wesleyan University is here.

Middletown /'mɪdltaun/ industrial city (steel, paper products, and aircraft parts) in SW Ohio, USA, on the Miami River, N of Cincinnati; population (1990) 46,000.

Mid Glamorgan /'mɪd glə'mɔːgən/ (Welsh *Morgannwg Ganol*) county in S Wales
area 1,020 sq km/394 sq mi
towns Cardiff (administrative headquarters), Porthcawl, Aberdare, Merthyr Tydfil, Bridgend, Pontypridd
features includes a small area of the former county of Monmouthshire to the E; mountains in the N; Caerphilly Castle, with its water defences
products the N was formerly a leading coal (Rhondda) and iron and steel area; Royal Mint at Llantrisant; agriculture in the S; Caerphilly mild cheese
population (1991) 536,500
languages 8% Welsh, English
famous people Geraint Evans.

Midi-Pyrénées /mɪ'di: ˌpɪrə'neɪ/ region of SW France, comprising the *départements* of Ariège, Aveyron, Haute-Garonne, Gers, Lot, Haute-Pyrénées, Tarn, and Tarn-et-Garonne

Mid Glamorgan

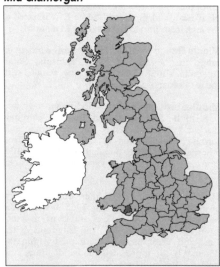

area 45,300 sq km/17,486 sq mi
population (1986) 2,355,000
towns capital Toulouse; Montauban, Cahors, Rodez, Lourdes
products fruit, wine, livestock
features several spa towns, winter resorts, and prehistoric caves
history occupied by the Basques since prehistoric times, this region once formed part of the prehistoric province of Gascony that was taken by the English 1154, recaptured by the French 1453, inherited by Henry of Navarre, and reunited with France 1607.

Midland /'mɪdlənd/ city in W Texas, USA, halfway between Fort Worth and El Paso; population (1990) 89,400. The city's economy depends on the oil companies located here after the discovery of oil 1923.

Midland /'mɪdlənd/ industrial city (chemicals and concrete) in central Michigan, USA, on the Tittabawassee River, NW of Saginaw; population (1990) 38,100. There are oil and gas wells.

Midlands /'mɪdləndz/ area of England corresponding roughly to the Anglo-Saxon kingdom of Mercia. The *E Midlands* comprises Derbyshire, Leicestershire, Northamptonshire, and Nottinghamshire. The *W Midlands* covers the former metropolitan county of West Midlands created from parts of Staffordshire, Warwickshire, and Worcestershire; and (often included) the *S Midlands* comprises Bedfordshire, Buckinghamshire, and Oxfordshire.

In World War II, the E Midlands was worked for oil, and substantial finds were made in the 1980s; the oilbearing E Midlands Shelf extends into Yorkshire and Lincolnshire.

Midlothian /mɪd'ləuðɪən/ former Scottish county S of the Firth of Forth, included 1975 in the region of Lothian; Edinburgh was the administrative headquarters.

Midway Islands /'mɪdweɪ/ two islands in the Pacific, 1,800 km/1,120 mi NW of Honolulu; area 5 sq km/2 sq mi; population (1980) 500. They were annexed by the USA 1867, and are now administered by the US Navy. The naval *Battle of Midway* 3–6 June 1942, between the USA and Japan, was a turning point in the Pacific in World War II; the US victory marked the end of Japanese expansion in the Pacific.

Midwest /ˌmɪd'west/ or *Middle West* large area of the N central USA. It is loosely defined, but is generally taken to comprise the states of Illinois, Iowa, Wisconsin, Minnesota, Nebraska, Kansas, Missouri, North Dakota, and South Dakota and the portions of Montana, Wyoming, and Colorado that lie E of the Rocky Mountains. Ohio, Michigan, and Indiana are often variously included, as well. Traditionally its economy is divided between agriculture and heavy industry. The main urban Midwest centre is Chicago.

Milan /mɪ'læn/ (Italian *Milano*) industrial city (aircraft, cars, locomotives, textiles), financial and cultural centre, capital of Lombardy, Italy; population (1988) 1,479,000.

features The Gothic cathedral, built about 1450, crowned with pinnacles, can hold 40,000 worshippers; the Pinacoteca di Brera art gallery; Leonardo da Vinci's *Last Supper* 1495–97 in the refectory of Sta Maria della Grazie; La Scala opera house (Italian *Teatro alla Scala*) 1778; an annual trade fair.

history Settled by the Gauls in the 5th century BC, it was conquered by the Roman consul Marcellus 222 BC to become the Roman city of *Mediolanum*. Under Diocletian, in AD 286 Milan was capital of the Western empire. Destroyed by Attila the Hun 452, and again by the Goths 539, the city regained its power through the political importance of its bishops. It became an autonomous commune 1045; then followed a long struggle for supremacy in Lombardy.

The city was taken by Frederick I (Barbarossa) 1162; only in 1176 were his forces finally defeated, at the battle of Legnano. Milanese forces were again defeated by the emperor at the battle of Cortenuova 1237. In the Guelph-Ghibelline struggle the Visconti family emerged at the head of the Ghibelline faction; they gained power 1277, establishing a dynasty which lasted until 1450 when Francesco Sforza seized control and became duke. The Sforza court marked the high point of Milan as a cultural and artistic centre. Control of the city passed to Louis XII of France 1499, and in 1540 it was annexed by Spain, beginning a long decline. The city was ceded to Austria by the Treaty of Utrecht 1714, and in the 18th century began a period of intellectual enlightenment.

Milan was in 1796 taken by Napoleon, who made it the capital of the Cisalpine Republic 1799, and in 1805 capital of the kingdom of Italy until 1814, when it reverted to the Austrians. In 1848, Milan rebelled unsuccessfully (the *Cinque Giornate*/Five Days), and in 1859 was joined to Piedmont.

Mildura /mɪl'dʊərə/ town in NW Victoria, Australia, on the Murray River, with food-processing industries; population (1985) 16,500.

Mile End /maɪl 'end/ area of the East End of London, England, in the district of Stepney, now part of the London borough of Tower Hamlets. Mile End Green (now Stepney Green) was the scene of Richard II's meeting with the rebel peasants 1381, and in later centuries was the exercise ground of the London 'trained bands', or militia.

Milford /'mɪlfəd/ industrial city (fabricated metal, writing pens, and electronics) in SW Connecticut, USA, situated by the Housatonic River and Long Island Sound, W of New Haven; population (1990) 49,900.

Milford Haven /'mɪlfəd 'heɪvən/ (Welsh *Aberdaugleddau*) seaport in Dyfed, SW Wales, on the estuary of the east and west Cleddau rivers; population (1985) 14,000. It has oil refineries, and a terminal for giant tankers linked by pipeline with Llandarcy, near Swansea.

Millville /'mɪlvɪl/ city in SW New Jersey, USA, on the Maurice River, SE of Philadelphia; popula-

tion (1990) 26,000. Products include vegetables, poultry, and glass.

Milton Keynes /'mɪltən 'kiːnz/ industrial new town (engineering, electronics) in Buckinghamshire, England; population (1989 est) 141,800. It was developed 1967 around the old village of the same name, following a grid design by Richard Llewelyn-Davies; it is the headquarters of the Open University.

Milwaukee /mɪl'wɔːki/ industrial port (meatpacking, brewing, engineering, machinery, electronic and electrical equipment, chemicals) in Wisconsin, USA, on Lake Michigan; population (1990) 628,100. The site was settled 1818 and drew a large influx of German immigrants, beginning in the 1840s.

Minas Gerais /'miːnəʒ ʒeˈraɪs/ state in SE Brazil; centre of the country's iron ore, coal, diamond and gold mining industries; area 587,172 sq km/226,710 sq mi; capital Belo Horizonte; population (1991) 16,956,900.

Mindanao /ˌmɪndəˈnau/ second-largest island of the Philippines
area 94,627 sq km/36,526 sq mi
towns Davao, Zamboanga
physical mountainous rainforest
features an isolated people, the Tasaday, were reputedly first seen by others 1971 (this may be a hoax). The active volcano Apo reaches 2,954 m/ 9,600 ft, and Mindanao is subject to severe earthquakes. There is a Muslim guerrilla resistance movement
products pineapples, coffee, rice, coconut, rubber, hemp, timber, nickel, gold, steel, chemicals, fertilizer
population (1980) 10,905,250.

Minden /'mɪndən/ industrial town (tobacco, food processing) of North Rhine–Westphalia, Germany, on the river Weser; population (1985) 80,000. The French were defeated here 1759 by an allied army from Britain, Hanover, and Brunswick, commanded by the duke of Brunswick.

Mindoro /mɪn'dɔːrəu/ island of the Philippine Republic, S of Luzon
area 10,347 sq km/3,995 sq mi
towns Calapan
features Mount Halcon 2,590 m/8,500 ft
population (1980) 500,000.

Minhow /ˌmɪn'hau/ name in use 1934–43 for ⋄Fuzhou, a town in SE China.

Minneapolis /ˌmɪni'æpəlɪs/ city in Minnesota, USA, forming with St Paul the Twin Cities area; population (1990) 368,400, metropolitan area 2,464,100. It is at the head of navigation of the Mississippi River. Industries include food processing and the manufacture of machinery, electrical and electronic equipment, precision instruments, transport machinery, and metal and paper products.

The powerful Cray computers are built here, used for long-range weather forecasting, space-

Minnesota

St Paul

craft design, and code-breaking. Shopping areas of the city centre are glass-covered against the difficult climate; there is an arts institute, a symphony orchestra, the University of Minnesota, and Tyrone Guthrie theatre. Minneapolis was incorporated as a village 1855.

Minnesota /ˌmɪnɪˈsəutə/ state in N midwest USA; nickname Gopher State/North Star State
area 218,700 sq km/84,418 sq mi
capital St Paul
towns Minneapolis, Duluth, Bloomington, Rochester
features sources of the Mississippi River and the Red River of the North; Voyageurs National Park near the Canadian border; Minnehaha Falls at Minneapolis; Mayo Clinic at Rochester; more than 15,000 lakes
products cereals, soya beans, livestock, meat and dairy products, iron ore (about two-thirds of US output), nonelectrical machinery, electronic equipment
population (1990) 4,375,100
famous people F Scott Fitzgerald, Hubert H Humphrey, Sinclair Lewis, Charles and William Mayo
history first European exploration, by French fur traders, in the 17th century; region claimed for France by Daniel Greysolon, Sieur Duluth, 1679; part E of Mississippi River ceded to Britain 1763 and to the USA 1783; part W of Mississippi passed to the USA under the Louisiana Purchase 1803; became a territory 1849; statehood achieved 1858.

Minorca /mɪˈnɔːkə/ (Spanish *Menorca*) second largest of the Balearic Islands in the Mediterranean
area 689 sq km/266 sq mi
towns Mahon, Ciudadela
products copper, lead, iron; tourism is important
population (1985) 55,500.

Minsk /mɪnsk/ or *Mensk* industrial city (machinery, textiles, leather, computers) and capital of Belarus; population (1987) 1,543,000.
   Minsk dates from the 11th century and has in turn been held by Lithuania, Poland, Sweden, and Russia before Belarus became an independent republic 1991. The city was devastated by Napoleon 1812 and heavily damaged by German forces 1944. Mass graves from between 1937 and 1941 of more than 30,000 victims of Joseph Stalin's terror were discovered in a forest outside Minsk 1989.

The town of partisans.

   On **Minsk** (occupied for 1,100 days by
   Germans during World War II)
   Russian saying

Miquelon Islands /ˌmiːkəlɒn/ small group of islands off the south coast of Newfoundland which, with St Pierre, form a French overseas *département*; see ⟡St Pierre and Miquelonq.
area 216 sq km/83 sq mi
products cod; silver fox and mink are bred
population (with St Pierre, 1982) 6,045.

Mirpur /ˌmɪəˈpuə/ district in SW Kashmir, Pakistan, between the Jhelum River and the Indian state of Jammu and Kashmir; capital Mirpur. Its products include cotton and grain.

Mirzapur /ˌmɪəzəˈpuə/ city of Uttar Pradesh, India, on the river Ganges; a grain and cotton market, with bathing sites and temples on the river; population (1981) 127,785.

Mishawaka /ˌmɪʃəˈwɔːkə/ industrial city (plastics, rubber, missiles, and vehicle and aircraft parts) in N central Indiana, USA, E of South Bend; population (1990) 42,600.

Miskolc /ˈmɪʃkɒlts/ industrial city (iron, steel, textiles, furniture, paper) in NE Hungary, on the river Sajo, 145 km/90 mi NE of Budapest; population (1988) 210,000.

Misr /ˈmɪsrə/ Egyptian name for ⟡Egypt and for ⟡Cairo.

Mission /ˈmɪʃn/ city in S Texas, USA, near the Rio Grande, W of McAllen; population (1990) 28,700. The economy is based on processing the area's citrus fruits and vegetables, and on nearby oil wells.

Mississippi /ˌmɪsɪˈsɪpi/ state in southeastern USA; nickname Magnolia State/Bayou State
area 123,600 sq km/47,710 sq mi
capital Jackson
towns Biloxi, Meridian, Hattiesburg
features rivers: Mississippi, Pearl, Big Black; Vicksburg National Military Park (Civil War site); Gulf Islands National Seashore; mansions and plantations, many in the Natchez area
products cotton, rice, soya beans, chickens, fish and shellfish, lumber and wood products, petroleum and natural gas, transportation equipment, chemicals
population (1990) 2,573,200
famous people Jefferson Davis, William Faulkner, Elvis Presley, Leontyne Price, Eudora Welty, Tennessee Williams, Richard Wright
history first explored by Hernando de Soto for Spain 1540; settled by the French 1699, the English 1763; ceded to USA 1798; statehood achieved 1817. After secession from the Union during the Civil War, it was readmitted 1870.

Mississippi

Mississippi /ˌmɪsɪ'sɪpɪ/ river in the USA, the main arm of the great river system draining the USA between the Appalachian and the Rocky mountains. The length of the Mississippi is 3,780 km/2,350 mi; with its tributary the Missouri 6,020 km/3,740 mi.

The Mississippi rises in the lake region of N Minnesota, with St Anthony Falls at Minneapolis. Below the tributaries Minnesota, Wisconsin, Des Moines, and Illinois, the confluence of the Missouri and Mississippi occurs at St Louis. Turning at the Ohio junction, it passes Memphis, and takes in the St Francis, Arkansas, Yazoo, and Red tributaries before reaching its delta on the Gulf of Mexico beyond New Orleans.

The Mississippi, ... with its paddle boats ferries and hoot owls, is the most haunted river in the world.

On the **Mississippi River** Cecil Beaton
It Gives Me Great Pleasure 1955

Missolonghi /ˌmɪsə'lɒŋgi/ (Greek *Mesolóngion*) town in W central Greece and Eubrea region, on the north shore of the Gulf of Patras; population (1981) 10,200. It was several times under siege by the Turks in the wars of 1822–26 and it was here that the British poet Byron died.

Missouri /mɪ'zuərɪ/ state in central USA; nickname Show Me State/Bullion State
area 180,600 sq km/69,712 sq mi
capital Jefferson City
towns St Louis, Kansas City, Springfield, Independence

Missouri

features rivers: Mississippi, Missouri; Pony Express Museum at St Joseph; birthplace of Jesse James; Mark Twain and Ozark state parks; Harry S Truman Library at Independence
products meat and other processed food, aerospace and transport equipment, lead, zinc
population (1990) 5,117,100
famous people George Washington Carver, T S Eliot, Jesse James, Joseph Pulitzer, Harry S Truman, Mark Twain
history explored by Hernando de Soto for Spain 1541; acquired by the USA under the Louisiana Purchase 1803; achieved statehood 1821, following the Missouri Compromise of 1820.

Missouri /mɪ'suərɪ/ major river in the central USA, a tributary of the Mississippi, which it joins N of St Louis; length 4,320 km/2,683 mi.

Since 1944 the muddy, turbulent river has been tamed by a series of locks and dams for irrigation and flood control.

Mitilíni /ˌmɪtɪ'liːni/ modern Greek name of ⊳Mytilene, a town on the island of Lesvos.

Mitylene /ˌmɪtɪ'liːni/ alternative spelling of ⊳Mytilene, a Greek town on the island of Lesvos.

Mizoram /ˌmaɪzə'ræm/ state of NE India
area 21,100 sq km/8,145 sq mi
capital Aizawl
products rice, hand-loom weaving
population (1991) 686,200
religion 84% Christian
history made a Union Territory 1972 from the Mizo Hills District of Assam. Rebels carried on a guerrilla war 1966–76, but in 1976 acknowledged Mizoram as an integral part of India. It became a state 1986.

Mmabatho /mə'baːtəʊ/ or *Sun City* capital of Bophuthatswana, South Africa; population (1985) 28,000. It is a casino resort frequented by many white South Africans.

MN abbreviation for ⊳*Minnesota*, a state of the USA.

Mizoram

MO abbreviation for ◊*Missouri*, a state of the USA.

Mobile /məʊ'biːl/ industrial city (meat-packing, paper, cement, clothing, chemicals) and only seaport in Alabama, USA; population (1990) 196,300. Founded 1702 by the French a little to the N of the present city, Mobile was capital of the French colony of Louisiana until 1763. It was then British until 1780, and Spanish to 1813.

Mobutu Sese Seko, Lake /mə'buːtu: 'seseɪ 'sekəʊ/ lake on the border of Uganda and Zaire in the Great Rift Valley; area 4,275 sq km/1,650 sq mi. The first European to see it was the British explorer Samuel Baker, who named it Lake Albert after the Prince Consort. It was renamed 1973 by Zaire's president Mobutu after himself.

Moçambique /ˌmuːsəm'biːkə/ Portuguese name for ◊Mozambique.

Modena /'mɒdɪnə/ city in Emilia, Italy, capital of the province of Modena, 37 km/23 mi NW of Bologna; population (1988) 177,000. Products include vehicles, glass, pasta, and sausages. It has a 12th-century cathedral, a 17th-century ducal palace, and a university (1683), known for its medical and legal faculties.

Modesto /mə'destəʊ/ city in central California, USA, on the Tuolumne River, SE of San Francisco, on the fringes of the San Joaquin Valley; population (1990) 164,700. It is an agricultural centre for wine, apricots, melons, beans, peaches, and livestock.

Mogadishu /ˌmɒgə'dɪʃuː/ or *Mugdisho* capital and chief port of Somalia; population (1988) 1,000,000. It is a centre for oil refining, food processing, and uranium mining. During the struggle to overthrow President Barre and the ensuing civil war 1991–92, much of the city was devastated and many thousands killed. In April 1992 the UN Security Council voted to send military observers to monitor a cease-fire in the city. In Sept the first UN peacekeeping troops were drafted in to the city.

It has mosques dating from the 13th century, and a cathedral built 1925–28.

Mogilev /ˌmɒgɪl'jɒf/ industrial city (tractors, clothing, chemicals, furniture) in Belarus, on the river Dneiper, 193 km/120 mi E of Minsk; population (1987) 359,000. It was annexed by Russia from Sweden 1772. Belarus became an independent republic 1991.

Mogok /'məʊgɒk/ village in Myanmar (Burma), 114 km/71 mi NNE of Mandalay, known for its ruby and sapphire mines.

Mojave Desert /mə'haːvi/ arid region in S California, USA, part of the Great Basin; area 38,500 sq km/15,000 sq mi.

The US military has appropriated thousands of square kilometres for bombing ranges and test stations, including Edwards Air Force Base, a landing place for space shuttles.

Mokha /'məʊkə/ or *Mocha* seaport of N Yemen near the mouth of the Red Sea, once famed for its coffee exports; population about 8,000.

Mold /məʊld/ (Welsh *Yr Wyddgrung*) market town in Clwyd, Wales, on the river Alyn; population (1989 est) 8,900. It is the administrative headquarters of Clywd. There are two theatres.

Moldavia /mɒl'deɪvɪə/ former principality in E Europe, on the river Danube, occupying an area divided today between Moldova (formerly a Soviet republic) and Romania. It was independent between the 14th and 16th centuries, when it became part of the Ottoman Empire. In 1861 Moldavia was united with its neighbouring principality Wallachia as Romania. In 1940 the eastern part, Bessarabia, became part of the USSR, whereas the western part remained in Romania.

Moldova Republic of
area 33,700 sq km/13,012 sq mi
capital Chişinău (Kishinev)
towns Tiraspol, Beltsy, Bendery
physical hilly land lying largely between the rivers Prut and Dniester; northern Moldova comprises the level plain of the Beltsy Steppe and uplands; the climate is warm and moderately continental
features Black Earth region
head of state Mircea Snegur from 1989
head of government Valerin Murovsky from 1992
political system emergent democracy
political parties Moldavian Popular Front (MRF), Romanian nationalist; Gagauz-Khalky People's Movement (GKPM), Gagauz separatist
products wine, tobacco, canned goods
population (1990) 4,400,000 (Moldavian 64%, Ukrainian 14%, Russian 13%, Gagauzi 4%, Bulgarian 2%)
language Moldavian, allied to Romanian
religion Russian Orthodox
chronology
1940 Bessarabia in the E became part of the Soviet Union whereas the W part remained in Romania.
1941 Bessarabia taken over by Romania-Germany.
1944 Red army reconquered Bessarabia.
1946–47 Widespread famine.
1988 A popular front, the Democratic Movement for Perestroika, campaigned for accelerated political reform.

1989 Jan–Feb: nationalist demonstrations in Chisinău. May: Moldavian Popular Front (MRF) established. July: former Communist Party deputy leader Mircea Snegur became head of state. Aug: Moldavian language granted official status, triggering clashes between ethnic Russians and Moldavians. Nov: Gagauz-Khalky People's Movement formed to campaign for Gagauz autonomy. *1990* Feb: MRF polled strongly in supreme soviet elections. June: economic and political sovereignty declared; renamed Republic of Moldova. Oct: Gagauzi held unauthorized elections to independent parliament; state of emergency declared after interethnic clashes. Trans-Dniester region declared its sovereignty. Nov: state of emergency declared in Trans-Dniester region after interethnic killings. *1991* March: Moldova boycotted the USSR's constitutional referendum. Aug: independence declared after abortive anti-Gorbachev coup; Communist Party outlawed. Dec: Moldova joined Commonwealth of Independent States. *1992* Admitted into United Nations and the Conference on Security and Cooperation in Europe; diplomatic recognition granted by USA. Possible union with Romania discussed. Trans-Dniester region fighting intensified; Russian peacekeeping force reportedly deployed after talks between Moldova and Russia. Andrei Sangheli became premier. *1993* Secessionist unrest continued in Gagauz and Trans-Dniester regions.

Moline /məʊˈliːn/ industrial city (farm implements, furniture, and metal products) in NW Illinois, USA, on the Mississippi River, W of Chicago; population (1990) 43,200.

Molise /mɒˈliːzeɪ/ mainly agricultural region of S central Italy, comprising the provinces of Campobasso and Isernia; area 4,400 sq km/1,698 sq mi; population (1990) 336,500. Its capital is Campobasso.

Molokai /ˌməʊləˈkaɪ/ mountainous island of Hawaii, USA, SE of Oahu
area 673 sq km/259 sq mi
features Kamakou 1,512 m/4,960 ft is the highest peak
population (1980) 6,049
history the island was the site of a leper colony organized 1873–89 by Belgian missionary Joseph De Veuster (Father Damien).

Molotov /ˈmɒlətɒf/ former name (1940–62) for the port of ◊Perm in Russia.

Moluccas /məʊˈlʌkəz/ another name for ◊Maluku, a group of Indonesian islands.

Mombasa /mɒmˈbæsə/ industrial port (oil refining, cement) in Kenya (serving also Uganda and Tanzania), built on Mombasa Island and adjacent mainland; population (1984) 481,000. It was founded by Arab traders in the 11th century and was an important centre for ivory and slave trading until the 16th century.

Mona /ˈməʊnə/ Latin name for ◊Anglesey, an island off the coast of Wales.

Monaco Principality of
area 1.95 sq km/0.75 sq mi
capital Monaco-Ville
towns Monte Carlo, La Condamine; heliport Fontvieille
physical steep and rugged; surrounded landwards by French territory; being expanded by filling in the sea
features aquarium and oceanographic centre; Monte Carlo film festival, motor races, and casinos; world's second-smallest state
head of state Prince Rainier III from 1949
head of government Jean Ausseil from 1986
political system constitutional monarchy under French protectorate
political parties National and Democratic Union (UND); Democratic Union Movement; Monaco Action; Monégasque Socialist Party
exports some light industry; economy dependent on tourism and gambling
currency French franc
population (1989) 29,000; growth rate –0.5% p.a.
languages French (official), English, Italian
religion Roman Catholic 95%
literacy 99% (1985)
chronology
1861 Became an independent state under French protection.
1918 France given a veto over succession to the throne.
1949 Prince Rainier III ascended the throne.
1956 Prince Rainier married US actress Grace Kelly.
1958 Birth of male heir, Prince Albert.
1959 Constitution of 1911 suspended.
1962 New constitution adopted.

Fantastic, giddy, dizzy, crazy, enchanting —supremely artificial and without the vegetation that softens similar extravagancies along the rest of the coast.

On **Monaco** William Samson *Blue Skies, Brown Studies* 1961

Monadnock /mə'nædnɒk/ mountain in New Hampshire, USA, 1,063 m/3,186 ft high. The term 'monadnock' is also used to mean any isolated hill or mountain.

Monaghan /'mɒnəhən/ (Irish **Mhuineachain**) county of the NE Republic of Ireland, province of Ulster; area 1,290 sq km/498 sq mi; products include cereals, linen, potatoes, and cattle; population (1991) 51,300. The county town is Monaghan. The county is low and rolling, and includes the rivers Finn and Blackwater.

Monastir /ˌmɒnə'stɪə/ resort town on the Mediterranean coast of Tunisia, 18 km/11 mi S of Sousse. It is the birthplace of former president Habib Bourguiba and summer residence of the president of Tunisia.

Monastir Turkish name for the town of ◊Bitolj in the Former Yugoslav Republic of Macedonia.

Mönchengladbach /ˌmʌnʃən'glædbæk/ industrial city in North Rhine–Westphalia, Germany, on the river Niers near Düsseldorf; industries include textiles, machinery, paper; population (1988) 255,000. It is the NATO headquarters for N Europe.

Mongolia State of
(**Outer Mongolia** until 1924; **People's Republic of Mongolia** until 1991)
area 1,565,000 sq km/604,480 sq mi
capital Ulaanbaatar
towns Darhan, Choybalsan
physical high plateau with desert and steppe (grasslands)
features Altai Mountains in SW; salt lakes; part of Gobi Desert in SE; the world's southernmost permafrost and northernmost desert
head of state Punsalmaagiyn Ochirbat from 1990
head of government Puntsagiyn Jasray from 1992
political system emergent democracy
political parties Mongolian People's Revolutionary Party (MPRP), reform-communist; Mongolian Democratic Party (MDP), main opposition party; Mongolian Democratic Union
exports meat and hides, minerals, wool, livestock,

grain, cement, timber
currency tugrik
population (1990 est) 2,185,000; growth rate 2.8% p.a.
life expectancy men 63, women 67 (1989)
languages Khalkha Mongolian (official), Chinese, Russian, and Turkic languages
religion officially none (Tibetan Buddhist Lamaism suppressed 1930s)
literacy 89% (1985)
GNP $3.6 bn; $1,820 per head (1986)
chronology
1911 Outer Mongolia gained autonomy from China.
1915 Chinese sovereignty reasserted.
1921 Chinese rule overthrown with Soviet help.
1924 People's Republic proclaimed.
1946 China recognized Mongolia's independence.
1966 20-year friendship, cooperation, and mutual-assistance pact signed with USSR. Relations with China deteriorated.
1984 Yumjaagiyn Tsedenbal, effective leader, deposed and replaced by Jambyn Batmonh.
1987 Soviet troops reduced; Mongolia's external contacts broadened.
1989 Further Soviet troop reductions.
1990 Democratization campaign launched by Mongolian Democratic Union. Punsalmaagiyn Ochirbat's MPRP elected in free multiparty elections. Mongolian script readopted.
1991 Massive privatization programme launched as part of move towards a market economy. The word 'Republic' dropped from country's name.
1992 Jan: New constitution introduced. Economic situation worsened. Puntsagiyn Jasray appointed new prime minister.

Mongolia, Inner /mɒŋ'gəʊlɪə/ (Chinese **Nei Mongol**) autonomous region of NE China from 1947
area 450,000 sq km/173,700 sq mi
capital Hohhot
features strategic frontier area with Russia; known for Mongol herders, now becoming settled farmers
physical grassland and desert
products cereals under irrigation; coal; reserves of rare earth oxides europium, and yttrium at Bayan Obo
population (1990) 21,457,000.

Monmouth /'mʌnməθ/ (Welsh **Trefynwy**) market town in Gwent, Wales, at the confluence of the rivers Wye and Monnow; population (1991) 75,000. There is some light industry. Henry V was born in the now ruined castle.

Monmouthshire /'mʌnməθʃə/ (Welsh **Sir Fynwy**) former county of Wales, which in 1974 became, minus a small strip on the border with Mid Glamorgan, the new county of **Gwent**.

Monroe /mən'rəʊ/ city in NE Louisiana, USA, on the Ouachita River, E of Shreveport; population (1990) 54,900. Industries include furniture, chemicals, paper, natural gas, and soya beans.

Monrovia /mɒn'rəʊvɪə/ capital and port of Liberia; population (1985) 500,000. Industries include rubber, cement, and petrol processing.

0 mls 500
0 km 1000
Ulaanbaatar
MONGOLIA
China
PACIFIC OCEAN

Montana

It was founded 1821 for slaves repatriated from the USA. Originally called **Christopolis**, it was renamed after US president James Monroe.

Mons /mɒnz/ (Flemish **Bergen**) industrial city (coalmining, textiles, sugar) and capital of the province of Hainaut, Belgium; population (1991) 91,700. The military headquarters of NATO is at nearby Chièvres-Casteau.

Montana /mɒn'tænə/ state in western USA, on the Canadian border; nickname Treasure State
area 318,100 sq km/147,143 sq mi
capital Helena
towns Billings, Great Falls, Butte
physical mountainous forests in the W, rolling grasslands in the E
features rivers: Missouri, Yellowstone, Little Bighorn; Glacier National Park on the Continental Divide and Yellowstone National Park; Museum of the Plains Indian; Custer Battlefield National Monument; hunting and ski resorts
products wheat (under irrigation), cattle, coal, copper, oil, natural gas, lumber, wood products
population (1990) 799,100
famous people Gary Cooper, Myrna Loy
history explored for France by Verendrye early 1740s; passed to the USA 1803 in the Louisiana Purchase; first settled 1809; W Montana obtained from Britain in the Oregon Treaty 1846; influx of gold-seeking immigrants mid-19th century; fierce Indian wars 1867–77, which included 'Custer's Last Stand' at the Little Bighorn with the Sioux; achieved statehood 1889.

Montauban /ˌmɒntəu'bɒn/ industrial town (porcelain, textiles) in the Midi-Pyrénées region, SW France, on the river Tarn; population (1982) 53,147. The classical painter Jean Auguste Ingres was born here.

Mont Blanc /ˌmɒm 'blɒŋ/ (Italian **Monte Bianco**) highest mountain in the Alps, between France and Italy; height 4,807 m/15,772 ft. It was first climbed 1786 by Jacques Balmat and Michel Paccard of Chamonix.

Mont Blanc is the monarch of mountains …

On **Mont Blanc** Lord Byron
Manfred 1816–17

Mont Cenis /ˌmɒn sə'niː/ pass in the Alps between Lyon, France, and Turin, Italy, at 2,082 m/ 6,831 ft.

Monte Bello Islands /'mɒnti 'beləu/ group of uninhabited islands in the Indian Ocean, off Western Australia, used by the UK for nuclear-weapons testing 1952; the largest of the group is Barrow Island.

Monte Carlo /'mɒnti 'kɑːləu/ town and luxury resort in the principality of Monaco, situated on a rocky promontory NE of Monaco town; population (1982) 12,000. It is known for its Casino (1878) designed by architect Charles Garnier, and the Monte Carlo car rally and Monaco Grand Prix.

Here in Monte Carlo, where we have to take amusement seriously, there is not much to laugh at.

On **Monte Carlo** Anthony Burgess in
Observer 28 May 1978

Monte Cristo /'mɒnti 'krɪstəu/ small uninhabited island 40 km/25 mi S of Elba, in the Tyrrhenian Sea; its name supplied a title for the French novelist Alexandre Dumas's hero in *The Count of Monte Cristo* 1844.

Montego Bay /mɒn'tiːgəu 'beɪ/ port and resort on the northwest coast of Jamaica; population (1982) 70,200. The fine beaches and climate attract tourists.

Montélimar /mɒn'teɪlɪmɑː/ town in Drôme district, France; known for the nougat to which its name is given; population (1990) 31,400.

Montenegro /ˌmɒntɪ'niːgrəu/ (Serbo-Croatian **Crna Gora**) constituent republic of Yugoslavia
area 13,800 sq km/5,327 sq mi
capital Titograd
town Cetinje
features smallest of the republics; Skadarsko Jezero (Lake Scutari) shared with Albania
physical mountainous
population (1986) 620,000, including 400,000 Montenegrins, 80,000 Muslims, and 40,000 Albanians
language Serbian variant of Serbo-Croat
religion Serbian Orthodox
famous people Milovan Djilas
history part of Serbia from the late 12th century, it became independent (under Venetian protection) after Serbia was defeated by the Turks 1389. It was forced to accept Turkish suzerainty in the late 15th century, but was never completely subdued by Turkey. It was ruled by bishop princes until 1851, when a monarchy was founded, and became a sovereign principality under the Treaty of Berlin 1878. The monarch used the title of king from 1910 with Nicholas I (1841–1921). Montenegro participated in the Balkan Wars 1912 and 1913. It was overrun by Austria in World War I, and in 1918 voted for the deposition of King Nicholas to become part of Serbia. In 1946 Montenegro became a republic of Yugoslavia.

In Jan 1989 the entire Communist Party leadership resigned after mass protests which secured the replacement of the party and state leaderships. The republic held multiparty elections for the first time in Dec 1990; the League of Communists of Montenegro remained in power. A staunch ally of Serbia, Montenegro sided with Serbia in the 1991 conflict with Slovenia and Croatia and approved a Serbian plan which called for a new, overwhelmingly Serb, Yugoslav federation Aug 1991. In a referendum March 1992 Montengrins voted to remain part of the Yugoslav federation; the referendum was boycotted by Montenegro's Muslim and Albanian communities.

Monterey /ˌmɒntəˈreɪ/ fishing port on Monterey Bay in California, USA, once the state capital; population (1990) 32,000. It is the setting for John Steinbeck's novels *Cannery Row* 1945 and *Tortilla Flat* 1935, dealing with migrant fruit workers.

Monterrey /ˌmɒntəˈreɪ/ industrial city (iron, steel, textiles, chemicals, food processing) in NE Mexico; population (1986) 2,335,000. It was founded 1597.

Montevideo /ˌmɒntɪvɪˈdeɪəʊ/ capital and chief port (grain, meat products, hides) of Uruguay, on the Río de la Plata; population (1985) 1,250,000. Industries include meat packing, tanning, footwear, flour milling, and textiles.

It was founded by the Spanish 1726 and has been the capital since 1830. There is a cathedral (1790–1804) on the Plaza Constitución and a university (1849). The park El Prado has museums of fine art and history.

Montgomery /mənˈgʌməri/ state capital of Alabama, USA; population (1990) 187,100. The *Montgomery Bus Boycott* 1955 began here when a black passenger, Rosa Parks, refused to give up her seat to a white person. Led by Martin Luther King, Jr, the boycott was a landmark in the civil-rights campaign.

Alabama's bus-segregation laws were nullified by the US Supreme Court 13 Nov 1956. Montgomery was the capital of the Confederacy in the first months of the American Civil War.

Montgomeryshire /mənˈgʌmrɪʃə/ (Welsh *Sir Drefaldwyn*) former county of N Wales, included in Powys 1974.

Montmartre /mɒmˈmɑːtrə/ district of Paris, France, dominated by the basilica of Sacré Coeur 1875. It is situated in the N of the city on a hill 120 m/400 ft. It is known for its nightlife and artistic associations and is a popular tourist site.

Montparnasse /ˌmɒmpɑːˈnæs/ district of Paris, France, formerly frequented by artists and writers. It is situated in the W of the city. The Pasteur Institute is also here.

Montpelier /mɒntˈpiːljə/ capital of Vermont, USA, in the Green Mountains in the central part of the state, on the Winooski River; population (1990) 8,200. Industries include granite, insurance, and tourism.

Montpellier /mɒmˈpeliɛɪ/ industrial city (electronics, medical research, engineering, textiles,

food processing, and a trade in wine and brandy), capital of Languedoc-Roussillon, France; population (1990) 210,900. It is the birthplace of the philosopher Auguste Comte.

The city is expanding, with a new music, exhibition, and convention centre, an underground railway, and a new town, which will eventually extend to the Mediterranean.

Montréal /ˌmɒntriˈɔːl/ inland port, industrial city (aircraft, chemicals, oil and petrochemicals, flour, sugar, brewing, meat packing) of Québec, Canada, on Montréal Island at the junction of the Ottawa and St Lawrence rivers; population (1986) 2,921,000.

features Mont Réal (Mount Royal, 230 m/753 ft) overlooks the city; an artificial island in the St Lawrence (site of the international exhibition 1967); three universities; except for Paris, the world's largest French-speaking city

history Jacques Cartier reached the site 1535, Samuel de Champlain established a trading post 1611, and the original Ville Marie (later renamed Montréal) was founded 1642 by Paul de Chomédy, Sieur de Maisonneuve (1612–1676). It was the last town surrendered by France to Britain 1760. Nevertheless, when troops of the rebel Continental Congress occupied the city 1775–76, the citizens refused to be persuaded (even by a visit from Benjamin Franklin) to join the future USA in its revolt against Britain.

Montreux /mɒnˈtrɜː/ winter resort in W Switzerland on Lake Geneva; population (1990) 19,900. It is the site of the island rock fortress of Chillon, where François Bonivard (commemorated by the poet Byron), prior of the Abbey of St Victor, was imprisoned 1530–36 for his opposition to the duke of Savoy. At the annual television festival (first held 1961), the premier award is the *Golden Rose of Montreux*.

Mont St Michel /mɒn sæm mɪˈʃel/ islet off the coast of NW France converted to a peninsula by an artificial causeway; it has a Benedictine monastery, founded 708.

Montserrat /ˌmɒntsəˈræt/ volcanic island in the West Indies, one of the Leeward group, a British crown colony; capital Plymouth; area 110 sq km/42 sq mi; population (1985) 12,000. Practically all buildings were destroyed by Hurricane Hugo Sept 1989.

Montserrat produces cotton, cotton-seed, coconuts, citrus and other fruits, and vegetables. Its first European visitor was Christopher Columbus 1493, who named it after the mountain in Spain. It was first colonized by English and Irish settlers who moved from St Christopher 1632. The island became a British crown colony 1871.

Montserrat /ˌmɒntsəˈræt/ (Spanish *monte serrado* 'serrated mountain') isolated mountain in NE Spain, height 1,240 m/4,070 ft, so called because its uneven outline of eroded pinnacles resembles the edge of a saw.

There is a monastery dating from 880. According to legend, St Peter hid an image of the

Virgin Mary on the mountain at the time of the Moorish invasion, which was later discovered by shepherds.

Monza /'mɒnzə/ town in N Italy, known for its motor-racing circuit; population (1988) 123,000. Once the capital of the Lombards, it preserves the Iron Crown of Lombardy in the 13th-century cathedral. Umberto I was assassinated here.

Moonie town in SE Queensland, W of Brisbane, the site of Australia's first commercial oil strike; population (1961) approximately 100.

Moorhead /'mɔːhed/ city in W Minnesota, USA, on the Red River, opposite Fargo, North Dakota; population (1990) 32,300. It is a centre for an agricultural region that produces sugar beet, potatoes, grain, and dairy products.

Moose Jaw /ˌmuːs 'dʒɔː/ town in S Saskatchewan, Canada, with grain elevators, extensive stockyards, petroleum refineries; population (1985) 35,500.

Moradabad /ˌmɔːrədə'bæd/ trading city in Uttar Pradesh, India, on the Ramganga River; population (1981) 348,000. It produces textiles and engraved brassware. It was founded 1625 by Rustan Khan, and the Great Mosque dates from 1631.

Moravia /mə'reɪvɪə/ (Czech *Morava*) district of central Europe, forming two regions of the Czech Republic:
South Moravia (Czech *Jihomoravský*)
area 15,030 sq km/5,802 sq mi
capital Brno
population (1991) 2,048,900.
North Moravia (Czech *Severomoravský*)
area 11,070 sq km/4,273 sq mi
capital Ostrava
population (1991) 1,961,500.
features (N and S) river Morava; 25% forested
products maize, grapes, wine in the S; wheat, barley, rye, flax, sugar beet in the N; coal and iron
history part of the Avar territory since the 6th century; conquered by Charlemagne's Holy Roman Empire. In 874 the kingdom of Great Moravia was founded by the Slavic prince Sviatopluk, who ruled until 894. It was conquered by the Magyars 906, and became a fief of Bohemia 1029. It was passed to the Habsburgs 1526, and became an Austrian crown land 1849. It was incorporated in the new republic of Czechoslovakia 1918, forming a province until 1949.

Moray Firth /'mʌri/ North Sea inlet in Scotland, between Burghead (Grampian Region) and Tarbat Ness (Highland Region), 38 km/15 mi wide at its entrance. The town of Inverness is situated at the head of the Firth.

Morayshire /'mʌrɪʃə/ former county of NE Scotland, divided 1975 between Highland Region (the SW section) and Grampian region (the NE); the county town was Elgin.

Morbihan, Gulf of /ˌmɔːbi'ɒn/ (Breton 'little sea') seawater lake in Brittany, W France, linked by a channel with the Bay of Biscay; area 104 sq km/40 sq mi. Morbihan gives its name to a *département*.

Mordovia /mɔː'dəuvɪə/ another name for ◊Mordvinia, an autonomous republic of Russia.

Mordvinia /mɔːd'vɪnɪə/ or *Mordovia* autonomous republic of central Russia
area 26,200 sq km/10,100 sq mi
capital Saransk
features river Sura on the E; forested in the W
products sugar beet, grains, potatoes; sheep and dairy farming; timber, furniture, and textiles
population (1986) 964,000
languages Russian, Mordvin
history Mordvinia was conquered by Russia during the 13th century. It was made an autonomous region 1930, and an Autonomous Soviet Socialist Republic 1934. It remained an autonomous republic of Russia after the dissolution of the Soviet Union 1991.

Morecambe /'mɔːkəm/ town and resort in Lancashire, England, on Morecambe Bay, conjoined with the port of Heysham, which has a ferry service to Ireland; joint population (1982) 43,000.

Morecambe Bay /'mɔːkəm 'beɪ/ inlet of the Irish Sea, between the Furness Peninsula (Cumbria) and Lancashire, England, with shallow sands. There are oil wells, and natural gas 50 km/30 mi offshore.

Morocco Kingdom of (*al-Mamlaka al-Maghrebia*)
area 458,730 sq km/177,070 sq mi (excluding Western Sahara)
capital Rabat
towns Marrakesh, Fez, Meknès; ports Casablanca, Tangier, Agadir
physical mountain ranges NE–SW; fertile coastal plains in W
features Atlas Mountains; the towns Ceuta (from 1580) and Melilla (from 1492) are held by Spain; tunnel crossing the Strait of Gibraltar to Spain proposed 1985
head of state Hassan II from 1961

head of government Mohamed Lamrani from 1992
political system constitutional monarchy
political parties Constitutional Union (UC), right-wing; National Rally of Independents (RNI), royalist; Popular Movement (MP), moderate socialist; Istiqlal, nationalist, right-of-centre; Socialist Union of Popular Forces (USFP), progressive socialist; National Democratic Party (PND), moderate, nationalist
exports dates, figs, cork, wood pulp, canned fish, phosphates
currency dirham
population (1990 est) 26,249,000; growth rate 2.5% p.a.
life expectancy men 62, women 65 (1989)
languages Arabic (official) 75%, Berber 25%, French, Spanish
religion Sunni Muslim 99%
literacy men 45%, women 22% (1985 est)
GNP $18.7 bn; $750 per head (1988)
chronology
1912 Morocco divided into French and Spanish protectorates.
1956 Independence achieved as the Sultanate of Morocco.
1957 Sultan restyled king of Morocco.
1961 Hassan II came to the throne.
1969 Former Spanish province of Ifni returned to Morocco.
1972 Major revision of the constitution.
1975 Western Sahara ceded by Spain to Morocco and Mauritania.
1976 Guerrilla war in Western Sahara with the Polisario Front. Sahrawi Arab Democratic Republic (SADR) established in Algiers. Diplomatic relations between Morocco and Algeria broken.
1979 Mauritania signed a peace treaty with Polisario.
1983 Peace formula for Western Sahara proposed by the Organization of African Unity (OAU); Morocco agreed but refused to deal directly with Polisario.
1984 Hassan signed an agreement for cooperation and mutual defence with Libya.
1987 Cease-fire agreed with Polisario, but fighting continued.
1988 Diplomatic relations with Algeria restored.
1989 Diplomatic relations with Syria restored.
1992 Mohamed Lamrani appointed prime minister; new constitution approved in national referendum.

Morocco is like a tree nourished by roots deep in the soil of Africa, which breathes through foliage rustling to the winds of Europe.

On **Morocco** King Hassan II of Morocco 1979

Moroni /mə'rəuni/ capital of the Comoros Republic, on Njazídja (Grande Comore); population (1980) 20,000. It has a small natural har-

bour from which coffee, cacao, and vanilla are exported.

Moscow /'mɒskəʊ/ (Russian *Moskva*) industrial city, capital of Russia and of the Moskva region, and formerly (1922–91) of the USSR, on the Moskva River 640 km/400 mi SE of St Petersburg; population (1987) 8,815,000. Its industries include machinery, electrical equipment, textiles, chemicals, and many food products.
features The 12th-century Kremlin (Citadel), at the centre of the city, is a walled enclosure containing a number of historic buildings, including three cathedrals, one of them the burial place of the tsars; the Ivan Veliki tower 90 m/300 ft, a famine-relief work commissioned by Boris Godunov 1600; various palaces, including the former imperial palace; museums; and the Tsar Kolokol, the world's largest bell (200 tonnes) 1735. The walls of the Kremlin are crowned by 18 towers and have five gates. Red Square, used for political demonstrations and processions, contains St Basil's Cathedral, the state department store GUM, and Lenin's tomb. The headquarters of the KGB, with Lubyanka Prison behind it, is in Dzerzhinsky Square; the underground railway was opened 1935. Institutions include Moscow University 1755 and People's Friendship University (for foreign students) 1953; the Academy of Sciences, which moved from Leningrad (now St Petersburg) 1934; Tretyakov Gallery of Russian Art 1856; Moscow State Circus. Moscow is the seat of the patriarch of the Russian Orthodox church. On the city outskirts is Star City (Zvezdnoy Gorodok), the space centre.
Moscow is the largest industrial centre of the Commonwealth of Independent States, linked with Stavropol by oil pipeline 480 km/300 mi, built 1957.
history Moscow, founded as the city-state of **Muscovy** 1127, was destroyed by the Mongols during the 13th century, but rebuilt 1294 by Prince Daniel (died 1303) as the capital of his principality. During the 14th century, it was under the rule of Alexander Nevski, Ivan I (1304–1341), and Dmitri Donskai (1350–1389), and became the foremost political power in Russia, and its religious capital.
It was burned 1571 by the khan of the Crimea, and ravaged by fire 1739, 1748, and 1753; in 1812 it was burned by its own citizens to save it from Napoleon's troops, or perhaps by accident. It became capital of the Russian Soviet Federated Social Republic (RSFSR) 1918, and of the Union of Soviet Socialist Republics (USSR) 1922. In World War II Hitler's troops were within 30 km/20 mi of Moscow on the NW by Nov 1941, but the stubborn Russian defence and severe winter weather forced their withdrawal in Dec.

Moscow is again, as the Russian poet said, the heart of Russia; even if the heart is broken.

On **Moscow** G K Chesterton *Generally Speaking* 1928

Moselle /məʊ'zel/ or **Mosel** river in W Europe some 515 km/320 mi long; it rises in the Vosges Mountains, France, and is canalized from Thionville to its confluence with the Rhine at Koblenz in Germany. It gives its name to the *départements* of Moselle and Meurthe-et-Moselle in France. Vineyards along the Moselle in Germany produce popular white wines.

Mosi-oa-tunya /'məʊsi 'əʊə 'tuːnjə/ African name for the ♢Victoria Falls of the Zambezi River.

Moskva /mʌsk'vaː/ Russian name for ♢Moscow, the capital of Russia.

Mosquito Coast /mə'skiːtəʊ/ Caribbean coast of Honduras and Nicaragua, characterized by swamp, lagoons, and tropical rainforest. The territory is inhabited by Miskito Indians, Garifunas, and Zambos, many of whom speak English. Between 1823 and 1860 Britain maintained a protectorate over the Mosquito Coast which was ruled by a succession of 'Mosquito Kings'.

The area was designated a biosphere reserve by the United Nations 1992, but the government sold the rights to exploit it (energy, road-building) to a US packaging giant (the Stone Container Corporation). The project threatens local people, whose numbers are diminishing.

Mostaganem /mə,stægə'nem/ industrial port (metal and cement) in NW Algeria, linked by pipeline with the natural gas fields at Hassi Messaoud; population (1982) 169,500. It was founded in the 11th century.

Mostar /'mɒstaː/ industrial town (aluminium, tobacco) in Bosnia-Herzegovina, known for its grapes and wines; population (1981) 110,000.

Mosul /'məʊsəl/ industrial city (cement, textiles) and oil centre in Iraq, on the right bank of the Tigris, opposite the site of ancient Nineveh; population (1985) 571,000.

Motherwell and Wishaw /'mʌðəwəl, 'wɪʃɔː/ industrial town (Ravenscraig iron and steel works, coal mines) in Strathclyde, Scotland, SE of Glasgow; population (1981) 68,000. The two burghs were amalgamated 1920.

Moulins /muː'læn/ capital of the *département* of Allier, Auvergne, central France; main industries are cutlery, textiles, and glass; population (1990) 23,400. Moulins was capital of the old province of Bourbonnais 1368–1527.

Moulmein /maul'meɪn/ port and capital of Mon state in SE Myanmar, on the Salween estuary; population (1983) 202,967. There is sawmilling and rice milling, and rice and tea are exported.

Mount Isa /'aɪzə/ mining town (copper, lead, silver, zinc) in NW Queensland, Australia; population (1984) 25,000.

Mount Lofty Range /'lɒfti/ mountain range in SE South Australia; Mount Bryan at 934 m/3,064 ft is the highest peak.

Mount Vernon /'vɜːnən/ village in Virginia, USA, on the Potomac River, where George Washington

lived 1752–99 and was buried on the family estate, now a national monument.

Mow Cop /'mau 'kɒp/ site in England of an open-air religious gathering 31 May 1807 that is considered to be the start of Primitive Methodism. Mow Cop is a hill at the southern end of the Pennines on the Cheshire–Staffordshire border and dominates the surrounding countryside. It remained a popular location for revivalist meetings.

Mozambique People's Republic of (*República Popular de Moçambique*)
area 799,380 sq km/308,561 sq mi
capital and chief port Maputo
towns Beira, Nampula
physical mostly flat tropical lowland; mountains in W
features rivers Zambezi, Limpopo; 'Beira Corridor' rail, road, and pipeline link with Zimbabwe
head of state and government Joaquim Alberto Chissano from 1986
political system emergent democratic republic
political parties National Front for the Liberation of Mozambique (Frelimo), Marxist-Leninist; Renamo, or Mozambique National Resistance (MNR), former rebel movement
exports prawns, cashews, sugar, cotton, tea, petroleum products, copra
currency metical (replaced escudo 1980)
population (1990 est) 14,718,000 (mainly indigenous Bantu peoples; Portuguese 50,000); growth rate 2.8% p.a.; nearly 1 million refugees in Malawi
life expectancy men 45, women 48 (1989)
languages Portuguese (official), 16 African languages
religion animist 60%, Roman Catholic 18%, Muslim 16%
literacy men 55%, women 22% (1985 est)
GDP $4.7 bn; $319 per head (1987)
chronology
1505 Mozambique became a Portuguese colony.
1962 Frelimo (liberation front) established.

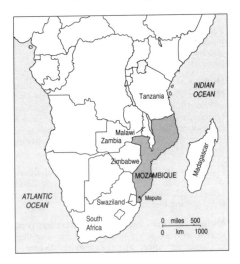

1975 Independence achieved from Portugal as a socialist republic, with Samora Machel as president and Frelimo as the sole legal party.
1977 Renamo resistance group formed.
1983 Re-establishment of good relations with Western powers.
1984 Nkomati accord of nonaggression signed with South Africa.
1986 Machel killed in air crash; succeeded by Joaquim Chissano.
1988 Tanzania announced withdrawal of its troops. South Africa provided training for Mozambican forces.
1989 Frelimo offered to abandon Marxist-Leninism; Chissano re-elected. Renamo continued attacks on government facilities and civilians.
1990 One-party rule officially ended. Partial cease-fire agreed.
1991 Peace talks resumed in Rome, delaying democratic process. Attempted antigovernment coup thwarted.
1992 Aug: peace accord agreed upon, but fighting continued. Oct: peace accord signed, but awaited ratification by government.

MS abbreviation for ◊*Mississippi*, a state of the USA.

MT abbreviation for ◊*Montana*, a state of the USA.

Mtwara /əm'twɑːrə/ deepwater seaport in S Tanzania, on Mtwara Bay; population (1978) 48,500. It was opened 1954.

Mukalla /muˈkælə/ seaport capital of the Hadhramaut coastal region of S Yemen; on the Gulf of Aden 480 km/300 mi E of Aden; population (1984) 158,000.

Mukden /ˈmukdən/ former name of ◊Shenyang, a city in China.

Mülheim an der Ruhr /ˈmjuːlhaɪm/ industrial city in North Rhine–Westphalia, Germany, on the river Ruhr; population (1988) 170,000. Industries include iron and steel, building materials, and construction.

Mulhouse /muˈluːz/ (German **Mülhausen**) industrial city (textiles, engineering, electrical goods) in Haut-Rhin *département*, Alsace, E France; population (1990) 109,900.

Mull /mʌl/ second largest island of the Inner Hebrides, Strathclyde, Scotland; area 950 sq km/367 sq mi; population (1981) 2,600. It is mountainous, and is separated from the mainland by the Sound of Mull. There is only one town, Tobermory. The economy is based on fishing, forestry, tourism, and some livestock.

Mullingar /ˌmʌlɪnˈgɑː/ county town and district of Westmeath, Republic of Ireland; population (1991) 30,300. It is a cattle market and trout-fishing centre.

Multan /ˌmulˈtɑːn/ industrial city (textiles, precision instruments, chemicals, pottery, jewellery) in Punjab province, central Pakistan, 205 km/190 mi SW of Lahore; population (1981) 732,000. It trades in grain, fruit, cotton, and wool. It is on a site inhabited since the time of Alexander the Great.

Mulu /ˈmuːluː/ mountainous region in N Borneo near the border with Sabah. Its limestone cave system, one of the largest in the world, was explored by a Royal Geographical Society Expedition 1978.

München /ˈmunʃən/ German name of ◊Munich, a city in Germany.

Muncie /ˈmʌnsi/ city in E central Indiana, NE of Indianapolis, USA; population (1990) 71,000. Industries include vehicle parts, livestock, dairy products, steel forgings, and wire. It was the subject of Robert and Helen Lynd's *Middletown*, a sociological study of a midwestern city.

Munich /ˈmjuːnɪk/ (German **München**) industrial city (brewing, printing, precision instruments, machinery, electrical goods, textiles), capital of Bavaria, Germany, on the river Isar; population (1986) 1,269,400.
features Munich owes many of its buildings and art treasures to the kings Ludwig I and Maximilian II of Bavaria. The cathedral is late 15th century. The Alte Pinakothek contains paintings by old masters, and the Neue Pinakothek, modern paintings; there is the Bavarian National Museum, the Bavarian State Library, and the Deutsches Museum (science and technology). The university, founded at Ingolstadt 1472, was transferred to Munich 1826; to the NE at Garching there is a nuclear research centre.
history Dating from the 12th century, Munich became the residence of the dukes of Wittelsbach in the 13th century, and the capital of independent Bavaria. It was the scene of the November revolution of 1918, the 'Soviet' republic of 1919, and the Hitler putsch of 1923. It became the centre of the Nazi movement, and the Munich Agreement of 1938 was signed there. When the 1972 Summer Olympics were held in Munich, a number of Israeli athletes were killed by guerrillas.

Munster /ˈmʌnstə/ southern province of the Republic of Ireland, comprising the counties of Clare, Cork, Kerry, Limerick, North and South Tipperary, and Waterford; area 24,140 sq km/9,318 sq mi; population (1991) 1,008,400.
It was a kingdom until the 12th century, and was settled in plantations by the English from 1586.

Münster /ˈmunstə/ industrial city (wire, cement, iron, brewing, and distilling) in North Rhine–Westphalia, NW Germany, formerly the capital of Westphalia; population (1988) 268,000. The Treaty of Westphalia was signed simultaneously here and at Osnabrück 1648, ending the Thirty Years' War.
Its university was founded 1773. Badly damaged in World War II, its ancient buildings, including the 15th-century cathedral and town hall, have been restored or rebuilt.

Murcia /'muəθiə/ autonomous region of SE Spain; area 11,300 sq km/4,362 sq mi; population (1986) 1,014,000. It includes the cities Murcia and Cartagena, and produces esparto grass, lead, zinc, iron, and fruit.

Murcia /'muəθiə/ industrial city (silk, metal, glass, textiles, pharmaceuticals), capital of the Spanish province of Murcia, on the river Segura; population (1991) 324,800. Murcia was founded 825 on the site of a Roman colony by 'Abd-ar-Rahman II, caliph of Córdoba. It has a university and 14th-century cathedral.

Murmansk /muə'mænsk/ seaport in NW Russia, on the Barents Sea; population (1987) 432,000. It is the largest city in the Arctic, Russia's most important fishing port, and the base of naval units and the icebreakers that keep the Northeast Passage open.

It is the centre of Russian Lapland and the only port on the Russian Arctic coast that is in use all year round. The Festival of the North in March marks the end of the two-month Arctic night.

Murray /'mʌri/ principal river of Australia, 2,575 km/ 1,600 mi long. It rises in the Australian Alps near Mount Kosciusko and flows W, forming the boundary between New South Wales and Victoria, and reaches the sea at Encounter Bay, South Australia. With its main tributary, the Darling, it is 3,750 km/2,330 mi long.

Its other tributaries include the Lachlan and the Murrumbidgee. The Dartmouth Dam (1979) in the Great Dividing Range supplies hydroelectric power and has drought-proofed the Murray River system, but irrigation (for grapes, citrus and stone fruits) and navigation schemes have led to soil salinization.

Murrumbidgee /,mʌrəm'bɪdʒi/ river of New South Wales, Australia; length 1,690 km/1,050 mi. It rises in the Australian Alps, flows N to the Burrinjuck reservoir, and then W to meet the river Murray.

Muscat /'mʌskæt/ or **Masqat** capital of Oman, E Arabia, adjoining the port of Matrah, which has a deep-water harbour; combined population (1982) 80,000. It produces natural gas and chemicals.

Muscat was the realisation of a pirate's lair as imagined by any schoolboy ... The place was clearly designed by Allah as a centre for piracy and slave-running.

On **Muscat** Cedric Belfrage *Away From It All* 1936

Muscat and Oman /'mʌskæt, əu'mɑːn/ former name of ⟡Oman, a country in the Middle East.

Musgrave Ranges /'mʌzgreɪv/ Australian mountain ranges on the border between South Australia

and the Northern Territory; the highest peak is Mount Woodruffe at 1,525 m/5,000 ft. The area is an Aboriginal reserve.

Muskegon /mə'skiːgən/ city in W Michigan, USA, on the Muskegon River where it enters into Lake Michigan, NW of Grand Rapids; population (1990) 40,300. Industries include heavy machinery, metal products, vehicle parts, and sporting goods.

Mustique /mu'stiːk/ island in the Caribbean; see ⟡St Vincent and the Grenadines.

Mutare /mu'tɑːri/ (formerly until 1982 *Umtali*) industrial town (vehicle assembly, engineering, tobacco, textiles, paper) in E Zimbabwe; population (1982) 69,621. It is the chief town of Manicaland province.

Myanmar Union of (*Thammada Myanmar Naingngandaw*) (formerly *Burma*, until 1989)
area 676,577 sq km/261,228 sq mi
capital and chief port Yangon (formerly Rangoon)
towns Mandalay, Moulmein, Pegu
physical over half is rainforest; rivers Irrawaddy and Chindwin in central lowlands ringed by mountains in N, W, and E
environment landslides and flooding during the rainy season (June–Sept) are becoming more frequent as a result of deforestation
features ruined cities of Pagan and Mingun
head of state and government Than Shwe from 1992
political system military republic
political parties National Unity Party (NUP), military-socialist ruling party; National League for Democracy (NLD), pluralist opposition grouping
exports rice, rubber, jute, teak, jade, rubies, sapphires
currency kyat
population (1990 est) 41,279,000; growth rate 1.9% p.a. (includes Shan, Karen, Raljome, Chinese, and Indian minorities)
life expectancy men 53, women 56 (1989)

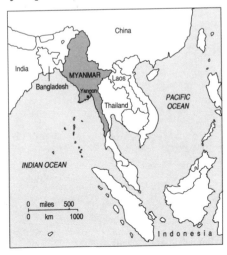

language Burmese
religions Hinayana Buddhist 85%, animist, Christian
literacy 66% (1989)
GNP $9.3 bn (1988); $210 per head (1989)
chronology
1886 United as province of British India.
1937 Became crown colony in the British Commonwealth.
1942–45 Occupied by Japan.
1948 Independence achieved from Britain. Left the Commonwealth.
1962 General Ne Win assumed power in army coup.
1973–74 Adopted presidential-style 'civilian' constitution.
1975 Opposition National Democratic Front formed.
1986 Several thousand supporters of opposition leader Suu Kyi arrested.
1988 Government resigned after violent demonstrations. General Saw Maung seized power in military coup Sept; over 1,000 killed.
1989 Martial law declared; thousands arrested including advocates of democracy and human

rights. Country renamed Myanmar and capital Yangon.
1990 Breakaway opposition group formed 'parallel government' on rebel-held territory.
1991 Martial law and human-rights abuses continued. Military offensives continued. Opposition leader, Aung San Suu Kyi, received Nobel Prize for Peace.
1992 Jan-April: Pogrom against Muslim community in Arakan province, W Myanmar, carried out with army backing. April: Saw Maung replaced by Than Shwe. Several political prisoners liberated. Sept: martial law lifted, but restrictions on political freedom remained.

Mysore /ˌmaɪˈsɔː/ or *Maisur* industrial city (engineering, silk) in Karnataka, S India, some 130 km/80 mi SW of Bangalore; population (1981) 476,000.

Mytilene /ˌmɪtɪˈliːni/ (modern Greek *Mitilíni*) port, capital of the Greek island of Lesvos (to which the name Mytilene is sometimes applied) and a centre of sponge fishing; population (1981) 24,000.

the state of Nagaland was established 1963 in response to demands for self-government, but fighting continued sporadically.

Nagasaki /ˌnæɡə'saːki/ industrial port (coal, iron, shipbuilding) on Kyushu Island, Japan; population (1990) 444,600. Nagasaki was the only Japanese port open to European trade from the 16th century until 1859. An atom bomb was dropped on it by the USA 9 Aug 1945.

Three days after Hiroshima, the second atom bomb was dropped here at the end of World War II. Of Nagasaki's population of 212,000, 73,884 were killed and 76,796 injured, not counting the long-term victims of radiation.

The Lourdes of the Atomic Age.

On **Nagasaki** Murray Sayle, quoted in *Spectator* 24 Sept 1980

Nablus /'naːbləs/ market town on the West Bank of the river Jordan, N of Jerusalem; the largest Palestinian town, after E Jerusalem, in Israeli occupation; population (1971) 64,000. Formerly Shechem, it was the ancient capital of Samaria, and a few Samaritans remain. The British field marshal Allenby's defeat of the Turks here 1918 completed the conquest of Palestine.

Nacala /nə'kaːlə/ seaport in Nampula province, N Mozambique; a major outlet for minerals. It is linked by rail with Malawi.

Naemen /'naːmən/ Flemish form of ◊Namur, a city in Belgium.

Nagaland /'naːgəlænd/ state of NE India, bordering Myanmar (Burma) on the E
area 16,721 sq km/6,456 sq mi
capital Kohima
products rice, tea, coffee, paper, sugar
population (1991) 1,215,600
history formerly part of Assam, the area was seized by Britain from Burma (now Myanmar) 1826. The British sent 18 expeditions against the Naga peoples in the N 1832–87. After India attained independence 1947, there was Naga guerrilla activity against the Indian government;

Nagorno-Karabakh /nə'gɔːnəu ˌkærəbæx/ autonomous region of Azerbaijan
area 4,400 sq km/1,700 sq mi
capital Stepanakert
products cotton, grapes, wheat; silk
population (1987) 180,000 (76% Armenian, 23% Azeri), the Christian Armenians forming an enclave within the predominantly Shi'ite Muslim Azerbaijan
history an autonomous protectorate after the Russian revolution 1917, Nagorno-Karabakh was annexed to Azerbaijan 1923 against the wishes of the largely Christian-Armenian population. Since the local, ethnic Armenian council declared its intention to transfer control of the region to Armenia 1989, the enclave has been racked by fighting between Armenian and Azeri troops, both attempting to assert control. Following a referendum and elections early Dec 1991, the 81-member parliament of Nagorno-Karabakh declared its independence. This was not recognized by Azerbaijan and conflict within the enclave worsened. By Feb 1992, the conflict had caused the loss of at least 1,000 lives (501 during 1991 alone) and the displacement of some 270,000 people, half of them Armenian and half Azeri. On 15 Aug 1992 the ethnic Armenian government resigned.

Nagoya /nə'gɔɪə/ industrial seaport (cars, textiles, clocks) on Honshu Island, Japan; population (1990) 2,154,700. It has a shogun fortress 1610 and a notable Shinto shrine, Atsuta Jingu.

Nagpur /ˌnæɡ'puə/ industrial city (textiles, metals) in Maharashtra, India, on the river Pench; population (1981) 1,298,000. Pharmaceuticals, cotton goods, and hosiery are produced, and oranges are traded. Nagpur was founded in the 18th century, and was the former capital of Berar and Madhya Pradesh states.

Naha /'naːhaː/ chief port on Okinawa Island, Japan; population (1990) 304,900. Industries include pottery and textiles; there is fishing.

Nairnshire /'neənʃə/ former county of Scotland, included 1975 in Highland Region. The county town was Nairn.

Nagaland

Kohima

INDIAN OCEAN

Nairobi /naɪˈrəubi/ capital of Kenya, in the central highlands at 1,660 m/5,450 ft; population (1985) 1,100,000. It has light industry and food processing and is the headquarters of the United Nations Environment Programme.

Nairobi was founded 1899. It has the International Louis Leakey Institute for African Prehistory 1977, and the International Primate Research Institute is nearby.

Najaf /ˈnædʒxf/ city near the Euphrates in Iraq, 144 km/90 mi S of Baghdad; population (1970) 179,000. The tomb of Ali, son-in-law of Muhammad, is a Shi'ite Muslim shrine.

Nakhichevan /ˌnæxɪtʃəˈvæn/ autonomous republic forming part of Azerbaijan, even though it is entirely outside the Azerbaijan boundary, being separated from it by Armenia; area 5,500 sq km/ 2,120 sq mi; population (1986) 272,000. Taken by Russia in 1828, it was annexed to Azerbaijan in 1924. Some 85% of the population are Muslim Azeris who maintain strong links with Iran to the south. Nakhichevan has been affected by the Armenia–Azerbaijan conflict; many Azeris have fled to Azerbaijan, and in Jan 1990 frontier posts and border fences with Iran were destroyed. In May 1992 Armenian forces made advances in the region, but Azeri forces had regained control by Aug. The republic has sought independence from Azerbaijan.

Nakhodka /nəˈxɒdkə/ Pacific port in E Siberia, Russia, on the Sea of Japan, E of Vladivostok; population (1985) 150,000. US-caught fish, especially pollock, is processed by Russian factory ships in a joint venture.

Nakuru, Lake /nəˈkuəruː/ salt lake in the Great Rift Valley, Kenya.

Namaqualand /næˈmɑːkwələænd/ or **Namaland** near-desert area on the southwest coast of Africa divided between Namibia and South Africa. **Great Namaqualand** is in Namibia, N of the Orange River, area 388,500 sq km/150,000 sq mi; sparsely populated by the Nama, a Hottentot people. **Little Namaqualand** is in Cape Province, South Africa, S of the Orange River, area 52,000 sq km/20,000 sq mi; copper and diamonds are mined here.

Namib Desert /ˈnɑːmɪb/ coastal desert region in Namibia between the Kalahari Desert and the Atlantic Ocean. Its sand dunes are among the tallest in the world, reaching heights of 370 m/ 1,200 ft.

Namibia Republic of (formerly **South West Africa**)
area 824,300 sq km/318,262 sq mi
capital Windhoek
towns Swakopmund, Rehoboth, Rundu
physical mainly desert
features Namib and Kalahari deserts; Orange River; Caprivi Strip links Namibia to Zambezi River; includes the enclave of Walvis Bay (area 1,120 sq km/432 sq mi)
head of state Sam Nujoma from 1990
head of government Hage Geingob from 1990
political system democratic republic

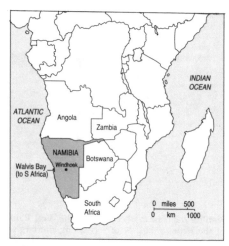

political parties South-West Africa People's Organization (SWAPO), socialist Ovambo-oriented; Democratic Turnhalle Alliance (DTA), moderate, multiracial coalition; United Democratic Front (UDF), disaffected ex-SWAPO members; National Christian Action (ACN), white conservative
exports diamonds, uranium, copper, lead, zinc
currency South African rand
population (1990 est) 1,372,000 (85% black African, 6% European)
life expectancy blacks 40, whites 69
languages Afrikaans (spoken by 60% of white population), German, English (all official), several indigenous languages
religion 51% Lutheran, 19% Roman Catholic, 6% Dutch Reformed Church, 6% Anglican
literacy whites 100%, nonwhites 16%
GNP $1.6 bn; $1,300 per head (1988)
chronology
1884 German and British colonies established.
1915 German colony seized by South Africa.
1920 Administered by South Africa, under League of Nations mandate, as British South Africa.
1946 Full incorporation in South Africa refused by United Nations (UN).
1958 South-West Africa People's Organization (SWAPO) set up to seek racial equality and full independence.
1966 South Africa's apartheid laws extended to the country.
1968 Redesignated Namibia by UN.
1978 UN Security Council Resolution 435 for the granting of full sovereignty accepted by South Africa and then rescinded.
1988 Peace talks between South Africa, Angola, and Cuba led to agreement on full independence for Namibia.
1989 Unexpected incursion by SWAPO guerrillas from Angola into Namibia threatened agreed independence. Transitional constitution created by elected representatives; SWAPO dominant party.
1990 Liberal multiparty 'independence' constitution adopted; independence achieved. Sam Nujoma elected president.

1991 Agreement on joint administration of Walvis Bay reached with South Africa, pending final settlement of dispute.
1992 Agreement on establishment of Walvis Bay Joint Administrative Body.

Nampo /ˌnæm'pəu/ (formerly to 1947 *Chinnampo*) city on the west coast of North Korea, 40 km/25 mi SW of Pyongan; population (1984) 691,000.

Namur /nə'mjuə/ (Flemish *Naemen*) industrial city (cutlery, porcelain, paper, iron, steel), capital of the province of Namur, in S Belgium, at the confluence of the Sambre and Meuse rivers; population (1991) 103,400. It was a strategic location during both world wars. The province of Namur has an area of 3,700 sq km/1,428 sq mi and a population (1991) of 423,300.

Nanaimo /næ'naɪməu/ coal-mining centre of British Columbia, Canada, on the east coast of Vancouver Island; population (1985) 50,500.

Nanchang /ˌnæn'tʃæŋ/ industrial city (textiles, glass, porcelain, soap), capital of Jiangxi province, China, about 260 km/160 mi SE of Wuhan; population (1989) 1,330,000.
Nanchang is a road and rail junction. It was originally a walled city built in the 12th century. The first Chinese Communist uprising took place here 1 Aug 1927.

Nancy /'nɒnsi/ capital of the *département* of Meurthe-et-Moselle and of the region of Lorraine, France, on the river Meurthe 280 km/175 mi E of Paris; population (1990) 102,400. Nancy dates from the 11th century.

Nanda Devi /nʌndə 'diːvi/ peak in the Himalayas, Uttar Pradesh, N India; height 7,817 m/25,645 ft. Until Kanchenjunga was absorbed into India, Nanda Devi was the country's highest mountain.

Nanga Parbat /'nʌŋgə 'pɑːbæt/ peak in the Himalayan Karakoram Mountains of Kashmir; height 8,126 m/26,660 ft.

Nanjing /ˌnæn'dʒɪŋ/ or *Nanking* capital of Jiangsu province, China, 270 km/165 mi NW of Shanghai; centre of industry (engineering, shipbuilding, oil refining), commerce, and communications; population (1989) 2,470,000. The bridge 1968 over the river Chang Jiang is the longest in China at 6,705 m/22,000 ft.
The city dates from the 2nd century BC, perhaps earlier. It received the name Nanjing ('southern capital') under the Ming dynasty (1368–1644) and was the capital of China 1368–1403, 1928–37, and 1946–49.

Nanking /ˌnæn'kɪŋ/ alternative name of ⟡Nanjing, a city in China.

Nanning /ˌnæn'nɪŋ/ industrial river port, capital of Guangxi Zhuang autonomous region, China, on the river You Jiang; population (1989) 1,050,00. It was a supply town during the Vietnam War and the Sino-Vietnamese confrontation 1979.

Nanshan Islands /næn'ʃæn/ Chinese name for the ⟡Spratly Islands.

Nantes /nɒnt/ industrial port in W France on the river Loire, capital of Pays de la Loire region;

industries include oil, sugar refining, textiles, soap, and tobacco; population (1990) 252,000. It has a cathedral 1434–1884 and a castle founded 938.

Nantucket /næn'tʌkɪt/ island and resort in Massachusetts, USA, south of Cape Cod, 120 sq km/46 sq mi. In the 18th–19th centuries, Nantucket was a whaling port; it is now a popular summer resort because of its excellent beaches.
The island was discovered 1602, settled 1659 by Quakers, and became part of Massachusetts 1692.

Napa /'næpə/ city in NW California on the Napa River, NE of San Francisco on San Pablo Bay. It is a major trading centre for wine; other products include fruits and clothing; population (1990) 61,800.

Napier /'neɪpiə/ wool port in Hawke Bay on the east coast of North Island, New Zealand; population (1991) 51,400.

Naples /'neɪpəlz/ (Italian *Napoli*) industrial port (shipbuilding, cars, textiles, paper, food processing) and capital of Campania, Italy, on the Tyrrhenian Sea; population (1988) 1,201,000. To the S is the Isle of Capri, and behind the city is Mount Vesuvius, with the ruins of Pompeii at its foot.
Naples is the third largest city of Italy, and as a port second in importance only to Genoa. Buildings include the royal palace, the San Carlo Opera House, the Castel Nuovo (1283), and the university (1224).
The city began as the Greek colony Neapolis in the 6th century BC and was taken over by Romans 326 BC; it became part of the Kingdom of the Two Sicilies 1140 and capital of the Kingdom of Naples 1282.

See Naples and die.

On **Naples** Italian proverb

Napoli /'nɑːpəli/ Italian form of ⟡Naples, a city in Italy.

Nara /'nɑːrə/ city in Japan, in the S of Honshu Island, the capital of the country 710–84; population (1990) 349,400. It was the birthplace of Japanese art and literature and has ancient wooden temples.

Narbonne /ˌnɑː'bɒn/ city in Aude *département*, S France; population (1990) 47,100. It was the chief town of S Gaul in Roman times and a port in medieval times.

Narmada River /nə'mɑːdə/ river that rises in the Maikala range in Madhya Pradesh state, central India, and flows 1,245 km/778 mi WSW to the Gulf of Khambat, an inlet of the Arabian Sea. Forming the traditional boundary between Hindustan and Deccan, the Narmada is a holy river of the Hindus.
India's Narmada Valley Project is one of the largest and most controversial river development

projects in the world. Between 1990 and 2040 it is planned to build 30 major dams, 135 medium-sized dams, and 3,000 smaller dams in a scheme that will involve moving 1 million of the valley's population of 20 million people.

Narragansett Bay /ˌnærəˈgænsɪt/ Atlantic inlet, Rhode Island, USA. Running inland for 45 km/ 28 mi, it encloses a number of islands.

Narvik /ˈnɑːvɪk/ seaport in Nordland county, N Norway, on Ofot Fjord, exporting Swedish iron ore; population (1991) 18,600. To secure this ore supply Germany seized Narvik April 1940. British, French, Polish, and Norwegian forces recaptured the port but had to abandon it 10 June to cope with the worsening Allied situation elsewhere in Europe.

Nashua /ˈnæʃuə/ city in S New Hampshire, USA, on the Nashua River where it meets the Merrimack River, just north of the Massachusetts border; population (1990) 79,600. Industries include electronics, asbestos, chemicals, and glass products.

Nashville /ˈnæʃvɪl/ port on the Cumberland River and capital of Tennessee, USA; population (1990) 488,300. It is a banking and commercial centre, and has large printing, music-publishing, and recording industries.

Most of the Bibles in the USA are printed here, and it is the hub of the country-music business.

Nashville dates from 1778, and the Confederate army was defeated here in 1864 in the American Civil War.

Nassau /ˈnæsɔː/ capital and port of the Bahamas, on New Providence Island; population (1980) 135,000.

English settlers founded it in the 17th century, and it was a supply base for Confederate blockade runners during the American Civil War.

Natal /nəˈtæl/ province of South Africa, NE of Cape Province, bounded on the east by the Indian Ocean
area 91,785 sq km/35,429 sq mi
capital Pietermaritzburg
towns Durban
physical slopes from the Drakensberg mountain range to a fertile subtropical coastal plain
features Ndumu Game Reserve, Kosi Bay Nature Reserve, Sodwana Bay National Park, Maple Lane Nature Reserve, and St Lucia National Park, which extends from coral reefs of the Indian Ocean N of Umfolozi River (whales, dolphins, turtles, crayfish), over forested sandhills to inland grasslands and swamps of Lake St Lucia, 324 sq km/125 sq mi (reedbuck, buffalo, crocodile, hippopotamus, black rhino, cheetah, pelican, flamingo, stork). It is under threat from titanium mining
products sugar cane, black wattle *Acacia mollissima*, maize, fruit, vegetables, tobacco, coal
population (1985) 2,145,000
history called Natal ('of [Christ's] birth') because Vasco da Gama reached it on Christmas Day 1497; part of the British Cape Colony from 1843 until 1856, when it was made into a separate colony. Zululand was annexed to Natal 1897, and the districts of Vrijheid, Utrecht, and part of Wakkerstroom were transferred from the Transvaal to Natal 1903; the colony became a part of the Union of South Africa 1910.

Natal /nəˈtæl/ industrial seaport (textiles, salt refining) in Brazil, capital of the state of Rio Grande do Norte; population (1980) 376,500. Natal was founded 1599 and became a city 1822.

Natchez /ˈnætʃɪz/ city in Mississippi, on the eastern bluffs above the Mississippi River; population (1990) 19,400. It has many houses of the antebellum period and was important in the heyday of steamboat traffic.

The Natchez Trace National Parkway is a restoration of the frontier road that followed Indian pathways and linked the city to Nashville, Tennessee.

Natron, Lake /ˈneɪtrən/ salt and soda lake in the Great Rift Valley, Tanzania; length 56 km/35 mi, width 24 km/15 mi.

Natural Bridge village in Virginia, USA, 185 km/ 115 mi W of Richmond. The nearby Cedar Creek is straddled by an arch of limestone, 66 m/215 ft high and 27 m/90 ft wide.

Nauru Republic of (*Naoero*)
area 21 sq km/8 sq mi
capital (seat of government) Yaren District
physical tropical island country in SW Pacific; plateau encircled by coral cliffs and sandy beaches
features lies just S of equator; one of three phosphate rock islands in the Pacific
head of state and government Bernard Dowiyogo from 1989
political system liberal democracy
political party Democratic Party of Nauru (DPN), opposition to government
exports phosphates
currency Australian dollar
population (1990 est) 8,100 (mainly Polynesian; Chinese 8%, European 8%); growth rate 1.7% p.a.
languages Nauruan (official), English
religion Protestant 66%, Roman Catholic 33%
literacy 99% (1988)
GNP $160 million (1986); $9,091 per head (1985)

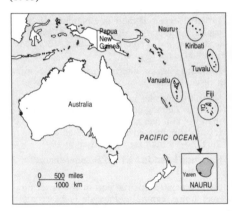

0   500 miles
0   1000 km

chronology
1888 Annexed by Germany.
1920 Administered by Australia, New Zealand, and UK until independence, except 1942–45, when it was occupied by Japan.
1968 Independence achieved from Australia, New Zealand, and UK with 'special member' Commonwealth status. Hammer DeRoburt elected president.
1976 Bernard Dowiyogo elected president.
1978 DeRoburt re-elected.
1986 DeRoburt briefly replaced as president by Kennan Adeang.
1987 DeRoburt re-elected; Adeang established the DPN.
1989 DeRoburt replaced by Kensas Aroi, who was later succeeded by Bernard Dowiyogo.

Navarre /nə'vɑː/ (Spanish *Navarra*) autonomous mountain region of N Spain
area 10,400 sq km/4,014 sq mi
capital Pamplona
features Monte Adi 1,503 m/4,933 ft; rivers: Ebro, Arga
population (1986) 513,000
history part of the medieval kingdom of Navarre. Estella, to the SW, where Don Carlos was proclaimed king 1833, was a centre of agitation by the Carlists.

Náxos /'næksɒs/ island of Greece, the largest of the Cyclades, area 453 sq km/175 sq mi. Known since early times for its wine, it was a centre for the worship of Bacchus, who, according to Greek mythology, found the deserted Ariadne on its shore and married her.

Nazareth /'næzərəθ/ town in Galilee, N Israel, SE of Haifa; population (1981) 64,000. According to the New Testament, it was the boyhood home of Jesus.

Nazca /'næskə/ town to the S of Lima, Peru, near a plateau that has geometric linear markings interspersed with giant outlines of birds and animals. The markings were made by American Indians, possibly in the 6th century AD, and their function is thought to be ritual rather than astronomical.

Naze, the /neɪz/ headland on the coast of Essex, England, S of the port of Harwich; also the English name for *Lindesnes*, a cape in S Norway.

NB abbreviation for ◊*New Brunswick*, a Canadian province.

NC abbreviation for ◊*North Carolina*, a state of the USA.

ND abbreviation for ◊*North Dakota*, a state of the USA.

N'djamena /ˌəndʒə'meɪnə/ capital of Chad, at the confluence of the Chari and Logone rivers, on the Cameroon border; population (1988) 594,000.
Founded 1900 by the French at the junction of caravan routes, it was used 1903–12 as a military centre against the kingdoms of central Sudan. Its name until 1973 was *Fort Lamy*.

Ndola /ən'dəʊlə/ mining centre and chief city of the Copperbelt province of central Zambia; population (1988) 442,700.

Nebraska

NE abbreviation for ◊*Nebraska*, a state of the USA.

Neagh, Lough /neɪ/ lake in Northern Ireland, 25 km/15 mi W of Belfast; area 396 sq km/153 sq mi. It is the largest lake in the British Isles.

Near East term used until the 1940s to describe the area of the Balkan states, Egypt, and SW Asia, now known as the Middle East.

Neath /niːθ/ (Welsh *Castell-nedd*) town in West Glamorgan, Wales, near the mouth of the river Neath; population (1984) 26,100. The Roman fort of Nidum was discovered nearby 1949; there are also remains of a Norman castle and abbey.

Nebr. abbreviation for ◊*Nebraska*, a state of the USA.

Nebraska /nə'bræskə/ state in central USA; nickname Cornhusker State/Blackwater State
area 200,400 sq km/77,354 sq mi
capital Lincoln
towns Omaha, Grand Island, North Platte
population (1990) 1,578,400
features Rocky Mountain foothills; tributaries of the Missouri; Boys' Town for the homeless, near Omaha; the ranch of Buffalo Bill; the only unicameral legislature
products cereals, livestock, processed foods, fertilizers, oil, natural gas
famous people Fred Astaire, William Jennings Bryan, Johnny Carson, Willa Cather, Henry Fonda, Harold Lloyd, Malcom X
history exploited by French fur traders in the early 1700s; ceded to Spain by France 1763; retroceded to France 1801; part of the Louisiana Purchase 1803; explored by Lewis and Clark 1804–06; first settlement at Bellevue 1823; became a territory 1854 and a state 1867 after the Union Pacific began its transcontinental railroad at Omaha 1865.

Needles, the /'niːdlz/ group of rocks in the sea near the Isle of Wight, S England.

Neenah /'niːnə/ city in E Wisconsin, USA, on the Fox River near Lake Winnebago, N of Oshkosh; population (1990) 23,200. To the E is Menasha, with which it forms one community. It is a processing and marketing centre for the area's agricultural products.

Negev /'negev/ desert in S Israel that tapers to the port of Eilat. It is fertile under irrigation, and minerals include oil and copper.

Negri Sembilan /'negri sem'bi:lən/ state of S Peninsular Malaysia; area 6,646 sq km/2,565 sq mi; population (1990) 723,800. It is mainly mountainous; products include rice and rubber. The capital is Seremban.

Nejd /nedʒd/ region of central Arabia consisting chiefly of desert; area about 2,072,000 sq km/ 800,000 sq mi. It forms part of the kingdom of Saudi Arabia and is inhabited by Bedouins. The capital is Riyadh.

Nepal Kingdom of (*Nepal Adhirajya*)
area 147,181 sq km/56,850 sq mi
capital Katmandu
towns Pátan, Moráng, Bhádgáon
physical descends from the Himalayan mountain range in N through foothills to the river Ganges plain in S
environment described as the world's highest rubbish dump, Nepal attracts 270,000 tourists, trekkers, and mountaineers each year. An estimated 1,100 lbs of rubbish is left by each expedition trekking or climbing in the Himalayas. Since 1952 the foothills of the Himalayas have been stripped of 40% of their forest cover
features Mount Everest, Mount Kangchenjunga; the only Hindu kingdom in the world; Lumbini, birthplace of the Buddha
head of state King Birendra Bir Bikram Shah Dev from 1972
head of government Girija Prasad Koirala from 1991
political system constitutional monarchy
political parties Nepali Congress Party (NCP), left-of-centre; United Nepal Communist Party (UNCP), Marxist-Leninist-Maoist; United Liberation Torchbearers; Democratic Front, radical republican
exports jute, rice, timber, oilseed
currency Nepalese rupee
population (1990 est) 19,158,000 (mainly known by name of predominant clan, the Gurkhas; the Sherpas are a Buddhist minority of NE Nepal); growth rate 2.3% p.a.
life expectancy men 50, women 49 (1989)
language Nepali (official); 20 dialects spoken

religion Hindu 90%; Buddhist, Muslim, Christian
literacy men 39%, women 12% (1985 est)
GNP $3.1 bn (1988); $160 per head (1986)
chronology
1768 Nepal emerged as unified kingdom.
1815–16 Anglo-Nepali 'Gurkha War'; Nepal became a British-dependent buffer state.
1846–1951 Ruled by the Rana family.
1923 Independence achieved from Britain.
1951 Monarchy restored.
1959 Constitution created elected legislature.
1960–61 Parliament dissolved by king; political parties banned.
1980 Constitutional referendum held following popular agitation.
1981 Direct elections held to national assembly.
1983 Overthrow of monarch-supported prime minister.
1986 New assembly elections returned a majority opposed to *panchayat* system of partyless government.
1988 Strict curbs placed on opposition activity; over 100 supporters of banned opposition party arrested; censorship imposed.
1989 Border blockade imposed by India in treaty dispute.
1990 *Panchayat* system collapsed after mass prodemocracy demonstrations; new constitution introduced; elections set for May 1991.
1991 NCP, led by Girija Prasad Koirala, won the general election.
1992 Communists led anti-government demonstrations in Katmandu and Pátan.

Ness, Loch /nes/ lake in Highland region, Scotland, forming part of the Caledonian Canal; 36 km/22.5 mi long, 229 m/754 ft deep. There have been unconfirmed reports of a *Loch Ness monster* since the 15th century.

The monster is worth £5 million a year to Scottish tourism.

Netherlands Kingdom of the (*Koninkrijk der Nederlanden*), popularly referred to as *Holland*
area 41,863 sq km/16,169 sq mi
capital Amsterdam

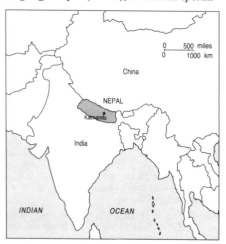

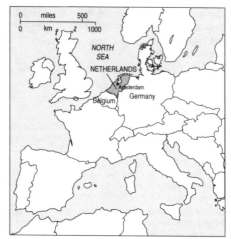

towns The Hague (seat of government), Utrecht, Eindhoven, Maastricht; chief port Rotterdam
physical flat coastal lowland; rivers Rhine, Scheldt, Maas; Frisian Islands
territories Aruba, Netherlands Antilles (Caribbean)
environment the country lies at the mouths of three of Europe's most polluted rivers, the Maas, Rhine, and Scheldt. Dutch farmers contribute to this pollution by using the world's highest quantities of nitrogen-based fertilizer per hectare/acre per year
features polders (reclaimed land) make up over 40% of the land area; dyke (*Afsluitdijk*) 32 km/20 mi long 1932 has turned the former Zuider Zee inlet into the freshwater IJsselmeer; Delta Project series of dams 1986 forms sea defence in Zeeland delta of the Maas, Scheldt, and Rhine
head of state Queen Beatrix Wilhelmina Armgard from 1980
head of government Ruud Lubbers from 1989
political system constitutional monarchy
political parties Christian Democratic Appeal (CDA), Christian, right-of-centre; Labour Party (PvdA), moderate, left-of-centre; People's Party for Freedom and Democracy (VVD), free enterprise, centrist
exports dairy products, flower bulbs, vegetables, petrochemicals, electronics
currency guilder
population (1990 est) 14,864,000 (including 300,000 of Dutch-Indonesian origin absorbed 1949–64 from former colonial possessions); growth rate 0.4% p.a.
life expectancy men 74, women 81 (1989)
language Dutch
religions Roman Catholic 40%, Protestant 31%
literacy 99% (1989)
GNP $223 bn (1988); $13,065 per head (1987)
chronology
1940–45 Occupied by Germany during World War II.
1947 Joined Benelux customs union.
1948 Queen Juliana succeeded Queen Wilhelmina to the throne.
1949 Became a founding member of North Atlantic Treaty Organization (NATO).
1953 Dykes breached by storm; nearly 2,000 people and thousands of cattle died in flood.
1958 Joined European Economic Community.
1980 Queen Juliana abdicated in favour of her daughter Beatrix.
1981 Opposition to cruise missiles averted their being sited on Dutch soil.
1989 Prime Minister Lubbers resigned; new Lubbers-led coalition elected.
1991 Treaty on political and monetary union signed by European Community (EC) members at Maastricht.

Netherlands Antilles /'neðələndz æn'tɪliːz/ two groups of Caribbean islands, part of the Netherlands with full internal autonomy, comprising Curaçao and Bonaire off the coast of Venezuela (Aruba is considered separately), and St Eustatius, Saba, and the southern part of St Maarten in the Leeward Islands, 800 km/500 mi to the NE

Netherlands Antilles

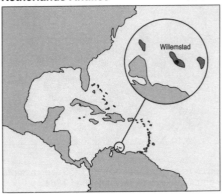

area 797 sq km/308 sq mi
capital Willemstad on Curaçao
products oil from Venezuela refined here; tourism is important
languages Dutch (official), Papiamento, English
population (1983) 193,000.

Netherlands East Indies former name of ⚬Indonesia (1798–1945).

Netzahualcóyotl /ˌnetsəˌwælkəˈjɒtl/ Mexican city lying to the S of Lake Texcoco, forming a suburb to the NE of Mexico City; population (1980) 1,341,200.

Neubrandenburg /nɔɪˈbrændənbɜːg/ former district of East Germany which, since 1990, has been absorbed into the state of Mecklenburg–West Pomerania, Germany.

Neuchâtel /ˌnɜːʃæˈtel/ (German *Neuenburg*) capital of Neuchâtel canton in NW Switzerland, on Lake Neuchâtel, W of Berne; population (1990) 32,800.

Neusiedler See /ˈnɔɪziːdləzeɪ/ (Hungarian *Fertö Tó*) shallow lake in E Austria and NW Hungary, SE of Vienna; area 152 sq km/60 sq mi; it is the only steppe lake in Europe.

Neuss /nɔɪs/ industrial city in North Rhine–Westphalia, Germany; population (1988) 144,000.

Nevada /nɪˈvɑːdə/ state in western USA; nickname Silver State/Sagebrush State
area 286,400 sq km/110,550 sq mi
capital Carson City
towns Las Vegas, Reno
population (1990) 1,201,800
physical Mojave Desert; lakes: Tahoe, Pyramid, Mead; mountains and plateaus alternating with valleys
features legal gambling and prostitution (in some counties); entertainment at Las Vegas and Reno casinos; Lehman Caves National Monument
products mercury, barite, gold
history explored by Kit Carson and John C Fremont 1843–45; ceded to the USA after the Mexican War 1848; first permanent settlement a Mormon trading post 1848. Discovery of

Nevada

Carson City

New Brunswick

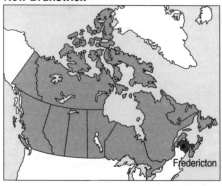

Fredericton

silver (the Comstock Lode) 1858 led to rapid population growth and statehood 1864.

Nevers /nə'veə/ industrial town in Burgundy, central France, at the meeting of the Loire and Nièvre rivers; capital of the former province of Nivernais and the modern *département* of Nièvre; population (1990) 43,900.

New Amsterdam /nju: 'æmstədæm/ town in Guyana, on the river Berbice, founded by the Dutch; population (1980) 25,000. Also a former name (1624–64) of New York.

Newark /'nju:ək/ largest city (industrial and commercial) of New Jersey, USA; population (1990) 275,200. Industries include electrical equipment, machinery, chemicals, paints, canned meats. The city dates from 1666, when a settlement called Milford was made on the site.

Newark /'nju:ək/ market town in Nottinghamshire, England; population (1988 est) 24,400. It has the ruins of a 12th-century castle in which King John died. The British Horological Institute is based here.

New Bedford /nju: 'bedfəd/ city in SE Massachusetts, USA, on the Acushnet River near Buzzards Bay, S of Boston; population (1990) 99,900. Industries include electronics, rubber and metal products, and fishing. During the 1800s it was a prosperous whaling town.

New Britain /nju: 'brɪtn/ largest island in the Bismarck Archipelago, part of Papua New Guinea; capital Rabaul; population (1985) 253,000.

New Britain /nju: 'brɪtn/ city in central Connecticut, USA, SW of Hartford and NE of Waterbury; population (1990) 75,400. Long an industrial city, its industries include tools, hardware, and household appliances.

New Brunswick /nju: 'brʌnzwɪk/ maritime province of E Canada
area 73,400 sq km/28,332 sq mi
capital Fredericton
towns St John, Moncton
features Grand Lake, St John River; Bay of Fundy
products cereals, wood, paper, fish, lead, zinc, copper, oil, natural gas
population (1991) 725,600; 37% French-speaking

history first reached by Europeans (Cartier) 1534; explored by Champlain 1604; remained a French colony as part of Nova Scotia until ceded to England 1713. After the American Revolution many United Empire Loyalists settled there, and it became a province of the Dominion of Canada 1867.

Newbury /'nju:bəri/ market town in Berkshire, England; population (1989 est) 28,100. It has a racecourse and training stables, and electronics industries. Nearby are Aldermaston and Harwell with nuclear research establishments, and RAF Greenham Common.

New Caledonia /nju: ˌkælɪ'dəuniə/ island group in the S Pacific, a French overseas territory between Australia and the Fiji Islands
area 18,576 sq km/7,170 sq mi
capital Nouméa
physical fertile, surrounded by a barrier reef
products nickel (the world's third largest producer), chrome, iron
currency CFP franc
population (1983) 145,300 (43% Kanak (Melanesian), 37% European, 8% Wallisian, 5% Vietnamese and Indonesian, 4% Polynesian)
language French (official)
religions Roman Catholic 60%, Protestant 30%
history New Caledonia was visited by Captain Cook 1774 and became French 1853. A general strike to gain local control of nickel mines 1974 was defeated. In 1981 the French socialist government promised moves towards independence. The 1985 elections resulted in control of most regions by Kanaks, but not the majority of seats. In 1986 the French conservative government reversed the reforms. The Kanaks boycotted a referendum Sept 1987 and a majority were in favour of remaining a French dependency. In 1989 the leader of the Socialist National Liberation front (the most prominent separatist group), Jean-Marie Tjibaou, was murdered.

Newcastle /'nju:ka:səl/ industrial port (iron, steel, chemicals, textiles, ships) in New South Wales, Australia; population (1986) 429,000. Coal was discovered nearby 1796. A penal settlement was founded 1804.

Newcastle-under-Lyme /'nju:ka:səl ʌndə 'laɪm/ industrial town (coal, bricks and tiles, clothing) in Staffordshire, England; population (1988 est) 117,500. Keele University is nearby.

Newcastle-upon-Tyne /'njuːkɑːsəl əpɒn 'taɪn/ industrial port (coal, shipbuilding, marine and electrical engineering, chemicals, metals), commercial and cultural centre, in Tyne and Wear, NE England, administrative headquarters of Tyne and Wear and Northumberland; population (1991 est) 263,000.

features Parts are preserved of a castle built by Henry II 1172–77 on the site of an older castle; the cathedral is chiefly 14th-century; there is a 12th-century church, and the Guildhall 1658. Newcastle is connected with the neighbouring town of Gateshead by several bridges. The headquarters of the Ministry of Social Security is here. The university (1962) is developing a centre for the computer-aided design of complex marine engineering projects.

history Chiefly known as a coaling centre, Newcastle first began to trade in coal in the 13th century. In 1826 ironworks were established by George Stephenson, and the first engine used on the Stockton and Darlington railway was made in Newcastle.

New Delhi /,njuː 'deli/ administrative centre and seat of government of India, adjacent to Old Delhi on the Yamuna River; see ▷Delhi.

New England /njuː 'ɪŋglənd/ region of northeastern USA, comprising the states of Maine, New Hampshire, Vermont, Massachusetts, Rhode Island, and Connecticut. It is a geographic region rather than a political entity, with an area of 172,681 sq km/66,672 sq mi. Boston is the principal urban centre of the region, and Harvard and Yale its major universities.

Originally settled by Pilgrims and Puritans from England, the area is still heavily forested and the economy relies on tourism and services as well as industry.

New England /njuː 'ɪŋglənd/ district of N New South Wales, Australia, especially the tableland area of Glen Innes and Armidale.

New Forest ancient forest in Hampshire, S England.

Newfoundland /'njuːfənlənd/ Canadian province on the Atlantic Ocean
area 405,700 sq km/156,600 sq mi
capital St John's
towns Corner Brook, Gander
physical Newfoundland island and Labrador on the mainland on the other side of the Straits of Belle Isle; rocky
features Grand Banks section of the continental shelf rich in cod; home of the Newfoundland and Labrador dogs
products newsprint, fish products, hydroelectric power, iron, copper, zinc, uranium, offshore oil
population (1991) 571,600
history colonized by Vikings about AD 1000; Newfoundland reached by the English, under the Italian navigator Giovanni Caboto, 1497. It was the first English colony, established 1583. French settlements made; British sovereignty was not recognized until 1713, although France retained

Newfoundland

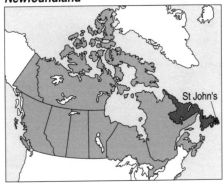

the offshore islands of St Pierre and Miquelon. Internal self-government was achieved 1855. In 1934, as Newfoundland had fallen into financial difficulties, administration was vested in a governor and a special commission. A 1948 referendum favoured federation with Canada and the province joined Canada 1949.

New Guinea /njuː 'gɪni/ island in the SW Pacific, N of Australia, comprising Papua New Guinea and the Indonesian province of West Irian (Irian Jaya area); total area about 885,780 sq km/342,000 sq mi. Part of the Dutch East Indies from 1828, West Irian was ceded by the United Nations to Indonesia 1963.

history The western half of New Guinea was annexed by the Dutch 1828. In 1884 the area of Papua on the southeast coast was proclaimed a protectorate by the British, and in the same year Germany took possession of the northeast quarter of New Guinea. Under Australian control 1914–21, German New Guinea was administered as a British mandate and then united with Papua 1945; Papua and New Guinea jointly gained full independence as Papua New Guinea 1975. The Dutch retained control over the western half of the island (West Irian) after Indonesia gained its independence 1949, but were eventually forced to transfer administrative reponsibility to Indonesia 1963.

Tension between Papua New Guinea and Indonesia has heightened as a result of a growing number of border incidents involving Indonesian troops and Irianese separatist guerrillas. At the same time large numbers of refugees have fled eastwards into Papua New Guinea from West Irian. Its tropical rainforest and the 0.5 million hunter-gatherers who inhabit it are under threat from logging companies and resettlement schemes.

Land of Apocalypse, where the earth dances, /The mountains speak, the doors of the spirit open,/And men are shaken by obscure trances.

On **New Guinea** James McAuley
A Vision of Ceremony 1956

New Hampshire

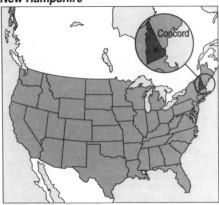

New Jersey

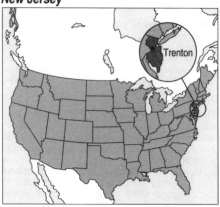

Newham /ˈnjuːəm/ inner borough of E Greater London, N of the Thames. It includes East and West Ham
features former residents include Dick Turpin and Gerard Manley Hopkins; former Royal Victoria and Albert and King George V docks
population (1991) 200,200.

New Hampshire /njuː ˈhæmpʃə/ state in northeastern USA; nickname Granite State
area 24,000 sq km/9,264 sq mi
capital Concord
towns Manchester, Nashua
population (1990) 1,109,300
features White Mountains, including Mount Washington, the tallest peak E of the Rockies (with its cog railway), and Mount Monadnock; the Connecticut River forms boundary with Vermont; earliest presidential-election party primaries every four years; no state income tax or sales tax; ski and tourist resorts
products dairy, poultry, fruits and vegetables; electrical and other machinery; pulp and paper
famous people Mary Baker Eddy, Robert Frost
history settled as a fishing colony near Rye and Dover 1623; separated from Massachusetts colony 1679. As leaders in the Revolutionary cause, its leaders received the honour of being the first to declare independence from Britain 4 July 1776. It became a state 1788, one of the original 13 states.

New Haven /njuː ˈheɪvən/ port town in Connecticut, USA, on Long Island Sound; population (1990) 130,500. *Yale University*, third oldest in the USA, was founded here 1701 and named after Elihu Yale (1648–1721), an early benefactor. New Haven was founded 1683 by English Protestants.

Newhaven /njuː ˈheɪvən/ port in E Sussex, SE England, with container facilities and cross-Channel services to Dieppe, France; population (1989 est) 11,200.

New Hebrides /njuː ˈhebrɪdiːz/ former name (until 1980) of ◊Vanuatu, a country in the S Pacific.

New Jersey /njuː ˈdʒɜːzi/ state in northeastern USA; nickname Garden State
area 20,200 sq km/7,797 sq mi

capital Trenton
towns Newark, Jersey City, Paterson, Elizabeth
population (1990) 7,730,200
features about 200 km/125 mi of seashore, including legalized gambling in Atlantic City and the Victorian beach resort of Cape May; Delaware Water Gap; Palisades along the west bank of the Hudson River; Princeton University; Morristown National Historic Park; Edison National Historic Site, Menlo Park; Walt Whitman House, Camden; Statue of Liberty National Monument (shared with New York); the Meadowlands stadium
products fruits and vegetables, fish and shellfish, chemicals, pharmaceuticals, soaps and cleansers, transport equipment, petroleum refining
famous people Stephen Crane, Thomas Edison, Thomas Paine, Paul Robeson, Frank Sinatra, Bruce Springsteen, Woodrow Wilson
history colonized in the 17th century by the Dutch (New Netherlands); ceded to England 1664; became a state 1787. It was one of the original 13 states.

New London /njuː ˈlʌndən/ port city in SE Connecticut, USA, on Long Island Sound at the mouth of the Thames River; population (1990) 28,500.

Newlyn /ˈnjuːlɪn/ seaport near Penzance, Cornwall, England, which gave its name to the *Newlyn School* of artists 1880–90, including Stanhope Forbes (1857–1947). The Ordnance Survey relates heights in the UK to mean sea level here.

Newmarket /ˈnjuːmɑːkɪt/ town in Suffolk, E England, a centre for horse racing since James I's reign, notably the 1,000 and 2,000 Guineas, the Cambridgeshire, and the Cesarewitch. It is the headquarters of the Jockey Club, and a bookmaker who is 'warned off Newmarket Heath' is banned from all British racecourses. The National Horseracing Museum (1983) and the National Stud are here.

New Mexico /njuː ˈmeksɪkəʊ/ state in southwestern USA; nickname Land of Enchantment
area 315,000 sq km/121,590 sq mi
capital Santa Fe

New Mexico

towns Albuquerque, Las Cruces, Roswell
population (1990) 1,515,100
physical more than 75% of the area lies
over 1,200 m/3,900 ft above sea level; plains,
mountains, caverns
features Great Plains; Rocky Mountains; Rio
Grande; Carlsbad Caverns, the largest known;
Los Alamos atomic and space research centre;
White Sands Missile Range (also used by space
shuttle); Kiowa Ranch, site of D H Lawrence's
stay in the Sangre de Christos Mountains; Taos
art colony; Santa Fe Opera Company; Navaho
and Hopi Indian reservations; White Sands and
Gila Cliff Dwellings national monuments
products uranium, potash, copper, oil, natural
gas, petroleum and coal products; sheep farming;
cotton; pecans; vegetables
famous people Billy the Kid, Kit Carson,
Georgia O'Keeffe
history explored by Francisco de Coronado for
Spain 1540–42; Spanish settlement 1598 on the
Rio Grande; Santa Fe founded 1610; most of
New Mexico ceded to the USA by Mexico 1848;
became a state 1912. The first atomic bomb,
a test device, was exploded in the desert near
Alamogordo 16 July 1945. Oil and gas development and tourism now contribute to the state
economy.

New Orleans /nju: 'ɔ:lɪnz/ commercial and industrial city (banking, oil refining, rockets) and
Mississippi river port in Louisiana, USA; population (1990) 496,900. It is the traditional
birthplace of jazz.

Founded by the French 1718, it still has
a distinctive French Quarter and Mardi Gras
celebrations. The Saturn rockets for Apollo
spacecraft are built here. Dixieland jazz exponents
still play at Preservation Hall. The Superdome
sports palace is among the world's largest
enclosed stadiums, and is adaptable to various
games and expected audience size.

New Plymouth /nju: 'plɪməθ/ port on the west
coast of North Island, New Zealand; population (1991) 48,400. It lies at the centre of
a dairy-farming region; Taranaki gas fields are
nearby.

Newport /'nju:pɔ:t/ (Welsh **Casnewydd**) seaport
and administrative headquarters in Gwent,
Wales, on the river Usk, NW of Bristol; population (1983) 130,200. There is a steelworks at
nearby Llanwern, and a high-tech complex at
Cleppa Park.

The Newport Transporter Bridge was built
1906.

Newport /'nju:pɔ:t/ river port, capital of the Isle of
Wight, England; population (1988 est) 25,700.
Charles I was imprisoned in nearby Carisbrooke
Castle.

Newport News /'nju:pɔ:t 'nju:z/ industrial city
(engineering, shipbuilding) and port of SE
Virginia, USA, at the mouth of the river James;
population (1990) 170,000. With neighbouring
Chesapeake, Norfolk, and Portsmouth, it forms
the Port of Hampton Roads, one of the chief US
ports. It is the site of one of the world's largest
shipyards.

New Rochelle /,nju: rə'ʃel/ residential suburb
of New York, USA, on Long Island Sound;
population (1990) 67,200.

New South Wales /'nju: sauθ 'weɪlz/ state of SE
Australia
area 801,600 sq km/309,418 sq mi
capital Sydney
towns Newcastle, Wollongong, Broken Hill
physical Great Dividing Range (including
Blue Mountains) and part of the Australian
Alps (including Snowy Mountains and Mount
Kosciusko); Riverina district, irrigated by the
Murray-Darling-Murrumbidgee river system;
other main rivers Lachlan, Macquarie-Bogan,
Hawkesbury, Hunter, Macleay, and Clarence
features a radio telescope at Parkes; Siding
Spring Mountain 859 m/2,817 ft, NW of Sydney,
with telescopes that can observe the central sector
of the Galaxy. Canberra forms an enclave within
the state, and New South Wales administers the
dependency of Lord Howe Island
products cereals, fruit, sugar, tobacco, wool,
meat, hides and skins, gold, silver, copper,
tin, zinc, coal; hydroelectric power from the
Snowy River
population (1987) 5,570,000; 60% in Sydney
history called New Wales by English explorer
Capt Cook, who landed at Botany Bay 1770
and thought that the coastline resembled that of
Wales. It was a convict settlement 1788–1850;
opened to free settlement by 1819; achieved self-
government 1856; and became a state of the
Commonwealth of Australia 1901. Since 1973

New South Wales

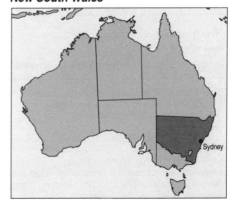

there has been decentralization to counteract the pull of Sydney, and the New England and Riverina districts have separate separatist movements.
New World the Americas, so called by the first Europeans who reached them.

The freedom of the New World is the hope of the universe.

On the **New World** attributed to
Simón Bolívar

New York /ˌnjuː ˈjɔːk/ largest city in the USA, industrial port (printing, publishing, clothing), cultural, financial, and commercial centre, in S New York State, at the junction of the Hudson and East rivers and including New York Bay. It comprises the boroughs of the Bronx, Brooklyn, Manhattan, Queens, and Staten Island; population (1990) 7,322,600, white 43.2%, black 25.2%, Hispanic 24.4%. New York is also known as the Big Apple.
features The Statue of Liberty stands on Liberty Island in the inner harbour of New York Bay. Manhattan skyscrapers include the twin towers of the World Trade Center (412 m/1,350 ft), the Art Deco Empire State Building (381 m/1,250 ft), and the Chrysler Building; the headquarters of the United Nations (UN) is also here. St Patrick's Cathedral is 19th-century Gothic. There are a number of art galleries, among them the Frick Collection, the Metropolitan Museum of Art (with a medieval crafts department, the Cloisters), the Museum of Modern Art, and the Guggenheim, designed by Frank Lloyd Wright. Columbia University (1754) is one of a number of institutions of higher education. Central Park is the largest park.
history The Italian navigator Giovanni da Verrazano (c. 1485–c. 1528) reached New York Bay 1524, and Henry Hudson explored it 1609. The Dutch established a settlement on Manhattan 1624, named *New Amsterdam* from 1626; this was captured by the English 1664 and renamed New York. During the War of Independence, British troops occupied New York 1776–84; it was the capital of the USA 1785–89. The five boroughs were linked 1898 to give the city its present extent.

New York is a catastrophe—but a magnificent catastrophe.

On **New York** Le Corbusier

New York /njuː ˈjɔːk/ state in northeastern USA; nickname Empire State/Excelsior State
area 127,200 sq km/49,099 sq mi
capital Albany
towns New York, Buffalo, Rochester, Yonkers, Syracuse
population (1990) 17,990,500
physical mountains: Adirondacks, Catskills; lakes: Champlain, Placid, Erie, Ontario; rivers: Mohawk, Hudson, St Lawrence (with Thou-

New York

sand Islands); Niagara Falls; Long Island; New York Bay
features West Point, site of the US Military Academy 1801; National Baseball Hall of Fame, Cooperstown; horse racing at Belmont, Aqueduct, Saratoga Springs; Washington Irving's home at Philipsburg Manor; Fenimore House (commemorating J F Cooper), Cooperstown; home of F D Roosevelt at Hyde Park, and the Roosevelt Library; home of Theodore Roosevelt, Oyster Bay; Statue of Liberty National Monument; Erie Canal
products dairy products, apples, clothing, periodical and book printing and publishing, electronic components and accessories, office machines and computers, communications equipment, motor vehicles and equipment, pharmaceuticals, aircraft and parts
famous people James Fenimore Cooper, George Gershwin, Washington Irving, Henry James, Herman Melville, Arthur Miller, Franklin D Roosevelt, Theodore Roosevelt, Walt Whitman
history explored by Giovanni da Verrazano for France 1524; explored by Samuel de Champlain for France and Henry Hudson for the Netherlands 1609; colonized by the Dutch from 1614; first permanent settlement at Albany (Fort Orange) 1624; Manhattan Island purchased by Peter Minuit 1625; New Amsterdam annexed by the English 1664. The first constitution was adopted 1777, when New York became one of the original 13 states.

New Zealand Dominion of
area 268,680 sq km/103,777 sq mi
capital and port Wellington
towns Hamilton, Palmerston North, Christchurch, Dunedin; port Auckland
physical comprises North Island, South Island, Stewart Island, Chatham Islands, and minor islands; mainly mountainous
overseas territories Tokelau (three atolls transferred 1926 from former Gilbert and Ellice Islands colony); Niue Island (one of the Cook Islands, separately administered from 1903: chief town Alafi); Cook Islands are internally self-governing but share common citizenship with New Zealand; Ross Dependency in Antarctica
features Ruapehu on North Island, 2,797 m/9,180 ft, highest of three active volcanoes; geysers and hot springs of the Rotorua district; Lake Taupo (616 sq km/238 sq mi), source of Waikato

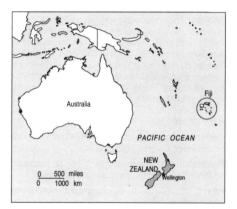

River; Kaingaroa state forest. On South Island are the Southern Alps and Canterbury Plains
head of state Elizabeth II from 1952 represented by governor general (Catherine Tizard from 1990)
head of government Jim Bolger from 1990
political system constitutional monarchy
political parties Labour Party, moderate, left-of-centre; New Zealand National Party, free enterprise, centre-right; Alliance Party, left-of-centre, ecologists
exports lamb, beef, wool, leather, dairy products, processed foods, kiwi fruit; seeds and breeding stock; timber, paper, pulp, light aircraft
currency New Zealand dollar
population (1990 est) 3,397,000 (European (mostly British) 87%; Polynesian (mostly Maori) 12%); growth rate 0.9% p.a.
life expectancy men 72, women 78 (1989)
languages English (official), Maori
religions Protestant 50%, Roman Catholic 15%
literacy 99% (1989)
GNP $37 bn; $11,040 per head (1988)
chronology
1840 New Zealand became a British colony.
1907 Created a dominion of the British Empire.
1931 Granted independence from Britain.
1947 Independence within the Commonwealth confirmed by the New Zealand parliament.
1972 National Party government replaced by Labour Party, with Norman Kirk as prime minister.
1974 Kirk died; replaced by Wallace Rowling.
1975 National Party returned, with Robert Muldoon as prime minister.
1984 Labour Party returned under David Lange.
1985 Non-nuclear military policy created disagreements with France and the USA.
1987 National Party declared support for the Labour government's non-nuclear policy. Lange re-elected. New Zealand officially classified as a 'friendly' rather than 'allied' country by the USA because of its non-nuclear military policy.
1988 Free-trade agreement with Australia signed.
1989 Lange resigned over economic differences with finance minister (he cited health reasons); replaced by Geoffrey Palmer.
1990 Palmer replaced by Mike Moore. Labour Party defeated by National Party in general election; Jim Bolger became prime minister.

1991 Formation of amalgamated Alliance Party set to challenge two-party system.
1992 Ban on visits by US warships lifted. Constitutional change agreed upon.

NF abbreviation for ▷*Newfoundland*, a Canadian province.

Ngorongoro Crater /əŋ‚ɡɒrəŋˈɡɒrəʊ/ crater in the Tanzanian section of the African Great Rift Valley notable for its large numbers of wildebeest, gazelle, and zebra.

NH abbreviation for ▷*New Hampshire*, a state of the USA.

Niagara Falls /naɪˈægərə/ two waterfalls on the Niagara River, on the Canada–USA border, between lakes Erie and Ontario and separated by Goat Island. The *American Falls* are 51 m/167 ft high, 330 m/1,080 ft wide; *Horseshoe Falls*, in Canada, are 49 m/160 ft high, 790 m/2,600 ft across.
    On the west bank of the river is *Niagara Falls*, a city in Ontario, Canada; population (1981) 71,000; on the east bank is *Niagara Falls*, New York State, USA; population (1990) 61,800. Their economy is based on hydroelectric generating plants, diversified industry, and tourism.

Fortissimo at last.

    On seeing **Niagara Falls** Gustav Mahler

Niamey /ˌnɪəˈmeɪ/ river port and capital of Niger; population (1983) 399,000. It produces textiles, chemicals, pharmaceuticals, and foodstuffs.

Nicaragua Republic of (*República de Nicaragua*)
area 127,849 sq km/49,363 sq mi
capital Managua
towns León, Granada; chief ports Corinto, Puerto Cabezas, El Bluff
physical narrow Pacific coastal plain separated from broad Atlantic coastal plain by volcanic mountains and lakes Managua and Nicaragua

features largest state of Central America and most thinly populated; Mosquito Coast, Fonseca Bay, Corn Islands
head of state and government Violeta Barrios de Chamorro from 1990
political system emergent democracy
political parties Sandinista National Liberation Front (FSLN), Marxist-Leninist; Democratic Conservative Party (PCD), centrist; National Opposition Union (UNO), loose, US-backed coalition
exports coffee, cotton, sugar, bananas, meat
currency cordoba
population (1990) 3,606,000 (mestizo 70%, Spanish descent 15%, Indian or black 10%); growth rate 3.3% p.a.
life expectancy men 61, women 63 (1989)
languages Spanish (official), Indian, English
religion Roman Catholic 95%
literacy 66% (1986)
GNP $2.1 bn; $610 per head (1988)
chronology
1838 Independence achieved from Spain.
1926–1933 Occupied by US marines.
1936 General Anastasio Somoza elected president; start of near-dictatorial rule by Somoza family.
1962 Sandinista National Liberation Front (FSLN) formed to fight Somoza regime.
1979 Somoza government ousted by FSLN.
1982 Subversive activity against the government by right-wing Contra guerrillas promoted by the USA. State of emergency declared.
1984 The USA mined Nicaraguan harbours.
1985 Denunciation of Sandinista government by US president Reagan. FSLN won assembly elections.
1987 Central American peace agreement cosigned by Nicaraguan leaders.
1988 Peace agreement failed. Nicaragua held talks with Contra rebel leaders. Hurricane left 180,000 people homeless.
1989 Demobilization of rebels and release of former Somozan supporters; cease-fire ended.
1990 FSLN defeated by UNO, a US-backed coalition; Violeta Barrios de Chamorro elected president. Antigovernment riots.
1992 June: US aid suspended because of concern over role of Sandinista in Nicaraguan government. Sept: around 16,000 made homeless by earthquake.

Nicaragua, Lake /ˌnɪkəˈrægwə/ lake in Nicaragua, the largest in Central America; area 8,250 sq km/3,185 sq mi.

Nice /niːs/ city on the French Riviera; population (1990) 345,700. Founded in the 3rd century BC, it repeatedly changed hands between France and the Duchy of Savoy from the 14th to the 19th century. In 1860 it was finally transferred to France. There is an annual Battle of Flowers, and chocolate and perfume are made. Chapels in the nearby village of Vence have been decorated by the artists Marc Chagall and Henri Matisse, and Nice has a Chagall museum.

Nicobar Islands /ˈnɪkəbɑː/ group of Indian islands, part of the Union Territory of Andaman and Nicobar Islands.

Nicosia /ˌnɪkəˈsiːə/ capital of Cyprus, with leather, textile, and pottery industries; population (1987) 165,000. Nicosia was the residence of Lusignan kings of Cyprus 1192–1475. The Venetians, who took Cyprus 1489, surrounded Nicosia with a high wall, which still exists; the city fell to the Turks 1571. It was again partly taken by the Turks in the invasion 1974.
The Greek and Turkish sectors are separated by the Attila Line.

Niederösterreich /ˈniːdərˌœːstəraɪʃ/ German name for the federal state of ◊Lower Austria.

Niedersachsen /ˈniːdəˌsæksən/ German name for the federal state of ◊Lower Saxony, Germany.

Nièvre /niˈeɪvrə/ river in central France, rising near Varzy and flowing 40 km/25 mi S to join the river Loire at Nevers; it gives its name to a *département*.

Niger /ˈnaɪdzə/ Republic of (*République du Niger*)
area 1,186,408 sq km/457,953 sq mi
capital Niamey
towns Zinder, Maradi, Tahoua
physical desert plains between hills in N and savanna in S; river Niger in SW, Lake Chad in SE
features part of the Sahara Desert and subject to Sahel droughts
head of state Mahamane Ousmane from 1993
head of government Mahamdou Issaufou from 1993
political system military republic
political parties Alliance for the Forces of Change (AFC), moderate, left of centre; National Movement for a Development Society (MNSD), right of centre
exports peanuts, livestock, gum arabic, uranium
currency franc CFA
population (1990 est) 7,691,000; growth rate 2.8% p.a.
life expectancy men 48, women 50 (1989)
languages French (official), Hausa, Djerma, and other minority languages
religions Sunni Muslim 85%, animist 15%
literacy men 19%, women 9% (1985 est)
GNP $2.2 bn; $310 per head (1987)
chronology
1960 Achieved full independence from France; Hamani Diori elected president.
1974 Diori ousted in army coup led by Seyni Kountché.

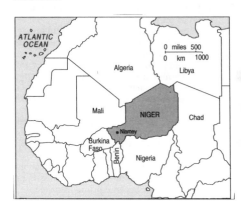

1977 Cooperation agreement signed with France.
1987 Kountché died; replaced by Col Ali Saibu.
1989 Ali Saibu elected president without opposition.
1990 Multiparty politics promised.
1991 Saibu stripped of executive powers; transitional government formed.
1992 Transitional government collapsed. Referendum endorsed adoption of multiparty politics.
1993 Mamahame Ousmane elected president in multiparty elections.

Niger third-longest river in Africa, 4,185 km/2,600 mi. It rises in the highlands bordering Sierra Leone and Guinea, flows NE through Mali, then SE through Niger and Nigeria to an inland delta on the Gulf of Guinea.

Nigeria Federal Republic of
area 923,773 sq km/356,576 sq mi
capital Abuja
towns Ibadan, Ogbomosho, Kano; ports Lagos, Port Harcourt, Warri, Calabar
physical arid savanna in N; tropical rainforest in S, with mangrove swamps along the coast; river Niger forms wide delta; mountains in SE
environment toxic waste from northern industrialized countries has been dumped in Nigeria
features harmattan (dry wind from the Sahara); rich artistic heritage, for example, Benin bronzes
head of state and government Sani Abacha from 1993
political system military republic pending promised elections
political parties Social Democratic Party (SDP), left of centre; National Republican Convention (NRC), right of centre
exports petroleum (largest oil resources in Africa), cocoa, peanuts, palm oil (Africa's largest producer), cotton, rubber, tin
currency naira
population (1991) 88,514,500 (Yoruba in W, Ibo in E, and Hausa-Fulani in N); growth rate 3.3% p.a.
life expectancy men 47, women 49 (1989)
languages English (official), Hausa, Ibo, Yoruba
media all radio and television stations and almost 50% of all publishing owned by the federal government or the Nigerian states
religions Sunni Muslim 50% (in N), Christian 40% (in S), local religions 10%

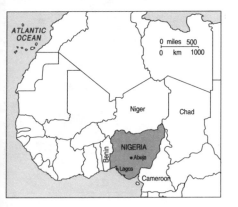

literacy men 54%, women 31% (1985 est)
GNP $78 bn (1987); $790 per head (1984)
chronology
1914 N Nigeria and S Nigeria united to become Britain's largest African colony.
1954 Nigeria became a federation.
1960 Independence achieved from Britain within the Commonwealth.
1963 Became a republic, with Nnamdi Azikiwe as president.
1966 Military coup, followed by a counter-coup led by General Yakubu Gowon. Slaughter of many members of the Ibo tribe in north.
1967 Conflict over oil revenues led to declaration of an independent Ibo state of Biafra and outbreak of civil war.
1970 Surrender of Biafra and end of civil war.
1975 Gowon ousted in military coup; second coup put General Olusegun Obasanjo in power.
1979 Shehu Shagari became civilian president.
1983 Shagari's government overthrown in coup by Maj-Gen Muhammadu Buhari.
1985 Buhari replaced in a bloodless coup led by Maj-General Ibrahim Babangida.
1989 Two new political parties approved. Babangida promised a return to pluralist politics.
1991 Nine new states created. Babangida confirmed his commitment to democratic rule.
1992 Multiparty elections won by Babangida's SDP.
1993 Results of presidential elections suspended by national commission, following complaints of ballot rigging. Interim government formed under Ernest Shonekan; Shonekan overthrown in military coup by General Sani Abacha.

Niigata /niːiɡɑːtə/ industrial port (textiles, metals, oil refining, chemicals) in Chubu region, Honshu Island, Japan; population (1990) 486,100.

Nijmegen /naimeiɡən/ industrial city (brewery, electrical engineering, leather, tobacco) in the E Netherlands, on the Waal River; population (1988) 241,000. The Roman *Noviomagus*, Nijmegen was a free city of the Holy Roman Empire and a member of the Hanseatic League.

Nikolayev /ˌnɪkəˈlaɪev/ port (with shipyards) and naval base on the Black Sea, Ukraine; population (1987) 501,000.

Nile /naɪl/ river in Africa, the world's longest, 6,695 km/4,160 mi. The *Blue Nile* rises in Lake Tana, Ethiopia, the *White Nile* at Lake Victoria; they join at Khartoum, Sudan. The river enters the Mediterranean Sea at a vast delta in N Egypt.
Its remotest headstream is the Luvironza, in Burundi. The Nile proper begins on leaving Lake Victoria above Owen Falls. From Lake Victoria it flows over rocky country, and there are many cataracts and rapids, including the Murchison Falls, until it enters Lake Mobutu (Albert). From here it flows across flat country and in places spreads out to form lakes. At Lake No it is joined by the Bahr el Ghazal, and from this point to Khartoum it is called the White Nile. At Khartoum it is joined by the Blue Nile, which rises in the Ethiopian highlands, and 320 km/200 mi below Khartoum it is joined by the Atbara. From Khartoum to Aswan there are six cataracts. The Nile is navigable to the

second cataract, a distance of 1,545 km/960 mi. The delta of the Nile is 190 km/120 mi wide. From 1982 Nile water has been piped beneath the Suez Canal to irrigate Sinai. The water level behind the Aswan Dam fell from 170 m/558 ft in 1979 to 150 m/492 ft in 1988, threatening Egypt's hydroelectric power generation. The 1988 water level was the lowest in a century.

As headstrong as an allegory on the banks of the Nile.

On the **Nile** Richard Sheridan
The Rivals 1775

Nîmes /niːm/ capital of Gard *département*, Languedoc-Roussillon, S France; population (1990) 133,600. Roman remains include an amphitheatre dating from the 2nd century and the Pont du Gard (aqueduct). The city gives its name to the cloth known as denim (*de Nîmes*).

Ningbo /ˌnɪŋˈbəu/ or *Ningpo* port and special economic zone in Zhejiang province, E China; industries include fishing, shipbuilding, and high-tech equipment; population (1984) 615,600. Already a centre of foreign trade under the Tang dynasty (618–907), it was one of the original treaty ports 1842.

Ningpo alternative name for ⬦Ningbo, a port in China.

Ningxia /ˈnɪŋʃiɑː/ or *Ningxia Hui* autonomous region (formerly *Ninghsia-Hui*) of NW China
area 170,000 sq km/65,620 sq mi
capital Yinchuan
physical desert plateau
products cereals and rice under irrigation; coal
population (1990) 4,655,000; including many Muslims and nomadic herders.

Nippon /ˈnɪpɒn/ English transliteration of the Japanese name for ⬦Japan.

Niterói /ˌniːtəˈrɔɪ/ or *Nictheroy* port and resort city in Brazil on the east shore of Guanabara Bay, linked by bridge with Rio de Janeiro; population (1991) 455,200.

Niue /ˈnjuːeɪ/ coral island in the S Pacific, W of the Cook Islands; overseas territory of New Zealand
area 260 sq km/100 sq mi
towns port Alofi
products coconuts, passion fruit, honey
population (1989) 2,300
history inhabited by warriors who stopped English explorer Capt Cook from landing 1774; British protectorate 1900; annexed by New Zealand 1901; attained self-government in free association with New Zealand (with which there is common citizenship).1974.

Nizhnevartovsk /ˌnɪʒni vɑːˈtɒfsk/ city in Tyumen region, N Siberia, Russia; population 240,000. The oilfield discovered 1965 at nearby Lake Samotlor is one of the largest in the world, and there are natural gas fields to the N. Poor technology prevented their efficient exploitation until the 1990s.

Nizhny Novgorod /ˈnɪʒni ˈnɒvgərɒd/ (formerly 1932–90 *Gorky*) city in central Russia; population (1987) 1,425,000. Cars, locomotives, and aircraft are manufactured here.

NJ abbreviation for ⬦*New Jersey*, a state of the USA.

NM abbreviation for ⬦*New Mexico*, a state of the USA.

Nord-Pas-de-Calais /ˈnɔː ˌpɑː də kæˈleɪ/ region of N France; area 12,400 sq km/4,786 sq mi; population (1990) 3,965,100. Its capital is Lille, and it consists of the *départements* of Nord and Pas-de-Calais.

   Pas-de-Calais is the French term for the Straits of Dover, between the English Channel and the North Sea.

Nore, the /nɔː/ sandbank at the mouth of the river Thames, England; site of the first lightship 1732.

Norfolk /ˈnɔːfək/ county on the east coast of England
area 5,360 sq km/2,069 sq mi
towns Norwich (administrative headquarters), King's Lynn; resorts: Great Yarmouth, Cromer, Hunstanton
physical rivers: Ouse, Yare, Bure, Waveney; the Norfolk Broads; Halvergate Marshes wildlife area
features traditional reed thatching; Grime's Graves (Neolithic flint mines); shrine of Our Lady of Walsingham, a medieval and present-day centre of pilgrimage; Blickling Hall (Jacobean); residence of Elizabeth II at Sandringham (built 1869–71)
products cereals, turnips, sugar beets, turkeys, geese, offshore natural gas
population (1991) 736,700
famous people Fanny Burney, John Sell Cotman, John Crome ('Old Crome'), Rider Haggard, Horatio Nelson, Thomas Paine.

Norfolk /ˈnɔːfək/ seaport in SE Virginia, USA, on the Atlantic Ocean at the mouth of the James and Elizabeth rivers; population (1990) 261,200.

Norfolk

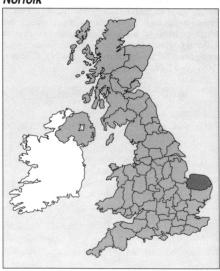

It is the headquarters of the US Navy's Atlantic fleet, and the home of 22 other Navy commands. Industries include shipbuilding, chemicals, and motor-vehicle assembly.

Norfolk Broads /ˈnɔːfək/ area of some 12 interlinked freshwater lakes in E England, created about 600 years ago by the digging-out of peat deposits; the lakes are used for boating and fishing.

Norfolk Island /ˈnɔːfək/ Pacific island territory of Australia, S of New Caledonia
area 40 sq km/15 sq mi
products citrus fruit, bananas; tourist industry
population (1986) 2,000
history reached by English explorer Capt Cook 1774: settled 1856 by descendants of the mutineers of the *Bounty* from Pitcairn Island; Australian territory from 1914: largely self-governing from 1979.

Norilsk /nəˈrɪlsk/ world's northernmost industrial city (nickel, cobalt, platinum, selenium, tellurium, gold, silver) in Siberian Russia; population (1987) 181,000. The permafrost is 300 m/1,000 ft deep, and the winter temperature may be 55°C/67°F.
It is the most polluted city in Russia, with more than 2.3 million tonnes of toxins emitted into the air every year, mostly from nickel mining and smelting. Norilsk was founded 1935 by Stalin as the administrative centre of the gulag network of prison camps.

Normal /ˈnɔːml/ city in central Illinois, USA, NE of Bloomington; population (1990) 40,000. It is a marketing centre for the livestock and grains produced in the surrounding area. Illinois State University is here.

Normandy /ˈnɔːməndi/ two regions of NW France: ◊Haute-Normandie and ◊Basse-Normandie. It was named after the Viking Norsemen (Normans), the people who conquered and settled in the area in the 9th century. As a French duchy it reached its peak under William the Conqueror and was renowned for its centres of learning established by Lanfranc and St Anselm. Normandy was united with England 1100–35. England and France fought over it during the Hundred Years' War, England finally losing it 1449 to Charles VII. In World War II the Normandy beaches were the site of the Allied invasion on D-day, 6 June 1944.
The main towns are Alençon, Bayeux, Caen, Cherbourg, Dieppe, Deauville, Lisieux, Le Havre, and Rouen. Features of Normandy include the painter Monet's restored home and garden at Giverny; Mont St Michel; Château Miromesnil, the birthplace of Guy de Maupassant; Victor Hugo's house at Villequier; and Calvados apple brandy.

Normandy has a bad reputation for rain and on that account is called the pot de chambre *of France.*

On **Normandy** R H Bruce Lockhart
My Europe 1952

Northallerton /nɔːˈθælətən/ market town, administrative headquarters of North Yorkshire, England; industries (tanning and flour milling); population (1985) 13,800.

North America third largest of the continents (including Greenland and Central America), and over twice the size of Europe
area 24,000,000 sq km/9,400,000 sq mi
largest cities (population over 1 million) Mexico City, New York, Chicago, Toronto, Los Angeles, Montréal, Guadalajara, Monterrey, Philadelphia, Houston, Guatemala City, Vancouver, Detroit, San Diego, Dallas
features Lake Superior (the largest body of fresh water in the world); Grand Canyon on the Colorado River; Redwood National Park, California has some of the world's tallest trees; San Andreas Fault, California; deserts: Death Valley, Mojave, Sonoran; rivers (over 1,600 km/1,000 mi) include Mississippi, Missouri, Mackenzie, Rio Grande, Yukon, Arkansas, Colorado, Saskatchewan-Bow, Columbia, Red, Peace, Snake
geographical extremities of mainland Boothia Peninsula in the Canadian Arctic in the N; SE Panama in the S; Cape Prince of Wales, Alaska, in the W; southeast coast of Labrador in the E
highest point Mount McKinley, Alaska (6,194 m/20,320 ft)
lowest point Badwater in Death Valley (–86 m/–282 ft)
physical occupying the northern part of the landmass of the western hemisphere between the Arctic Ocean and the tropical southeastern tip of the isthmus that joins Central America to South America. In Canada and the USA, the Great Plains of the interior separate mountain belts to the E (Appalachians, Laurentian Highlands) and W (Rocky Mountains, Coast Mountains, Cascade Range, Sierra Nevada). The western range extends S into Mexico as the Sierra Madre. The Mississippi river system drains from the central Great Plains into the Gulf of Mexico.
Low coastal plains on the Atlantic coast are indented by the Gulf of St Lawrence, Bay of Fundy, Delaware Bay, Chesapeake Bay; the St Lawrence and Great Lakes form a rough crescent (with Lake Winnipeg, Lake Athabasca, the Great Bear, and the Great Slave lakes) around the exposed rock of the great Canadian/Laurentian shield, into which Hudson Bay breaks from the north; Greenland (the largest island in the world next to Australia) is a high, ice-covered plateau with a deeply indented coastline of fjords
products with abundant resources and an ever-expanding home market, the USA's fast-growing industrial and technological strength has made it less dependent on exports and a dominant economic power throughout the continent. Canada is the world's leading producer of nickel, zinc, uranium, potash, and linseed, and the world's second largest producer of asbestos, silver, titanium, gypsum, sulphur, and molybdenum; Mexico is the world's leading producer of silver and the fourth largest oil producer; the USA is the world's leading producer of salt and the second

largest producer of oil and cotton; nearly 30% of the world's beef and veal is produced in North America
population (1990 est) 395 million, rising to an estimated 450 million by the year 2000; annual growth rate from 1980 to 1985: Canada 1.08%, USA 0.88%, Mexico 2.59%, Honduras 3.39%; the native American Indian, Inuit, and Aleut peoples are now a minority within a population predominantly of European immigrant origin. Many Africans were brought in as part of the slave trade
languages English predominates in Canada, USA, and Belize; Spanish is the chief language of the countries of Latin America and a sizeable minority in the USA; French is spoken by about 25% of the population of Canada, and by people of the French *département* of St Pierre and Miquelon; indigenous non-European minorities, including the Inuit of Arctic Canada, the Aleuts of Alaska, North American Indians, and the Maya of Central America, have their own languages and dialects
religions Christian and Jewish religions predominate; 97% of Latin Americans, 47% of Canadians, and 21% of those living in the USA are Roman Catholic.

Northampton /nɔː'θæmptən/ county town of Northamptonshire, England; population (1991) 178,200. Boots and shoes (of which there is a museum) are still made, but engineering has superseded them as the chief industry; there is also food processing and brewing.

Northamptonshire /nɔː'θæmptənʃə/ county in central England
area 2,370 sq km/915 sq mi
towns Northampton (administrative headquarters), Kettering
features rivers Welland and Nene; Canons Ashby, Tudor house, home of the Drydens for 400 years; churches with broached spires

Northamptonshire

North Carolina

Raleigh

products cereals, cattle
population (1991) 568,900
famous people John Dryden, Richard III, Robert Browne.

Northants abbreviation for *Northamptonshire*.

North Brabant /'nɔːθ brə'bænt/ (Dutch *Noord Brabant*) southern province of the Netherlands, lying between the river Maas (Meuse) and Belgium
area 4,940 sq km/1,907 sq mi
capital 's-Hertogenbosch
towns Breda, Eindhoven, Tilburg
physical former heathland is now under mixed farming
products brewing, engineering, microelectronics, textile manufacture
population (1991) 2,209,000.

North Cape (Norwegian *Nordkapp*) cape in the Norwegian county of Finnmark; the most northerly point of Europe.

North Carolina /ˌnɔːθ ˌkærə'laɪnə/ state in eastern USA; nickname Tar Heel State/Old North State
area 136,400 sq km/52,650 sq mi
capital Raleigh
towns Charlotte, Greensboro, Winston-Salem
features Appalachian Mountains (including Blue Ridge and Great Smoky mountains), site of Fort Raleigh on Roanoke Island, Wright Brothers National Memorial at Kitty Hawk, the Research Triangle established 1956 (Duke University, University of North Carolina, and North Carolina State University) for high-tech industries, Cape Hatteras and Cape Lookout national seashores
products tobacco, corn, soya beans, livestock, poultry, textiles, clothing, cigarettes, furniture, chemicals, machinery
population (1990) 6,628,600
famous people Billy Graham, O Henry, Jesse Jackson, Thomas Wolfe
history after England's Roanoke Island colony was unsuccessful 1585 and 1587, permanent settlement was made 1663; it was one of the original 13 states 1789.

Northd abbreviation for ◊*Northumberland*, an English county.

North Dakota /ˌnɔːθ də'kəʊtə/ state in northern USA; nickname Flickertail State/Sioux State
area 183,100 sq km/70,677 sq mi
capital Bismarck

North Dakota

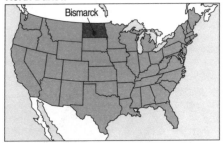

Northern Territory

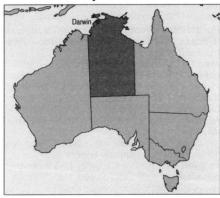

towns Fargo, Grand Forks, Minot
features fertile Red River valley, Missouri Plateau; Garrison Dam on the Missouri River; Badlands, so called because the pioneers had great difficulty in crossing them (also site of Theodore Roosevelt's Elkhorn Ranch); International Peace Garden, on Canadian border; 90% of the land is cultivated
products cereals, meat products, farm equipment, oil, coal
population (1990) 638,800
famous people Maxwell Anderson, Louis L'Amour
history explored by La Verendrye's French Canadian expedition 1738–40; acquired by the USA partly in the Louisiana Purchase 1803 and partly by treaty with Britain 1813. The earliest settlement was Pembina 1812, by Scottish and Irish families. North Dakota became a state 1889, attracting many German and Norwegian settlers.

North Downs line of chalk hills in SE England, running from Salisbury Plain across Hampshire, Surrey, and Kent to the cliffs of South Foreland. They face the South Downs across the Weald of Kent and Sussex, and are much used for sheep pasture.

North-East Frontier Agency former name (until 1972) for ▷Arunachal Pradesh, a territory of India.

North-East India area of India (Meghalaya, Assam, Mizoram, Tripura, Manipur, Nagaland, and Arunachal Pradesh) linked with the rest of India only by a narrow corridor. There is opposition to immigration from Bangladesh and rest of India, and demand for secession.

Northeast Passage sea route from the N Atlantic, around Asia, to the N Pacific, pioneered by Swedish explorer Nils Nordenskjöld 1878–79 and developed by the USSR in settling N Siberia from 1935. Russia owns offshore islands and claims it as an internal waterway; the USA claims that it is international.

Northern Areas districts N of Azad Kashmir, directly administered by Pakistan but not merged with it. India and Azad Kashmir each claim them as part of disputed Kashmir. They include Baltistan, Gilgit, Skardu, and Hunza (an independent principality for 900 years until 1974).

Northern Ireland see ▷Ireland, Northern.

Northern Rhodesia /ˈnɔːðən rəʊˈdiːʃə/ former name (until 1964) of ▷Zambia, a country in Africa.

Northern Territory territory of Australia
area 1,346,200 sq km/519,633 sq mi
capital Darwin (chief port)
towns Alice Springs
features mainly within the tropics, although with wide range of temperature; very low rainfall, but artesian bores are used; Macdonnell Ranges (Mount Zeil 1,510 m/4,956 ft); Cocos and Christmas Islands included in the territory 1984; 50,000–60,000-year-old rock paintings of animals, birds, and fish in Kakadu National Park
products beef cattle, prawns, bauxite (Gove), gold and copper (Tennant Creek), uranium (Ranger)
population (1987) 157,000
government there is an administrator and a legislative assembly, and the territory is also represented in the federal parliament
history originally part of New South Wales, it was annexed 1863 to South Australia but from 1911 until 1978 (when self-government was introduced) was under the control of the Commonwealth of Australia government. Mineral discoveries on land occupied by Aborigines led to a royalty agreement 1979.

North Holland (Dutch *Noord Holland*) low-lying coastal province of the Netherlands occupying the peninsula jutting northwards between the North Sea and the IJsselmeer
area 2,670 sq km/1,031 sq mi
population (1991) 2,397,000
capital Haarlem
towns Amsterdam, Hilversum, Den Helder, the cheese centres Alkmaar and Edam
physical most of the province is below sea level, protected from the sea by a series of sand dunes and artificial dykes
products bulbs, grain, and vegetables
history once part of the former county of Holland that was divided into two provinces (North and South) 1840.

North Island smaller of the two main islands of New Zealand.

North Korea see ▷Korea, North.

North Pole the northern point where an imaginary line penetrates the Earth's surface by the axis about which it revolves; see also ⟡ Arctic.

North Rhine–Westphalia /nɔːθ 'raɪn westˈfeɪliə/ (German *Nordrhein-Westfalen*) administrative *Land* of Germany
area 34,100 sq km/13,163 sq mi
capital Düsseldorf
towns Cologne, Essen, Dortmund, Duisburg, Bochum, Wuppertal, Bielefeld, Bonn, Gelsenkirchen, Münster, Mönchengladbach
features valley of the Rhine; Ruhr industrial district
products iron, steel, coal, lignite, electrical goods, fertilizers, synthetic textiles
religion 53% Roman Catholic, 42% Protestant
population (1988) 16,700,000.

North Sea sea to the E of Britain and bounded by the coasts of Belgium, the Netherlands, Germany, Denmark, and Norway; area 523,000 sq km/ 202,000 sq mi; average depth 55 m/180 ft, greatest depth 660 m/2,165 ft. In the NE it joins the Norwegian Sea, and in the S it meets the Strait of Dover. It has fisheries, oil, and gas.

Despite the 1972 Oslo Treaty, Britain still dumps the largest quantity of sewage sludge in the North Sea. In 1987, Britain dumped more than 4,700 tonnes of sewage sludge into the North Sea; Britain is also responsible for the highest input of radioactive isotopes, atmospheric nitrogen, and hydrocarbons. In 1991 the North Sea received more than 2 million tonnes of liquid chemical waste. The effects of pollution are most noted along the east coast of the North Sea.

Continued investment in North Sea oil and gas fields came under threat 1992 due to uncertainty over the future of coal pits and government support for gas-fired power stations. In 1992 1,300 jobs were lost at the Ardersier oil-fabrication yard in N Scotland.

A gradual lowering of the British coastline since the Ice Age has meant a gradual rise in the level of the North Sea, producing floods in 1881, 1928, 1953, and 1978. It is believed that the melting of the polar ice cap as a result of the greenhouse effect will increase this rise in sea level.

North Uist /ˈjuːɪst/ island of the Outer Hebrides, Scotland. Lochmaddy is the main port of entry. It produces tweeds and seaweed, and crofting is practised.

Northumberland /nɔːˈθʌmbələnd/ county in N England
area 5,030 sq km/1,942 sq mi
towns Newcastle-upon-Tyne (administrative headquarters), Berwick-upon-Tweed, Hexham
features Cheviot Hills; rivers: Tweed, upper Tyne; Northumberland National Park in the west; Holy Island; the Farne island group; part of Hadrian's Wall and Housestead's Fort; Alnwick and Bamburgh castles; Thomas Bewick museum; large moorland areas are used for military manoeuvres; Longstone Lighthouse from which Grace Darling rowed to the rescue is no longer inhabited, the crew having been replaced by an automatic light; wild white cattle of Chillingham
products sheep
population (1991) 300,600

Northumberland

famous people Thomas Bewick, Jack Charlton, Grace Darling.

North-West Frontier Province province of Pakistan; capital Peshawar; area 74,500 sq km/ 28,757 sq mi; population (1985) 12,287,000. It was a province of British India 1901–47. It includes the strategic Khyber Pass, the site of constant struggle between the British Raj and the Pathan warriors. In the 1980s it had to accommodate a stream of refugees from neighbouring Afghanistan.

Northwest Passage Atlantic–Pacific sea route around the north of Canada. Canada, which owns offshore islands, claims it as an internal waterway; the USA insists that it is an international waterway and sent an icebreaker through without permission 1985.

Early explorers included Englishmen Martin Frobisher and, later, John Franklin, whose failure to return 1847 led to the organization of 39 expeditions in the next ten years. John Ross reached Lancaster Sound 1818 but mistook a bank of cloud for a range of mountains and turned back. R McClune explored the passage 1850–53 although he did not cover the whole route by sea. The polar explorer Roald Amundsen was the first European to sail through 1903–06.

Northwest Territories /nɔːθwest ˈtentəriz/ territory of Canada
area 3,426,300 sq km/1,322,552 sq mi
capital Yellowknife
physical extends to the North Pole, to Hudson's Bay in the E, and in the W to the edge of the Canadian Shield
features Mackenzie River; lakes: Great Slave, Great Bear; Miles Canyon
products oil, natural gas, zinc, lead, gold, tungsten, silver
population (1991) 54,000; over 50% native peoples (Indian, Inuit)
history the area was the northern part of Rupert's Land, bought by the Canadian government from

Northwest Terriitories

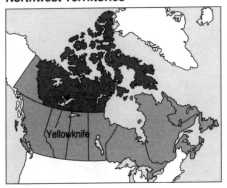

the Hudson's Bay Company 1869. An act of 1952 placed the Northwest Territories under a commissioner acting in Ottawa under the Ministry of Northern Affairs and Natural Resources. In 1990 territorial control of over 350,000 sq km/135,000 sq mi of the Northwest Territories was given to the Inuit, and in 1992 the creation of an Inuit autonomous homeland, Nunavut, was agreed in a regional referendum.

North Yorkshire /ˈjɔːkʃə/ county in NE England
area 8,320 sq km/3,212 sq mi
towns Northallerton (administrative head-quarters), York; resorts: Harrogate, Scarborough, Whitby
features England's largest county; including part of the Pennines, the Vale of York, and the Cleveland Hills and North Yorkshire Moors, which form a national park (within which is Fylingdales radar station to give early warning—4 minutes—of nuclear attack); and Rievaulx Abbey; Yorkshire Dales National Park (including Swaledale, Wensleydale, and Bolton Abbey in Wharfedale); rivers: Derwent, Ouse; Fountains Abbey near Ripon, with Studley Royal Gardens;

North Yorkshire

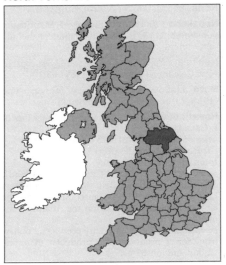

York Minster; Castle Howard, designed by Vanbrugh, has Britain's largest collection of 18th–20th-century costume; largest accessible cavern in Britain, the Battlefield Chamber, Ingleton
products cereals, wool and meat from sheep, dairy products, coal, electrical goods
population (1991) 698,000
famous people Alcuin, W H Auden, Guy Fawkes.

Norwalk /ˈnɔːwɔːk/ city in SW Connecticut, USA, on the Norwalk River where it flows into Long Island Sound, NE of Stamford; population (1990) 78,300. Industries include electronic equipment, clothing, hardware, and furniture.

Norway Kingdom of (*Kongeriket Norge*)
area 387,000 sq km/149,421 sq mi (includes Svalbard and Jan Mayen)
capital Oslo
towns Bergen, Trondheim, Stavanger
physical mountainous with fertile valleys and deeply indented coast; forests cover 25%; extends N of Arctic Circle
territories dependencies in the Arctic (Svalbard and Jan Mayen) and in Antarctica (Bouvet and Peter I Island, and Queen Maud Land)
environment an estimated 80% of the lakes and streams in the southern half of the country have been severely acidified by acid rain
features fjords, including Hardanger and Sogne, longest 185 km/115 mi, deepest 1,245 m/4,086 ft; glaciers in north; midnight sun and northern lights
head of state Harald V from 1991
head of government Gro Harlem Brundtland from 1990
political system constitutional monarchy
political parties Norwegian Labour Party (DNA), moderate left of centre; Conservative Party, progressive, right of centre; Christian People's Party (KrF), Christian, centre-left; Centre Party (SP), left of centre, rural-oriented
exports petrochemicals from North Sea oil and gas, paper, wood pulp, furniture, iron ore and other minerals, high-tech goods, sports goods, fish
currency krone
population (1990 est) 4,214,000; growth rate 0.3% p.a.
life expectancy men 73, women 80 (1989)

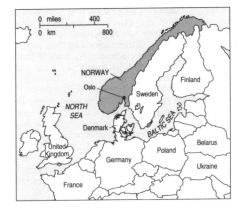

languages Norwegian (official); there are Saami (Lapp) and Finnish-speaking minorities
religion Evangelical Lutheran (endowed by state) 94%
literacy 100% (1989)
GNP $89 bn (1988); $13,790 per head (1984)
chronology
1814 Became independent from Denmark; ceded to Sweden.
1905 Links with Sweden ended; full independence achieved.
1940–45 Occupied by Germany.
1949 Joined North Atlantic Treaty Organization (NATO).
1952 Joined Nordic Council.
1957 King Haakon VII succeeded by his son Olaf V.
1960 Joined European Free Trade Association (EFTA).
1972 Accepted into membership of European Economic Community; application withdrawn after a referendum.
1988 Gro Harlem Brundtland awarded Third World Prize.
1989 Jan P Syse became prime minister.
1990 Brundtland returned to power.
1991 King Olaf V died; succeeded by his son Harald V.
1992 Defied whaling ban and resumed whaling industry.

Norwegian Sea /nɔːˈwiːdʒən ˈsiː/ part of the Arctic Ocean.

Norwich /ˈnɒrɪdʒ/ cathedral city in Norfolk, E England; population (1991) 121,000. Industries include shoes, clothing, chemicals, confectionery, engineering, printing, and insurance. It has a Norman castle, a 15th-century Guildhall, medieval churches, Tudor houses, and a Georgian Assembly House.
  The cathedral was founded 1096. The castle has a collection of paintings by the Norwich School (John Sell Cotman and John Crome). The University of East Anglia (1963) has the Sainsbury Centre for Visual Arts on its campus. The Sainsbury Laboratory (1987), in association with the John Innes Institute, was founded here to study the molecular causes of disease.

Norwich /ˈnɔːwɪtʃ/ city in SE Connecticut, USA, at the confluence of the Yantic and Quinebaug rivers that form the Thames River, N of New London; seat of New London County; population (1990) 37,400. Industries include leather, paper, and metal products; electronic equipment; and clothing. In the 1700s it was a shipping and shipbuilding centre.

Nottingham /ˈnɒtɪŋəm/ industrial city (engineering, coalmining, bicycles, textiles, knitwear, pharmaceuticals, tobacco, lace, electronics) and administrative headquarters of Nottinghamshire, England; population (1991) 261,500.
  Features include the university (1881), the Playhouse (opened 1963), and the Theatre Royal. Nearby are Newstead Abbey, home of Byron, and D H Lawrence's home at Eastwood.

Nottinghamshire

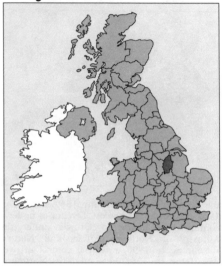

Nottinghamshire /ˈnɒtɪŋəmʃə/ county in central England
area 2,160 sq km/834 sq mi
towns Nottingham (administrative headquarters), Mansfield, Worksop
features river Trent; the remaining areas of Sherwood Forest (home of Robin Hood), formerly a royal hunting ground, are included in the 'Dukeries'; Cresswell Crags (remains of prehistoric humans); D H Lawrence commemorative walk from Eastwood (where he lived) to Old Brinsley Colliery
products cereals, cattle, sheep, light engineering, footwear, limestone, ironstone, oil
population (1991) 980,600
famous people William Booth, D H Lawrence, Alan Sillitoe
history in World War II Nottinghamshire produced the only oil out of U-boat reach, and drilling revived in the 1980s.

Notts abbreviation for *Nottinghamshire*.

Nouakchott /ˌnuːækˈʃɒt/ capital of Mauritania; population (1985) 500,000. Products include salt, cement, and insecticides.

Nouméa /nuːˈmeɪə/ port on the southwest coast of New Caledonia; population (1989) 65,100.

Nova Lisboa /ˈnəʊvə lɪzˈbəʊə/ former name (1928–73) for ⬦Huambo, a town in Angola.

Nova Scotia /ˈnəʊvə ˈskəʊʃə/ maritime province of E Canada
area 55,500 sq km/21,423 sq mi
capital Halifax (chief port)
towns Dartmouth, Sydney
features Cabot Trail (Cape Breton Island); Alexander Graham Bell Museum; Fortress Louisbourg; Strait of Canso Superport, the largest deepwater harbour on the Atlantic coast of North America
products coal, gypsum, dairy products, poultry, fruit, forest products, fish products (including scallop and lobster)

Nova Scotia

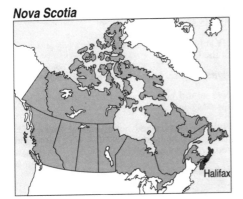

Halifax

physical comprising a peninsula with a highly indented coastline extending SE from New Brunswick into the Atlantic Ocean, and Cape Breton Island which is linked to the mainland by the Canso Causeway
population (1991) 897,500
history Nova Scotia was visited by the Italian navigator Giovanni Caboto 1497. A French settlement was established 1604, but expelled 1613 by English colonists from Virginia. The name of the colony was changed from *Acadia* to Nova Scotia 1621. England and France contended for possession of the territory until Nova Scotia (which then included present-day New Brunswick and Prince Edward Island) was ceded to Britain 1713; Cape Breton Island remained French until 1763. Nova Scotia was one of the four original provinces of the Dominion of Canada.

Novaya Zemlya /ˈnɔuviə ˈzemliə/ Arctic island group off NE Russia; area 81,279 sq km/ 31,394 sq mi; population, a few Samoyed. It is rich in birds, seals, and walrus.

Novgorod /ˈnɒvɡərɒd/ industrial city (chemicals, engineering, clothing, brewing) on the Volkhov River, NW Russia; a major trading city in medieval times; population (1987) 228,000.
Novgorod was the original capital of the Russian state, founded 862. In 912 the capital moved to Kiev, but Novgorod continued to flourish until the 13th century as a major commercial centre for trade with Scandinavia, the Byzantine empire, and the Muslim world. The city's prosperity declined from the 14th century, and it came under the control of Ivan the Great III 1478 and was sacked by Ivan the Terrible 1570.

Who can resist God and Lord Novgorod the Great?

On **Novgorod** Russian proverb

Novi Sad /ˈnɒvi ˈsɑːd/ industrial and commercial city (pottery and cotton), capital of the autonomous province of Vojvodina in N Serbia, Yugoslavia, on the river Danube; population (1981) 257,700. Products include leather, textiles, and tobacco.

Novokuznetsk /ˌnɒvəkuzˈnetsk/ industrial city (steel, aluminium, chemicals) in the Kuznetsk Basin, S central Russia; population (1987) 589,000. It was called *Stalinsk* 1932–61.

Novorossiisk /ˌnɒvərɒˈsiːsk/ Black Sea port and industrial city (cement, metallurgy, food processing) in Russia; population (1987) 179,000.

Novosibirsk /ˌnɒvəsɪˈbɪəsk/ industrial city (engineering, textiles, chemicals, food processing) in W Siberian Russia, on the river Ob; population (1987) 1,423,000. Winter lasts eight months here.
At *Akademgorodok* ('Science City'), population 25,000, advanced research is carried on into Siberia's local problems.

Nowa Huta /ˈnɔuvə ˈhuːtə/ industrial suburb of Kraków, on the Vistula River. It is the centre of Poland's steel industry.
Because of excessive pollution, the area around Nowa Huta has been declared one of Poland's 27 ecologically endangered areas. In 1981 an aluminium plant was closed after the dumping of toxic waste led to severe illnesses. The amount of soot in the air is 35 times greater than that judged dangerous to health.

NS abbreviation for ⟡*Nova Scotia*, a Canadian province.

NSW abbreviation for ⟡*New South Wales*, an Australian state.

Nukua'lofa /ˌnuːkuəˈlɒfə/ capital and port of Tonga on Tongatapu Island; population (1986) 29,000.

Nullarbor Plain /ˈnʌləbɔː/ (Latin *nullus arbor* 'no tree') arid coastal plateau area divided between Western and South Australia; there is a network of caves beneath it. Atom-bomb experiments were carried out in the 1950s at Maralinga, an area in the NE bordering on the Great Victoria Desert.

Nunavut (Inuit 'our land') semi-autonomous Inuit homeland in Northwest Territories, Canada, extending over 2,000,000 sq km/ 772,000 sq mi. Its creation (not due to take effect until 1999) was narrowly approved in a regional plebiscite May 1992 after representatives of the 17,000 Inuit had negotiated hunting, fishing, and mineral rights in the area, as well as outright ownership of 350,000 sq km/135,100 sq mi.
Creation of the homeland was opposed by Dene Indians from the western Arctic, where 74% voted against (the Dene Indians claimed that the homeland cut across their traditional hunting grounds). In the eastern Arctic where most Inuit live, 84% voted in favour.

Nuneaton /nʌnˈiːtn/ market town in Warwickshire, England, on the river Anker, NE of Coventry; industries include ceramics, tiles and bricks; population (1986 est) 71,300.

Nuremberg /ˈnjuərəmbɜːg/ (German *Nürnberg*) industrial city (electrical and other machinery, precision instruments, textiles, toys) in Bavaria, Germany; population (1988) 467,000. From 1933 the Nuremberg rallies were held here, and in 1945 the Nuremberg trials of war criminals.
Created an imperial city 1219, it has an 11th–16th-century fortress and many medieval

buildings (restored after destruction of 75% of the city in World War II), including the home of the 16th-century composer Hans Sachs, where the Meistersingers met.

Nusa Tenggara /ˈnuːsə teŋˈɡɑːrə/ or *Lesser Sunda Island* volcanic archipelago in Indonesia, including Bali, Lombok, and Timor; area 73,144 sq km/28,241 sq mi. The islands form two provinces of Indonesia: *Nusu Tenggara Barat*, population (1980) 2,724,500; and *Nusu Tenggara Timur*, population (1980) 2,737,000.

Nuuk /nuːk/ Greenlandic for ◊Godthaab, the capital of Greenland.

NV abbreviation for ◊*Nevada*, a state of the USA.

NY abbreviation for ◊*New York*, a city and state of the USA.

Nyasa /niˈæsə/ former name for Lake ◊Malawi.

Nyasaland /niˈæsəlænd/ former name (until 1964) for ◊Malawi.

Nyíregyháza /ˈniːredʒˌhɑːzə/ market town in E Hungary; population (1988) 119,000. It trades in tobacco and vegetables.

NZ abbreviation for ◊*New Zealand*.

machinery, plastics, chemicals) and coalmining city in the Ruhr Valley, North Rhine–Westphalia, Germany; population (1988) 222,000.

Oberösterreich /ˈəubərˌɜːstəraɪʃ/ German name for the federal state of ◊Upper Austria.

Ocala /əuˈkælə/ city in N central Florida, USA, SE of Gainesville; seat of Marion County; population (1990) 42,000. It is a marketing and shipping centre for the citrus, poultry, cotton, and tobacco products grown in the surrounding area. Tourism is also vital to the economy.

Oceania /ˌəuʃiˈɑːniə/ general term for the islands of the central and S Pacific, including Australia, New Zealand, and the eastern half of New Guinea; although not strictly a continent, Oceania is often referred to as such to facilitate handling of global statistics
area 8,500,000 sq km/3,300,000 sq mi (land area)
largest cities (population over 500,000) Sydney, Melbourne, Brisbane, Perth, Adelaide, Auckland
features the Challenger Deep in the Mariana Trench −11,034 m/−36,201 ft is the greatest known depth of sea in the world; Ayers Rock in Northern Territory, Australia, is the world's largest monolith; the Great Barrier Reef is the longest coral reef in the world; Mount Kosciusko 2,229 m/7,316 ft in New South Wales is the highest peak in Australia; Mount Cook 3,764 m/ 21,353 ft is the highest peak in New Zealand; the Murray River in SE Australia is the longest river (2,590 km/1,609 mi)
highest point Mount Wilhelm, Papua New Guinea (4,509 m/14,793 ft)
lowest point Lake Eyre, South Australia (−16 m/ −52 ft)
physical stretching from the Tropic of Cancer in the N to the southern tip of New Zealand, Oceania can be broadly divided into groups of volcanic and coral islands on the basis of the ethnic origins of their inhabitants: Micronesia (Guam, Kiribati, Mariana, Marshall, Caroline Islands), Melanesia (Papua New Guinea, Vanuatu, New Caledonia, Fiji, Solomon Islands) and Polynesia (Tonga, Samoa, Line Islands, Tuvalu, French Polynesia, Pitcairn); Australia (the largest island in the world) occupies more than 90% of the land surface
products with a small home market, Oceania has a manufacturing sector dedicated to servicing domestic requirements and a large export-oriented sector, 70% of which is based on exports of primary agricultural or mineral products. Australia is a major producer of bauxite, nickel, silver, cobalt, gold, iron ore, diamonds, lead, and uranium; New Caledonia is a source of cobalt, chromite, and nickel; Papua New Guinea produces gold and copper. Agricultural products include coconuts, copra, palm oil, coffee, cocoa, phosphates (Nauru), rubber (Papua New Guinea), 40% of the world's wool (Australia, New Zealand); New Zealand and Australia are, respectively, the world's second and third largest producers of mutton and lamb; fishing and tourism are also major industries
population 26 million, rising to 30 million by

Oahu /əuˈɑːhuː/ island of Hawaii, USA, in the N Pacific
area 1,525 sq km/589 sq mi
towns Honolulu (state capital)
physical formed by two extinct volcanoes
features Waikiki beach; Pearl Harbor naval base; Diamond Head; punchbowl craters
products sugar, pineapples; tourism is a major industry
population (1988 est) 838,500.

Oakland /ˈəuklənd/ industrial port (vehicles, textiles, chemicals, food processing, shipbuilding) in California, USA, on the east coast of San Francisco Bay; population (1990) 372,200. It is linked by bridge (1936) with San Francisco. A major earthquake 1989 buckled the bay bridge and an Oakland freeway section, causing more than 60 deaths.

Oak Ridge /ˈəuk ˈrɪdʒ/ town in Tennessee, eastern USA, on the river Clinch, noted for the Oak Ridge National Laboratory (1943), which manufactures plutonium for nuclear weapons; population (1990) 27,300. The community was founded 1942 as part of the Manhattan Project to develop an atomic bomb; by the end of World War II its population was more than 75,000. Ownership of the community passed to the residents in the late 1950s.

Oaxaca /wəˈhɑːkə/ capital of a state of the same name in the Sierra Madre del Sur mountain range, central Mexico; population (1990) 212,900. Industries include food processing, textiles, and handicrafts.

Ob /ɒb/ river in Asian Russia, flowing 3,380 km/ 2,100 mi from the Altai Mountains through the W Siberian Plain to the Gulf of Ob in the Arctic Ocean. With its main tributary, the **Irtysh**, it is 5,600 km/3,480 mi long.
    Although frozen for half the year, and subject to flooding, the Ob is a major transport route. Novosobirsk and Barnaul are the main ports.

Oban /ˈəubən/ seaport and resort in Strathclyde, W Scotland; population (1981) 8,100.

Obeid, El /ˈəubeɪd/ see ◊El Obeid, a city in Sudan.

Oberhausen /ˈəubəˌhauzən/ industrial (metals,

the year 2000; annual growth rate from 1980 to 1985 1.5%; Australia accounts for 65% of the population

languages English, French (French Polynesia, New Caledonia, Wallis and Fatuna, Vanuatu); a wide range of indigenous Aboriginal, Maori, Melanesian, Micronesian, and Polynesian languages and dialects (over 700 in Papua New Guinea) are spoken

religions predominantly Christian; 30% of the people of Tonga adhere to the Free Wesleyan Church; 70% of the people of Tokelau adhere to the Congregational Church; French overseas territories are largely Roman Catholic.

Ocean Island /ˈəuʃən/ another name for Banaba, an island belonging to ◊Kiribati.

Ocussi Ambeno /ɒˈkuːsi mˈbeɪnəu/ port on the north coast of Indonesian West Timor, until 1975 an exclave of the Portuguese colony of East Timor. The port is an outlet for rice, copra, and sandalwood.

Odense /ˈəudənsə/ industrial port (shipbuilding, electrical goods, glass, textiles) on the island of Fyn, Denmark; population (1990) 176,100. It is the birthplace of Hans Christian Andersen.

Oder /ˈəudə/ (Polish *Odra*) European river flowing N from the Czech Republic to the Baltic Sea (the Neisse River is a tributary); length 885 Km/550 mi.

Odessa /ɒˈdesə/ seaport in Ukraine, on the Black Sea, capital of Odessa region; population (1989) 1,115,000. Products include chemicals, pharmaceuticals, and machinery.

Odessa was founded by Catherine II 1795 near the site of an ancient Greek settlement. Occupied by Germany 1941–44, it suffered severe damage under the Soviet scorched-earth policy and from German destruction.

Odessa /ɒuˈdesə/ city in W Texas, USA, SW of Big Springs; seat of Ector County; population (1990) 89,700. It grew up among the oilfields located here. Industries include petroleum products; it is also a livestock processing and shipping centre.

Offaly /ˈɒfəli/ county of the Republic of Ireland, in the province of Leinster, between Galway on the W and Kildare on the E; area 2,000 sq km/772 sq mi; population (1991) 58,500.

Towns include the county town of Tullamore. Features include the rivers Shannon (along the western boundary), Brosna, Clodagh, and Broughill and the Slieve Bloom mountains in the SE.

Offa's Dyke /ˈɒfəz/ defensive earthwork along the Welsh border, of which there are remains from the mouth of the river Dee to that of the river Severn. It represents the boundary secured by Offa's wars with Wales.

The dyke covered a distance of 240 km/149 mi, of which 130 km/81 mi are still standing.

Offenbach am Main /ˈɒfənbæx æm ˈmaɪn/ city in Hessen, Germany; population (1988) 111,900. It faces Frankfurt on the opposite side of the river Main.

Ogaden /ˌɒgəˈden/ desert region in Harar province, SE Ethiopia, that borders on Somalia. It is a desert plateau, rising to 1,000 m/3,280 ft, inhabited mainly by Somali nomads practising arid farming. The area became one of five new autonomous provinces created in Ethiopia 1987.

A claim to the area was made by Somalia in the 1960s, resulting in guerrilla fighting and major Somali advances during 1977. By 1980 Ethiopia, backed by the USSR and Cuba, was again in virtual control of the area, but armed clashes continued. In 1988 diplomatic relations were restored between Ethiopia and Somalia and troops were withdrawn from their shared border. Internal troubles in Somalia 1990 created a large refugee population in E Ogaden.

Ogallala Aquifer /ˌəugəˈlælə/ the largest source of groundwater in the USA, stretching from S South Dakota to NW Texas.

The over exploitation of this water resource resulted in the loss of more than 18% of the irrigated farmland of Oklahoma and Texas in the period 1940–90.

Ogbomosho /ˌɒgbəˈməuʃəu/ city and commercial centre in W Nigeria, 80 km/50 mi NE of Ibadan; population (1981) 590,600.

Ogden /ˈɒgdən/ city in N Utah, USA, on the Weber and Ogden rivers, N of Salt Lake City; population (1990) 63,900. It is a railway, trading, and military supply centre; Hill Air Force Base is nearby.

Ogun /ˈəugun/ state of SW Nigeria; area 16,762 sq km/6,474 sq mi; capital Abeokuta; population (1988) 3,397,900.

OH abbreviation for ◊*Ohio*, a state of the USA.

Ohio /əuˈhaɪəu/ state in N central USA; nickname Buckeye State

area 107,100 sq km/41,341 sq mi

capital Columbus

towns Cleveland, Cincinnati, Dayton, Akron, Toledo, Youngstown, Canton

features Ohio River; Lake Erie; Serpent Mound, a 1.3-m/ 4-ft embankment, 405 m/1,330 ft long and about 5 m/18 ft across (built by Hopewell Indians about 2nd–1st centuries BC)

products coal, cereals, livestock, dairy foods, machinery, chemicals, steel, motor vehicles, automotive and aircraft parts, rubber products, office equipment, refined petroleum

Ohio

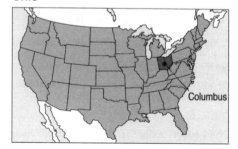

Columbus

population (1990) 10,847,100
famous people Thomas Edison, John Glenn,
Paul Newman, General Sherman, Orville
Wright; six presidents (Garfield, Grant, Harding,
Harrison, Hayes, and McKinley)
history explored for France by La Salle 1669;
ceded to Britain by France 1763; first settled at
Marietta (capital of the Northwest Territory) by
Europeans 1788; became a state 1803.

Ohio /əʊˈhaɪəʊ/ river in the USA, 1,580 km/980
mi long; it is formed by the union of the
Allegheny and Monongahela at Pittsburgh,
Pennsylvania, and flows SW until it joins the river
Mississippi at Cairo, Illinois.

Ohrid, Lake /ˈɒxrɪd/ lake on the frontier between
Albania and the Former Yugoslav Republic of
Macedonia; area 350 sq km/135 sq mi.

Oise /waːz/ European river that rises in the
Ardennes plateau, Belgium, and flows SW
through France for 300 km/186 mi to join the
Seine about 65 km/40 mi below Paris. It gives its
name to a French *département* in Picardie.

OK abbreviation for ♦*Oklahoma*, a state of the
USA.

Okavango Swamp /ˌɒkəˈvæŋgəʊ/ marshy area in
NW Botswana, fed by the **Okavango River**,
which rises in Angola and flows SE about 1,600
km/1,000 mi. It is an important area for wildlife.

Okayama /ˌɒkəˈjɑːmə/ industrial port (textiles,
cotton) in the SW of Honshu Island, Japan;
population (1990) 593,700.

Okeechobee /ˌəʊkiˈtʃəʊbi/ lake in the N
Everglades, Florida, USA; 65 km/40 mi long and
40 km/25 mi wide. It is the largest lake in the
southern USA, about 1,800 sq km/700 sq mi.

Okefenokee /ˌəʊkɪfɪˈnəʊki/ swamp in SE Georgia
and NE Florida, USA, rich in alligators, bears,
deer, and birds. Much of its 1,700 sq km/660 sq
mi forms a natural wildlife refuge. It is drained by
the St Marys and Suwannee rivers.

Okhotsk, Sea of /əʊˈxɒtsk/ arm of the N Pacific
Ocean between the Kamchatka Peninsula and
Sakhalin and bordered to the S by the Kuril
Islands; area 937,000 sq km/361,700 sq mi. It is
free of ice only in summer, and is often fogbound.

Okinawa /ˌɒkɪˈnɑːwə/ group of islands, forming
part of the Japanese Ryukyu Islands in the W
Pacific; the largest island is Okinawa
area 2,250 sq km/869 sq mi
capital Naha
features Okinawa, the largest island of the group
(area 1,176 sq km/453 sq mi; population (1990)
105,852), has a large US military base
population (1990) 3,145,500
history virtually all buildings were destroyed in
World War II. The principal island, Okinawa,
was captured by the USA in the **Battle of
Okinawa** 1 April–21 June 1945, with 47,000 US
casualties (12,000 dead) and 60,000 Japanese
(only a few hundred survived as prisoners).
During the invasion over 150,000 Okinawans,
mainly civilians, died; many of them were

Oklahoma

massacred by Japanese forces. The island was
returned to Japan 1972.

Oklahoma /ˌəʊkləˈhəʊmə/ state in S central USA;
nickname Sooner State
area 181,100 sq km/69,905 sq mi
capital Oklahoma City
towns Tulsa, Lawton, Norman, Enid
features Arkansas, Red, and Canadian rivers;
Wichita and Ozark mountain ranges; the high
plains have Indian reservations (Cherokee,
Chickasaw, Choctaw, Creek, and Seminole);
American Indian Hall of Fame; Chicasaw
National Recreation Area
products cereals, peanuts, cotton, livestock, oil,
natural gas, helium, machinery and other metal
products
population (1990) 3,145,600
famous people John Berryman, Ralph Ellison,
Woody Guthrie, Mickey Mantle, Will Rogers,
Jim Thorpe
history explored for Spain by Francisco de
Coronado 1541; most acquired by the USA from
France with the Louisiana Purchase 1803.
   Part of the present state formed the Territory of
Oklahoma from 1890, and was thrown open to
settlers with lotteries and other hurried methods of
distributing land. Together with what remained of
Indian Territory, it became a state 1907.

Oklahoma City /ˌəʊkləˈhəʊmə/ industrial city (oil
refining, machinery, aircraft, telephone equipment), capital of Oklahoma, USA, on the
Canadian River; population (1990) 444,700. On
22 April 1889, a tent city of nearly 10,000
inhabitants was set up overnight as the area was
opened to settlement. In 1910 Oklahoma City
had 64,000 people and became the state capital.

Oldenburg /ˈəʊldənbɜːg/ industrial city in Lower
Saxony, Germany, on the river Hunte;
population (1988) 140,400. It is linked by river
and canal to the Ems and Wieser rivers.

Oldham /ˈəʊldəm/ industrial city in Greater
Manchester, England; population (1991)
211,400. Industries include textiles and textile
machinery, plastics, electrical goods, and
electronic equipment.

Olduvai Gorge /ˈɒlduvaɪ ˈgɔːdʒ/ deep cleft in the
Serengeti steppe, Tanzania, where Louis and
Mary Leakey found prehistoric stone tools in the
1930s. They discovered Pleistocene remains of
prehumans and gigantic animals 1958–59.

The gorge has given its name to the **Olduvai culture**, a simple stone-tool culture of prehistoric hominids, dating from 2–0.5 million years ago. The Pleistocene remains include sheep the size of a carthorse, pigs as big as rhinoceroses, and a gorilla-sized baboon. The skull of an early hominid (1.75 million years old), *Australopithecus boisei* (its massive teeth earned it the nickname 'Nutcracker Man') was also found here, as well as remains of *Homo habilis* and primitive types of *Homo erectus*.

Old World the continents of the eastern hemisphere, so called because they were familiar to Europeans before the Americas.

Olmos /'ɒlmɒs/ small town on the edge of the Sechura Desert, NW Peru. It gives its name to the large-scale **Olmos Project** which began 1926 in an attempt to irrigate the desert plain and increase cotton and sugar-cane production.

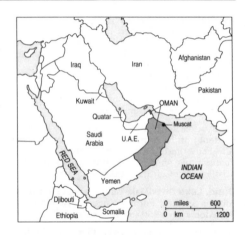

Olney /'əʊlni/ small town in Buckinghamshire, England, where every Shrove Tuesday local women run a pancake race.

Olomouc /'ɒləmaʊts/ industrial city in the Czech Republic, at the confluence of the Bystrice and Morava rivers; population (1991) 105,700. Industries include sugar refining, brewing, and metal goods.

Olsztyn /'ɒlʃtɪn/ formerly **Allenstein** industrial town in NE Poland, at the centre of the Mazurian Lakes region; population (1990) 162,900. It was founded 1334 and was formerly in East Prussia.

Olympia /ə'lɪmpɪə/ capital of Washington, USA, located in the west central part of the state, on the Deschutes River near Puget Sound; population (1990) 30,800. It is a deep-water port; fishing and tourism are important to the economy.

Olympus /ə'lɪmpəs/(Greek **Olimbos**) any of several mountains in Greece and elsewhere, one of which is **Mount Olympus** in N Thessaly, Greece, 2,918 m/9,577 ft high. In ancient Greece it was considered the home of the gods.

Omagh /əʊ'mɑː/ county town of Tyrone, Northern Ireland, on the river Strule, 48 km/30 mi S of Londonderry; population (1981) 14,600. Industries include footwear and engineering; there is salmon fishing.

Omaha /'əʊməhɑː/ city in E Nebraska, USA, on the Missouri River; population (1990) 335,800. It is a livestock-market centre, with food-processing and meat-packing industries.

Oman Sultanate of (*Saltanat 'Uman*)
area 272,000 sq km/105,000 sq mi
capital Muscat
towns Salalah, Nizwa
physical mountains to N and S of a high arid plateau; fertile coastal strip
features Jebel Akhdar highlands; Kuria Muria islands; Masirah Island is used in aerial reconnaissance of the Arabian Sea and Indian Ocean; exclave on Musandam Peninsula controlling Strait of Hormuz
head of state and government Qaboos bin Said from 1970

political system absolute monarchy
political parties none
exports oil, dates, silverware, copper
currency rial Omani
population (1990 est) 1,305,000; growth rate 3.0% p.a.
life expectancy men 55, women 58 (1989)
languages Arabic (official), English, Urdu, other Indian languages
religions Ibadhi Muslim 75%, Sunni Muslim, Shi'ite Muslim, Hindu
literacy 20% (1989)
GNP $7.5 bn (1987); $5,070 per head (1988)
chronology
1951 The Sultanate of Muscat and Oman achieved full independence from Britain. Treaty of Friendship with Britain signed.
1970 After 38 years' rule, Sultan Said bin Taimur replaced in coup by his son Qaboos bin Said. Name changed to Sultanate of Oman.
1975 Left-wing rebels in south defeated.
1982 Memorandum of Understanding with UK signed, providing for regular consultation on international issues.
1985 Diplomatic ties established with USSR.
1991 Sent troops to Operation Desert Storm, as part of coalition opposing Iraq's occupation of Kuwait.

Omdurman /ˌɒmdə'mɑːn/ city in Sudan, on the White Nile, a suburb of Khartoum; population (1983) 526,000. It was the residence of the Sudanese sheik known as the Mahdi 1884–98.

Omsk /ɒmsk/ industrial city (agricultural and other machinery, food processing, sawmills, oil refining) in Russia, capital of Omsk region, W Siberia; population (1987) 1,134,000. Its oil refineries are linked with Tuimazy in the Bashkir republic by a 1,600-km/1,000-mi pipeline.

Onega, Lake /əʊ'neɪgə/ second-largest lake in Europe, NE of St Petersburg, partly in Karelia, Russia; area 9,600 sq km/3,710 sq mi. The **Onega Canal**, along its S shore, is part of the Mariinsk system linking St Petersburg with the river Volga.

Oneida /əʊ'naɪdə/ town in New York State, USA, named after the Oneida people (a nation of the

Ontario

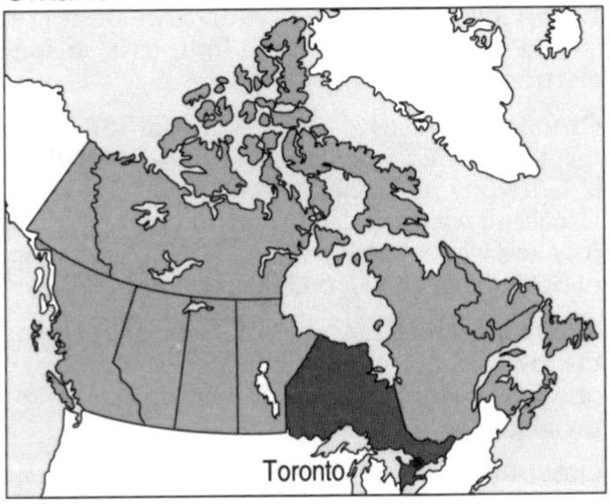

Iroquois confederacy). From 1848 the ***Oneida Community***, a religious sect, practised a form of 'complex marriage' until its dissolution 1879.

Ont. abbreviation for ⇨***Ontario***, a Canadian province.

Ontario /ɒn'teəriəu/ province of central Canada
area 1,068,600 sq km/412,480 sq mi
capital Toronto
towns Hamilton, Ottawa (federal capital), London, Windsor, Kitchener, St Catharines, Oshawa, Thunder Bay, Sudbury
features Black Creek Pioneer Village; Niagara Falls; richest, chief manufacturing, most populated, and leading cultural province of English-speaking Canada
products nickel, iron, gold, forest products, motor vehicles, iron, steel, paper, chemicals, copper, uranium
population (1986) 9,114,000
history first explored by the French in the 17th century, it came under British control 1763 (Treaty of Paris).

An attempt 1841 to form a merged province with French-speaking Québec failed, and Ontario became a separate province of Canada 1867. Under the protectionist policies of the new federal government, Ontario gradually became industrialized and urban. Since World War II, more than 2 million immigrants, chiefly from Europe, have settled in Ontario.

Ontario, Lake /ɒn'teəriəu/ smallest and easternmost of the Great Lakes, on the US–Canadian border; area 19,200 sq km/7,400 sq mi. It is connected to Lake Erie by the Welland Canal and the Niagara River, and drains into the St Lawrence River. Its main port is Toronto.

It has the merit, from the shore, of producing
a slight ambiguity of vision. It is the sea,
and yet just not the sea.

On **Lake Ontario** Henry James
Portraits of Places 1883

Oostende /əust'endə/ Flemish form (meaning 'east end') of ⇨Ostend.

Opole /ɒ'pəuleɪ/ industrial town in S Poland, on the river Oder; population (1990) 128,400. It is an agricultural centre; industries include textiles, chemicals, and cement.

Oporto /ə'pɔːtəu/ alternative form of ⇨Porto, a city in Portugal.

OR abbreviation for ⇨***Oregon***, a state of the USA.

Oradea /ɒ'rɑːdiə/ or ***Oradea-Mare*** industrial city in Romania, on the river Koös; population (1985) 208,500. Industries include agricultural machinery, chemicals, non-ferrous metallurgy, leather goods, printing, glass, textiles, clothing, and brewing.

Created the seat of a bishopric by St Ladislas 1083, Oradea was destroyed by the Turks 1241 and rebuilt. Many of its buildings date from the reign of Maria Theresa in the 18th century. It was ceded by Hungary to Romania 1919 and held by Hungary 1940–45.

Oran /ɔː'rɑːn/ (Arabic ***Wahran***) seaport in Algeria; population (1983) 663,500. Products include iron, textiles, footwear, and processed food; the port trades in grain, wool, vegetables, and native esparto grass.

Oran was part of the Ottoman Empire, except when it was under Spanish rule 1509–1708 and 1732–91. It was occupied by France 1831. After the surrender of France to Germany 1940, the French warships in the naval base of Mers-el-Kebir nearby were put out of action by the British navy to prevent them from falling into German hands.

Orange /'ɒrɪndʒ/ river in South Africa, rising on the Mont aux Sources in Lesotho and flowing W to the Atlantic Ocean; length 2,100 km/1,300 mi. It runs along the southern boundary of the Orange Free State and was named 1779 after William of Orange. Water from the Orange is diverted via the Orange-Fish River Tunnel (1975) to irrigate the semi-arid E Cape Province.

Orange /ɒ'rɒnʒ/ town in France, N of Avignon; population (1990) 28,100. It has the remains of a Roman theatre and arch. It was a medieval principality from which came the European royal house of Orange.

Orange /'ɒrɪndʒ/ town in New South Wales, Australia, 200 km/125 mi NW of Sydney; population (1987) 32,500. There is a woollen-textile industry based on local sheep flocks, and fruit is grown.

Orange County /'ɒrɪndʒ/ metropolitan area of S California, USA; area 2,075 sq km/801 sq mi; it adjoins Los Angeles County; population (1980) 1,932,700. Industries include aerospace and electronics. Oranges and strawberries are grown. Disneyland is here, and Santa Ana is the chief town.

Orange Free State /'ɒrɪndʒ ˌfriː 'steɪt/ province of the Republic of South Africa
area 127,993 sq km/49,405 sq mi
capital Bloemfontein

towns Springfontein, Kroonstad, Bethlehem, Harrismith, Koffiefontein
features plain of the High Veld; Lesotho forms an enclave on the Natal–Cape Province border
products grain, wool, cattle, gold, oil from coal, cement, pharmaceuticals
population (1987) 1,863,000; 82% ethnic Africans
history original settlements from 1810 were complemented by the Great Trek, and the state was recognized by Britain as independent 1854. Following the South African, or Boer, War 1899–1902, it was annexed by Britain until it entered the union as a province 1910.

Orasul Stalin /ˈɔːrəsuːl ˈstɑːlɪn/ name 1948–56 of the Romanian town ⟫Braşov.

Ordzhonikidze former name (1954–91) of ⟫Vladikavkaz, the capital of North Ossetia in SW Russia.

Oregon /ˈɒrɪɡən/ state in northwestern USA, on the Pacific coast; nickname Beaver State
area 251,500 sq km/97,079 sq mi
capital Salem
towns Portland, Eugene
population (1990) 2,842,300
features fertile Willamette river valley; rivers: Columbia, Snake; Crater Lake, deepest in the USA (589 m/1,933 ft); mountains: Coast and Cascades; Oregon Dunes National Recreation Area, on Pacific coast
products wheat, livestock, timber, electronics
famous people Linus Pauling
history settled 1811 by the Pacific Fur Company, Oregon Territory included Washington until 1853; Oregon became a state 1859. The Oregon Trail (3,200 km/2,000 mi from Independence, Missouri, to the Columbia river) was the pioneer route across the USA 1841–60.

Oregon ... a pleasant, homogeneous, self-contained state, filled with pleasant, homogeneous, self-contained people ...

On **Oregon** Arthur M Schlesinger Jr. *Robert Kennedy and His Times* 1978

Orel /aˈrjɔl/ industrial city in Russia, capital of Orel region, on the river Oka, 320 km/200 mi SSW of Moscow; population (1987) 335,000. Industries include engineering, textiles, and foodstuffs.

Oregon

Orem city in N central Utah, USA, SE of Salt Lake City; population (1990) 67,500. It was settled by Mormons 1861. Industries include electronics and steel.

Orenburg /ˈɒrənbɜːɡ/ city in S central Russia, on the Ural River; population (1989) 547,000. It is a trading and mining centre and capital of Orenburg region. It dates from the early 18th century and was called **Chkalov** 1938–57 in honour of Soviet aviator Valeri Chkalov (1904–1938).

Orense /ɒˈrenseɪ/ town in NW Galicia, Spain, on the river Miño; population (1991) 107,200. It produces textiles, furniture, food products, and metal goods.

Öresund /ˌɜːrəˈsund/ strait between Sweden and Denmark; in English it is called the Sound.

Orinoco /ˌɒrɪˈnəʊkəʊ/ river in N South America, flowing for about 2,400 km/1,500 mi through Venezuela and forming for about 320 km/200 mi the boundary with Colombia; tributaries include the Guaviare, Meta, Apure, Ventuari, Caura, and Caroni.

It is navigable by large steamers for 1,125 km/700 mi from its Atlantic delta; rapids obstruct the upper river.

Orissa /ɒˈrɪsə/ state of NE India
area 155,800 sq km/60,139 sq mi
capital Bhubaneswar
towns Cuttack, Rourkela
features mainly agricultural; Chilka Lake with fisheries and game; temple of Jagannath or Juggernaut at Puri
products rice, wheat, oilseed, sugar, timber, chromite, dolomite, graphite, iron
population (1991) 31,512,000
language Oriya (official)
religion 90% Hindu
history administered by the British 1803–1912 as a subdivision of Bengal, it joined with Bihar to become a province. In 1936 Orissa became a separate province, and in 1948–49 its area was doubled before its designation as a state 1950.

Orissa

Orizaba /ˌɒrɪ'sɑːbə/ industrial city (brewing, paper, and textiles) and resort in Veracruz state, Mexico; population (1980) 115,000. An earthquake severely damaged it 1973.

Orizaba /ˌɒrɪ'sɑːbə/ Spanish name for ▷Citlaltépetl, a mountain in Mexico.

Orkney Causeway /'ɔːkni/ or **Churchill Barriers** construction in N Scotland put up in World War I, completed 1943, joining four of the Orkney Islands. It was built to protect the British fleet from intrusion through the eastern entrances to Scapa Flow. The Orkney Causeway links the east mainland with the islands of Lambholm, Glimsholm, Burray, and South Ronaldsay.

Orkney Islands /'ɔːkni/ island group off the northeast coast of Scotland
area 970 sq km/375 sq mi
towns Kirkwall (administrative headquarters), on Mainland (Pomona)
features comprises about 90 islands and islets, low-lying and treeless; mild climate owing to the Gulf Stream; Skara Brae, a well-preserved Neolithic village on Mainland. Population, long falling, has in recent years risen as the islands' remoteness from the rest of the world attracts new settlers. Scapa Flow, between Mainland and Hoy, was a naval base in both world wars, and the German fleet scuttled itself here 21 June 1919
products fishing and farming, wind power (Burgar Hill has the world's most productive wind-powered generator; a 300 KW wind turbine with blades 60 m/197 ft diameter, capable of producing 20% of the islands' energy needs)
population (1989) 19,400
famous people Edwin Muir, John Rae
history Harald I (Fairhair) of Norway conquered the islands 876; they were pledged to James III of Scotland 1468 for the dowry of Margaret of Denmark and annexed by Scotland (the dowry unpaid) 1472.

Orkney Islands

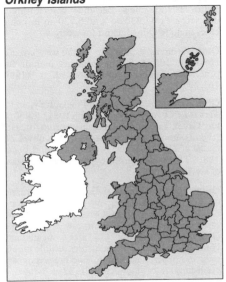

Orkneys, South /'ɔːkniz/ islands in the British Antarctic Territory; see ▷South Orkney Islands.

Orlando /ɔː'lændəu/ industrial city in Florida, USA; population (1990) 164,700. It is a winter resort and tourist centre, with Walt Disney World and the Epcot Center nearby. Electronic and aerospace equipment are manufactured in the city, and citrus-fruit products are processed here. Educational institutions include the University of Central Florida. Orlando was settled 1843.
The city was named 1857 after Orlando Reeves, a soldier killed in a clash with Indians.

Orléans /ɔː'liɒnz/ industrial city of France, on the river Loire; 115 km/70 mi SW of Paris; population (1990) 108,000. It is the capital of Loiret *département*. Industries include engineering and food processing.
Orléans, of pre-Roman origin and formerly the capital of the old province of Orléanais, is associated with Joan of Arc, who liberated it from English rule 1429.

Orly /'ɔːli/ suburb of Paris in the *département* of Val-de-Marne; population (1990) 21,800. Orly international airport is the busiest in France.

Ormuz /'ɔːmuz/ alternative name for the Iranian island, ▷Hormuz.

Orne /ɔːn/ French river rising E of Sées and flowing NW, then NE to the English Channel below Caen; 152 km/94 mi long. A ship canal runs alongside it from Caen to the sea at Ouistreham. The Orne gives its name to a *département* in Normandy; population (1990) 293,200.

Orontes /ɒ'rɒntiːz/ (Arabic *'Asi*) river flowing from Lebanon through Syria and Turkey to the Mediterranean Sea and used mainly for irrigation; length 400 km/250 mi.

Orsk /ɔːsk/ industrial city in S central Russia, at the junction of the Or and Ural rivers; population (1987) 273,000. Industries include mining, oil refining, locomotives, and aluminium. Its refineries are fed by a pipeline from Guriev. The town was originally a fortress.

Orvieto /ˌɔːvi'etəu/ town in Umbria, Italy, NE of Lake Bolsena, population (1981) 22,800. Built on the site of *Volsinii*, an Etruscan town destroyed by the Romans 280 BC, Orvieto has many Etruscan remains. The name is from Latin *Urbs Vetus* meaning 'old town'.

Osaka /əu'sɑːkə/ industrial port (iron, steel, shipbuilding, chemicals, textiles) on Honshu Island, Japan; population (1990) 2,623,800, metropolitan area 8,000,000. It is the oldest city of Japan and was at times the seat of government in the 4th–8th centuries.
Lying on a plain sheltered by hills and opening onto Osaka Bay, Osaka is honeycombed with waterways. It is a tourist centre for Kyoto and the Seto Inland Sea and is linked with Tokyo by fast electric train (travelling up to 200 kph/124 mph). An underground shopping and leisure centre (1951) has been used as a model for others throughout Japan. It was a mercantile centre in

the 18th century, and in the 20th century set the pace for Japan's revolution based on light industries.

In trade it is a Chicago. In situation it is a Venice.

On **Osaka** John Foster Fraser *Round the World on a Wheel* 1899

Oshkosh /ˈɒʃkɒʃ/ city in E central Wisconsin, USA, where the Fox River flows into Lake Winnebago, NW of Milwaukee; seat of Winnebago County; population (1990) 55,000. Industries include clothing, machinery, lumber, and electronics.

Oshogbo /ɒˈʃɒgbəʊ/ city and trading centre on the river Osua, in W Nigeria, 200 km/125 mi NE of Lagos; population (1986) 405,000. It is the capital of Osua State. Industries include cotton and brewing.

Osijek /ˈɒsiek/ (German *Esseg*) industrial port in Croatia, on the river Drava; population (1981) 158,800. Industries include textiles, chemicals, and electrical goods.

Oslo /ˈɒzləʊ/ capital and industrial port (textiles, engineering, timber) of Norway; population (1991) 461,600. The first recorded settlement was made in the 11th century by Harald III, but after a fire 1624, it was entirely replanned by Christian IV and renamed **Christiania** 1624–1924.

The port is built at the head of Oslo fjord, which is kept open in winter by icebreakers. There is a Viking museum, the 13th-century Akershus Castle, a 17th-century cathedral, and the National Gallery, which includes many paintings by Edvard Munch.

Osnabrück /ˌɒznəˈbrʊk/ industrial city in Lower Saxony, Germany; 115 km/71 mi W of Hanover; population (1988) 154,000. Industries include engineering, iron, steel, textiles, clothing, paper, and food processing. The Treaty of Westphalia was signed at Osnabrück and Münster 1648, ending the Thirty Years' War.

Osnabrück bishopric was founded by Charlemagne 783. A type of rough fabric, *osnaburg*, was originally made here.

Ossa /ˈɒsə/ mountain in Thessaly, Greece; height 1,978 m/6,490 ft. In mythology, two of Poseidon's giant sons were said to have tried to dislodge the gods from Olympus by piling nearby Mount Pelion on top of Ossa to scale the great mountain.

Ossa, Mount /ˈɒsə/ the highest peak on the island of Tasmania, Australia; height 1,617 m/5,250 ft.

Ossetia /ɒˈsiːʃə/ region in the Caucasus, on the border between Russia and Georgia. It is inhabited by the Ossets, who speak the Iranian language Ossetic, and who were conquered by the Russians 1802. Some live in *North Ossetia*, an autonomous republic in southwestern Russia; area 8,000 sq km/3,088 sq mi; population (1989)

634,000; capital Vladikavkaz (formerly Ordzhonikidze). The rest live in *South Ossetia*, an autonomous region of the Georgian republic, population (1989) 99,000; capital Tshkinvali. The region has been the scene of Osset–Georgian interethnic conflict since 1989.

The Ossets demanded that South Ossetia be upgraded to an autonomous republic as a preliminary to reunification with North Ossetia, and in Sept 1990 declared the region independent of the USSR (a claim rejected by President Gorbachev). Separatist moves were violently suppressed by Georgian authorities and states of emergency imposed Dec 1990 in the capital, Tskhinvali, and in the town of Dzhava, following clashes between police and Ossets. During 1991 several hundred lives were claimed in interethnic gun battles. The violence escalated June 1992, when armed South Ossetian nationalists launched a drive to unite the two regions, spreading violence to North Ossetia and heightening tensions between Georgia and Russia. A peace accord reached 24 June had little effect, and from July Georgian and Russian peace-keeping troops were deployed.

Ostend /ɒstˈend/ (Flemish *Oostende*) seaport and pleasure resort in W Flanders, Belgium; 108 km/67 mi NW of Brussels; population (1989) 68,400. There are large docks, and the Belgian fishing fleet has its headquarters here. There are ferry links to Dover and Folkestone, England.

Ostrava /ˈɒstrəvə/ industrial city (iron works, furnaces, coal, chemicals) in the Czech Republic, capital of Severomoravsky region, NE of Brno; population (1991) 327,600.

Oswestry /ˈɒzwəstri/ market town in Shropshire, England; population (1981) 13,100. Industries include agricultural machinery and plastics. It is named after St Oswald, killed here 642.

Otago /əʊˈtɑːgəʊ/ peninsula and coastal plain on South Island, New Zealand, constituting a district; area 64,230 sq km/25,220 sq mi; chief cities include Dunedin and Invercargill.

Otaru /əʊˈtɑːruː/ fishing port on the west coast of Hokkaido Island, Japan; industries include fish processing, paper, sake; population (1990) 163,200.

Otranto /ɒˈtræntəʊ/ seaport in Puglia, Italy, on the *Strait of Otranto*; population (1981) 5,000. It has Greek and Roman remains, a ruined castle (the inspiration for Horace Walpole's novel *The Castle of Otranto* 1764), and a cathedral begun 1080. The port is linked by ferry with the island of Corfu.

Ottawa /ˈɒtəwə/ capital of Canada, in E Ontario, on the hills overlooking the Ottawa River and divided by the Rideau Canal into the Upper (western) and Lower (eastern) towns; population (1986) 301,000, metropolitan area (with adjoining Hull, Québec) 819,000. Industries include timber, pulp and paper, engineering, food processing, and publishing. It was founded 1826–32 as Bytown, in honour of John By (1781–1836), whose army engineers were building the Rideau Canal. It was renamed 1854 after the Outaouac Indians.

Features include the National Museum, National Art Gallery, Observatory, Rideau Hall (the governor general's residence), and the National Arts Centre 1969 (with an orchestra and English/French theatre). In 1858 it was chosen by Queen Victoria as the country's capital.

A sub-Arctic lumber village converted by royal mandate into a political cockpit.

On **Ottawa** Goldwin Smith about 1880

Otztal Alps /ˈɜːtstɑːl/ range of the Alps in Italy and Austria, rising to 3,774 m/12,382 ft at Wildspitze, Austria's second highest peak.

Ouagadougou /ˌwæɡəˈduːɡuː/ capital and industrial centre of Burkina Faso; population (1985) 442,000. Products include textiles, vegetable oil, and soap. The city has the palace of Moro Naba, emperor of the Mossi people, a neo-Romanesque cathedral, and a central avenue called the Champs Elysées. It was the capital of the Mossi empire from the 15th century.

Oudenaarde /ˈuːdənɑːd/ town of E Flanders, W Belgium, on the river Scheldt, 28 km/18 mi SSW of Ghent; population (1991) 27,200. It is a centre of tapestry-making and carpet-weaving. Oudenaarde was the site of the victory by the British, Dutch, and Austrians over the French 1708 during the War of the Spanish Succession.

Oudh /aud/ region of N India, now part of Uttar Pradesh. An independent kingdom before it fell under Mogul rule, Oudh regained independence 1732–1856, when it was annexed by Britain. Its capital was Lucknow, centre of the Indian Mutiny 1857–58. In 1877 it was joined with Agra, from 1902 as the United Provinces of Agra and Oudh, renamed Uttar Pradesh 1950.

Ouessant /ˈwesɒn/ French form of ▷Ushant, an island W of Brittany.

Oujda /uːˈdʒɑː/ industrial and commercial city (lead and coalmining) in N Morocco, near the border with Algeria; population (1982) 471,000. It trades in wool, grain, and fruit.

Oulu /ˈəuluː/ (Swedish *Uleåborg*) industrial port (saw mills, tanneries, shipyards) in W Finland, on the Gulf of Bothnia; population (1990) 101,400. It was originally a Swedish fortress 1375.

Ouse /uːz/ (Celtic 'water') any of several British rivers. The **Great Ouse** rises in Northamptonshire and winds its way across 250 km/160 mi to enter the Wash N of King's Lynn. A large sluice across the Great Ouse, near King's Lynn, was built as part of extensive flood-control works 1959. The **Little Ouse** flows for 38 km/24 mi along part of the Norfolk/Suffolk border and is a tributary of the Great Ouse. The Yorkshire **Ouse** is formed by the junction of the Ure and Swale near Boroughbridge and joins the river Trent to form the Humber. The Sussex

Ouse rises between Horsham and Cuckfield and flows through the South Downs to enter the English Channel at Newhaven.

The Sussex Ouse derives its name from the town of Lewes, the river being known in early charters as *aqua de Lewes*, later misunderstood as 'de l'Ouse'.

Ovamboland /əuˈvæmbəulænd/ region of N Namibia stretching along the Namibia–Angola frontier; the scene of conflict between SWAPO guerrillas and South African forces in the 1970s and 1980s.

Ovens River /ˈʌvənz/ river in Victoria, Australia, a tributary of the Murray.

Overijssel /ˌəuvərˈaisəl/ province of the E central Netherlands
area 3,340 sq km/1,289 sq mi
capital Zwolle
towns Enschede, Hengelo, Deventer
physical it is generally flat and contains the rivers IJssel and Vecht
products livestock, dairy products, textiles
population (1991) 1,026,300
history ruled by the bishops of Utrecht during the Middle Ages, Overijssel was sold to Charles V of Spain 1527. Joining the revolt against Spanish authority, it became one of the United Provinces of the Netherlands 1579.

Oviedo /ˌɒviˈeidəu/ industrial city (textiles, metal goods, pharmaceuticals, matches, chocolate, sugar) and capital of Asturias region, Spain, 25 km/16 mi S of the Bay of Biscay; population (1991) 203,200.

Owen Falls /ˈəuin/ waterfall in Uganda on the White Nile, 4 km/2.5 mi below the point at which the river leaves Lake Victoria. A dam, built 1949–60, provides hydroelectricity for Uganda and Kenya and helps to control the flood waters.

Owensboro /ˈəuinzbʌrə/ city in NW Kentucky, USA, on the Ohio River, SW of Louisville; seat of Davies County; population (1990) 53,500. Industries include bourbon (whisky), electronics, tobacco, and steel.

Oxford /ˈɒksfəd/ university city and administrative centre of Oxfordshire in S central England, at the confluence of the rivers Thames and Cherwell; population (1991) 109,000.

Oxford University has 40 colleges, the oldest being University College (1249). Other notable buildings are the Bodleian Library (1488), the Ashmolean Museum (1683), and Christopher Wren's Sheldonian Theatre (1664–68).

Industries include motor vehicles at Cowley, steel products, electrical goods, paper, publishing, and English language schools. Tourism is important.

One of the supreme gratifications of travel, the perfect prose of Gothic.

On **Oxford** Henry James *English Hours* 1905

Oxfordshire

Oxfordshire /ˈɒksfədʃə/ county in S central England
area 2,610 sq km/1,007 sq mi
towns Oxford (administrative headquarters), Abingdon, Banbury, Henley-on-Thames, Witney, Woodstock

features river Thames and tributaries; Cotswolds and Chiltern Hills; Vale of the White Horse (chalk hill figure at Uffington, 114 m/374 ft long); Oxford University; Blenheim Palace, Woodstock (started 1705 by Vanbrugh with help from Nicholas Hawksmoor, completed 1722); Europe's major fusion project JET (Joint European Trust) at the UK Atomic Energy Authority's fusion laboratories at Culham
products cereals, cars, paper, bricks, cement
population (1991) 553,800
famous people William Davenant, Flora Thompson, Winston Churchill.

Oxnard /ˈɒksnɑːd/ city in SW California, USA, NW of Los Angeles; population (1990) 142,200. Industries include paper products, aircraft parts, and oil refining.

Oxus /ˈɒksəs/ ancient name of ◊Amu Darya, a river in Russia.

Ozark Mountains /ˈəʊzɑːk/ area in the USA (shared by Arkansas, Illinois, Kansas, Mississippi, Oklahoma) of ridges, valleys, and streams; highest point only 700 m/2,300 ft; area 130,000 sq km/50,000 sq mi. This heavily forested region between the Missouri and Arkansas rivers has agriculture and lead and zinc mines.

166,242,500 sq km/64,170,000 sq mi; average depth 4,188 m/13,749 ft; greatest depth of any ocean 11,034 m/36,210 ft in the Mariana Trench.

That great sea, miscalled the Pacific.

On the **Pacific Ocean** Charles Darwin
*Journal...During the Voyage
...of HMS Beagle* 1832–36

Padang /'pɑːdæŋ/ port on the west coast of Sumatra, Indonesia; population (1980) 481,000. The Dutch secured trading rights here 1663. The port trades in copra, coffee, and rubber.

Paderborn /,pɑːdə'bɔːn/ market town in North Rhine–Westphalia, Germany; population (1988) 110,000. Industries include leather goods, metal products, and precision instruments. It was the seat of a bishopric in Charlemagne's time and later became a member of the Hanseatic League.

Padua /'pædjuə/(Italian **Padova**) city in N Italy, 45 km/28 mi W of Venice; population (1988) 224,000. The astronomer Galileo taught at the university, founded 1222.

The 13th-century Palazzo della Ragione, the basilica of Sant' Antonio, and the botanical garden laid out 1545 are notable. Padua is the birthplace of the Roman historian Livy and the painter Andrea Mantegna.

Pagalu /pə'gɑːluː/ former name (1973–79) of ◊Annobón, an island in Equatorial Guinea.

Pago Pago /'pɑːŋgəu 'pɑːŋgəu/ chief port of American Samoa on the island of Tutuila; population (1980) 3,060. Formerly a naval coaling station, it was acquired by the USA under a commercial treaty with the local king 1872.

Pahang /pə'hʌŋ/ state of E Peninsular Malaysia; capital Kuantan; area 36,000 sq km/13,896 sq mi; population (1980) 799,000. It is mountainous and forested and produces rubber, tin, gold, and timber. There is a port at Tanjung Gelang. Pahang is ruled by a sultan.

Pahsien /,pɑː'ʃjen/ alternative name of ◊*Chongqing*, a port in SW China.

Pakistan Islamic Republic of
area 796,100 sq km/307,295 sq mi; one-third of Kashmir under Pakistani control
capital Islamabad
towns Karachi, Lahore, Rawalpindi, Peshawar
physical fertile Indus plain in E, Baluchistan plateau in W, mountains in N and NW
environment about 68% of irrigated land is waterlogged or suffering from salinization
features the 'five rivers' (Indus, Jhelum, Chenab, Ravi, and Sutlej) feed the world's largest irrigation system; Tarbela (world's largest earthfill dam); K2 mountain; Khyber Pass; sites of the Indus Valley civilization
head of state Ghulam Ishaq Khan from 1988
head of government Benazir Bhutto from 1993
political system emergent democracy

PA abbreviation for the state of ◊*Pennsylvania*, a state of the USA.

Paarl /pɑːl/ town on the Great Berg River, Cape Province, South Africa; population (1980) 71,300. It is the centre of a wine-producing area, 50 km/31 mi NE of Cape Town. Nelson Mandela served the last days of his imprisonment at the Victor Vester prison near here.

Pacaraima, Sierra /,pækə'raɪmə/ mountain range along the Brazil–Venezuela frontier, extending into Guyana; length 620 km/385 mi; highest point **Mount Roraima**, a plateau about 50 sq km/20 sq mi, 2,810 m/9,222 ft above sea level, surrounded by cliffs 300 m/1,000 ft high, at the conjunction of the three countries. Formed 300 million years ago, it has unique fauna and flora, because of its isolation, consisting only of grasses, bushes, flowers, insects, and small amphibians.

Pacific Islands /pə'sɪfɪk 'aɪləndz/ former (1947–1990) UN Trust Territory in the W Pacific captured from Japan during World War II. The territory comprised over 2,000 islands and atolls and was assigned to the USA 1947. The islands were divided into four governmental units: the Northern Marianas (except Guam) which became a self-governing commonwealth in union with the USA 1975; the Marshall Islands, the Federated States of Micronesia, and the Republic of Belau (formerly Palau) became self-governing 1979-80, signing agreements of free association with the USA during the 1980s. In Dec 1990 the United Nations Security Council voted to dissolve its trusteeship over the islands with the exception of Belau.

The first experience can never be repeated. The first love, the first sun-rise, the first South Sea Island, are memories apart, and touched a virginity of sense.

On the **Pacific Islands** R L Stevenson
In the South Seas 1889

Pacific Ocean world's largest ocean, extending from Antarctica to the Bering Strait; area

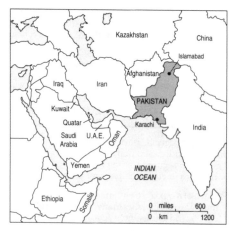

political parties Pakistan People's Party (PPP), moderate, Islamic, socialist; Islamic Democratic Alliance (IDA), including the Pakistan Muslim League (PML), Islamic conservative; Mohajir National Movement (MQM), Sind-based *mohajir* (Muslims previously living in India) settlers
exports cotton textiles, rice, leather, carpets
currency Pakistan rupee
population (1990 est) 113,163,000 (Punjabi 66%, Sindhi 13%); growth rate 3.1% p.a.
life expectancy men 54, women 55 (1989)
languages Urdu and English (official); Punjabi, Sindhi, Pashto, Baluchi, other local dialects
religion Sunni Muslim 75%, Shi'ite Muslim 20%, Hindu 4%
literacy men 40%, women 19% (1985 est)
GDP $39 bn (1988); $360 per head (1984)
chronology
1947 Independence achieved from Britain, Pakistan formed following partition of British India.
1956 Proclaimed a republic.
1958 Military rule imposed by General Ayub Khan.
1969 Power transferred to General Yahya Khan.
1971 Secession of East Pakistan (Bangladesh). After civil war, power transferred to Zulfiqar Ali Bhutto.
1977 Bhutto overthrown in military coup by General Zia ul-Haq; martial law imposed.
1979 Bhutto executed.
1981 Opposition Movement for the Restoration of Democracy formed. Islamization process pushed forward.
1985 Nonparty elections held, amended constitution adopted, martial law and ban on political parties lifted.
1988 Zia introduced Islamic legal code, the Shari'a. He was killed in a military plane crash Aug. Benazir Bhutto elected prime minister.
1989 Pakistan rejoined the Commonwealth.
1990 Army mobilized in support of Muslim separatists in Indian Kashmir. Bhutto dismissed on charges of incompetence and corruption. Islamic Democratic Alliance (IDA), led by Nawaz Sharif, won general election.
1991 Shari'a bill enacted; privatization and economic deregulation programme launched.
1992 Floods devastated N of country. Pakistan

elected to UN Security Council 1993–95.
1993 Sharif resigned; Benazir Bhutto elected prime minister.

Palau /pə'laʊ/ former name (until 1981) of the Republic of ◊Belau.

Paldiski /'pɑːldɪski/ small, ice-free port in Estonia; a naval base 40 km/25 mi W of Tallinn at the entrance to the Gulf of Finland.

Palembang /pə'lembæŋ/ oil-refining city in Indonesia, capital of S Sumatra province; population (1980) 786,000. Palembang was the capital of a sultanate when the Dutch established a trading station here 1616.

Palermo /pə'leəməʊ/ capital and seaport of Sicily; population (1988) 729,000. Industries include shipbuilding, steel, glass, and chemicals. It was founded by the Phoenicians in the 8th century BC.

Palestine /'pælɪstaɪn/ (Arabic *Falastin* 'Philistine') geographical area at the E end of the Mediterranean Sea, also known as the Holy Land because of its historic and symbolic importance for Jews, Christians, and Muslims. In ancient times Palestine extended E of the river Jordan, though today it refers to the territory of the State of Israel and the two Israeli-occupied territories of the West Bank and the Gaza Strip. Early settlers included the Canaanites, Hebrews, and Philistines. Over the centuries it became part of the Egyptian, Assyrian, Babylonian, Macedonian, Ptolemaic, Seleucid, Roman, Byzantine, Arab, and Ottoman empires.

Many Arabs refuse to recognize a Jewish state in Palestine, where for centuries Arabs constituted the majority of the population.

Palk Strait /pɔːlk/ channel separating SE India from the island of Sri Lanka; it is 53 km/33 mi at the widest point.

Palma /'pælmə/ (Spanish *Palma de Mallorca*) industrial port (textiles, cement, paper, pottery), resort, and capital of the Balearic Islands, Spain, on Majorca; population (1991) 308,600. Palma was founded 276 BC as a Roman colony. It has a Gothic cathedral, begun 1229.

Palma, La /'pælmə/ one of the Canary Islands, Spain
area 730 sq km/282 sq mi
capital Santa Cruz de la Palma
features forested
products wine, fruit, honey, silk; tourism is important
population (1981) 77,000.

Palmas, Las /'pælməs/ port in the Canary Islands; see ◊Las Palmas.

Palm Beach luxurious winter resort in Florida, USA, on an island between Lake Worth and the Atlantic; population (1990) 9,800.

Palmerston North /'pɑːməstən 'nɔːθ/ town on the southwest coast of North Island, New Zealand; population (1991) 70,200. Industries include textiles, dairy produce, and electrical goods.

Palm Springs /'pɑːm 'sprɪŋz/ resort and spa in S

California, USA, about 160 km/100 mi E of Los Angeles; population (1990) 40,200.

Palmyra /pæl'maɪrə/ ancient city and oasis in the desert of Syria, about 240 km/150 mi NE of Damascus. Palmyra, the biblical *Tadmor*, was flourishing by about 300 BC. It was destroyed AD 272 after Queen Zenobia had led a revolt against the Romans. Extensive temple ruins exist, and on the site is a village called Tadmor.

Palmyra /pæl'maɪrə/ coral atoll 1,600 km/1,000 mi SW of Hawaii, in the Line Islands, S Pacific, purchased by the USA from a Hawaiian family 1979 for the storage of highly radioactive nuclear waste from 1986.

Palo Alto /'pæləu 'æltəu/ city in California, USA, situated SE of San Francisco at the centre of the high-tech region known as Silicon Valley; population (1990) 55,900. It is the site of Stanford University.

Pamirs /pə'mɪəz/ central Asian plateau mainly in Tajikistan, but extending into China and Afghanistan, traversed by mountain ranges. Its highest peak is Kommunizma Pik (Communism Peak 7,495 m/24,600 ft) in the Akademiya Nauk range.

Pamplona /pæm'pləunə/ industrial city (wine, leather, shoes, textiles) in Navarre, N Spain, on the Arga River; population (1986) 184,000. A pre-Roman town, it was rebuilt by Pompey 68 BC, captured by the Visigoths 476, sacked by Charlemagne 778, became the capital of Navarre, and was taken by the Duke of Wellington in the Peninsular War 1813. An annual running of bulls takes place in the streets in July as part of the fiesta of San Fermin, a local patron saint.

Panama Republic of (*República de Panamá*)
area 77,100 sq km/29,768 sq mi
capital Panama City
towns Cristóbal, Balboa, Colón, David
physical coastal plains and mountainous interior; tropical rainforest in E and NW; Pearl Islands in Gulf of Panama
features Panama Canal; Barro Colorado Island in Gatún Lake (reservoir supplying the canal), a

tropical forest reserve since 1923; Smithsonian Tropical Research Institute
head of state and government Guillermo Endara from 1989
political system emergent democratic republic
political parties Democratic Revolutionary Party (PRD), right-wing; Labour Party (PALA), right-of-centre; Panamanian Republican Party (PPR), right-wing; Nationalist Liberal Republican Movement (MOLIRENA), left-of-centre; Authentic Panamanian Party (PPA), centrist; Christian Democratic Party (PDC), centre-left
exports bananas, petroleum products, copper, shrimps, sugar
currency balboa
population (1990 est) 2,423,000 (70% mestizo (mixed race), 14% W Indian, 10% European descent, 6% Indian (Cuna, Choco, Guayami)); growth rate 2.2% p.a.
life expectancy men 71, women 75 (1989)
languages Spanish (official), English
religions Roman Catholic 93%, Protestant 6%
literacy 87% (1989)
GNP $4.2 bn (1988); $1,970 per head (1984)
chronology
1821 Achieved independence from Spain; joined Gran Colombia.
1903 Full independence achieved on separation from Colombia.
1974 Agreement to negotiate full transfer of the Panama Canal from the USA to Panama.
1977 USA–Panama treaties transferred the canal to Panama, effective from 1999, with USA guaranteeing its protection and an annual payment.
1984 Nicolás Ardito Barletta elected president.
1985 Barletta resigned; replaced by Eric Arturo del Valle.
1987 General Noriega resisted calls for his removal, despite suspension of US military and economic aid.
1988 Del Valle replaced by Manuel Solis Palma. Noriega, charged with drug smuggling by the USA, declared a state of emergency.
1989 Opposition won election; Noriega declared results invalid; Francisco Rodríguez sworn in as president; coup attempt against Noriega failed; Noriega declared head of government by assembly; 'state of war' with the USA announced; US invasion deposed Noriega; Guillermo Endara installed as president; Noriega sought asylum in Vatican embassy; later surrendered and taken to US for trial.
1991 Attempted antigovernment coup foiled. Army abolished.
1992 Noriega found guilty of drug offences. Constitutional reform referendum voted down overwhelmingly.

Panama Canal /ˌpænə'mɑː/ canal across the Panama isthmus in Central America, connecting the Pacific and Atlantic oceans; length 80 km/50 mi, with 12 locks. Built by the USA 1904–14 after an unsuccessful attempt by the French, it was formally opened 1920. The *Panama Canal Zone* was acquired 'in perpetuity' by the USA 1903, comprising land extending about 5 km/3 mi on either side of the canal. The zone passed to

Panama 1979, and control of the canal itself was ceded to Panama by the USA Jan 1990 under the terms of the Panama Canal Treaty 1977. The Canal Zone has several US military bases.

It is the greatest liberty that Man has ever taken with Nature.

On the **Panama Canal** James Bryce
South America 1912

Panama City /ˌpænəˈmɑː/ capital of the Republic of Panama, near the Pacific end of the Panama Canal; population (1990) 584,800. Products include chemicals, plastics, and clothing. An earlier Panama, to the NE, founded 1519, was destroyed 1671, and the city was founded on the present site 1673.

Pan-American Highway road linking the USA with Central and South America; length 25,300 km/15,700 mi. Starting from the US-Canadian frontier (where it links with the Alaska Highway), it runs through San Francisco, Los Angeles, and Mexico City to Panama City, then down the west side of South America to Valparaiso, Chile, where it crosses the Andes and goes to Buenos Aires, Argentina. The road was first planned 1923.

Panay /pɑːˈnaɪ/ one of the Philippine islands, lying between Mindoro and Negros
area 11,515 sq km/4,446 sq mi
capital Iloilo
features mountainous, 2,215 m/7,265 ft in Madiaás
products rice, sugar, pineapples, bananas, copra, copper
history seized by Spain 1569; occupied by Japan 1942–45.

Panipat /pɑːˈnɪpət/ market town in Punjab, India, near the river Yamuna; population (1981 est) 137,900. It was the scene of three decisive battles: 1526, when Babur (1483–1530), great-grandson of Tamerlane, defeated the emperor of Delhi and founded the Mogul empire; 1556, won by his descendant Akbar; 1761, when the Marathas were defeated by Ahmad Shah Durrani of Afghanistan.

Panjshir Valley /ˈpʌndʒɪə/ valley of the river Panjshir, which rises in the Panjshir range to the N of Kabul, E Afghanistan. It was the chief centre of mujaheddin rebel resistance against the Soviet-backed Najibullah government in the 1980s.

Pantanal /ˌpæntəˈnɑːl/ large area of swampland in the Mato Grosso of SW Brazil, occupying 220,000 sq km/84,975 sq mi in the upper reaches of the Paraguay River; one of the world's great wildlife refuges of which 1,370 sq km/530 sq mi were designated a national park 1981.

Pantelleria /ˌpæntelɪˈrɪə/ volcanic island in the Mediterranean, 100 km/62 mi SW of Sicily and part of that region of Italy

area 115 sq km/45 sq mi
town Pantelleria
products sheep, fruit, olives, capers
population (1981) 7,800
history Pantelleria has drystone dwellings dating from prehistoric times. The Romans called it *Cossyra* and sent people into exile there. Strategically placed, the island has been the site of many battles. It was strongly fortified by Mussolini in World War II but surrendered to the Allies 11 June 1943.

Papeete /ˌpɑːpiˈeɪti/ capital and port of French Polynesia on the northwest coast of Tahiti; population (1983) 79,000. Products include vanilla, copra, and mother-of-pearl.

Paphos /ˈpæfɒs/ resort town on the southwest coast of Cyprus; population (1985) 23,200. It was the capital of Cyprus in Roman times and the legendary birthplace of the goddess Aphrodite, who rose out of the sea. Archaeological remains include the 2,300-year-old underground 'Tombs of the Kings', the Roman villa of Dionysus, and a 7th-century Byzantine castle.

Papua /ˈpɑːpuə/ original name of the island of New Guinea, but latterly its southeastern section, now part of Papua New Guinea.

Papua New Guinea
area 462,840 sq km/178,656 sq mi
capital Port Moresby (on E New Guinea)
towns Lae, Rabaul, Madang
physical mountainous; includes tropical islands of New Ireland, New Britain, and Bougainville; Admiralty Islands, D'Entrecasteaux Islands, and Louisiade Archipelago
features one of world's largest swamps on SW coast; world's largest butterfly, orchids; Sepik River
head of state Elizabeth II, represented by governor general
head of government Paias Wingti from 1992
political system liberal democracy
political parties Papua New Guinea Party (Pangu Pati: PP), urban-and coastal-oriented nationalist; People's Democratic Movement (PDM), 1985 breakaway from the PP; National Party (NP), highlands-based; Melanesian Alliance (MA), Bougainville-based autonomy; People's Progress Party (PPP), conservative

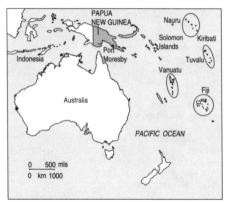

exports copra, coconut oil, palm oil, tea, copper, gold, coffee
currency kina
population (1989 est) 3,613,000 (Papuans, Melanesians, Negritos, various minorities); growth rate 2.6% p.a.
life expectancy men 53, women 54 (1987)
languages English (official); pidgin English, 715 local languages
religions Protestant 63%, Roman Catholic 31%, local faiths
literacy men 55%, women 36% (1985 est)
GNP $2.5 bn; $730 per head (1987)
chronology
1883 Annexed by Queensland; became the Australian Territory of Papua.
1884 NE New Guinea annexed by Germany; SE claimed by Britain.
1914 NE New Guinea occupied by Australia.
1921–42 Held as a League of Nations mandate.
1942–45 Occupied by Japan.
1975 Independence achieved from Australia, within the Commonwealth, with Michael Somare as prime minister.
1980 Julius Chan became prime minister.
1982 Somare returned to power.
1985 Somare challenged by deputy prime minister, Paias Wingti, who later formed a five-party coalition government.
1988 Wingti defeated on no-confidence vote and replaced by Rabbie Namaliu, who established a six-party coalition government.
1989 State of emergency imposed on Bougainville in response to separatist violence.
1991 Peace accord signed with Bougainville secessionists. Economic boom as gold production doubled. Wiwa Korowi elected as new governor general. Deputy prime minister, Ted Diro, resigned, having been found guilty of corruption.
1992 April: killings by outlawed Bougainville secessionists reported. July: Wingti elected premier.

Pará /pə'rɑː/ alternative name of the Brazilian port ♢Belém.

Paracels /,pærə'selz/ (Chinese **Xisha**; Vietnamese **Hoang Sa**) group of about 130 small islands in the S China Sea. Situated in an oil-bearing area, they were occupied by China following a skirmish with Vietnam 1974.

Paraguay Republic of (*República del Paraguay*)
area 406,752 sq km/157,006 sq mi
capital Asunción
towns Puerto Presidente Stroessner, Pedro Juan Caballero; port Concepción
physical low marshy plain and marshlands; divided by Paraguay River; Paraná River forms SE boundary
features Itaipú dam on border with Brazil; Gran Chaco plain with huge swamps
head of state and government General Andrés Rodríguez from 1989
political system emergent democratic republic
political parties National Republican Association (Colorado Party), right-of-centre; Liberal Party (PL), right-of-centre; Radical Liberal Party (PLR), centrist

exports cotton, soya beans, timber, vegetable oil, maté
currency guaraní
population (1990 est) 4,660,000 (95% mixed Guarani Indian–Spanish descent); growth rate 3.0% p.a.
life expectancy men 67, women 72 (1989)
languages Spanish 6% (official), Guarani 90%
religion Roman Catholic 97%
literacy men 91%, women 85% (1985 est)
GNP $7.4 bn; $1,000 per head (1987)
chronology
1811 Independence achieved from Spain.
1865–70 War with Argentina, Brazil, and Uruguay; much territory lost.
1932–35 Territory won from Bolivia during the Chaco War.
1940–48 Presidency of General Higinio Morínigo.
1948–54 Political instability; six different presidents.
1954 General Alfredo Stroessner seized power.
1989 Stroessner ousted in coup led by General Andrés Rodríguez. Rodríguez elected president; Colorado Party won the congressional elections.
1991 Colorado Party successful in assembly elections.

Paramaribo /,pærə'mærɪbəu/ port and capital of Surinam, South America, 24 km/15 mi from the sea on the river Suriname; population (1980) 193,000. Products include coffee, fruit, timber, and bauxite. It was founded by the French on an Indian village 1540, made capital of British Surinam 1650, and placed under Dutch rule 1816–1975.

Paraná /,pærə'nɑː/ industrial port (flour mills, meat canneries) and capital of Entre Rios province in E Argentina, on the Paraná River, 560 km/350 mi NW of Buenos Aires; population (1991) 276,000.

Paraná /,pærə'nɑː/ river in South America, formed by the confluence of the Río Grande and Paranaiba; the Paraguay joins it at Corrientes, and it flows into the Río de la Plata with the Uruguay; length 4,500 km/2,800 mi. It is used for hydroelectric power by Argentina, Brazil, and Paraguay.

Paris /'pæris/ port and capital of France, on the river Seine; *département* in the Ile de France region; area 105 sq km/40.5 sq mi; population (1990) 2,175,200. Products include metal, leather, and luxury goods and chemicals, glass, and tobacco.

features the river Seine is spanned by 32 bridges, the oldest is the Pont Neuf 1578. Churches include Notre Dame cathedral built 1163–1250; the Invalides, housing the tomb of Napoleon; the Gothic Sainte-Chapelle; and the 19th-century basilica of Sacré-Coeur, 125 m/410 ft high. Notable buildings include the Palais de Justice, the Hôtel de Ville, and the Luxembourg Palace and Gardens. The former palace of the Louvre (with its glass pyramid entrance by I M Pei 1989) is one of the world's major art galleries; the Musée d'Orsay 1986 has Impressionist and other paintings from the period 1848–1914; the Pompidou Centre (Beaubourg) 1977 exhibits modern art. Other landmarks are the Tuileries Gardens, the Place de la Concorde, the Eiffel Tower, and the Champs-Elysées avenue leading to the Arc de Triomphe. Central Paris was replanned in the 19th century by Baron Haussmann. To the W is the Bois de Boulogne and, beyond the river, La Défense business park with the Grande Arche 1989 by Danish architect Johan Otto von Sprekelsen; Montmartre is in the N of the city; the university, founded about 1150, is on the Left Bank.

history Paris, the Roman *Lutetia*, capital of the Parisii, a Gaulish people, was occupied by Julius Caesar 53 BC. The Merovingian king Clovis made Paris the capital in about AD 508, and the city became important under the Capetian kings 987–1328. Paris was occupied by the English 1420–36, and was besieged by Henry IV 1590–94. The Bourbon kings did much to beautify the city. Napoleon I added new boulevards, bridges, and triumphal arches, as did Napoleon III. Paris was the centre of the revolutions of 1789–94, 1830, and 1848. It was besieged by Prussia 1870–71, and by government troops during the Commune period (local socialist government) March–May 1871. During World War I it suffered from air raids and bombardment, and in World War II it was occupied by German troops June 1940–Aug 1944.

If you are lucky enough to have lived in Paris as a young man, then wherever you go for the rest of your life it stays with you, for Paris is a moveable feast.

On **Paris** Ernest Hemingway
A Moveable Feast 1964

Paris-Plage /pæ'ri: 'plɑːʒ/ resort in Nord-Pas-de-Calais region, N France, adjoining Le Touquet.

Parkersburg /'pɑːkəzbɜːg/ city in NW West Virginia, USA, where the Little Kanawha River flows into the Ohio River, N of Charleston; population (1990) 33,900. Industries include chemicals, glassware, paper, and plastics.

Parma /'pɑːmə/ city in Emilia-Romagna, N Italy; population (1988) 175,000. Industries include food processing, textiles, and engineering. Founded by the Etruscans, it was the capital of the duchy of Parma 1545–1860. It has given its name to Parmesan cheese.

Parnassus /pɑː'næsəs/ mountain in central Greece, height 2,457 m/8,064 ft, revered by the ancient Greeks as the abode of Apollo and the Muses. The sacred site of Delphi lies on its southern flank.

Parramatta /,pærə'mætə/ river inlet, western arm of Sydney Harbour, New South Wales, Australia. It is 24 km/15 mi long and is lined with industrial suburbs of Sydney: Balmain, Drummoyne, Concord, Parramatta, Ermington and Rydalmere, Ryde, and Hunter's Hill.

Pasadena /,pæsə'diːnə/ city in SW California, USA, part of Greater ⟶Los Angeles; population (1990) 131,600. Products include electronic equipment and precision instruments.

On 1 Jan the East–West football game is held here in the 85,000-seat Rose Bowl. The California Institute of Technology (Caltech) owns the Hale Observatories (which include the Mount Palomar telescope) and is linked with the Jet Propulsion Laboratories.

Pascagoula /,pæskə'guːlə/ city in SE Mississippi, USA, at the mouth of the Pascagoula River, E of Biloxi; population (1990) 25,900. Industries include fishing, shipbuilding, paper, petroleum, and chemicals. A French fort 1718, it was at other times owned by Britain, Spain, and the free state of West Florida.

Pasco /'pæskəu/ city in SE Washington, USA, on the Columbia River, seat of Franklin County; population (1990) 20,300.

Pas-de-Calais /,pɑːdə'kæleɪ/ French name for the Strait of Dover and also the French *département* bordering it, of which Arras is the capital and Calais the chief port. See also ⟶Nord-Pas-de-Calais.

Passaic /pə'seɪk/ city in NW New Jersey, USA, on the Passaic River, N of Jersey City; population (1990) 58,000. Products include television cables, chemicals, plastics, pharmaceuticals, and clothing.

Passau /'pæsau/ town in SE Bavaria, Germany, at the junction of the rivers Inn and Ilz with the Danube. The Treaty of Passau 1552 between Maurice, Elector of Saxony, and the future emperor Ferdinand I allowed the Lutherans full religious liberty and prepared the way for the Peace of Augsburg.

Patagonia /,pætə'gəuniə/ geographic area of South America, S of latitude 40° S, with sheep farming, and coal and oil resources. Sighted by Ferdinand

Magellan 1520, it was claimed by both Argentina and Chile until divided between them 1881.

I believe that we have here the secret of the persistence of Patagonian images ... it is that nature in these desolate scenes ... moves us more deeply than in others.

On **Patagonia** W H Hudson *Idle Days in Patagonia* 1893

Patiala /ˌpʌtiˈɑːlə/ city in E Punjab, India; population (1981) 206,254. Industries include textiles and metalwork.

Patmos /ˈpætmɒs/ Greek island in the Aegean Sea, one of the Dodecanese; the chief town is Hora. St John is said to have written the New Testament Book of Revelation while in exile here.

Patna /ˈpætnə/ capital of Bihar state, India, on the river Ganges; population (1981) 916,000. It has remains of a hall built by the emperor Asoka in the 3rd century BC.

Patras /ˈpætrəs/ (Greek *Pátrai*) industrial city (hydroelectric installations, textiles, paper) in the NW Peloponnese region, Greece, on the Gulf of Patras; population (1981) 141,500. The ancient Patrai is the only one of the 12 cities of the ancient Greek province of Achaea to survive.

Pau /pəʊ/ industrial city (electrochemical and metallurgical products) and resort, capital of Pyrénées-Atlantiques *département* in Aquitaine, SW France, near the Spanish border; population (1990) 83,900. It is the centre of the Basque area of France, and the site of fierce guerrilla activity.

Pawtucket /pəˈtʌkɪt/ city in NE Rhode Island, USA, on the Blackstone River, NE of Providence; population (1990) 72,600. Industries include textiles, thread, machinery, and metal and glass products. It was the home of the first US water-powered cotton mill 1790.

Paysandú /ˌpaɪsænˈduː/ city in Uruguay, capital of Paysandú department, on the river Uruguay; population (1985) 74,000. Tinned meat is the main product. The city dates from 1772 and is linked by a bridge built 1976 with Puerto Colón in Argentina.

Pays de la Loire /peiˈiː də lɑː ˈlwɑː/ agricultural region of W France, comprising the *départements* of Loire-Atlantique, Maine-et-Loire, Mayenne, Sarthe, and Vendée; capital Nantes; area 32,100 sq km/12,391 sq mi; population (1986) 3,018,000. Industries include shipbuilding and wine production.

Peace /piːs/ river formed in British Columbia, Canada, by the union at Finlay Forks of the Finlay and Parsnip rivers and flowing through the Rocky Mountains and across Alberta to join the river Slave just N of Lake Athabasca; length 1,600 km/ 1,000 mi.

Peak District /piːk/ tableland of the S Pennines in

NW Derbyshire, England. It is a tourist region and a national park (1951). The highest point is Kinder Scout, 636 m/2,088 ft.

Pearl Islands group of some 180 islands in the Gulf of Panama, Central America. The main islands are San Miguel (the largest), San José, and Pedro González. There is pearl fishing.

Pechenga /ˈpetʃɪŋgə/ (Finnish *Petsamo*) ice-free fishing port in Murmansk, NW Russia, on the Barents Sea. Russia ceded Pechenga to Finland 1920 but recovered it under the 1947 peace treaty.

Pechora /pɪˈtʃɔːrə/ river in N Russia, rising in the N Urals. It transports coal, timber, and furs (June–Sept) to the Barents Sea, 1,800 km/ 1,125 mi to the N.

Pécs /peɪtʃ/ city in SW Hungary, the centre of a coalmining area on the Croatian frontier; population (1988) 182,000. Industries include metal, leather, and wine. The town dates from Roman times and was under Turkish rule 1543–1686.

Peeblesshire /ˈpiːbəlzʃə/ former county of S Scotland, included from 1975 in Borders Region. Peebles was the county town.

Peel /piːl/ fishing port in the Isle of Man, UK, 19 km/12 mi NW of Douglas.

Pegu /peˈguː/ city in S Myanmar on the river Pegu, NE of Yangon; population (1983) 254,762. It was founded 573 and is the site of the celebrated Shwemawdaw pagoda.

Peiping /ˌpeɪˈpɪŋ/ name (meaning 'northern peace') 1928–49 of ▷Beijing in China.

Peipus, Lake /ˈpaɪpəs/ (Estonian *Peipsi*, Russian *Chudskoye*) lake forming the boundary between Estonia and Pskov'oblast', an administrative region of Russia. Alexander Nevski defeated the Teutonic Knights on its frozen surface 1242.

Peking /ˈpiːˈkɪŋ/ alternative transcription of ▷Beijing, the capital of China.

Pelée, Mont /pəˈleɪ/ volcano on the island of Martinique in the West Indies; height 1,350 m/ 4,428 ft. It destroyed the town of St Pierre during its eruption 1902.

Pelion /ˈpiːlɪən/ mountain in Thessaly, Greece, near Mount Ossa; height 1,548 m/5,079 ft. In Greek mythology it was the home of the centaurs, creatures half-human and half-horse.

Peloponnese /ˌpeləpəˈniːs/ (Greek *Peloponnesos*) peninsula forming the southern part of Greece; area 21,549 sq km/8,318 sq mi; population (1991) 1,077,000.

It is joined to the mainland by the narrow isthmus of Corinth and is divided into the nomes (administrative areas) of Argolis, Arcadia, Achaea, Elis, Corinth, Lakonia, and Messenia, representing its seven ancient states.

Pemba /ˈpembə/ coral island in the Indian Ocean, 48 km/30 mi NE of Zanzibar, and forming with it part of Tanzania

area 984 sq km/380 sq mi
capital Chake Chake
products cloves, copra
population (1985) 257,000.

Pembroke /'pembruk/ (Welsh *Penfro*) seaport and engineering centre in Dyfed, Wales; population (1981) 15,600. Henry VII was born in Pembroke Castle.

Pembrokeshire /'pembrukʃə/ (Welsh *Sir Benfro*) former extreme southwestern county of Wales, which became part of Dyfed 1974; the county town was Haverfordwest.

Penang /pɪ'næŋ/ (Malay *Pulau Pinang*) state in W Peninsular Malaysia, formed of Penang Island, Province Wellesley, and the Dindings on the mainland; area 1,030 sq km/398 sq mi; capital Penang (George Town); population (1990) 1,142,200. Penang Island was bought by Britain from the ruler of Kedah 1785; Province Wellesley was acquired 1800.

Penarlâg /ˌpenɑː'lɑːg/ Welsh name of ⟩ Hawarden, a town in Clwyd, Wales.

Pennines /'penaɪnz/ mountain system, 'the backbone of England', broken by a gap through which the river Aire flows to the E and the Ribble to the W; length (Scottish border to the Peaks in Derbyshire) 400 km/250 mi.
Britain's first long-distance footpath was the *Pennine Way*, opened 1965.

Pennsylvania /ˌpensl'veɪnɪə/ state in northeastern USA; nickname Keystone State
area 117,400 sq km/45,316 sq mi
capital Harrisburg
towns Philadelphia, Pittsburgh, Erie, Allentown, Scranton
features Allegheny Mountains; rivers: Ohio, Susquehanna, Delaware; Independence National Historic Park, Philadelphia; Valley Forge National Historic Park; Gettysburg Civil War battlefield; Pennsylvania Dutch country; Poconos resort region
products hay, cereals, mushrooms, cattle, poultry, dairy products, cement, coal, steel, petroleum products, pharmaceuticals, motor vehicles and equipment, electronic components, textiles
population (1990) 11,881,600
famous people Marian Anderson, Stephen Foster, Benjamin Franklin, George C Marshall, Robert E Peary, Gertrude Stein, John Updike

Pennsylvania

Harrisburg

history founded and named by William Penn 1682, following a land grant by Charles II. The Declaration of Independence was proclaimed in Philadelphia, and many important Revolutionary War battles were fought here 1777–78. It was one of the original 13 states.

Pensacola /ˌpensə'kəulə/ port in NW Florida, USA, on the Gulf of Mexico, with a large naval air-training station; population (1990) 58,200. Industries include chemicals, synthetic fibres, and paper. Pensacola was founded by the Spanish 1696.

Pentland Firth /'pentlənd 'fɜːθ/ channel separating the Orkney Islands from N Scotland.

Penza /'penzə/ industrial city (sawmills, bicycles, watches, calculating machines, textiles) in W central Russia, capital of Penza region, 560 km/350 mi SE of Moscow, at the junction of the Penza and Sura rivers; population (1987) 540,000. It was founded as a fort 1663.

Penzance /pen'zæns/ seaport for the Scilly Isles and resort in Cornwall, SW England, on Mount's Bay; population (1981) 19,600. It now incorporates the seaport of Newlyn.

Peoria /pi'ɔːrɪə/ city in central Illinois, USA, on the Illinois River; a transport, mining, and agricultural centre; population (1990) 113,500. Fort Crève Coeur was built here by the French explorer Robert Cavalier de la Salle 1680 and became a trading centre. The first US settlers arrived 1818, and the town was known as Fort Clark until 1825.

Perak /'peərə/ state of W Peninsular Malaysia; capital Ipoh; area 21,000 sq km/8,106 sq mi; population (1990) 2,222,200. It produces tin and rubber. The government is a sultanate. The other principal town is Taiping.

Pereira /pə'reərə/ capital of Risaralda department, central Colombia, situated at an altitude of 1,463 m/4,800 ft, overlooking the fertile Cauca Valley, W of Bogotá; population (1985) 390,000. Founded 1863, the city has developed into a centre for the national coffee and cattle industries.

Périgueux /ˌperɪ'gɜː/ capital of Dordogne *département*, Aquitaine, France, on the river Isle, 127 km/79 mi ENE of Bordeaux; population (1990) 32,800. It is a trading centre for wine and truffles. The Byzantine cathedral dates from 984.

Perim /'perɪm/ island in the strait of Bab-el-Mandeb, the southern entrance to the Red Sea; part of Yemen; area 13 sq km/5 sq mi.

Perlis /'pɜːlɪs/ border state of Peninsular Malaysia, NW Malaysia; capital Kangar; area 800 sq km/309 sq mi; population (1990) 187,700. It produces rubber, rice, coconuts, and tin. Perlis is ruled by a raja. It was transferred by Siam to Britain 1909.

Perm /pɜːm/ industrial city (shipbuilding, oil refining, aircraft, chemicals, sawmills), and capital of Perm region, N Russia, on the Kama near the Ural Mountains; population (1987) 1,075,000. It was called Molotov 1940–57.

Pernambuco /ˌpɜːnæmˈbuːkəʊ/ state of NE Brazil, on the Atlantic coast
area 98,281 sq km/37,946 sq mi
capital Recife (former name Pernambuco)
features highlands; the coast is low and humid
population (1985) 6,776,000.

Perpignan /ˌpɜːpiːnˈjɒŋ/ market town (olives, fruit, wine), resort, and capital of the Pyrénées-Orientales *département* of France, just off the Mediterranean coast, near the Spanish border; population (1990) 108,000. Overlooking Perpignan is the castle of the counts of Roussillon.

Persian Gulf or *Arabian Gulf* large shallow inlet of the Arabian Sea; area 233,000 sq km/90,000 sq mi. It divides the Arabian peninsula from Iran and is linked by the Strait of Hormuz and the Gulf of Oman to the Arabian Sea. Oilfields surround it in the Gulf States of Bahrain, Iran, Iraq, Kuwait, Oman, Qatar, Saudi Arabia, and the United Arab Emirates.

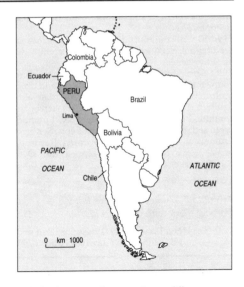

No arm of the sea has been, or is of greater interest alike to the geologist and archaeologist, the historian and geographer, the merchant and the student of strategy ...

On the **Persian Gulf**
Arnold Wilson *The Persian Gulf* 1928

Perth /pɜːθ/ industrial town in Tayside, E Scotland, on the river Tay; population (1989 est) 44,000. It was the capital of Scotland from the 12th century until James I of Scotland was assassinated here 1437.

Perth /pɜːθ/ capital of Western Australia, with its port at nearby Fremantle on the Swan River; population (1990) 1,190,100. Products include textiles, cement, furniture, and vehicles. It was founded 1829 and is the commercial and cultural centre of the state.

Perthshire /ˈpɜːθʃə/ former inland county of central Scotland, of which the major part was included in Tayside 1975, the southwestern part being included in Central Region; Perth was the administrative headquarters.

Peru Republic of (*República del Perú*)
area 1,285,200 sq km/496,216 sq mi
capital Lima, including port of Callao
towns Arequipa, Iquitos, Chiclayo, Trujillo
physical Andes mountains NW–SE cover 27% of Peru, separating Amazon river-basin jungle in NE from coastal plain in W; desert along coast N–S
environment an estimated 3,000 out of the 8,000 sq km/3,100 sq mi of coastal lands under irrigation are either waterlogged or suffering from saline water. Only half the population have access to clean drinking water
features Lake Titicaca; Atacama Desert; Nazca lines, monuments of Machu Picchu, Chan Chan, Charín de Huantar
head of state and government Alberto Fujimori from 1990

political system democratic republic
political parties American Popular Revolutionary Alliance (APRA), moderate, left-wing; United Left (IU), left-wing; Change 90, centrist
exports coca, coffee, alpaca, llama and vicuña wool, fish meal, lead (largest producer in South America), copper, iron, oil
currency new sol
population (1990 est) 21,904,000 (46% Indian, mainly Quechua and Aymara; 43% mixed Spanish–Indian descent); growth rate 2.6% p.a.
life expectancy men 61, women 66
languages Spanish 68% (both official), Aymara 3%
religion Roman Catholic 90%
literacy men 91%, women 78% (1985 est)
GNP $19.6 bn (1988); $940 per head (1984)
chronology
1824 Independence achieved from Spain.
1849–74 Some 80,000–100,000 Chinese labourers arrived in Peru to fill menial jobs such as collecting guano.
1902 Boundary dispute with Bolivia settled.
1927 Boundary dispute with Colombia settled.
1942 Boundary dispute with Ecuador settled.
1948 Army coup, led by General Manuel Odría, installed a military government.
1963 Return to civilian rule, with Fernando Belaúnde Terry as president.
1968 Return of military government in a bloodless coup by General Juan Velasco Alvarado.
1975 Velasco replaced, in a bloodless coup, by General Morales Bermúdez.
1980 Return to civilian rule, with Fernando Belaúnde as president.
1981 Boundary dispute with Ecuador renewed.
1985 Belaúnde succeeded by Social Democrat Alan García Pérez.
1987 President García delayed the nationalization of Peru's banks after a vigorous campaign against the proposal.
1988 García pressured to seek help from the International Monetary Fund.
1989 Mario Vargas Llosa entered presidential

race; his Democratic Front won municipal elections Nov. **1990** Alberto Fujimori defeated Vargas Llosa in presidential elections. Assassination attempt on president failed. **1992** April: Fujimori sided with army to avert coup, announcing crackdown on rebels and drug traffickers. USA suspended humanitarian aid. Sendero Luminoso ('Shining Path') terrorists continued campaign of violence. May: Fujimori promised return to democracy. Sept: Sendero Luminoso leader, Abimael Guzman Reynoso, arrested; extremist attacks stepped up. Oct: Reynoso received life sentence.

I would not change my native land/
For rich Peru with all her gold.

On **Peru** Isaac Watts 'Praise for Birth and Education in a Christian Land' 1753

Perugia /pə'ruːdʒə/ capital of Umbria, Italy, 520 m/1,700 ft above the river Tiber, about 137 km/85 mi N of Rome; population (1988) 148,000. Its industries include textiles, liqueurs, and chocolate. One of the 12 cities of the ancient country of Etruria, it surrendered to Rome 309 BC. There is a university, founded 1276; a municipal palace, built 1281; and a 15th-century cathedral.

Pescadores /ˌpeskə'dɔːrɪz/ (Chinese **Penghu**) group of about 60 islands off Taiwan, of which they form a dependency; area 130 sq km/50 sq mi.

Pescara /pe'skɑːrə/ town in Abruzzi, E Italy, at the mouth of the Pescara River, on the Adriatic coast; population (1988) 131,000. Hydroelectric installations supply Rome with electricity. It is linked by ferry to Split in Croatia.

Peshawar /pə'ʃauə/ capital of North-West Frontier Province, Pakistan, 18 km/11 mi E of the Khyber Pass; population (1981) 555,000. Products include textiles, leather, and copper.

Petaluma /ˌpetə'luːmə/ city in NW California, USA, N of San Francisco, on the Petaluma River; population (1990) 43,200. It is an agricultural centre for poultry and dairy products.

Peterborough /'piːtəbərə/ city in Cambridgeshire, England, noted for its 12th-century cathedral; population (1991) 148,800, one of the fastest growing cities in Europe. It has an advanced electronics industry. Nearby Flag Fen disclosed 1985 a well-preserved Bronze Age settlement of 660 BC. Peterborough was designated a new town 1967.

Peterhead /ˌpiːtə'hed/ industrial seaport (fishing, shipbuilding, light engineering, whisky distilling, woollens) in Grampian Region, Scotland, 54 km/ 33 mi NE of Aberdeen; population (1986 est) 18,150. James Edward Stuart, the Old Pretender, landed here 1715. The harbour is used by service industries for North Sea oil.

Peter I Island /'piːtə/ uninhabited island in the Bellingshausen Sea, Antarctica, belonging to Norway since 1931; area 180 sq km/69 sq mi.

Peterlee /ˌpiːtə'liː/ new town in County Durham, England, established 1948; population (1981) 22,750. It was named after Peter Lee, first Labour chair of a county council.

Petersburg /'piːtəzbɜːg/ city in SE Virginia, USA, on the Appomattox River, S of Richmond; population (1990) 38,400. Industries include tobacco products, textiles, leather products, boat building, and chemicals. It was the site of Fort Henry 1646, as well as of War of American Independence and Civil War battles.

Petra /'petrə/ (Arabic **Wadi Musa**) ancient city carved out of the red rock at a site in Jordan, on the eastern slopes of the Wadi el Araba, 90 km/56 mi S of the Dead Sea. An Edomite stronghold and capital of the Nabataeans in the 2nd century, it was captured by the Roman emperor Trajan 106 and destroyed by the Arabs in the 7th century. It was forgotten in Europe until 1812 when the Swiss traveller Jacob Burckhardt (1818–1897) came across it.

Match me such marvel save in Eastern
clime/A rose-red city 'half as old as time'!

On **Petra** John William Burgon *Petra* 1845

Petrograd /'petrəgræd/ former name (1914–24) of ⬦St Petersburg, a city in Russia.

Petropavlovsk /ˌpetrəupæv'lɒvsk/ industrial city (flour, agricultural machinery, leather) in N Kazakhstan, on the Ishim River, the Trans-Siberian railway, and the Transkazakh line, opened 1953; population (1987) 233,000. It was founded as a Russian fortress 1782.

Petropavlovsk-Kamchatskiy /ˌpetrəupæv'lɒvsk kæm'tʃætski/ Pacific seaport and naval base on the east coast of the Kamchatka peninsula, Russia; population (1987) 252,000.

Petrópolis /pe'trɒpəlɪs/ hill resort in SE Brazil, founded by Pedro II; population (1991) 294,200.

Petrovsk /pɪ'trɒvsk/ former name (until 1921) of the Russian port ⬦Makhachkala.

Petrozavodsk /ˌpetrəuzə'vɒdsk/ industrial city (metal goods, cement, prefabricated houses, sawmills), and capital of the autonomous republic of Karelia, Russia, on the west shore of Lake Onega; population (1987) 264,000. Peter the Great established the township 1703 as an ironworking centre; it was named Petrozavodsk 1777.

Petsamo /'petsəməu/ Finnish name of the Murmansk port ⬦Pechenga.

Pevensey /'pevənsi/ English village in Sussex, 8 km/5 mi NE of Eastbourne, the site of the Norman king William the Conqueror's landing 1066. The walls remain of the Roman fortress of Anderida, later a Norman castle, which was prepared against German invasion in World War II.

Pforzheim /'pfɔːtshaɪm/ city in Baden-Württemberg, Germany, 26 km/16 mi SE of Karlsruhe; there are goldsmith industries; population (1988) 105,000. It was a Roman settlement, and the residence of the margraves (princes) of Baden 1300–1565.

Phil. abbreviation for *Philadelphia*, a state of the USA.

Philadelphia /ˌfɪlə'delfiə/ ('the city of brotherly love') industrial city and port on the Delaware River in Pennsylvania, USA; population (1990) 1,585,600, metropolitan area 5,899,300. Products include refined oil, chemicals, textiles, processed food, printing and publishing. Founded 1682, it was the first capital of the USA 1790–1800.

Philae /'faɪliː/ island in the river Nile, Egypt, above the first rapids, famed for the beauty of its temple of Isis (founded about 350 BC and in use until the 6th century AD). In 1977 the temple was re-erected on the nearby island of Agilkia above the flooding caused by the Aswan Dam.

Philae as it is, is perhaps the one perfect thing in the world ...

On **Philae** Wilfred Blunt 1895

Philippeville /'fɪlɪpvɪl/ former name (until 1962) of the Algerian port of ⬦Skikda.

Philippines Republic of the (*Republika ng Pilipinas*)
area 300,000 sq km/115,800 sq mi
capital Manila (on Luzon)
towns Quezon City (Luzon), Zamboanga (Mindanao); ports Cebu, Davao (on Mindanao), and Iloilo
physical comprises over 7,000 islands; volcanic mountain ranges traverse main chain N–S; 50% still forested. The largest islands are Luzon 108,172 sq km/41,754 sq mi and Mindanao

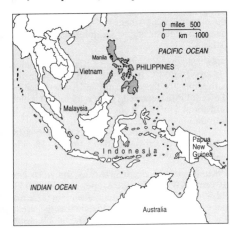

94,227 sq km/36,372 sq mi; others include Samar, Negros, Palawan, Panay, Mindoro, Leyte, Cebu, and the Sulu group
environment cleared for timber, tannin, and the creation of fish ponds, the mangrove forest was reduced from an area of 5,000 sq km/1,930 sq mi to 380 sq km/146 sq mi between 1920 and 1988
features Luzon, site of Clark Field, US air base used as a logistical base in Vietnam War; Subic Bay, US naval base; Pinatubo volcano (1,759 m/5,770 ft); Mindanao has active volcano Apo (2,954 m/9,690 ft) and mountainous rainforest
head of state and government Fidel Ramos from 1992
political system emergent democracy
political parties People's Power, including the PDP–Laban Party and the Liberal Party, centrist pro-Aquino; Nationalist Party, Union for National Action (UNA), and Grand Alliance for Democracy (GAD), conservative opposition groupings; Mindanao Alliance, island-based decentralist body
exports sugar, copra (world's largest producer) and coconut oil, timber, copper concentrates, electronics, clothing
currency peso
population (1990 est) 66,647,000 (93% Malaysian); growth rate 2.4% p.a.
life expectancy men 63, women 69 (1989)
languages Tagalog (Filipino, official); English and Spanish
religions Roman Catholic 84%, Protestant 9%, Muslim 5%
literacy 88% (1989)
GNP $38.2 bn; $667 per head (1988)
chronology
1542 Named the Philippines (Filipinas) by Spanish explorers.
1565 Conquered by Spain.
1898 Ceded to the USA after Spanish–American War.
1935 Granted internal self-government.
1942–45 Occupied by Japan.
1946 Independence achieved from USA.
1965 Ferdinand Marcos elected president.
1983 Opposition leader Benigno Aquino murdered by military guard.
1986 Marcos overthrown by Corazon Aquino's People's Power movement.
1987 'Freedom constitution' adopted, giving Aquino mandate to rule until June 1992; People's Power won majority in congressional elections. Attempted right-wing coup suppressed. Communist guerrillas active. Government in rightward swing.
1988 Land Reform Act gave favourable compensation to large estate-holders.
1989 Referendum on southern autonomy failed; Marcos died in exile; Aquino refused his burial in Philippines. Sixth coup attempt suppressed with US aid; Aquino declared state of emergency.
1990 Seventh coup attempt survived by President Aquino.
1991 June: eruption of Mount Pinatubo, hundreds killed. USA agreed to give up Clark Field airbase but keep Subic Bay naval base for ten more years. Sept: Philippines Senate voted to urge withdrawal of all US forces. US renewal of

Subic Bay lease rejected. Nov: Imelda Marcos returned.
1992 Fidel Ramos elected to replace Aquino.

Three hundred years in the convent and fifty years in Hollywood.

On the **Philippines** a common saying used to describe the Philippines' Spanish and US colonial heritage

---

Phnom Penh /'nɒm 'pen/ capital of Cambodia, on the Mekong River, 210 km/130 mi NW of Saigon; population (1989) 800,000. Industries include textiles and food-processing.
On 17 April 1975 the entire population (about 3 million) was forcibly evacuated by the Khmer Rouge communist movement; survivors later returned.

Phoenix /'fi:nɪks/ capital of Arizona, USA; industrial city (steel, aluminium, electrical goods, food processing) and tourist centre on the Salt River; population (1990) 983,400.

Phoenix Islands /'fi:nɪks/ group of eight islands in the South Pacific, included in Kiribati; total land area 18 sq km/11 sq mi; population (1990) 45. Drought has rendered them all uninhabitable.

Piacenza /,pɪə'tʃentsə/ industrial city (agricultural machinery, textiles, pottery) in Emilia-Romagna, N Italy, on the river Po, 65 km/40 mi SE of Milan; population (1988) 105,000. The Roman *Placentia*, Piacenza dates from 218 BC and has a 12th-century cathedral.

Picardy /'pɪkədi/ (French *Picardie*) region of N France, including Aisne, Oise, and Somme *départements*
area 19,400 sq km/7,488 sq mi
products chemicals and metals
population (1986) 1,774,000
history in the 13th century the name Picardy was used to describe the feudal smallholdings N of Paris added to the French crown by Philip II. During the Hundred Years' War the area was hotly contested by France and England, but it was eventually occupied by Louis XI 1477. Picardy once more became a major battlefield in World War I.

Picton /'pɪktən/ small port at the northeast extremity of South Island, New Zealand, with a ferry to Wellington, North Island.

Piedmont /'pi:dmɒnt/ (Italian *Piemonte*) region of N Italy, bordering Switzerland to the N and France to the W, and surrounded, except to the E, by the Alps and the Apennines; area 25,400 sq km/9,804 sq mi; population (1990) 4,356,200. Its capital is Turin, and towns include Alessandria, Asti, Vercelli, and Novara. It also includes the fertile Po river valley. Products include fruit, grain, cattle, cars, and textiles. The movement for the unification of Italy started in the 19th century in Piedmont, under the house of Savoy.

Pierre /pɪə/ capital of South Dakota, USA, located in the central part of the state, on the Missouri River, near the geographical centre of North America; population (1990) 12,900. Industries include tourism and grain and dairy products. As Fort Pierre in the early 1800s, it served as a fur-trading post; in the late 1800s it was a supply centre for gold miners.

Pietermaritzburg /,pi:tə'mærɪtsbɜ:g/ industrial city (footwear, furniture, aluminium, rubber, brewing), and capital, from 1842, of Natal, South Africa; population (1985) 192,400. Founded 1838 by Boer trekkers from the Cape, it was named after their leaders, Piet Retief and Gert Maritz, who were killed by Zulus.

Pikes Peak /,paɪks 'pi:k/ mountain in the Rampart of the Rocky Mountains, Colorado, USA; height 4,300 m/14,110 ft. It has commanding views, accessible by cog railway and road. Pikes Peak was discovered 1806 by Zebulon Pike and first scaled 1820.

Pilsen /'pɪlzən/ German form of ⬦Plzeň, a Czech town.

Pinatubo, Mount /,pɪnə'tu:bəu/ active volcano on Luzon Island, the Philippines, 88 km/55 mi N of Manila. Dormant for 600 years, it erupted June 1991, killing 343 people and leaving as many as 200,000 homeless. Surrounding rice fields were covered with 3 m/10 ft of volcanic ash.

Pindus Mountains /'pɪndəs/ (Greek *Pindhos Oros*) range in NW Greece and Albania, between Epirus and Thessaly; highest point Smolikas, 2,633 m/8,638 ft.

Pine Bluff /'paɪn blʌf/ city in SE Arkansas, USA, on the Arkansas River, SE of Little Rock, seat of Jefferson County; population (1990) 57,100. Industries include paper, cotton, grain, and furniture.

Piraeus /paɪ'ri:əs/ port of both ancient and modern Athens and main port of Greece, on the Gulf of Aegina; population (1981) 196,400. Constructed as the port of Athens about 493 BC, it was linked with that city by the Long Walls, a fortification protecting the approaches to Athens comprising three walls built 496–456 BC. After the destruction of Athens by Sulla 86 BC, Piraeus declined. Piraeus is now an industrial suburb of Athens.

Pisa /'pi:zə/ city in Tuscany, Italy; population (1988) 104,000. It has an 11th–12th-century cathedral. Its famous campanile, the Leaning Tower of Pisa (repaired 1990) is 55 m/180 ft high and about 5 m/16.5 ft out of perpendicular. It has foundations only about 3 m/10 ft deep.
Pisa was a maritime republic in the 11th–12th centuries. The university dates from 1338. The scientist Galileo was born here.

Pistoia /pɪ'stɔɪə/ city in Tuscany, Italy, 16 km/10 mi NW of Florence; population (1982) 92,500. Industries include steel, small arms, paper, pasta, and olive oil. Pistoia was the site of the Roman rebel Catiline's defeat 62 BC. It is surrounded by walls (1302) and has a 12th-century cathedral.

Pitcairn Islands /ˈpɪtkeən/ British colony in Polynesia, 5,300 km/3,300 mi NE of New Zealand
area 27 sq km/10 sq mi
capital Adamstown
features the uninhabited Henderson Islands, an unspoiled coral atoll with a rare ecology, and tiny Ducie and Oeno islands, annexed by Britain 1902
products fruit and souvenirs to passing ships
population (1990) 52
language English
government the governor is the British high commissioner in New Zealand
history settled 1790 by nine mutineers from the British ship the Bounty together with some Tahitians; their occupation remained unknown until 1808.

Pittsburgh /ˈpɪtsbɜːɡ/ industrial city (machinery, chemicals) in the NE USA and the nation's largest inland port, where the Allegheny and Monongahela rivers join to form the Ohio River in Pennsylvania; population (1990) 369,900, metropolitan area 2,242,800.
    Established by the French as Fort Duquesne 1750, the site was taken by the British 1758 and renamed Fort Pitt. The main growth of the city's iron and steel industry came after 1850.

Pittsfield /ˈpɪtsfiːld/ city in W central Massachusetts, USA, on the Housatonic River, just E of the New York border; population (1990) 48,600. Industries include electronics and tourism. Herman Melville wrote Moby Dick 1851 at his home here.

Piura /ˈpjuərə/ capital of the department of the same name in the arid NW of Peru, situated on the Piura River, 160 km/100 mi SW of Punta Pariñas; population (1981) 186,000. It is the westernmost point in South America and was founded 1532 by the Spanish conquistadors left behind by Francisco Pizarro. Cotton is grown in the surrounding area.

Plata, Río de la /læ ˈplɑːtə/ or *River Plate* estuary in South America into which the rivers Paraná and Uruguay flow; length 320 km/200 mi and width up to 240 km/150 mi. The basin drains much of Argentina, Bolivia, Brazil, Uruguay, and Paraguay, which all cooperate in its development.

Plenty, Bay of /ˈplenti/ broad inlet on the northeast coast of North Island, New Zealand, with the port of Tauranga. One of the first canoes bringing Maori immigrants landed here about 1350.

Pleven /ˈplevən/ industrial town (textiles, machinery, ceramics) in N Bulgaria; population (1990) 168,000. In the Russo-Turkish War 1877, Pleven surrendered to the Russians after a siege of five months.

Ploeşti /plɔɪˈeʃt/ industrial city (textiles, paper, petrochemicals; oil centre) in SE Romania; population (1985) 234,000.

Plovdiv /ˈplɒvdɪv/ industrial city (textiles, chemicals, leather, tobacco) in Bulgaria, on the river Maritsa; population (1990) 379,100. Conquered by Philip of Macedon in the 4th century BC, it was

known as **Philippopolis** ('Philip's city'). It was capital of Roman Thrace.

Plymouth /ˈplɪməθ/ city and seaport in Devon, England, at the mouth of the river Plym, with dockyard, barracks, and a naval base at Devonport; population (1981) 243,900.
    The city rises N of the Hoe headland where tradition has it that Francis Drake played bowls before leaving to fight the Spanish Armada. The city centre was reconstructed after heavy bombing in World War II.

Plynlimon /plɪnˈlɪmən/ mountain in Powys, Wales, with three summits; the highest is 752 m/2,468 ft.

Plzeň /ˈpɪlzən/ (German *Pilsen*) industrial city (heavy machinery, cars, beer) in the Czech Republic, at the confluence of the Radbuza and Mze rivers, capital of Západočeský (West Bohemia) region; 84 km/52 mi SW of Prague; population (1991) 173,100.

Pnom Penh /nɒm ˈpen/ alternative form of ◊Phnom Penh, the capital of Cambodia.

Po /pəʊ/ longest river in Italy, flowing from the Cottian Alps to the Adriatic Sea; length 668 km/415 mi. Its valley is fertile and contains natural gas. The river is heavily polluted with nitrates, phosphates, and arsenic.

Pobeda, Pik /pəbˈjedə/ highest peak in the Tian Shan mountain range on the Kyrgyz-Chinese border, at 7,439 m/24,406 ft.

Podgorica /ˈpɒdɡərɪtsə/ former name (until 1946) of ◊Titograd, a city in Montenegro, Yugoslavia.

Podolsk /pəˈdɒlsk/ industrial city (oil refining, machinery, cables, cement, ceramics) in Russia, 40 km/25 mi SW of Moscow; population (1987) 209,000.

Pointe-Noire /pwænt ˈnwɑː/ chief port of the Congo, formerly (1950–58) the capital; population (1984) 297,000. Industries include oil refining and shipbuilding.

Poitiers /ˈpwɒtieɪ/ capital of Poitou-Charentes, W France; population (1990) 82,500. Products include chemicals and clothing.
    The Merovingian king Clovis defeated the Visigoths under Alaric here 507; Charles Martel stemmed the Saracen advance 732, and Edward the Black Prince of England defeated the French 1356.

Poitou-Charentes /pwɑːˈtuː ʃæˈrɒnt/ region of W central France, comprising the *départements* of Charente, Charente-Maritime, Deux-Sèvres, and Vienne
area 25,800 sq km/9,959 sq mi
capital Poitiers
products dairy products, wheat, chemicals, metal goods; brandy is made at Cognac
population (1986) 1,584,000
history once part of the Roman province of Aquitaine, this region was captured by the Visigoths in the 5th century and taken by the Franks AD 507. The area was contested by the English and French until the end of the Hundred Years' War 1453, when it was incorporated into France by Charles II.

Poland Republic of (*Polska Rzeczpospolita*)
area 127,886 sq km/49,325 sq mi
capital Warsaw
towns Lódź, Kraków, Wroclaw, Poznań, Katowice, Bydgoszcz, Lublin; ports Gdańsk, Szczecin, Gdynia
physical part of the great plain of Europe; Vistula, Oder, and Neisse rivers; Sudeten, Tatra, and Carpathian mountains on S frontier
environment atmospheric pollution derived from coal (producing 90% of the country's electricity), toxic waste from industry, and lack of sewage treatment have resulted in the designation of 27 ecologically endangered areas. Half the country's lakes have been seriously contaminated and three-quarters of its drinking water does not meet official health standards
features last wild European bison (only in protected herds)
head of state Lech Wałesa from 1990
head of government Waldemar Pawlak from 1993
political system emergent democratic republic
political parties Democratic Union, centrist, ex-Solidarity; Democratic Left Alliance, ex-communist; Centre Alliance, right of centre, Walesa-linked; Social Democratic Party of the Polish Republic, 1990 successor to Polish United Workers' Party (PUWP), social democratic; Union of Social Democrats, radical breakaway from PUWP formed 1990; Solidarność (Solidarity) Parliamentary Club (OKP), anti-communist coalition
exports coal, softwood timber, chemicals, machinery, ships, vehicles, meat, copper
currency zloty
population (1990 est) 38,363,000; growth rate 0.6% p.a.
life expectancy men 66, women 74 (1989)
languages Polish (official), German
religion Roman Catholic 95%
literacy 98% (1989)
GNP $276 billion (1988); $2,000 per head (1986)
chronology
1918 Poland revived as independent republic.

1939 German invasion and occupation.
1944 Germans driven out by Soviet forces.
1945 Polish boundaries redrawn at Potsdam Conference.
1947 Communist people's republic proclaimed.
1956 Poznań riots. Wladyslaw Gomulka installed as Polish United Workers' Party (PUWP) leader.
1970 Gomulka replaced by Edward Gierek after Gdańsk riots.
1980 Solidarity emerged as a free trade union following Gdańsk disturbances.
1981 Martial law imposed by General Jaruzelski.
1983 Martial law ended.
1984 Amnesty for political prisoners.
1985 Zbigniew Messner became prime minister.
1987 Referendum on economic reform rejected.
1988 Solidarity-led strikes and demonstrations called off after pay increases. Messner resigned; replaced by the reformist Mieczyslaw F Rakowski.
1989 Solidarity relegalized. April: new 'socialist pluralist' constitution formed. June: widespread success for Solidarity in assembly elections, the first open elections in 40 years. July: Jaruzelski elected president. Sept: 'Grand coalition', first non-Communist government since World War II formed; economic restructuring undertaken on free-market lines; W Europe and US create $1 billion aid package.
1990 Jan: PUWP dissolved; replaced by Social Democratic Party and breakaway Union of Social Democrats. Lech Wałesa elected president.
1991 Oct: Multiparty general election produced inconclusive result. Five-party centre-right coalition formed under Jan Olszewski. Treaty signed agreeing to withdrawal of Soviet troops.
1992 June: Olszewski ousted on vote of no confidence; succeeded by Waldemar Pawlak. July: Hanna Suchocka replaced Pawlak.
1993 Suchocka lost vote of confidence. Oct: inconclusive general election; Pawlak appointed premier, heading coalition government.

The soul of Poland is indestructible ...
She will rise again like a rock, which may
for a spell be submerged by a tidal wave,
but which remains a rock.

On **Poland** Winston Churchill, speech in the House of Commons 2 Aug 1939

Poltava /pɒl'tɑːvə/ industrial city (machinery, foodstuffs, clothing) in Ukraine, capital of Poltava region, on the river Vorskla; population (1987) 309,000. Peter the Great defeated Charles XII of Sweden here 1709.

Polynesia /ˌpɒlɪ'niːzɪə/ islands of Oceania E of 170° E latitude, including Hawaii, Kiribati, Tuvalu, Fiji, Tonga, Tokelau, Samoa, Cook Islands, and French Polynesia.

Pomerania /ˌpɒmə'reɪnɪə/ (Polish *Pomorze*, German *Pommern*) region along the southern shore of the Baltic Sea, including the island of Rügen, divided between Poland and (west of the Oder–Neisse line) East Germany 1945–90, and the

Federal Republic of Germany after reunification 1990. The chief port is Gdańsk. It was formerly a province of Germany.

An independent Slavic duchy in the 11th century, Pomerania was taken by Poland in the 12th century and divided into the principalities of West Pomerania and East Pomerania (or Pomerelia). West Pomerania became part of the Holy Roman Empire, while East Pomerania remained part of Poland until 1772, when it was ceded to Prussia.

Pomfret /pʌmfrɪt/ old form of ⟡Pontefract, a town in West Yorkshire, England.

Pommern /pɒmən/ German form of ⟡Pomerania, a region of N Europe, now largely in Poland.

Pomorze /pɑ'mɒʒeɪ/ Polish form of ⟡Pomerania, a region of N Europe, now largely in Poland.

Pompano Beach /ˌpɒmpənəu 'biːtʃ/ city in SE Florida, USA, N of Fort Lauderdale, on the Atlantic Ocean; population (1990) 72,400. Tourism and fruit processing are important to the economy.

Ponce /pɒnseɪ/ major city and industrial port (iron, textiles, sugar, rum) in S Puerto Rico, population (1980) 161,739. The settlement, established in the late 17th century, was named after the Spanish explorer Juan Ponce de León.

Pondicherry /ˌpɒndɪ'tʃeri/ union territory of SE India; area 480 sq km/185 sq mi; population (1991) 789,400. Its capital is Pondicherry, and products include rice, peanuts, cotton, and sugar. Pondicherry was founded by the French 1674 and changed hands several times among the French, Dutch, and British before being returned to France 1814 at the close of the Napoleonic wars. Together with Karaikal, Yanam, and Mahé (on the Malabar Coast) it formed a French colony until 1954 when all were transferred to the government of India; since 1962 they have formed the Union Territory of Pondicherry. Languages spoken include French, English, Tamil, Telegu, and Malayalam.

Ponta Delgada /pɒntə del'gɑːdə/ port, resort, and chief commercial centre of the Portuguese Azores Islands, on São Miguel; population (1981) 22,200.

Pontefract /pɒntɪfrækt/ town in Wakefield borough, West Yorkshire, N England, 34 km/21 mi SW of York; population (1981) 32,000. Products are coal and confectionery (liquorice Pontefract cakes). Features include the remains of the Norman castle where Richard II died.

Polynesia

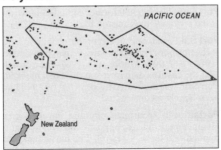

Pontiac /pɒntiæk/ motor-manufacturing city in Michigan, USA, 38 km/24 mi NW of Detroit; population (1990) 71,200.

Pontine Marshes /pɒntaɪn/ formerly malarial marshes in the Lazio region of Italy, near the coast 40 km/25 mi SE of Rome. Roman attempts to drain them were unsuccessful, and it was not until 1926, under Mussolini's administration, that they were brought into cultivation. Products include cereals, fruit and wine, and sugar beet.

Pontypool /ˌpɒntə'puːl/ (Welsh *Pontypŵl*) industrial town in Torfaen district, Gwent, SE Wales, on the Afon Llwyd, 15 km/9 mi N of Newport; population (1981) 36,800. Products include coal, iron and steel goods, tinplate, glass, synthetic textiles, and scientific instruments.

Pontypridd /ˌpɒntə'priːð/ industrial town (coal mining, chain and cable works, light industry on the Treforest trading estate) in Taff-Ely district, Mid Glamorgan, S Wales; population (1990 est) 33,600.

Poole /puːl/ industrial town (chemicals, engineering, boatbuilding, confectionery, pottery from local clay) and yachting centre on Poole Harbour, Dorset, S England, 8 km/5 mi W of Bournemouth; population (1991) 130,400.

The first Scout camp was held 1907 on Brownsea Island in the harbour; the island is now owned by the National Trust. Furzey Island, also within the harbour, is part of Wytch Farm, Britain's largest onshore oil development.

Pool Malebo /puːl mə'liːbəu/ lake on the border between the Congo Republic and Zaire, formed by a widening of the Zaïre River, 560 km/350 mi from its mouth.

Poona /puːnə/ former English spelling of ⟡Pune, a city in India; after independence 1947 the form Poona was gradually superseded by Pune.

Popocatépetl /ˌpɒpə,kætə'petl/ (Aztec 'smoking mountain') volcano in central Mexico, 50 km/30 mi SE of Mexico City; 5,340 m/17,526 ft. It last erupted 1920.

Pori /pɔːri/ (Swedish *Björneborg*) ice-free industrial port (nickel and copper refining, sawmills, paper, textiles) on the Gulf of Bothnia, SW Finland; population (1990) 76,400. A deepwater harbour was opened 1985.

Port Adelaide /pɔːt 'ædɪleɪd/ industrial port (cement, chemicals) in South Australia, on Gulf St Vincent, 11 km/7 mi NW of Adelaide; population (1985) 37,000.

Port Arthur /pɔːt 'ɑːθə/ industrial deepwater port (oil refining, shipbuilding, brass, chemicals) in Texas, USA, 24 km/15 mi SE of Beaumont; population (1990) 58,700. Founded 1895, it gained importance with the discovery of petroleum near Beaumont 1901.

There are great open spaces everywhere. Everything is planned to be a mile away from everything else.

On **Port Arthur** (and **Fort William**) Stephen Leacock *My Discovery of the West* 1937

Port Arthur /pɔːt 'ɑːθə/ former name (until 1905) of the port and naval base of Lüshun in NE China, now part of ▷Lüda.

Port Augusta /ɔː'ɡʌstə/ port (trading in wool and grain) in South Australia, at the head of Spencer Gulf, NNW of Adelaide; population (1985) 17,000. It is a base for the Royal Flying Doctor Service.

Port-au-Prince /,pɔːtəu'prɪns/ capital and industrial port (sugar, rum, textiles, plastics) of Haiti; population (1982) 763,000.
Founded by the French 1749, it was destroyed by earthquakes 1751 and 1770.

Port Darwin /,pɔːt 'dɑːwɪn/ port serving Darwin, the capital of Northern Territory, Australia.

Port Elizabeth /ɪ'lɪzəbəθ/ industrial port (engineering, steel, food processing) in Cape Province, South Africa, about 710 km/440 mi E of Cape Town on Algoa Bay; population (1980) 492,140.

Porterville /'pɔːtəvɪl/ city in S central California, USA, N of Bakersfield; population (1990) 29,600. Industries include citrus fruits and olive oil.

Port Harcourt /'hɑːkɔːt/ port (trading in coal, palm oil, and groundnuts) and capital of Rivers state in SE Nigeria, on the river Bonny in the Niger delta; population (1983) 296,200. It is also an industrial centre producing refined mineral oil, sheet aluminium, tyres, and paints.

Port Kelang /kə'læŋ/ (formerly until 1971 *Port Swettenham*) Malaysian rubber port on the Strait of Malacca, 40 km/25 mi SW of Kuala Lumpur; population (1980) 192,080.

Portland /'pɔːtlənd/ industrial port (aluminium, paper, timber, lumber machinery, electronics) and capital of Multnomah County, NW Oregon, USA; on the Columbia River, 173 km/108 mi from the sea, at its confluence with the Willamette River; population (1990) 437,300.

Portland /'pɔːtlənd/ industrial port and largest city of Maine, USA, on Casco Bay, SE of Sebago Lake; population (1990) 64,400.
It is the birthplace of the poet Henry Wadsworth Longfellow.

Portland, Isle of /'pɔːtlənd/ rocky peninsula off Dorset, S England, joined to the mainland by the bank of shingle, Chesil bank.
Portland Castle was built by Henry VIII 1520; Portland harbour is a naval base; building stone is still quarried.

Port Laoise /,pɔːt'liːʃə/ or *Portlaoighise* (formerly *Maryborough*) county town of County Laois, Republic of Ireland, 80 km/50 mi WSW of Dublin; population (1990 est) 9,500. It has woollen and flour-milling industries, and is the site of a top-security prison.

Port Louis /pɔːt 'luːi/ capital of Mauritius, on the island's northwest coast; population (1987) 139,000. Exports include sugar, textiles, watches, and electronic goods.

Port Mahón /pɔːt mɑː'ɒn/ or *Maó* port serving the capital Mahón on the Spanish island of Minorca (second in size of the Balearic Islands, after Majorca). The largest natural port in the Mediterranean, Mahón was occupied and colonized by the great seafaring powers of antiquity (the Phoenicians and the Romans), and later the Dutch, the French, and the English during the 17th–19th centuries.

Portmeirion /pɔːt'meəriən/ holiday resort in Gwynedd, Wales, built by the architect Clough Williams-Ellis in Italianate fantasy style; it was the setting of the 1967 cult television series *The Prisoner*.

Port Moresby /'mɔːzbi/ capital and port of Papua New Guinea on the south coast of New Guinea; population (1987) 152,000.

Pôrto /'portu/ (English *Oporto*) industrial city (textiles, leather, pottery) in Portugal, on the river Douro, 5 km/3 mi from its mouth; population (1984) 327,000. It exports port wine; the suburb Vila Nova de Gaia on the south bank of the Douro is known for its port lodges.
Pôrto is the second largest city in Portugal and has a 12th-century cathedral.

Pôrto Alegre /pɔːtəu æleɪɡri/ port and capital of Rio Grande do Sul state, S Brazil; population (1991) 1,254,600. It is a freshwater port for ocean-going vessels, and is Brazil's major commercial centre.

Port-of-Spain /'pɔːt əv 'speɪn/ port and capital of Trinidad and Tobago, on the island of Trinidad; population (1988) 58,000. It has a cathedral (1813–28) and the San Andres Fort (1785).

Porto Novo /'pɔːtəu 'nəuvəu/ capital of Benin, W Africa; population (1982) 208,258. It was a former Portuguese centre for the slave and tobacco trade with Brazil and became a French protectorate 1863.

Porto Rico /'pɔːtəu 'riːkəu/ name until 1932 of ▷Puerto Rico, a US island in the Caribbean.

Port Phillip Bay /'fɪlɪp/ inlet off Bass Strait, Victoria, Australia, on which the city of Melbourne stands.

Port Pirie /'pɪri/ industrial port (smelting of ores from the Broken Hill mines, and chemicals) in South Australia; population (1985) 16,030.

Port Rashid /ræ'ʃiːd/ port serving Dubai in the United Arab Emirates.

Port Royal /,pɔːt 'rɔɪəl/ former capital of Jamaica, at the entrance to Kingston harbour.

Port Said /saɪd/ port in Egypt, on reclaimed land at the northern end of the Suez Canal; population (1983) 364,000. During the 1967 Arab-Israeli War the city was damaged and the canal blocked; Port Said was evacuated by 1969 but by 1975 had been largely reconstructed.

Portsmouth /'pɔːtsməθ/ city and naval port in Hampshire, England, opposite the Isle of Wight; population (1991) 174,700. The naval dockyard was closed 1981 although some facilities remain.

It was already a port in the days of King Alfred, but in 1194 Richard I recognized its strategic importance and created a settlement on Portsea Island. The Tudor warship *Mary Rose* and Admiral Horatio Nelson's flagship, HMS *Victory*, are exhibited here. The novelist Charles Dickens was born in the Portsmouth suburb of Landport.

Portsmouth /ˈpɔːtsməθ/ port and independent city in SE Virginia, USA, on the Elizabeth River, seat of a US navy yard and training centre; population (1990) 103,900. Manufactured goods include electronic equipment, chemicals, clothing, and processed food.

Portsmouth /ˈpɔːtsməθ/ port in Rockingham County, SE New Hampshire, USA, on the estuary of the Piscataqua River; the state's only seaport; population (1990) 25,900. The nearby US Navy Yard (on Seavy's Island) dates from the 1790s and specializes in submarine construction and maintenance. Founded 1623, Portsmouth was the state capital 1679–1775. The John Paul Jones House is here. The treaty ending the Russo-Japanese War was signed here 1905.

Port Sunlight /pɔːt ˈsʌnlaɪt/ model village built 1888 by W H Lever (1851–1925) for workers at the Lever Brothers soap factory at Birkenhead, near Liverpool, NW England. Designed for a population of 3,000, and covering an area of 353 ha/130 acres, it includes an art gallery, church, library, and social hall.

Port Swettenham /ˈswetnəm/ former name of ◊Port Kelang, a port in Peninsular Malaysia.

Port Talbot /ˈtɔːlbət/ industrial port (tinplate and steel strip mill) in West Glamorgan, Wales; population (1988 est) 49,000.

Portugal Republic of (*República Portuguesa*)
area 92,000 sq km/35,521 sq mi (including the Azores and Madeira)
capital Lisbon
towns Coimbra, ports Pôrto, Setúbal
physical mountainous in N, plains in S
features rivers Minho, Douro, Tagus (Tejo), Guadiana; Serra da Estrêla mountains
head of state Mario Alberto Nobre Lopes Soares from 1986
head of government Aníbal Cavaco Silva from 1985
political system democratic republic
political parties Social Democratic Party (PSD), moderate, left of centre; Socialist Party (PS), progressive socialist; Democratic Renewal Party (PRD), centre left; Democratic Social Centre Party (CDS), moderate, left of centre
exports wine, olive oil, resin, cork, sardines, textiles, clothing, pottery, pulpwood
currency escudo
population (1990 est) 10,528,000; growth rate 0.5% p.a.
life expectancy men 71, women 78 (1989)
language Portuguese
religion Roman Catholic 97%
literacy men 89%, women 80% (1985)
GNP $33.5 bn (1987); $2,970 per head (1986)

chronology
1928–68 Military dictatorship under António de Oliveira Salazar.
1968 Salazar succeeded by Marcello Caetano.
1974 Caetano removed in military coup led by General Antonio Ribeiro de Spínola. Spínola replaced by General Francisco da Costa Gomes.
1975 African colonies became independent.
1976 New constitution, providing for return to civilian rule, adopted. Minority government appointed, led by Socialist Party leader Mario Soares.
1978 Soares resigned.
1980 Francisco Balsemão formed centre-party coalition after two years of political instability.
1982 Draft of new constitution approved, reducing powers of presidency.
1983 Centre-left coalition government formed.
1985 Aníbal Cavaco Silva became prime minister.
1986 Mario Soares elected first civilian president in 60 years. Portugal joined European Community.
1988 Portugal joined Western European Union.
1989 Constitution amended to allow major state enterprises to be denationalized.
1991 Mario Soares re-elected president; Social Democrats (PSD) majority slightly reduced in assembly elections.

Portuguese East Africa /ˌpɔːtjuˈgiːz/ former name of ◊Mozambique in SE Africa.

Portuguese Guinea /ˈgɪni/ former name of ◊Guinea-Bissau in W Africa.

Portuguese West Africa former name of ◊Angola in SW Africa.

Posen /ˈpəʊzən/ German form of ◊Poznań, a city in Poland.

Potchefstroom /ˈpɒtʃəfstrəʊm/ oldest town in the Transvaal, South Africa on the river Mooi, founded by Boers (descendants of Dutch settlers) trekking from the Cape 1838. It is the centre of a large cattle-rearing area.

Potomac /pəˈtəʊmək/ river in W Virginia, Virginia, and Maryland states, USA, rising in the Allegheny Mountains, and flowing SE through

Powys

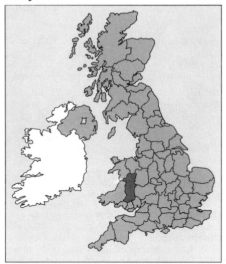

Washington DC, into Chesapeake Bay. It is formed by the junction of the N Potomac, about 153 km/95 mi long, and S Potomac, about 209 km/130 mi long, and is itself 459 km/285 mi long.

Potosí /ˌpɒtəʊˈsiː/ town in SW Bolivia, on the Cerro de Potosí slopes at 4,020 m/13,189 ft; it is one of the highest towns in the world; population (1988) 114,100. Silver, tin, lead, and copper are mined here.

It was founded by Spaniards 1545; during the 17th and 18th centuries it was the chief silver-mining town and foremost city in South America.

Potsdam /ˈpɒtsdæm/ capital of the state of Brandenburg, Germany, on the river Havel SW of Berlin; population (1986) 140,000. Products include textiles, pharmaceuticals, and electrical goods. A leading garrison town and Prussian military centre, Potsdam was restored to its position as capital of Brandenburg with the reunification of Germany 1990. The New Palace 1763–70 and Sans Souci were both built by Frederick the Great, and Hitler's Third Reich was proclaimed in the garrison church 21 March 1933. The Potsdam Conference took place here 1945.

Poughkeepsie /pəˈkɪpsi/ city in SE New York, USA, on the Hudson River, N of New York City; population (1990) 28,900. Products include chemicals, ball bearings, and cough drops. Vassar College is here. Settled by the Dutch 1687, it was the temporary capital of New York 1717.

Poverty Bay /ˈpɒvəti ˈbeɪ/ inlet on the east coast of North Island, New Zealand, on which the port of Gisborne stands. The English explorer Capt James Cook made his first landing here 1769.

> *The Bay ... I have named Poverty Bay because it afforded us not one thing we wanted.*
>
> On **Poverty Bay** Capt James Cook *Journal* 11 October 1769

Powys /ˈpəʊɪs/ county in central Wales
area 5,080 sq km/1,961 sq mi
towns Llandrindod Wells (administrative headquarters)
features Brecon Beacons National Park; Black Mountains; rivers: Wye, Severn, which both rise on Plynlimon in Dyfed; Lake Vyrnwy, artificial reservoir supplying Liverpool and Birmingham; alternative technology centre near Machynlleth
products agriculture, dairy cattle, sheep
population (1991) 117,500
languages 20% Welsh, English
famous people George Herbert, Robert Owen.

Poznań /ˈpɒznæn/ (German *Posen*) industrial city (machinery, aircraft, beer) in W Poland; population (1985) 553,000. Founded 970, it was settled by German immigrants 1253 and passed to Prussia 1793; it was restored to Poland 1919.

Pozzuoli /ˌpɒtsuːˈəʊli/ port in Campania, S Italy, W of Naples; population (1981) 71,000. It is shaken by some 25 earthquakes a day, 60% of its buildings are uninhabitable, and an eventual major disaster seems inevitable.

Prague /prɑːg/ (Czech *Praha*) city and capital of the Czech Republic on the river Vltava; population (1991) 1,212,000. Industries include cars, aircraft, chemicals, paper and printing, clothing, brewing, and food processing. It became the capital 1918.

> *... the town that conceived the Reformation and hatched the Thirty Years' War.*
>
> On **Prague** Jerome K Jerome *Three Men on the Bummel* 1900

Praha /ˈprɑːha/ Czech name for ◊Prague, the capital of the Czech Republic.

Praia /ˈpraɪa/ port and capital of the Republic of Cape Verde, on the island of São Tiago (Santiago); population (1980) 37,500. Industries include fishing and shipping.

Prato /ˈprɑːtəʊ/ industrial town (woollens) in Tuscany, central Italy; population (1988) 165,000. The 12th-century cathedral has works of art by Donatello, Filippo Lippi, and Andrea della Robbia.

Pressburg /ˈpresbʊək/ German name of ◊Bratislava, the capital of the Slovak Republic.

Preston /ˈprestən/ industrial town (textiles, chemicals, electrical goods, aircraft, and shipbuilding), and administrative headquarters of Lancashire, NW England, on the river Ribble, 34 km/21 mi S of Lancaster; population (1990 est) 125,800. Oliver Cromwell defeated the Royalists at Preston 1648. It is the birthplace of Richard Arkwright, inventor of cotton-spinning machinery.

Prestwick /ˈprestwɪk/ town in Strathclyde, SW Scotland; population (1985) 13,532. Industries include engineering and aerospace engineering. There is an international airport, which is linked with a free port.

Pretoria /prɪˈtɔːriə/ administrative capital of the Republic of South Africa from 1910 and capital of Transvaal province from 1860; population (1985) 741,300. Industries include engineering, chemicals, iron, and steel. Founded 1855, it was named after Boer leader Andries Pretorius (1799–1853).

Pribilof Islands /ˈprɪbɪlɒf/ group of four islands in the Bering Sea, of volcanic origin, 320 km/200 mi SW of Bristol Bay, Alaska, USA. Named after Gerasim Pribilof, who reached them 1786, they were sold by Russia to the USA 1867 with Alaska, of which they form part. They were made a fur-seal reservation 1868.

Primorye /priːˈmɔːriei/ territory of SE Siberian Russia, on the Sea of Japan; area 165,900 sq km/64,079 sq mi; population (1985) 2,136,000; capital Vladivostok. Timber and coal are produced.

Prince Edward Island /ˈedwəd/ province of E Canada
area 5,700 sq km/2,200 sq mi
capital Charlottetown
features Prince Edward Island National Park; Summerside Lobster Carnival
products potatoes, dairy products, lobsters, oysters, farm vehicles
population (1991) 129,900
history first recorded visit by Cartier 1534, who called it Isle St-Jean; settled by French; taken by British 1758; annexed to Nova Scotia 1763; separate colony 1769; renamed after Prince Edward of Kent, father of Queen Victoria 1798; settled by Scottish 1803; joined Confederation 1873.
　　In the late 1980s, there was controversy about whether to build a bridge to the mainland.

Prince Rupert /ˈruːpət/ fishing port at the mouth of the Skeena River in British Columbia, Canada, on Kaien Island, west side of Tsimpsean peninsula; population (1983) 16,786.

Princeton /ˈprɪnstən/ borough in Mercer County, W central New Jersey, USA, 80 km/50 mi SW of New York; population (1990) 13,200. It is the seat of Princeton University, founded 1746 at Elizabethtown and moved to Princeton 1756.
　　It is the site of an important battle of the War of American Independence 1777.

Prince Edward Island

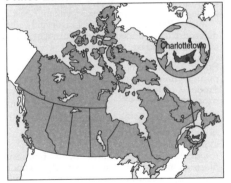

Princetown /ˈprɪnstaun/ village on the W of Dartmoor, Devon, SW England, containing Dartmoor prison, opened 1809.

Prince William Sound /prɪns ˈwɪljəm/ channel in the Gulf of Alaska, extending 200 km/125 mi NW from Kayak Island. In March 1989 the oil tanker *Exxon Valdez* ran aground here, spilling 12 million gallons of crude oil in one of the world's greatest oil-pollution disasters.

Pripet /ˈpriːpɪt/ (Russian *Pripyat*) river in E Europe, a tributary of the river Dnieper, which it joins 80 km/50 mi above Kiev, Ukraine, after a course of about 800 km/500 mi. The *Pripet marshes* near Pinsk were of strategic importance in both world wars.

Priština /ˈpriʃtɪna/ capital of Kosovo autonomous province, S Serbia, Yugoslavia; population (1981) 216,000. Once capital of the medieval Serbian empire, it is now a trading centre.

Prokopyevsk /prəˈkɒpjefsk/ chief coalmining city of the Kuznetsk Basin, Siberian Russia, on the river Aba; population (1987) 278,000.

Provence-Alpes-Côte d'Azur /prəˈvɒns ælps ˌkəut dæˈzjuə/ region of SE France, comprising the *départements* of Alpes-de-Haute-Provence, Hautes-Alpes, Alpes-Maritimes, Bouches-du-Rhône, Var, and Vaucluse; area 31,400 sq km/12,120 sq mi; capital Marseille; population (1986) 4,059,000. The *Côte d'Azur*, on the Mediterranean, is a tourist centre. Provence was an independent kingdom in the 10th century, and the area still has its own language, Provençal.

Providence /ˈprɒvɪdəns/ industrial seaport (jewellery, silverware, textiles and textile machinery, watches, chemicals, meatpacking) and capital of Rhode Island, USA, on Narragansett Bay and the Providence River, 27 mi/43 km from the Atlantic Ocean; population (1990) 160,700.
　　Providence was founded 1636 by Roger Williams, who had been banished from Plymouth colony for his religious beliefs.

Provo /ˈprəuvəu/ city in N central Utah, USA, on the Provo River, SE of Salt Lake City; seat of Utah County; population (1990) 86,800. Industries include iron and steel, food processing, and electronics. Brigham Young University is here.

Prudhoe Bay /ˈprʌdəu ˈbeɪ/ bay in N Alaska; a pipeline links oil fields with the Gulf of Alaska to the S.

Prussia /ˈprʌʃə/ N German state 1618–1945 on the Baltic coast. It was an independent kingdom until 1867, when it became, under Otto von Bismarck, the military power of the North German Confederation and part of the German Empire 1871 under the Prussian king Wilhelm I. West Prussia became part of Poland under the Treaty of Versailles, and East Prussia was largely incorporated into the USSR after 1945.

Prut /pruːt/ river that rises in the Carpathian Mountains of SW Ukraine, and flows 900 km/565 mi to meet the Danube at Reni. For part of its course it follows the eastern frontier of Romania.

Przemysl /ˈpʃemɪsuː/ industrial city (timber, ceramics, flour milling, tanning, distilling, food processing, gas, engineering) in SE Poland; population (1991) 68,500.
history Founded in the 8th century, it belonged alternately to Poland and Kiev in the 10th–14th centuries. An Austrian territory 1722–1919, it was a frontier fortress besieged by Soviet troops Sept 1914–March 1915 and was occupied by the Germans June 1941–July 1944.

Pskov /pskɒf/ industrial city (food processing, leather) in Russia, on the Velikaya River, SW of St Petersburg; population (1987) 202,000. Dating from 965, it was independent 1348–1510.

Puebla (de Zaragoza) /ˈpweblə deɪ ˌsærəˈɡɒsə/ industrial city (textiles, sugar refining, metallurgy, hand-crafted pottery and tiles) and capital of Puebla state, S central Mexico; population (1986) 1,218,000. Founded 1535 as *Pueblo de los Angeles*, it was later renamed after General de Zaragoza, who defeated the French here 1862.

Puebla was the only Mexican town in which it seemed to me possible to live with some happiness. It had more than the usual beauty: it had grace.

On **Puebla** Graham Greene 'The Lawless Roads' 1939

Pueblo /ˈpweblaʊ/ city in S central Colorado, USA, on the Arkansas River, SE of Colorado Springs; population (1990) 98,600. Industries include steel, coal, lumber, and livestock and other agricultural products.

Puerto Rico /ˈpweətəʊ ˈriːkəʊ/ the Commonwealth of; easternmost island of the Greater Antilles, situated between the US Virgin Islands and the Dominican Republic
area 9,000 sq km/3,475 sq mi
capital San Juan

Punjab

towns ports Mayagüez, Ponce
features volcanic mountains run E–W; the islands of Vieques and Culebra belong to Puerto Rico
exports sugar, tobacco, rum, pineapples, textiles, plastics, chemicals, processed foods
currency US dollar
population (1990) 3,522,000
languages Spanish and English (official)
religion Roman Catholic
government under the constitution of 1952, similar to that of the USA, with a governor elected for four years, and a legislative assembly with a senate and house of representatives
history visited 1493 by Columbus; annexed by Spain 1509; ceded to the USA after the Spanish-American War 1898; achieved commonwealth status with local self-government 1952.
 This was confirmed in preference to independence by a referendum 1967, but there is both an independence movement and one preferring incorporation as a state of the USA. Although legislation in favour of a further referendum was proposed and discussed 1990–91, it was later shelved.

It is a kind of lost love-child, born to the Spanish Empire and fostered by the United States.

On **Puerto Rico** Nicholas Wollaston
Red Rumba 1962

Puerto Sandino /ˈpweətəʊ sænˈdiːnaʊ/ major port on the Pacific west coast of Nicaragua, known as *Puerto Somoza* until 1979.

Puget Sound /ˈpjuːdʒxt/ inlet of the Pacific Ocean on the west coast of Washington State, USA.

Puglia (English *Apulia*) region of Italy, the southeastern 'heel'; area 19,300 sq km/7,450 sq mi; capital Bari; population (1990) 4,081,500. Products include wheat, grapes, almonds, olives, and vegetables. The main industrial centre is Taranto.

Pula /ˈpuːlə/ commercial and naval port in W Croatia, on the Adriatic coast; population (1981) 77,278. A Roman naval base, *Colonia Pietas Julia*, it was seized by Venice 1148, passed to Austria 1815, to Italy 1919, to Yugoslavia 1947, and in 1991 became part of independent Croatia.
 It has a Roman theatre, and a castle and cathedral constructed under Venetian rule. There is an annual film festival.

Pune /ˈpuːnə/ formerly *Poona* city in Maharashtra, India; population (1985) 1,685,000. Products include chemicals, rice, sugar, cotton, paper, and jewellery.

Punjab /ˌpʌnˈdʒɑːb/ state of NW India
area 50,400 sq km/19,454 sq mi
capital Chandigarh
towns Amritsar, Jalandhar, Faridkot, Ludhiana
features mainly agricultural, crops chiefly under irrigation; longest life expectancy rates in India

PYRENEES

(59 for women, 64 for men); Harappa has ruins from the Indus Valley civilization 2500 to 1600 BC
population (1991) 20,190,800
language Punjabi
religion 60% Sikh, 30% Hindu; there is friction between the two groups.

Punjab /ˌpʌn'dʒɑːb/ state of NE Pakistan
area 205,344 sq km/79,263 sq mi
capital Lahore
features wheat cultivation (by irrigation)
population (1981) 47,292,000
languages Punjabi, Urdu
religion Muslim.

Punta Arenas /'puntə ə'reməs/ (Spanish 'sandy point', former name *Magallanes*) seaport (trading in meat, wool, and oil) in Chile, capital of Magallanes province, on Magellan Strait, southernmost town on the American mainland; population (1990) 120,000.

Purbeck, Isle of /'pɜːbek/ peninsula in the county of Dorset, S England. Purbeck marble and china clay are obtained from the 'isle', which includes Corfe Castle and Swanage.

Puri /'puəri/ town in Orissa, E India, with a statue of Jagganath or Vishnu, one of the three gods of Hinduism, dating from about 318, which is annually taken in procession on a large vehicle (hence the word 'juggernaut' used for a very large lorry). Devotees formerly threw themselves beneath its wheels.

Pusan /ˌpuː'sæn/ or *Busan* chief industrial port (textiles, rubber, salt, fishing) of South Korea; population (1985) 3,797,600. It was invaded by the Japanese 1592 and opened to foreign trade 1883.

Pushkin /'puʃkɪn/ town NW of St Petersburg, Russia; population (1985) 91,000. Founded by Peter the Great as Tsarskoe Selo ('tsar's village') 1708, it has a number of imperial summer palaces, restored after German troops devastated the town 1941–44. In the 1920s it was renamed Detskoe Selo ('children's village'), but since 1937 it has been known as Pushkin, after the poet Aleksandr Pushkin, who was educated at the school that is now a museum commemorating him.

Puy, Le see ▷Le Puy, a town in France.

Pwllheli /puɬ'heli/ resort in Gwynedd, Wales, on Cardigan Bay; the Welsh National Party, Plaid Cymru, was founded here 1925.

Pyongyang /ˌpjɒŋ'jæŋ/ capital and industrial city (coal, iron, steel, textiles, chemicals) of North Korea; population (1984) 2,640,000.

Pyrenees /ˌpɪrə'niːz/ (French *Pyrénées*; Spanish *Pirineos*) mountain range in SW Europe between France and Spain; length about 435 km/ 270 mi; highest peak Aneto (French Néthon) 3,404 m/11,172 ft. Andorra is entirely within the range. Hydroelectric power has encouraged industrial development in the foothills.

Qatar State of (*Dawlat Qatar*)
area 11,400 sq km/4,402 sq mi
capital and chief port Doha
town Dukhan, centre of oil production
physical mostly flat desert with salt flats in S
features negligible rain and surface water; only
3% is fertile, but irrigation allows self-sufficiency
in fruit and vegetables; extensive oil discoveries
since World War II
head of state and government Sheik Khalifa
bin Hamad al-Thani from 1972
political system absolute monarchy
political parties none
exports oil, natural gas, petrochemicals, fertilizers, iron, steel
currency riyal
population (1990 est) 498,000 (half in Doha;
Arab 40%, Indian 18%, Pakistani 18%); growth
rate 3.7% p.a.
life expectancy men 68, women 72 (1989)
languages Arabic (official), English
religion Sunni Muslim 95%
literacy 60% (1987)
GNP $5.9 bn (1983); $35,000 per head
chronology
1916 Qatar became a British protectorate.
1970 Constitution adopted, confirming the
emirate as an absolute monarchy.
1971 Independence achieved from Britain.
1972 Emir Sheik Ahmad replaced in blood-

less coup by his cousin, Crown Prince Sheik
Khalifa.
1991 Forces joined UN coalition in Gulf War
against Iraq.

Qattara Depression /kə'tɑːrə/ tract of the Western Desert, Egypt, up to 125 m/400 ft below sea
level; area 20,000 sq km/7,500 sq mi. Its very
soft sand makes it virtually impassable to vehicles,
and it protected the left flank of the Allied armies
before and during the battle of Alamein 1942.

Qingdao /ˌtʃɪŋ'dau/ or *Tsingtao* industrial port
(brewing) and summer resort in Shandong
province, E China; population (1984) 1,229,500.

Qinghai /ˌtʃɪŋ'haɪ/ or *Tsinghai* province of NW
China
area 721,000 sq km/278,306 sq mi
capital Xining
features mainly desert, with nomadic herders
products oil, livestock, medical products
population (1990) 4,457,000; minorities
include 900,000 Tibetans (mostly nomadic herders); Tibetan nationalists regard the province as
being under colonial rule.

Qisarya /kiː'sɑːriə/ Mediterranean port N of Tel
Aviv-Jaffa, Israel; there are underwater remains
of Herod the Great's port of Caesarea.

Qld abbreviation for ▷*Queensland*, an
Australian state.

Qom /kum/ or *Qum* holy city of Shi'ite Muslims,
in central Iran, 145 km/90 mi S of Tehran;
population (1986) 551,000. The Islamic academy of Madresseh Faizieh 1920 became the
headquarters of Ayatollah Khomeini.

Quai d'Orsay /'keɪ dɔː'seɪ/ part of the left bank
of the river Seine in Paris, where the French
Foreign Office and other government buildings
are situated. The name has become synonymous
with the Foreign Office itself.

Que. abbreviation for ▷*Québec*, a Canadian
province.

Québec /kwɪ'bek/ province of E Canada
area 1,540,700 sq km/594,710 sq mi
capital Quebec
towns Montréal, Laval, Sherbrooke, Verdun,
Hull, Trois-Rivières
features immense water-power resources (for
example, the James Bay project)
products iron, copper, gold, zinc, cereals, potatoes, paper, textiles, fish, maple syrup (70% of
world's output)
population (1991) 6,811,800
language French (the only official language
since 1974, although 17% speak English). Language laws 1989 prohibit the use of English on
street signs
history known as New France 1534–1763; captured by the British and became province of
Québec 1763–90, Lower Canada 1791–1846,
Canada East 1846–67; one of the original provinces 1867. Nationalist feelings 1960s (despite
existing safeguards for Québec's French-derived
civil law, customs, religion, and language) were
encouraged by French president de Gaulle's
exclamation '*Vive le Québec libre/Long live free
Québec*' on a visit to the province, and led to

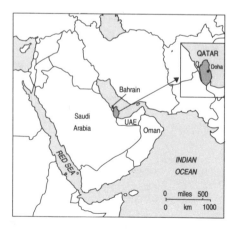

Québec

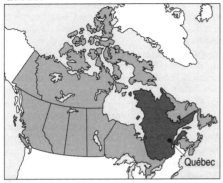

Québec

the foundation of the Parti Québecois by René Lévesque 1968.

The Québec Liberation Front (FLQ) separatists had conducted a bombing campaign in the 1960s and fermented an uprising 1970; Parti Québecois won power 1976; a referendum on 'sovereignty-association' (separation) was defeated 1980. In 1982, when Canada severed its last legal ties with the UK, Québec opposed the new Constitution Act as denying the province's claim to an absolute veto over constitutional change. Robert Bourassa and Liberals returned to power 1985 and enacted restrictive English-language legislation. The right of veto was proposed for all provinces of Canada 1987, but the agreement failed to be ratified by its 1990 deadline and support for independence grew. The Parti Québecois was defeated by the Liberal Party 1989.

In a referendum 1992 Quebec's French-speaking population rejected constitutional reforms giving greater autonomy to the province, on the grounds that the reforms were insufficient to protect them from losing their identity within the English-speaking majority.

Québec /kwɪˈbek/ capital and industrial port (textiles, leather, timber, paper, printing, and publishing) of Québec province, on the St Lawrence River, Canada; population (1986) 165,000, metropolitan area 603,000.

The city was founded by the French explorer Samuel de Champlain 1608, and was a French colony 1608–1763. The British, under General Wolfe, captured Québec 1759 after a battle on the nearby Plains of Abraham; both Wolfe and the French commander Montcalm were killed. Québec is a centre of French culture, there are two universities, Laval 1663 (oldest in North America) and Québec 1969. Its picturesque old town survives below the citadel about 110 m/ 360 ft above the St Lawrence River.

The impression made upon the visitor by this Gibraltar of America ... is at once unique and lasting.

On the city of **Québec** Charles Dickens
American Notes 1842

Queen Charlotte Islands /ˈʃɑːlət/ archipelago about 160 km/100 mi off the coast of British Columbia, W Canada, of which it forms part; area 9,790 sq km/3,780 sq mi; population 2,500. Graham and Moresby are the largest of about 150 islands. There are timber and fishing industries.

Queen Maud Land /kwiːn ˈmɔːd/ region of Antarctica W of Enderby Land, claimed by Norway since 1939.

Queens /kwinz/ mainly residential borough and county at the west end of Long Island, New York City, USA; population (1980) 1,891,300.

Queen's County /kwiːnz/ former name (until 1920) of ▷Laois, a county in the Republic of Ireland.

Queensland /ˈkwiːnzlænd/ state in NE Australia
area 1,727,200 sq km/666,699 sq mi
capital Brisbane
towns Townsville, Toowoomba, Cairns
features Great Dividing Range, including Mount Bartle Frere 1,657 m/5,438 ft; Great Barrier Reef (collection of coral reefs and islands about 2,000 km/1,250 mi long, off the east coast); Gold Coast, 32 km/20 mi long, S of Brisbane; Mount Isa mining area; Sunshine Coast, a 100-km/60-mi stretch of coast N of Brisbane, between Rainbow Beach and Bribie Island, including the resorts of Noosa Heads, Coolum Beach, and Caloundra
products sugar, pineapples, beef, cotton, wool, tobacco, copper, gold, silver, lead, zinc, coal, nickel, bauxite, uranium, natural gas
population (1987) 2,650,000
history part of New South Wales until 1859, when it became self-governing. In 1989 the ruling National Party was defeated after 32 years in power and replaced by the Labor Party.

Queenstown /ˈkwiːnztaun/ former name (1849–1922) of ▷Cóbh, a port in the Republic of Ireland.

Quemoy /keˈmɔɪ/ island off the southeast coast of China, and administered, along with the island

Queensland

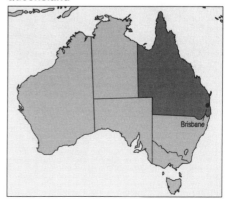

of Matsu, by Taiwan. Quemoy: area 130 sq km/ 50 sq mi; population (1982) 57,847. Matsu: 44 sq km/17 sq mi; population (1982) 11,000. China claims Quemoy; the USA supported Taiwan until 1979, eight years after Taiwan had ceased to be a member of the United Nations (UN). When the islands were shelled from the mainland 1960, the USA declared they would be defended if attacked. Since 1986 the Taiwanese government has encouraged the growth of tourism on Quemoy, and in 1992 ended 43 years of martial-law rule.

A dagger pointed at the heart of communism.

On **Quemoy** Chinese saying

Quetta /'kwetə/ summer resort and capital of Baluchistan, W Pakistan; population (1981) 281,000. It was linked to Shikarpur by a gas pipeline 1982.

Quezon City /'keɪsɒn 'sɪti/ former capital of the Philippines 1948–76, northeastern part of metropolitan Manila (the present capital), on Luzon Island; population (1990) 1,166,800. It was named after the Philippines' first president, Manuel Luis Quezon (1878–1944).

Qufu /,tʃuː'fuː/ or *Chufu* town in Shandong province, China; population 27,000. Qufu was capital of the state of Lu during the Zhou dynasty (1122–221 BC). It is the birthplace of Kong Zi (Confucius) and the site of the Great Temple of Confucius. The tomb of Confucius is to the N of the temple.

Quiberon /,kiːbə'rɒn/ peninsula and coastal town in Brittany, NW France. In 1759 the British admiral Hawke defeated a French fleet (under Conflans) in Quiberon Bay.

Quimper /kæm'peə/ town in Brittany, NW France, on the river Odet; a centre for the manufacture of decorative pottery since the 16th century; population (1990) 62,500. There is a fine 15th-century Gothic cathedral.

Quintana Roo /kɪn'tɑːnə 'rəuəu/ state in SE Mexico, E of the Yucatán peninsula; population (1989) 414,300. There are Maya remains at Tulum; Cancun is a major resort and free port.

Quito /'kiːtəu/ capital and industrial city (textiles, chemicals, leather, gold, silver) of Ecuador, about 3,000 m/9,850 ft above sea level; population (1986) 1,093,300. It was an ancient settlement, taken by the Incas about 1470 and by the Spanish 1534. It has a temperate climate all year round.

Qum /kuːm/ alternative spelling of ⟡Qom, a city of Iran.

QwaQwa /'kwɑːkwɑː/ black homeland of South Africa that achieved self-governing status 1974; population (1985) 181,600.

R

Rajasthan

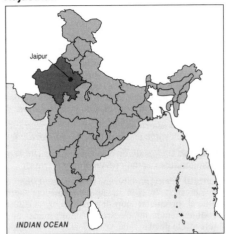

Rabat /rə'bɑːt/ capital of Morocco, industrial port (cotton textiles, carpets, leather goods) on the Atlantic coast, 177 km/110 mi W of Fez; population (1982) 519,000, Rabat-Salé 842,000. It is named after its original *ribat* or fortified monastery.

Rabaul /rɑː'baul/ largest port (trading in copra and cocoa) of Papua New Guinea, on the volcanic island of New Britain, SW Pacific; population (1980) 14,954. It was destroyed by British bombing after its occupation by the Japanese in 1942, but was rebuilt.

Racine /rə'siːn/ city and port of entry in SW Wisconsin, USA, on the Root River where it flows into Lake Michigan; population (1990) 84,300. Industries include automotive parts, farm machinery, and wax products.

Radium Hill /'reɪdiəm/ mining site SW of Broken Hill, New South Wales, Australia, formerly a source of radium and uranium.

Radnorshire /'rædnəʃə/ (Welsh *Sir Faesyfed*) former border county of Wales, merged with Powys 1974. Presteign was the county town.

Radom /'rɑːdɒm/ industrial city (flour-milling, brewing, tobacco, leather, bicycles, machinery; iron works) in Poland, 96 km/60 mi S of Warsaw; population (1990) 228,500. Radom became Austrian 1795 and Russian 1825 and was returned to Poland 1919.

Ragusa /rə'guːzə/ town in Sicily, Italy, 54 km/34 mi SW of Syracuse; there are textile industries; population (1981) 64,492. It stands over 450 m/1,500 ft above the river Ragusa, and there are ancient tombs in caves nearby.

Ragusa /rə'guːzə/ Italian name (until 1918) for the Croatian town of ▷Dubrovnik. Its English name was *Arrogosa*, from which the word 'argosy' is derived, because of the town's fame for its trading fleets while under Turkish rule in the 16th century.

Rainier, Mount /rə'nɪə/ mountain in the Cascade Range, Washington State, USA; 4,392 m/14,415 ft, crowned by 14 glaciers and carrying dense forests on its slopes. It is a quiescent volcano. Mount Rainier national park was dedicated 1899.

Rajasthan /ˌrɑːdʒə'stɑːn/ state of NW India
area 342,200 sq km/132,089 sq mi
capital Jaipur
features includes the larger part of the Thar Desert, where India's first nuclear test was carried out; in the SW is the Ranthambhor wildlife reserve, formerly the private hunting ground of the maharajahs of Jaipur, and rich in tiger, deer, antelope, wild boar, crocodile, and sloth bear
products oilseed, cotton, sugar, asbestos, copper, textiles, cement, glass
population (1991) 43,880,600
languages Rajasthani, Hindi
religions 90% Hindu, 3% Muslim
history formed 1948; enlarged 1956.

Rajshahi /rɑːdʒ'ʃɑːhi/ capital of Rajshahi region, W Bangladesh; population (1981) 254,000. It trades in timber and vegetable oil.

Raleigh /'rɔːli/ industrial city (food processing, electrical machinery, textiles) and capital of North Carolina, USA; population (1990) 208,000.

Raleigh, Fort /'rɔːli/ site of the first English settlement in America, at the north end of Roanoke Island, North Carolina, USA, to which in 1585 Walter Raleigh sent 108 colonists from Plymouth, England, under his cousin Richard Grenville. In 1586 Francis Drake took the dissatisfied survivors back to England. The outline fortifications are preserved.

Ramat Gan /'rɑːmɑːt gɑːn/ industrial city (textiles, food processing) in W Israel, NE of Tel Aviv; population (1987) 116,000. It was established 1921.

Rambouillet /ˌrɒmbuː'jeɪ/ town in the S of the forest of Rambouillet, SW of Paris, France; population (1990) 25,300. The former royal château is now the presidential summer residence. A breed of fine-woolled sheep is named after the town.

Ramsgate /ˈræmzgeɪt/ seaside resort and cross-Channel port (ferry to Dunkirk, France) in the Isle of Thanet, Kent, SE England; population (1981) 39,600. There is a maritime museum. The architect Pugin built his home here, and is buried in the church next door (St Augustine's).

Rance /rɑːns/ river in Brittany, NW France, flowing into the English Channel between Dinard and St Malo, where a dam built 1960–67 (with a lock for ships) uses the 13-m/44-ft tides to feed the world's first successful tidal power station.

Rand /rænd/ shortened form of ♢Witwatersrand, a mountain ridge in Transvaal, South Africa.

Rangoon /ˌræŋˈɡuːn/ former name (until 1989) of ♢Yangon, the capital of Myanmar (Burma).

Rantoul /rænˈtuːl/ city in E Illinois, USA, S of Chicago. It is a trading centre for agricultural products; population (1990) 17,200. Products include electronic equipment and motorcycle parts.

Rapallo /rəˈpæləʊ/ port and winter resort in Liguria, NW Italy, 24 km/15 mi SE of Genoa on the Gulf of Rapallo; population (1981) 29,300. Treaties were signed here 1920 (settling the common frontiers of Italy and Yugoslavia) and 1922 (cancelling German and Russian counterclaims for indemnities for World War I).

Rapa Nui /ˈrɑːpə ˈnuːi/ another name for ♢Easter Island, an island in the Pacific.

Ras el Khaimah /ræs æl ˈxaɪmə/ or *Ra's al Khaymah* emirate on the Persian Gulf; area 1,690 sq km/652 sq mi; population (1980) 73,700. Products include oil, pharmaceuticals, and cement. It is one of the seven members of the United Arab Emirates.

Rathlin /ˈræθlɪn/ island off the N Irish coast, in Antrim; St Columba founded a church here in the 6th century, and in 1306 Robert Bruce hid here after his defeat by the English at Methven.

Ratisbon /ˈrætɪzbɒn/ English name for the German city of ♢Regensburg.

Ravenna /rəˈvenə/ historical city and industrial port (petrochemical works) in Emilia-Romagna, Italy; population (1988) 136,000. It lies in a marshy plain and is known for its Byzantine churches with superb mosaics.
history Ravenna was a Roman port and naval station. It was capital of the W Roman emperors 404–93, of Theodoric the Great 493–526, and later of the Byzantine exarchs (bishops) 539–750. The British poet Byron lived for some months in Ravenna, home of Countess Guiccioli, during the years 1819–21.

Ravenna itself preserves perhaps more of the old Italian manners than any City in Italy … They make love a good deal, – and assassinate a little.

On **Ravenna** Lord Byron, letter of 1819

Ravi /ˈrɑːvi/ river in the Indian subcontinent, a tributary of the Indus. It rises in India, forms the boundary between India and Pakistan for some 110 km/70 mi, and enters Pakistan above Lahore, the chief town on its 725-km/450-mi course. It is an important source of water for the Punjab irrigation canal system.

Rawalpindi /rɔːlˈpɪndi/ city in Punjab province, Pakistan, in the foothills of the Himalayas; population (1981) 928,400. Industries include oil refining, iron, chemicals, and furniture.

Reading /ˈredɪŋ/ industrial town (biscuits, electronics) on the river Thames; administrative headquarters of Berkshire, England; population (1991) 122,600. It is an agricultural and horticultural centre, and was extensively rebuilt after World War II. There is a university (1892).

Reading /ˈredɪŋ/ industrial city (textiles, special steels) in E Pennsylvania, USA; population (1990) 78,400.

Recife /reˈsiːfə/ industrial seaport (cotton textiles, sugar refining, fruit canning, flour milling) and naval base in Brazil, at the mouth of the river Capibaribe; capital of Pernambuco state; population (1991) 1,335,700. It was founded 1504.

Recklinghausen /ˈreklɪŋˌhauzən/ industrial town (coal, iron, chemicals, textiles, engineering) in North Rhine–Westphalia, Germany, 24 km/15 mi NW of Dortmund; population (1988) 118,000. It is said to have been founded by Charlemagne.

Redbridge /ˈredbrɪdʒ/ borough of NE Greater London. It includes Ilford, Wanstead, and Woodford, and parts of Chigwell and Dagenham
features part of Epping Forest; Hainault Forest
population (1991) 220,600.

Redditch /ˈredɪtʃ/ industrial town (needles, fishing tackle, car and aircraft components, cycles, motorcycles, electrical equipment) in Hereford and Worcester, England; population (1989 est) 78,000.
It was developed from 1965 as a new town to take Birmingham's overspill.

Redoubt, Mount /rɪˈdaʊt/ active volcanic peak rising to 3,140 m/10,197 ft, W of Cook inlet in S Alaska, USA. There were eruptions in 1966 and 1989.

Red River /red/ name of two rivers in the USA. (1) The **Red River of the South** is a western tributary of the Mississippi River, 1,638 km/1,018 mi long; so called because of the reddish soil sediment it carries. The stretch that forms the Texas–Oklahoma border is called Tornado Alley because of the storms caused by the collision in spring of warm air from the Gulf of Mexico with cold fronts from the N. The largest city on the river is Shreveport, Louisiana. (2) The **Red River of the North**, about 877 km/ 545 mi long, runs from North Dakota into Manitoba, Canada, and through Winnipeg, emptying into Lake Winnipeg. The fertile soil of the river valley produces large yields of wheat and other crops.

Red River red river in N Vietnam, 500 km/310 mi long, that flows into the Gulf of Tonkin. Its extensive delta is a main centre of population.

Redruth /ˌredˈruːθ/ town in Cornwall, SW England, part of the combined town of Camborne-Redruth.

Red Sea submerged section of the Great Rift Valley (2,000 km/1,200 mi long and up to 320 km/200 mi wide). Egypt, Sudan, and Ethiopia (in Africa) and Saudi Arabia (Asia) are on its shores.

Regensburg /ˈreɪɡənsbuək/ (English **Ratisbon**) city in Bavaria, Germany, on the river Danube at its confluence with the Regen, 100 km/63 mi NE of Munich; population (1988) 124,000. It has many medieval buildings, including a Gothic cathedral 1275–1530.

history Regensburg stands on the site of a Celtic settlement dating from 500 BC. It became the Roman **Castra Regina** AD 179, the capital of the Eastern Frankish Empire, a free city 1245, and seat of the German *Diet* (parliament) 16th century–1806. It was included in Bavaria 1810.

Reggio di Calabria /ˈredʒəu diː kəˈlæbriə/ industrial centre (farm machinery, olive oil, perfume) of Calabria, S Italy; population (1988) 179,000. It was founded by Greeks about 720 BC.

Reggio nell'Emilia /ˈredʒəu nel eˈmiːljə/ chief town of the province of the same name in Emilia-Romagna region, N Italy; population (1987) 130,000. It was here in 1797 that the Congress of the cities of Emilia adopted the tricolour flag that was later to become the national flag of Italy.

Regina /rəˈdʒaɪnə/ industrial city (oil refining, cement, steel, farm machinery, fertilizers), and capital of Saskatchewan, Canada; population (1986) 175,000. It was founded 1882 as **Pile O'Bones**, and renamed in honour of Queen Victoria of England.

Rehoboth Gebeit /rɪˈhəubəθ/ district of Namibia to the S of Windhoek; area 32,168 sq km/12,420 sq mi; chief town Rehoboth. The area is occupied by the Basters, a mixed race of European-Nama descent.

Reigate /ˈraɪɡɪt/ town in Surrey, England, at the foot of the North Downs; population (1988 est) 50,550. With Redhill it forms a residential suburb of London.

Reims /riːmz, French ræns/ (English **Rheims**) capital of Champagne-Ardenne region, France; population (1990) 185,200. It is the centre of the champagne industry and has textile industries. It was known in Roman times as **Durocorturum**. From 987 all but six French kings were crowned here. Ceded to England 1420 under the Treaty of Troyes, it was retaken by Joan of Arc, who had Charles VII consecrated in the 13th-century cathedral. In World War II, the German High Command formally surrendered here to US general Eisenhower 7 May 1945.

Remscheid /ˈremʃaɪt/ industrial city in North Rhine–Westphalia, Germany, where stainless-steel implements are manufactured; population (1988) 121,000.

Renfrew /ˈrenfruː/ town on the Clyde, in Strathclyde, 8 km/5 mi NW of Glasgow, Scotland; population (1981) 21,396. It was formerly the county town of Renfrewshire.

Renfrewshire /ˈrenfruːʃə/ former county of W central Scotland, bordering the Firth of Clyde. It was merged with the region of Strathclyde 1975. The county town was Renfrew.

Rennes /ren/ industrial city (oil refining, chemicals, electronics, cars) and capital of Ille-et-Vilaine *département*, W France, at the confluence of the Ille and Vilaine, 56 km/35 mi SE of St Malo; population (1990) 203,500. It was the old capital of Brittany. Its university specializes in Breton culture.

Reno /ˈriːnəu/ city in Nevada, USA, known for gambling and easy divorces; population (1990) 133,850. Products include building materials and electronic equipment.

Réunion /ˌreɪuːnˈjɒŋ/ French island of the Mascarenes group, in the Indian Ocean, 650 km/400 mi E of Madagascar and 180 km/110 mi SW of Mauritius
area 2,512 sq km/970 sq mi
capital St Denis
physical forested, rising in Piton de Neiges to 3,069 m/10,072 ft
features administers five uninhabited islands, also claimed by Madagascar
products sugar, maize, vanilla, tobacco, rum
population (1990) 597,800
history explored by Portuguese (the first European visitors) 1513; annexed by Louis XIII of France 1642; overseas *département* of France 1946; overseas region 1972.

Reus /ˈreɪus/ industrial city with an international airport in Catalonia, E Spain, 10 km/6 mi NW of Tarragona; population (1991) 87,700. Products include textiles, flowers, dried fruit, and vegetables. The architect Antonio Gaudí was born here.

Reval /ˈreɪvæl/ former name of the Estonian port of ◊Tallinn.

Reykjavik /ˈreɪkjəviːk/ capital (from 1918) and chief port of Iceland, on the southwest coast;

Réunion

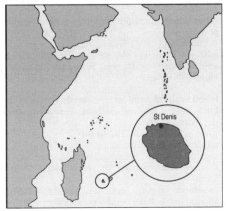

population (1988) 93,000. Fish processing is the main industry. Reykjavik is heated by underground mains fed by volcanic springs. It was a seat of Danish administration from 1801 to 1918.

Rheims /riːmz/ English version of ⬦Reims, a city in France.

Rheinland-Pfalz /ˈraɪnlænt ˈpfælts/ German name for the ⬦Rhineland-Palatinate, a region of Germany.

Rhine /raɪn/ (German **Rhein**, French **Rhin**) European river rising in Switzerland and reaching the North Sea via Germany and the Netherlands; length 1,320 km/820 mi. Tributaries include the Moselle and the Ruhr. The Rhine is linked with the Mediterranean by the Rhine–Rhône Waterway, and with the Black Sea by the Rhine–Main–Danube Waterway. It is the longest, and the dirtiest, river in Europe.

The river was severely polluted by a chemical factory fire 1986, as a result of which 30 tonnes of pesticides entered the river, rendering it lifeless for 100–200 km/60–120 mi. The governments of the five countries through which the Rhine flows announced a major clean-up campaign 1987.

The **Lorelei** is a rock in the river in Rhineland-Palatinate, Germany, with a remarkable echo; the German poet Brentano gave currency to the legend of a siren who lured sailors to death with her song, also the subject of a poem by Heinrich Heine.

The principal artery of capitalism in Europe.

On the **Rhine** Richard West *Victory in Vietnam* 1974

Rhineland-Palatinate /ˈraɪnlænd pəˈlætɪnət/ (German **Rheinland-Pfalz**) administrative region (German *Land*) of Germany
area 19,800 sq km/7,643 sq mi
capital Mainz
towns Ludwigshafen, Koblenz, Trier, Worms
physical wooded mountain country, river valleys of Rhine and Moselle
products wine (75% of German output), tobacco, chemicals, machinery, leather goods, pottery
population (1992) 3,702,000
history formed 1946 of the Rhenish Palatinate and parts of Hessen, Rhine province, and Hessen-Nassau.

Rhode Island /rəud ˈaɪlənd/ smallest state of the USA, in New England; nickname Little Rhody or the Ocean State
area 3,100 sq km/1,197 sq mi
capital Providence
towns Cranston, Newport, Woonsocket
features Narragansett Bay runs inland 45 km/28 mi
products apples, potatoes, poultry (Rhode Island Reds), dairy products, jewellery (30% of

Rhode Island

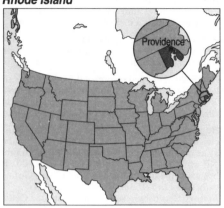

the workforce), textiles, silverware, machinery, rubber, plastics, electronics
population (1990) 1,003,500
history founded 1636 by Roger Williams, exiled from Massachusetts Bay colony for religious dissent; one of the original 13 states.

Rhodes /rəudz/ (Greek **Ródhos**) Greek island, largest of the Dodecanese, in the E Aegean Sea
area 1,412 sq km/545 sq mi
capital Rhodes
products grapes, olives
population (1981) 88,000
history settled by Greeks about 1000 BC; the Colossus of Rhodes (fell 224 BC) was one of the Seven Wonders of the World; held by the Knights Hospitallers of St John 1306–1522; taken from Turkish rule by the Italian occupation 1912; ceded to Greece 1947.

Rhodes, where history lies sleeping.

On **Rhodes** Freya Stark *Beyond Euphrates* 1951

Rhodesia /rəuˈdiːʃə/ former name of ⬦Zambia (Northern Rhodesia) and ⬦Zimbabwe (Southern Rhodesia), in S Africa.

Rhodope Mountains /ˈrɒdəpi/ range of mountains on the frontier between Greece and Bulgaria, rising to 2,925 m/9,497 ft at Musala.

Rhondda /ˈrɒnðə/ industrial town in Mid Glamorgan, Wales; population (1981) 81,725. Light industries have replaced coalmining, formerly the main source of employment. The closure of the Maerdy mine (opened 1875) in 1990 ended mining in the valley; its coal powered 90% of the Royal Navy's ships in World War I. The Rhondda Heritage Park recreates a 1920s-style mining village for visitors.

Rhône /rəun/ river of S Europe; length 810 km/500 mi. It rises in Switzerland and flows through Lake Geneva to Lyon in France, where at its confluence with the Saône the upper limit of navigation is reached. The river turns due S, passes

Vienne and Avignon, and takes in the Isère and other tributaries. Near Arles it divides into the *Grand* and *Petit Rhône*, flowing respectively SE and SW into the Mediterranean W of Marseille. Here it forms a two-armed delta; the area between the tributaries is the marshy region known as the Camargue.

The Rhône is harnessed for hydroelectric power, the chief dam being at Genissiat in Ain *département*, constructed 1938–48. Between Vienne and Avignon the Rhône flows through a major wine-producing area. The river gives its name to a *département*.

The Rhône is the river of Angels.

On the **Rhône** Robert Louis Stevenson,
letter of 1886

Rhône-Alpes /ˈrəun ˈælps/ region of E France in the upper reaches of the Rhône; area 43,700 sq km/16,868 sq mi; population (1992) 5,344,000. It consists of the *départements* of Ain, Ardèche, Drôme, Isère, Loire, Rhône, Savoie, and Haute-Savoie. The chief town is Lyon. There are several wine-producing areas, including Chenas, Fleurie, and Beaujolais. Industrial products include chemicals, textiles, and motor vehicles.

Rhyl /rɪl/ seaside holiday resort in Clwyd, N Wales; population (1980) 23,000. Products include furniture; tourism is important.

RI abbreviation for ▷*Rhode Island*, a state of the USA.

Ribble /ˈrɪbəl/ river in N England; length 120 km/ 75 mi. From its source in the Pennine hills, North Yorkshire, it flows S and SW past Preston, Lancashire, to join the Irish Sea.

Richborough /ˈrɪtʃbərə/ (Roman *Rutupiae*) former port in Kent, England; now marooned in salt marshes, it was militarily reactivated in both world wars.

Richland /ˈrɪtʃlənd/ city in SE Washington, USA, on the Columbia River, NW of Walla Walla; population (1990) 33,300. It is a centre for research for the US Department of Energy and a major producer of plutonium for nuclear weapons. It grew as a residential community for employees of the Hanford Engineer Works that helped to develop the atomic bomb from 1943.

Richmond /ˈrɪtʃmənd/ capital of Virginia, USA; population (1990) 203,000. It is the centre of the Virginian tobacco trade. It was the capital of the Confederacy 1861–65. A museum commemorates the writer Edgar Allan Poe, who grew up here.

Richmond /ˈrɪtʃmənd/ town in North Yorkshire, England; population (1981) 7,600. It has a theatre built 1788.

Richmond-upon-Thames /ˈrɪtʃmənd əpɒn ˈtemz/ borough of SW Greater London

features **Hampton** Garrick Villa; Old Court House (the architect Christopher Wren's last home), Faraday House; Hampton Court Palace and Bushy Park; *Kew* outhoused departments of the Public Record Office; Kew Palace (former royal residence), within the Royal Botanic Gardens; *Richmond* gatehouse of former Richmond Palace (see Henry VIII and Elizabeth I), Richmond Hill and Richmond Park (including White Lodge, home of the Royal Ballet School); Ham House (17th century); *Teddington* highest tidal point of the Thames; National Physical Laboratory; *Twickenham* Kneller Hall (Royal Military School of Music); Marble Hill House (Palladian home of the Duchess of Suffolk, mistress of George II); Strawberry Hill (home of Horace Walpole); Twickenham Rugby Ground; Alexander Pope is buried in the church *population* (1991) 154,600.

Rif, Er /rɪf/ mountain range about 290 km/180 mi long on the Mediterranean seaboard of Morocco.

Rift Valley, Great /rɪft/ volcanic valley formed 10–20 million years ago by a crack in the Earth's crust and running about 8,000 km/5,000 mi from the Jordan Valley through the Red Sea to central Mozambique in SE Africa. It is marked by a series of lakes, including Lake Turkana (formerly Lake Rudolf), and volcanoes, such as Mount Kilimanjaro.

At some points its traces have been lost by erosion, but elsewhere, as in S Kenya, cliffs rise thousands of metres.

Riga /ˈriːgə/ capital and port of Latvia; population (1987) 900,000. A member of the Hanseatic League from 1282, Riga has belonged in turn to Poland 1582, Sweden 1621, and Russia 1710.

It was the capital of independent Latvia 1918–40 and was occupied by Germany 1941–44, before being annexed by the USSR. It again became independent Latvia's capital 1991.

Rigi /ˈriːgi/ mountain in central Switzerland, between lakes Lauerz, Lucerne, and Zug; height 1,800 m/5,908 ft. The cogwheel train to the top was the first in Europe.

Rijeka /riˈekə/ (Italian *Fiume*) industrial port (oil refining, distilling, paper, tobacco, chemicals) in Croatia; population (1983) 193,044. It has changed hands many times and, after being seized by the Italian nationalist Gabriele d'Annunzio 1919, was annexed by Italy 1924. It was ceded back to Yugoslavia 1949; in 1991 it became part of newly independent Croatia.

Rimini /ˈrɪmɪni/ industrial port (pasta, footwear, textiles, furniture) and holiday resort in Emilia-Romagna, Italy; population (1988) 131,000.
history Its name in Roman times was *Ariminum*, and it was the terminus of the Flaminian Way from Rome. In World War II it formed the eastern strongpoint of the German 'Gothic' defence line and was badly damaged in the severe fighting Sept 1944, when it was taken by the Allies.

Rineanna /ˈrɪnjənə/ Irish name of Shannon Airport, County Clare, in the Republic of Ireland.

Rio de Janeiro /ˈriːəu də ʒəˈnɪərəu/ port and resort in E Brazil; population (1991) 5,487,300. The name (Portuguese 'river of January') commemorates the arrival of Portuguese explorers 1 Jan 1502, but there is in fact no river. Sugar Loaf Mountain stands at the entrance to the harbour. Rio was the capital of Brazil 1763–1960.

Some colonial churches and other buildings survive; there are modern boulevards, including the Avenida Rio Branco, and Copacabana is a luxurious beachside suburb. The city is the capital of the state of Rio de Janeiro, which has a population of 13,267,100 (1987 est).

It is hard for man to make any city worthy of such surroundings as Nature has given to Rio.

On **Rio de Janeiro** James Bryce
South America 1912

Río de Oro /ˈriːəu deɪ ˈɔːrəu/ former district in the S of the province of Spanish Sahara. See ◊Western Sahara.

Rio Grande /ˈriːəu ˈgrænd, ˈgrændi/ river rising in the Rocky Mountains in S Colorado, USA, and flowing S to the Gulf of Mexico, where it is reduced to a trickle by irrigation demands on its upper reaches; length 3,050 km/1,900 mi. Its last 2,400 km/1,500 mi form the US–Mexican border (Mexican name **Río Bravo del Norte**).

Rio Grande do Norte /ˈriːuː ˈgrʌndi duː ˈnɔːti/ state of NE Brazil; area 53,000 sq km/20,460 sq mi; population (1990) 2,522,700. Its capital is Natal. Industries include agriculture and textiles; there is offshore oil.

Rio Grande do Sul /ˈriːuˌgrʌndiduˈsuːl/ most southerly state of Brazil, on the frontier with Argentina and Uruguay; area 282,184 sq km/108,993 sq mi; population (1990) 9,348,300. Its capital is Pôrto Alegre. Industries include agriculture and cattle, food processing, and textiles.

Rioja, La see ◊La Rioja, a region of Spain.

Riom /riˈɒm/ town on the river Ambène, in the Puy-de-Dôme *département* of central France. It was the scene in World War II of the 'war guilt' trials of several prominent Frenchmen by the Vichy government Feb–April 1942. The accused included the former prime ministers Blum and Daladier, and General Gamelin. The occasion turned into a wrangle over the reasons for French unpreparedness for war, and, at the German dictator Hitler's instigation, the court was dissolved. The defendants remained in prison until released by the Allies 1945.

Río Muni /ˈriːəu ˈmuːni/ the mainland portion of ◊Equatorial Guinea.

Río Negro /ˈriːəu ˈneɪgrəu/ river in South America, rising in E Colombia and joining the Amazon at Manáus, Brazil; length 2,250 km/1,400 mi.

Río Tinto /ˈriːəu ˈtɪntəu/ town in Andalusia, Spain;

population (1983) 8,400. Its copper mines, first exploited by the Phoenicians, are now almost worked out.

Ripon /ˈrɪpən/ city and market town in North Yorkshire, England; population (1987 est) 13,100. There is a cathedral 1154–1520; and the nearby 12th-century ruins of Fountains Abbey are among the finest monastic ruins in Europe.

Riva del Garda /ˈriːvə del ˈgɑːdə/ town on Lake Garda, Italy, where the Prix Italia broadcasting festival has been held since 1948.

Riverina /ˌrɪvəˈriːnə/ district of New South Wales, Australia, between the Lachlan and Murray rivers, through which runs the Murrumbidgee. On fertile land, artificially irrigated from the three rivers, wool, wheat, and fruit are produced.

Riverside /ˈrɪvəsaɪd/ city in California, USA, on the Santa Ana River, E of Los Angeles; population (1990) 226,500. It was founded 1870. It is the centre of a citrus-growing district and has a citrus research station; the seedless orange was developed at Riverside 1873.

Riviera /ˌrɪviˈeərə/ the Mediterranean coast of France and Italy from Marseille to La Spezia. The most exclusive stretch of the Riviera, with the finest climate, is the Côte d'Azur, from Menton to St Tropez, which includes Monaco.

Riyadh /ˈriːæd/ (Arabic *Ar Riyād*) capital of Saudi Arabia and of the Central Province, formerly the sultanate of Nejd, in an oasis, connected by rail with Dammam on the Arabian Gulf; population (1986) 1,500,000.

Outside the city are date gardens irrigated from deep wells. There is a large royal palace and an Islamic university 1950.

One does not often see a whole city being built at one moment, but this is what is happening here.

On **Riyadh** Arnold Toynbee
East to West 1958

Roanoke /ˈrəuənəuk/ (American Indian 'shell money') industrial city (railway repairs, chemicals, steel goods, furniture, textiles) in Virginia, USA, on the Roanoke River; population (1980) 100,500. Founded 1834 as **Big Lick**, it was a small village until 1881 when the repair shops of the Virginia Railway were set up here, after which it developed rapidly.

Robben Island /ˈrɒbɪn/ island in Table Bay, Cape Town, South Africa. It is used by the South African government to house political prisoners.

Rochdale /ˈrɒtʃdeɪl/ industrial town (textiles, machinery, asbestos) in Greater Manchester, England, on the river Roch 16 km/10 mi NE of Manchester; population (1991) 196,400. The so-called Rochdale Pioneers founded the first Co-operative Society in England, in Toad Lane, Rochdale, 1844. The singer Gracie Fields was born here and a theatre is named after her.

Rochefort /ˌrəuʃ'fɔ:/ industrial port (metal goods, machinery) in W France, SE of La Rochelle and 15 km/9 mi from the mouth of the river Charente; population (1990) 26,900. The port dates from 1666 and it was from here that Napoleon embarked for Plymouth on the *Bellerophon* on his way to final exile 1815.

Rochelle, La /rɒ'ʃel/ see ◊La Rochelle, a port in W France.

Rochester /'rɒtʃɪstə/ industrial city in New York, USA, on the Genesee River, S of Lake Ontario; population (1990) 231,600. Its manufactured products include photographic equipment and optical and other precision instruments. It was the birthplace of the Xerox copier, and the world headquarters of the Eastman Kodak Company are here.

Rochester /'rɒtʃɪstə/ commercial centre with dairy and food-processing industries in Minnesota, USA; population (1990) 70,745. Rochester is the home of the Mayo Clinic, part of a medical centre established 1889.

Rochester upon Medway /'rɒtʃɪstə, 'medweɪ/ city in Kent, England; population (1983, with Chatham and Strood) 146,200. It has a 12th-century Norman castle keep (the largest in England), a 12th–15th-century cathedral, and many timbered buildings. The Dickens Centre 1982 commemorates the town's links with the novelist Charles Dickens.

Rockall /rɒkɔ:l/ British islet in the Atlantic, 24 m/80 ft across and 22 m/65 ft high, part of the Hatton-Rockall bank, and 370 km/230 mi W of North Uist in the Hebrides. The bank is part of a fragment of Greenland that broke away 60 million years ago. It is in a potentially rich oil/gas area. A party of British marines landed 1955 formally to annex Rockall, but Denmark, Iceland, and Ireland challenge Britain's claims for mineral, oil, and fishing rights. The *Rockall Trough* between Rockall and Ireland, 250 km/155 mi wide and up to 3,000 m/10,000 ft deep, forms an ideal marine laboratory.

Rockford /'rɒkfəd/ city in N Illinois, USA, on the Rock River, NW of Chicago; population (1990) 139,400. Industries include automotive parts, furniture, machine tools, and food products.

Rockhampton /ˌrɒk'hæmptən/ port in E Queensland, Australia, on the Fitzroy River; population (1986) 55,700. It is the centre of a beef-producing area.

Rock Hill /ˌrɒk 'hɪl/ city in N central South Carolina, USA, S of Charlotte, North Carolina; population (1990) 41,600. Industries include chemicals, textiles, and paper products.

Rock Island /rɒk aɪlənd/ city in NW Illinois, USA, on the Mississippi River, W of Chicago; population (1990) 40,600. Industries include rubber and electronics. A US government arsenal is located here.

Rocky Mountains /'rɒki/ or *Rockies* largest North American mountain system. It extends from the junction with the Mexican plateau, northwards through the W central states of the USA, through Canada to S Alaska. The highest mountain is Mount McKinley (6,194 m/20,320 ft).

Many large rivers rise in the Rocky Mountains, including the Missouri. The Rocky Mountain National Park (1915) in Colorado has more than 100 peaks over 3,350 m/11,000 ft; Mount Logan on the Canadian–Alaskan border is 5,951 m/19,524 ft. In the 1980s computer techniques enabled natural gas in large quantities to be located in the W Rockies.

Ródhos /'rɒðɒs/ Greek name for the island of ◊Rhodes.

Roeselare /'ru:sələ:rə/ (French *Roulers*) textile town in West Flanders province, NW Belgium; population (1991) 52,900. It was a major German base in World War I.

Roma /'rəumə/ town in SE Queensland, linked by rail and gas pipeline to Brisbane; population (1985) 6,500.

Romagna /rəu'mɑ:njə/ area of Italy on the Adriatic coast, under papal rule 1278–1860 and now part of the region of Emilia-Romagna.

Romania
area 237,500 sq km/91,699 sq mi
capital Bucharest
towns Braşov, Timişoara, Cluj–Napoca, Iaşi; ports Galaţi, Constanta, Brăila
physical mountains surrounding a plateau, with river plains S and E
environment although sulphur-dioxide levels are low, only 20% of the country's rivers can provide drinkable water
features Carpathian Mountains, Transylvanian Alps; river Danube; Black Sea coast; mineral springs
head of state Ion Iliescu from 1989
head of government Theodor Stolojan from 1991
political system emergent democratic republic
political parties National Salvation Front (NSF), reform socialist; Civic Alliance (CA), right-of-centre; Convention for Democracy, umbrella organization for right-of-centre National Liberal Party (NLP) and ethnic-Hungarian Magyar Democratic Union (UDM)

exports petroleum products and oilfield equipment, electrical goods, cars, cereals
currency leu
population (1990 est) 23,269,000 (Romanians 89%, Hungarians 7.9%, Germans 1.6%); growth rate 0.5% p.a.
life expectancy men 67, women 73 (1989)
languages Romanian (official), Hungarian, German
media television is state-run; there are an estimated 900 newspapers and magazines, but only the progovernment papers have adequate distribution and printing facilities
religions Romanian Orthodox 80%, Roman Catholic 6%
literacy 98% (1988)
GNP $151 bn (1988); $6,400 per head
chronology
1944 Pro-Nazi Antonescu government overthrown.
1945 Communist-dominated government appointed.
1949 New Soviet-style constitution adopted. Joined Comecon.
1952 Second new Soviet-style constitution.
1955 Romania joined Warsaw Pact.
1958 Soviet occupation forces removed.
1965 New constitution adopted.
1974 Ceauşescu created president.
1985–86 Winters of austerity and power cuts.
1987 Workers demonstrated against austerity programme.
1988–89 Relations with Hungary deteriorated over 'systematization programme'.
1989 Announcement that all foreign debt paid off. Razing of villages and building of monuments to Ceauşescu; Communist orthodoxy reaffirmed; demonstrations violently suppressed; massacre in Timisoara; army joined uprising; heavy fighting; bloody overthrow of Ceauşescu regime in 'Christmas Revolution'; Ceauşescu and wife tried and executed; estimated 10,000 dead in civil warfare; power assumed by new military-dissident-reform communist National Salvation Front, headed by Ion Iliescu.
1990 Securitate replaced by new Romanian Intelligence Service (RIS); religious practices resumed; mounting strikes and protests against effects of market economy.
1991 April: treaty on cooperation and good neighbourliness signed with USSR. Aug: privatization law passed. Sept: prime minister Petre Roman resigned following riots, succeeded by Theodor Stolojan heading a new cross-party coalition government. Dec: new constitution endorsed by referendum.
1992 President Iliescu re-elected.

No country is kinder to the wanderer who has good legs.

On **Romania** Walter Starkie
Raggle-Taggle 1933

Rome /rəum/ (Italian **Roma**) capital of Italy and of Lazio region, on the river Tiber, 27 km/17 mi from the Tyrrhenian Sea; population (1987) 2,817,000.
Rome has few industries but is an important cultural, road, and rail centre. A large section of the population finds employment in government offices. Remains of the ancient city include the Forum, Colosseum, and Pantheon.
features E of the river are the seven hills on which Rome was originally built (Quirinal, Aventine, Caelian, Esquiline, Viminal, Palatine, and Capitol); to the W are the quarter of Trastevere, the residential quarters of the Prati, and the Vatican. Among Rome's buildings are Castel Sant' Angelo (the mausoleum of the emperor Hadrian) and baths of Caracalla. Among the Renaissance palaces are the Lateran, Quirinal (with the Trevi fountain nearby), Colonna, Borghese, Barberini, and Farnese. There are a number of churches of different periods; San Paolo was founded by the emperor Constantine on St Paul's grave. The house where the English poet Keats died is near the Piazza di Spagna.
history traditionally founded 753 BC, Rome became a republic 510 BC. From then, its history is one of almost continual expansion until the murder of Julius Caesar and foundation of the empire under Augustus and his successors. At its peak under Trajan, the Roman Empire stretched from Britain to Mesopotamia and the Caspian Sea. Under Diocletian, the empire was divided into two parts—East and West—although temporarily reunited under Constantine, the first emperor formally to adopt Christianity. The end of the Roman Empire is generally dated by the sack of Rome by the Goths AD 410, or by the deposition of the last emperor in the west AD 476. The Eastern Empire continued until 1453 at Constantinople.
After the deposition of the last emperor, Romulus Augustulus, 476, the papacy became the real ruler of Rome and from the 8th century was recognized as such. As a result of the French Revolution, Rome temporarily became a republic 1798–99, and was annexed to the French Empire 1808–14, until the pope returned on Napoleon's fall. During the 1848–49 revolution, a republic was established under Mazzini's leadership, but, in spite of Garibaldi's defence, was overthrown by French troops.
In 1870 Rome became the capital of Italy, the pope retiring into the Vatican until 1929 when the Vatican City was recognized as a sovereign state. The occupation of Rome by the Fascists 1922 marked the beginning of Mussolini's rule, but in 1943 Rome was occupied by Germany and then captured by the Allies 1944.

While stands the Coliseum, Rome shall stand; /When falls the Coliseum, Rome shall fall; /And when Rome falls—the World.

On **Rome** Lord Byron *Childe Harold's Pilgrimage* 1812

Rome /rəum/ city in central New York, USA, on the Mohawk River, NW of Albany; population (1990) 44,400. Industries include copper and brass products, paint, and household appliances. Construction of the Erie Canal began here 1817.

Romney Marsh /'rɒmni 'mɑːʃ/ stretch of drained marshland on the Kent coast, SE England, between Hythe and Rye, used for sheep pasture. The seaward point is Dungeness. Romney Marsh was reclaimed in Roman times.

New Romney, formed by the amalgamation of Romney, one of the Cinque Ports, with Littlestone and Greatstone, is now more than a mile from the sea; population about 5,000.

Romsey /'rʌmzi/ market town in Hampshire, S England; population (1990 est) 14,700. The fine Norman church of Romsey Abbey (founded by Edward the Elder) survives, as does King John's Hunting Box of about 1206 (now a museum); nearby Broadlands was the seat of Earl Mountbatten and Lord Palmerston.

Roncesvalles /rɒnsəvælz, Spanish rɒnθez'væljes/ village of N Spain, in the Pyrenees 8 km/5 mi S of the French border, the scene of the defeat of the rearguard of Charlemagne's army under Roland who was killed 778.

Rondônia /rɒn'dəuniə/ state in NW Brazil; the centre of Amazonian tin and gold mining and of experiments in agricultural colonization; area 243,044 sq km/93,876 sq mi; population (1991) 1,373,700. Known as the Federal Territory of *Guaporé* until 1956, it became a state 1981.

Roodepoort-Maraisburg /'ruːdəpuət məˈreɪsbɜːg/ goldmining town in Transvaal, South Africa, 15 km/9 mi W of Johannesburg, at an altitude of 1,745 m/5,725 ft; population (1986) 141,764. Leander Starr Jameson and his followers surrendered here 1896 after an attempt to overthrow the government.

Roraima, Mount /rɔːˈraɪmə/ plateau in the Pacaraima mountain range in South America, rising to 2,810 m/9,222 ft on the Brazil–Guyana–Venezuela frontier.

Rosario /rəuˈsɑːriəu/ industrial river port (sugar refining, meat packing, maté processing) in Argentina, 280 km/175 mi NW of Buenos Aires, on the river Paraná; population (1991) 1,078,400. It was founded 1725.

Roscoff /'rɒskɒf/ port in N Brittany, France, with a ferry link to Plymouth in England; population (1982) 4,000.

Roscommon /rɒs'kɒmən/ (originally Ros-Comain, 'wood around a monastery') county of the Republic of Ireland in the province of Connacht
area 2,460 sq km/950 sq mi
towns Roscommon (county town)
physical bounded on the E by the river Shannon; lakes: Gara, Key, Allen; rich pastures
features remains of a castle put up in the 13th century by English settlers
population (1991) 51,900.

Roseau /rəuˈzəu/ formerly *Charlotte Town* capital of Dominica, West Indies, on the SW coast of the island; population (1981) 20,000.

Roseires, Er /rɒ'seəres/ port at the head of navigation of the Blue Nile in Sudan. A hydroelectric scheme here provides the country with 70% of its electrical power.

Roskilde /'rɒskɪlə/ port at the southern end of Roskilde Fjord, Zealand, Denmark; population (1990) 49,100. It was the capital of the country from the 10th century until 1443.

Ross and Cromarty /rɒs, 'krɒməti/ former county of Scotland. In 1975 Lewis, in the Outer Hebrides, became part of the Western Isles, and the mainland area was included in Highland Region.

Ross Dependency /rɒs/ all the Antarctic islands and territories between 160° E and 150° W longitude and S of 60° S latitude; it includes Edward VII Land, Ross Sea and its islands, and parts of Victoria Land
area 450,000 sq km/173,700 sq mi
features the **Ross Ice Shelf** (or Ross Barrier), a permanent layer of ice across the Ross Sea about 425 m/1,400 ft thick
population a few scientific bases with about 250 staff members, 12 of whom are present during winter
history given to New Zealand 1923. It is probable that marine organisms beneath the ice shelf had been undisturbed from the Pleistocene period until drillings were made 1976.

Ross Island /rɒs/ either of two islands in Antarctica:
Ross Island in the Weddell Sea, discovered 1903 by the Swedish explorer Nils Nordenskjöld, area about 3,885 sq km/1,500 sq mi;
Ross Island in the Ross Sea, discovered 1841 by the British explorer James Ross, area about 6,475 sq km/2,500 sq mi, with the research stations Scott Base (New Zealand) and McMurdo (USA). Mount Erebus (3,794 m/12,520 ft) is the world's southernmost active volcano; its lake of molten lava may provide a window on the magma beneath the Earth's crust that fuels volcanoes.

Rosslare /ˌrɒs'leə/ port in County Wexford, Republic of Ireland, 15 km/9 mi SE of Wexford; population (1980 est) 600. It was founded by the English 1210 and has been the Irish terminus of the ferry route from Fishguard from 1906.

Ross Sea /rɒs/ Antarctic inlet of the S Pacific. See also ▷Ross Dependency and ▷Ross Island.

Rostock /'rɒstɒk/ industrial port (electronics, fish processing, ship repair) in the state of Mecklenburg-West Pomerania, Germany, on the river Warnow 13 km/8 mi S of the Baltic; population (1990) 250,000.
Founded 1189, in the 14th century Rostock became a powerful member of the Hanseatic League. It was rebuilt in the 1950s and was capital of an East German district of the same name 1952–90.

Rostov-on-Don /'rɒstɒv ɒn 'dɒn/ industrial port (shipbuilding, tobacco, cars, locomotives, textiles) in SW Russia, capital of Rostov region, on the river

Don, 23 km/14 mi E of the Sea of Azov; population (1987) 1,004,000. Rostov dates from 1761 and is linked by river and canal with Volgograd on the river Volga.

Rosyth /rə'saɪθ/ naval base and dockyard used for nuclear submarine refits, in Fife, Scotland, built 1909 on the northern shore of the Firth of Forth; population (1988 est) 11,500.

Rota /'rəʊtə/ port and naval base, 57 km/35 mi NW of Cadiz, Spain; population (1991) 27,100.

Rothamsted /'rɒθəmsted/ agricultural research centre in Hertfordshire, England, NW of St Albans.

Rothenburg /'rəʊtnbuək/ town in Bavaria, Germany, 65 km/40 mi W of Nuremberg; population (1978) 13,000. It is known for its medieval buildings, churches, and walls.

Rotherham /'rɒðərəm/ industrial town (pottery, glass, coal) in South Yorkshire, England, on the river Don, NE of Sheffield; population (1988 est) 251,800.

Rotherhithe Tunnel /'rɒðəhaɪð/ road tunnel extending 1,481 m/4,860 ft under the river Thames E of Wapping, London, connecting Rotherhithe with Shadwell. It was built 1904–08 to a design by Maurice Fitzmaurice.

Rotorua /,rəʊtə'ruːə/ town with medicinal hot springs and other volcanic activity in North Island, New Zealand, near Lake Rotorua; population (1991) 53,600.

Spirit of Picnics, here should thy temple be.

On **Rotorua** Charles Heaphy *Expedition to the South West of Nelson* 1846

Rotterdam /'rɒtədæm/ industrial port (brewing, distilling, shipbuilding, sugar and petroleum refining, margarine, tobacco) in the Netherlands and one of the foremost ocean cargo ports in the world, in the Rhine-Maas delta, linked by canal 1866–90 with the North Sea; population (1991) 582,266.

Rotterdam dates from the 12th century or earlier, but the centre was destroyed by German air attack 1940, and rebuilt; its notable art collections were saved. The philosopher Erasmus was born here.

Roubaix /ruː'beɪ/ town in Nord *département*, N France, adjacent to Lille; population (1990) 98,200. It is a major centre of French woollen textile production.

Rouen /ruː'ɒŋ/ industrial port (cotton textiles, electronics, distilling, oil refining) on the river Seine, in Seine-Maritime *département*, central N France; population (1990) 105,500.
history Rouen was the capital of Normandy from 912. Lost by King John 1204, it returned briefly to English possession 1419–49; Joan of Arc was burned in the square 1431. The novelist Flaubert was born here, and the hospital where his father was chief surgeon is now a Flaubert museum.

Roulers /ruː'leɪ/ French name of ▷Roeselare, a town in Belgium.

Rovaniemi /'rɒvənjeɪmi/ capital of Lappi province, N Finland, and chief town of Finnish Lapland, situated just S of the Arctic Circle; population (1990) 33,500. After World War II the town was rebuilt by the architect Alvar Aalto, who laid out the main streets in the form of a reindeer's antlers.

Roxburgh /'rɒksbərə/ former border county of Scotland, included 1975 in Borders Region. Jedburgh was the county town.

RSFSR abbreviation for *Russian Soviet Federal Socialist Republic*, the largest republic of the former Soviet Union; renamed the ▷Russian Federation on achieving independence 1991.

Ruahine /,ruːə'hiːni/ mountain range in North Island, New Zealand. It stretches NE from the Manawater Gorge to the headwaters of the Ngaruro River and lies within the 900 sq km/347 sq mi Ruahine State Forest Park designated 1976. The highest point is Mangaweka, 1,733 m/5,686 ft.

Ruapehu /,ruːə'peɪhuː/ volcano in New Zealand, SW of Lake Taupo; the highest peak in North Island, 2,797 m/9,175 ft.

Rub' al Khali /'rub æl 'kɑːli/ (Arabic 'empty quarter') vast sandy desert in S Saudi Arabia; area 650,000 sq km/250,000 sq mi. The British explorer Bertram Thomas (1892–1950) was the first European to cross it 1930–31.

Ruda Slaska town in Silesia, Poland, with metallurgical industries, created 1959 by a merger of Ruda and Nowy Butom; population (1984) 163,000. Silesia's oldest mine is nearby.

Rudolf, Lake /'ruːdɒlf/ former name (until 1979) of Lake ▷Turkana in E Africa.

Rugby /'rʌgbi/ market town and railway junction in Warwickshire, England; population (1989 est) 60,000. Rugby School 1567 established its reputation under headmaster Thomas Arnold. Rugby football originated here.

Rügen /'ruːgən/ Baltic island in the state of Mecklenburg-West Pomerania, Germany; area 927 sq km/358 sq mi. It is a holiday centre, linked by causeway to the mainland; chief town Bergen, main port Sassnitz. As well as tourism, there is agriculture and fishing, and chalk is mined. Rügen was annexed by Denmark 1168, Pomerania 1325, Sweden 1648, and Prussia 1815.

Ruhr /ruə/ river in Germany; it rises in the Rothaargebirge Mountains and flows W to join the Rhine at Duisburg. The *Ruhr Valley* (228 km/142 mi) is a metropolitan industrial area (petrochemicals, cars; iron and steel at Duisburg and Dortmund). It was formerly a coalmining centre.

The area was occupied by French and Belgian troops 1923–25 in an unsuccessful attempt to force Germany to pay reparations laid down in the Treaty of Versailles. During World War II the Ruhr district was severely bombed. Allied control of the area from 1945 came to an end with the setting-up of the European Coal and Steel Community 1952.

Rum /rʌm/ or **Rhum** island of the Inner Hebrides, Highland Region, Scotland, area 110 sq km/42 sq mi, a nature reserve from 1957. Askival is 810 m/2,658 ft high.

Rumania /ruːˈmeɪnɪə/ alternative spelling of ◊Romania.

Rumeila /ruˈmeɪlə/ oilfield straddling the frontier between Iraq and Kuwait. Kuwait's extraction of oil from this field was a contributory factor leading up to the Iraqi invasion of 1991 and the subsequent Gulf War 1992.

Rum Jungle /rʌm 'dʒʌŋgl/ uranium-mining centre in the NW of Northern Territory, Australia.

Runcorn /ˈrʌŋkɔːn/ industrial town (chemicals) in Cheshire, England, 24 km/15 mi up the Mersey estuary from Liverpool; population (1985 est) 64,600. As a new town it has received Merseyside overspill from 1964.

Rupert's Land /ˈruːpəts lænd/ area of N Canada, of which the English Royalist general and admiral Prince Rupert (1619–1682) was the first governor. Granted to the Hudson's Bay Company 1670, it was later split among Québec, Ontario, Manitoba, and the Northwest Territories.

Ruse /ˈruːseɪ/ (anglicized name **Rustchuk**) Danube port in Bulgaria, linked by rail and road bridge with Giurgiu in Romania; population (1990) 209,800.

Rushmore, Mount /ˈrʌʃmɔː/ mountain in the Black Hills, South Dakota, USA; height 1,890 m/6,203 ft. On its granite face are carved giant portrait heads of presidents Washington, Jefferson, Lincoln, and Theodore Roosevelt. The sculptor was Gutzon Borglum.

Russia /ˈrʌʃə/ originally the prerevolutionary Russian Empire (until 1917), now accurately restricted to the ◊Russian Federation.

Russian Far East geographical, not administrative, division of Asiatic Russia, on the Pacific coast. It includes the Amur, Lower Amur, Kamchatka, and Sakhalin regions, and Khabarovsk and Maritime territories.

Russian Federation formerly (until 1991) Russian Soviet Federal Socialist Republic (RSFSR)
area 17,075,500 sq km/6,591,100 sq mi
capital Moscow
towns St Petersburg (Leningrad), Nizhny-Novgorod (Gorky), Rostov-on-Don, Samara (Kuibyshev), Tver (Kalinin), Volgograd, Vyatka (Kirov), Ekaterinburg (Sverdlovsk)
physical fertile Black Earth district; extensive forests; the Ural Mountains with large mineral resources
features the heavily industrialized area around Moscow; Siberia; includes 16 autonomous republics (capitals in parentheses): Bashkir (Ufa); Buryat (Ulan-Ude); Checheno-Ingush (Grozny); Chuvash (Cheboksary); Dagestan (Makhachkala); Kabardino-Balkar (Nalchik); Kalmyk (Elista); Karelia (Petrozavodsk); Komi (Syktyvkar); Mari (Yoshkar-Ola); Mordovia (Saransk); Vladikavkaz (formerly Ordzhonikidze); Tatarstan (Kazan); Tuva (Kizyl); Udmurt (Izhevsk); Yakut (Yakutsk)
head of state Boris Yeltsin from 1990/91
head of government Viktor Chernomyrdin from 1992
political system emergent democracy
political parties Democratic Russia, liberal-radical, pro-Yeltsin; Congress of Civil and Patriotic Groups (CCDG), right-wing; Civic Union, right of centre; Nashi (Ours), far-right, Russian imperialist coalition; Communist (Bolshevik) Party, Stalinist-communist
products iron ore, coal, oil, gold, platinum, and other minerals, agricultural produce
currency rouble
population (1990) 148,000,000 (82% Russian, Tatar 4%, Ukrainian 3%, Chuvash 1%)
language Great Russian
religion traditionally Russian Orthodox
chronology
1945 Became a founding member of United Nations (UN).
1988 Aug: Democratic Union formed in Moscow as political party opposed to totalitarianism. Oct: Russian-language demonstrations in Leningrad, tsarist flag raised.
1989 March: Boris Yeltsin elected to USSR Congress of People's Deputies. Sept: conservative-nationalist Russian United Workers' Front established in Sverdlovsk.
1990 May: anticommunist May Day protests in Red Square, Moscow; Yeltsin narrowly elected RSFSR president by Russian parliament. June: economic and political sovereignty declared; Ivan Silaev became Russian prime minister. Aug: Tatarstan declared sovereignty. Dec: rationing introduced in some cities; private land ownership allowed.
1991 June: Yeltsin directly elected president under a liberal-radical banner. July: Yeltsin issued a decree to remove Communist Party cells from workplaces; sovereignty of the Baltic republics recognized by the republic. Aug: Yeltsin stood out against abortive anti-Gorbachev coup, emerging as key power-broker within Soviet Union; national

guard established and pre-revolutionary flag restored. Sept: Silaev resigned as Russian premier. Nov: Yeltsin named prime minister; Soviet and Russian Communist Parties banned; Yeltsin's government gained control of Russia's economic assets and armed forces. Oct: Checheno-Ingush declared its independence. Dec: Yeltsin negotiated formation of new confederal Commonwealth of Independent States; admitted into United Nations as newly independent state; independence recognized by USA and European Community. *1992* Jan: admitted into Conference on Security and Cooperation in Europe; assumed former USSR's permanent seat on UN Security Council; prices freed. Feb: demonstrations in Moscow and other cities as living standards plummeted. June: Yeltsin–Bush summit meeting. March: 18 out of 20 republics signed treaty agreeing to remain within loose Russian Federation; Tatarstan and Checheno-Ingush refused to sign. Dec: Victor Chernomyrdin elected prime minister; new constitution agreed in referendum. START II arms-reduction agreement signed with USA. *1993* March: Power struggle between Yeltsin and Congress of People's Deputies. Yeltsin declared temporary presidential 'special rule' pending referendum. April: Referendum gave vote of confidence in Yeltsin's presidency but did not support constitutional change. Sept–Oct: Yeltsin dissolved parliament, calling legislative elections for Dec. Attempted coup, led by conservative opponents, foiled by troops loyal to Yeltsin. Coup leaders arrested.

Ruthenia /ru:'θi:niə/ or *Carpathian Ukraine* region of central Europe, on the southern slopes of the Carpathian Mountains, home of the Ruthenes or Russniaks. Dominated by Hungary from the 10th century, it was part of Austria-Hungary until World War I. In 1918 it was divided between Czechoslovakia, Poland, and Romania; independent for a single day in 1938, it was immediately occupied by Hungary, captured by the USSR 1944, and 1945–47 was incorporated into Ukraine Republic, USSR. Ukraine became an independent republic 1991.

Rutland /'rʌtlənd/ formerly the smallest English county, part of Leicestershire since 1974.

Ruwenzori /,ru:ən'zɔ:ri/ mountain range on the frontier between Zaire and Uganda, rising to 5,119 m/16,794 ft at Mount Stanley.

Rwanda Republic of (*Republika y'u Rwanda*)
area 26,338 sq km/10,173 sq mi
capital Kigali
towns Butare, Ruhengeri
physical high savanna and hills, with volcanic mountains in NW
features part of lake Kivu; highest peak Mount Karisimbi 4,507 m/14,792 ft; Kagera River (whose headwaters are the source of the Nile) and National Park
head of state Maj-Gen Juvenal Habyarimana from 1973
head of government Agathe Uwilingyimana

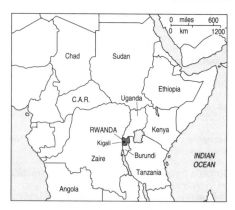

political system one-party military republic
political party National Revolutionary Movement for Development (MRND), nationalist, socialist
exports coffee, tea, pyrethrum
currency franc
population (1990 est) 7,603,000 (Hutu 90%, Tutsi 9%, Twa 1%); growth rate 3.3% p.a.
life expectancy men 49, women 53 (1989)
languages Kinyarwanda, French (official); Kiswahili
religions Roman Catholic 54%, animist 23%, Protestant 12%, Muslim 9%
literacy men 50% (1989)
GNP $2.3 bn (1987); $323 per head (1986)
chronology
1916 Belgian troops occupied Rwanda; League of Nations mandated Rwanda and Burundi to Belgium as Territory of Ruanda-Urundi.
1959 Interethnic warfare between Hutu and Tutsi.
1962 Independence from Belgium achieved, with Grégoire Kayibanda as president.
1972 Renewal of interethnic fighting.
1973 Kayibanda ousted in a military coup led by Maj-Gen Juvenal Habyarimana.
1978 New constitution approved; Rwanda remained a military-controlled state.
1980 Civilian rule adopted.
1988 Refugees from Burundi massacres streamed into Rwanda.
1990 Government attacked by Rwandan Patriotic Front (FPR), a Tutsi military-political organization based in Uganda.
1992 Peace accord made with FPR.
1993 Power-sharing agreement with government repudiated by FPR. Peace accord formally signed.

Ryazan /rɪ'zæn/ industrial city (agricultural machinery, leather, shoes) dating from the 13th century, capital of Ryazan region, W Russia, on the river Oka SE of Moscow; population (1987) 508,000.

Rybinsk /'rɪbɪnsk/ port and industrial city (engineering) in NW Russia, on the river Volga, NE of Moscow; population (1987) 254,000. In 1984 it was renamed *Andropov*, commemorating the death that year of the president of the USSR. It reverted to its former name March 1989.

Ryde /raɪd/ English resort on the northeast coast of the Isle of Wight, on the Solent opposite Portsmouth, with which there is ferry and hovercraft connection; population (1988 est) 27,100.

Rye /raɪ/ town in East Sussex, England; population (1985) 4,500. It was formerly a flourishing port (and one of the Cinque Ports), but silt washed down by the river Rother has left it 3 km/2 mi inland. The novelist Henry James lived here; another writer, E F Benson (who was mayor of Rye 1934–37), later lived in James's house.

Ryukyu Islands /riˈuːkjuː/ southernmost island group of Japan, stretching towards Taiwan and including Okinawa, Miyako, and Ishigaki
area 2,254 sq km/870 sq mi
capital Naha, on Okinawa
features 73 islands, some uninhabited; subject to typhoons

products sugar, pineapples, fish
population (1985) 1,179,000
history originally an independent kingdom; ruled by China from the late 14th century until seized by Japan 1609 and controlled by the Satsuma feudal lords until 1868, when the Japanese government took over. Chinese claims to the islands were relinquished 1895. In World War II the islands were taken by USA 1945; northernmost group, Oshima, restored to Japan 1953, the rest 1972.

The Ryukyu Islands, the tail to the rocket of Japan ...

On the **Ryukyu Islands** George Woodcock
Asia, Gods and Cities 1966

SA abbreviation for ◊*South Africa*; ◊*South Australia*, an Australian state.

Saar /sɑː/ (French *Sarre*) river in W Europe; it rises in the Vosges Mountains in France, and flows 240 km/149 mi N to join the river Moselle in Germany. Its valley has many vineyards.

Saarbrücken /zɑːˈbrukən/ city on the river Saar, Germany; population (1988) 184,000. It is situated on a large coalfield, and is an industrial centre (engineering, optical equipment). It has been the capital of Saarland since 1919.

Saarland /ˈsɑːlænd, German ˈzɑːlænt/ (French *Sarre*) *Land* (state) of Germany
area 2,570 sq km/992 sq mi
capital Saarbrücken
features one-third forest; crossed NW–S by the river Saar
products cereals and other crops; cattle, pigs, poultry. Former flourishing coal and steel industries survive only by government subsidy
population (1988) 1,034,000
history in 1919, the Saar district was administered by France under the auspices of the League of Nations; a plebiscite returned it to Germany 1935; Hitler gave it the name Saarbrücken. Part of the French zone of occupation 1945, it was included in the economic union with France 1947. It was returned to Germany 1957.

Sabah /ˈsɑːbə/ self-governing state of the federation of Malaysia, occupying NE Borneo, forming (with Sarawak) East Malaysia
area 73,613 sq km/28,415 sq mi
capital Kota Kinabalu (formerly Jesselton)
physical chiefly mountainous (highest peak Mount Kinabalu 4,098 m/13,450 ft) and forested
products hardwoods (25% of the world's supplies), rubber, fish, cocoa, palm oil, copper, copra, and hemp
population (1990) 1,470,200, of which the Kadazans form the largest ethnic group at 30%; also included are 250,000 immigrants from Indonesia and the Philippines
languages Malay (official) and English
religions Sunni Muslim and Christian (the Kadazans, among whom there is unrest about increasing Muslim dominance)
government consists of a constitutional head of state with a chief minister, cabinet, and legislative assembly
history in 1877–78 the Sultan of Sulu made concessions to the North Borneo Company, which was eventually consolidated with Labuan as a British colony 1946, and became the state of Sabah within Malaysia 1963. The Philippines advanced territorial claims on Sabah 1962 and 1968 on the grounds that the original cession by the Sultan was illegal, Spain having then been sovereign in the area.

Sachsen /ˈzæksən/ German form of ◊Saxony, a former kingdom and state of Germany.

Sacramento /ˌsækrəˈmentəu/ industrial port and capital (since 1854) of California, USA, 130 km/80 mi NE of San Francisco; population (1990) 369,400, metropolitan area 1,481,100. It stands on the Sacramento River, which flows 615 km/382 mi through Sacramento Valley to San Francisco Bay. Industries include the manufacture of detergents and jet aircraft and food processing, including almonds, peaches, and pears.

It was founded as **Fort Sutter** 1848 on land bought by John Sutter 1839. Its old town has been restored. A deepwater port channel to San Francisco Bay was completed 1963.

Safi /sæˈfiː/ Atlantic port in Tensift province, NW Morocco; population (1981) 256,000. It exports phosphates and has fertilizer plants, sardine factories, and boat-building yards.

Sagamihara /səˌɡɑːmɪˈhɑːrə/ city on the island of Honshu, Japan, with a large silkworm industry; population (1990) 531,600.

Sagarmatha /ˌsæɡəˈmɑːtə/ Nepalese name for Mount ◊Everest, 'the Goddess of the Universe', and the official name of the 1,240-sq-km/476-sq-mi Himalayan national park established 1976.

Saginaw /ˈsæɡɪnɔː/ city and port in E Michigan, USA, on the Saginaw River, near Lake Huron, NW of Flint; population (1990) 69,500. Industries include automotive parts, metal products, salt, coal, and sugar beet. Saginaw was a lumber centre until the late 19th century.

Saguenay /ˌsæɡəˈneɪ/ river in Québec, Canada, used for hydroelectric power as it flows from Lac St Jean SE to the St Lawrence estuary; length 765 km/475 mi.

Sahara /səˈhɑːrə/ largest desert in the world, occupying 5,500,000 sq km/2,123,000 sq mi of N Africa from the Atlantic to the Nile, covering: W Egypt; part of W Sudan; large parts of Mauritania, Mali, Niger, and Chad; and southern parts of Morocco, Algeria, Tunisia, and Libya. Small areas in Algeria and Tunisia are below sea level, but it is mainly a plateau with a central mountain system, including the Ahaggar Mountains in Algeria, the Air Massif in Niger, and the Tibesti Massif in Chad, of which the highest peak is Emi Koussi, 3,415 m/11,208 ft. The area of the Sahara expanded by 650,000 sq km/251,000 sq mi 1940–90, but reafforestation is being attempted in certain areas.

Oases punctuate the caravan routes, now modern roads. Resources include oil and gas in the N. Satellite observations have established a pattern below the surface of dried-up rivers that existed 2 million years ago. Cave paintings confirm that 4,000 years ago running rivers and animal life existed.

Sahel /'sɑːhel/ (Arabic *sahil* 'coast') marginal area to the S of the Sahara, from Senegal to Somalia, where the desert is gradually encroaching. The desertification is partly due to climatic change but has also been caused by the pressures of a rapidly expanding population, which have led to overgrazing and the destruction of trees and scrub for fuelwood. In recent years many famines have taken place in the area.

Saida /'saɪdə/ (ancient *Sidon*) port in Lebanon; population (1980) 24,740. It stands at the end of the Trans-Arabian oil pipeline from Saudi Arabia. Sidon was the chief city of Phoenicia, a bitter rival of Tyre about 1400–701 BC, when it was conquered by Sennacherib. Later a Roman city, it was taken by the Arabs AD 637 and fought over during the Crusades.

Saigon /ˌsaɪ'ɡɒn/ former name (until 1976) of ◊ Ho Chi Minh City, Vietnam.

St Albans /sənt 'ɔːlbənz/ city in Hertfordshire, England, on the river Ver, 40 km/25 mi NW of London; population (1981) 55,700. The cathedral was founded 793 in honour of St Alban; nearby are the ruins of the Roman city of Verulamium on Watling Street.

St Andrews /sənt 'ændruːz/ town at the eastern tip of Fife, Scotland, 19 km/12 mi SE of Dundee; population (1987 est) 13,800. Its university (1411) is the oldest in Scotland, and the Royal and Ancient Club (1754) is the ruling body of golf.

There are four golf courses, all municipal; the Old Course dates from the 16th century. One of the earliest patrons was Mary Queen of Scots. St Andrews has been used to stage the British Open 24 times between 1873 and 1984.

St Augustine /seɪnt 'ɔːɡəstiːn/ port and holiday resort in Florida, USA; population (1990) 11,700. Founded by the Spanish 1565, the oldest permanent settlement in the USA, it was burned by the English sea captain Francis Drake 1586 and ceded to the USA 1821. It includes the oldest house (late 16th century) and oldest masonry fort (Castillo de San Marcos 1672) in the continental USA.

St Austell /sənt 'ɔːstəl/ market town in Cornwall, England, 22 km/14 mi NE of Truro; population (1985 est) 39,200 (with Fowey, with which it is administered). It is the centre of the China clay area, which supplies the Staffordshire potteries.

St Bernard Passes /sənt 'bɜːnəd/ the passes through the Alps (the *Great St Bernard Pass* and *Little St Bernard Pass*).

St Christopher (St Kitts)–Nevis Federation of *area* 269 sq km/104 sq mi (St Christopher 176 sq

km/68 sq mi, Nevis 93 sq km/36 sq mi)
capital Basseterre (on St Christopher)
towns Charlestown (largest on Nevis)
physical both islands are volcanic
features fertile plains on coast; black beaches
head of state Elizabeth II from 1983 represented by governor general
head of government Kennedy Simmonds from 1980
political system federal constitutional monarchy
political parties People's Action Movement (PAM), centre-right; Nevis Reformation Party (NRP), Nevis-separatist; Labour Party, moderate, left-of-centre
exports sugar, molasses, electronics, clothing
currency E Caribbean dollar
population (1990 est) 45,800; growth rate 0.2% p.a.
life expectancy men 69/women 72
language English
media no daily newspaper; two weekly papers, published by the governing and opposition party respectively—both receive advertising support from the government
religions Anglican 36%, Methodist 32%, other protestant 8%, Roman Catholic 10% (1985 est)
literacy 90% (1987)
GNP $40 million (1983); $870 per head
chronology
1871–1956 Part of the Leeward Islands Federation.
1958–62 Part of the Federation of the West Indies.
1967 St Christopher, Nevis, and Anguilla achieved internal self-government, within the British Commonwealth, with Robert Bradshaw, Labour Party leader, as prime minister.
1971 Anguilla returned to being a British dependency.
1978 Bradshaw died; succeeded by Paul Southwell.
1979 Southwell died; succeeded as prime minister by Lee L Moore.
1980 Coalition government led by Kennedy Simmonds.
1983 Full independence achieved within the Commonwealth.

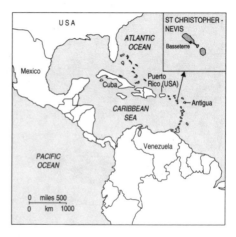

1984 Coalition government re-elected.
1989 Prime Minister Simmonds won a third successive term.

St-Cloud /sæŋ 'kluː/ town in the Ile de France region, France; population (1990) 28,700. The château of St-Cloud, linked with Marie Antoinette and Napoleon, was demolished 1781, but the park remains. It is the site of the Sèvres porcelain factory.

St David's /sənt 'deɪvɪdz/ (Welsh *Tyddewi*) small city in Dyfed, Wales. Its cathedral, founded by St David, was rebuilt 1180–1522.

St-Denis /ˌsæn də'niː/ industrial town, a northern suburb of Paris, France; population (1990) 90,800. The French philosopher and theologian Peter Abelard was a monk at the 12th-century Gothic abbey, which contains many tombs of French kings.

St Elias Mountains /ˌseɪnt ɪ'laɪəs/ mountain range on the Alaska–Canada border. Its highest peak, Mount Logan 6,050 m/19,850 ft, is Canada's highest mountain.

Saint-Etienne /ˌsænt et'jen/ city in S central France, capital of Loire *département*, Rhônes-Alpes region; population (1990) 201,600. Industries include the manufacture of aircraft engines, electronics, and chemicals, and it is the site of a school of mining, established 1816.

St Gall /sæŋ 'gæl/ (German *Sankt Gallen*) town in NE Switzerland; population (1990) 73,400. Industries include natural and synthetic textiles. It was founded in the 7th century by the Irish missionary St Gall, and the Benedictine abbey library has many medieval manuscripts.

St George's /sənt 'dʒɔːdʒɪz/ port and capital of Grenada, on the SW coast; population (1986) 7,500, urban area 29,000. It was founded 1650 by the French.

St George's Channel /sənt 'dʒɔːdʒɪz/ stretch of water between SW Wales and SE Ireland, linking the Irish Sea with the Atlantic Ocean. It is 160 km/100 mi long and 80–150 km/50–90 mi wide. It is

St Helena

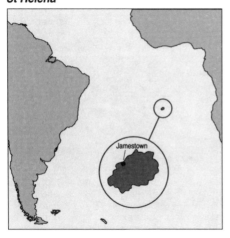

also the name of a channel between New Britain and New Ireland, Papua New Guinea.

St Gotthard Pass /sənt 'gɒtəd/ pass through the Swiss Alps, at an altitude of 2,000 m/6,500 ft. A rail tunnel is planned, running 48 km/30 mi through the St Gotthard massif, to be built 1990–2010.

St Helena /ˌsent ɪ'liːnə/ British island in the S Atlantic, 1,900 km/1,200 mi W of Africa, area 122 sq km/47 sq mi; population (1987) 5,600. Its capital is Jamestown, and it exports fish and timber. Ascension and Tristan da Cunha are dependencies.
St Helena became a British possession 1673, and a colony 1834. Napoleon died in exile here 1821.

St Helens /sənt 'helənz/ town in Merseyside, England, 19 km/12 mi NE of Liverpool, and connected to the river Mersey by canal; population (1991) 175,300. It is a leading centre for the manufacture of sheet glass.

St Helens, Mount /seɪnt 'helənz/ volcanic mountain in Washington State, USA. When it erupted 1980 after being quiescent since 1857, it devastated an area of 600 sq km/230 sq mi and its height was reduced from 2,950 m/9,682 ft to 2,560 m/8,402 ft.

St Helier /sənt 'heliə/ resort and capital of Jersey, Channel Islands; population (1988 est) 82,500. The 'States of Jersey', the island legislature, sits here in the *salle des états*.

St Ives /sənt 'aɪvz/ fishing port and resort in Cornwall; population (1981 est) 11,000. Its artists' colony, founded by Walter Sickert and James Whistler, later included Naum Gabo, Barbara Hepworth (a museum and sculpture gardens commemorate her), and Ben Nicholson.

Saint John /seɪnt 'dʒɒn/ largest city of New Brunswick, Canada, on the Saint John River; population (1986) 121,000. It is a fishing port and has shipbuilding, timber, fish-processing, petroleum, refining, and textile industries. Founded by the French as *Saint-Jean* 1635, it was taken by the British 1758.

St John's /seɪnt 'dʒɒnz/ capital and chief port of Newfoundland, Canada; population (1986) 96,000, urban area 162,000. The main industry is fish processing; other products include textiles, fishing equipment, furniture, and machinery.
It was founded by English navigator Humphrey Gilbert 1582. The inventor Guglielmo Marconi's first transatlantic radio message was received on Signal Hill 1901. Memorial University was founded 1925.

St John's /sənt 'dʒɒnz/ port and capital of Antigua and Barbuda, on the northwest coast of Antigua; population (1982) 30,000. It exports rum, cotton, and sugar.

St John's Wood /sənt ˌdʒɒnz 'wud/ residential suburb of NW London. It is the site of Lord's cricket ground, headquarters of the Marylebone Cricket Club (MCC).

St Kilda /sənt 'kɪldə/ group of three mountainous islands, the most westerly of the Outer Hebrides, 200 km/124 mi W of the Scottish mainland; area 16 sq km/6 sq mi. They were populated from prehistory until 1930, and are now a nature reserve.

St Kitts-Nevis /sənt 'kɪts 'niːvɪs/ contracted form of ◊St Christopher–Nevis.

St Lawrence /seɪnt 'lɒrəns/ river in E North America. From ports on the Great Lakes it forms, with linking canals (which also give great hydro-electric capacity to the river), the St Lawrence Seaway for oceangoing ships, ending in the Gulf of St Lawrence. It is 745 mi/1,200 km long and is icebound for four months each year.

St Leonards /'lenədz/ seaside town near Hastings, England.

St-Lô /sæn 'ləu/ market town in Normandy, France, on the river Vire; population (1990) 22,800. In World War II it was almost entirely destroyed 10–18 July 1944, when US forces captured it from the Germans.

St Louis /seɪnt 'luːɪs/ city in Missouri, USA, on the Mississippi River; population (1990) 396,700, metropolitan area 2,444,100. Its products include aerospace equipment, aircraft, vehicles, chemicals, electrical goods, steel, and beer. Founded as a French trading post 1764, it passed to the USA 1803 under the Louisiana Purchase. The Gateway Arch 1965 is a memorial by US architect Eliel Saarinen to the pioneers of the West.

St Lucia
area 617 sq km/238 sq mi
capital Castries
towns Vieux-Fort, Soufrière
physical mountainous island with fertile valleys; mainly tropical forest
features volcanic peaks; Gros and Petit Pitons
head of state Elizabeth II from 1979 represented by governor general
head of government John Compton from 1982
political system constitutional monarchy
political parties United Workers' Party (UWP),

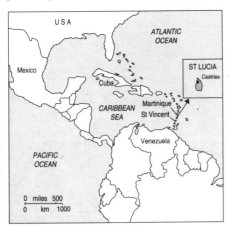

moderate, left-of-centre; St Lucia Labour Party (SLP), moderate, left-of-centre; Progressive Labour Party (PLP), moderate, left-of-centre
exports coconut oil, bananas, cocoa, copra
currency E Caribbean dollar
population (1990 est) 153,000; growth rate 2.8% p.a.
life expectancy men 68, women 73 (1989)
languages English; French patois
media two independent biweekly newspapers
religion Roman Catholic 90%
literacy 78% (1989)
GNP $166 million; $1,370 per head (1987)
chronology
1814 Became a British crown colony following Treaty of Paris.
1967 Acquired internal self-government as a West Indies associated state.
1979 Independence achieved from Britain within the Commonwealth, with John Compton, leader of the UWP, as prime minister. Allan Louisy, leader of the SLP, replaced Compton as prime minister.
1981 Louisy resigned; replaced by Winston Cenac.
1982 Compton returned to power
1987 Compton re-elected with reduced majority.
1991 Integration with Windward Islands proposed.
1992 UWP won general election.

St-Malo /ˌsæm mɑːˈləu/ seaport and resort in the Ille-et-Vilaine *département*, W France, on the Rance estuary; population (1990) 49,300. It took its name from the Welshman Maclou, who was bishop here in about 640.

St Michael's Mount /sənt 'maɪkəlz/ island in Mount's Bay, Cornwall, England, linked to the mainland by a causeway.

St-Nazaire /ˌsæn næˈzeə/ industrial seaport in the Loire-Atlantique *département*, Pays de la Loire region, France; population (1990) 66,100. It stands at the mouth of the river Loire and in World War II was used as a German sub-marine base. Industries include shipbuilding, engineering, and food canning.

St-Omer /ˌsæn əuˈmeə/ town in Pas-de-Calais *département*, France, 42 km/26 mi SE of Calais; population (1985) 15,500. In World War I it was the site of British general headquarters 1914–16.

St Paul /seɪnt 'pɔːl/ capital and industrial city of Minnesota, USA, adjacent to Minneapolis; population (1990) 272,200. Industries include electronics, publishing and printing, chemicals, refined petroleum, machinery, and processed food.

St Peter Port /sənt 'piːtə 'pɔːt/ only town of Guernsey, Channel Islands; population (1986 est) 16,100.

St Petersburg /'piːtəzbɜːg/ capital of the St Peters-burg region, Russia, at the head of the Gulf of Finland; population (1989 est) 5,023,500. Industries include shipbuilding, machinery, chemicals, and textiles. It was renamed *Petrograd* 1914 and was called *Leningrad* from 1924 until 1991, when its original name was restored.

Built on a low and swampy site, St Petersburg is split up by the mouths of the river Neva, which connects it with Lake Ladoga. The climate is severe. The city became a seaport when it was linked with the Baltic by a ship canal built 1875–93. It is also linked by canal and river with the Caspian and Black seas, and in 1975 a seaway connection was completed via lakes Onega and Ladoga with the White Sea near Belomorsk, allowing naval forces to reach the Barents Sea free of NATO surveillance.

features St Petersburg is notable for its wide boulevards and the scale of its architecture. Most of its fine baroque and classical buildings of the 18th and early 19th centuries survived World War II. Museums include the Winter Palace, occupied by the tsars until 1917, the Hermitage, the Russian Museum (formerly Michael Palace), and St Isaac's Cathedral. The oldest building in St Petersburg is the fortress of St Peter and St Paul, on an island in the Neva, now a political prison. The university was founded 1819.

history Saint Petersburg was founded as an outlet to the Baltic 1703 by Peter the Great, who took up residence here 1712. It was capital of the Russian Empire 1709–1918 and the centre of all the main revolutionary movements from the Decembrist revolt 1825 up to the 1917 revolution.

During the German invasion in World War II the city withstood siege and bombardment 30 Aug 1941–27 Jan 1944. Over 100,000 bombs were dropped by the Luftwaffe, and 150,000–200,000 shells were fired, but most deaths (estimate 1.3–1.5 million) resulted from famine and the cold. Soviet counter attacks began 1943, but the siege was not completely lifted until Jan 1944.

In June 1991, the city's electors voted by 55% to 43% to restore the Tsarist (1703–1914) designation, St Petersburg. This vote received parliamentary sanction Sept 1991. In the previous month, the city's mayor, Anatoly Sobchak, and thousands of citizens had resisted an attempted coup to oust Soviet leader Gorbachev.

(St Petersburg is) Russian, but it is not Russia.

On **St Petersburg** Tsar Nicholas I

St Petersburg /'piːtəzbɜːg/ seaside resort and industrial city (space technology), W Florida, USA; population (1990) 238,600. It is across Tampa Bay from Tampa.

St Pierre and Miquelon /,sæmpi'eə 'miːkəlɒn/ territorial dependency of France, eight small islands off the south coast of Newfoundland, Canada
area St Pierre group 26 sq km/10 sq mi; Miquelon-Langlade group 216 sq km/83 sq mi
capital St Pierre
features the last surviving remnant of France's North American empire
products fish
currency French franc

population (1987) 6,300
language French
religion Roman Catholic
government French-appointed commissioner and elected local council; one representative in the National Assembly in France
history settled 17th century by Breton and Basque fisherfolk; French territory 1816–1976; overseas *département* until 1985; violent protests 1989 when France tried to impose its claim to a 320-km/200-mi fishing zone around the islands; Canada maintains that there is only a 19-km/12-mi zone.

St-Quentin /,sæŋ kɒn'tæn/ town on the river Somme, Picardie, N France; population (1990) 62,100. It was the site of a Prussian defeat of the French 1871 and was almost obliterated in World War I. It is linked by canal to the industrial centres of Belgium and Germany. Its traditional textile production has been replaced by chemicals and metalworks.

St-Tropez /,sæntrəu'peɪ/ resort and fishing port on the French Côte d'Azur; population (1985) 6,250. It became popular as a resort in the 1960s.

St Vincent /sənt 'vɪnsənt/ cape of the Algarve region of SW Portugal off which England defeated the French and Spanish fleets 1797.

St Vincent and the Grenadines

area 388 sq km/150 sq mi, including islets of the Northern Grenadines 43 sq km/17 sq mi
capital Kingstown
towns Georgetown, Chateaubelair
physical volcanic mountains, thickly forested
features Mustique, one of the Grenadines, a holiday resort; Soufrière volcano
head of state Elizabeth II from 1979 represented by governor general
head of government James Mitchell from 1984
political system constitutional monarchy
political parties New Democratic Party (NDP), moderate, left of centre; St Vincent Labour Party (SVLP), moderate, left of centre
exports bananas, taros, sweet potatoes, arrowroot, copra
currency E Caribbean dollar

population (1990 est) 106,000; growth rate –4% p.a.
life expectancy men 69, women 74 (1989)
languages English; French patois
media government-owned radio station; two privately owned weekly newspapers, subject to government pressure
religions 47% Anglican, 28% Methodist, 13% Roman Catholic
literacy 85% (1989)
GNP $188 million; $1,070 per head (1987)
chronology
1783 Became a British crown colony.
1958–62 Part of the West Indies Federation.
1969 Achieved internal self-government.
1979 Achieved full independence from Britain within the Commonwealth, with Milton Cato as prime minister.
1984 James Mitchell replaced Cato.
1989 Mitchell decisively re-elected.
1991 Integration with Windward Islands proposed.

St Vincent Gulf /sənt 'vɪnsənt/ inlet of the Southern Ocean on which Adelaide, South Australia, stands. It is named after Adam John Jervis, 1st Earl of St Vincent (1735–1823).

Sakai /sɑːˈkaɪ/ city on the island of Honshu, Japan; population (1990) 807,900. Industries include engineering, aluminium, and chemicals.

Sakhalin /ˌsæxəˈliːn/ (Japanese **Karafuto**) island in the Pacific, N of Japan, that since 1947, with the Kurils, forms a region of Russia; capital Yuzhno-Sakhalinsk (Japanese **Toyohara**); area 74,000 sq km/28,564 sq mi; population (1981) 650,000, including aboriginal Ainu and Gilyaks. There are two parallel mountain ranges, rising to over 1,525 m/5,000 ft, which extend throughout its length, 965 km/600 mi.

The economy is based on dairy farming, leguminous crops, oats, barley, and sugar beet. In the milder south, there is also timber, rice, wheat, fish, some oil, and coal. The island was settled by both Russians and Japanese from the 17th century. In 1875 the south was ceded by Japan to Russia, but Japan regained it 1905, only to cede it again 1945. It has a missile base.

Sakkara /səˈkɑːrə/ or **Saqqara** village in Egypt, 16 km/10 mi S of Cairo, with 20 pyramids, of which the oldest (third dynasty) is the 'Step Pyramid' designed by Imhotep, whose own tomb here was the nucleus of the Aesklepieion, a centre of healing in the ancient world.

Salado /səˈlɑːdəʊ/ two rivers of Argentina, both rising in the Andes Mountains, and about 1,600 km/1,000 mi long. **Salado del Norte**, or **Juramento**, flows from the Andes to join the Paraná River; the **Salado del Sud**, or **Desaguadero**, joins the Colorado River and flows into the Atlantic S of Bahia Blanca.

Salamanca /ˌsæləˈmæŋkə/ city in Castilla-León, W Spain, on the river Tormes, 260 km/162 mi NW of Madrid; population (1986) 167,000. It produces pharmaceuticals and wool. Its university was founded about 1230. It has a superbly

designed square, the Plaza Mayor.

Salamis /ˈsæləmɪs/ island off Piraeus, the port of Athens, Greece; area 101 sq km/39 sq mi; population (1981) 19,000. The town of Salamis, on the west coast, is a naval station.

Salang Highway /ˈsɑːlæŋ/ the main N–S route between Kabul, capital of Afghanistan, and Tajikistan; length 422 km/264 mi. The high-altitude **Salang Pass** and **Salang Tunnel** cross a natural break in the Hindu Kush Mountains about 100 km/60 mi N of Kabul. This supply route was a major target of the Mujaheddin resistance fighters during the Soviet occupation of Afghanistan.

Sale /seɪl/ residential suburb of Manchester, England; population (1981) 58,000.

Sale /seɪl/ town in Victoria, Australia, linked by canal via the Gippsland Lake to Bass Strait; population (1981) 13,000. It has benefited from the Strait deposits of oil and natural gas, and the brown coal to the S. The town was named after the British general Robert Sale (1782–1845).

Salem /ˈseɪləm/ industrial city (iron mining, textiles) in Tamil Nadu, India; population (1981) 515,000.

Salem /ˈseɪləm/ city in NW Oregon, USA, settled about 1840 and made state capital 1859; population (1990) 107,800. It processes timber into wood products and has a prosperous fruit- and vegetable-canning industry.

Salem /ˈseɪləm/ city and manufacturing centre in Massachusetts, USA, 24 km/15 mi NE of Boston; population (1990) 38,100.
It was the site of witch trials 1692, which ended in the execution of 19 people.

Salerno /səˈleənəʊ/ port in Campania, SW Italy, 48 km/30 mi SE of Naples; population (1988) 154,000. It was founded by the Romans about 194 BC, destroyed by Charlemagne, and sacked by Holy Roman Emperor Henry VI 1194. The temple ruins of the ancient Greek city of Paestum, with some of the earliest Greek paintings known, are nearby. Salerno has had a university (1150–1817, revived 1944) and medical school since medieval times.

Salford /ˈsɔːlfəd/ industrial city in Greater Manchester, England, on the river Irwell; population (1991) 217,900. Industries include engineering, electrical goods, textiles, and chemicals.

Salinas /səˈliːnəs/ city in W central California, USA, S of San Jose; population (1990) 108,800. Fruits and vegetables, such as lettuce, are the economy's mainstay.

Salisbury /ˈsɔːlzbəri/ city and market town in Wiltshire, England, 135 km/84 mi SW of London; population (1983 est) 40,000. Salisbury is an agricultural centre, and industries include brewing and carpet manufacture (in nearby Wilton). The cathedral of St Mary, built 1220–66, is an example of Early English architecture; its decorated spire 123 m/404 ft is the highest in England; its clock (1386) is one of the oldest still working. The cathedral library contains one of only four copies of the *Magna Carta*.

Another name for Salisbury is *New Sarum*, Sarum being a medieval Latin corruption of the ancient Romano-British name Sorbiodonum. *Old Sarum*, on a 90-m/300-ft hill to the N, was deserted when New Sarum was founded 1220 but was later again inhabited; it was brought within the town boundary 1953.

Salisbury /ˈsɔːlzbəri/ former name (until 1980) of ▷Harare, the capital of Zimbabwe.

Salisbury Plain /ˈsɔːlzbəri/ area of open downland 775 sq km/300 sq mi between Salisbury and Devizes in Wiltshire, England. It rises to 235 m/770 ft in Westbury Down. Stonehenge stands on Salisbury Plain. For many years it has been a military training area.

Salonika /ˌsælɒˈnaɪkə, səˈlɒnɪkə/ English name for ▷Thessaloníki, a port in Greece.

Salop /ˈsæləp/ abbreviation and former official name (1972–80) for ▷Shropshire, a county in England.

Salt Lake City /ˈsɔːlt ˌleɪk ˈsɪti/ capital of Utah, USA, on the river Jordan, 18 km/11 mi SE of the Great Salt Lake; population (1990) 159,900.

Founded 1847, it is the headquarters of the Mormon church of Jesus Christ of the Latter Day Saints. Mining, construction, and other industries are being replaced by high technology.

Salton Sea /ˈsɔːltən/ brine lake in SE California, USA, area 650 sq km/250 sq mi, accidentally created in the early 20th century during irrigation works from the Colorado River. It is used to generate electricity.

Salvador /ˌsælvəˈdɔː/ port and naval base in Bahia state, NE Brazil, on the inner side of a peninsula separating Todos Santos Bay from the Atlantic Ocean; population (1991) 2,075,400. Products include cocoa, tobacco, and sugar. Founded 1510, it was the capital of Brazil 1549–1763.

Salvador, El /elˈsælvədɔː/ republic in Central America; see ▷El Salvador.

Salween /ˈsælwiːn/ river rising in E Tibet and flowing 2,800 km/1,740 mi through Myanmar (Burma) to the Andaman Sea; it has many rapids.

Salzburg /ˈsæltsbɜːɡ/ capital of the state of Salzburg, W Austria, on the river Salzach, in W Austria; population (1981) 139,400. The city is dominated by the Hohensalzburg fortress. It is the seat of an archbishopric founded by St Boniface about 700 and has a 17th-century cathedral. Industries include stock rearing, dairy farming, forestry, and tourism.

It is the birthplace of the composer Wolfgang Amadeus Mozart and an annual music festival has been held here since 1920.

Salzburg /ˈsæltsbɜːɡ/ federal province of Austria; area 7,200 sq km/2,779 sq mi; population (1987) 462,000. Its capital is Salzburg.

Salzgitter /ˈzæltsˌɡɪtə/ city in Lower Saxony, Germany; population (1988) 105,000. Industries

include iron and steel, shipbuilding, vehicles, and machine tools.

Samara /səˈmɑːrə/ capital of Kuibyshev region, W central Russia, and port at the junction of the rivers Samara and Volga, situated in the centre of the fertile middle Volga plain; population (1987) 1,280,000. Industries include aircraft, locomotives, cables, synthetic rubber, textiles, fertilizers, petroleum refining, and quarrying. It was called *Kuibyshev* 1935–91, reverting to its former name Jan 1991.

The city was provisional capital of the USSR 1941–43.

Samarkand /ˌsæmɑːˈkænd/ city in E Uzbekistan, capital of Samarkand region, near the river Zerafshan, 217 km/135 mi E of Bukhara; population (1987) 388,000. Industries include cotton-ginning, silk manufacture, and engineering.

Samarkand was the capital of the empire of Tamerlane, the 14th-century Mongol ruler, who is buried here, and was once a major city on the Silk Road. It was occupied by the Russians in 1868 but remained a centre of Muslim culture until the Russian Revolution.

For lust of knowing what should not be known,/We take the golden road to Samarkand.

On **Samarkand** James Elroy Flecker *Hassan* 1913

Samarra /səˈmærə/ ancient town in Iraq, on the river Tigris, 105 km/65 mi NW of Baghdad; population (1970) 62,000. Founded 836 by the Abbasid Caliph Motassim, it was the Abbasid capital until 892 and is a place of pilgrimage for Shi'ite Muslims. It is one of the largest archaeological sites in the world, and includes over 6,000 separate sites. The best preserved palace is Qasr al-Ashiq, built entirely of brick 878–882.

Samoa /səˈməʊə/ volcanic island chain in the SW Pacific. It is divided into Western Samoa and American Samoa.

In the South Seas the Creator seems to have laid himself out to show what he can do.

On **Samoa** Rupert Brooke *Letters from America, 1913* 1916

Samoa, American /səˈməʊə/ group of islands 4,200 km/2,610 mi S of Hawaii, administered by the USA
area 200 sq km/77 sq mi
capital Fagatogo on Tutuila
features five volcanic islands, including Tutuila, Tau, and Swain's Island, and two coral atolls. National park (1988) includes prehistoric village of Saua, virgin rainforest, flying foxes
exports canned tuna, handicrafts

currency US dollar
population (1990) 46,800
languages Samoan and English
religion Christian
government as a non-self-governing territory of the USA, under Governor A P Lutali, it is constitutionally an unincorporated territory of the USA, administered by the Department of the Interior
history the islands were acquired by the USA Dec 1899 by agreement with Britain and Germany under the Treaty of Berlin. A constitution was adopted 1960 and revised 1967.

Samoa, Western Independent State of (*Samoa i Sisifo*)
area 2,830 sq km/1,093 sq mi
capital Apia (on Upolu island)
physical comprises South Pacific islands of Savai'i and Upolu, with two smaller tropical islands and islets; mountain ranges on main islands
features lava flows on Savai'i
head of state King Malietoa Tanumafili II from 1962
head of government Tofilau Eti Alesana from 1988
political system liberal democracy
political parties Human Rights Protection Party (HRPP), led by Tofilau Eti Alesana; the Va'ai Kolone Group (VKG); Christian Democratic Party (CDP), led by Tupua Tamasese Efi. All 'parties' are personality-based groupings
exports coconut oil, copra, cocoa, fruit juice, cigarettes, timber
currency talà
population (1989) 169,000; growth rate 1.1% p.a.
life expectancy men 64, women 69 (1989)
languages English, Samoan (official)
religions Protestant 70%, Roman Catholic 20%
literacy 90% (1989)
GNP $110 million (1987); $520 per head
chronology
1899–1914 German protectorate.
1920–61 Administered by New Zealand.
1959 Local government elected.
1961 Referendum favoured independence.
1962 Independence achieved within the Commonwealth, with Fiame Mata Afa Mulinu'u as prime minister.

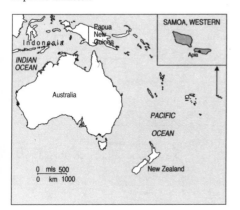

1975 Mata Afa died.
1976 Tupuola Taisi Efi became first nonroyal prime minister.
1982 Va'ai Kolone became prime minister; replaced by Tupuola Efi. Assembly failed to approve budget; Tupuola Efi resigned; replaced by Tofilau Eti Alesana.
1985 Tofilau Eti resigned; head of state invited Va'ai Kolone to lead the government.
1988 Elections produced a hung parliament, with first Tupuola Efi as prime minister and then Tofilau Eti Alesana.
1990 Universal adult suffrage introduced.
1991 Tofilau Eti Alesana re-elected. Fiame Naome became first woman in cabinet.

Samos /'seɪmɒs, Greek 'sɑːmɒs/ Greek island in the Aegean Sea, off the W coast of Turkey; area 476 sq km/184 sq mi; capital Limén Vathéos; population (1981) 31,600. Mountainous but fertile, it produces wine (muscat) and olive oil. The mathematician Pythagoras was born here. The town of Teganion is on the site of the ancient city of Samos, which was destroyed by Darius I of Persia.

Samsun /sæm'suːn/ Black Sea port and capital of a province of the same name in N Turkey; situated at the mouth of the Murat river in a tobacco-growing area; population (1990) 303,900. It is the site of the ancient city of Amisus.

San'a /sæ'nɑː/ capital of Yemen, SW Arabia, 320 km/200 mi N of Aden; population (1986) 427,000. A walled city, with fine mosques and traditional architecture, it is rapidly being modernized. Weaving and jewellery are local handicrafts.

San Andreas fault /,sæn æn'dreɪəs/ geological fault line stretching for 1,125 km/700 mi NW–SE through the state of California, USA.
   Two sections of the Earth's crust meet at the San Andreas fault, and friction is created as the coastal Pacific plate moves NW, rubbing against the American continental plate, which is moving slowly SE. The friction caused by this tectonic movement gives rise to periodic earthquakes, theatening the nearby cities of Los Angeles and San Francisco.

San Angelo /sæn 'ændʒələʊ/ city in W central Texas, USA, where the North and Middle Concho rivers meet, SW of Abilene; population (1990) 84,500. Industries include wool, food processing, oil, livestock, and clay products.

San Antonio /,sæn æn'təʊniəʊ/ city in S Texas, USA; population (1990) 935,900. It is a commercial and financial centre; industries include aircraft maintenance, oil refining, and meat packing.
   Founded 1718, it grew up round the site of the Alamo fort.

... you have the sensation in San Antonio by day of the world being deliciously excluded.

On **San Antonio** Graham Greene
'The Lawless Roads' 1939

San Bernardino /'sæn ˌbɜːnəˈdiːnəʊ/ city in California, USA, 80 km/50 mi E of Los Angeles; population (1990) 164,200. Products include processed food, steel, and aerospace and electronic equipment. It was founded 1851 by Mormons.

San Cristóbal /ˌsæn krɪsˈtəʊbæl/ capital of Tachira state, W Venezuela, near the Colombian border; population (1989) 355,900. It was founded by Spanish settlers 1561 and stands on the Pan-American Highway.

Sandhurst /'sændhɜːst/ small town in Berkshire, England. The Royal Military Academy (the British military officer training college), founded 1799, is nearby.

San Diego /ˌsæn diˈeɪɡəʊ/ city and military and naval base in California, USA; population (1990) 1,110,500, metropolitan area 2,498,000. It is an important Pacific Ocean fishing port. Manufacturing includes aerospace and electronic equipment, metal fabrication, printing and publishing, seafood canning, and shipbuilding.

   Tijuana adjoins San Diego across the Mexican border.

Sandwich /'sænwɪtʃ/ resort and market town in Kent, England; population (1981) 4,200. It has many medieval buildings and was one of the Cinque Ports, but recession of the sea has left the harbour useless since the 16th century.

Sandwich Islands /'sænwɪtʃ/ former name of ◊Hawaii, a group of islands in the Pacific.

San Francisco /ˌsæn frænˈsɪskəʊ/ chief Pacific port of the USA, in California; population (1990) 724,000, metropolitan area of San Francisco and Oakland 3,686,600. The city stands on a peninsula, on the south side of the Golden Gate Strait, spanned 1937 by the world's second longest single-span bridge, 1,280 m/4,200 ft. The strait gives access to San Francisco Bay. Manufactured goods include textiles, fabricated metal products, electrical equipment, petroleum products, chemicals, and pharmaceuticals.

   history In 1578 Sir Francis Drake's flagship, the *Golden Hind*, stopped near San Francisco on its voyage around the world. A Spanish fort and mission were established 1776. The original Spanish village was called Yerba Buena; its name was changed to San Francisco 1846. In the same year the town was occupied during the war with Mexico. When gold was discovered 1848 its population increased from about 800 to about 25,000 in two years. In 1906 the city was almost destroyed by an earthquake and subsequent fire that killed 452 people. It was the site of the drawing up of the United Nations Charter 1945 and of the signing of the peace treaty between the Allied nations and Japan 1951. Another earthquake (6.9 on the Richter scale) rocked the city 1989.

San José /'sæn həʊˈzeɪ/ capital of Costa Rica; population (1989) ·284,600. Products include coffee, cocoa, and sugar cane. It was founded 1737 and has been the capital since 1823.

San José /'sæn həʊˈzeɪ/ city in Santa Clara Valley, California, USA; population (1990) 782,200. It is the centre of the valley, nickname 'Silicon Valley', the site of many high-technology electronic firms turning out semiconductors and other computer components. There are also electrical, aerospace, missile, rubber, metal, and machine industries, and it is a commercial and transportation centre for orchard crops and wines produced in the area.

San Juan /'sæn 'wɑːn/ capital of Puerto Rico; population (1990) 437,750. It is a port and industrial city. Products include chemicals, pharmaceuticals, machine tools, electronic equipment, textiles, plastics, and rum.

San Luis Potosí /'sæn luːˈiːs ˌpɒtəʊˈsiː/ silver-mining city and capital of San Luis Potosí state, central Mexico; population (1986) 602,000. Founded 1586 as a Franciscan mission, it became the colonial administrative headquarters and has fine buildings of the period.

San Marino Republic of (*Repubblica di San Marino*)

area 61 sq km/24 sq mi
capital San Marino
towns Serravalle (industrial centre)
physical on the slope of Mount Titano
features surrounded by Italian territory; one of the world's smallest states
head of state and government two captains regent, elected for a six-month period
political system direct democracy
political parties San Marino Christian Democrat Party (PDCS), right-of-centre; Democratic Progressive Party (PDP), Socialist Unity Party (PSU), and Socialist Party (PSS), all three left-of-centre
exports wine, ceramics, paint, chemicals, building stone
currency Italian lira
population (1990 est) 23,000; growth rate 0.1% p.a.
life expectancy men 70, women 77
language Italian
religion Roman Catholic 95%
literacy 97% (1987)
chronology
1862 Treaty with Italy signed; independence

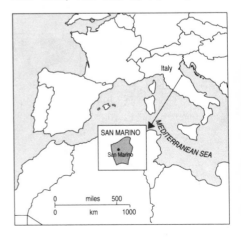

recognized under Italy's protection.
1947–86 Governed by a series of left-wing and centre-left coalitions.
1986 Formation of Communist and Christian Democrat 'grand coalition'.
1992 Joined the United Nations.

San Pedro Sula /sæn ˈpedrəʊ ˈsuːlə/ main industrial and commercial city in NW Honduras, the second-largest city in the country; population (1989) 300,900. It trades in bananas, coffee, sugar, and timber and manufactures textiles, plastics, furniture, and cement.

San Salvador /sæn ˈsælvədɔː/ capital of El Salvador 48 km/30 mi from the Pacific Ocean, at the foot of San Salvador volcano (2,548 m/ 8,360 ft); population (1984) 453,000. Industries include food processing and textiles. Since its foundation 1525, it has suffered from several earthquakes.

San Sebastián /sæn sɪˈbæstiən/ port and resort in the Basque Country, Spain; population (1986) 180,000. It was formerly the summer residence of the Spanish court.

Santa Ana /sæntə ˈænə/ commercial city in NW El Salvador, the second largest city in the country; population (1980) 205,000. It trades in coffee and sugar.

Santa Barbara /ˈbɑːbərə/ town in S California, USA; population (1990) 85,571. It is the site of a campus of the University of California. The Santa Ynez Mountains are to the N. Vandenburg air force base is 80 km/50 mi to the NW.

Santa Cruz /sæntə kruːz/ city in W central California at the northern end of Monterey Bay, SW of San Jose; population (1990) 49,000. A division of the University of California is here. Industries include tourism, food processing, fishing, and electronics.

Santa Cruz de la Sierra /sæntə ˈkruːz ˌdelə siˈerə/ capital of Santa Cruz department in E Bolivia, the second-largest city in the country; population (1988) 615,100. Sugar cane and cattle were the base of local industry until newly discovered oil and natural gas led to phenomenal growth.

Santa Cruz de Tenerife /sæntə ˈkruːz də ˌtenəˈriːf/ capital of Tenerife and of the Canary Islands; population (1991) 192,000. It is a fuelling port and cable centre. Industry also includes oil refining, pharmaceuticals, and trade in fruit. Santa Cruz was bombarded by the British admirals Blake 1657 and Nelson 1797 (the action in which he lost his arm).

Santa Fe /sæntə ˈfeɪ/ capital of New Mexico, USA, on the Santa Fe River, 65 km/40 mi W of Las Vegas; population (1990) 55,900, many Spanish-speaking. A number of buildings date from the Spanish period, including a palace 1609–10; the cathedral 1869 is on the site of a monastery built 1622. Santa Fe produces American Indian jewellery and textiles; its chief industry is tourism.

Santa Fé /sæntə ˈfeɪ/ capital of Santa Fé province, Argentina, on the Salado River 153 km/95 mi N of Rosario; population (1980) 287,000. It has shipyards and exports timber, cattle, and wool. It

was founded 1573, and the 1853 constitution was adopted here.

Santa Maria /sæntə məˈriːə/ city in SW California, USA, NW of Santa Barbara; population (1990) 61,284. Industries include oil, food processing, and dairy products.

Santander /ˌsæntænˈdeə/ port on the Bay of Biscay, Cantabria, N Spain; population (1986) 189,000. Industries include chemicals, textiles, vehicles, and shipyards. It was sacked by the French marshal Nicolas Soult 1808 and was largely rebuilt after a fire 1941. Palaeolithic cave wall paintings of bison, wild boar, and deer were discovered at the nearby **Altamira** site 1879.

Santa Rosa /sæntə ˈrəʊzə/ city in NW California, USA, N of San Francisco; population (1990) 113,300. Industries include wine, fruit, chemicals, and clothing.

Santiago /ˌsæntiˈɑːgəʊ/ capital of Chile; population (1990) 4,385,500. Industries include textiles, chemicals, and food processing. It was founded 1541 and is famous for its broad avenues.

Santiago de Compostela /ˌsæntiˈɑːgəʊ deɪ ˌkɒmpɒsˈtelə/ city in Galicia, Spain; population (1991) 105,500. The 11th-century cathedral was reputedly built over the grave of Sant Iago el Mayor (St James the Great), patron saint of Spain, and was a world-famous centre for medieval pilgrims.

Santiago de Cuba /ˌsæntiˈɑːgəʊ deɪ ˈkuːbə/ port on the south coast of Cuba; population (1989) 405,400. Products include sugar, rum, and cigars.

Santiago de los Caballeros /ˌsæntiˈɑːgəʊ deɪ lɒs ˌkæbælˈjeərɒs/ second largest city in the Dominican Republic; population (1982) 395,000. It is a processing and trading centre for sugar, coffee, and cacao.

Santo Domingo /ˈsæntəʊ dəˈmɪŋgəʊ/ capital and chief sea port of the Dominican Republic; population (1982) 1,600,000. Founded 1496 by Bartolomeo, brother of Christopher Columbus, it is the oldest colonial city in the Americas. Its cathedral was built 1515–40.

Santos /ˈsæntɒs/ coffee-exporting port in SE Brazil, 72 km/45 mi SE of São Paulo; population (1991) 546,600. The Brazilian footballer Pelé played here for many years.

Saône /səʊn/ river in E France, rising in the Vosges Mountains and flowing 480 km/300 mi to join the Rhône at Lyon.

São Paulo /saʊm ˈpaʊləʊ/ city in Brazil, 72 km/ 45 mi NW of its port Santos; population (1991) 9,700,100, metropolitan area 15,280,000. It is 900 m/3,000 ft above sea level, and 2°S of the Tropic of Capricorn. It is South America's leading industrial city, producing electronics, steel, and chemicals; it has meat-packing plants and is the centre of Brazil's coffee trade. It originated as a Jesuit mission 1554.

The locomotive that pulls the rest of Brazil.

On **São Paulo** local saying

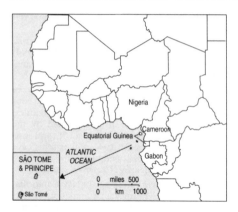

São Tomé /ˌsaun tɒˈmeɪ/ port and capital of São Tomé e Príncipe, on the northeast coast of São Tomé island, Gulf of Guinea; population (1984) 35,000. It exports cocoa and coffee.

São Tomé e Príncipe Democratic Republic of
area 1,000 sq km/386 sq mi
capital São Tomé
towns Santo Antonio, Santa Cruz
physical comprises two main islands and several smaller ones, all volcanic; thickly forested.
head of state and government Miguel Trovoada from 1991
political system emergent democratic republic
political parties Movement for the Liberation of São Tomé e Príncipe (MLSTP), nationalist socialist; Democratic Convergence Party–Reflection Group (PCD–EM), centre left
exports cocoa, copra, coffee, palm oil and kernels
currency dobra
population (1990 est) 125,000; growth rate 2.5% p.a.
life expectancy men 62, women 62
languages Portuguese (official), Fang (Bantu)
religions Roman Catholic 80%, animist
literacy men 73%, women 42% (1981)
GNP $32 million (1987); $384 per head (1986)
chronology
1471 Discovered by Portuguese.
1522–1973 A province of Portugal.
1973 Achieved internal self-government.
1975 Independence achieved from Portugal, with Manuel Pinto da Costa as president.
1984 Formally declared a nonaligned state.
1987 Constitution amended.
1988 Unsuccessful coup attempt against da Costa.
1990 New constitution approved.
1991 First multiparty elections held; Miguel Trovoada replaced Pinto da Costa.

Sapporo /səˈpɒːrəʊ/ capital of Hokkaido prefecture, Japan; population (1990) 1,671,800. Industries include rubber and food processing. It is a winter sports centre and was the site of the 1972 Winter Olympics. Giant figures are sculpted in ice at the annual snow festival.
   The university was founded 1876 as the Sapporo Agricultural College.

Saragossa /ˌsærəˈgɒsə/ English spelling of ◊Zaragoza, a city in Spain.

Sarajevo /ˌsærəˈjeɪvəu/ capital of Bosnia-Herzegovina; population (1982) 449,000. Industries include engineering, brewing, chemicals, carpets, and ceramics. Since April 1992 the city has been the target of a siege by Serb militia units in their fight to carve up the newly independent republic.
   A Bosnian, Gavrilo Princip, assassinated Archduke ◊Franz Ferdinand here 1914, thereby precipitating World War I. In Jan 1992 the city was attacked by Serb militia units opposed to Bosnia-Herzegovina's moves towards independence, and following EC recognition of the republic's independence in April, the city was under constant siege. Fighting intensified in July with fresh Serb offensives endangering a mounting UN effort to deliver relief supplies.

Sarasota /ˌsærəˈsəutə/ city in SW Florida, USA, on the Gulf of Mexico, S of Tampa; population (1990) 51,000. It is a resort town specializing in food processing and electronics research. It is the winter home of Ringling Brothers and Barnum and Bailey Circus.

Saratoga Springs /ˈsærətəugə ˈsprɪŋz/ city and spa in New York State, USA; population (1990) 25,000. In 1777 the British general John Burgoyne was defeated in two engagements nearby during the War of American Independence.

Saratov /səˈrɑːtɒf/ industrial port (chemicals, oil refining) on the river Volga in W central Russia; population (1987) 918,000. It was established in the 1590s as a fortress to protect the Volga trade route.

Sarawak /səˈrɑːwæk/ state of Malaysia, on the northwest corner of the island of Borneo
area 124,400 sq km/48,018 sq mi
capital Kuching
physical mountainous; the rainforest, which may be 10 million years old, contains several thousand tree species. A third of all its plant species are endemic to Borneo. 30% of the forest was cut down 1963–89; timber is expected to run out 1995–2001
products timber, oil, rice, pepper, rubber, and coconuts
population (1991) 1,669,000; 24 ethnic groups make up almost half this number
history Sarawak was granted by the Sultan of Brunei to English soldier James Brooke 1841, who became 'Rajah of Sarawak'. It was a British protectorate from 1888 until captured by the Japanese in World War II. It was a crown colony 1946–63, when it became part of Malaysia.

Sardinia /sɑːˈdɪnɪə/ (Italian *Sardegna*) mountainous island, special autonomous region of Italy; area 24,100 sq km/9,303 sq mi; population (1990) 1,664,400. Its capital is Cagliari, and it exports cork and petrochemicals. It is the second largest Mediterranean island and includes Costa Smeralda (Emerald Coast) tourist area in the NE and *nuraghi* (fortified Bronze Age dwellings). After centuries of foreign rule, it became linked 1720 with Piedmont, and this dual kingdom became the basis of a united Italy 1861.

Sardinia

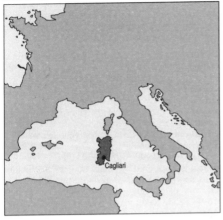

Sargasso Sea /sɑːˈgæsəu/ part of the N Atlantic (between 40° and 80°W and 25° and 30°N) left static by circling ocean currents, and covered with floating weed *Sargassum natans*.

Sark /sɑːk/ one of the Channel Islands, 10 km/ 6 mi E of Guernsey; area 5 sq km/2 sq mi; there is no town or village. It is divided into Great and Little Sark, linked by an isthmus, and is of great natural beauty. The Seigneurie of Sark was established by Elizabeth I, the ruler being known as Seigneur/Dame, and has its own parliament, the Chief Pleas. There is no income tax, and cars are forbidden; immigration is controlled.

Sarum /ˈseərəm/ former settlement from which the modern city of Salisbury, Wiltshire, England, developed.

Sasebo /ˈsɑːsebəu/ seaport and naval base on the west coast of Kyushu Island, Japan; population (1990) 244,700.

Sask. abbreviation for *Saskatchewan*, a Canadian province.

Saskatchewan /sæˈskætʃəwən/ (Cree *Kis-is-ska-tche-wan* 'swift flowing') province of W Canada
area 652,300 sq km/251,788 sq mi
capital Regina

Saskatchewan

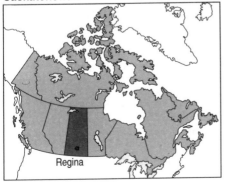

towns Saskatoon, Moose Jaw, Prince Albert
physical prairies in the S; to the N, forests, lakes, and subarctic tundra; Prince Albert National Park
products more than 60% of Canada's wheat; oil, natural gas, uranium, zinc, potash (world's largest reserves), copper, helium (the only western reserves outside the USA)
population (1991) 995,300
history once inhabited by Indians speaking Athabaskan, Algonquin, and Sioux languages, who depended on caribou and moose in the N and buffalo in the S. French trading posts established about 1750; owned by Hudson's Bay Company, first permanent settlement 1774; ceded to Canadian government 1870 as part of Northwest Territories; became a province 1905.

The Lord said 'let there be wheat' and Saskatchewan was born.

On **Saskatchewan** Stephen Leacock
My Discovery of America 1937

Saskatoon /ˌsæskəˈtuːn/ largest city in Saskatchewan, Canada; population (1986) 177,641. Industries include cement, oil refining, chemicals, metal goods, and processed foods. The University of Saskatchewan is here. Saskatoon was settled 1882.

Sassari /ˈsæsəri/ capital of the province of the same name, in the northwest corner of Sardinia, Italy; population (1987) 121,000. Every May the town is the scene of the Sardinian Cavalcade, the greatest festival on the island.

Saudi Arabia Kingdom of (*al-Mamlaka al-'Arabiya as-Sa'udiya*)
area 2,200,518 sq km/849,400 sq mi
capital Riyadh
towns Mecca, Medina, Taif; ports Jidda, Dammam
physical desert, sloping to the Persian Gulf from a height of 2,750 m/9,000 ft in the W
environment oil pollution caused by the Gulf War 1990–91 has affected 460 km/285 mi of the Saudi coastline, threatening desalination

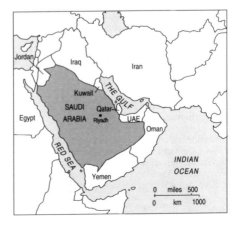

plants and damaging the wildlife of saltmarshes, mangrove forest, and mudflats
features Nafud Desert in N and the Rub'al Khali (Empty Quarter) in S, area 650,000 sq km/250,000 sq mi; with a ban on women drivers, there are an estimated 300,000 chauffeurs
head of state and government King Fahd Ibn Abdul Aziz from 1982
political system absolute monarchy
political parties none
exports oil, petroleum products
currency rial
population (1990 est) 16,758,000 (16% nomadic); growth rate 3.1% p.a.
life expectancy men 64, women 67 (1989)
language Arabic
religion Sunni Muslim; there is a Shi'ite minority
literacy men 34%, women 12% (1980 est)
GNP $70 bn (1988); $6,170 per head (1988)
chronology
1926–32 Territories united and kingdom established.
1953 King Ibn Saud died and was succeeded by his eldest son, Saud.
1964 King Saud forced to abdicate; succeeded by his brother, Faisal.
1975 King Faisal assassinated; succeeded by his half-brother, Khalid.
1982 King Khalid died; succeeded by his brother, Crown Prince Fahd.
1987 Rioting by Iranian pilgrims caused 400 deaths in Mecca; diplomatic relations with Iran severed.
1990 Iraqi troops invaded and annexed Kuwait and massed on Saudi Arabian border. King Fahd called for help from US and UK forces.
1991 King Fahd provided military and financial assistance in Gulf War. Calls from religious leaders for 'consultative assembly' to assist in government of kingdom. Saudi Arabia attended Middle East peace conference.
1992 Formation of a 'consultative council' seen as possible move towards representative government.

Sault Ste Marie /'suː seɪnt məˈriː/ twin industrial ports on the Canadian/US border, one in Ontario and one in Michigan; population (1981) 82,902 and (1990) 14,700, respectively. They stand at the falls (French *sault*) in St Mary's River, which links Lakes Superior and Huron. The falls are bypassed by canals. Industries include steel, pulp, and agricultural trade.

Saumur /səʊˈmjʊə/ town in Maine-et-Loire *département*, France, on the river Loire; population (1990) 31,900. The area produces sparkling wines. The cavalry school, founded 1768, has since 1942 also been a training school for the French armed forces.

Savannah /səˈvænə/ city and port of Georgia, USA, 29 km/18 mi from the mouth of the Savannah River; population (1990) 137,600. Founded 1733, Savannah was the first city in the USA to be laid out in geometrically regular blocks.

It exports cotton, and produces cottonseed oil, fertilizers, and machinery. The *Savannah*, the

first steam-powered ship to cross the Atlantic, was built here; most of the 25-day journey, in 1819, was made under sail. The first nuclear-powered merchant ship, launched by the USA 1959, was given the same name.

Saxony /'sæksəni/ (German *Sachsen*) administrative *Land* (state) of Germany
area 17,036 sq km/6,580 sq mi
capital Dresden
towns Leipzig, Chemnitz, Zwickau
physical on the plain of the river Elbe N of the Erzgebirge mountain range
products electronics, textiles, vehicles, machinery, chemicals, coal
population (1990) 5,000,000
history conquered by Charlemagne 792, Saxony became a powerful medieval German duchy. The electors of Saxony were also kings of Poland 1697–1763. Saxony was part of East Germany 1946–90, forming a region with Anhalt.
Saxony takes its name from the early Saxon inhabitants whose territories originally reached as far W as the Rhine. The duchy of Saxony was divided 1260 but reconstituted 1424 when a new electorate embracing Thuringia, Meissen, and Wittenberg was formed. The northern part of Saxony became a province of Prussia 1815, its king having sided with Napoleon.
In 1946 Saxony was joined with Anhalt as a region of East Germany and in 1952 it was split into the districts of Leipzig, Dresden, and Chemnitz (later named Karl-Marx-Stadt). The state of Saxony was restored 1990 following German reunification and the abolition of the former districts of East Germany.

Saxony-Anhalt /'sæksəni 'ænhælt/ administrative *Land* (state) of Germany
area 20,450 sq km/10,000 sq mi
capital Magdeburg
towns Halle, Dessau
products chemicals, electronics, rolling stock, footwear, cereals, vegetables
population (1990) 3,000,000
history Anhalt became a duchy 1863 and a member of the North German Confederation 1866. Between 1946 and 1990 it was joined to the former Prussian province of Saxony as a region of East Germany.
The territory of Anhalt, named after the medieval castle of Anhalt, was divided and reunited many times before becoming a duchy. In 1952, as part of the East German region of Saxony-Anhalt, it was divided into the districts of Halle and Magdeburg. After the reunification of Germany 1990, Saxony-Anhalt was reconstituted as one of the five new *Länder* of the Federal Republic.

Sayan Mountains /saɪˈæn/ range in E Siberian Russia, on the Mongolian border; the highest peak is Munku Sardik 3,489 m/11,451 ft. The mountains have coal, gold, silver, graphite, and lead resources.

SC abbreviation for ♢*South Carolina*, a state of the USA.

Scafell Pike /ˌskɔːˈfel/ highest mountain in England, 978 m/3,210 ft. It is in Cumbria in the

Lake District and is separated from Scafell (964 m/3,164 ft) by a ridge called Mickledore. The summit of Scafell Pike was presented to the National Trust by the third Lord Leconfield, as a war memorial, 1919.

Scandinavia /ˌskændɪ'neɪvɪə/ peninsula in NW Europe, comprising Norway and Sweden; politically and culturally it also includes Denmark, Iceland, the Faroe Islands, and Finland.

Scapa Flow /'skɑːpə 'fləʊ/ expanse of sea in the Orkney Islands, Scotland, until 1957 a base of the Royal Navy. It was the main base of the Grand Fleet during World War I and in 1919 was the scene of the scuttling of 71 surrendered German warships.

Scarborough /'skɑːbərə/ spa and holiday resort in North Yorkshire, England; population (1985) 50,000. A ruined Norman castle overlooks the town, which is a touring centre for the Yorkshire Moors.

Schaffhausen /'ʃæf,haʊzən/ town in N Switzerland; population (1990) 34,200. Industries include the manufacture of watches, chemicals, and textiles. The Rhine falls here in a series of cascades 60 m/197 ft high.

Scheldt /skelt/ (Dutch *Schelde*; French *Escaut*) river rising in Aisne *département*, N France, and flowing 400 km/250 mi to join the North Sea S of Walcheren in the Netherlands. Antwerp is the chief town on the Scheldt.

Schenectady /skə'nektədɪ/ industrial city on the Mohawk River, New York State, USA; population (1990) 65,600. It dates from 1662 and has long been a producer of electrical goods.

Scheveningen /'sxeɪfənɪŋə/ seaside resort and northern suburb of The Hague in the Netherlands. There is a ferry link with Great Yarmouth, England.

Schiedam /ˌsxiː'dæm/ port in Zuid-Holland province, SW Netherlands, on the river Meuse, 5 km/3 mi W of Rotterdam; population (1991) 70,200. It is famous for its gin.

Schleswig-Holstein /'ʃlezwɪg 'hɒlstaɪn/ *Land* (state) of Germany
area 15,700 sq km/6,060 sq mi
capital Kiel
towns Lübeck, Flensburg, Schleswig
features river Elbe, Kiel Canal, Heligoland
products shipbuilding, mechanical and electrical engineering, food processing
population (1988) 2,613,000
religions 87% Protestant; 6% Catholic
history Schleswig (Danish *Slesvig*) and Holstein were two duchies held by the kings of Denmark from 1460, but were not part of the kingdom; a number of the inhabitants were German, and Holstein was a member of the Confederation of the Rhine formed 1815. Possession of the duchies had long been disputed by Prussia, and when Frederick VII of Denmark died without an heir 1863, Prussia, supported by Austria, fought and defeated the Danes 1864, and in 1866 annexed

the two duchies. A plebiscite held 1920 gave the northern part of Schleswig to Denmark, which made it the province of Haderslev and Aabenraa; the rest, with Holstein, remained part of Germany.

Schwarzwald /'ʃvɑːtsvɛlt/ German name for the ▷Black Forest, a coniferous forest in Germany.

Schwerin /ʃve'riːn/ capital of Mecklenburg–West Pomerania, administrative region (*Land*) of Germany, on the western shore of the lake of Schwerin; population (1990) 130,000. Products include machinery and chemicals. Formerly the capital of Mecklenburg and earlier of the old republic of Mecklenburg–Schwerin, Schwerin became capital of Mecklenburg–West Pomerania with the reunification of Germany 1990.

Schwyz /ʃviːts/ capital of Schwyz canton, Switzerland; population (1990) 12,700. Schwyz was one of the three original cantons of the Swiss Confederation 1291, which gave its name to the whole country about 1450.

Scilly, Isles of /'sɪlɪ/ or *Scilly Isles/Islands*, or *Scillies* group of 140 islands and islets lying 40 km/25 mi SW of Land's End, England; administered by the Duchy of Cornwall; area 16 sq km/6.3 sq mi; population (1981) 1,850. The five inhabited islands are *St Mary's*, the largest, on which is Hugh Town, capital of the Scillies; *Tresco*, the second largest, with subtropical gardens; *St Martin's*, noted for beautiful shells; *St Agnes*; and *Bryher*.

Products include vegetables and flowers, and tourism is important. The islands have remains of Bronze Age settlements. The numerous wreck sites off the islands include many of Sir Cloudesley Shovell's fleet 1707. The Scilly Islands are an important birdwatching centre with breeding sea birds in the summer and rare migrants in the spring and autumn.

Scone /skuːn/ site of ancient palace where most of the Scottish kings were crowned on the Stone of Destiny (now in the Coronation Chair at Westminster, London). The village of Scone is in Tayside, Scotland, N of Perth.

Scotland /'skɒtlənd/ the northernmost part of Britain, formerly an independent country, now part of the UK
area 78,470 sq km/30,297 sq mi
capital Edinburgh
towns Glasgow, Dundee, Aberdeen
features the Highlands in the N (with the Grampian Mountains); central Lowlands, including valleys of the Clyde and Forth, with most of the country's population and industries; Southern Uplands (including the Lammermuir Hills); and islands of the Orkneys, Shetlands, and Western Isles; the world's greatest concentration of nuclear weapons are at the UK and US bases on the Clyde, near Glasgow; 8,000-year-old pinewood forests once covered 1,500,000 hectares/3,706,500 acres, now reduced to 12,500 hectares/30,900 acres
industry electronics, marine and aircraft engines, oil, natural gas, chemicals, textiles, clothing, printing, paper, food processing, tourism

Scotland: regions

regions	administrative headquarters	area in sq km
Borders	Newtown St Boswells	4,662
Central	Stirling	2,590
Dumfries and Galloway	Dumfries	6,475
Fife	Glenrothes	1,308
Grampian	Aberdeen	8,550
Highland	Inverness	26,136
Lothian	Edinburgh	1,756
Strathclyde	Glasgow	13,856
Tayside	Dundee	7,668
island authorities		
Orkney	Kirkwall	974
Shetland	Berwick	1,427
Western Islands	Stornoway	2,901
		78,303

currency pound sterling
population (1988 est) 5,094,000
languages English; Scots, a lowland dialect (derived from Northumbrian Anglo-Saxon); Gaelic spoken by 1.3%, mainly in the Highlands
religions Presbyterian (Church of Scotland), Roman Catholic
famous people Robert Bruce, Walter Scott, Robert Burns, Robert Louis Stevenson, Adam Smith
government Scotland sends 72 members to the UK Parliament at Westminster. Local government is on similar lines to that of England, but there is a differing legal system (Scots law). There is a movement for an independent or devolved Scottish assembly.

What I think is that it has suffered in the past, and is suffering now, from too much England.

On **Scotland** A G Macdonell
My Scotland 1937

Scranton /ˈskræntən/ industrial city on the Lackawanna River, Pennsylvania, USA; population (1990) 81,800. Anthracite coal is mined nearby, but production has declined sharply, and the city now manufactures such products as electronic equipment, fabricated metal, clothing, plastic goods, and printed materials.

Scunthorpe /ˈskʌnθɔːp/ industrial town in Humberside, England, 39 km/24 mi W of Grimsby; population (1987 est) 61,500. It has one of Europe's largest iron and steel works, which has been greatly expanded with help from the European Community.

SD abbreviation for ◊*South Dakota*, a state of the USA.

Seaham /ˈsiːəm/ seaport in Durham, England, 8 km/5 mi S of Sunderland; population (1987 est) 22,600. Coal mines and engineering were developed from the 19th century. The poet Byron married Anne Isabella Milbanke at Seaham Hall nearby.

Seaside /ˈsiːsaɪd/ city in W central California, USA, on the southern shore of Monterey Bay, S of San Francisco; population (1990) 38,900. Industries include fruit processing.

Seattle /siˈætl/ port (grain, timber, fruit, fish) of the state of Washington, USA, situated between Puget Sound and Lake Washington; population (1990) 516,300, metropolitan area (with Everett) 2,559,200. It is a centre for the manufacture of jet aircraft (Boeing), and also has shipbuilding, food processing, and paper industries.
First settled 1851, as the nearest port for Alaska, Seattle grew in the late 19th century under the impetus of the gold rush. It is named after the Indian Sealth.

Sebastopol /sɪˈbæstəpɒl/ alternative spelling of ◊Sevastopol, a port in Ukraine.

Secunderabad /səˈkʌndərəbæd/ northern suburb of Hyderabad city, Andhra Pradesh, India, separated from the rest of the city by the Hussain Sagar Lake; population (1981) 144,287. Formerly a separate town, it was founded as a British army cantonment, with a parade ground where 7,000 troops could be exercised.

Sedan /sɪˈdæn/ town on the river Meuse, in Ardennes *département*, NE France; population (1990) 22,400. Industries include textiles and dyestuffs; the town's prosperity dates from the 16th–17th centuries, when it was a Huguenot centre. In 1870 Sedan was the scene of Napoleon III's surrender to Germany during the Franco-Prussian War. It was the focal point of the German advance into France 1940.

Seeland /ˈzeɪlænt/ German form of ◊Sjælland, the main island of Denmark.

Segovia /sɪˈɡəʊviə/ town in Castilla-León, central Spain; population (1991) 58,000. Thread, fertilizer, and chemicals are produced. It has a Roman aqueduct with 118 arches in current use, and the Moorish alcázar (fortress) was the palace of the monarchs of Castile. Isabella of Castile was crowned here 1474.

Seikan Tunnel /ˈseɪkæn/ the world's longest underwater tunnel, opened 1988, linking the Japanese islands of Hokkaido and Honshu, which are separated by the Tsungaru Strait; length 51.7 km/32.3 mi.

Seine /seɪn/ French river rising on the Langres plateau NW of Dijon, and flowing 774 km/472 mi NW to join the English Channel near Le Havre, passing through Paris and Rouen.

Sekondi-Takoradi /ˌsekənˈdiː ˌtɑːkəˈrɑːdi/ seaport of Ghana; population (1982) 123,700. The old port was founded by the Dutch in the 16th century. Takoradi has an artificial harbour, opened 1928, and railway engineering, boat building, and cigarette manufacturing industries.

Selangor /sə'læŋə/ state of the Federation of Malaysia; area 7,956 sq km/3,071 sq mi; population (1990) 1,978,000. It was under British protection from 1874 and was a federated state 1895–1946. The capital was transferred to Shah Alam from Kuala Lumpur 1973. Klang is the seat of the sultan and a centre for rubber-growing and tin-mining; Port Klang (formerly Port Keland and, in 1971, Port Swettenham) exports tin.

Selborne /'selbɔːn/ village in Hampshire, S England, 8 km/5 mi SE of Alton. Gilbert White, author of *The Natural History of Selborne* 1789, was born here. The Selborne Society (founded 1885) promotes the study of wildlife.

Selby /'selbi/ town on the river Ouse, North Yorkshire, England; population (1981) 10,700. The nearby Selby coalfield, discovered 1967, consists of 2,000 million tonnes of pure coal.

Selkirkshire /'selkɜːkʃə/ former inland county of Scotland, included in Borders Region 1975.

Semarang /sə'mɑːræŋ/ port in N Java, Indonesia; population (1980) 1,027,000. There is a shipbuilding industry, and exports include coffee, teak, sugar, tobacco, kapok, and petroleum from nearby oilfields.

Semipalatinsk /ˌsemɪpə'lætɪnsk/ town in NE Kazakhstan, on the river Irtysh; population (1987) 330,000. It was founded 1718 as a Russian frontier post and moved to its present site 1776. Industries include meat-packing, tanning, and flour-milling, and the region produces nickel and chromium. The Kyzyl Kum atomic-weapon-testing ground is nearby.

Sendai /'sendaɪ/ city in Tōhoku region, NE Honshu Island, Japan; population (1990) 918,400. Industries include metal goods (a metal museum was established 1975), electronics, textiles, pottery, and food processing. It was a feudal castle town from the 16th century.

Senegal Republic of (*République du Sénégal*)
area 196,200 sq km/75,753 sq mi
capital and chief port Dakar
towns Thiès, Kaolack
physical plains rising to hills in SE; swamp and tropical forest in SW
features river Senegal; the Gambia forms an enclave within Senegal
head of state and government Abdou Diouf from 1981
political system emergent socialist democratic republic
political parties Senegalese Socialist Party (PS), democratic socialist; Senegalese Democratic Party (PDS), left-of-centre
exports peanuts, cotton, fish, phosphates
currency franc CFA
population (1990 est) 7,740,000; growth rate 3.1% p.a.
life expectancy men 51, women 54 (1989)
languages French (official); African dialects are spoken
religions Muslim 80%, Roman Catholic 10%, animist

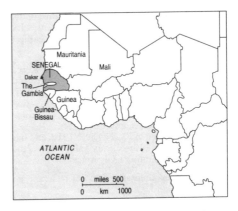

literacy men 37%, women 19% (1985 est)
GNP $2 bn (1987); $380 per head (1984)
chronology
1659 Became a French colony.
1854–65 Interior occupied by French.
1902 Became a territory of French West Africa.
1959 Formed the Federation of Mali with French Sudan.
1960 Independence achieved from France, but withdrew from the federation. Léopold Sédar Senghor, leader of the Senegalese Progressive Union (UPS), became president.
1966 UPS declared the only legal party.
1974 Pluralist system re-established.
1976 UPS reconstituted as Senegalese Socialist Party (PS). Prime Minister Abdou Diouf nominated as Senghor's successor.
1980 Senghor resigned; succeeded by Diouf. Troops sent to defend Gambia.
1981 Military help again sent to Gambia.
1982 Confederation of Senegambia came into effect.
1983 Diouf re-elected. Post of prime minister abolished.
1988 Diouf decisively re-elected.
1989 Violent clashes between Senegalese and Mauritanians in Dakar and Nouakchott killed more than 450 people; over 50,000 people repatriated from both countries. Senegambia federation abandoned.
1991 Constitutional changes outlined.
1992 Diplomatic links with Mauritania re-established.

Senegal /ˌsenɪ'gɔːl/ river in W Africa, formed by the confluence of the Bafing and Bakhoy rivers and flowing 1,125 km/700 mi NW and W to join the Atlantic Ocean near St Louis, Senegal. In 1968 the Organization of Riparian States of the River Senegal (Guinea, Mali, Mauritania, and Senegal) was formed to develop the river valley, including a dam for hydroelectric power and irrigation at Joina Falls in Mali; its headquarters is in Dakar. The river gives its name to the Republic of Senegal.

Sennar /se'nɑː/ town about 260 km/160 mi SE of Khartoum, on the Blue Nile, Sudan Republic; population (1972) 10,000. Nearby is the Sennar Dam 1926, part of the Gezira irrigation scheme.

Sens /sɒns/ town in Yonne *département*, Burgundy, France; population (1990) 27,800. Its 12th–16th-century cathedral is one of the earliest in the Gothic style in France.

Seoul /səul/ or *Sŏul* capital of South Korea (Republic of Korea), near the Han River, and with its chief port at Inchon; population (1985) 10,627,800. Industries include engineering, textiles, food processing, electrical and electronic equipment, chemicals, and machinery.

It was the capital of Korea 1392–1910, and has universities and a 14th-century palace. It was the site of the 1988 Summer Olympics.

Serang /sə'ræŋ/ alternative form of ◊Ceram, an Indonesian island.

Serbia /'sɜːbiə/ (Serbo-Croatian *Srbija*) constituent republic of Yugoslavia, which includes Kosovo and Vojvodina
area 88,400 sq km/34,122 sq mi
capital Belgrade
physical fertile Danube plains in the N, mountainous in the S
features includes the autonomous provinces of *Kosovo*, capital Priština, of which the predominantly Albanian population demands unification with Albania, and *Vojvodina*, capital Novi Sad, largest town Subotica, with a predominantly Serbian population
population (1986) 9,660,000
language the Serbian variant of Serbo-Croatian
religion Serbian Orthodox
history The Serbs settled in the Balkans in the 7th century and became Christians in the 9th century. They were united as one kingdom about 1169; the Serbian hero Stephan Dushan (1331–1355) founded an empire covering most of the Balkans. After their defeat at Kosovo 1389 they came under the domination of the Turks, who annexed Serbia 1459. Uprisings 1804–16, led by Kara George and Milosh Obrenovich, forced the Turks to recognize Serbia as an autonomous principality under Milosh. The assassination of Kara George on Obrenovich's orders gave rise to a long feud between the two houses. After a war with Turkey 1876–78, Serbia became an independent kingdom. On the assassination of the last Obrenovich 1903 the Karageorgevich dynasty came to the throne.

The two Balkan Wars 1912–13 greatly enlarged Serbia's territory at the expense of Turkey and Bulgaria. Serbia's designs on Bosnia-Herzegovina, backed by Russia, led to friction with Austria, culminating in the outbreak of war 1914. Serbia was overrun 1915–16 and was occupied until 1918, when it became the nucleus of the new kingdom of the Serbs, Croats, and Slovenes, and subsequently Yugoslavia. Rivalry between Croats and Serbs continued within the republic. During World War II Serbia was under a puppet government set up by the Germans; after the war it became a constituent republic of Yugoslavia.

From 1986 Slobodan Milosević as Serbian party chief and president waged a populist campaign to end the autonomous status of the provinces of Kosovo and Vojvodina. Despite a violent Albanian backlash in Kosovo 1989–90 and

growing pressure in Croatia and Slovenia to break away from the federation, Serbia formally annexed Kosovo Sept 1990. Milosević was re-elected by a landslide majority Dec 1990, but in March 1991 there were anticommunist and anti-Milosević riots in Belgrade. The 1991 civil war in Yugoslavia arose from the Milosević nationalist government attempting the forcible annexation of Serb-dominated regions in Croatia, making use of the largely Serbian federal army. In Oct 1991 Milosević renounced territorial claims on Croatia pressured by threats of European Community (EC) and United Nations (UN) sanctions, but the fighting continued until a cease-fire was agreed Jan 1992. EC recognition of Slovenia's and Croatia's independence in Jan 1992 and Bosnia-Herzegovina's in April left Serbia dominating a greatly reduced 'rump' Yugoslavia. A successor Yugoslavia, announced by Serbia and Montenegro April 1992, was rejected by the USA and EC because of concerns over serious human rights violations in Kosovo and Serbia's continued attempted partition of Bosnia-Herzegovina. In March 1992, and again in June, thousands of Serbs marched through Belgrade, demanding the ousting of President Milosević and an end to the war in Bosnia-Herzegovina.

Sergiyev Posad (formerly 1930–92 *Zagorsk*) town 70 km/45 mi NE of Moscow; population (1989) 115,000. Paint, optical instruments, building materials, textiles, and furniture are produced. The Moscow Theological Academy is here.

The town developed in the 14th century around a centre of pilgrimage, the Trinity Monastery of St Sergius, and gained city status 1919.

Seringapatam /sə,rɪŋɡəpə'tæm/ town in Karnataka, India, on an island in the Cauvery River. It was the capital of Mysore state 1610–1799, when it was taken from the sultan of Mysore, Tipu Sahib, by the British general Charles Cornwallis.

Sète /seɪt/ town on the Mediterranean coast of France, in Hérault *département*, SW of Montpellier; population (1990) 41,916. It is a seaport and handles fish, wine, brandy, and chemicals. It was founded 1666 as an outlet to the Canal du Midi.

Seto Naikai /'setəu 'naɪkəɪ/ (Japanese 'inland sea') narrow body of water almost enclosed by the Japanese islands of Honshu, Shikoku, and Kyushu. It is both a transport artery and a national park (1934) with 3,000 islands.

Sevastopol /sɪ'væstəpəl/ or *Sebastopol* Black Sea port, resort, and fortress in the Crimea, Ukraine; population (1987) 350,000. It was the base of the former Soviet Black Sea fleet. It also has shipyards and a wine-making industry. Founded by Catherine II 1784, it was successfully besieged by the English and French in the Crimean War (Oct 1854–Sept 1855), and in World War II by the Germans (Nov 1941–July 1942), but was retaken by Soviet forces 1944.

Sevenoaks /'sevənəuks/ town in Kent, England. It lies 32 km/20 mi SE of London, population (1980) 19,000. Nearby are the 17th-century houses of Knole and Chevening. Most of its seven oak trees were blown down in the 1987 gale.

Severn /'sevən/ river of Wales and England, rising on the northeast side of Plynlimmon, N Wales, and flowing 338 km/210 mi through Shrewsbury, Worcester, and Gloucester to the Bristol Channel. The *Severn bore* is a tidal wave up to 2 m/6 ft high.

S England and S Wales are linked near Chepstow by a rail tunnel (1873–85) over the Severn, and a road bridge (1966). A second road bridge is under construction.

Seveso /sɪ'veɪsəu/ town in Lombardy, Italy, site of a factory manufacturing the herbicide hexachlorophene. In 1976 one of the by-products that escaped in a cloud contaminated the area, resulting in severe skin disorders and deformed births.

Seville /sɪ'vɪl/ (Spanish *Sevilla*) city in Andalusia, Spain, on the Guadalquivir River, 96 km/60 mi N of Cadiz; population (1991) 683,500. Products include machinery, spirits, porcelain, pharmaceuticals, silk, and tobacco.

Formerly the centre of a Moorish kingdom, it has a 12th-century alcazar, or fortified palace, and a 15th–16th-century Gothic cathedral. Seville was the birthplace of the artists Murillo and Velázquez. The international trade fair Expo 92 celebrated the 500th anniversary of the discovery of the New World.

Seville lights up for a feast-day as a face lights up with a smile.

On **Seville** Arthur Symons *Cities* 1903

Sèvre /'seɪvrə/ either of two French rivers from which the *département* of Deux Sèvres takes its name. The *Sèvre Nantaise* joins the Loire River at Nantes; the *Sèvre Niortaise* flows into the Bay of Biscay.

Seychelles Republic of
area 453 sq km/175 sq mi
capital Victoria (on Mahé island)
towns Cascade, Port Glaud, Misere
physical comprises two distinct island groups, one concentrated, the other widely scattered, totalling over 100 islands and islets
features Aldabra atoll, containing world's largest tropical lagoon; the unique 'double coconut' (*coco de mer*); tourism is important
head of state and government France-Albert René from 1977
political system one-party socialist republic
political party Seychelles People's Progressive Front (SPPF), nationalist socialist
exports copra, cinnamon
currency Seychelles rupee
population (1990) 71,000; growth rate 2.2% p.a.
life expectancy 66 years (1988)
languages creole (Asian, African, European mixture) 95%, English, French (all official)

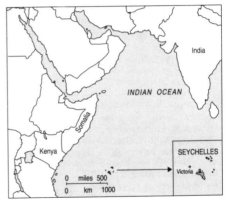

religion Roman Catholic 90%
literacy 80% (1989)
GNP $175 million; $2,600 per head (1987)
chronology
1744 Became a French colony.
1794 Captured by British.
1814 Ceded by France to Britain; incorporated as a dependency of Mauritius.
1903 Became a separate British colony.
1975 Internal self-government agreed.
1976 Independence achieved from Britain as a republic within the Commonwealth, with James Mancham as president.
1977 Albert René ousted Mancham in an armed coup and took over presidency.
1979 New constitution adopted; SPPF as sole legal party.
1981 Attempted coup by South African mercenaries thwarted.
1984 René re-elected.
1987 Coup attempt foiled.
1989 René re-elected.
1991 Multiparty politics promised.
1992 Mancham returned from exile. Constitutional commission elected.

Sfax /sfæks/ (Arabic *Safaqis*) port and second largest city in Tunisia; population (1984) 232,000. It is the capital of Sfax district, on the Gulf of Gabès, and lies about 240 km/150 mi SE of Tunis. Products include leather, soap, and carpets; there are also salt works and phosphate workings nearby. Exports include phosphates, olive oil, dates, almonds, esparto grass, and sponges.

's-Gravenhage /sˌxrɑːvən'hɑːxə/ Dutch name for The ♢Hague, a city in the Netherlands.

Shaanxi /ʃɑːn'ʃiː/ or *Shensi* province of NW China
area 195,800 sq km/75,579 sq mi
capital Xian
physical mountains; Huang He Valley, one of the earliest settled areas of China
products iron, steel, mining, textiles, fruit, tea, rice, wheat
population (1990) 32,882,000.

Shaba /'ʃɑːbə/ (formerly until 1972 *Katanga*), region of Zaire; area 496,965 sq km/191,828 sq mi;

population (1984) 3,874,000. Its main town is Lubumbashi, formerly Elisabethville.

Shache /ʃɑːˈtʃeɪ/ alternative name for ⟡Yarkand, a city in China.

Shaftesbury /ˈʃɑːftsbəri/ market town and agricultural centre in Dorset, England, 30 km/19 mi SW of Salisbury; population (1989 est) 7,200. King Alfred is said to have founded an abbey on the site 880; Canute died at Shaftesbury 1035.

Shakhty /ˈʃæxti/ town in the Donets Basin region of Russia, 80 km/50 mi NE of Rostov; population (1987) 225,000. Industries include anthracite mining, stone quarrying, textiles, leather, and metal goods. It was known as *Aleksandrovsk Grushevskii* until 1921.

Shandong /ʃænˈdʌŋ/ or *Shantung* province of NE China
area 153,300 sq km/59,174 sq mi
capital Jinan
towns ports: Yantai, Weihai, Qingdao, Shigiusuo
features crossed by the Huang He River and the Grand Canal; Shandong Peninsula
products cereals, cotton, wild silk, varied minerals
population (1990) 84,393,000.

Shanghai /ʃæŋˈhaɪ/ port on the Huang-pu and Wusong rivers, Jiangsu province, China, 24 km/15 mi from the Chang Jiang estuary; population (1986) 6,980,000, the largest city in China. The municipality of Shanghai has an area of 5,800 sq km/2,239 sq mi and a population of 13,342,000. Industries include textiles, paper, chemicals, steel, agricultural machinery, precision instruments, shipbuilding, flour and vegetable-oil milling, and oil refining. It handles about 50% of China's imports and exports.

Shanghai is reckoned to be the most densely populated area in the world, with an average of 6 sq m/65 sq ft of living space and 2.2 sq m/2.6 sq yd of road per person.
features Notable buildings include the Jade Buddha Temple 1882, the former home of the revolutionary Sun Yat-sen, the house where the First National Congress of the Communist Party of China met secretly in 1921, and the house, museum, and tomb of the writer Lu Xun.
history Shanghai was a city from 1360 but became significant only after 1842, when the treaty of Nanking opened it to foreign trade. In the 1920s, about 40,000 foreigners lived in Shanghai, more than 2% of the total population; in 1925 the electorate consisted of 2,742 voters, of which 1,000 were British. The Municipal Council issued a de facto unilateral declaration of independence 1925. The international settlement then developed, which remained the commercial centre of the city after the departure of European interests 1943–46.

As Vincent Sheean once said, Shanghai is the city par excellence of two things, money and the fear of losing it.

On **Shanghai** John Gunther
Inside Asia 1939

Shannon /ˈʃænən/ longest river in Ireland, rising in County Cavan and flowing 386 km/240 mi through loughs Allen and Ree and past Athlone, to reach the Atlantic Ocean through a wide estuary below Limerick. It is also the greatest source of electric power in the republic, with hydroelectric installations at and above Ardnacrusha, 5 km/3 mi N of Limerick.

Shansi /ʃænˈsiː/ alternative transliteration of Chinese province of ⟡Shanxi.

Shantou /ʃænˈtau/ or *Swatow* port and industrial city in SE China; population (1984) 746,400. It was opened as a special foreign trade area 1979.

Shantung /ʃænˈtʌŋ/ alternative transliteration of the Chinese province of ⟡Shandong.

Shanxi /ʃænˈʃiː/ or *Shansi* or *Shensi* province of NE China
area 157,100 sq km/60,641 sq mi
capital Taiyuan
features a drought-ridden plateau, partly surrounded by the Great Wall
products coal, iron, fruit
population (1990) 28,759,000
history saw the outbreak of the Boxer Rebellion 1900.

Sharjah /ˈʃɑːdʒə/ or *Shariqah* third largest of the seven member states of the United Arab Emirates, situated on the Arabian Gulf NE of Dubai; area 2,600 sq km/1,004 sq mi; population (1985) 269,000. Since 1952 it has included the small state of Kalba. In 1974 oil was discovered offshore. Industries include ship repair, cement, paint, and metal products.

Sharon /ˈʃeərən/ coastal plain in Israel between Haifa and Tel Aviv, and a sub-district of Central district; area 348 sq km/134 sq mi; population (1983) 190,400. It has been noted since ancient times for its fertility.

Sharon /ˈʃærən/ city in W Pennsylvania, USA, on the Shenango River, N of Pittsburgh, near the Ohio border; population (1990) 17,500. Industries include steel products and electronics.

Sharpeville /ˈʃɑːpvɪl/ black township in South Africa, 65 km/40 mi S of Johannesburg and N of Vereeniging; 69 people were killed here when police fired on a crowd of anti-apartheid demonstrators 21 March 1960.

The massacre took place during a campaign launched by the Pan-Africanist Congress against the pass laws (laws requiring nonwhite South Africans to carry identity papers). On the anniversary of the massacre in 1985, during funerals of people who had been killed protesting against unemployment, 19 people were shot by the police at Langa near Port Elizabeth.

Shasta, Mount /ˈʃæstə/ dormant volcano rising to a height of 4,317 m/14,162 ft in the Cascade Range, N California, USA.

Shatt-al-Arab /ʃæt æl ˈærəb/ (Persian *Arvand*) waterway formed by the confluence of the rivers Euphrates and Tigris; length 190 km/120 mi to the Persian Gulf. Basra, Khorramshahr, and Abadan stand on it.

Its lower reaches form a border of disputed demarcation between Iran and Iraq. In 1975 the two countries agreed on the deepest water line as the frontier, but Iraq repudiated this 1980; the dispute was a factor in the Iran–Iraq war 1980–88.

Sheba /ˈʃiːbə/ ancient name for S ◊Yemen (Sha'abijah). It was once renowned for gold and spices. According to the Old Testament, its queen visited Solomon; until 1975 the Ethiopian royal house traced its descent from their union.

Sheboygan /ʃɪˈbɔɪɡən/ city in E Wisconsin, USA, on Lake Michigan, N of Milwaukee; population (1990) 50,000. Industries include wood, food, plastic, and enamel products.

Sheerness /ʃɪəˈnes/ seaport and resort on the Isle of Sheppey, Kent, England; population (1989 est) 12,000. Situated at the confluence of the rivers Thames and Medway, it was originally a fortress 1660, and was briefly held by the Dutch admiral de Ruyter 1667. It was a royal dockyard until 1960.

Sheffield /ˈʃefiːld/ industrial city on the river Don, South Yorkshire, England; population (1991 est) 499,700. From the 12th century, iron smelting was the chief industry, and by the 14th century, Sheffield cutlery, silverware, and plate were made. During the Industrial Revolution the iron and steel industries developed rapidly. It now produces alloys and special steels, cutlery of all kinds, permanent magnets, drills, and precision tools. Other industries include electroplating, type-founding, and the manufacture of optical glass.
The parish church of St Peter and St Paul (14th–15th centuries) is the cathedral of Sheffield bishopric established 1914. Mary Queen of Scots was imprisoned at Sheffield 1570–84, part of the time in the Norman castle, which was captured by the Parliamentarians 1644 and subsequently destroyed. There are two art galleries; Cutlers' Hall; Ruskin museum, opened 1877 and revived 1985; and two theatres, the Crucible 1971 and the Lyric Theatre, designed by W R Sprague 1897; the restored Lyceum theatre reopened 1990; there is also a university (1905) and a polytechnic (1969). The city is a touring centre for the Peak District. The headquarters of the National Union of Mineworkers are in Sheffield.

Shenandoah /ʃenənˈdəuə/ river in Virginia, USA, 89 km/55 mi long, a tributary of the Potomac, which it joins at Harper's Ferry, West Virginia.
The Union general Philip Sheridan laid waste the Shenandoah Valley in the American Civil War.

Shensi /ʃenˈsiː/ alternative transcription of the Chinese province of ◊Shanxi.

Shenyang /ʃenˈjæŋ/ industrial city and capital of Liaoning province, China; population (1990) 4,500,000. It was the capital of the Manchu emperors 1644–1912; their tombs are nearby.
Historically known as **Mukden**, it was taken from Russian occupation by the Japanese in the Battle of Mukden 20 Feb–10 March 1905, and was again taken by the Japanese 1931.

Shenzen /ʃʌnˈdzʌn/ special economic zone established 1980 opposite Hong Kong on the coast of Guangdong province, S China. Its status provided much of the driving force of its spectacular development in the 1980s when its population rose from 20,000 in 1980 to 600,000 in 1989. Part of the population is 'rotated': newcomers from other provinces return to their homes after a few years spent learning foreign business techniques.

Sheppey /ˈʃepi/ island off the north coast of Kent, England; area 80 sq km/31 sq mi; population about 27,000. Situated at the mouth of the river Medway, it is linked with the mainland by Kingsferry road and rail bridge over the river Swale, completed 1960. The resort and port of Sheerness is here.

Sherman /ˈʃɜːmən/ city in NE Texas, USA, N of Dallas; population (1990) 31,600. It is a processing and shipping centre for agricultural products; textiles, electronics, and machinery are manufactured.

's-Hertogenbosch /seə,təuxənˈbɒs/ or **Den Bosch** (French **Bois-le-Duc**) capital of North Brabant province, the Netherlands, at the confluence of the Aa and Dommel rivers, 45 km/28 mi SE of Utrecht; population (1991) 92,000. It has a Gothic cathedral and was the birthplace of the painter Hieronymus Bosch.

Sherwood Forest /ˈʃɜːwud/ hilly stretch of parkland in W Nottinghamshire, England, area about 520 sq km/200 sq mi. Formerly a royal forest, it is associated with the legendary outlaw Robin Hood.

Shetland Islands /ˈʃetlənd/ islands off the north coast of Scotland, beyond the Orkneys
area 1,400 sq km/541 sq mi
towns Lerwick (administrative headquarters), on Mainland, largest of 19 inhabited islands

Shetland Islands

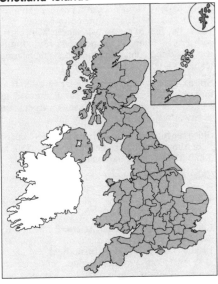

physical over 100 islands including Muckle Flugga (latitude 60° 51′ N) the northernmost of the British Isles
products processed fish, handknits from Fair Isle and Unst, miniature ponies. Europe's largest oil port is Sullom Voe, Mainland
population (1988 est) 22,900
language dialect derived from Norse, the islands having been a Norse dependency from the 8th century until 1472.

Shihchiachuang alternative transliteration of the city of ▷Shijiazhuang in China.

Shijiazhuang /ˌʃiːˌdʒɪəˈdʒwæŋ/ or **Shihchiachuang** city and major railway junction in Hebei province, China; population (1989) 1,300,000. Industries include textiles, chemicals, printing, and light engineering.

Shikoku /ʃiːˈkəukuː/ smallest of the four main islands of Japan, S of Honshu, E of Kyushu; area 18,800 sq km/7,257 sq mi; population (1986) 4,226,000; chief town Matsuyama. Products include rice, wheat, soya beans, sugar cane, orchard fruits, salt, and copper.
It has a mild climate, and annual rainfall in the S can reach 266 cm/105 in. The highest point is Mount Ishizuchi (1,980 m/6,498 ft). A suspension bridge links Shikoku to Awajishima Island over the Naruto whirlpool in the Seto Naikai (Inland Sea).

Shillelagh /ʃɪˈleɪlə/ village in County Wicklow, Republic of Ireland, which gives its name to a rough cudgel of oak or blackthorn. The district was once covered by the Shillelagh Wood, which supplied oak roofing for St Patrick's cathedral in Dublin.

Shillong /ʃɪˈlɒŋ/ capital of Meghalaya state, NE India; population (1981) 109,244. It trades in rice, cotton, and fruit. It was the former capital of Assam.

Shimonoseki /ˌʃɪmənəuˈseki/ seaport in the extreme SW of Honshu Island, Japan; population (1990) 262,600. It was opened to foreign trade 1890. The first of the Sino-Japanese Wars ended with a treaty signed at Shimonoseki 1895. Industries include fishing, shipbuilding, engineering, textiles, and chemicals.

Shiraz /ʃɪəˈræz/ ancient walled city of S Iran, the capital of Fars province; population (1986) 848,000. It is known for its wines, carpets, and silverwork and for its many mosques.

Shiré Highlands /ˈʃɪəreɪ/ upland area of S Malawi, E of the Shiré River; height up to 1,750 m/5,800 ft. Tea and tobacco are grown.

Shizuoka /ˌʃiːzuːˈəukə/ town in Chubo region, Honshu Island, Japan; population (1990) 472,200. Industries include metal and food processing, especially tea.

Shkodër /ˈʃkəudə/ (Italian *Scutari*) town on the river Bojana, NW Albania, SE of Lake Shkodër, 19 km/12 mi from the Adriatic Sea; population (1983) 71,000. Products include woollens and cement. During World War I it was occupied

by Austria 1916–18, and during World War II by Italy.

Sholapur /ˌʃəuləˈpuə/ town in Maharashtra state, India; population (1981) 514,860. Industries include textiles, leather goods, and chemicals.

Shreveport /ˈʃriːvpɔːt/ port on the Red River, Louisiana, USA; population (1990) 198,500. Industries include oil, natural gas, steel, telephone equipment, glass, and timber. It was founded 1836 and named after Henry Shreeve, a riverboat captain who cleared a giant logjam.

Shrewsbury /ˈʃrəuzbəri/ market town on the river Severn, Shropshire, England; population (1991) 90,900. It is the administrative headquarters of the county. To the E is the site of the Roman city of Viroconium (larger than Pompeii). In the 5th century, as *Pengwern*, Shrewsbury was capital of the kingdom of Powys, which later became part of Mercia. In the battle of Shrewsbury 1403, Henry IV defeated the rebels led by Hotspur (Sir Henry Percy).
The city declined an invitation 1539, at the dissolution of the monasteries, to become a cathedral city.

Shropshire /ˈʃrɒpʃə/ county in W England. Sometimes abbreviated to *Salop*, it was officially known by this name from 1974 until local protest reversed the decision 1980
area 3,490 sq km/1,347 sq mi
towns Shrewsbury (administrative headquarters), Telford, Oswestry, Ludlow
physical bisected, on the Welsh border, NW–SE by the river Severn; Ellesmere, the largest of several lakes; the Clee Hills rise to about 610 m/1,800 ft in the SW
features Ironbridge Gorge open-air museum of industrial archaeology, with the Iron Bridge (1779)
products chiefly agricultural: sheep and cattle
population (1991) 401,600

Shropshire

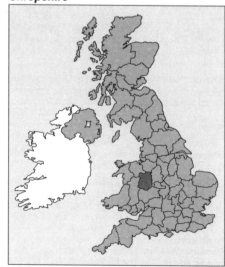

famous people Charles Darwin, A E Housman, Wilfred Owen, Gordon Richards

history Shropshire became a county in the 10th century, as part of the kingdom of Mercia in its defence against the Danes. During the Middle Ages, it was part of the Welsh Marches and saw much conflict between the lords of the Marches and the Welsh.

Siachen Glacier /si'ætʃen/ Himalayan glacier at an altitude of 5,236 m/17,000 ft in the Karakoram Mountains of N Kashmir. Occupied by Indian forces 1984, the glacier has been the focal point of a territorial dispute between India and Pakistan since independence 1947. Three wars in 1947, 1965, and 1971 resulted in the establishment of a temporary boundary between the two countries through the province of Jammu and Kashmir, but the accords failed to define a frontier in the farthest reaches of N Kashmir. Pakistan responded to the 1984 Indian action by sending troops to the heights of the nearby Baltoro Glacier.

Sialkot /si'ælkɒt/ city in Punjab province, E Pakistan; population (1981) 302,000. Industries include the manufacture of surgical and sports goods, metalware, carpets, textiles, and leather goods.

Sian /si'æn/ alternative transliteration of ◊Xian, a city in China.

Siberia /saɪ'bɪərɪə/ Asian region of Russia, extending from the Ural Mountains to the Pacific Ocean

area 12,050,000 sq km/4,650,000 sq mi

towns Novosibirsk, Omsk, Krasnoyarsk, Irkutsk

features long and extremely cold winters

products hydroelectric power from rivers Lena, Ob, and Yenisei; forestry; mineral resources, including gold, diamonds, oil, natural gas, iron, copper, nickel, cobalt

history overrun by Russia in the 17th century, Siberia was used from the 18th century to exile political and criminal prisoners. The first *Trans-Siberian Railway* 1892–1905 from St Petersburg (via Omsk, Novosibirsk, Irkutsk, and Khabarovsk) to Vladivostok, approximately 8,700 km/5,400 mi, began to open it up. A popular front was formed 1988, campaigning for ecological and political reform.

> '*It's a place where 100 roubles is not money, 1,000 kilometres is no distance, and half a litre of vodka is no drink.*'
>
> On **Siberia** a joke Siberians tell against themselves

Sichuan /,sɪtʃ'wɑːn/ or *Szechwan* province of central China

area 569,000 sq km/219,634 sq mi

capital Chengdu

towns Chongqing

features surrounded by mountains, it was the headquarters of the Nationalist government 1937–45, and China's nuclear research cen-

tres are here. It is China's most populous administrative area

products rice, coal, oil, natural gas

population (1990) 107,218,000.

Sicily /'sɪsəli/ (Italian *Sicilia*) the largest Mediterranean island, an autonomous region of Italy; area 25,700 sq km/9,920 sq mi; population (1990) 5,196,800. Its capital is Palermo, and towns include the ports of Catania, Messina, Syracuse, and Marsala. It exports Marsala wine, olives, citrus, refined oil and petrochemicals, pharmaceuticals, potash, asphalt, and marble. The region also includes the islands of Lipari, Egadi, Ustica, and Pantelleria. Etna, 3,323 m/10,906 ft high, is the highest volcano in Europe; its last major eruption was in 1971.

Conquered by most of the major powers of the ancient world, Sicily flourished under the Greeks who colonized the island during the 8th–5th centuries BC. It was invaded by Carthage and became part of the Roman empire 241 BC–AD 476. In the Middle Ages it was ruled successively by the Arabs; the Normans 1059–1194, who established the *Kingdom of the Two Sicilies* (that is, Sicily and the southern part of Italy); German emperors; and then the Angevins, until the popular revolt known as the *Sicilian Vespers* 1282. Spanish rule was invited and continued in varying forms, with a temporary displacement of the Spanish Bourbons by Napoleon, until Garibaldi's invasion 1860 resulted in the two Sicilies being united with Italy 1861.

Sidi Barrâni /'sidi bə'rɑːni/ coastal settlement in Egypt, about 370 km/230 mi W of Alexandria. It was the scene of heavy fighting 1940–42 during World War II.

Sidi-Bel-Abbès /'sidi 'bel æ'bes/ trading city in Algeria; population (1983) 187,000. Because of its strategic position, it was the headquarters of the French Foreign Legion until 1962.

Sidon /'saɪdn/ ancient name for the port of ◊Saida, Lebanon.

Siegen /'ziːgən/ city in North Rhine–Westphalia, Germany; population (1988) 107,000. There are iron and steel industries. It was once the seat of the Princes of Nassau-Orange and today is the cultural and economic centre of the Siegerland. The artist Rubens was born here.

Siena /si'enə/ city in Tuscany, Italy; population (1985) 60,670. Founded by the Etruscans, it has medieval architecture by Pisano and Donatello, including a 13th-century Gothic cathedral, and many examples of the Sienese school of painting that flourished from the 13th to the 16th centuries. The *Palio* ('banner', in reference to the prize) is a dramatic and dangerous horse race in the main square, held annually (2 July and 16 August) since the Middle Ages.

Sierra Leone Republic of

area 71,740 sq km/27,710 sq mi

capital Freetown

towns Koidu, Bo, Kenema, Makeni

physical mountains in E; hills and forest; coastal mangrove swamps

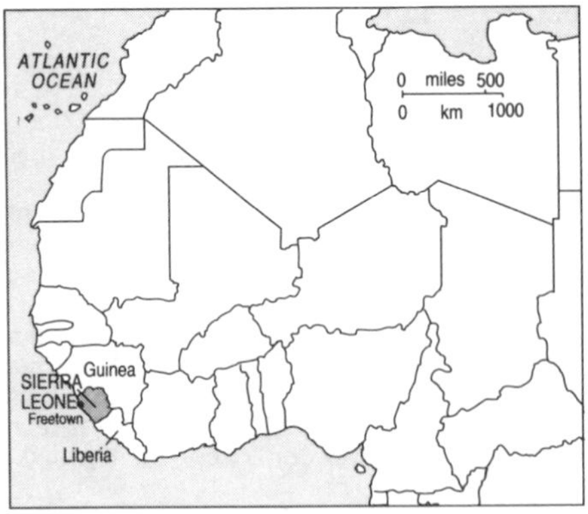

features hot and humid climate (3,500 mm/138 in rainfall p.a.)
head of state and government military council headed by Capt Valentine Strasser from 1992
political system transitional
political parties All People's Congress (APC), moderate socialist; United Front of Political Movements (UNIFORM), centre-left
exports palm kernels, cocoa, coffee, ginger, diamonds, bauxite, rutile
currency leone
population (1990 est) 4,168,000; growth rate 2.5% p.a.
life expectancy men 41, women 47 (1989)
languages English (official), local languages
media no daily newspapers; 13 weekly papers, of which 11 are independent but only one achieves sales of over 5,000 copies
religions animist 52%, Muslim 39%, Protestant 6%, Roman Catholic 2% (1980 est)
literacy men 38%, women 21% (1985 est)
GNP $965 million (1987); $320 per head (1984)
chronology
1808 Became a British colony.
1896 Hinterland declared a British protectorate.
1961 Independence achieved from Britain within the Commonwealth, with Milton Margai, leader of Sierra Leone People's Party (SLPP), as prime minister.
1964 Milton succeeded by his half-brother, Albert Margai.
1967 Election results disputed by army, who set up a National Reformation Council and forced the governor general to leave.
1968 Army revolt made Siaka Stevens, leader of the All People's Congress (APC), prime minister.
1971 New constitution adopted, making Sierra Leone a republic, with Stevens as president.
1978 APC declared only legal party. Stevens sworn in for another seven-year term.
1985 Stevens retired; succeeded by Maj-Gen Joseph Momoh.
1989 Attempted coup against Momoh foiled.
1991 Referendum endorsed multiparty politics.
1992 Military take-over; President Momoh fled. National Provisional Ruling Council (NPRC) established under Capt Valentine Strasser.

Sierra Madre /si'erə 'mɑːdreɪ/ chief mountain system of Mexico, consisting of three ranges, enclosing the central plateau of the country; highest point Pico de Orizaba 5,700 m/18,700 ft. The Sierra Madre del Sur ('of the south') runs along the SW Pacific coast.

Sierra Nevada /si'erə nɪ'vɑːdə/ mountain range of S Spain; highest point Mulhacén 3,481 m/11,425 ft.

Sierra Nevada /si,erə nɪ'vɑːdə/ mountain range in E California; highest point Mount Whitney 4,418 m/14,500 ft. The Sierra Nevada includes the King's Canyon, Sequoia, and Yosemite Valley national parks.

About 640 km/400 mi in length, the Sierra Nevada separates California from the rest of the continent. In 1848 settlers found gold in its western foothills, touching off the great 1849 gold rush. Silver mines have been opened on its east side.

Si-Kiang /'ʃiː ki'æŋ/ alternative transliteration of ⇨Xi Jiang, a Chinese river.

Sikkim /'sɪkɪm/ or ***Denjong*** state of NE India; formerly a protected state, it was absorbed by India 1975, the monarchy being abolished. China does not recognize India's sovereignty
area 7,300 sq km/2,818 mi
capital Gangtok
features Mount Kangchenjunga; wildlife including birds, butterflies, and orchids
products rice, grain, tea, fruit, soya beans, carpets, cigarettes, lead, zinc, copper
population (1991) 403,600
languages Bhutia, Lepecha, Khaskura (Nepali)
religions Mahayana Buddhism, Hinduism
history ruled by the Namgyol dynasty from the 14th century to 1975, when the last chogyal, or king, was deposed. Allied to Britain 1886, Sikkim became a protectorate of India 1950 and a state of India 1975.

Silicon Valley nickname given to Santa Clara County, California, since the 1950s the site of many high-technology electronic firms, whose prosperity is based on the silicon chip.

Sikkim

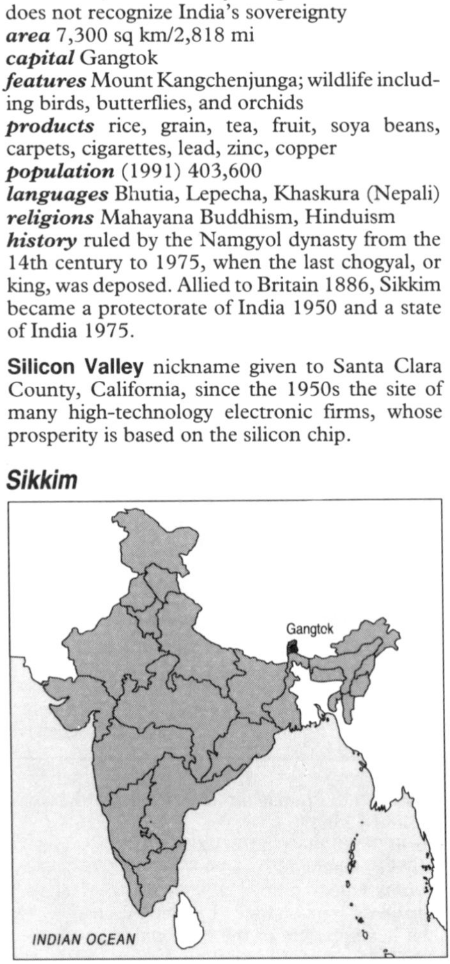

Simferopol /ˌsɪmfəˈrəupɒl/ city in Ukraine; population (1987) 338,000. Industries include the manufacture of soap and tobacco.
It is on the site of the Tatar town of *Ak-Mechet*, conquered by the Russians 1783 and renamed.

Similaun Glacier glacier in the Tyrolean Alps on the Austria/Italy frontier. In 1991 the oldest intact human body was discovered here, having been preserved in ice for over 5,300 years.

Simla /ˈsɪmlə/ capital of Himachal Pradesh state, India, 2,300 m/7,500 ft above sea level, population (1980) 70,604. It was the summer administrative capital of British India 1864–1947.

Simplon /ˈsæmplɒn/ (Italian *Sempione*) Alpine pass Switzerland–Italy. The road was built by Napoleon 1800–05, and the Simplon Tunnel 1906, 19.8 km/12.3 mi, is one of Europe's longest.

Simpson Desert /ˈsɪmpsən/ desert area in Australia, chiefly in Northern Territory; area 145,000 sq km/56,000 sq mi.
The desert was named after a president of the South Australian Geographical Society who financed its exploration.

Sinai /ˈsaɪnaɪ/ Egyptian peninsula, at the head of the Red Sea; area 65,000 sq km/25,000 sq mi. Resources include oil, natural gas, manganese, and coal; irrigation water from the river Nile is carried under the Suez Canal.
Sinai was occupied by Israel 1967–82. After the Battle of Sinai 1973, Israel began a gradual withdrawal from the area, under the disengagement agreement 1975, and the Camp David peace treaty 1979, and restored the whole of Sinai to Egyptian control by April 1982.

Sinai, Mount /ˈsaɪnaɪ/ (Arabic *Gebel Mûsa*) mountain near the tip of the Sinai Peninsula; height 2,285 m/7,500 ft. According to the Old Testament this is where Moses received the Ten Commandments from God.

Among all the stupendous works of Nature, not a place can be selected more fitted for the exhibition of Almighty power …

On **Mount Sinai** John Lloyd Stephens
Incidents of Travel in Egypt 1837

Sind /sɪnd/ province of SE Pakistan, mainly in the Indus delta
area 140,914 sq km/54,393 sq mi
capital and chief seaport Karachi
population (1981) 19,029,000
languages 60% Sindi; others include Urdu, Punjabi, Baluchi, Pashto
features Sukkur Barrage, which enables water from the Indus River to be used for irrigation
history annexed 1843, it became a province of British India, and part of Pakistan on independence. There is agitation for its creation as a separate state, Sindhudesh.

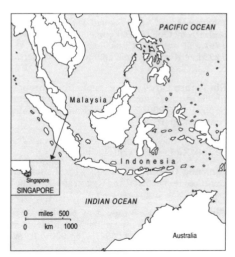

Singapore Republic of
area 622 sq km/240 sq mi
capital Singapore City
towns Jurong, Changi
physical comprises Singapore Island, low and flat, and 57 small islands
features Singapore Island is joined to the mainland by causeway across Strait of Johore; temperature range 21°–34°C/69°–93°F
head of state Wee Kim Wee from 1985
head of government Goh Chok Tong from 1990
political system liberal democracy with strict limits on dissent
political parties People's Action Party (PAP), conservative; Workers' Party (WP), socialist; Singapore Democratic Party (SDP), liberal
exports electronics, petroleum products, rubber, machinery, vehicles
currency Singapore dollar
population (1990 est) 2,703,000 (Chinese 75%, Malay 14%, Tamil 7%); growth rate 1.2% p.a.
life expectancy men 71, women 77 (1989)
languages Malay (national tongue), Chinese, Tamil, English (all official)
religions Buddhist, Taoist, Muslim, Hindu, Christian
literacy men 93%, women 79% (1985 est)
GDP $19.9 bn (1987); $7,616 per head
chronology
1819 Singapore leased to British East India Company.
1858 Placed under crown rule.
1942 Invaded and occupied by Japan.
1945 Japanese removed by British forces.
1959 Independence achieved from Britain; Lee Kuan Yew became prime minister.
1963 Joined new Federation of Malaysia.
1965 Left federation to become independent.
1984 Opposition made advances in parliamentary elections.
1986 Opposition leader convicted of perjury and prohibited from standing for election.
1988 Ruling conservative party elected to all but one of available assembly seats; increasingly authoritarian rule.

1990 Lee Kuan Yew resigned as prime minister; replaced by Goh Chok Tong.
1991 PAP and Goh Chok Tong re-elected.
1992 Lee Kuan Yew surrendered PAP leadership to Goh Chok Tong.

Singapore City /ˌsɪŋə'pɔː/ capital of Singapore, on the southeast coast of the island of Singapore; population (1980) 2,413,945. It is an oil refining centre and port.

Sining /ˌʃiː'nɪŋ/ alternative transliteration of the city of ◊Xining in W central China.

Sinkiang-Uighur /ˌʃɪnki'æŋ 'wiːgə/ alternative transliteration of ◊Xinjiang Uygur, an autonomous region of NW China.

Sinuiju /ˌsɪnwiː'dʒuː/ capital of North Pyongan province, near the mouth of the Yalu River, North Korea; population (1984) 754,000. It was founded 1910.

Sioux City /ˌsuː 'sɪti/ city in NW Iowa, USA, on the Missouri River, near Iowa's border with Nebraska and South Dakota; population (1990) 80,500. Industries include food processing, fabricated metals, fertilizer, and meatpacking. It is the head of navigation for the Missouri River.

Sioux Falls /ˌsuː 'fɔːlz/ largest city in South Dakota, USA; population (1990) 100,800. Its industry (electrical goods and agricultural machinery) is powered by the Big Sioux River over the Sioux Falls 30 m/100 ft.

Sirte, Gulf of /'sɜːti/ gulf off the coast of Libya, on which the port of Benghazi stands. Access to the gulf waters has been a cause of dispute between Libya and the USA.

Six Counties the six counties that form Northern Ireland: Antrim, Armagh, Down, Fermanagh, Derry, and Tyrone.

Sjælland /'ʃelənd/ or **Zealand** main island of Denmark, on which Copenhagen is situated; area 7,000 sq km/2,700 sq mi; population (1970) 2,130,000. It is low-lying with an irregular coastline. The chief industry is dairy farming.

Skagerrak /'skægəræk/ arm of the North Sea between the south coast of Norway and the north coast of Denmark. In May 1916 it was the scene of the inconclusive Battle of Jutland between British and German fleets.

Skåne /'skɔːnə/ or **Scania** area of S Sweden. It is a densely populated and fertile agricultural region, comprising the counties of Malmöhus and Kristianstad. Malmö and Hälsingborg are leading centres. It was under Danish rule until ceded to Sweden 1658.

Skegness /ˌskeg'nes/ holiday resort on the coast of Lincolnshire, England; population (1985) 14,600. It was the site of the first Butlin holiday camp.

Skelmersdale /'skelməzdeɪl/ town in Lancashire, N England, W of Wigan; population (1985) 41,800. It was developed as a 'new town' from

1962, with many light industries, including electronics, engineering, and textiles.

Skiddaw /'skɪdɔː/ mountain (930 m/3,052 ft) in Cumbria, England, in the Lake district, N of Keswick.

Skikda /'skɪkdɑː/ trading port in Algeria; population (1983) 141,000. Products include wine, citrus, and vegetables. It was founded by the French 1838 as **Philippeville** and renamed after independence 1962.

Skipton /'skɪptən/ industrial (engineering) town in North Yorkshire, England; population (1981) 13,246.

Skopje /'skɒpjeɪ/ industrial city and capital of the Former Yugoslav Republic of Macedonia; population (1981) 506,547. Industries include iron, steel, chromium mining, and food processing.

It stands on the site of an ancient town destroyed by an earthquake in the 5th century and was taken in the 13th century by the Serbian king Milutin, who made it his capital. Again destroyed by an earthquake 1963, Skopje was rebuilt on a safer site nearby. It is an Islamic centre.

Skye /skaɪ/ largest island of the Inner Hebrides, Scotland; area 1,740 sq km/672 sq mi; population (1987) 8,100. It is separated from the mainland by the Sound of Sleat. The chief port is Portree. The economy is based on crofting, tourism, and livestock.

A privately financed toll bridge to the island is due to be completed 1995.

Bonnie Prince Charlie (Charles Edward Stuart) took refuge here after the Battle of Culloden 1746.

Skyros /'skaɪrɒs/ or **Skiros** Greek island, the largest of the northern Sporades; area 210 sq km/81 sq mi; population (1981) 2,750. It is known for its furniture and weaving. The English poet Rupert Brooke is buried here.

Slavonia /slə'vəʊniə/ region of E Croatia bounded by the Sava, Drava, and Danube rivers; Osijek is the largest town. Slavonia was the scene of fierce fighting between Croatian forces and Serb-dominated Yugoslav federal troops 1991–92. Following Croatia's declaration of independence from Yugoslavia 1991, Eastern and Western Slavonia declared themselves autonomous provinces of Serbia. After the cease-fire 1992, 10,000 UN troops were deployed in E and W Slavonia and contested Krajina.

Sligo /'slaɪgəʊ/ county in the province of Connacht, Republic of Ireland, situated on the Atlantic coast of NW Ireland; area 1,800 sq km/695 sq mi; population (1991) 54,700. The county town is Sligo; there is livestock and dairy farming.

Slough /slaʊ/ industrial town (pharmaceuticals, electronics, engineering, chocolate manufacture) in Berkshire, England, near Windsor; population (1981) 97,000. The home of astronomer William Herschel is now a museum.

Slovak Republic Slovak Republic (*Slovenská Republika*)
area 49,035 sq km/18,940 sq mi
capital Bratislava
towns Košice, Nitra, Prešov, Banská Bystrica
physical W range of the Carpathian Mountains including Tatra and Beskids in N; Danube plain in S; numerous lakes and mineral springs
features fine beech and oak forests with bears and wild boar
head of state Michal Kovac from 1993
head of government Vladimir Meciar from 1993
political system emergent democracy
political parties Civic Democratic Union (CDU), centre-left; Movement for a Democratic Slovakia (HZDS), centre-left, nationalist; Christian Democratic Movement (KDH), right of centre; Slovak National Party, nationalist; Party of the Democratic Left, left-wing, ex-communist; Coexistence and Hungarian Christian Democratic Movement, both representing Hungarian minority
exports iron ore, copper, mercury, magnesite, armaments, chemicals, textiles, machinery
currency new currency based on koruna
population (1991) 5,268,900 (with Hungarian and other minorities); growth rate 0.4% p.a.
life expectancy men 68, women 75
languages Slovak (official)
religions Roman Catholic (over 50%), Lutheran, Reformist, Orthodox
literacy 100%
GDP $10,000 million (1990); $1,887 per head
chronology
906–1918 Under Magyar domination.
1918 Independence achieved from Austro-Hungarian Empire; Slovaks joined Czechs in forming Czechoslovakia as independent nation.
1948 Communists assumed power in Czechoslovakia.
1968 Slovak Socialist Republic created under new federal constitution.
1989 Prodemocracy demonstrations in Bratislava; new political parties formed, including Slovak-based People Against Violence (PAV);

Communist Party stripped of powers. Dec: new government formed, including former dissidents; political parties legalized; Václav Havel appointed president.
1991 Evidence of increasing Slovak separatism. March: PAV splinter group formed under Slovak premier Vladimir Meciar. April: Meciar dismissed, replaced by Jan Carnogursky; pro-Meciar rallies held in Bratislava.
1992 March: PAV renamed Civic Democratic Union (CDU). June: Havel resigned following Slovak gains in assembly elections. Aug: agreement on creation of separate Czech and Slovak states.
1993 Jan: Slovak Republic became sovereign state, with Meciar, leader of the MFDS, as prime minister. Feb: Michal Kovak became president.

Slovenia Republic of
area 20,251 sq km/7,817 sq mi
capital Ljubljana
towns Maribor, Kranj, Celji; chief port: Koper
physical mountainous; Sava and Drava rivers
head of state Milan Kucan from 1990
head of government Janez Drnovsek from 1992
political system emergent democracy
political parties Christian Democratic Party, right of centre; People's Party, right of centre; Liberal Democratic Party, left of centre; Democratic Party, left of centre
products grain, sugarbeet, livestock, timber, cotton and woollen textiles, steel, vehicles
currency tolar
population (1990) 2,000,000 (Slovene 91%, Croat 3%, Serb 2%)
languages Slovene, resembling Serbo-Croat, written in Roman characters
religion Roman Catholic
chronology
1918 United with Serbia and Croatia.
1929 The kingdom of Serbs, Croats, and Slovenes took the name of Yugoslavia.
1945 Became a constituent republic of Yugoslav Socialist Federal Republic.
mid-1980s The Slovenian Communist Party liberalized itself and agreed to free elections. Yugoslav counterintelligence (KOV) began repression.
1989 Jan: Social Democratic Alliance of Slovenia

launched as first political organization independent of Communist Party. Sept: constitution changed to allow secession from federation. **1990** April: nationalist DEMOS coalition secured victory in first multiparty parliamentary elections; Milan Kucan became president. July: sovereignty declared. Dec: independence overwhelmingly approved in referendum. **1991** June: independence declared; 100 killed after federal army intervened; cease-fire brokered by European Community. July: cease-fire agreed between federal troops and nationalists. Oct: withdrawal of Yugoslav army completed. Dec: DEMOS coalition dissolved. **1992** Jan: EC recognized Slovenia's independence. April: Janez Drnovsek appointed prime minister; independence recognized by USA. May: admitted into United Nations and Conference on Security and Cooperation in Europe. Dec: Liberal Democrats and Christian Democrats won assembly elections; Kucan re-elected president.

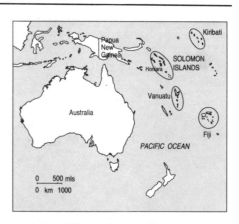

Smithfield /ˈsmɪθfiːld/ site of a meat market from 1868 and poultry and provision market from 1889, in the City of London, England. Formerly an open space, it was the scene of the murder of Wat Tyler, leader of the Peasants' Revolt 1381, and the execution of many Protestant martyrs in the 16th century. The annual Bartholomew Fair was held here 1614–1855.

Smolensk /sməˈlensk/ city on the river Dnieper, W Russia; population (1987) 338,000. Industries include textiles, distilling, and flour milling. It was founded 882 as the chief town of a Slavic tribe and was captured by Napoleon 1812.

Snaefell /ˌsneɪˈfel/ highest mountain in the Isle of Man, 620 m/2,035 ft.

Snake /sneɪk/ tributary of the Columbia River, in northwestern USA; length 1,670 km/1,038 mi. It flows 65 km/40 mi through Hell's Canyon, one of the deepest gorges in the world.

Snowdon /ˈsnəʊdn/(Welsh **Y Wyddfa**) highest mountain in Wales, 1,085 m/3,560 ft above sea level. It consists of a cluster of five peaks. At the foot of Snowdon are the Llanberis, Aberglaslyn, and Rhyd-ddu passes. A rack railway ascends to the summit from Llanberis. **Snowdonia**, the surrounding mountain range, was made a national park 1951. It covers 2,188 sq km/845 sq mi of mountains, lakes, and forest land.

Snowy Mountains /ˈsnəʊi/ range in the Australian Alps, chiefly in New South Wales, near which the **Snowy River** rises; both river and mountains are known for a hydroelectric and irrigation system.

Society Islands /səˈsaɪəti/ (French **Archipel de la Société**) archipelago in French Polynesia, divided into the Windward Islands and the Leeward Islands; area 1,685 sq km/650 sq mi; population (1983) 142,000. The administrative headquarters is Papeete on Tahiti. The **Windward Islands** (French **Iles du Vent**) have an area of 1,200 sq km/460 sq mi and a population (1983) of 123,000. They comprise Tahiti, Moorea (area 132 sq km/51 sq mi; population 7,000), Maio (or Tubuai Manu; 9 sq km/3.5 sq mi;

population 200), and the smaller Tetiaroa and Mehetia. The **Leeward Islands** (French **Iles sous le Vent**) have an area of 404 sq km/156 sq mi and a population of 19,000. They comprise the volcanic islands of Raiatea (including the main town of Uturoa), Huahine, Bora-Bora, Maupiti, Tahaa, and four small atolls. Claimed by France 1768, the group became a French protectorate 1843 and a colony 1880.

Socotra /səʊˈkəʊtrə/ Yemeni island in the Indian Ocean; capital Tamridah; area 3,500 sq km/1,351 sq mi. Under British protection from 1886, it became part of South Yemen 1967.

Sofia /ˈsəʊfiə/ or **Sofiya** capital of Bulgaria since 1878; population (1990) 1,220,900. Industries include textiles, rubber, machinery, and electrical equipment. It lies at the foot of the Vitosha Mountains.

Sogne Fjord /ˈsɒŋnə ˈfiːɔːd/ longest and deepest fjord in Norway, stretching 204 km/127 mi from inland Skjolden to its mouth on the west coast at Solund; it is 1,245 m/4,080 ft deep.

Soho /ˈsəʊhəʊ/ district of London, England, which houses the offices of publishing, film, and recording companies; restaurants; nightclubs; and a decreasing number of sex shops.

Soissons /ˈswæsɒn/ market town in Picardie region, N France; population (1990) 32,100. The chief industry is metallurgy. In 486 the Frankish king Clovis defeated the Gallo-Romans here, ending their rule in France.

Sokoto /ˈsəʊkətəʊ/ state in Nigeria, established 1976; capital Sokoto; area 102,500 sq km/39,565 sq mi; population (1984) 7,609,000. It was a Fula sultanate from the 16th century until occupied by the British 1903.

Solent, the /ˈsəʊlənt/ channel between the coast of Hampshire, England, and the Isle of Wight. It is now a yachting centre.

Solingen /ˈzəʊlɪŋən/ city in North Rhine–Westphalia, Germany; population (1988) 158,000. It was once a major producer of swords and today makes high-quality steel for razor blades and cutlery.

Solomon Islands
area 27,600 sq km/10,656 sq mi
capital Honiara (on Guadalcanal)
towns Gizo, Yandina
physical comprises all but the northernmost islands (which belong to Papua New Guinea) of a Melanesian archipelago stretching nearly 1,500 km/900 mi. The largest is Guadalcanal (area 6,500 sq km/2,510 sq mi); others are Malaita, San Cristobal, New Georgia, Santa Isabel, Choiseul; mainly mountainous and forested
features rivers ideal for hydroelectric power
head of state Elizabeth II represented by governor general
head of government Solomon Mamaloni from 1989
political system constitutional monarchy
political parties People's Alliance Party (PAP), centre-left; Solomon Islands United Party (SIUPA), right of centre
exports fish products, palm oil, copra, cocoa, timber
currency Solomon Island dollar
population (1990 est) 314,000 (Melanesian 95%, Polynesian 4%); growth rate 3.9% p.a.
life expectancy men 66, women 71
languages English (official); there are some 120 Melanesian dialects
religions Anglican 34%, Roman Catholic 19%, South Sea Evangelical 17%
literacy 60% (1989)
GNP $141 million; $420 per head (1987)
chronology
1893 Solomon Islands placed under British protection.
1978 Independence achieved from Britain within the Commonwealth, with Peter Kenilorea as prime minister.
1981 Solomon Mamaloni of the People's Progressive Party (PPP) replaced Kenilorea as prime minister.
1984 Kenilorea returned to power, heading a coalition government.
1986 Kenilorea resigned after allegations of corruption; replaced Ezekiel Alebua.
1988 Kenilorea elected deputy prime minister. Joined Vanuatu and Papua New Guinea to form the Spearhead Group, aiming to preserve Melanesian culture and secure independence for the French territory of New Caledonia.
1989 Solomon Mamaloni, now leader of the People's Action Party (PAP), elected prime minister; formed PAP-dominated coalition.
1990 Mamaloni resigned PAP leadership, but continued as head of government of national unity.

Solway Firth /ˈsɒlweɪ/ inlet of the Irish Sea, formed by the estuaries of the rivers Eden and Esk, at the western end of the border between England and Scotland.

Somalia Somali Democratic Republic (*Jamhuriyadda Dimugradiga Somaliya*)
area 637,700 sq km/246,220 sq mi
capital Mogadishu
towns Hargeisa, Kismayu, port Berbera
physical mainly flat, with hills in N
environment destruction of trees for fuel and by livestock has led to an increase in desert area

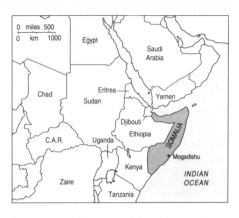

features occupies a strategic location on the Horn of Africa
head of state and government Ali Mahdi Mohammed from 1991
political system one-party socialist republic
political party Somali Revolutionary Socialist Party (SRSP), nationalist, socialist
exports livestock, skins, hides, bananas, fruit
currency Somali shilling
population (1990 est) 8,415,000 (including 350,000 refugees in Ethiopia and 50,000 in Djibouti); growth rate 3.1% p.a.
life expectancy men 53, women 53 (1989)
languages Somali, Arabic (both official), Italian, English
religion Sunni Muslim 99%
literacy 40% (1986)
GNP $1.5 bn; $290 per head (1987)
chronology
1884–87 British protectorate of Somaliland established.
1889 Italian protectorate of Somalia established.
1960 Independence achieved from Italy and Britain.
1963 Border dispute with Kenya; diplomatic relations broken with Britain.
1968 Diplomatic relations with Britain restored.
1969 Assassination of President Shermarke; army coup led by Maj-Gen Mohamed Siad Barre; constitution suspended, Supreme Revolutionary Council set up; name changed to Somali Democratic Republic.
1978 Defeated in eight-month war with Ethiopia. Armed insurrection began in north.
1979 New constitution for socialist one-party state adopted.
1982 Antigovernment Somali National Movement formed. Oppressive countermeasures by government.
1987 Barre re-elected president.
1989 Dissatisfaction with government and increased guerrilla activity in north.
1990 Civil war intensified. Constitutional reforms promised.
1991 Mogadishu captured by rebels; Barre fled; Ali Mahdi Mohammed named president; free elections promised. Secession of NE Somalia, as the Somaliland Republic, announced. Cease-fire signed, but later collapsed. Thousands of casualties as a result of heavy fighting in capital.

1992 Relief efforts to ward off impending famine severely hindered by unstable political situation; relief convoys hijacked by 'warlords'. Dec: UN peacekeeping troops, mainly US Marines, sent in to protect relief operations; dominant warlords agreed truce.
1993 March: leaders of armed factions agreed to federal system of government, based on 18 autonomous regions. Peacekeeping troops engaged in intermittent combat throughout year, with mounting casualties. US announced date of March 1994 for withdrawal of most of its troops.

Somerset /'sʌməset/ county in SW England
area 3,460 sq km/1,336 sq mi
towns administrative headquarters Taunton; Wells, Bridgwater, Glastonbury, Yeovil
physical rivers Avon, Parret, and Exe; marshy coastline on the Bristol Channel; Mendip Hills (including Cheddar Gorge and Wookey Hole, a series of limestone caves where Stone Age flint implements and bones of extinct animals have been found); the Quantock Hills; Exmoor
products engineering, dairy products, cider
population (1991) 459,100
famous people Ernest Bevin, Henry Fielding, John Pym.

Somme /sɒm/ river in N France, on which Amiens and Abbeville stand; length 240 km/150 mi. It rises in Aisne *département* and flows W through Somme *département* to the English Channel.

Soochow /ˌsuːˈtʃaʊ/ alternative transliteration of the Chinese city of ◊Suzhou.

Sosnowiec /sɒ'snɒvjets/ chief city of the Darowa coal region in the Upper Silesian province of Katowice, S Poland; population (1990) 259,300.

Sound, the /saʊnd/ (Swedish and Danish **Øresund**) strait dividing SW Sweden from Denmark and linking the Kattegat strait and the Baltic Sea; length 113 km/70 mi; width between 5–60 km/3–37 mi.

Somerset

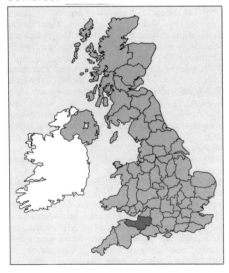

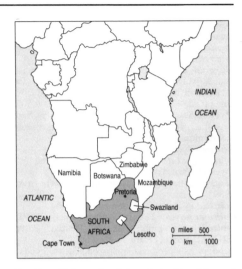

South Africa /'saʊθə'merɪkə/ Republic of (*Republiek van Suid-Afrika*)
area 1,223,181 sq km/472,148 sq mi (includes Walvis Bay and independent black homelands)
capital and port Cape Town (legislative), Pretoria (administrative), Bloemfontein (judicial)
towns Johannesburg; ports Durban, Port Elizabeth, East London
physical southern end of large plateau, fringed by mountains and lowland coastal margin
territories Marion Island and Prince Edward Island in the Antarctic
features Drakensberg Mountains, Table Mountain; Limpopo and Orange rivers; the Veld and the Karoo; part of Kalahari Desert; Kruger National Park
head of state and government F W de Klerk from 1989
political system racist, nationalist republic, restricted democracy
political parties White: National Party (NP), right of centre, racist; Conservative Party of South Africa (CPSA), extreme right, racist; Democratic Party (DP), left of centre, multiracial. Coloureds: Labour Party of South Africa, left of centre; People's Congress Party, right of centre. Indian: National People's Party, right of centre; Solidarity Party, left of centre
exports maize, sugar, fruit, wool, gold (world's largest producer), platinum, diamonds, uranium, iron and steel, copper; mining and minerals are largest export industry, followed by arms manufacturing
currency rand
population (1990 est) 39,550,000 (73% black: Zulu, Xhosa, Sotho, Tswana; 18% white: 3% mixed, 3% Asian); growth rate 2.5% p.a.
life expectancy whites 71, Asians 67, blacks 58
languages Afrikaans and English (both official), Bantu
religions Dutch Reformed Church 40%, Anglican 11%, Roman Catholic 8%, other Christian 25%, Hindu, Muslim
literacy whites 99%, Asians 69%, blacks 50% (1989)
GNP $81 bn; $1,890 per head (1987)

chronology
1910 Union of South Africa formed from two British colonies and two Boer republics.
1912 African National Congress (ANC) formed.
1948 Apartheid system of racial discrimination initiated by Daniel Malan, leader of National Party (NP).
1955 Freedom Charter adopted by ANC.
1958 Malan succeeded as prime minister by Hendrik Verwoerd.
1960 ANC banned.
1961 South Africa withdrew from Commonwealth and became a republic.
1962 ANC leader Nelson Mandela jailed.
1964 Mandela, Walter Sisulu, Govan Mbeki, and five other ANC leaders sentenced to life imprisonment.
1966 Verwoerd assassinated; succeeded by B J Vorster.
1976 Soweto uprising.
1977 Death in custody of Pan African Congress activist Steve Biko.
1978 Vorster resigned and was replaced by Pieter W Botha.
1984 New constitution adopted, giving segregated representation to Coloureds and Asians and making Botha president. Nonaggression pact with Mozambique signed but not observed.
1985 Growth of violence in black townships.
1986 Commonwealth agreed on limited sanctions. US Congress voted to impose sanctions. Some major multinational companies closed down their South African operations.
1987 Government formally acknowledged the presence of its military forces in Angola.
1988 Botha announced 'limited constitutional reforms'. South Africa agreed to withdraw from Angola and recognize Namibia's independence as part of regional peace accord.
1989 Botha gave up NP leadership and state presidency. F W de Klerk became president. ANC activists released; beaches and public facilities desegregated. Elections held in Namibia to create independence government.
1990 ANC ban lifted; Nelson Mandela released from prison. NP membership opened to all races. ANC leader Oliver Tambo returned. Daily average of 35 murders and homicides recorded.
1991 Mandela and Zulu leader Mangosuthu Buthelezi urged end to fighting between ANC and Inkatha. Mandela elected ANC president. Revelations of government support for Inkatha threatened ANC cooperation. De Klerk announced repeal of remaining apartheid laws. South Africa readmitted to international sport; USA lifted sanctions. PAC and Buthelezi withdrew from negotiations over new constitution.
1992 Constitution leading to all-races majority rule approved by whites-only referendum. Massacre of civilians at black township of Boipatong near Johannesburg by Inkatha, aided and abetted by police, threatened constitutional talks.
1993 Feb: de Klerk and Nelson Mandela agreed to formation of government of national unity after free elections. April: ANC leader Chris Hani assassinated by white extremist. July: riots in townships followed announcement of April 1994 date for nonracial elections.

South America fourth largest of the continents, nearly twice as large as Europe (13% of the world's land surface), extending S from Central America *area* 17,864,000 sq km/6,900,000 sq mi *largest cities* (population over 3.5 million) Buenos Aires, São Paulo, Rio de Janeiro, Bogotá, Santiago, Lima, Caracas *features* Lake Titicaca (the world's highest navigable lake); La Paz (highest capital city in the world); Atacama Desert; Inca ruins at Machu Picchu; rivers include the Amazon (world's largest and second longest), Parana, Madeira, São Francisco, Purus, Paraguay, Orinoco, Araguaia, Negro, Uruguay *geographical extremities of mainland* Point Gallinas in Colombia in the N; Cape Froward on the Brunswick peninsula, S Chile, in the S; Point Pariñas, Peru, in the W; Point Coqueiros, just N of Recife, Brazil, in the E *physical* occupying the southern part of the landmass of the western hemisphere, the South American continent stretches from the Caribbean coast of Colombia to Cape Horn at the southern tip of Horn Island, which lies adjacent to Tierra del Fuego. Five-sixths of the continent lies in the southern hemisphere and two-thirds within the tropics. South America can be divided into the following physical regions: 1) the Andes mountain system; 2) the uplifted remains of the old continental mass in the Brazilian Highlands and Guiana Highlands; 3) the plain of the Orinoco River, lying between the Venezuelan Andes and the Guiana Highlands; 4) the tropical Amazon Plain; 5) the Pampa-Chaco plain of Argentina, Paraguay, and Bolivia; and 6) the Patagonian Plateau in the S *products* produces 44% of the world's coffee (Brazil, Colombia), 22% of its cocoa (Brazil), 35% of its citrus fruit, meat (Argentina, Brazil), soya beans (Argentina, Brazil), cotton (Brazil), linseed (Argentina); Argentina is the world's second largest producer of sunflower seed; Brazil is the world's largest producer of bananas, its second largest producer of tin, and its third largest producer of manganese, tobacco, and mangoes; Peru is the world's second largest producer of silver; Chile is the world's largest producer of copper *population* (1988) 285 million, rising to 550 million (est) by the year 2000; annual growth rate from 1980 to 1985 2.3% *languages* Spanish, Portuguese (chief language in Brazil), Dutch (Surinam), French (French Guiana), Amerindian languages; Hindi, Javanese, and Chinese spoken by descendants of Asian immigrants to Surinam and Guyana; a variety of Creole dialects spoken by those of African descent *religions* 90–95% Roman Catholic; local animist beliefs among Amerindians; Hindu and Muslim religions predominate among the descendants of Asian immigrants in Surinam and Guyana.

Southampton /saʊθˈhæmptən/ port in Hampshire, S England; population (1981) 204,600. Industries include engineering, chemicals, plastics, flour-milling, and tobacco; it is also a passenger and container port. There is a university (1952).
The *Mayflower* set sail from here en route to North America 1620, as did the *Titanic* on its fateful maiden voyage 1912.

South Australia

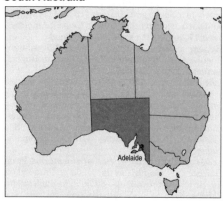

South Arabia, Federation of /'sauθ ə'reɪbɪə/ former grouping (1959–67) of Arab emirates and sheikdoms, joined by Aden 1963. The western part of the area was claimed by Yemen, and sporadic fighting and terrorism from 1964 led to British withdrawal 1967 and the proclamation of the Republic of South Yemen (part of united Yemen from 1990).

South Australia state of the Commonwealth of Australia
area 984,000 sq km/379,824 sq mi
capital Adelaide (chief port)
towns Whyalla, Mount Gambier
features Murray Valley irrigated area, including wine-growing Barossa Valley; lakes: Eyre, Torrens; mountains: Mount Lofty, Musgrave, Flinders; parts of the Nullarbor Plain, and Great Victoria and Simpson deserts; experimental rocket range in the arid N at Woomera
products meat and wool (80% of area cattle and sheep grazing), wines and spirits, dried and canned fruit, iron (Middleback Range), coal (Leigh Creek), copper, uranium (Roxby Downs), oil and natural gas in the NE, lead, zinc, iron, opals, household and electrical goods, vehicles
population (1987) 1,388,000; 1% Aborigines
history possibly known to the Dutch in the 16th century; surveyed by Dutch navigator Abel Tasman 1644; first European settlement 1834; province 1836; became a state 1901. In 1963 British nuclear tests were made at Maralinga, in which Aborigines were said to have died.

South Bank area of London S of the river Thames, the site of the Festival of Britain 1951, and now a cultural centre. Buildings include the Royal Festival Hall 1951 (Robert Matthew and Leslie Martin) and the National Theatre 1976 (Denys Lasdun), all connected by a series of walkways.

South Bend /'sauθ 'bend/ city on the St Joseph River, N Indiana, USA; population (1990) 105,500. Industries include the manufacture of agricultural machinery, cars, and aircraft equipment.

South Carolina /,sauθ ,kærə'laɪnə/ state in southeastern USA; nickname Palmetto State

South Carolina

area 80,600 sq km/31,112 sq mi
capital Columbia
towns Charleston, Greenville-Spartanburg
population (1990) 3,486,700
features large areas of woodland; subtropical climate in coastal areas; antebellum Charleston; Myrtle Beach and Hilton Head Island ocean resorts
products tobacco, soya beans, lumber, textiles, clothing, paper, wood pulp, chemicals, nonelectrical machinery, primary and fabricated metals
famous people John C Calhoun, 'Dizzy' Gillespie
history first Spanish settlers 1526; Charles I gave the area (known as Carolina) to Robert Heath (1575–1649), attorney general 1629; Declaration of Independence, one of the original 13 states 1776; joined the Confederacy 1860; readmitted to Union 1868.

South Dakota /'sauθ də'kəutə/ state in western USA; nickname Coyote or Sunshine State
area 199,800 sq km/77,123 sq mi
capital Pierre
cities Sioux Falls, Rapid City, Aberdeen
physical Great Plains; Black Hills (which include granite Mount Rushmore, on whose face giant relief portrait heads of former presidents Washington, Jefferson, Lincoln, and T Roosevelt are carved); Badlands
products cereals, hay, livestock, gold (second-largest US producer), meat products
population (1990) 696,000
famous people Crazy Horse, Sitting Bull, Ernest O Lawrence
history claimed by French 18th century; first white settlements 1794; became a state 1889.

South Dakota

South Glamorgan

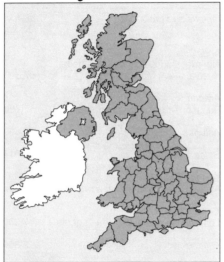

South Downs line of chalk hills in SE England, running across Sussex to Beachy Head. They face the North Downs and are used as sheep pasture.

South East Cape southernmost point of Australia, in Tasmania.

Southend-on-Sea /ˌsauθ'end ɒn 'siː/ resort in Essex, England; population (1988 est) 165,400. Industries include light engineering and boat-building. The shallow water of the Thames estuary enabled the building of a pier 2 km/1.25 mi long.

Southern and Antarctic Territories French overseas territory created 1955. It comprises the islands of *St Paul* and *Amsterdam* (67 sq km/ 26 sq mi); the *Kerguelen* and *Crozet* Islands (7,515 sq km/2,901 sq mi); and *Adélie Land* on Antarctica (432,000 sq km/165,500 sq mi). All are uninhabited, except for research stations.

Southern Ocean corridor linking the Pacific, Atlantic, and Indian oceans, all of which receive cold polar water from the world's largest ocean surface current, the Antarctic Circumpolar current, which passes through the Southern Ocean.

Southern Uplands one of the three geographical divisions of Scotland, occupying most of the hilly Scottish Borderland to the S of a geological fault line that stretches from Dunbar on the North Sea to Girvan on the Firth of Clyde. The Southern Uplands, largely formed by rocks of the Silurian and Ordovician age, are intersected by the broad valleys of the Nith and Tweed rivers.

South Georgia /'sauθ 'dʒɔːdʒə/ island in the S Atlantic, a British crown colony administered with the South Sandwich Islands; area 3,757 sq km/1,450 sq mi. South Georgia lies 1,300 km/800 mi SE of the Falkland Islands, of which it was a dependency until 1985. The British Antarctic Survey has a station on nearby Bird

Island. The island was visited by Capt James Cook 1775. The explorer Edward Shackleton is buried here. The chief settlement, Grytviken, was established as a whaling station 1904 and abandoned 1966; it was reoccupied by a small military garrison after the Falklands War 1982.

Not a tree or shrub was to be seen ... our Botanists found here only three plants.

On **South Georgia** Capt James Cook
Journal 1775

South Glamorgan /'sauθ glə'mɔːgən/ (Welsh *De Morgannwg*) county in S Wales
area 416 sq km/161 sq mi
towns Cardiff (administrative headquarters), Barry, Penarth
features mixed farming in the fertile Vale of Glamorgan; Welsh Folk Museum at St Fagans, near Cardiff
products dairy farming, industry (steel, plastics, engineering) in the Cardiff area
population (1991) 383,300
language English; 6% are Welsh-speaking
famous people Sarah Siddons, Shirley Bassey, R S Thomas.

South Holland (Dutch *Zuid Holland*) low-lying coastal province of the Netherlands, between the provinces of North Holland and Zeeland
area 2,910 sq km/1,123 sq mi
capital The Hague
towns Rotterdam, Dordrecht, Leiden, Delft, Gouda
products bulbs, horticulture, livestock, dairy products, chemicals, textiles
population (1991) 3,245,300
history once part of the former county of Holland, which was divided into two provinces 1840.

South Island larger of the two main islands of New Zealand.

South Korea see ◊Korea, South.

Southland Plain /'sauθlænd/ plain on S South Island, New Zealand, on which Invercargill stands. It is an agricultural area with sheep and dairy farming.

South Orkney Islands /'sauθ 'ɔːkni/ group of barren, uninhabited islands in British Antarctic Territory, SE of Cape Horn; area 622 sq km/ 240 sq mi. They were discovered by the naval explorer Capt George Powell 1821. Argentina, which lays claim to the islands, maintained a scientific station here 1976–82.

South Sandwich Islands /'sauθ 'sænwɪtʃ/ actively volcanic uninhabited British Dependent Territory; area 337 sq km/130 sq mi. Along with South Georgia, 750 km/470 mi to the NW, it is administered from the Falkland Islands. They were claimed by Capt Cook 1775 and named after his patron John Montagu, the 4th Earl of Sandwich. The islands were annexed by the UK 1908 and 1917. They were first formally claimed by Argentina 1948. In Dec 1976, 50 Argentine 'scientists' landed on Southern Thule and were removed June 1982. Over 21 million penguins breed on Zavadovski Island.

South Shetland Islands /'sauθ 'ʃetləndz/ archipelago of 12 uninhabited islands in the South Atlantic, forming part of British Antarctic Territory; area 4,622 sq km/1,785 sq mi.

South Shields /'sauθ 'ʃiːldz/ manufacturing port in Tyne and Wear, England, on the Tyne estuary, E of Gateshead; population (1981) 87,000. Products include electrical goods, cables, and chemicals.

South, the historically, the states of the USA bounded on the N by the Mason-Dixon Line, the Ohio River, and the E and N borders of Missouri, with an agrarian economy based on plantations worked by slaves, and which seceded from the Union 1861, beginning the American Civil War, as the Confederacy. The term is now loosely applied in a geographical and cultural sense, with Texas often regarded as part of the Southwest rather than the South.

The countries of the Third World are sometimes referred to as the South.

South Uist /'sauθ 'juːɪst/ island in the Outer Hebrides, Scotland, separated from North Uist by the island of Benbecula. There is a guided-missile range here.

Southwark /'sʌðək/ borough of S London, England; population (1986) 215,000. It is the site of the Globe Theatre (built on Bankside 1599 by Burbage, Shakespeare, and others, burned down 1613, now being rebuilt) the 12th-century Southwark Cathedral; the George Inn (last galleried inn in London); the Imperial War Museum; Dulwich College and Picture Gallery, and the Horniman Museum. The Romans built a large baths complex here about AD 120; excavations have revealed wall paintings of a high standard.

South Yorkshire

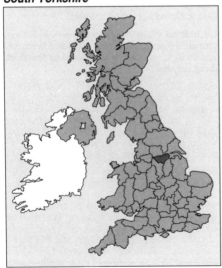

Spain: territorial divisions

regions and provinces	area in sq km
Andalusia Almería, Cádiz, Córdoba, Granada, Huelva, Jaén, Málaga, Sevilla	87,300
Aragon Huesca, Teruel, Zaragoza	47,700
Asturias	10,600
Balearic Islands	5,014
Basque Country Alava, Guipúzcoa, Vizcaya	7,300
Canary Islands Las Palmas, Santa Cruz de Tenerife	7,300
Cantabria	5,300
Castilla–La Mancha Albacete, Ciudad Real, Cuenca, Guadalajara, Toledo	79,200
Castilla–León Avila, Burgos, León, Palencia, Salamanca, Segovia, Soria, Valladolid, Zamora	94,100
Catalonia Barcelona, Gerona, Lérida, Tarragona	31,900
Extremadura Badajoz, Cáceres	41,600
Galicia La Coruña, Lugo, Orense, Pontevedra	29,400
Madrid	8,000
Murcia	11,300
Navarra	10,400
La Rioja	5,000
Valencian Community Alicante, Castellón, Valencia	23,300
Ceuta	18
Melilla	14
	499,732

South West Africa /'sauθ ˌwest 'æfrɪkə/ former name (until 1968) of ◊Namibia.

South Yorkshire /'sauθ 'jɔːkʃə/ metropolitan county of England, created 1976, originally administered by an elected council; its powers reverted to district councils from 1986 *area* 1,560 sq km/602 sq mi

towns Barnsley, Sheffield, Doncaster
features river Don; part of Peak District
National Park
products metal work, coal, dairy, sheep, arable
farming
population (1991) 1,269,300
famous people Ian Botham, Arthur Scargill.

Sovetsk /sɒv'jetsk/ town in Kaliningrad region, W
Russia. In 1807 Napoleon signed peace treaties
with Prussia and Russia here. Until 1945 it was
known as *Tilsit* and was part of East Prussia.

Soviet Central Asia former name (until 1991)
of the ◊Central Asian Republics.

Soviet Far East former name (until 1991) of
a geographical division of Asiatic Russia, now
known as the ◊Russian Far East.

Soviet Union /'səuviət 'juːniən/ alternative name
for the former ◊Union of Soviet Socialist
Republics (USSR).

Soweto /sə'weɪtəu/ (acronym for *So*uth *We*st
*To*wnship) racially segregated urban settlement
in South Africa, SW of Johannesburg; population
(1983) 915,872. It has experienced civil unrest
because of the apartheid regime.

It began as a shanty town in the 1930s and is
now the largest black city in South Africa, but
until 1976 its population could have status only
as temporary residents, serving as a workforce
for Johannesburg. There were serious riots June
1976, sparked by a ruling that Afrikaans be used
in African schools there. Reforms followed, but
riots flared up again 1985 and have continued into
the 1990s.

Spa /spɑː/ town in Liège province, Belgium; population (1991) 10,100. A health resort since the
14th century for its mineral springs, it has given
its name to similar centres elsewhere.

Spaghetti Junction nickname for a complex
system of motorway flyovers and interchanges at
Gravelly Hill, N Birmingham, in the W Midlands
of England.

Spain (*España*)
area 504,750 sq km/194,960 sq mi
capital Madrid
towns Zaragoza, Seville, Murcia, Córdoba; ports
Barcelona, Valencia, Cartagena, Málaga, Cádiz,
Vigo, Santander, Bilbao
physical central plateau with mountain ranges;
lowlands in S
territories Balearic and Canary Islands; in N
Africa: Ceuta, Melilla, Alhucemas, Chafarinas Is,
Peñón de Vélez
features rivers Ebro, Douro, Tagus, Guadiana,
Guadalquivir; Iberian Plateau (Meseta);
Pyrenees, Cantabrian Mountains, Andalusian
Mountains, Sierra Nevada
head of state King Juan Carlos I from 1975
head of government Felipe González Márquez
from 1982
political system constitutional monarchy
political parties Socialist Workers' Party
(PSOE), democratic socialist; Popular Alliance
(AP), centre-right; Christian Democrats (DC),
centrist; Liberal Party (PL), left-of-centre
exports citrus fruits, grapes, pomegranates, veg-

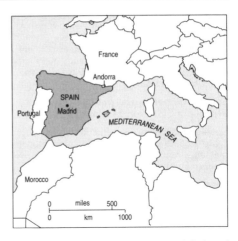

etables, wine, sherry, olive oil, canned fruit and
fish, iron ore, cork, vehicles, textiles, petroleum
products, leather goods, ceramics
currency peseta
population (1990 est) 39,623,000; growth rate
0.2% p.a.
life expectancy men 74, women 80 (1989)
languages Spanish (Castilian, official), Basque,
Catalan, Galician, Valencian, Majorcan
religion Roman Catholic 99%
literacy 97% (1989)
GNP $288 bn (1987); $4,490 per head (1984)
chronology
1936–39 Civil war; General Francisco Franco
became head of state and government; fascist party Falange declared only legal political
organization.
1947 General Franco announced restoration of
the monarchy after his death, with Prince Juan
Carlos as his successor.
1975 Franco died; succeeded as head of state by
King Juan Carlos I.
1978 New constitution adopted with Adolfo
Suárez, leader of the Democratic Centre Party, as
prime minister.
1981 Suárez resigned; succeeded by Leopoldo
Calvo Sotelo. Attempted military coup thwarted.
1982 PSOE, led by Felipe González, won
a sweeping electoral victory. Basque separatist
organization (ETA) stepped up its guerrilla
campaign.
1985 ETA's campaign spread to holiday resorts.
1986 Referendum confirmed NATO membership. Spain joined the European Economic
Community.
1988 Spain joined the Western European Union.
1989 PSOE lost seats to hold only parity after
general election. Talks between government and
ETA collapsed and truce ended.
1992 ETA's 'armed struggle' resumed.

*Spain flourished as a province, and has
declined as a kingdom.*

On **Spain** Edward Gibbon *The Decline and
Fall of the Roman Empire* 1776–88

Spalato /'spɑːlətəu/ Italian name for ▷Split, a port in Croatia.

Spalding /'spɔːldɪŋ/ market town on the river Welland, in Lincolnshire, England; population (1985 est) 19,000. There are bulb farms and a flower festival in May.

Spanish Guinea /'spænɪʃ 'gɪnɪ/ former name of the Republic of ▷Equatorial Guinea.

Spanish Main common name for the Caribbean Sea in the 16th–17th centuries, but more properly the South American mainland between the river Orinoco and Panama.

Spanish Sahara /'spænɪʃ sə'hɑːrə/ former name for ▷Western Sahara.

Spanish Town /'spænɪʃ taun/ town in Middlesex county, Jamaica; population (1982) 89,000. Founded by Diego Columbus about 1525, it was the capital of Jamaica 1535–1871.

Spartanburg /'spɑːtnbɜːg/ city in NW South Carolina, USA, NW of Columbia, in the foothills of the Blue Ridge Mountains; population (1990) 43,500. It is an agricultural centre. Its industries include food products, furniture, textiles, paper, and plumbing supplies.

Spey /speɪ/ river in Highland and Grampian regions, Scotland, rising SE of Fort Augustus, and flowing 172 km/107 mi to the Moray Firth between Lossiemouth and Buckie. It has salmon fisheries at its mouth.

Speyer /'ʃpaɪə/ (English **Spires**) ancient city on the Rhine, in Rhineland-Palatinate, Germany, 26 km/16 mi S of Mannheim; population (1983) 43,000. It was at the **Diet of Spires** 1529 that Protestantism received its name.

Spice Islands /spaɪs/ former name of the ▷Moluccas, a group of islands in the Malay Archipelago.

Spinans Hill prehistoric site on a hill near Baltinglass, County Wicklow in the Republic of Ireland, 48 km/30 miles SW of Dublin. Discovered 1992, it is the largest Bronze Age fortified hill fort yet to be found in the British Isles, covering an area of 130 hectares/320 acres.

Spires /spɪə or 'spaɪəz/ English name for the German city of ▷Speyer.

Spitalfields /'spɪtlfiːldz/ district in the Greater London borough of Tower Hamlets. It was once the home of Huguenot silk weavers.

Spithead /,spɪt'hed/ roadstead (partly sheltered anchorage) between the mainland of England and the Isle of Wight. The name is often applied to the entire eastern area of the Solent.

Spitsbergen /'spɪts,bɜːgən/ mountainous island with a deeply indented coastline in the Arctic Ocean, the main island in the Norwegian archipelago of Svalbard, 657 km/408 mi N of Norway; area 39,043 sq km/15,075 sq mi. Fishing, hunting, and coal mining are the chief economic activities. The Norwegian Institute of Polar Research operates an all-year scientific station on the west coast. Mount Newton rises to 1,713 m/5,620 ft.

Split /splɪt/ (Italian **Spalato**) port in Croatia, on the Adriatic coast; population (1981) 236,000. Industries include engineering, cement, and textiles. Split was bombed during 1991 as part of the Yugoslavia's blockade of the Croatian coast.

The Roman emperor Diocletian retired here 305.

Spokane /spəu'kæn/ city on the Spokane River, E Washington, USA; population (1990) 177,200. It is situated in a mining, timber, and rich agricultural area, and is the seat of Gonzaga University (1887).

Spoleto /spə'leɪtəu/ town in Umbria, central Italy; population (1985) 37,000. There is an annual opera and drama festival (June–July) established by Gian Carlo Menotti. It was a papal possession 1220–1860 and has Roman remains and medieval churches.

Sporades /'spɒrədiːz/ Greek island group in the Aegean Sea. The chief island of the **Northern Sporades** is Skyros. The **Southern Sporades** are more usually referred to as the Dodecanese.

Spratly Islands /'sprætli/ (Chinese **Nanshan Islands**) group of small islands, coral reefs, and sandbars dispersed over a distance of 965 km/600 mi in the South China Sea.

Used as a submarine base by the Japanese during World War II, the islands are claimed in whole or part by the People's Republic of China, Taiwan, Malaysia, Vietnam (which calls the islands **Truong Sa**), and the Philippines (which calls them **Kalayaan**). The islands are of strategic importance, commanding the sea passage from Japan to Singapore, and in 1976 oil was discovered.

Springdale /'sprɪŋdeɪl/ city in NW Arkansas, USA, NW of Little Rock; population (1990) 29,900. Industries include food and livestock processing.

Springfield /'sprɪŋfiːld/ city in Massachusetts, USA; population (1990) 157,000. It was the site (1794–1968) of the US arsenal and armoury, known for the Springfield rifle.

Springfield /'sprɪŋfiːld/ city and agricultural centre in Missouri, USA; population (1990) 140,500. Industries include electronic equipment and processed food.

Springfield /'sprɪŋfiːld/ capital and agricultural and mining centre of Illinois, USA; population (1990) 105,200. President Abraham Lincoln was born and is buried here.

Springfield /'sprɪŋfiːld/ city in W Oregon on the Willamette River, E of Eugene; population (1990) 45,000. Industries include lumber, animal feeds, and agricultural products.

Springs /sprɪŋz/ city in Transvaal, South Africa, 40 km/25 mi E of Johannesburg; population (1980) 154,000. It is a mining centre, producing gold, coal, and uranium.

Sri Lanka Democratic Socialist Republic of (*Prajathanrika Samajawadi Janarajaya Sri Lanka*) (until 1972 *Ceylon*)
area 65,600 sq km/25,328 sq mi
capital and chief port Colombo
towns Kandy; ports Jaffna, Galle, Negombo, Trincomalee
physical flat in N and around the coast; hills and mountains in S and central interior
features Adam's Peak (2,243 m/7,538 ft); ruined cities of Anuradhapura, Polonnaruwa
head of state Ranasinghe Premadasa from 1989
head of government Dingiri Banda Wijetunge from 1989
political system liberal democratic republic
political parties United National Party (UNP), right-of-centre; Sri Lanka Freedom Party (SLFP), left-of-centre; Democratic United National Front (DUNF), centre-left; Tamil United Liberation Front (TULF), Tamil autonomy; Eelam People's Revolutionary Liberation Front (EPLRF), Indian-backed Tamil-secessionist 'Tamil Tigers'
exports tea, rubber, coconut products, graphite, sapphires, rubies, other gemstones
currency Sri Lanka rupee
population (1990 est) 17,135,000 (Sinhalese 74%, Tamils 17%, Moors 7%); growth rate 1.8% p.a.
life expectancy men 67, women 72 (1989)
languages Sinhala, Tamil, English
religions Buddhist 69%, Hindu 15%, Muslim 8%, Christian 7%
literacy 87% (1988)
GNP $7.2 bn; $400 per head (1988)
chronology
1802 Ceylon became a British colony.
1948 Ceylon achieved independence from Britain within the Commonwealth.
1956 Sinhala established as the official language.
1959 Prime Minister Solomon Bandaranaike assassinated.
1972 Socialist Republic of Sri Lanka proclaimed.
1978 Presidential constitution adopted by new Jayawardene government.
1983 Tamil guerrilla violence escalated; state of emergency imposed.

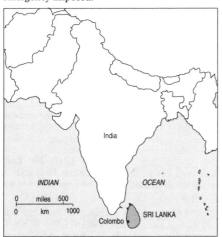

1987 President Jayawardene and Indian prime minister Rajiv Gandhi signed Colombo Accord. Violence continued despite cease-fire policed by Indian troops.
1988 Left-wing guerrillas campaigned against Indo-Sri Lankan peace pact. Prime Minister Ranasinghe Premadasa elected president.
1989 Premadasa became president; D B Wijetunge, prime minister. Leaders of the TULF and terrorist People's Liberation Front assassinated.
1990 Indian peacekeeping force withdrawn. Violence continued.
1991 March: defence minister, Ranjan Wijeratne, assassinated; Sri Lankan army killed 2,552 Tamil Tigers at Elephant Pass. October: impeachment motion against President Premadasa failed. Dec: new party, the Democratic United National Front, formed by former members of UNP.

Srinagar /srɪˈnʌgə/ summer capital of the state of Jammu and Kashmir, India; population (1981) 520,000. It is a beautiful resort, intersected by waterways, and has carpet, papier mâché, and leather industries.

Staffa /ˈstæfə/ uninhabited island in the Inner Hebrides, W of Mull. It has a rugged coastline and many caves, including Fingal's Cave.

Staffordshire /ˈstæfədʃə/ county in W central England
area 2,720 sq km/1,050 sq mi
towns Stafford (administrative headquarters), Stoke-on-Trent
features largely flat, comprising the Vale of Trent and its tributaries; Cannock Chase; Keele University 1962; Staffordshire bull terriers
products coal in the N; china and earthenware in the Potteries and the upper Trent basin
population (1991) 1,020,300
famous people Arnold Bennett, Peter de Wint, Robert Peel.

Staffs abbreviation for *Staffordshire*, an English county.

Stalin /ˈstɑːlɪn/ former name (1949–56) of the port of ⟡Varna, Bulgaria.

Stalingrad /ˈstɑːlɪŋgræd/ former name (1925–61) of the Russian city of ⟡Volgograd.

Stalinsk /ˈstɑːlɪnsk/ former name (1932–61) of ⟡Novokuznetsk, a city in Russia.

Stamboul /ˌstæmˈbuːl/ old part of the Turkish city of ⟡Istanbul, the area formerly occupied by the ancient Greek city of Byzantium.

Stamford /ˈstæmfəd/ city in SW Connecticut, USA, on Long Island Sound, NE of New York City; population (1990) 108,100. Industries include computers, hardware, rubber, plastics, and pharmaceuticals.

Stanley /ˈstænlɪ/ town on E Falkland, capital of the Falkland Islands; population (1986) 1,200. After changing its name only once between 1843 and 1982, it was renamed five times in the space of six weeks during the Falklands War April–June 1982.

Stanley Falls /ˈstænlɪ/ former name (until 1972) of ⟡Boyoma Falls, on the Zaïre River.

Staffordshire

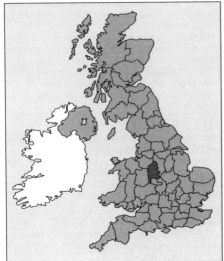

Stanley Pool /'stænli/ former name (until 1972) of ⟡Pool Malebo, on the Zaïre River.

Stanleyville /'stænlivil/ former name (until 1966) of the Zairean port of ⟡Kisangani.

Stansted /'stænsted/ London's third international airport, in Essex, England. The passenger terminal opened March 1991, designed by Norman Foster, is the centrepiece of a £400 million development, which took the airport's annual capacity to 8 million passengers.

Staten Island /'stætn/ island in New York harbour, part of New York City, USA, comprising the county of Richmond and, since 1975, the borough of Staten Island; area 155 sq km/60 sq mi.

Stavanger /stə'væŋə/ seaport and capital of Rogaland county, SW Norway, population (1991) 98,200. It has fish-canning, oil, and ship-building industries. It was founded in the 8th century and has a 12th-century cathedral.

Stavropol /'stævrəpɒl/ territory of the Russian Federation, lying N of the Caucasus Mountains; area 80,600 sq km/31,128 sq mi; population (1985) 2,715,000. The capital is Stavropol. Irrigated land produces grain and sheep are also reared. There are natural gas deposits.

Stavropol /'stævrəpɒl/ (formerly 1935–43 *Voroshilovsk*) town SE of Rostov, in the N Caucasus, SW Russia; population (1987) 306,000. Founded 1777 as a fortress town, it is now a market centre for an agricultural area and makes agricultural machinery, textiles, and food products.

Steelville town in Missouri, USA, on the edge of the Ozark Mountains; founded 1835 as a staging post. On completion of the 1990 census, it was calculated to be the demographic centre of the USA.

Steep Point /sti:p/ the westernmost extremity of Australia, in Western Australia, NW of the Murchison River.

Steiermark /'ʃtaɪəmɑːk/ German name for ⟡Styria, a province of Austria.

Stellenbosch /,steləm'bɒs/ town in Cape Province, South Africa; population (1985) 43,000. It is the centre of a wine-producing district. It was founded 1679, and is the oldest European settlement in South Africa after Cape Town.

Stepney /'stepni/ district of London, now part of the borough of Tower Hamlets, N of the Thames, and E of the City of London.

Sterea Ellas-Evvoia /'steriə 'elæs 'eviə/ region of central Greece and Euboea, occupying the southern part of the Greek mainland between the Ionian and Aegean seas and including the island of Euboea; area 24,391 sq km/9,421 sq mi; population (1991) 1,235,600. The chief city is Athens.

Stettin /ʃte'ti:n/ German name for the Polish city of ⟡Szczecin.

Steubenville /'stju:bənvɪl/ city in E Ohio, USA, on the Ohio River, S of Youngstown, near the West Virginia border; population (1990) 22,100. Industries include steel, coal, paper, and chemicals. Originally *Fort Steuben*, it was built 1786 to protect government land agents from the Indians.

Stevenage /'sti:vənɪdʒ/ town in Hertfordshire, England, 45 km/28 mi N of London; population (1985 est) 75,700. Dating from medieval times, in 1946 Stevenage was the first place chosen for development as a new town.

Stewart Island /'stju:ət/ volcanic island divided from South Island, New Zealand, by the Foveaux Strait; area 1,750 sq km/676 sq mi; population (1981) 600. Industries include farming, fishing, and granite quarrying. Oban is the main settlement.

Stirling /'stɜːlɪŋ/ administrative headquarters of Central Region, Scotland, on the river Forth; population (1981) 39,000. Industries include the manufacture of agricultural machinery, textiles, and carpets. The castle, which guarded a key crossing of the river, predates the 12th century and was long a Scottish royal residence. William Wallace won a victory at Stirling bridge 1297. Edward II of England (in raising a Scottish siege of the town) went into battle at Bannockburn 1314 and was defeated by Robert I (the Bruce).

Stirlingshire /'stɜːlɪŋʃə/ former county of Scotland. In 1975 most of it was merged with Central Region, but a southwestern section, including Kilsyth, went to Strathclyde. The area lay between the Firth of Forth and Loch Lomond, and included the Lennox Hills and the fringe of the Highlands. The county town was Stirling.

Stockholm /'stɒkhəum/ capital and industrial port of Sweden; population (1990) 674,500. It is built on a number of islands. Industries include engineering, brewing, electrical goods, paper, textiles, and pottery.

A network of bridges links the islands and the mainland; an underground railway was completed 1957. The 18th-century royal palace stands on the site of the 13th-century fortress that defended the trading settlements of Lake Mälar, around which the town first developed. The old town is well preserved and has a church (1264). The town hall was designed by Ragnar Östberg 1923. Most of Sweden's educational institutions are in Stockholm (including the Nobel Institute). The warship *Wasa* (built for King Gustavus Adolphus, 69 m/75 yd long and 52 m/57 yd high), which sank in the harbour 1628, was raised 1961 and is preserved in a museum.

Stockport /'stɒkpɔːt/ town in Greater Manchester, England; population (1981) 289,000. The rivers Tame and Goyt join here to form the Mersey. Industries include electronics, chemicals, engineering, and still some cotton textiles.

Stockton /'stɒktən/ industrial river port (agricultural machinery, food processing) on the San Joaquin River in California, USA; population (1990) 210,900.

Stockton-on-Tees /'stɒktən ɒn 'tiːz/ town and port on the river Tees, Cleveland, NE England; population (1991) 170,200. There are shipbuilding, steel, and chemical industries, and it was the starting point for the world's first passenger railway 1825.

It has the oldest railway-station building in the world, and there are many Georgian buildings.

Stoke-on-Trent /'stəuk ɒn 'trent/ city in Staffordshire, England, on the river Trent; population (1991) 244,800. It is the heart of the Potteries and a major ceramic centre. Other industries include steel, chemicals, engineering machinery, paper, rubber, and coal.

Stoke was formed 1910 from Burslem, Hanley, Longton, Stoke-upon-Trent, Fenton, and Tunstall. The ceramics factories of Minton and Wedgwood are here.

The Gladstone Pottery Museum is the only working pottery museum.

Stoke Poges /'stəuk 'pəudʒɪz/ village in Buckinghamshire, England, 3 km/2 mi N of Slough, which inspired Thomas Gray to write his 'Elegy in a Country Churchyard'; the poet is buried here.

Stornoway /'stɔːnəweɪ/ port on the island of Lewis in the Outer Hebrides, Scotland; population (1989 est) 8,400. It is the administrative centre for the Western Isles. The economy is based on fishing, tourism, tweeds, and offshore oil. Stornoway was founded by James VI of Scotland (James I of England).

Stourbridge /'stauəbrɪdʒ/ market town in West Midlands, England, on the river Stour, SW of Birmingham; population (1981) 54,700. Industries include the manufacture of glass and bricks.

Straits Settlements /'streɪts 'setlmənts/ former province of the East India Company 1826–58, a British crown colony 1867–1946; it comprised Singapore, Malacca, Penang, Cocos Islands, Christmas Island, and Labuan.

Stranraer /stræn'rɑː/ port in Dumfries and Galloway Region, Scotland; population (1981) 10,200. There is a ferry service to Larne in Northern Ireland.

Strasbourg /'stræzbuəg/ city on the river Ill, in Bas-Rhin *département*, capital of Alsace, France; population (1990) 255,900. Industries include car manufacture, tobacco, printing and publishing, and preserves. The Council of Europe meets here, and sessions of the European Parliament alternate between Strasbourg and Luxembourg.

Seized by France 1681, it was surrendered to Germany 1870–1919 and 1940–44. It has a 13th-century cathedral.

Stratford /'strætfəd/ port and industrial town in SW Ontario, Canada; population (1981) 26,000. It is the site of a Shakespeare festival.

Stratford-upon-Avon /'strætfəd əpɒn 'eɪvən/ market town on the river Avon, in Warwickshire, England; population (1986 est) 20,900. It is the birthplace of William Shakespeare.

The Royal Shakespeare Theatre 1932 replaced an earlier building 1877–79 that burned down 1926. Shakespeare's birthplace contains relics of his life and times. His grave is in the parish church; his wife Anne Hathaway's cottage is nearby. It receives over 2 million tourists a year.

Strathclyde /ˌstræθ'klaɪd/ region of Scotland
area 13,900 sq km/5,367 sq mi
towns Glasgow (administrative headquarters), Paisley, Greenock, Kilmarnock, Clydebank, Hamilton, Coatbridge, Prestwick
features includes some of Inner Hebrides; river Clyde; part of Loch Lomond; Glencoe, site of the massacre of the Macdonald clan; Breadalbane; islands: Arran, Bute, Mull

Strathclyde

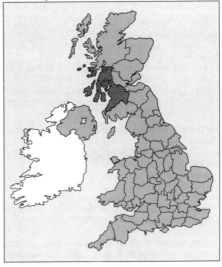

products dairy, pig, and poultry products; shipbuilding; engineering; coal from Ayr and Lanark; oil-related services
population (1991) 2,218,200, half the population of Scotland
famous people William Burrell, James Keir Hardie, David Livingstone.

Stretford /ˈstretfəd/ town in Greater Manchester, England; population (1981) 48,000. It includes the Old Trafford cricket ground. There are engineering, chemical, and textile industries.

Stromboli /ˈstrɒmbəli/ Italian island in the Tyrrhenian Sea, one of the Lipari Islands; area 12 sq km/5 sq mi. It has an active volcano, 926 m/3,039 ft high. The island produces Malmsey wine and capers.

Stuttgart /ˈʃtʊtɡɑːt/ capital of Baden-Württemberg, on the river Neckar, Germany; population (1988) 565,000. Industries include the manufacture of vehicles and electrical goods, foodstuffs, textiles, papermaking and publishing; it is a fruit-growing and wine-producing centre. There are two universities. Stuttgart was founded in the 10th century.

It is the headquarters of the US European Command (Eucom). The philosopher Hegel was born here.

Styria /ˈstɪriə/ (German **Steiermark**) Alpine province of SE Austria; area 16,400 sq km/6,330 sq mi; population (1987) 1,181,000. Its capital is Graz, and its industries include iron, steel, lignite, vehicles, electrical goods, and engineering. An independent state from 1056 until it passed to the Habsburgs in the 13th century, it was annexed by Germany 1938.

Subotica /ˈsubətitsə/ largest town in Vojvodina, NW Serbia, Yugoslavia; population (1981) 155,000. Industries include chemicals and electrical machinery.

Suceava /ˌsuːtʃiˈɑːvə/ capital of Suceava county, N Romania; population (1985) 93,000. Industries include textiles and lumber. It was a former centre of pilgrimage and capital of Moldavia 1388–1564.

Sucre /ˈsuːkreɪ/ legal capital and judicial seat of Bolivia; population (1988) 95,600. It stands on the central plateau at an altitude of 2,840 m/9,320 ft.

The city was founded 1538, its cathedral dates from 1553, and the University of San Francisco Xavier (1624) is probably the oldest in South America. The first revolt against Spanish rule in South America began here 25 May 1809.

Sudan Democratic Republic of (*Jamhuryat es-Sudan*)
area 2,505,800 sq km/967,489 sq mi
capital Khartoum
towns Omdurman, Juba, Wadi Medani, al-Obeid, Kassala, Atbara, al-Qadarif, Kosti; chief port Port Sudan
physical fertile valley of river Nile separates Libyan Desert in W from high rocky Nubian Desert in E
environment the building of the Jonglei Canal to supply water to N Sudan and Egypt threatens the grasslands of S Sudan

features Sudd swamp; largest country in Africa
head of state and government General Omar Hassan Ahmed el-Bashir from 1989
political system military republic
political parties New National Umma Party (NNUP), Islamic, nationalist; Democratic Unionist Party (DUP), moderate, nationalist; National Islamic Front, Islamic, nationalist
exports cotton, gum arabic, sesame seed, peanuts, sorghum
currency Sudanese pound
population (1990 est) 25,164,000; growth rate 2.9% p.a.
life expectancy men 51, women 55 (1989)
languages Arabic 51% (official), local languages
religions Sunni Muslim 73%, animist 18%, Christian 9% (in south)
literacy 30% (1986)
GNP $8.5 bn (1988); $330 per head (1988)
chronology
1820 Sudan ruled by Egypt.
1885 Revolt led to capture of Khartoum by self-proclaimed Mahdi.
1896–98 Anglo-Egyptian offensive led by Lord Kitchener subdued revolt.
1899 Sudan administered as an Anglo-Egyptian condominium.
1955 Civil war between Muslim north and non-Muslim south broke out.
1956 Sudan achieved independence from Britain and Egypt as a republic.
1958 Military coup replaced civilian government with Supreme Council of the Armed Forces.
1964 Civilian rule reinstated.
1969 Coup led by Col Gaafar Mohammed Nimeri established Revolutionary Command Council (RCC); name changed to Democratic Republic of Sudan.
1970 Union with Egypt agreed in principle.
1971 New constitution adopted; Nimeri confirmed as president; Sudanese Socialist Union (SSU) declared only legal party.
1972 Proposed Federation of Arab Republics, comprising Sudan, Egypt, and Syria, abandoned. Addis Ababa conference proposed autonomy for southern provinces.
1974 National assembly established.
1983 Nimeri re-elected. Shari'a (Islamic law) introduced.
1985 Nimeri deposed in a bloodless coup led

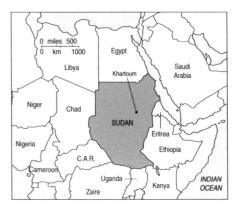

by General Swar al-Dahab; transitional military council set up. State of emergency declared.
1986 More than 40 political parties fought general election; coalition government formed.
1987 Virtual civil war with Sudan People's Liberation Army (SPLA).
1988 Al-Mahdi formed a new coalition. Another flare-up of civil war between north and south created tens of thousands of refugees. Floods made 1.5 million people homeless. Peace pact signed with SPLA.
1989 Sadiq al-Mahdi overthrown in coup led by General Omar Hassan Ahmed el-Bashir.
1991 Federal system introduced, with division of country into nine states.
1993 March: SPLA leaders announced unilateral cease-fire in ten years' war with government in Khartoum. Peace talks began.

Sudbury /ˈsʌdbəri/ city in Ontario, Canada; population (1986) 149,000. A buried meteorite here yields 90% of the world's nickel.

Sudetenland /suːˈdeɪtnlænd/ mountainous region of the Czech Republic. As part of Czechoslovakia, the region was annexed by Germany under the Munich Agreement 1938; it was returned to Czechoslovakia 1945.

Suez /ˈsuːɪz/ (Arabic *El Suweis*) port at the Red Sea terminus of the Suez Canal; population

(1985) 254,000. Industries include oil refining and the manufacture of fertilizers. It was reconstructed 1979, after the Arab-Israeli Wars.

Suez Canal /ˈsuːɪz/ artificial waterway, 160 km/100 mi long, from Port Said to Suez, linking the Mediterranean and Red seas, separating Africa from Asia, and providing the shortest eastwards sea route from Europe. It was opened 1869, nationalized 1956, blocked by Egypt during the Arab-Israeli War 1967, and not reopened until 1975.

The French Suez Canal Company was formed 1858 to execute the scheme of Ferdinand de Lesseps. The canal was opened 1869. In 1875 British prime minister Disraeli acquired a major shareholding for Britain from the khedive of Egypt. The 1888 Convention of Constantinople opened it to all nations. The Suez Canal was administered by a company with offices in Paris controlled by a council of 33 (10 of them British) until 1956 when it was forcibly nationalized by President Nasser of Egypt. The new Damietta port complex on the Mediterranean at the mouth of the canal was inaugurated 1986. The port is designed to handle 16 million tonnes of cargo.

It cannot be made, it shall not be made; but if it were made there would be a war between England and France for the possession of Egypt.

On the **Suez Canal** Lord Palmerston 1851

Suffolk /ˈsʌfək/ county of E England
area 3,800 sq km/1,467 sq mi
towns Ipswich (administrative headquarters), Bury St Edmunds, Lowestoft, Felixstowe
physical low undulating surface and flat coastline; rivers: Waveney, Alde, Deben, Orwell, Stour; part of the Norfolk Broads
features Minsmere marshland bird reserve, near Aldeburgh; site of Sutton Hoo (7th-century ship-burial); site of Sizewell B, Britain's first pressurized-water nuclear reactor plant (under construction)
products cereals, sugar beet, working horses (Suffolk punches), fertilizers, agricultural machinery
population (1991) 629,900
famous people John Constable, Thomas Gainsborough, Elizabeth Garrett Anderson, Benjamin Britten, George Crabbe.

Sukkur /suˈkuə/ or *Sakhar* port in Sind province, Pakistan, on the river Indus; population (1981) 191,000. The Sukkur River–Lloyd Barrage 1928–32 lies to the W.

Sulawesi /ˌsuːləˈweɪsi/ formerly *Celebes* island in E Indonesia, one of the Sunda Islands; area (with dependent islands) 190,000 sq km/73,000 sq mi; population (1980) 10,410,000. It is mountainous and forested and produces copra and nickel.

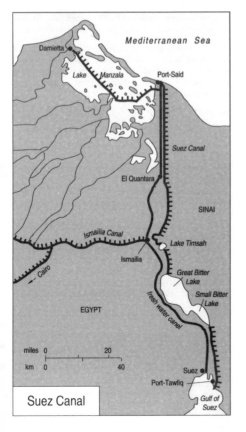

Mediterranean Sea
Damietta
Lake Manzala
Port-Said
Suez Canal
El Quantara
SINAI
Ismailia Canal
Lake Timsah
Ismailia
Cairo
Great Bitter Lake
Small Bitter Lake
EGYPT
fresh water canal
miles 0 · · · 20
km 0 · · · 40
Suez
Port-Tawfiq
Gulf of Suez

Suez Canal

Suffolk

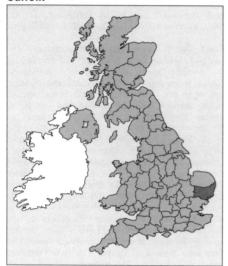

Sulu Archipelago /'suːluː ˌɑːkɪ'peləgəu/ group of about 870 islands off SW Mindanao in the Philippines, between the Sulawesi and Sulu seas; area 2,700 sq km/1,042 sq mi; population (1989) 12,507,700. The capital is Jolo, on the island (the largest) of the same name. Until 1940 the islands were an autonomous sultanate.

Sumatra /suː'mɑːtrə/ or **Sumatera** second largest island of Indonesia, one of the Sunda Islands; area 473,600 sq km/182,800 sq mi; population (1989) 36,882,000. East of a longitudinal volcanic mountain range is a wide plain; both are heavily forested. Products include rubber, rice, tobacco, tea, timber, tin, and petroleum. The main towns are Palembang, Padang, and Benkuelen.

A Hindu empire was founded in the 8th century, but Islam was introduced by Arab traders from the 13th century and by the 16th century was adopted throughout the island. Northern Sumatra is rapidly being industrialized, and the Asakan River (rising in Lake Toba) was dammed for power 1974.

Sunbelt /'sʌnbelt/ popular name for a region of the USA, S of Washington, DC, between the Pacific and Atlantic coasts, so-called because of its climate. The largest city in the Sunbelt is Los Angeles.

Sunbury-on-Thames /'sʌnbəri ɒn 'temz/ market town and boating centre in Surrey, SE England, on the river Thames; population (1981) 28,200.

Sun City /'sʌn 'sɪti/ alternative name for ⇨Mmabatho, a resort in Bophuthatswana, South Africa.

Sunda Islands /'sʌndə/ islands W of the Moluccas, in the Malay Archipelago, the greater number belonging to Indonesia. They are so named because they lie largely on the Indonesian extension of the Sunda continental shelf. The **Greater Sundas** include Borneo, Java (including the small island of Madura), Sumatra, Sulawesi, and Belitung. The **Lesser Sundas** (Indonesian *Nusa Tenggara*) are all Indonesian and include Bali, Lombok, Flores, Sumba, Sumbawa, and Timor.

Sunderland /'sʌndələnd/ port in Tyne and Wear, NE England; population (1983 est) 299,400. Industries were formerly only coalmining and shipbuilding but have now diversified to electronics, glass, and furniture. There is a polytechnic and a civic theatre, the Sunderland Empire.

Sundsvall /'sundsvæl/ port in E Sweden; population (1990) 93,800. It has oil, timber, and wood-pulp industries.

Sungari /'suŋgəri/ river in Manchuria, NE China, that joins the Amur River on the Siberian frontier; length 1,300 km/800 mi.

Sunshine Coast chain of sandy beaches on the coast of Queensland, Australia, stretching for about 100 km/60 mi from Bribie Island, N of Brisbane, to Rainbow Beach. It includes the resorts of Noosa Heads and Caloundra.

Superior, Lake /suː'pɪəriə/ largest and deepest of the Great Lakes of North America, and the second largest lake in the world; area 83,300 sq km/32,200 sq mi.

Sur /suə/ or **Soûr** Arabic name for the Lebanese port of ⇨Tyre.

Surabaya /ˌsuərə'baɪə/ port on the island of Java, Indonesia; population (1980) 2,028,000. It has oil refineries and shipyards and is a naval base.

Surat /su'rɑːt/ city in Gujarat, W India, at the mouth of the Tapti River; population (1981) 913,000. The chief industry is textiles. The first East India Company trading post in India was established here 1612.

Surinam Republic of (*Republiek Suriname*)
area 163,820 sq km/63,243 sq mi
capital Paramaribo
towns Nieuw Nickerie, Brokopondo, Nieuw Amsterdam
physical hilly and forested, with flat and narrow coastal plain
features Suriname River
head of state and government Ronald Venetiaan from 1991
political system emergent democratic republic
political parties Party for National Unity and Solidarity (KTPI)*, Indonesian, left-of-centre; Surinam National Party (NPS)*, Creole, left-of-centre; Progressive Reform Party (VHP)*, Indian, left-of-centre; New Front For Democracy (NF) *members of Front for Democracy and Development (FDD)
exports alumina, aluminium, bauxite, rice, timber
currency Surinam guilder
population (1990 est) 408,000 (Hindu 37%, Creole 31%, Javanese 15%); growth rate 1.1% p.a.
life expectancy men 66, women 71 (1989)
languages Dutch (official), Sranan (creole), English, others

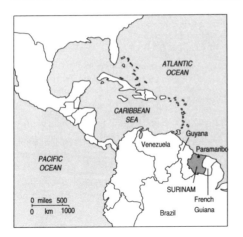

religion Christian 30%, Hindu 27%, Muslim 20%
literacy 65% (1989)
GNP $1.1 bn (1987); $2,920 per head (1985)
chronology
1667 Became a Dutch colony.
1954 Achieved internal self-government as Dutch Guiana.
1975 Independence achieved from the Netherlands, with Dr Johan Ferrier as president and Henck Arron as prime minister; 40% of the population emigrated to the Netherlands.
1980 Arron's government overthrown in army coup; Ferrier refused to recognize military regime; appointed Dr Henk Chin A Sen to lead civilian administration. Army replaced Ferrier with Dr Chin A Sen.
1982 Army, led by Lt Col Desi Bouterse, seized power, setting up a Revolutionary People's Front.
1985 Ban on political activities lifted.
1986 Antigovernment rebels brought economic chaos to Surinam.
1987 New constitution approved.
1988 Ramsewak Shankar elected president.
1989 Bouterse rejected peace accord reached by President Shankar with guerrilla insurgents, vowed to continue fighting.
1990 Shankar deposed in army coup.
1991 Johan Kraag became interim president. New Front for Democracy won assembly majority. Ronald Venetiaan elected president.
1992 Peace accord with guerrilla groups.

Surrey /'sʌri/ county in S England
area 1,660 sq km/641 sq mi
towns Kingston upon Thames (administrative headquarters), Guildford, Woking
features rivers: Thames, Mole, Wey; hills: Box and Leith; North Downs; Runnymede, Thameside site of the signing of Magna Carta; Yehudi Menuhin School; Kew Palace and Royal Botanic Gardens; in 1989 it was the most affluent county in Britain—average income 40% above national average
products vegetables, agricultural products, service industries
population (1991) 997,000

famous people Eric Clapton, John Galsworthy, Aldous Huxley, Laurence Olivier.

Surtsey /'sɜːtsi/ volcanic island 20 km/12 mi SW of Heimaey in the Westman Islands of Iceland. The island was created by an underwater volcanic eruption Nov 1963.

Susa /'suːzə/ (French *Sousse*) port and commercial centre in NE Tunisia; population (1984) 83,500. It was founded by the Phoenicians and has Roman ruins.

Susquehanna /ˌsʌskwɪˈhænə/ river rising in central New York State, USA, and flowing 715 km/ 444 mi to Chesapeake Bay. It is used for hydroelectric power. On the strength of its musical name, Samuel Coleridge planned to establish a communal settlement here with his fellow poet Robert Southey.

Sussex /'sʌsɪks/ former county of England, on the south coast, now divided into ⟠East Sussex and ⟠West Sussex.
  According to tradition, the Saxon Ella landed here 477, defeated the inhabitants, and founded the kingdom of the South Saxons, which was absorbed by Wessex 825.

Sutherlandshire /'sʌðələndʃə/ former county of Scotland, merged 1975 with Highland Region. Dornoch was the administrative headquarters.

Sutlej /'sʌtlɪdʒ/ river in Pakistan, a tributary of the river Indus; length 1,370 km/851 mi.

Sutton /'sʌtn/ borough of S Greater London; population (1981) 168,000. It was the site of Nonsuch Palace built by Henry VIII, which was demolished in the 17th century.

Sutton Coldfield /'sʌtn 'kəʊldfiːld/ residential part of the West Midlands conurbation around Birmingham, England; population (1981) 86,500.

Surrey

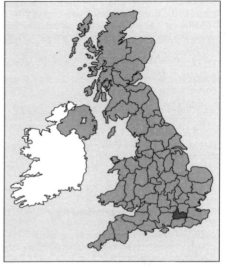

Sutton-in-Ashfield /'sʌtn ɪn 'æʃfiːld/ town in Nottinghamshire, England; population (1990 est) 40,200. It has coal, hosiery, and plastics industries.

Suva /'suːvə/ capital and industrial port of Fiji, on Viti Levu; population (1981) 68,000. It produces soap and coconut oil.

Suzhou /ˌsuːˈdʒəu/ or **Soochow** (formerly 1912–49 **Wuhsien**) city S of the Yangtze river delta and E of the Grand Canal, in Jiangsu province, China; population (1983) 670,000. It has embroidery and jade-carving traditions and Shizilin and Zhuozheng gardens. The city dates from about 1000 BC, and the name Suzhou from the 7th century AD; it was reputedly visited by the Venetian Marco Polo.

Svalbard /'svɑːlbɑː/ Norwegian archipelago in the Arctic Ocean. The main island is Spitsbergen; other islands include North East Land, Edge Island, Barents Island, and Prince Charles Foreland.
area 62,000 sq km/23,938 sq mi
towns Longyearbyen on Spitsbergen
features weather and research stations; wildlife includes walrus and polar bear; fossil palms show that it was in the tropics 40 million years ago
products coal, phosphates, asbestos, iron ore, and galena—all mined by Russia, Kazakhstan, and Norway
population (1982) 4,000; 62% Russian, 36% Norwegian
history under the **Svalbard Treaty** 1925, Norway has sovereignty, but allows free scientific and economic access to others.

Sverdlovsk /sviəd'lɒvsk/ former name (1924–91) of ♢Ekaterinburg, a town in Russia.

Swanage /'swɒnɪdʒ/ town on the Isle of Purbeck, Dorset, England.

Swansea /'swɒnzi/ (Welsh **Abertawe**) port and administrative headquarters of West Glamorgan, S Wales, at the mouth of the river Tawe where it meets the Bristol Channel; population (1981) 168,000. It has oil refineries and metallurgical industries and manufactures stained glass (since 1936).
It is the vehicle-licensing centre of the UK.

Swatow /ˌswɑːˈtau/ another name for the Chinese port of ♢Shantou.

Swaziland Kingdom of (*Umbuso weSwatini*)
area 17,400 sq km/6,716 sq mi
capital Mbabane
towns Manzini, Big Bend
physical central valley; mountains in W (Highveld); plateau in E (Lowveld and Lubombo plateau)
features landlocked enclave between South Africa and Mozambique
head of state and government King Mswati III from 1986
political system near-absolute monarchy
political party Imbokodvo National Movement (INM), nationalist monarchist

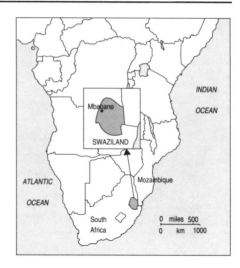

exports sugar, canned fruit, wood pulp, asbestos
currency lilangeni
population (1990 est) 779,000; growth rate 3% p.a.
life expectancy men 47, women 54 (1989)
languages Swazi 90%, English (both official)
religions Christian 57%, animist
literacy men 70%, women 66% (1985 est)
GNP $539 million; $750 per head (1987)
chronology
1903 Swaziland became a special High Commission territory.
1967 Achieved internal self-government.
1968 Independence achieved from Britain, within the Commonwealth, as the Kingdom of Swaziland, with King Sobhuza II as head of state.
1973 The king suspended the constitution and assumed absolute powers.
1978 New constitution adopted.
1982 King Sobhuza died; his place was taken by one of his wives, Dzeliwe, until his son, Prince Makhosetive, reached the age of 21.
1983 Queen Dzeliwe ousted by another wife, Ntombi.
1984 After royal power struggle, it was announced that the crown prince would become king at 18.
1986 Crown prince formally invested as King Mswati III.
1987 Power struggle developed between advisory council Liqoqo and Queen Ntombi over accession of king. Mswati dissolved parliament; new government elected with Sotsha Dlamini as prime minister.
1991 Calls for democratic reform.
1992 Food shortages suffered as result of widespread drought in S Africa.

Sweden Kingdom of (*Konungariket Sverige*)
area 450,000 sq km/173,745 sq mi
capital Stockholm
towns Göteborg, Malmö, Uppsala, Norrköping, Västerås
physical mountains in W; plains in S; thickly forested; more than 20,000 islands off the Stockholm coast

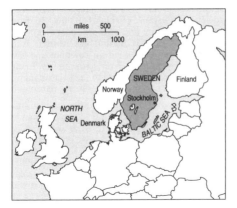

environment of the country's 90,000 lakes,
20,000 are affected by acid rain; 4,000 are so
severely acidified that no fish are thought to
survive in them
features lakes, including Vänern, Vättern,
Mälaren, Hjälmaren; islands of Öland and
Gotland; wild elk
head of state King Carl XVI Gustaf from 1973
head of government Carl Bildt from 1991
political system constitutional monarchy
political parties Social Democratic Labour
party (SAP), moderate, left-of-centre; Moderate
Party, right-of-centre; Liberal Party, centre-left;
Centre Party, centrist; Christian Democratic
Party, Christian, centrist; Left (Communist)
Party, European, Marxist; Green, ecological
exports aircraft, vehicles, ballbearings, drills,
missiles, electronics, petrochemicals, textiles,
furnishings, ornamental glass, paper, iron and
steel
currency krona
population (1990 est) 8,407,000 (including
17,000 Saami (Lapps) and 1.2 million postwar
immigrants from Finland, Turkey, Yugoslavia,
Greece, Iran, other Nordic countries); growth
rate 0.1% p.a.
life expectancy men 74, women 81 (1989)
languages Swedish; there are Finnish-and
Saami-speaking minorities
religion Lutheran (official) 95%
literacy 99% (1989)
GNP $179 bn; $11,783 per head (1989)
chronology
12th century United as an independent nation.
1397–1520 Under Danish rule.
1914–45 Neutral in both world wars.
1951–76 Social Democratic Labour Party (SAP)
in power.
1969 Olof Palme became SAP leader and prime
minister.
1971 Constitution amended, creating a single-
chamber Riksdag, the governing body.
1975 Monarch's last constitutional powers
removed.
1976 Thorbjörn Fälldin, leader of the Centre
Party, became prime minister, heading centre-
right coalition.
1982 SAP, led by Palme, returned to power.
1985 SAP formed minority government, with
Communist support.

1986 Olof Palme murdered. Ingvar Carlsson
became prime minister and SAP party leader.
1988 SAP re-elected with reduced majority;
Green Party gained representation in Riksdag.
1990 SAP government resigned. Sweden to apply
for European Community (EC) membership.
1991 Formal application for EC membership
submitted. Election defeat for SAP; Carlsson
resigned. Coalition government formed; Carl
Bildt became new prime minister.
1992 Across-parties agreement to solve economic
problems.

Swindon /'swɪndən/ town in Wiltshire, 124 km/
77 mi W of London, England; population (1985
est) 129,300. The site of a major railway engineering works 1841–1986, the town has diversified
since 1950 into heavy engineering, electronics,
and electrical manufacture.

Switzerland Swiss Confederation (German
Schweiz, French *Suisse*, Romansch *Svizzera*)
area 41,300 sq km/15,946 sq mi
capital Bern
towns Zürich, Geneva, Lausanne; river port
Basel (on the Rhine)
physical most mountainous country in Europe (Alps and Jura mountains); highest peak
Dufourspitze 4,634 m/15,203 ft in Apennines
environment an estimated 43% of coniferous
trees, particularly in the central alpine region,
have been killed by acid rain, 90% of which comes
from other countries. Over 50% of bird species
are classified as threatened
features winter sports area of the upper valley
of the river Inn (Engadine); lakes Maggiore,
Lucerne, Geneva, Constance
head of state and government René Felber
from 1992
government federal democratic republic
political parties Radical Democratic Party
(FDP), radical, centre-left; Social Democratic
Party (SPS), moderate, left-of-centre; Christian
Democratic Party (PDC), Christian, moderate,
centrist; People's Party (SVP), centre-left; Liberal
Party (PLS), federalist, centre-left; Green Party,
ecological

exports electrical goods, chemicals, pharmaceuticals, watches, precision instruments, confectionery
currency Swiss franc
population (1990 est) 6,628,000; growth rate 0.2% p.a.
life expectancy men 74, women 82 (1989)
languages German 65%, French 18%, Italian 12%, Romansch 1% (all official)
religions Roman Catholic 50%, Protestant 48%
literacy 99% (1989)
GNP $111 bn (1988); $26,309 per head (1987)
chronology
1648 Became independent of the Holy Roman Empire.
1798–1815 Helvetic Republic established by French revolutionary armies.
1847 Civil war resulted in greater centralization.
1874 Principle of the referendum introduced.
1971 Women given the vote in federal elections.
1984 First female cabinet minister appointed.
1986 Referendum rejected proposal for membership of United Nations.
1989 Referendum supported abolition of citizen army and military service requirements.
1991 18-year-olds allowed to vote for first time in national elections. Four-party coalition remained in power.
1992 René Felber elected president with Adolf Ogi as vice president. Decision to apply for full European Community (EC) membership.

That country which a Philosopher would perhaps prefer to the rest of Europe.

On **Switzerland** Edward Gibbon, letter of 1775

Sydney /'sɪdni/ capital and port of New South Wales, Australia; population (1990) 3,656,900. Industries include engineering, oil refining, electronics, scientific equipment, chemicals, clothing, and furniture. It is a financial centre, and has three universities. The 19th-century Museum of Applied Arts and Sciences is the most popular museum in Australia.

Originally a British penal colony 1788, Sydney developed rapidly following the discovery of gold in the surrounding area. The main streets still follow the lines of the original wagon tracks, and the Regency Bligh House survives. Modern landmarks are the harbour bridge (single span 503.5 m/1,652 ft) 1923–32, Opera House 1959–73, and Centre Point Tower 1980.

God made the harbour, and that all right, but Satan made Sydney.

On **Sydney** local saying

Syktyvkar /ˌsɪktɪf'kɑ:/ capital of the autonomous republic of Komi, N central Russia; population (1987) 224,000. Industries include timber, paper, and tanning. It was founded 1740 as a Russian colony.

Sylhet /sɪl'het/ capital of Sylhet region, NE Bangladesh; population (1981) 168,000. It is a tea-growing centre and also produces rice, jute, and sugar. There is natural gas nearby. It is the former capital of a Hindu kingdom and was conquered by Muslims in the 14th century. In the 1971 civil war, which led to the establishment of Bangladesh, it was the scene of heavy fighting.

Syracuse /'saɪrəkjuːz/ (Italian *Siracusa*) industrial port (chemicals, salt) in E Sicily; population (1988) 124,000. It has a cathedral and remains of temples, aqueducts, catacombs, and an amphitheatre. Founded 734 BC by the Corinthians, it became a centre of Greek culture under the elder and younger Dionysius. After a three-year siege it was taken by Rome 212 BC. In AD 878 it was destroyed by the Arabs, and the rebuilt town came under Norman rule in the 11th century.

Syracuse /'sɪrəkjuːs/ industrial city on Lake Onondaga, in New York State, USA; population (1990) 163,900. Industries include the manufacture of electrical and other machinery, paper, and food processing. There are canal links with the Great Lakes, and the Hudson and St Lawrence rivers.

Syracuse was settled in the 1780s on the site of a former Iroquois capital and developed as a salt-mining centre.

Syria Syrian Arab Republic (*al-Jamhuria al-Arabya as-Suriya*)
area 185,200 sq km/71,506 sq mi
capital Damascus
towns Aleppo, Homs, Hama; chief port Latakia
physical mountains alternate with fertile plains and desert areas; Euphrates River
features Mount Hermon, Golan Heights; crusader castles (including Krak des Chevaliers); Phoenician city sites (Ugarit), ruins of ancient Palmyra
head of state and government Hafez al-Assad from 1971
political system socialist republic
political parties National Progressive Front (NPF), pro-Arab, socialist; Communist Action Party, socialist
exports cotton, cereals, oil, phosphates, tobacco
currency Syrian pound

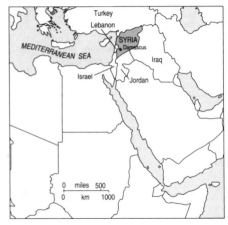

population (1990 est) 12,471,000; growth rate 3.5% p.a.
life expectancy men 67, women 69 (1989)
languages Arabic 89% (official), Kurdish 6%, Armenian 3%
religions Sunni Muslim 74%; ruling minority Alawite, and other Islamic sects 16%; Christian 10%
literacy men 76%, women 43% (1985 est)
GNP $17 bn (1986); $702 per head
chronology
1946 Achieved full independence from France.
1958 Merged with Egypt to form the United Arab Republic (UAR).
1961 UAR disintegrated.
1967 Six-Day War resulted in the loss of territory to Israel.
1970–71 Syria supported Palestinian guerrillas against Jordanian troops.
1971 Following a bloodless coup, Hafez al-Assad became president.
1973 Israel consolidated its control of the Golan Heights after the Yom Kippur War.
1976 Substantial numbers of troops committed to the civil war in Lebanon.
1978 Assad re-elected.
1981–82 Further military engagements in Lebanon.
1982 Islamic militant uprising suppressed; 5,000 dead.
1984 Presidents Assad and Gemayel approved plans for government of national unity in Lebanon.
1985 Assad secured the release of 39 US hostages held in an aircraft hijacked by extremist Shi'ite group, Hezbollah. Assad re-elected.
1987 Improved relations with USA and attempts to secure the release of Western hostages in Lebanon.
1989 Diplomatic relations with Morocco restored. Continued fighting in Lebanon; Syrian forces reinforced in Lebanon; diplomatic relations with Egypt restored.
1990 Diplomatic relations with Britain restored.
1991 Syria fought against Iraq in Gulf War. President Assad agreed to US Middle East peace plan. Assad re-elected as president.
1992 European Community (EC) blocked aid to Syria, due to poor human rights record.

... a symbol of the permanent physical conditions that run throughout history; the permanent geographical limits of human settlement, government and war.

On **Syria** Hillaire Belloc *Places* 1942

Szczecin /'ʃtʃetʃiːn/ (German *Stettin*) industrial (shipbuilding, fish processing, synthetic fibres, tools, iron) port on the river Oder, in NW Poland; population (1990) 413,400.

A Hanseatic port from 1278, it was Swedish from 1648 until 1720, when it was taken by Prussia. It was Germany's chief Baltic port until captured by the Russians 1945, and came under Polish administration. Catherine the Great of Russia was born here.

Szechwan /,seɪtʃ'wɑːn/ alternative spelling for the central Chinese province of ⇔Sichuan.

Szeged /'seged/ port on the river Tisza and capital of Csongrad county, S Hungary; population (1988) 188,000. The chief industry is textiles, and the port trades in timber and salt.

Székesfehérvár /'seɪkeʃ,feheəvɑː/ industrial city (metal products) in W central Hungary; population (1988) 113,000. It is a market centre for wine, tobacco, and fruit.

Tabah /'tɑːbə/ or **Taba** small area of disputed territory, 1 km/0.6 mi long, between Eilat (Israel) to the E and the Sinai Desert (Egypt) to the W on the Red Sea. Under an Anglo-Egyptian-Turkish agreement 1906, the border ran through Tabah; under a British survey of 1915 headed by T E Lawrence (of Arabia), who made 'adjustments' allegedly under British government orders, runs to the E. Taken by Israel 1967, Tabah was returned to Egypt 1989.

Table Bay /'teɪbəl 'beɪ/ inlet on the southwest coast of the Cape of Good Hope, South Africa, on which Cape Town stands. It is overlooked by **Table Mountain** (highest point Maclear's Beacon 1,087 m/3,568 ft), the cloud that often hangs above it being known as the 'tablecloth'.

Tabora /tə'bɔːrə/ trading centre in W Tanzania; population (1978) 67,400. It was founded about 1820 by Arab traders of slaves and ivory.

Tabriz /tæ'briːz/ city in NW Iran; population (1986) 972,000. Industries include metal casting, carpets, cotton, and silk textiles.

Tacna /'tæknə/ city in S Peru; population (1988) 138,000. It is undergoing industrialization. In 1880 Chile defeated a combined Peruvian–Bolivian army nearby and occupied Tacna until 1929.

Tacoma /tə'kəumə/ port in Washington State, USA, on Puget Sound, 40 km/25 mi S of Seattle; population (1990) 176,700. It is a lumber and shipping centre, with fishing and boat-building industries. Industries include primary metals, wood and paper products, chemicals, and processed foods.

Founded 1868, the city developed after being chosen as the terminus of the North Pacific Railroad 1873.

Taegu /,teɪ'guː/ third largest city in South Korea, situated between Seoul and Pusan; population (1990) 2,228,800. Nearby is the Haeinsa Temple, one of the country's largest monasteries and repository of the *Triptaka Koreana*, a collection of 80,000 wood blocks on which the Buddhist sculptures are carved. Grain, fruit, textiles, and tobacco are produced.

Taejon /,teɪ'dʒɒn/ (Korean 'large rice paddy') capital of South Chungchong province, central South Korea; population (1990) 1,062,100. Korea's tallest standing Buddha and oldest wooden building are found NE of the city at Popchusa in the Mount Songnisan National Park.

Taganrog /,tægən'rɒg/ port in the northeast corner of the Sea of Azov, SW Russia, W of Rostov; industries include iron, steel, metal goods, aircraft, machinery, and shoes; population (1987) 295,000. A museum commemorates the playwright Anton Chekhov, who was born here.

Tagus /'teɪgəs/ (Spanish *Tajo*, Portuguese *Tejo*) river rising in Aragon, Spain, and reaching the Atlantic Ocean at Lisbon, Portugal; length 1,007 km/626 mi. At Lisbon it is crossed by the April 25 (formerly Salazar) Bridge, so named in honour of the 1974 revolution. The *Tagus-Segura* irrigation scheme serves the rainless Murcia/Alicante region for early fruit and vegetable growing.

Tahiti /tə'hiːti/ largest of the Society Islands, in French Polynesia; area 1,042 sq km/402 sq mi; population (1983) 116,000. Its capital is Papeete. Tahiti was visited by Capt James Cook 1769 and by Admiral Bligh of the *Bounty* 1788. It came under French control 1843 and became a colony 1880.

The island to which every voyager has offered up his tribute of imagination.

On **Tahiti** Charles Darwin 1830s

Taipei /,taɪ'peɪ/ or **Taibei** capital and commercial centre of Taiwan; population (1990) 2,719,700. Industries include electronics, plastics, textiles, and machinery. The National Palace Museum 1965 houses the world's greatest collection of Chinese art, taken there from the mainland 1948.

Taiwan Republic of China (*Chung Hua Min Kuo*)
area 36,179 sq km/13,965 sq mi
capital Taipei
towns ports Kaohsiung, Keelung
physical island (formerly Formosa) off People's Republic of China; mountainous, with lowlands in W
environment industrialization has taken its toll: an estimated 30% of the annual rice crop is dangerously contaminated with mercury, cadmium, and other heavy metals
features Penghu (Pescadores), Jinmen (Quemoy), Mazu (Matsu) islands
head of state Lee Teng-hui from 1988
head of government Hau Pei-tsun from 1990
political system emergent democracy
political parties Nationalist Party of China (Kuomintang: KMT), anticommunist, Chinese nationalist; Democratic Progressive Party (DPP), centrist-pluralist, pro-self-determination grouping; Workers' Party (Kuntang), left-of-centre
exports textiles, steel, plastics, electronics, foodstuffs

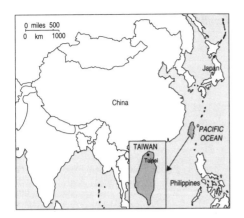

currency New Taiwan dollar
population (1990) 20,454,000 (84%
Taiwanese, 14% mainlanders); growth rate
1.4% p.a.
life expectancy 70 men, 75 women (1986)
language Mandarin Chinese (official); Taiwan,
Hakka dialects
religions officially atheist; Taoist, Confucian,
Buddhist, Christian
literacy 90% (1988)
GNP $119.1 bn; $6,200 per head (1988)
chronology
1683 Taiwan (Formosa) annexed by China.
1895 Ceded to Japan.
1945 Recovered by China.
1949 Flight of Nationalist government to Taiwan
after Chinese communist revolution.
1954 US-Taiwanese mutual defence treaty.
1971 Expulsion from United Nations.
1972 Commencement of legislature elections.
1975 President Chiang Kai-shek died; replaced
as Kuomintang leader by his son, Chiang
Ching-kuo.
1979 USA severed diplomatic relations and
annulled 1954 security pact.
1986 Democratic Progressive Party (DPP)
formed as opposition to the nationalist
Kuomintang.
1987 Martial law lifted; opposition parties
legalized; press restrictions lifted.
1988 President Chiang Ching-kuo died; replaced
by Taiwanese-born Lee Teng-hui.
1989 Kuomintang (KMT) won assembly elections.
1990 Formal move towards normalization of
relations with China. Hau Pei-tsun became prime
minister.
1991 President Lee Teng-hui declared end
to state of civil war with China. Constitution amended. KMT won landslide victory in
assembly elections.
1992 Diplomatic relations with South Korea
broken.

Taiyuan /ˌtaɪjuˈɑːn/ capital of Shanxi province, on
the river Fen He, NE China; industries include
iron, steel, agricultural machinery, and textiles;
population (1989) 1,900,000. It is a walled city,

founded in the 5th century AD, and is the seat of
Shanxi University.

Ta'iz /taːˈɪz/ city in N Yemen, at the centre
of a coffee-growing region; population (1980)
119,500. Cotton, leather, and jewellery are also
produced.

Tajikistan Republic of
area 143,100 sq km/55,251 sq mi
capital Dushanbe
towns Khodzhent (formerly Leninabad),
Kurgan-Tyube, Kulyab
physical mountainous, more than half of its
territory lying above 3,000 m/10,000 ft; huge
mountain glaciers which are the source of many
rapid rivers
features Pik Kommuniza (Communism Peak);
health resorts and mineral springs
head of state Akbasho Iskandrov from 1992
head of government Abdumalik Abdullajanov
from 1992
political system emergent democracy
political parties Socialist (formerly Communist) Party of Tajikistan (SPT); Democratic Party;
Islamic Revival Party
products fruit, cereals, cotton, cattle, sheep,
silks, carpets, coal, lead, zinc, chemicals, oil, gas
population (1990) 5,300,000 (63% Tajik, 24%
Uzbek, 8% Russian, 1% Tatar, 1% Kyrgyz, 1%
Ukrainian)
language Tajik, similar to Farsi (Persian)
religion Sunni Muslim
chronology
1921 Part of Turkestan Soviet Socialist Autonomous Republic.
1929 Became a constituent republic of USSR.
1990 Ethnic Tajik/Armenian conflict in
Dushanbe resulted in rioting against Communist
Party of Tajikistan (TCP); state of emergency and
curfew imposed.
1991 Jan: curfew lifted in Dushanbe. March:
maintenance of Union endorsed in referendum.

Aug: President Makhkamov forced to resign after failed anti-Gorbachev coup; TCP broke links with Moscow. Sept: declared independence; Nabiyev elected president; TCP renamed Socialist Party of Tajikistan (SPT); state of emergency declared. Dec: joined new Commonwealth of Independent States (CIS).
1992 Jan: admitted into Conference for Security and Cooperation in Europe (CSCE). Nabiyev temporarily ousted; state of emergency lifted. Feb: joined the Muslim Economic Cooperation Organization (ECO). March: admitted into United Nations (UN); US diplomatic recognition achieved. May: coalition government formed. Sept: Nabiyev forced to resign; replaced by Akbasho Iskandrov

Tajo /'tɑːxəu/ Spanish name for the river ▷Tagus.

Takao /tæ'kau/ Japanese name for ▷Kaohsiung, a city on the west coast of Taiwan.

Takoradi /ˌtɑːkə'rɑːdi/ port in Ghana, administered with ▷Sekondi.

Talavera de la Reina /ˌtælə'veərə delə 'reɪnə/ town in Castilla-Léon, central Spain, on the river Tagus, 120 km/75 mi SW of Madrid; population (1991) 68,600. It produces soap, pharmaceuticals, and textiles. Spanish and British forces defeated the French here in the Peninsular War 1809.

Talcahuano /ˌtælkə'wɑːnəu/ port and chief naval base in Biobio region, Chile; industries include oil refining and timber; population (1987) 231,000.

Talien /'tælien/ part of the port of ▷Lüda, China.

Tallahassee /ˌtælə'hæsi/ (Cree Indian 'old town') capital of Florida, USA; an agricultural and lumbering centre; population (1990) 124,800. The Spanish explorer Hernando de Soto founded an Indian settlement here 1539, and the site was chosen as the Florida territorial capital '1821.
During the Civil War, Tallahassee was the only Confederate capital E of the Mississippi River not captured by Union troops, and the city still has many pre-Civil War mansions.

Tallinn /'tælɪn/ (German *Reval*) naval port and capital of Estonia; population (1987) 478,000. Industries include electrical and oil-drilling machinery, textiles, and paper. Founded 1219, it was a member of the Hanseatic League; it passed to Sweden 1561 and to Russia 1750. Vyshgorod castle (13th century) and other medieval buildings remain. It is a yachting centre.

Tamale /tə'mɑːli/ town in NE Ghana; population (1982) 227,000. It is a commercial centre, dealing in rice, cotton, and peanuts.

Tamar /'teɪmə/ river rising in N Cornwall, England, and flowing to Plymouth Sound; for most of its 97 km/60 mi length it forms the Devon–Cornwall border.

Tamar /'teɪmɑː/ river flowing into Bass Strait, Tasmania, formed by the union of the North and South Esk; length 65 km/40 mi.

Tambov /tæm'bɒv/ city in W central Russia; population (1987) 305,000. Industries include

Tamil Nadu

engineering, flour milling, and the manufacture of rubber and synthetic chemicals.

Tamil Nadu /'tæmɪl nɑː'duː/ (formerly until 1968 *Madras State*) state of SE India
area 130,100 sq km/50,219 sq mi
capital Madras
products mainly industrial: cotton, textiles, silk, electrical machinery, tractors, rubber, sugar refining
population (1991) 55,638,300
language Tamil
history the present state was formed 1956. Tamil Nadu comprises part of the former British Madras presidency (later province) formed from areas taken from France and Tipu Sahib, the sultan of Mysore, in the 18th century, which became a state of the Republic of India 1950. The NE was detached to form Andhra Pradesh 1953; in 1956 other areas went to Kerala and Mysore (now Karnataka), and the Laccadive Islands (now Lakshadweep) became a separate Union Territory.

Tampa /'tæmpə/ port and resort on Tampa Bay in W Florida, USA; population (1990) 280,000. Industries include fruit and vegetable canning, shipbuilding, and the manufacture of fertilizers, clothing, beer, and cigars.
Tampa was settled 1823, and a fort was built the next year that was taken from Confederate forces by Union troops in the Civil War.

Tampere /'tæmpəreɪ/ (Swedish *Tammerfors*) city in SW Finland; population (1990) 172,600, metropolitan area 258,000. Industries include textiles, paper, footwear, and turbines. It is the second largest city in Finland.

Tampico /tæm'piːkəu/ port on the Rio Pánuco, 10 km/6 mi from the Gulf of Mexico, in Tamaulipas state, Mexico; population (1980) 268,000. Industries include oil refining and fishing.

Tamworth /'tæmwɜːθ/ town in Staffordshire, England, on the river Tame, NE of Birmingham;

industries include engineering, paper, and clothing; population (1989 est) 65,000.

Tamworth /'tæmwɜːθ/ dairying centre with furniture industry in New South Wales, Australia, on the river Peel; population (1984) 34,000.

Tana /'tɑːnə/ lake in Ethiopia, 1,800 m/5,900 ft above sea level; area 3,600 sq km/1,390 sq mi. It is the source of the Blue Nile.

Tananarive /təˌnænə'riːv/ former name for ▷Antananarivo, the capital of Madagascar.

Tanga /'tæŋgə/ seaport and capital of Tanga region, NE Tanzania, on the Indian Ocean; population (1978) 103,000. The port trades in sisal, fruit, cocoa, tea, and fish.

Tanganyika /ˌtæŋgən'jiːkə/ former British colony in E Africa, which now forms the mainland of ▷Tanzania.

Tanganyika, Lake /ˌtæŋgən'jiːkə/ lake 772 m/ 2,534 ft above sea level in the Great Rift Valley, E Africa, with Zaire to the W, Zambia to the S, and Tanzania and Burundi to the E. It is about 645 km/400 mi long, with an area of about 31,000 sq km/12,000 sq mi, and is the deepest lake (1,435 m/4,710 ft) in Africa. The mountains around its shores rise to about 2,700 m/8,860 ft. The chief ports on the lake are Bujumbura (Burundi), Kigoma (Tanzania), and Kalémié (Zaire).

Tangier /tæn'dʒɪə/ or *Tangiers* or *Tanger* port in N Morocco, on the Strait of Gibraltar; population (1982) 436,227. It was a Phoenician trading centre in the 15th century BC. Captured by the Portuguese 1471, it passed to England 1662 as part of the dowry of Catherine of Braganza, but was abandoned 1684, and later became a lair of Barbary Coast pirates. From 1923 Tangier and a small surrounding enclave became an international zone, administered by Spain 1940–45. In 1956 it was transferred to independent Morocco and became a free port 1962.

Tangier the White ... posted like a sentinel on the most northern part of Africa.

On **Tangier** Pierre Loti *Morocco* 1914

Tangshan /ˌtæŋ'ʃæn/ industrial city in Hebei province, China; population (1986) 1,390,000. Almost destroyed by an earthquake 1976, with 200,000 killed, it was rebuilt on a new site, coal seams being opened up under the old city.

Tannu-Tuva /'tænuː 'tuːvə/ former independent republic in NE Asia; see ▷Tuva.

Tanzania United Republic of (*Jamhuri ya Muungano wa Tanzania*)
area 945,000 sq km/364,865 sq mi
capital Dodoma (since 1983)
towns Zanzibar Town, Mwanza; chief port and former capital Dar es Salaam

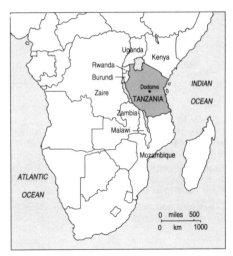

physical central plateau; lakes in N and W; coastal plains; lakes Victoria, Tanganyika, and Niasa
environment the black rhino faces extinction as a result of poaching
features comprises islands of Zanzibar and Pemba; Mount Kilimanjaro, 5,895 m/19,340 ft, the highest peak in Africa; Serengeti National Park, Olduvai Gorge; Ngorongoro Crater, 14.5 km/9 mi across, 762 m/2,500 ft deep
head of state and government Ali Hassan Mwinyi from 1985
political system one-party socialist republic
political party Revolutionary Party of Tanzania (CCM), African, socialist
exports coffee, cotton, sisal, cloves, tea, tobacco, cashew nuts, diamonds
currency Tanzanian shilling
population (1990 est) 26,070,000; growth rate 3.5% p.a.
life expectancy men 49, women 54 (1989)
languages Kiswahili, English (both official)
religions Muslim 35%, Christian 35%, traditional 30%
literacy 85% (1987)
GNP $4.9 bn; $258 per head (1987)
chronology
16th–17th centuries Zanzibar under Portuguese control.
1890–1963 Zanzibar became a British protectorate.
1920–46 Tanganyika administered as a British League of Nations mandate.
1946–62 Tanganyika came under United Nations (UN) trusteeship.
1961 Tanganyika achieved independence from Britain, within the Commonwealth, with Julius Nyerere as prime minister.
1962 Tanganyika became a republic with Nyerere as president.
1964 Tanganyika and Zanzibar became the United Republic of Tanzania with Nyerere as president.
1967 East African Community (EAC) formed. Arusha Declaration.
1977 Revolutionary Party of Tanzania (CCM)

proclaimed the only legal party. EAC dissolved.
1978 Ugandan forces repulsed after crossing into
Tanzania.
1979 Tanzanian troops sent to Uganda to help
overthrow the president, Idi Amin.
1985 Nyerere retired from presidency but stayed
on as CCM leader; Ali Hassan Mwinyi became
president.
1990 Nyerere surrendered CCM leadership;
replaced by President Mwinyi.
1992 CCM agreed to abolish one-party rule. East
African cooperation pact with Kenya and Uganda
to be re-established.

Taormina /ˌtɑːɔːˈmiːnə/ coastal resort in E Sicily, at
the foot of Mount Etna; population (1985) 9,000.
It has an ancient Greek theatre.

Taperinha archaeological site on the Amazon
River, E of Santarem, Brazil. Its discovery 1991
provided evidence that an ancient New World
civilization, predating Mexican and Andean
cultures, existed 6,000–8,000 years ago.

Tara Hill /ˈtɑːrə/ ancient religious and political
centre in County Meath, S Ireland. It was the site
of a palace and coronation place of many Irish
kings, abandoned in the 6th century. St Patrick,
patron saint of Ireland, preached here.

Taranaki /ˌtærəˈnæki/ peninsula in North Island,
New Zealand, dominated by Mount Egmont;
volcanic soil makes it a rich dairy-farming area,
and cheese is manufactured here.

Taranto /təˈræntəu/ naval base and port in Puglia
region, SE Italy; population (1988) 245,000. It
is an important commercial centre, and its steelworks are part of the new industrial complex
of S Italy. It was the site of the ancient Greek
Tarentum, founded in the 8th century BC by
Sparta, and was captured by the Romans 272 BC.

Tarawa /təˈrɑːwə/ port and capital of Kiribati;
population (1990) 28,800. Mother-of-pearl and
copra are exported.

Tarbes /tɑːb/ capital of Hautes-Pyrénées
département, SW France, a tourist centre for the
Pyrenees; population (1990) 50,200. It belonged
to England 1360–1406.

Taree /ˌtɑːˈriː/ town in a dairying area of NE
New South Wales, Australia; population (1981)
16,000.

Tarim Basin /ˌtɑːˈriːm/ (Chinese *Tarim Pendi*)
internal drainage area in Xinjiang Uygur province, NW China, between the Tien Shan and
Kunlun mountains; area about 900,000 sq km/
350,000 sq mi. It is crossed by the river Tarim
He and includes the lake of Lop Nur. The
Taklimakan desert lies to the S of the Tarim He.

Tarn /tɑːn/ river in SW France, rising in the
Cévennes Mountains and flowing 350 km/217 mi
to the Garonne River. It cuts picturesque gorges
in the limestone plateaus of the Lozère and
Aveyron *départements*.

Tarragona /ˌtærəˈɡəunə/ port in Catalonia, Spain;
population (1986) 110,000. Industries include
petrochemicals, pharmaceuticals, and electrical
goods. It has a cathedral and Roman remains,
including an aqueduct and amphitheatre.

Tarrasa /təˈrɑːsə/ town in Catalonia, NE Spain;
population (1991) 153,500. Industries include
textiles and fertilizers.

Tarsus /ˈtɑːsəs/ city in İçel province, SE Turkey,
on the river Pamuk; population (1980) 121,000.
Formerly the capital of the Roman province of
Cilicia, it was the birthplace of St Paul.

Tartu /ˈtɑːtuː/ city in Estonia; population (1981)
107,000. Industries include engineering and food
processing. Once a stronghold of the Teutonic
Knights, it was taken by Russia 1558 and then
held by Sweden and Poland but returned to
Russian control 1704.

Tas. abbreviation for ◊ *Tasmania*, an island off
Australia.

Tashkent /ˌtæʃˈkent/ capital of Uzbekistan; population (1990) 2,100,000. Industries include the
manufacture of mining machinery, chemicals,
textiles, and leather goods. Founded in the 7th
century, it was taken by the Turks in the 12th
century and captured by Tamerlane 1361. In
1865 it was taken by the Russians. It was severely
damaged by an earthquake 1966.
   A temporary truce between Pakistan and India
over Kashmir was established at the Declaration
of Tashkent 1966.

Tasmania /tæzˈmeɪnɪə/ former name (1642–1856)
Van Diemen's Land island off the south coast
of Australia; a state of the Commonwealth of
Australia;
area 67,800 sq km/26,171 sq mi
capital Hobart
towns Launceston (chief port)
features an island state (including small islands
in the Bass Strait, and Macquarie Island);
Franklin River, a wilderness area saved from
a hydroelectric scheme 1983, which also has
a prehistoric site; unique fauna including the
Tasmanian devil
products wool, dairy products, apples and other
fruit, timber, iron, tin, coal, copper, silver
population (1987) 448,000
history the first European to visit here was Abel
Tasman 1642; the last of the Tasmanian Aboriginals died 1876. Tasmania joined the Australian
Commonwealth as a state 1901.

Tasmania

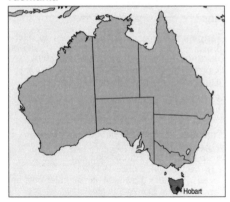

Tasman Sea /ˈtæzmən/ part of the Pacific Ocean between SE Australia and NW New Zealand. It is named after the Dutch explorer Abel Tasman.

Tatarstan /ˌtɑːtəˈstɑːn/ formerly *Tatar Autonomous Republic* autonomous republic of E Russia
area 68,000 sq km/26,250 sq mi
capital Kazan
products oil, chemicals, textiles, timber
population (1986) 3,537,000 (48% Tatar, 43% Russian)
history a territory of Volga-Kama Bulgar state from the 10th century when Islam was introduced; conquered by the Mongols 1236; the capital of the powerful Khanate of Kazan until conquered by Russia 1552; an autonomous republic from 1920. In recent years the republic (mainly Muslim and an important industrial and oil-producing area) has seen moves towards increased autonomy. In Aug 1990 the republic's assembly upgraded Tatarstan to full republic status, proclaiming its economic and political 'sovereignty', and in April 1991 there were popular demonstrations in support of this action. In June 1991 it refused to participate in the Russian presidential election, and in March 1992 declined to be party to a federal treaty, signed in Moscow by 18 of Russia's other 20 main political subdivisions. A referendum 21 March 1992 favoured Tatarstan becoming a sovereign state within Russia.

Tatra Mountains /ˈtɑːtrə/ range in central Europe, extending for about 65 km/40 mi along the Polish–Slovak border; the highest part of the central Carpathian Mountains.

Taunton /ˈtɔːntən/ market town and administrative headquarters of Somerset, England; population (1987 est) 53,300. Industries include cider, leather, optical instruments, and light engineering; there is a weekly cattle market. The Elizabethan hall survives, in which Judge Jeffreys held the Bloody Assizes 1685 after the Duke of Monmouth's rebellion.

Taunus Mountains /ˈtaunəs/ mountain range in Hessen, Germany; the Grosser Feldberg, 881 m/2,887 ft, is the highest peak in the Rhenish uplands. There are several mineral spas, including Wiesbaden, Bad Nauheim, and Bad Soden.

Taupo /ˈtaupəu/ largest lake in New Zealand, in a volcanic area of hot springs; area 620 sq km/239 sq mi. It is the source of the Waikato River.

Tauranga /tauˈræŋə/ port in North Island, New Zealand; population (1991) 70,700. It exports citrus fruit, dairy produce, and timber.

Taurus Mountains /ˈtɔːrəs/ (Turkish *Toros Dağlari*) mountain range in S Turkey, forming the southern edge of the Anatolian plateau and rising to over 3,656 m/12,000 ft.

Tavistock /ˈtævɪstɒk/ market town 24 km/15 mi N of Plymouth, Devon, England; population (1981) 9,300.

Tay /teɪ/ longest river in Scotland; length 189 km/118 mi. Rising in NW Central region, it flows

Tayside

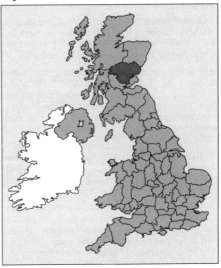

NE through Loch Tay, then E and SE past Perth to the Firth of Tay, crossed at Dundee by the Tay Bridge, before joining the North Sea. The Tay has salmon fisheries; its main tributaries are the Tummel, Isla, and Earn.

Tayside /ˈteɪsaɪd/ region of Scotland
area 7,700 sq km/2,973 sq mi
towns Dundee (administrative headquarters), Perth, Arbroath, Forfar
features river Tay; Grampian Mountains; Lochs Tay and Rannoch; hills: Ochil and Sidlaw; vales of the North and South Esk
products beef and dairy products, soft fruit from the fertile Carse of Gowrie (SW of Dundee)
population (1991) 385,300
famous people J M Barrie, John Buchan, Princess Margaret.

Tbilisi /dbɪˈliːsi/ (formerly *Tiflis*) capital of Georgia; population (1987) 1,194,000; industries include textiles, machinery, ceramics, and tobacco. Dating from the 5th century, it is a centre of Georgian culture, with fine medieval churches. Anti-Russian demonstrations were quashed here by troops 1981 and 1989; the latter clash followed rejected demands for autonomy from the Abkhazia enclave, and resulted in 19 or more deaths from poison gas (containing chloroacetophenone) and 100 injured. In Dec 1991 at least 50 people were killed as well-armed opposition forces attempted to overthrow President Gamsakhurdia, eventually forcing him to flee.

... Tiflis is a very civilised town, closely imitating St Petersburg and doing it rather well.

On **Tbilisi** (Tiflis) Leo Tolstoy 1851

Teddington /ˈtedɪŋtən/ part of Twickenham, in the Greater London borough of Richmond-upon-

Thames; site of the National Physical Laboratory, established 1900.

Tees /tiːz/ river flowing from the Pennines in Cumbria, England, to the North Sea via Tees Bay in Cleveland; length 130 km/80 mi. It is polluted with industrial waste, sewage, and chemicals.

Teesside /'tiːzsaɪd/ industrial area at the mouth of the river Tees, Cleveland, NE England; population (1981) 382,700. Industries include high technology, capital-intensive steelmaking, chemicals, an oil-fuel terminal, and the main North Sea natural-gas terminal. Middlesbrough is a large port.

Tegucigalpa /teɪˌɡuːsɪˈɡælpə/ capital of Honduras; population (1989) 608,000. Industries include textiles and food-processing. It was founded 1524 as a gold- and silver-mining centre.

Tehran /ˌteəˈrɑːn/capital of Iran; population (1986) 6,043,000. Industries include textiles, chemicals, engineering, and tobacco. It was founded in the 12th century and made the capital 1788 by Muhammad Shah. Much of the city was rebuilt in the 1920s and 1930s. Tehran is the site of the Gulistan Palace (the former royal residence). There are three universities; the Shahyad Tower is a symbol of modern Iran.

Teignmouth /'tɪnməθ/ port and resort in S Devon, England, at the mouth of the river Teign; population (1985 est) 13,500.

Tejo /'tʌʒuː/ Portuguese name for the river ◊Tagus.

Tel Aviv /'telə'viːv/ officially **Tel Aviv-Jaffa** city in Israel, on the Mediterranean coast; population (1987) 320,000. Industries include textiles, chemicals, sugar, printing, and publishing. Tel Aviv was founded 1909 as a Jewish residential area in the Arab town of Jaffa, with which it was combined 1949; their ports were superseded 1965 by Ashdod to the S.

Tema /'tiːmə/ port in Ghana; population (1982) 324,000. It has the largest artificial harbour in Africa, opened 1962, as well as oil refineries and a fishing industry.

Temple /'templ/ industrial city (building materials, steel, furniture, railway supplies) in central Texas, USA, S of Waco; population (1990) 46,100.

Temuco /te'muːkəʊ/ market town and capital of Araucanía region, S Chile, situated to the N of the Lake District; population (1987) 218,000. Founded 1881, it is a market town for the Mapuche Indians and a centre for coal, steel, and textile production.

Tenerife /ˌtenəˈriːf/ largest of the Canary Islands, Spain; area 2,060 sq km/795 sq mi; population (1981) 557,000. **Santa Cruz** is the main town, and **Pico de Teide** is an active volcano.

Tennessee /ˌtenəˈsiː/ state in E central USA; nickname Volunteer State
area 109,200 sq km/42,151 sq mi
capital Nashville
towns Memphis, Knoxville, Chattanooga, Clarksville

Tennessee

features Tennessee Valley Authority; Great Smoky Mountains National Park; Grand Old Opry, Nashville; Beale Street Historic District and Graceland, estate of Elvis Presley, Memphis; research centres, including Oak Ridge National Laboratory
products cereals, cotton, tobacco, soya beans, livestock, timber, coal, zinc, copper, chemicals
population (1990) 4,877,200
famous people Davy Crockett, David Farragut, W C Handy, Cordell Hull, Andrew Jackson, Andrew Johnson, Dolly Parton, John Crowe Ransom, Bessie Smith
history settled by Europeans 1757; became a state 1796. Tennessee was deeply divided in the Civil War; the battles of Shiloh, Murfreesboro, Chattanooga, and Nashville were fought here.

Teplice /'teplɪtseɪ/ industrial city (peat- and lignite-mining, glass, porcelain, cement, paper) and spa in N Bohemia, Czech Republic; population (1991) 127,800.

Terengganu alternative spelling of ◊Trengganu, a state in Peninsular Malaysia.

Terni /'teəni/ industrial city in the valley of the river Nera, Umbria region, central Italy; population (1987) 111,000. The nearby Marmore Falls, the highest in Italy, were created by the Romans in order to drain the Rieti marshes.

Terre Adélie /'teər ædeɪ'li/ French name for ◊Adélie Land, Antarctica.

Terre Haute /'teər 'həʊt/ city in W Indiana, USA, on the Wabash River; population (1990) 57,500. Industries include plastics, chemicals, and glass.

Tetuán /te'twɑːn/ or **Tétouan** town in NE Morocco, near the Mediterranean coast, 64 km/40 mi SE of Tangier; population (1982) 372,000. Products include textiles, leather, and soap. It was settled by Moorish exiles from Spain in the 16th century.

Texarkana /ˌteksɑːˈkænə/ twin cities that straddle the Texas–Arkansas border, USA; population (1990) 22,600 (Texas), 31,700 (Arkansas). Industries include furniture, lumber, cotton, and sand and gravel.

Texas /'teksəs/ state in southwestern USA; nickname Lone Star State
area 691,200 sq km/266,803 sq mi
capital Austin
towns Houston, Dallas-Fort Worth, San Antonio, El Paso, Corpus Christi, Lubbock

Texas

Austin

features rivers: Rio Grande, Red; arid Staked Plains, reclaimed by irrigation; the Great Plains; Gulf Coast resorts; Lyndon B Johnson Space Center, Houston; Alamo, San Antonio; Big Bend and Guadalupe Mountains national parks
products rice, cotton, sorghum, wheat, hay, livestock, shrimps, meat products, lumber, wood and paper products, petroleum (nearly one-third of US production), natural gas, sulphur, salt, uranium, chemicals, petrochemicals, nonelectrical machinery, fabricated metal products, transportation equipment, electric and electronic equipment
population (1990) 16,986,500
famous people James Bowie, George Bush, Buddy Holly, Sam Houston, Howard Hughes, Lyndon Johnson, Janis Joplin, Katherine Anne Porter, Patrick Swayze, Tina Turner
history settled by the Spanish 1682; part of Mexico 1821–36; Santa Anna massacred the Alamo garrison 1836, but was defeated by Sam Houston at San Jacinto the same year; Texas became an independent republic 1836–45, with Houston as president; in 1845 it became a state of the USA. Texas is the only state in the USA to have previously been an independent republic.

Other states were carved or born, /Texas grew from hide to horn.

On **Texas** Bert Hart Nance about 1930

Texas City /ˌteksəs ˈsɪti/ city in SE Texas, USA, on Galveston Bay, SE of Houston; population (1990) 40,800. Industries include tin, chemicals, oil, and grains.

Texel /ˈtesəl/ or *Tessel* largest and southernmost of the Frisian Islands, in North Holland province, the Netherlands; area 190 sq km/73 sq mi; population (1991) 12,700. Texel sheep are kept for their wool and cheese. The island is a good breeding ground for birds. Den Burg is the chief settlement.

Thailand Kingdom of (*Prathet Thai* or *Muang-Thai*)
area 513,115 sq km/198,108 sq mi
capital and chief port Bangkok
towns Chiangmai, Nakhon Sawan river port

physical mountainous, semi-arid plateau in NE, fertile central region, tropical isthmus in S
environment tropical rainforest was reduced to 18% of the land area 1988 (from 93% in 1961); logging was banned by the government 1988
features rivers Chao Phraya, Mekong, Salween; ancient ruins of Sukhothai and Ayurrhaya
head of state King Bhumibol Adulyadej from 1946
head of government Chuan Leekpai from 1992
political system military-controlled emergent democracy
political parties New Aspiration Party; Samakkhi Tham (Justice and Unity) Party, right-of-centre, airforce-linked; Palang Dharma, anti-corruption; Social Action Party (Kij Sangkhom), right-of-centre; Thai Nation (Chart Thai), conservative, pro-business; Liberal Democratic Party
exports rice, textiles, rubber, tin, rubies, sapphires, maize, tapioca
currency baht
population (1990 est) 54,890,000 (Thai 75%, Chinese 14%); growth rate 2% p.a.
life expectancy men 62, women 68 (1989)
languages Thai and Chinese (both official); regional dialects
religions Buddhist 95%, Muslim 4%
literacy 89% (1988)
GNP $52 bn (1988); $771 per head (1988)
chronology
1782 Siam absolutist dynasty commenced.
1896 Anglo-French agreement recognized Siam as independent buffer state.
1932 Constitutional monarchy established.
1939 Name of Thailand adopted.
1941–44 Japanese occupation.
1947 Military seized power in coup.
1972 Withdrawal of Thai troops from South Vietnam.
1973 Military government overthrown.
1976 Military reassumed control.
1980 General Prem Tinsulanonda assumed power.
1983 Civilian government formed; martial law maintained.
1988 Prime Minister Prem resigned; replaced by

Chatichai Choonhavan.
1989 Thai pirates continued to murder, pillage, and kidnap Vietnamese 'boat people' at sea.
1991 Military seized power in coup. Interim civilian government formed under Anand Panyarachun. 50,000 demonstrated against new military-oriented constitution.
1992 March: general election produced five-party coalition; Narong Wongwan named premier but removed a month later. April: appointment of General Suchinda Kraprayoon as premier provoked widespread riots. May: Suchinda forced to stand down. June: Anand made interim prime minister. Sept: new coalition government led by Chuan Leekpai.

Thames /temz/ river in S England; length 338 km/ 210 mi. It rises in the Cotswold Hills above Cirencester and is tidal as far as Teddington.
Below London there is protection from flooding by means of the Thames Barrier. The headstreams unite at Lechlade.
Tributaries from the N are the Windrush, Evenlode, Cherwell, Thame, Colne, Lea, and Roding; and from the S, Kennet, Loddon, Wey, Mole, Darent, and Medway. Above Oxford the river is sometimes poetically called the *Isis*.

Sweete Themmes runne softly, till I end my song.

On the river **Thames** Edmund Spenser
Prothalamium 1596

Thames, Firth of /temz/ inlet between Auckland and the Coromandel Peninsula, New Zealand.

Thames Tunnel /temz/ tunnel extending 365 m/ 1,200 ft under the river Thames, London, linking Rotherhithe with Wapping; the first underwater tunnel in the world. It was designed by Marc Isambard Brunel and was completed 1843. Originally intended as a road tunnel, it remained a pedestrian tunnel, for lack of funds, until the 1860s, when it was converted into a railway tunnel for the East London Railway. Today it carries underground trains.

Thanet, Isle of /ˈθænɪt/ northeast corner of Kent, England, bounded by the North Sea and the river Stour. It was an island until the 16th century, and includes the coastal resorts of Broadstairs, Margate, and Ramsgate.

Thar Desert /tɑː/ or *Indian Desert* desert on the borders of Rajasthan and Pakistan; area about 250,000 sq km/96,500 sq mi.

Thessaloníki /ˌθesəlɒˈniːki/ (English *Salonika*) port in Macedonia, NE Greece, at the head of the Gulf of Thessaloníki, the second largest city of Greece; population (1981) 706,200. Industries include textiles, shipbuilding, chemicals, brewing, and tanning. It was founded from Corinth by the Romans 315 BC as *Thessalonica* (to whose inhabitants St Paul addressed two epistles), captured by the Saracens AD 904 and by the Turks 1430, and restored to Greece 1912.

Thessaly /ˈθesəli/ (Greek *Thessalia*) region of E central Greece, on the Aegean coast; area 13,904 sq km/5,367 sq mi; population (1991) 731,200. It is a major area of cereal production. Thessaly was an independent state in ancient Greece and later formed part of the Roman province of Macedonia. It was Turkish from the 14th century until incorporated in Greece 1881.

Thetford /ˈθetfəd/ market town in Norfolk, England; population (1987 est) 19,900. There is light industry and printing. It is the birthplace of the political pamphleteer Thomas Paine.

Thetford Mines /ˈθetfəd/ site of the world's largest asbestos deposits, 80 km/50 mi S of Québec, Canada; discovered 1876.

Thibodaux /ˌtɪbəˈdəu/ town in SE Louisiana, USA, SW of New Orleans; population (1980) 15,810. It is an agricultural centre for sugar, dairy products, vegetables, and cotton.

Thimbu /ˈθɪmbuː/ or *Thimphu* capital since 1962 of the Himalayan state of Bhutan; population (1987) 15,000. There is a 13th-century fortified monastery, Tashichoedzong, and the Memorial Charter to the Third King (1974).

Thousand Islands group of about 1,500 islands on the border between Canada and the USA in the upper St Lawrence River.

Thrace /θreɪs/ (Greek *Thráki*) ancient empire (6000 BC–AD 300) in the Balkans, SE Europe, formed by parts of modern Greece and Bulgaria. It was held successively by the Greeks, Persians, Macedonians, and Romans.
The area was divided 1923 into western Thrace (the Greek province of Thráki) and eastern Thrace (European Turkey). The heart of the ancient Thracian Empire was Bulgaria, where since 1945 there have been tomb finds of gold and silver dishes, drinking vessels, and jewellery with animal designs. The legend of Orpheus and the cult of Dionysus were both derived by the Greeks from Thrace. The area was conquered by Persia 6th–5th centuries BC and by Macedonia 4th–2nd centuries BC. From AD 46 it was a Roman province, then part of the Byzantine Empire, and Turkish from the 15th century until 1878; it was then subject to constant dispute until after World War I.

Three Mile Island island in the Shenandoah River near Harrisburg, Pennsylvania, USA, site of a nuclear power station which was put out of action following a major accident March 1979. Opposition to nuclear power in the USA was reinforced after this accident and safety standards reassessed.

Three Rivers English name for the Canadian port of ▷Trois-Rivières.

Thule /ˈθjuːli/ Greek and Roman name for the northernmost land known. It was applied to the Shetlands, the Orkneys, and Iceland, and by later writers to Scandinavia.

Thunder Bay /ˈθʌndə ˈbeɪ/ city and port on Lake Superior, Ontario, Canada, formed by the union of Port Arthur and its twin city of Fort William to the S; industries include shipbuilding, timber, paper, wood pulp, and export of wheat; population (1986) 122,000.

Thuringia /θju'rɪndʒiə/ administrative *Land* (state) of Germany;
area 15,482 sq km/5,980 sq mi
capital Erfurt
towns Weimar, Gera, Jena, Eisenach
products machine tools, optical instruments, steel, vehicles, ceramics, electronics, glassware, timber
population (1990) 2,500,000
history an historic, densely forested region of Germany that became a province 1918 and a region of East Germany 1946. It was split into the districts of Erfurt, Gera, and Suhl 1952 but reconstituted as a state following German reunification 1990.

Thursday Island /'θɜːzdeɪ/ island in Torres Strait, Queensland, Australia; area 4 sq km/1.5 sq mi; chief centre Port Kennedy. It is a centre of the pearl-fishing industry.

Thurso /'θɜːsəu/ port in Highland Region, Scotland. It is the mainland terminus of the steamer service to the Orkneys, and the experimental atomic station of Dounreay lies to the W.

Tianjin /ˌtjen'dʒɪn/ or **Tientsin** port and industrial and commercial city in Hubei province, central China; population (1989) 5,620,000. The special municipality of Tianjin has an area of 4,000 sq km/1,544 sq mi and a population (1990) of 8,788,000. Its handmade silk and wool carpets are renowned. Dagan oilfield is nearby. Tianjin was opened to foreign trade 1860 and occupied by the Japanese 1937.

Tian Shan /ti'en 'ʃɑːn/ (Chinese *Tien Shan*) mountain system in central Asia. *Pik Pobedy* on the Xinjiang-Kyrgyz border is the highest peak at 7,440 m/24,415 ft.

Tiber /'taɪbə/ (Italian *Tevere*) river in Italy on which Rome stands; length from its source in the Apennines to the Tyrrhenian Sea 400 km/250 mi.

Tiberias, Lake /taɪ'bɪəriæs/ or *Sea of Galilee* lake in N Israel, 210 m/689 ft below sea level, into which the river Jordan flows; area 170 sq km/66 sq mi. The first Israeli kibbutz (cooperative settlement) was founded nearby 1909.

Tibesti Mountains /tɪ'besti/ range in the central Sahara, N Chad; the highest peak is *Emi Koussi* at 3,415 m/11,208 ft.

Tibet /tɪ'bet/ autonomous region of SW China (Pinyin form *Xizang*);
area 1,221,600 sq km/471,538 sq mi
capital Lhasa
features Tibet occupies a barren plateau bounded S and SW by the Himalayas and N by the Kunlun Mountains, traversed W to E by the Bukamagna, Karakoram, and other mountain ranges, and having an average elevation of 4,000–4,500 m/13,000–15,000 ft. The Sutlej, Brahmaputra, and Indus rivers rise in Tibet, which has numerous lakes, many of which are salty. The yak is the main domestic animal
government Tibet is an autonomous region of China, with its own People's Government and

People's Congress. The controlling force in Tibet is the Communist Party of China, represented locally by First Secretary Wu Jinghua from 1985. Tibetan nationalists regard the province as being under colonial rule
products wool, borax, salt, horn, musk, herbs, furs, gold, iron pyrites, lapis lazuli, mercury, textiles, chemicals, agricultural machinery
population (1991) 2,190,000 including 2,090,000 Tibetan nationalists (94.5%); many Chinese have settled in Tibet; 2 million Tibetans live in China outside Tibet
religion traditionally Lamaist (a form of Mahāyāna Buddhism)
history Tibet was an independent kingdom from the 5th century AD. It came under nominal Chinese rule about 1700. Independence was regained after a revolt 1912. China regained control 1951 when the historic ruler and religious leader, the Dalai Lama, was driven from the country and the monks (who formed 25% of the population) were forced out of the monasteries. Between 1951 and 1959 the Chinese People's Liberation Army (PLA) controlled Tibet, although the Dalai Lama returned as nominal spiritual and temporal head of state. In 1959 a Tibetan uprising spread from bordering regions to Lhasa and was supported by Tibet's local government. The rebellion was suppressed by the PLA, prompting the Dalai Lama and 9,000 Tibetans to flee to India. The Chinese proceeded to dissolve the Tibet local government, abolish serfdom, collectivize agriculture, and suppress Lamaism. In 1965 Tibet became an autonomous region of China. Chinese rule continued to be resented, however, and the economy languished.

From 1979, the leadership in Beijing adopted a more liberal and pragmatic policy towards Tibet. Traditional agriculture, livestock, and trading practices were restored (under the 1980 slogan 'relax, relax, and relax again'), a number of older political leaders and rebels were rehabilitated or pardoned, and the promotion of local Tibetan cadres was encouraged. In addition, a somewhat more tolerant attitude towards Lamaism was adopted (temples damaged during the 1965–68 Cultural Revolution were being repaired) and attempts made to persuade the Dalai Lama to return from exile.

Pro-independence demonstrations erupted in Lhasa in Sept-Oct 1987, repeatedly throughout 1988, and in March 1989 and were forcibly suppressed by Chinese troops.

Tien Shan /'tjen 'ʃɑːn/ Chinese form of ▷Tian Shan, a mountain system of central Asia.

Tientsin /ˌtjen'tsɪn/ alternative form of ▷Tianjin, an industrial city in NE China.

Tierra del Fuego /ti'eərə del 'fweɪgəu/ island group divided between Chile and Argentina. It is separated from the mainland of South America by the Strait of Magellan, and Cape Horn is at the southernmost point. The chief town, Ushuaia, Argentina, is the world's most southerly town. Industries include oil and sheep farming.

To the S of the main island is *Beagle Channel* (named after the ship of the scientist Charles

Darwin's voyage), with three islands at the eastern end, finally awarded 1985 to Chile rather than Argentina.

Tiflis /'tɪflɪs/ former name (until 1936) of the city of ◊Tbilisi in Georgia.

Tigré /'tiːɡreɪ/ or *Tigray* region in the northern highlands of Ethiopia; area 65,900 sq km/25,444 sq mi. The chief town is Mekele. The region had an estimated population of 2.4 million in 1984, at a time when drought and famine were driving large numbers of people to fertile land in the S or into neighbouring Sudan. Since 1978 a guerrilla group known as the Tigré People's Liberation Front (TPLF) has been fighting for regional autonomy. In 1989 government troops were forced from the province, and the TPLF advanced towards Addis Ababa, playing a key role in the fall of the Ethiopian government May 1991.

Tigris /'taɪɡrɪs/ (Arabic *Shatt Dijla*) river flowing through Turkey and Iraq, joining the Euphrates above Basra, where it forms the Shatt-al-Arab; length 1,600 km/1,000 mi.

Tijuana /tɪ'wɑːnə/ city and resort in NW Mexico; population (1990) 742,700. It is known for horse races and casinos. San Diego adjoins it across the US border.

Tilbury /'tɪlbəri/ port in Essex, England, on the north bank of the Thames; population (1981) 12,000. Greatly extended 1976, it became London's largest container port. It dates from Roman times.

Tilsit /'tɪlzɪt/ former name (until 1945) of the Russian town of ◊Sovetsk.

Timaru /'tɪməruː/ (Maori 'place of shelter') industrial port and resort on South Island, New Zealand; population (1991) 27,700. Industries include flour milling, deep freezing, pottery, and brewing.

Timbuktu /,tɪmbʌk'tuː/ or *Tombouctou* town in Mali; population (1976) 20,500. A camel caravan centre from the 11th century on the fringe of the Sahara, since 1960 it has been surrounded by the southward movement of the desert, and the former canal link with the river Niger is dry. Products include salt.

To that impracticable place Timbuctoo, where geography finds no one to oblige her/ With such a chart as may be safely stuck to.

On **Timbuktu** Lord Byron
Don Juan 1819–24

Timişoara /,tɪmɪ'ʃwɑːrə/ capital of Timiş county, W Romania; population (1985) 319,000. Industries include electrical engineering, chemicals, pharmaceuticals, textiles, food processing, metal, and footwear. The revolt against the Ceauşescu regime began here Dec 1989 when demonstrators prevented the arrest and deportation of a popular Protestant minister who was promoting the rights of ethnic Hungarians. This soon led to large prodemocracy rallies.

Timor /'tiːmɔː/ largest and most easterly of the Lesser Sunda Islands, part of Indonesia; area 33,610 sq km/12,973 sq mi. *West Timor* (capital Kupang) was formerly Dutch and was included in Indonesia on independence. ◊*East Timor* (capital Dili), an enclave on the northwest coast, and the islands of Atauro and Jaco formed an overseas province of Portugal until it was seized by Indonesia 1975. The annexation is not recognized by the United Nations, and guerrilla warfare by local people seeking independence continues. Since 1975 over 500,000 Timorese have been killed by Indonesian troops or have resettled in West Timor, according to Amnesty International. Products include coffee, maize, rice, and coconuts.

Tindouf /tɪn'duːf/ Saharan oasis in the Aïn-Sefra region of Algeria, crossed by the Agadir–Dakar desert route. There are large iron deposits in the area; the oasis acted as a base for exiled Polisario guerrillas of the Western Sahara.

Tintagel /tɪn'tædʒəl/ village resort on the coast of N Cornwall, England. There are castle ruins, and legend has it that King Arthur was born and held court here.

Tipperary /,tɪpə'reəri/ county in the Republic of Ireland, province of Munster, divided into north and south regions. *North Tipperary*: administrative headquarters Nenagh; area 2,000 sq km/772 sq mi; population (1991) 57,800.
South Tipperary: administrative headquarters Clonmel; area 2,260 sq km/872 sq mi; population (1991) 74,800. It includes part of the Golden Vale, a dairy-farming region.

Tirana /tɪ'rɑːnə/ or *Tiranë* capital (since 1920) of Albania; population (1990) 210,000. Industries include metallurgy, cotton textiles, soap, and cigarettes. It was founded in the early 17th century by Turks when part of the Ottoman Empire. Although the city is now largely composed of recent buildings, some older districts and mosques have been preserved.

Tîrgu Mureş /'tɜːɡuː 'muəreʃ/ city in Transylvania, Romania, on the river Mureş, 450 km/280 mi N of Bucharest; population (1985) 157,400. With a population comprising approximately equal numbers of ethnic Hungarians and Romanians, the city was the scene of rioting between the two groups following Hungarian demands for greater autonomy 1990.

Tirol /tɪ'rəʊl/ federal province of Austria; area 12,600 sq km/4,864 sq mi; population (1989) 619,600. Its capital is Innsbruck, and it produces diesel engines, optical instruments, and hydroelectric power. Tirol was formerly a province (from 1363) of the Austrian Empire, divided 1919 between Austria and Italy (see ◊Trentino–Alto Adige).

Mountains holding up cups of snow to the fiery sun, who glares on them in vain.

On the **Tirol** George Meredith,
letter of 1861

Tiruchirapalli /ˌtɪrətʃɪˈrɑːpəli/ former name **Trichinopoly** ('town of the sacred rock') city in Tamil Nadu, India; chief industries are cotton textiles, cigars, and gold and silver filigree; population (1981) 362,000. It is a place of pilgrimage and was the capital of Tamil kingdoms during the 10th to 17th centuries.

Tisza /ˈtiːsə/ tributary of the river Danube, rising in Ukraine and flowing through Hungary to Yugoslavia; length 967 km/601 mi.

Titicaca /ˌtɪtɪˈkɑːkə/ lake in the Andes, 3,810 m/12,500 ft above sea level and 1,220 m/4,000 ft above the tree line; area 8,300 sq km/3,200 sq mi, the largest lake in South America. It is divided between Bolivia (port at Guaqui) and Peru (ports at Puno and Huancane). It has enormous edible frogs, and is one of the few places in the world where reed boats are still made (Lake Tana in Ethiopia is another).

Titograd /ˈtiːtəʊɡræd/ (formerly until 1948 *Podgorica*) capital of Montenegro, Yugoslavia; population (1981) 132,300. Industries include metalworking, furniture-making, and tobacco. It was damaged in World War II and after rebuilding was renamed in honour of Marshal Tito. It was the birthplace of the Roman emperor Diocletian.

Titusville /ˈtaɪtəsvɪl/ town in E Florida, USA, on the Indian River, E of Orlando; population (1990) 39,400. Industries include citrus fruits and sport fishing. The Kennedy Space Center is nearby.

Tivoli /ˈtɪvəli/ town NE of Rome, Italy; population (1981) 52,000. It has remains of Hadrian's villa, with gardens; and the Villa d'Este with Renaissance gardens laid out 1549 for Cardinal Ippolito d'Este.

Tlemcen /tlemˈsen/ (Roman *Pomaria*) town in NW Algeria; population (1983) 146,000. Carpets and leather goods are made, and there is a 12th-century mosque.

TN abbreviation for ◊*Tennessee*, a state of the USA.

Tobago /təˈbeɪɡəʊ/ island in the West Indies; part of the republic of Trinidad and Tobago.

Tobolsk /təˈbɒlsk/ river port and lumber centre at the confluence of the Tobol and Irtysh rivers in N Tyumen, W Siberia, Russia; population (1985) 75,000. It was founded by Cossacks 1587; Tsar Nicholas II was exiled here 1917.

Tobruk /təˈbruk/ Libyan port; population (1984) 94,000. There is a naval ship repair industry. Occupied by Italy 1911, it was taken by Britain 1941 during World War II, and unsuccessfully besieged by Axis forces April–Dec 1941. It was captured by Germany June 1942 after the retreat of the main British force to Egypt, and this precipitated the replacement of Auchinleck by Montgomery as British commander.

Togliatti /tɒlˈjæti/ or *Tolyatti* (formerly *Stavropol*) port on the river Volga, W central Russia; population (1989) 630,000. Industries include engineering and food processing. The

city was relocated in the 1950s after a flood and renamed after the Italian communist Palmiro Togliatti.

Togo Republic of (*République Togolaise*)
area 56,800 sq km/21,930 sq mi
capital Lomé
towns Sokodé, Kpalimé
physical two savanna plains, divided by range of hills NE–SW; coastal lagoons and marsh
environment the homes of thousands of people in Keto were destroyed by coastal erosion as a result of the building of the Volta dam
features Mono Tableland, Oti Plateau, Oti River
head of state President Eyadéma
head of government Jospeh Kokou Koffigoh from 1991
political system transitional
political parties Rally of the Togolese People (RPT), centrist nationalist; Alliance of Togolese Democrats (ADT), left-of-centre; Togolese Movement for Democracy (MDT), left of centre
exports phosphates, cocoa, coffee, coconuts
currency franc CFA
population (1990 est) 3,566,000; growth rate 3% p.a.
life expectancy men 53, women 57 (1989)
languages French (official), Ewe, Kabre
religions animist 46%, Catholic 28%, Muslim 17%, Protestant 9%
literacy men 53%, women 28% (1985 est)
GNP $1.3 bn (1987); $240 per head (1985)
chronology
1885–1914 Togoland was a German protectorate until captured by Anglo-French forces.
1922 Divided between Britain and France under League of Nations mandate.
1946 Continued under United Nations trusteeship.
1956 British Togoland integrated with Ghana.
1960 French Togoland achieved independence from France as the Republic of Togo with Sylvanus Olympio as head of state.
1963 Olympio killed in a military coup. Nicolas Grunitzky became president.
1967 Grunitzky replaced by Lt-Gen Etienne Gnassingbé Eyadéma in bloodless coup.
1973 Assembly of Togolese People (RPT) formed as sole legal political party.
1975 EEC Lomé convention signed in Lomé,

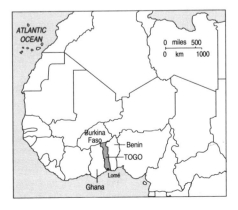

establishing trade links with developing countries. **1979** Eyadéma returned in election. Further EEC Lomé convention signed.
1986 Attempted coup failed.
1991 Eyadéma legalized opposition parties. National conference elected Joseph Kokou Koffigoh head of interim government; troops loyal to Eyadéma failed to reinstate him.
1992 Overwhelming referendum support for multiparty politics.

Tōhoku /təuˈhəuku:/ mountainous region of N Honshu Island, Japan; area 66,971 sq km/ 25,867 sq mi; population (1988) 9,745,000. Timber, fruit, fish, and livestock are produced. The chief city is Sendai. Aomori in the NE is linked to Hakodate on the island of Hokkaido by the **Seikan Tunnel**, the world's longest underwater tunnel.

Tokelau /ˈtəukəlau/ (formerly **Union Islands**) overseas territory of New Zealand, 480 km/ 300 mi N of Western Samoa, comprising three coral atolls: Atafu, Fakaofo, and Nukunonu; area 10 sq km/4 sq mi; population (1986) 1,700. The islands belong to the Polynesian group. Their resources are small, and until 1975 many of the inhabitants settled in New Zealand, which has administered them since 1926 when they were separated from the British Gilbert and Ellice islands colony.

Tokyo /ˈtəukiəu/ capital of Japan, on Honshu Island; population (1990) 8,163,100, metropolitan area over 12 million. The Sumida River delta separates the city from its suburb of Honjo. It is Japan's main cultural and industrial centre (engineering, chemicals, textiles, electrical goods). Founded in the 16th century as **Yedo** (or **Edo**), it was renamed when the emperor moved his court there from Kyoto 1868. An earthquake 1923 killed 58,000 people and destroyed much of the city, which was again severely damaged by Allied bombing in World War II. The subsequent rebuilding has made it into one of the world's most modern cities.
Features include the Imperial Palace, National Diet (parliament), Asakusa Kannon Temple (7th century, rebuilt after World War II), National Theatre, National Museum and other art collections, Tokyo University 1877, Tokyo Disneyland, and the National Athletic Stadium.

Frequent outbreaks of fire constitute one of the greatest perils of life in Japan. In Tokyo, burning houses are given the poetic name of the Flower of Edo.

On **Tokyo** James Kirkup *Heaven Hell and Hare-kiri* 1974

Toledo /tɒˈleɪdəu/ city on the river Tagus, Castilla–La Mancha, central Spain; population (1982) 62,000. It was the capital of the Visigoth kingdom 534–711, then became a Moorish city, and was the Castilian capital 1085–1560.
In the 12th century, Toledo had a flourishing steel industry and a school of translators, run by Archbishop Raymond (1125–1151), writing Latin versions of Arabic philosophical works. The painter El Greco worked here from about 1575 (his house and garden are preserved), and the local landscape is the setting of Cervantes' novel *Don Quixote*.

Toledo /təˈliːdəu/ port on Lake Erie, Ohio, USA, at the mouth of the Maumee River; population (1990) 332,900. Industries include food processing and the manufacture of vehicles, electrical goods, and glass. A French fort was built 1700, but permanent settlement did not begin until after the War of 1812.

Tomsk /tɒmsk/ city on the river Tom, W central Siberia; industries include synthetic fibres, timber, distilling, plastics, and electrical motors; population (1987) 489,000. It was formerly a gold-mining town and the administrative centre of much of Siberia.

Tonga Kingdom of *(Pule'anga Fakatu'i 'o Tonga)* or **Friendly Islands**
area 750 sq km/290 sq mi
capital Nuku'alofa (on Tongatapu island)
towns Pangai, Neiafu
physical three groups of islands in SW Pacific, mostly coral formations, but actively volcanic in W
features of 170 islands in the Tonga group, 36 are inhabited
head of state King Taufa'ahau Tupou IV from 1965
head of government Baron Vaea from 1991
political system constitutional monarchy
political parties none
currency Tongan dollar or pa'anga
population (1988) 95,000; growth rate 2.4% p.a.
life expectancy men 69, women 74 (1989)
languages Tongan (official), English
religions Wesleyan 47%, Roman Catholic 14%, Free Church of Tonga 14%, Mormon 9%, Church of Tonga 9%
literacy 93% (1988)
GNP $65 million (1987); $430 per head
chronology
1831 Tongan dynasty founded by Prince Taufa'ahau Tupou.
1900 Became a British protectorate.
1965 Queen Salote died; succeeded by her son,

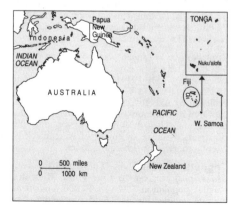

King Taufa'ahau Tupou IV.
1970 Independence achieved from Britain within the Commonwealth.
1990 Three prodemocracy candidates elected. Calls for reform of absolutist power.

Tongariro /ˌtɒŋəˈrɪərəʊ/ volcanic peak at the centre of North Island, New Zealand. Considered sacred by the Maori, the mountain was presented to the government by chief Te Heuheu Tukino IV 1887. It was New Zealand's first national park and the fourth to be designated in the world.

Tonkin /ˌtɒnˈkɪn/ or **Tongking** former region of Vietnam, on the China Sea; area 103,500 sq km/ 39,951 sq mi. Under Chinese rule from 111 BC, Tonkin became independent AD 939 and remained self-governing until the 19th century. A part of French Indochina 1885–1946, capital Hanoi, it was part of North Vietnam from 1954 and was merged into Vietnam after the Vietnam War.

Tonkin, Gulf of /ˌtɒnˈkɪn/ part of the South China Sea, with oil resources. China and Vietnam disagree over their respective territorial boundaries in the area.

Tonle Sap /ˈtɒnli ˈsæp/ or **Great Lake** lake on a tributary of the Mekong River, W Cambodia; area ranging from 2,600 sq km/1,000 sq mi to 6,500 sq km/2,500 sq mi at the height of the monsoon. During the June–Nov wet season it acts as a natural flood reservoir.

Toowoomba /təˈwʊmbə/ town and commercial and industrial centre (coal-mining, iron-working, engineering, clothing) in the Darling Downs, SE Queensland, Australia; population (1987) 79,000.

Topeka /təˈpiːkə/ capital of Kansas, USA; population (1990) 119,900. It is a centre for agricultural trade, and its products include processed food, printed materials, and rubber and metal products.

Torbay /ˌtɔːˈbeɪ/ district in S Devon, England; population (1981) 116,000. It was created 1968 by the union of the seaside resorts of Paignton, Torquay, and Brixham.

Torgau /ˈtɔːɡaʊ/ town in Leipzig county, E Germany; population 20,000. In 1760, during the Seven Years' War, Frederick II of Prussia defeated the Austrians nearby, and in World War II the US and Soviet forces first met here.

Torino /tɒˈriːnəʊ/ Italian name for the city of ⬦Turin.

Torness /ˌtɔːˈnes/ site of an advanced gas-cooled nuclear reactor 7 km/4.5 mi SW of Dunbar, East Lothian, Scotland. It started to generate power 1987.

Toronto /təˈrɒntəʊ/ (North American Indian 'place of meeting') known until 1834 as **York**. Port and capital of Ontario, Canada, on Lake Ontario; metropolitan population (1985) 3,427,000. It is Canada's main industrial and commercial centre (banking, shipbuilding, cars, farm machinery, food processing, publishing) and

also a cultural centre, with theatres and a film industry. A French fort was established 1749, and the site became the provincial capital 1793.

> *... no Canadian city can ever be anything better or different. If they are good they become Toronto.*
>
> On **Toronto** Rupert Brooke, letter of 1913

Torquay /ˌtɔːˈkiː/ resort in S Devon, England, part of the district of Torbay. It is a sailing centre and has an annual regatta in August.

Torremolinos /ˌtɒrɪməˈliːnɒs/ tourist resort on the Costa del Sol between Málaga and Algeciras in Andalucia, S Spain; population (1991) 31,700. There is a wine museum and a modern congress and exhibition centre.

Torrens /ˈtɒrənz/ salt lake 8 m/25 ft below sea level in E South Australia; area 5,800 sq km/ 2,239 sq mi. It is reduced to a marsh in dry weather.

Torreón /ˌtɒriˈɒn/ industrial and agricultural city in Coahuila state, N Mexico, on the river Nazas at an altitude of 1,127 m/3,700 ft; population (1986) 730,000. Before the arrival of the railway 1907 Torreón was the largest of the three Laguna cotton-district cities (with Gómez Palacio and Ciudad Lerdo). Since then it has developed as a major thoroughfare and commercial centre.

Torres Strait /ˈtɒrɪs/ channel separating New Guinea from Australia, with scattered reefs; width 130 km/80 mi. The first European to sail through it was the Spanish navigator Luis Vaez de Torres 1606.

Torres Vedras /ˈtɒrɪs ˈveɪdrəs/ town in Portugal, 40 km/25 mi N of Lisbon, where the fortifications known as the **lines of Torres Vedras** were built by the British commander Wellington 1810, during the Peninsular War 1808–14, when the British, Spanish, and Portuguese joined forces against the French.

Tortuga /tɔːˈtuːɡə/ (French **La Tortue** 'turtle') island off the north coast of Haiti; area 180 sq km/ 69 sq mi. It was a pirate lair during the 17th century.

Toruń /ˈtɒrʊn/ (German **Thorn**) industrial river port (electronics, fertilizers, synthetic fibres) in N Poland, on the river Vistula; population (1990) 262,200. It was founded by the Teutonic Knights 1230 and is the birthplace of the astronomer Copernicus.

Toscana /tɒˈskɑːnə/ Italian name for the region of ⬦Tuscany.

Tottenham /ˈtɒtənəm/ district of the Greater London borough of Haringey.

Toulon /tuːˈlɒn/ port and capital of Var **département**, SE France, on the Mediterranean Sea, 48 km/30 mi SE of Marseille; population (1990) 170,200. It is the chief Mediterranean naval station of France. Industries include oil refining, chemicals, furniture, and clothing. Toulon was the Roman *Telo Martius* and was made a port by Henry IV. It was occupied by the British 1793, and Napoleon first distinguished himself in driving them out. In World War II the French fleet was scuttled here to avoid its passing to German control.

Toulouse /tuː'luːz/ capital of Haute-Garonne *département*, SW France, on the river Garonne, SE of Bordeaux; population (1990) 365,900. The chief industries are textiles and aircraft construction (Concorde was built here). It was the capital of the Visigoths and later of Aquitaine 781–843. *features* Toulouse is known as 'la ville rose' (the red city) because most of the buildings are made from red brick. It has a fine 12th–13th-century Romanesque church (St Sernin), the church of the Jacobins (belonging to a monastery founded 1216), and the 11th–17th-century cathedral of St Etienne. The main square is the Place du Capitole. The university was founded 1229. *history* Founded by the Romans, Toulouse was the Visigoth capital in the 5th century, then capital of Aquitaine 781–843, and later of Languedoc. The cultural centre of medieval France in the 12th–13th centuries, it was captured 1218 by Simon de Montfort in the pope's crusade against the Albigensian heretics (the Count of Toulouse had been accused of complicity in the murder of a papal legate). The Duke of Wellington repulsed the French marshal Soult here 1814 in the Peninsular War. During the 20th century the city has expanded as a major European centre of scientific research (aerospace, electronics, data processing, agriculture).

Touraine /tuə'rein/ former province of W central France, now part of the *départements* of Indre-et-Loire and Vienne; capital Tours.

Tourcoing /tuə'kwæŋ/ town in Nord *département*, France, part of metropolitan Lille; population (1990) 94,400. It is situated near the Belgian border, and has been a textile centre since the 12th century.

Tournai /tuə'nei/ (Flemish *Doornik*) town in Hainaut province, Belgium, on the river Scheldt; population (1991) 67,700. Industries include carpets, cement, and leather. It stands on the site of a Roman relay post and has an 11th-century Romanesque cathedral.

Tours /tuə/ industrial city (chemicals, textiles, machinery) and capital of the Indre-et-Loire *département*, W central France, on the river Loire; population (1990) 133,400. It has a 13th–15th-century cathedral. An ancient city and former capital of Touraine, it was the site of the French defeat of the Arabs 732 under Charles Martel. Tours became the French capital for four days during World War II.

Tower Hamlets /'tauə 'hæmləts/ borough of E Greater London; population (1988 est) 161,800. It includes: the Tower of London, and the World Trade Centre in former St Katharine's Dock; the *Isle of Dogs* bounded on three sides by the

Thames, including the former India and Millwall docks (redevelopment includes Canary Wharf and Docklands, Billingsgate fish market, removed here 1982, and the Docklands light railway, linking the isle with the City); *Limehouse district*, the main centre of 18th-and 19th-century shipbuilding, which in the 1890s became a focal point for Chinese sailors operating out of the West India Dock; *Spitalfields district* which derives its name from the Hospital of St Mary founded there 1197; silk was woven by exiled Flemish and French Huguenots 16th–17th centuries; the industry collapsed mid-19th century; *Bethnal Green* which has a Museum of Childhood; *Wapping* which has replaced Fleet Street as the centre of the newspaper industry. Mile End Green (later Stepney Green) was where Richard II met the rebels of the 1381 Peasant's Revolt.

Townsville /'taunzvil/ port on Cleveland Bay, N Queensland, Australia; population (1987) 108,000. It is the centre of a mining and agricultural area and exports meat, wool, sugar, and minerals, including gold and silver.

Trabzon /'træbzɒn/ (formerly *Trebizond*) port on the Black Sea, NE Turkey, 355 km/220 mi SW of Batum; population (1990) 143,900. Its exports include fruit, tobacco, and hides.

Trail /treil/ mining centre in British Columbia, Canada, on the Columbia River; population (1981) 10,000. It has lead, zinc, and copper industries.

Trans-Amazonian Highway /,trænz, æmə'zəuniən/ or *Transamazonica* road in Brazil, linking Recife in the E with the provinces of Rondonia, Amazonas, and Acre in the W. Begun as part of the Brazilian National Integration Programme (PIN) 1970, the Trans-Amazonian Highway was designed to enhance national security, aid the industrial development of the N of Brazil, and act as a safety valve for the overpopulated coastal regions.

Transcaucasia /,trænzkɔː'keiziə/ geographical region S of the Caucasus Mountains, which includes Armenia, Azerbaijan, and Georgia. It formed the *Transcaucasian Republic* 1922, but was broken up 1936 into three separate Soviet republics. All three republics became independent 1991.

Trans-Dniester region of NE Moldova, lying between the river Dniester and the Ukraine; it is largely inhabited by ethnic Slavs (Russians and Ukrainians). In Oct 1990, Slav separatists unilaterally declared a breakaway republic, fearing a resurgence of ethnic Romanian nationalism as Moldova moved toward independence; a state of emergency was declared Nov 1990 after ethnic clashes left six dead. The violence escalated May 1992 in response to mounting fears of reunification with Romania and by July 1992 hundreds had died in the fighting. By Aug a cease-fire was in place and a Russian peacekeeping force reportedly deployed in the region.

Transjordan /trænz'dʒɔːdn/ former name (1923–49) of the Hashemite kingdom of ◊Jordan.

Transkei /ˌtrænsˈkaɪ/ largest of South Africa's Bantustans, or homelands, extending NE from the Great Kei River, on the coast of Cape Province, to the border of Natal; area 43,808 sq km/ 16,910 sq mi; population (1985) 3,000,000, including small white and Asian minorities. It became self-governing 1963, and achieved full 'independence' 1976. Its capital is Umtata, and it has a port at Mnganzana. It is one of the two homelands of the Xhosa people (the other is Ciskei), and products include livestock, coffee, tea, sugar, maize, and sorghum. It is governed by a military council since a 1987 coup (military leader Maj-Gen H B Holomisa from 1987).

Trans-Siberian Railway /ˌtrænssaɪˈbɪərɪən/ railway line connecting the cities of European Russia with Omsk, Novosibirsk, Irkutsk, and Khabarovsk, and terminating at Vladivostok on the Pacific. It was built 1891–1905; from Leningrad to Vladivostok is about 8,700 km/ 5,400 mi. A 3,102 km/1,928 mi northern line was completed 1984 after ten years' work.

Transvaal /ˈtrænzvɑːl/ province of NE South Africa, bordering Zimbabwe to the N; area 262,499 sq km/101,325 sq mi; population (1985) 7,532,000. Its capital is Pretoria, and towns include Johannesburg, Germiston, Brakpan, Springs, Benoni, Krugersdorp, and Roodepoort. Products include diamonds, coal, iron ore, copper, lead, tin, manganese, meat, maize, tobacco, and fruit.
    The main rivers are the Vaal and Limpopo with their tributaries. Swaziland forms an enclave on the Natal border. It was settled by *Voortrekkers*, Boers who left Cape Colony in the Great Trek from 1831. Independence was recognized by Britain 1852, until the settlers' difficulties with the conquered Zulus led to British annexation 1877. It was made a British colony after the South African War 1899–1902, and in 1910 became a province of the Union of South Africa.

Transylvania /ˌtrænsɪlˈveɪnɪə/ mountainous area of central and NW Romania, bounded to the S by the Transylvanian Alps (an extension of the Carpathian Mountains), formerly a province, with its capital at Cluj. It was part of Hungary from about 1000 until its people voted to unite with Romania 1918. It is the home of the vampire legends.

Poverty reigns in Transylvania, but it is a distinguished poverty.

On **Transylvania** Walter Starkie
Raggle-Taggle 1933

Trapani /ˈtrɑːpəni/ port and naval base in NW Sicily, about 48 km/30 mi N of Marsala; population (1981) 72,000. It trades in wine, salt, and fish.

Trebizond /ˈtrebɪzɒnd/ former English name of ♢Trabzon, a city in Turkey.

Trengganu /treŋˈgɑːnuː/ or *Terengganu* state of E Peninsular Malaysia; capital Kuala Trengganu; area 13,000 sq km/5,018 sq mi; population (1990) 752,000. Its exports include copra, black pepper, tin, and tungsten; there are also fishing and offshore oil industries.

Trent /trent/ third longest river of England; length 275 km/170 mi. Rising in the S Pennines, it flows first S and then NE through the Midlands to the Humber estuary and out into the North Sea. It is navigable by barge for nearly 160 km/100 mi.

Trentino–Alto Adige /trenˈtiːnəʊ ˈæltəʊ ˈædɪdʒeɪ/ autonomous region of N Italy, comprising the provinces of Bolzano and Trento; capital Trento; chief towns Trento in the Italian-speaking southern area, and Bolzano-Bozen in the northern German-speaking area of South Tirol (the region

Transylvania

was Austrian until ceded to Italy 1919); area 13,600 sq km/5,250 sq mi; population (1990) 891,400.

Trento /'trentəu/ capital of Trentino–Alto Adige region, Italy, on the Adige River; population (1988) 101,000. Industries include the manufacture of electrical goods and chemicals. The Council of Trent was held here 1545–63.

Trenton /'trentən/ capital and industrial city (metalworking, ceramics) of New Jersey, USA, on the Delaware River; population (1990) 88,700. It was first settled by Quakers 1679; George Washington defeated the British here 1776. It became the state capital 1790.

Trèves /trev/ French name for ♢Trier, a city in Germany.

Treviso /tre'vi:zəu/ city in Veneto, NE Italy; population (1981) 88,000. Its industries include the manufacture of machinery and ceramics. The 11th-century cathedral has an altarpiece by Titian.

Tribal Areas, Federally Administered part of the mountainous frontier of NW Pakistan with Afghanistan, comprising the districts of Malakand, Mohmand, Khyber, Kurram, and Waziristan, administered directly from Islamabad; area 27,219 sq km/10,507 sq mi; population (1985) 2,467,000. The chief towns are Wana, Razmak, and Miram Shah.

Trichinopoly /ˌtrɪkɪ'nɒpəli/ former name for ♢Tiruchirapalli, a city in India.

Trier /triə/ (French *Trèves*) city in Rhineland-Palatinate, Germany; population (1984) 95,000. It is a centre for the wine trade. Once the capital of the Treveri, a Celto-Germanic tribe, it became known as *Augusta Treverorum* under the Roman emperor Augustus about 15 BC and was the capital of an ecclesiastical principality during the 14th–18th centuries. Karl Marx was born here.

Trieste /tri'est/ port on the Adriatic coast, opposite Venice, in Friuli-Venezia-Giulia, Italy; population (1988) 237,000, including a large Slovene minority. It is the site of the International Centre for Theoretical Physics, established 1964. *history* Trieste was under Austrian rule from 1382 (apart from Napoleonic occupation 1809–14) until transferred to Italy 1918. It was claimed after World War II by Yugoslavia, and the city and surrounding territory were divided 1954 between Italy and Yugoslavia.

Triglav /'tri:glau/ mountain in Slovenia, rising to 2,863 m/9,393 ft, the highest peak in the Julian Alps.

Trincomalee /ˌtrɪŋkəmə'li:/ port in NE Sri Lanka; population (1981) 45,000. It was an early Tamil settlement, and a British naval base until 1957.

Trinidad /ˌtrɪnɪ'dæd/ town in Beni region, N Bolivia, near the river Mamoré, 400 km/250 mi NE of La Paz; population (1980) 36,000. It is built on an artificial earth mound, above flood level, the work of a little-known early American Indian people.

Trinidad and Tobago Republic of
area Trinidad 4,828 sq km/1,864 sq mi and Tobago 300 sq km/116 sq mi
capital Port-of-Spain
towns San Fernando, Arima, Scarborough (Tobago)
physical comprises two main islands and some smaller ones; coastal swamps and hills E–W
features Pitch Lake, a self-renewing source of asphalt used by 16th-century explorer Walter Raleigh to repair his ships
head of state Noor Hassanali from 1987
head of government Patrick Manning from 1991
political system democratic republic
political parties National Alliance for Reconstruction (NAR), nationalist, left-of-centre; People's National Movement (PNM), nationalist, moderate, centrist
exports oil, petroleum products, chemicals, sugar, cocoa
currency Trinidad and Tobago dollar
population (1990 est) 1,270,000 (40% African descent, 40% Indian, 16% European, 2% Chinese and others), 1.2 million on Trinidad; growth rate 1.6% p.a.
life expectancy men 68, women 72 (1989)
languages English (official), Hindi, French, Spanish
media freedom of press guaranteed by constitution and upheld by government; there are two independent morning newspapers and several weekly tabloids
religions Roman Catholic 32%, Protestant 29%, Hindu 25%, Muslim 6%
literacy 97% (1988)
GNP $4.5 bn; $3,731 per head (1987)
chronology
1888 Trinidad and Tobago united as a British colony.
1956 People's National Movement (PNM) founded.
1959 Achieved internal self-government, with PNM leader Eric Williams as chief minister.
1962 Independence achieved from Britain, within the Commonwealth.
1976 Became a republic, with Ellis Clarke as president and Williams as prime minister.

1981 Williams died and was succeeded by George Chambers, with Arthur Robinson as opposition leader.
1986 NAR, headed by Arthur Robinson, won general election.
1987 Noor Hassanali became president.
1990 Attempted antigovernment coup defeated.
1991 General election saw victory for PNM, with Patrick Manning as prime minister.

Tripoli /'trɪpəli/ (Arabic *Tarabolus al-Gharb*) capital and chief port of Libya, on the Mediterranean coast; population (1982) 980,000. Products include olive oil, fruit, fish, and textiles.
history Tripoli was founded about the 7th century BC by Phoenicians from Oea (now Tripoli in Lebanon). It was a base for Axis powers during World War II. In 1986 it was bombed by the US Air Force in retaliation for international guerrilla activity.

Tripoli /'trɪpəli/ (Arabic *Tarabolus esh-sham*) port in N Lebanon, 65 km/40 mi NE of Beirut; population (1980) 175,000. There is oil refining. It stands on the site of the Phoenician city of Oea.

Tripolitania /ˌtrɪpəlɪ'teɪnɪə/ former province of Libya, stretching from Cyrenaica in the E to Tunisia in the W. It came under Turkish rule in the 16th century; Italy captured it from Turkey 1912, and the British captured it from Italy 1942 and controlled it until it was incorporated into the newly independent United Kingdom of Libya, established 1951. In 1963 Tripolitania was subdivided into administrative divisions.

Tripura /'trɪpurə/ state of NE India since 1972, formerly a princely state, between Bangladesh and Assam
area 10,500 sq km/4,053 sq mi
capital Agartala
features agriculture on a rotation system in the rainforest, now being superseded by modern methods
products rice, cotton, tea, sugar cane; steel, jute
population (1991) 2,744,800
language Bengali
religion Hindu.

Tristan da Cunha /'trɪstən də 'kuːnjə/ group of islands in the S Atlantic, part of the British dependency of St Helena;
area 110 sq km/42 sq mi
features comprises four islands: Tristan, Gough, Inaccessible, and Nightingale. Tristan consists of a single volcano 2,060 m/6,761 ft; it is an important meteorological and radio station
government administrator and island council
products crayfish
currency pound sterling
population (1990) 299
language English
history the first European to visit the then uninhabited islands was the Portuguese admiral after whom they are named, 1506; they were annexed by Britain 1816. Believed to be extinct, the Tristan volcano erupted 1961 and the population was evacuated, but in 1963 they chose to return.

Trivandrum /trɪ'vændrəm/ capital of Kerala, SW

India; population (1981) 483,000. It has chemical, textile, and rubber industries. Formerly the capital of the princely state of Travancore, it has many palaces, an old fort, and a shrine.

Trobriand Islands /'trəubriənd/ group of coral islands in the Solomon Sea, forming part of the province of Milne Bay, Papua New Guinea; chief town Losuia; area 440 sq km/170 sq mi.

Trois-Rivières /'trwɑ: riːv'jeə/ port on the St Lawrence River, at the point where the St Maurice River enters the St Lawrence, Québec, Canada; population (1986) 129,000. The chief industry is the production of newsprint. It was founded by the French explorer Samuel de Champlain 1634.

Tromsø /'trɒmsɜ:/ fishing port and the largest town in NW Norway, on Tromsø Island; population (1991) 51,300. A church was founded here in the 13th century and the town grew up around it. It is used as a base for Arctic expeditions.

Trondheim /'trɒndhaɪm/ fishing port in Norway; population (1990) 137,800. It has canning, textile, margarine, and soap industries. It was the medieval capital of Norway, and Norwegian kings are crowned in the cathedral (1066–93). Trondheim was occupied by the Germans 1940–45.

Trossachs /'trɒsəks/ woodland glen between lochs Katrine and Achray in Central Region, Scotland, 3 km/2 mi long. Featured in the novels of Walter Scott, it has become a favoured tourist spot.

Trowbridge /'trəubrɪdʒ/ market town in Wiltshire, England, 12 km/7 mi SE of Bath; population (1981) 23,000. Its industries include dairy produce, bacon, ham, and wool.

Troy /trɔɪ/ city in E New York, USA, E of Albany on the east bank of the Hudson River; seat of Rensselear County, incorporated 1816; population (1990) 54,300. Industries include clothing,

Tripura

abrasives, metals, paper, car and railway parts, and processed foods.

A Mohegan Indian fortress, it was explored by Henry Hudson 1609, granted by the Dutch East India Company as a patroonship to Kiliaen Van Rensselaer, and founded as a town 1786.

Troyes /trwɑː/ industrial town (textiles and food processing) in Champagne-Ardenne, NE France; population (1990) 60,800. It was the capital of the medieval province of Champagne. The *Treaty of Troyes* 1420 made Henry V of England heir to the French crown.

Trucial States /'truːʃəl 'steɪts/ former name (until 1971) of the ⟡United Arab Emirates. It derives from the agreements made with Britain 1820 to ensure a truce in the area and to suppress piracy and slavery.

Trujillo /truːˈxiːəu/ city in NW Peru, with its port at Salaverry; population (1988) 491,000. Industries include engineering, copper, sugar milling, and vehicle assembly.

Truk /trʌk/ group of about 55 volcanic islands surrounded by a coral reef in the E Caroline islands of the W Pacific, forming one of the four states of the Federated States of Micronesia. Fish and copra are the main products.

The lagoon area (one of the largest in the world) has an area of 2,130 sq km/822 sq mi. The main island is Moen. More than 60 ships of the Japanese fleet, destroyed during World War II, lie at the bottom of the lagoon.

Truong Sa /ˌtruːɒŋ ˈsɑː/ one of the Spratly Islands in the South China Sea.

Truro /'truərəu/ city in Cornwall, England, and administrative headquarters of the county; population (1982) 16,000.

Truro was the traditional meeting place of the Stannary (local parliament), and was formerly a centre for the nearby tin-mining industry. The cathedral, designed by J L Pearson (1817–1897) dates from 1880–1910, and the museum and art gallery has works by John Opie.

Tsaritsyn /tsɑːˈrɪtsɪn/ former name (until 1925) of ⟡Volgograd, a city in Russia.

Tsavo /'tsɑːvəu/ national park in SE Kenya, established 1948. One of the world's largest, it occupies 20,821 sq km/8,036 sq mi.

Tselinograd /tseˈlɪnəgræd/ (formerly until 1961 *Akmolinsk*) commercial and industrial city in N Kazakhstan, on the river Ishim, situated at a railway junction; population (1983) 253,000. It produces agricultural machinery, textiles, and chemicals.

Tsinan /ˌtsiːˈnæn/ alternative transliteration of ⟡Jinan, the capital of Shandong province, E China.

Tsingtao /ˌtsɪŋˈtau/ alternative transliteration of ⟡Qingdao, a port in E China.

Tsumeb /'tsuːmeb/ principal mining centre (diamonds, copper, lead, zinc) of N Namibia, NW of Grootfontein; population 13,500.

Tsushima /'tsuːʃimɑ/ Japanese island between Korea and Japan in *Tsushima Strait*; area 702 sq km/271 sq mi; population (1990) 59,300. The Russian fleet was destroyed by the Japanese

here 27 May 1905 in the Russo-Japanese War, and 12,000 Russians were killed. The chief settlement is Izuhara.

Tuamotu Archipelago /ˌtuːəˈmautuː/ two parallel ranges of 78 atolls, part of French Polynesia; area 690 sq km/266 sq mi; population (1983) 11,800, including the Gambier Islands to the E. The atolls stretch 2,100 km/1,300 mi N and E of the Society Islands. The administrative headquarters is Apataki. The largest atoll is Rangiroa, the most significant is Hao; they produce pearl shell and copra. Mururoa and Fangataufa atolls to the SE have been a French nuclear test site since 1966. Spanish explorers landed 1606, and the islands were annexed by France 1881.

Tübingen /'tjuːbɪŋən/ town in Baden-Württemberg, Germany, on the river Neckar, 30 km/19 m S of Stuttgart; population (1985) 75,000. Industries include paper, textiles, and surgical instruments. The town dates from the 11th century; the university was established 1477.

Tubuai Islands /ˌtuːbuːˈaɪ/ or *Austral Islands* chain of volcanic islands and reefs 1,300 km/800 mi long in French Polynesia, S of the Society Islands; area 148 sq km/57 sq mi; population (1983) 6,300. The main settlement is Mataura on Tubuai. They were visited by Capt Cook 1777 and annexed by France 1880.

Tucson /'tuːsɒn/ resort city in the Sonora Desert in SE Arizona, USA; population (1990) 405,400. It stands 760 m/2,500 ft above sea level, and the Santa Catalina Mountains to the NE rise to about 2,750 m/9,000 ft. Industries include aircraft, electronics, and copper smelting. Tucson passed from Mexico to the USA 1853 and was the territorial capital 1867–77.

Tucumán /ˌtukuːˈmɑːn/ or *San Miguel de Tucumán* capital of Tucumán province, NW Argentina, on the Rio Sali, in the foothills of the Andes; population (1991) 473,000. Industries include sugar mills and distilleries. Founded 1565, Tucumán was the site of the signing of the Argentine declaration of independence from Spain 1816.

Tula /'tuːlɑ/ city in Russia, on the river Upa, 193 km/121 mi S of Moscow; population (1987) 538,000. Industries include engineering and metallurgy. It was the site of the government ordnance factory, founded 1712 by Peter the Great.

Tulare /tuˈleəri/ town in S central California, USA, S of Fresno, in the San Joaquin Valley; population (1990) 33,200. Industries include food processing, wine, and dairy products. Tulare was built 1871 as a division headquarters for the Southern Pacific Railroad.

Tulsa /'tʌlsə/ city in NE Oklahoma, USA, on the Arkansas River, NE of Oklahoma City; population (1990) 367,300. It is an oil-producing and aerospace centre; other industries include mining, machinery, metal, and cement.

Tunbridge Wells, Royal /'tʌnbrɪdʒ 'welz/ spa town in Kent, SE England, with iron-rich springs discovered 1606; population (1985) 98,500. There is an expanding light industrial estate. The

Pantiles or shopping parade (paved with tiles in the reign of Queen Anne), was a fashionable resort; the town has been named 'Royal' since 1909 after visits by Queen Victoria.

Tunbs, the /tumbz/ two islands in the Strait of Hormuz, formerly held by Ras al Khaimah and annexed from other Gulf states by Iran 1971; their return to their former owners was an Iraqi aim in the Iran–Iraq War.

Tungurahua /ˌtuŋguˈrɑːwə/ active volcano in the Andes of central Ecuador, the tenth highest peak in Ecuador; height 5,016 m/16,456 ft. Its last major eruption was in 1886.

Tunis /ˈtjuːnɪs/ capital and chief port of Tunisia; population (1984) 597,000. Industries include chemicals and textiles. Founded by the Arabs, it was captured by the Turks 1533, then occupied by the French 1881 and by the Axis powers 1942–43. The ruins of ancient Carthage are to the NE.

Tunisia Tunisian Republic (*al-Jumhuriya at-Tunisiya*)
area 164,150 sq km/63,378 sq mi
capital and chief port Tunis
towns ports Sfax, Sousse, Bizerta
physical arable and forested land in N graduates towards desert in S
features fertile island of Jerba, linked to mainland by causeway (identified with island of lotus-eaters); Shott el Jerid salt lakes; holy city of Kairouan, ruins of Carthage
head of state and government Zine el-Abidine Ben Ali from 1987
political system emergent democratic republic
political party Constitutional Democratic Rally (RCD), nationalist, moderate, socialist
exports oil, phosphates, chemicals, textiles, food, olive oil
currency dinar
population (1990 est) 8,094,000; growth rate 2% p.a.
life expectancy men 68, women 71 (1989)
languages Arabic (official), French
media publications must be authorized; the offence of defamation is used to protect members of the government from criticism
religions Sunni Muslim 95%; Jewish, Christian

literacy men 68%, women 41% (1985 est)
GNP $9.6 bn (1987); $1,163 per head (1986)
chronology
1883 Became a French protectorate.
1955 Granted internal self-government.
1956 Independence achieved from France as a monarchy, with Habib Bourguiba as prime minister.
1957 Became a republic with Bourguiba as president.
1975 Bourguiba made president for life.
1985 Diplomatic relations with Libya severed.
1987 Bourguiba removed Prime Minister Rashed Sfar and appointed Zine el-Abidine Ben Ali. Ben Ali declared Bourguiba incompetent and seized power.
1988 Constitutional changes towards democracy announced. Diplomatic relations with Libya restored.
1989 Government party, RDC, won all assembly seats in general election.
1991 Opposition to US actions during the Gulf War. Crackdown on religious fundamentalists.

Tunja /ˈtuŋhɑː/ capital of the Andean department of Boyacá, E central Colombia; population (1985) 93,800. Formerly the seat of the Chibcha Indian kings, the Spanish built a city here 1539. In 1818 Simón Bolívar defeated Spanish Royalists near Tunja.

Turin /tjuˈrɪn/ (Italian *Torino*) capital of Piedmont, NW Italy, on the river Po; population (1988) 1,025,000. Industries include iron, steel, cars, silk and other textiles, fashion goods, chocolate, and wine. There is a university, established 1404, and a 15th-century cathedral. Features include the Palazzo Reale (Royal Palace) 1646–58 and several gates to the city. It was the first capital of united Italy 1861–64.

Turin became important after the union of Savoy and Piedmont 1416. Its growth as a major city dates from 1559 when Emanuele Filiberto chose it as the capital of the House of Savoy, with Italian as the official language. In 1706 Prince Eugène defeated a French army besieging the city, thus ensuring the survival of the Savoy duchy.

Turkana, Lake /tɜːˈkɑːnə/ (formerly until 1979 *Lake Rudolf*) lake in the Great Rift Valley,

375 m/1,230 ft above sea level, with its northernmost end in Ethiopia and the rest in Kenya; area 9,000 sq km/3,475 sq mi. It is saline, and shrinking by evaporation. Its shores were an early human hunting ground, and valuable remains have been found that are accurately datable because of undisturbed stratification.

Turkestan /ˌtɜːkɪˈstɑːn/ area of central Asia divided among Kazakhstan, Kyrgyzstan, Tajikistan, Turkmenistan, Uzbekistan, Afghanistan, and China (part of Xinjiang Uygur).

Turkey Republic of (*Türkiye Cumhuriyeti*)
area 779,500 sq km/300,965 sq mi
capital Ankara
towns ports Istanbul and Izmir
physical central plateau surrounded by mountains
environment only 0.3% of the country is protected by national parks and reserves compared with a global average of 7% per country
features Bosporus and Dardanelles; Mount Ararat; Taurus Mountains in SW (highest peak Kaldi Dag, 3,734 m/12,255 ft); sources of rivers Euphrates and Tigris in E; archaeological sites include Çatal Hüyük, Ephesus, and Troy; rock villages of Cappadocia; historic towns (Antioch, Iskenderun, Tarsus)
head of state Suleyman Demirel from 1993
head of government Tansu Ciller from 1993
political system democratic republic
political parties Motherland Party (ANAP), Islamic, nationalist, right of centre; Social Democratic Populist Party (SDPP), moderate, left of centre; True Path Party (TPP), centre-right
exports cotton, yarn, hazelnuts, citrus, tobacco, dried fruit, chromium ores
currency Turkish lira
population (1990 est) 56,549,000 (Turkish 85%, Kurdish 12%); growth rate 2.1% p.a.
life expectancy men 63, women 66 (1989)
languages Turkish (official), Kurdish, Arabic
religion Sunni Muslim 98%
literacy men 86%, women 62% (1985)
GNP $112.5 bn (1992)
chronology
1919–22 Turkish War of Independence provoked by Greek occupation of Izmir. Mustafa Kemal (Atatürk), leader of nationalist congress, defeated Italian, French, and Greek forces.
1923 Treaty of Lausanne established Turkey as independent republic under Kemal. Westernization began.
1950 First free elections; Adnan Menderes became prime minister.
1960 Menderes executed after military coup by General Cemal Gürsel.
1965 Suleyman Demirel became prime minister.
1971 Army forced Demirel to resign.
1973 Civilian rule returned under Bulent Ecevit.
1974 Turkish troops sent to protect Turkish community in Cyprus.
1975 Demirel returned to head of a right-wing coalition.
1978 Ecevit returned, as head of coalition, in the face of economic difficulties and factional violence.

1979 Demeril returned. Violence grew.
1980 Army took over, and Bulent Ulusu became prime minister. Harsh repression of political activists attracted international criticism.
1982 New constitution adopted.
1983 Ban on political activity lifted. Turgut Özal became prime minister.
1987 Özal maintained majority in general election.
1988 Improved relations and talks with Greece.
1989 Turgut Özal elected president; Yildirim Akbulut became prime minister. Application to join European Community (EC) rejected.
1991 Mesut Yilmaz became prime minister. Conflict with Kurdish minority continued. Coalition government formed under Suleyman Demirel after inconclusive election result.
1992 Earthquake claimed thousands of lives.
1993 Özal died and was succeeded by Demirel. Tansu Ciller became prime minister.

Turkmenistan Republic of
area 488,100 sq km/188,406 sq mi
capital Ashgabat
towns Chardzhov, Mary (Merv), Nebit-Dag, Krasnovodsk
physical some 90% of land is desert including the Kara Kum 'Black Sands' desert (area 310,800 sq km/120,000 sq mi)
features on the edge of the Kara Kum desert is the Altyn Depe, 'golden hill', site of a ruined city with a ziggurat, or stepped pyramid; river Amu Darya; rich deposits of petroleum, natural gas, sulphur, and other industrial raw materials
head of state Saparmurad Niyazov from 1991
head of government Sakhat Muradov from 1992
political system socialist pluralist
products silk, karakul, sheep, astrakhan fur, carpets, chemicals, rich deposits of petroleum, natural gas, sulphur, and other industrial raw materials
population (1990) 3,600,000 (Turkmen 72%, Russian 10%, Uzbek 9%, Kazakh 3%, Ukrainian 1%)
language West Turkic, closely related to Turkish

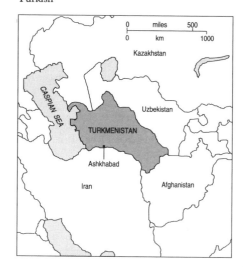

religion Sunni Muslim
chronology
1921 Part of Turkestan Soviet Socialist Autonomous Republic.
1925 Became a constituent republic of USSR.
1990 Aug: economic and political sovereignty declared.
1991 Jan: Communist Party leader Niyazov became state president. March: endorsed maintenance of the Union in USSR referendum. Aug: President Niyazov initially supported attempted anti-Gorbachev coup. Oct: independence declared. Dec: joined new Commonwealth of Independent States (CIS).
1992 Jan: admitted into Conference for Security and Cooperation in Europe (CSCE). Feb: joined the Muslim Economic Cooperation Organization (ECO). March: admitted into United Nations (UN); US diplomatic recognition achieved. May: new constitution adopted.

Turks and Caicos Islands /tɜːks, ˈkeɪkɒs/ British crown colony in the West Indies, the southeastern archipelago of the Bahamas
area 430 sq km/166 sq mi
capital Cockburn Town on Grand Turk
features a group of some 30 islands, of which six are inhabited. The largest is the uninhabited *Grand Caicos*; others include *Grand Turk* (population 3,100), *South Caicos* (1,400), *Middle Caicos* (400), *North Caicos* (1,300), *Providenciales* (1,000), and *Salt Cay* (300); since 1982 the Turks and Caicos have developed as a tax haven
government governor, with executive and legislative councils (chief minister from 1987 Michael John Bradley, Progressive National Party)
exports crayfish and conch (flesh and shell)
currency US dollar
population (1980) 7,500, 90% of African descent
languages English, French Creole
religion Christian
history secured by Britain 1766 against French and Spanish claims, the islands were a Jamaican dependency 1873–1962, and in 1976 attained internal self-government.

Turku /ˈtuəkuː/ (Swedish *Åbo*) port in SW Finland, near the mouth of the river Aura, on the Gulf of Bothnia; population (1990) 159,200. Industries include shipbuilding, engineering, textiles, and food processing. It was the capital of Finland until 1812.

Tuscaloosa /ˌtʌskəˈluːsə/ city in W central Alabama, USA, on the Black Warrior River, SW of Birmingham; population (1990) 77,800. Industries include chemicals, tyres, paper, and lumber. It was originally founded by Creek Indians 1809.

Tuscany /ˈtʌskəni/ (Italian *Toscana*) region of N central Italy, on the west coast; area 23,000 sq km/8,878 sq mi; population (1990) 3,562,500. Its capital is Florence, and towns include Pisa, Livorno, and Siena. The area is mainly agricultural, with many vineyards, such as in the Chianti hills; it also has lignite and iron mines and marble

quarries (Carrara marble is from here). The Tuscan dialect has been adopted as the standard form of Italian. Tuscany was formerly the Roman *Etruria*, and inhabited by Etruscans around 500 BC. In medieval times the area was divided into small states, united under Florentine rule during the 15th–16th centuries. It became part of united Italy 1861.

Tuva /ˈtuːvə/ (Russian *Tuvinskaya*) autonomous republic (administrative unit) of Russia, NW of Mongolia;
area 170,500 sq km/65,813 sq mi
capital Kyzyl
features good pasture; gold, asbestos, cobalt
population (1986) 284,000
history part of Mongolia until 1911 and declared a Russian protectorate 1914; after the 1917 revolution it became the independent Tannu-Tuva republic 1920, until incorporated in the USSR as an autonomous region 1944. It was made the Tuva Autonomous Republic 1961.

Tuvalu South West Pacific State of (formerly *Ellice Islands*)
area 25 sq km/9.5 sq mi
capital Funafuti
physical nine low coral atolls forming a chain of 579 km/650 mi in the SW Pacific
features maximum height above sea level 6 m/20 ft; coconut palms are main vegetation
head of state Elizabeth II from 1978 represented by governor general
head of government Bikenibeu Paeniu from 1989
political system liberal democracy
political parties none; members are elected to parliament as independents
exports copra, handicrafts, stamps
currency Australian dollar
population (1990 est) 9,000 (Polynesian 96%); growth rate 3.4% p.a.
life expectancy 60 men, 63 women (1989)
languages Tuvaluan, English
religion Christian (Protestant)
literacy 96% (1985)
GDP (1983) $711 per head
chronology
1892 Became a British protectorate forming part of the Gilbert and Ellice Islands group.

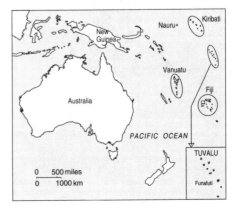

1916 The islands acquired colonial status.
1975 The Ellice Islands were separated from the Gilbert Islands.
1978 Independence achieved from Britain within the Commonwealth with Toaripi Lauti as prime minister.
1981 Tomasi Puapua replaced Lauti as premier.
1986 Islanders rejected proposal for republican status.
1989 Bikenibeu Paeniu elected new prime minister.

Tver /tveǝ/ (formerly 1932–90 *Kalinin*) city of NW Russia, capital of Tver region, a transport centre on the river Volga, 160 km/100 mi NW of Moscow; population (1987) 447,000.

Tweed /twiːd/ river rising in SW Borders Region, Scotland, and entering the North Sea at Berwick-upon-Tweed, Northumberland; length 156 km/ 97 mi.

Twickenham /'twɪkǝnǝm/ district in the Greater London borough of Richmond-upon-Thames.

TX abbreviation for ◊ *Texas*, a state of the USA.

Tyburn /'taɪbǝn/ stream in London, England, near which (at the junction of Oxford Street and Edgware Road) Tyburn gallows stood from the 12th century until 1783. The Tyburn now flows underground.

Tyler /'taɪlǝ/ city in NE Texas, USA, SE of Dallas; population (1990) 75,508. Industries include oil, roses, vegetables, furniture, and plastics.

Tyne /taɪn/ river of NE England formed by the union of the North Tyne (rising in the Cheviot Hills) and South Tyne (rising in Cumbria) near Hexham, Northumberland, and reaching the North Sea at Tynemouth; length 72 km/45 mi. Kielder Water (1980) in the N Tyne Valley is Europe's largest artificial lake, 12 km/7.5 mi long and 0.8 km/0.5 mi wide, and supplies the industries of Tyneside, Wearside, and Teesside.

Tyne and Wear /taɪn, wɪǝ/ metropolitan county in NE England, created 1974, originally admin-

Tyne and Wear

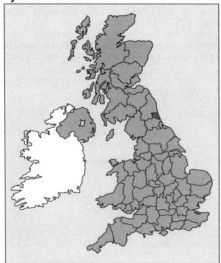

istered by an elected metropolitan council; its powers reverted to district councils 1986
area 540 sq km/208 sq mi
towns Newcastle-upon-Tyne, South Shields, Gateshead, Sunderland
features bisected by the rivers Tyne and Wear; includes part of Hadrian's Wall; Newcastle and Gateshead, linked with each other and with the coast on both sides by the Tyne and Wear Metro (a light railway using existing suburban lines, extending 54 km/34 mi)
products once a centre of heavy industry, it is now being redeveloped and diversified
population (1991) 1,087,000
famous people Thomas Bewick, Robert Stephenson, Harry Patterson ('Jack Higgins').

Tynemouth /'taɪnmauθ, 'tɪnmǝθ/ port and resort in Tyne and Wear, England; population (1985) 9,442.

Tyre /taɪǝ/ (Arabic *Sur* or *Soûr*) town in SW Lebanon, about 80 km/50 mi S of Beirut, formerly a port until its harbour silted up; population (1980 est) about 14,000. It stands on the site of the ancient city of the same name, a seaport of Phoenicia.
history Built on the mainland and two small islands, the city was a commercial centre, known for its purple dye. Besieged and captured by Alexander the Great 333–332 BC, it came under Roman rule 64 BC and was taken by Arab forces AD 638. The Crusaders captured it 1124, and it never recovered from the destruction it suffered when retaken by the Arabs 1291. In the 1970s it became a Palestinian guerrilla stronghold and was shelled by Israel 1979.

Tyre the crowning city, whose merchants are princes.

On **Tyre** the Bible, Isaiah 23:8

Tyrol variant spelling of ◊ Tirol, a state of Austria.

Tyrone /tɪ'rǝun/ county of Northern Ireland;
area 3,160 sq km/1,220 sq mi
towns Omagh (county town), Dungannon, Strabane, Cookstown
features rivers: Derg, Blackwater, Foyle; Lough Neagh
products mainly agricultural
population (1981) 144,000.

Tyrrhenian Sea /tɪ'riːniǝn/ arm of the Mediterranean Sea surrounded by mainland Italy, Sicily, Sardinia, Corsica, and the Ligurian Sea. It is connected to the Ionian Sea through the Straits of Messina. Islands include Elba, Ustica, Capri, Stromboli, and the Lipari Islands.

Tyumen /tju:'men/ oldest town in Siberia, central Russia (founded 1586), on the river Nitsa; population (1987) 456,000. Industries include oil refining, machine tools, and chemicals.

Tywi /'taui/ or *Towy* river in Dyfed, SW Wales; length 108 km/68 mi. It rises in the Cambrian Mountains of central Wales, flowing SW to enter Carmarthen Bay.

U

Ubangi-Shari /uːˈbæŋgi ˈʃɑːri/ former name for the ▷Central African Republic.

Udaipur /uːˈdaɪpuə/ or **Mecvar** industrial city (cotton, grain) in Rajasthan, India, capital of the former princely state of Udaipur; population (1981) 232,588. It was founded 1568 and has several palaces (two on islands in a lake) and the Jagannath Hindu temple 1640.

Udine /uːˈdɪneɪ/ industrial city (chemicals, textiles) NE of Venice, Italy; population (1984) 101,000. Udine was the capital of Friuli in the 13th century and passed to Venice 1420.

Udmurt /ˈudmuət/ (Russian **Udmurtskaya**) autonomous republic in the W Ural foothills, central Russia
area 42,100 sq km/16,200 sq mi
capital Izhevsk
products timber, flax, potatoes, peat, quartz
population (1985) 1,559,000 (58% Russian, 33% Udmurt, 7% Tatar)
history conquered in the 15th–16th centuries; constituted the Votyak Autonomous Region 1920; name changed to Udmurt 1932; Autonomous Republic 1934; part of the independent republic of Russia from 1991.

Ufa /uːˈfɑː/ industrial city (engineering, oil refining, petrochemicals, distilling, timber) and capital of the Bashkir autonomous republic, central Russia, on the river Bielaia, in the W Urals; population (1987) 1,092,000. It was founded by Russia 1574 as a fortress.

Uganda Republic of
area 236,600 sq km/91,351 sq mi
capital Kampala
towns Jinja, M'Bale, Entebbe, Masaka
physical plateau with mountains in W; forest and grassland; arid in NE
features Ruwenzori Range (Mount Margherita, 5,110 m/16,765 ft); national parks with wildlife (chimpanzees, crocodiles, Nile perch to 70 kg/160 lb); Owen Falls on White Nile where it leaves Lake Victoria; Lake Albert in W
head of state and government Yoweri Museveni from 1986
political system emergent democratic republic
political parties National Resistance Move-

ment (NRM), left-of-centre; Democratic Party (DP), centre-left; Conservative Party (CP), centre-right; Uganda People's Congress (UPC), left-of-centre; Uganda Freedom Movement (UFM), left-of-centre
exports coffee, cotton, tea, copper
currency Uganda new shilling
population (1990 est) 17,593,000 (largely the Baganda, from whom the country is named; also Langi and Acholi, some surviving Pygmies); growth rate 3.3% p.a.
life expectancy men 49, women 51 (1989)
languages English (official), Kiswahili, Luganda, and other African languages
religions Roman Catholic 33%, Protestant 33%, Muslim 16%, animist
literacy men 70%, women 45% (1985 est)
GNP $3.6 bn (1987); $220 per head
chronology
1962 Independence achieved from Britain within the Commonwealth with Milton Obote as prime minister.
1963 Proclaimed a federal republic with King Mutesa II as president.
1966 King Mutesa ousted in coup led by Obote, who ended the federal status and became executive president.
1969 All opposition parties banned after assassination attempt on Obote.
1971 Obote overthrown in army coup led by Maj-Gen Idi Amin Dada; ruthlessly dictatorial regime established; nearly 49,000 Ugandan Asians expelled; over 300,000 opponents of regime killed.
1978 Amin forced to leave country by opponents backed by Tanzanian troops. Provisional government set up with Yusuf Lule as president. Lule replaced by Godfrey Binaisa.
1978–79 Fighting broke out against Tanzanian troops.
1980 Binaisa overthrown by army. Elections held and Milton Obote returned to power.
1985 After opposition by National Resistance Army (NRA), and indiscipline in army, Obote ousted by Brig Tito Okello; power-sharing agreement entered into with NRA leader Yoweri Museveni.
1986 Agreement ended; Museveni became president, heading broad-based coalition government.
1992 Announcement made that East African

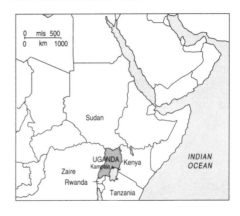

cooperation pact with Kenya and Tanzania would be revived.

Uist /ˈjuːɪst/ two small islands in the Outer Hebrides, Scotland: North Uist and South Uist.

Ujiji /uːˈdʒiːdʒi/ port on Lake Tanganyika, Tanzania, where Henry Stanley found David Livingstone 1871; population (1970) 17,000. It was originally an Arab trading post for slaves and ivory.

Ujung Pandang /uːdʒuŋ pænˈdæŋ/ formerly (until 1973) *Macassar* or *Makassar* chief port (trading in coffee, rubber, copra, and spices) on Sulawesi, Indonesia, with fishing and food-processing industries; population (1980) 709,000.

UK abbreviation for the ◊*United Kingdom*.

Ukraine
area 603,700 sq km/233,089 sq mi
capital Kiev
towns Kharkov, Donetsk, Odessa, Dnepro-petrovsk, Lugansk (Voroshilovgrad), Lviv (Lvov), Mariupol (Zhdanov), Krivoi Rog, Zaporozhye
physical Russian plain; Carpathian and Crimean Mountains; rivers: Dnieper (with the Dnieper dam 1932), Donetz, Bug
features Askaniya-Nova Nature Reserve (established 1921); health spas with mineral springs
head of state and government Leonid Kravchuk (president from 1990; in charge of government from 1993)
political system emergent democracy
political party Ukrainian People's Movement (Rukh), umbrella nationalist grouping, with three leaders
products grain, coal, oil, various minerals
currency grivna
population (1990) 51,800,000 (Ukrainian 73%, Russian 22%, Byelorussian 1%, Russian-speaking Jews 1%—some 1.5 million have emigrated to the USA, 750,000 to Canada)
language Ukrainian (Slavonic)
religions traditionally Ukrainian Orthodox; also Ukrainian Catholic
chronology
1918 Independent People's Republic proclaimed.

1920 Conquered by Soviet Red Army.
1921 Poland allotted charge of W Ukraine.
1932–33 Famine caused the deaths of more than 7.5 million people.
1939 W Ukraine occupied by Red Army.
1941–44 Under Nazi control; Jews massacred at Babi Yar; more than 5 million Ukrainians and Ukrainian Jews deported and exterminated.
1944 Soviet control re-established.
1945 Became a founder member of the United Nations.
1946 Ukrainian Uniate Church proscribed and forcibly merged with Russian Orthodox Church.
1986 April: Chernobyl nuclear disaster.
1989 Ukrainian People's Movement (Rukh) established. Ban on Ukrainian Uniate Church lifted.
1990 July: voted to proclaim sovereignty; former Communist Party (CP) leader Leonid Kravchuk indirectly elected president; sovereignty declared.
1991 Aug: demonstrations against the abortive anti-Gorbachev coup; independence declared, pending referendum; CP activities suspended. Oct: voted to create independent army. Dec: Kravchuk popularly elected president; independence overwhelmingly endorsed in referendum; joined new Commonwealth of Independent States; independence acknowledged by USA and European Community.
1992 Jan: admitted into Conference on Security and Cooperation in Europe (CSCE); pipeline deal with Iran to end dependence on Russian oil; prices freed. Feb: prices 'temporarily' re-regulated. March: agreed tactical arms shipments to Russia suspended. May: Crimean sovereignty declared, but subsequently rescinded. Aug: joint control of Black Sea fleet agreed with Russia.
1993 Kravchuk eliminated post of prime minister.

Ulaanbaatar /ˈuːlɑːn ˈbɑːtɔː/ or *Ulan Bator*; (formerly until 1924 *Urga*) capital of the Mongolian Republic; a trading centre producing carpets, textiles, vodka; population (1991) 575,000.

Ulan Bator /ˈuːlɑːn ˈbɑːtɔː/ alternative spelling of Ulaanbaatar, the capital of Mongolia.

Ulan-Ude /uˈlɑːn uˈdeɪ/ (formerly until 1934 *Verkhne-Udinsk*) industrial city (sawmills, cars, glass) and capital of the autonomous republic of Buryat in SE Russia, on the river Ibla and the Trans-Siberian railway; population (1987) 351,000. It was founded as a Cossack settlement in the 1660s.

Uleåborg /ˈuːlioˌbɔrjə/ Swedish name for the Finnish port of ◊Oulu.

Ulm /ʊlm/ industrial city (vehicles, agricultural machinery, precision instruments, textiles) in Baden-Württemberg, Germany, on the river Danube; population (1988) 101,000. Its Gothic cathedral has the highest stone spire ever built (161 m/528 ft) escaped damage in World War II when two-thirds of Ulm was destroyed. It was a free imperial city from the 14th century to 1802. The physicist and mathematician Albert Einstein was born here.

Ulsan /ˌʊlˈsæn/ industrial city (vehicles, ship-building, oil refining, petrochemicals) in South

Kyongsang province, SE South Korea; population (1985) 551,000.

Ulster /ˈʌlstə/ former kingdom in Northern Ireland, annexed by England 1461, from Jacobean times a centre of English, and later Scottish, settlement on land confiscated from its owners; divided 1921 into Northern Ireland (counties Antrim, Armagh, Down, Fermanagh, Londonderry, and Tyrone) and the Republic of Ireland (counties Cavan, Donegal, and Monaghan).

Ulster will fight; Ulster will be right.

On **Ulster** Lord Randolph Churchill, letter of 1886

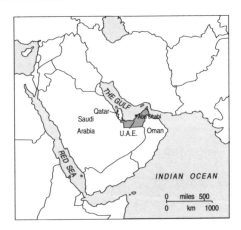

Ulundi /uˈlundi/ capital of the black African homeland Kwa Zulu in Natal, South Africa.

Umbria /ˈʌmbriə/ mountainous region of Italy in the central Apennines, including the provinces of Perugia and Terni; area 8,500 sq km/3,281 sq mi; population (1990) 822,700. Its capital is Perugia, and the river Tiber rises in the region. Industries include wine, grain, olives, tobacco, textiles, chemicals, and metalworking. This is the home of the Umbrian school of artists, including Raphael.

Umm al Qaiwain /ˈum æl kaɪˈwaɪn/ one of the ▷United Arab Emirates.

Umtali /umˈtɑːli/ former name (until 1982) for the town of ▷Mutare in Zimbabwe.

Umtata /umˈtɑːtə/ capital of the South African Bantu homeland of Transkei; population (1976) 25,000.

Ungava /ʌnˈgɑːvə/ district in N Québec and Labrador, Canada, E of Hudson Bay; area 351,780 sq mi/911,110 sq km. It has large deposits of iron ore.

Union of Soviet Socialist Republics (USSR) former country in N Asia and E Europe that reverted to independent states 1991; see ▷Armenia, ▷Azerbaijan, ▷Belarus, ▷Estonia, ▷Georgia, ▷Kazakhstan, ▷Kyrgyzstan, ▷Latvia, ▷Lithuania, ▷Moldova, ▷Russian Federation, ▷Tajikistan, ▷Turkmenistan, ▷Ukraine, and ▷Uzbekistan.

United Arab Emirates (UAE) (*Ittihad al-Imarat al-Arabiyah*) federation of the emirates of Abu Dhabi, Ajman, Dubai, Fujairah, Ras al Khaimah, Sharjah, Umm al Qaiwain
total area 83,657 sq km/32,292 sq mi
capital Abu Dhabi
towns (chief port) Dubai
physical desert and flat coastal plain; mountains in E
features linked by dependence on oil revenues
head of state and of government Sheik Sultan Zayed bin al-Nahayan of Abu Dhabi from 1971
political system absolutism
political parties none

exports oil, natural gas, fish, dates
currency UAE dirham
population (1990 est) 2,250,000 (10% nomadic); growth rate 6.1% p.a.
life expectancy men 68, women 72 (1989)
languages Arabic (official), Farsi, Hindi, Urdu, English
religions Muslim 96%, Christian, Hindu
literacy 68% (1989)
GNP $22 bn (1987); $11,900 per head
chronology
1952 Trucial Council established.
1971 Federation of Arab Emirates formed; later dissolved. Six Trucial States formed United Arab Emirates, with ruler of Abu Dhabi, Sheik Zayed, as president.
1972 The seventh state joined.
1976 Sheik Zayed threatened to relinquish presidency unless progress towards centralization became more rapid.
1985 Diplomatic and economic links with USSR and China established.
1987 Diplomatic relations with Egypt restored.
1990–91 Iraqi invasion of Kuwait opposed; UAE fights with UN coalition.
1991 Bank of Commerce and Credit International (BCCI), controlled by Abu Dhabi's ruler, collapsed.

United Arab Republic union formed 1958, broken 1961, between Egypt and Syria. Egypt continued to use the name after the breach until 1971.

United Kingdom of Great Britain and Northern Ireland (UK)
area 244,100 sq km/94,247 sq mi
capital London
towns Birmingham, Glasgow, Leeds, Sheffield, Liverpool, Manchester, Edinburgh, Bradford, Bristol, Belfast, Newcastle-upon-Tyne, Cardiff
physical became separated from European continent about 6000 BC; rolling landscape, increasingly mountainous towards the N, with Grampian Mountains in Scotland, Pennines in N England, Cambrian Mountains in Wales; rivers include Thames, Severn, and Spey
territories Anguilla, Bermuda, British Antarctic Territory, British Indian Ocean Territory, British

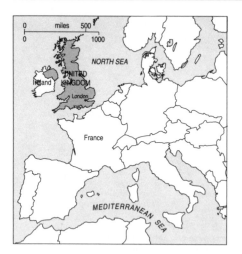

1707 Act of Union between England and Scotland under Queen Anne.
1721 Robert Walpole unofficially first prime minister, under George I.
1783 Loss of North American colonies that form USA; Canada retained.
1801 Act of Ireland united Britain and Ireland.
1819 Peterloo massacre: cavalry charged a meeting of supporters of parliamentary reform.
1832 Great Reform Bill became law, shifting political power from upper to middle class.
1838 Chartist working-class movement formed.
1846 Corn Laws repealed by Robert Peel.
1851 Great Exhibition in London.
1867 Second Reform Bill, extending the franchise, introduced by Disraeli and passed.
1906 Liberal victory; programme of social reform.
1911 Powers of House of Lords curbed.
1914 Irish Home Rule Bill introduced.
1914–18 World War I.
1916 Lloyd George became prime minister.
1920 Home Rule Act incorporated NE of Ireland (Ulster) into the United Kingdom of Great Britain and Northern Ireland.
1921 Ireland, except for Ulster, became a dominion (Irish Free State, later Eire, 1937).
1924 First Labour government led by Ramsay MacDonald.
1926 General Strike.
1931 Coalition government; unemployment reached 3 million.
1939 World War II began.
1940 Winston Churchill became head of coalition government.
1945 Labour government under Clement Attlee; welfare state established.
1951 Conservatives under Winston Churchill defeated Labour.
1956 Suez Crisis.
1964 Labour victory under Harold Wilson.
1970 Conservatives under Edward Heath defeated Labour.
1972 Parliament prorogued in Northern Ireland; direct rule from Westminster began.
1973 UK joined European Economic Community.
1974 Three-day week, coal strike; Wilson replaced Heath.
1976 James Callaghan replaced Wilson as prime minister.
1977 Liberal–Labour pact.
1979 Victory for Conservatives under Margaret Thatcher.
1981 Formation of Social Democratic Party (SDP). Riots occurred in inner cities.
1982 Unemployment over 3 million. Falklands War.
1983 Thatcher re-elected.
1984–85 Coal strike, the longest in British history.
1986 Abolition of metropolitan counties.
1987 Thatcher re-elected for third term.
1988 Liberals and most of SDP merged into the Social and Liberal Democrats, leaving a splinter SDP. Inflation and interest rates rose.
1989 The Green Party polled 2 million votes in the European elections.
1990 Riots as poll tax introduced in England.

Virgin Islands, Cayman Islands, Falkland Islands, Gibraltar, Hong Kong (until 1997), Montserrat, Pitcairn Islands, St Helena and Dependencies (Ascension, Tristan da Cunha), Turks and Caicos Islands
environment an estimated 67% (the highest percentage in Europe) of forests have been damaged by acid rain
features milder climate than N Europe because of Gulf Stream; considerable rainfall. Nowhere more than 120 km/74.5 mi from sea; indented coastline, various small islands
head of state Elizabeth II from 1952
head of government John Major from 1990
political system liberal democracy
political parties Conservative and Unionist Party, right-of-centre; Labour Party, moderate, left-of-centre; Social and Liberal Democrats, centre-left; Scottish National Party (SNP), Scottish nationalist; Plaid Cymru (Welsh Nationalist Party), Welsh nationalist; Official Ulster Unionist Party (OUP), Northern Ireland moderate right-of-centre; Democratic Unionist Party (DUP), Northern Ireland, right-of-centre; Social Democratic Labour Party (SDLP), Northern Ireland, moderate, left-of-centre; Ulster People's Unionist Party (UPUP), Northern Ireland, militant right-of-centre; Sinn Féin, Northern Ireland, pro-united Ireland; Green Party, ecological
exports cereals, rape, sugar beet, potatoes, meat and meat products, poultry, dairy products, electronic and telecommunications equipment, engineering equipment and scientific instruments, oil and gas, petrochemicals, pharmaceuticals, fertilizers, film and television programmes, aircraft
currency pound sterling
population (1990 est) 57,121,000 (81.5% English, 9.6% Scottish, 1.9% Welsh, 2.4% Irish, 1.8% Ulster); growth rate 0.1% p.a.
religion Christian (55% Protestant, 10% Roman Catholic); Muslim, Jewish, Hindu, Sikh
life expectancy men 72, women 78 (1989)
languages English, Welsh, Gaelic
literacy 99% (1989)
GNP $758 bn; $13,329 per head (1988)

Troops sent to the Persian Gulf following Iraq's invasion of Kuwait. British hostages held in Iraq, later released. Britain joined European exchange rate mechanism (ERM). Thatcher replaced by John Major as Conservative leader and prime minister. *1991* British troops took part in US-led war against Iraq under United Nations umbrella. Severe economic recession and rising unemployment. *1992* Recession continued. April: Conservative Party won fourth consecutive general election, but with reduced majority. John Smith replaced Neil Kinnock as Labour leader. Sept: sterling devalued and UK withdrawn from ERM. Oct: coal mine closure programme encountered massive public opposition; later reviewed. Major's popularity at unprecedentedly low rating. Nov: government motion in favour of ratification of Maastricht Treaty narrowly passed. Revelations of past arms sales to Iraq implicated senior government figures, including the prime minister. *1993* Recession continued. July: Maastricht Treaty ratified by parliament. Nov: Government engaged in talks with Northern Ireland politicians. Dec: peace proposal, the Downing Street Declaration, issued jointly with Irish government.

United Provinces of Agra and Oudh /ɑːgrə, aud/ former province of British India, which formed the major part of the state of Uttar Pradesh; see also ◊Agra, ◊Oudh.

United States of America
area 9,368,900 sq km/3,618,770 sq mi
capital Washington DC
towns New York, Los Angeles, Chicago, Philadelphia, Detroit, San Francisco, Washington, Dallas, San Diego, San Antonio, Houston, Boston, Baltimore, Phoenix, Indianapolis, Memphis, Honolulu, San José
physical topography and vegetation from tropical (Hawaii) to arctic (Alaska); mountain ranges parallel with E and W coasts; the Rocky

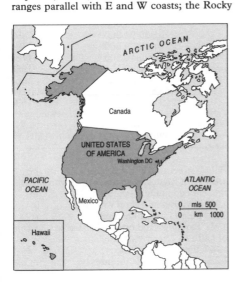

Mountains separate rivers emptying into the Pacific from those flowing into the Gulf of Mexico; Great Lakes in N; rivers include Hudson, Mississippi, Missouri, Colorado, Columbia, Snake, Rio Grande, Ohio
environment the USA produces the world's largest quantity of municipal waste per person (850 kg/1,900 lb)
features see individual states
territories the commonwealths of Puerto Rico and Northern Marianas; Guam, the US Virgin Islands, American Samoa, Wake Island, Midway Islands, Belau, and Johnston and Sand Islands
head of state and government Bill Clinton from 1993
political system liberal democracy
political parties Democratic Party, liberal centre; Republican Party, centre-right
currency US dollar
population (1990 est) 250,372,000 (white 80%, black 12%, Asian/Pacific islander 3%, American Indian, Inuit, and Aleut 1%, Hispanic [included in above percentages] 9%); growth rate 0.9% p.a.
life expectancy men 72, women 79 (1989)
languages English, Spanish
religions Christian 86.5% (Roman Catholic 26%, Baptist 19%, Methodist 8%, Lutheran 5%), Jewish 1.8%, Muslim 0.5%, Buddhist and Hindu less than 0.5%
literacy 99% (1989)
GNP $5,880.7 bn (1990)
chronology
1776 Declaration of Independence.
1787 US constitution drawn up.
1789 Washington elected as first president.
1803 Louisiana Purchase.
1812-14 War with England, arising from commercial disputes caused by Britain's struggle with Napoleon.
1819 Florida purchased from Spain.
1836 The battle of the Alamo, Texas, won by Mexico.
1841 First wagon train left Missouri for California.
1846 Mormons, under Brigham Young, founded Salt Lake City, Utah.
1846-48 Mexican War resulted in cession to USA of Arizona, California, part of Colorado and Wyoming, Nevada, New Mexico, Texas, and Utah.
1848-49 California gold rush.
1860 Lincoln elected president.
1861-65 Civil War between North and South.
1865 Slavery abolished. Lincoln assassinated.
1867 Alaska bought from Russia.
1890 Battle of Wounded Knee, the last major battle between American Indians and US troops.
1898 War with Spain ended with the Spanish cession of Philippines, Puerto Rico, and Guam; it was agreed that Cuba be independent. Hawaii annexed.
1917-18 USA entered World War I.
1919-1921 Wilson's 14 Points became base for League of Nations.
1920 Women achieved the vote.
1924 American Indians made citizens by Congress.
1929 Wall Street stock-market crash.

United States of America

state	capital	area in sq km/sq mi	date of joining the Union
Alabama	Montgomery	134,700/52,000	1819
Alaska	Juneau	1,531,100/591,200	1959
Arizona	Phoenix	294,100/113,600	1912
Arkansas	Little Rock	137,800/53,200	1836
California	Sacramento	411,100/158,700	1850
Colorado	Denver	269,700/104,100	1876
Connecticut	Hartford	13,000/5,000	1788
Delaware	Dover	5,300/2,000	1787
Florida	Tallahassee	152,000/58,700	1845
Georgia	Atlanta	152,600/58,900	1788
Hawaii	Honolulu	16,800/6,500	1959
Idaho	Boise	216,500/83,600	1890
Illinois	Springfield	146,100/56,400	1818
Indiana	Indianapolis	93,700/36,200	1816
Iowa	Des Moines	145,800/56,300	1846
Kansas	Topeka	213,200/82,300	1861
Kentucky	Frankfort	104,700/40,400	1792
Louisiana	Baton Rouge	135,900/52,500	1812
Maine	Augusta	86,200/33,300	1820
Maryland	Annapolis	31,600/12,200	1788
Massachusetts	Boston	21,500/8,300	1788
Michigan	Lansing	151,600/58,500	1837
Minnesota	St Paul	218,700/84,400	1858
Mississippi	Jackson	123,600/47,700	1817
Missouri	Jefferson City	180,600/69,700	1821
Montana	Helena	381,200/147,200	1889
Nebraska	Lincoln	200,400/77,400	1867
Nevada	Carson City	286,400/110,600	1864
New Hampshire	Concord	24,000/9,300	1788
New Jersey	Trenton	20,200/7,800	1787
New Mexico	Santa Fé	315,000/121,600	1912
New York	Albany	127,200/49,100	1788
North Carolina	Raleigh	136,400/52,700	1789
North Dakota	Bismarck	183,100/70,700	1889
Ohio	Columbus	107,100/41,400	1803
Oklahoma	Oklahoma City	181,100/69,900	1907
Oregon	Salem	251,500/97,100	1859
Pennsylvania	Harrisburg	117,400/45,300	1787
Rhode Island	Providence	3,100/1,200	1790
South Carolina	Columbia	80,600/31,100	1788
South Dakota	Pierre	199,800/77,100	1889
Tennessee	Nashville	109,200/42,200	1796
Texas	Austin	691,200/266,900	1845
Utah	Salt Lake City	219,900/84,900	1896
Vermont	Montpelier	24,900/9,600	1791
Virginia	Richmond	105,600/40,800	1788
Washington	Olympia	176,700/68,200	1889
West Virginia	Charleston	62,900/24,300	1863
Wisconsin	Madison	145,500/56,200	1848
Wyoming	Cheyenne	253,400/97,800	1890
District of Columbia	Washington	180/70	
	Total	9,391,880/3,625,870	

1933 F D Roosevelt's New Deal to alleviate the Depression put into force.
1941–45 The Japanese attack on Pearl Harbor Dec 1941 precipitated US entry into World War II.
1945 USA ended war in the Pacific by dropping A-bombs on Hiroshima and Nagasaki, Japan.
1950–53 US involvement in Korean War. McCarthy anticommunist investigations

(HUAC) became a 'witch hunt'.
1954 Civil Rights legislation began with segregation ended in public schools.
1957 Civil Rights bill on voting.
1958 First US satellite in orbit.
1961 Bay of Pigs abortive CIA-backed invasion of Cuba.
1963 President Kennedy assassinated; L B Johnson assumed the presidency.

1964–68 'Great Society' civil-rights and welfare measures in the Omnibus Civil Rights bill.
1964–75 US involvement in Vietnam War.
1965 US intervention in Dominican Republic.
1969 US astronaut Neil Armstrong was the first human on the Moon.
1973 OPEC oil embargo almost crippled US industry and consumers. Inflation began.
1973–74 Watergate scandal began in effort to re-elect Richard Nixon and ended just before impeachment; Nixon resigned as president; replaced by Gerald Ford, who 'pardoned' Nixon.
1975 Final US withdrawal from Vietnam.
1979 US–Chinese diplomatic relations normalized.
1979–80 Iranian hostage crisis; relieved by Reagan's concessions.
1981 Space shuttle mission was successful.
1983 US invasion of Grenada.
1986 'Irangate' scandal over secret US government arms sales to Iran, with proceeds to antigovernment Contra guerrillas in Nicaragua.
1987 Reagan and Gorbachev (for USSR) signed intermediate-range nuclear forces treaty. Wall Street stock-market crash caused by programme trading.
1988 USA became world's largest debtor nation, owing $532 billion. George Bush elected president.
1989 Bush met Gorbachev at Malta, end to Cold War declared; large cuts announced for US military; USA invaded Panama; Noriega taken into custody.
1990 Bush and Gorbachev met again. Nelson Mandela freed in South Africa, toured USA. US troops sent to Middle East following Iraq's invasion of Kuwait.
1991 Jan–Feb: US-led assault drove Iraq from Kuwait in Gulf War. US support was given to the USSR during the dissolution of communism and the recognition of independence of the Baltic republics. July: Strategic Arms Reduction Treaty (START) signed at US–Soviet summit in Moscow.
1992 Bush's popularity slumped as economic recession continued. Widespread riots in Los Angeles. US troops sent to Somalia as part of UN peacekeeping force. Nov: Bill Clinton won presidential elections for the Democrats.
1993 Jan: Clinton inaugurated. July: air strike on Baghdad, Iraq. Oct: date of March 1994 given for withdrawal of US troops from Somalia.

She of the open soul and open door, With room about her hearth for all mankind.

On the **United States of America**

J R Lowell

Unzen /ˈunzen/ active volcano on the Shimbara peninsula, Kyushu Island, Japan, opposite the city of Kumamoto. Its eruption June 1991 led to the evacuation of 10,000 people.

Upper Austria (German *Oberösterreich*) mountainous federal province of Austria, drained by the river Danube; area 12,000 sq km/4,632 sq mi; population (1987) 1,294,000. Its capital is Linz. In addition to wine, sugar beet and grain, there are reserves of oil. Manufactured products include textiles, chemicals, and metal goods.

Upper Volta /ˈvɒltə/ former name (until 1984) of ◊Burkina Faso.

Uppsala /ʊpˈsɑːlə/ city in Sweden, NW of Stockholm; population (1990) 167,500. Industries include engineering and pharmaceuticals. The university was founded 1477; there are Viking relics and a Gothic cathedral.

Ural Mountains /ˈjʊərəl/ (Russian *Ural'skiy Khrebet*) mountain system running from the Arctic to the Caspian Sea, traditionally separating Europe from Asia. The highest peak is Naradnaya, 1,894 m/6,214 ft. It has vast mineral wealth. The middle Urals is one of the most industrialized regions of Russia. Perm, Chelyabinsk, Ekaterinburg (Sverdlovsk), Magnitogorsk, and Zlatoust are major industrial centres.

Urbana /ɜːˈbænə/ industrial city (food processing, electronics, and metal products) in E central Illinois, USA, E of Springfield; population (1990) 36,300. The University of Illinois 1867 is here.

Urga /ˈɜːgə/ former name (until 1924) of ◊Ulaanbaatar, the capital of Mongolia.

Uruguay Oriental Republic of (*República Oriental del Uruguay*)
area 176,200 sq km/68,031 sq mi
capital Montevideo
towns Salto, Paysandú
physical grassy plains (pampas) and low hills
features rivers Negro, Uruguay, Río de la Plata
head of state and government Luis Lacalle Herrera from 1989
political system democratic republic

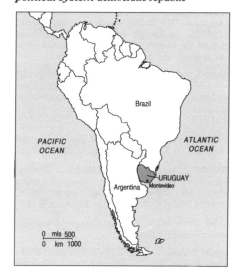

political parties Colorado Party (PC), progressive, centre-left; National (Blanco) Party (PN), traditionalist, right of centre; Amplio Front (FA), moderate, left-wing
exports meat and meat products, leather, wool, textiles
currency nuevo peso
population (1990 est) 3,002,000 (Spanish, Italian; mestizo, mulatto, black); growth rate 0.7% p.a.
life expectancy men 68, women 75 (1989)
language Spanish
media the Ministry of Defence controls broadcasting licences
religion Roman Catholic 66%
literacy 96% (1984)
GNP $7.5 bn; $2,470 per head (1988)
chronology
1825 Independence declared from Brazil.
1836 Civil war.
1930 First constitution adopted.
1966 Blanco party in power, with Jorge Pacheco Areco as president.
1972 Colorado Party returned, with Juan Maria Bordaberry Arocena as president.
1976 Bordaberry deposed by army; Dr Méndez Manfredini became president.
1984 Violent antigovernment protests.
1985 Agreement reached between the army and political leaders for return to constitutional government. Colorado Party won general election; Dr Julio Maria Sanguinetti became president.
1986 Government of national accord established under President Sanguinetti's leadership.
1989 Luis Lacalle Herrera elected president.

Urumchi alternative spelling of Urumqi, a city in China.

Urumqi /u'ruːmtʃiː/ or *Urumchi* industrial city and capital of Xinjiang Uygur autonomous region, China, at the northern foot of the Tian Shan Mountains; population (1989) 1,110,000. It produces cotton textiles, cement, chemicals, iron, and steel.

Ushant /ˈʌʃənt/ (French *Ouessant*) French island 18 km/11 mi W of Brittany, area 15 sq km/9 sq mi, off which the British admiral Richard Howe defeated the French navy 1794 on 'the Glorious First of June'. The chief town is Lampaul.

Ushuaia /uːˈswaɪə/ southernmost town in the world, at the tip of Tierra del Fuego, Argentina, less than 1,000 km/620 mi from Antarctica; population (1991) 29,700. It is a free port and naval base.

Usküb /ˈuskuːb/ Turkish name of ◊Skopje, capital of the Former Yugoslav Republic of Macedonia.

Usküdar /ˌuskuːˈdɑː/ suburb of Istanbul, Turkey; formerly a separate town, which under the name *Scutari* was the site of the hospital set up by Florence Nightingale during the Crimean War.

USSR abbreviation for the former ◊*Union of Soviet Socialist Republics*.

Ussuri /uˈsuəri/ river in E Asia, tributary of the Amur. Rising N of Vladivostok and joining the

Utah

Amur S of Khabarovsk, it forms part of the border between Russia and the Chinese province of Heilongjiang. There were military clashes 1968–69 over the sovereignty of Damansky Island (Chenpao).

Ust-Kamenogorsk /ˈuːst kəˌmenəˈɡɔːsk/ river port and chief centre of the nuclear industry in Russia, situated in the Altai Mountains, on the river Irtysh; population (1987) 321,000.

UT abbreviation for ◊Utah, a state of the USA.

Utah /ˈjuːtɑː/ state in western USA; nickname Beehive State/Mormon State
area 219,900 sq km/84,881 sq mi
capital Salt Lake City
towns Provo, Ogden
physical Colorado Plateau to the E, mountains in centre, Great Basin to the W, Great Salt Lake
features Great American Desert; Colorado river system; Dinosaur and Rainbow Bridge national monuments; five national parks: the Arches, Bryce Canyon, Canyonlands, Capitol Reef, Zion; auto racing at Bonneville Salt Flats; Mormon temple and tabernacle, Salt Lake City
products wool, gold, silver, copper, coal, salt, steel
population (1990) 1,722,850
famous people Brigham Young
history explored first by Franciscan friars for Spain 1776; Great Salt Lake discovered by US frontier scout Jim Bridger 1824; part of the area ceded by Mexico 1848; developed by Mormons, still by far the largest religious group in the state; territory 1850, but not admitted to statehood until 1896 because of Mormon reluctance to relinquish plural marriage.

Utica /ˈjuːtɪkə/ industrial city (engine parts, clothing) in central New York State, USA; population (1990) 68,600. Utica was an important textile centre from about 1850 to 1950. The first Woolworth store was opened here 1879.
    The settlement 1773 was on the site of an Iroquois centre and a British fort.

Utrecht /ˈjuːtrekt/ province of the Netherlands lying SE of Amsterdam, on the Kromme Rijn (Crooked Rhine)
area 1,330 sq km/513 sq mi
capital Utrecht
towns Amersfoort, Zeist, Nieuwegeun, Veenendaal
products chemicals, livestock, textiles, electrical goods

Uttar Pradesh

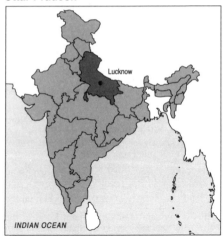

population (1991) 1,026,800
history ruled by the bishops of Utrecht in
the Middle Ages, the province was sold to the
emperor Charles V of Spain 1527. It became a
centre of Protestant resistance to Spanish rule
and, with the signing of the Treaty of Utrecht,
became one of the seven United Provinces of the
Netherlands 1579.

Uttar Pradesh /ˈutə prəˈdeʃ/ state of N India
area 294,400 sq km/113,638 sq mi
capital Lucknow
towns Kanpur, Varanasi, Agra, Allahabad,
Meerut
features most populous state; Himalayan peak
Nanda Devi 7,817 m/25,655 ft
population (1991) 138,760,400
famous people Indira Gandhi, Ravi Shankar
language Hindi
religions 80% Hindu, 15% Muslim
history formerly the heart of the Mogul Empire
and generating point of the Indian Mutiny 1857
and subsequent opposition to British rule; see also
⟡Agra and ⟡Oudh.

---

*... the centre of the stage on which the
drama of Indian history has been played ...*

On **Uttar Pradesh** Ved Mehta
Portrait of India 1970

Uzbekistan Republic of
area 447,400 sq km/172,741 sq mi
capital Tashkent
towns Samarkand, Bukhara, Namangan
physical oases in the deserts; rivers: Amu Darya,
Syr Darya; Fergana Valley; rich in mineral
deposits

features more than 20 hydroelectric plants; three
natural gas pipelines
head of state Islam Karimov from 1990
head of government Abdul Hashim Mutalov
from 1991
political system socialist pluralist
political parties Uzbekistan Socialist (formerly
Communist) Party, reform-socialist; Democratic
Party, tolerated opposition
products rice, dried fruit, vines (all grown by
irrigation); cotton, silk
population (1990) 20,300,000 (71% Uzbek, 8%
Russian, 5% Tajik, 4% Kazakh)
language Uzbek, a Turkic language
religion Sunni Muslim
chronology
1921 Part of Turkestan Soviet Socialist Autonomous Republic.
1925 Became constituent republic of the USSR.
1944 Some 160,000 Meskhetian Turks forcibly
transported from their native Georgia to
Uzbekistan by Stalin.
1989 June: Tashlak, Yaipan, and Ferghana were
the scenes of riots in which Meskhetian Turks
were attacked; 70 killed and 850 wounded.
1990 June: economic and political sovereignty
declared.
1991 March: Uzbek supported 'renewed federation' in USSR referendum. Aug: anti-
Gorbachev coup in Moscow initially accepted by
President Karimov; later, Karimov resigned from
Soviet Communist Party (CPSU) Politburo;
Uzbek Communist Party (UCP) broke with
CPSU; pro-democracy rallies dispersed by militia; independence declared. Dec: joined new
Commonwealth of Independent States (CIS).
1992 Jan: admitted into Conference on Security
and Cooperation in Europe (CSCE); violent
food riots in Tashkent. March: joined the
United Nations (UN); US diplomatic recognition
achieved.

VA abbreviation for ⬦ *Virginia*, a state of the USA.

Vaal /vɑːl/ river in South Africa, the chief tributary of the Orange River. It rises in the Drakensberg mountain range and for much of its course of 1,200 km/750 mi it separates Transvaal from Orange Free State.

Vadodara /wɔ'dəudərə/ (formerly until 1976 *Baroda*) industrial city (metal goods, chemicals, jewellery, textiles) and rail junction in Gujarat, W India; population (1981) 744,881. Until 1947 it was capital of the princely state of Baroda. It has Lakshmi Vilas Palace, Pratap Vilas Palace (now the Railway Staff College), and several multi-level step wells (baoli).

Vaduz /fæ'duts/ capital of the European principality of Liechtenstein; industries include engineering and agricultural trade; population (1984) 5,000.

Valdai Hills /væl'daɪ/ small forested plateau in Russia, between St Petersburg and Moscow, where the Volga and W Dvina rivers rise. The Viking founders of the Russian state used it as a river route centre to reach the Baltic, Black, Caspian, and White seas. From the 15th century it was dominated by Moscow.

Valdivia /væl'diːviə/ industrial port (shipbuilding, leather, beer, soap) and resort in Chile; population (1990) 113,500. It was founded 1552 by the Spanish conquistador Pedro de Valdivia (c. 1500–1554), conqueror of Chile.

Valence /væ'lɒns/ market town and capital of Drôme *département*, SE France, on the river Rhône; population (1990) 65,000. Industries include electrical goods and components for aerospace. It is of pre-Roman origin and has a Romanesque cathedral consecrated 1095.

Valencia /vɔ'lensiɔ/ industrial city (wine, fruit, chemicals, textiles, ship repair) in Valencia region, E Spain; population (1991) 777,400. The Community of Valencia, consisting of Alicante, Castellón, and Valencia, has an area of 23,300 sq km/8,994 sq mi and a population of 3,772,000.

Valencia was ruled by El Cid 1094–99, after he recaptured it from the Moors. There is a cathedral of the 13th–15th centuries and a university founded 1500.

Valencia /vɔ'lensiɔ/ industrial city (textiles, leather, sugar) and agricultural centre in Carabobo state, N Venezuela, on the Cabriales River; population (1989) 1,227,472. It is 478 m/ 1,569 ft above sea level and was founded 1555.

Valenciennes /ˌvælɒnsi'en/ industrial town in Nord *département*, NE France, near the Belgian border, once known for its lace; population (1990) 39,300. It became French 1678.

Valladolid /ˌvæljədəu'liːð/ industrial town (food processing, vehicles, textiles, engineering), and capital of Valladolid province, Spain; population (1991) 345,300.

It was the capital of Castile and Leon in the 14th–15th centuries, then of Spain until 1560. The Catholic monarchs Ferdinand and Isabella were married at Valladolid 1469. The explorer Christopher Columbus died here, and the home of the writer Miguel de Cervantes is preserved. It has a university founded 1346 and a cathedral 1595.

Valle d'Aosta /'væleɪ dɑː'ɒstə/ autonomous region of NW Italy; area 3,300 sq km/1,274 sq mi; population (1990) 116,000, many of whom are French-speaking. It produces wine and livestock. Its capital is Aosta.

Vallejo /vɔ'leɪəu/ industrial city (fruit and flour processing and petroleum refining) in NW California, USA, on San Pablo Bay, NE of Berkeley; population (1990) 109,200. It was California's capital 1852–53.

Valletta /vɔ'letə/ capital and port of Malta; population (1987) 9,000; urban area 101,000.

It was founded 1566 by the Knights of St John of Jerusalem and named after their grand master Jean de la Valette (1494–1568), who fended off a Turkish siege May–Sept 1565. The 16th-century palace of the grand masters survives. Malta was formerly a British naval base and came under heavy attack in World War II.

Valley of Ten Thousand Smokes valley in SW Alaska, on the Alaska Peninsula, where in 1912 Mount Katmai erupted in one of the largest volcanic explosions ever known, though without loss of human life since the area was uninhabited. It was dedicated as the Katmai National Monument 1918. Thousands of fissures on the valley floor continue to emit steam and gases.

Valona /vɔ'ləunə/ Italian form of ⬦ Vlorë, a port in Albania.

Valparaíso /ˌvælpɔ'raɪzəu/ industrial port (sugar, refining, textiles, chemicals) in Chile; capital of Valparaíso province, on the Pacific; population (1990) 276,800. Founded 1536, it was occupied by the English naval adventurers Francis Drake 1578 and John Hawkins 1595, pillaged by the Dutch 1600, and bombarded by Spain 1866; it has also suffered from earthquakes.

Van /vɑːn/ city in Turkey on a site on *Lake Van* that has been inhabited for more than 3,000 years; population (1990) 153,100. It is a commercial centre for a fruit- and grain-producing area.

Vancouver /væn'kuːvə/ industrial city (oil refining, engineering, shipbuilding, aircraft, timber, pulp and paper, textiles, fisheries) in Canada, its chief Pacific seaport, on the mainland of British Columbia; population (1986) 1,381,000.

Vancouver is situated on Burrard Inlet, at the mouth of the Fraser River. George Vancouver took possession of the site for Britain 1792. It was settled by 1875, under the name of *Granville*, and was renamed when it became a city 1886, having been reached by the Canadian Pacific Railroad. In 1989 it had an ethnic Chinese population of 140,000, and this was rapidly augmented by thousands of immigrants from Hong Kong.

Vancouver /væn'kuːvə/ city in SW Washington, USA, on the Columbia River, N of Portland, Oregon; population (1990) 46,400. It is a manufacturing and shipping centre for agriculture and timber. It began as a trading post for the Hudson's Bay Company 1825.

Vancouver Island /væn'kuːvə/ island off the west coast of Canada, part of British Columbia
area 32,136 sq km/12,404 sq mi
towns Victoria, Nanaimo, Esquimalt (naval base)
products coal, timber, fish
history visited by British explorer Capt Cook 1778; surveyed 1792 by Capt George Vancouver.

Van Diemen's Land /væn 'diːmənz/ former name (1642–1855) of ◊Tasmania, Australia. It was named by Dutch navigator Abel Tasman after the governor general of the Dutch East Indies, Anthony van Diemen. The name Tasmania was used from the 1840s and became official 1855.

Vänern, Lake /venən/ largest lake in Sweden, area 5,550 sq km/2,140 sq mi. Karlstad, Vänersborg, Lidköping, and Mariestad are on its banks.

Vannin, Ellan /'elən 'vænɪn/ Gaelic name for the Isle of ◊Man.

Vanuatu Republic of (*Ripablik Blong Vanuatu*)
area 14,800 sq km/5,714 sq mi
capital Vila (on Efate)
towns Luganville (on Espíritu Santo)
physical comprises around 70 islands, including Espíritu Santo, Malekula, and Efate; densely forested, mountainous
features three active volcanoes
head of state Fred Timakata from 1989
head of government Maxime Carlot from 1991
political system democratic republic
political parties Union of Moderate Parties (UMP), Francophone centrist; Vanuatu National United Party (VNUP), formed by Walter Lini; Vanua'aku Pati (VP), Anglophone centrist; Melanesian Progressive Party (MPP), Melanesian centrist; Fren Melanesian Party
exports copra, fish, coffee, cocoa
currency vatu
population (1989) 152,000 (90% Melanesian);

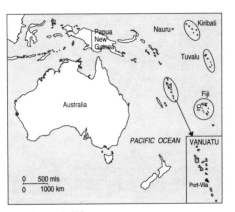

growth rate 3.3% p.a.
life expectancy men 67, women 71 (1989)
languages Bislama 82%, English, French (all official)
literacy 53%
religion Presbyterian 40%, Roman Catholic 16%, Anglican 14%, animist 15%
GDP $125 million (1987); $927 per head
chronology
1906 Islands jointly administered by France and Britain.
1975 Representative assembly established.
1978 Government of national unity formed, with Father Gerard Leymang as chief minister.
1980 Revolt on the island of Espíritu Santo delayed independence but it was achieved within the Commonwealth, with George Kalkoa (adopted name Sokomanu) as president and Father Walter Lini as prime minister.
1988 Dismissal of Lini by Sokomanu led to Sokomanu's arrest for treason. Lini reinstated.
1989 Sokomanu sentenced to six years' imprisonment; succeeded as president by Fred Timakata.
1991 Lini voted out by party members; replaced by Donald Kalpokas. General election produced UMP–VNUP coalition under Maxime Carlot.

Var /vɑː/ river in S France, rising in the Maritime Alps and flowing generally SSE for 134 km/84 mi into the Mediterranean near Nice. It gives its name to a *département* in the Provence-Alpes-Côte d'Azur region.

Varanasi /və'rɑːnəsi/ or *Benares* holy city of the Hindus in Uttar Pradesh, India, on the river Ganges; population (1981) 794,000. There are 1,500 golden shrines, and a 5 km/3 mi frontage to the Ganges with sacred stairways (ghats) for purification by bathing.

At the burning ghats, the ashes of the dead are scattered on the river to ensure a favourable reincarnation.

Varna /vɑːnə/ port in Bulgaria, on an inlet of the Black Sea; population (1990) 320,600. Industries include shipbuilding and the manufacture of chemicals.

Varna was a Greek colony in the 6th century BC and part of the Ottoman Empire 1391–1878; it was renamed *Stalin* 1949–56.

Vatican City State (*Stato della Città del Vaticano*)
area 0.4 sq km/109 acres
physical forms an enclave in the heart of Rome, Italy
features Vatican Palace, official residence of the pope; basilica and square of St Peter's; churches in and near Rome, the pope's summer villa at Castel Gandolfo; the world's smallest state
head of state and government John Paul II from 1978
political system absolute Catholicism
currency Vatican City lira; Italian lira
population (1985) 1,000
languages Latin (official), Italian
religion Roman Catholic
chronology
1929 Lateran Treaty recognized sovereignty of the pope.
1947 New Italian constitution confirmed the sovereignty of the Vatican City State.
1978 John Paul II became the first non-Italian pope for more than 400 years.
1985 New concordat signed under which Roman Catholicism ceased to be Italy's state religion.

Vaucluse /vəʊ'kluːz/ mountain range in SE France, part of the Provence Alps, E of Avignon, rising to 1,242 m/4,075 ft. It gives its name to a *département*. The Italian poet Petrarch lived in the Vale of Vaucluse 1337–53.

Venda /'vendə/ Black National State from 1979, near the Zimbabwe border, in South Africa
area 6,500 sq km/2,510 sq mi
capital Thohoyandou
towns MaKearela
features homeland of the Vhavenda people
government military council since a coup 1990 (military leader Ramushwana from 1990)
products coal, copper, graphite, stone
population (1980) 343,500
languages Luvenda, English.

Vendée /vɒn'deɪ/ river in W France that rises near the village of La Châtaigneraie and flows 72 km/45 mi to join the Sèvre Niortaise 11 km/7 mi E of the Bay of Biscay.

Veneto /'venətəʊ/ region of NE Italy, comprising

the provinces of Belluno, Padova (Padua), Treviso, Rovigo, Venezia (Venice), and Vicenza; area 18,400 sq km/7,102 sq mi; population (1990) 4,398,100. Its capital is Venice, and towns include Padua, Verona, and Vicenza. The Veneto forms part of the N Italian plain, with the delta of the river Po; it includes part of the Alps and Dolomites, and Lake Garda. Products include cereals, fruit, vegetables, wine, chemicals, ships, and textiles.

Venezia /ve'netsiə/ Italian form of ◊Venice, a city, port, and naval base on the Adriatic Sea.

Venezuela Republic of (*República de Venezuela*)
area 912,100 sq km/352,162 sq mi
capital Caracas
towns Barquisimeto, Valencia; port Maracaibo
physical Andes Mountains and Lake Maracaibo in NW; central plains (llanos); delta of river Orinoco in E; Guiana Highlands in SE
features Angel Falls, world's highest waterfall
head of state and of government Carlos Andrés Pérez from 1988
government federal democratic republic
political parties Democratic Action Party (AD), moderate, left-of-centre; Social Christian Party (COPEI), Christian centre-right; Movement towards Socialism (MAS), left-of-centre
exports coffee, timber, oil, aluminium, iron ore, petrochemicals
currency bolívar
population (1990 est) 19,753,000 (mestizos 70%, white (Spanish, Portuguese, Italian) 20%, black 9%, amerindian 2%); growth rate 2.8% p.a.
life expectancy men 67, women 73 (1989)
religions Roman Catholic 96%, Protestant 2%
languages Spanish (official), Indian languages 2%
literacy 88% (1989)
GNP $47.3 bn (1988); $2,629 per head (1985)
chronology
1961 New constitution adopted, with Rómulo Betancourt as president.
1964 Dr Raúl Leoni became president.
1969 Dr Rafael Caldera became president.
1974 Carlos Andrés Pérez became president.
1979 Dr Luis Herrera became president.
1984 Dr Jaime Lusinchi became president; social

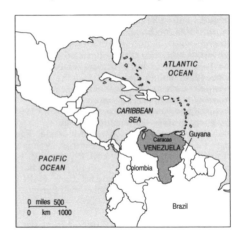

pact established between government, trade unions, and business; national debt rescheduled. *1987* Widespread social unrest triggered by inflation; student demonstrators shot by police. *1988* Carlos Andrés Pérez elected president. Payments suspended on foreign debts. *1989* Economic austerity programme enforced by $4.3 billion loan from International Monetary Fund. Price increases triggered riots; 300 people killed. Feb: martial law declared. May: General strike. Elections boycotted by opposition groups. *1991* Protests against austerity programme. *1992* Attempted anti-government coups failed. Pérez promised constitutional changes.

Venice /'venɪs/ (Italian **Venezia**) city, port, and naval base on the northeast coast of Italy; population (1990) 79,000. It is the capital of Veneto region. The old city is built on piles on low-lying islands in a salt-water lagoon, sheltered from the Adriatic Sea by the Lido and other small strips of land. There are about 150 canals crossed by some 400 bridges. Apart from tourism (it draws 8 million tourists a year), industries include glass, jewellery, textiles, and lace. Venice was an independent trading republic from the 10th century, ruled by a doge, or chief magistrate, and was one of the centres of the Italian Renaissance.

features It is now connected with the mainland and its industrial suburb, Mestre, by road and rail viaduct. The Grand Canal divides the city and is crossed by the Rialto Bridge; transport is by traditional gondola or *vaporetto* (water bus).

St Mark's Square has the 11th-century Byzantine cathedral of San Marco, the 9th–16th-century campanile (rebuilt 1902), and the 14th–15th-century Gothic Doge's Palace (linked to the former state prison by the 17th-century Bridge of Sighs). The nearby Lido is a bathing resort. The *Venetian School* of artists includes the Bellinis, Carpaccio, Giorgione, Titian, Tintoretto, and Veronese. The Venetian Carnival is held annually at the end of February, with spectacular costumes and masks.

history In 1991 archaeologist Ernesto Canal established that the city was founded by the Romans in the 1st century AD; it was previously thought to have been founded by mainlanders fleeing from the Barbarians in AD 421. Venice became a wealthy independent trading republic in the 10th century, stretching by the mid-15th century to the Alps and including Crete. It was governed by an aristocratic oligarchy, the Council of Ten, and a senate, which appointed the doge 697–1797. Venice helped defeat the Ottoman Empire in the naval battle of Lepanto 1571 but the republic was overthrown by Napoleon 1797. It passed to Austria 1815 but finally became part of the kingdom of Italy 1866.

This is the city of mazes. You may set off from the same place to the same place every day and never go by the same route.

On **Venice** Jeanette Winterson
The Passion 1987

Vent, Iles du /iːl djuː 'vɒn/ French name for the ◊Windward Islands, part of the Society Islands in French Polynesia. The Leeward Islands are known as the *Iles sous le Vent*.

Ventura /venˈtjʊərə/ city in SW California, USA, on the Pacific Ocean, NW of Los Angeles; population (1980) 83,475. Industries include oil and agricultural products, such as citrus fruits and lima beans.

Veracruz /ˌverəˈkruːz/ port (trading in coffee, tobacco, and vanilla) in E Mexico, on the Gulf of Mexico; population (1980) 305,456. Products include chemicals, sisal, and textiles. It was founded by the Spanish conquistador Hernando Cortés as Villa Nueva de la Vera Cruz ('new town of the true cross') on a nearby site 1519 and transferred to its present site 1599.

Vermont /vɜːˈmɒnt/ state in northeastern USA; nickname Green Mountain State
area 24,900 sq km/9,611 sq mi
capital Montpelier
towns Burlington, Rutland, Barre
features brilliant autumn foliage and winter sports; Green Mountains; Lake Champlain
products apples, maple syrup, dairy products, china clay, granite, marble, slate, business machines, paper and allied products; tourism is important
population (1990) 562,800
famous people Chester A Arthur, Calvin Coolidge, John Dewey
history explored by the Frenchman Samuel de Champlain from 1609; settled 1724; became a state 1791.

The first French settlement was at Fort Ste Anne 1666; first English settlers 1724 at Fort Drummer (now Brattleboro). England controlled the area from 1763 after the French and Indian War. The Green Mountain Boys, organized 1764 to protect Vermont from New York's territorial claims, captured Ticonderoga and Crown Point from the British 1775.

Verona /vəˈrəʊnə/ industrial city (printing, paper, plastics, furniture, pasta) in Veneto, Italy, on the

Vermont

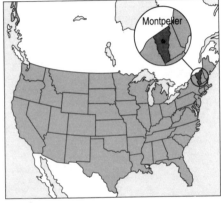

Adige River; population (1988) 259,000. It also trades in fruit and vegetables.

Its historical sights include one of the largest Roman amphitheatres in the world; Castelvecchio, the 14th-century residence of the Scaligers, lords of Verona; the tomb of Juliet; and a 12th-century cathedral.

Versailles /veə'saɪ/ city in N France, capital of Les Yvelines *département*, on the outskirts of Paris; population (1990) 91,000. It grew up around the palace of Louis XV. Within the palace park are two small châteaux, Le Grand Trianon and Le Petit Trianon, built for Louis XIV (by Jules-Hardouin Mansart (1646–1708)) and Louis XV (by Jacques Gabriel (1698–1782)) respectively.

Vesuvius /vɪ'suːvɪəs/ (Italian *Vesuvio*) active volcano SE of Naples, Italy; height 1,277 m/4,190 ft. In 79 BC it destroyed the cities of Pompeii, Herculaneum, and Oplonti.

VI abbreviation for ◊ *Virgin Islands*, in the West Indies; ◊ *Vancouver Island*, off the coast of Canada.

Viborg /'viːbɔː/ industrial town (brewing, engineering, textiles, tobacco) in Jutland, Denmark; population (1990) 39,400. It is also the Swedish name for ◊Vyborg, a port and naval base in Russia.

Vicenza /vɪ'tʃentsə/ city in Veneto region, NE Italy, capital of Veneto province, manufacturing textiles and musical instruments; population (1988) 110,000.

It has a 13th-century cathedral and many buildings by Palladio, including the Teatro Olimpico 1583.

Victoria /vɪk'tɔːrɪə/ state of SE Australia
area 227,600 sq km/87,854 sq mi
capital Melbourne
towns Geelong, Ballarat, Bendigo
physical part of the Great Dividing Range, running E–W and including the larger part of the Australian Alps; Gippsland lakes; shallow lagoons on the coast; the mallee shrub region
products sheep, beef cattle, dairy products, tobacco, wheat, vines for wine and dried fruit, orchard fruits, vegetables, gold, brown coal (Latrobe Valley), oil and natural gas (Bass Strait)

Victoria

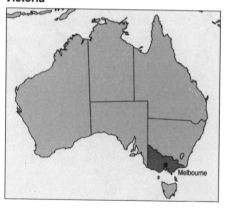

Melbourne

population (1987) 4,184,000; 70% in the Melbourne area
history annexed for Britain by Capt Cook 1770; settled in the 1830s; after being part of New South Wales became a separate colony 1851, named after the queen; became a state 1901.

Victoria /vɪk'tɔːrɪə/ industrial port (shipbuilding, chemicals, clothing, furniture) on Vancouver Island, capital of British Columbia, Canada; population (1986) 66,303.

It was founded as Fort Victoria 1843 by the Hudson's Bay Company. Its university was founded 1964.

Victoria /vɪk'tɔːrɪə/ port and capital of the Seychelles, on Mahé Island; population (1987) 24,300. Industries include copra, vanilla, and cinnamon.

Victoria /vɪk'tɔːrɪə/ city in S Texas, USA, on the Guadalupe River, near the Gulf of Mexico; population (1990) 55,100. It is a transportation centre for oil, natural gas, chemicals, and dairy products.

Victoria /vɪk'tɔːrɪə/ district of Hong Kong, rising to 554 m/1,800 ft at Victoria Park.

Victoria Falls /vɪk'tɔːrɪə/ or *Mosi-oa-tunya* waterfall on the river Zambezi, on the Zambia–Zimbabwe border. The river is 1,700 m/5,580 ft wide and drops 120 m/400 ft to flow through a 30-m/100-ft wide gorge.

The falls were named after Queen Victoria by the Scottish explorer David Livingstone 1855.

Victoria, Lake /vɪk'tɔːrɪə/ or *Victoria Nyanza* largest lake in Africa; area over 69,400 sq km/26,800 sq mi; length 410 km/255 mi. It lies on the equator at an altitude of 1,136 m/3,728 ft, bounded by Uganda, Kenya, and Tanzania. It is a source of the river Nile.

The British explorer John Speke named it after Queen Victoria 1858.

Vienna /vi'enə/ (German *Wien*) capital of Austria, on the river Danube at the foot of the Wiener Wald (Vienna Woods); population (1986) 1,481,000. Industries include engineering and the production of electrical goods and precision instruments.

The United Nations city (1979) houses the United Nations Industrial Development Organization (UNIDO) and the International Atomic Energy Agency (IAEA).

features Renaissance and Baroque architecture; the Hofburg (former imperial palace); the 18th-century royal palaces of Schönbrunn and Belvedere, with formal gardens; the Steiner House 1910 by Adolf Loos; and several notable collections of paintings. Vienna is known for its theatre and opera. Sigmund Freud's home is a museum, and there is a university (1365).

history Vienna was the capital of the Austro-Hungarian Empire 1278–1918 and the commercial centre of E Europe. The old city walls were replaced by a wide street, the Ringstrasse, 1860. After much destruction in World War II the city was divided into US, British, French, and Soviet occupation zones 1945–55. Vienna is associated with J Strauss waltzes, as well as the music of Haydn, Mozart, Beethoven, and Schubert and the development of atonal music. Also figuring in Vienna's cultural history were the Vienna Sezession group of painters and the philosophical Vienna Circle; psychoanalysis originated here.

The nostalgic city with a streak of gentle hopelessness, where Freud discovered sex.

On **Vienna** Alan Whicker 1980

Vientiane /vi,enti'ɑ:n/ (Lao *Vieng Chan*) capital and chief port of Laos on the Mekong River; population (1985) 377,000. Noted for its pagodas, canals, and houses on stilts, it is a trading centre for forest products and textiles. The Temple of the Heavy Buddha, the Pratuxai triumphal arch, and the Black Stupa are here. The Great Sacred Stupa to the NE of the city is the most important national monument in Laos.

Vietnam Socialist Republic of (*Công Hòa Xã Hôi Chu Nghĩa Viêt Nam*)
area 329,600 sq km/127,259 sq mi
capital Hanoi
towns ports Ho Chi Minh City (formerly Saigon), Da Nang, Haiphong
physical Red River and Mekong deltas, centre of cultivation and population; tropical rainforest; mountainous in N and NW
environment during the Vietnam War an estimated 2.2 million hectares/5.4 million acres of forest were destroyed. The country's National Conservation Strategy is trying to replant 500 million trees each year
features Karst hills of Halong Bay, Cham Towers
head of state Vo Chi Cong from 1987
head of government Vo Van Kiet from 1991
political system communism
political party Communist Party
exports rice, rubber, coal, iron, apatite
currency dong

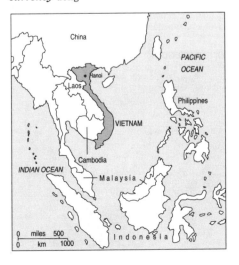

population (1990 est) 68,488,000 (750,000 refugees, majority ethnic Chinese left 1975–79, some settled in SW China, others fled by sea—the 'boat people'—to Hong Kong and elsewhere); growth rate 2.4% p.a.
life expectancy men 62, women 66 (1989)
languages Vietnamese (official), French, English, Khmer, Chinese, local
media independent newspapers prohibited by law 1989; central government approval is required for appointment of editors
religions Buddhist, Taoist, Confucian, Christian
literacy 78% (1989)
GNP $12.6 bn; $180 per head (1987)
chronology
1945 Japanese removed from Vietnam at end of World War II.
1946 Commencement of Vietminh war against French.
1954 France defeated at Dien Bien Phu. Vietnam divided along 17th parallel.
1964 US troops entered Vietnam War.
1973 Paris cease-fire agreement.
1975 Saigon captured by North Vietnam.
1976 Socialist Republic of Vietnam proclaimed.
1978 Admission into Comecon. Vietnamese invasion of Cambodia.
1979 Sino-Vietnamese border war.
1986 Retirement of 'old guard' leaders.
1987–88 Over 10,000 political prisoners released.
1988–89 Troop withdrawals from Cambodia continued.
1989 'Boat people' leaving Vietnam murdered and robbed at sea by Thai pirates. Troop withdrawal from Cambodia completed. Hong Kong forcibly repatriated some Vietnamese refugees.
1991 Vo Van Kiet replaced Do Muoi as prime minister. Cambodia peace agreement signed. Relations with China normalized.

Vigo /'vi:gəʊ/ industrial port (oil refining, leather, paper, distilling) and naval station on Vigo Bay, Galicia, NW Spain; population (1991) 276,700.

Viipuri /'vi:puri/ Finnish name of ◊Vyborg, a port and naval base in the Russia.

Vila /'vi:lə/ or **Port-Vila** port and capital of Vanuatu, on the SW of Efate Island; population (1988) 15,000.

Vilnius /'vɪlnius/ capital of Lithuania; population (1987) 566,000. Industries include engineering and the manufacture of textiles, chemicals, and foodstuffs.
   From a 10th-century settlement, Vilnius became the Lithuanian capital 1323 and a centre of Polish and Jewish culture. It was then Polish from 1386 until the Russian annexation 1795. Claimed by both Poland and Lithuania after World War I, it was given to Poland 1921, occupied by the USSR 1939, and immediately transferred to Lithuania. The city was the focal point of Lithuania's agitation for independence from the USSR 1989–91, and became the country's capital when independence was achieved 1991.

Vineland /'vaɪnlənd/ industrial city (foundry products, glassware, chemicals, and vegetables) in

Virginia

SW New Jersey, USA, N of Millville; population (1990) 54,800.

Vinson Massif /'vɪnsən mæ'siːf/ highest point in Antarctica, rising to 5,140 m/16,863 ft in the Ellsworth Mountains.

Virginia /və'dʒɪnɪə/ state in eastern USA; nickname Old Dominion
area 105,600 sq km/40,762 sq mi
capital Richmond
towns Norfolk, Virginia Beach, Newport News, Hampton, Chesapeake, Portsmouth
features Blue Ridge Mountains, which include the Shenandoah National Park; Arlington National Cemetery; Mount Vernon (home of George Washington 1752–99); Monticello (Thomas Jefferson's home near Charlottesville); Stratford Hall (Robert E. Lee's birthplace at Lexington); Williamsburg restoration; Jamestown and Yorktown historic sites
products sweet potatoes, maize, tobacco, apples, peanuts, coal, ships, lorries, paper, chemicals, processed food, textiles
population (1990) 6,187,400.

The good Old Dominion, the blessed mother of us all.

On **Virginia** Thomas Jefferson,
letter of 1791

Virginia Beach /və,dʒɪnjə 'biːtʃ/ town and resort in SE Virginia, USA, on Chesapeake Bay, E of Norfolk; population (1990) 393,100. The colonists who settled Jamestown first landed here 1607.

Virgin Islands /'vɜːdʒɪn/ group of about 100 small islands, northernmost of the Leeward Islands in the Antilles, West Indies. Tourism is the main industry.

They comprise the *US Virgin Islands* St Thomas (with the capital, Charlotte Amalie), St Croix, St John, and about 50 small islets; area 350 sq km/135 sq mi; population (1990) 101,800; and the *British Virgin Islands* Tortola (with the capital, Road Town), Virgin Gorda, Anegada, and Jost van Dykes, and about 40 islets; area 150 sq km/58 sq mi; population (1987) 13,250.

The US Virgin Islands were purchased from Denmark 1917, and form an 'unincorporated territory'. The British Virgin Islands were taken over from the Dutch by British settlers 1666, and have partial internal self-government.

Visalia /vaɪ'seɪljə/ city in central California, USA, in the San Joaquin Valley, SE of Fresno; population (1990) 75,600. It is an agricultural centre for grapes, citrus fruits, and dairy products.

Vistula /'vɪstjulə/ (Polish *Wisła*) river in Poland that rises in the Carpathian Mountains and runs SE to the Baltic Sea at Gdańsk; length 1,090 km/677 mi. It is heavily polluted, carrying into the Baltic every year large quantities of industrial and agricultural waste, including phosphorus, oil, nitrogen, mercury, cadmium, and zinc.

Vitebsk /'viːtebsk/ industrial city (glass, textiles, machine tools, shoes) in NE Belarus, on the Dvina River; population (1987) 347,000. Vitebsk dates from the 10th century and has been Lithuanian, Russian, and Polish.

Do you think I have come so far to conquer these miserable huts?

On **Vitebsk** attributed to Napoleon,
before the march to Moscow 1812

Vitoria /vɪ'tɔːrɪə/ capital of Alava province, in the Basque country, N Spain; population (1991) 208,600. Products include motor vehicles, agricultural machinery, and furniture.

Vittorio Veneto /vɪ'tɔːrɪəu 'venɪtəu/ industrial town (motorcycles, agricultural machinery, furniture, paper, textiles) in Veneto, NE Italy; population (1981) 30,000. It was the site of the final victory of Italy and its allies over Austria Oct 1918.

Vizcaya /vɪ'skaɪə/ Basque form of ⟡Biscay, a bay in the Atlantic Ocean off France and Spain. It is also the name of one of the three Spanish Basque provinces.

Vladikavkaz (former name 1954–91 *Ordzhonikidze*) capital of the autonomous region of North Ossetia, Russia, on the river Terek in the Caucasus Mountains; population (1989) 300,000. Metal products, vehicles, and textiles are produced. Vladikavkaz was founded 1784 as a frontier fortress.

Virgin Islands

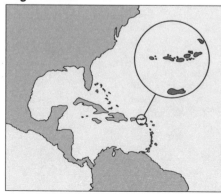

Vladivostok /ˌvlædɪˈvɒstɒk/ port (naval and commercial) in E Siberian Russia, at the Amur Bay on the Pacific coast; population (1987) 615,000. It is kept open by icebreakers during winter. Industries include shipbuilding and the manufacture of precision instruments.

It was established 1860 as a military port. It is the administrative centre of the Far East Science Centre 1969, with subsidiaries at Petropavlovsk, Khabarovsk, and Magadan.

Vlissingen /ˈflɪsɪŋə/ (English *Flushing*) port on Walcheren Island, Zeeland, SW Netherlands; population (1987) 44,900. It stands at the entrance to the Scheldt estuary, one of the principal sea routes into Europe. Industries include fishing, shipbuilding, and petrochemicals, and there is a ferry service to Harwich, England.

Admiral de Ruyter was born here and is commemmorated in the Jacobskerk. In 1944 bombing of the dams of the island of Walcheren submerged it, but it was subsequently reclaimed from the sea.

Vlorë /ˈvlɔːrə/ port and capital of Vlorë province, SW Albania, population (1980) 58,000. A Turkish possession from 1464, it was the site of the declaration of independence by Albania 1912.

Vojvodina /ˌvɔɪvəˈdiːnə/ autonomous area in N Serbia, Yugoslavia; area 21,500 sq km/ 8,299 sq mi; population (1986) 2,050,000, including 1,110,000 Serbs and 390,000 Hungarians. Its capital is Novi Sad. In Sept 1990 Serbia effectively stripped Vojvodina of its autonomous status, causing anti-government and anti-communist riots in early 1991.

Volga /ˈvɒlgə/ longest river in Europe; 3,685 km/ 2,290 mi, 3,540 km/2,200 mi of which are navigable. It drains most of the central and eastern parts of European Russia, rises in the Valdai plateau, and flows into the Caspian Sea 88 km/ 55 mi below the city of Astrakhan.

Volgograd /ˈvɒlgəɡræd/ (formerly until 1925 *Tsaritsyn* and 1925–61 *Stalingrad*) industrial city (metal goods, machinery, sawmills, oil refining) in SW Russia, on the river Volga; population (1987) 988,000.

Its successful defence 1942–43 against Germany was a turning point in World War II.

Volta /ˈvɒltə/ main river in Ghana, about 1,600 km/ 1,000 mi long, with two main upper branches, the Black and White Volta. It has been dammed to provide power.

Volta, Upper /ˈvɒltə/ former name (until 1984) of ◊ Burkina Faso.

Vorarlberg /ˈfɔːrˌɑːlbɜːɡ/ ('in front of the Arlberg') Alpine federal province of W Austria draining into the river Rhine and Lake Constance; area 2,600 sq km/1,004 sq mi; population (1987) 314,000. Its capital is Bregenz. Industries include forestry and dairy farming.

Voronezh /vəˈrɒneʃ/ industrial city (chemicals, construction machinery, electrical equipment) and capital of the Voronezh region of Russia, S of Moscow on the Voronezh River; population (1987) 872,000. There has been a town on the site since the 11th century.

Voroshilovgrad /ˌvɒrəˈʃiːlɒfɡræd/ former name (1935–58; 1970–89) of ◊ Lugansk, a city in the Ukraine.

Vosges /vəuʒ/ mountain range in E France, rising in the Ballon de Guebwiller to 1,422 m/4,667 ft and forming the western edge of the Rhine rift valley. It gives its name to a *département*.

VT abbreviation for ◊ *Vermont*, a state of the USA.

Vukovar /ˈvuːkəuvɑː/ river port in Croatia at the junction of the rivers Vuka and Danube, 32 km/20 mi SE of Osijek; population (1981) 81,200. Industries include foodstuffs manufacture, fishing, and agricultural trade. In 1991 the town resisted three months of siege by the Serb-dominated Yugoslav army before capitulating.

It suffered the severest damage inflicted to any European city since the bombing of Dresden during World War II.

Vyborg /ˈviːbɔːɡ/ (Swedish *Viborg*, Finnish *Viipuri*) port (trading in timber and wood products) and naval base in E Karelia, NW Russia, on the Gulf of Finland, 112 km/70 mi NW of St Petersburg; population (1973) 51,000. Products include electrical equipment and agricultural machinery. Founded by the Swedes 1293, it was part of Finland until 1940.

WA abbreviation for ▷ *Washington*, a state of the USA; ▷ *Western Australia*, an Australian state.

Waco /'weɪkəʊ/ city in E central Texas, USA, on the Brazos River, S of Fort Worth; population (1990) 103,600. It is an agricultural shipping centre for cotton, grain, and livestock; industries include aircraft parts, glass, cement, tyres, and textiles.

Waddenzee /'wɒdnzeɪ/ European estuarine area (tidal flats, salt marshes, islands, and inlets) N of the Netherlands and Germany, and W of Denmark; area 10,000 sq km/4,000 sq mi. It is the nursery for the North Sea fisheries, but the ecology is threatened by tourism and other development.

Wadi Halfa /'wɒdi 'haɪfə/ frontier town in Sudan, NE Africa, on Lake Nuba (the Sudanese section of Lake Nasser, formed by the Nile dam at Aswan, Egypt), which partly flooded the archaeological sites here).

Wagga Wagga /'wɒgə 'wɒgə/ agricultural town in SE New South Wales, Australia; population (1985) 49,500.

Waikato /waɪ'kætəʊ/ river on North Island, New Zealand, 355 km/220 mi long; Waikato is also the name of the dairy area the river traverses; chief town Hamilton.

Wairarapa /ˌwaɪrə'ræpə/ area of North Island, New Zealand, round *Lake Wairarapa*, specializing in lamb and dairy farming; population (1986) 39,600. The chief market centre is Masterton.

Wairau /'waɪrau/ river in N South Island, New Zealand, flowing 170 km/105 m NE to Cook Strait.

Waitaki /waɪ'tæki/ river in SE South Island, New Zealand, that flows 215 km/135 mi to the Pacific. The Benmore hydroelectric installation has created an artificial lake.

Wakefield /'weɪkfiːld/ industrial city (chemicals, machine tools), administrative headquarters of West Yorkshire, England, on the river Calder, S of Leeds; population (1991) 306,300. The Lancastrians defeated the Yorkists here 1460 during the Wars of the Roses.

Wake Islands /weɪk/ small Pacific atoll comprising three islands 3,700 km/2,300 mi W of Hawaii, under US Air Force administration since 1972; area 8 sq km/3 sq mi; population (1980) 300. It was discovered by Capt William Wake 1841, annexed by the USA 1898, and uninhabited until 1935 when it was made an air staging point, with a garrison. It was occupied by Japan 1941–45.

Walcheren /'vɑːlkərən/ island in Zeeland province, the Netherlands, in the estuary of the river Scheldt
area 200 sq km/80 sq mi
capital Middelburg
towns Vlissingen (Flushing)
features flat and for the most part below sea level
products dairy, sugar beet, and other root vegetables
history a British force seized Walcheren 1809; after 7,000 of the garrison of 15,000 had died of malaria, the remainder were withdrawn. It was flooded by deliberate breaching of the dykes to drive out the Germans 1944–45, and in 1953 by abnormally high tides.

Wales /weɪlz/ (Welsh *Cymru*) Principality of; constituent part of the UK, in the W between the British Channel and the Irish Sea
area 20,780 sq km/8,021 sq mi
capital Cardiff
towns Swansea, Wrexham, Newport, Carmarthen
features Snowdonia Mountains (Snowdon 1,085 m/3,561 ft, the highest point in England and Wales) in the NW and in the SE the Black Mountains, Brecon Beacons, and Black Forest ranges; rivers Severn, Wye, Usk, and Dee
exports traditional industries (coal and steel) have declined, but varied modern and high-technology ventures are being developed; Wales has the largest concentration of Japanese-owned plants in the UK. It also has the highest density of sheep in the world and a dairy industry; tourism is important
currency pound sterling
population (1988 est) 2,857,000
languages Welsh 19% (1981), English
religions Nonconformist Protestant denominations; Roman Catholic minority
Wales: counties

county	administrative headquarters	area in sq km
Clwyd	Mold	2,420
Dyfed	Carmarthen	5,770
Gwent	Cwmbran	1,380
Gwynedd	Caernarvon	3,870
Mid Glamorgan	Cardiff	1,020
Powys	Llandrindod Wells	5,080
South Glamorgan	Cardiff	420
West Glamorgan	Swansea	820
		20,780

government returns 38 members to the UK Parliament.

Wallis and Futuna /'wɒlɪs, fuːˈtjuːnə/ two island groups in the SW Pacific, an overseas territory of France; area 367 sq km/143 sq mi; population (1990) 13,700. They produce copra, yams and bananas. Discovered by European sailors in the 18th century, the islands became a French protectorate 1842 and an overseas territory 1961.

Wallsend /'wɔːlzend/ town in Tyne and Wear, NE England, on the river Tyne at the east end of Hadrian's Wall; population (1981) 51,200. Industries include shipbuilding, engineering, and coalmining.

Wall Street /wɔːl/ street in Manhattan, New York, on which the stock exchange is situated, and a synonym for stock dealing in the USA. Its name derives from a stockade erected 1653.

Walsall /'wɔːlsəl/ industrial town (castings, tubes, electrical equipment, leather goods) in the West Midlands, England, 13 km/8 mi NW of Birmingham; population (1981) 265,900.

Waltham Forest borough of N Greater London *population* (1991) 203,400.

Walvis Bay /'wɔːlvɪs 'beɪ/ chief port serving Namibia, SW Africa; population (1980) 26,000. It has a fishing industry with allied trades. It has been a detached part (area 1,100 sq km/425 sq mi) of Cape Province from 1884 (administered solely by South Africa 1922–92; from 1992 jointly by South Africa and Namibia).

The South African and Namibian governments agreed Sept 1991 on joint control of the disputed port of Walvis Bay, pending a final settlement of the future of the disputed territory. In 1992 it was agreed to establish a joint administrative body.

Wandsworth borough of SW central Greater London, S of the Thames
features Wandsworth Park; Putney Heath
population (1991) 237,500.

Wanganui /ˌwɒŋəˈnuːi/ port (textiles, clothing) in SW North Island, New Zealand, at the mouth of the Wanganui River; population (1991) 41,200.

Wankie /'wæŋki/ name until 1982 of ◊Hwange, a town and national park in Zimbabwe.

Wapping /'wɒpɪŋ/ district of the Greater London borough of Tower Hamlets; situated between the Thames and the former London Docks. Since the 1980s it has become a centre of the UK newspaper industry.

Warner Robins /ˌwɔːnə 'rɒbɪnz/ city in central Georgia, USA, S of Macon; population (1990) 43,700. Industries include nuts, fruits, and aeroplane parts.

Warren /'wɒrən/ industrial city (steel and iron products, tools, and paint) in NE Ohio, USA, on the Mahoning River, NW of Youngstown; population (1990) 50,800.

Warrington /'wɒrɪŋtən/ industrial town (metal goods, chemicals, brewing) in Cheshire, NW England, on the river Mersey; population (1989 est) 188,000. A trading centre since Roman times, it was designated a new town 1968.

Warrnambool /'wɔːnəmbuːl/ port near the mouth of Hopkins River, SW Victoria, Australia; population (1981) 22,000. A tourist centre, it also manufactures textiles and dairy products.

Warrumbungle Range /ˌwɒrəmˈbʌŋgəl/ (Aboriginal 'broken-up small mountains') mountain range of volcanic origin in New South Wales, Australia. Siding Spring Mountain, 859 m/2,819 ft, is the site of an observatory; the Breadknife is a 90-m/300-ft high rock only 1.5 m/5 ft wide; the highest point is Mount Exmouth, 1,228 m/4,030 ft.

Warsaw /'wɔːsɔː/ (Polish *Warszawa*) capital of Poland, on the river Vistula; population (1990) 1,655,700. Industries include engineering, food processing, printing, clothing, and pharmaceuticals.
history Founded in the 13th century, it replaced Kraków as capital 1595. Its university was founded 1818. Between the mid nineteenth century and 1940, a third of the population were Jews. It was taken by the Germans 27 Sept 1939, and 250,000 Poles were killed during two months of street fighting that started 1 Aug 1944. It was finally liberated 17 Jan 1945. The old city was virtually destroyed in World War II but has been reconstructed. The physicist and chemist Marie Curie was born here.

Warwick /'wɒrɪk/ market town, administrative headquarters of Warwickshire, England; population (1981) 21,900. Industries include carpets and engineering. Founded 914, it has many fine medieval buildings including a 14th-century castle.

Warwickshire /'wɒrɪkʃə/ county in central England
area 1,980 sq km/764 sq mi

Warwickshire

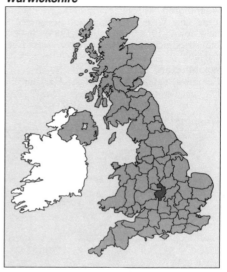

towns Warwick (administrative headquarters), Leamington, Nuneaton, Rugby, Stratford-upon-Avon
features Kenilworth and Warwick castles; remains of the 'Forest of Arden' (portrayed by Shakespeare in *As You Like It*); site of the Battle of Edgehill; annual Royal Agricultural Show held at Stoneleigh
products mainly agricultural, engineering, textiles
population (1991) 477,000
famous people Rupert Brooke, George Eliot, William Shakespeare.

Wash, the /wɒʃ/ bay of the North Sea between Norfolk and Lincolnshire, England. The rivers Nene, Ouse, Welland, and Witham drain into the Wash. King John lost his baggage and treasure in crossing it 1216.

… the sink of more than thirteen counties.

On **the Wash** Daniel Defoe *Tour through the Whole Island of Great Britain* 1724–27

Washington /'wɒʃɪŋtən/ state in northwestern USA; nickname Evergreen State/Chinook State
area 176,700 sq km/68,206 sq mi
capital Olympia
towns Seattle, Spokane, Tacoma
features Columbia River; national parks: Olympic (Olympic Mountains), Mount Rainier (Cascade Range), North Cascades; 90 dams
products apples and other fruits, potatoes, livestock, fish, timber, processed food, wood products, paper and allied products, aircraft and aerospace equipment, aluminium
population (1990) 4,866,700 (including 1.4% Indians, mainly of the Yakima people)
famous people Bing Crosby, Jimi Hendrix, Mary McCarthy, Theodore Roethke
history explored by Spanish, British, and Americans in the 18th century; settled from 1811; became a territory 1853 and a state 1889.

Washington /'wɒʃɪŋtən/ town on the river Wear, Tyne and Wear, NE England, designated a 'new town' 1964; population (1985) 56,000. Industries include textiles, electronics, and car assembly.

Washington, DC /'wɒʃɪŋtən/ (District of Columbia) national capital of the USA, on the Potomac River

Washington

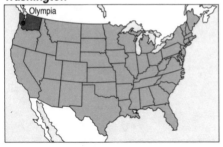

Olympia

area 180 sq km/69 sq mi
capital the District of Columbia covers only the area of the city of Washington
features designed by a French engineer, Pierre l'Enfant (1754–1825). Among buildings of architectural note are the Capitol, the Pentagon, the White House, and the Lincoln Memorial. The National Gallery has a fine collection of paintings; libraries include the Library of Congress, the National Archives, and the Folger Shakespeare Library. The Smithsonian Institution is here
population (1990) 606,900 (metropolitan area, extending outside the District of Columbia, 3 million)
history the District of Columbia, initially land ceded from Maryland and Virginia, was established by Act of Congress 1790–91, and was first used as the seat of Congress 1800. The right to vote in national elections was not granted to residents until 1961. Local self-rule began 1975.

Washington isn't a city, it's an abstraction.

On **Washington, DC** Dylan Thomas 1950

Waterbury /'wɔːtəbəri/ city in W Connecticut, USA, on the Naugatuck River; population (1990) 109,000. Products include clocks, watches, brass and copper ware, and plastics. It was founded 1674.

Waterford /'wɔːtəfəd/ county in Munster province, Republic of Ireland; area 1,840 sq km/710 sq mi; population (1991) 91,600. The county town is Waterford. The county includes the rivers Suir and Blackwater, and the Comeragh and Monavallagh mountain ranges in the N and centre. Products include cattle, beer, whisky, and glassware.

Waterford /'wɔːtəfəd/ port and county town of County Waterford, SE Republic of Ireland, on the Suir; population (1991) 40,300. Handmade Waterford crystal glass (34% lead content instead of the normal 24%) was made here until 1851 and again from 1951.

Watford /'wɒtfəd/ industrial town (printing, engineering, and electronics) in Hertfordshire, SE England; population (1986) 77,000. It is a dormitory town for London.

Waziristan /wə,zɪərɪ'stɑːn/ mountainous territory in Pakistan, on the border with Afghanistan, inhabited by Waziris and Mahsuds.

Weald, the /wiːld/ (Old English 'forest') area between the North and South Downs, England, once thickly wooded, and forming part of Kent, Sussex, Surrey, and Hampshire. Now an agricultural area, it produces fruit, hops, and vegetables. In the Middle Ages its timber and iron ore made it the industrial heart of England.

Its oaks were used in shipbuilding. The name often refers only to the area of Kent SW of the greensand ridge running from Hythe to Westerham.

Wear /wɪə/ river in NE England; length 107 km/ 67 mi. From its source in the Pennines it flows E, past Durham to meet the North Sea at Sunderland.

Weddell Sea /'wedl/ arm of the S Atlantic Ocean that cuts into the Antarctic continent SE of Cape Horn; area 8,000,000 sq km/3,000,000 sq mi. Much of it is covered with thick pack ice for most of the year.

It is named after the British explorer James Weddell.

Weihai /ˌweɪ'haɪ/ commercial port (textiles, rubber articles, matches, soap, vegetable oils) in Shandong, China; population about 220,000. It was leased to Britain 1898–1930, during which time it was a naval and coaling station. It was occupied by Japan 1938–45.

Weimar /'vaɪmɑː/ town in the state of Thuringia, Germany, on the river Elm; population (1990) 80,000. Products include farm machinery and textiles. It was the capital of the grand duchy of Saxe-Weimar 1815–1918; in 1919 the German National Assembly drew up the constitution of the new Weimar Republic here. The writers Goethe, Schiller, and Herder and the composer Liszt lived in the town. The former concentration camp of Buchenwald is nearby.

It is a very queer, little town, although called the 'Athens of Germany' on account of the great poets who have lived here.

On **Weimar** G H Lewes, letter of 1854

Weirton /'wɪətn/ industrial city (coal, steel, chemicals, and cement) in West Virginia, USA, on the Ohio River, W of Pittsburgh; population (1990) 22,100.

Welland Ship Canal /'welənd/ Canadian waterway, part of the St Lawrence Seaway, linking Lake Erie to Lake Ontario.

West Bengal

Calcutta

INDIAN OCEAN

Wellington /'welɪŋtən/ capital and industrial port (woollen textiles, chemicals, soap, footwear, bricks) of New Zealand on North Island on the Cook Strait; population (1991) 149,600; urban area 324,800. The harbour was sighted by Capt James Cook 1773.

Founded 1840 by Edward Gibbon Wakefield as the first settlement of the New Zealand Company, it has been the seat of government since 1865, when it replaced Auckland. Victoria University was founded 1897. A new assembly hall (designed by the British architect Basil Spence and popularly called 'the beehive' because of its shape) was opened 1977 alongside the original parliament building.

Wells /welz/ market and cathedral town in Somerset, SW England; population (1989 est) 10,000. Industries include printing and the manufacture of animal foodstuffs. The cathedral, built near the site of a Saxon church in the 12th and 13th centuries, has a west front with 386 carved figures. Wells was made the seat of a bishopric about 900 (Bath and Wells from 1244) and has a bishop's palace.

Welwyn Garden City /'welɪn/ industrial town (chemicals, electrical engineering, clothing, food) in Hertfordshire, England, 32 km/20 mi N of London; population (1981) 40,500. It was founded as a garden city 1919–20 by Ebenezer Howard, and designated a new town 1948.

Wembley /'wembli/ district of the Greater London borough of Brent, site of Wembley Stadium.

Wenchow /ˌwen'tʃau/ alternative transcription of the Chinese town Wenzhou.

Wenzhou /ˌwen'dʒəu/ industrial port (textiles, medicine) in Zhejiang, SE China; population (1984) 519,000. It was opened to foreign trade 1877 and is now a special economic zone.

Wesermünde /ˌveɪzə'mundə/ name until 1947 of ¢Bremerhaven, a port in Germany.

West Bank area (5,879 sq km/2,270 sq mi) on the west bank of the river Jordan; population (1988) 866,000. The West Bank was taken by the Jordanian army 1948 at the end of the Arab-Israeli war that followed the creation of the state of Israel, and was captured by Israel during the Six-Day War 5–10 June 1967. The continuing Israeli occupation and settlement of the area has created tensions with the Arab population.

In 1988 King Hussein announced that Jordan was cutting 'legal and administrative ties' with the West Bank, leaving responsibility for Arabs in the region to the Palestine Liberation Organization (PLO) (which was already the position in practice).

West Bengal /ben'gɔːl/ state of NE India
area 87,900 sq km/33,929 sq mi
capital Calcutta
towns Asansol, Durgapur
physical occupies the west part of the vast alluvial plain created by the rivers Ganges and Brahmaputra, with the Hooghly River; annual rainfall in excess of 250 cm/100 in

Western Australia

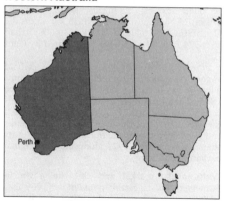

products rice, jute, tea, coal, iron, steel, cars, locomotives, aluminium, fertilizers
population (1991) 67,982,700
history created 1947 from the former British province of Bengal, with later territories added: Cooch Behar 1950, Chandernagore 1954, and part of Bihar 1956.

West Bromwich /'brɒmɪdʒ/ industrial town (metalworking, springs, tubes) in West Midlands, England, NW of Birmingham; population (1981) 92,000.

Western Australia /ɒ'streɪlɪə/ state of Australia
area 2,525,500 sq km/974,843 sq mi
capital Perth
towns main port Fremantle, Bunbury, Geraldton, Kalgoorlie-Boulder, Albany
features largest state in Australia; Monte Bello Islands; rivers Fitzroy, Fortescue, Gascoyne, Murchison, Swan; northwest coast subject to hurricanes (willy-willies); Lasseter's Reef
products wheat, fresh and dried fruit, meat and dairy products, natural gas (northwest shelf) and oil (Canning Basin), iron (the Pilbara), copper, nickel, uranium, gold, diamonds
population (1987) 1,478,000
history a short-lived convict settlement at King George Sound 1826; first non-convict settlement founded at Perth 1829 by Capt James Stirling (1791–1865); self-government 1890; became a state 1901,

Western Isles island area of Scotland, comprising the Outer Hebrides (Lewis, Harris, North and South Uist, and Barra)
area 2,900 sq km/1,120 sq mi
towns Stornoway on Lewis (administrative headquarters)
features divided from the mainland by the Minch channel; Callanish monolithic circles of the Stone Age on Lewis
products Harris tweed, sheep, fish, cattle
population (1991) 29,100
famous people Flora MacDonald.

Western Provinces in Canada, the provinces of Alberta, British Columbia, Manitoba, and Saskatchewan.

Western Isles

Western Sahara /sə'hɑːrə/ (formerly *Spanish Sahara*) disputed territory in NW Africa bounded to the N by Morocco, to the W and S by Mauritania, and to the E by the Atlantic Ocean
area 266,800 sq km/103,011 sq mi
capital Laâyoune (Arabic *El Aaiún*)
towns Dhakla
features electrically monitored fortified wall enclosing the phosphate area
exports phosphates
currency dirham
population (1988) 181,400; another estimated 165,000 live in refugee camps near Tindouf, SW Algeria. Ethnic composition: Sawrawis (traditionally nomadic herders)
language Arabic
religion Sunni Muslim
government administered by Morocco
history this 1,000-km-long Saharan coastal region, which during the 19th century separated

Western Sahara

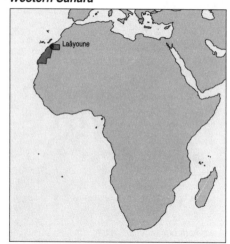

French-dominated Morocco and Mauritania, was designated a Spanish 'sphere of influence' 1884 because it lies opposite the Spanish-ruled Canary Islands. On securing its independence 1956, Morocco laid claim to and invaded this 'Spanish Sahara' territory, but was repulsed. Moroccan interest was rekindled from 1965, following the discovery of rich phosphate resources at Bou-Craa, and within Spanish Sahara a pro-independence nationalist movement developed, spearheaded by the Popular Front for the Liberation of Saguia al Hamra and Río de Oro (Polisario), established 1973.

partition After the death of the Spanish ruler General Franco, Spain withdrew and the territory was partitioned between Morocco and Mauritania. Polisario rejected this partition, declared their own independent Saharan Arab Democratic Republic (SADR), and proceeded to wage a guerrilla war, securing indirect support from Algeria and, later, Libya. By 1979 they had succeeded in their struggle against Mauritania, which withdrew from their southern sector and concluded a peace agreement with Polisario, and in 1982 the SADR was accepted as a full member of the Organization of African Unity (OAU). By the end of 1990, 70 countries had granted diplomatic recognition to the SADR.

defensive wall Morocco, which occupied the Mauritanian-evacuated zone, still retained control over the bulk of the territory, including the key towns and phosphate mines, which they protected with an 'electronic defensive wall' 4,000 km/ 2,500 mi long, completed 1987. From the mid-1980s this wall was gradually extended outwards as Libya and Algeria reduced their support for Polisario and drew closer to Morocco. In 1988, Morocco and the Polisario Front agreed to United Nations-sponsored plans for a cease-fire and a referendum in Western Sahara, based on 1974 voting rolls, to decide the territory's future. However, divisions persisted during 1989 and 1990 over the terms of the referendum and sporadic fighting continued.

Western Samoa see ♢Samoa, Western.

West Germany see ♢Germany, West.

West Glamorgan /gləˈmɔːgən/ (Welsh *Gorllewin Morgannwg*) county in SW Wales
area 817 sq km/315 sq mi
towns Swansea (administrative headquarters), Margam, Port Talbot, Neath
features Gower Peninsula
products tinplate, copper, steel, chemicals
population (1991) 357,800
languages 16% Welsh, English
famous people Richard Burton, Anthony Hopkins, Dylan Thomas.

West Indies /ˈɪndiz/ archipelago of about 1,200 islands, dividing the Atlantic from the Gulf of Mexico and the Caribbean. The islands are divided into:
Bahamas;
Greater Antilles Cuba, Hispaniola (Haiti, Dominican Republic), Jamaica, and Puerto Rico;
Lesser Antilles Aruba, Netherlands Antilles,

West Glamorgan

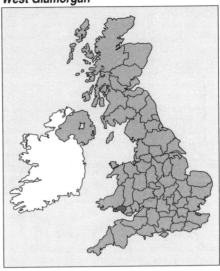

Trinidad and Tobago, the Windward Islands (Grenada, Barbados, St Vincent, St Lucia, Martinique, Dominica, Guadeloupe), the Leeward Islands (Montserrat, Antigua, St Christopher (St Kitts)–Nevis, Barbuda, Anguilla, St Martin, British and US Virgin Islands), and many smaller islands.

The West Indies I behold/Like the Hesperides of old/ – Trees of life with fruits of gold.

On the **West Indies** James Montgomery
'A Voyage Round the World' 1841

West Irian /ˈɪriən/ former name of ♢Irian Jaya, a province of Indonesia.

West Lothian /ˈləʊðiən/ former county of central Scotland, bordering the southern shore of the Firth of Forth; from 1975 included (except for the Bo'ness area, which went to Central Region) in Lothian Region.

Westman Islands /ˈwestmən/ small group of islands off the south coast of Iceland. In 1973 volcanic eruption caused the population of 5,200 to be temporarily evacuated, and added 2.5 sq km/ 1 sq mi to the islands' area. Heimaey, the largest of the islands, is one of Iceland's chief fishing ports.

Westmeath /ˌwestˈmiːð/ inland county of Leinster province, Republic of Ireland
area 1,760 sq km/679 sq mi
town Mullingar (county town)
physical rivers: Shannon, Inny, Brosna; lakes: Ree, Sheelin, Ennell
products agricultural and dairy products, limestone, textiles
population (1991) 61,900.

West Midlands /ˈmɪdləndz/ metropolitan county in central England, created 1974, originally

West Midlands

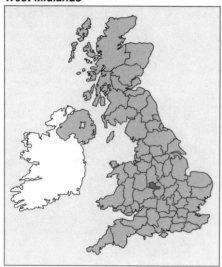

West Sussex

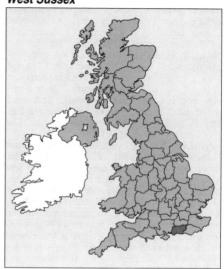

administered by an elected council; its powers reverted to district councils from 1986
area 900 sq km/347 sq mi
towns Birmingham, Coventry
products industrial goods
population (1991) 2,500,400
famous people Edward Burne-Jones, Neville Chamberlain, Philip Larkin.

Westminster, City of /'west,mɪnstə/ borough of central Greater London, England, on the north bank of the Thames between Kensington and the City of London; population (1991) 181,500. It encompasses Bayswater, Belgravia, Mayfair, Paddington, Pimlico, Soho, St John's Wood, and Westminster.
Bayswater is a residential and hotel area N of Kensington Gardens;
Belgravia bounded to the N by Knightsbridge, has squares laid out 182–30 by Thomas Cubitt;
Mayfair between Oxford Street and Piccadilly, includes Park Lane and Grosvenor Square (with the US embassy);
Paddington includes Little Venice on the Grand Union Canal;
Pimlico has the Tate Gallery (Turner collection, British, and modern art);
Soho has many restaurants and a Chinese community around Gerrard Street. It was formerly known for strip clubs and sex shops;
St John's Wood has Lord's cricket ground and the studios at 11 Abbey Road where the Beatles recorded their music;
Westminster encompasses Buckingham Palace (royal residence), Green Park, St James's Park and St James's Palace (16th century), Marlborough House, Westminster Abbey, Westminster Hall (1097–1401), the Houses of Parliament with Big Ben, Whitehall (government offices), Downing Street (homes of the prime minister at number 10 and the chancellor of the Exchequer at number 11), Hyde Park with the Albert Memorial

opposite the Royal Albert Hall, Trafalgar Square with the National Gallery and National Portrait Gallery.

Westmorland /'westmələnd/ former county in the Lake District, England, part of Cumbria from 1974.

Weston-super-Mare /'westən ˌsuːpə 'meə/ seaside resort and town in Avon, SW England, on the Bristol Channel; population (1984) 170,500. Industries include plastics and engineering.

West Palm Beach /ˌwest paːm 'biːtʃ/ town and resort on the southeast coast of Florida, USA, on the lagoon Lake Worth, N of Miami; population (1990) 67,600. Industries include transistors, aircraft parts, building materials, and citrus fruits; tourism is important to the economy.

West Sussex /sʌsɪks/ county on the south coast of England
area 2,020 sq km/780 sq mi
towns Chichester (administrative headquarters), Crawley, Horsham, Haywards Heath, Shoreham (port); resorts: Worthing, Littlehampton, Bognor Regis
physical the Weald, South Downs; rivers: Arun, West Rother, Adur
features Arundel and Bramber castles; Goodwood, Petworth House (17th century); Wakehurst Place, where the Royal Botanic Gardens, Kew, has additional grounds; the Weald and Downland Open Air Museum at Singleton
population (1991) 692,800
famous people William Collins, Richard Cobden, Percy Bysshe Shelley.

West Virginia /vəˈdʒɪnɪə/ state in E central USA; nickname Mountain State
area 62,900 sq km/24,279 sq mi
capital Charleston
towns Huntington, Wheeling
physical Allegheny Mountains; Ohio River

West Virginia

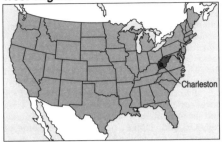

features port of Harper's Ferry, restored as when John Brown seized the US armoury 1859
products apples, maize, poultry, dairy and meat products, coal, natural gas, oil, chemicals, synthetic fibres, plastics, steel, glass, pottery
population (1990) 1,793,500
famous people Pearl S. Buck, Thomas 'Stonewall' Jackson, Walter Reuther, Cyrus Vance
history first settlement at Mill Creek 1731 by Morgan Morgan, who came from Delaware; German settlements 1730s; coal discovered on the Coal River 1742; industrial development early 19th century. On the secession of Virginia from the Union 1861, west Virginians dissented and formed a new state 1863. Industrial expansion was accompanied by labour strife in the early 20th century.

West Yorkshire /'jɔːksə/ metropolitan county in NE England, created 1976, originally administered by an elected metropolitan council; its powers reverted to district councils from 1986
area 2,040 sq km/787 sq mi
towns Wakefield, Leeds, Bradford, Halifax, Huddersfield
features Ilkley Moor, Haworth Moor, Haworth Parsonage; part of the Peak District National Park
products coal, woollen textiles
population (1988 est) 2,056,500
famous people the Brontës, David Hockney, Henry Moore, J B Priestley.

Wexford /'weksfəd/ county in the Republic of Ireland, province of Leinster
area 2,350 sq km/907 sq mi
towns Wexford (county town), Rosslare
products fish, livestock, oats, barley, potatoes
population (1991) 102,000.

Wexford /'weksfəd/ seaport and county town of Wexford, Republic of Ireland; population (1991) 9,500. Products include textiles, cheese, and agricultural machinery. It was founded by the Danes in the 9th century and devastated by Oliver Cromwell 1649.

Weymouth /'weiməθ/ seaport and resort in Dorset, S England; population (1989 est) 62,500. It is linked by ferry to France and the Channel Islands. Weymouth, dating from the 10th century, was the first place in England to suffer from the Black Death 1348. It was popularized as a bathing resort by George III.

Wheeling /'wiːlɪŋ/ city in NW West Virginia, USA, on the Ohio River, SW of Pittsburgh, Pennsylvania; population (1990) 34,900. Industries include coal and natural-gas processing, iron and steel, textiles, glass, pottery, paper, and chemicals. Fort Henry, site of the last battle 1782 of the American War of Independence, is here.

Whipsnade /'(h)wɪpsneɪd/ zoo in Bedfordshire, England, 5 km/3 mi S of Dunstable, opened 1931, where wild animals and birds are bred and exhibited in conditions resembling their natural state.

Whitby /'(h)wɪtbi/ port and resort in North Yorkshire, England, on the North Sea coast; population (1981) 13,400. Industries include boatbuilding, fishing, and plastics. Remains of a Benedictine abbey built 1078 survive on the site of the original foundation by St Hilda 657, which was destroyed by the Danes 867. Capt Cook's ship *Resolution* was built in Whitby, where he had served his apprenticeship, and he sailed from here on his voyage to the Pacific 1768.

Whitehall /'(h)waɪt'hɔːl/ street in central London, England, between Trafalgar Square and the Houses of Parliament, with many government offices and the Cenotaph war memorial.

Whitehaven /'(h)waɪt,heɪvən/ town and port in Cumbria, NW England, on the Irish sea coast; population (1981) 27,700. Industries include chemicals and printing. Britain's first nuclear power station was sited at Calder Hall to the SE, where there is also a plant for reprocessing spent nuclear fuel at Sellafield.

Whitehorse /'(h)waɪthɔːs/ capital of Yukon Territory, Canada; population (1986) 15,199. Whitehorse is on the NW Highway. It replaced Dawson as capital 1953.

White Russia English translation of ◊Belarus.

West Yorkshire

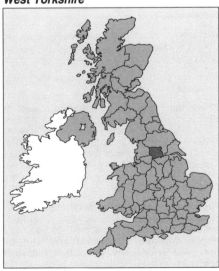

White Sea (Russian *Beloye More*) gulf of the Arctic Ocean, on which the port of Archangel stands. There is a warship construction base, including nuclear submarines, at Severodvinsk. The North Dvina and Onega rivers flow into it, and there are canal links with the Baltic, Black, and Caspian seas.

Whitstable /'(h)wɪtstəbəl/ resort in Kent, SE England, at the mouth of the river Swale, noted for its oysters; population (1985 est) 27,300.

Whyalla /(h)waɪˈælə/ port and industrial city (iron and steel) in South Australia; population (1985) 30,000.

WI abbreviation for ◊*West Indies*, an archipelago of islands between the Gulf of Mexico and the Caribbean, and the Atlantic Ocean; ◊*Wisconsin*, a state of the USA.

Wichita /'wɪtʃɪtɔː/ industrial city (oil refining, aircraft, motor vehicles) in S Kansas, USA; population (1990) 304,000. Wichita was founded about 1867.

Wichita Falls /ˌwɪtʃətɔː 'fɔːlz/ city in N Texas, USA, on the Wichita River, S of the Oklahoma border; population (1990) 96,300. It is an important petroleum-processing centre. Other industries include leather goods, textiles, foodstuffs, electronics, and pharmaceutical goods.

Wick /wɪk/ fishing port and industrial town (shipping, distilleries, North Sea oil) in NE Scotland, in Highland Region; population about 8,000. Air services to the Orkneys and Shetlands operate from here.

Wicklow /'wɪkləu/ county in the Republic of Ireland, province of Leinster
area 2,030 sq km/784 sq mi
towns Wicklow (county town)
physical Wicklow Mountains; rivers: Slane, Liffey
features the village of Shillelagh gave its name to rough cudgels of oak or blackthorn made there
population (1991) 97,300.

Wicklow /'wɪkləu/ port and county town of County Wicklow, Republic of Ireland; population (1991) 5,800.

Wien /viːn/ German name for ◊Vienna, the capital of Austria.

Wiesbaden /'viːsbɑːdn/ spa town and capital of Hessen, Germany, on the river Rhine 20 km/ 12 mi W of Frankfurt; population (1988) 267,000. Products include cement, plastics, wines, and spirits; most of the German sparkling wine cellars are in this area. Wiesbaden was the capital of the former duchy of Nassau from the 12th century until 1866.

Wigan /'wɪgən/ industrial town (food processing, engineering, paper) in Greater Manchester, NW England; population (1988 est) 307,600. The *Wigan Alps* are a recreation area with ski slopes and water sports created from industrial dereliction including colliery spoil heaps.

Wight, Isle of /waɪt/ island and county in S England
area 380 sq km/147 sq mi
towns Newport (administrative headquarters), resorts: Ryde, Sandown, Shanklin, Ventnor
features the *Needles*, a group of pointed chalk rocks up to 30 m/100 ft high in the sea to the W; the *Solent*, the sea channel between Hampshire and the island (including the anchorage of *Spithead* opposite Portsmouth, used for naval reviews); *Cowes*, venue of Regatta Week and headquarters of the Royal Yacht Squadron; Osborne House, near Cowes, a home of Queen Victoria, for whom it was built 1845; Farringford, home of the poet Alfred Tennyson, near Freshwater
products chiefly agricultural; tourism
population (1991) 126,600
famous people Thomas Arnold, Robert Hooke
history called *Vectis* ('separate division') by the Romans, who conquered it AD 43. Charles I was imprisoned 1647–48 in Carisbrooke Castle, now ruined.

Wigtown /'wɪgtaun/ former county of SW Scotland extending to the Irish Sea, merged 1975 in Dumfries and Galloway. The administrative headquarters was Wigtown.

Wilhelmshaven /ˌvɪlhelms'haːfən/ North Sea industrial port, resort, and naval base in Lower Saxony, Germany, on Jade Bay; population (1983) 99,000. Products include chemicals, textiles, and machinery.

Wilkes Barre /'wɪlks bæri/ industrial city (furniture, textiles, wire, tobacco products, and heavy machinery) in NE Pennsylvania, USA, on the Susquehanna River, SW of Scranton; population (1990) 47,500.

Williamsburg /'wɪljəmzbɜːg/ historic city in Virginia, USA; population (1990) 11,500. Founded 1632, capital of the colony of Virginia 1699–1779, much of it has been restored to its 18th-century appearance. The College of William and Mary (1693) is one of the oldest in the USA.

Williamsport /'wɪljəmzpɔːt/ industrial city (electronics, plastics, metals, lumber, textiles, and aircraft parts) in N central Pennsylvania, USA, on the Susquehanna River, N of Harrisburg; population (1990) 31,900. It is the birthplace of Little League baseball (1939).

Wilmington /'wɪlmɪŋtən/ industrial port and city (chemicals, textiles, shipbuilding, iron and steel goods; headquarters of Du Pont enterprises) in Delaware, USA; population (1990) 71,500. Founded by Swedish settlers as *Fort Christina* 1638, it was taken from the Dutch and renamed by the British 1664.

Wilmington /'wɪlmɪŋtən/ port and industrial city (textiles, tobacco, lumber, and chemicals) in SE North Carolina, USA, on the Cape Fear River, near the Atlantic Ocean; population (1990) 55,500. Tourism is important to the economy.

Wiltshire

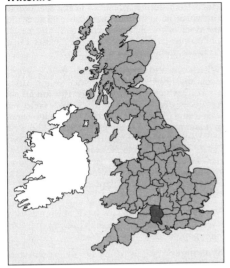

Wilton /'wɪltən/ market town in Wiltshire, S England, outside Salisbury; population (1981) 4,000. It has manufactured carpets since the 16th century. Wilton House, the seat of the earls of Pembroke, was built from designs by Holbein and Inigo Jones, and is associated with Sir Philip Sidney and Shakespeare.

Wilts abbreviation for *Wiltshire*.

Wiltshire /'wɪltʃə/ county in SW England
area 3,480 sq km/1,343 sq mi
towns Trowbridge (administrative headquarters), Salisbury, Swindon, Wilton
physical Marlborough Downs; Savernake Forest; rivers: Kennet, Wylye, Salisbury and Bristol Avons; Salisbury Plain
features Salisbury Plain, a military training area used since Napoleonic times; Longleat House (Marquess of Bath); Wilton House (Earl of Pembroke); Stourhead, with 18th-century gardens; Neolithic Stonehenge, Avebury stone circle
products wheat, cattle, pig and sheep farming, rubber, engineering
population (1990) 553,300
famous people Isaac Pitman, William Talbot, Christopher Wren.

Wimbledon /'wɪmbəldən/ district of the Greater London borough of Merton, headquarters of the All-England Lawn Tennis and Croquet Club.

Winchester /'wɪntʃɪstə/ cathedral city and administrative headquarters of Hampshire, on the river Itchen; population (1984) 93,000. Tourism is important, and there is also light industry. Originally a Roman town, Winchester was capital of the Anglo-Saxon kingdom of Wessex, and later of England. The cathedral is the longest medieval church in Europe and was remodelled from Norman-Romanesque to Perpendicular Gothic under the patronage of William of Wykeham (founder of Winchester College 1382), who is

buried there, as are Saxon kings, St Swithun, and the writers Izaac Walton and Jane Austen.

A tribal centre of the Britons under the name *Caer Gwent* and later one of the largest Roman settlements in Britain as *Venta Belgarum*, the town become capital of Wessex 519 and under Alfred the Great and Canute was the seat of government. In 827 Egbert was crowned first king of all England here. Under William the Conqueror, Winchester was declared dual capital of England with London.

A medieval 'reconstruction' of Arthur's Round Table is preserved in the 13th-century hall (all that survives) of the castle.

Windermere /'wɪndəmɪə/ largest lake in England, in Cumbria, 17 km/10.5 mi long and 1.6 km/1 mi wide.

Windhoek /'wɪndhuk/ capital of Namibia; population (1988) 115,000. It is just N of the Tropic of Capricorn, 290 km/180 mi from the west coast.

Windsor /'wɪnzə/ town in Berkshire, S England, on the river Thames; population (1981) 28,000. It is the site of Windsor Castle, a royal residence, and Eton College (public school, 1540) and has a 17th-century guildhall designed by Christopher Wren.

Thy Forests, Windsor! and thy green Retreats, /At once the Monarch's and the Muse's Seats.

On **Windsor** Alexander Pope
Windsor Forest 1713

Windsor /'wɪnzə/ industrial lake port (car engines, pharmaceuticals, iron and steel goods, paint, bricks) in Ontario, SE Canada, opposite Detroit, Michigan, USA; population (1986) 254,000. It was founded as a Hudson's Bay Company post 1853.

Windward Islands /'wɪndwəd/ islands in the path of the prevailing wind, notably: *West Indies* see ▷Antilles; ▷*Cape Verde Islands*; and ▷*French Polynesia* (Tahiti, Moorea, Makatea).

Winnipeg /'wɪnɪpeg/ capital and industrial city (processed foods, textiles, transportation, and transportation equipment) in Manitoba, Canada, on the Red River, S of Lake Winnipeg; population (1986) 623,000. Established as Winnipeg 1870 on the site of earlier forts, the city expanded with the arrival of the Canadian Pacific Railroad 1881.

With the thermometer at 30 below zero and the wind behind him, a man walking on Main Street in Winnipeg knows which side of him is which.

On **Winnipeg** Stephen Leacock *My Discovery of the West* 1937

Wisconsin

Winnipeg, Lake /ˈwɪnɪpeg/ lake in S Manitoba, Canada, draining much of the Canadian prairies; area 24,500 sq km/9,460 sq mi.

Winston-Salem /ˌwɪnstən ˈseɪləm/ industrial city (tobacco products, textiles, clothing, and furniture) in N central North Carolina, USA, NE of Charlotte; population (1990) 143,500. Wake Forest University 1834 is here.

Winter Haven /ˌwɪntə ˈheɪvn/ city in central Florida, USA, E of Lakeland; a centre for citrus-fruit processing and shipping; population (1990) 24,700. Other industries include tourism, cigars, and alcohol.

Winterthur /ˈvɪntətuə/ Swiss town and spa NE of Zürich; population (1990) 85,700. Manufacturing includes engines and textiles.

Wisconsin /wɪˈskɒnsɪn/ state in N central USA; nickname Badger State
area 145,500 sq km/56,163 sq mi
capital Madison
cities Milwaukee, Green Bay, Racine
features lakes: Superior, Michigan; Mississippi River; Door peninsula
products leading US dairy state; maize, hay, industrial and agricultural machinery, engines and turbines, precision instruments, paper products, cars and lorries, plumbing equipment
population (1990) 4,891,800
famous people Edna Ferber, Harry Houdini, Joseph McCarthy, Spencer Tracy, Orson Welles, Thornton Wilder, Frank Lloyd Wright
history explored by Jean Nicolet for France 1634; originally settled near Ashland by the French; passed to Britain 1763; included in USA 1783. Wisconsin became a territory 1836 and a state 1848.

Wisconsin has always boasted of raising the largest crops of talking humanity.

On **Wisconsin** Bayard Taylor *Eldorado, or Adventures in the Path of Empire* 1850

Witten /ˈvɪtn/ city in North Rhine–Westphalia, Germany, on the Ruhr River; population (1988) 102,000. Industries include steel and chemicals.

Wittenberg /ˈvɪtnbeək/ town in the state of Saxony-Anhalt, Germany, on the river Elbe, SW of Berlin; population (1981) 54,000. Wittenberg university was founded 1502, but transferred to Halle 1815. Luther preached in the Stadtkirche (in which he is buried), nailed his 95 theses to the door of the Schlosskirche 1517, and taught philosophy at the university. The artists Lucas Cranach, father and son, lived here.

Witwatersrand /wɪtˈwɔːtəzrænd/ or *the Rand* economic heartland of S Transvaal, South Africa. Its reef, which stretches nearly 100 km/60 mi, produces over half the world's gold. Gold was first found here 1854. The chief city of the region is Johannesburg. Forming a watershed between the Vaal and the Olifant rivers, the Rand comprises a series of parallel ranges which extend 100 km/60 mi E–W and rise to 1,525–1,830 m/5,000–6,000 ft above sea level. Gold occurs in reefs that are mined at depths of up to 3,050 m/10,000 ft.

Wolfsburg /ˈvɒlfsbuək/ town NE of Brunswick in Germany, chosen 1938 as the Volkswagen factory site; population (1988) 122,000.

Wollongong /ˈwuləŋgɒŋ/ industrial city (iron, steel) in New South Wales, Australia, 65 km/40 mi S of Sydney; population (1985, with Port Kembla) 238,000.

Wolverhampton /ˌwulvəˈhæmptən/ industrial town (metalworking, chemicals, tyres, aircraft, commercial vehicles) in West Midlands, England, 20 km/12 mi NW of Birmingham; population (1991) 239,800.

Woolwich /ˈwulɪdʒ/ district in London, England, cut through by the river Thames, the northern section being in the borough of Newham and the southern in Greenwich. There is a ferry service and the Thames Barrier (a flood barrier, 1984) is here. The Royal Arsenal, an ordnance depot from 1518, was closed down 1967.

Woonsocket /wuːnˈsɒkɪt/ industrial city (rubber, chemicals, woollen goods) in N Rhode Island, USA, on the Blackstone River, NW of Providence; population (1990) 43,900.

Worcester /ˈwustə/ cathedral city with industries (gloves, shoes, Worcester sauce; Royal Worcester porcelain from 1751) in Hereford and Worcester, W central England, administrative headquarters of the county, on the river Severn; population (1991) 81,000. The cathedral dates from the 13th and 14th centuries. The birthplace of the composer Elgar at nearby Broadheath is a museum. At the **Battle of Worcester** 1651 Oliver Cromwell defeated Charles I.

Worcester /ˈwustə/ industrial port (textiles, engineering, printing) in central Massachusetts, USA, on the Blackstone River; population (1990) 169,800. It was permanently settled 1713.

Worcestershire /ˈwustəʃə/ former Midland county of England, merged 1974 with Herefordshire in the new county of Hereford and Worcester, except for a small projection in the N, which went to West Midlands. Worcester was the county town.

Worcs abbreviation for *Worcestershire*.

Worksop /'wɜːksɒp/ market and industrial town (coal, glass, chemicals, engineering) in Nottinghamshire, central England, on the river Ryton; population (1988 est) 36,200. Mary Queen of Scots was imprisoned at Worksop Manor (burned 1761).

Worms /wɜːmz,German vɔːms/ industrial town in Rhineland-Palatinate, Germany, on the river Rhine; population (1984) 73,000. The vineyards of the Liebfrauenkirche produced the original Liebfraumilch wine; it is now produced by many growers around Worms. The Protestant reformer Martin Luther appeared before the *Diet* (Assembly) *of Worms* 1521 and was declared an outlaw by the Roman Catholic church.

Worthing /'wɜːðɪŋ/ seaside resort in West Sussex, England, at the foot of the South Downs; population (1989 est) 97,700. Industries include electronics, engineering, plastics, and furniture. There are traces of prehistoric and Roman occupation in the vicinity.

Wrexham /'reksəm/ (Welsh **Wrecsam**) town in Clwyd, NE Wales, 19 km/12 mi SW of Chester; population (1983) 40,000. Industries include coal, electronics, and pharmaceuticals. It is the seat of the Roman Catholic bishopric of Menevia (Wales). Elihu Yale, benefactor of Yale University, died in Wrexham and is buried in the 15th-century church of St Giles.

Wrocław /'vrɒtslɑːf/ industrial river port in Poland, on the river Oder; population (1990) 643,200. Under the German name of **Breslau**, it was the capital of former German Silesia. Industries include shipbuilding, engineering, textiles, and electronics.

Wuchang /,wuː'tʃæŋ/ former city in China; amalgamated with ⟡Wuhan.

Wuhan /,wuː'hæn/ river port and capital of Hubei province, China, at the confluence of the Han and Chang Jiang rivers, formed 1950 as one of China's greatest industrial areas by the amalgamation of Hankou, Hanyang, and Wuchang; population (1989) 3,710,000. It produces iron, steel, machine tools, textiles, and fertilizer.

A centre of revolt in both the Taiping Rebellion 1851–65 and the 1911 revolution, it had an anti-Mao revolt 1967 during the Cultural Revolution.

Wuhsien /,wuː'ʃi'en/ alternative transliteration for ⟡Suzhou, a city in China.

Wuppertal /'vupətɑːl/ industrial town in North Rhine–Westphalia, Germany, 32 km/20 mi E of Düsseldorf; population (1988) 374,000. Industries include textiles, plastics, brewing, and electronics. It was formed 1929 (named 1931) by uniting Elberfeld (13th century) and Barmen (11th century).

Württemberg /'vɜːtəmbɜːg/ former kingdom (1805–1918) in SW Germany that joined the German Reich 1870. Its capital was Stuttgart. Divided 1946 between the administrative West German *Länder* of Württemberg-Baden and Württemberg-Hohenzollern, from 1952 it was part of the *Land* of Baden-Württemberg.

Würzburg /'vɜːtsbɜːg/ industrial town (engineering, printing, wine, brewing) in NW Bavaria, Germany; population (1988) 127,000. The bishop's palace was decorated by the Italian Rococo painter Tiepolo.

WV abbreviation for ⟡ *West Virginia*, a state of the USA.

WY abbreviation for ⟡ *Wyoming*, a state of the USA.

Wye /waɪ/ (Welsh **Gwy**) river in Wales and England; length 208 km/130 mi. It rises on Plynlimmon, NE Dyfed, flows SE and E through Powys, and Hereford and Worcester, then follows the Gwent–Gloucestershire border before joining the river Severn S of Chepstow.

Other rivers of the same name in the UK are found in Buckinghamshire (15 km/9 mi) and Derbyshire (32 km/20 mi).

Wyoming /waɪ'əʊmɪŋ/ state in western USA:
nickname Equality State
area 253,400 sq km/97,812 sq mi
capital Cheyenne
cities Casper, Laramie
features Rocky Mountains; national parks: Yellowstone (including the geyser Old Faithful), Grand Teton
products oil, natural gas, sodium salts, coal, uranium, sheep, beef
population (1990) 453,600
famous people Buffalo Bill, Jackson Pollock
history acquired by USA from France as part of the Louisiana Purchase 1803; Fort Laramie, a trading post, settled 1834; women achieved the vote 1869; became a state 1890.

Wyoming

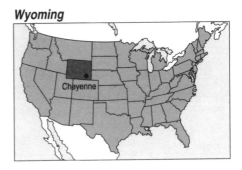

South China Sea; length 1,900 km/1,200 mi. Guangzhou lies on the northern arm of its delta, and Hong Kong island at its mouth. The name means 'west river'.

Xingú /ʃɪŋˈguː/ region in Pará, Brazil, crossed by branches of the Xingú River which flows for 1,900 km/1,200 mi to the Amazon Delta. In 1989 Xingú Indians protested at the creation of a vast, intrusive lake for the Babaquara and Kararao dams of the Altamira complex.

Xining /ʃiːˈnɪŋ/ or *Sining* industrial city (chemicals, textiles, machinery, processed foods) and capital of Qinghai province, China, on the Xining River; population (1989) 640,000. For centuries it was a major trading centre on the caravan route to Tibet.

Xinjiang Uygur /ʃɪndʒiˈæŋ ˈwiːguə/ or *Sinkiang Uighur* autonomous region of NW China
area 1,646,800 sq km/635,665 sq mi
capital Urumqi
features largest of Chinese administrative areas; Junggar Pendi (Dzungarian Basin) and Tarim Pendi (Tarim Basin, which includes Lop Nor, China's nuclear testing ground, although the research centres were moved to the central province of Sichuan 1972) separated by the Tian Shan Mountains
products cereals, cotton, fruit in valleys and oases; uranium, coal, iron, copper, tin, oil
population (1990) 15,156,000; the region has 13 recognized ethnic minorities, the largest being 6 million Uigurs (Muslim descendants of Turks)
religion 50% Muslim
history under Manchu rule from the 18th century. Large sections were ceded to Russia 1864 and 1881; China has raised the question of their return and regards the frontier between Xinjiang Uygur and Tajikistan, which runs for 480 km/ 300 mi, as undemarcated.

Xizang /ʃiːˈzæŋ/ Chinese name for ⟡Tibet, an autonomous region of SW China from 1965.

Xochimilco /ˌsɒtʃɪˈmɪlkəu/ lake about 11 km/7 mi SE of Mexico City, Mexico, which features floating gardens, all that remains of an ancient water-based agricultural system.

Xiamen /ʃiˌɑːˈmʌn/ formerly (until 1979) *Amoy* port on Ku Lang Island in Fujian province, SE China; population (1984) 533,000. Industries include textiles, food products, and electronics. It was one of the original five treaty ports used for trade under foreign control 1842–1943 and a special export-trade zone from 1979.

Xian /ʃiːˈæn/ industrial city and capital of Shaanxi province, China; population (1989) 2,710,000. It produces chemicals, electrical equipment, and fertilizers.

It was the capital of China under the Zhou dynasty (1126–255 BC); under the Han dynasty (206 BC–AD 220), when it was called *Changan* ('long peace'); and under the Tang dynasty 618–906, as *Siking* ('western capital'). The Manchus called it *Sian* ('western peace'), now spelled Xian. It reverted to Changan 1913–32, was Siking 1932–43, and again Sian from 1943. It was here that the imperial court retired after the Boxer Rebellion 1900.

Its treasures include the 600-year-old Ming wall; the pottery soldiers buried to protect the tomb of the first Qin emperor, Shi Huangdi; Big Wild Goose Pagoda, one of the oldest in China; and the Great Mosque 742.

Xi Jiang /ʃiː dʒiˈæŋ/ or *Si-Kiang* river in China, that rises in Yunnan province and flows into the

Yakima /'jækɪmɔ:/ city in S central Washington, USA, on the Yakima River, SE of Seattle; population (1990) 54,800. It is an agricultural processing centre for sugar beet, apples, hops, and livestock, and produces cider and flour

Yakut /jæ'kut/ (Russian *Yakutskaya*) autonomous republic in Siberian Russia
area 3,103,000 sq km/1,197,760 sq mi
capital Yakutsk
features one of world's coldest inhabited places; river Lena
products furs, gold, natural gas, some agriculture in the S
population (1986) 1,009,000 (50% Russians, 37% Yakuts)
history the nomadic Yakuts were conquered by Russia in the 17th century; Yakut was a Soviet republic 1922–91. It remained an autonomous republic within the Russian Federation after the collapse of the Soviet Union 1991, since when it has agitated for greater independence.

Yakutia is the enormous backyard of Russia.

On **Yakut** Erik de Mauny
Russian Prospect 1969

Yakutsk /jæ'kutsk/ capital of Yakut autonomous republic, Russia, on the river Lena; population (1987) 220,000. Industries include timber, tanning, and brick-making. It is the coldest point of the Arctic in NE Siberia, average winter temperature –50°C/–68°F, and has an institute for studying the permanently frozen soil area (permafrost). The lowest temperature ever recorded was in Yakutia, –70°C/–126°F.

Yalu /'ja:lu:/ river forming the northern boundary between North Korea and Jilin and Liaoning provinces (Manchuria) in China; length 790 km/491 mi. It is only navigable near the mouth and is frozen Nov to March.

Yamal Peninsula /jə'ma:l/ peninsula in NW Siberian Russia, with gas reserves estimated at 6 trillion cu m/212 trillion cu ft; supplies are piped to W Europe.

Yamoussoukro /jæmu:'su:krəu/ capital of Ivory Coast; population (1986) 120,000. The economy is based on tourism and agricultural trade. It replaced Abidja'n as capital 1983, but was not officially recognized as such until 1992.
A Roman Catholic basilica (said to be the largest church in the world) was completed 1989.

Yamuna /'jæmunə/ alternative name for the ⇨Jumna River in India.

Yan'an /jæn'æn/ or *Yenan* industrial city in Shaanxi province, central China; population (1984) 254,000. The Long March ended here Jan 1937, and it was the communist headquarters 1936–47 (the caves in which Mao lived are preserved).

Yangon /jæŋ'gɒn/ (formerly 1755–1989 *Rangoon*) capital and chief port of Myanmar (Burma) on the Yangon River, 32 km/20 mi from the Indian Ocean; population (1983) 2,459,000. Products include timber, oil, and rice. The city *Dagon* was founded on the site AD 746; it was given the name Rangoon (meaning 'end of conflict') by King Alaungpaya 1755.

Yangtze-Kiang /'jæŋktsi ki'æŋ/ alternative transcription of ⇨Chang Jiang, the longest river in China.

Yangzhou /jæŋ 'dʒəu/ or *Yangchow* canal port in Jiangsu province, E China, on the Chang Jiang River; population (1984) 382,000. Among its features are gardens and pavilions. It is an artistic centre for crafts, jade carving, and printing.

Yantai /jæn'taɪ/ (formerly *Chefoo*) ice-free port in Shandong province, E China; population (1984) 700,000. It is a special economic zone; its industries include tourism, wine, and fishing.

Yaoundé /ja:'undeɪ/ capital of Cameroon, 210 km/130 mi E of the port of Douala; population (1984) 552,000. Industry includes tourism, oil refining, and cigarette manufacturing.
Established by the Germans as a military port 1899, it became capital of French Cameroon 1921.

Yarkand /ja:'kænd/ or *Shache* walled city in the Xinjiang Uygur region of China, in an oasis of the Tarim Basin, on the caravan route to India and W Russia; population (1985) 100,000. It is a centre of Islamic culture.

Yarmouth /'ja:məθ/ or *Great Yarmouth* holiday resort and port in Norfolk, England, at the mouth of the river Yare; population (1988 est) 89,000. Formerly a fishing town, it is now a base for North Sea oil and gas.

Yaroslavl /jærə'sla:vəl/ industrial city (textiles, rubber, paints, commercial vehicles) in W central Russia, capital of Yaroslavl region, on the river Volga, 250 km/155 mi NE of Moscow; population (1987) 634,000.

Yazd /ja:zd/ or *Yezd* silk-weaving town in central Iran, an oasis on a trade route; population (1986) 231,000.

Yellowknife /'jeləunaɪf/ capital of Northwest Territories, Canada, on the northern shore of Great

Slave Lake; population (1986) 11,753. It was founded 1935 when gold was discovered in the area and became the capital 1967.

Yellow River English name for the ⟡Huang He River, China.

Yellow Sea /'jeləu/ gulf of the Pacific Ocean between China and Korea; area 466,200 sq km/ 180,000 sq mi. It receives the Huang He (Yellow River) and Chang Jiang.

Yellowstone National Park /'jeləustəun/ largest US nature reserve, established 1872, on a broad plateau in the Rocky Mountains, chiefly in NW Wyoming, but also in SW Montana and E Idaho; area 8,983 sq km/3,469 sq mi. The park contains more than 3,000 geysers and hot springs, including periodically erupting Old Faithful. It is one of the world's greatest wildlife refuges. Much of the park was ravaged by forest fires 1988.

Yemen Republic of (*al Jamhuriya al Yamaniya*)
area 531,900 sq km/205,367 sq mi
capital San'a
towns Ta'iz; and chief port Aden
physical hot moist coastal plain, rising to plateau and desert
features once known as *Arabia felix* because of its fertility, includes islands of Perim (in strait of Bab-el-Mandeb, at S entrance to Red Sea), Socotra, and Kamaran
head of state and of government Ali Abdullah Saleh from 1990
political system authoritarian republic
political parties Yemen Socialist Party (YSP), Democratic Unionist Party, National Democratic Front, Yemen Reform Group
exports cotton, coffee, grapes, vegetables
currency rial
population (1990 est) 11,000,000; growth rate 2.7% p.a.
life expectancy men 47, women 50
language Arabic
religions Sunni Muslim 63%, Shi'ite Muslim 37%
literacy men 20%, women 3% (1985 est)
GNP $4.9 bn (1983); $520 per head
chronology
1918 Yemen became independent.
1962 North Yemen declared the Yemen Arab Republic (YAR), with Abdullah al-Sallal as president. Civil war broke out between royalists and republicans.
1967 Civil war ended with the republicans victorious. Sallal deposed and replaced by Republican Council. The People's Republic of South Yemen was formed.
1970 People's Republic of South Yemen renamed People's Democratic Republic of Yemen.
1971–72 War between South Yemen and the YAR; union agreement signed but not kept.
1974 Ibrahim al-Hamadi seized power in North Yemen and Military Command Council set up.
1977 Hamadi assassinated and replaced by Ahmed ibn Hussein al-Ghashmi.
1978 Constituent People's Assembly appointed in North Yemen and Military Command Council dissolved. Ghashmi killed by envoy from South

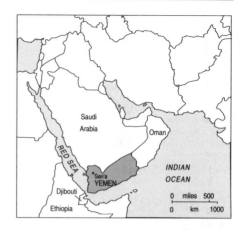

Yemen; succeeded by Ali Abdullah Saleh. War broke out again between the two Yemens. South Yemen president deposed and Yemen Socialist Party (YSP) formed with Abdul Fattah Ismail as secretary general, later succeeded by Ali Nasser Muhammad.
1979 Cease-fire agreed with commitment to future union.
1983 Saleh elected president of North Yemen for a further five-year term.
1984 Joint committee on foreign policy for the two Yemens met in Aden.
1985 Ali Nasser Muhammad re-elected secretary general of the YSP in South Yemen; removed his opponents. Three bureau members killed.
1986 Civil war in South Yemen; Ali Nasser dismissed. New administration under Haydar Abu Bakr al-Attas.
1988 President Saleh re-elected in North Yemen.
1989 Draft constitution for single Yemen state published.
1990 Border between two Yemens opened; countries formally united 22 May as Republic of Yemen.
1991 New constitution approved.

The Cuba of the Arabian peninsula.

On **Yemen** (South) Richard Nixon
The Real War 1980

Yemen, North /'jemən/ former country in SW Asia. It was united with South Yemen 1990 as the Republic of Yemen.

Yemen, South /'jemən/ former country in SW Asia. It was united with North Yemen 1990 as the Republic of Yemen.

Yenan /jʌn'æn/ alternative transcription of ⟡Yan'an, a city in China.

Yenisei /jenɪ'seɪ/ river in Asian Russia, rising in the Tuva region and flowing across the Siberian plain into the Arctic Ocean; length 4,100 km/ 2,550 mi.

Yerevan /jeri'væn/ industrial city (tractor parts, machine tools, chemicals, bricks, bicycles, wine, fruit canning) and capital of Armenia, a few miles N of the Turkish border; population (1987) 1,168,000. It was founded in the 7th century and was alternately Turkish and Persian from the 15th century until ceded to Russia 1828. Its university was founded 1921. Armenia became an independent republic 1991.

The city has seen mounting inter-ethnic violence and Armenian nationalist demonstrations since 1988, fanned by the Nagorno-Karabakh dispute.

Yezd /jezd/ alternative name for the Iranian town of ⟡Yazd.

Yezo /'jezəu/ former name (until 1868) of ⟡Hokkaido, the northernmost of the four main islands of Japan.

Yichang /ji:'tʃæŋ/ port at the head of navigation of the Chang Jiang River, Hubei province, China; population (1982) 175,000.

Yinchuan /jɪn'tʃwɑːn/ capital of Ningxia autonomous region, NW China, on the Huang He River; population (1989) 576,000. It is a trading centre for the Ningxia plain, producing textiles and coal.

Yogyakarta /jɒgjə'kɑːtə/ city in Java, Indonesia, capital 1945–1949; population (1980) 399,000. The Buddhist pyramid shrine to the NW at Borobudur (122 m/400 ft square) was built AD 750–850.

Yokohama /jəukəu'hɑːmə/ Japanese port on Tokyo Bay; population (1990) 3,220,350. Industries include shipbuilding, oil refining, engineering, textiles, glass, and clothing.

In 1859 it was the first Japanese port opened to foreign trade. From then it grew rapidly from a small fishing village to become the chief centre of trade with Europe and the USA. Almost destroyed in an earthquake 1923, it was again rebuilt after World War II.

Yokosuka /jəukəu'suːkə/ Japanese seaport and naval base (1884) on Tokyo Bay, S of Yokohama; population (1990) 433,400.

Yonkers /'jɒŋkəz/ city in Westchester County, New York, USA, on the Hudson River, just N of the Bronx, New York City; population (1990) 188,300. Products include machinery, processed foods, chemicals, clothing, and electric and electronic equipment. Yonkers was a Dutch settlement from about 1650.

Yonne /jɒn/ French river, 290 km/180 mi long, rising in central France and flowing N into the Seine; it gives its name to a *département* in Burgundy region.

York /jɔːk/ cathedral and industrial city (railway rolling stock, scientific instruments, sugar, chocolate, and glass) in North Yorkshire, N England; population (1991) 100,600. The city is visited by 3 million tourists a year.

features The Gothic York Minster contains medieval stained glass; the south transept was severely damaged by fire 1984, but has been restored. Much of the 14th-century city wall survives, with four gates or 'bars', as well as the medieval streets collectively known as the Shambles (after the slaughterhouse). The Jorvik Viking Centre, opened 1984 after excavation of a site at Coppergate, contains wooden remains of Viking houses. There are fine examples of 17th- to 18th-century domestic architecture; the Theatre Royal, site of a theatre since 1765; the Castle Museum; the National Railway Museum; and the university 1963.

history Traditionally the capital of the N of England, the city became from AD 71 the Roman fortress of **Eboracum**. Recent excavations of the Roman city have revealed the fortress, baths, and temples to Serapis and Mithras. The first bishop of York (Paulinus) was consecrated 627 in the wooden church that preceded York Minster. Paulinus baptized King Edwin there 627, and York was created an archbishopric 732. In the 10th century it was a Viking settlement. During the Middle Ages its commercial prosperity depended on the wool trade. An active Quaker element in the 18th and 19th centuries included the Rowntree family that founded the chocolate factory.

York /jɔːk/ city in S Pennsylvania, USA, SE of Harrisburg; population (1990) 42,100. It is an agricultural processing centre for the area and manufactures paper products, building materials, and heavy machinery. The Articles of Confederation were adopted here during the Continental Congress 1777–78.

Yorks. abbreviation for **Yorkshire**, an English county.

Yorkshire /'jɔːkʃə/ former county in NE England on the North Sea divided administratively into north, east, and west ridings (thirds), but reorganized to form a number of new counties 1974: the major part of **Cleveland** and **Humberside**, **North Yorkshire**, **South Yorkshire**, and **West Yorkshire**. Small outlying areas also went to Durham, Cumbria, Lancashire, and Greater Manchester.

Yosemite /jəu'semɪti/ area in the Sierra Nevada, E California, USA, a national park from 1890; area 3,079 sq km/1,189 sq mi. It includes Yosemite Gorge, Yosemite Falls (739 m/2,425 ft in three leaps) with many other lakes and waterfalls, and groves of giant sequoia trees.

Youngstown /'jʌŋstaun/ industrial city (fabricated metals) in E Ohio, USA, on the Mahoning River; population (1990) 95,700. Youngstown was laid out 1797.

Ysselmeer alternative spelling of ⟡IJsselmeer, a lake in the Netherlands.

Yuba City /'juːbə 'sɪti/ city in N central California, USA, on the Feather River, N of San Francisco; population (1990) 27,400. It is an agricultural trading centre for nuts, fruits, rice, and dairy products.

Yucatán /juːkə'tɑːn/ peninsula in Central America, divided among Mexico, Belize, and Guatemala; area 180,000 sq km/70,000 sq mi. Tropical crops are grown. It is inhabited by Maya Indians and contains the remains of their civilization.

There are ruins at Chichén Itzá and Uxmal. The Mexican state of Yucatán has an area of 38,402 sq km/14,823 sq mi, and a population (1980) of 1,035,000. Its capital is Mérida.

Yugoslavia

area 58,300 sq km/22,503 sq mi
capital Belgrade
towns Kraljevo, Leskovac, Pristina, Novi Sad, Titograd
head of state Dobrica Cosic from 1992
head of government Radoje Kontic from 1993
political system socialist pluralist republic
political parties Socialist Party of Serbia, ex-communist; Serbian Renaissance Movement, pro-monarchist; Democratic Party, Serbia-based, liberal, free-market; Montenegro League of Communists; Democratic Coalition of Muslims; Alliance of Reform Forces, Montenegro-based
exports machinery, electrical goods, chemicals, clothing, tobacco
currency dinar
population (1990) 10,500,000
life expectancy men 69, women 75 (1989)
languages Serbian variant of Serbo-Croatian, Slovenian
religion Eastern Orthodox 41% (Serbs), Roman Catholic 12% (Croats), Muslim 3%
literacy 90% (1989)
GNP $154.1 bn; $6,540 per head (1988)
chronology
1918 Creation of Kingdom of the Serbs, Croats, and Slovenes.
1929 Name of Yugoslavia adopted.
1941 Invaded by Germany.
1945 Yugoslav Federal Republic formed under leadership of Tito; communist constitution introduced.
1948 Split with USSR.
1953 Self-management principle enshrined in constitution.
1961 Nonaligned movement formed under Yugoslavia's leadership.
1974 New constitution adopted.
1980 Tito died; collective leadership assumed power.

1988 Economic difficulties: 1,800 strikes, 250% inflation, 20% unemployment. Ethnic unrest in Montenegro and Vojvodina; party reshuffled and government resigned.
1989 Reformist Croatian Ante Marković became prime minister. Twenty-nine died in ethnic riots in Kosovo province, protesting against Serbian attempt to end autonomous status of Kosovo and Vojvodina; state of emergency imposed. May: inflation rose to 490%; tensions with ethnic Albanians rose.
1990 Multiparty systems established in Serbia and Croatia.
1991 June: Slovenia and Croatia declared independence, resulting in clashes between federal and republican armies; Slovenia accepted European Community (EC)-sponsored peace pact. Fighting continued in Croatia. Dec: President Stipe Mesic and Prime Minister Ante Marković resigned.
1992 Jan: EC-brokered cease-fire established in Croatia; EC and USA recognized Slovenia's and Croatia's independence. Bosnia-Herzegovina and Macedonia declared independence. April: Bosnia-Herzegovina recognized as independent by EC and USA amid increasing ethnic hostility. New Federal Republic of Yugoslavia (FRY) proclaimed by Serbia and Montenegro but not recognized externally. May: Western ambassadors left Belgrade. International sanctions imposed against Serbia and Montenegro. Hostilities continued. Jun: Dobrica Cosic became president. Sept: UN membership suspended. Dec: Slobodan Milosevic re-elected Serbian president.
1993 Macedonia recognized as independent under name of Former Yugoslav Republic of Macedonia.

Yukon /'juːkɒn/ territory of NW Canada
area 483,500 sq km/186,631 sq mi
capital Whitehorse
towns Dawson, Mayo
features named after its chief river, the Yukon; includes the highest point in Canada, Mount Logan, 6,050 m/19,850 ft; Klondike Gold Rush International Historical Park, which extends into Alaska
products gold, silver, lead, zinc, oil, natural gas, coal
population (1991) 26,500
history settlement dates from the gold rush

Yukon

1896–1910, when 30,000 people moved to the Klondike river valley (silver is now worked there). It became separate from the Northwest Territories 1898, with Dawson as the capital 1898–1951. Construction of the Alcan Highway during World War II helped provide the basis for further development.

This is the Law of the Yukon, that only the Strong shall thrive;/ That surely the Weak shall perish, and only the Fit survive.

On the **Yukon** Robert Service *Songs of a Sourdough* 1907

Yukon River /ˈjuːkɒn/ river in North America, 3,185 km/1,979 mi long, flowing from Lake Tagish in Yukon Territory into Alaska, where it empties into the Bering Sea.

Yungning /juŋˈnɪŋ/ alternative transcription of ▷Nanning, a Chinese port.

Yunnan /juːˈnæn/ province of SW China, adjoining Myanmar (Burma), Laos, and Vietnam
area 436,200 sq km/168,373 sq mi
capital Kunming
physical rivers: Chang Jiang, Salween, Mekong; crossed by the Burma Road; mountainous and well forested
products rice, tea, timber, wheat, cotton, rubber, tin, copper, lead, zinc, coal, salt
population (1990) 36,973,000.

Yuzovka /ˈjuːzəfkə/ former name (1872–1924) for the town of ▷Donetsk, Ukraine, named after the Welshman John Hughes who established a metallurgical factory there in the 1870s.

Z

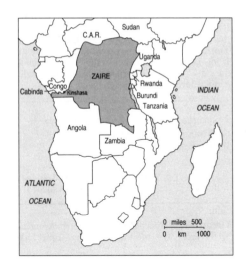

Zaandam /ˌzɑːnˈdæm/ industrial port (timber, paper) in North Holland province, the Netherlands, on the river Zaan, NW of Amsterdam, since 1974 included in the municipality of Zaanstad.

Zaanstad /ˈzɑːnstæd/ industrial town in W Netherlands which includes the port of Zaandam; population (1991) 130,700.

Zabrze /ˈzæbʒeɪ/ industrial city (coalmining, iron, chemicals) in Silesia, S Poland; formerly (until 1945) the German town of Hindenburg; population (1985) 198,000.

Zadar /ˈzædɑː/ (Italian *Zara*) port and resort in Croatia; population (1981) 116,000. The city was sacked by the army of the Fourth Crusade 1202, which led to the Crusade being excommunicated by Pope Innocent III. It was alternately held and lost by the Venetian republic from the 12th century until its seizure by Austria 1813. It was the capital of Dalmatia 1815–1918 and part of Italy 1920–47, when it became part of Yugoslavia; it now belongs to independent Croatia.

Zagorsk /zəˈɡɔːsk/ former name (1930–91) of ⏲Sergiyev Posad, a town in Russia.

Zagreb /ˈzɑːɡreb/ industrial city (leather, linen, carpets, paper, and electrical goods) and capital of Croatia, on the Sava River; population (1981) 1,174,512. Zagreb was a Roman city (*Aemona*) and has a Gothic cathedral. Its university was founded 1874. The city was damaged by bombing Oct 1991 during the Croatian civil war.

Zaire Republic of (*République du Zaire*) (formerly *Congo*)
area 2,344,900 sq km/905,366 sq mi
capital Kinshasa
towns Lubumbashi, Kananga, Kisangani; ports Matadi, Boma
physical Zaïre River basin has tropical rainforest and savanna; mountains in E and W
features lakes Tanganyika, Mobutu Sese Seko, Edward; Ruwenzori mountains
head of state Mobutu Sese Seko Kuku Ngbendu wa Zabanga from 1965
head of government Etienne Tshisekedi from 1992
political system socialist pluralist republic
political parties Popular Movement of the

Revolution (MPR), African socialist; numerous new parties registered 1991
exports coffee, copper, cobalt (80% of world output), industrial diamonds, palm oil
currency zaïre
population (1990 est) 35,330,000; growth rate 2.9% p.a.
life expectancy men 51, women 54 (1989)
languages French (official), Swahili, Lingala, other African languages; over 300 dialects
religions Christian 70%, Muslim 10%
literacy men 79%, women 45% (1985 est)
GNP $5 bn (1987); $127 per head
chronology
1908 Congo Free State annexed to Belgium.
1960 Independence achieved from Belgium as Republic of the Congo. Civil war broke out between central government and Katanga province.
1963 Katanga war ended.
1967 New constitution adopted.
1970 Col Mobutu elected president.
1971 Country became the Republic of Zaire.
1972 The MPR became the only legal political party. Katanga province renamed Shaba.
1974 Foreign-owned businesses and plantations seized by Mobutu and given in political patronage.
1977 Original owners of confiscated properties invited back. Mobutu re-elected; Zairians invaded Shaba province from Angola, repulsed by Belgian paratroopers.
1978 Second unsuccessful invasion from Angola.
1988 Potential rift with Belgium avoided.
1990 Mobutu announced end of ban on multiparty politics, following internal dissent.
1991 Multiparty elections promised for 1992; after antigovernment riots, Mobutu agreed to share power with opposition. Etienne Tshisekedi appointed premier. Oct: Tshisekedi dismissed.
1992 Army mutiny crushed. Aug: Tshisekedi reinstated against Mobuto's wishes; interim opposition parliament formed. Oct: renewed rioting.

Zaïre River /zɑːˈɪə/ (formerly until 1971 *Congo*) second longest river in Africa, rising near the Zambia–Zaire border (and known as the *Lualaba River* in the upper reaches) and flowing 4,500 km/2,800 mi to the Atlantic Ocean, running in a great curve that crosses the equator twice, and discharging a volume of water second only to the Amazon. The chief tributaries are the Ubangi, Sangha, and Kasai.

Navigation is interrupted by dangerous rapids up to 160 km/100 mi long, notably from the Zambian border to Bukama; below Kongolo, where the gorge known as the Gates of Hell is located; above Kisangani, where the Stanley Falls are situated; and between Kinshasa and Matadi.

Boma is a large port on the estuary; Matadi is a port 80 km/50 mi from the Atlantic, for ocean-going ships; and at Pool Malebo (formerly Stanley Pool), a widening of the river 560 km/350 mi from its mouth which encloses the marshy island of Bamu, are Brazzaville on the western shore and Kinshasa on the southwestern. The Inga Dam supplies Matadi and Kinshasa with electricity.

history The mouth of the Zaïre was seen by the Portuguese navigator Diego Cão 1482, but the vast extent of its system became known to Europeans only with the explorations of David Livingstone and Henry Stanley. Its navigation from source to mouth was completed by the expedition 1974 led by the English explorer John Blashford-Snell (1936–), supported by President Mobutu.

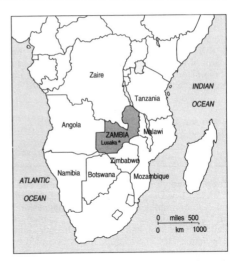

... resembling an immense snake uncoiled, with its head in the sea, its body at rest curving afar over a vast country, and its tail lost in the depths of the land.

On the **Zaïre River** Joseph Conrad
Heart of Darkness 1902

Zákinthos /ˈzækɪnθɒs/ or *Zante* southernmost of the Ionian Islands, Greece; area 410 sq km/158 sq mi; population (1981) 30,000. Products include olives, currants, grapes, and carpets.

Zambezi /zæmˈbiːzi/ river in central and SE Africa; length 2,650 km/1,650 mi from NW Zambia through Mozambique to the Indian Ocean, with a wide delta near Chinde. Major tributaries include the Kafue in Zambia. It is interrupted by rapids, and includes on the Zimbabwe–Zambia border the Victoria Falls (Mosi-oa-tunya) and Kariba Dam, which forms the reservoir of Lake Kariba with large fisheries.

Zambia Republic of
area 752,600 sq km/290,579 sq mi
capital Lusaka
towns Kitwe, Ndola, Kabwe, Chipata, Livingstone
physical forested plateau cut through by rivers
features Zambezi River, Victoria Falls, Kariba Dam
head of state and government Frederick Chiluba from 1991
political system socialist pluralist republic

political parties United National Independence Party (UNIP), African socialist; Movement for Multiparty Democracy (MMD); National Democratic Alliance (Nada)
exports copper, cobalt, zinc, emeralds, tobacco
currency kwacha
population (1990 est) 8,119,000; growth rate 3.3% p.a.
life expectancy men 54, women 57 (1989)
language English (official); Bantu dialects
religions Christian 66%, animist, Hindu, Muslim
literacy 54% (1988)
GNP $2.1 bn (1987); $304 per head (1986)
chronology
1899–1924 As Northern Rhodesia, under administration of the British South Africa Company.
1924 Became a British protectorate.
1964 Independence achieved from Britain, within the Commonwealth, as the Republic of Zambia with Kenneth Kaunda as president.
1972 United National Independence Party (UNIP) declared the only legal party.
1976 Support for the Patriotic Front in Rhodesia declared.
1980 Unsuccessful coup against President Kaunda.
1985 Kaunda elected chair of the African Front Line States.
1987 Kaunda elected chair of the Organization of African Unity (OAU).
1988 Kaunda re-elected unopposed for sixth term.
1990 Multiparty system announced for 1991.
1991 MMD won landslide election victory; Frederick Chiluba became president.
1992 Food and water shortages caused by severe drought.

Zante /ˈzænti/ Italian name for the Ionian island of ♢Zákinthos, Greece.

Zanzibar /ˌzænzɪˈbɑː/ island region of Tanzania
area 1,658 sq km/640 sq mi (80 km/50 mi long)
towns Zanzibar
products cloves, copra
population (1985) 571,000

history settled by Arab traders in the 7th century; occupied by the Portuguese in the 16th century; became a sultanate in the 17th century; under British protection 1890–1963. Together with the island of Pemba, some nearby islets, and a strip of mainland territory, it became a republic 1963. It merged with Tanganyika as Tanzania 1964.

When the flute is played in Zanzibar all Africa east of the Lakes must dance.

On **Zanzibar** Arab proverb

Zaporozhye /ˌzæpəˈrɒʒji/ (formerly until 1921 *Aleksandrovsk*) industrial city (steel, chemicals, aluminium goods, pig iron, magnesium) in Ukraine, on the river Dnieper; capital of Zaporozhye region and site of the Dnieper Dam; population (1987) 875,000. It was occupied by Germany 1941–43.

Zara /ˈzɑːrə/ Italian name for ♢Zadar, a port on the Adriatic coast of Croatia.

Zaragoza /ˌsærəˈɡɒsə/ (English *Saragossa*) industrial city (iron, steel, chemicals, plastics, canned food, electrical goods) in Aragon, Spain; population (1991) 614,400. The medieval city walls and bridges over the river Ebro survive, and there is a 15th-century university.
history A Celtic settlement known as *Salduba* was captured by the Romans in the 1st century BC, they named it *Caesarea Augusta*, after their leader; later it was captured by Visigoths and Moors and was taken 1118 by Alfonso the Warrior, King of Navarre and Aragon, after a nine-month siege. It remained the capital of Aragon until the end of the 15th century. From June 1808 to Feb 1809, in the Peninsular War, it resisted a French siege. Maria Augustin (died 1859), known as the 'Maid of Zaragoza', became a national hero for her part in the defence; her story is told in Byron's *Childe Harold* 1812–18.

Zealand /ˈziːlənd/ another name for ♢*Sjælland*, the main island of Denmark, and for ♢*Zeeland*, a province of the SW Netherlands.

Zeebrugge /ˈziːbruɡə/ small Belgian ferry port on the North Sea, linked to Bruges by a canal (built 1896–1907), 14 km/9 mi long. In March 1987 it was the scene of a disaster in which over 180 passengers lost their lives when the car ferry *Herald of Free Enterprise* put to sea from Zeebrugge with its car-loading doors still open.

Zeeland /ˈziːlənd/ province of the SW Netherlands
area 1,790 sq km/691 sq mi
capital Middelburg
towns Vlissingen, Terneuzen, Goes
features mostly below sea level, Zeeland is protected by a system of dykes
products cereals, potatoes
population (1991) 357,500
history disputed by the counts of Flanders and Holland during the Middle Ages, Zeeland was annexed to Holland 1323 by Count William III.

Zelenograd /ˌzelɪnəˈɡræd/ city on the Skhodnia

River, 37 km/23 mi NW of Moscow, Russia; population (1989) 158,000. Construction began 1960 and it achieved city status 1963. It is a centre for the microelectronics industry; construction materials and fruit and vegetables are also produced.

Zermatt /ˈzɜːmæt, German tseəˈmæt/ ski resort in the Valais canton, Switzerland, at the foot of the Matterhorn; population (1985) 3,700.

Zetland /ˈzetlənd/ official form until 1974 of ♢Shetland, a group of islands off N Scotland.

Zhangjiakou /ˌdʒæŋdʒiəˈkəu/ or *Changchiakow* historic town and trade centre in Hebei province, China, 160 km/100 mi NW of Beijing, on the Great Wall; population (1980) 1,100,000. Zhangjiakou is on the border of Inner Mongolia (its Mongolian name is *Kalgan*, 'gate') and on the road and railway to Ulaanbaatar in Mongolia. It developed under the Manchu dynasty, and was the centre of the tea trade from China to Russia.

Zhdanov /ˈʒdɑːnɒv/ former name (1948–89) of ♢Mariupol, a port in Ukraine.

Zhejiang /ˌdʒɜːdʒiˈæŋ/ or *Chekiang* province of SE China
area 101,800 sq km/39,295 sq mi
capital Hangzhou
features smallest of the Chinese provinces; the base of the Song dynasty 12th–13th centuries; densely populated
products rice, cotton, sugar, jute, maize; timber on the uplands
population (1990) 41,446,000.

Zhengzhou /ˌdʒʌŋˈdʒəu/ or *Chengchow* industrial city (light engineering, cotton textiles, foods) and capital (from 1954) of Henan province, China, on the Huang He River; population (1989) 1,660,000.
In the 1970s the earliest city found in China, from 1500 BC, was excavated near the walls of Zhengzhou. The Shaolin temple, where the martial art of kung fu originated, is nearby.

Zhitomir /ʒɪˈtəumɪə/ capital of Zhitomir region in W Ukraine, W of Kiev; population (1987) 287,000. It is a timber and grain centre and has furniture factories. Zhitomir dates from the 13th century.

Zhonghua Renmin Gonghe Guo /ˌdʒɒŋˈhwɑː ˌrenˈmɪn ˌɡɒŋhɑːˈɡwəu/ Chinese for People's Republic of ♢China.

Zian alternative spelling of ♢Xian, a city in China.

Zimbabwe Republic of
area 390,300 sq km/150,695 sq mi
capital Harare
towns Bulawayo, Gweru, Kwekwe, Mutare, Hwange
physical high plateau with central high veld and mountains in E; rivers Zambezi, Limpopo
features Hwange National Park, part of Kalahari Desert; ruins of Great Zimbabwe
head of state and government Robert Mugabe from 1987
political system effectively one-party socialist republic

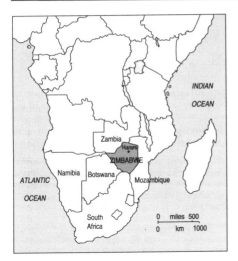

political party Zimbabwe African National Union–Patriotic Front (ZANU–PF), African socialist
exports tobacco, asbestos, cotton, coffee, gold, silver, copper
currency Zimbabwe dollar
population (1990 est) 10,205,000 (Shona 80%, Ndbele 19%; about 100,000 whites); growth rate 3.5% p.a.
life expectancy men 59, women 63 (1989)
languages English (official), Shona, Sindebele
religions Christian, Muslim, Hindu, animist
literacy men 81%, women 67% (1985 est)
GNP $5.5 bn (1988); $275 per head (1986)
chronology
1889–1923 As Southern Rhodesia, under administration of British South Africa Company.
1923 Became a self-governing British colony.
1961 Zimbabwe African People's Union (ZAPU) formed, with Joshua Nkomo as leader.
1962 ZAPU declared illegal.
1963 Zimbabwe African National Union (ZANU) formed, with Robert Mugabe as secretary general.
1964 Ian Smith became prime minister. ZANU banned. Nkomo and Mugabe imprisoned.
1965 Smith declared unilateral independence.
1966–68 Abortive talks between Smith and UK prime minister Harold Wilson.
1974 Nkomo and Mugabe released.
1975 Geneva conference set date for constitutional independence.
1979 Smith produced new constitution and established a government with Bishop Abel Muzorewa as prime minister. New government denounced by Nkomo and Mugabe. Conference in London agreed independence arrangements (Lancaster House Agreement).
1980 Independence achieved from Britain, with Robert Mugabe as prime minister.
1981 Rift between Mugabe and Nkomo.
1982 Nkomo dismissed from the cabinet, leaving the country temporarily.

1984 ZANU–PF party congress agreed to create a one-party state in future.
1985 Relations between Mugabe and Nkomo improved. Troops sent to Matabeleland to suppress rumoured insurrection; 5,000 civilians killed.
1986 Joint ZANU–PF rally held amid plans for merger.
1987 White-roll seats in the assembly were abolished. President Banana retired; Mugabe combined posts of head of state and prime minister with the title executive president.
1988 Nkomo returned to the cabinet and was appointed vice president.
1989 Opposition party, the Zimbabwe Unity Movement, formed by Edgar Tekere; draft constitution drawn up, renouncing Marxism–Leninism; ZANU and ZAPU formally merged.
1990 ZANU–PF re-elected. State of emergency ended. Opposition to creation of one-party state.
1992 United Front formed to oppose ZANU–PF. March: Mugabe declared dire drought and famine situation a national disaster.

Zlatoust /ˌzlætəʊˈuːst/ industrial city (metallurgy) in Chelyabinsk region, central Russia, in the S Ural Mountains; population (1987) 206,000. It was founded 1754 as an iron- and copper-working settlement, destroyed 1774 by a peasant uprising, but developed as an armaments centre from the time of Napoleon's invasion of Russia.

Zomba /ˈzɒmbə/ former capital of Malawi, 32 km/ 20 mi W of Lake Shirwa; population (1985) 53,000. Industries included cement and fishing tackle. It was replaced by Lilongwe as capital 1975 but remains the university town.

Zuider Zee /ˈzaɪdə ˈziː, Dutch ˈzaudə ˈzeɪ/ former sea inlet in the NW Netherlands, closed off from the North Sea by a 32-km/20-mi dyke 1932; much of it has been reclaimed as land. The remaining lake is called the IJsselmeer.

Zululand /ˈzuːluːlænd/ region in Natal, South Africa, largely corresponding to the Black National State KwaZulu. It was formerly a province, annexed to Natal 1897.

Zürich /ˈzjʊərɪk/ financial centre and industrial city (machinery, electrical goods, textiles) on Lake Zürich; population (1990) 341,300. Situated at the foot of the Alps, it is the capital of Zürich canton and the largest city in Switzerland. The university was refounded 1833.

Zutphen /ˈzʌtfən/ town in Gelderland province, the Netherlands; population (1991) 31,000.

Zwickau /ˈtsvɪkaʊ/ coalmining and industrial town (vehicles, textiles) SW of Chemnitz in the state of Saxony, Germany, on the river Mulde; population (1986) 121,000. It is the birthplace of the composer Robert Schumann.

Zwolle /ˈzwɒlə/ capital of Overijssel province, the Netherlands; a market town with brewing, distilling, butter making, and other industries; population (1991) 95,600.

GLOSSARY OF
GEOGRAPHICAL TERMS

ablation loss of snow and ice from a glacier by melting and evaporation. It is the opposite of accumulation. If total ablation exceeds total accumulation for a particular glacier, then the glacier will retreat, and vice versa.

abrasion the effect of corrasion, a type of erosion in which rock fragments scrape and grind away a surface. The rock fragments may be carried by rivers, wind, ice, or the sea. Striations, or grooves, on rock surfaces are a common effect, caused by the scratching of rock debris embedded in glacier ice.

abyssal plain broad expanse of sea floor lying 3–6 km/2–4 mi below sea level. Abyssal plains are found in all the major oceans, and they extend from bordering continental rises to mid-oceanic ridges.

abyssal zone dark ocean area 2,000–6,000 m/6,500–19,500 ft deep; temperature 4°C/39°F. Three-quarters of the area of the deep ocean floor lies in the abyssal zone. The region above is the bathyal zone; the region below, the hadal zone.

accumulation addition of snow and ice to a glacier. It is the opposite of ablation. Snow is added through snowfall and avalanches, and is gradually compressed to form ice as the glacier progresses.

acid rain acidic rainfall, thought to be caused mainly by the release into the atmosphere of sulphur dioxide (formed from the burning of fossil fuels) and nitrogen oxides (contributed from industrial activities and from car exhaust fumes). Acid rain is linked with damage to and the death of forests; it also results in damage to buildings.

afforestation planting of trees in areas that have not previously held forests. Trees may be planted to provide timber and wood pulp, to provide firewood in countries where this is an energy source, to bind soil together and prevent erosion, and to act as windbreaks.

agglomeration the clustering of activities or people at specific points or areas, for example at a route centre where lines of communication meet.

agribusiness commercial farming on an industrial scale, often financed by companies whose main interests lie outside agriculture. Agribusiness farms are mechanized, large in size, highly structured, and reliant on chemicals.

aid money or resources given or lent on favourable terms to developing countries. Short-term aid, usually food and medicine, is given to relieve conditions in emergencies such as famine. Long-term aid is intended to promote economic development and improve the quality of life, for example, by funding irrigation, education, and communications programmes.

alluvial fan roughly triangular sedimentary formation found at the base of slopes. An alluvial fan results when a sediment-laden stream or river rapidly deposits its load of gravel and silt as its speed is reduced on entering a plain.

alluvium fine silty material deposited by a river where the water's velocity is low -for example, on the inside bend of a meander. Alluvium periodically deposited by floodwater creates a floodplain.

anabatic wind warm wind that blows uphill in steep-sided valleys in the early morning. As the sides of a valley warm up in the morning the air above is also warmed and rises up the valley to give a gentle breeze.

anemometer device for measuring wind speed. A cup-type anemometer consists of cups at the ends of arms, which rotate when the wind blows. The speed of rotation indicates the wind speed in kilometres per hour or knots.

Antarctic Circle imaginary line that encircles the South Pole at latitude 66° 32′ S. The line encompasses the continent of Antarctica and the Antarctic Ocean.

anticline a fold in the rocks of the Earth's crust in which the layers or beds bulge upwards to form an arch (seldom preserved intact).

anticyclone area of high atmospheric pressure caused by descending air, which becomes warm and dry. Winds radiate from a calm centre, taking a clockwise direction in the northern hemisphere and an anticlockwise direction in the southern hemisphere. Anticyclones are characterized by clear weather and the absence of rain and violent winds.

antipodes places at opposite points on the globe.

aquifer any rock formation containing water. The rock of an aquifer must be porous and permeable (full of interconnected holes) so that it can absorb water. Aquifers supply artesian wells, and are actively sought in arid areas as sources of drinking and irrigation water.

arable farming cultivation of crops, as opposed to the keeping of animals. Crops may be cereals, vegetables, or plants for producing oils or cloth.

archipelago group of islands, or an area of sea containing a group of islands. The islands of an archipelago are usually volcanic in origin, and they sometimes represent the tops of peaks in areas around continental margins flooded by the sea.

Arctic Circle imaginary line that encircles the North Pole at latitude 66° 32′ N. Within this line there is at least one day in the summer during which the Sun never sets, and at least one day in the winter during which the Sun never rises.

arête sharp narrow ridge separating two glacier troughs. The typical U-shaped cross sections of glacier troughs give arêtes very steep sides. Arêtes are common in glaciated mountain regions.

arid zone infertile area with small, infrequent rainfall that rapidly evaporates because of high temperatures.

artesian well well that is supplied with water rising from an underground water-saturated rock layer (aquifer). Such a well may be drilled into an aquifer that is confined by impermeable rocks both above and below. The water rises under its own pressure.

asthenosphere division of the Earth's structure lying beneath the lithosphere, at a depth of approximately 70 km/45 mi to 260 km/160 mi. It is thought to be the soft, partially molten layer of the mantle on which the rigid plates of the Earth's surface move to produce the motions of plate tectonics.

atmosphere the mixture of gases that surrounds the Earth; it is prevented from escaping by the pull of the Earth's gravity.

atoll continuous or broken circle of coral reef and low coral islands surrounding a lagoon.

attrition process by which particles of rock being transported by river, wind, or sea are rounded and gradually reduced in size by being struck against one another.

aurora coloured light in the night sky near the Earth's magnetic poles, called *aurora borealis*, 'northern lights', in the northern hemisphere and *aurora australis* in the southern hemisphere. An aurora is usually in the form of a luminous arch with its apex towards the magnetic pole followed by arcs, bands, rays, curtains, and coronas, usually green but often showing shades of blue and red, and sometimes yellow or white.

avalanche fall of a mass of snow and ice down a steep slope. Avalanches occur because of the unstable nature of snow masses in mountain areas.

badlands barren landscape cut by erosion into a maze of ravines, pinnacles, gullies, and sharp-edged ridges.

bar deposit of sand or silt formed in a river channel, or a long sandy ridge running parallel to a coastline. Coastal bars can extend across estuaries to form bay bars.

barometer instrument that measures atmospheric pressure as an indication of weather. Most often used are the mercury barometer and the aneroid barometer.

barrier island long island of sand, lying offshore and parallel to the coast. Most barrier islands are derived from marine sands piled up by shallow longshore currents that sweep sand parallel to the

seashore. Others are derived from former spits, connected to land and built up by drifted sand, that were later severed from the mainland.

barrier reef coral reef that lies offshore, separated from the mainland by a shallow lagoon.

batholith large, irregular, deep-seated mass of igneous rock, usually granite, with an exposed surface of more than 100 sq km/40 sq mi. The mass forms by the intrusion or upswelling of magma (molten rock) through the surrounding rock. Batholiths form the core of all major mountain ranges.

beach strip of land bordering the sea, normally consisting of boulders and pebbles on exposed coasts or sand on sheltered coasts. It is usually defined by the high-and low-water marks.

bearing the direction of a fixed point from a point of observation on the Earth's surface, expressed as an angle from the north. Bearings are taken by a compass and measured in degrees, given as three-digit numbers increasing clockwise. For instance, north is 000°, south is 180°, and southwest is 225°.

Beaufort scale system of recording wind velocity. It is a numerical scale ranging from 0 to 17, calm being indicated by 0 and a hurricane by 12; 13–17 indicate degrees of hurricane force.

bed a single sedimentary rock unit with a distinct set of physical characteristics or contained fossils, readily distinguishable from those of beds above and below.

bergschrund deep crevasse that may be found at the head of a glacier.

berm on a beach, a ridge of sand or pebbles running parallel to the water's edge, formed by the action of the waves on beach material. Sand and pebbles are deposited at the farthest extent of swash (advance of water) on that particular beach.

biofuel any solid, liquid, or gaseous fuel produced from organic matter, either directly from plants or indirectly from industrial, commercial, domestic, or agricultural wastes.

biological weathering form of weathering caused by the activities of living organisms —for example, the growth of roots or the burrowing of animals. Tree roots are probably the most significant agents of biological weathering as they are capable of prising apart rocks by growing into cracks and joints.

biosphere or *ecosphere* the region of the Earth's surface (both land and water), together with the atmosphere above it, that can be occupied by living organisms.

bog type of wetland where decomposition is slowed down and dead plant matter accumulates as peat. Bogs develop under conditions of low temperature, high acidity, low nutrient supply, stagnant water, and oxygen deficiency.

bore surge of tidal water up an estuary or a river, caused by the funnelling of the rising tide by a narrowing river mouth. A very high tide, possibly fanned by wind, may build up when it is held back by a river current in the river mouth. The result is a broken wave, a metre or a few feet high, that rushes upstream.

Bouguer anomaly an increase in the Earth's gravity observed near a mountain or dense rock mass. This is due to the gravitational force exerted by the rock mass.

boulder clay another name for ⟡till, a type of glacial deposit.

braiding the subdivision of a river into several channels caused by deposition from islets in the channel. Braided channels are common in meltwater streams.

butte steep-sided, flat-topped hill, formed in horizontally layered sedimentary rocks, largely in arid areas. A large butte with a pronounced tablelike profile is a mesa.

cacao tropical American evergreen tree *Theobroma cacao* of the Sterculia family, now also cultivated in W Africa and Sri Lanka. Its seeds are cocoa beans, from which cocoa and chocolate are prepared.

caldera very large basin-shaped crater. Calderas are found at the tops of volcanoes, where the original peak has collapsed into an empty chamber beneath. The basin, many times larger than the original volcanic vent, may be flooded, producing a crater lake, or the flat floor may contain a number of small volcanic cones, produced by volcanic activity after the collapse.

canyon deep, narrow valley or gorge running through mountains. Canyons are formed by stream down-cutting, usually in arid areas, where the stream or river receives water from outside the area.

cartography art and practice of drawing maps.

cash crop crop grown solely for sale rather than for the farmer's own use, for example, coffee, cotton, or sugar beet. Many Third World countries grow cash crops to meet their debt repayments rather than grow food for their own people.

catch crop crop such as turnip that is inserted between two principal crops in a rotation in order to provide some quick livestock feed or soil improvement at a time when the land would otherwise be lying idle.

catchment area area from which water is collected by a river and its tributaries.

cave roofed-over cavity in the Earth's crust usually produced by the action of underground water or by waves on a seacoast. Caves of the former type commonly occur in areas underlain by limestone, where the rocks are soluble in water.

cavitation erosion of rocks caused by the forcing of air into cracks. Cavitation results from the pounding of waves on the coast and the swirling of turbulent river currents, and exerts great pressure, eventually causing rocks to break apart.

Celsius scale temperature scale, previously called Centigrade, in which the range from freezing to boiling of water is divided into 100 degrees, freezing point being 0 degrees and boiling point 100 degrees.

census official gathering of information about the population in a particular area. The data collected is used by government departments in planning for the future in such areas as health, education, transport, and housing. In the UK a census has been conducted every ten years since 1801.

cereal grass grown for its edible, nutrient-rich, starchy seeds. The term refers primarily to wheat, oats, rye, and barley, but may also refer to corn, millet, and rice. If all the world's cereal crop were consumed as whole-grain products directly by humans, everyone could obtain adequate protein and carbohydrate; however, a large proportion of cereal production in affluent nations is used as animal feed to boost the production of meat, dairy products, and eggs.

chalk soft, fine-grained, whitish rock composed of calcium carbonate, $CaCO_3$, extensively quarried for use in cement, lime, and mortar, and in the manufacture of cosmetics and toothpaste.

chemical weathering form of weathering brought about by a chemical change in the rocks affected. Chemical weathering involves the 'rotting', or breakdown, of the minerals within a rock, and usually produces a claylike residue (such as china clay and bauxite). Some chemicals are dissolved and carried away from the weathering source.

city important urban settlement. In the past, a town in Britain needed either a cathedral or a royal charter before it could be called a city, but in modern-day usage a city is a settlement with a population of more than 150,000.

climate weather conditions at a particular place over an extended period of time. The primary factors that determine the variations of climate over the surface of the Earth are: (a) the effect of latitude and the tilt of the Earth's axis to the plane of the orbit about the Sun (66.5°); (b) the large-scale movements of different wind belts over the Earth's surface; (c) the temperature difference between land and sea; (d) contours of the ground; and (e) location of the area in relation to ocean currents. Catastrophic variations to climate may be caused by the impact of another planetary body, or by clouds resulting from volcanic activity. The most important local or global metereological changes brought about by human activity are those linked with ozone depleters and the greenhouse effect.

clint one of a number of flat-topped limestone blocks that make up a limestone pavement. Clints are separated from each other by enlarged joints called grykes.

cloud water vapour condensed into minute water particles that float in masses in the atmosphere. Clouds, like fogs or mists, which occur at lower levels, are formed by the cooling of air containing water vapour, which generally condenses around tiny dust particles.

coal black or blackish mineral substance formed from the compaction of ancient plant matter in tropical swamp conditions. It is used as a fuel and in the chemical industry.

coastal erosion erosion of the land by the constant battering of the sea's waves. This produces two effects. The first is a hydraulic effect, in which the force of the wave compresses air pockets in coastal rocks and cliffs, and the air then expands explosively. The second is the effect of abrasion, in which rocks and pebbles are flung against the cliffs, wearing them away.

coffee tropical evergreen shrub of the genus *Coffea* whose beanlike seeds are roasted and ground to produce a drink of the same name. Coffee grows best on frost-free hillsides with moderate rainfall. The world's largest producers are Brazil, Colombia, and the Ivory Coast.

collective farming system in which a group of farmers pool their land, domestic animals, and agricultural implements, retaining as private property enough only for the members' own requirements. The profits are divided among the members.

combe or *coombe* steep-sided valley found on the scarp slope of a chalk escarpment.

compass any instrument for finding direction. The most commonly used is a magnetic compass with a needle which points to the magnetic north, from which true north can be found from tables of magnetic correction.

confluence point at which two rivers join.

coniferous forest forest consisting of evergreen trees such as pines and firs. Most conifers grow quickly and can tolerate poor soil, steep slopes, and short growing seasons. Coniferous forests are often planted in afforestation schemes. Conifers also grow in woodland.

conservation action taken to protect and preserve the natural world, usually from pollution, overexploitation, and other harmful features of human activity.

constructive margin in plate tectonics, a region in which two plates are moving away from each other. Magma, or molten rock, escapes to the surface along this margin to form new crust, usually in the form of a ridge. Over time, as more and more magma reaches the surface, the sea floor spreads.

continent any one of the seven large land masses of the Earth, as distinct from the oceans. They are Asia, Africa, North America, South America, Europe, Australia, and Antarctica. Continents are constantly moving and evolving. A continent does not end at the coastline; its boundary is the edge of the shallow continental shelf, which may extend several hundred kilometres or miles out to sea.

continental drift theory that, about 200 million years ago, the Earth consisted of a single large continent (Pangaea) that subsequently broke apart to form the continents known today. Such vast continental movements could not be satisfactorily explained until the study of plate tectonics in the 1960s.

continental rise the portion of the ocean floor rising gently from the abyssal plain towards the steeper continental slope. The continental rise is a depositional feature formed from sediments transported down the slope mainly by turbidity currents.

continental shelf gently sloping submarine plain extending into the ocean from a continent. The plain has a gradient of less than 1°; when the angle of the sea bed is 1°–5°, it is known as the continental slope.
This change usually occurs several hundred kilometres away from the land.

continental slope sloping, submarine portion of a continent. It extends downwards at an angle of 1°–5° from the continental margin at the edge of the continental shelf. In some places continental slopes extend directly to the ocean deeps or abyssal plain. In others they grade into the gentler continental rises that in turn grade into the abyssal plains.

contour line drawn on a map to join points of equal height. Contours are drawn at regular height intervals; for example, every 10 m/33 ft. The closer together the lines are, the steeper the slope. Contour patterns can be used to interpret the relief of an area and to identify land forms.

conurbation or *metropolitan area* large continuous built-up area formed by the joining together of several urban settlements.

cooperative farming system in which individual farmers pool their resources (excluding land) to buy commodities such as seeds and fertilizers, and services such as marketing.

coordinated universal time (UTC) measurement of time, based on uniform atomic time. It replaced Greenwich Mean Time (GMT) 1986.

cordillera group of mountain ranges and their valleys, all running in a specific direction, formed by the continued convergence of two tectonic plates along a line.

core the innermost part of the Earth. It is divided into an inner core, the upper boundary of which is 1,700 km/1,060 mi from the centre, and an outer core, 1,820 km/1,130 mi thick. Both parts are thought to consist of iron-nickel alloy, with the inner core being solid and the outer core being semisolid. The temperature may be 3,000°C/5,400°F.

Coriolis effect the effect of the Earth's W–E rotation upon the atmosphere and upon all objects on the Earth's surface. In the northern hemisphere it causes winds, ocean currents, and aircraft to be deflected to the right (clockwise); in the southern hemisphere it causes deflection to the left (anticlockwise).

corn main cereal crop of a region—for example, wheat in the UK, oats in Scotland and Ireland, maize in the USA.

corrasion the grinding away of solid rock surfaces by particles carried by water, ice, and wind. It is generally held to be the most significant form of erosion. As the eroding particles are carried along they become eroded themselves due to the process of attrition.

corrie (Welsh *cwm*; French, North American *cirque*) Scottish term for a steep-walled hollow in the mountainside of a glaciated area representing the source of a melted glacier. The weight of the ice has ground out the bottom and worn back the sides.

cotton tropical and subtropical herbaceous plant of the genus *Gossypium* of the mallow family Malvaceae. Fibres surround the seeds inside the ripened fruits, or bolls, and these are spun into yarn for cloth.

country park pleasure ground or park, often located near an urban area, providing facilities for the public enjoyment of the countryside.

cover crop crop that is inserted between two principal crops to prevent soil erosion or leaching, or to provide 'green manure' (natural enrichment of the soil, by the ploughing in of the growing crop).

crag in previously glaciated areas, a large lump of rock that a glacier has been unable to wear away. As the glacier passed up and over the crag, weaker rock on the far side was largely protected from erosion and formed a tapering ridge, or tail, of debris.

crater bowl-shaped depression, usually round and with steep sides. Craters are formed by explosive events such as the eruption of a volcano or by the impact of a meteorite.

craton or *shield* core of a continent, a vast tract of highly deformed metamorphic rock around which the continent has been built. Intense mountain-building periods shook these shield areas in Precambrian times before stable conditions set in.

crevasse deep crack in the surface of a glacier; it can reach several metres in depth. Crevasses often occur where a glacier flows over the break of a slope, because the upper layers of ice are unable to stretch and cracks result. Crevasses may also form at the edges of glaciers owing to friction with the bedrock.

croft small farm in the Highlands of Scotland, traditionally farming common land collectively. Today, although grazing land is still shared, arable land is typically enclosed.

crop rotation system of regularly changing the crops grown on a piece of land. The crops are grown in a particular order to utilize and add to the nutrients in the soil and to prevent the build-up of insect and fungal pests.

crust the outermost part of the structure of Earth, consisting of two distinct parts, the oceanic crust and the continental crust. The *oceanic crust* is on average about 10 km/6.2 mi thick and consists mostly of basaltic types of rock. By contrast, the *continental crust* is largely made of granite and is more complex in its structure. Because of the movements of plate tectonics, the oceanic crust is in no place older than about 200 million years. However, parts of the continental crust are over 3 billion years old.

current flow of a body of water or air, or of heat, moving in a definite direction. There are three

basic types of oceanic current: *drift currents* are broad and slow-moving; *stream currents* are narrow and swift-moving; and *upwelling currents* bring cold, nutrient-rich water from the ocean bottom.

cyclone alternative name for a ◊depression, an area of low atmospheric pressure.

dam structure built to hold back water in order to prevent flooding, to supply water for irrigation and storage, and to provide hydroelectric power.

deciduous forest woodland area consisting of broad-leaved trees (such as oak) which shed their leaves in winter to reduce water loss and conserve energy. Broad-leaved trees grow slowly, reaching maturity 100–200 years after being planted, thus limiting their economic value.

deforestation the cutting down of forest without planting new trees to replace those lost or allowing the forest to regenerate itself naturally.

delta tract of land at a river's mouth, composed of silt deposited as the water slows on entering the sea. The shape of the Nile delta is like the Greek letter *delta* Δ, and thus gave rise to the name.

demography the study of population.

depression or *cyclone* or *low* region of low atmospheric pressure. A depression forms as warm, moist air from the tropics mixes with cold, dry polar air, producing warm and cold boundaries (fronts) and unstable weather—low cloud and drizzle, showers, or fierce storms. Depressions tend to travel eastwards and can remain active for several days. A severe depression that forms in the tropics is called a tropical cyclone or hurricane.

desert arid area without sufficient rainfall and, consequently, vegetation to support human life. The term includes the ice areas of the polar regions (known as cold deserts). Almost 33% of the Earth's land surface is desert, and this proportion is increasing.

destructive margin in plate tectonics, a region on the Earth's crust in which two plates are moving towards one another. Usually one plate (the denser of the two) is forced to dive below the other into what is called the subduction zone. The descending plate melts to form a body of magma, which may then rise to the surface through cracks and faults to form volcanoes. If the two plates consist of more buoyant continental crust, subduction does not occur. Instead, the crust crumples gradually to form fold mountains.

dew precipitation in the form of moisture that collects on the ground. It forms after the temperature of the ground has fallen below the dew point of the air in contact with it. As the temperature falls during the night, the air and its water vapour become chilled, and condensation takes place on the cooled surfaces.

dew point temperature at which the air becomes saturated with water vapour. At temperatures below the dew point, the water vapour condenses out of the air as droplets. If the droplets are large they become deposited on the ground as dew; if small they remain in suspension in the air and form mist or fog.

diagenesis or *lithification* physical and chemical changes by which a sediment becomes a sedimentary rock. The main processes involved include compaction of the grains, and the cementing of the grains together by the growth of new minerals deposited by percolating groundwater.

diapirism geological process in which a particularly light rock, such as rock salt, punches upwards through the heavier layers above. The resulting structure is called a salt dome, and oil is often trapped in the curled-up rocks at each side.

dip the angle and direction in which a bed of rock is plunging. Rocks that are dipping have usually been affected by folding.

dispersed settlement settlement made up of buildings scattered over a wide area.

distributary river that has branched away from a main river. Distributaries are most commonly found on a delta, where the very gentle gradient and large amounts of silt deposited encourage channels to split.

diurnal temperature the range in temperature over a 24-hour period.

doldrums area of low atmospheric pressure along the equator, in the intertropical convergence zone where the NE and SE trade winds converge. The doldrums are characterized by calm or very light westerly winds, during which there may be sudden squalls and stormy weather.

drainage basin or *catchment area* the area of land drained by a river and its tributaries.

drought period of prolonged dry weather. The area of the world subject to serious droughts is increasing because of destruction of forests, overgrazing, and poor agricultural practices.

drumlin long streamlined hill created in formerly glaciated areas. Rocky debris (till) is gathered up by the glacial icesheet and moulded to form an egg-shaped mound, 8–60 m/26–197 ft in height and 0.5–1 km/0.3–0.6 mi in length. Drumlins commonly occur in groups on the floors of glacial troughs, producing a 'basket-of-eggs' landscape.

dune mound or ridge of wind-drifted sand. Loose sand is blown and bounced along by the wind, up the windward side of a dune. The sand particles then fall to rest on the lee side, while more are blown up from the windward side. In this way a dune moves gradually downwind.

dust devil small dust storm caused by intense local heating of the ground in desert areas. The air swirls upwards, carrying fine particles of dust with it.

dyke sheet of igneous rock created by the intrusion of magma (molten rock) across layers of pre-existing rock. It may form a ridge when exposed on the surface. A dyke is also a human-made structure used in flood control along coastlines.

Earth third planet from the Sun. It is almost spherical, flattened slightly at the poles, and is composed of three concentric layers: the core, the mantle, and the crust. About 70% of the surface (including the north and south polar icecaps) is covered with water. The Earth is surrounded by a life-supporting atmosphere and is the only planet on which life is known to exist.

earthquake shaking of the Earth's surface as a result of the sudden release of stresses built up in the Earth's crust. Most earthquakes occur along faults (fractures or breaks) in the crust. Plate tectonic movements generate the major proportion: as two plates move past each other they can become jammed and deformed, and a series of shock waves (seismic waves) occur when they spring free.

earth sciences scientific study of the planet Earth as a whole, a synthesis of several traditional subjects such as geology, meteorology, oceanography, geophysics, geochemistry, and palaeontology.

ecology the study of the relationship between organisms and the environments in which they live, including all living and nonliving components.

ecosystem ecological unit made up of living organisms and the nonliving, or physical, environment with which they interact. Ecosystems are fragile because the relationships between components of the unit are so finely balanced: the alteration or removal of any one species can cause the whole ecosystem to collapse.

Ekman spiral effect application of the Coriolis effect to ocean currents, whereby the currents flow at an angle to the winds that drive them.

environment the conditions affecting a particular organism, including physical surroundings, climate, and influences of other living organisms. In common usage, 'the environment' often means the total global environment, without reference to any particular organism.

epicentre the point on the Earth's surface immediately above the seismic focus of an earthquake. Most damage usually takes place at an earthquake's epicentre.

epoch subdivision of a geological period in the geological time scale. Epochs are sometimes given their own names (such as the Palaeocene, Eocene, Oligocene, Miocene, and Pliocene epochs comprising the Tertiary period), or they are referred to as the late, early, or middle portions of a given period (as the Late Cretaceous or the Middle Triassic epoch).

equator imaginary line, 40,092 km/24,901.8 mi long, encircling the broadest part of the Earth, and representing 0° latitude. It is divided into 360° of longitude. The equator divides the Earth into two halves, called the northern and the southern hemispheres. The *celestial equator* is the circle in which the plane of the Earth's equator intersects the celestial sphere (an imaginary sphere surrounding the Earth on which the celestial bodies appear to lie).

equinox the points in spring and autumn at which the Sun's path, the ecliptic, crosses the celestial equator, so that the day and night are of approximately equal length. The *vernal equinox* occurs about 21 March and the *autumnal equinox* about 23 Sept.

era any of the major divisions of geological time, each including several periods, but smaller than an eon. The currently recognized eras all fall within the Phanerozoic eon—or the vast span of time, starting about 590 million years ago, when fossils are found to become abundant. The eras in ascending order are the Palaeozoic, Mesozoic, and Cenozoic. We are living in the Recent epoch of the Quaternary period of the Cenozoic era.

erosion wearing away of the Earth's surface, caused by the breakdown and transportation of particles of rock or soil (by contrast, weathering does not involve transportation). Agents of erosion include the sea, rivers, glaciers, and wind. People also contribute to erosion by bad farming practices and the cutting down of forests, which can lead to the formation of dust bowls.

erratic a displaced rock that has been transported by a glacier or some other natural force to a site of different geological composition.

escarpment or *cuesta* large ridge created by the erosion of folded rocks. It has one steep side (scarp) and one gently sloping side (dip). Escarpments are common features of chalk landscapes.

esker winding steep-walled ridge of sands and gravels formed beneath a glacier. It represents the course of a subglacial river channel.

estuary river mouth widening into the sea, where fresh water mixes with salt water and tidal effects are felt.

eutrophication excessive enrichment of rivers, lakes, and shallow sea areas, primarily by nitrate fertilizers and detergents in municipal sewage. These encourage the growth of algae and bacteria which use up the oxygen in the water, thereby making it uninhabitable for fishes and other animal life.

exfoliation form of physical weathering brought about by extreme changes of temperature. Exfoliation may cause the outer layer of a rock surface to break away—a process that is sometimes called 'onion-skin weathering'.

exosphere the uppermost layer of the atmosphere. It is an ill-defined zone above the thermosphere, beginning at about 700 km/435 mi and fading off into the vacuum of space. The gases are extremely thin, with hydrogen as the main constituent.

extensive agriculture farming system where the area of the farm is large but there are low inputs (such as labour or fertilizers). Extensive farming generally gives rise to lower yields per hectare than intensive agriculture.

extrusive rock or *volcanic rock* igneous rock formed on the surface of the Earth. It is usually fine-grained. The magma (molten rock) that cools to form extrusive rock may reach the surface through a crack or through the vent of a volcano.

facies any assemblage of mineral, rock, or fossil features that reflect the environment in which rock was formed.

Fahrenheit scale temperature scale which was commonly used in English-speaking countries up until the 1970s, after which the Celsius scale was generally adopted, in line with the rest of the world. In the Fahrenheit scale intervals are measured in degrees (°F), with water freezing at 32° and boiling at 212°; °F = (°C × 9/5) + 32.

famine severe shortage of food affecting a large number of people. Famine may be the result of drought, of population pressure, or of civil war.

fault a fracture in the Earth's crust along which the two sides have moved as a result of differing strains in the adjacent rock bodies. Displacement of rock masses horizontally or vertically along a fault may be microscopic, or it may be massive, causing major earthquakes.

firn or *névé* snow that has lain on the ground for a full calendar year. Firn is common at the tops of mountains. After many years, compaction turns firn into ice and a glacier forms.

fjord or *fiord* narrow sea inlet enclosed by high cliffs, formed when an overdeepened glacial valley is drowned by a rise in sea-level. At the mouth of the fjord there is a characteristic lip causing a shallowing of the water. This is due to reduced glacial erosion at this point.

flax any plant of the genus *Linum*, family Linaceae. The species *L. usitatissimum* is the cultivated strain; linen is produced from the fibre in its stems. The seeds yield linseed oil, used in paints and varnishes.

flood plain area of periodic flooding along the course of river valleys. When river discharge exceeds the capacity of the channel, water rises over the channel banks and floods the adjacent low-lying lands.

focus the point within the Earth's crust at which an earthquake originates.

fog cloud that collects at the surface of the Earth, composed of water vapour that has condensed on particles of dust in the atmosphere, caused by the air temperature falling below dew point. Usually, fog is formed by the meeting of two currents of air, one cooler than the other, or by warm air flowing over a cold surface. Sea fogs commonly occur where warm and cold currents meet and the air above them mixes.

fold a bend in beds or layers of rock. If the bend is arched in the middle it is called an anticline; if it sags downwards in the middle it is called a syncline.

forestry the science of forest management. Recommended forestry practice aims at multipurpose crops, allowing the preservation of varied plant and animal species as well as human uses (lumbering, recreation).

fossil fuel fuel, such as coal, oil, and natural gas, formed from the fossilized remains of plants that lived hundreds of millions of years ago. Fossil fuels are a nonrenewable resource and will eventually run out.

fringing reef coral reef that is attached to the coast with or without an intervening lagoon.

front the boundary between two air masses of different temperature or humidity. A *cold front* marks the line of advance of a cold air mass from below, as it displaces a warm air mass; a *warm front* marks the advance of a warm air mass as it rises up over a cold one.

frontal rainfall rainfall associated with the meeting of air masses at fronts.

frost condition of the weather that occurs when the air temperature is below freezing, $0°C/32°F$. Water in the atmosphere is deposited as ice crystals on the ground or exposed objects.

geochemistry science of chemistry as it applies to geology. It deals with the relative and absolute abundances of the chemical elements and their isotopes in the Earth, and also with the chemical changes that accompany geologic processes.

geodesy methods of surveying the Earth for making maps and correlating geological, gravitational, and magnetic measurements. Geodesic surveys are now commonly made by using radio signals and laser beams from orbiting satellites.

geography science of the Earth's surface; its topography, climate, and physical conditions, and how these factors affect people and society.

geological time time scale embracing the history of the Earth from its physical origin to the present day. Geological time is divided into eras (Precambrian, Palaeozoic, Mesozoic, Cenozoic), which in turn are divided into periods, epochs, ages, and finally chrons.

geology science of the Earth, its origin, composition, structure, and history.

geomorphology branch of geology that deals with the nature and origin of surface landforms such as mountains, valleys, plains, and plateaus.

geophysics branch of geology using physics to study the Earth's surface, interior, and atmosphere. Studies also include winds, weather, tides, earthquakes, volcanoes, and their effects.

geyser natural spring that intermittently discharges an explosive column of steam and hot water into the air.

glacial deposition the laying-down of rocky material by a glacier. When ice melts, it deposits the material that it has been carrying. The material dumped on the valley floor forms a deposit called till or boulder clay.

glacial trough or *U-shaped valley* steep-sided, flat-bottomed valley formed by the erosive action of a glacier and the debris carried by it.

glacier tongue of ice, originating in mountains in snowfields above the snowline, which moves slowly downhill and is constantly replenished from its source. The scenery produced by the erosive action of glaciers is characteristic and includes glacial troughs, corries, arêtes, and pyramidal peaks.

gorge narrow steep-sided valley (or canyon) that may or may not have a river at the bottom. A gorge may be formed as a waterfall retreats upstream, or it may be caused by rejuvenation, when a river begins to cut downwards into its channel once again. Gorges are common in limestone country, where they are often formed by the collapse of the roofs of underground caverns.

gravimetry study of the Earth's gravitational field. Small variations in the gravitational field can be caused by varying densities of rocks and structure beneath the surface. Such variations are measured by a gravimeter. Gravimetry is used by geologists to map the subsurface features of the Earth's crust.

great circle circle drawn on a sphere such that the diameter of the circle is a diameter of the sphere. On the Earth, all meridians of longitude are half great circles; among the parallels of latitude, only the equator is a great circle.

greenhouse effect in the Earth's atmosphere, the trapping of solar radiation, which, absorbed by the Earth and re-emitted from the surface, is prevented from escaping by various gases in the air, principally carbon dioxide, methane, and chlorofluorocarbons. The result is a rise in the Earth's temperature; in a garden greenhouse, the glass walls have the same effect. Fossil-fuel consumption and forest fires are the main causes of carbon dioxide build-up; methane is a byproduct of agriculture (rice, cattle, sheep). Water vapour is another greenhouse gas. The United Nations Environment Programme estimates an increase in average world temperatures of $1.5°C/2.7°F$ with a consequent rise of 20 cm/7.7 in in sea level by the year 2025.

Greenwich Mean Time (GMT) local time on the zero line of longitude (the Greenwich meridian), which passes through the Old Royal Observatory at Greenwich, London. It was replaced 1986 by coordinated universal time (UTC), but continued to be used to measure longitudes and the world's standard time zones.

groundnut another name for ◊peanut.

groundwater water collected underground in porous rock strata; it emerges at the surface as springs and streams. Recent research suggests that usable groundwater amounts to more than 90% of all the fresh water on Earth; keeping such supplies free of pollutants is of critical environmental concern.

gryke enlarged joint that separates blocks of limestone (clints) in a limestone pavement.

gyre circular surface rotation of ocean water in each major sea (a type of current). Gyres are large and permanent, and occupy the northern and southern halves of the three major oceans. They move clockwise in the northern hemisphere and anticlockwise in the southern hemisphere.

habitat localized environment in which an organism lives and which provides for all (or almost all) of its needs. Habitats range from the inorganic or physical, such as the Arctic ice cap, a cliff face, or a cave, to more complex habitats, such as a woodland or a forest floor. Most habitats provide a home for many species of plants and animals.

hail precipitation in the form of pellets of ice (hailstones). It is caused by the circulation of moisture in strong convection currents, usually associated with thunderstorms.

hamlet small rural settlement that is more than just an isolated dwelling but not large enough to be a village. Typically it has 11–100 people and half a dozen houses.

hanging valley valley that joins a larger glacial trough at a higher level than the trough floor. A river or stream flowing along the hanging valley often forms a waterfall as it enters the trough.

harmattan a dry and dusty northeast wind that blows over W Africa.

high-yield variety crop that has been specially bred or selected to produce more than the natural varieties of the same species.

honeypot site area that is of special interest or appeal to tourists. At peak times honeypot sites may become crowded and congested, and noise and litter may eventually spoil such areas.

hot spot a hypothetical region of high thermal activity in the Earth's mantle. It is believed to be the origin of many chains of ocean islands.

humus component of soil consisting of decomposed or partly decomposed organic matter, dark in colour and usually richer towards the surface. It is an important source of minerals in soil fertility.

hurricane or *tropical cyclone* or *typhoon* in tropical regions, an intense depression. Hurricanes originate between 5° and 20° N or S of the equator when the surface temperature of the ocean is above 27°C/80°F. A central calm area, called the eye, is surrounded by inwardly spiralling winds of up to 320 kph/200 mph. A hurricane is accompanied by lightning and torrential rain, and can cause extensive damage.

hydroelectric power (HEP) electricity generated by moving water. In a typical HEP scheme, water stored in a reservoir, often created by damming a river, is piped into water turbines which are coupled to electricity generators.

hydrography study and charting of Earth's surface waters in seas, lakes, and rivers.

hydrological cycle or *water cycle* the cycle by which water moves between the Earth's surface and its atmosphere. It is a complex system involving a number of physical processes (such as evaporation, precipitation, and throughflow) and stores (such as rivers, oceans, and soil).

hydrology study of the location and movement of inland water, both frozen and liquid, above and below ground. It is applied to major civil engineering projects such as irrigation schemes, dams, and hydroelectric power, and in planning water supply.

hydrothermal vein a crack in rock filled with minerals precipitated through the action of circulating high-temperature fluids. Igneous activity often gives rise to the circulation of heated fluids that migrate outwards and move through the surrounding rock. When such solutions carry metallic ions, ore-mineral deposition occurs in the new surroundings on cooling.

hydrothermal vent hot fissure in the ocean floor, known as a smoker.

ice age any period of glaciation occurring in the Earth's history, but particularly that in the Pleistocene epoch, immediately preceding historic times. On the North American continent, glaciers reached as far south as the Great Lakes, and an ice sheet spread over N Europe, leaving its remains as far south as Switzerland. There were several glacial advances separated by interglacial stages during which the ice melted and temperatures were higher than today.

iceberg floating mass of ice, about 80% of which is submerged, rising sometimes to 100 m/ 300 ft above sea level. Glaciers that reach the coast become extended into a broad foot; as this enters the sea, masses break off and drift towards temperate latitudes, becoming a danger to shipping.

ice cap body of ice that is larger than a glacier but smaller than an ice sheet. Such ice masses cover mountain ranges or small islands. Glaciers often originate from ice caps.

ice sheet body of ice that covers a large land mass or continent; it is larger than an ice cap. Today there are two ice sheets, covering much of Antarctica and Greenland. About 96% of all present-day ice is in the form of ice sheets.

igneous rock rock formed from cooling magma or lava, and solidifying from a molten state. Igneous rocks that crystallize below the Earth's surface are called plutonic or intrusive, depending on the depth of formation. They have large crystals produced by slow cooling; examples include dolerite and granite. Those extruded at the surface are called extrusive or volcanic. Rapid cooling results in small crystals; basalt is an example.

inselberg prominent steep-sided hill of resistant solid rock, such as granite, rising out of a plain, usually in a tropical area. Its rounded appearance is caused by so-called onion-skin weathering, in which the surface is eroded in successive layers.

insolation the amount of solar radiation (heat energy from the Sun) that reaches the Earth's surface. Insolation varies with season and latitude, being greatest at the equator and least at the poles.

intensive agriculture farming system where large quantities of inputs, such as labour or fertilizers, are involved over a small area of land. Market gardening is an example. Yields are often much higher than those obtained from extensive agriculture.

International Date Line (IDL) imaginary line that approximately follows the 180° line of longitude. The date is put forward a day when crossing the line going west, and back a day when crossing the line going east.

intertropical convergence zone (ITCZ) area of heavy rainfall found in the tropics and formed as two air masses converge and rise to form cloud and rain. The ITCZ is responsible for most of the rain that falls in Africa.

intrusive rock igneous rock formed within the Earth. Magma, or molten rock, cools slowly at these depths to form coarse-grained rocks, such as granite, with large crystals. A mass of intrusive rock is called an intrusion.

ionosphere ionized layer of Earth's outer atmosphere (60–1,000 km/38–620 mi) that contains sufficient free electrons to modify the way in which radio waves are propagated, for instance by reflecting them back to Earth. The ionosphere is thought to be produced by absorption of the Sun's ultraviolet radiation.

island arc curved chain of islands produced by volcanic activity at a destructive margin (where one tectonic plate slides beneath another).

isostasy the theoretical balance in buoyancy of all parts of the Earth's crust, as though they were floating on a denser layer beneath. High mountains, for example, have very deep roots, just as an iceberg floats with most of its mass submerged.

isotherm line on a map linking all places having the same temperature at a given time.

isthmus narrow strip of land joining two larger land masses.

jet stream narrow band of very fast wind (velocities of over 150 kph/95 mph) found at altitudes of 10–16 km/6–10 mi in the upper troposphere or lower stratosphere. Jet streams usually occur about the latitudes of the Westerlies (35°–60°).

joint vertical crack in a rock, formed by compression, usually several metres in length. Joints are common in limestone and granite, and the weathering of joints in these rocks is responsible for the formation of features such as limestone pavements and tors.

kame fluvioglacial feature, usually in the form of a mound or ridge, formed by the deposition of rocky material carried by a stream of glacial meltwater. Kames are commonly laid down in front of or at the edge of a glacier (kame terrace), and are associated with the disintegration of glaciers at the end of an ice age.

karst landscape characterized by remarkable surface and underground forms, created as a result of the action of water on permeable limestone.

katabatic wind cool wind that blows down a valley on calm clear nights. When the sky is clear, heat escapes rapidly from ground surfaces, and the air above the ground becomes chilled. The cold dense air moves downhill, forming a wind that tends to blow most strongly just before dawn.

kettle hole pit or depression created by glacial activity. A kettle hole is formed when a block of ice from a receding glacier becomes isolated and buried in glacial debris (till). As the block melts the till collapses to form a hollow, which may become filled with water to form a kettle lake or pond.

K-T boundary geologists' shorthand for the boundary between the rocks of the Cretaceous and the Tertiary periods. It marks the extinction of the dinosaurs and in many places reveals a layer of iridium, possibly deposited by a meteorite that may have caused the extinction by its impact.

laccolith intruded mass of igneous rock that forces apart two strata and forms a round lens-shaped mass many times wider than thick. The overlying layers are often pushed upward to form a dome.

lagoon coastal body of shallow salt water, usually with limited access to the sea. The term is normally used to describe the shallow sea area cut off by a coral reef or barrier islands.

lahar mudflow formed of a fluid mixture of water and volcanic ash. During a volcanic eruption melting ice may combine with ash to form a powerful flow capable of causing great destruction.

lake body of still water lying in depressed ground without direct communication with the sea. Lakes are common in formerly glaciated regions, along the courses of slow rivers, and in low land near the sea. The main classifications are by origin: *glacial lakes*, formed by glacial scouring; *barrier lakes*, formed by landslides and glacial moraines; *crater lakes*, found in volcanoes; and *tectonic lakes*, occurring in natural fissures.

land breeze gentle breeze blowing from the land towards the sea and affecting coastal areas. It forms at night in the summer or autumn and tends to be cool.

latitude and longitude imaginary lines used to locate position on the globe. Lines of latitude are drawn parallel to the equator, with 0° at the equator and 90° at the north and south poles. Lines of longitude are drawn at right angles to these, with 0° (the Prime Meridian) passing through Greenwich, London.

lava molten rock that erupts from a volcano and cools to form extrusive, or volcanic, igneous rock. Lava that is high in silica is viscous and sticky and does not flow far. Low-silica lava can flow for long distances.

leaching process by which substances are washed out of the soil. Fertilizers leached out of the soil drain into rivers, lakes, and ponds and cause water pollution. In tropical areas, leaching of the soil after the destruction of forests removes scarce nutrients and leads to a dramatic loss of soil fertility.

lead ore any of several minerals from which lead is extracted. The main primary ore is galena or lead sulphite, PbS.

levee naturally formed raised bank along the side of a river channel. When a river overflows its banks, the rate of flow in the flooded area is less than that in the channel, and silt is deposited. After the waters have withdrawn the silt is left as a bank that grows with successive floods.

limestone pavement bare rock surface resembling a block of chocolate, found on limestone plateaus. It is formed by the weathering of limestone into individual upstanding blocks (clints) separated from each other by joints (grykes).

limnology study of lakes and other bodies of open fresh water, in terms of their plant and animal biology, and their physical properties.

linear development or *ribbon development* housing that has grown up along a route such as a road.

lithification another term for ⟳diagenesis.

lithosphere topmost layer of the Earth's structure, forming the jigsaw of plates that take part in the movements of plate tectonics. The lithosphere comprises the crust and a portion of the upper mantle. It is regarded as being rigid and moves about on the semi-molten asthenosphere. The lithosphere is about 75 km/47 mi thick.

livestock farming the rearing or keeping of animals for their meat, dairy products, and other products such as skins.

loam type of fertile soil, a mixture of sand, silt, clay, and organic material. It is porous, which allows for good air circulation and retention of moisture.

lode geologic deposit rich in certain minerals, generally consisting of a large vein or set of veins containing ore minerals.

loess fertile yellow soil derived from glacial meltwater deposits and accumulated by wind in periglacial regions during the ice ages.

longitude see ⟳latitude and longitude.

longshore drift the movement of material along a beach. When a wave breaks obliquely, pebbles are carried up the beach in the direction of the wave (swash). The wave draws back at right angles to the beach (backwash), carrying some pebbles with it. In this way, material moves in a zigzag fashion along a beach. Longshore drift is responsible for the formation of spits.

maelstrom whirlpool off the Lofoten Islands, Norway, also known as the Moskenesstraumen, which gave its name to whirlpools in general.

magma molten rock material beneath the Earth's surface from which igneous rocks are formed. Lava is magma that has reached the surface and solidified, losing some of its components on the way.

mangrove swamp muddy swamp found on tropical coasts and estuaries, characterized by dense thickets of mangrove trees. These low trees are adapted to live in creeks of salt water.

mantle intermediate zone of the Earth between the crust and the core. It is thought to consist of silicate minerals such as olivine and spinel.

map diagrammatic representation of an area—for example, part of the Earth's surface or the distribution of the stars. Modern maps of the Earth are made using satellites in low orbit to take a series of overlapping stereoscopic photographs from which a three-dimensional image can be prepared.

maritime climate climate typical of coastal areas that is characterized by mild and damp conditions because of the nearness of the sea.

market gardening farming system that specializes in the commercial growing of vegetables, fruit, or flowers. It is an intensive agriculture with crops often being grown inside greenhouses on small farms.

market town settlement with a permanent or periodic market (once a week, for instance). This acts as a selling point for goods produced in the surrounding area.

marsh low-lying wetland. Freshwater marshes are common wherever groundwater, surface springs, streams, or run-off causes frequent flooding or more or less permanent shallow water. Near the sea, a salt marsh may form on the sheltered side of sand and shingle spits. Marshes are alkaline.

mass movement downhill movement of surface materials under their own weight (by gravity). Some types of mass movement are very rapid—for example, landslides, whereas others, such as soil creep, are slow. Water often plays a significant role as it acts as a lubricant, enabling material to move.

meander loop-shaped curve in a river flowing across flat country. The current is fastest on the outside of the curve where it cuts into the bank; on the inside of the curve the current is slow and deposits any transported material. In this way the river changes its course across the flood plain.

Mediterranean climate climate characterized by hot dry summers and warm wet winters. Mediterranean zones are situated in either hemisphere on the western side of continents, between latitudes of 30° and 60°.

Mercalli scale scale used to measure the intensity of an earthquake, based on observed phenomena, and varying from place to place with the same earthquake.

meridian half a great circle drawn on the Earth's surface passing through both poles and thus through all places with the same longitude. Terrestrial longitudes are usually measured from the Greenwich meridian (the zero line of longitude).

mesa flat-topped, steep-sided plateau, consisting of horizontal weak layers of rock topped by a resistant formation. A small mesa is called a butte.

mesosphere layer in the Earth's atmosphere above the stratosphere and below the thermosphere. It lies between about 50 km/31 mi and 80 km/50 mi above the ground.

metamorphic rock rock altered in structure and composition by pressure, heat, or chemically active fluids after original formation. (If heat is sufficient to melt the original rock, technically it becomes an igneous rock upon cooling.)

meteorology scientific observation and study of the atmosphere, so that weather can be accurately forecast. Data from meteorological stations and weather satellites are collated by computer at central agencies, and forecast and weather maps based on current readings are issued at regular intervals.

metropolitan area another name for ⟳conurbation.

midnight sun the constant appearance of the Sun (within the Arctic and Antarctic circles) above the horizon during the summer.

mineral naturally formed inorganic substance with a particular chemical composition. Minerals are the constituents of rocks. In more general usage, a mineral is any substance economically valuable for mining -including coal and oil, which are of organic origin.

mist low cloud caused by the condensation of water vapour in the lower part of the atmosphere. Mist is less thick than fog, visibility being 1–2 km/ 0.6–1.2 mi.

mixed farming system of farming which combines both the cultivation of crops and the keeping of animals.

Mohorovičić discontinuity also *Moho* or *M-discontinuity* boundary that separates the Earth's crust and mantle, marked by a rapid increase in the speed of earthquake waves. It follows the variations in the thickness of the crust and is found approximately 32 km/20 mi below the continents and about 10 km/6 mi below the oceans.

monsoon wind pattern that brings seasonally heavy rain to S Asia; it blows towards the sea in winter and towards the land in summer. The monsoon may cause destructive flooding all over India and SE Asia from April to Sept.

moraine rocky debris or till carried along and deposited by a glacier. Material eroded from the side of a glaciated valley and carried along the glacier's edge is called *lateral moraine*; that worn from the valley floor and carried along the base of the glacier is called *ground moraine*. Rubble dropped at the foot of a melting glacier is called *terminal moraine*. A linear ridge of rocky debris running along the centre of a glacier (commonly formed by the joining of two lateral moraines when two glaciers merge) is called *medial moraine*.

mountain natural upward projection of the Earth's surface, higher and steeper than a hill. The process of mountain building (orogeny) consists of volcanism, folding, faulting, and thrusting, resulting from the collision and welding together of two tectonic plates.

mudflow downhill movement of muddy sediment containing a large proportion of water. Mudflows can be fast and destructive.

national park land set aside and conserved for public enjoyment. The first was Yellowstone National Park, USA, established 1872. National parks include not only the most scenic places, but also places distinguished for their historic, prehistoric, or scientific interest, or for their superior recreational assets.

natural gas mixture of flammable gases found in the Earth's crust, often in association with oil. Natural gas is usually transported from its source by pipeline, although it may be liquefied for transport and storage, and is therefore often used in remote areas where other fuels are scarce and expensive.

nature reserve area set aside to protect a habitat and the wildlife that lives within it, with only restricted admission for the public. A nature reserve often provides a sanctuary for rare species. The world's largest is Etosha Reserve, Namibia; area 99,520 sq km/38,415 sq mi.

new town centrally planned urban area. In the UK, new towns were partly designed to accommodate the overspill from large cities and towns with provision for housing, employment, and other amenities after World War II, when the population was rapidly expanding and city centres had either decayed or been destroyed. In 1976 the policy of building new towns was abandoned.

nodule a lump of mineral or other matter found within rocks or formed on the seabed surface; mining technology is being developed to exploit them.

nomadic pastoralism farming system where animals (cattle, goats, camels) are taken to different locations in order to find fresh pastures. It is practised in the developing world.

nucleated settlement settlement where the buildings are grouped or clustered around a central point, or nucleus. For example, a water supply such as a spring might form the central point for the inititial development of a settlement.

nuée ardente glowing white-hot cloud of ash and gas emitted by a volcano during a violent eruption.

nunatak mountain peak protruding through an ice sheet.

oasis area of land made fertile by the presence of water near the surface in an otherwise arid region. The occurrence of oases affects the distribution of plants, animals, and people in the desert regions of the world.

occluded front weather front formed when a cold front catches up with a warm front. It brings cloud and rain as air is forced to rise upwards along the front, cooling and condensing as it does so.

ocean great mass of salt water. Strictly speaking there are three oceans—the Atlantic, Indian, and Pacific—to which the Arctic is often added. They cover approximately 70% or 363,000,000 sq km/ 140,000,000 sq mi of the total surface area of the Earth. Water levels recorded in the world's oceans have shown an increase of 10–15 cm/4–6 in over the past 100 years.

oceanography study of the oceans, their origin, composition, structure, history, and wildlife (seabirds, fish, plankton, and other organisms). Much oceanography uses computer simulations to plot the possible movements of the waters, and many studies are carried out by remote sensing.

ocean ridge mountain range on the seabed indicating the presence of a constructive margin. Ocean ridges consist of many segments offset along faults, and can rise thousands of metres above the surrounding seabed.

ocean trench deep trench in the seabed indicating the presence of a destructive margin. The subduction or dragging downwards of one plate of the lithosphere beneath another means that the ocean floor is pulled down.

oil crop plant from which vegetable oils are pressed from the seeds. Cool temperate areas grow oilseed rape; warm temperate regions produce sunflowers, olives, and soya beans; tropical regions produce peanuts, palm oil, and coconuts.

oil, crude thick flammable mineral oil found underground in permeable rocks; see ⟡petroleum.

ooze sediment of fine texture consisting mainly of organic matter found on the ocean floor at depths greater than 2,000 m/6,600 ft. Several kinds of ooze exist, each named after its constituents.

organic farming farming without the use of synthetic fertilizers (such as nitrates and phosphates) or pesticides (herbicides, insecticides, and fungicides) or other agrochemicals (such as hormones, growth stimulants, or fruit regulators).

orogeny or *orogenesis* the formation of mountains. It is closely associated with the movements of the rigid plates making up the Earth's crust (described by plate tectonics). Where two plates collide crustal rocks become folded and lifted to form chains of fold mountains.

orographic rainfall rainfall that occurs when an airstream is forced to rise over a mountain range. The air becomes cooled and precipitation takes place. The orographic effect can sometimes occur in large cities, when air rises over tall buildings.

outwash sands and gravels deposited by streams of meltwater (water produced by the melting of a glacier). Such material may be laid down ahead of the glacier's snout to form a large flat expanse called an outwash plain.

oxbow lake curved lake found on the flood plain of a river. Oxbows are caused by the loops of meanders being cut off at times of flood and the river subsequently adopting a shorter course. In the USA, the term bayou is often used.

ozone layer layer of highly reactive pale-blue gas, a form of oxygen, in the upper atmosphere. It protects life on Earth from ultraviolet rays, a cause of skin cancer. At lower atmospheric levels it is an air pollutant and contributes to the greenhouse effect. A continent-sized hole has formed over Antarctica as a result of damage to the ozone layer, caused in part by the emission of chlorofluorocarbons (CFCs).

palaeomagnetism science of the reconstruction of the Earth's ancient magnetic field and the former positions of the continents from the evidence of remanent magnetization in ancient rocks; that is, traces left by the Earth's magnetic field in igneous rocks before they cool.

palaeontology the study of ancient life that encompasses the structure of ancient organisms and their environment, evolution, and ecology, as revealed by their fossils. The practical aspects of palaeontology are based on using the presence of different fossils to date particular rock strata and to identify rocks that were laid down under particular conditions.

pampas /'pæmpəz/ flat, treeless, Argentine plains, lying between the Andes Mountains and the Atlantic Ocean, and rising gradually from the coast to the lower slopes of the mountains.

pastoral farming rearing or keeping of animals in order to obtain meat or other products, such as milk, skins, and hair. Animals can be kept in one place or periodically moved (nomadic pastoralism).

peanut or *groundnut* or *monkey nut* South American annual plant *Arachis hypogaea*, family Leguminosae. The nuts are a staple food in many tropical countries. They yield a valuable edible oil and are also used to make oilcake for cattle.

peat fibrous organic substance found in bogs and formed by the incomplete decomposition of plants such as sphagnum moss. Peat has been dried and used as fuel from ancient times.

pedology the study of soil. It has a practical application in agriculture, where knowledge of the type of soil is important in relation to the cultivation of specific crops.

permafrost condition in which a deep layer of soil does not thaw out during the summer but remains at below 0°C/32°F for at least two years, despite thawing of the soil above. It is claimed that 26% of the world's land surface is permafrost.

petroleum or *crude oil* natural mineral oil, a thick greenish-brown flammable liquid formed underground by the decomposition of organic matter. Oil may flow naturally from wells under gas pressure from above or water pressure from below, or it may require pumping to bring it to the surface.

petrology branch of geology that deals with the study of rocks, their mineral compositions, and their origins.

physical weathering or *mechanical weathering* form of weathering responsible for the mechanical breakdown of rocks but involving no chemical change. Processes involved include freeze-thaw (the alternate freezing and melting of ice in rock cracks) and exfoliation (the alternate expansion and contraction of rocks in response to extreme changes in temperature).

pingo landscape feature of tundra terrain consisting of a hemispherical mound about 30 m/100 ft high, covered with soil that is cracked at the top. The core consists of ice, probably formed from the water of a former lake.

plain or *grassland* large area of flat land, usually covered with grass. Plains cover a large proportion of the Earth's surface, especially between the deserts of the tropics and the rainforests of the equator, and have rain in one season only.

plantation large farm or estate where commercial production of one crop is carried out. Plantations are usually owned by large companies, often multinational corporations, and run by an estate manager.

plate according to plate tectonics, one of a number of slabs of solid rock, about a hundred kilometres thick and often several thousands of kilometres across, making up the Earth's surface.

plateau elevated area of fairly flat land, or a mountainous region in which the peaks are at the same height.

plate tectonics theory that explains the formation of the major physical features of the Earth's surface. The Earth's outermost layer, or lithosphere, is regarded as being divided into a number of rigid plates up to 100 km/62 mi thick, which move relative to each other. Their movement may be due to convection currents within the semisolid mantle beneath. Major landforms occur at the margins of the plates—for example, volcanoes, young fold mountains, ocean trenches, and ocean ridges.

playa temporary lake in a region of interior drainage. Such lakes are common features in arid desert basins fed by intermittent streams. The streams bring dissolved salts to the lakes, and when the lakes shrink during dry spells, the salts precipitate as evaporite deposits.

plunge pool deep pool at the bottom of a waterfall. It is formed by the hydraulic action of the water as it crashes down onto the river bed from a height.

plutonic rock igneous rock derived from magma that has cooled and solidified deep in the crust of the Earth; granites and gabbros are examples of plutonic rocks.

polar reversal changeover in polarity of the Earth's magnetic poles. Studies of the magnetism retained in rocks at the time of their formation have shown that in the past the Earth's north magnetic pole repeatedly became the south magnetic pole, and vice versa.

polder area of flat reclaimed land that used to be covered by a river, lake, or the sea. Polders have been artificially drained and protected from flooding by building dykes.

pole either of the geographic north and south points of the axis about which the Earth rotates. The magnetic poles are the points towards which a freely suspended magnetic needle will point; however, they vary continually.

pollution contamination of the environment caused by human activities. It frequently takes the form of chemicals added to the land, water, or air as a by-product of industry, traffic, or agriculture. Pollutants may enter the food chain and be passed on from one organism to another. They are frequently harmful and may have side effects such as acid rain.

population the number of people inhabiting a country, region, area, or town. Population statistics are derived from many sources, such as the registration of births and deaths, and from censuses.

potato perennial plant *Solanum tuberosum*, family Solanaceae, with edible tuberous roots that are rich in starch.

pothole small hollow in the rock bed of a river. Potholes are formed by the erosive action of rocky material carried by the river, and are commonly found along the river's upper course, where it tends to flow directly over solid bedrock. In caving, a pothole is a vertical hole in rock caused by water descending a crack.

prairie the central North American plain, formerly grass-covered, extending over most of the region between the Rocky Mountains on the W and the Great Lakes and Ohio River on the E.

precipitation meteorological term for water that falls to the Earth from the atmosphere. It includes rain, snow, sleet, hail, dew, and frost.

prevailing wind the direction from which the wind most commonly blows in a locality.

pulse crop such as peas and beans. Pulses are grown primarily for their seeds, which provide a concentrated source of vegetable protein, and make a vital contribution to human diets in poor countries where meat is scarce, and among vegetarians.

pyroclastic deposit deposit made up of fragments of rock, ranging in size from fine ash to large boulders, ejected during an explosive volcanic eruption.

rain form of precipitation in which separate drops of water fall to the Earth's surface from clouds. The drops are formed by the accumulation of fine droplets that condense from water vapour in the air. The condensation is usually brought about by rising and subsequent cooling of air.

rainforest dense forest found on or near the equator where the climate is hot and wet. Rainforests provide the bulk of the oxygen needed globally for plant and animal respiration and harbour at least 40% of the Earth's species of plants and animals. The forests are being destroyed at an increasing rate as their valuable timber is harvested and land cleared for agriculture, causing problems of deforestation, soil erosion, flooding, extinction of species, and displacement of tribal people.

rainshadow zone in the lee of a mountain range that receives less rainfall than the windward side.

raised beach beach that has been raised above the present-day shoreline and is therefore no longer washed by the sea. It is an indication of a fall in sea level or of a rise in land level.

ranching commercial form of pastoral farming that involves extensive use of large areas of land for grazing cattle or sheep.

reef underwater ridge of rock, sand, or coral rising near to the surface of the sea.

regolith the surface layer of loose material that covers most rocks. It consists of eroded rocky material, volcanic ash, river alluvium, vegetable matter, or a mixture of these known as soil.

relief the rise and fall in the level of a landscape. An area can be said to have undulating or flat relief. All areas have relief.

resort town urban settlement whose main function is to cater for tourism of one type or another.

Coastal resorts offer seaside activities; alpine resorts offer skiing. Resorts may also have an attractive climate or scenery, or historic interest.

resource commodity that can be used to satisfy human needs. Resources can be categorized into human resources, such as labour, supplies, and skills, and natural resources, such as climate, fossil fuels, and water.

ria long narrow sea inlet, usually branching and surrounded by hills. A ria is deeper and wider towards its mouth, unlike a fjord. It is formed by the flooding of a river valley due to either a rise in sea level or a lowering of a landmass.

ribbon development another term for ◊linear development.

rice principal cereal of the wet regions of the tropics; derived from grass of the species *Oryza sativa*, probably native to India and SE Asia. It is unique among cereal crops in that it is grown standing in water. The yield is very large, and rice is said to be the staple food of one-third of the world population.

Richter scale scale based on measurement of seismic waves, used to determine the magnitude of an earthquake at its epicentre.

rift valley valley formed by the subsidence of a block of the Earth's crust between two or more parallel faults. Rift valleys are steep-sided and form where the crust is being pulled apart.

river long water course that flows down a slope along a channel. It originates at a point called its source, and enters a sea or lake at its mouth. Along its length it may be joined by smaller rivers called tributaries.

river terrace part of an old floodplain that has been left perched on the side of a river valley. It results from rejuvenation, a renewal in the erosive powers of a river. River terraces are very fertile and are often used for farming. They are also commonly chosen as sites for human settlement because they are safe from flooding.

roche moutonnée outcrop of tough bedrock having one smooth side and one jagged side, found on the floor of a glacial trough. As the glacier moved over its surface, ice and debris eroded the upstream side and rendered it smooth. On the downstream side fragments of rock were plucked away by the ice, causing it to become jagged.

rock constituent of the Earth's crust, composed of mineral particles and/or materials of organic origin consolidated into a hard mass as igneous, sedimentary, or metamorphic rocks.

root crop plant cultivated for its swollen edible root (which may or may not be a true root). Potatoes are the major temperate root crop. The major tropical root crops are cassava, yams, and sweet potatoes. Root crops are second in importance only to cereals as human food.

rubber coagulated latex of a variety of plants, mainly from the New World. Most important is Para rubber, which derives from the tree *Hevea brasiliensis* of the spurge family.

rural depopulation loss of people from remote country areas to cities. The population left behind will be increasingly aged and agriculture declines.

salt marsh wetland with characteristic vegetation that is tolerant of sea water. Salt marshes develop around estuaries and on the sheltered side of sand and shingle spits. They usually have a network of creeks and drainage channels by which tidal waters enter and leave the marsh.

sand loose grains of rock, sized 0.02–2.00 mm/0.0008–0.0800 in in diameter, consisting chiefly of quartz, but owing their varying colour to mixtures of other minerals.

sandbar ridge of sand built up by the currents across the mouth of a river or bay. A sandbar may be entirely underwater or it may form an elongated island that breaks the surface. A sandbar stretching out from a headland is a sand spit.

satellite town new town planned and built to serve a particular local industry, or as a dormitory or overspill town for people who work in nearby urban areas.

savanna or *savannah* extensive open tropical grasslands, with scattered trees and shrubs.

scale on a map, the distance between two places compared with the distance in the real world. The scale may be shown as a representative fraction (such as 1:50,000), a scale line, or a statement (such as 'two centimetres equal one kilometre').

scarp and dip the two slopes formed when a sedimentary bed outcrops as a landscape feature. The scarp is the slope that cuts across the bedding plane; the dip is the opposite slope which follows the bedding plane. The scarp is usually steep, while the dip is a gentle slope.

scree pile of rubble and sediment that forms an ascending slope at the foot of a mountain range or cliff.

sea breeze gentle coastal wind blowing off the sea towards the land. It is most noticeable in summer.

seafloor spreading growth of the ocean crust outwards (sideways) from ocean ridges. The concept of seafloor spreading has been combined with that of continental drift and incorporated into plate tectonics.

season period of the year having a characteristic climate. The change in seasons is mainly due to the change in attitude of the Earth's axis in relation to the Sun, and hence the position of the Sun in the sky at a particular place. In temperate latitudes four seasons are recognized: spring, summer, autumn, and winter. Tropical regions have two seasons—the wet and the dry. Monsoon areas around the Indian Ocean have three seasons: the cold, the hot, and the rainy.

sediment any loose material that has been deposited from suspension in water, ice, or air, generally as the water current or wind speed decreases. Typical sediments are, in order of increasing coarseness, clay, mud, silt, sand, gravel, pebbles, cobbles, and boulders.

sedimentary rock rock formed by the accumulation and cementing together of deposits laid down by water, wind, or ice—for example, limestone, shale, and sandstone. Sedimentary rocks cover more than two-thirds of the Earth's surface.

seismograph instrument used to record the activity of an earthquake.

seismology study of earthquakes and how their shock waves travel through the Earth. By examining the global pattern of waves produced by an earthquake, seismologists can deduce the nature of the materials through which they have passed.

settlement place where people live, varying in size from isolated dwellings to the largest cities.

shelf sea relatively shallow sea, usually no deeper than 200 m/650 ft, overlying the continental shelf around the coastlines.

shield alternative name for ◊craton, the ancient core of a continent.

shifting cultivation farming system where farmers move on from one place to another. The most common form is *slash-and-burn agriculture*: land is cleared by burning, so that crops can be grown. After a few years, soil fertility is reduced and the land is abandoned. A new area is cleared while the old land recovers its fertility.

sial the substance of the Earth's continental crust. The name is derived from *si*lica and *al*umina, its two main chemical constituents.

sill sheet of igneous rock created by the intrusion of magma (molten rock) between layers of pre-existing rock.

sima the substance of the Earth's oceanic crust. The name is derived from *si*lica and *ma*gnesia, its two main chemical constituents.

sleet precipitation consisting of a mixture of water and ice, formed from melted falling snow or hail. In North America the term refers to precipitation in the form of ice pellets smaller than 5 mm/0.2 in.

smoker or *hydrothermal vent* crack in the ocean floor, associated with an ocean ridge, through which hot, mineral-rich ground water erupts into the sea, forming thick clouds of suspended material.

snow precipitation in the form of soft, white, crystalline flakes caused by the condensation in air of excess water vapour below freezing point. Light reflecting in the crystals, which have a basic hexagonal geometry, gives snow its white appearance.

soil loose covering of broken rocky material and decaying organic matter overlying the bedrock of the Earth's surface.

soil creep gradual movement of soil down a slope. As each soil particle is dislodged by a raindrop it moves slightly further downhill. This eventually results in a mass downward movement of soil on the slope.

solstice either of the points at which the Sun is farthest north or south of the celestial equator each year. The *summer solstice*, when the Sun is farthest north, occurs around 21 June; the *winter solstice* around 22 December.

spa town town with a spring, the water of which, it is claimed, has the power to cure illness and restore health. Spa treatment involves drinking and bathing in the naturally mineralized spring water.

speleology scientific study of caves, their origin, development, physical structure, flora, fauna, folklore, exploration, mapping, photography, cave-diving, and rescue work.

spit ridge of sand or shingle projecting from the land into a body of water. It is deposited by a current carrying material from one direction to another across the mouth of an inlet (longshore drift). Deposition in the brackish water behind a spit may result in the formation of a salt marsh.

spring a natural flow of water from the ground, formed where the water table meets the ground's surface. The source of water is rain that has percolated through the overlying rocks. During its underground passage, the water may have dissolved mineral substances, which may then be precipitated at the spring.

spring line geological feature where water springs up in several places along the edge of a permeable rock escarpment.

spur ridge of rock jutting out into a valley or plain. In mountainous areas rivers often flow around interlocking spurs because they are not powerful enough to erode through them. Spurs may be eroded away by large and powerful glaciers to form truncated spurs.

stack isolated pillar of rock that has become separated from a headland by coastal erosion. It is usually formed by the collapse of an arch.

stalactite and stalagmite cave structures formed by the deposition of calcite dissolved in ground water. *Stalactites* grow downwards from the roofs or walls whereas *stalagmites* grow upwards from the cave floor. They may meet to form a continuous column from floor to ceiling.

steppe the temperate grasslands of Europe and Asia. Arable and pastoral farming are carried out there. Sometimes the term refers to other temperate grasslands and semi-arid desert edges.

strata (singular *stratum*) layers or beds of sedimentary rock.

stratigraphy branch of geology that deals with the sequence of formation of sedimentary rock layers and the conditions under which they were formed.

stratosphere that part of the atmosphere 10–40 km/6–25 mi from Earth, where the temperature slowly rises from a low of –55°C/–67°F to around 0°C/32°F. The air is rarefied and at around 25 km/15 mi much ozone is concentrated.

striation scratch formed by the movement of a glacier over a rock surface. Striations are caused by the scraping of rocky debris embedded in the base of the glacier, and provide an useful indicator of the direction of ice flow in past ice ages.

stromatolite mound produced in shallow water by mats of algae that trap mud particles. Another mat grows on the trapped mud layer and this traps another layer of mud and so on. Stromatolites are uncommon today but their fossils are among the earliest evidence for living things—over 2,000 million years old.

subduction zone in plate tectonics, a region in which one plate descends below another. Subduction zones are a feature of destructive margins; most are marked by ocean trenches.

subsistence farming farming when the produce is enough to feed only the farmer and family and there is no surplus to sell.

surveying the accurate measuring of the Earth's crust, or of land features or buildings. It is used to establish boundaries, and to evaluate the topography for engineering work. The measurements used are both linear and angular, and geometry and trigonometry are applied in the calculations.

swallow hole hole, often found in limestone areas, through which a surface stream disappears underground. It will usually lead to an underground network of caves.

syncline a fold in the rocks of the Earth's crust in which the layers or beds dip inwards and downwards, thus forming a troughlike structure with a sag in the middle.

synoptic chart weather chart in which symbols are used to represent the weather conditions experienced over an area at a particular time. Synoptic charts appear on television and in newspaper forecasts, although the symbols used may differ.

taiga or *boreal forest* Russian name for the forest zone south of the tundra, found across the northern hemisphere. Dense forests of conifers (spruces and hemlocks), birches, and poplars occupy glaciated regions punctuated with cold lakes, streams, bogs, and marshes.

tea evergreen shrub *Camellia sinensis*, family Theaceae, of which the fermented, dried leaves are infused to make a beverage of the same name. Producers today include Africa, South America, Georgia, Azerbaijan, Indonesia, and Iran.

tectonics the study of the movements of rocks on the Earth's surface. On a small scale tectonics involves the formation of folds and faults, but on a large scale plate tectonics deals with the movement of the Earth's surface as a whole.

tenant farming system whereby farmers rent their holdings from a landowner in return for the use of agricultural land.

terrace farming farming on steep slopes that are terraced to produce a series of flat fields, stepped in appearance. This flat land can then be cultivated.

thermosphere layer in the Earth's atmosphere above the mesosphere and below the exosphere. Its lower level is about 80 km/50 mi above the ground, but its upper level is undefined. The ionosphere is located in the thermosphere. In the thermosphere the temperature rises with increasing height to several thousand degrees Celsius. However, because of the thinness of the air, very little heat is actually present.

throughflow the seepage, or percolation, of water through soil. In the hydrological, or water, cycle it is one of the processes responsible for the movement of water from the land to the oceans.

tidal energy renewable energy derived from the tides. If water is trapped at a high level during high tide, perhaps by means of a barrage across an estuary, it may then be gradually released and its energy exploited to drive turbines and generate electricity.

tide rise and fall of sea level due to the gravitational forces of the Moon and Sun. High water occurs at an average interval of 12 hr 24 min 30 sec. The highest or *spring tides* are at or near new and full Moon; the lowest or *neap tides* when the Moon is in its first or third quarter. Some seas, such as the Mediterranean, have very small tides.

till or *boulder clay* deposit of clay, mud, gravel, and boulders left by a glacier. It is unsorted, with all sizes of fragments mixed up together, and does not form clear layers or beds.

tombolo spit, or ridge of sand or shingle, that connects the mainland to an island.

topography the surface shape and composition of the landscape, comprising both natural and artificial features. Topographical features include the relief and contours of the land; the distribution of mountains, valleys, and human settlements; and the patterns of rivers, roads, and railways.

tor isolated mass of rock, usually granite, left upstanding on a hilltop after the surrounding rock has been broken down. Weathering takes place along the joints in the rock, reducing the outcrop into a mass of rounded blocks. The term is used more generally to mean a high, often bare and rocky, hill.

tornado extremely violent revolving storm with swirling, funnel-shaped clouds, caused by a rising column of warm air propelled by strong wind. A tornado can rise to a great height, but with a diameter of only a few hundred metres or yards or less. Tornadoes move with wind speeds of 160–480 kph/100–300 mph, destroying everything in their path.

town settlement intermediate in size between a village and a city. Towns in the UK typically have a population of 10,000–90,000.

trade wind prevailing wind that blows towards the equator from the NE and SE. Trade winds are caused by hot air rising at the equator and the consequent movement of air from north and south to take its place. The winds are deflected towards the west because of the Earth's W–E rotation.

tributary river that joins a larger river.

tropics the area between the tropics of Cancer and Capricorn, defined by the parallels of latitude approximately 23°30′ N and S of the equator. They are the limits of the area of Earth's surface in which the Sun can be directly overhead. The mean monthly temperature is over 20°C/68°F.

troposphere lower part of the Earth's atmosphere extending about 10.5 km/6.5 mi from the Earth's surface, in which temperature decreases with height to about −60°C/−76°F except in local layers of temperature inversion.

tsunami giant wave generated by an undersea earthquake or volcanic eruption. In the open ocean it may take the form of several successive waves, in excess of a metre in height. In the coastal shallows tsunamis slow down and build up, producing towering waves that can sweep inland and cause great loss of life and property.

tundra region of high latitude almost devoid of trees, resulting from the presence of permafrost. The vegetation consists mostly of grasses, sedges, heather, mosses, and lichens.

typhoon violently revolving storm in the W Pacific Ocean.

unconformity a break in the sequence of sedimentary rocks. It is usually seen as an eroded surface, with the beds above and below lying at different angles. An unconformity represents an ancient land surface, where exposed rocks were worn down by erosion and later covered in a renewed cycle of deposition.

U-shaped valley another term for a ⟳glacial trough, a valley formed by a glacier.

varve a pair of thin sedimentary beds, one coarse and one fine, representing a cycle of thaw followed by an interval of freezing, in lakes of glacial regions.

veldt subtropical grassland in South Africa, equivalent to the pampas of South America.

vent hole in a volcano through which magma, or molten rock, rises. If the magma is particularly thick it may clog up the vent, causing a tremendous build-up of pressure, which will be released as a very violent eruption.

village rural settlement intermediate in size between a hamlet and a town (population about 200–3,000). The term may also apply to village-style parts of larger urban areas.

volcanic rock igneous rock formed at the surface of the Earth. It is usually fine-grained, unlike the more coarse-grained intrusive (under the surface) types of igneous rocks. Volcanic rock can be either *lava* (solidified magma) or a *pyroclastic deposit* (fragmentary lava or ash), such as tuff (volcanic ash that has fused to form rock).

volcano vent in the Earth's crust from which molten rock, lava, ashes, and gases are ejected. Usually it is cone-shaped with a pitlike opening at the top called the crater. Some volcanoes eject the material with explosive violence; others are quiet and the lava simply rises into the crater and flows over the rim.

V-shaped valley river valley with a V-shaped cross-section. Such valleys are usually found near the source of a river, where the steeper gradient means that erosion cuts downwards more than it does sideways. However, a V-shaped valley may also be formed in the lower course of a river when its powers of downward erosion become renewed by a fall in sea level, a rise in land level, or the capture of another river.

wadi in arid regions of the Middle East, a steep-sided valley containing an intermittent stream that flows in the wet season.

water cycle the natural circulation of water through the biosphere. Water is lost from the Earth's surface to the atmosphere either by evap-

oration from the surface of lakes, rivers, and oceans, or through the transpiration of plants. This atmospheric water forms clouds that condense to deposit moisture on the land and sea as rain or snow. The water that collects on land flows to the ocean in streams and rivers.

waterfall cascade of water in a river or stream. It occurs when the water flows over a bed of rock that resists erosion; weaker rocks downstream are worn away, creating a steep, vertical drop and a plunge pool into which the water falls.

watershed the boundary between two drainage basins, usually a ridge of high ground.

water table the upper level of ground water. Water that is above the water table will drain downwards; a spring forms where the water table cuts the surface of the ground. The water table rises and falls in response to rainfall and the rate at which water is extracted, for example, for irrigation.

wave-cut platform gently sloping rock surface found at the foot of a coastal cliff. Covered by water at high tide but exposed at low tide, it represents the last remnant of an eroded headland.

wave power power obtained by harnessing the energy of water waves. A number of wave-power devices have been advanced (such as the duck—a floating boom whose segments bob up and down with the waves, driving a generator), but few have yet proved economical.

weather day-to-day variation of climatic and atmospheric conditions at any one place, or the state of these conditions at a place at any one time. Such conditions include humidity, precipitation, temperature, cloud cover, visibility, and wind.

wheat cereal plant derived from the wild *Triticum*, a grass native to the Middle East. It is the chief cereal used in breadmaking and is widely cultivated in temperate climates suited to its growth. Wheat is killed by frost, and damp renders the grain soft, so warm, dry regions produce the most valuable grain.

wind the horizontal movement of air across the surface of the Earth. Wind is caused by the movement of air from areas of high atmospheric pressure (anticyclones) to areas of low pressure (depressions). Very strong winds can cause a great deal of damage, killing people and blowing down buildings and crops.

wind energy energy derived from the wind. Wind turbines are aerodynamically advanced windmills that drive electricity generators when their blades are spun by the wind. Wind energy is beginning to be used to produce electricity on a large scale.

woodland area in which trees grow more or less thickly; generally smaller than a forest. Temperate climates, with four distinct seasons a year, tend to support a mixed woodland habitat, with some conifers but mostly broad-leaved and deciduous trees, shedding their leaves in autumn and regrowing them in spring.

yardang ridge formed by wind erosion from a dried-up riverbed or similar feature.

40-566-1